CONTROL SYSTEM ANALYSIS AND DESIGN

CONTROL SYSTEM ANALYSIS AND DESIGN

SECOND EDITION

A K Tripathi
Assistant Professor
Department of Electronics Engineering (Degree Division)
Institute of Engineering and Rural Technology
Allahabad, UP

Dinesh Chandra
Professor
Department of Electrical Engineering
MNNIT, Allahabad, UP

Rajeev Gupta
Assistant Professor
Department of Electronics and Communication Engineering
MNNIT, Allahabad, UP

Publishing Globally

NEW AGE INTERNATIONAL (P) LIMITED, PUBLISHERS
LONDON • NEW DELHI • NAIROBI
Bangalore • Chennai • Cochin • Guwahati • Hyderabad • Kolkata • Lucknow • Mumbai
Visit us at **www.newagepublishers.com**

Published by New Age International (P) Ltd., Publishers
First Edition: 2009
Second Edition: 2015
Reprint: 2018

GLOBAL OFFICES

- **New Delhi** **NEW AGE INTERNATIONAL (P) LIMITED, PUBLISHERS**
7/30 A, Daryaganj, New Delhi-110002, (INDIA)
Tel.: (011) 23253771, 23253472, **Telefax:** 23267437, 43551305
E-mail: contactus@newagepublishers.com • Visit us at www.newagepublishers.com
- **London** **NEW AGE INTERNATIONAL (UK) LTD.**
27 Old Gloucester Street, London, WC1N 3AX, UK
E-mail: info@newacademicscience.co.uk • Visit us at www.newacademicscience.co.uk
- **Nairobi** **NEW AGE GOLDEN (EAST AFRICA) LTD.**
Ground Floor, Westlands Arcade, Chiromo Road (Next to Naivas Supermarket)
Westlands, Nairobi, KENYA, **Tel.:** 00-254-713848772, 00-254-725700286
E-mail: kenya@newagepublishers.com

BRANCHES

- **Bangalore** 37/10, 8th Cross (Near Hanuman Temple), Azad Nagar, Chamarajpet, Bangalore- 560 018
Tel.: (080) 26756823, **Telefax:** 26756820, **E-mail: bangalore@newagepublishers.com**
- **Chennai** 26, Damodaran Street, T. Nagar, Chennai-600 017, **Tel.:** (044) 24353401, **Telefax:** 24351463
E-mail: chennai@newagepublishers.com
- **Cochin** CC-39/1016, Carrier Station Road, Ernakulam South, Cochin-682 016, **Tel.:** (0484) 2377303, **Telefax:** 4051304
E-mail: cochin@newagepublishers.com
- **Guwahati** Hemsen Complex, Mohd. Shah Road, Paltan Bazar, Near Starline Hotel, Guwahati-781 008,
Tel.: (0361) 2513881, **Telefax:** 2543669, **E-mail: guwahati@newagepublishers.com**
- **Hyderabad** 105, 1st Floor, Madhiray Kaveri Tower, 3-2-19, Azam Jahi Road, Near Kumar Theater, Nimboliadda
Kachiguda, Hyderabad-500 027, **Tel.:** (040) 24652456, **Telefax:** 24652457
E-mail: hyderabad@newagepublishers.com
- **Kolkata** RDB Chambers (Formerly Lotus Cinema) 106A, 1st Floor, S N Banerjee Road, Kolkata-700 014
Tel.: (033) 22273773, **Telefax:** 22275247, **E-mail: kolkata@newagepublishers.com**
- **Lucknow** 16-A, Jopling Road, Lucknow-226 001, **Tel.:** (0522) 2209578, 4045297, **Telefax:** 2204098
E-mail: lucknow@newagepublishers.com
- **Mumbai** 142C, Victor House, Ground Floor, N.M. Joshi Marg, Lower Parel, Mumbai-400 013
Tel.: (022) 24927869, **Telefax:** 24915415, **E-mail: mumbai@newagepublishers.com**
- **New Delhi** 22, Golden House, Daryaganj, New Delhi-110 002, **Tel.:** (011) 23262368, 23262370, **Telefax:** 43551305
E-mail: sales@newagepublishers.com

ISBN: 978-81-224-3885-7

C-17-11-10838

Printed in India at A. P. Offset, Delhi.
Typeset at In-house, New Delhi.

NEW AGE INTERNATIONAL (P) LIMITED, PUBLISHERS
7/30 A, Daryaganj, New Delhi-110002
Visit us at **www.newagepublishers.com**
(CIN: U74899DL1966PTC004618)

PREFACE

Stepping into modern technological developments requires that the future engineers be aptly equipped with knowledge of fundamental concepts and techniques which will enable them to be at natural ease in analysing and designing the control systems. In fact, the control system is an extraordinarily rich subject of Electronics, Electrical and Mechanical Engineering with diverse applications. Keeping this in view, this book is intended to be used as text for an introductory course in Control Systems with the prerequisite that the reader should have an introductory knowledge of differential equations, matrix algebra and Laplace transform.

The book starts with an introduction to the basic concepts of control systems in Chapter 1. System, system classification, system modeling and need for the control in a system, have been discussed. This chapter also familiarises the reader with evolution of transfer function model and establishing the correspondence of time functions with individual partial fraction terms of transfer function. Chapter 2 identifies the standard test signals and applies them to the system of order one, order two and higher order to evaluate responses both in transients and steady state.

Chapter 3 is designed to familiarise the reader with block diagram manipulation, signal flow graphs and use of Mason's gain rule. Chapter 4 discusses system stability. The determination of root distribution through Routh's test, testing of adjustable systems and axis shift to evaluate relative stability are thoroughly treated.

Chapter 5 aims at thorough understanding of root locus for positive parameter variation, negative parameter variation and multiple parameter variation. The root locus for systems of forms other than those with unity subtractive feedback type, is also discussed.

Chapter 6 is devoted to all important frequency response analysis. There is wide speculation among students that Nyquist plot and Bode plot are the most tedious segments of control system analysis to learn. Considerable thought and effort have been devoted to ease out learning them. The step-by-step approach for sketching Nyquist plot for minimum phase systems, non-minimum phase systems and those with poles on imaginary axis, is expected to fascinate the reader. A similar approach is demonstrated to treat Bode plots as well. Interpreting stability from Nyquist and Bode plots is still more dexterously treated. This chapter begins with identifying the strengths and weaknesses of frequency response approach and culminates in finding the transfer function model from Bode plot. The frequency response plots for systems with delay, are also well treated. In fact, the entire text has been developed exactly in the manner in which the author has learnt and has been teaching in class.

Chapter 7 is entirely devoted to state space analysis including asymptotic stability, controllability and observability. The state space design including pole placement and observer design is also covered. Finding transfer function and time response through state space approach, are also included.

Chapter 8 covers the analysis and design of industrial controllers: P, PI, PD and PID. The step-by-step procedure is used while treating the design of these controllers from root locus perspective. The analysis of phase lag, phase lead and lag-lead compensators is included. The design of these compensators through root locus and Bode plots is thoroughly treated using step-by-step approach. Evolution of design strategies has been categorically explained with manual sketches. The abstraction through computer computations is avoided.

On successive demand of students, Chapter 9 lays stress on Control System Components of wider applications including stepper motors and encoders that are quite pervasive in robotics and design of digital control systems.

The idea of writing a textbook on control systems, was conceived when Dr. H. Kar and we were enjoying a cup of tea in department of electronics, MNNIT Allahabad. It is he who gave me a tip of developing the text exactly in the manner we teach in the class. Probably our classroom teaching, for which he would have got feedback from students, fascinated him. The text development begins with aim to bring even the challenging control concepts within grasp of average students. Step-by-step approach is adapted throughout the text. All significant points concerning a particular concept have been put together in the form of tables and 'note the following'. A potential care has been exercised in developing the example problems. Each chapter includes sufficient number of multiple-choice questions designed to test the student rigorously on the material covered.

The authors would like to express their sincere appreciation to professors, Rajeev Tripathi (Dean Academic, MNNIT, Allahabad), Neeraj Shukla (Academic Coordinator, IERT, Allahabad), S.K. Shukla (Department of Mathematics, IERT) and Vikash Choubey (Department of IP, IERT) who made valuable suggestions and constructive comments in revision routine. Appreciation is also due to my students Anand Agrawal, Nishu, Rupam and Sarala who helped in identifying the corrections in first edition and made significant contribution in revising, identifying corrections and adding the problems at the end of the chapters.

The authors are pleased to acknowledge their indebtedness to Dr. H. Kar and Dr. M.M. Dwivedi who reviewed the entire script. Their most gracious encouragements and valued constructive criticism are most sincerely appreciated. Sincere thanks is due to Mr. Satya Prakash for his invaluable contribution towards reading some portion of text and consistent encouragement to complete the task.

The authors express deep appreciation and thanks to Kailash Nath Prajapati for his most skillful services in the preparation of script. He tirelessly typed the manuscript multiple times and developed the figures with keen interest and utmost care. The authors would like to thank Mr. Bhupendra Pandey, an educator and motivator for his consistent persuasion to complete the task. Authors also thank all colleagues and students for their assistance.

The encouragement, patience, technical support and enthusiasm provided by the publishers have been, in deed, crucial in making this text a reality.

The authors highly appreciate your comments and suggestions so as to upkeep the vision to the goal of producing best possible text on Control System.

A.K. Tripathi
Dinesh Chandra
Rajeev Gupta

CONTENTS

PREFACE v

1 Introductory Control Concepts 1–52

1.1 Introduction 1
- Open loop control system 2
- Closed loop control system (feedback control system) 2

1.2 Effects of Feedback 3

1.3 Systems Modelling 10
- Electrical systems 10
- Mechanical system 11

1.4 Mathematical Modelling of Translational/Rotational Mechanical System 13

1.5 Analogous Systems 18
- Force-voltage (*f-v*) analogy 18
- Procedure to construct analogous electrical system using *f-v* analogy 20
- Procedure to construct analogous electrical system using *f-i* analogy 21

1.6 Gear Train and its Electrical Analog 22
- Electrical analog of gear train 24

1.7 Transfer Function 26
- Transfer function model of system 27
- Zero state response 29
- Zero input response 30
- Transfer function of a system with multi inputs and multi outputs 30

1.8 Classification of Systems 32
- Linear and non-linear systems 32
- Time varying and time invariant systems 32
- Continuous and discrete systems 33
- Deterministic and stochastic systems 33
- System with memory and without memory 33
- Causal system 33

Problems and Solutions 34

Drill Problems 41
Multiple Choice Questions 43
Answers and Hints to Multiple Choice Questions 50

2 Control System Analysis in Time Domain **53–114**

2.1 Introduction 53
2.2 Standard Test Signals 54
2.3 First Order System 55
- Unit step response of first order system 55
- Unit ramp response of first order system 57
- Unit impulse response of first order system 58
- Illustrative examples 59

2.4 Second Order System 60
- Unit step response of a general second order control system 60
- Transient response specifications 62
- The characteristic equation roots and corresponding time response 65
- Unit impulse response of a general second order system 70

2.5 Steady State Performance of Linear Control System 72
- Type number of a system 73
- Steady state error due to step input 74
- Steady state error due to ramp input (velocity input) 75
- Steady state error due to parabolic input (acceleration input) 76
- Drawbacks of approach based on error constant evaluation 78

2.6 The Error Series 78
2.7 Higher Order Systems 79
- Technique to cast away insignificant poles 80

2.8 Effect of Adding Poles and Zeros to Transfer Functions 81
- Addition of a zero to closed loop transfer function 81
- Addition of a zero to open loop transfer function 82
- Addition of a pole to open loop transfer function 82
- Addition of pole to closed loop transfer function 83

Problems and Solutions 84
Drill Problems 95
Multiple Choice Questions 97
Answers and Hints to Multiple Choice Questions 110

3 Block Diagrams and Signal Flow Graph **115–162**

3.1 Block Diagrams 115
- Block diagram development 116
- Block diagram reduction 117

3.2 Signal Flow Graph 119
- Important definitions 120
- Construction of signal flow graph from block diagram 123
- Construction of signal flow graph for electrical network 123
- Mason's gain rule 124

Problems and Solutions 125
Drill Problems 148
Multiple Choice Questions 154
Answers and Hints to Multiple Choice Questions 160

4 System Stability **163–205**

4.1 Introduction 163
- Asymptotic stability 163
- Impulse response stability 163
- Bounded input bounded output (BIBO) stability 164

4.2 Coefficient Test for Stability 166
4.3 Routh's Stability Test 167
4.4 Left Column Zero of Array 169
4.5 Premature Termination of Array 171
4.6 Relative Stability Analysis 175
4.7 Routh's Stability Test in Control System Analysis 176
Problems and Solutions 177
Drill Problems 194
Multiple Choice Questions 197
Answers and Hints to Multiple Choice Questions 202

5 Root Locus **206–277**

5.1 Introduction 206
5.2 Root Locus for Feedback Systems 208
- Graphical evaluation of angle and magnitude of G(s) H(s) 210

5.3 Root Locus Construction 213
- Some root locus plots 226

5.4 Root Loci for Systems with Other Forms 228
5.5 Root Loci for Systems with Positive Feedback 230
5.6 Root Locus of a G(s) H(s) Product with Pole-zero Cancellation 234
5.7 Effects of Adding Poles and Zeros to the Product G(s) H(s) on Shape of Root Locus 235
- Addition of poles 235
- Addition of zeros 236

5.8 Effects of Delay on Root Locus 237

5.9 Root Contours (Multiple Parameter Variation) 239
Problems and Solutions 241
Drill Problems 265
Multiple Choice Questions 268
Answers and Hints to Multiple Choice Questions 275

6 Frequency Response Analysis 278–438

6.1 Introduction 278
- Strengths of frequency response approach 278
- Weaknesses of frequency response approach 279

6.2 Frequency Response 279
- Steady state sinusoidal response 279
- Frequency response evaluation 281
- Graphical evaluation of frequency response 281

6.3 Correlation between Time Response and Frequency Response 282
- Frequency response specifications 283

6.4 Graphical Representation of Frequency Response 287
- Polar plots 287
- Polar plot construction 287
- Effects of addition of poles and zeros to G(s) on the shape of polar plots 302
- Addition of poles at origin 302
- Addition of finite non-origin poles 304
- Addition of zeros at origin 305

6.5 The Nyquist Stability Criterion 305
- Mapping 306
- Nyquist stability criterion 310
- Nyquist procedure for minimum phase system 312
- Relative stability using Nyquist procedure 313
- The measures of relative stability: Gain margin and phase margin 315
- Generating complete Nyquist plot and interpretting stability 321

6.6 Bode Plot 327
- The product terms of G($j\omega$) H($j\omega$) 327
- Poles and zeros at origin of s plane 328
- Real axis poles or zeros 331
- Complex conjugate poles or zeros 336
- Sketching magnitude (dB) plot 336
- Sketching phase plot 337
- Gain margin and phase margin from Bode plot 339
- Relationship between Bode magnitude (dB) plot and number type of a system 341
- Evaluation of K_p from Bode plot 341

- Evaluation of K_v from Bode plot 342
- Evaluation of K_a from Bode plot 343
- Irrational transmittances 349
- All pass systems 349
- Effect of variation in gain K on Bode plot 350
- Effect of presence of delay in system on Bode plot 351
- Finding transfer function models 352
- Imaginary axis zeros and poles 354

6.7 Closed Loop Frequency Response of Unity Feedback System 355
- Constant magnitude loci (M circles) 357

Problems and Solutions 364
Drill Problems 406
Multiple Choice Questions 411
Answers and Hints to Multiple Choice Questions 431

7 State Space Analysis and Design 439–531

7.1 Introduction 439
7.2 State Space Representation 440
- State 440
- State vector 440
- State space 440
- Phase variable form of state model 445
- Dual phase variable form of state model 447
- State model for systems with single input and multiple outputs 448
- State model for systems with multiple inputs and single output 449
- State model for systems with multiple inputs and multiple outputs 451
- Other ways of modelling 451
- State space model using canonical variables 455

7.3 Modelling Electrical and Mechanical Systems 459
7.4 Finding Transfer Function from State Space Model 461
7.5 Finding Time Response from State Model 464
- First order systems 464
- State transition matrix 465
- Properties of STM 466
- Higher order systems 467

7.6 Controllability and Observability 468
- Test of state controllability for diagonal systems 469
- Test of output controllability 472
- Test of observability for diagonal systems 472
- Causes of uncontrollability and/or unobservability 473
- Principle of duality 475

7.7 Finding Decoupled State Equations (Diagonalisation) 476
7.8 State Feedback and Pole Placement 481
7.9 Observer Design 484
Problems and Solutions 488
Drill Problems 517
Multiple Choice Questions 522
Answers and Hints to Multiple Choice Questions 528

8 Control System Design 532–635

8.1 Introduction 532
8.2 Controller Configurations 532
8.3 Industrial Automatic Controllers 534
- Two position/on-off control 534
- Proportional control 535
- Integral control 536
- Proportional *plus* Integral control (PI) 537
- Proportional *plus* Derivative control (PD) 539
- Proportional *plus* Integral *plus* Derivative control (PID) 541

8.4 Generating Hardware for Industrial Controllers 543
- PI controller 543
- PD controller 545
- PID controller 545

8.5 The Compensator Elements 546
- Phase lead compensator 546
- Phase lag-compensator 551
- Lag-lead compensator 555

8.6 Root Locus Design 557
- Cascade compensator design for improving steady state performance 559
- PI compensator design 559
- Lag-compensator design 561
- Cascade compensator design for improving transient response 564
- PD compensator design 565
- Lead compensator design 568
- PID controller design 571
- Lag-lead compensator design 574

8.7 Frequency Response Design 577
- Lag-compensator design 578
- Lead compensator design 582
- Phase lag-lead compensator design 586

8.8 Rate Feedback Compensator Design 590
- Minor loop feedback compensation 593

Problems and Solutions 596
Drill Problems 626
Multiple Choice Questions 629
Answers and Hints to Multiple Choice Questions 634

9 Control System Components **636–698**

9.1 Introduction 636
9.2 Potentiometer 637
- Potentiometer performance indices 641
- Merits and demerits of potentiometer 642
- Types of potentiometer 642

9.3 Synchros 643
- Synchro construction and operation 643
- Synchro transmitter (ST) 643
- Synchro control transformer (CT) 644

9.4 Servo Motor 647
- DC servo motor 647
- Armature controlled DC servo motor 647
- Field controlled DC servo motor 651
- Comparing armature controlled and field controlled DC servo motors 653
- AC servo motor 653

9.5 AC Tachometer 657
- DC tachometer 657
- AC tachometer 658

9.6 Servo Amplifier 659
- Amplidyne 660

9.7 Stepper Motor 663
9.8 Optical Encoder 673
- Incremental encoder 673
- Absolute encoder 675

9.9 Mechanical Arrangements to Convert Rotational Motion into Corresponding Linear Motion 676
9.10 Belt or Chain Drive and Lever System 677
Problems and Solutions 678
Drill Problems 690
Multiple Choice Questions 694
Answers and Hints to Multiple Choice Questions 698

Index **699–705**

1

INTRODUCTORY CONTROL CONCEPTS

1.1 INTRODUCTION

In recent years, the control systems have greatly influenced the development and advancement of modern life and technology. They have become an integrated part of our everyday life too. Some of the widely pervasive examples in day to day life, are automatic washers and dryers, microwave ovens, pollution controls and economic regulation. The control systems appear in almost everything, right from simple electronic household products to aeroplanes and spacecrafts. They have significantly advanced into robotics, space vehicle systems including successful lunar soft landing, intercontinental missile guidance systems, high speed rail systems and most recently magnetic levitation rail systems. They have also become essential in industrial operations in the sense of controlling pressure, temperature, humidity, viscosity and flow in process industries. Thus the control systems have become interdisciplinary today cutting across all specialized engineering fields.

The current chapter is devoted to understand a system, need for control in a system and mathematical modelling of systems. The **system** is any interconnection of components to achieve desired objective. A system is called so provided it produces response of particular interest or delivers energy in some useful form. It is well known that any system cannot create energy by itself. The energy must either be inherent in the system itself or be supplied by an external source. The energy inherent in the system is modelled as initial condition (I.C.) thereof. A typical system with input, output and initial condition is shown in Fig. 1.1.

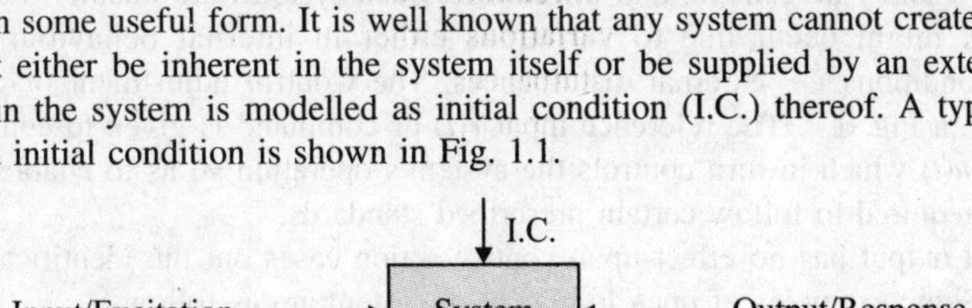

Fig. 1.1: A system

A system excited by only initial conditions is said to be **autonomous** and a system having external input together with or without initial condition is said to be **non autonomous**.

Experimental investigation and engineering work often involve forcing a parameter which is otherwise undetermined or does not have the desired value, to track a reference to a specified degree of accuracy. Here a common approach is to use a controller to manipulate response of system as desired. Some physical process takes place in the system of interest. Under ideal conditions, in absence of control, response of system would be constant. However, in reality, because of external disturbances like change in temperature and other environmental parameters, because of noise intrinsic to the system and because of changes in internal parameters of the system, the response changes with time. The overall response is sum of pure (steady) response and disturbances. Because of fluctuating nature of disturbances, the system response becomes free running.

The system together with appropriate control strategy constitutes a control system. Although a control system can assume different shapes depending on control situation but common to them all is their function to manage the system's operation so that the overall response approximates the commanded behaviour. The controller generates actuating signals to be inputed to the system to produce desired output as shown in Fig. 1.2. Some of the system inputs are accessible and some are generally not available. The inaccessible inputs are often disturbances to the system. The double lines in the figure indicate that several signals of each type are involved. Systems with one input and one output are called single input single output (SISO) systems and those with more than one input and one output, are called multi input multi output (MIMO) systems.

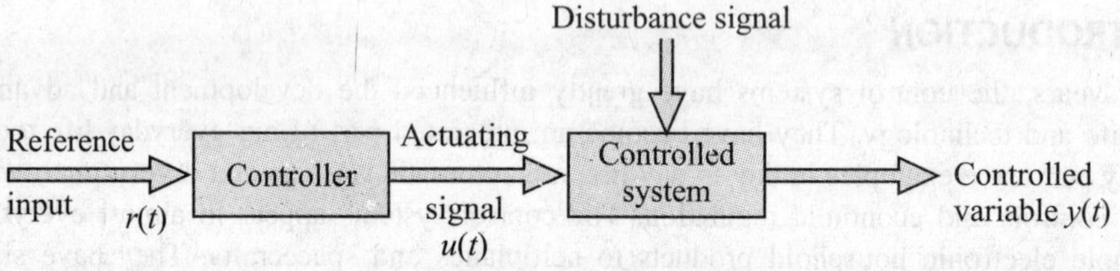

Fig. 1.2: Open loop control system

The control systems are broadly classified into the following:

Open loop control system

The control system wherein the control inputs are not influenced by system outputs is termed as open loop control system as shown in Fig. 1.2. Such systems have no feedback around them. The open loop control system are the **simplest** and **economical**. Yet they are not generally preferred because they are usually **inaccurate** and **unreliable**. Such systems are unable to adapt to variations in response that might occur due to variations either in internal behaviour of system or in environmental conditions *i.e.* external disturbances. The Control adjustment of open loop control system is shown in Fig. 1.2. The reference input $r(t)$ or command is given to controller to generate actuating signal $u(t)$ which in turn controls the system's operation so as to enable it to produce the response that is required to follow certain prescribed standards.

The fact that output has no effect upon control action eases out the identification of open loop control system. Some examples of open loop control system are as follows:

(*i*) Execution of programme on computer.

(*ii*) Separately excited d.c. generator.

(*iii*) Washing machine

(*iv*) Toaster to produce desired darkness of toasted bread.

(*v*) Traffic Control System.

Closed loop control system (feedback control system)

A link or feedback path established between input and output provides more satisfactory control. Note that such a link is absent in open loop control system.

Closed loop control systems are class of control systems wherein, for an accurate control, the actual output $y(t)$ is fedback, compared with reference input $r(t)$ or set point (desired output) and error signal proportional to difference between these two, is sent to the controller to correct the error $e(t)$. The functional layout of closed loop control system is shown in Fig. 1.3.

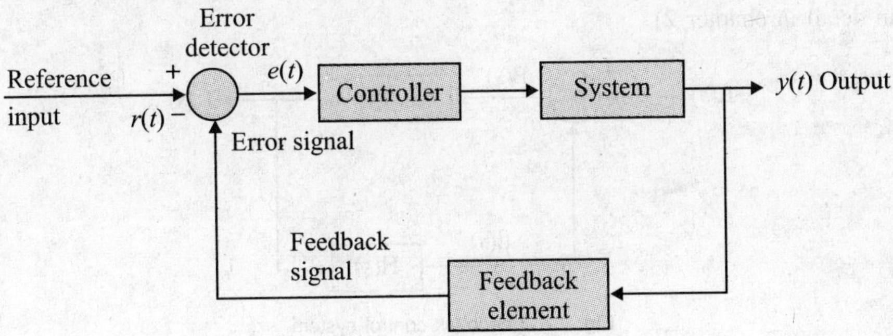

Fig. 1.3: Closed loop control system

Some examples of closed loop control systems are as follows:

(*i*) Human being (biological system).

(*ii*) Automobile driving system.

(*iii*) Constant d.c. voltage generator.

(*iv*) Automatic editing of text by computer.

The feedback action introduced in control system makes the response of the system relatively insensitive to external disturbances or internal variations in system parameters. It is therefore possible to use relatively less accurate and inexpensive components to achieve the desired result. This is impossible in case of open loop configuration. From the view point of stability (to be discussed in detail in chapter 4) the open loop systems are easier to build since stability does not pose any big problem. But stability is always a major problem in closed loop case since such systems sometimes tend to overcorrect the error that might result in instability.

Open loop systems are advised in cases where input signals are known ahead of time and where there is no unpredictable variations in system parameters. The closed loop systems become advantageous in a situation where unpredictable external disturbances are anticipated.

A feedback control system where controlled outputs are mechanical positions are derivatives thereof is called **servo system**.

1.2 EFFECTS OF FEEDBACK

A feedback appropriately established in a control system, brings about improvement in system performance. Some of the significant changes that are brought about by feedback in performance characteristics of the system, are as follows:

(*a*) Overall gain

(*b*) Stability

(*c*) Sensitivity

(*d*) Disturbances.

(a) Overall gain

Consider a model of feedback control system shown in Fig. 1.4 where R(*s*) is input, C(*s*) is output, E(*s*) is error variable and β(*s*) is feedback variable, all are transform variables. (The term transfer function is dealt in detail in section 1.8 of this chapter and block diagram algebra is discussed in detail in chapter 2).

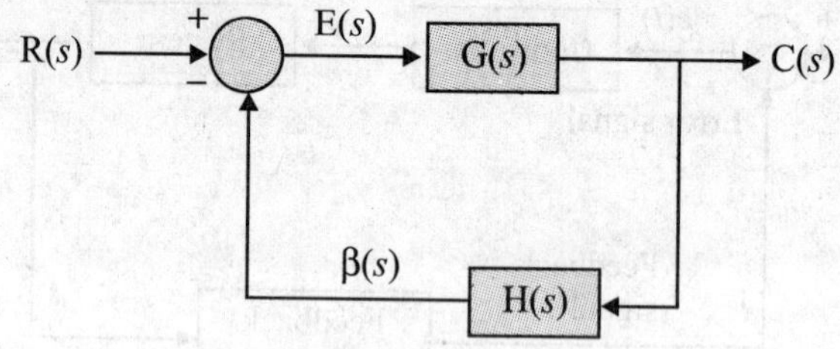

Fig. 1.4: Feedback control system

From Fig. 1.4, we have

$$C(s) = E(s) \cdot G(s)$$

where
$$E(s) = R(s) - \beta(s)$$

and
$$\beta(s) = C(s)\, H(s)$$

then
$$C(s) = [R(s) - C(s)\, H(s)]\, G(s)$$

and
$$\frac{C(s)}{R(s)} = \frac{G(s)}{1 + G(s)\, H(s)} = T(s) \qquad ...(1.1)$$

G(*s*) = C(*s*)/E(*s*) is called forward path transfer function, H(*s*) feedback path transfer function, G(*s*)H(*s*) = β(*s*)/E(*s*) open loop transfer function and T(*s*) = G(*s*)/[1 + G(*s*)H(*s*)] closed loop transfer function.

The configuration of Fig. 1.4 has negative feedback. It is obvious from (1.1) that generally gain of closed loop configuration is reduced by a factor of (1+ G(*s*)H(*s*)) as compared to open loop system as shown in Fig. 1.5 where G(*s*) = C(*s*)/R(*s*). However, depending on whether feedback is negative or positive in nature (1 + G(*s*)H(*s*)) may be greater than one or less than one and overall gain may decrease or increase. Since G(*s*) and H(*s*) are frequency dependent, it is also possible that overall gain may increase in a particular range of frequencies and decrease in some other range of frequencies.

(b) Stability

In a non rigorous sense, a system is said to be stable if it provides finite response for finite input. Any system providing infinite response for finite input, is said to be unstable. An unstable system is of no practical interest. In the expression for overall gain (1.1) if G(*s*) H(*s*) = – 1, the overall gain will turn out to be infinite, meaning infinite response and original stable system might turn out to be unstable. Stability in detail will be discussed later in chapter 4.

(c) Sensitivity

The prime objective of feedback in control system is to reduce sensitivity of system to parameter variations or to improve accuracy of system. The term sensitivity is truly the measure of effectiveness of feedback in reducing the influence of parameter variations on system performance.

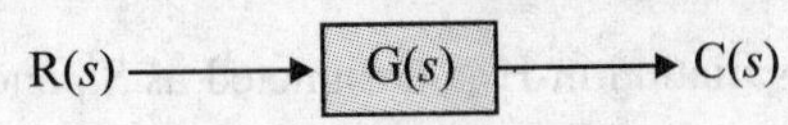

Fig. 1.5: Open loop system

Consider open loop configuration as shown in Fig. 1.5, where

$$C(s) = R(s)\,G(s)$$

Due to parameter variation let $G(s)$ change to $G(s) + \Delta G(s)$ where it is assumed that $|\Delta G(s)| << |G(s)|$. Correspondingly let $C(s)$ change to $C(s) + \Delta C(s)$, then

$$C(s) + \Delta C(s) = [G(s) + \Delta G(s)]\,R(s) = G(s).R(s) + \Delta G(s)\,R(s)$$

and
$$\Delta C(s) = \Delta G(s)\cdot R(s) \qquad ...(1.2)$$

Now consider closed loop configuration of Fig.1.4 let $G(s)$ change to $G(s) + \Delta G(s)$ again due to parameter variation and correspondingly $C(s)$ to $C(s) + \Delta C(s)$. Assuming $|G(s)| >> |\Delta G(s)|$ and ignoring change $\Delta G(s)$ in denominator, equation (1.1) takes the form as follows:

$$C(s) + \Delta C(s) = \frac{G(s)\,R(s)}{1 + G(s)\,H(s)} + \frac{\Delta G(s)\cdot R(s)}{1 + G(s)\,H(s)}$$

and
$$\Delta C(s) = \frac{\Delta G(s)\,R(s)}{1 + G(s)\,H(s)} \qquad ...(1.3)$$

Comparing (1.2) and (1.3) it is obvious that change in response $\Delta C(s)$ for the same change in $G(s)$ in closed loop system is reduced by a factor of $(1 + G(s)\,H(s))$ as compared to change in response in open loop system. **Note** that the factor $1 + G(s)\,H(s)$ is much greater than unity in most practical systems.

Recalling equation (1.1) as

$$T = \frac{C}{R} = \frac{G}{1 + GH}\text{; (the variable } s \text{ is omitted for ease)}$$

the sensitivity of T to variation in G is designated as S_G^T and defined as:

$$S_G^T \triangleq \frac{\%\text{ change in T}}{\%\text{ change in G}} = \frac{\partial T/T}{\partial G/G} = \frac{\partial T}{\partial G} \times \frac{G}{T}$$

where
$$\frac{\partial T}{\partial G} = \frac{(1 + GH) - GH}{(1 + GH)^2} = \frac{1}{(1 + GH)^2},$$

So,
$$S_G^T = \frac{1}{(1 + GH)^2} \times \frac{G\,(1 + GH)}{G}$$

or
$$S_G^T = \frac{1}{1 + GH} \qquad ...(1.4)$$

Similarly, the sensitivity of open loop system where $T = G$, is

$$S_G^T = \frac{\partial T}{\partial G} \times \frac{G}{T} = 1 \qquad ...(1.5)$$

Note (1.4) and (1.5) to conclude that sensitivity is reduced by factor of $(1 + GH)$ in closed loop compared to open loop. But this improvement in sensitivity is achieved at the cost of loss in system gain by the same factor.

Similarly, sensitivity of T to variation in H is designated as S_H^T and defined as;

$$S_H^T \triangleq \frac{\%\text{ change in T}}{\%\text{ change in H}} = \frac{\partial T/T}{\partial H/H} = \frac{\partial T}{\partial H} \times \frac{H}{T}$$

where
$$\frac{\partial T}{\partial H} = \frac{\partial}{\partial H}[G(1+GH)^{-1}] = -G(1+GH)^{-2}G = -\frac{G^2}{(1+GH)^2}$$

So,
$$S_H^T = \frac{-G^2}{(1+GH)^2} \times \frac{H}{G} \times (1+GH)$$

$$S_H^T = \frac{-GH}{1+GH} \qquad ...(1.6)$$

Note (1.4) and (1.6) to conceive the following significant points

(*i*) For large values of GH, S_H^T approaches unity whereas large GH is desirable so as to keep S_G^T small.

(*ii*) Since $S_H^T\big|_{GH>>1} \rightarrow 1$, variation in H directly affects the response of the system and in general more than that due to variation in G, there stands a need to choose accurate feedback components that do not vary with environmental changes. In closed loop systems, H(*s*) is usually a sensor and constituted of elements operating at low power. G(*s*) is constituted of high power operating elements. Therefore, selection of accurate H(*s*) at low power is far less costly than selection of G(*s*) at high power to meet prescribed system specifications.

In order to have still more insight into what has been explained just above, consider the feedback system representing an amplifier over certain range of frequencies as shown in Fig. 1.6.

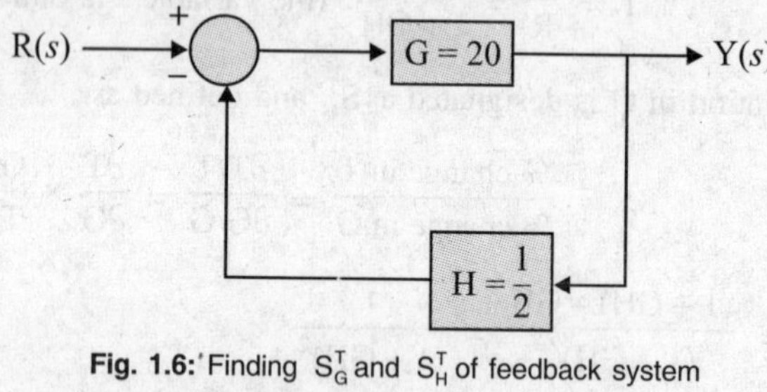

Fig. 1.6: Finding S_G^T and S_H^T of feedback system

Use (1.1), (1.4) and (1.6) to get the following

$$T = \frac{G}{1+GH} = \frac{20}{1+20\times\frac{1}{2}} = \frac{20}{11}$$

$$S_G^T = \frac{1}{1+GH} = \frac{1}{11}$$

$$S_H^T = \frac{-GH}{1+GH} = \frac{-10}{11}$$

This is to say that, with feedback, the closed loop transfer function T changes only 1/11 as much with small changes in G as it would without feedback.

Also changes in T with changes in H is much more than that with changes in G. The minus sign in S_H^T indicates that T increases with decrease in H and vice versa.

The sensitivity of T to variations in any other system parameter can also be computed using exactly similar computational routine for example, consider the configuration shown in Fig. 1.7.

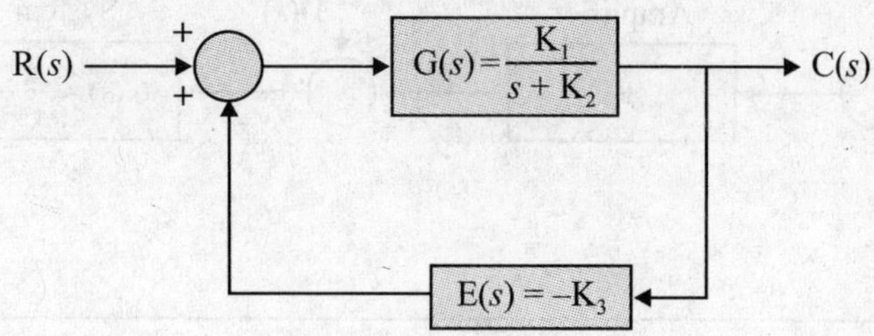

Fig. 1.7: Finding sensitivity of feedback system to changes in system parameters

Let us evaluate, $S_{K_1}^T$, $S_{K_2}^T$ and $S_{K_3}^T$ *i.e.,* sensitivities of T = C/R to small changes in K_1, K_2 and K_3 about their nominal values $K_1 = 1$, $K_2 = 2$ and $K_3 = 3$.

Use (1.1) to compute

$$T(s) = \frac{C(s)}{R(s)} = \frac{G}{1+GH} = \frac{K_1}{s + K_2 + K_1K_3}$$

$$S_{K_1}^T = \frac{\partial T}{\partial K_1} \times \frac{K_1}{T} = \frac{s + K_2}{s + K_2 + K_1K_3}$$

$$S_{K_2}^T = \frac{-K_2}{s + K_2 + K_1K_3}$$

and
$$S_{K_3}^T = \frac{-K_1K_3}{s + K_2 + K_1K_3}$$

Substituting nominal values of K_1, K_2 and K_3, we have

$$S_{K_1}^T = \frac{s+2}{s+5}$$

$$S_{K_2}^T = \frac{-2}{s+5}$$

$$S_{K_3}^T = \frac{-3}{s+5}$$

Note that sensitivities are, in general, functions of complex variable *s*. So, there stands a need to compute sensitivities over entire frequency range in which the input has significant frequency components. For example,

$$\left|S_{K_2}^T\right| = \frac{2}{\sqrt{25+\omega^2}}$$

monotonically decreases from 0.4 at $\omega = 0$ to 0 at $\omega = \infty$.

(d) Disturbances

The control systems are often subjected to unwanted inaccessible disturbance signals. For example, sudden gusts of wind tending to change dynamics of radar antenna, noise generated in electronic amplifiers, etc. Let us investigate how feedback assists in mitigating the effect of these disturbances on system response. Consider the system of Fig. 1.8.

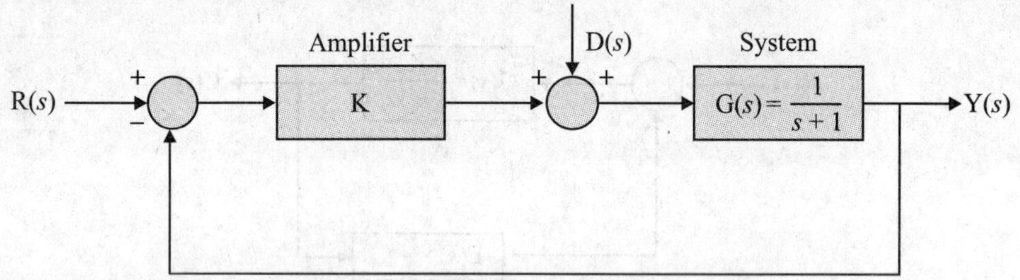

Fig. 1.8: Effect of disturbance on feedback system

Using the procedure as explained to derive equation (1.1), the transfer function relating Y(s) to D(s) assuming R(s) = 0, is as follows:

$$T_D(s) = \frac{Y(s)}{D(s)} = \frac{\dfrac{1}{(s+1)}}{1+\dfrac{K}{s+1}} = \frac{1}{s+1+K} \qquad \text{...(1.7)}$$

The response Y(s) is related to D(s) as

$$Y(s) = \frac{D(s)}{s+1+K}$$

It is easy to see that system response due to disturbance can be made arbitrarily small by choosing K sufficiently large.

Consider the following illustrative examples for exposure to computational routine of sensitivity.

Example 1: A negative feedback control system has forward path transfer function $G(s) = \dfrac{K}{s\,(s+1)}$ and feedback path transfer function H(s) = 5. Determine sensitivity of closed loop transfer function with respect to G and H at ω = 1 rad./sec. Assume K = 10 (nominal value).

Solution:

$$S_G^T = \frac{1}{1+GH} = \frac{1}{1+\dfrac{5K}{s\,(s+1)}} = \frac{s\,(s+1)}{s^2+s+5K}$$

$$S_G^T(j\omega) = \frac{-\omega^2+j\omega}{-\omega^2+j\omega+5K}$$

$$\left|S_G^T(j\omega)\right|_{K=10} = \sqrt{\frac{\omega^2+\omega^4}{(50-\omega^2)^2+\omega^2}}$$

$$\left|S_G^T(j\omega)\right|_{\omega=1} = 0.02885$$

$$S_H^T(s) = \frac{-GH}{1+GH} = \frac{-\dfrac{5K}{s(s+1)}}{1+\dfrac{5K}{s(s+1)}} = \frac{-5K}{s^2+s+5K}$$

$$S_H^T(j\omega) = \frac{-5K}{(5K-\omega^2)+j\omega}$$

$$\left|S_H^T(j\omega)\right|_{\substack{\omega=1\\K=10}} = \frac{50}{\sqrt{(50-1)^2+1^2}} = 1.020196.$$

Example 2: A closed loop configuration is given in Fig. 1.9 where $\beta_1 = 4$ and $\beta_2 = 9$. Calculate S_α^T; $T = \dfrac{Y(s)}{X(s)}$. Find α such that S_α^T equals 0.2 in steady state.

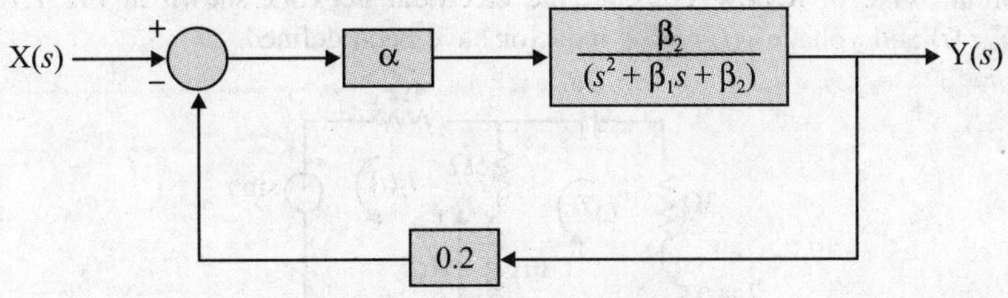

Fig. 1.9: Closed loop configuration

Solution:

$$T(s) = \frac{\dfrac{\alpha\beta_2}{(s^2+\beta_1 s+\beta_2)}}{1+\dfrac{0.2\,\alpha\beta_2}{(s^2+\beta_1 s+\beta_2)}} = \frac{\alpha\beta_2}{s^2+\beta_1 s+\beta_2+0.2\,\alpha\beta_2}$$

Substituting given values of β_1 and β_2 ;

$$T(s) = \frac{9\alpha}{s^2+4s+9+1.8\alpha}$$

$$S_\alpha^T = \frac{\partial T}{\partial \alpha} \times \frac{\alpha}{T}$$

$$= \frac{9(s^2+4s+9+1.8\alpha) - 9\alpha\times 1.8}{(s^2+4s+9+1.8\alpha)^2} \times \frac{\alpha(s^2+4s+9+1.8)}{9\alpha}$$

$$= \frac{(s^2+4s+9)}{(s^2+4s+9+1.8\alpha)}$$

Under steady state conditions ($s = 0$)

$$S_{\alpha}^{T}\Big|_{s=0} = \frac{9}{9 + 1.8\alpha} = 0.2 \text{ (given)}$$

So, $\alpha = 20.$

1.3 SYSTEMS MODELLING

The first step involved in design and analysis of a continuous dynamical system is to develop differential equation model for the system. The Laplace transform of differential equation (explained little later) will yield transfer function model of system, of course all initial conditions are assumed to be zero therein. The components of which the control system is constituted, are usually electrical, electronic, mechanical or electromechanical in nature.

Electrical systems

It is deemed that the reader is well conversant with Kirchoff's two laws: KCL and KVL by which models of electrical networks are governed. So, electrical networks are not dealt in detail here. However for the sake of review, consider the electrical network shown in Fig 1.10 where loop currents $i_1(t)$, $i_2(t)$ and voltage $v(t)$ across inductor have been defined.

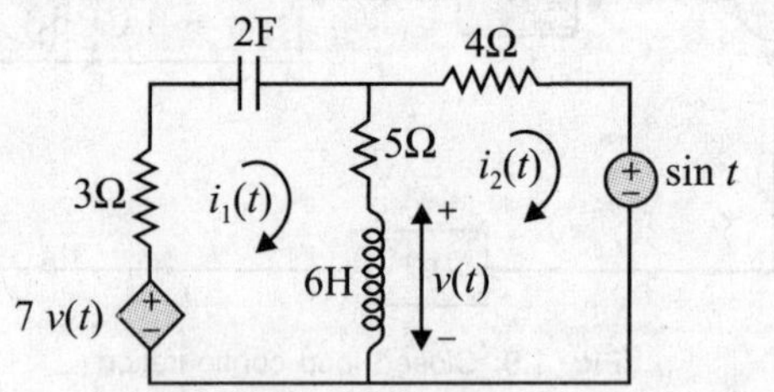

Fig. 1.10: Electrical network

The simultaneous loop equations obtained using KVL are as follows:

$$7\,v(t) - 3i_1(t) - \frac{1}{2}\int_0^t i_1(t)\,dt - 5[i_1(t) - i_2(t)] - 6\frac{d}{dt}[i_1(t) - i_2(t)] = 0; \qquad (i_1 \text{ loop})$$

$$6\frac{d}{dt}[i_1(t) - i_2(t)] - 5\,[i_2(t) - i_1(t)] - 4i_2(t) - \sin t = 0 \qquad (i_2 \text{ loop})$$

$$6\frac{d}{dt}\,[i_1(t) - i_2(t)] = V(t)$$

The Laplace transform of above equations while initial conditions assumed to be zero, are as follows:

$$7V(s) - \left[6s + 8 + \frac{1}{2s}\right] I_1(s) + 6(s + 5)\,I_2(s) = 0$$

$$6\,(s + 5)\,I_1(s) - 6\,(s + 9)\,I_2(s) - \frac{1}{s^2 + 1} = 0$$

$$6sI_1(s) - 6sI_2(s) - V(s) = 0$$

The electrical system models are, in general constituted of resistors, inductors capacitors, current sources and voltage sources. These elements together with voltage current relations therein are listed in Table 1.1.

TABLE 1.1: Electrical elements and voltage current relations

Electrical elements	*Symbol*	*Voltage-current relation*
Independent voltage source	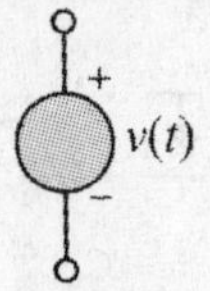	$v(t)$ is expressed as a function of time
Independent current source	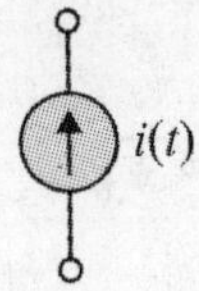	$i(t)$ is expressed as a function of time
Dependent voltage source		$v(t)$ is expressed as a function of other voltage or current in the network
Dependent current source		$i(t)$ is expressed as a function of other voltage or current in the network
Resistor		$v = iR$
Inductor		$v = L\frac{di}{dt}$ $i = \frac{1}{L}\int v\,dt$
Capacitor	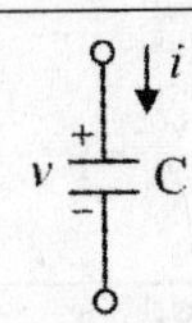	$v = \frac{1}{C}\int i\,dt$ $i = C\frac{dv}{dt}$

Mechanical system

The dynamics of a mechanical system may be translational or rotational or both. The variables that describe translational dynamics are displacement (x), velocity ($\dot{x}$) and acceleration ($\ddot{x}$). Mass, spring and damper are three mechanical elements whose symbols and force position relationships are as given in Table 1.2.

TABLE 1.2: Mechanical elements and force-position relationship

Mechanical elements	*Symbol*	*Force-position relationship*
Mass	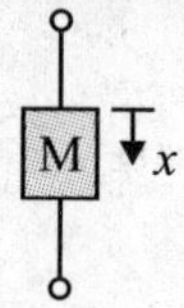	Inertial force $F_m = M\ddot{x}$ M = mass
Damper/Dashpot	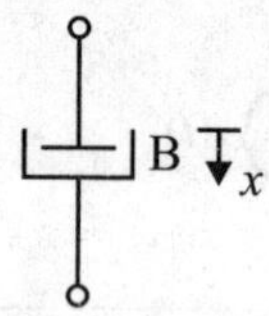	Damping force $F_B = B\dot{x}$ B = Damping coefficient
Spring	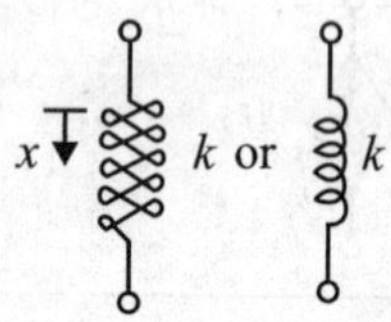	Spring force $F_k = kx$ k = Spring constant

Similarly rotational dynamics is characterised by three variables θ (angular displacement), $\dot{\theta}$ (angular velocity) and $\ddot{\theta}$ (angular acceleration). The torque-angular position relationship and symbols of these three elements are given in Table 1.3.

TABLE 1.3: Mechanical elements and torque-position relationship

Mechanical elements	*Symbol*	*Torque-angular position relation*
Moment of Inertia		Inertial torque $\tau_J = J\ddot{\theta}$ J = moment of inertial (MI)
Damper/Dashpot	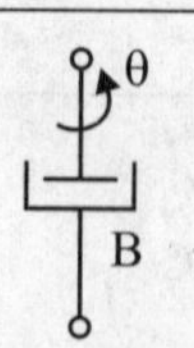	Damping torque $\tau_B = B\dot{\theta}$ B = Damping coefficient
Spring	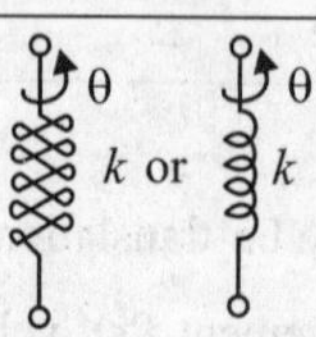	Spring torque $\tau_k = k\theta$ k = Spring constant

1.4 MATHEMATICAL MODELLING OF TRANSLATIONAL/ROTATIONAL MECHANICAL SYSTEM

A differential equation or a set of differential equation(s) characterising the translational/rotational dynamics of systems is/are called Mathematical model. The steps involved in modelling mechanical system are outlined below.

1. Identify various displacements (linear/angular) from equilibrium position in mechanical system. Each displacement corresponds to a node in mechanical network. All points on a rigid mass are considered as same node and one terminal of mass is always connected to the ground, reason being the velocity (or displacement) of a mass is always referred to earth.
2. Draw mechanical network by connecting together terminals of elements which change the same displacement.
3. Write force/torque balance equations for each node by applying D′ Alembert's principle of mechanics. The D′ Alembert's principle of mechanics equates the algebraic sum of forces at a node to zero in mechanical network and is analogous to KCL that equates algebraic sum of currents at a node to zero in electrical network.
4. The corresponding algebraic equations as a function of complex variable s are obtained by taking Laplace transform of differential equations obtained using steps 1 to 3. By little manipulation of algebraic equations so obtained, one can easily develop transfer function model of system relating output variable to input variable.

The following examples will illustrate the procedure.

Example 3: Write simultaneous differential equations for the translational mechanical system of Fig. 1.11 and Laplace transform the equations assuming all initial conditions to be zero.

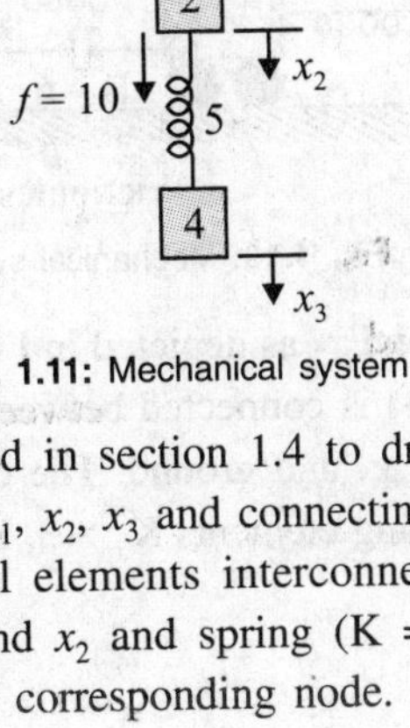

Fig. 1.11: Mechanical system

Solution: Use the procedure outlined in section 1.4 to draw mechanical network as shown in Fig. 1.12 by first choosing three nodes x_1, x_2, x_3 and connecting the masses between ground and the corresponding node. Fit the mechanical elements interconnecting two nodes in the network for example damper (B = 7) between x_1 and x_2 and spring (K = 5) between x_2 and x_3. Then fit the remaining elements between ground and corresponding node.

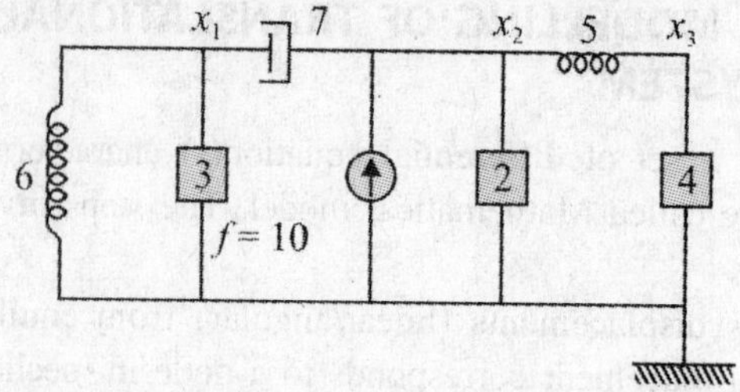

Fig. 1.12: Mechanical network of system of Fig. 1.11

While equating the algebraic sum of internal forces to that of externally applied forces, write differential equations describing the system dynamics as follows.

$$3\ddot{x}_1 + 6x_1 + 7(\dot{x}_1 - \dot{x}_2) = 0; \qquad \text{(node } x_1\text{)}$$

or $$3\ddot{x}_1 + 7\dot{x}_1 + 6x_1 - 7\dot{x}_2 = 0 \qquad \text{...(1.8)}$$

$$2\ddot{x}_2 + 7(\dot{x}_2 - \dot{x}_1) + 5(x_2 - x_3) = 10; \qquad \text{(node } x_2\text{)}$$

or $$2\ddot{x}_2 + 7\dot{x}_2 + 5x_2 - 7\dot{x}_1 - 5x_3 = 10 \qquad \text{...(1.9)}$$

and $$4\ddot{x}_3 + 5(x_3 - x_2) = 0; \qquad \text{(node } x_3\text{)}$$

or $$4\ddot{x}_3 + 5x_3 - 5x_2 = 0 \qquad \text{...(1.10)}$$

Laplace transform of (1.9), (1.10) and (1.11) yields respectively

$$(3s^2 + 7s + 6)\, X_1(s) - 7s\, X_2(s) = 0 \qquad \text{...(1.11)}$$

$$(2s^2 + 7s + 5)\, X_2(s) - 7s\, X_1(s) - 5X_3(s) = \frac{10}{s} \qquad \text{...(1.12)}$$

$$(4s^2 + 5)\, X_3(s) - 5X_2(s) = 0 \qquad \text{...(1.13)}$$

Example 4: Develop differential equation model for mechanical system shown in Fig. 1.13.

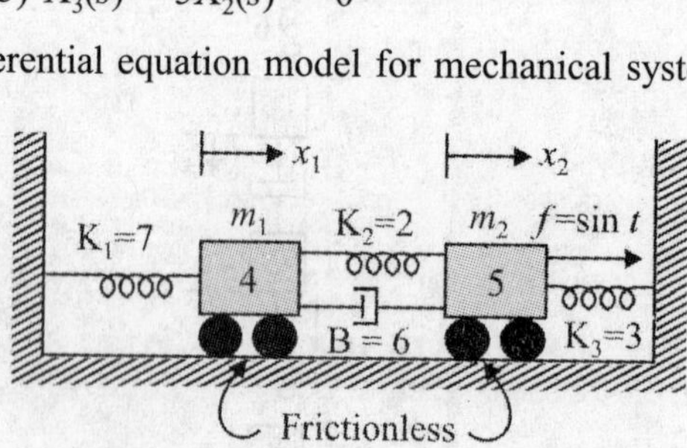

Fig. 1.13: Mechanical system

Solution: Choose two nodes x_1 and x_2 as depicted in Fig. 1.13 to draw the mechanical network as shown in Fig.1.14. The mass ($m_1 = 4$) is connected between node x_1 and ground. Similarly, the mass ($m_2 = 5$) is connected between node x_2 and ground. The elements B = 6 and $K_2 = 2$ are connected between nodes x_1 and x_2. The remaining elements $K_3 = 3$, $K_1 = 7$ and $f = \sin t$ are connected between ground and corresponding nodes.

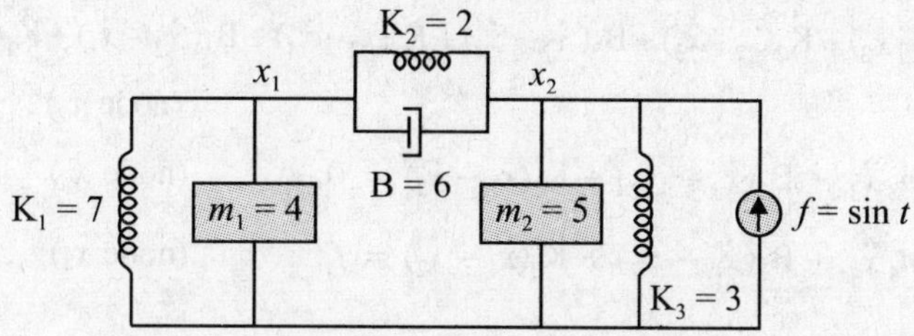

Fig. 1.14: Mechanical network of system of Fig. 1.13

While equating the algebraic sum of internal forces to that of externally applied forces the differential equations for nodes x_1 and x_2 are written as follows:

$$m_1 \ddot{x}_1 + B(\dot{x}_1 - \dot{x}_2) + K_2(x_1 - x_2) + K_1 x_1 = 0 \qquad \text{(node } x_1\text{)}$$

and $$m_2 \ddot{x}_2 + B(\dot{x}_2 - \dot{x}_1) + K_2(x_2 - x_1) + K_3 x_2 = \sin t \qquad \text{(node } x_2\text{)}$$

Substitute given values to get

$$4\ddot{x}_1 + 6\dot{x}_1 + 9x_1 - 6\dot{x}_2 - 2x_2 = 0$$

and $$5\ddot{x}_2 + 6\dot{x}_2 + 5x_2 - 2x_1 - 6\dot{x}_1 = \sin t$$

Example 5: The human body is often modelled by springs, masses and dampers. For the model of seated body with applied force f, Fig. 1.15, find the system equations.

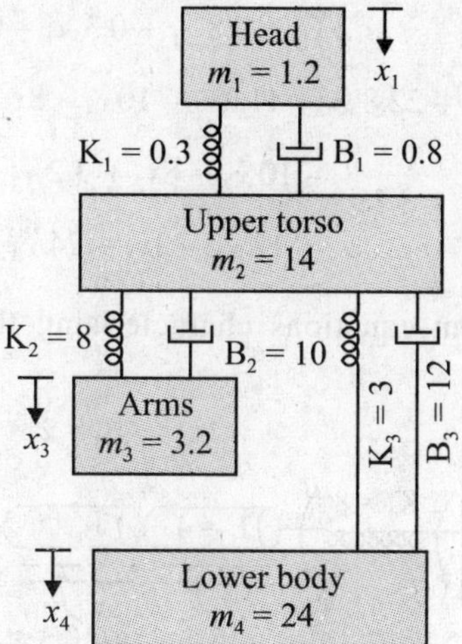

Fig. 1.15: Model of seated human body

Solution: Having identified four displacements of head, upper torso, arms and lower body in mechanical system of Fig. 1.15, four nodes x_1, x_2, x_3 and x_4 are correspondingly chosen and the mechanical network as explained in example 4 is drawn in Fig. 1.16. The system equations can be now written as follows:

$$m_1 \ddot{x}_1 + B_1(\dot{x}_1 - \dot{x}_2) + K_1(x_1 - x_2) = 0 \qquad \text{(node } x_1\text{)} \qquad ...(1.14)$$

$$m_2\ddot{x}_2+B_2(\dot{x}_2-\dot{x}_3)+K_2(x_2-x_3)+B_3(\dot{x}_2-\dot{x}_4)+K_3(x_2-x_4)+B_1(\dot{x}_2-\dot{x}_1)+K_1(x_2-x_1)=0$$

(node x_2) ...(1.15)

$$m_3\ddot{x}_3 + B_2(\dot{x}_3 - \dot{x}_2) + K_2(x_3 - x_2) = 0; \quad \text{(node } x_3) \quad ...(1.16)$$

$$m_4\ddot{x}_4 + B_3(\dot{x}_4 - \dot{x}_2) + K_3(x_4 - x_2) = f; \quad \text{(node } x_4) \quad ...(1.17)$$

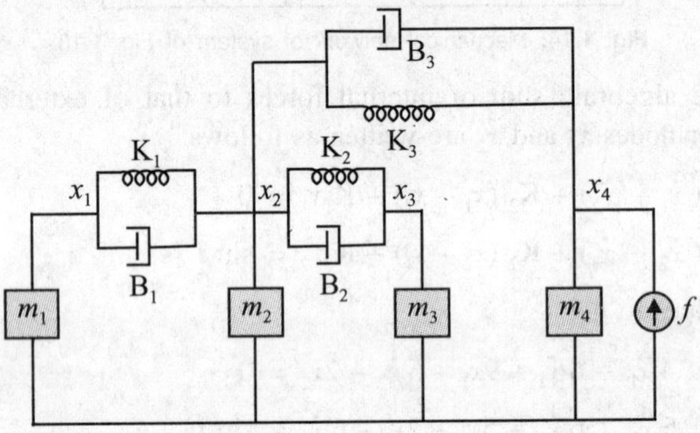

Fig. 1.16: Mechanical network of system of Fig. 1.15

Fitting element values and rearranging (1.14), (1.15), (1.16) and (1.17), the following equations result respectively.

$$1.2\ddot{x}_1 + 0.8\dot{x}_1 + 0.3x_1 - 0.8\dot{x}_2 - 0.3x_2 = 0 \quad ...(1.18)$$

$$-0.8\dot{x}_1 - 0.3x_1 + 14\ddot{x}_2 + 22.8\dot{x}_2 + 11.3x_2 - 10\dot{x}_3 - 8x_3 - 12\dot{x}_4 - 3x_4 = 0 \quad ...(1.19)$$

$$-10\dot{x}_2 - 8x_2 + 3.2\ddot{x}_3 + 10\dot{x}_3 + 8x_3 = 0 \quad ...(1.20)$$

$$-12\dot{x}_2 - 3x_2 + 24\ddot{x}_4 + 12\dot{x}_4 + 3x_4 = 5t^2 \quad ...(1.21)$$

Example 6: Find the system equations characterising the mechanical system shown in Fig. 1.17.

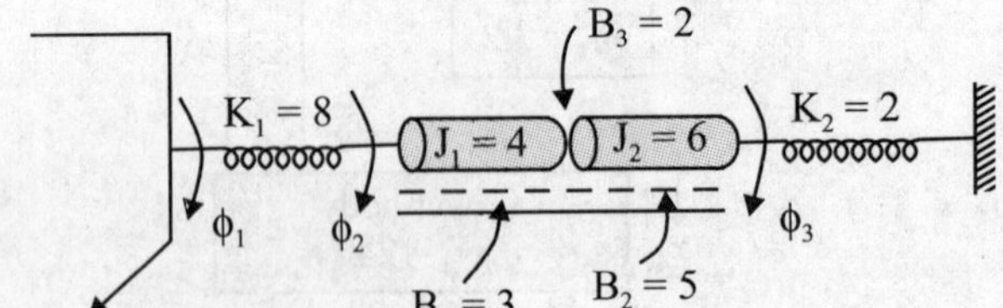

Fig. 1.17: Mechanical system of example 6

Solution: For the system shown in Fig. 1.17, the corresponding mechanical network is drawn in Fig. 1.18. ϕ_1, ϕ_2 and ϕ_3 identified as three nodes in mechanical network are three angular displacements from equilibrium. J_1, J_2 are connected between ground and corresponding nodes ϕ_2 and ϕ_3 respectively. The remaining self explanatory connections are also shown in Fig. 1.18.

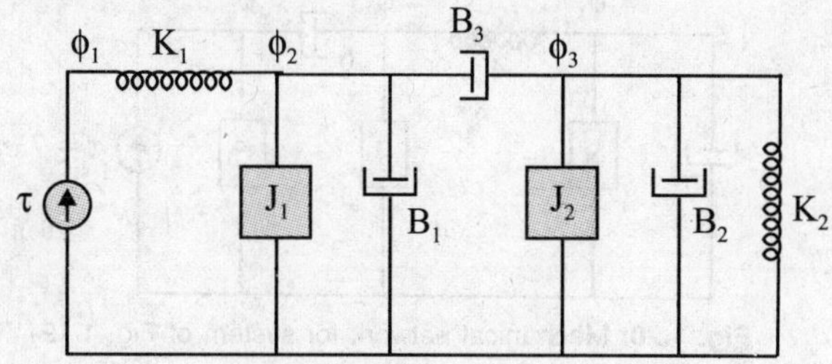

Fig. 1.18: Mechanical network for system of Fig. 1.17

Use mechanical network to write system equations as follows:

$$K_1(\phi_1 - \phi_2) = \tau; \qquad \text{(node } \phi_1) \qquad ...(1.22)$$

$$J_1\ddot{\phi}_2 + B_1\dot{\phi}_2 + B_3(\dot{\phi}_2 - \dot{\phi}_3) + K_1(\phi_2 - \phi_1) = 0; \qquad \text{(node } \phi_2) \qquad ...(1.23)$$

$$J_2\ddot{\phi}_3 + B_2\dot{\phi}_3 + K_2\phi_3 + B_3(\dot{\phi}_3 - \dot{\phi}_2) = 0; \qquad \text{(node } \phi_3) \qquad ...(1.24)$$

Fit the values of elements and rearrange to get the following equations describing the dynamics of given system.

$$8\phi_1 - 8\phi_2 = 10\, e^{-t/2} \qquad ...(1.25)$$

$$-8\phi_1 + 4\ddot{\phi}_2 + 3\dot{\phi}_2 + 2\dot{\phi}_2 + 8\phi_2 - 2\dot{\phi}_3 = 0 \qquad ...(1.26)$$

$$-2\dot{\phi}_2 + 6\ddot{\phi}_3 + 5\dot{\phi}_3 + 2\phi_3 + 2\dot{\phi}_3 = 0 \qquad ...(1.27)$$

Example 7: Write simultaneous equations in *s* domain for rotational system shown in Fig. 1.19. Assume all initial conditions to be zero.

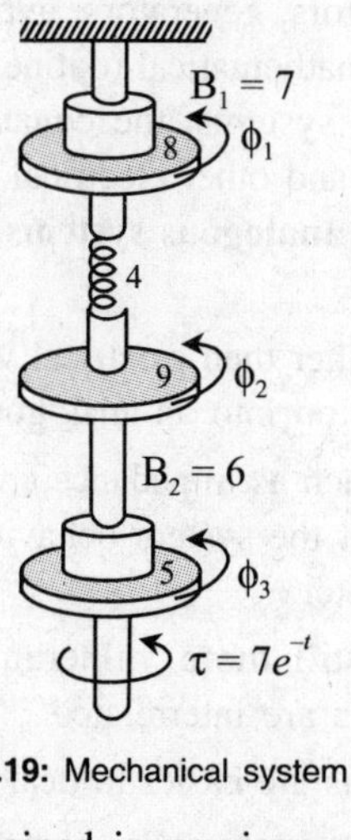

Fig. 1.19: Mechanical system

Solution: Using the procedure as explained in previous examples, three angular displacements ϕ_1, ϕ_2 and ϕ_3 have been identified as depicted in Fig. 1.19 and the corresponding network is drawn in Fig. 1.20.

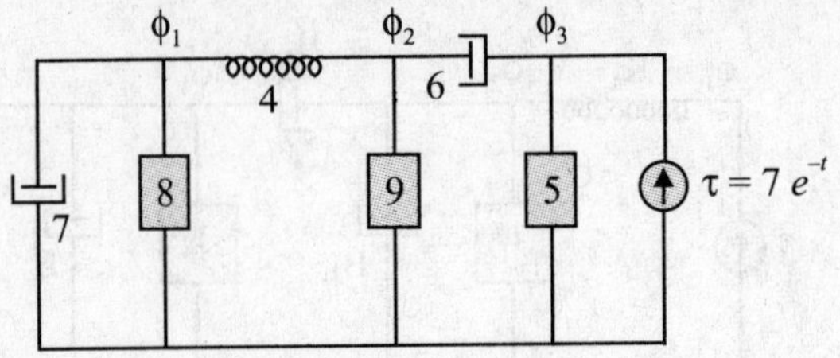

Fig. 1.20: Mechanical network for system of Fig. 1.19

The differential equations are as follows

$$8\ddot{\phi}_1 + 7\dot{\phi}_1 + 4(\phi_1 - \phi_2) = 0 \qquad ...(1.28)$$

$$9\ddot{\phi}_2 + 4(\phi_2 - \phi_1) + 6(\dot{\phi}_2 - \dot{\phi}_3) = 0 \qquad ...(1.29)$$

$$5\ddot{\phi}_3 + 6(\dot{\phi}_3 - \dot{\phi}_2) = 7e^{-t} \qquad ...(1.30)$$

The Laplace transform (with zero initial conditions) of (1.28), (1.29) and (1.30) gives system equations in s domain as follows.

$$(8s^2 + 7s + 4)\phi_1(s) - 4\phi_2(s) = 0 \qquad ...(1.31)$$

$$-4\phi_1(s) + (9s^2 + 6s + 4)\phi_2(s) - 6s\phi_3(s) = 0 \qquad ...(1.32)$$

$$-6s\phi_2(s) + (5s^2 + 6s)\phi_3(s) = \frac{7}{(s+1)} \qquad ...(1.33)$$

1.5 ANALOGOUS SYSTEMS

Numerous electromechanical devices are often encountered in engineering applications. Some of the examples are solenoids, actuators, motors, generators, gyroscopes, accelerometers and loud speakers.

In analysis of linear systems, the mathematical routine to obtain solution to given set of equations, does not depend upon what physical systems, the equations represent. The two different physical systems (for example one mechanical and other electrical in nature) which can be described by same set of differential equations are called **analogous systems** and will produce same response to a given excitation.

While dealing with the systems other than electrical we have following distinct advantages if we transform the system under consideration into an analogous electrical network.

1. Electrical network theory such as impedance concept and various network theorems can be applied to study and predict the system behaviour such as resonance, pass band, damping coefficient, time constant, etc.
2. This technique becomes still more vulnerable when we deal with electromechanical systems wherein phenomena are interrelated.
3. In general, electrical systems are easier to deal with experimentally.

Force-voltage (*f-v*) analogy

Consider a mechanical system and an electrical system shown in Figs. 1.21(*a*) and 1.21(*b*) respectively. The corresponding mechanical network for mechanical system of Fig. 1.21(*a*) is drawn in Fig. 1.21(*c*) where node x represents displacement of mass.

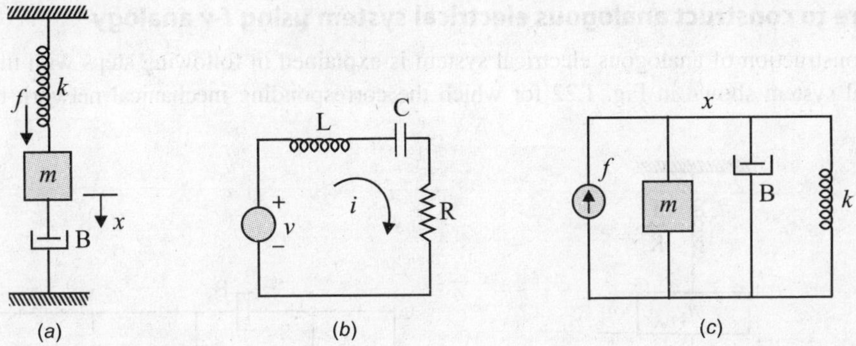

Fig. 1.21: (*a*) Mechanical system, (*b*) Analogous electrical system, (*c*) Mechanical network

The differential equation describing the dynamics of mechanical system can be written as

$$m\ddot{x} + B\dot{x} + kx = f \qquad ...(1.34)$$

where x is displacement

and that for electrical system as

$$L\frac{di}{dt} + \frac{1}{C}\int i\,dt + Ri = v \qquad ...(1.35)$$

where i = current. The charge q and current i bear the relationship

$i = \dot{q}$ using which (1.35) takes the following form.

$$L\ddot{q} + R\dot{q} + \frac{1}{C}\cdot q = v \qquad ...(1.36)$$

Compare (1.34) and (1.36) to note that they are of identical form. Therefore the mechanical and electrical systems as shown in Fig. 1.21 (a) and 1.21 (b) respectively are analogous to each other. The quantities in corresponding position of equations (1.34) and (1.36), termed as analogous quantities are given in Table 1.4.

TABLE 1.4: Analogous quantities in force voltage analogy

Translational mechanical system	*Rotational mechanical system*	*Electrical system*
Force (f)	Torque (τ)	Votlages (v)
Mass (m)	Moment of inertial (J)	Inductance (L)
Damping coefficient (B)	Damping coefficient (B)	Resistance (R)
Spring constant (k)	Spring constant (k)	Reciprocal of Capacitance (1/C)
Linear displacement (x)	Angular displacement (θ)	Charge (q)
Linear velocity ($\dot{x}$)	Angular Velocity ($\dot{\theta}$)	Current (i)

Procedure to construct analogous electrical system using *f-v* analogy

The construction of analogous electrical system is explained in following steps with the help of a mechanical system shown in Fig. 1.22 for which the corresponding mechanical network is drawn in Fig. 1.23.

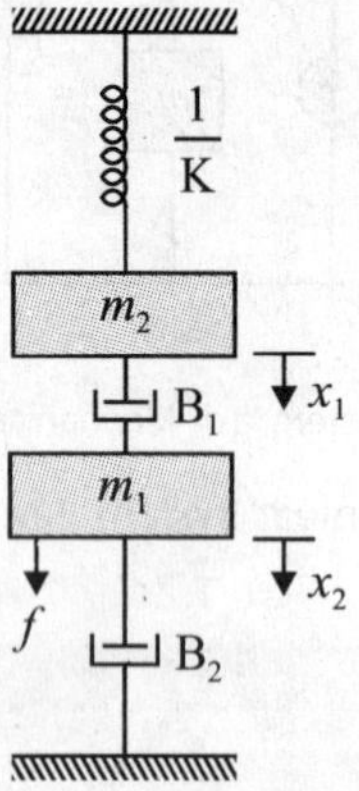

Fig. 1.22: Mechanical system

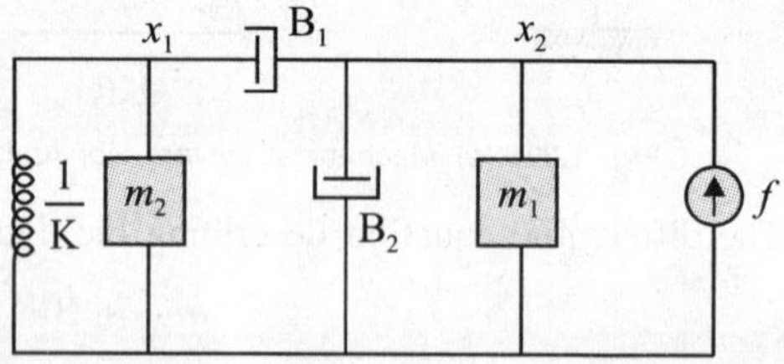

Fig. 1.23: Mechanical network of system of Fig. 1.22

1. Each node in mechanical network of Fig. 1.23 corresponds to a loop in electrical system. The nodes x_1 and x_2 correspond to loop 1 and loop 2 respectively as demonstrated in Fig. 1.24.
2. Identify the mechanical elements connected to each node. The corresponding loop will contain their electrical analogs, for example loop 1 will contain three electrical analogs R_1, L_2 and C for their mechanical counter elements B_1 , m_2 and $1/k$ respectively. Similarly, loop 2 will contain R_1, R_2, L_1 and voltage source v for B_1, B_2, m_1 and force f respectively.
3. Identify those mechanical elements interconnecting two nodes. The respective electrical analogous will be connected in series while being common to the two loops corresponding to these two nodes for example R_1 analogous to B_1 will be common to loop 1 and loop 2.
4. Identify those mechanical elements connected distinctly to each node. The respective electrical analogous will be connected in series and will constitute a distinct segment of corresponding loop for example C and L_2 are in series and are part of only loop 1. Similarly L_1, R_2 and v are in series and appear in only loop 2.

   The complete analogous electrical system is shown in Fig. 1.24.

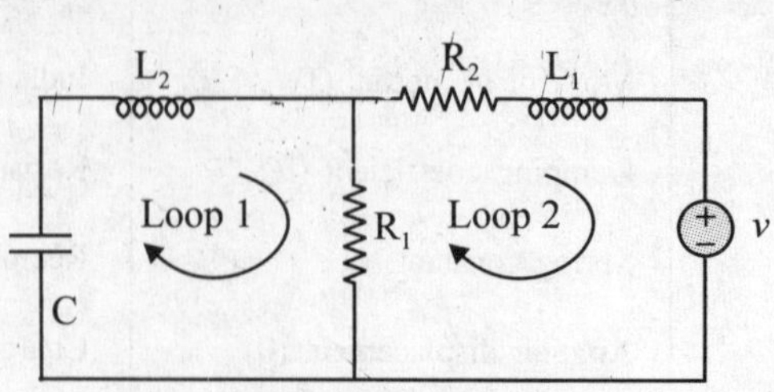

Fig. 1.24: Electrical $(f - v)$ analog

Force-current (*f-i*) analogy: Consider the electrical system shown in Fig. 1.25.

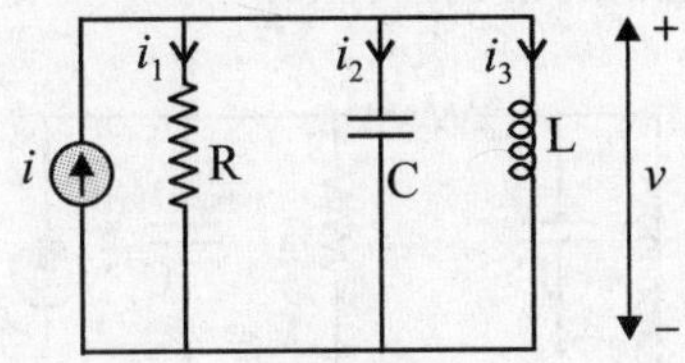

Fig. 1.25

KCL yields

$$i = i_1 + i_2 + i_3 \qquad \text{...(1.37)}$$

which in terms of node voltage v takes the following form.

$$\frac{v}{R} + \frac{1}{L}\int v\,dt + C\dot{v} = i \qquad \text{...(1.38)}$$

It is known that voltage v and magnetic flux ϕ are related as:

$$v = \dot{\phi} \qquad \text{...(1.39)}$$

use equation (1.39) to arrange equation (1.38) as follows:

$$C\ddot{\phi} + \frac{1}{R}\dot{\phi} + \frac{1}{L}\phi = i \qquad \text{...(1.40)}$$

Compare (1.34) and (1.40) to note that both have identical form and therefore mechanical system of Fig. 1.21 (*a*) and electrical system of Fig. 1.25 are analogous. The analogous quantities therein are listed in Table 1.5.

TABLE 1.5: Analogous quantities in force current analogy

Translational mechanical system	*Rotational mechanical system*	*Electrical system*
Force (f)	Torque (τ)	Current (i)
Mass (m)	Moment of inertial (J)	Capacitance (c)
Damping coefficient (B)	Damping coefficient (B)	Conductance (G)
Spring constant (k)	Spring constant (k)	Reciprocal of inductance (1/L)
Linear displacement (x)	Angular displacement (θ)	Magnetic flux (ϕ)
Linear velocity ($\dot{x}$)	Angular velocity ($\dot{\theta}$)	Voltages (v)

Procedure to construct analogous electrical system using *f-i* analogy

Consider mechanical system as shown in Fig. 1.22. The current interest is to construct analogous electrical system. Each node in mechanical network as shown in Fig. 1.23 drawn for mechanical system of Fig. 1.22, corresponds to a junction/node in electrical analog as well. Mere replacement of excitation sources and passive elements in mechanical network by their analogous electrical counterparts, will yield analogous electrical system. For example Analogous electrical system for mechanical system of Fig. 1.22 is drawn in Fig. 1.26.

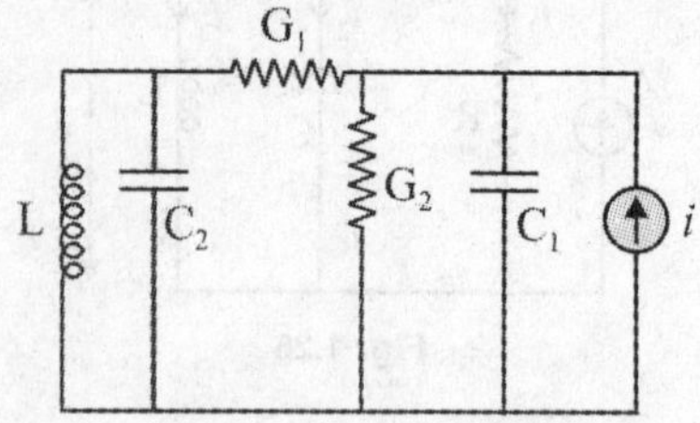

Fig. 1.26: Electrical system using *f-i* analogy

1.6 GEAR TRAIN AND ITS ELECTRICAL ANALOG

Gear Trains: The gear train plays similar role of exchanging energy from one part to another in mechanical system as a transformer does in electrical system. A situation is usually experienced in mechanical systems where a servo motor operating at high speed but low torque is to drive a load with high torque and low speed. Then the requirement of torque magnification and speed regulation is met by gear train. The two gears labelled primary and secondary coupled together are shown in Fig. 1.27.

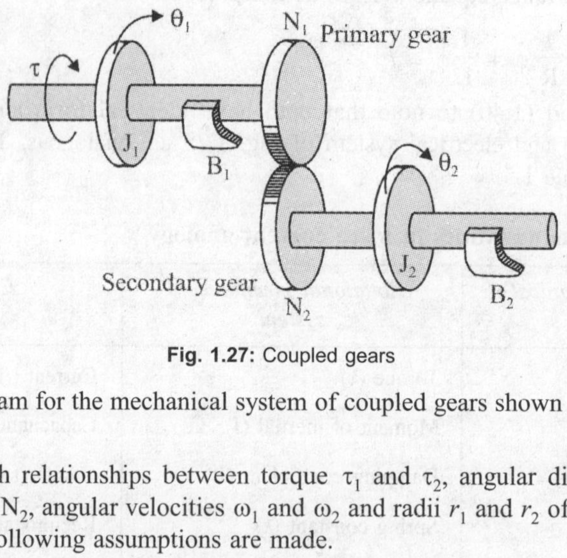

Fig. 1.27: Coupled gears

The free body diagram for the mechanical system of coupled gears shown in Fig 1.27, is drawn in Fig. 1.28.

In order to establish relationships between torque τ_1 and τ_2, angular displacements θ_1 and θ_2, number of teeth N_1 and N_2, angular velocities ω_1 and ω_2 and radii r_1 and r_2 of primary and secondary gears respectively, the following assumptions are made.

1. The number of teeth on the surface of gears is proportional to radii so that

$$r_1 N_2 = r_2 N_1 \qquad ...(1.41)$$

2. The distance travelled along the surface of each gear is same so that

$$\theta_1 r_1 = \theta_2 r_2 \qquad ...(1.42)$$

3. Assume no loss so that work done by each gear is same and

$$\tau_1 \theta_1 = \tau_2 \theta_2 \qquad ...(1.43)$$

The equations (1.41), (1.42) and (1.43) yield the following relationship

$$\frac{\tau_1}{\tau_2} = \frac{\theta_2}{\theta_1} = \frac{N_1}{N_2} = \frac{\omega_2}{\omega_1} = \frac{r_1}{r_2} \qquad ...(1.44)$$

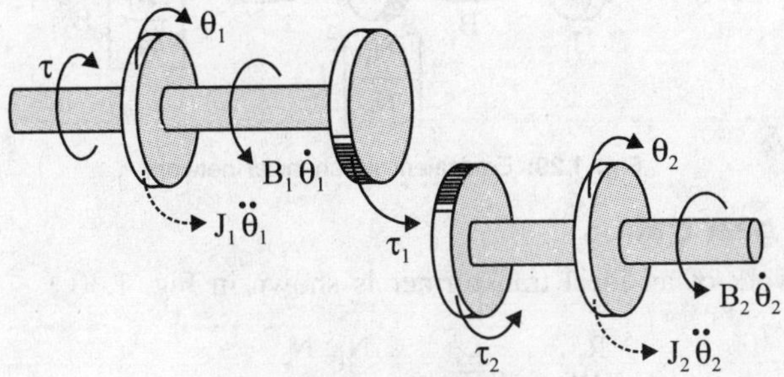

Fig. 1.28: Free body diagram

The rotational dynamics of mechanical network of Fig. 1.28 is described by following equations:

$$J_1\ddot{\theta}_1 + B_1\dot{\theta}_1 + \tau_1 = \tau \qquad ...(1.45)$$

$$J_2\ddot{\theta}_2 + B_2\dot{\theta}_2 + \tau_2 = 0 \qquad ...(1.46)$$

use (1.44) to have gear relations as

$$\tau_2 = \frac{N_2}{N_1}\tau_1 \quad \text{and} \quad \theta_2 = -\frac{N_1}{N_2}\theta_1 \qquad ...(1.47)$$

Substitute these values in equation (1.45) to get

$$\tau_1 = \left(\frac{N_1}{N_2}\right)^2 (J_2\ddot{\theta}_1 + B_2\dot{\theta}_1)$$

Substitute this value of τ_1 in (1.46) to get

$$J_1\ddot{\theta}_1 + B_1\dot{\theta}_1 + \left(\frac{N_1}{N_2}\right)^2 [J_2\ddot{\theta}_1 + B_2\dot{\theta}_1] = \tau \qquad ...(1.48)$$

or

$$\tau = J_e\ddot{\theta}_1 + B_e\dot{\theta}_1 \qquad ...(1.49)$$

where

$$J_e = J_1 + \left(\frac{N_1}{N_2}\right)^2 J_2$$

and

$$B_e = B_1 + \left(\frac{N_1}{N_2}\right)^2 B_2$$

J_e and B_e may be regarded as equivalent inertia and friction referred to primary gear. An equivalent mechanical network satisfying (1.48) is shown in Fig. 1.29.

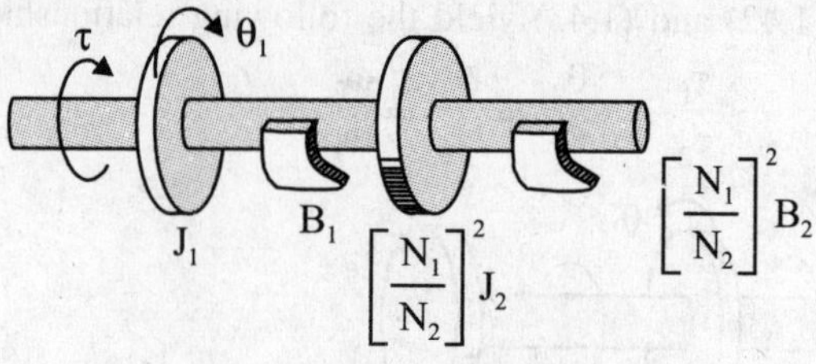

Fig. 1.29: Equivalent mechanical network

Electrical analog of gear train

The electrical network of an ideal transformer is shown in Fig. 1.30.

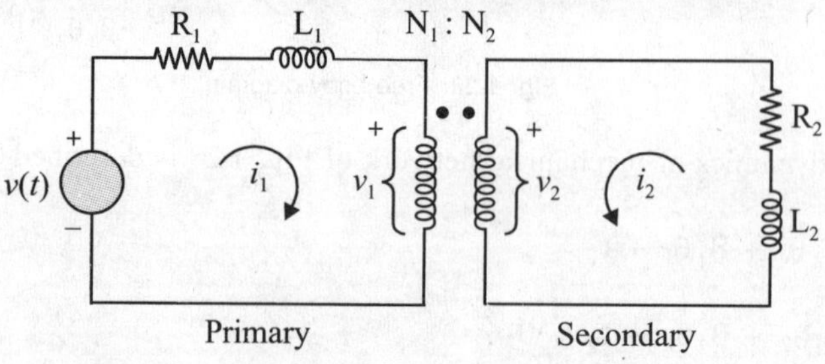

Fig. 1.30: Electrical network of an ideal transformer

The loop equations can be written as

$$L_1 \frac{di_1}{dt} + R_1 i_1 + v_1 = v$$

$$L_2 \frac{di_2}{dt} + R_2 i_2 + v_2 = 0$$

But $\quad v_2 = \frac{N_2}{N_1} v_1 \quad \text{and} \quad i_2 = -\frac{N_1}{N_2} i_1$

Substituting v_2, i_2 and eliminating v_1, we get

$$\left[L_1 + \left(\frac{N_1}{N_2}\right)^2 L_2\right] \frac{di_1}{dt} + \left[R_1 + \left(\frac{N_1}{N_2}\right)^2 R_2\right] i_2 = v \qquad ...(1.50)$$

The equations (1.49) and (1.50) are analogous with following equivalent quantities:

$$i_1 \longleftrightarrow \dot{\theta}_1$$

$$i_2 \longleftrightarrow \dot{\theta}_2$$

$$v \longleftrightarrow \tau$$

$$L_1 \longleftrightarrow J_1$$

$$R_1 \longleftrightarrow B_1$$

$$L_2 \longleftrightarrow J_2$$

$$R_2 \longleftrightarrow B_2$$

The equivalent network with secondary load reflected to primary satisfying (1.50) is shown in Fig. 1.31. A careful observation of Fig. 1.29 and Fig. 1.31, reveals that the transformer is an electrical analog of coupled gears.

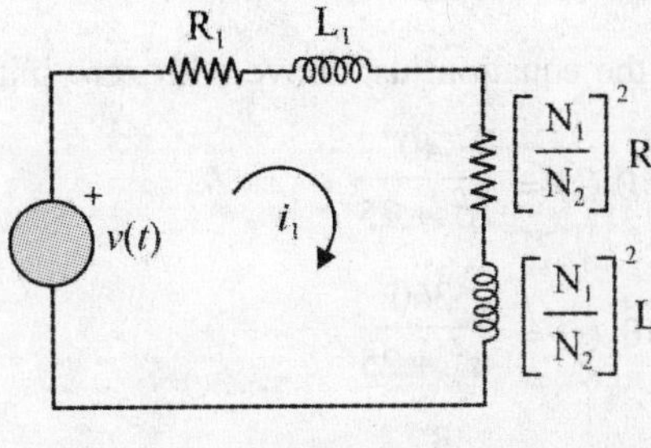

Fig. 1.31: Equivalent network with secondary load reflected to primary

Example 8: Write differential equations for rotational mechanical system shown in Fig. 1.32 (*a*) and then Laplace transform them assuming zero initial conditions.

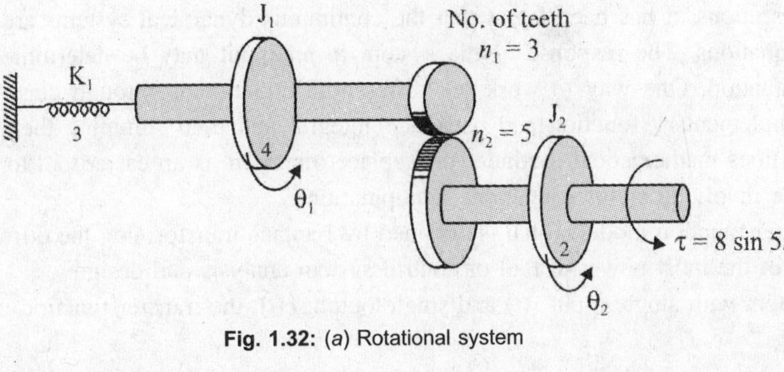

Fig. 1.32: (*a*) Rotational system

Solution: The equivalent rotational system with components $K_1 = 3$ and $J_1 = 4$ referred to side of applied torque τ, is shown in Fig. 1.32 (*b*).

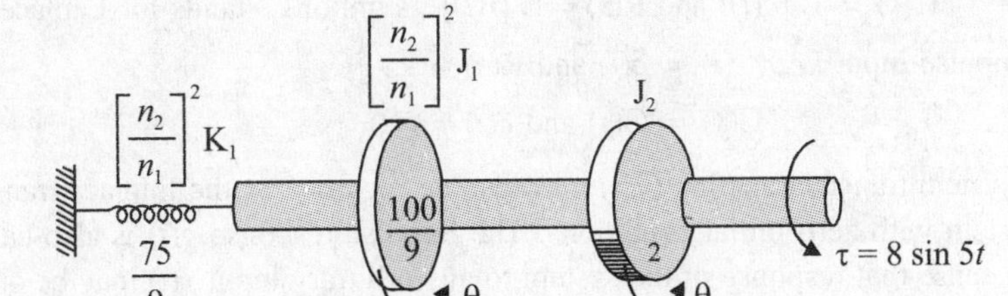

Fig. 1.32: (*b*) Equivalent rotational system

Using d'Alembert's Law, the differential equation describing the system dynamics is written as follows:

$$\left(\frac{n_2}{n_1}\right)^2 K_1\theta_2 + J_2\ddot{\theta}_2 + \left(\frac{n_2}{n_1}\right)^2 J_1\ddot{\theta}_2 = \tau$$

Substitute the component values to get

$$2\ddot{\theta}_2 + \frac{25}{9} \times 4\ddot{\theta}_2 + \frac{25}{9} \times 3\theta_2 = 8 \sin 5t$$

The Laplace transform of the equation just above with zero initial conditions, yields

$$\left(2 + \frac{100}{9}\right)s^2\theta_2(s) + \frac{75}{9}\,\theta_2(s) = \frac{40}{s^2 + 25}$$

or
$$(118s^2 + 75)\theta_2(s) = \frac{360}{s^2 + 25}$$

where
$$\theta_2(s) = -\frac{3}{5}\theta_1(s)$$

1.7 TRANSFER FUNCTION

In preceding sections, it has been learnt that the continuous dynamical systems are characterized by differential equations. The response of the system to an input may be determined by solving the differential equation. One way to work out differential equation is through classical technique by obtaining complementary function and particular integral and then summing them. But it involves some what tedious mathematical routine. The Laplace transform is an easier tool to solve differential equation while involving simpler algebraic manipulations.

The transfer function model which is obtained by Laplace transforming the differential equations, becomes one of the most powerful tool of control system analysis and design.

For a system with single input $r(t)$ and single output $c(t)$, the transfer function is defined as

$$G(s) = \left.\frac{C(s)}{R(s)}\right|_{\text{all initial conditions} = 0} \qquad ...(1.51)$$

where $C(s) = L[c(t)]$ and $R(s) = L[r(t)]$; symbol L stands for Laplace transform

For unit impulse input *i.e.* $r(t) = \delta(t)$ and $R(s) = 1$.

$$C(s) = G(s) \text{ and } c(t) = g(t)$$

Thus, the system transfer function G(*s*) may also be regarded as the Laplace transform of impulse response of system with zero initial conditions. The impulse response $g(t)$ is also called as **weighing function** in the sense that response of the system to any arbitrary input $x(t)$ may be obtained by taking convolution of $g(t)$ with the input $x(t)$. Thus,

$$c(t) = g(t) * x(t) = \int_0^\infty g(t-\tau)\,x(\tau)\,d\tau\,; \; * \text{ stands for convolution} \qquad ...(1.52)$$

Note the following important points regarding transfer function of a system:

1. The concept of transfer function applies to only linear time invariant systems. In general it is not applicable to non linear systems, although it can be extended to certain class of non linear control systems with certain assumptions made.

2. The transfer function is expressed in terms of system parameters and is property of system itself. It is independent of the input or driving function.
3. All initial conditions of system are set to zero *i.e.* the system is assumed to be in possession of no energy inherent in itself.
4. The dynamics of continuous time system is described by algebraic equation in complex variable s. The highest power of s in denominator of G(s) is equal to the order of highest derivative term of output. If this highest power equals n, the system is said to be an nth **order system**. It is also worth noting that the discrete time systems are modelled by difference equation(s). The transfer function of a discrete time system is function of z when z transform is used.
5. In systems described by linear, constant coefficient integro differential equations, every Laplace transformed signal is related to every other such signal by a transfer function. To avoid confusion the term **transfer function** is reserved to describe input-output relation and **transmittance** is used to denote the similar relation between a pair of signals other than input and output.
6. The denominator polynomial of transfer function is called **characteristic polynomial** and roots of characteristic polynomial *i.e.* solution of **characteristic equation** which is actually characteristic polynomial equated to zero, are called system's **characteristic roots**. Later we shall see that these roots play an important role in determining the stability of system.

Transfer function model of system

In order to develop transfer function model of a given system, with input $r(t)$ and output $c(t)$, the following steps are used.

1. Write differential equation of given system.
2. Laplace transform the system equations with zero initial conditions.
3. Obtain the ratio of output C(s) to input R(s) to get transfer function model.

For example, consider the system described by differential equation

$$\ddot{c}(t) + 5\dot{c}(t) + 6c(t) = \dot{r}(t) + 7r(t) \qquad ...(1.53)$$

The Laplace transform of (1.53) with zero initial conditions yields.

$$s^2C(s) + 5s\,C(s) + 6\,C(s) = s\,R(s) + 7\,R(s)$$

and

$$G(s) = \frac{C(s)}{R(s)} = \frac{s+7}{s^2+5s+6} \qquad ...(1.54)$$

The correspondence of time functions with individual partial fraction terms of transfer function

The nature of time function (exponential/sinusoidal) corresponding to each partial fraction expansion term of G(s) depends upon location of characteristic root in s plane and upon whether or not the characteristic root is repeated. These time functions may also be regarded as impulse responses which are inverse Laplace transform of G(s) = C(s) for R(s) = 1. Table 1.6 shows typical impulse responses associated with various root locations.

TABLE 1.6: Impulse responses corresponding to various characteristic root locations

Roots locations(s) on the complex plane	*Impulse response*	*Nature of response*
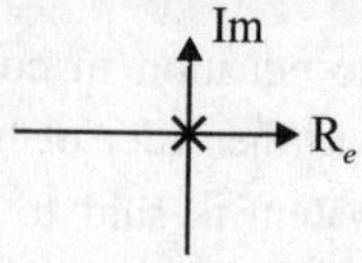	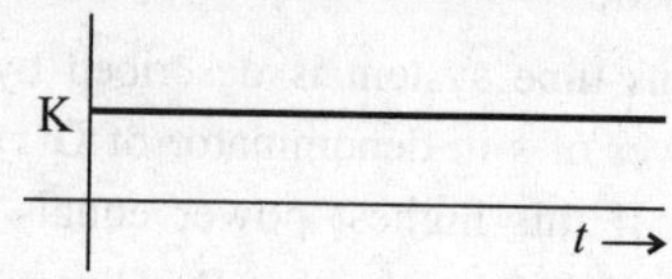	Constant K
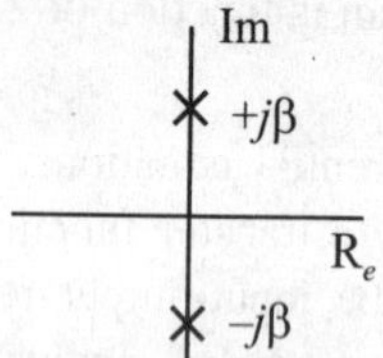	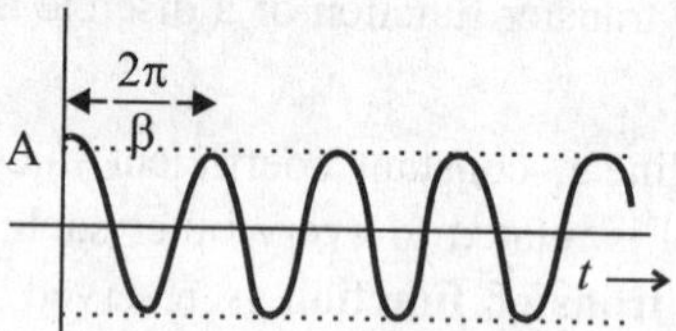	Sinusoid with radian frequency β, A $\cos(\beta t + \theta)$
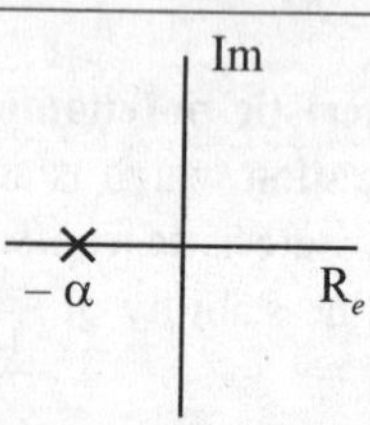	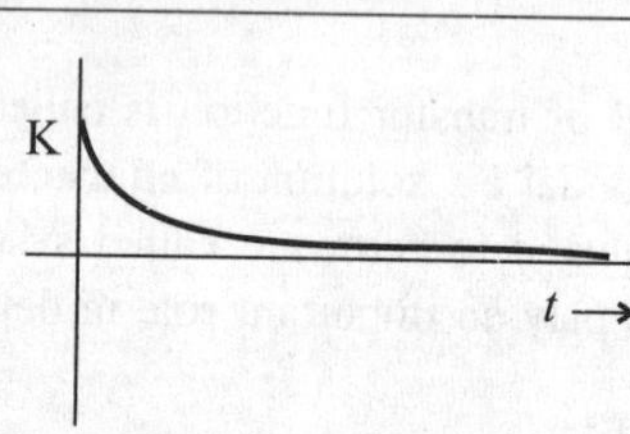	Decaying exponential $Ke^{-\alpha t}$
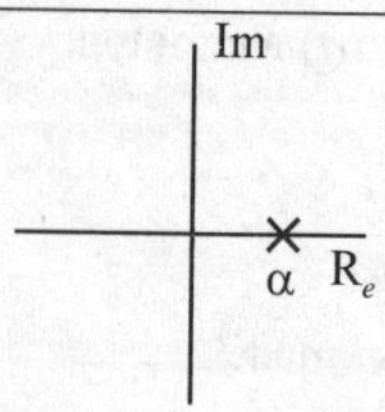	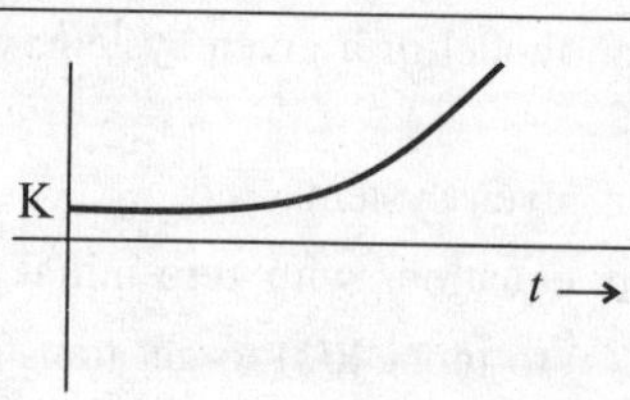	Growing exponential $Ke^{\alpha t}$
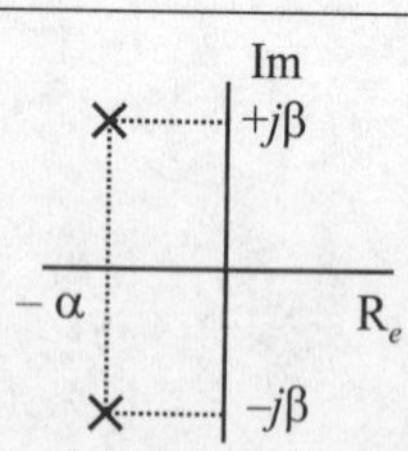	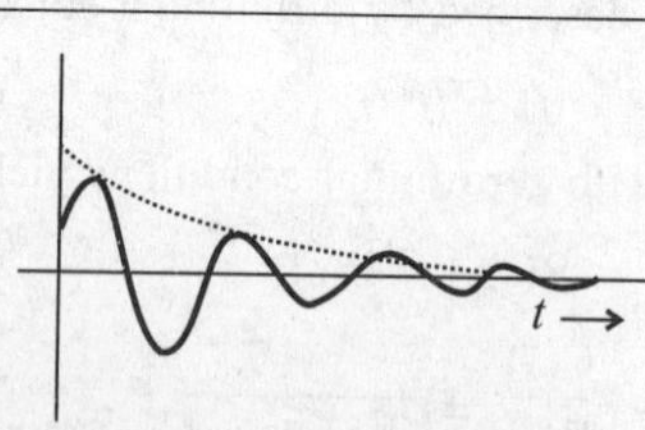	Exponentially decaying sinusoid $Ae^{-\alpha t}\cos(\beta t + \theta)$ Exponential constant is α and sinusoidal radian frequency is β
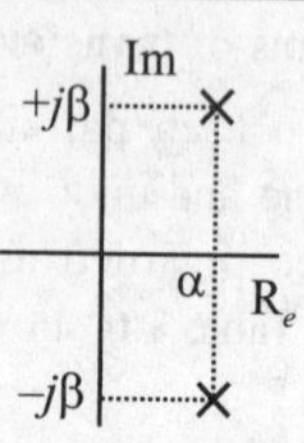	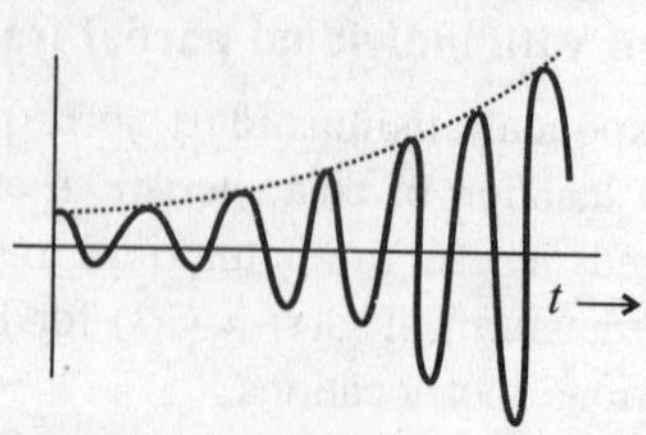	Exponentially growing sinusoid $Ae^{\alpha t}\cos(\beta t + \theta)$

(*Contd.*)

Roots locations(s) on the complex plane	*Impulse response*	*Nature of response*
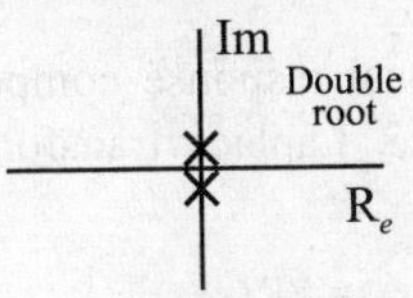	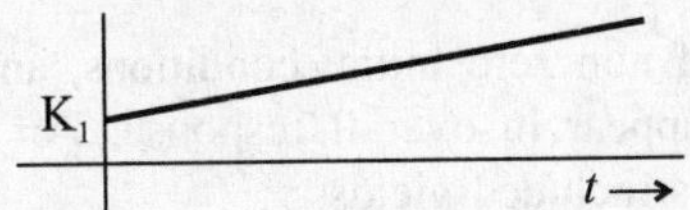	Constant plus a constant times t, $K_1 + K_2 t$
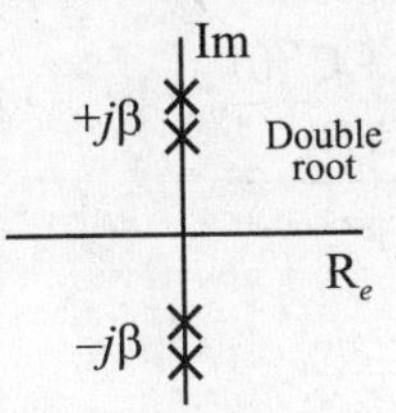	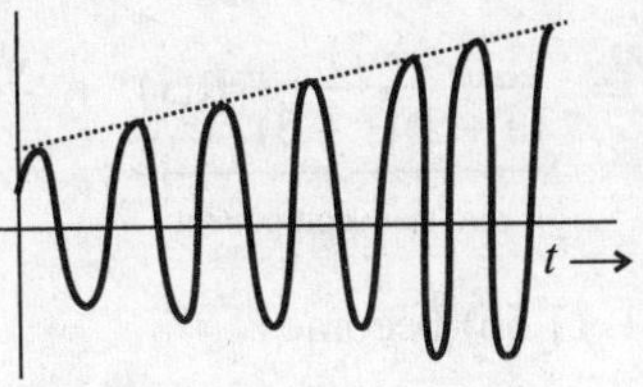	Sinusoidal with radian frequency β plus t times sinusoidal $A_1 \cos(\beta t + \theta_1) + A_2\, t \cos(\beta t + \theta_2)$.
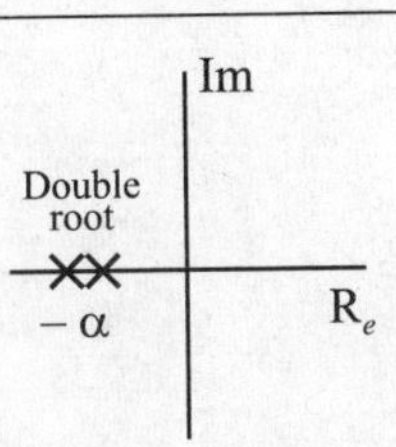	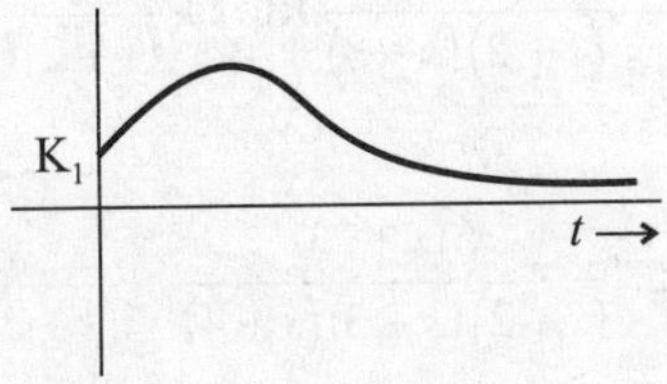	Decaying exponential plus t times a decaying exponential, $K_1 e^{-\alpha t} + K_2 t e^{-\alpha t}$
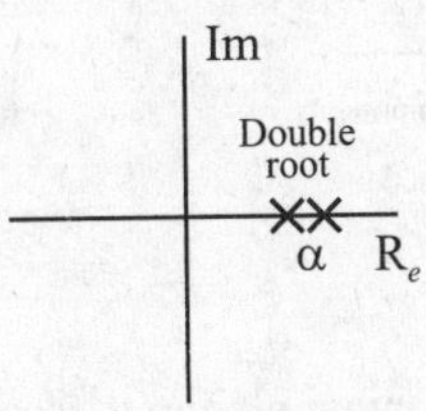	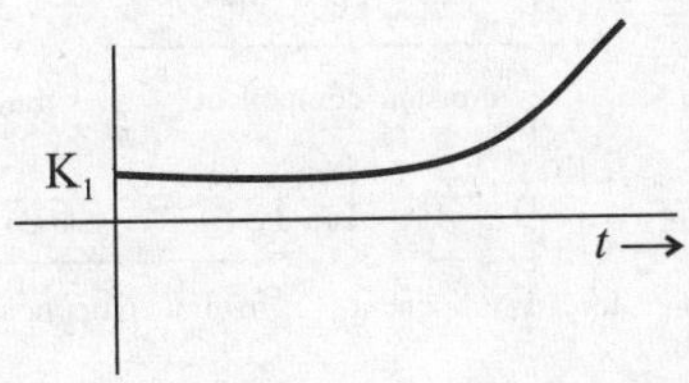	Growing exponential plus t times, growing exponential, $K_1 e^{\alpha t} + K_2 t e^{\alpha t}$

Zero state response

The zero state response is defined as the system output when all the initial conditions are zero for example, consider system described by equation (1.54) where

$$r(t) = 6\,u(t)\,e^{-4t} \text{ and } R(s) = \frac{6}{s+4}$$

then
$$C(s)\big|_{\text{zero state}} = \frac{6\,(s+7)}{(s^2+5s+6)\,(s+4)} = \frac{\alpha_1}{(s+2)} + \frac{\alpha_2}{(s+3)} + \frac{\alpha_3}{(s+4)}$$

where
$$\alpha_1 = \left.\frac{6\,(s+7)}{(s+4)\,(s+3)}\right|_{s=-2} = 15$$

$$\alpha_2 = \left.\frac{6\,(s+7)}{(s+2)\,(s+4)}\right|_{s=-3} = -24$$

$$\alpha_3 = \left.\frac{6\,(s+7)}{(s+2)\,(s+3)}\right|_{s=-4} = 9$$

and $$c(t)\Big|_{\text{zero state}} = 15e^{-2t} - 24e^{-3t} + 9e^{-4t} \qquad ...(1.55)$$

Zero input response

If the system is in possession of non zero initial conditions, an additional response component called as zero input response will appear in overall response. For example, Laplace transform of equation (1.53) with initial conditions included yields.

$$[s^2C(s) - s\,C(0) - C'(0)] + 5[s\,C(s) - C(0)] + 6C(s) = sR(s) + 7R(s)$$

$$C(s) = \underbrace{\frac{(s+7)}{(s+2)(s+3)}R(s)}_{\text{zero state component}} + \underbrace{\frac{(s+5)\,C(0) + C'(0)}{(s+2)\,(s+3)}}_{\text{zero input component}} \qquad ...(1.56)$$

Assuming $c'(0) = 0$ and $c(0) = 1$, (1.56) becomes.

$$C(s) = \frac{(s+7)}{(s+2)(s+3)}R(s) + \frac{(s+5)}{(s+2)(s+3)}$$

Now if $r(t) = 6\,u(t)\,e^{-4t}$, then

$$C(s) = \frac{6(s+7)}{(s+2)(s+3)(s+4)} + \frac{s+5}{(s+2)(s+3)}$$

$$c(t) = \underbrace{15e^{-2t} - 24e^{-3t} + 9e^{-4t}}_{\text{zero state component}} + \underbrace{3e^{-2t} - 2e^{-3t}}_{\text{zero input component}} \qquad ...(1.57)$$

$$= \underbrace{9e^{-4t}}_{\text{forced component}} + \underbrace{18e^{-2t} - 26e^{-3t}}_{\text{natural component}}$$

In overall response $c(t)$ of equation (1.57), the response component containing input is called **forced response** and remainder of response is called **natural response**. It is obvious from equation (1.57) that both the zero input and zero state response components generally contribute to natural response of system.

It is important to note that the control engineer seldom explicitly computes natural response of system due to their dependence on the specific initial conditions, which are often unknown or of little concern. It is usually sufficient to know the nature of natural response and to know that the system is stable as stability will guarantee the decay of these terms to zero with time. Also an unstable system can never be made stable by any choice of initial conditions. However if desired, the initial conditions may be considered to be system inputs and transfer functions can be found that relate the outputs to initial condition inputs as illustrated just before.

Transfer function of a system with multi inputs and multi outputs

If a system has multiple inputs $r_1(t)$, $r_2(t)$ .... and multiple outputs $c_1(t)$, $c_2(t)$, ..., there is a transfer function which relates each one of outputs to each one of inputs, while all other inputs are zero;

$$G_{ij}(s) = \frac{C_i(s)}{R_j(s)}\Bigg|_{\substack{\text{all initial conditions are zero and} \\ \text{all inputs except } R_j \text{ are zero}}}$$

where i = output number

j = input number.

In general for a system with p inputs and q outputs, the outputs are given by

$$C_1(s) = G_{11}(s)R_1(s) + G_{12}(s)R_2(s) + \dots + G_{1p}(s)R_p(s)$$

$$C_2(s) = G_{21}(s)R_1(s) + G_{22}(s)R_2(s) + \dots + G_{2p}(s)R_p(s)$$

$$\vdots$$

$$C_q(s) = G_{q1}(s)R_1(s) + G_{q2}(s)R_2(s) + \dots + G_{qp}(s)R_p(s)$$

or

$$\begin{bmatrix} C_1(s) \\ C_2(s) \\ \vdots \\ \vdots \\ C_q(s) \end{bmatrix} = \begin{bmatrix} G_{11}(s) & G_{12}(s) & \dots & G_1p(s) \\ G_{21}(s) & G_{22}(s) & \dots & G_2p(s) \\ & & \vdots & \\ & & \vdots & \\ G_{q1}(s) & G_{q2}(s) & \dots & G_{qp}(s) \end{bmatrix} \begin{bmatrix} R_1(s) \\ R_2(s) \\ \vdots \\ \vdots \\ R_p(s) \end{bmatrix} \quad \text{...(1.58)}$$

To illustrate, let a system with two inputs and two outputs be described by following differential equations.

$$\left.\begin{aligned} \frac{d^3 y_1}{dt^3} + 7\frac{d^2 y_1}{dt^2} + 6\frac{dy_1}{dt} + y_1 &= \frac{d^2 r_1}{dt^2} + 3r_1 + r_2 \\ \frac{d^3 y_2}{dt^3} + 7\frac{d^2 y_2}{dt^2} + 6\frac{dy_2}{dt} + y_2 &= 4\frac{dr_2}{dt} \end{aligned}\right\} \quad \text{...(1.59)}$$

The Laplace transform of equations (1.59) with zero initial conditions gives

$$\left.\begin{aligned} s^3 Y_1(s) + 7s^2 Y_1(s) + 6sY_1(s) + Y_1(s) &= s^2 R_1(s) + 3R_1(s) + R_2(s) \\ \text{and} \quad s^3 Y_2(s) + 7s^2 Y_2(s) + 6sY_2(s) + Y_2(s) &= 4s\,R_2(s) \end{aligned}\right\} \quad \text{...(1.60)}$$

To get $\dfrac{Y_1(s)}{R_1(s)}$ and $\dfrac{Y_2(s)}{R_1(s)}$ set $R_2(s) = 0$

then $$\frac{Y_1(s)}{R_1(s)} = \frac{s^2 + 3}{s^3 + 7s^2 + 6s + 1}$$

and $$\frac{Y_2(s)}{R_1(s)} = 0$$

To get $\dfrac{Y_2(s)}{R_2(s)}$ and $\dfrac{Y_1(s)}{R_2(s)}$ set $R_1(s) = 0$

then $$\frac{Y_1(s)}{R_2(s)} = \frac{1}{s^3 + 7s^2 + 6s + 1}$$

and $$\frac{Y_2(s)}{R_2(s)} = \frac{4s}{s^3 + 7s^2 + 6s + 1}.$$

1.8 CLASSIFICATION OF SYSTEMS

The systems are classified as follows:

1. Linear and non-linear systems

The mathematical model of a system is said to be linear if it obeys the laws of superposition (additivity) and homogeneity (scalar multiplicativity). These laws imply that if system model has zero state responses $y_1(t)$ and $y_2(t)$ for any two inputs $x_1(t)$ and $x_2(t)$ respectively, then the system response to linear combination of these inputs $\alpha_1 x_1(t) + \alpha_2 x_2(t)$ is given by linear combination of individual responses $\alpha_1 y_1(t) + \alpha_2 y_2(t)$ where α_1 and α_2 are constants. The systems which do not obey any or both the laws are classified as non linear systems.

For example system of Fig. 1.33 described by mathematical model,

$$y = \alpha x$$

is linear and system of Fig. 1.34 whose input-output relationship is given by

$$y = \alpha_1 x + \alpha_2$$

is non-linear as one can verify that it does not obey the laws of superposition and homogeneity.

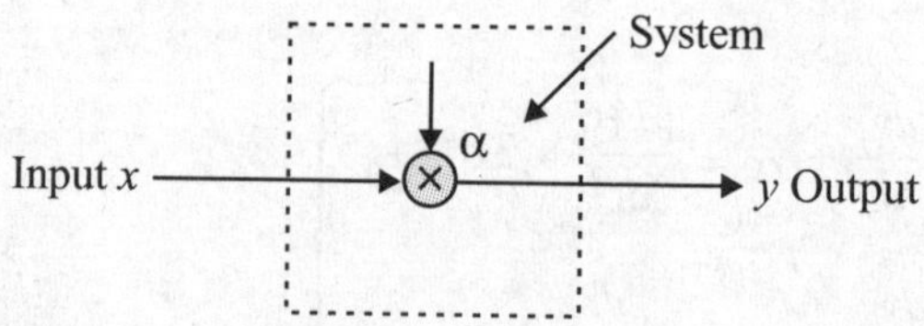

Fig. 1.33: Linear system

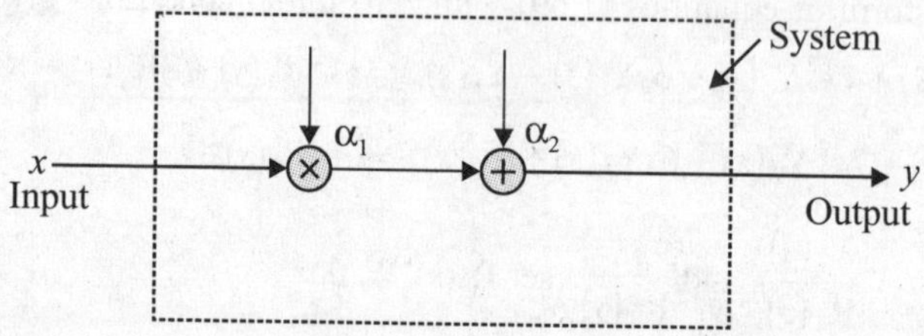

Fig. 1.34: Non-linear system

2. Time varying and time invariant systems

This classification is based on whether parameters of a system remain functions of time or independent of time variable during operation. It is known that continuous dynamical systems are characterised by differential equation(s). A dynamical system is classified as time invariant if coefficients of differential equation(s) characterising the system, are constants and time varying if any one or more than one or all coefficients are functions of time. For, example

$$\alpha_1 \ddot{y}(t) + \alpha_2 \dot{y}(t) + \alpha_3 y = x(t) \text{ is time invariant.}$$

whereas,

$$a(t)\ddot{y}(t) + b(t)\dot{y}(t) + c(t)y(t) = x(t) \text{ is time varying.}$$

3. Continuous and discrete systems

If the signals at input terminal, output terminal and at all other points within the system are functions of continuous time variable t, then system is said to be continuous and if these signals are either a pulse train or a digital code, the system is referred to as discrete system. The discrete systems are sub classified as sampled data systems and digital systems. Signals are in the form of pulses in sampled data system and these are digitally coded (*e.g.*; binary code) in digital systems.

The continuous systems are characterised by differential equations and discrete systems are characterised by difference equations. For example

$$\ddot{y} + \alpha\,\dot{y} + \beta y \;=\; x \text{ is continuous sytem}$$

but $$y\,(k+2) + \alpha\, y\,(k+1) + \beta\, y\,(k) \;=\; x(k) \text{ is discrete system.}$$

4. Deterministic and stochastic systems

If the coefficients of differential/difference equations are deterministic in nature, the system is said to be deterministic and if they are probabilistic in nature (random variable), it is said to stochastic system.

5. System with memory and without memory

A system is said to be memoryless if its current response depends on only current excitation. If its current response depends not only on current excitation to it but also on past history, then it is said to be system with memory. For example ;

$$y\,(t) \;=\; r\,(t) + \alpha_1\, r^2(t) \text{ is memoryless system.}$$

A resistor is another example of memoryless system ; with $x(t)$ and $y(t)$ respectively considered as input current and output voltage, the input output relationship is given by

$$y(t) \;=\; \text{R}\; x(t) \text{ where R = resistance}$$

A memoryless system whose input and output are identical is called **identity system**. The input-output relationship for a continuous identity system is $y(t) = x(t)$

$y(t) = r(t) + \alpha_1\, r^2(t-2)$ is an example of system with memory. A capacitor C is another example of this class as the voltage $v_c(t)$ in terms of current $i(t)$ is given by

$$v_c(t) \;=\; \frac{1}{\text{C}}\int_{-\infty}^{t} i(\tau)\,dc$$

A system with memory, physically signifies that the system has capability of storing the energy.

6. Causal system

A system is said to be causal if its output at any time depends only on inputs at present time and in the past. Causal systems are also called as non-anticipative systems in the sense that system output does not anticipate future values of input. For example, RC circuit is causal as voltage across capacitor depends on present and past values of source voltage whereas system described by equation

$$y(t) \;=\; x(t) + x(t+1)$$

is not causal. While investigating the condition of causality, one is required to be bit careful in checking the input-output relationship for all times. For example, consider a system

$$y(t) \;=\; x(t)\cos(t+1)$$

One may be temped to conclude that it is not causal. But it is causal system as $\cos(t + 1)$ is only a time varying function and if expressed as $g(t) = \cos(t + 1)$ then $y(t) = g(t)\, x(t)$. Now it is obvious that current output is influenced by only current input.

PROBLEMS AND SOLUTIONS

P1.1: Fig. P1.1 shows the block diagram of a control system. The system in block A has an impulse response $h_A(t) = e^{-t}\, u(t)$. The system in block B has an impulse response $h_B(t) = e^{-2t}\, u(t)$. The block 'k' amplifies its input by a factor k. For the overall system with input $x(t)$ and output $y(t)$, find

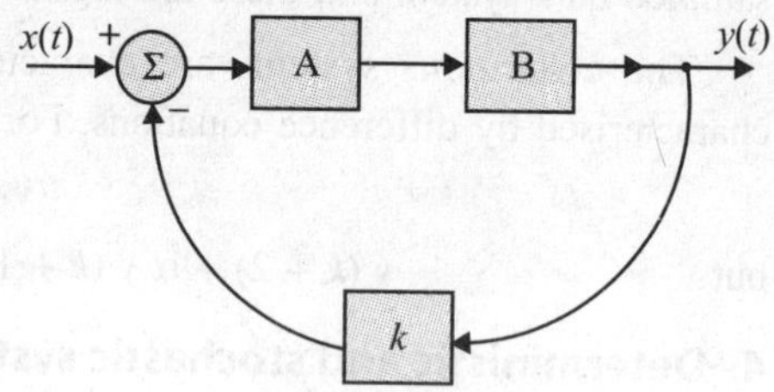

Fig. P1.1

(*a*) the transfer function $\dfrac{Y(s)}{X(s)}$ when $k = 1$

(*b*) the impulse response when $k = 0$.

Solution:

(*a*)
$$H_A(s) = \frac{1}{s+1}, \qquad H_B(s) = \frac{1}{s+2}$$

$$\left.\frac{Y(s)}{X(s)}\right|_{k=1} = \left.\frac{H_A(s)\cdot H_B(s)}{1 + k\, H_A(s)\, H_B(s)}\right|_{k=1} = \frac{1}{s^2 + 3s + 3}$$

(*b*)
$$\left.\frac{Y(s)}{X(s)}\right|_{k=0} = \left.\frac{H_A(s)\cdot H_B(s)}{1 + k\, H_A(s)\, H_B(s)}\right|_{k=0} = H_A(s)H_B(s) = \frac{1}{(s+1)(s+2)}$$

$$\text{Impulse response} = L^{-1}\left[\,Y(s)\Big|_{X(s)=1}\right] = e^{-t} - e^{-2t}$$

P1.2: A speed control system of an engine is shown in Fig. P1.2. Determine:

(*a*) Sensitivity of closed loop system to changes in engine gain k_1 and tachometer feedback gain k_2.

(*b*) Steady state speed for reference speed = 50 km/hr, $k_1 = 100$, $k_2 = 1$ and $d(t) = 0$

(*c*) Steady state speed when there is 5 % change in k_1 and other values as in part (*b*).

(*d*) Steady state speed when there is 5 % change in k_2 and other values as in part (*b*).

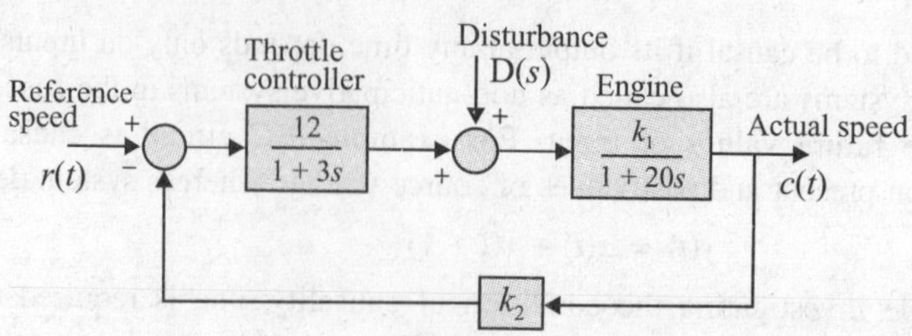

Fig. P1.2

Solution:

(*a*) $$T(s)\big|_{D(s)=0} = \frac{C(s)}{R(s)} = \frac{12k_1}{(1+3s)(1+20s)+12k_1k_2} = \frac{12k_1}{60s^2+23s+12k_1k_2+1}$$

$$S_{k_1}^{T} = \frac{\partial T}{\partial k_1}\times\frac{k_1}{T} = \frac{60s^2+23s+1}{60s^2+23s+1+12k_1k_2}$$

$$S_{k_2}^{T} = \frac{\partial T}{\partial k_2}\times\frac{k_2}{T} = \frac{-12k_1k_2}{60s^2+23s+1+12k_1k_2}$$

(*b*) $$T(s)\big|_{s=0} = \frac{12k_1}{1+12k_1k_2} = \frac{1200}{1201}$$

$$\text{Steady state speed} = c(\infty) = T(s)\times R(s)\big|_{s=0}$$

$$= \frac{1200}{1201}\times 50 = 49.958 \text{ km/hr}$$

(*c*) $$\frac{\Delta T(s)}{T(s)} = S_{k_1}^{T(s)}\times\frac{\Delta k_1}{k_1}$$

$$\Delta T(s)\big|_{s=0} = T(0)\cdot S_{k_1}^{T(0)}\cdot\frac{\Delta k_1}{k_1} = \frac{1200}{1201}\times\frac{1}{1201}\times 0.05$$

and $$\Delta C(s)\big|_{s=0} = \frac{1200\times 0.05}{1201\times 1201}\times 50 = 2.079\times 10^{-3} \cong 0.002$$

$$c(\infty) = 50 + 0.002 = 50.002 \text{ km/hr}$$

(*d*) $$\frac{\Delta T(s)}{T(s)} = S_{k_2}^{T(s)}\times\frac{\Delta k_2}{k_2}$$

$$\Delta T(s)\big|_{s=0} = T(0)\cdot S_{k_2}^{T(0)}\cdot\frac{\Delta k_2}{k_2} = \frac{1200}{1201}\times\frac{-1200}{1201}\times 0.05 = -0.0499$$

$$\Delta C(s)\big|_{s=0} = -0.0499\times 50 = -2.495$$

$$\text{Steady state speed} = C(s)\big|_{s=0} = 50 - 2.495 = 47.50 \text{ km/hr}$$

P1.3: The forward path transfer function of a position control system with velocity feedback is given by

$$G(s) = \frac{K}{s(s+p)}$$

Determine the sensitivity of the transfer function of the closed loop system to changes in K, p and α for transfer function of the feedback path $H(s) = (1 + \alpha s)$. The nominal values of the parameters are $K = 12$, $p = 3$ and $\alpha = 0.14$.

Solution: $$T(s) = \frac{K}{s^2+(p+\alpha K)s+K}$$

$$S_K^T = \frac{\partial T}{\partial K} \times \frac{K}{T} = \frac{s(s+p)}{s^2 + (p+\alpha K)s + K}$$

$$S_P^T = \frac{\partial T}{\partial p} \times \frac{p}{T} = \frac{-sp}{s^2 + (p+\alpha K)s + K}$$

$$S_\alpha^T = \frac{\partial T}{\partial \alpha} \times \frac{\alpha}{T} = \frac{-K\alpha s}{s^2 + (p+\alpha K)s + K}$$

Substitute nominal values of parameters K = 12, p = 3 and α = 0.14 to get

$$S_K^T = \frac{s(s+3)}{s^2 + 4.68s + 12}$$

$$S_P^T = \frac{-3s}{s^2 + 4.68s + 12}$$

$$S_\alpha^T = \frac{-1.68s}{s^2 + 4.68s + 12}$$

P1.4: Write differential equations governing the behaviour of mechanical system shown in Fig. P1.4 and obtain analogous electrical networks based on;

(*a*) *f-v* analogy

(*b*) *f-i* analogy.

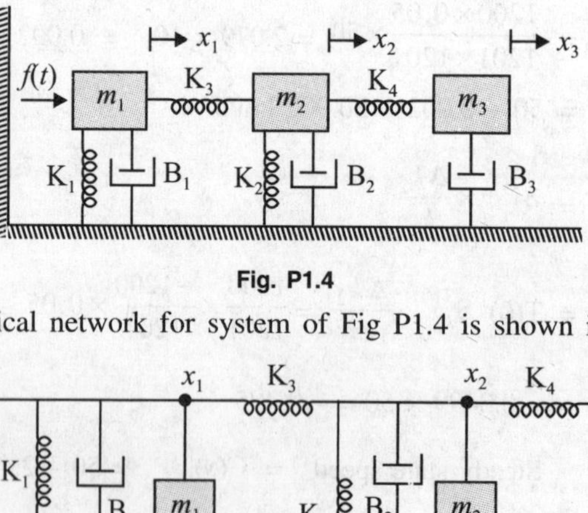

Fig. P1.4

Solution: The mechanical network for system of Fig P1.4 is shown in Fig. P 1.4 (*a*).

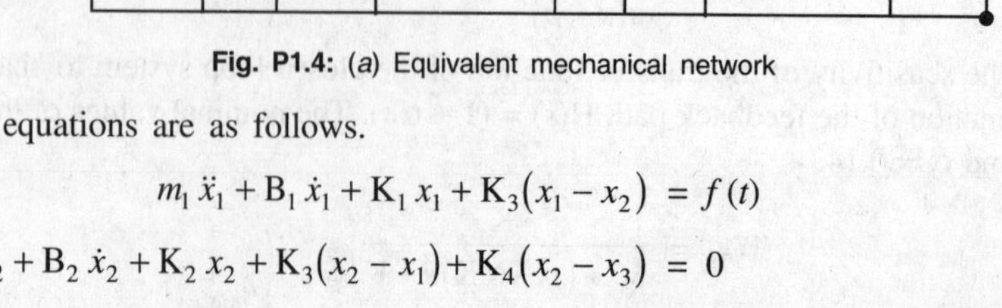

Fig. P1.4: (*a*) Equivalent mechanical network

The system equations are as follows.

$$m_1 \ddot{x}_1 + B_1 \dot{x}_1 + K_1 x_1 + K_3(x_1 - x_2) = f(t)$$

$$m_2 \ddot{x}_2 + B_2 \dot{x}_2 + K_2 x_2 + K_3(x_2 - x_1) + K_4(x_2 - x_3) = 0$$

and $$m_3 \ddot{x}_3 + B_3 \dot{x}_3 + K_4(x_3 - x_2) = 0$$

(*a*) *f-v* analogous electrical network will consist of 3 loops as described below. The elements marked (*) are common elements between two loops. Loop 1 corresponding to node x_1 contains

$v(t)$ for $f(t)$
C_1 for $1/K_1$
R_1 for B_1
L_1 for m_1
and C_3 for $1/K_3$*

Loop 2 corresponding to node x_2 contains

C_3 for $1/K_3$*
C_2 for $1/K_2$
R_2 for B_2
L_2 for m_2
and C_4 for $1/K_4$*

Loop 3 corresponding to node x_3 contains

L_3 for m_3
R_3 for B_3
and C_4 for $1/K_4$*

C_3 is common in loops 1 and 2. C_4 is common in loops 2 and 3. *f–v* analogous network is drawn in Fig. P 1.4 (*b*).

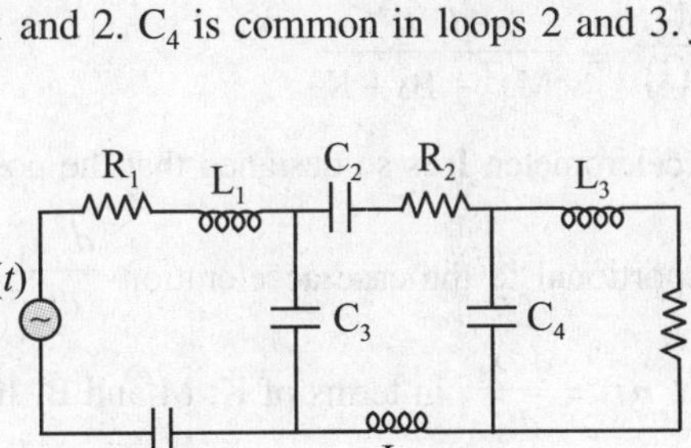

Fig. P1.4: (*b*) *f–v* analog

(*a*) The force-current analogous network is drawn in Fig. P1.4(*c*).

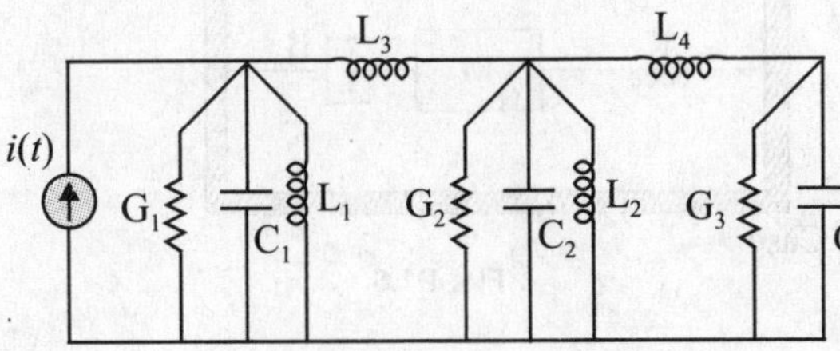

Fig. P1.4: (*c*) *f–i* analog

P1.5: Find transmittance $\frac{X(s)}{F(s)}$ for the system shown in Fig. P1.5

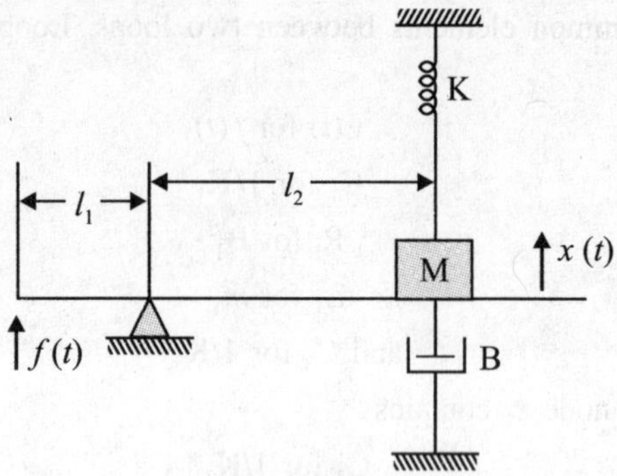

Fig. P 1.5

Solution: Assuming $f^*(t)$ is force on mass M in the direction of displacement $x(t)$, the system equations can be written as

$$f(t) \times l_1 = -f^*(t)\, l_2$$

$$f^*(t) = (-l_1/l_2) \,.\, f(t)$$

and $$M\ddot{x}(t) + B\dot{x}(t) + Kx(t) = f^*(t)$$

Laplace transforming with zero initial conditions yields

$$Ms^2X(s) + BsX(s) + KX(s) = F^*(s)$$

or $$\frac{X(s)}{F(s)} = \frac{-l_1/l_2}{Ms^2 + Bs + K}$$

P1.6: Fig. P 1.6 shows an accelerometer. It is so designed that the position x_2 of mass with respect to the case is approximately proportional to the case acceleration $\frac{d^2x_1}{dt_2}$. Find the transmittance that relates x_2 to the case acceleration, $r(t) = \frac{d^2x_1}{dt^2}$ in terms of K, M and B. If M = 0.2, select K and B so that the characteristic polynomial has both roots at – 80.

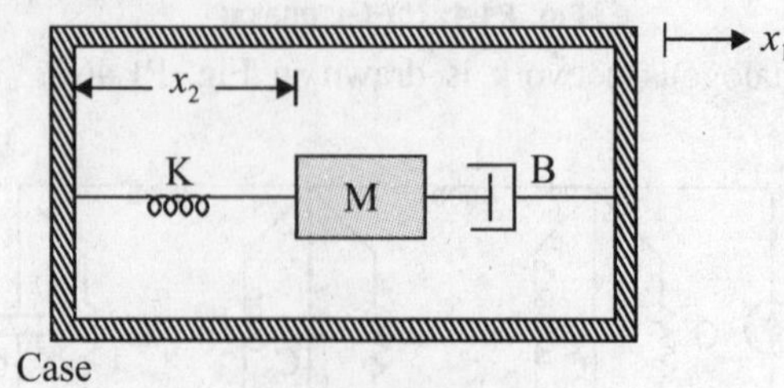

Fig. P1.6

Solution:

Assume x_0 = displacement of mass M relative to inertial space

Given x_1 = displacement of case relative to inertial space

and x_2 = displacement of mass M relative to case

so, $x_2 = x_0 - x_1$

System equation can be written as

$$M\ddot{x}_0 + B(\dot{x}_0 - \dot{x}_1) + K(x_0 - x_1) = 0$$

$$M(\ddot{x}_2 + \ddot{x}_1) + B\dot{x}_2 + Kx_2 = 0$$

$$M\ddot{x}_2 + B\dot{x}_2 + Kx_2 = -M\ddot{x}_1 = -Mr(t)$$

Laplace transforming

$$Ms^2 X_2(s) + Bs X_2(s) + KX_2(s) = -MR(s)$$

and

$$\frac{X_2(s)}{R(s)} = \frac{-M}{Ms^2 + Bs + K} = -\frac{1}{s^2 + \frac{B}{M}s + \frac{K}{M}}$$

The characteristic polynomial is

$$P(s) = s^2 + \frac{B}{0.2}s + \frac{K}{0.2} = s^2 + 5Bs + 5K$$

whose roots $s_1, s_2 = \dfrac{-5B \pm \sqrt{(5B)^2 - 20K}}{2} = -2.5B \pm \sqrt{\dfrac{25B^2}{4} - 5K}$

are required to lie at – 80

So, $2.5B = 80$ gives $B = 32$

and $\dfrac{25B^2}{4} = 5K$ gives $K = 1280$

P1.7: Obtain transfer function of each of two mechanical systems given in Fig. P 1.7, where x_i represents input displacement and x_0 represents output displacement.

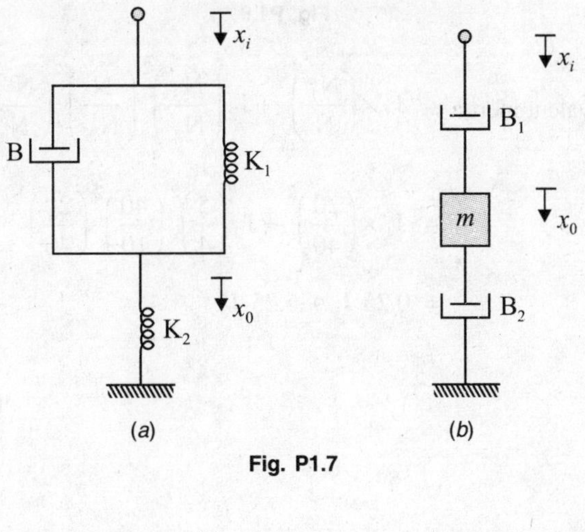

Fig. P1.7

Solution:

(*a*) System equation corresponding to node x_0 is

$$K_2 x_0 + K_1(x_0 - x_i) + B(\dot{x}_0 - \dot{x}_i) = 0$$

whose Laplace transform is

$$K_2X_0(s) + K_1X_0(s) - K_1X_i(s) + Bs[X_0(s) - X_i(s)] = 0$$

$$\frac{X_0(s)}{X_i(s)} = \frac{Bs + K_1}{Bs + K_1 + K_2}$$

(*b*) The system equation corresponding to node x_0 is

$$m\ddot{x}_0 + B_2\dot{x}_0 + B_1(\dot{x}_0 - \dot{x}_i) = 0$$

whose Laplace transform is

$$ms^2 X_0(s) + B_2 s X_0(s) + B_1 s[X_0(s) - X_i(s)] = 0$$

$$\frac{X_0(s)}{X_i(s)} = \frac{B_1 s}{ms^2 + s(B_1 + B_2)} = \frac{B_1}{ms + (B_1 + B_2)}$$

P1.8: Find equivalent inertia referred to shaft shown in Fig. P1.8.

Given $\frac{N_3}{N_4} = \frac{5}{4}$

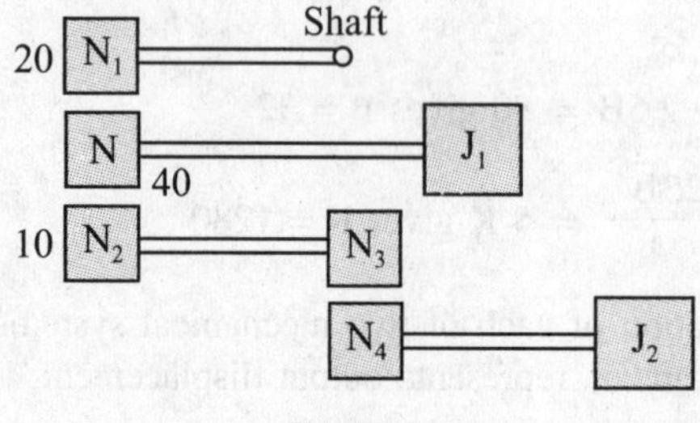

Fig. P1.8

Solution: Equivalent inertia $= J_1 \times \left(\frac{N_1}{N}\right)^2 + J_2\left(\frac{N_3}{N_4}\right)^2\left(\frac{N}{N_2}\right)^2\left(\frac{N_1}{N}\right)^2$

$$= J_1 \times \left(\frac{20}{40}\right)^2 + J_2\left(\frac{5}{4}\right)^2\left(\frac{40}{10}\right)^2\left(\frac{20}{40}\right)^2$$

$$= 0.25\ J_1 + 6.25\ J_2.$$

DRILL PROBLEMS

D1.1: Write simultaneous Laplace transformed differential equations for mechanical system shown in Fig. D1.1. Assume zero initial conditions.

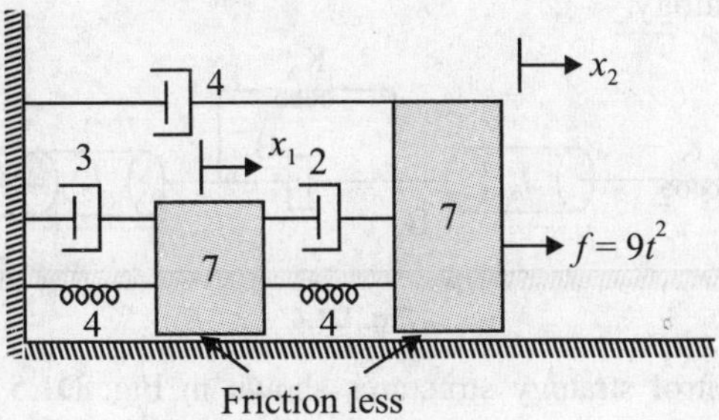

Fig. D1.1

Ans. $7s^2X_1(s) + 5sX_1(s) + 8X_1(s) - 2sX_2(s) - 4X_2(s) = 0$

$7s^2X_2(s) + 6sX_2(s) + 4X_2(s) - 2sX_1(s) - 4X_1(s) = 18/s^3$

D1.2: Show that the systems in Fig. D1.2 (*a*). and D1.2 (*b*) are analogous.

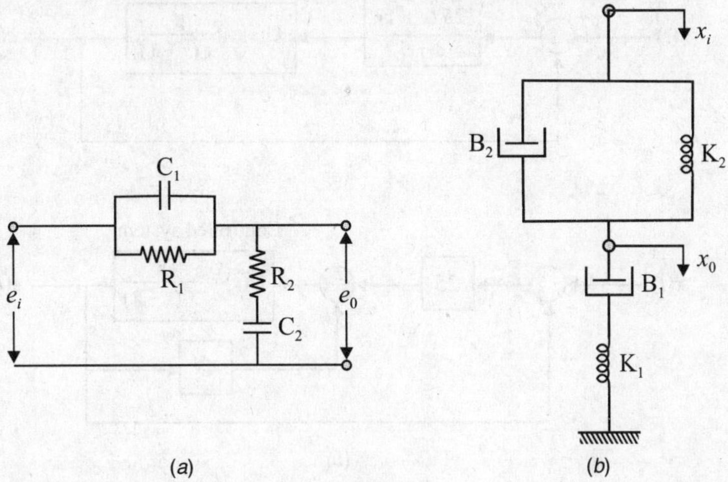

Fig. D1.2

D1.3: Fig. D1.3 shows a schematic diagram of a seismograph. x_i and x_o represent displacements of case and mass *m* respectively with respect to inertial space. If *y* is displacement of mass *m* with respect to case, find transmittance $Y(s)/X_i(s)$.

Ans. $-\dfrac{s^2}{s^2 + \dfrac{B}{m}s + \dfrac{K}{m}}$.

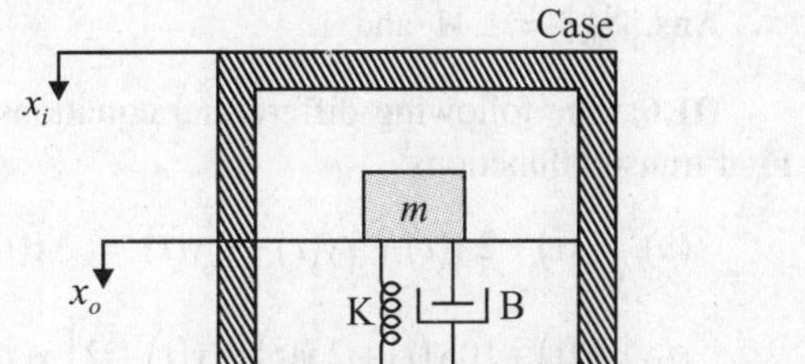

Fig. D1.3

D1.4: For the rotational system shown in Fig. D1.4, find Laplace transformed system equations and draw electrical analogs using

(*a*) Force-voltage analogy

(*b*) Force-current analogy.

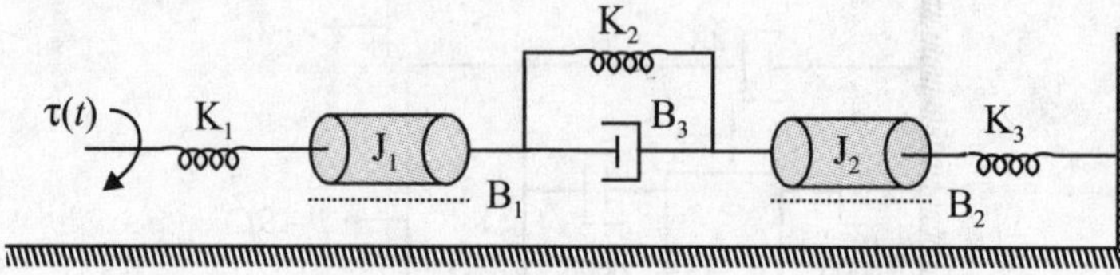

Fig. D1.4

D1.5: Consider the control strategy structures shown in Fig. D1.5 (*a*) and (*b*). Given nominal value of K = 1, show that both have same transfer function C(*s*)/X(*s*). Evaluate $\left|S_K^T\right|$ for both at ω = 5 rad/sec and comment on result. Does information about sensitivity provide any clue to the selection of a particular control strategy?

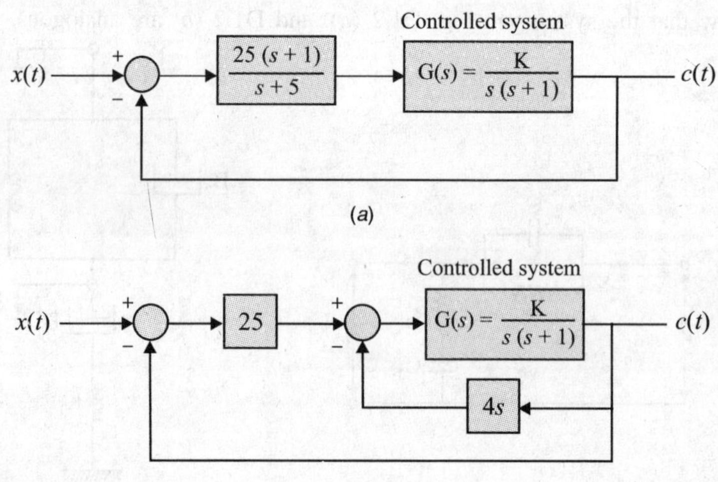

(*a*) (*b*)

Fig. D1.5

Ans. $\left|S_K^T\right|$ = 1.41 and 1.

D1.6: The following differential equations represent LTI systems with input *x*(*t*) and output *y*(*t*). Find transfer functions.

(*a*) $\dddot{y}(t)+2\ddot{y}(t)+5\dot{y}(t)+6y(t) = 3\dot{x}(t)+x(t)$

(*b*) $\ddddot{y}(t)+10\dddot{y}(t)+2\ddot{y}(t)+\dot{y}(t)+2\int_0^t y(\tau)d\tau = \dot{x}(t)+2x(t)$

Ans. (*a*) $\dfrac{3s+1}{s^3+2s^2+5s+6}$ (*b*) $\dfrac{s(s+2)}{s^4+10s^3+2s^2+s+2}$

D1.7: The dynamics of a multi input and multi output system with inputs $x_1(t)$ and $x_2(t)$ and outputs $y_1(t)$ and $y_2(t)$ is described by following set of differential equations.

$$\ddot{y}_1(t) + 2\dot{y}_1(t) + 3y_2(t) = x_1(t) + x_2(t)$$

$$\ddot{y}_2(t) + 3\dot{y}_1(t) + y_1(t) - y_2(t) = x_2(t) + \dot{x}_1(t)$$

Find transfer functions:

(*a*) $\left.\dfrac{Y_1(s)}{X_1(s)}\right|_{x_2 = 0}$ (*b*) $\left.\dfrac{Y_2(s)}{X_1(s)}\right|_{x_2 = 0}$ (*c*) $\left.\dfrac{Y_1(s)}{X_2(s)}\right|_{x_1 = 0}$ (*d*) $\left.\dfrac{Y_2(s)}{X_2(s)}\right|_{x_1 = 0}$

D1.8: Determine which of the following properties hold and which do not hold for each of systems with input $x(t)$ and output $y(t)$ given below.

Properties	**Systems**
(*a*) Causal	(*i*) $y(t) = t^2 x(t-1)$
(*b*) linear	(*ii*) $y(t) = [\cos 3t]\, x(t)$
(*c*) time invariant	(*iii*) $y(t) = \dot{x}(t)$
(*d*) memory less	(*iv*) $y(t) = \begin{cases} 0 & ; t < 0 \\ x(t) + x(t-2) & ; t \geq 0 \end{cases}$

MULTIPLE CHOICE QUESTIONS

M1.1: As compared to a closed-loop system, an open-loop system is:

(*a*) more stable as well as more accurate

(*b*) less stable as well as less accurate

(*c*) more stable but less accurate

(*d*) less stable but more accurate.

M1.2: Consider the following statements regarding the advantages of closed-loop negative feedback control system over open-loop system:

1. The overall reliability of the closed-loop system is more than that of open-loop system
2. The transient response in the closed-loop system decays more quickly than in the open-loop system
3. In an open-loop system, closing of the loop increases the overall gain of the system
4. In the closed-loop system, the effect of variation of component parameters on its performance is reduced.

Of these statements:

(*a*) 1 and 3 are correct (*b*) 1, 2 and 4 are correct

(*c*) 2 and 4 are correct (*d*) 3 and 4 are correct.

M1.3: Consider the following statements regarding a linear system $y = f(x)$:

1. $f(x_1 + x_2) = f(x_1) + f(x_2)$
2. $f[x(t + T)] = f[x(t)] + f[x(T)]$
3. $f(Kx) = K f(x)$

Of these statements:

(*a*) 1 , 2 and 3 are correct (*b*) 1 and 2 are correct

(*c*) 2 alone is correct (*d*) 1 and 3 are correct.

M1.4: Consider the systems shown in Fig. 1 and Fig. 2. If the forward path gain is reduced by 10% in each system, then the variation in C_1 and C_2 will be respectively:

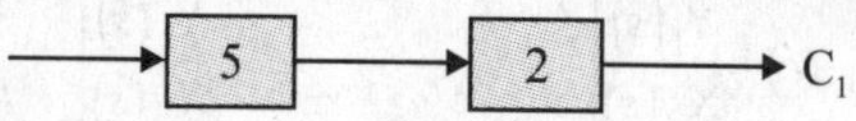

Fig. 1

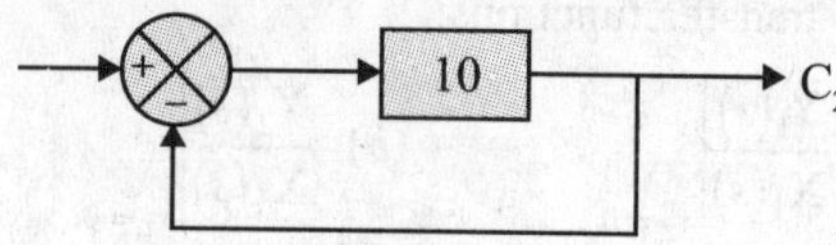

Fig. 2

(*a*) 10% and 10% (*b*) 2% and 10%

(*c*) 5% and 1% (*d*) 10% and 1%.

M1.5: In the system shown below, to eliminate the effect of disturbance D(*s*) on C(*s*), the transfer function $G_d(s)$ should be:

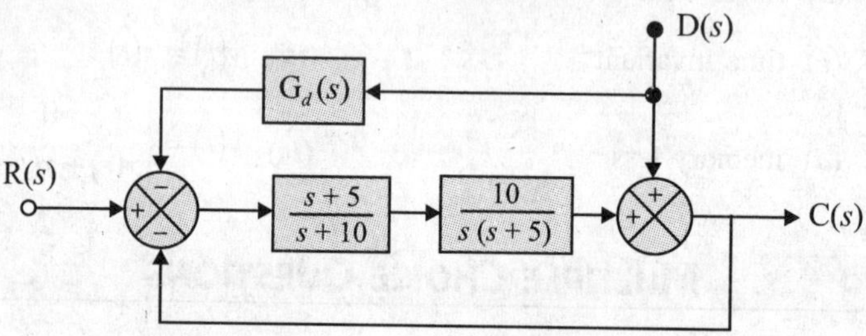

(*a*) $\dfrac{(s+10)}{10}$ (*b*) $\dfrac{s(s+10)}{10}$ (*c*) $\dfrac{10}{(s+10)}$ (*d*) $\dfrac{10}{s(s+10)}$

M1.6: A closed-loop system is shown below. The noise transfer function $\dfrac{C_n(s)}{N(s)}$ [$C_n(s)$ = output corresponding to noise input N(*s*)] is approximately:

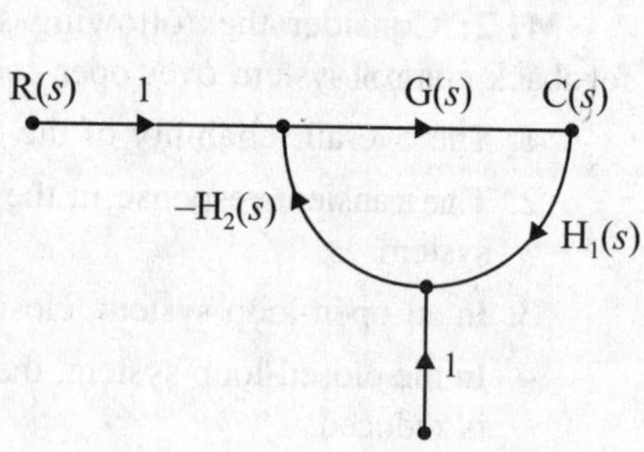

(*a*) $\dfrac{1}{G(s)H_1(s)}$ for $|G(s)H_1(s)H_2(s)| << 1$

(*b*) $-\dfrac{1}{H_1(s)}$ for $|G(s)H_1(s)H_2(s)| >> 1$

(*c*) $-\dfrac{1}{H_1(s)H_2(s)}$ for $|G(s)H_1(s)H_2(s)| >> 1$

(*d*) $-\dfrac{1}{G(s)H_1(s)H_2(s)}$ for $|G(s)H_1(s)H_2(s)| << 1$.

M1.7: A linear system, initially at rest, is subjected to an input signal;

$$r(t) = 1 - e^{-t} \ (t \geq 0)$$

The response of the system for $t \geq 0$ is given by

$$c(t) = 1 - e^{-2t}$$

The transfer function of the system is:

(*a*) $\dfrac{(s+2)}{(s+1)}$ (*b*) $\dfrac{(s+1)}{(s+2)}$ (*c*) $\dfrac{(s+1)^2}{s(s+2)}$ (*d*) $\dfrac{1}{2}\dfrac{(s+1)}{(s+2)}$.

M1.8: The impulse response of a system is $5e^{-10t}$. Its step response is equal to

(*a*) $0.55\ e^{-10t}$ (*b*) $5\ (1 - e^{-10t})$

(*c*) $0.5\ (11 - e^{-10t})$ (*d*) $10\ (1 - e^{-10t})$

M1.9: Given: $KK_t = 99$; $s = j1\,rad/s$: the sensitivity of the closed-loop system shown below, to variation in parameter K is approximately:

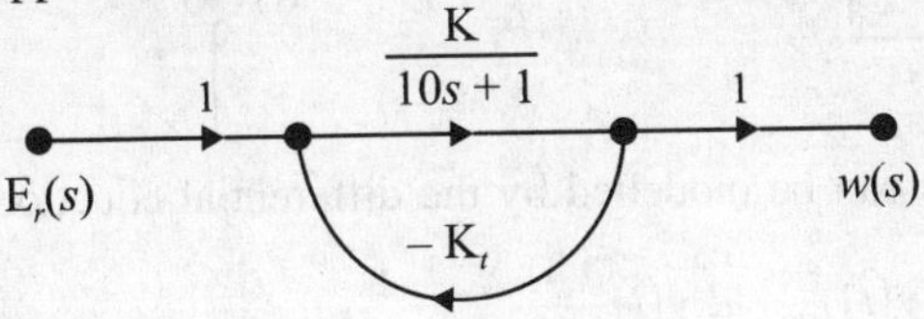

(*a*) 0.01 (*b*) 0.1 (*c*) 1.0 (*d*) 10

M1.10: Which one of the following systems is open-loop?

(*a*) The respiratory system of man

(*b*) A system for controlling the movement of the slide of a copying milling machine

(*c*) A thermostatic control

(*d*) Traffic light control.

M1.11: Match List-I (Physical action or activity) with List-II (Category of system) and select the correct answer using the codes given below the Lists:

List I	List II
A. Human respiration system	1. Man-made control system
B. Pointing of an object with a finger	2. Natural including biological control system
C. A man driving a car	3. Control system which is man-made and natural
D. A thermostatically controlled room heater	

Codes:	A	B	C	D
(*a*)	2	2	3	1
(*b*)	3	1	2	1
(*c*)	3	2	2	3
(*d*)	2	1	3	3

M1.12: Feedback control systems are

(*a*) insensitive to both forward and feedback path parameter changes

(*b*) less sensitive to feedback path parameter changes than to forward path parameter changes

(*c*) less sensitive to forward path parameter changes than to feedback path parameter changes

(*d*) equally sensitive to forward and feedback path parameter changes.

M1.13: Consider the following systems:

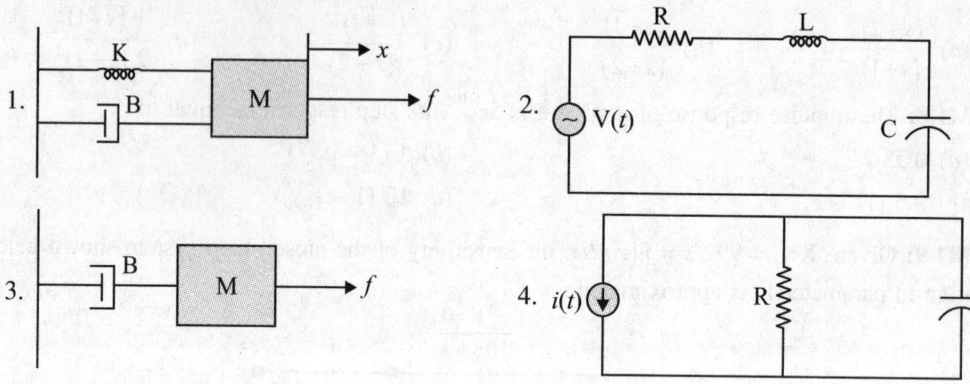

Which of these systems can be modelled by the differential equation of form

$$a_2 \frac{d^2 y(t)}{dt^2} + a_1 \frac{d\,y(t)}{dt} + a_0\,y(t) \;=\; x\,(t)$$

(*a*) 1 and 2 (*b*) 1 and 3

(*c*) 2 and 4 (*d*) 1, 2 and 4

M1.14: Which one of the following input-output relationship is that of a linear system

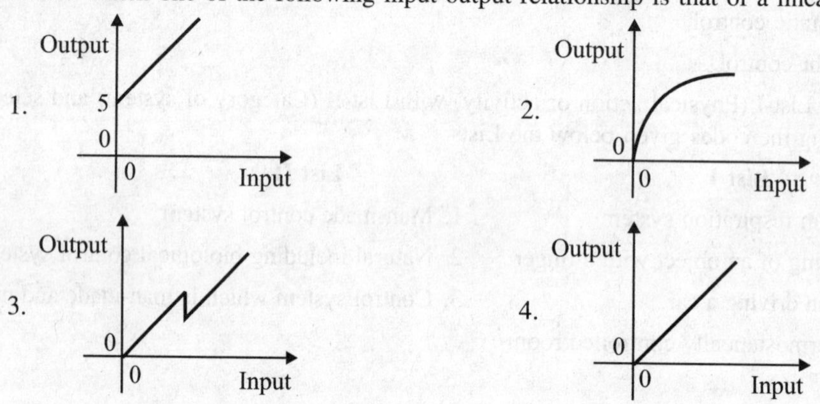

M1.15: Which one of the following pairs is NOT correctly matched; (input = $x(t)$ and output = $y(t)$)

(*a*) Unstable system .... $\dfrac{dy(t)}{dt} - 0.1\,y(t) = x(t)$ (*b*) Nonlinear system .... $\dfrac{dy(t)}{dt} + 2t^2\,y(t) = x(t)$

(*c*) Noncausal system .... $y(t) = x\,(t + 2)$ (*d*) Nondynamic system .... $y(t) = 3\,x^2(t)$.

M1.16: Consider the mechanical system shown below. If the system is set into motion by unit impulse force, the equation of the resulting oscillation will be:

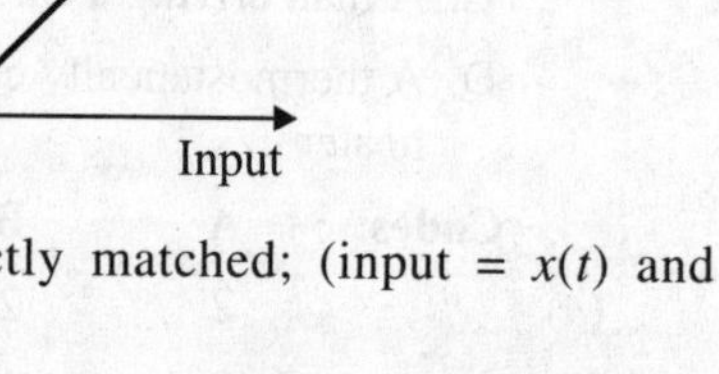

(*a*) $x(t) = \sin t$ (*b*) $x(t) = \sqrt{2}\,\sin t$

(*c*) $x(t) = \dfrac{1}{2}\sin 2t$ (*d*) $x(t) = \sin \sqrt{2}\,t$

M1.17: The open loop DC gain of unity negative feedback system with closed-loop transfer function $\frac{s+4}{s^2+7s+13}$ is

(*a*) $\frac{4}{13}$ (*b*) $\frac{4}{9}$ (*c*) 4 (*d*) 13

M1.18: Choose the correct matching:

(P) $a_1 \frac{d^2 y}{d x^2} + a_2 y \frac{d y}{d x} + a_3 y = a_4$

(Q) $a_1 \frac{d^3 y}{d x^3} + a_2 y = a_3$

(R) $a_1 \frac{d^2 y}{d x^2} + a_2 x \frac{d y}{d x} + a_3 x^2 y = 0$

(1) Non-linear differential equation
(2) Linear differential equation with constant coefficient
(3) Linear homogeneous differential equation
(4) Non-linear homogeneous differential equation
(5) Non-linear first order differential equation.

(*a*) P-1, Q-2, R-4 (*b*) P-1, Q-5, R-2
(*c*) P-2, Q-1, R-4 (*d*) P-5, Q-2, R-1.

M1.19: The transfer function of the system shown below is

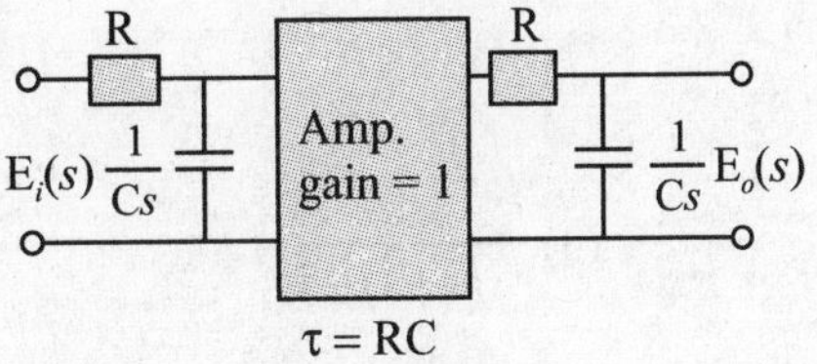

(*a*) $\frac{1}{1+\tau s}$ (*b*) $\frac{1}{(1+\tau s)^2}$

(*c*) $\frac{\tau s}{1+\tau s}$ (*d*) $\frac{\tau s}{(1+\tau s)^2}$

M1.20: Consider a system shown below:

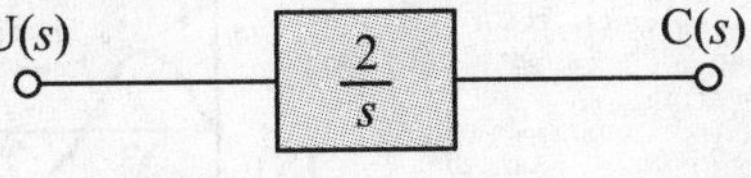

If the system is disturbed so that $c(0) = 1$, then $c(t)$ for a unit step input will be

(*a*) $1 + t$ (*b*) $1 - t$ (*c*) $1 + 2t$ (*d*) $1 - 2t$

M1.21: The impulse response of an initially relaxed linear system is $e^{-2t}u(t)$. To produce a response of $te^{-2t}u(t)$, the input must be equal to

(*a*) $2\,e^{-t}u(t)$ (*b*) $(1/2)\,e^{-2t}u(t)$ (*c*) $e^{-2t}u(t)$ (*d*) $e^{-t}u(t)$

M1.22: The unit impulse response of a unity feedback control system is given by;

$$c(t) = -te^{-t} + 2e^{-t}, \ (t \geq 0)$$

The open loop transfer function is equal to

(*a*) $\dfrac{(s+1)}{(s+2)^2}$ (*b*) $\dfrac{2s+1}{s^2}$ (*c*) $\dfrac{s+1}{(s+1)^2}$ (*d*) $\dfrac{s+1}{s^2}$

M1.23: The system has a transfer function $P(s) = \dfrac{2}{s+2}$. The gain for $\omega = 2$ rad/sec will be

(*a*) 0.707 (*b*) 0.666 (*c*) 0.5 (*d*) 0.25

M1.24: A linear time-invariant system initially at rest, when subjected to a unit-step input gives a response $y(t) = te^{-t}$; $t > 0$. The transfer function of the system is

(*a*) $\dfrac{1}{(s+1)^2}$ (*b*) $\dfrac{1}{s(s+1)^2}$ (*c*) $\dfrac{s}{(s+1)^2}$ (*d*) $\dfrac{1}{s(s+1)}$

M1.25: Match the following transfer functions and impulse responses:

Transfer function **Impulse responses**

(1) $\dfrac{s}{s+1}$ (P) 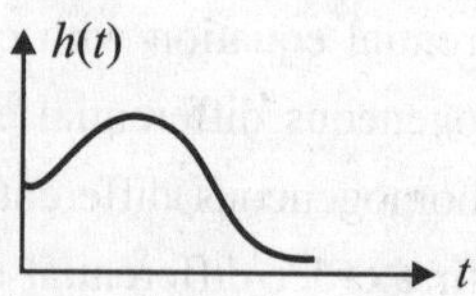

(2) $\dfrac{s}{(s+1)^2}$ (Q) 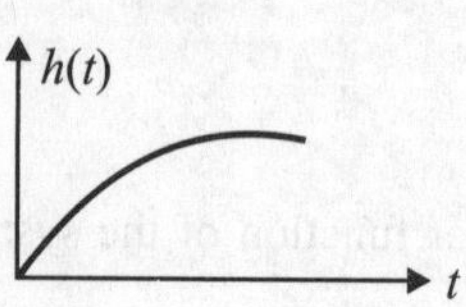

(3) $\dfrac{1}{s(s+1)+1}$ (R) 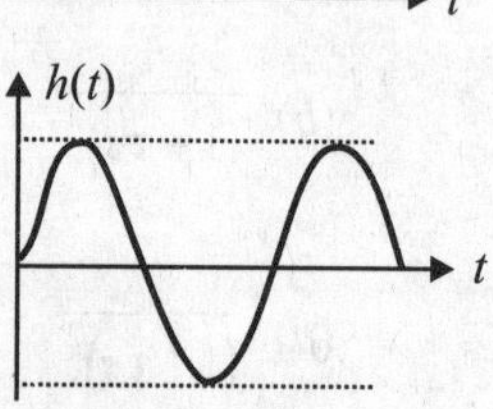

(4) $\dfrac{s}{s^2+1}$ (S) 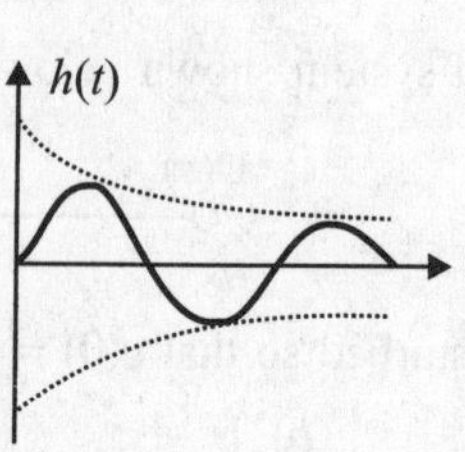

(*a*) 1-P, 2-Q, 3-S, 4-R (*b*) 1-P, 2-Q, 3-R, 4-S

(*c*) 1-Q, 2-P, 3-S, 4-R (*d*) 1-R, 2-Q, 3-S, 4-P.

M1.26: The output $y(t)$ of figure shown below with three inputs u_1, u_2 and u_3 using superposition, is

(*a*) $y(t) = 5\ (\cos t - 2 \sin 2t - t^2)$

(*b*) $y(t) = 5\ (\cos t + 2 \sin 2t - t^2)$

(*c*) $y(t) = 5\ (\cos t - 2 \sin 2t + t^2)$

(*d*) $y(t) = 5\ (\cos t + 2 \sin 2t + t^2)$

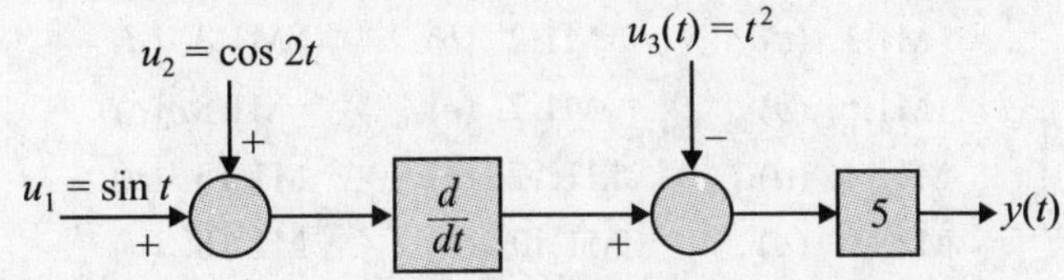

M1.27: The unit step response of a system is

$$y(t) = 1 - \frac{7}{3}e^{-t} + \frac{3}{2}e^{-2t} - \frac{1}{6}e^{-4t}$$

The transfer function model of the system is

(*a*) $\dfrac{(s+8)}{(s+1)(s+2)(s+4)}$ (*b*) $\dfrac{(s+4)}{(s+1)(s+2)(s+8)}$

(*c*) $\dfrac{(s+1)}{(s+2)(s+4)(s+8)}$ (*d*) $\dfrac{1}{(s+1)(s+2)(s+4)}$

M1.28: The figure shown below represents system I and system II. Now, consider following statements in respect of these systems when $K_1 = K_2 = 100$ nominally.

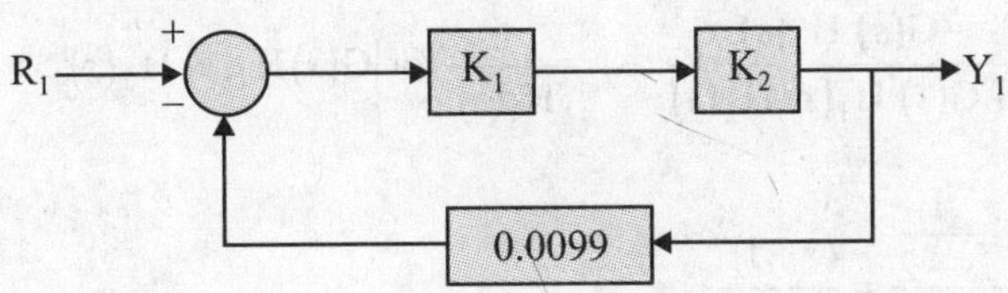

System I

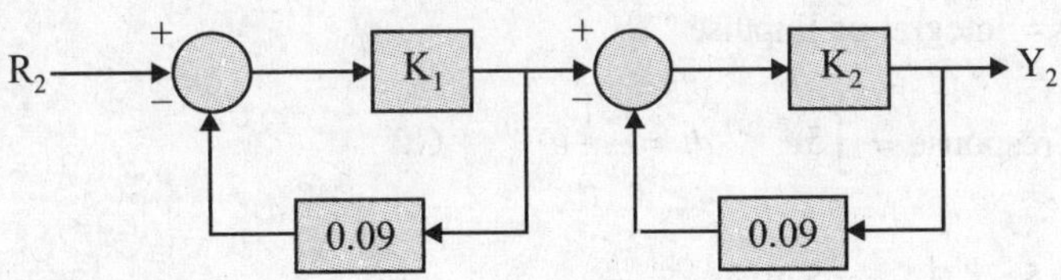

System II

1. System I and system II both have same transfer function model.
2. The response of system II is **ten** times more sensitive to variation in K_1 than is the response of system I.

Of these

(*a*) 1 and 2 both are correct (*b*) only 1 is correct

(*c*) 1 and 2 both are incorrect (*d*) only 2 is correct.

ANSWERS

M1.1. (*c*)	M1.2. (*b*)	M1.3. (*d*)	M1.4. (*d*)	M1.5. (*b*)
M1.6. (*b*)	M1.7. (*c*)	M1.8. (*c*)	M1.9. (*b*)	M1.10. (*d*)
M1.11. (*a*)	M1.12. (*c*)	M1.13. (*a*)	M1.14. (*d*)	M1.15. (*c*)
M1.16. (*a*)	M1.17. (*b*)	M1.18. (*a*)	M1.19. (*b*)	M1.20. (*c*)
M1.21. (*c*)	M1.22. (*b*)	M1.23. (*a*)	M1.24. (*c*)	M1.25. (*c*)
M1.26. (*a*)	M1.27. (*a*)	M1.28. (*a*)		

Important Hints

M1.4: OL → variation is directly reflected in response

CL → variation is reduced by a factor of (1 + GH).

M1.5: $\frac{C(s)}{D(s)} = \left[1 - \frac{10G_d(s)}{s(s+10)}\right] \Big/ \left[1 + \frac{10}{s(s+10)}\right]$

So that $\frac{C(s)}{D(s)} = 0, \; G_d(s) = \frac{s(s+10)}{10}$

M1.6: $\frac{C(s)}{N(s)} = \frac{-G(s)\,H_2(s)}{1+G(s)\,H_1(s)\,H_2(s)} = -\frac{1}{H_1(s)}$ for $|G(s)H_1(s)\,H_2(s)| >> 1$

M1.7: $\frac{C(s)}{R(s)} = \frac{1-\frac{1}{s+2}}{1-\frac{1}{s+1}} = \frac{(s+1)^2}{s(s+2)}$

M1.8: Step = Integral of impulse

step response = $\int_0^t 5e^{-10t}\,dt = -\frac{1}{2}e^{-10t} + C$

since $5e^{-10t}\big|_{t=0} = 5$ gives

$$\left(-\frac{1}{2}e^{-10t} + C\right)\Big|_{t=0} = 5$$

C = 5.5. ∴ Step response = 0.5 [11 – e^{-10t}]

M1.9: $T(s) = \frac{K}{10s+1+KK_t}$; $S_K^T = \frac{(10s+1)}{10(s+10)}$ and $\left|S_K^T\right|_{s=j1} = 0.1$

M1.16: $\dfrac{X(s)}{F(s)} = \dfrac{1}{s^2+1}$

M1.17: $\dfrac{G(s)}{1+G(s)} = \dfrac{s+4}{s^2+7s+13}$; $G(s) = \dfrac{s+4}{s^2+6s+9}$; $G(s)\Big|_{s=0} = \dfrac{4}{9}$

M1.20: $\dfrac{C(s)}{U(s)} = \dfrac{2}{s}$ where $U(s) = \dfrac{1}{s} \Rightarrow C(s) = \dfrac{2}{s^2} \Rightarrow c(t) = 2t$

but $c(0) = 1$; overall $c(t) = 1 + 2t$

M1.21: $G(s)$ = L [impulse response] = $\dfrac{1}{s+2}$; L stands for Laplace transform

$C(s) = G(s)\, R(s) = \dfrac{1}{(s+2)^2} \Rightarrow R(s) = \dfrac{C(s)}{G(s)} = \dfrac{1}{(s+2)}$

$r(t) = e^{-2t}u(t)$

M1.22: CLTF = $-\dfrac{1}{(s+1)^2} + \dfrac{2}{(s+1)} = \dfrac{2s+1}{(s+1)^2}$

but CLTF with unity feedback = $\dfrac{G(s)}{1+G(s)} = \dfrac{2s+1}{s^2+2s+1}$; $G(s)$ is OLTF

OLTF = $\dfrac{2s+1}{s^2}$

M1.23: $\left|P(j\omega)\right| = \dfrac{2}{\sqrt{4+\omega^2}}\Bigg|_{\omega=2} = \dfrac{1}{\sqrt{2}}$

M1.24: $Y(s) = \dfrac{1}{(s+1)^2} = G(s)\, R(s)$

$G(s) = \dfrac{Y(s)}{R(s)} = \dfrac{s}{(s+1)^2}$

M1.26: $y_1\Big|_{u_2 = u_3 = 0} = 5\dfrac{d}{dt}\sin t = 5\cos t$

$y_2\Big|_{u_1 = u_3 = 0} = 5\dfrac{d}{dt}(\cos 2t) = -10\sin 2t$

$y_3\Big|_{u_1 = u_2 = 0} = -5t^2$

$y = y_1 + y_2 + y_3$

M1.27: Impulse response $= \dfrac{d}{dt}\left[1 - \dfrac{7}{3}e^{-t} + \dfrac{3}{2}e^{-2t} - \dfrac{1}{6}e^{-4t}\right]$

$$= \frac{7}{3}e^{-t} - 3e^{-2t} + \frac{2}{3}e^{-4t}$$

$$\text{Transfer function} = \frac{7}{3}\cdot\frac{1}{(s+1)} - 3\cdot\frac{1}{(s+2)} + \frac{2}{3}\cdot\frac{1}{(s+4)}$$

$$= \frac{(s+8)}{(s+1)\,(s+2)\,(s+4)}$$

M1.28: $T_1 = \dfrac{Y_1}{R_1} = \dfrac{K_1K_2}{1+0.0099K_1K_2} = \dfrac{100\times 100}{1+0.0099\times 100\times 100} = 100$

$$T_2 = \frac{Y_2}{R_2} = \left[\frac{K_1}{1+0.09K_1}\right]\left[\frac{K_2}{1+0.09K_2}\right] = \left(\frac{100}{1+0.09\times 100}\right)\left(\frac{100}{1+0.09\times 100}\right) = 100$$

$$S_{K_1}^{T_1} = \frac{\partial T_1}{\partial K_1}\times\frac{K_1}{T_1} = \frac{1}{1+0.0099K_1K_2} = \frac{1}{1+0.0099\times 100\times 100} = 0.01$$

$$S_{K_1}^{T_2} = \frac{\partial T_2}{\partial K_1}\times\frac{K_1}{T_2} = \frac{1}{1+0.09K_1} = \frac{1}{1+0.09\times 100} = 0.1$$

2

CONTROL SYSTEM ANALYSIS IN TIME DOMAIN

2.1 INTRODUCTION

The time is often obvious choice of control engineers as independent variable for the analysis of a control system. To study the behavioural features of a control system under dynamic and steady conditions, the very first step involved, is development of mathematical model of system. The mathematical model of a continuous system is a differential equation or a set of differential equations. It is well known that solution of a differential equation consists of two parts.

(*i*) The complementary function which reveals the transient or dynamic behaviour of the system.

(*ii*) Particular integral which provides the information about steady state behaviour of the system.

The entire time domain response $y(t)$ of a control system is divided into two parts: the transient response $y_t(t)$ and the steady state response $y_s(t)$.

Thus $$y(t) = y_t(t) + y_s(t)$$

It is worth noting that $y_t(t)$ decays to zero as time becomes very large *i.e.*

$$\underset{t \to \infty}{\text{lt}} \; y_t(t) = 0$$

and steady state part of total response, persists after transient part has died out. Nevertheless, the steady state part may still vary but it will always do so in a fixed pattern, *e.g.* sinusoidally for sinusoidal input to linear system.

Note that the nature of transient behaviour while being independent of type of input, depends only upon system components and their layout whereas the steady state behaviour depends not only on system components and their arrangement but also on type of input. The commonly used standard test input signals are those of impulse, step, ramp and parabolic.

2.2 STANDARD TEST SIGNALS

(a) Step displacement input (Step function)

This type of input abruptly changes from one level (usually zero) to another level R in zero time as shown in Fig. 2.1(*a*) and is mathematically modelled as

$$r(t) = R; \quad t \geq 0$$
$$= 0; \text{ elsewhere}$$

$$L[r(t)] = R(s) = \frac{R}{s}$$

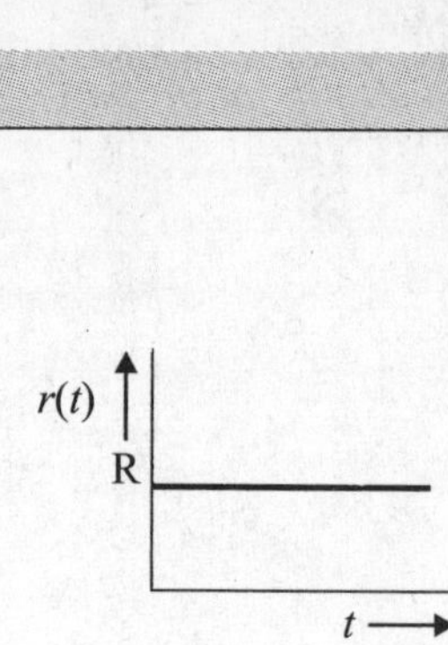

Fig. 2.1: (*a*) Step function

(b) Step velocity input (Ramp function)

This signal begins at zero level and increases linearly with time as shown in Fig. 2.1(*b*).

$$r(t) = Rt; \quad t \geq 0$$
$$= 0; \text{ elsewhere}$$

$$L[r(t)] = R(s) = \frac{R}{s^2}$$

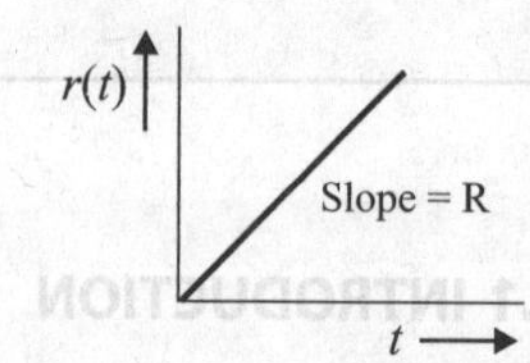

Fig. 2.1: (*b*) Ramp function

(c) Step acceleration input (Parabolic function)

This signal is one order faster than ramp as shown in Fig. 2.1(*c*) and mathematically represented as:

$$r(t) = \frac{Rt^2}{2}; \quad t \geq 0$$
$$= 0; \text{ elsewhere}$$

$$L[r(t)] = R(s) = \frac{R}{s^3}$$

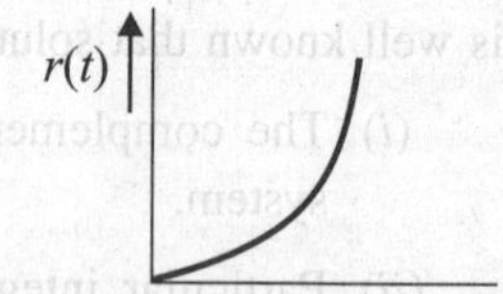

Fig. 2.1: (*c*) Parabolic

(d) Impulse function

The impulse function $\delta(t)$ is derivative of step function *i.e.*

$$\delta(t) = \frac{d}{dt} u(t) \quad \text{or} \quad u(t) = \int_0^t \delta(\tau)\, d\tau$$

but slope (derivative) of $u(t)$ is 0 except at $t = 0$ where it is ∞. Thus $\delta(t)$ is defined as a function having zero value for all values of t except at $t = 0$, additionally satisfying the relationship

$$\int_{-\infty}^{+\infty} \delta(t)\, dt = 1 \quad \text{or} \quad \int_{0^-}^{0^+} \delta(t)\, dt = 1$$

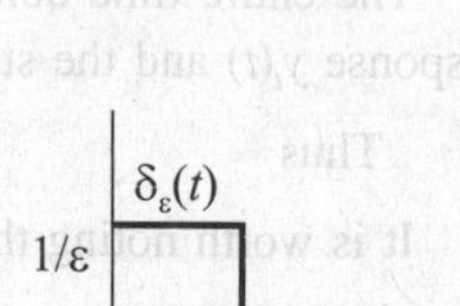

Fig. 2.1: (*d*) Pulse of duration ε

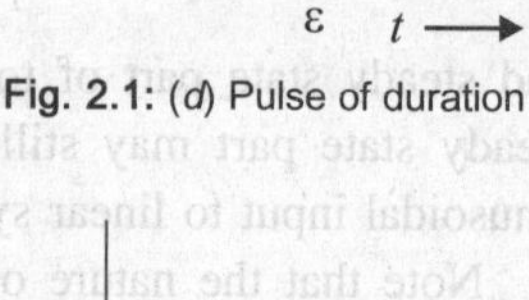

Fig. 2.1: (*e*) Non-unity shifted impulse function

It may be noted that an impulse having infinite magnitude and zero duration does not occur in physical systems. In practice, a pulse input with a very short duration compared with the significant time constant of the system, can be treated as an impulse. Let us define $\delta_\varepsilon(t)$ as shown in Fig. 2.1(*d*).

As $\varepsilon \to 0$, $\delta_\varepsilon(t)$ becomes $\delta(t)$ of zero width and infinite magnitude. In general, a non-unity impulse occurring at $t = t_0$ may be written as

$$k\delta\,(t - t_0) = k\frac{d}{dt}u\,(t - t_0)$$

where k is called weight or strength of impulse. See Fig. 2.1(*e*).

It may be noted that the nature of transient response can be revealed by any of test signal as it is independent of type of input, it is quite adequate to derive information about it for only one of the standard input for which the step input is usually used. The obvious choice for step input is because

(*i*) this is the worst kind of input, a system may be subjected during its entire operational tenure.

(*ii*) it becomes a direct investigation of quickness of system in responding to abrupt changes.

(*iii*) the spectrum of step signal contains wide band of frequencies due to the jump discontinuity and it is equivalent to simultaneous application of numerous sinusoidal signals.

2.3 FIRST ORDER SYSTEM

Let us consider the first order system shown in Fig. 2.2. A physical example of such a system could be a thermal system, an RC circuit, etc.

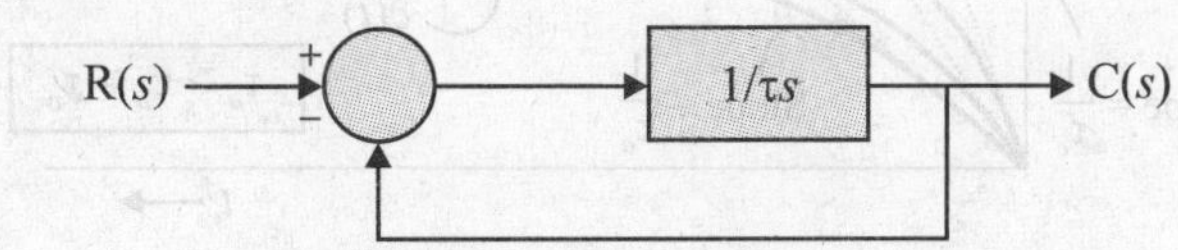

Fig. 2.2: Block diagram of a first-order system

The input-output relationship is given by

$$\frac{C(s)}{R(s)} = \frac{1}{s\tau + 1} \quad \text{...(2.1)}$$

The cross multiplication yields $\tau s\,C(s) + C(s) = R(s)$

whose inverse Laplace yields

$$\tau\dot{c}(t) + c(t) = r(t) \quad \text{...(2.2)}$$

Equation (2.2) provide some relevant information about dynamical behaviour of system. If system under consideration is position control system and $c(t)$ = mechanical position then $\dot{c}(t)$ obviously denotes velocity. Thus a first order system involves only velocity in its dynamics. Intuitively higher order dynamical systems will involve some more complex dynamics. For example, a second order system will include acceleration also. Dynamics of third order system will have time variation of acceleration and so on.

Unit step response of first order system

Consider the system shown in Fig. 2.2. With $r(t) = u(t)$ or $R(s) = 1/s$, (2.1) takes the form

$$C(s) = \frac{1}{s(\tau s+1)} = \frac{\alpha_1}{s} + \frac{\alpha_2}{\tau s+1} \qquad ...(2.3)$$

where $$\alpha_1 = \left.\frac{1}{\tau s+1}\right|_{s=0} = 1$$

and $$\alpha_2 = \left.\frac{1}{s}\right|_{s=-1/\tau} = -\tau$$

$$C(s) = \frac{1}{s} - \frac{\tau}{\tau s+1} \Rightarrow c(t) = 1 - e^{-t/\tau} \qquad ...(2.4)$$

Therefore $$c(0) = c(t)\big|_{t=0} = 0 \text{ and } c(t)\big|_{t=\infty} = c(\infty) = 1$$

Thus response rises from $c(0) = 0$ to $c(\infty) = 1$ exponentially. Figure 2.3 shows typical unit step response curves for three different values of τ.

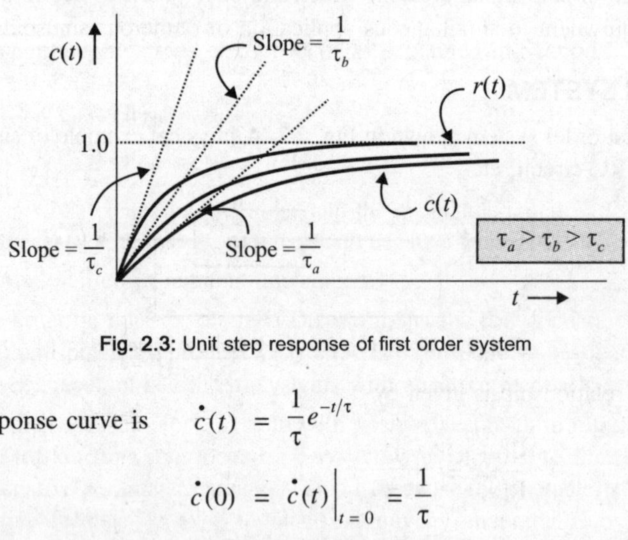

Fig. 2.3: Unit step response of first order system

The slope of response curve is $$\dot{c}(t) = \frac{1}{\tau}e^{-t/\tau}$$

The initial slope $$\dot{c}(0) = \dot{c}(t)\Big|_{t=0} = \frac{1}{\tau}$$

and terminal slope is $$\dot{c}(\infty) = \dot{c}(t)\Big|_{t=\infty} = 0$$

i.e., slope of response curve decreases monotonically from $\frac{1}{\tau}$ at $t = 0$ to 0 at $t = \infty$.

This τ, called as time constant, is an important parameter of the system. τ is indicative of how fast the system tends to reach the final value. A large time constant τ_a corresponds to sluggish and smaller time constant τ_b to faster response. The system with time constant τ_c exhibits still faster response as shown in Fig. 2.3.

Also $$c(t)\big|_{t=\tau} = 1 - e^{-1} = 0.632 \qquad ...(2.5)$$

The time constant, alternatively according to equation (2.5), may also be defined as time taken by exponential response to reach from 0 to 63.2% of final value.

The exponential response reaches from 0 to 63.2% of final value in one time constant. The response reaches 86.5% of final value in two time constants. At $t = 3\tau, 4\tau, 5\tau$, the response reaches 95%, 98.2% and 99.3% of final value. Though mathematical routine suggests arrival of steady state only after an infinite time, but a practically reasonable time to reach steady state within 2% tolerance band is four time constants as shown in Fig. 2.4.

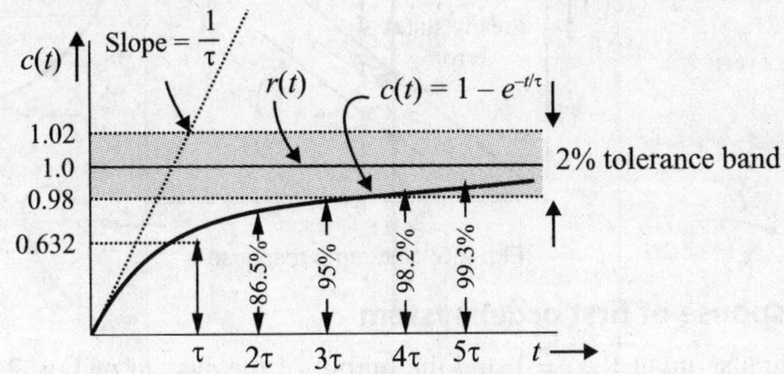

Fig. 2.4: Exponential response curve

At steady state $\quad c(\infty) = c(t)\big|_{t=\infty} = 1$

and error
$$e(t) = r(t) - c(t) = 1 - [1 - e^{-t/\tau}] = e^{-t/\tau} \qquad \text{...(2.6)}$$

Steady state error for unit step input $= e(\infty) = e(t)\big|_{t=\infty} = 0$ *i.e.* the system tracks unit step input with zero steady state error. An important conclusion one can derive from this analysis is that any change in τ will bring about change in only transient behaviour of system leaving steady state behaviour unchanged.

Unit ramp response of first order system

Since L [Unit ramp function] $= \dfrac{1}{s^2}$

$$C(s) \text{ for } R(s) = \frac{1}{s^2} \text{ is } \frac{1}{s^2(\tau s+1)} = \frac{1}{s^2} - \frac{\tau}{s} + \frac{\tau^2}{\tau s+1}$$

and
$$c(t) = L^{-1}[C(s)] = t - \tau + \tau e^{-t/\tau} \qquad \text{...(2.7)}$$

where
$$c(t)\big|_{t=0} = c(0) = 0 \text{ and } c(t)\big|_{t=\infty} = c(\infty) = \infty$$

The error
$$e(t) = r(t) - c(t) = t - [t - \tau + \tau e^{-t/\tau}] = \tau[1 - e^{-t/\tau}] \qquad \text{...(2.8)}$$

The initial error $e(0)$ and terminal error $e(\infty)$ take the following forms

$$e(0) = e(t)\big|_{t=0} = 0 \text{ and } e(\infty) = e(t)\big|_{t=\infty} = \tau$$

Slope of response curve is $\dot{c}(t) = 1 - e^{-t/\tau}$

and
$$\dot{c}(t)\Big|_{t=0} = \dot{c}(0) = 0 \text{ and } \dot{c}(\infty) = \dot{c}(t)\Big|_{t=\infty} = 1$$

Thus it may be seen, any change in τ will bring about change in not only transient behaviour of system but also in steady state error. The smaller the value of time constant τ, the smaller is the steady state error and faster is dynamics. (See Fig. 2.5)

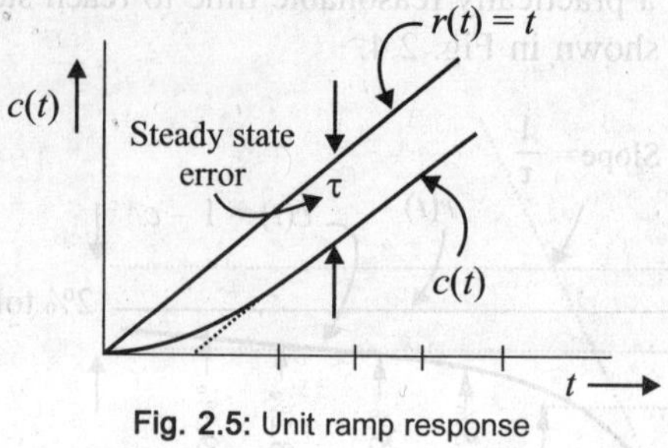

Fig. 2.5: Unit ramp response

Unit impulse response of first order system

For the unit impulse input R(s) = 1 and the output of the system of Fig. 2.2 can be obtained as

$$C(s) = \frac{1}{\tau s+1} \quad \text{and} \quad c(t) = \frac{1}{\tau}e^{-t/\tau} \qquad ...(2.9)$$

where
$$c(0) = c(t)\big|_{t=0} = \frac{1}{\tau} \quad \text{and} \quad c(\infty) = c(t)\big|_{t=\infty} = 0$$

suggest that the impulse response decays form $\frac{1}{\tau}$ at $t = 0$ to zero at $t = \infty$ with an intermediate value $c(t)\big|_{t=\tau} = \frac{1}{\tau}e^{-1} = 0.368/\tau$. The response given by (2.9) is shown in Fig. 2.6.

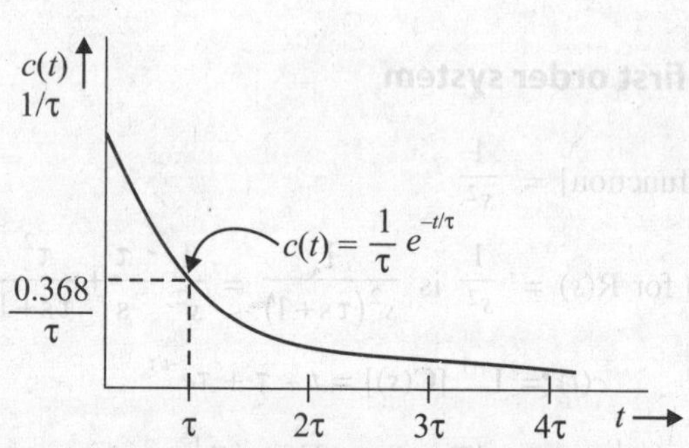

Fig. 2.6: Unit impulse response

From foregoing analysis, the following important property for a LTI system is observed.

1. $c(t)\big|_{\text{for ramp input}} = t - \tau + \tau e^{-t/\tau}$
2. Unit step input = derivative of unit ramp input.
3. Derivative of unit ramp response = $1 - e^{-t/\tau}$ = unit step response.
4. Unit impulse input = derivative of unit step input.
5. Derivative of unit step response = $\frac{1}{\tau}e^{-t/\tau}$ = unit impulse response.

From above illustration one can derive the following conclusions:

(*i*) Response to the derivative of an input can be obtained by differentiating the response of the system to original input.

(*ii*) Response to the integral of an input can be obtained by integrating the response of the system to original input and by evaluating of the integration constants from initial conditions.

(*iii*) The first derivative of parabolic function is ramp function. The first derivative of ramp function is step function. The first derivative of step function is similarly the impulse function.

Note that linear time varying systems and non linear systems do not possess this property.

Illustrative Examples

1. The impulse response of a initially relaxed system is $5e^{-10t}$. Compute its step response.

Solution: Impulse input = Derivative of step unit

Step input = Integral of impulse input

response to step input = Integral of response to impulse input.

$$\therefore \quad \text{Step response} = \int_0^t 5e^{-10t}\,dt = -\frac{5}{10}e^{-10t}\Big|_0^t$$

$$= -0.5\left[e^{-10t} - 1\right] = 0.5 - 0.5\,e^{-10t}$$

2. When the input to a system was withdrawn at $t = 0$, its output was found to decrease exponentially from 1000 units to 500 units in 1.386 sec. What is time constant of system?

Solution: Response with exponential decay is given as;

$$c(t) = 1000\,e^{-t/\tau}$$

$$c(t)\big|_{t=1.386} = 500 \quad \Rightarrow \quad \tau = 1.9995 \cong 2.0$$

3. A first order system and its response to a step input are shown in Fig. 2.7, compute system parameters α and k.

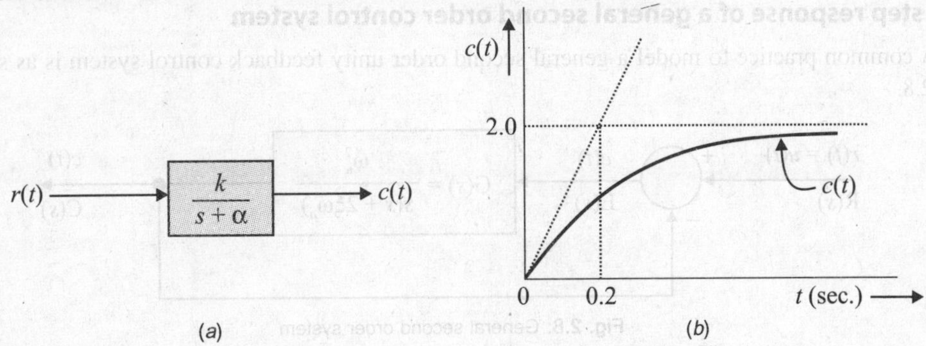

Fig. 2.7: (*a*) System, (*b*) Step response

Solution: Given $\dfrac{C(s)}{R(s)} = \dfrac{k}{s+\alpha}$

and $$R(s) = \frac{2}{s}$$

$$C(s) = \frac{2k}{s(s+\alpha)} = \frac{\alpha_1}{s} + \frac{\alpha_2}{s+\alpha}$$

where $$\alpha_1 = \left.\frac{2k}{(s+\alpha)}\right|_{s=0} = \frac{2k}{\alpha}$$

$$\alpha_2 = \left.\frac{2k}{s}\right|_{s=-\alpha} = -\frac{2k}{\alpha}$$

$$\therefore \quad C(s) = \frac{2k/\alpha}{s} - \frac{2k/\alpha}{s+\alpha}$$

Inverse Laplace yields $$c(t) = \frac{2k}{\alpha} - \frac{2k}{\alpha}e^{-\alpha t} = \frac{2k}{\alpha}[1 - e^{-\alpha t}]$$

$$c(\infty) = c(t)\big|_{t=\infty} = \frac{2k}{\alpha}$$

$$\dot{c}(t) = 2k[e^{-\alpha t}]$$

$$\text{Initial slope} = \dot{c}(t)\big|_{t=0} = 2k$$

From given response $\frac{2k}{\alpha} = 2$ and $2k = 10$ solving which we have

$$k = 5 \quad \text{and} \quad \alpha = 5.$$

2.4 SECOND ORDER SYSTEM

Although a practical system, rarely turns out to be of order two, yet it does not become irrelevant to control system rather provides very good base for analysis and design of system of higher order.

Unit step response of a general second order control system

A common practice to model a general second order unity feedback control system is as shown in Fig. 2.8.

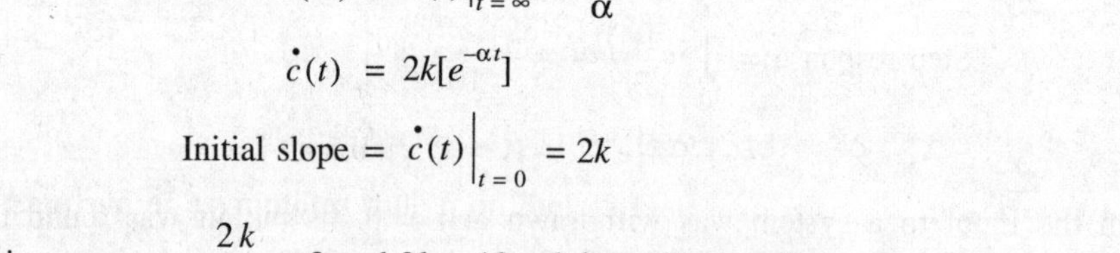

Fig. 2.8: General second order system

The response C(s) is given by

$$C(s) = \frac{\omega_n^2}{s^2 + 2\xi\omega_n s + \omega_n^2} \times R(s) \qquad \text{...(2.10)}$$

With $R(s) = \frac{1}{s}$, (2.10) takes the following form

$$C(s) = \frac{k_1}{s} + \frac{k_2 s + k_3}{s^2 + 2\xi\omega_n s + \omega_n^2} \quad ...(2.11)$$

where k_1, k_2 and k_3 are partial coefficients. It is easy to verify that $k_1 = 1$, $k_2 = -1$, and $k_3 = -2\xi\omega_n$.

Substitution of k_1, k_2 and k_3 in (2.11) yields

$$C(s) = \frac{1}{s} - \left[\frac{s + 2\xi\omega_n}{(s+\xi\omega_n)^2 + \omega_n^2(1-\xi^2)}\right]$$

$$= \frac{1}{s} - \left[\frac{s + \xi\omega_n}{(s+\xi\omega_n)^2 + \omega_n^2(1-\xi^2)} + \frac{\xi\omega_n}{(s+\xi\omega_n)^2 + \omega_n^2(1-\xi^2)}\right] \quad ...(2.12)$$

Let $\xi\omega_n = \sigma$ and $\omega_n^2(1 - \xi^2) = \omega_d^2$ then (2.12) may be written as

$$C(s) = \frac{1}{s} - \left[\frac{s+\sigma}{(s+\sigma)^2 + \omega_d^2} + \frac{\xi}{\sqrt{1-\xi^2}} \frac{\omega_d}{(s+\sigma)^2 + \omega_d^2}\right] \quad ...(2.13)$$

Inverse Laplace transform of (2.13) gives

$$c(t) = 1 - \left[e^{-\sigma t}\cos\omega_d t + \frac{\xi}{\sqrt{1-\xi^2}} e^{-\sigma t}\sin\omega_d t\right]$$

$$= 1 - e^{-\sigma t}\left[\cos\omega_d t + \frac{\xi}{\sqrt{1-\xi^2}}\sin\omega_d t\right] \quad ...(2.14)$$

$$= 1 - \frac{e^{-\sigma t}}{\sqrt{1-\xi^2}}\left[\sqrt{1-\xi^2}\cos\omega_d t + \xi\sin\omega_d t\right]$$

$$c(t) = 1 - \frac{e^{-\sigma t}}{\sqrt{1-\xi^2}} \sin(\omega_d t + \phi) \quad ...(2.15)$$

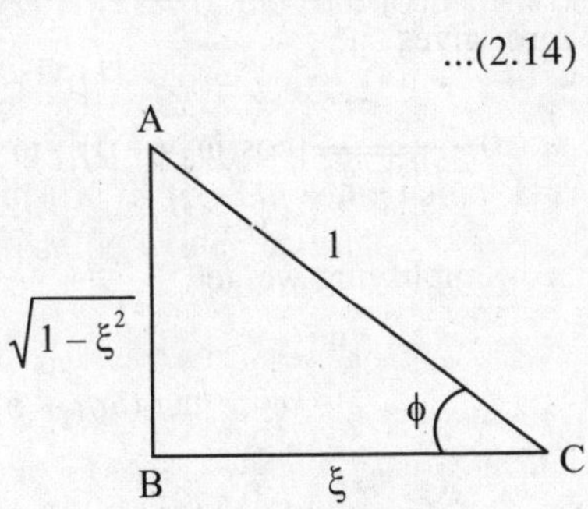

Fig. 2.9

where $\phi = \tan^{-1}\left[\frac{\sqrt{1-\xi^2}}{\xi}\right]$; refer Fig. 2.9

It may be noted that the term σ appears in exponential term of unit step response (2.15) and therefore indicates the rate of decay of response $c(t)$. Alternatively it provides information about damping of system and is called **damping constant** or **damping factor**, the inverse of which is proportional to time constant of system. In expression for damping factor $\sigma = \xi\omega_n$, ξ is called **damping ratio** and ω_n the **natural undamped frequency**. The term $\omega_d = \omega_n\sqrt{1-\xi^2}$ is called **damped frequency** or **conditional frequency**.

It is further obvious from (2.15) that the dynamics of the system to unit step input, will be exponential as well as sinusoidal in nature. For a typical second order system the typical response is shown in Fig. 2.10.

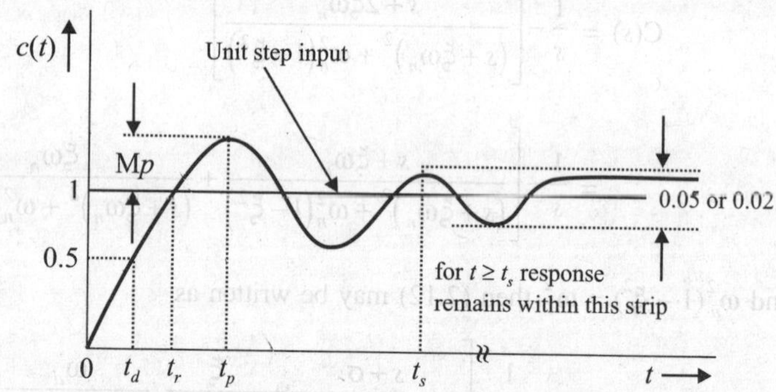

Fig. 2.10: Transient response specifications

In characterising the transient response characteristics of a control system to a unit step input, it is common practice to specify the following.

Transient response specifications

1. Peak time (t_p)

The peak time is the time required for the response to reach the peak of the first overshoot. Refer Fig. 2.10 to note that t_p can be evaluated by equation $\dot{c}(t) = 0$. Differentiating (2.15) and equating it to zero, gives

$$0 - \frac{e^{-\sigma t}}{\sqrt{1-\xi^2}}[\cos(\omega_d t + \phi)] \cdot \omega_d + \frac{\sigma e^{-\sigma t}}{\sqrt{1-\xi^2}} \sin(\omega_d + \phi) = 0$$

Simplifying we get

$$\tan(\omega_d t + \phi) = \frac{\omega_d}{\sigma} = \frac{\sqrt{1-\xi^2}}{\xi}$$

or $$\tan(\omega_d t + \phi) = \tan\phi$$

Since $$\tan(n\pi + \phi) = \tan\phi$$

$$\omega_d t = n\pi$$

and $$t = \frac{n\pi}{\omega_n\sqrt{1-\xi^2}} \quad \text{...(2.16)}$$

Equation (2.16) gives the time at which extremum (maxima or minima) of response occurs.

In general n = 1, 3, 5, 7, ... correspond to overshoots and

n = 2, 4, 6, 8, ... correspond to undershoots.

It is important to note the following:

(*a*) *t* will assume a real value only for $0 < \xi < 1$. Though it shall assume real value for some –ve values of ξ also, but it is not of practical interest as –ve damping will make the response grow without bound *i.e.* it will lead to unstable system.

(*b*) The maximum overshoot occurs at $n = 1$ for $0 < \xi < 1$ *i.e.*

$$t_p = \frac{\pi}{\omega_n\sqrt{1-\xi^2}}$$

(*c*) For $\xi \geq 1$, maxima of $c(t) = c(t)\big|_{t=\infty} = c(\infty)$ = steady state response.

(*d*) The unit step response $c(t)$ is not periodic for $\xi \neq 0$. But the undershoots and overshoots do occur at periodic intervals with period $T = \pi/\omega_d$. See Fig. 2.11

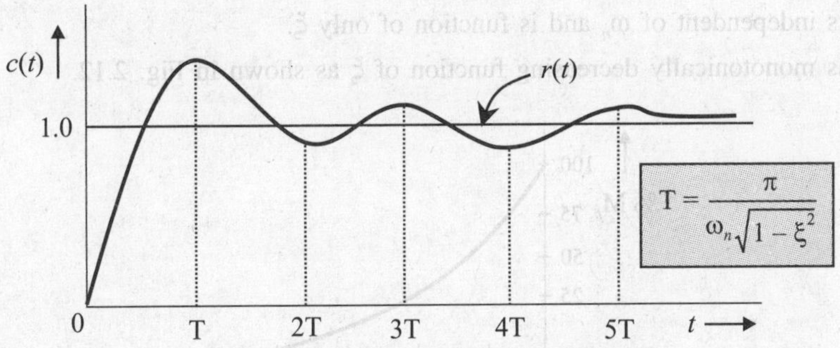

Fig. 2.11: Periodic overshoots and undershoots

2. Maximum overshoot (M_p)

In unit step response $c(t)$

let c_{max} = maximum value of $c(t)$

c_{ss} = steady state value of $c(t)$

Then maximum overshoot $M_p = c_{max} - c_{ss}$

M_p is usually expressed in percentage as ;

$$\%\, M_p = \frac{\text{max.overshoot}}{c_{ss}} \times 100$$

M_p is often recognised as measure of relative stability of system. A system with large overshoot is usually undesirable. It is usually seen that maxima occurs at first overshoot. However, for some system it might occur at later peak. A negative overshoot may also occur in case the system transfer function has an odd number of zeros in right half of *s*-plane.

Substituting (2.16) in (2.15) we get the magnitudes of overshoots and undershoots as follows.

$$c(t)\Big|_{\text{max/min}} = 1 - \frac{1}{\sqrt{1-\xi^2}} e^{-\frac{n\pi\xi}{\sqrt{1-\xi^2}}} \sin(n\pi + \phi)\,;\ \sin\,(n\pi + \phi) = (-1)^n \sin\phi$$

$$= 1-(-1)^n \frac{1}{\sqrt{1-\xi^2}} \left[e^{-n\pi\xi/\sqrt{1-\xi^2}}\right] \sqrt{1-\xi^2}\ ;\ \sin\phi = \sqrt{1-\xi^2}$$

$$= 1-(-1)^n e^{-n\pi\xi/\sqrt{1-\xi^2}}\quad ;\ n = 1, 2, 3, ...$$

The peak overshoot occurs for $n = 1$ *i.e.*

$$M_p = c(t)\Big|_{\max} - 1 = e^{-\pi\xi/\sqrt{1-\xi^2}} \qquad ...(2.17)$$

and

$$\%\ M_p = 100\, e^{-\pi\xi/\sqrt{1-\xi^2}} \qquad ...(2.18)$$

It is worth noting here that

(*i*) M_p is independent of ω_n and is function of only ξ.

(*ii*) M_p is monotonically decreasing function of ξ as shown in Fig. 2.12.

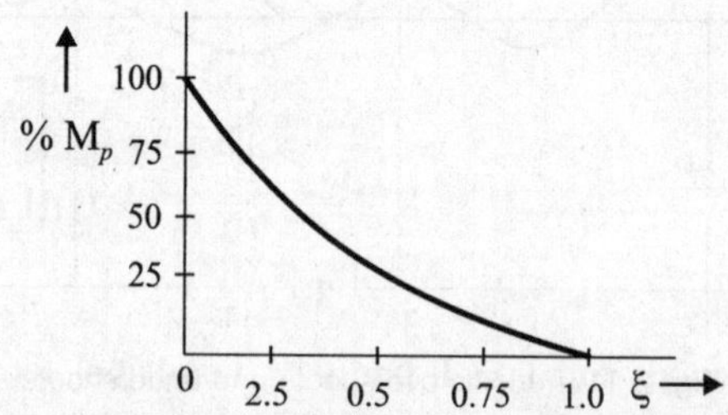

Fig. 2.12: Peak overshoot as function of ξ

3. Settling time (t_s)

The settling time t_s is defined as time required for the step response to decrease and stay within specified tolerance band (usually 2% or 5% of its final value). Though it is bit difficult to find exact expression for t_s but the approximate expression can be obtained by considering only exponential decay and assuming ξ to be arbitrarily small.

For ± 2% tolerance band

$$e^{-\sigma t_s} = 0.02 \quad \text{and} \quad t_s \cong \frac{4}{\xi\omega_n} \qquad ...(2.19)$$

and for ± 5 % tolerance band

$$e^{-\sigma t_s} = 0.05 \quad \text{and} \quad t_s \cong \frac{3}{\xi\omega_n} \qquad ...(2.20)$$

Note that settling time is inversely proportional to the product of damping ratio and undamped natural frequency of system. Since ξ is usually selected for the requirement of permissible M_p, the settling time primarily becomes function of ω_n. Thus duration over which the transients persist, can easily be varied by adjusting ω_n keeping maximum overshoot invariant. It is evident that large ω_n will result in rapid response.

4. Rise time (t_r)

The rise time t_r is defined as time required for the step response to reach from 0% to 90% of its final value for over damped systems and from 0% to 100% for underdamped systems.

Alternatively t_r is also equal to reciprocal of slope of step response at the instant, response is half the final value. *i.e.*;

$$t_r = \frac{1}{\dot{c}(t)\big|_{t=t_d}}$$

Assuming system response to be underdamped nature, t_r may be obtained by following equation.

$$c(t)\big|_{t=t_r} = 1$$

Use (2.15) to get $\sin(\omega_d t_r + \phi) = 0$

or $\omega_d t_r + \phi = \pi$

and $$\boxed{t_r = \frac{\pi - \phi}{\omega_n\sqrt{1-\xi^2}}}$$

where $$\phi = \tan^{-1}\left(\frac{\sqrt{1-\xi^2}}{\xi}\right) \quad \text{for } 0<\xi<1 \qquad ...(2.21)$$

Note the following;

(*i*) Keeping ω_n fixed, as ξ increases, t_r increases. Keeping ξ fixed as ω_n increases t_r decreases.

(*ii*) Improvement in damping of system consequently increases rise time of system.

5. Delay time (t_d)

The delay time t_d is defined as time required for the step response to reach 50% of its final value in first attempt.

Note that if one specifies the values of t_d, t_r, t_p, t_s and M_p, then the shape of response curve is virtually determined. See Fig. 2.10.

The characteristic equation roots and corresponding time response

Consider a general closed loop transfer function as follows.

$$\frac{C(s)}{R(s)} = \frac{G(s)}{1+G(s)H(s)} = \frac{N(s)}{D(s)} \qquad ...(2.22)$$

The denominator polynomial D(s) is called characteristic polynomial and when equated to zero *i.e.*

$$1 + G(s)\,H(s) = 0$$

is called **characteristic equation** of system. The roots of this characteristic equation are called **characteristic roots** or closed loop poles whose location in *s*-plane uniquely specifies the nature of transient response of the system.

Consider a general second order unity feedback control system (Fig. 2.8), the general characteristic equation is given as

$$s^2 + 2\xi\omega_n s + \omega_n^2 = 0 \qquad \text{...(2.23)}$$

The roots of (2.23) are $s_1, s_2 = -\xi\omega_n \pm j\omega_n\sqrt{1-\xi^2}$ as shown in Fig. 2.13 ...(2.24)

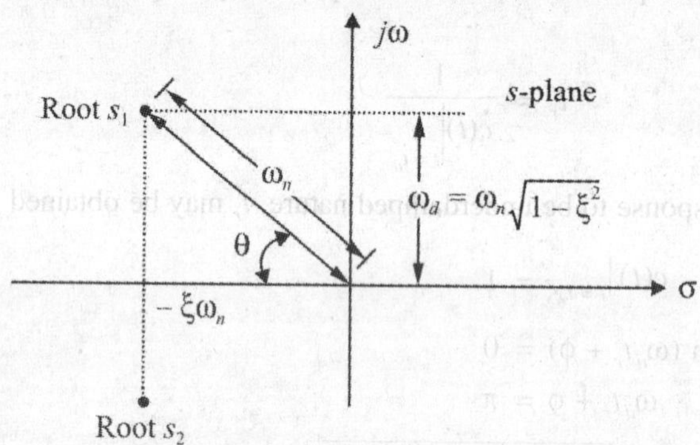

Fig. 2.13: Characteristic roots

Note the following from location of complex conjugate roots s_1, s_2 on *s*-plane (Fig. 2.13).

(*i*) ω_n = radial distance between origin of *s*-plane and root s_1.

(*ii*) damping factor σ = real part of the root.

(*iii*) damped frequency ω_d = imaginary part of the root.

(*iv*) damping ratio $\xi = \cos\theta$ = cosine of angle between radial line and –ve real axis.

(*v*) constant ξ loci can be drawn as shown in Fig. 2.14.

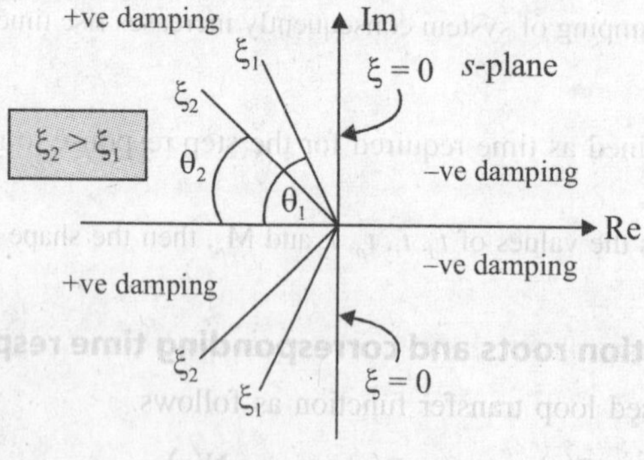

Fig. 2.14: Constant ξ loci

Note that left half of *s*-plane corresponds to positive damping ($\xi > 0$), right half of *s*-plane to negative damping ($\xi < 0$) and imaginary axis to no damping ($\xi = 0$). Positive damping will permit the unit step response $c(t)$ given by Equation (2.15) to settle to final steady state value due to term $e^{-\sigma t}$, but the negative damping will permit the response to grow with time due to term $e^{\sigma t}$ and lead to unstable system. $\xi = 0$ obviously leads the system to sustained oscillations.

(*vi*) Constant ω_n loci, constant ω_d loci and constant damping factor $\sigma(=\xi\omega_n)$ loci are drawn in Figs. 2.15, 2.16 and 2.17 respectively.

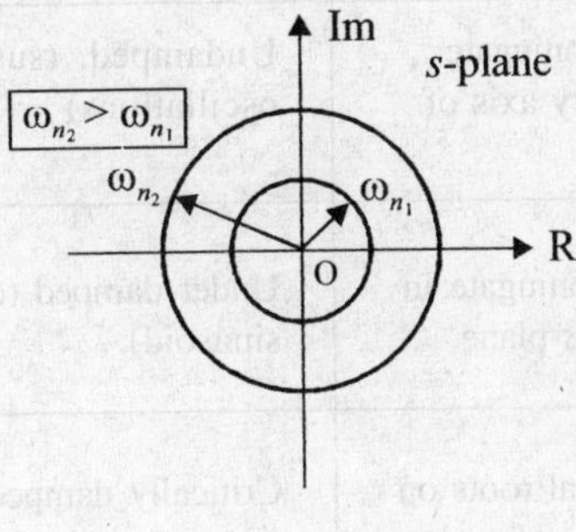

Fig. 2.15: Constant ω_n loci

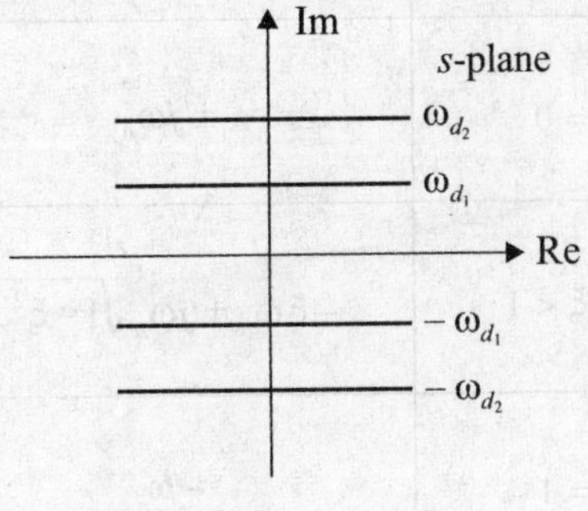

Fig. 2.16: Constant ω_d loci

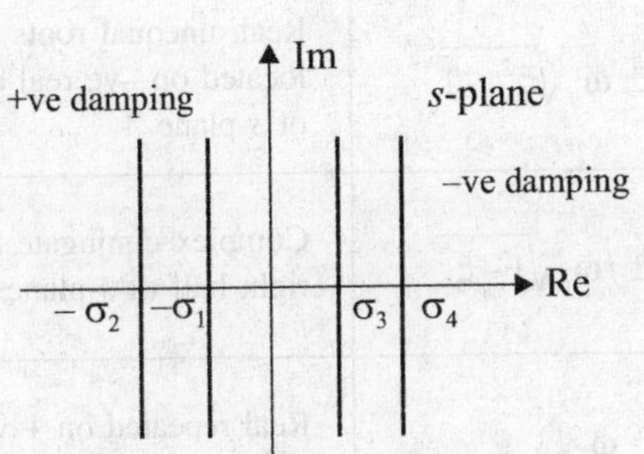

Fig. 2.17: Constant damping factor loci

(*vii*) Keeping ω_n constant and varying ξ from $-\infty$ to $+\infty$, the locus of roots of characteristic equation (2.23) is drawn in Fig. 2.18.

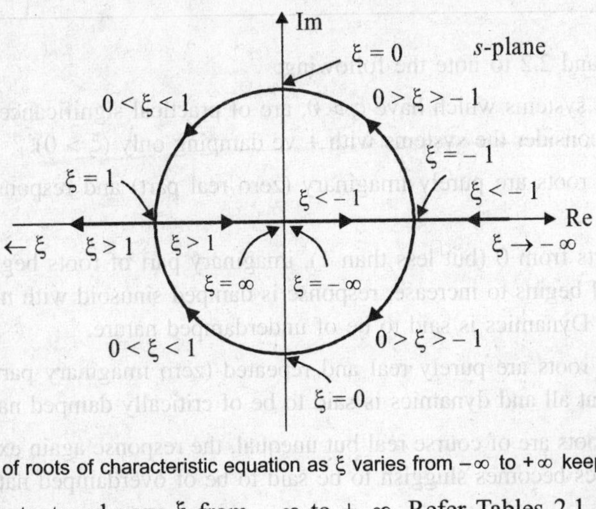

Fig. 2.18: Locus of roots of characteristic equation as ξ varies from $-\infty$ to $+\infty$ keeping ω_n constant

Let us keep ω_n constant and vary ξ from $-\infty$ to $+\infty$. Refer Tables 2.1 and 2.2 for complete insight into system dynamics against variation in ξ.

TABLE 2.1:

Range of values of ξ	*Characteristic roots s_1, s_2*	*Nature of roots*	*Nature of system dynamics*
$\xi = 0$	$\pm j\omega_n$	Complex conjugate on imaginary axis of *s*-plane	Undamped, (sustained oscillations)
$0 < \xi < 1$	$-\xi\omega_n \pm j\omega_n\sqrt{1-\xi^2}$	Complex conjugate in left half of *s*-plane	Under damped (damped sinusoid)
$\xi = 1$	$-\omega_n$	Repeated real roots on –ve real axis of *s*-plane	Critically damped (exponentially decaying)
$\xi > 1$	$-\xi\omega_n \pm \omega_n\sqrt{\xi^2-1}$	Real, unequal roots located on –ve real axis of *s*-plane	Overdamped (slower exponential decay)
$\xi < 0$	$+\xi\omega_n \pm j\omega_n\sqrt{1-\xi^2}$	Complex conjugate in right half of *s*-plane	Negatively damped, exponentially growing sinusoidal without bound
$\xi = -1$	ω_n	Real repeated on + ve real axis of *s*-plane	Exponentially growing
$\xi < -1$	$\xi\omega_n \pm \omega_n\sqrt{\xi^2-1}$	Real, unequal roots on +ve real axis of *s*-plane	Exponentially growing

Scan Tables 2.1 and 2.2 to note the following:

(*i*) Only those systems which have $\xi > 0$, are of practical significance and therefore we shall, here after consider the systems with + ve damping only ($\xi > 0$).

(*ii*) For $\xi = 0$, roots are purely imaginary (zero real part) and response breaks into sustained oscillation.

(*iii*) As ξ departs from 0 (but less than 1), imaginary part of roots begins to decrease and real part thereof begins to increase, response is damped sinusoid with more damping for higher value of ξ. Dynamics is said to be of underdamped nature.

(*iv*) For $\xi = 1$, roots are purely real and repeated (zero imaginary part), response exhibits no oscillation at all and dynamics is said to be of critically damped nature.

(*v*) For $\xi > 1$, roots are of course real but unequal, the response again exhibits no oscillation but the dynamics becomes sluggish to be said to be of overdamped nature.

(*vi*) One can easily conclude that imaginary part of root contributes sinusoidal oscillation to system dynamics where as real part of the root is responsible for damping in system. Larger is the imaginary part, more oscillatory is the dynamics and larger is the real part, more is damping in system *i.e.* less oscillatory is the dynamics.

TABLE 2.2: Typical step response for corresponding location of roots of characteristic equation

Location of roots in s-plane	*Typical time response*
$0 < \xi < 1$ 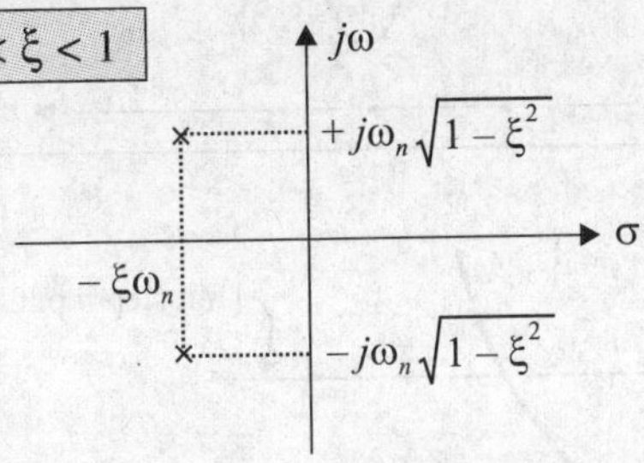	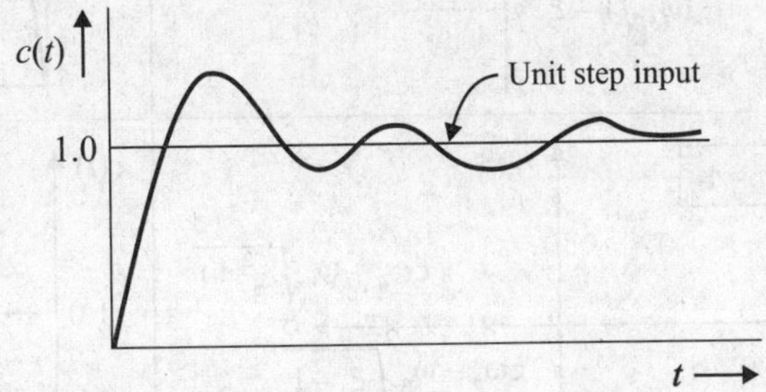
$\xi = 0$ 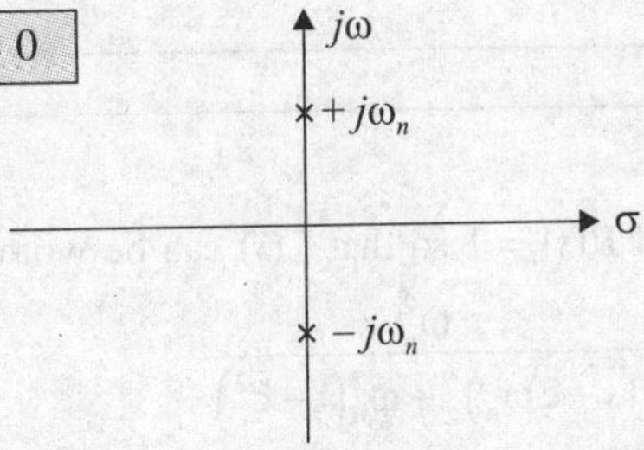	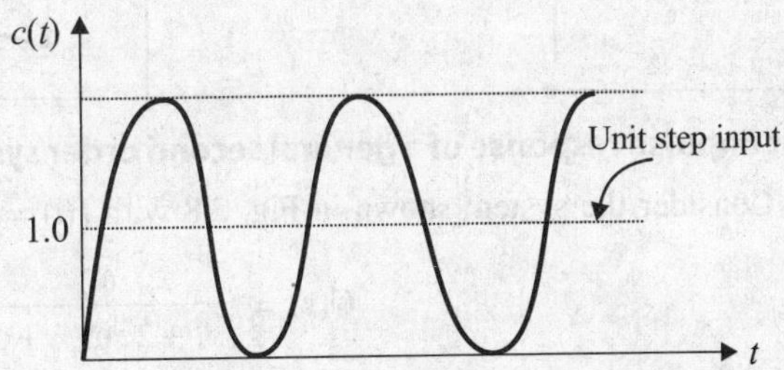
$\xi = 1$ 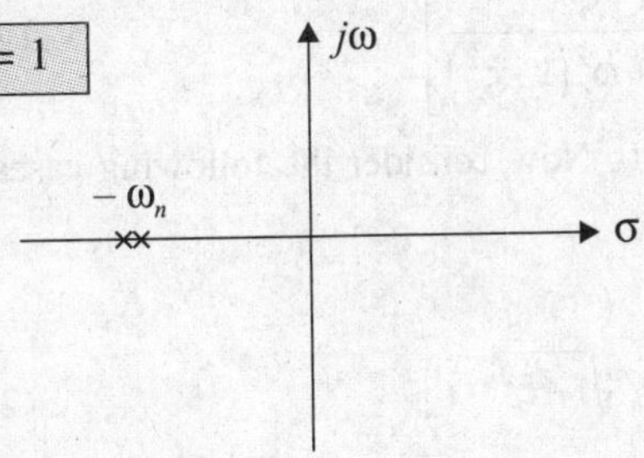	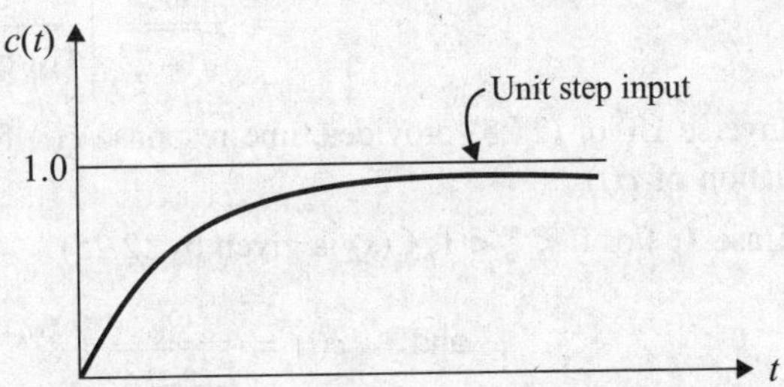
$\xi > 1$ 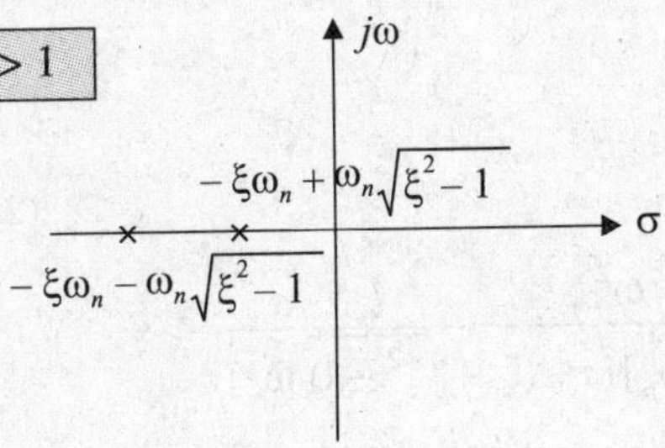	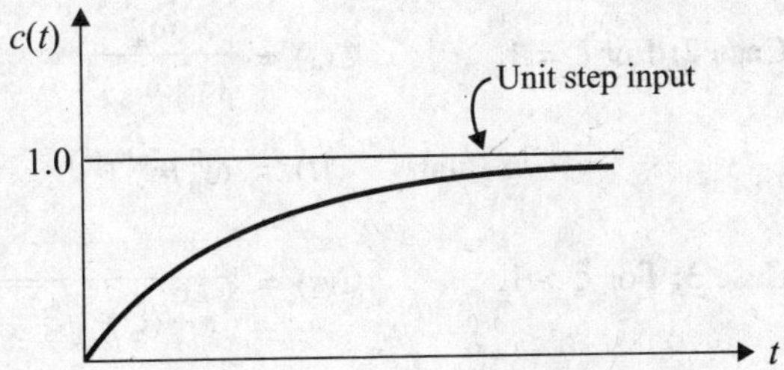

Location of roots in s-plane	*Typical time response*
$0 > \xi > -1$ 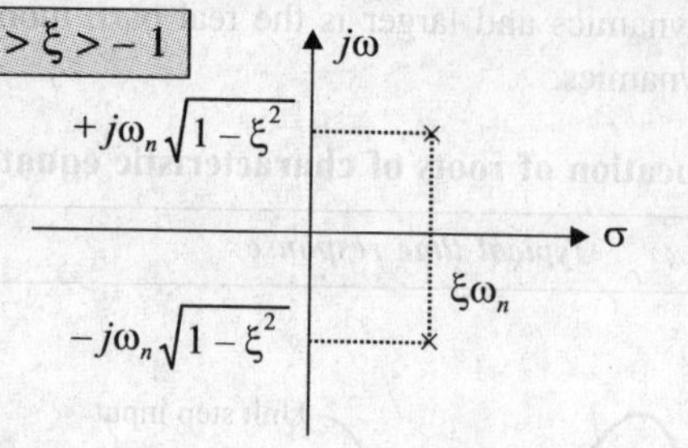	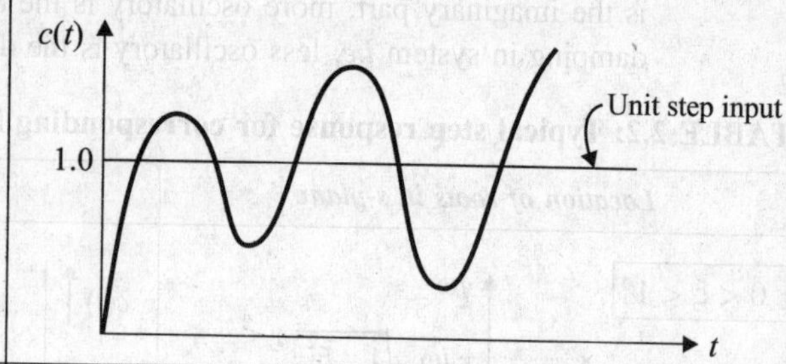
$\xi < -1$ 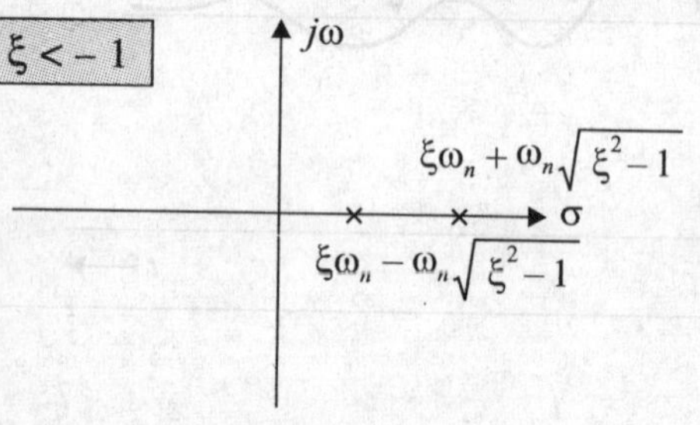	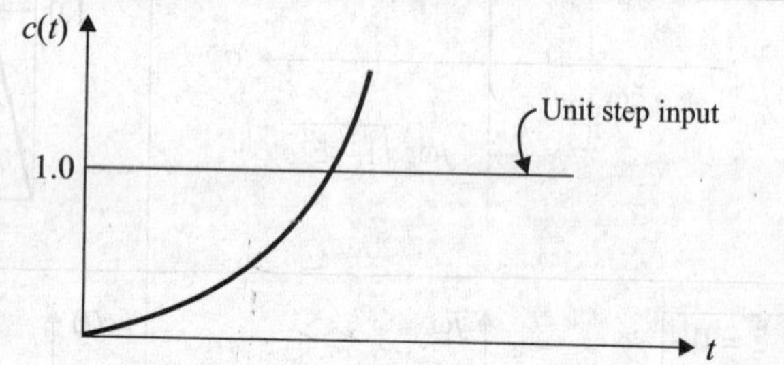

Unit impulse response of a general second order system

Consider the system shown in Fig. 2.8 with $r(t) = \delta(t)$ or $R(s) = 1$ so that $C(s)$ can be written as

$$C(s) = \frac{\omega_n^2}{s^2 + 2\xi\omega_n s + \omega_n^2} = \frac{\omega_n^2}{(s+\xi\omega_n)^2 + \omega_n^2(1-\xi^2)}$$

$$= \frac{\omega_n}{\sqrt{1-\xi^2}}\left[\frac{\omega_n\sqrt{1-\xi^2}}{(s+\xi\omega_n)^2 + \omega_n^2(1-\xi^2)}\right] \quad ...(2.25)$$

Inverse LT of (2.25) provides time response $c(t)$ for $t > 0$. Now consider the following cases for evaluation of $c(t)$.

Case 1: For $0 < \xi < 1$, $C(s)$ is given by (2.25)

and $$c(t) = \frac{\omega_n}{\sqrt{1-\xi^2}}\, e^{-\xi\omega_n t} \sin\left(\omega_n\sqrt{1-\xi^2}\cdot t\right) \quad ...(2.26)$$

Case 2: For $\xi = 1$, $$C(s) = \frac{\omega_n^{\,2}}{(s+\omega_n)^2}$$

and $$c(t) = \omega_n^2\, t e^{-\omega_n t} \quad ...(2.27)$$

Case 3: For $\xi > 1$, $$C(s) = \frac{\omega_n^2}{[s + (\xi + \sqrt{\xi^2-1})\,\omega_n][s + (\xi - \sqrt{\xi^2-1})\,\omega_n]}$$

and $$c(t) = \frac{\omega_n}{2\sqrt{\xi^2-1}}\left[e^{-\left(\xi-\sqrt{\xi^2-1}\right)\omega_n t} - e^{-\left(\xi+\sqrt{\xi^2-1}\right)\omega_n t}\right] \quad ...(2.28)$$

The typical responses for all above three cases are plotted in Fig. 2.19.

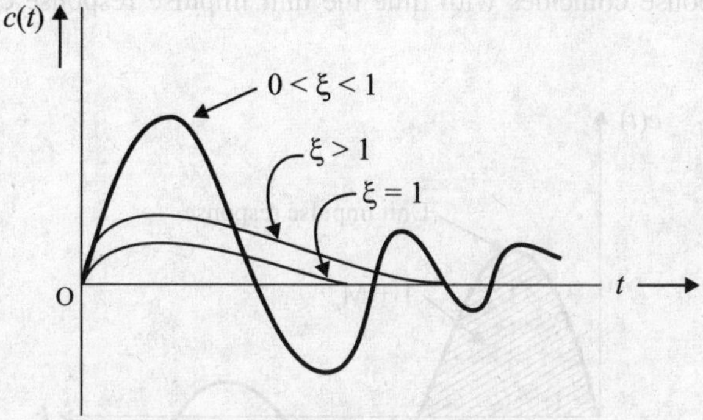

Fig. 2.19: Typical impulse response of second order system

From the foregoing analysis, the following points are worth noting ;

(*i*) Since unit impulse signal is time derivative of unit step signal, (2.26), (2.27) and (2.28) could be directly obtained from unit step response (2.15) by merely differentiating it so that mathematical routine in *s* domain could be avoided.

(*ii*) The unit impulse response is never negative in case its dynamics is of critically damped or overdamped nature.

(*iii*) If underdamped dynamics is possessed by unit impulse response, it assumes both +ve and –ve values and oscillates about zero mark.

(*iv*) The maximum overshoot in underdamped system when impulse excited, can be obtained by

$$\dot{c}(t)\Big|_{t = t_{max}} = 0$$

Differentiating 2.26 and equating it to zero yields

$$\tan\left[\omega_n \sqrt{1-\xi^2}\cdot t\right]\Big|_{t=t_{max}} = \frac{\sqrt{1-\xi^2}}{\xi}$$

and

$$t_{max} = \frac{\tan^{-1}\dfrac{\sqrt{1-\xi^2}}{\xi}}{\omega_n\sqrt{1-\xi^2}} \quad \text{for } 0 < \xi < 1 \qquad ...(2.29)$$

(*v*) The maximum overshoot in underdamped dynamics can be obtained by substituting (2.29) in (2.26) as follows.

$$c(t)_{max} = \omega_n \exp\left[-\frac{\xi}{\sqrt{1-\xi^2}}\tan^{-1}\left(\frac{\sqrt{1-\xi^2}}{\xi}\right)\right] \quad \text{for } 0 < \xi < 1 \qquad ...(2.30)$$

(*vi*) The maximum overshoot for unit step response can be obtained from unit impulse response by computing area under unit impulse response curve from time $t = 0$ to the time first zero appears, as step signal is integral of impulse signal. See Fig. 2.20 to note that the shaded area equals $1 + M_p$, where M_p = max. overshoot in unit step response. The peak time t_p for unit step response coincides with time the unit impulse response crosses the time axis for the first time.

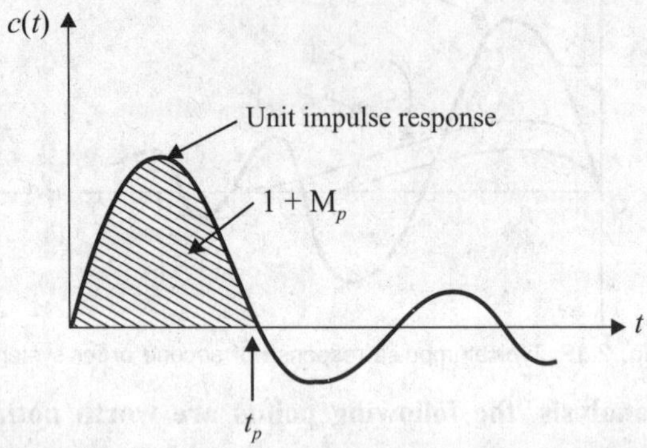

Fig. 2.20: M_p and t_p of unit step response from unit impulse response

2.5 STEADY STATE PERFORMANCE OF LINEAR CONTROL SYSTEM

As discussed earlier, the steady state portion of the time response is that part which remains after transients have died out. A control system designed only for dynamic precision may not suffice in a typical control situation. Then the additional requirement felt here is that the steady state controlled response must not be in substantial deviation from its desired value. This deviation, called as steady state error may be caused by imperfection in system components such as static friction, backslash, amplifier drift, ageing or deterioration, etc. The steady state error is indicative of accuracy of controlled system when subjected to a specific type of input. It is expected of control system that its response follows this input accurately with zero error. But it seldom agrees exactly with the reference input. Thus it is inevitable in almost every control situation. However the concerted effort of designer is to keep it within certain tolerable limits.

Consider a control configuration shown in Fig. 2.21.

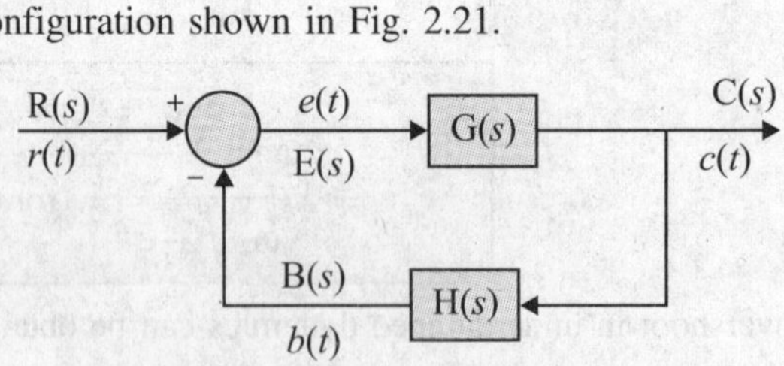

Fig. 2.21: General feedback configuration

The steady state error $e(t)$ in a unity feedback control system where reference input $r(t)$ and controlled output $c(t)$ are of same dimension meaning $H(s) = 1$ is defined as

$$e(t) = r(t) - c(t)$$

If $r(t)$ and $c(t)$ are not of same dimension, *i.e.* $H(s) \neq 1$, then

$$e(t) = r(t) - b(t) \qquad \text{...(2.31)}$$

or

$$E(s) = R(s) - B(s) = R(s) - H(s)\, C(s) = R(s) - H(s)\, [E(s) \cdot G(s)]$$

and

$$E(s) = \frac{R(s)}{1+G(s)H(s)}$$

One can use Final Value Theorem of Laplace transform to evaluate steady state error (e_{ss}) as

$$e_{ss} = \lim_{t \to \infty} e(t) = \lim_{s \to o} sE(s) = \lim_{s \to o} \left[\frac{sR(s)}{1+G(s)H(s)} \right] \quad ...(2.32)$$

Note that steady state error depends on:

(*i*) input R(*s*)

(*ii*) open loop system structure G(*s*) H(*s*).

(*iii*) type number of open loop transfer function.

Type number of a system

In general open loop transfer function G(*s*) H(*s*) can be written in following two forms:

(*a*) **Time constant form:**

$$G(s)\, H(s) = \frac{K \prod^{i} (T_{Z_i} s+1)}{s^r \prod^{j} (T_{p_j} s+1)} \quad ...(2.33)$$

(*b*) **Pole-zero form:**

$$G(s)\, H(s) = \frac{K^* \prod^{i} (s+z_i)}{s^r \prod^{j} (s+p_j)} \quad ...(2.34)$$

The gains in above two forms are related as

$$K = \frac{K^* \prod^{i} (z_i)}{s^r \prod^{j} (p_j)} \quad ...(2.35)$$

The term s^r, appearing in denominator of both the forms which corresponds to the number of integrations involved in system or number of poles at origin, classifies the system on the basis of type. A system is called type 0, type 1, type 2, ... if $r = 0$, $r = 1$, $r = 2$ ...., respectively.

Note the following from foregoing discussion:

(*i*) The type number of system is different from that of order of system.

(*ii*) As type number increases you will see little later that steady state accuracy improves but the stability which is characterisation of transient behaviour of system, worsens.

(*iii*) In practice, it is rare to have a system of type 3 or higher because it becomes difficult to design stable systems having more than two integrations in feed forward path.

(*iv*) In steady state evolution routine that shall follow, we shall use time constant form (2.33) wherein each term in numerator and denominator except the term s^r, approaches unity as s approaches zero. Then the open loop gain K will have direct relationship with steady state error and the term s^r will play a dominant role in determination of steady state error.

Steady state error due to step input

Let the system of Fig. 2.21 be excited by step input of magnitude R *i.e.*

$$R(t) = R\,u(t) \quad \text{and} \quad R(s) = \frac{R}{s}$$

Use equation (2.32) to find

$$e_{ss} = \lim_{s\to 0} \frac{R}{1+G(s)H(s)} = \frac{R}{1+\lim_{s\to 0} G(s)H(s)}$$

The step error constant or position error constant K_p is defined as

$$K_p \triangleq \lim_{s\to 0} G(s)H(s) \qquad \text{...(2.36)}$$

then

$$e_{ss} = \frac{R}{1+K_p} \qquad \text{...(2.37)}$$

Note the following:

(*i*) for type 0 system

$$K_p = \lim_{s\to 0} \frac{K\cdot \prod^{i} (T_{Z_i} s+1)}{\prod^{j} (T_{p_j} s+1)} = K$$

and

$$e_{ss} = \frac{R}{1+K} \qquad \text{...(2.38)}$$

(*ii*) for system of type 1 or higher

$$K_p = \lim_{s\to o} \frac{K\cdot \prod^{i} (T_{Z_i} s+1)}{s^r\cdot \prod^{j} (T_{p_j} s+1)} = \infty \quad \text{for } r \geq 1$$

and

$$e_{ss} = 0$$

(*iii*) If system does not involve any integration in forward path (system with no pole at origin), it suffers from steady state error given by equation (2.38) and is shown in Fig. 2.22. This error may be tolerable if K is large but it will be relatively difficult to achieve reasonable relative stability with large K.

(*iv*) If zero steady state error due to step input is desired, the type of system must be one or higher.

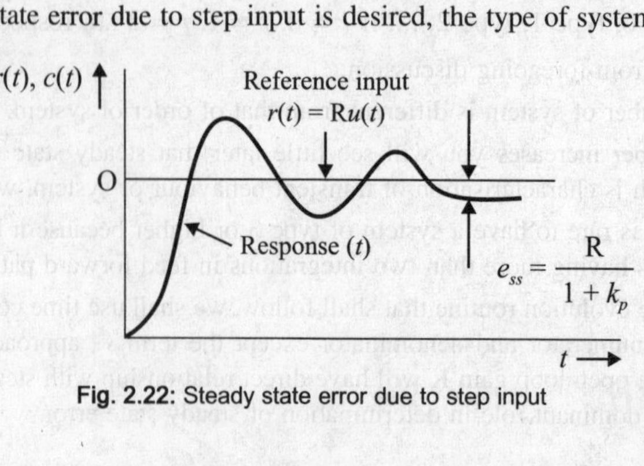

Fig. 2.22: Steady state error due to step input

Steady state error due to ramp input (Velocity input)

Let the system of Fig. 2.21 be excited by ramp input given by

$$R(t) = Rt\,u(t);\ R = \text{real constant}$$

The Laplace transform of $r(t)$ is

$$R(s) = R/s^2$$

Use equation (2.32) to get

$$e_{ss} = \underset{s\to 0}{\text{lt}} \frac{R/s}{1+G(s)H(s)} = \frac{R}{\underset{s\to 0}{\text{lt}}\ sG(s)H(s)}$$

The ramp error constant or velocity error constant K_V is defined as

$$K_V \overset{\Delta}{=} \lim_{s\to 0} sG(s)H(s)$$

Then $$e_{ss} = \frac{R}{K_V} \qquad ...(2.39)$$

Note the following:

(*i*) For type 0 system

$$K_V = \underset{s\to 0}{\text{lt}}\ s\cdot\frac{K\cdot\prod^{i}(T_{Z_i}s+1)}{\prod^{j}(T_{p_j}s+1)} = 0$$

and $e_{ss} = \infty$. *i.e.* type 0 system is incapable of following a ramp input in steady state.

(*ii*) For type 1 system

$$K_V = \underset{s\to 0}{\text{lt}} \frac{K\cdot\prod^{i}(T_{Z_i}s+1)}{\prod^{j}(T_{p_j}s+1)} = K$$

and $e_{ss} = \frac{R}{K}$ *i.e.* a type 1 system is capable of following a ramp input with finite error. A typical response is shown in Fig. 2.23.

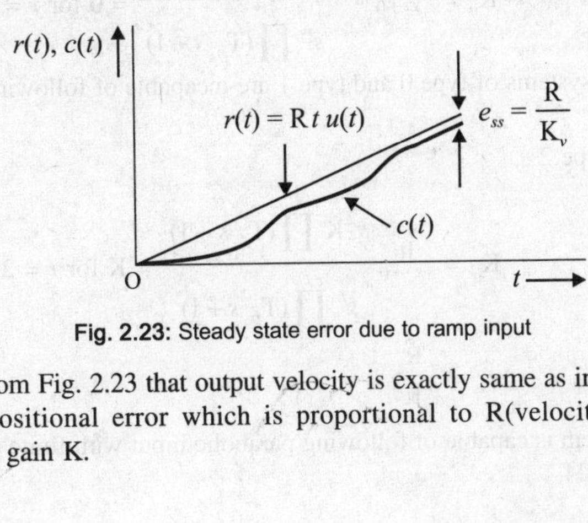

Fig. 2.23: Steady state error due to ramp input

It is obvious from Fig. 2.23 that output velocity is exactly same as input velocity in steady state, but there is positional error which is proportional to R(velocity of input) and inversely proportional to gain K.

(*iii*) For system of type 2 or higher

$$K_V = \underset{s\to 0}{\text{lt}} \frac{sK\prod^{i}(T_{Z_i}s+1)}{s^r\prod^{j}(T_{p_j}s+1)} = \infty \text{ for } r \geq 2$$

and $e_{ss} = 0$ *i.e.* the system of type 2 or higher is capable of following the ramp input with zero error in steady state.

Steady state error due to parabolic input (Acceleration input)

Let the system of Fig. 2.21 be excited by parabolic input given by

$$r(t) = \frac{Rt^2}{2}u(t) \qquad ...(2.40)$$

whose Laplace transform is

$$R(s) = \frac{R}{s^3}$$

use equation (2.32) to get

$$e_{ss} = \underset{s\to 0}{\text{lt}} \frac{R/s^2}{1+G(s)H(s)} = \frac{R}{\underset{s\to 0}{\text{lt}}\, s^2 G(s)H(s)}$$

The parabolic error constant or acceleration error constant is defined as

$$K_a \triangleq \lim_{s\to 0} s^2 G(s)H(s) \qquad ...(2.41)$$

then
$$e_{ss} = \frac{R}{K_a} \qquad ...(2.42)$$

Note the following:

(*i*) For type 0 and type 1 system

$$K_a = \underset{s\to 0}{\text{lt}}\ s^2 \frac{K\prod^{i}(T_{Z_i}s+1)}{s^r\prod^{j}(T_{p_j}s+1)} = 0 \text{ for } r = 0 \text{ and } r = 1$$

and $e_{ss} = \infty$ *i.e.* systems of type 0 and type 1 are incapable of following parabolic input in steady state.

(*ii*) For system of type 2

$$K_a = \underset{s\to 0}{\text{lt}} \frac{s^2K\prod^{i}(T_{Z_i}s+1)}{s^r\prod^{j}(T_{p_j}s+1)} = K \text{ for } r = 2$$

$$e_{ss} = \frac{R}{K} \qquad ...(2.43)$$

i.e. a type 2 system is capable of following parabolic input with finite error. A typical response is shown in Fig. 2.24.

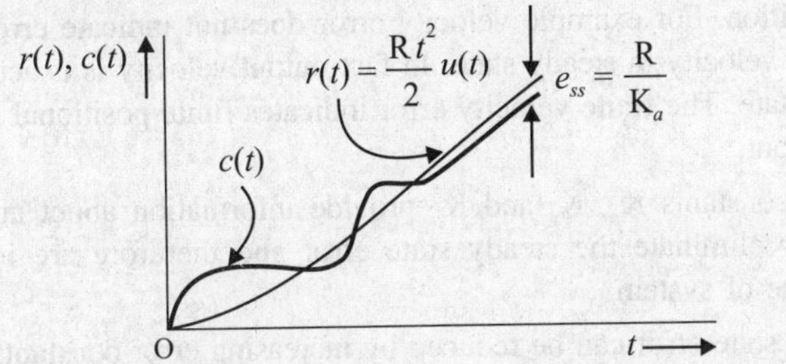

Fig. 2.24: Steady state error due to parabolic input

Note that output acceleration is exactly same as input acceleration in steady state but there is positional error which is proportional to R and inversely proportional to gain K.

(*iii*) For system of type 3 or higher

$$K_a = \underset{s \to 0}{\text{lt}}\ s^2 \frac{K \cdot \prod^{i} (T_{Z_i} s + 1)}{s^r \prod^{j} (T_{p_j} s + 1)} = \infty \text{ for } r \geq 3 \qquad \text{...(2.44)}$$

and $e_{ss} = 0$ i.e. the system of type 3 or higher, is capable of following parabolic input with zero error in steady state.

Table 2.3 summarizes the steady state error for type 0, type 1 and type 2 systems when they are subjected to various inputs.

TABLE 2.3: Steady state error in terms of gain K

Type of system (r)	*Error constants*			*Step input* $r(t) = R\,u(t)$	*Ramp Input* $r(t) = Rt\,u(t)$	*Parabolic input* $r(t) = \frac{Rt^2}{2} u(t)$
	K_p	K_v	K_a	$e_{ss} = \frac{R}{1+K_p}$	$e_{ss} = \frac{R}{K_v}$	$e_{ss} = R/K_a$
0	K	0	0	$\frac{R}{1+K}$	∞	∞
1	∞	K	0	0	$\frac{R}{K}$	∞
2	∞	∞	K	0	0	$\frac{R}{K}$
≥ 3	∞	∞	∞	0	0	0

Scan Table 2.3 and **note** the following points:

(*i*) For type 0, type 1 and type 2 systems, the steady state error assumes finite values only on diagonal, above the diagonal they are infinite and below the diagonal they are zero.

(*ii*) The terms position error, velocity error and acceleration error mean steady state error in output position. For example velocity error does not indicate error between input velocity and output velocity in steady state. In fact output velocity is exactly same as input velocity in steady state. The finite velocity error indicates finite positional difference after transients have died out.

(*iii*) The error constants K_p, K_v and K_a provide information about ability of system to either mitigate or eliminate the steady state error and therefore are indicative of steady state performance of system.

(*iv*) The steady state error can be reduced by increasing error constants which obviously means increasing K. However the gain K cannot be increased arbitrarily as it will worsen the transient behaviour which is also desired to remain within acceptable limits.

(*v*) In order to bring about an improvement in steady state performance of system, the designer can also increase type of system by adding one or more integrations in forward path. But this poses the problem of stability. The design of satisfactorily stable system with more than two integrations in forward path, is usually an uphill task due to which K_a becomes less important than K_v.

(*vi*) If input to the system is linear combination of some or all the test inputs (step, ramp, parabolic), then the steady state error can be computed by summing the errors contributed by individual inputs.

Drawbacks of approach based on error constant evaluation

Having discussed in detail the steady state evaluation routine using error constant approach, let us discuss the drawbacks therein.

(*i*) Error constant approach is not applicable to any arbitrary input *e.g.*; one cannot compute it if input is a sinusoid. Steady state error can be computed only if input is step, ramp or parabolic.

(*ii*) Since the evaluation of error constants involves use of final value theorem (FVT) of Laplace transform, it becomes imperative to first investigate whether s E(s) has any pole on $j\omega$ axis or in right half of s-plane. If so FVT cannot be applied.

(*iii*) K_p, K_v, K_a are strictly defined for a system with configuration shown in Fig. 2.21. If the system configuration is different, this approach might fail. In such a situation the usual strategy used for steady state performance evaluation is to identify the error signal and apply final value theorem.

(*iv*) The steady state error obtained through error constant approach, assumes either zero value or a finite non zero value or infinity. When it is infinite, it is due to the fact that error continues to grow with time. Then the information about how the error varies with time, is not available in this approach.

Now, we shall introduce a steady state evaluation routine in the following section wherein the drawbacks stated above are almost alleviated.

2.6 THE ERROR SERIES

In this approach of steady state evaluation, the dynamic error is expressed in terms of dynamic error coefficients. The error series is given as

$$e(t) = C_0\, r(t) + C_1\, \dot{r}(t) + \frac{C_2}{2!}\, \ddot{r}(t) + \frac{C_3}{3!}\, \dddot{r}(t) + ...$$

where $r(t)$ is input and C_0, C_1, C_2, ... are called dynamic error coefficients. To evaluate these dynamic error coefficients, let us express the error function E(s) of Fig. 2.21 as

$$\frac{E(s)}{R(s)} = W(s) = \frac{1}{1+G(s)\,H(s)}$$

then
$$C_0 = \underset{s\to 0}{\text{lt}}\, W(s)$$

$$C_1 = \underset{s\to 0}{\text{lt}}\, \frac{d}{ds} W(s)$$

$$C_2 = \underset{s\to 0}{\text{lt}}\, \frac{d^2}{ds^2} W(s)$$

$$\vdots$$

$$C_n = \underset{s\to 0}{\text{lt}}\, \frac{d^n}{ds^n} W(s)$$

2.7 HIGHER ORDER SYSTEMS

We have discussed in detail the dynamical behaviour of second order system in section 2.4. It is also known that the most of the practical control systems are seldom of order two. For the systems of order higher than two, we no longer can use damping ratio ξ and natural undamped frequency ω_n that are strictly defined for second order system, for the purpose of investigating the dynamical behaviour. Let us explore the possibility of approximating high order systems by lower order (for example of order 2) systems in so far evaluation of dynamical behaviour is concerned.

In high order systems (assuming all poles lying in left of s-plane), the poles located far away from imaginary axis, exhibit fast decaying dynamics and those located near imaginary axis exhibit relatively slower dynamics of decaying nature. See Figs. 2.25 and 2.26.

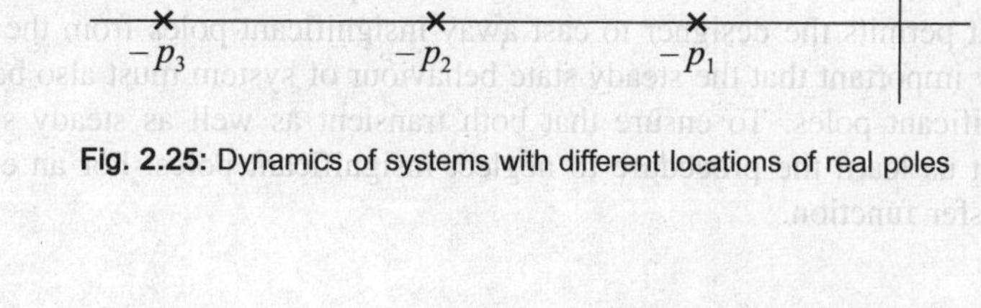

Fig. 2.25: Dynamics of systems with different locations of real poles

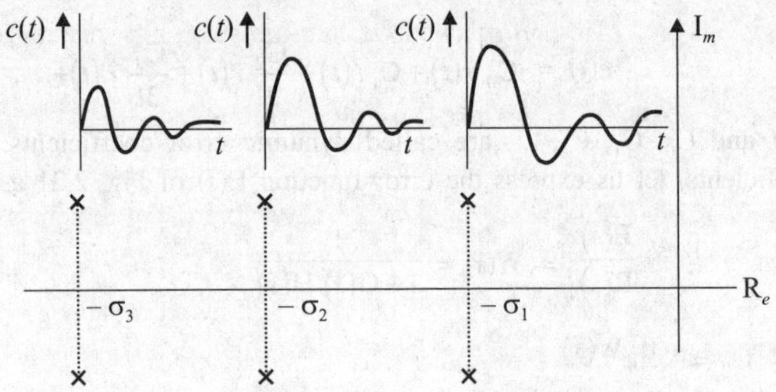

Fig. 2.26: Dynamics of system with different locations of complex conjugate poles

Figure 2.25 shows typical unit impulse responses $c(t)$ corresponding to real poles located at $-p_1$, $-p_2$ and $-p_3$. It can be observed that one located at $-p_1$ has slower dynamics and one located at $-p_3$ has fastest dynamics. Fig. 2.26 shows typical impulse response corresponding to pair of complex conjugate poles having real parts $-\sigma_1$, $-\sigma_2$, $-\sigma_3$. A pair with real part equal to $-\sigma_1$ has slower dynamics and a pair with real part equal to $-\sigma_3$ has fastest dynamics.

Though all the poles do contribute to consolidate the total transient response of the system but those near imaginary axis contribute significantly and those far away from imaginary axis make negligibly small contribution. The poles with significant contribution are called dominant poles and those with negligibly small contribution are called insignificant poles.

Now the question is how far away from imaginary axis the poles should be located so that it can be called insignificant. No separatrix can be drawn to this effect. As thumb rule, it is usually accepted that if the magnitude of real part of pole is at least 5 to 10 times that of a dominant pole or a pair of complex dominant poles, then the pole may be regarded as insignificant in so far as transient response is concerned See Fig. 2.27.

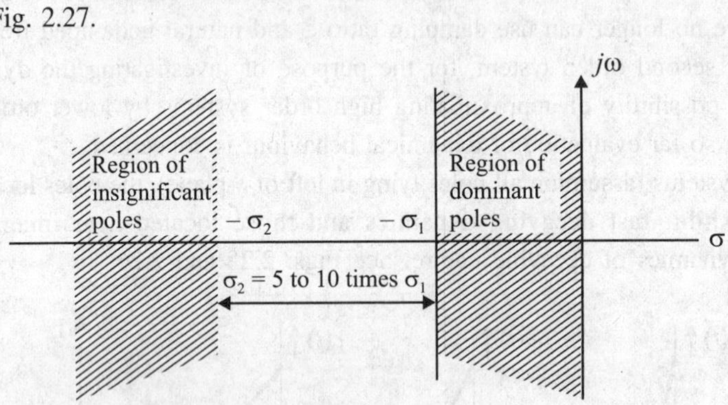

Fig. 2.27: Region of dominant and insignificant poles in complex plane

Technique to cast away insignificant poles

The transient response of system without insignificant poles will be almost same as that with insignificant poles. It permits the designer to cast away insignificant poles from the transfer function but it will be equally important that the steady state behaviour of system must also be preserved while casting away insignificant poles. To ensure that both transient as well as steady state response are almost preserved, let us learn the procedure to neglect insignificant poles. For an example; consider the closed loop transfer function.

$$\frac{C(s)}{R(s)} = \frac{10}{(s+5.5)(s^2+1.55s+1.75)}$$

The pole at $s = -5.5$ can be neglected. To do this write equation as

$$\frac{C(s)}{R(s)} = \frac{10}{5.5\,(1+s/5.5)\,(s^2+1.55s+1.75)} \quad ...(2.45)$$

Now approximate second order system is

$$\frac{10/5.5}{(s^2+1.55s+1.75)} = \frac{1.82}{s^2+1.55s+1.75} \quad ...(2.46)$$

Here we considered closed loop transfer function.

A similar technique can be used to approximate open loop transfer function also.

2.8 EFFECT OF ADDING POLES AND ZEROS TO TRANSFER FUNCTIONS

In this section we shall see how transient response of closed loop system gets modified when we add poles and zeros to open loop and closed loop transfer functions.

Addition of a zero to closed loop transfer function

Let us add a zero at $s = -1/\alpha$ to closed loop transfer function of Fig. 2.8 so that

$$T(s) = \frac{C(s)}{R(s)} = \frac{\omega_n^2(1+\alpha s)}{s^2+2\xi\omega_n s+\omega_n^2} = \frac{\omega_n^2}{s^2+2\xi\omega_n s+\omega_n^2} + \frac{\alpha\omega_n^2 s}{s^2+2\xi\omega_n s+\omega_n^2} \quad ...(2.47)$$

Let the 1st term in Equation (2.47) correspond to $c_1(t)$ then the total response $c(t)$ may be written as

$$c(t) = c_1(t) + \alpha\dot{c}_1(t) \quad ...(2.48)$$

which means total response is sum of $c_1(t)$ and a term proportional to its first derivative. Figure 2.28 shows the constituent responses $c_1(t), \alpha_1\,\dot{c}_1(t)$ and overall time response $c(t)$.

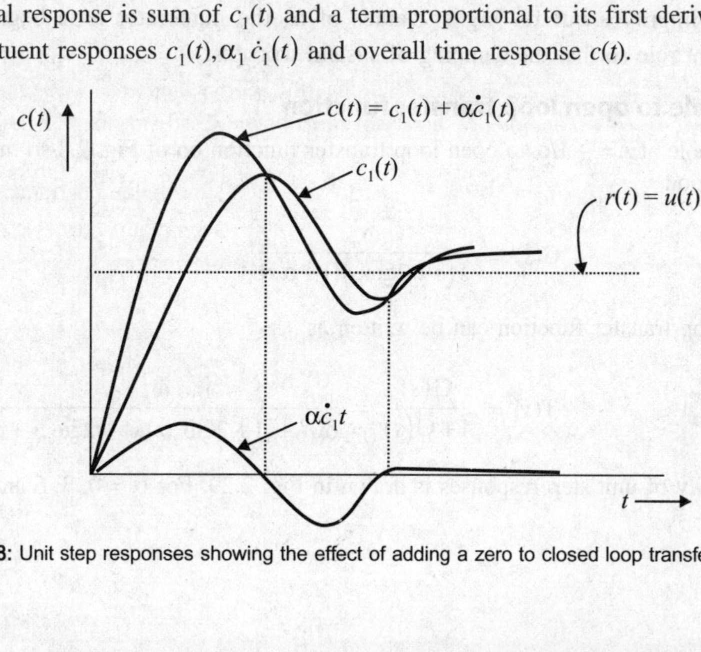

Fig. 2.28: Unit step responses showing the effect of adding a zero to closed loop transfer function

Note that the addition of a zero to closed loop transfer function poses the following changes in step response.

(*a*) Rise time decreases.

(*b*) Peak overshoot increases.

(*c*) The peak overshoot further increases as the value of α increases, *i.e.* placing zero closer to imaginary axis will correspond to larger peak overshoot. However the system continues to have dynamics of decaying nature (stable system) provided $\xi > 0$.

Addition of a zero to open loop transfer function

Let us add a zero at $s = -1/\alpha$ to open loop transfer function of Fig. 2.8 so that

$$G(s) = \frac{\omega_n^2(1+\alpha s)}{s(s+2\xi\omega_n)}$$

then closed loop transfer function will be

$$T(s) = \frac{C(s)}{R(s)} = \frac{\omega_n^2(1+\alpha s)}{s^2 + s(2\xi\omega_n + \alpha\omega_n^2) + \omega_n^2} \quad ...(2.49)$$

It is obvious from closed loop transfer function that the term $(1 + \alpha s)$ appears in numerator but the denominator also contains the term α. The following observations can be made.

(*a*) Since α appears in coefficient of *s* in denominator, its obvious effect is that it improves the damping *i.e.* reduces peak overshoot.

(*b*) The appearance of term $(1 + \alpha s)$ in numerator will have an obvious effect of increasing the peak over shoot.

(*c*) From observations (*a*) and (*b*) one can conclude that for smaller value of α, damping of system improves but for higher values of α, the numerator term begins to play more dominant role so that the damping of system worsens.

Addition of a pole to open loop transfer function

Let us add a pole at $s = -1/\alpha$ to open loop transfer function on of Fig. 2.8 so that modified open loop transfer function

$$G(s) = \frac{\omega_n^2}{s(s+2\xi\omega_n)(1+\alpha s)}$$

then the closed loop transfer function can be written as

$$T(s) = \frac{G(s)}{1+G(s)} = \frac{\omega_n^2}{\alpha s^3 + (1+2\xi\omega_n\alpha)s^2 + 2\xi\omega_n s + \omega_n^2} \quad ...(2.50)$$

A typical family of unit step responses is drawn in Fig. 2.29. For $\alpha = 0, 3, 6$ and $\xi = 1$, $\omega_n = 1$.

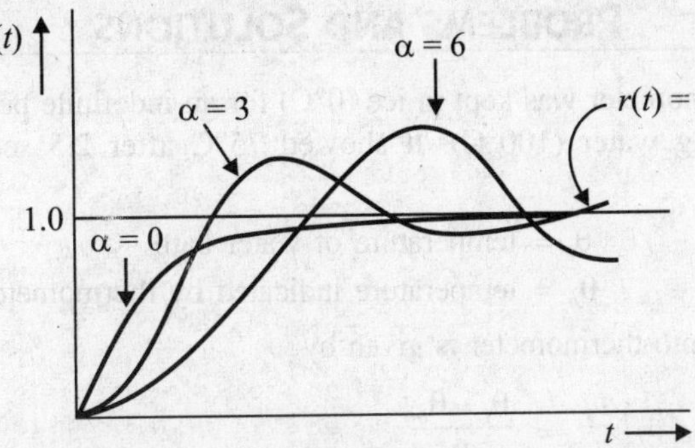

Fig. 2.29: Unit step responses showing the effect of adding a pole to open loop transfer function

One can make the following observations from Fig. 2.29.

(*a*) as α increases *i.e.* as pole gets closer to imaginary axis in *s*-plane, the peak overshoot goes on increasing.

(*b*) as α increases, the rise time of step response also goes on increasing.

Addition of pole to closed loop transfer function

Let us add a pole at $s = -1/\alpha$ to the closed loop transfer function of Fig. 2.8 so that

$$T(s) = \frac{C(s)}{R(s)} = \frac{\omega_n^2}{\left(s^2 + 2\xi\omega_n s + \omega_n^2\right)(1 + \alpha s)} \quad ...(2.51)$$

A typical family of unit step responses is drawn in Fig. 2.30 for $\xi = 0.5$, $\omega_n = 1$ and $\alpha = 0, 0.5, 1$

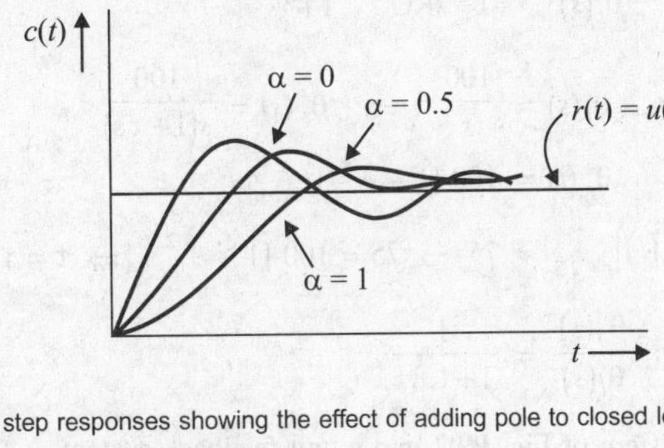

Fig. 2.30: Unit step responses showing the effect of adding pole to closed loop transfer function

One can make the following observations from Fig. 2.30.

(*a*) as α increases, peak overshoot decreases.

(*b*) as α increases, rise time increases.

(*c*) as far as peak overshoot is concerned, addition of a pole to closed loop transfer function poses the opposite effect to that of addition of a pole to open loop transfer function.

PROBLEMS AND SOLUTIONS

P2.1: A mercury thermometer was kept in ice (0°C) for an indefinite period. It was removed and immediately put in boiling water (100°C). It showed 75°C after 2.5 seconds. Evaluate transfer function of thermometer.

Solution: Assuming θ_i = temperature of water bath

and θ_o = temperature indicated by thermometer,

the rate of flow of heat q into thermometer is given by

$$\frac{dq}{dt} = \frac{\theta_i - \theta_o}{R}$$

where R = thermal resistance of thermometer wall.

The indicated temperature rises at the rate

$$\frac{d\theta_o}{dt} = \frac{1}{C}\frac{dq}{dt}$$

where C = thermal capacity of thermometer.

Then $$\dot{\theta}_o = \frac{1}{C}\left[\frac{\theta_i - \theta_o}{R}\right] \text{ whose LT yields}$$

$$s\,\theta_o(s) = \frac{1}{RC}\left[\theta_i(s) - \theta_o(s)\right]$$

$$\frac{\theta_o(s)}{\theta_i(s)} = \frac{1}{1+sRC} = \frac{1}{1+\tau s};\ \tau = RC \text{ (time constant)}$$

Given $$\theta_i(s) = \frac{100}{s} \Rightarrow \theta_o(s) = \frac{100}{s(1+\tau s)}$$

and $$\theta_o(t) = 100\,[1 - e^{-t/\tau}]$$

But $$\theta_o(t)\big|_{t=2.5} = 75° \Rightarrow 75 = 100\,[1 - e^{-2.5/\tau}] \Rightarrow \tau = 1.8$$

$$\therefore \quad \frac{\theta_o(s)}{\theta_i(s)} = \frac{1}{1+1.8s}$$

P2.2: Transform the system of Fig. P2.2 into a unit feedback system.

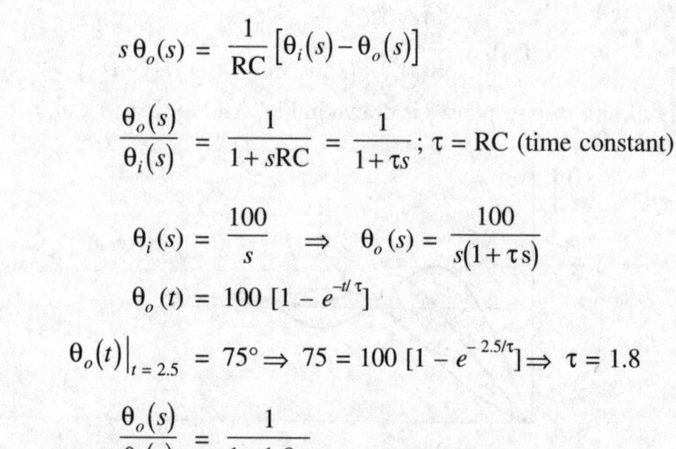

Fig. P2.2

Solution: $$T(s) = \frac{C(s)}{R(s)} = \frac{5(6s+11)}{0.5s^3 + s^2 + 90s + 165} = \frac{N(s)}{D(s)} \text{ (let)}$$

The forward transmittance in unity feedback structure G(*s*)

$$= \frac{N(s)}{D(s) - N(s)} = \frac{5(6s+11)}{0.5s^3 + s^2 + 60s + 110}$$

In fact, a system with forward transmittance G(*s*), in unity feedback configuration has closed loop transfer function.

$$T(s) = \frac{G(s)}{1+G(s)} = \frac{N(s)}{D(s)}$$

Then $$G(s)\,D(s) = N(s) + N(s)\,G(s)$$

or $$G(s) = \frac{N(s)}{D(s) - N(s)}$$

The configuration of unity feedback system is shown below in Fig. P2.2(*a*):

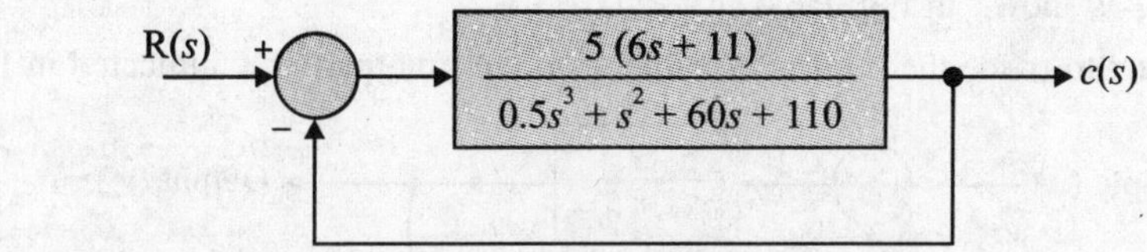

Fig. P2.2(*a*)

P2.3: Block diagram model of a position control system is shown in Fig. P2.3:

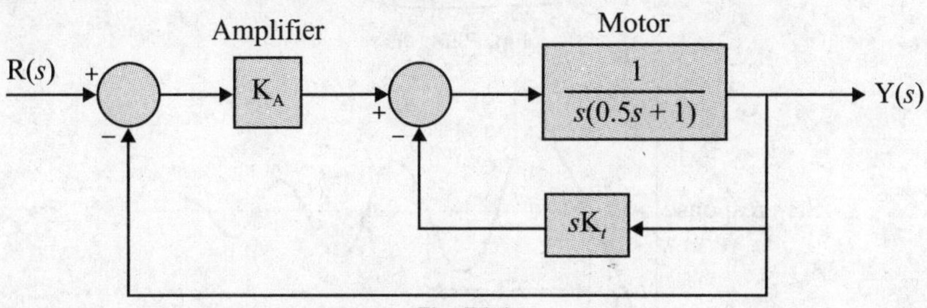

Fig. P2.3

(*a*) In absence of derivative feedback ($K_t = 0$), determine damping ratio of the system for amplifier gain $K_A = 5$. Also find the steady state error to unit ramp input.

(*b*) Find suitable values of the parameters K_A and K_t so that damping ratio of the system is increased to 0.7 without affecting the steady state error as obtained in part (*a*).

Solution:

(*a*) Ch. Equation: $$s^2 + 2s + 2K_A = 0$$

Put $K_A = 5$ and compare with general characteristics equation $s^2 + 2\xi\omega_n s + \omega_n^2 = 0$ to get damping ratio $\xi = 0.316$

Forward transmittance $$G(s) = \frac{5}{s(0.5s+1)}$$

$$K_v = \lim_{s \to 0} sG(s) = 5 \quad \text{and} \quad e_{ss} = 0.2$$

(*b*) Ch. Equation: $s^2 + 2(1 + K_t)\,s + 2K_A = 0$ gives $\omega_n = \sqrt{2K_A}$ and

$$\xi = \frac{1+K_t}{\sqrt{2K_A}} = 0.7$$

$$K_V = \lim_{s\to 0} s\left[\frac{K_A}{s(0.5s+1+K_t)}\right] = \frac{K_A}{1+K_t}$$

$$e_{ss} = \frac{1+K_t}{K_A} = 0.2$$

Solve these two equations to get $K_A = 24.5$ and $K_t = 3.9$.

P2.4: The block diagram of a feedback system is shown in Fig. P2.4 (*a*).

(*a*) Find the closed loop transfer function.

(*b*) Find the minimum value of G for which the step response of the system would exhibit an overshoot, as shown in Fig. P2.4 (*b*)

(*c*) For G equal to twice the minimum value, find the time period T indicated in Fig. P2.4 (*b*).

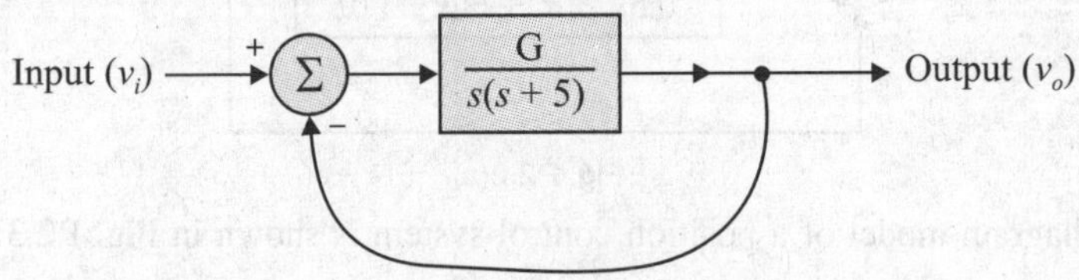

Fig. P2.4 (*a*)

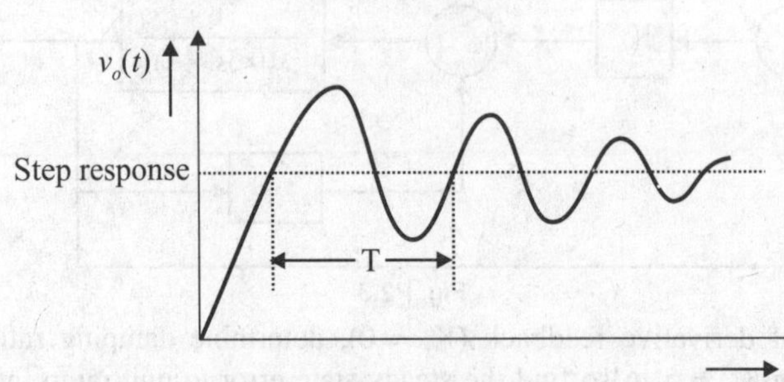

Fig. P2.4 (*b*)

Solution:

(*a*) $$\frac{V_o(s)}{V_i(s)} = \frac{G}{s^2+5s+G}$$

(*b*) System should be underdamped to exhibit an overshoot and $0 < \xi < 1$.

$$\xi = \frac{5}{2\sqrt{G}} < 1 \text{ gives } G > 6.25$$

$$G_{min} = 6.25 + \in$$

(*c*) Ch. Equation for $2G_{min}$ is

$$s^2 + 5s + 12.5 = 0$$

Comparing with general equation

$$s^2 + 2\xi\omega_n s + \omega_n^2 = 0 \text{ gives}$$

$$\omega_n = \sqrt{12.5}, \quad \xi = \frac{5}{2\times\sqrt{12.5}}$$

and

$$\omega_d = \omega_n\sqrt{1-\xi^2} = \sqrt{12.5}\sqrt{1-\frac{25}{50}} = 2.5$$

$$T = \frac{2\pi}{\omega_d} = 2.5 \text{ sec}$$

P 2.5: Consider the system shown in Fig. P2.5. Determine the value of *a* such that the damping ratio is 0.5. Also obtain the values of the rise time t_r and maximum overshoot M_P in its step response.

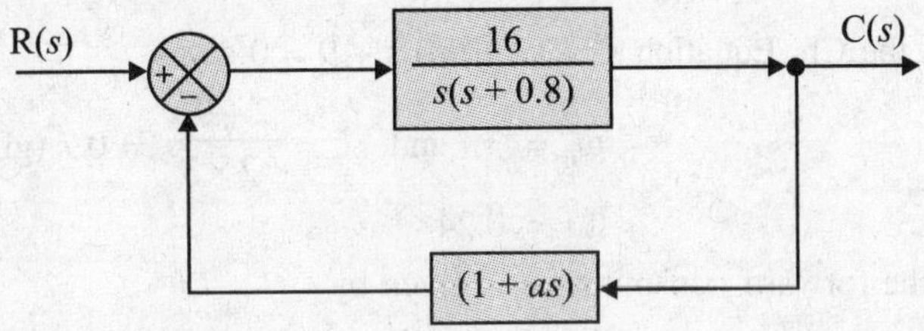

Fig. P2.5

Solution:

Ch. Equation: $s^2 + (0.8 + 16a)\, s + 16$

Compare it with general equation to get $\xi = 0.1 + 2a$

and $a\big|_{\xi = 0.5} = 0.2$

$$t_r = \frac{\pi - \phi}{\omega_d} = \frac{\pi - \tan^{-1}\left(\sqrt{1-\xi^2}/\xi\right)}{\omega_n\sqrt{1-\xi^2}} = 0.605 \text{ sec}$$

$$M_p = e^{-\pi\xi/\sqrt{1-\xi^2}} = 0.163$$

$$\% M_p = 16.3\%$$

P 2.6: Consider the system shown in Fig. P2.6:

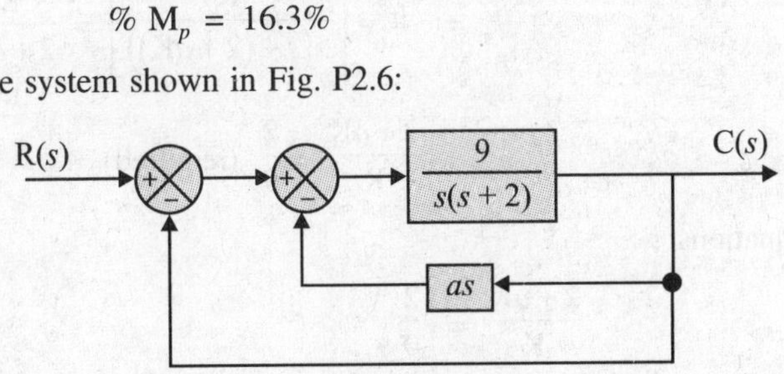

Fig. P2.6

(*i*) In the absence of derivative feedback ($a = 0$) determine the damping factor and natural frequency. Also determine the steady state error resulting from a unit ramp input.

(*ii*) Determine the derivative feedback constant a which will increase the damping factor of the system to 0.7. What is the steady state error to unit ramp input with this setting of the derivative feedback constant.

(*iii*) Illustrate how the steady state error of the system with the derivative feedback to unit - ramp can be reduced to the same value as in part (*i*) while the damping factor is maintained at 0.7.

Solution:

(*i*) Ch. Equation: $s^2 + 2s + 9 = 0$ gives $\xi = 0.33$, $\omega_n = 3$.

$$k_v = \lim_{s \to 0} s\,G(s) = \frac{9}{2}$$

$$e_{ss} = \frac{1}{k_v} = \frac{2}{9}$$

(*ii*) With $a \neq 0$, the Ch. Equation $s^2 + (2 + 9a)\,s + 9 = 0$ gives

$$\omega_n = 3 \quad \text{and} \quad \xi = \frac{2+9a}{2\times 3} = 0.7 \text{ (given)}$$

$$a = 0.24$$

with this setting of a the forward transmittance is given by

$$G^*(s) = \frac{9}{s(s+4.16)}$$

$$k_v^* = \operatorname*{lt}_{s\to 0} sG^*(s) = \frac{9}{4.16}$$

and
$$e_{ss}^* = \frac{4.16}{9} = 0.46$$

(*iii*) The steady state error can be decreased while maintaining $\xi = 0.7$ by changing forward gain of 9 to another value K. Then Ch. Equation $s^2 + (2 + a\mathrm{K})\,s + \mathrm{K} = 0$

gives
$$2\,\xi\omega_n = 2 + a\mathrm{K}$$

or
$$2 \times 0.7\,\omega_n = (2 + a\mathrm{K})$$

Also
$$k_v = \operatorname*{lt}_{s\to 0} s\left[\frac{\mathrm{K}}{s(s+(2+a\mathrm{K}))}\right] = \frac{\mathrm{K}}{2+a\mathrm{K}}$$

$$e_{ss} = \frac{2+a\mathrm{K}}{\mathrm{K}} = \frac{2}{9} \text{ (required)}$$

and solving the two equations

$$\frac{2+a\mathrm{K}}{\mathrm{K}} = \frac{2}{9}$$

and
$$1.4\,\sqrt{\mathrm{K}} = (2+a\mathrm{K}) \quad \text{we get}$$

$$a = 0.245$$

$$\mathrm{K} = 8.82$$

Alternatively, an amplifier with gain K_A can be placed between two summing blocks. Then Ch. Equation: $s^2 + (2 + 9a)\, s + 9\, K_A = 0$ gives

$$2\xi\omega_n = 2 + 9a \quad \text{and} \quad \xi = \frac{2+9a}{2\times 3.\sqrt{K_A}} = 0.7$$

$$k_v = \frac{9\,K_A}{(2+9a)} \quad \text{and} \quad e_{ss} = \frac{2+9a}{9\,K_A} = \frac{2}{9}$$

Solving these two equations

$$a = 0.\,757$$

$$K_A = 4.41$$

The second solution although requires smaller gain but needs additional hardware in the form of amplifier.

P 2.7: For the system shown in Fig. P2.7, determine characteristic equation. Hence find the following when excitation is a unit step ;

(*a*) undamped natural frequency
(*b*) damped frequency of oscillation
(*c*) damping ratio and damping factor
(*d*) maximum overshoot
(*e*) settling time.
(*f*) number of cycles before the response is settled within 2 % of its final value.
(*g*) Time interval after which maxima & minima will occur.

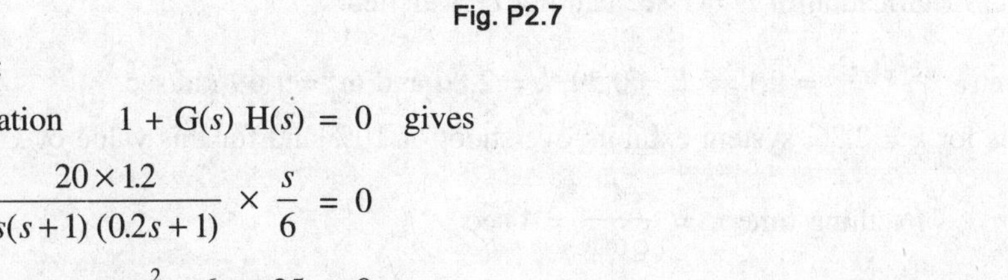

Fig. P2.7

Solution:

Ch. Equation $1 + G(s)\, H(s) = 0$ gives

$$1+\frac{20\times 1.2}{s(s+1)\,(0.2s+1)} \times \frac{s}{6} = 0$$

or $$s^2 + 6s + 25 = 0$$

On comparison with general Ch. Equation $s^2 + 2\xi\omega_n s + \omega_n^2 = 0$, one can find

(*a*) undamped natural frequency $\omega_n = 5$ rad/sec and $\xi = 0.6$

(*b*) damped frequency of oscillation $\omega_d = \omega_n\sqrt{1-\xi^2} = 4$ rad/sec

(*c*) damping ratio $\xi = 0.6$ and damping factor $\sigma = \xi\omega_n = 3$.

(*d*) max. overshoot = $e^{-\pi\xi/\sqrt{1-\xi^2}}$ = 0.09478 or 9.48%

(*e*) settling time $t_s = \dfrac{4}{\xi\omega_n}$ = 1.33 sec

(*f*) time period of occurrence of optima = $\dfrac{2\pi}{\omega_d}$ = 1.57 sec.

number of cycles within $t_s = \dfrac{1.33}{1.57} = 0.85$

(*g*) $t_p = \dfrac{\pi}{\omega_d} = \dfrac{\pi}{4} = 0.785\,\text{sec}$

P2.8: The open loop transfer function of a unity feedback system is

$$G(s) = \frac{k}{s(s+2)}$$

It is specified that the response of the system to a unit step input should have maximum overshoot of 10 % and the settling time should be less than one second.

(*a*) Is it possible to satisfy both the specification simultaneously by adjusting *k* ?

(*b*) If not, determine values of *k* that will satisfy the first specification. What will be the settling time and the time to reach the first peak for this case ?

Solution: Ch. equation $s^2 + 2s + k = 0$ gives

$$\omega_n = \sqrt{k}\,, \quad \xi = \frac{1}{\sqrt{k}}\,, \quad \sigma = \xi\omega_n = 1$$

(*a*) For $k > 1$, $\xi < 1$ and system exhibits underdamped dynamics. Since $\xi\omega_n$ remains constant equal to 1, settling time t_s remains constant equal to 4 sec.

For $k < 1$, $\xi > 1$ and system exhibits overdamped dynamics.

For $k = 1$, $\xi = 1$ and system exhibits critically damped dynamics.

Thus specification of $t_s = 1$ sec can not be satisfied.

(*b*) Given $e^{-\pi\xi/\sqrt{1-\xi^2}} = 0.1 \Rightarrow \xi = 0.59$, $k = 2.86$ and $\omega_n = 1.69$ rad/sec.

Thus for $k = 2.86$, system exhibits overshoot of 10% and for this value of *k*

$$\text{settling time } t_s = \frac{4}{\xi\omega_n} = 4 \text{ sec}$$

and
$$t_p = \frac{\pi}{\omega_n\sqrt{1-\xi^2}} = 2.3 \text{ sec}$$

P 2.9: The open loop transfer function of a unity feedback system is $G(s) = \dfrac{\alpha}{s(1+\beta s)}$. For this system overshoot reduces from 0.6 to 0.2 due to change in α only. Show that

$$\frac{\beta\alpha_1 - 1}{\beta\alpha_2 - 1} \cong 43$$

where α_1 and α_2 are values of α for 0.6 and 0.2 overshoot respectively.

Solution: Ch. equation $s^2 + \frac{s}{\beta} + \frac{\alpha}{\beta} = 0$ gives $\omega_n = \sqrt{\alpha/\beta}$ and $\xi = \frac{1}{2\sqrt{\alpha\beta}}$

Assuming $\xi = \xi_1$ for $M_p = 0.6$ and $\xi = \xi_2$ for $M_p = 0.2$,

$$0.6 = e^{-\pi\xi_1/\sqrt{1-\xi_1^2}}$$

and $$0.2 = e^{-\pi\xi_2/\sqrt{1-\xi_2^2}} \text{ gives}$$

$$\xi_1 = 0.1602 = \frac{1}{2\sqrt{\alpha_1\beta}} \Rightarrow \beta\alpha_1 = 9.7412$$

$$\xi_2 = 0.4559 = \frac{1}{2\sqrt{\alpha_2\beta}} \Rightarrow \beta\alpha_2 = 1.2028$$

then $$\frac{\beta\alpha_1 - 1}{\beta\alpha_2 - 1} \cong 43$$

P2.10: Fig. P2.10 (*a*). shows a mechanical vibratory system. When 8.9 N force is applied to the system, the mass exhibits the dynamics as shown in Fig. P2.10(*b*). Compute *m*, *f* and *k* of the system.

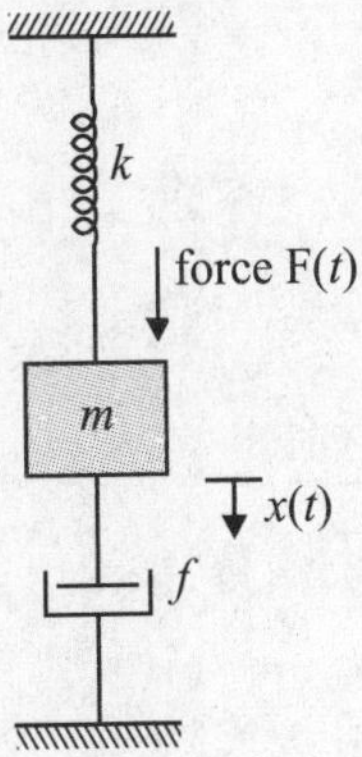

Fig. P2.10 (*a*)

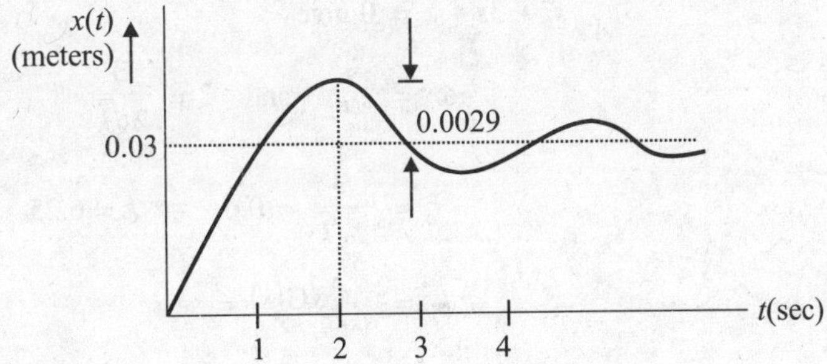

Fig. P2.10 (*b*)

Solution: For system of Fig. P2.10 (*a*), one can get

$$\frac{X(s)}{F(s)} = \frac{1}{Ms^2 + fs + k}$$

where $$F(s) = \frac{8.9}{s}$$

$$x(\infty) = \lim_{s\to 0} s\left[\frac{1}{Ms^2 + fs + k}\right] \times \frac{8.9}{s} = \frac{8.9}{k} = 0.03 \text{ (given)}$$

$$k = \frac{8.9}{0.03} = 296.67$$

For this value of k, Ch. equation $s^2 + \frac{f}{m}s + \frac{296.67}{m} = 0$ gives

$$\omega_n = \sqrt{296.67/m} \quad \text{and} \quad \xi = \frac{f}{2\sqrt{296.67\,m}} = \frac{f}{34.45\sqrt{m}}$$

Given $t_p = 2 = \dfrac{\pi}{\omega_n\sqrt{1-\xi^2}}$ and $e^{-\pi\xi/\sqrt{1-\xi^2}} = 0.0029$

Solving these two equations after putting values of ω_n and ξ in terms of m and f, we get

$$m = 77.3 \text{ kg}$$

and

$$f = 181.8 \text{ Nm}^{-1}\text{s}$$

P2.11: The forward transfer function of a unity feedback type 1, second order system has a pole at $s = -3$. The gain k is adjusted for the damping ratio of 0.6. Compute steady state error if this system is excited by $r(t) = 1 + 5t$.

Solution: For given information about type of system and location of system pole, forward path transfer function can be written as

$$G(s) = \frac{k}{s(s+3)}$$

Ch. equation for unity feedback system

$$s^2 + 3s + k = 0 \text{ gives}$$

$$\omega_n = \sqrt{k} \quad \text{and} \quad \xi = \frac{3}{2\sqrt{k}}$$

but

$$\xi = \frac{3}{2\sqrt{k}} = 0.6 \quad \Rightarrow k = 6.25$$

$$k_p = \lim_{s\to 0} sG(s) = \infty$$

$$K_v = \lim_{s\to 0} sG(s) = \frac{6.25}{3} = 2.083$$

$$e(\infty) = \frac{1}{1+k_p} + \frac{5}{k_v} = 0 + \frac{5}{2.083} = 2.4$$

P2.12: The open loop transfer function of unity feedback system is given by

$$G(s) = \frac{40}{s(0.2s+1)}$$

Determine steady state error as a function of time for the input given by $r(t) = (3 + 4t)t$.

Solution: $e(t) = c_0 r(t) + c_1 \dot{r}(t) + c_2 \ddot{r}(t) + ...$

where $\dot{r}(t) = 3 + 8t$

and $\ddot{r}(t) = 8$

Higher order derivatives are obviously zero.

$$W(s) = \frac{E(s)}{R(s)} = \frac{1}{1+G(s)\,H(s)} = \frac{s(0.2s+1)}{s(0.2s+1)+40}$$

$$C_0 = \underset{s\to 0}{\text{lt}}\ W(s) = 0$$

$$C_1 = \underset{s\to 0}{\text{lt}} \frac{d}{ds} W(s) = \underset{s\to 0}{\text{lt}} \left[\frac{16s+40}{(0.2s^2+s+40)^2}\right] = 0.025$$

$$C_2 = \underset{s\to 0}{\text{lt}} \frac{d^2}{ds^2} W(s) = \underset{s\to 0}{\text{lt}} \left[\frac{-4.8s^2 - 48s + 560}{(0.2s+s+40)^3}\right] = 0.00875$$

$$e(t) = 0.025\ (3 + 8t) + 0.00875 \times 8 = 0.2t + 0.145$$

P2.13: The closed loop transfer function of a system is given by

$$T(s) = \frac{C(s)}{R(s)} = \frac{k(s+z)}{s^2+4s+8}$$

where k and z are adjustable.

(*a*) If $r(t) = t$, find values of k and z so that steady state error is zero.

(*b*) For the values of k and z obtained in part (*a*), find $e(\infty)$ for input $r(t) = \frac{1}{2}t^2$.

Solution:

(*a*) $$E(s) = R(s) - C(s) = R(s) - R(s)\,T(s) = R(s)\,[1 - T(s)]$$

$$e(\infty) = \underset{s\to 0}{\text{lt}}\ sE(s) = \underset{s\to 0}{\text{lt}}\left[sR(s)(1-T(s))\right]$$

where $R(s) = \frac{1}{s^2}$

$\therefore$ $$e(\infty) = \underset{s\to 0}{\text{lt}}\left[\frac{1-T(s)}{s}\right]$$

It is obvious that $e(\infty) = \infty$ unless $T(0) = 1$ which means $\frac{kz}{8} = 1$

If $T(0) = 1$, use of L' Hospital's rule gives

$$e(\infty) = \underset{s\to 0}{\text{lt}}\left[-\frac{dT(s)}{ds}\right] = 0 \text{ (given)}$$

$$\Rightarrow \qquad \underset{s\to 0}{\text{lt}}\left[-\frac{k\left(s^2+4s+8\right)-k(s+z)(2s+4)}{\left(s^2+4s+8\right)^2}\right] = 0$$

$$z = 2 \quad \text{and} \quad k = \frac{8}{z} = 4$$

(*b*) Putting values of *k* and *z* of part (*a*)

$$\frac{C(s)}{R(s)} = T(s) = \frac{4(s+2)}{\left(s^2+4s+8\right)}$$

where $\qquad R(s) = \frac{1}{s^3}$

$$e(\infty) = \underset{s\to 0}{\text{lt}}\left[sR(s)(1-T(s))\right] = \underset{s\to 0}{\text{lt}}\left[\frac{1-T(s)}{s^2}\right]$$

$$= \underset{s\to 0}{\text{lt}}\left[\frac{1}{s^2+4s+8}\right] = 0.125$$

P2.14: The closed loop transfer function of a position control system is given by

$$\frac{C(s)}{R(s)} = T(s) = \frac{k(s+z)(s+4)}{(s+p)\left(s^2+6s+25\right)}$$

where *k*, *p* and *z* are adjustable. Is it possible to select them so that system exhibits zero steady state error for step, ramp and parabolic inputs ? If possible find steady state error for input $r(t) = t^3/6$.

Solution: $\qquad e(\infty) = \underset{s\to 0}{\text{lt}}\left[sR(s)(1-T(s))\right]$

$$= \underset{s\to 0}{\text{lt}}\left[sR(s)\left(\frac{(s+p)\left(s^2+6s+25\right)-K(s+z)(s+4)}{(s+p)\left(s^2+6s+25\right)}\right)\right]$$

For $e(\infty)$ to be zero for $R(s) = \frac{1}{s^3}$ *i.e.* parabolic input, the expression $(s + p)(s^2 + 6s + 25) - k(s + z)(s + 4)$ must be equal to s^3 or coefficients of term s^2, s and constant of expression must be equal to zero, *i.e.*

$$6 + p - k = 0$$

$$6p + 25 - k(z + 4) = 0$$

$$25p - 4kz = 0$$

Solving these equations, we get

$$k = \frac{106}{17}, \quad p = \frac{4}{17} \quad \text{and} \quad z = \frac{25}{106}$$

Then $$1 - \mathrm{T}(s) = \frac{s^3}{s^3 + \frac{106}{17}s^2 + \frac{449}{17}s + \frac{100}{17}}$$

and $e(\infty)$ for $r(t) = \frac{t^3}{6}$ or $\mathrm{R}(s) = \frac{1}{s^4}$ is given by

$$e(\infty) = \underset{s \to 0}{\mathrm{lt}} \left[(s)\left(\frac{1}{s^4}\right)\left(\frac{s^3}{s^3 + \frac{106}{17}s^2 + \frac{449}{17}s + \frac{100}{17}}\right)\right] = 0.17$$

DRILL PROBLEMS

D 2.1: Identify order and type of systems shown in Fig. D2.1 (*a*) and (*b*).

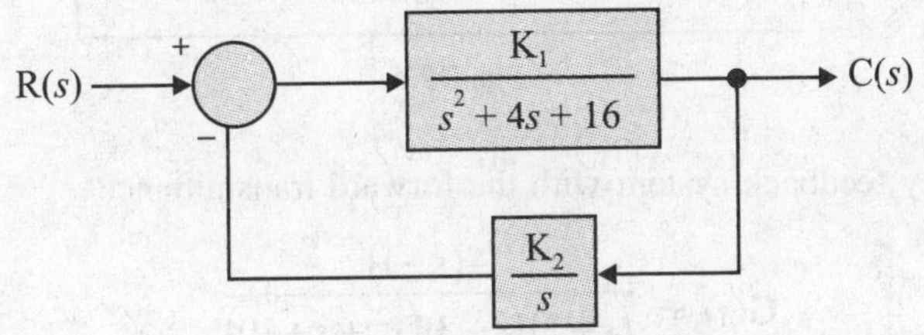

Fig. D2.1 (*a*)

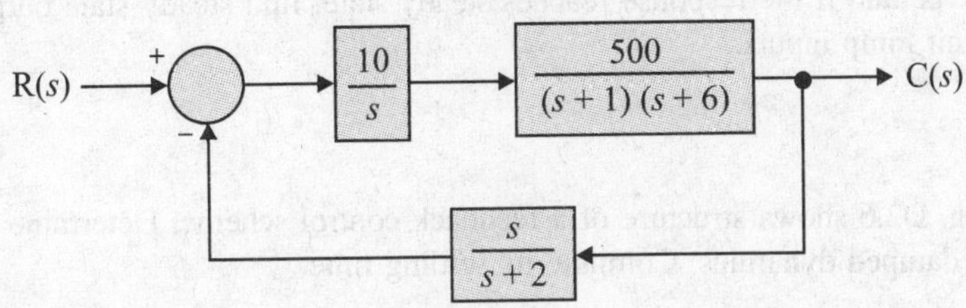

Fig. D2.1 (*b*)

Ans. (*a*) type = 1, order = 3, (*b*) type = 0, order = 4

D 2.2: Determine which of the following second order systems are underdamped, which are critically damped and which are overdamped.

(*a*) $\mathrm{T}(s) = \frac{9s^2 + 3s + 10}{s^2 + 5s + 2}$ (*b*) $\mathrm{T}(s) = \frac{s^2 - 2s}{s^2 + 6s + 9}$

(*c*) $\mathrm{T}(s) = \frac{64}{3s^2 + 4s + 5}$ (*d*) $\mathrm{T}(s) = \frac{19s - 20}{s^2 + s + 100}$

(*e*) $\mathrm{T}(s) = \frac{s^2 + 2s + 100}{s^2 + 7s + 49}$

Ans. (*a*) overdamped, (*b*) critically damped, (*c*), (*d*), (*e*) underdamped.

D2.3: Find the constant k for which the system with transfer function T(s) has the given second order response property.

(*a*) $T(s) = \dfrac{10}{s^2 + 40s + k}, \quad \xi = 0.7$

Ans. 816

(*b*) $T(s) = \dfrac{s^2 - 6}{ks^2 + s + 6}, \quad \omega_n = 0.7$

Ans. 12.24.

D2.4: For the system shown in Fig. D2.4, find k for which damping ratio of overall system is 0.7. For this value of K find undamped natural frequency.

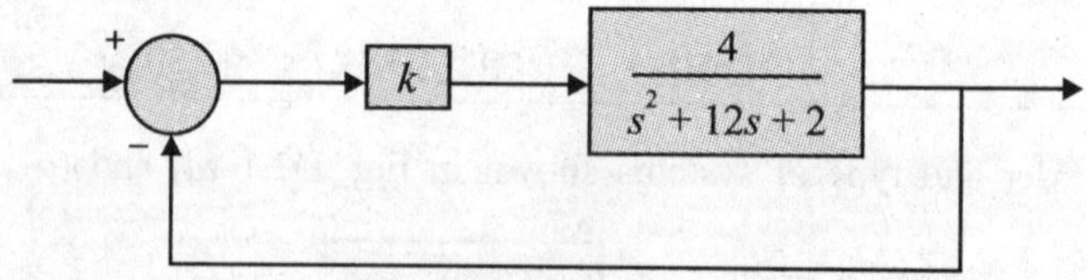

Fig. D2.4

Ans. 17.87, 8.57.

D2.5: For the unity feedback system with the forward transmittance

$$G(s) = \frac{4(s+1)}{(s+2)(s+3)(s^2+s+10)}$$

Find system type and if the response reaches steady state, find steady state output-input errors to unit step and to unit ramp inputs.

Ans. 0, $\dfrac{15}{16}$, ∞.

D2.6: The Fig. D2.6 shows structure of a feedback control scheme. Determine T_d so that system exhibits critically damped dynamics. Compute its settling time.

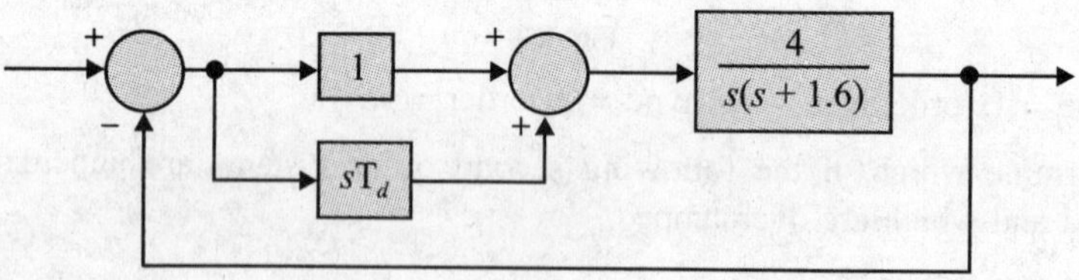

Fig. D2.6

Ans. $T_d = 0.6$, $t_s = 2$ sec.

D2.7: Fig. D2.7 shows a servo system. Determine

(*a*) Characteristics equation

(*b*) Undamped frequency of oscillation

(*c*) Damping ratio

(*d*) Damping factor

(*e*) Peak overshoot

(*f*) First undershoot

(*g*) Time interval after which optima occur

(*h*) Settling time

(*i*) No. of cycles completed before response settles within 2% of final value.

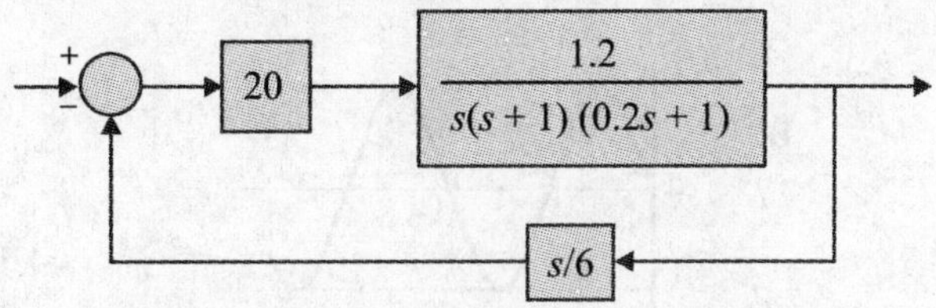

Fig. D2.7

Ans. $s^2 + 6s + 25 = 0$, $\omega_n = 5$ rad/sec, $\xi = 0.6$, $\omega_d = 4$, $M_p = 9.5\%$, 0.94%, 0.785 sec, 1.33 sec; 0.85 cycles.

D2.8: Compute static error coefficients for a unity feedback system with $G(s) = \dfrac{12}{s(s+6)}$. If input $r(t) = 4 + 3t$, find steady state error. For this system if steady state error is to be reduced to 10% of existing value, what should be percentage change in gain ?

Ans. $k_p = \infty$, $k_v = 2$, $e(\infty) = 1.5$, 900% change.

D2.9: The open loop transfer function of a unity feedback control system is given by

$$G(s) = \frac{k}{s(sT+1)}$$

(*a*) By what factor gain k should be multiplied so that damping ratio increases from 0.2 to 0.8 ?

(*b*) By what factor the time constant T should be multipled so that damping ratio reduces from 0.6 to 0.3 ?

Ans. $\dfrac{1}{16}$, 4.

D2.10: The dynamics of a servo mechanism is characterised by

$$\ddot{y} + 4.8\dot{y} = 144\, e$$

where $e = c - 0.5\, y$ is error signal.

(*a*) Sketch block diagram of system

(*b*) Find damping ratio, damped and undamped frequency of oscillations.

Ans. 0.281, 8.48 rad/sec, 8.14 rad/sec.

MULTIPLE CHOICE QUESTIONS

M2.1: Match List-I with List-II and select the correct answer using the codes given below the lists:

List I (Response to a unit step input)	**List II (Location of poles in the s-plane)**
A. 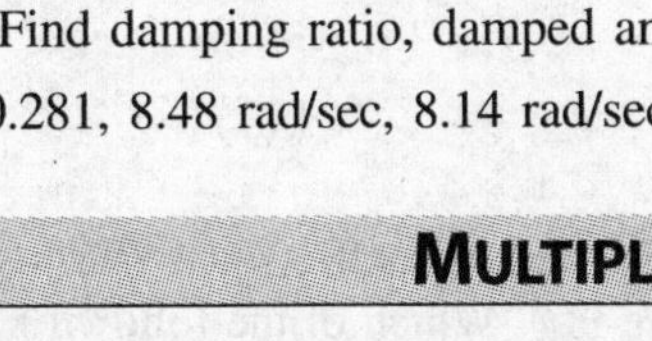	1. One at the origin

B. 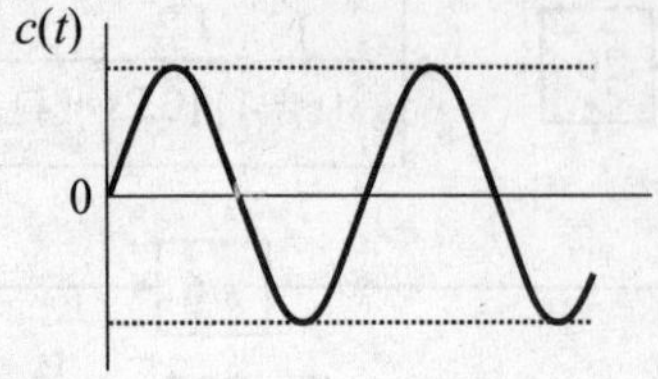

2. Two identical roots on the negative real axis

C. 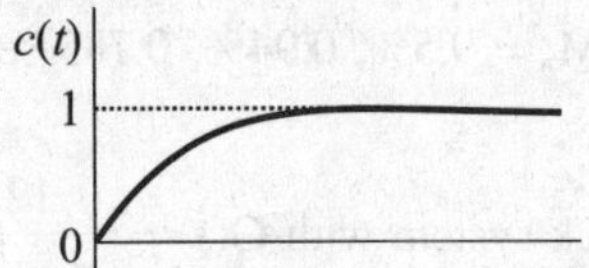

3. Two on the imaginary axis

D. 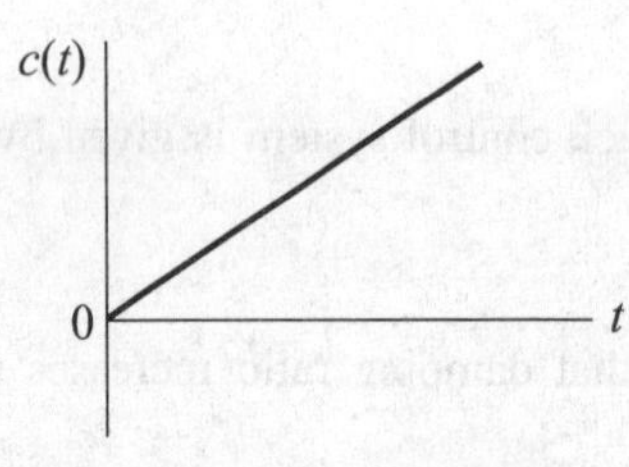

4. One on the positive real axis

Codes:	**A**	**B**	**C**	**D**
(*a*)	4	3	2	1
(*b*)	3	4	1	2
(*c*)	3	4	2	1
(*d*)	4	3	1	2.

M 2.2: The open-loop transfer function of a unity feedback control system is given by $G(s) = \dfrac{K}{s(s+1)}$. If the gain K is increased to infinity, then the damping ratio will tend to become:

(*a*) $\dfrac{1}{\sqrt{2}}$ (*b*) 1 (*c*) 0 (*d*) ∞

M2.3: In the given figure, spring constant is K, viscous friction coefficient is B, mass is M and the system output motion is $y(t)$ corresponding to input force $F(t)$. Which of the following parameters relate to the above system:

1. Time constant $= \dfrac{1}{M}$
2. Damping coefficient $= \dfrac{B}{2\sqrt{KM}}$
3. Natural frequency of oscillation $= \sqrt{\dfrac{K}{M}}$

(*a*) 1, 2 and 3 (*b*) 1 and 2 (*c*) 2 and 3 (*d*) 1 and 3

M2.4: The system shown in the given figure has a unit step input ;

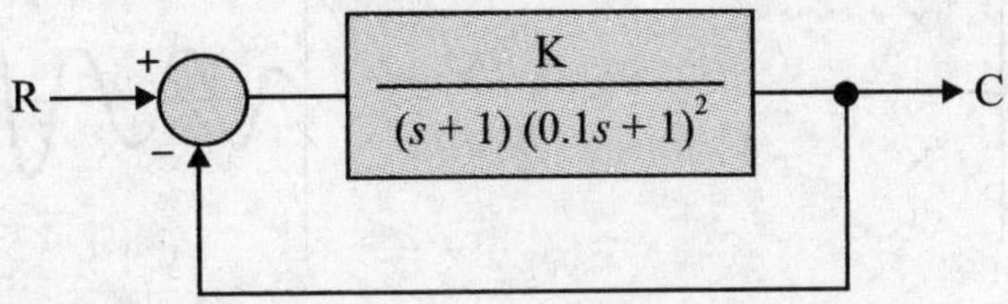

In order that the steady state error is 0.1, the value of K required is:

(*a*) 0.1 (*b*) 0.9 (*c*) 1.0 (*d*) 9.0

M2.5: The system shown in the given figure has a second order response with a damping ratio of 0.6 and a frequency of damped oscillations of 10 rad/sec. The values of K_1 and K_2 are respectively :

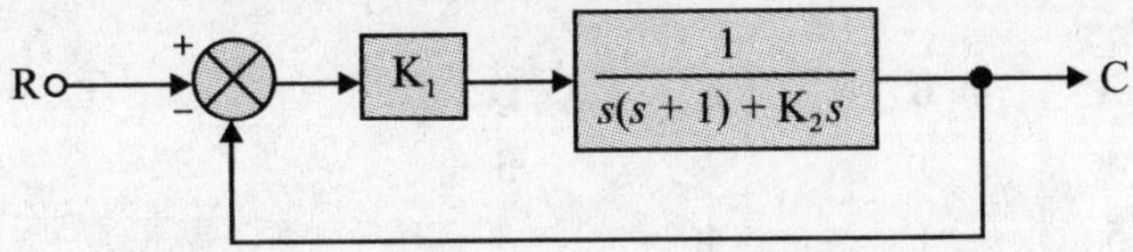

(*a*) 12.5 and 15 (*b*) 156.25 and 15 (*c*) 156.25 and 14 (*d*) 12.5 and 14

M2.6: Match List-I with List-II and select the correct answer using the codes given below the lists:

List-I (Location of poles on s-plane)

List-II (Type of response)

A. 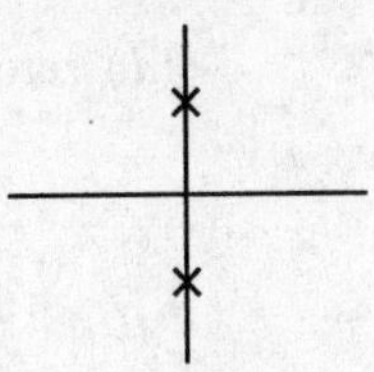

1. 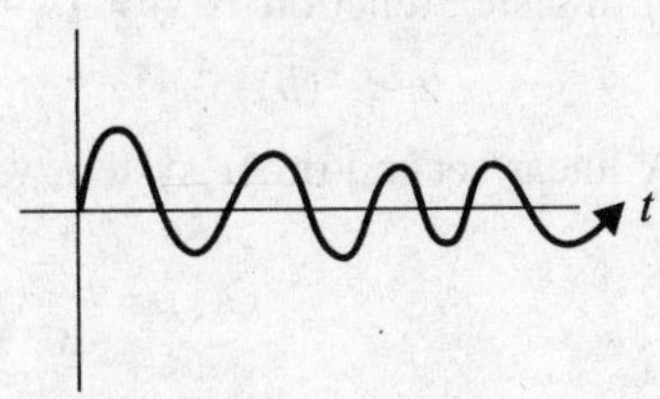

B. 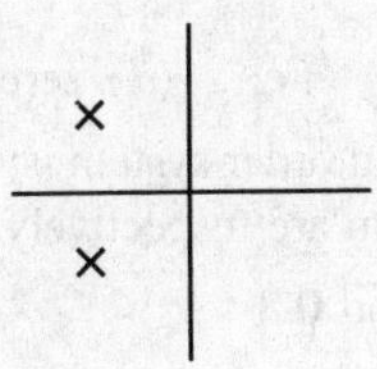

2. 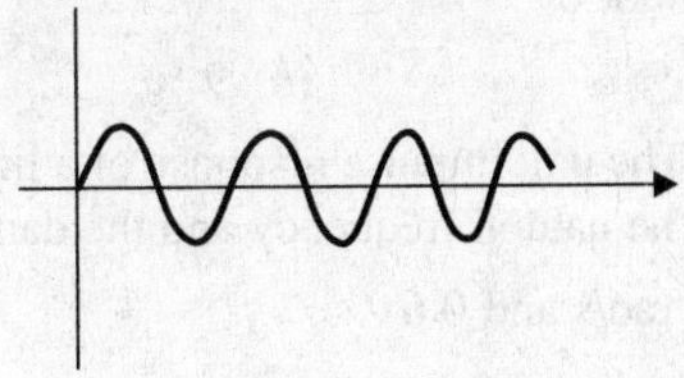

C. 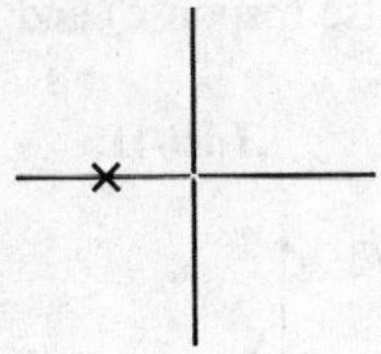

3. 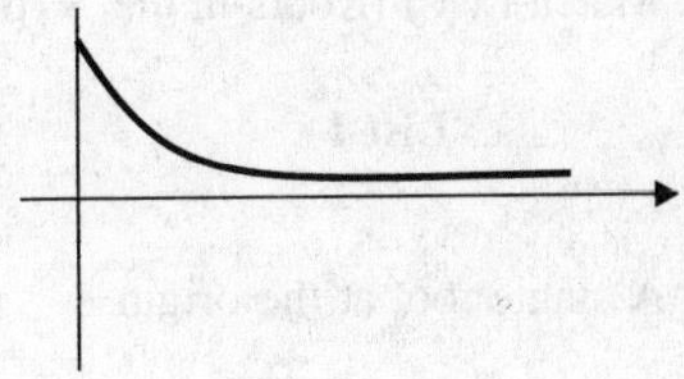

D. 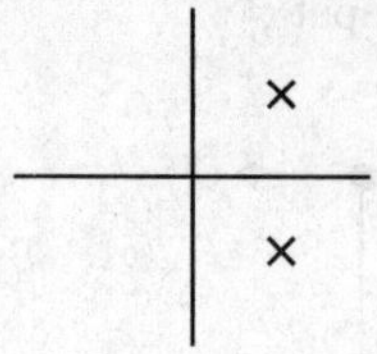

4. 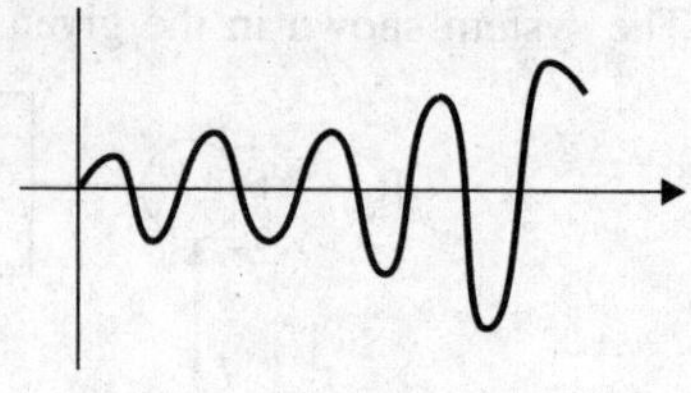

5. 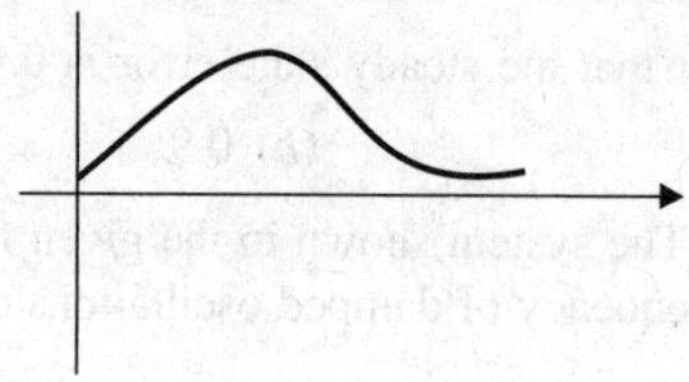

Codes:	A	B	C	D
(*a*)	4	1	2	3
(*b*)	5	1	4	3
(*c*)	2	3	5	4
(*d*)	2	1	3	4

M 2.7: For a unit step input, a system with forward path transfer function $G(s) = \dfrac{20}{s^2}$ and feedback path transfer function $H(s) = (s + 5)$, has a steady state output of:

(*a*) 20 (*b*) 5 (*c*) 0.2 (*d*) zero

M 2.8: A linear second-order system with transfer function:

$$G(s) = \frac{49}{s^2 + 16s + 49}$$

is initially at rest and is subjected to a step input signal. The response of the system will exhibit a peak overshoot of:

(*a*) 16 % (*b*) 9 % (*c*) 2 % (*d*) zero

M2.9: The unit impulse response of a linear time-invariant second order system is; $g(t) = 10\,e^{-8t}$ sin $6t$ ($t < 0$). The natural frequency and the damping factor of the system are respectively:

(*a*) 10 rad/s and 0.6 (*b*) 10 rad/s and 0.8

(*c*) 6 rad/s and 0.6 (*d*) 6 rad/s and 0.8

M2.10: Match List-I (Roots in the 's' plane) with List-II (Impulse response) and select the correct answer:

List-I	List-II
A. A single root at the origin	(1) 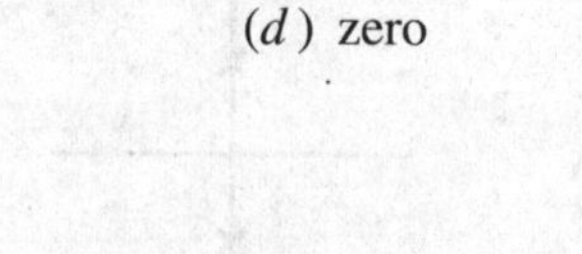

B. A single root on the negative real axis (2)

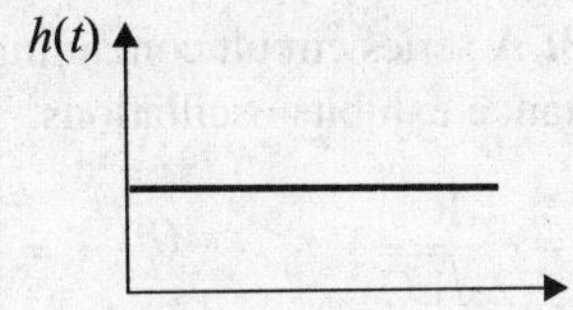

C. Two imaginary roots (3)

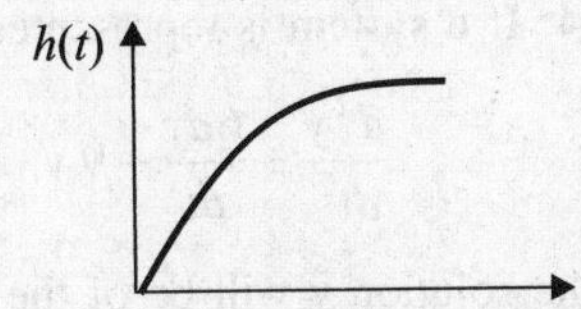

D. Two complex roots in the right half plane (4)

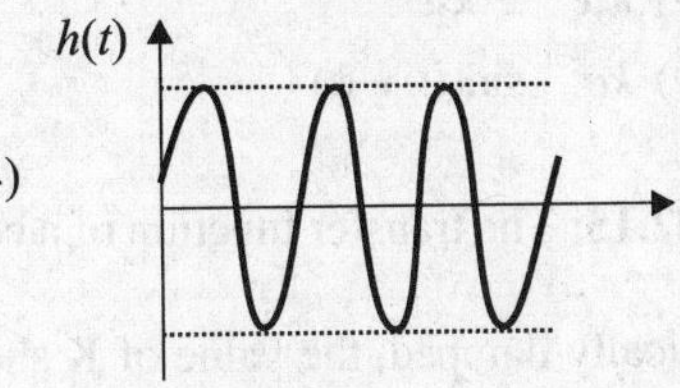

(5)

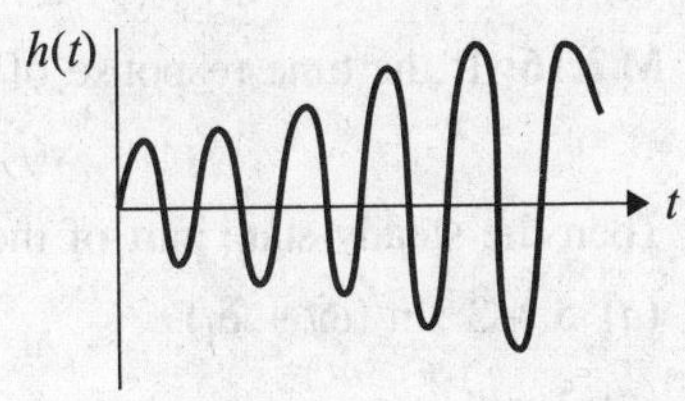

Codes:	A	B	C	D
(*a*)	2	1	5	4
(*b*)	3	2	4	5
(*c*)	3	2	5	4
(*d*)	2	1	4	5

M2.11: For a second order system, if both the roots of the characteristic equation are real and equal then the value of damping ratio will be:

(*a*) less then unity (*b*) equal to zero

(*c*) equal to unity (*d*) greater than unity

M2.12: $[-a \pm jb]$ are the complex conjugate roots of the characteristic equation of a second order system. Its damping coefficient and natural frequency will be respectively:

(*a*) $\dfrac{b}{\sqrt{a^2+b^2}}$ and $\sqrt{a^2+b^2}$ (*b*) $\dfrac{a}{\sqrt{a^2+b^2}}$ and a^2+b^2

(*c*) $\dfrac{a}{\sqrt{a^2+b^2}}$ and $\sqrt{a^2+b^2}$ (*d*) $\dfrac{a}{\sqrt{a^2+b^2}}$ and a^2+b^2

M2.13: A series circuit containing R, L and C is excited by a step voltage input. The voltage across the capacitance exhibits oscillations. The damping coefficient (ratio) of this circuit is given by:

(*a*) $\zeta = \dfrac{R}{2\sqrt{LC}}$ (*b*) $\zeta = \dfrac{R}{LC}$ (*c*) $\zeta = \dfrac{R}{2\sqrt{C/L}}$ (*d*) $\zeta = \dfrac{R}{2\sqrt{L/C}}$

M2.14: If a system is represented by the differential equation ;

$$\frac{d^2y}{dt^2} + \frac{6dy}{dt} + 9y = 0,$$

then the solution y will be of the form:

(*a*) $k_1e^{-t} + k_2e^{-9t}$ (*b*) $(k_1 + k_2t)\,e^{-3t}$
(*c*) $ke^{-3t}\sin(t + \phi)$ (*d*) $(k_1 + k_2t)\,e^{3t}$

M2.15: The transfer function of a control system is given as ; $T(s) = \dfrac{K}{s^2 + 4s + K}$. For this system to be critically damped, the value of K should be:

(*a*) 1 (*b*) 2 (*c*) 3 (*d*) 4

M 2.16: If the time response of a system is given by the following equation;

$$y(t) = 5 + 3\sin(\omega t + \delta_1) + e^{-3t}\sin(\omega t + \delta_2) + e^{-5t}$$

Then the steady state part of the above response is given by:

(*a*) $5 + 3\sin(\omega t + \delta_1)$ (*b*) $5 + 3\sin(\omega t + \delta_1) + e^{-3t}\sin(\omega t + \delta_2)$
(*c*) $5 + e^{-5t}$ (*d*) 5

M2.17: The transfer function of a system is $\dfrac{10}{1+s}$. When operated as a unity feedback system, the steady state error to a unit step input will be:

(*a*) zero (*b*) 1/11 (*c*) 10 (*d*) infinity

M2.18: A unity feedback second order control system is characterised by $G(s) = \dfrac{K}{s(Js + B)}$ where J = moment of inertia, K = system gain, B = Viscous damping coefficient. The transient response specification which is NOT affected by variation of system gain is the:

(*a*) peak overshoot (*b*) rise time
(*c*) settling time (*d*) damped frequency of oscillations

M2.19: The open loop transfer function of a unity feedback system is given by $\dfrac{K}{s(s+1)}$. If the value of gain K is such that the system is critically damped, the closed loop poles of the system will lie at:

(*a*) – 0.5 and – 0.5 (*b*) $\pm j\,0.5$
(*c*) 0 and – 1 (*d*) $0.5 \pm j\,0.5$

M2.20: For what values of 'a' does the system shown in figure has a zero steady state error [*i.e.*, $\lim_{t \to \infty} e(t)$] for a step input:

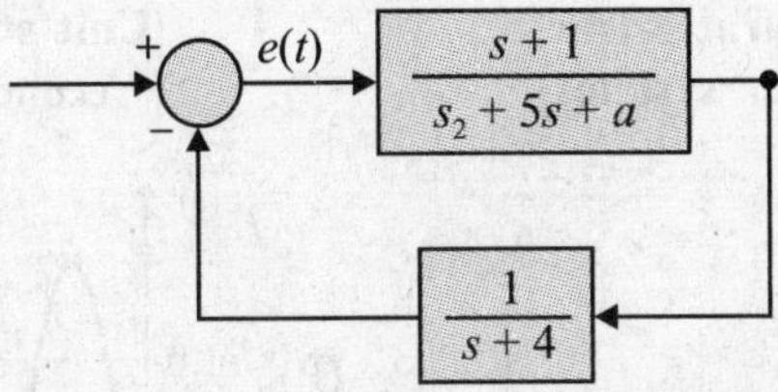

(*a*) $a = 0$ (*b*) $a = 1$ (*c*) $a < 4$ (*d*) For no value of a

M 2.21: Match the following transfer functions and impulse responses and choose the correct answer:

Transfer function		**Impulse responses**
(1) $\dfrac{s}{s+1}$	(P)	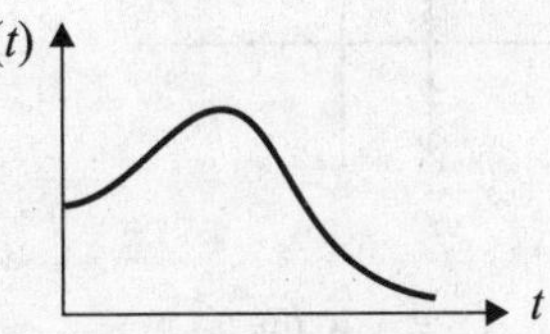
(2) $\dfrac{s}{(s+1)^2}$	(Q)	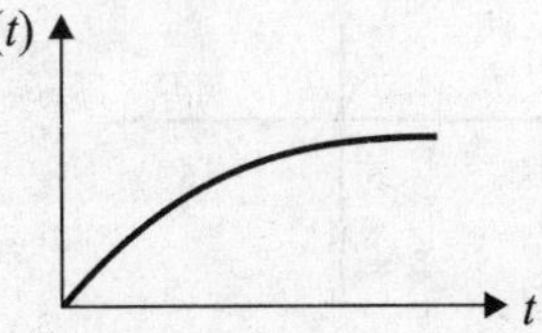
(3) $\dfrac{s}{s(s+1)+1}$	(R)	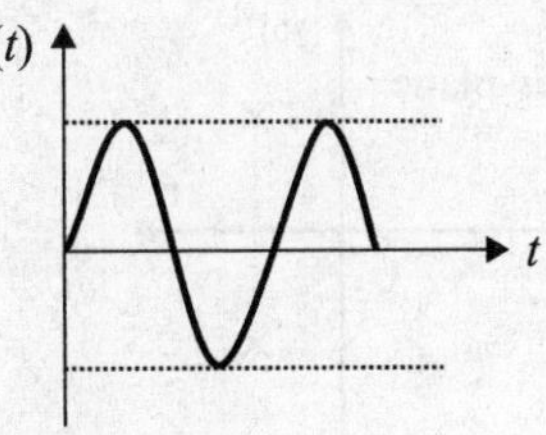
(4) $\dfrac{1}{s^2+1}$	(S)	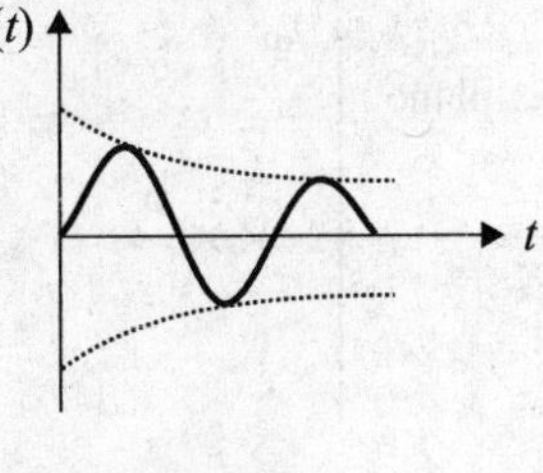

(*a*) 1-P, 2-Q, 3-S, 4-R (*b*) 1-P, 2-Q, 3-R, 4-S
(*c*) 1-Q, 2-P, 3-S, 4-R (*d*) 1-P, 2-R, 3-S, 4-Q

M 2.22: Match the following items from List-I and List-II and choose the correct answer:

List-I
Root locations of the characteristic equations of second order systems

(1) 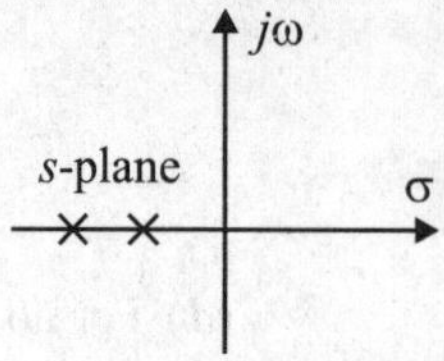

(2) 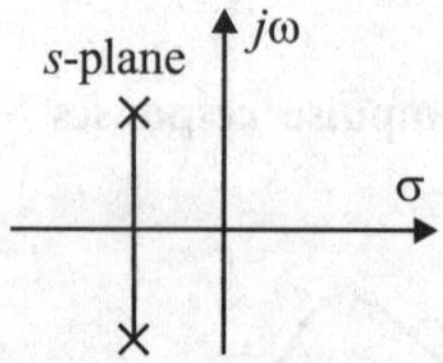

(3) 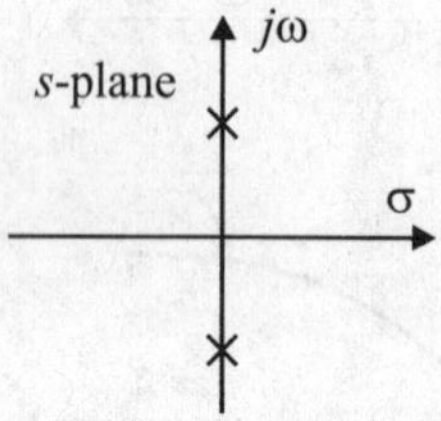

(4) 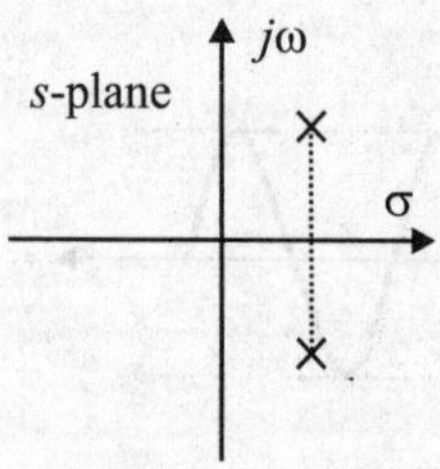

(5) 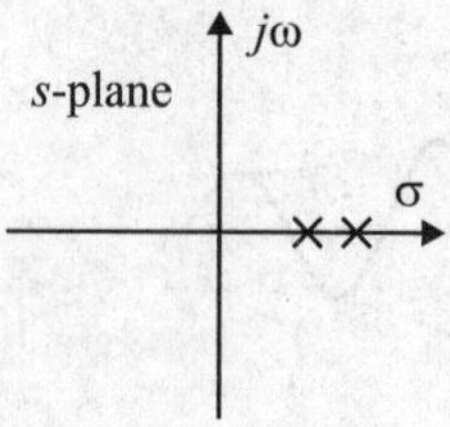

List-II
Unit step responses of second order systems

(P) 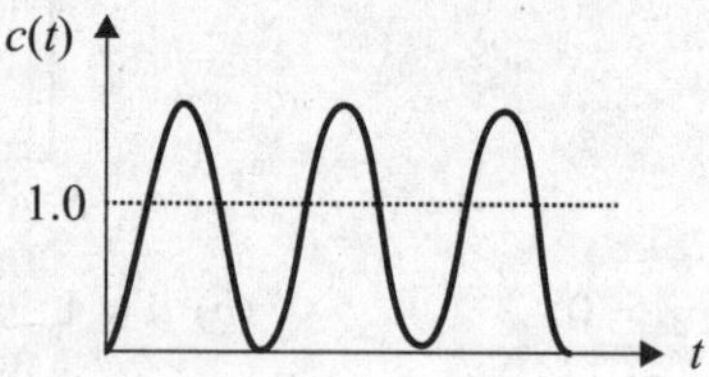

(Q) 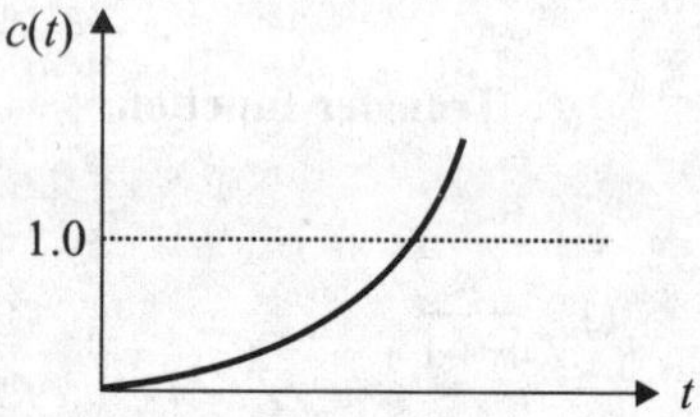

(R) 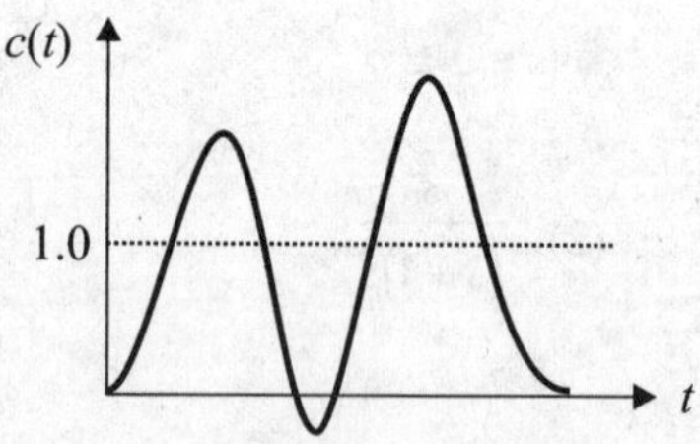

(S) 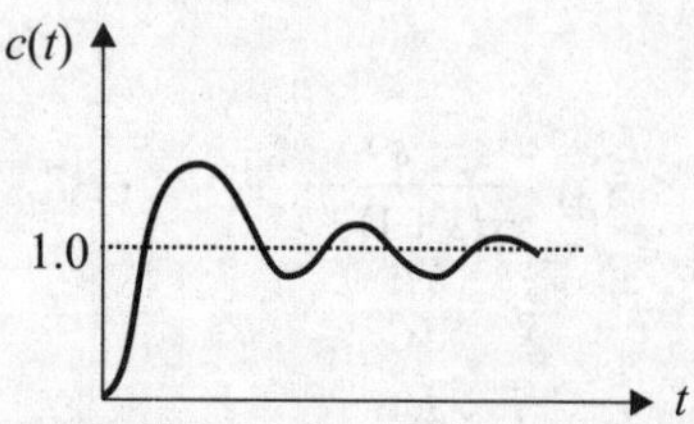

(T) 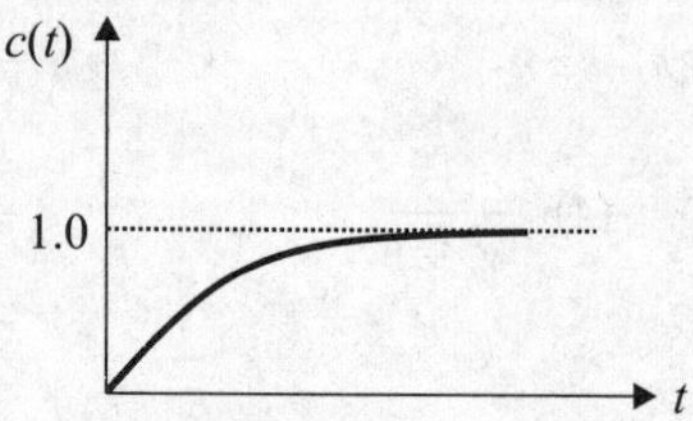

(*a*) 1-T, 2-S, 3-P, 4-R, 5-Q
(*b*) 1-Q, 2-S, 3-P, 4-R, 5-T
(*c*) 1-T, 2-R, 3-P, 4-S, 5-Q
(*d*) 1-Q, 2-S, 3-R, 4-P, 5-T

M2.23: For the system shown in figure with a damping ratio ζ of 0.7 and an undamped natural frequency ω_n of 4 rad/sec, the values of K and a are:

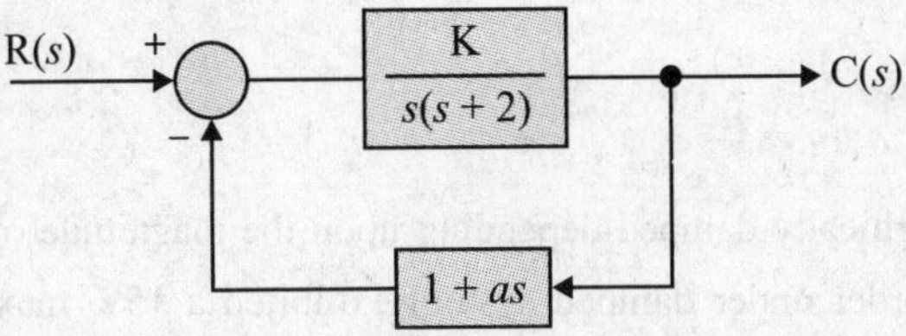

(*a*) K = 4, a = 0.35 (*b*) K = 8, a = 0.455

(*c*) K = 16, a = 0.225 (*d*) K = 64, a = 0.9

M2.24: Consider the following expressions which indicate the step or impulse response of an initially relaxed control system:

1. $(5 - 4e^{-2t})\, u(t)$
2. $(e^{-2t} + 5)\, u(t)$
3. $\delta(t) + 8e^{-2t}\, u(t)$
4. $\delta(t) + 4e^{-2t}\, u(t)$

Those which correspond to the step and impulse response of the same system include:

(*a*) 1 and 3 (*b*) 1 and 4 (*c*) 2 and 4 (*d*) none of these

M2.25: Consider the systems with the following open loop transfer functions:

1. $\dfrac{36}{s(s+3.6)}$ 2. $\dfrac{100}{s(s+5)}$ 3. $\dfrac{6.25}{s(s+4)}$

The correct sequence of these systems in increasing order of the time taken for the unit step response to settle is:

(*a*) 1 , 2 , 3 (*b*) 3, 1, 2 (*c*) 2, 3, 1 (*d*) 3, 2, 1

M2.26: Match List-I with List-II and select the correct answer using the codes given below the lists:

List-I (Characteristics equations)	**List-II (Nature of damping)**
A. $s^2 + 15s + 26.25$	1. Undamped
B. $s^2 + 5s + 6.25$	2. Underdamped
C. $s^2 + 20.25$	3. Critically damped
D. $s^2 + 4.5s + 42.45$	4. Overdamped

Codes:	**A**	**B**	**C**	**D**
(*a*)	1	2	3	4
(*b*)	2	3	1	4
(*c*)	4	3	1	2
(*d*)	1	2	4	3

M2.27: A step function voltage is applied to an RLC series circuit having R = 2Ω, L = 1H and C = 1F. The transient current response of the circuit would be:

(*a*) overdamped

(*b*) critically damped

(*c*) underdamped

(*d*) over, under or critically damped depending upon the magnitude of the step voltage.

M2.28: A second order under damped system exhibited a 15% maximum overshoot on being excited by a step input $r(t) = 2u(t)$, and then attained a steady state value of 2 (see figures given). If, at $t = t_0$, the input were changed to a unit step $r(t) = u(t)$, then its time response $c(t)$ would be similar to:

(*a*) 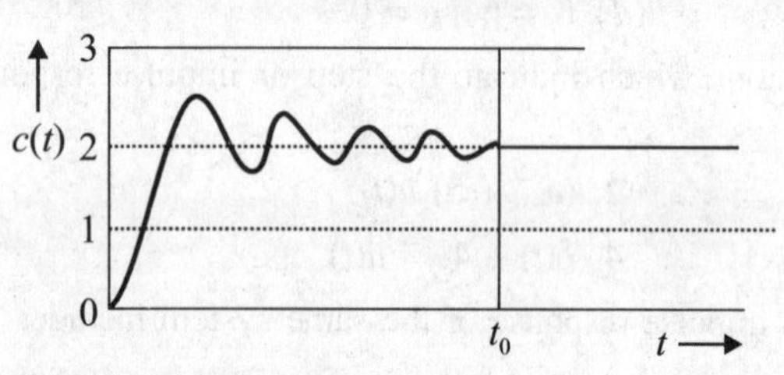

(*b*) 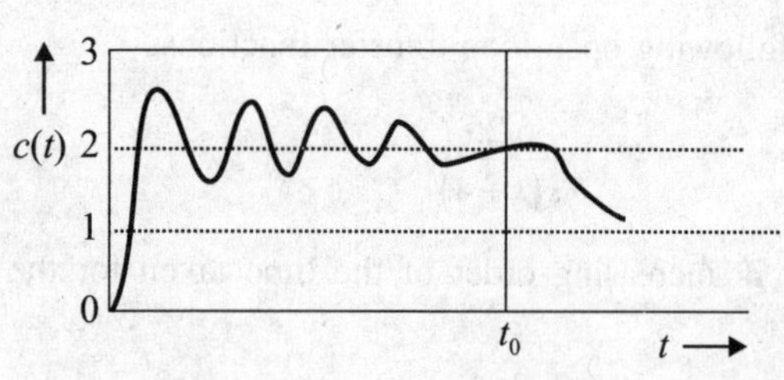

(*c*) 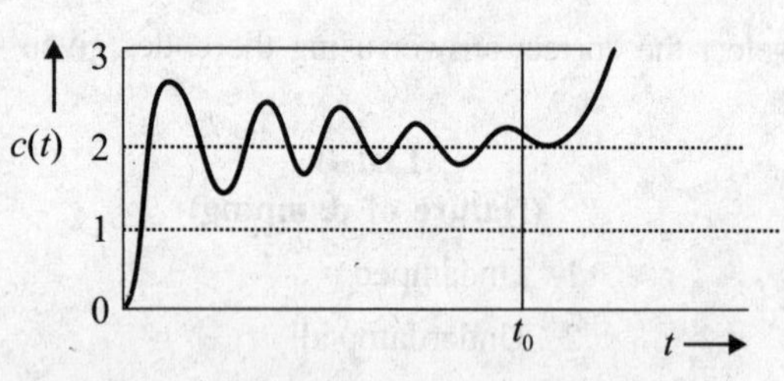

(*d*) 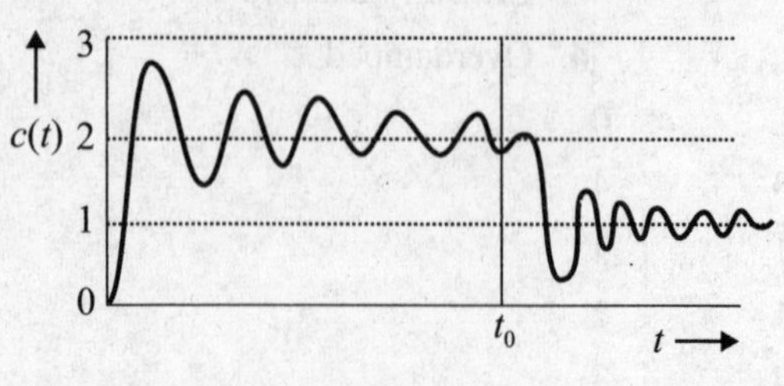

M2.29: If the closed loop transfer function T(*s*) of a unity negative feedback system is given by;

$$T(s) = \frac{a_{n-1}s + a_n}{s^n + a_1 s^{n-1} + \ldots\ldots + a_{n-1}s + a_n}$$

then the steady state error for a unit ramp input is:

(*a*) $\frac{a_n}{a_{n-1}}$ (*b*) $\frac{a_n}{a_{n-2}}$ (*c*) $\frac{a_{n-1}}{a_{n-2}}$ (*d*) zero

M2.30: A unity feedback control system has a forward path transfer function; $G(s) = \frac{10(1+4s)}{s^2(1+s)}$, the system is subjected to an input $r(t) = 1 + t + \frac{t^2}{2}$ $(t \geq 0)$, the steady state error of the system will be:

(*a*) zero (*b*) 10 (*c*) 0.1 (*d*) infinity

M2.31: The system shown in figure below is excited by unit step input and is required to exhibit output equal to input in steady state. The design effort is made to ensure that peak over shoot and 2% settling time do not exceed 5% and 1 second respectively. For successful design the numerical values of *a*, *b* and K respectively are

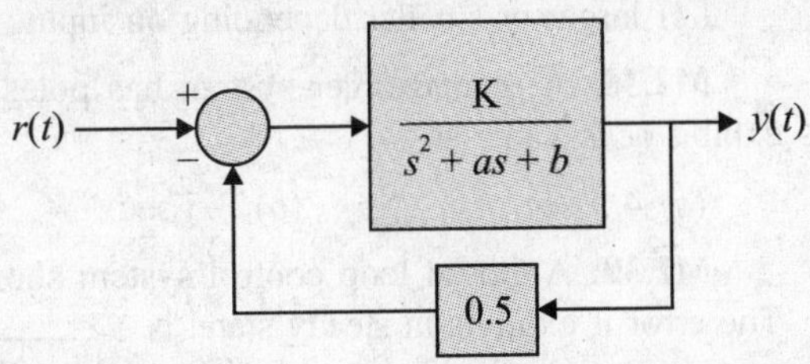

(*a*) 16, 16.25 and 32.5

(*b*) 8, 32.5 and 16.25

(*c*) 8, 16.25 and 32.5

(*d*) 16, 32.5 and 16.25

M2.32: A continuous system has transfer function with a zero at $s = -1$, a pole at $s = -2$ and gain factor = 2. The unit step response generated by this system is

(*a*) $1 + e^{-t}$ (*b*) $1 - e^{-2t}$ (*c*) $1 - e^{-t}$ (*d*) $1 + e^{-2t}$

M2.33: The unit ramp response of a system having transfer function $G(s) = \frac{s+1}{s+2}$, is

(*a*) $\frac{1}{2}\left(1 - e^{-2t}\right) - t$ (*b*) $\frac{1}{4} + \frac{1}{4}e^{-2t} - \frac{1}{2}t$

(*c*) $\frac{1}{4}\left(1 - e^{-2t}\right) + t$ (*d*) $\frac{1}{4} - \frac{1}{4}e^{-2t} + \frac{1}{2}t$

M2.34: If two identical first order stable low pass filters are cascaded non-interactively, then the unit step response of the composite filter will be

(*a*) critically damped (*b*) overdamped (*c*) underdamped (*d*) oscillatory

M2.35: Consider the system shown in figure below. The value of *K* that contributes steady state error of 20% to a unit step input, is

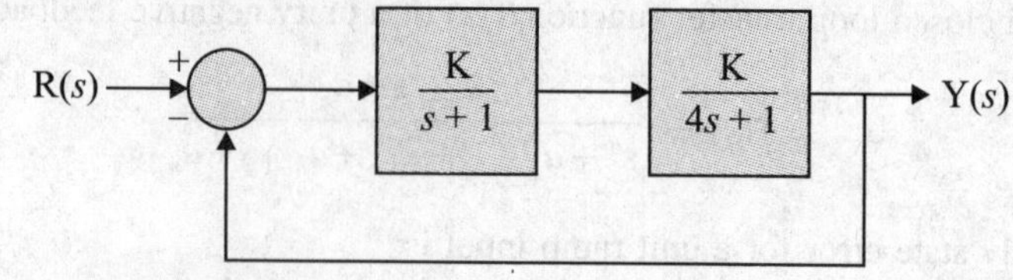

(*a*) 2 (*b*) 100 (*c*) 20 (*d*) 4

M2.36: When a unit step input is applied, a second order underdamped system exhibits a peak over shoot of M_p at $t = t_p$. If another step input equal in magnitude to peak overshoot M_p is applied at $t = t_p$, then the system will settle at

(*a*) $1 + M_p$ (*b*) $1 - M_p$ (*c*) M_p (*d*) 1.0

M2.37: A third order system is approximated to an equivalent second order system. The rise time of the lower order system will be

(*a*) same as that of original system for any input.

(*b*) smaller than that of original system for any input.

(*c*) larger than that of original system for any input.

(*d*) larger or smaller depending on input.

M2.38: A second order system has poles at $s = -1 \pm j$. The step response of this system will exhibit peak value at

(*a*) 4.5 sec (*b*) 3.5 sec (*c*) 3.14 sec (*d*) 1 sec.

M2.39: A closed loop control system shown in figure below, is excited by a unit step function. The error it exhibits in steady state, is

(*a*) −1.0

(*b*) −0.5

(*c*) 0.5

(*d*) 0

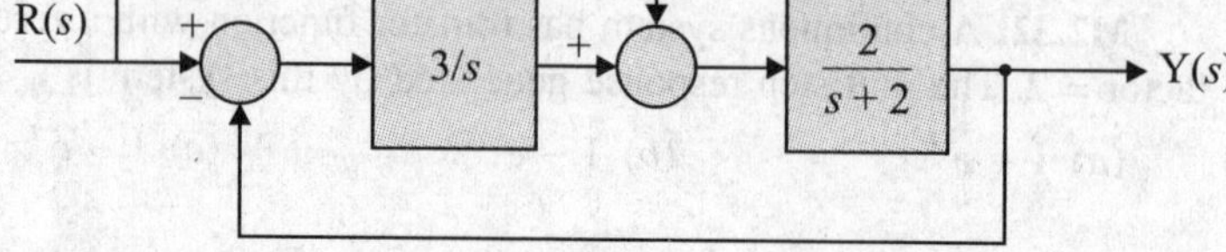

M2.40: The unit step response of a unity feedback system with open loop transfer function

$G(s) = \dfrac{K}{(s+1)(s+2)}$ is shown in the figure below. The value of K, is

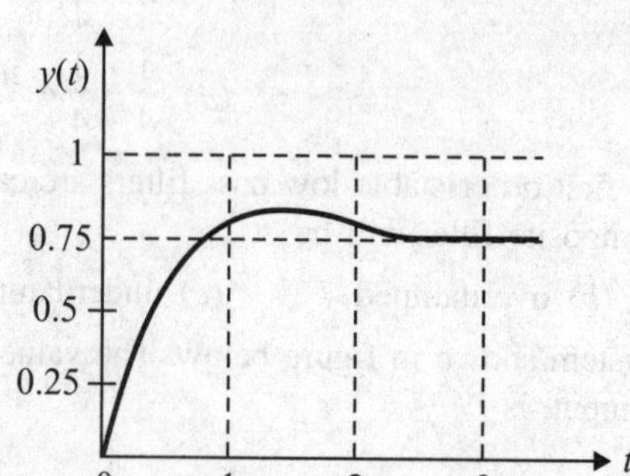

(*a*) 0.5 (*b*) 2 (*c*) 4 (*d*) 6

M2.41: The system shown in the figure below is excited by unit step input. The system is stable and $K_p = 4$, $K_i = 10$, $\omega = 500$, $\xi = 0.7$. The steady state value of $z(t)$, is

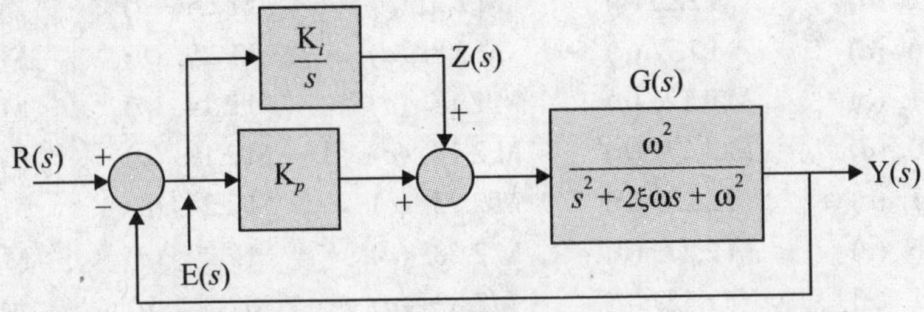

(*a*) 1 (*b*) 0.25 (*c*) 0.1 (*d*) 0

M2.42: A two loop position control system is shown below.

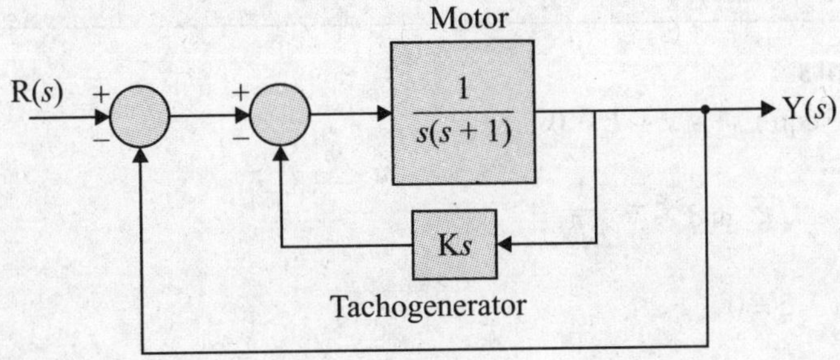

The gain K of tachogenerator influences mainly the

(*a*) peak overshoot.

(*b*) natural frequency of oscillation.

(*c*) phase shift of closed loop transfer function at very low frequencies ($\omega \to 0$).

(*d*) phase shift of closed loop transfer function at very high frequencies ($\omega \to \infty$).

ANSWERS

M 2.1. (*a*)	M 2.2. (*c*)	M 2.3. (*c*)	M 2.4. (*d*)	M 2.5. (*c*)
M 2.6. (*d*)	M 2.7. (*c*)	M 2.8. (*d*)	M 2.9. (*b*)	M 2.10. (*d*)
M 2.11. (*c*)	M 2.12. (*c*)	M 2.13. (*d*)	M 2.14. (*b*)	M 2.15. (*d*)
M 2.16. (*a*)	M 2.17. (*b*)	M 2.18. (*c*)	M 2.19. (*a*)	M 2.20. (*a*)
M 2.21. (*c*)	M 2.22. (*a*)	M 2.23. (*c*)	M 2.24. (*a*)	M 2.25. (*c*)
M 2.26. (*c*)	M 2.27. (*b*)	M 2.28. (*d*)	M 2.29. (*d*)	M 2.30. (*c*)
M2.31. (*c*)	M2.32. (*d*)	M2.33. (*d*)	M2.34. (*a*)	M2.35. (*a*)
M2.36. (*a*)	M2.37. (*c*)	M2.38. (*c*)	M2.39. (*d*)	M2.40. (*d*)
M2.41. (*a*)	M2.42. (*a*)			

Important Hints

M 2.2: Ch. eqn.: $s^2 + s + k = 0$

$$\omega_n = \sqrt{K} \text{ and } \xi = \frac{1}{2\sqrt{K}}$$

$$\underset{K \to \infty}{\text{lt}} \xi = 0$$

M 2.3: $M\ddot{y} + B\dot{y} + Ky = F(t)$

$$\text{Ch. eqn.} = s^2 + \frac{B}{M}s + \frac{K}{M} = 0$$

$$\omega_n = \sqrt{K/M}, \ \xi = \frac{B}{2\sqrt{KM}}$$

M 2.4: $K_p = \underset{s \to 0}{\text{lt}} G(s)\,H(s) = K$

$$e(\infty) = \frac{1}{1+K} = 0.1 \quad \Rightarrow K = 9$$

M 2.5: Ch. eqn. $= s^2 + (1 + K_2)s + K_1 = 0$

$$\omega_n = \sqrt{K_1}, \ \xi = \frac{1+K_2}{2\sqrt{K_1}} = 0.6 \text{ and } \omega_n\sqrt{1-\xi^2} = 10.$$

$$K_1 = 156.25, \quad K_2 = 14$$

M 2.7: $C(s) = \dfrac{20}{s(s^2 + 20s + 100)}$

$$c(\infty) = \underset{s \to 0}{\text{lt}} sC(s) = \frac{1}{5}$$

M2.8: $\omega_n = 7,\ \xi = \frac{16}{14},\ \xi > 1 \Rightarrow$ over damped system.

System does not exhibit over shoot.

M2.9: $\left.\begin{aligned} \xi\omega_n &= 8 \\ \omega_n\sqrt{1-\xi^2} &= 6 \end{aligned}\right\} \Rightarrow \begin{aligned} \xi &= 0.8 \\ \omega_n &= 10 \end{aligned}$

M2.12: Ch. eqn. $= (s + a + jb)\ (s + b - jb) = 0$

$= s^2 + 2as + a^2 + b^2$

$\Rightarrow \omega_n = \sqrt{a^2 + b^2},\quad \xi = \frac{a}{\sqrt{a^2 + b^2}}$

M2.13: Ch. Eqn. $= s^2 + s\frac{R}{L} + \frac{1}{LC} \quad \Rightarrow \xi = \frac{R}{2\sqrt{L/C}}$

M2.14: $G(s) = \frac{1}{s^2 + 6s + 9} = \frac{1}{(s+3)^2} = \frac{k_1}{s+3} + \frac{k_2}{(s+3)^2}$ and $g(t) = (k_1 + k_2 t)e^{-3t}$

M2.16: Exponential terms will decay to 0 as $t \to \infty$.

M2.17: $K_p = \underset{s\to 0}{\text{lt}}\ G(s) = 10$ and $e(\infty) = \frac{1}{1 + K_p}$

M2.18: Ch. Eqn $= s^2 + \frac{B}{J}s + \frac{K}{J} = 0$

$\xi\omega_n = \frac{B}{2J}$ independent of K.

M2.19: Ch. Eqn. $= s^2 + s + K = 0$

$K = \frac{1}{4}$ for critical damping and roots $s_1,\ s_2 = -0.5$

M2.20: $G(s)\ H(s) = \frac{s+1}{(s+4)\left(s^2 + 5s + a\right)}$

For $a = 0$, type of system = 1

M2.23: Ch. eqn. $\Rightarrow 1 + G(s)\ H(s) = 0$

$\Rightarrow s^2 + (2 + aK)s + K = 0$

$\Rightarrow \omega_n^2 = K = 16$

$2\xi\omega_n = 5.6 = (2 + 16a)$

$\Rightarrow a = 0.225$

M 2.24: (3) is derivative of (1).

M 2.27: Ch. Eqn. $\Rightarrow s^2 + s\frac{R}{L} + \frac{1}{LC} = 0$

$$\Rightarrow s^2 + 2s + 1 = 0$$

$$\Rightarrow (s+1)^2 = 0$$

M 2.28: Transient response does not depend upon nature of excitation.

M 2.29: OLTF = $\frac{a_{n-1}s + a_n}{s^n + a_1 s^{n-1} + \dots\dots + a_{n-2}s^2}$; System is of type 2.

M 2.30: $K_p = \lim_{s\to 0} G(s) = \infty, \quad K_v = \lim_{s\to 0} s\,G(s) = \infty$

$$K_a = \lim_{s\to 0} s^2 G(s) = 10$$

$$e_{ss} = \frac{1}{1+K_p} + \frac{1}{K_v} + \frac{1}{K_a} = 0.1$$

M2.31: $\frac{Y(s)}{R(s)} = \frac{K}{s^2 + as + b + 0.5K}$

$$y(\infty) = \operatorname{lt}_{s\to 0} \left.\frac{sKR(s)}{s^2 + as + b + 0.5K}\right|_{R(s) = 1/s} = \frac{K}{b + 0.5K}$$

and $\frac{K}{b + 0.5K} = 1 \to b = 0.5K$

Also $e^{-\pi\xi/\sqrt{1-\xi^2}} = 0.05 \Rightarrow \xi = 0.7$

$$\frac{4}{\xi\omega_n} = 1 \Rightarrow \omega_n = 5.7$$

Compare with general second order underdamped system dynamics to get

$$K = \omega_n^2 = 32.5$$

$$a = 2\xi\omega_n = 8$$

$$b = 0.5K = 16.25$$

M2.32: $$\frac{Y(s)}{R(s)} = \frac{2(s+1)}{(s+2)} \Rightarrow Y(s)\big|_{R(s)=1/s} = \frac{2(s+1)}{s(s+2)} \Rightarrow y(t) = 1+e^{-2t}$$

M2.33: $$Y(s)\big|_{R(s)=1/s^2} = G(s)\,R(s) = \frac{s+1}{s^2(s+2)} \text{ and } y(t) = \frac{1}{4}-\frac{1}{4}e^{-2t}+\frac{1}{2}t$$

M2.34: The composite second order filter will have equal roots.

M2.35: $$K_p = \underset{s\to 0}{\text{lt}}\ G(s) = \underset{s\to 0}{\text{lt}}\ \frac{K}{(s+1)}\cdot\frac{K}{(4s+1)} = K^2$$

$$e(\infty) = \frac{1}{1+K_p} = \frac{1}{1+K^2} \text{ and } \frac{1}{1+K^2} = 0.2 \text{ gives } K = 2$$

M2.36: Nature of system dynamics is independent of input.

M2.37: Pole corresponding to faster dynamics, is discarded.

M2.38: Characteristic equation: $s^2 + 2s + 2 = 0$

$$\omega_n = \sqrt{2} \text{ and } \xi = \frac{1}{\sqrt{2}}$$

$$t_\rho = \frac{\pi}{\omega_n\sqrt{1-\xi^2}} = \pi$$

M2.39: $e(\infty)$ can not be evaluated through error constant approach. Instead,

$$\frac{Y(s)}{R(s)} = \frac{2(s+3)}{s^2+2s+6} \text{ and } Y(s)\big|_{R(s)=1/s} = \frac{2(s+3)}{s(s^2+2s+6)}$$

$$y(\infty) = \underset{s\to 0}{\text{lt}}\ sY(s) = 1 \text{ and } e(\infty) = y(\infty)-1 = 0$$

M2.40: $$\frac{Y(s)}{R(s)} = \frac{G(s)}{1+G(s)} = \frac{K}{(s+1)(s+2)+K}$$

and $$Y(s)\big|_{R(s)=\frac{1}{s}} = \frac{K}{s[(s+1)(s+2)+K]}$$

$$y(\infty) = \underset{s\to 0}{\text{lt}}\ sY(s) = \frac{K}{2+K} \text{ and } \frac{K}{2+K} = 0.75 \Rightarrow K=6$$

M2.41: $$Z(s) = E(s)\cdot\frac{K_i}{s} \text{ and } E(s) = R(s) - E(s)\left(K_p + \frac{K_i}{s}\right)G(s)$$

or $$E(s) = \frac{R(s)}{1+\left(sK_p + K_i\right)\dfrac{G(s)}{s}} = \frac{sR(s)}{s+\left(sK_p + K_i\right)G(s)}$$

$\therefore$ $$Z(s) = \frac{s\cdot R(s)\left(K_i/s\right)}{s+\left(sK_p + K_i\right)G(s)}$$

and $$Z(s)\Big|_{R(s)=\frac{1}{s}} = \frac{K_i}{s\left[s+\left(sK_p + K_i\right)G(s)\right]}$$

$$z(\infty) = \underset{s\to 0}{\text{lt}}\ sZ(s) = \underset{s\to 0}{\text{lt}}\ \frac{K_i}{s+\left(sK_p + K_i\right)G(s)}$$

$$= \underset{s\to 0}{\text{lt}}\ \frac{1}{G(s)} = 1$$

M2.42: Characteristic equation is $s^2 + (1 + K)s + 1 = 0$

K influences damping ratio and peak overshoot. ω_n is independent of K.

3

BLOCK DIAGRAMS AND SIGNAL FLOW GRAPH

3.1 BLOCK DIAGRAMS

A control system usually consists of a number of components and multiple feedback loops. Block diagrams are often used by control engineers to describe the composition and interconnection of a system. This is due to simplicity and versatility with which the block diagram can be used together with transfer function in order to depict input-output relationship throughout the system. A block diagram has following three basic elements.

(*i*) **Block:** A block is used to indicate a proportional relationship between two Laplace transformed signals. The proportionality function, also called as transmittance, that describes the relationship between incoming and outgoing signals, is indicated within the block as shown in Fig. 3.1 (*a*).

(*ii*) **Summing point:** A summing point indicates addition and subtraction of signals. At a summing point there can be any number of incoming signals but only one outgoing signal. The algebraic signs involved in summation are indicated next to arrow head for each incoming signal as shown in Fig. 3.1 (*b*).

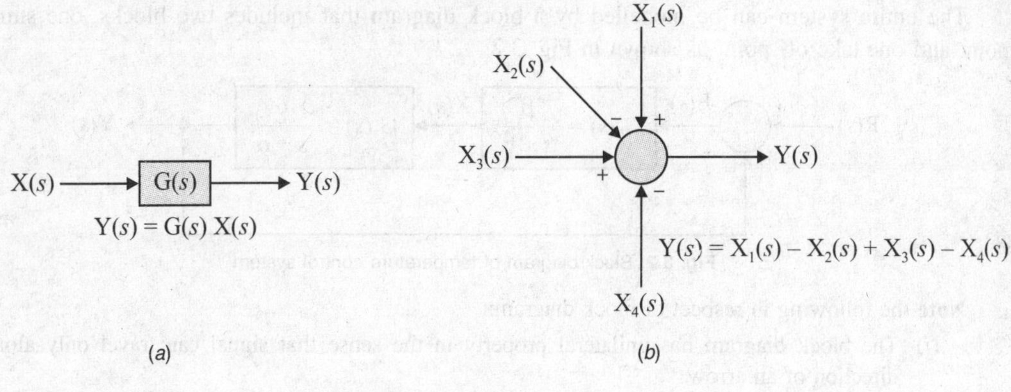

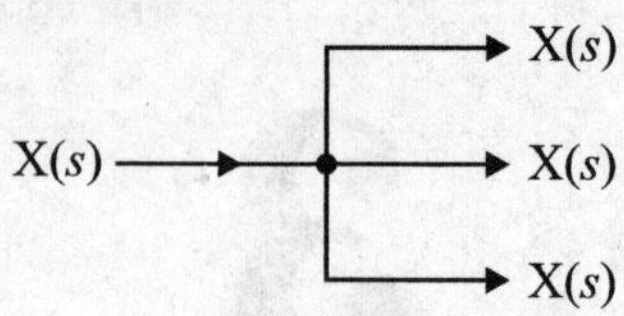

(c)

Fig. 3.1: Elements of block diagram. (*a*) Block (*b*) Summing point (*c*) Pick off point

(*iii*) **Pick off point:** A junction, also called as pick off point or take off point, indicates same signal being sent to several places as shown in Fig. 3.1 (*c*).

Block diagram development

Consider an example of **temperature control system** wherein x represents heat inputed and y, the output temperature satisfying the relationship

$$\dot{y} + \alpha y = \alpha x$$

provides the transfer function

$$G_1(s) = \left.\frac{Y(s)}{X(s)}\right|_{\text{initial conditions} = 0} = \frac{\alpha}{s+\alpha}$$

If the system incorporates comparison of desired temperature (***r***) with actual temperature (***y***), producing error (***e***), then

$$e = r - y$$

and
$$E(s) = R(s) - Y(s)$$

If the system includes oven to modify heat supplied (x) depending upon error (e) satisfying the relationship

$$\dot{x} + \beta x = \beta e$$

then it provides the transfer function

$$G_2(s) = \left.\frac{X(s)}{E(s)}\right|_{\text{initial conditions} = 0} = \frac{\beta}{s+\beta}$$

The entire system can be modelled by a block diagram that includes two blocks, one summing point and one take off point as shown in Fig. 3.2.

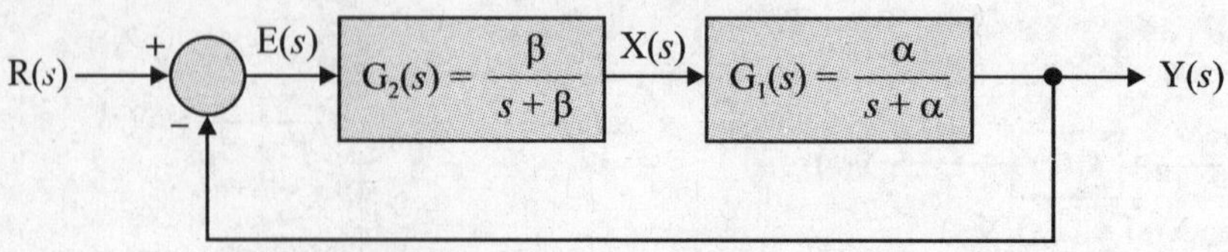

Fig. 3.2: Block diagram of temperature control system

Note the following in respect of block diagram:

(*i*) The block diagram has unilateral property in the sense that signal can travel only along the direction of an arrow.

(*ii*) The block diagram for a given system is not unique. The blocks constituting a system can be arranged depending on nature of analysis. But overall relationship between input and output is unique.

(*iii*) The block diagram does not have information about physical structure of system.

(*iv*) Individual as well as overall performance of system can be studied with the help of transfer functions entered in block diagram.

(*v*) While using block diagram reduction rules, it is assumed that blocks are non interactory contributing no loading error.

Block diagram reduction

The block diagram reduction means simplifying the diagram consisting of many blocks and feedback loops to an extent where it is a single block providing a transfer function that relates the output to the input. The rules that are followed in doing so, are listed in Table 3.1.

TABLE 3.1: Block diagram reduction rules

	Rule	*Original diagram*	*Equivalent diagram*
1.	Associative law for summing points with no additional block or connection in between	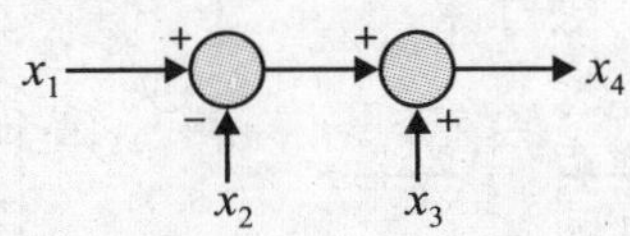	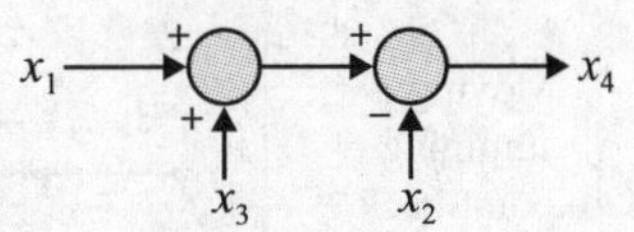
2.	Blocks in cascade (or series) with no additional connection in between	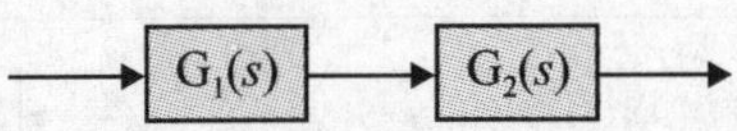	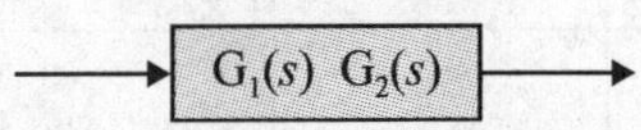
3.	Blocks in tandem (or parallel)	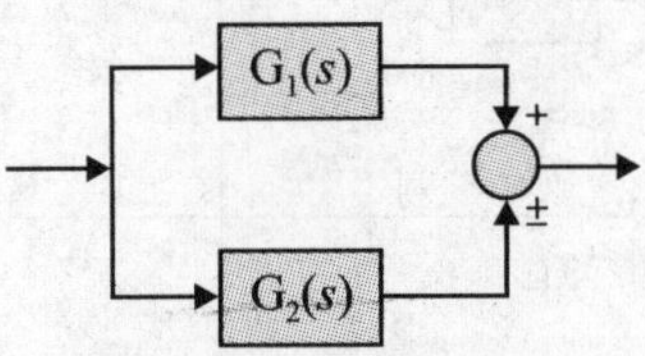	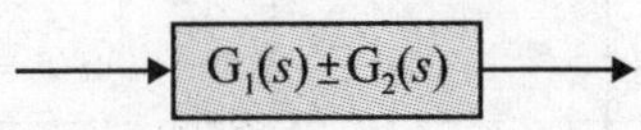
4.	Two blocks in feedback configuration	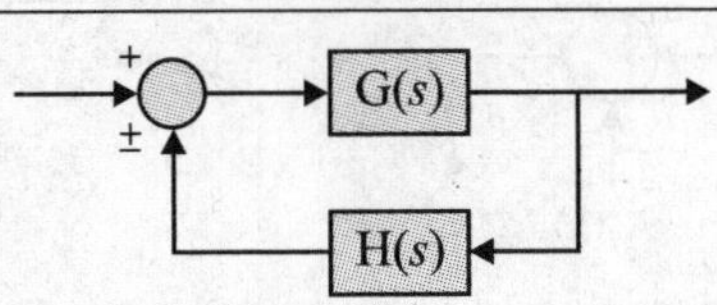	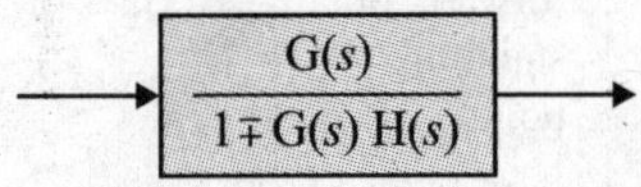
5.	Insertion / removal of unity gain		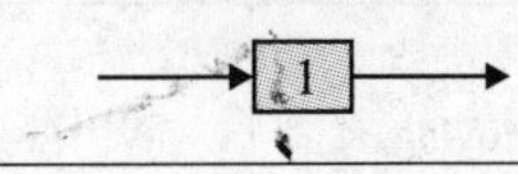
6.	Changing a summer sign	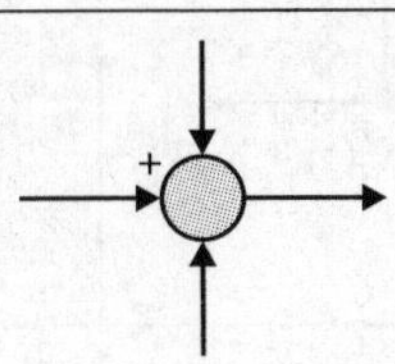	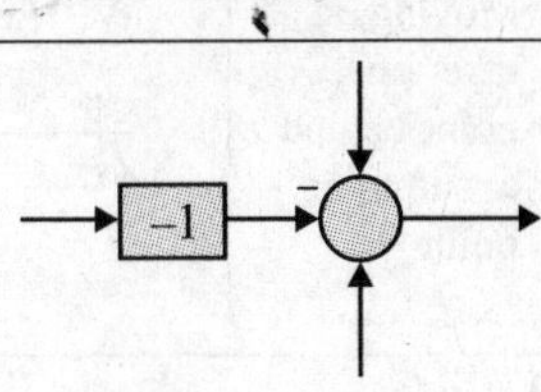

	Rule	*Original diagram*	*Equivalent diagram*
7.	Moving a take off point back	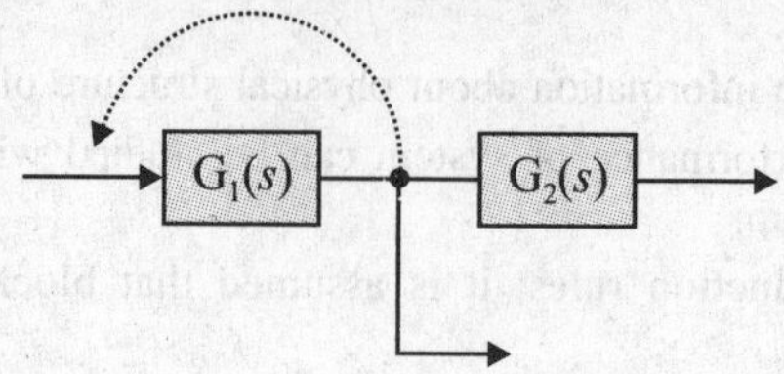	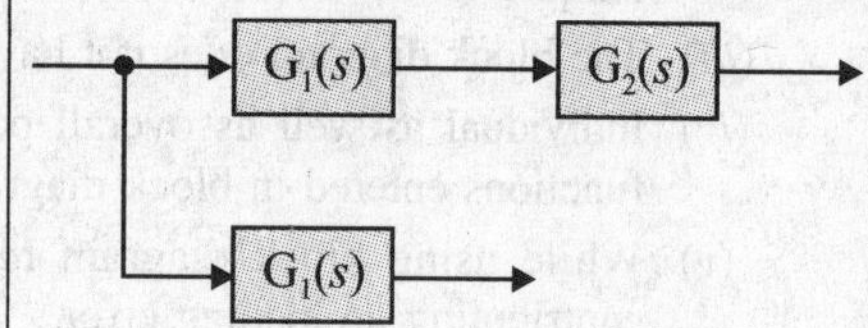
8.	Moving a takeoff point forward	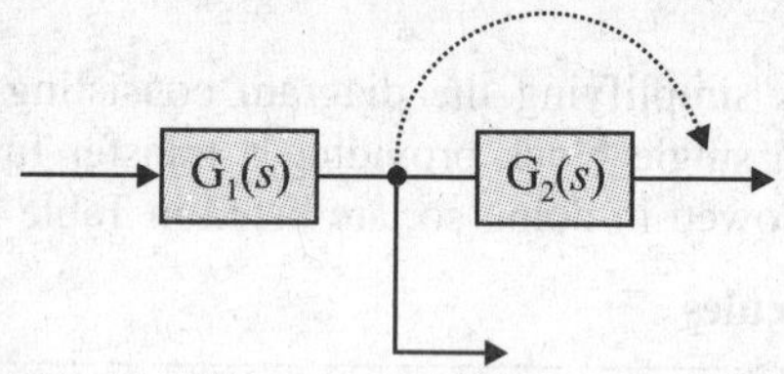	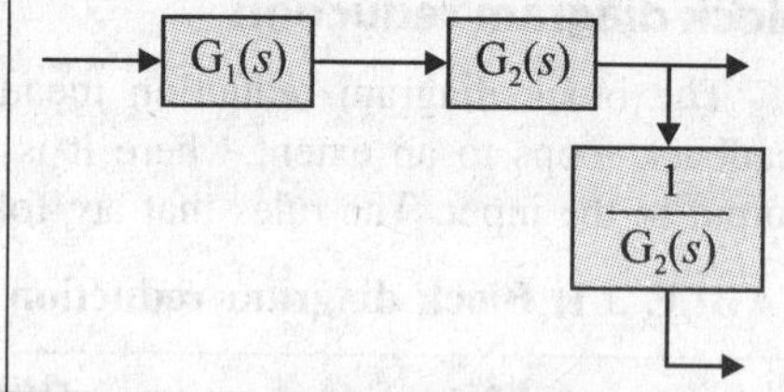
9.	Moving a summing point back	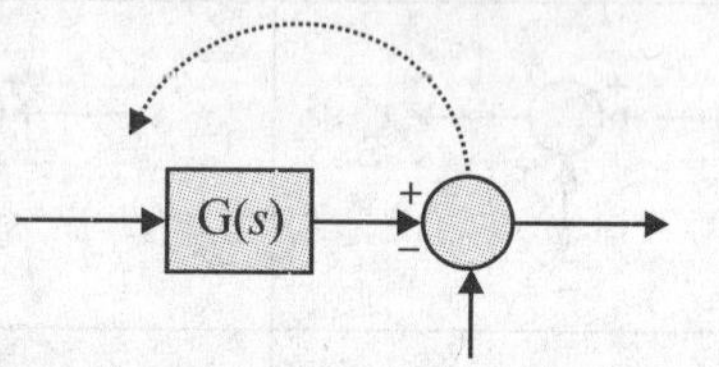	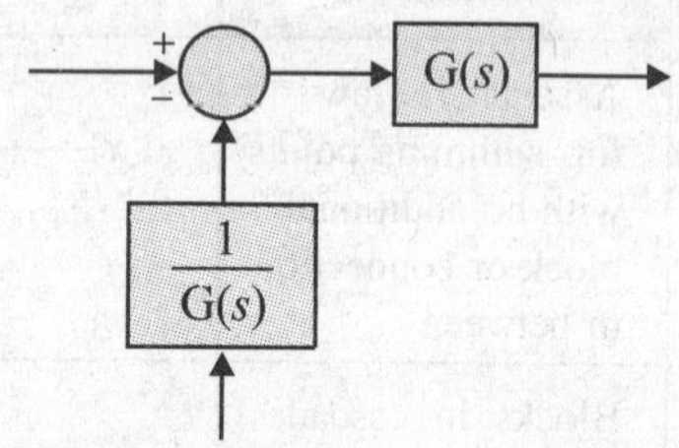
10.	Moving a summing point forward	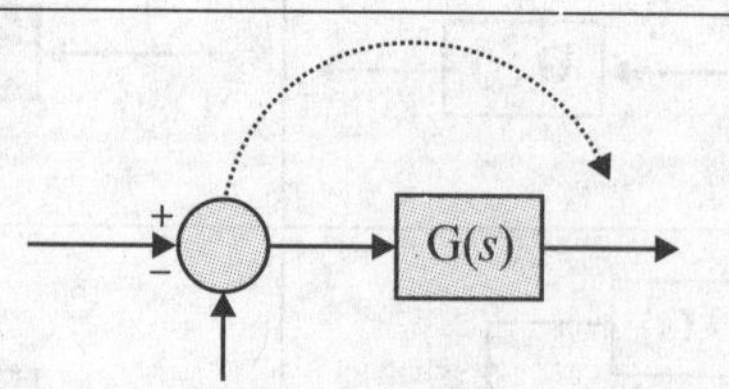	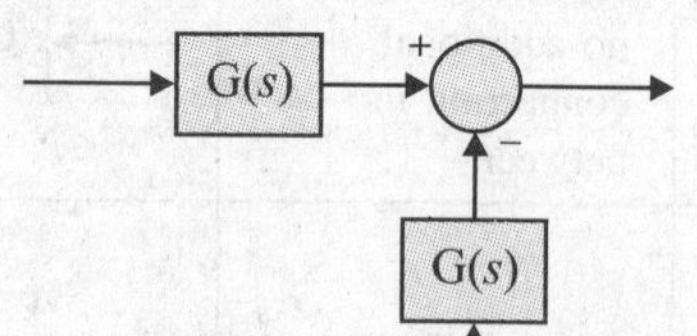
11.	Moving a take off point forward of a summing point	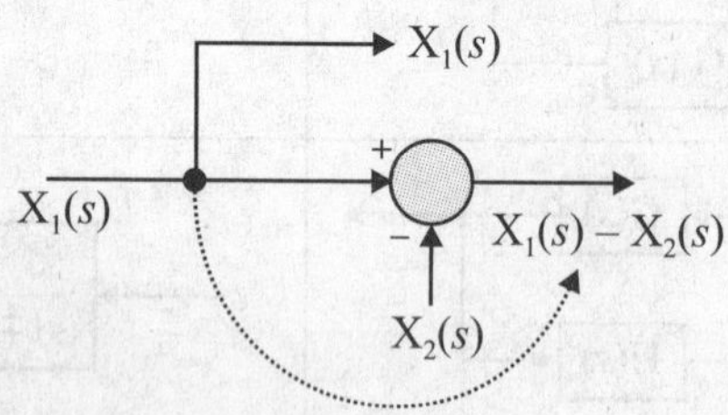	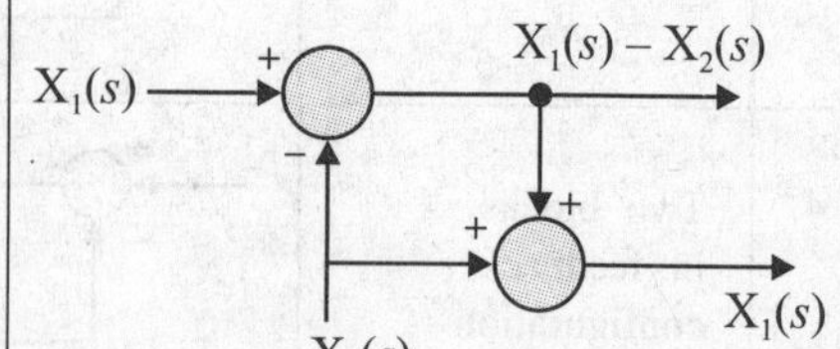
12.	Moving a take off point behind a summing point	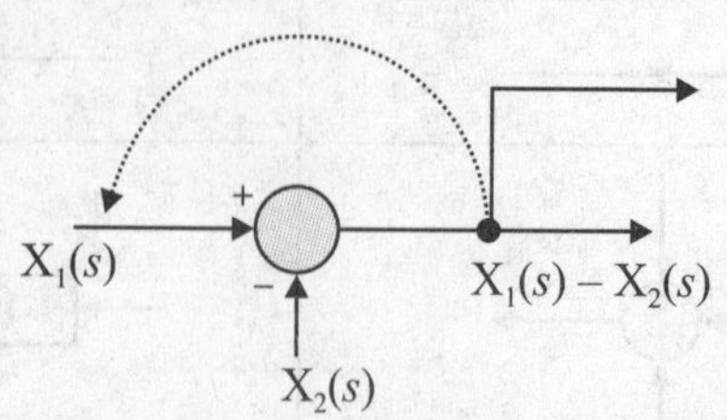	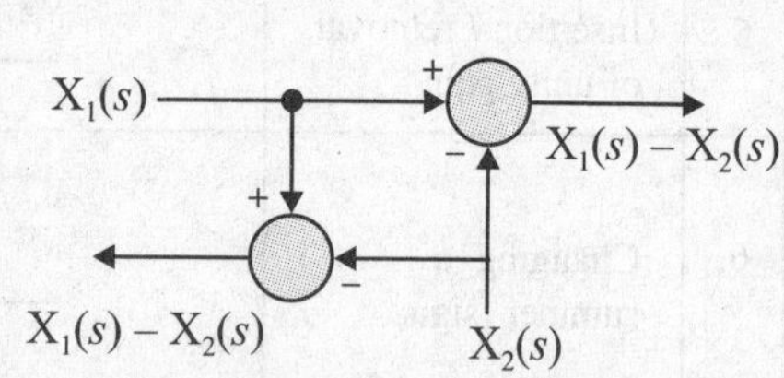

3.2 SIGNAL FLOW GRAPH

The block diagram algebra is a convenient way of depicting the relations between Laplace transformed signals in a system, but this technique involves a tedious computational routine and becomes time consuming for the larger systems. The signal flow graph turns out to be a better option then. The signal flow graph contains essentially the same information as the block diagram. The advantage herein is use of Mason's gain formula (to be introduced later in this section) that gives relation between system variables in one step avoiding multiple steps of block reduction. Thus a signal flow graph may be regarded as simplified version of a block diagram.

The signal flow graph is a graphical representation of the variables of a set of linear algebraic equations describing a system. The significant elements constituting a signal flow graph are nodes and branches. The nodes are connected by directed branches. Each node represents a system variable. Each branch connecting two nodes acts as a signal multiplier. The direction of signal flow is indicated by an arrow placed thereon and multiplication or transmittance factor is written along the branch. The value of signal at any node is sum of signals coming into the node from the branches.

Let us consider the following Laplace transformed equations to illustrate construction of the signal flow graph.

$$X_1(s) = \frac{1}{s} X_2(s) - 4X_1(s)$$

$$X_2(s) = \frac{s}{s+3} X_3(s) + 2R_2(s)$$

$$X_3(s) = \frac{2}{s+2} X_1(s) - 3X_2(s) + 2R_1(s)$$

$$Y(s) = -3X_1(s)$$

The signal flow graph for above equations, is constructed step by step as shown in Fig. 3.3 (*a*), (*b*), (*c*) and (*d*). Combining all these, Fig. 3.3 (*e*) depicts the complete signal flow graph.

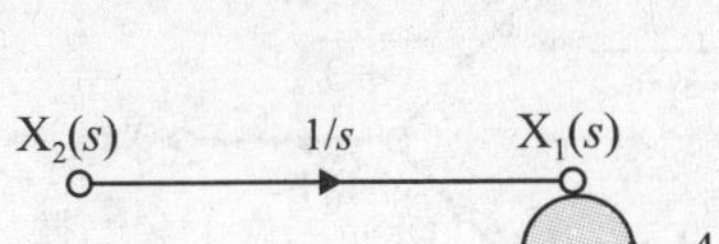

(*a*)

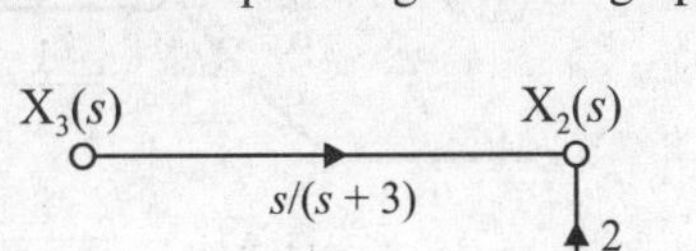

(*b*)

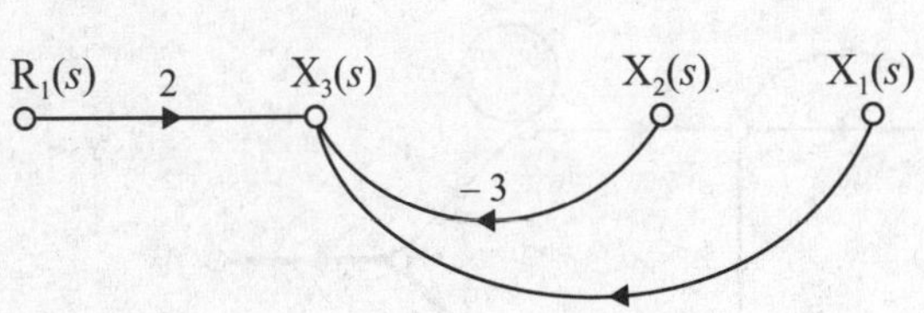

(*c*)

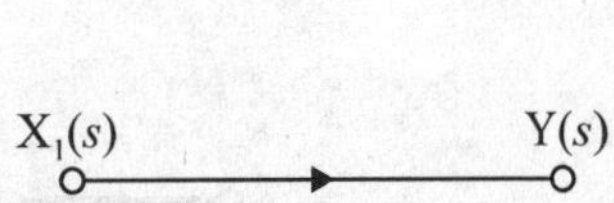

(*d*)

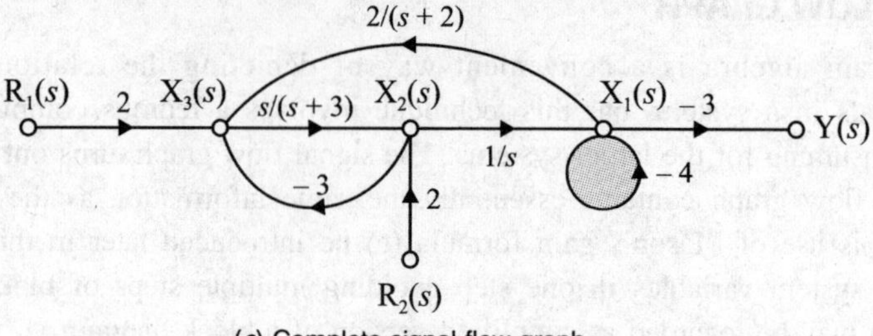

(e) Complete signal flow graph

Fig. 3.3: Step by step construction of signal flow graph

Note the following regarding signal flow graph:

(i) A signal flow graph is applicable to linear time invariant systems only.

(ii) For a given system, the signal flow graph is not unique.

(iii) The system for which the signal flow graph is to be drawn must be characterised by equations in algebraic form.

(iv) Signals travel along the branches only in the direction indicated by arrows on the branches.

(v) Each node represents a variable and branch indicates the functional dependence of one variable on the other for example. The branch directed from node x_i to x_j represents the dependence of variable x_j on x_i, but not the reverse.

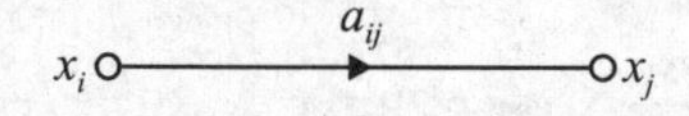

$$x_j = a_{ij} x_i$$

(vi) A node adds the signals of all incoming branches and transmits this sum to all outgoing branches.

Important definitions

Nodes and branches have been defined. The following terms will also be useful in explaining the application of Mason's gain rule. Consider the signal flow graph of Fig. 3.4.

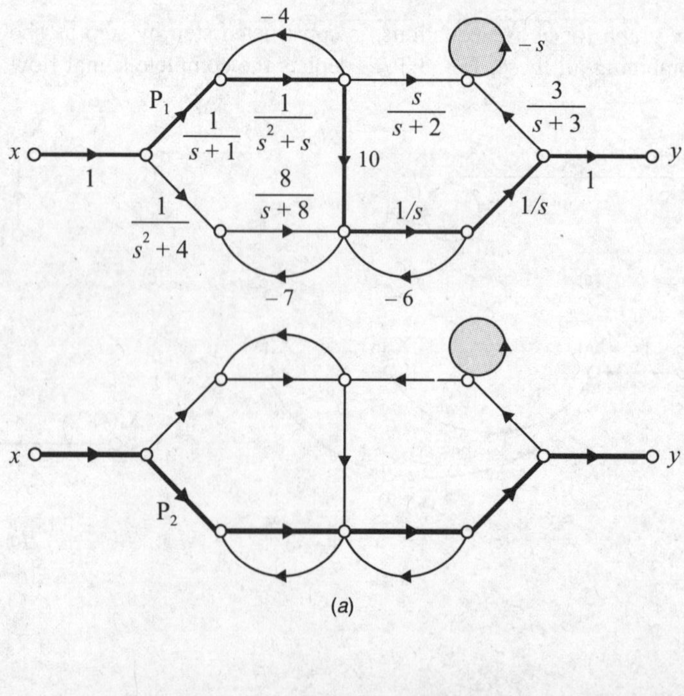

(a)

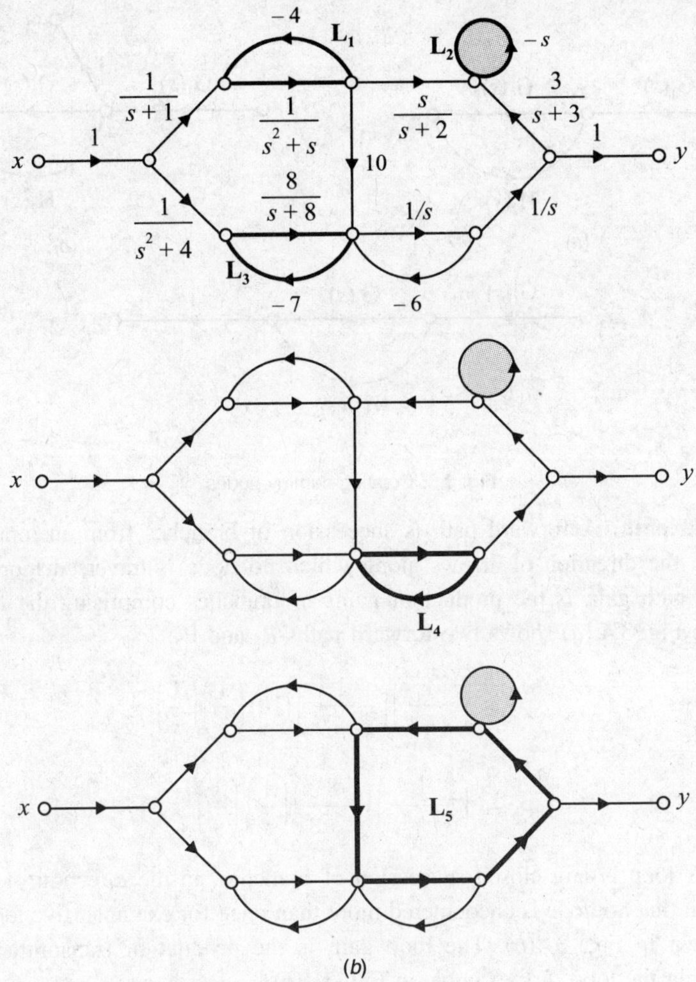

(b)

Fig. 3.4: (a) Forward paths, (b) Loops

(a) Input node (source): The node having only outgoing branches is known as source or input node, for example the node labelled x in Fig. 3.4 is input node.

(b) Output node (sink): The node having only incoming branches is known as sink or output node, for example node labelled y in Fig. 3.4 is output node. However, there may be a signal flow graph Fig. 3.5(a) where this condition is not satisfied by any node. In such a case it is customary (although not absolutely necessary) to bring output signal out of signal flow graph with unity transmittance, for example. x_2 or x_3 can be made output node via a branch with unity transmittance as shown in Fig. 3.5 (b) and (c) respectively. The new nodes created in such a way are called **dummy nodes**.

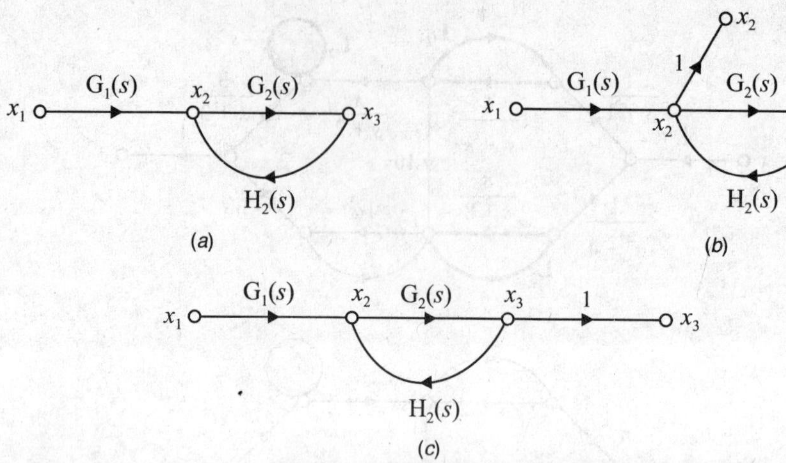

Fig. 3.5: Creating dummy nodes

(*c*) **Forward path:** A forward path is succession of branches from an input node to output node, in the direction of arrows along which no node is traversed more than once. The forward path gain is the product of gains of branches comprising the forward path. For example Fig. 3.4 (*a*) shows two forward paths P_1 and P_2:

$$P_1 = \left(\frac{1}{s+1}\right)\left(\frac{1}{s^2+s}\right)(10)\left(\frac{1}{s}\right)\left(\frac{1}{s}\right)$$

$$P_2 = \left(\frac{1}{s^2+4}\right)\left(\frac{8}{s+8}\right)\left(\frac{1}{s}\right)\left(\frac{1}{s}\right)$$

(*d*) **Loop:** A loop is any closed succession of branches, in the direction of arrows, with the condition that no node is encountered more than once for example, five loops from L_1 to L_5 are shown in Fig. 3.4(*b*). The loop gain is the product of transmittances of branches comprising the loop. For example in Fig. 3.4(*b*)

$$L_1 = \frac{-4}{s^2+s}, \qquad L_2 = -s, \qquad L_3 = \frac{-56}{s+8}$$

$$L_4 = \frac{-6}{s} \quad \text{and} \quad L_5 = (10)\left(\frac{1}{s}\right)\left(\frac{1}{s}\right)\left(\frac{3}{s+3}\right)\left(\frac{s}{s+2}\right)$$

A loop consisting of only one node is called as self loop, for example, L_2 in Fig. 3.4 (*b*) is **self loop**.

Two loops are said to be touching if they have any node in common, else, they are non touching. Similarly, a loop and a forward path are said to be touching if they have any node in common. For example, in Fig. 3.4 the two non touching loops are L_1 and L_2, L_1 and L_3, L_1 and L_4, L_2 and L_3, L_2 and L_4 and the three non touching loops are L_1, L_2, L_3 and L_1, L_2, L_4.

Construction of signal flow graph from block diagram

The construction of signal flow graph from system equations has been illustrated just above. Here we present a systematic approach to construct signal flow graph from block diagram with the help of an example. The following steps are involved.

(*i*) Identify each summing and take off point. If summing and take off points are near each other, represent them by separate nodes. However two summing points or two take off points near each other with no additional block or connection between them, may be represented by a single node.

(*ii*) Represent them by a distinct node in signal flow graph.

(*iii*) Connect the nodes by branches instead of blocks, indicating block transfer functions as gains of corresponding branches and show input and output nodes.

An example: The block diagram is shown in Fig. 3.6(*a*) and corresponding signal flow graph constructed using above steps, is shown in Fig. 3.6(*b*). The three take off points have been represented by nodes x_1, x_3 and x_5 and the three summing points have been represented by nodes x_2, x_4 and x_6. Input and output nodes are also shown. The nodes are inter connected by branches, the arrows according to direction of flow of signals in the block diagram are placed on branches and transmittances within the blocks are shown as gain along the branches.

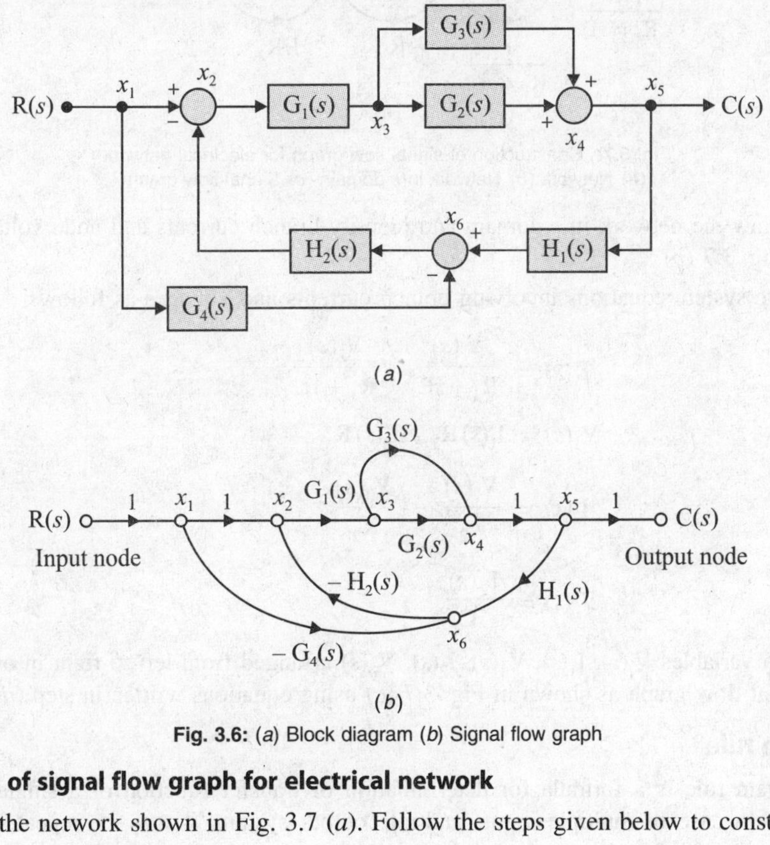

Fig. 3.6: (*a*) Block diagram (*b*) Signal flow graph

Construction of signal flow graph for electrical network

Consider the network shown in Fig. 3.7 (*a*). Follow the steps given below to construct the signal flow graph.

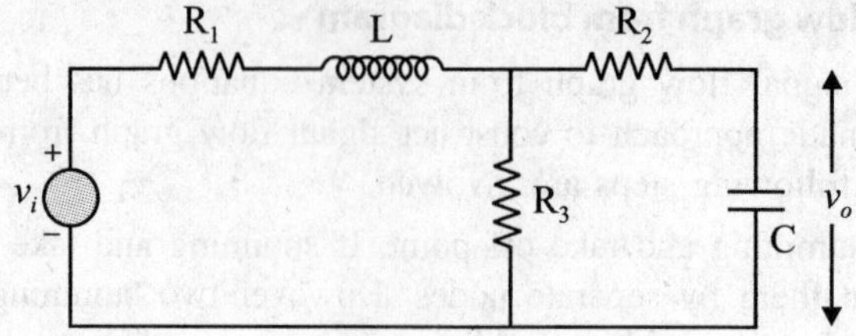

(a)

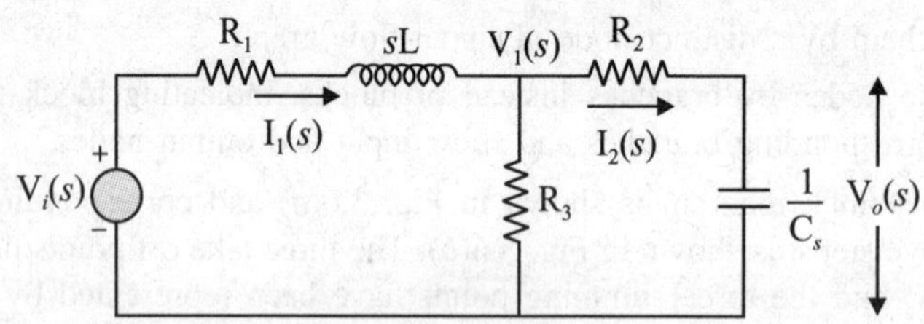

(b)

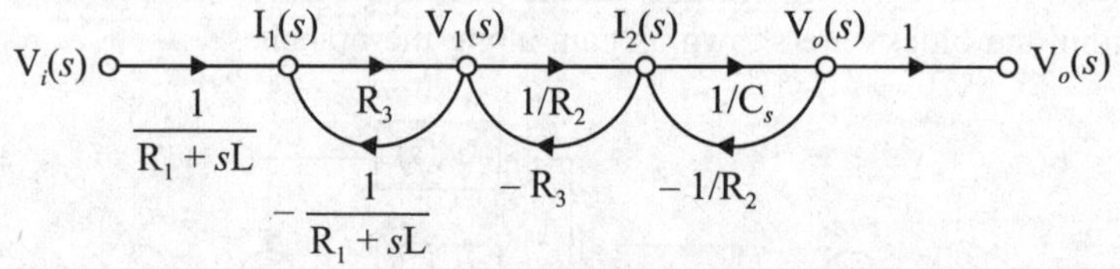

(c)

Fig. 3.7: Construction of signal flow graph for electrical network (*a*) Network (*b*) Network in *s* domain (*c*) Signal flow graph

(*i*) Redraw the network in *s* domain and identify branch currents and node voltages as shown in Fig. 3.7 (*b*).

(*ii*) Write system equations involving branch currents and voltages as follows:

$$I_1(s) = \frac{V_i(s)}{R_1 + sL} - \frac{V_1(s)}{R_1 + sL}$$

$$V_1(s) = I_1(s)R_3 - I_2(s)R_3$$

$$I_2(s) = \frac{V_1(s)}{R_2} - \frac{V_0(s)}{R_2}$$

$$V_o(s) = \frac{I_2(s)}{sC}$$

(*iii*) With variables $V_i(s)$, $I_1(s)$, $V_1(s)$, $I_2(s)$, $V_o(s)$ arranged from left to right in order, construct signal flow graph as shown in Fig. 3.7 (*c*) using equations written in step (*ii*).

Mason's gain rule

Mason's gain rule is a formula for determination of transfer function of a single input, single output system. In case of multiple input, multiple output system, it may be repeatedly applied to determine each transfer function.

The general gain formula is as follows. The transfer function of a single input, single output signal flow graph is

$$T(s) = \sum_{i=1}^{n} \frac{P_i \Delta_i}{\Delta}$$

where n = Total number of forward paths between input and output.

P_i = Path gain or transmittance of ith forward path between input and output

Δ = **Determinant** of a signal flow graph

= 1 – (sum of all loop gains) + (sum of gain products of all possible combinations of two non touching loops) – (sum of gain products of all possible combinations of three non touching loops) + ...

Δ_i = **Co-factor** of ith forward path and it is determinant of signal flow graph formed by deleting all loops touching i^{th} forward path.

For example, consider signal flow graph shown in Fig. 3.4

$$\Delta = 1 - (L_1 + L_2 + L_3 + L_4 + L_5) + (L_1L_2 + L_1L_3 + L_1L_4 + L_2L_3 + L_2L_4) - (L_1L_2L_3 + L_1L_2L_4)$$

$$\Delta_1 = 1 - L_2 = 1 + s$$

$$\Delta_2 = 1 - (L_1 + L_2) + L_1L_2 = 1 + \frac{4}{s^2+s} + s + \frac{4s}{s^2+s}$$

and $$T(s) = \frac{P_1\Delta_1 + P_2\Delta_2}{\Delta}$$

PROBLEMS AND SOLUTIONS

P3.1: Reduce the block diagram shown in Fig. P3.1 to obtain system transfer function.

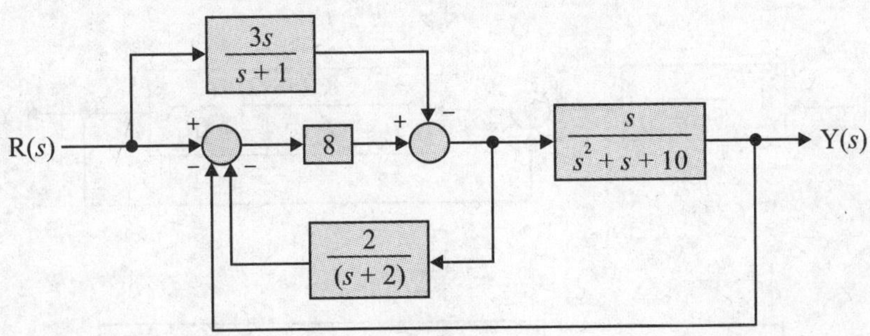

Fig. P3.1

Solution:

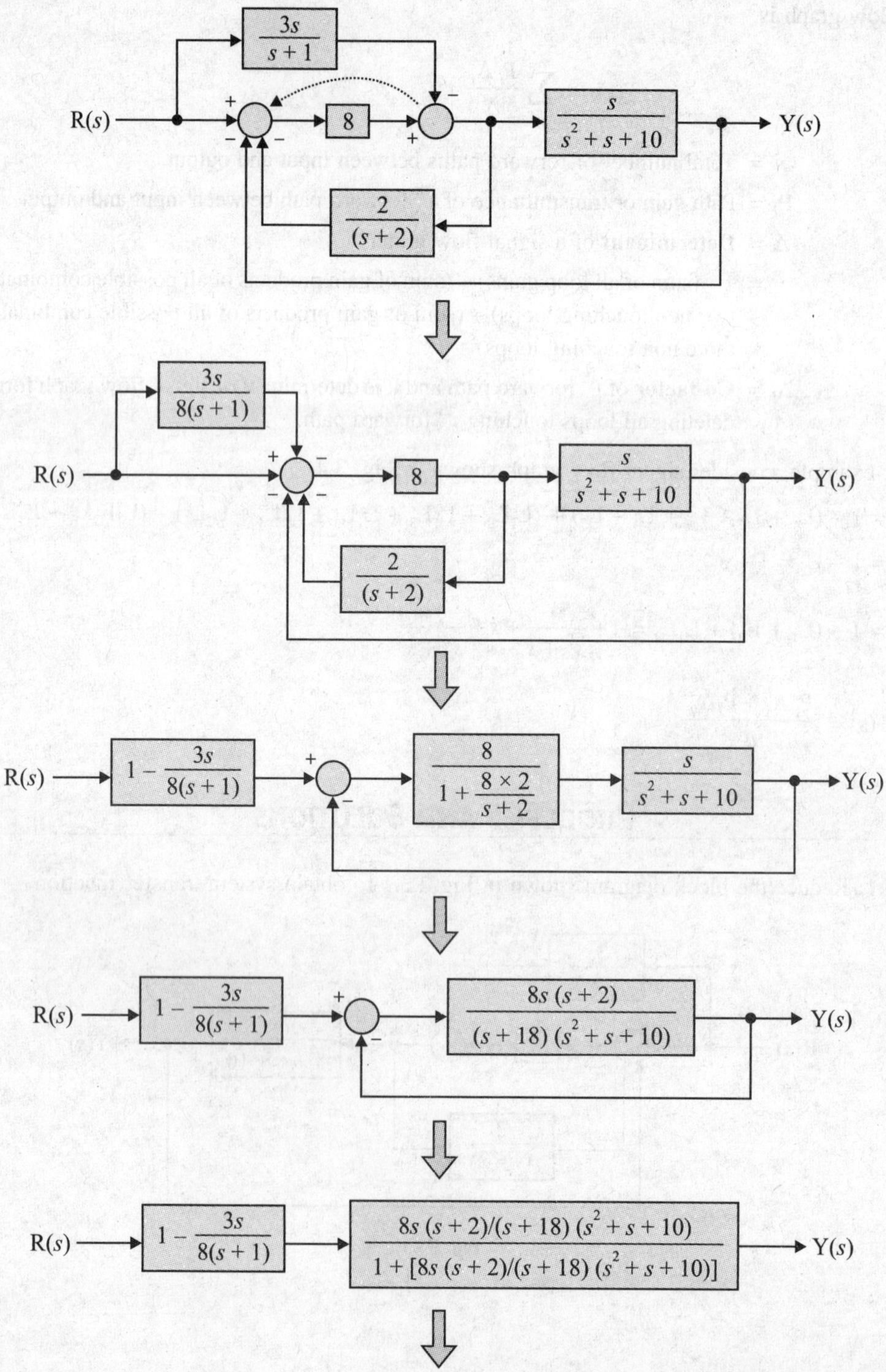

3s / (s + 1)
R(s)
8
s / (s^2 + s + 10)
Y(s)
2 / (s + 2)
3s / 8(s + 1)
8 / (1 + 8 × 2 / (s + 2))
1 − 3s / 8(s + 1)
8s (s + 2) / (s + 18) (s^2 + s + 10)
8s (s + 2)/(s + 18) (s^2 + s + 10) / 1 + [8s (s + 2)/(s + 18) (s^2 + s + 10)]

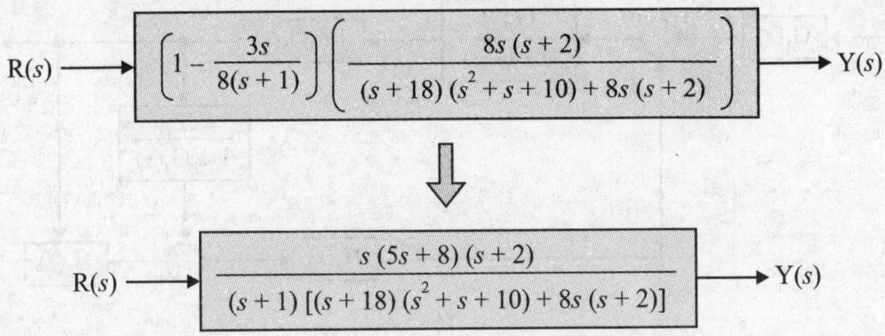

P3.2: Obtain overall transfer function for diagram shown in Fig. P3.2.

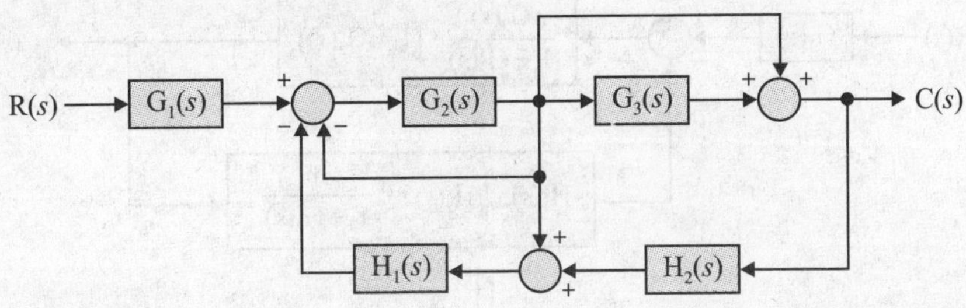

Fig. P3.2

Solution:

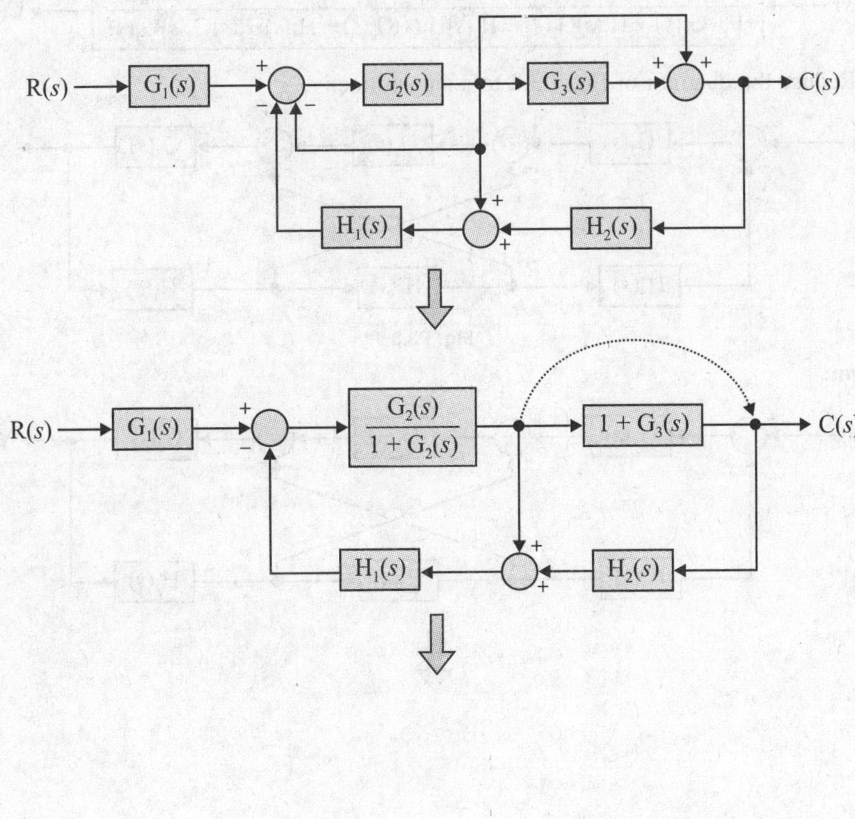

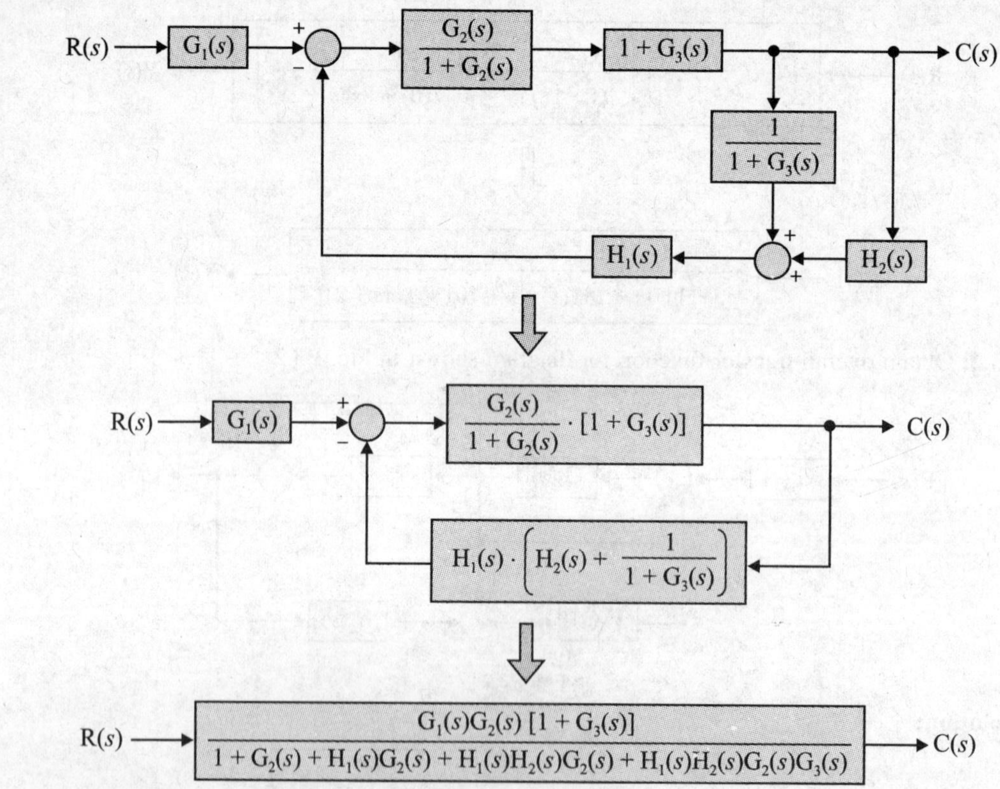

P3.3: Reduce the diagram of Fig. P3.3 to a single block.

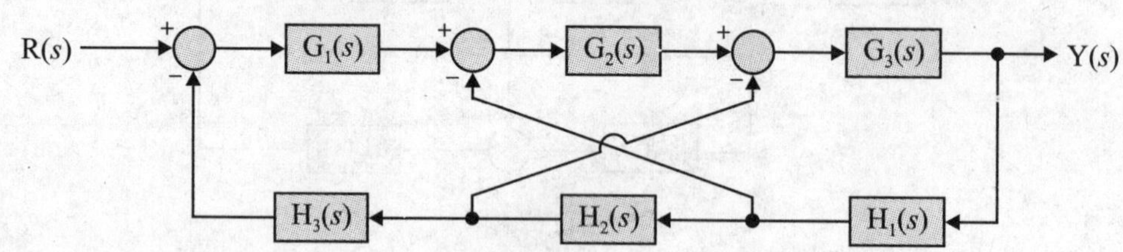

Fig. P3.3

Solution:

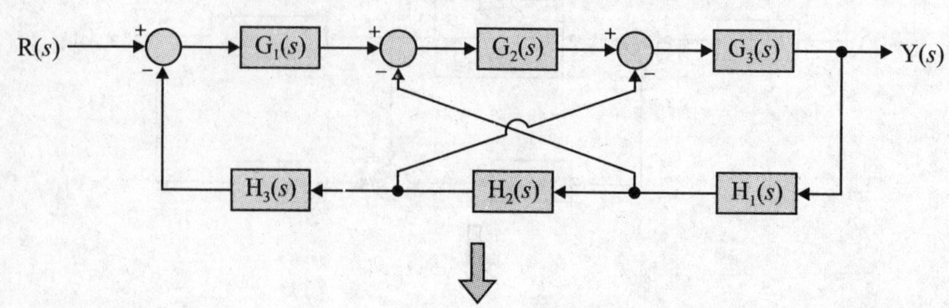

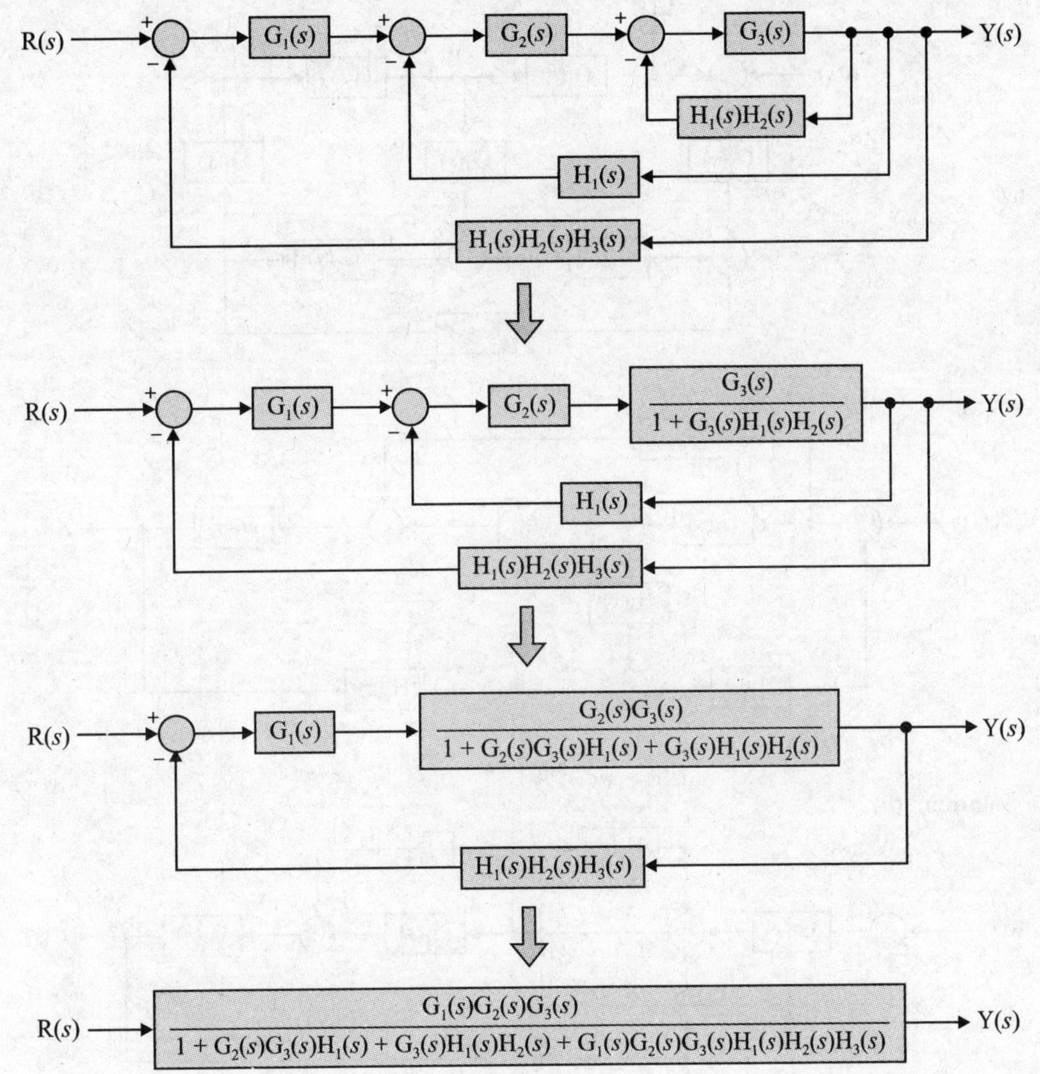

P3.4: Find closed loop transfer function of systems shown in Fig. P3.4. (*a*), (*b*) and (*c*).

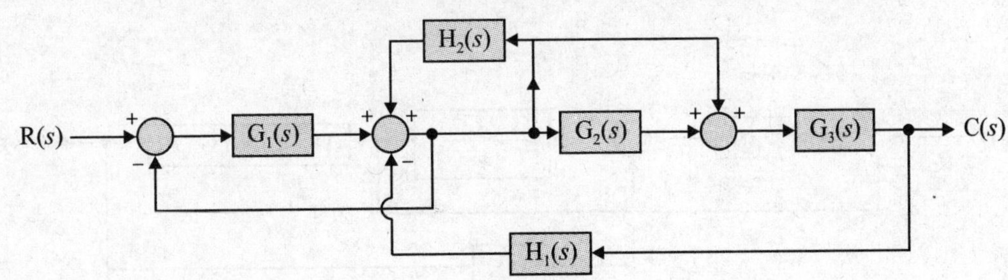

(*a*)

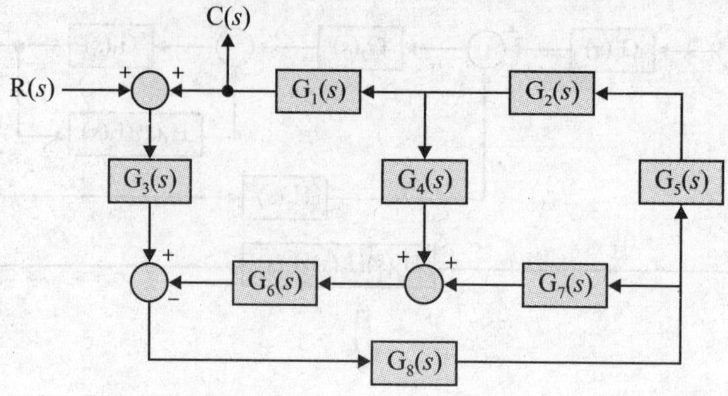

(b)

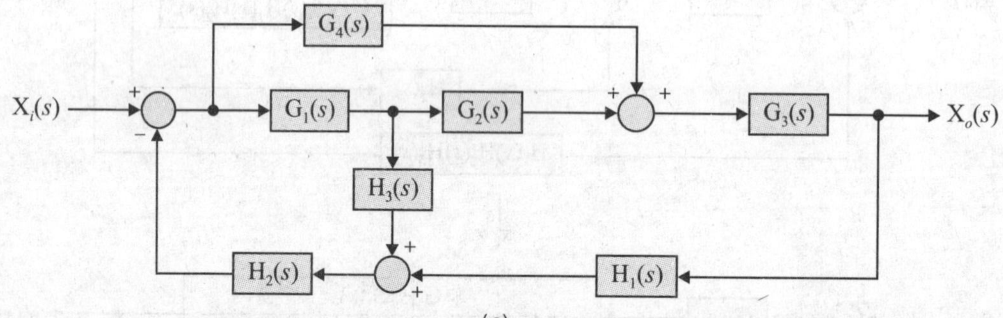

(c)

Fig. P3.4

Solution: (*a*)

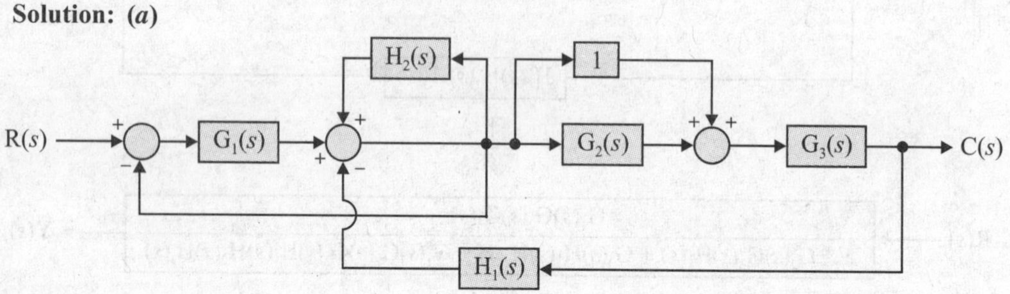

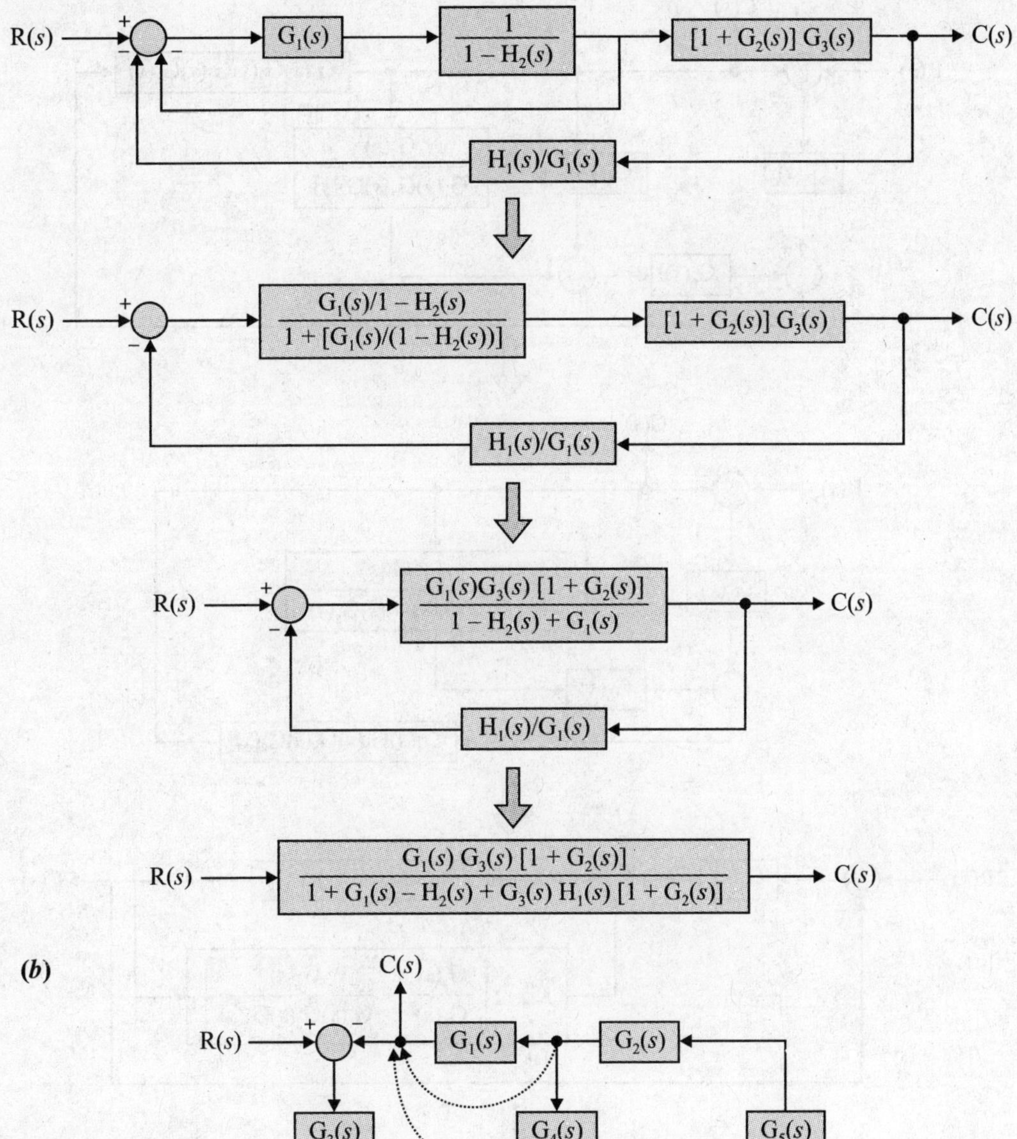
R(s)
+
−
−
G1(s)
1 / (1 − H2(s))
[1 + G2(s)] G3(s)
C(s)
H1(s)/G1(s)
R(s)
+
−
G1(s)/1 − H2(s) / 1 + [G1(s)/(1 − H2(s))]
[1 + G2(s)] G3(s)
C(s)
H1(s)/G1(s)
R(s)
+
−
G1(s)G3(s) [1 + G2(s)] / 1 − H2(s) + G1(s)
C(s)
H1(s)/G1(s)
R(s)
G1(s) G3(s) [1 + G2(s)] / 1 + G1(s) − H2(s) + G3(s) H1(s) [1 + G2(s)]
C(s)

(b)

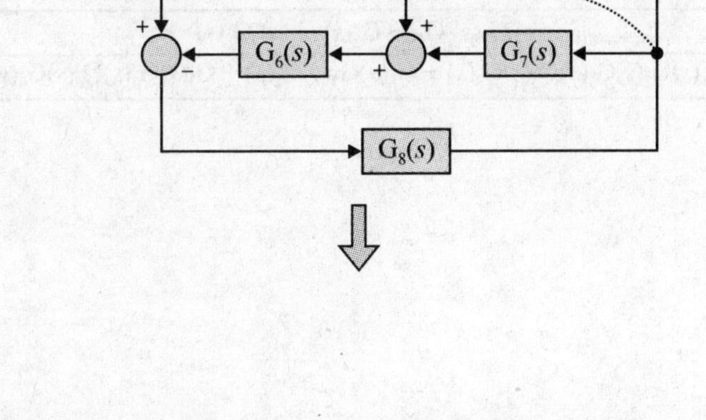
C(s)
R(s)
+
−
G1(s)
G2(s)
G3(s)
G4(s)
G5(s)
+
−
G6(s)
+
+
G7(s)
G8(s)

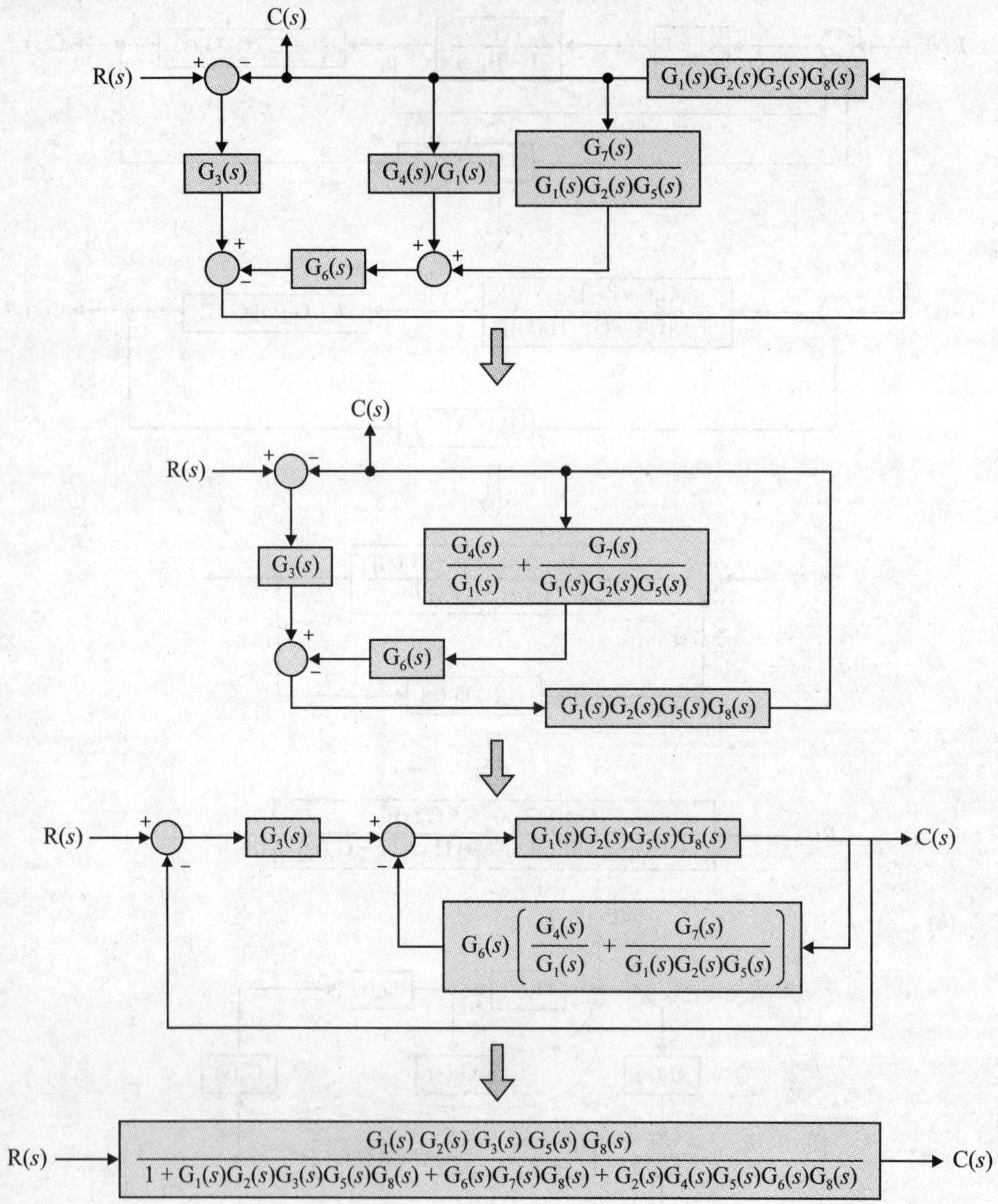

C(s)
R(s)
+
−
G1(s)G2(s)G5(s)G8(s)
G3(s)
G4(s)/G1(s)
G7(s) / G1(s)G2(s)G5(s)
G6(s)
G4(s)/G1(s) + G7(s)/G1(s)G2(s)G5(s)
G6(s) [G4(s)/G1(s) + G7(s)/G1(s)G2(s)G5(s)]
G1(s) G2(s) G3(s) G5(s) G8(s) / 1 + G1(s)G2(s)G3(s)G5(s)G8(s) + G6(s)G7(s)G8(s) + G2(s)G4(s)G5(s)G6(s)G8(s)

(c)

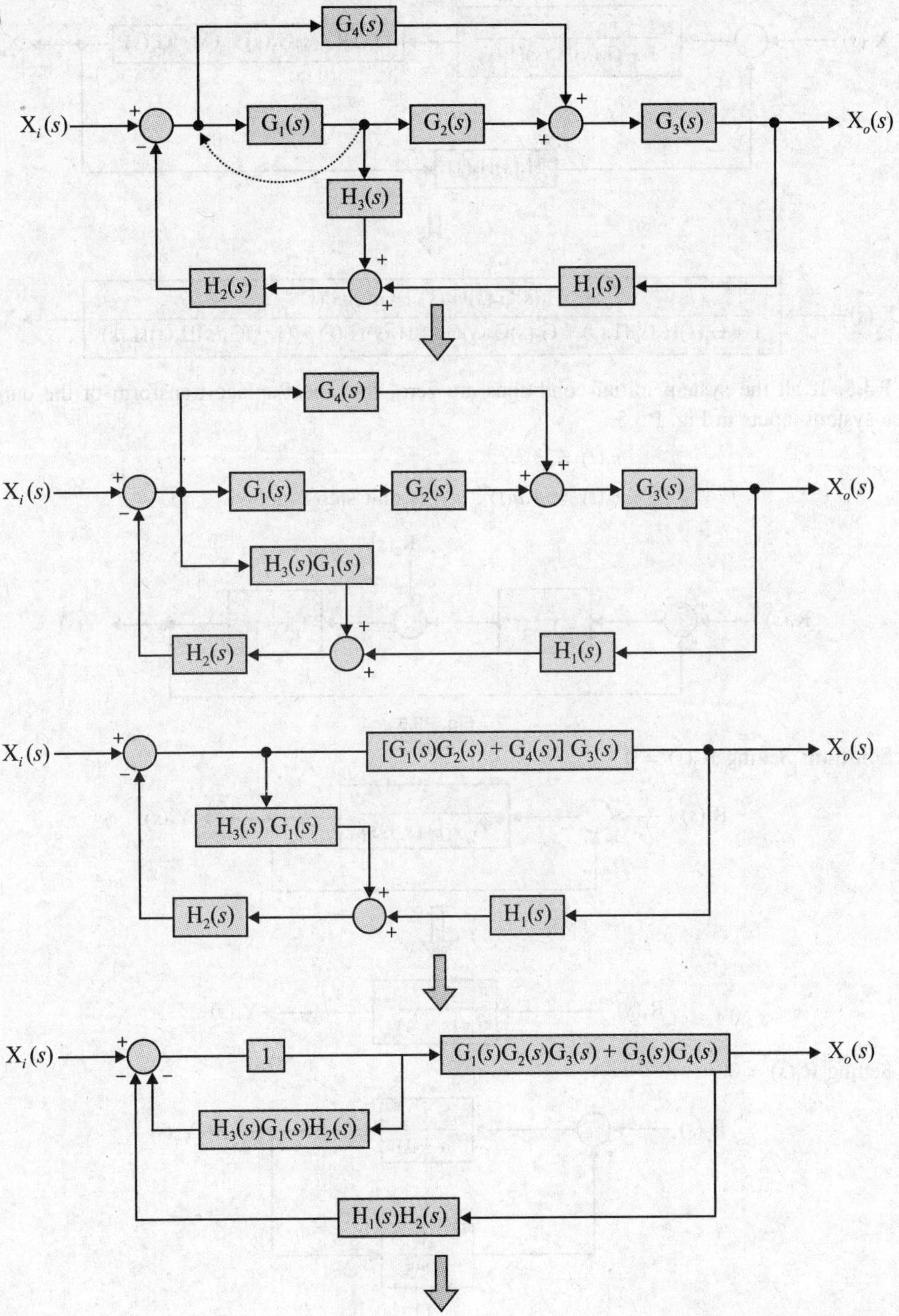
G₄(s)
Xᵢ(s)
G₁(s)
G₂(s)
G₃(s)
Xₒ(s)
H₃(s)
H₂(s)
H₁(s)
G₄(s)
Xᵢ(s)
G₁(s)
G₂(s)
G₃(s)
Xₒ(s)
H₃(s)G₁(s)
H₂(s)
H₁(s)
Xᵢ(s)
[G₁(s)G₂(s) + G₄(s)] G₃(s)
Xₒ(s)
H₃(s) G₁(s)
H₂(s)
H₁(s)
Xᵢ(s)
1
G₁(s)G₂(s)G₃(s) + G₃(s)G₄(s)
Xₒ(s)
H₃(s)G₁(s)H₂(s)
H₁(s)H₂(s)

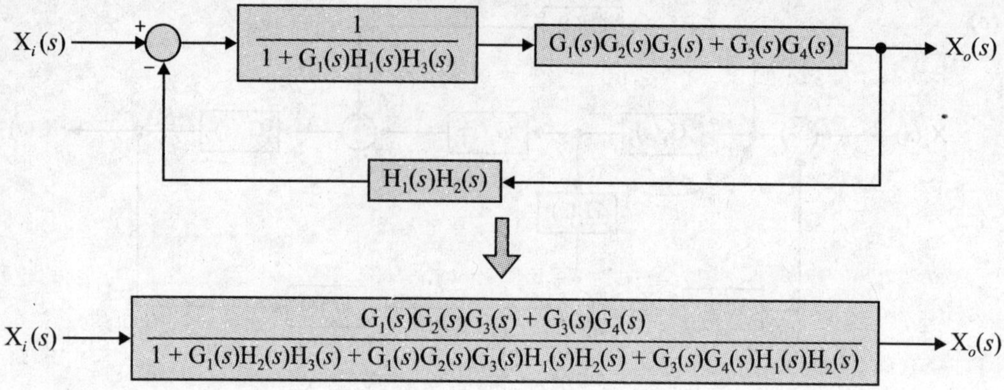

P 3.5: If all the system initial conditions are zero, find the Laplace transform of the output for given system inputs in Fig. P3.5.

$$r_1(t) = 3\,e^{-t}$$

$$r_2(t) = 4\,u(t)\ ;\ u(t) \text{ is unit step function.}$$

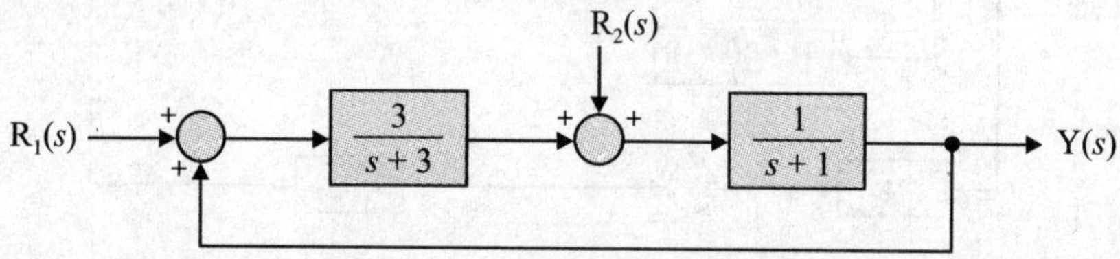

Fig. P3.5

Solution: Setting $R_2(s) = 0$

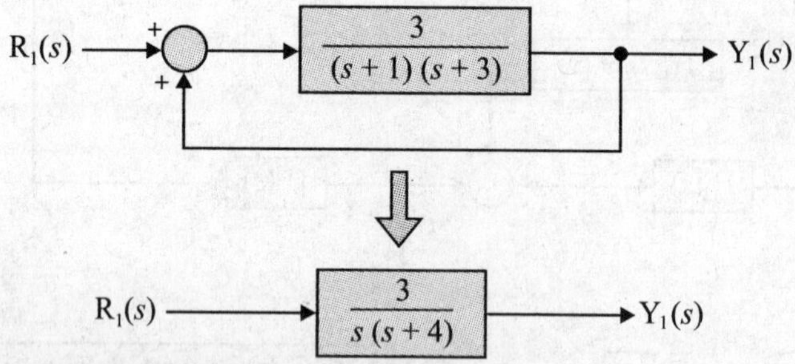

Setting $R_1(s) = 0$

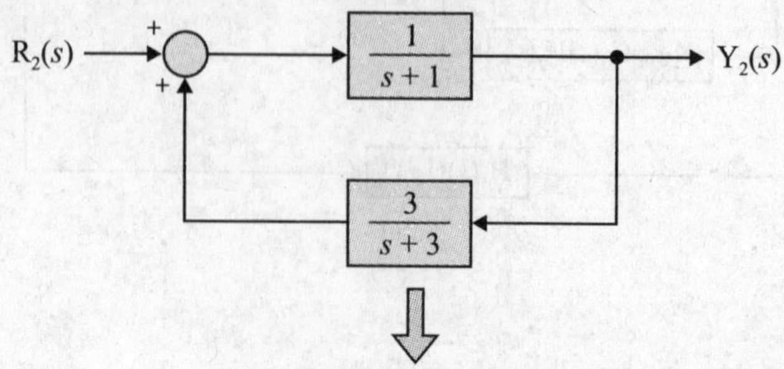

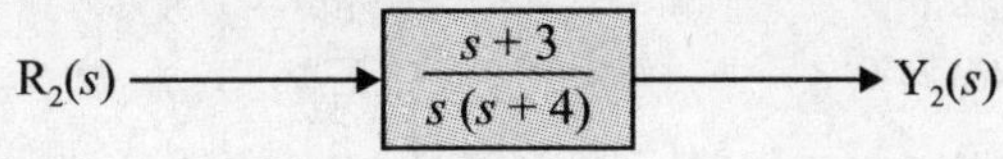

$$Y(s) = Y_1(s) + Y_2(s)$$

$$= \frac{3}{s(s+4)} R_1(s) + \frac{s+3}{s(s+4)} R_2(s)$$

$$= \left[\frac{3}{s(s+4)}\right]\left[\frac{3}{(s+1)}\right] + \left[\frac{s+3}{s(s+4)}\right]\left[\frac{4}{s}\right]$$

$$= \frac{4s^2 + 25s + 12}{s^2(s+1)(s+4)}$$

P 3.6: Find $\frac{C_1(s)}{R_1(s)}$ and $\frac{C_2(s)}{R_2(s)}$ for the system shown in Fig. P3.6 where α_1, α_2 and α_3 are constants.

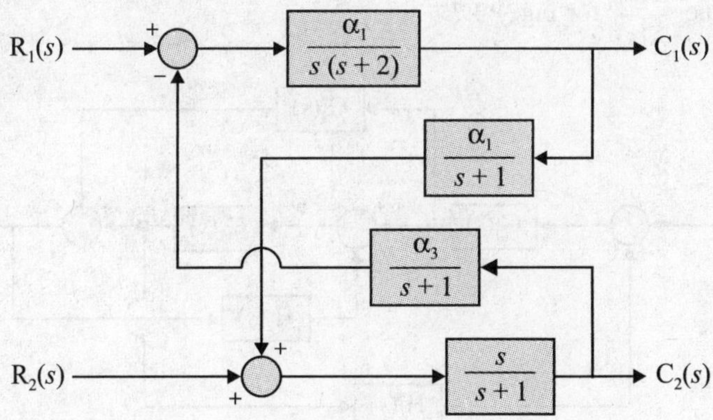

Fig. P3.6

Solution: To find $\frac{C_1(s)}{R_1(s)}$, set $R_2(s) = 0$

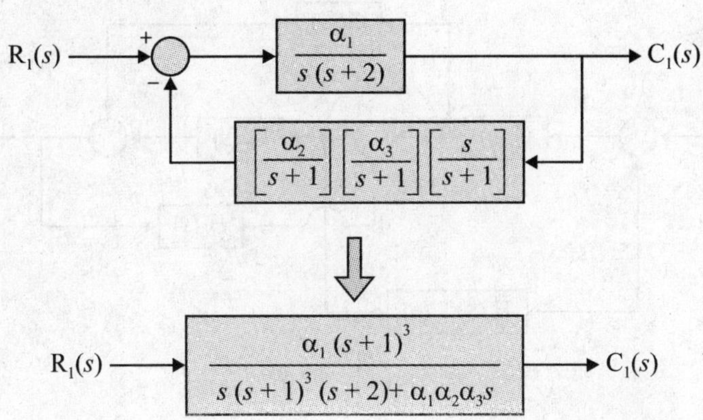

To find $\frac{C_2(s)}{R_2(s)}$, set $R_1(s) = 0$

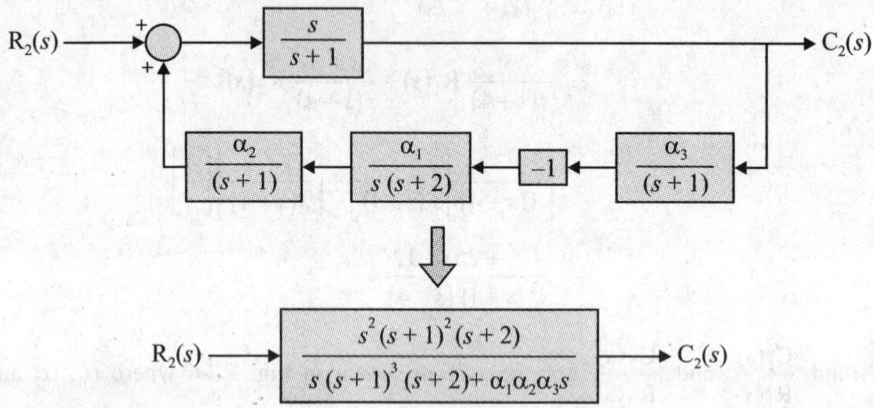

P3.7: Determine $\frac{Y(s)}{X(s)}$ for Fig. P3.7.

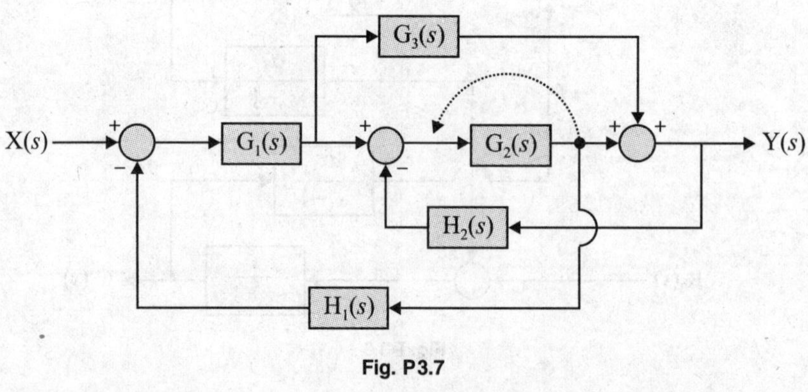

Fig. P3.7

Solution: Moving take off point as shown by dashed line in Fig. P 3.7, the following block diagram results.

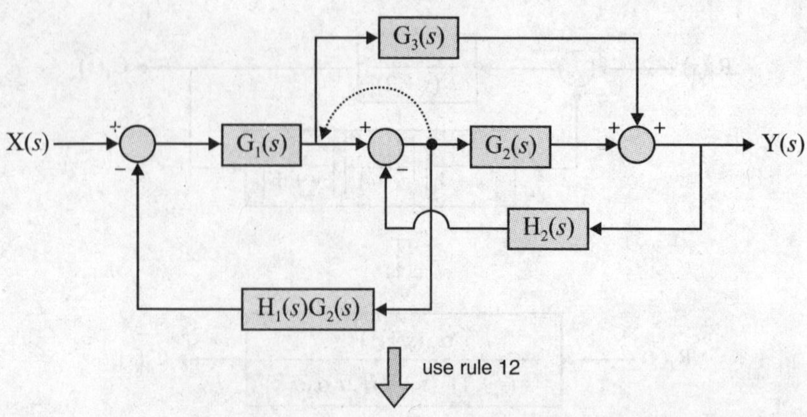

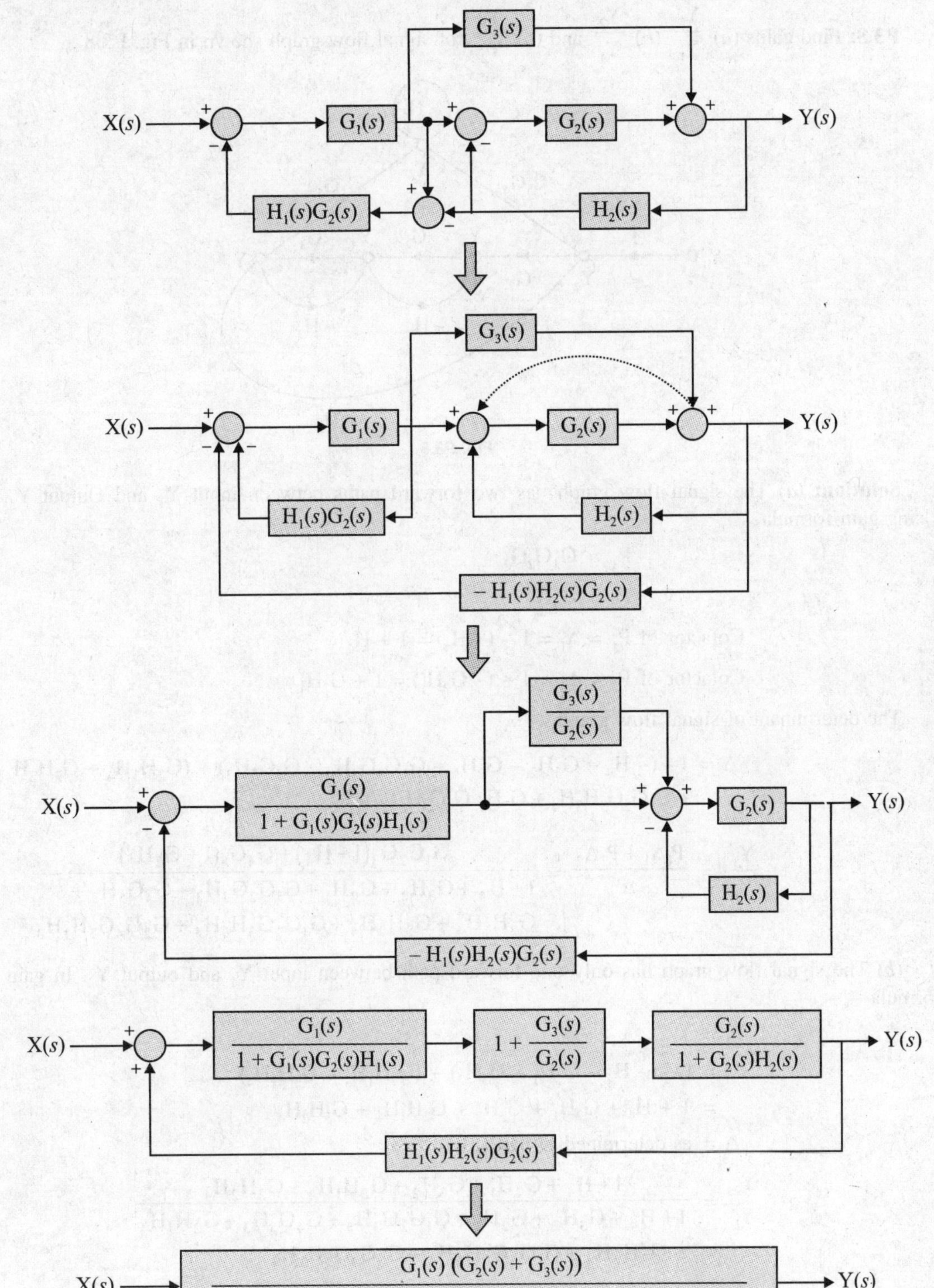
X(s)
Y(s)
G1(s)
G2(s)
G3(s)
H1(s)G2(s)
H2(s)
– H1(s)H2(s)G2(s)
G3(s)/G2(s)
G1(s) / 1 + G1(s)G2(s)H1(s)
1 + G3(s)/G2(s)
G2(s) / 1 + G2(s)H2(s)
H1(s)H2(s)G2(s)
G1(s) (G2(s) + G3(s)) / 1 + G2(s)H2(s) + G1(s)G2(s)H1(s) – G1(s)G2(s)G3(s)H1(s)H2(s)

P3.8: Find gains (*a*) $\frac{Y_5}{Y_1}$ (*b*) $\frac{Y_2}{Y_1}$ and (*c*) $\frac{Y_5}{Y_2}$ for signal flow graph shown in Fig. P3.8

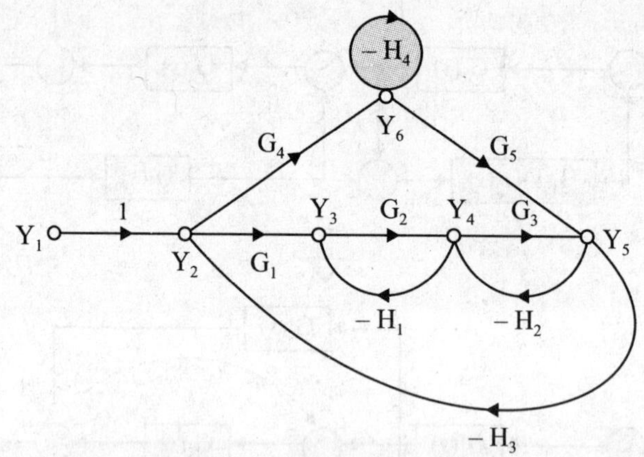

Fig. P3.8

Solution: (*a*) The signal flow graph has two forward paths between input Y_1 and Output Y_5. Using gain formula

$$P_1 = G_1G_2G_3$$

$$P_2 = G_4G_5$$

$$\text{Cofactor of } P_1 = \Delta_1 = 1 - (-H_4) = 1 + H_4$$

$$\text{Cofactor of } P_2 = \Delta_2 = 1 - (-G_2H_1) = 1 + G_2H_1$$

The determinant of signal flow graph

$$\Delta = 1 - (-H_4 - G_3H_2 - G_2H_1 - G_1G_2G_3H_3 - G_4G_5H_3) + (G_2H_1H_4 + G_3H_2H_4 + G_1G_2G_3H_3H_4 + G_2H_1\,G_4G_5H_3)$$

$$\frac{Y_5}{Y_1} = \frac{P_1\Delta_1 + P_2\Delta_2}{\Delta} = \frac{G_1G_2G_3(1+H_4)+G_4G_5(1+G_2H_1)}{1+H_4+G_3H_2+G_2H_1+G_1G_2G_3H_3+G_4G_5H_3+G_2H_1H_4+G_3H_2H_4+G_1G_2G_3H_3H_4+G_2G_4G_5H_1H_3}$$

(*b*) The signal flow graph has only one forward path between input Y_1 and output Y_2. In gain formula

$$P_1 = 1$$

$$\Delta_1 = 1 - (-H_4 - G_2H_1 - G_3H_2) + (G_2H_1H_4 - G_3H_2H_4)$$

$$= 1 + H_4 + G_2H_1 + G_3H_2 + G_2H_1H_4 + G_3H_2H_4$$

$$\Delta = \text{as determined in part (a)}$$

$$\frac{Y_2}{Y_1} = \frac{1+H_4+G_2H_1+G_3H_2+G_2H_1H_4+G_3H_2H_4}{1+H_4+G_3H_2+G_2H_1+G_1G_2G_3H_3+G_4G_5H_3+G_2H_1H_4+G_3H_2H_4+G_1G_2G_3H_3H_4+G_2G_4G_5H_1H_3}$$

(*c*) $$\frac{Y_5}{Y_2} = \frac{Y_5}{Y_1} \times \frac{Y_1}{Y_2} = \frac{G_1G_2G_3(1+H_4)+G_4G_5(1+G_2H_1)}{1+H_4+G_2H_1+G_3H_2+G_2H_1H_4+G_3H_2H_4}$$

P 3.9: Determine following transfer functions for signal flow graph shown in Fig. P 3.9.

(*a*) $T_{11}(s) = \dfrac{Y_1(s)}{R_1(s)}$

(*b*) $T_{21}(s) = \dfrac{Y_2(s)}{R_1(s)}$

(*c*) $T_{12}(s) = \dfrac{Y_1(s)}{R_2(s)}$

(*d*) $T_{22}(s) = \dfrac{Y_2(s)}{R_2(s)}$

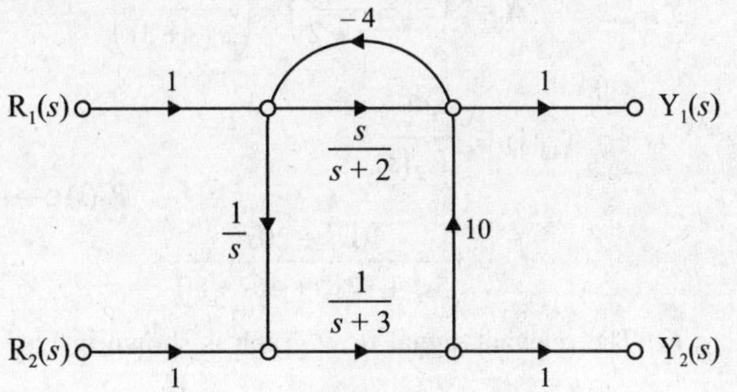

Fig. P 3.9

Solution: (*a*) The relevant signal flow graph is shown in Fig. P3.9 (*a*). It has two forward paths

$$P_1 = \frac{s}{s+2}, \quad \Delta_1 = 1$$

$$P_2 = \frac{10}{s(s+3)}, \quad \Delta_2 = 1$$

$$\Delta = 1 - \left(-\frac{40}{s(s+3)} - \frac{4s}{(s+2)}\right)$$

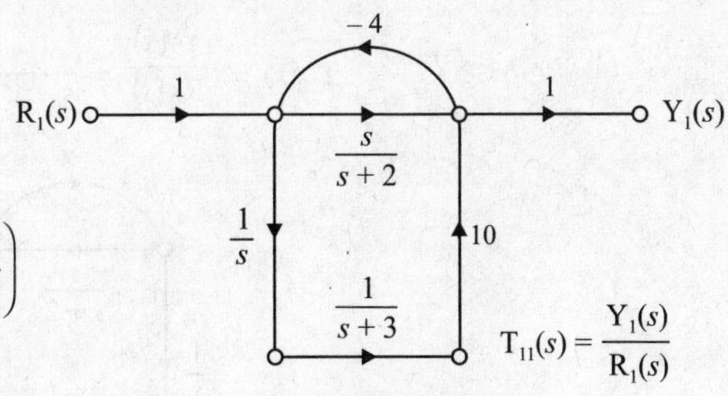

Fig. P 3.9 (*a*)

and $$T_{11}(s) = \frac{Y_1(s)}{R_1(s)}$$

$$= \frac{P_1\Delta_1 + P_2\Delta_2}{\Delta} = \frac{s^3+3s^2+10s+20}{5s^3+17s^2+46s+80}$$

(*b*) The relevant signal flow graph is shown in Fig. P3.9 (*b*). It has only one forward path.

$$P_1 = \frac{1}{s(s+3)}, \quad \Delta_1 = 1$$

$$\Delta = 1 - \left(-\frac{4s}{s+2}\right) - \left(\frac{-40}{s(s+3)}\right)$$

$$T_{21}(s) = \frac{Y_2(s)}{R_1(s)} = \frac{s+2}{5s^3+17s^2+46s+80}$$

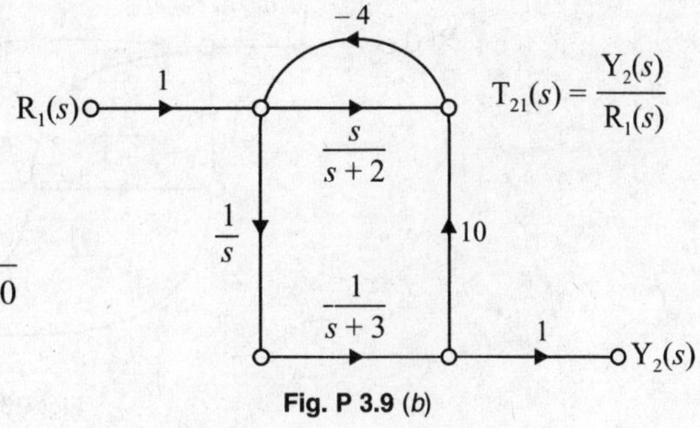

Fig. P 3.9 (*b*)

(*c*) The relevant signal flow graph is shown in Fig. P3.9 (*c*) It has only one forward path.

$$P_1 = \frac{10}{s+3}, \quad \Delta_1 = 1$$

$$\Delta = 1 - \left(-\frac{4s}{s+2}\right) - \left(\frac{-40}{s(s+3)}\right)$$

$$T_{12}(s) = \frac{Y_1(s)}{R_2(s)}$$

$$= \frac{10s^2 + 20s}{5s^3 + 17s^2 + 46s + 80}$$

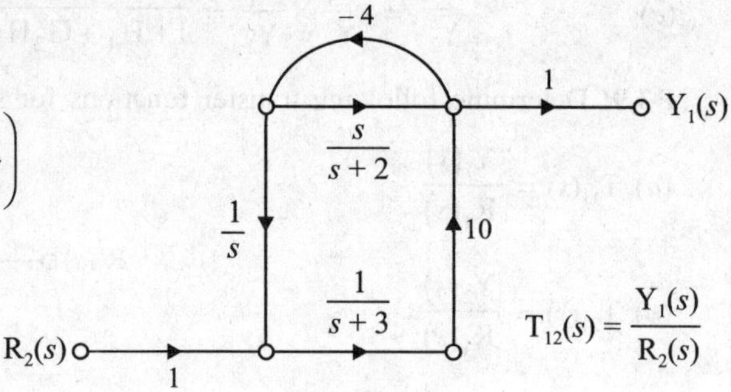

Fig. P 3.9 (*c*)

(*d*) The relevant signal flow graph is shown in Fig. P3.9 (*d*). It has only one forward path.

$$P_1 = \frac{1}{s+3}, \qquad \Delta_1 = 1 - \left(\frac{-4s}{s+2}\right)$$

$$\Delta = 1 - \left(-\frac{4s}{s+2}\right) - \left(\frac{-40}{s(s+3)}\right)$$

$$T_{22}(s) = \frac{Y_2(s)}{R_2(s)} = \frac{5s^2 + 2s}{5s^3 + 17s^2 + 46s + 80}$$

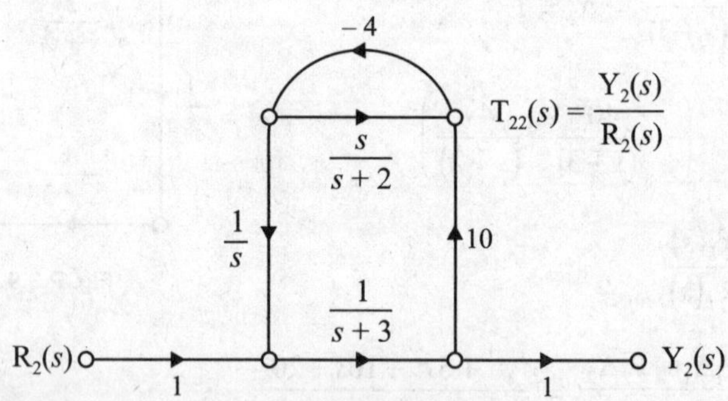

Fig. P3.9 (*d*)

P3.10: Use Mason's gain rule to find transfer function of system shown in Fig. P3.10.

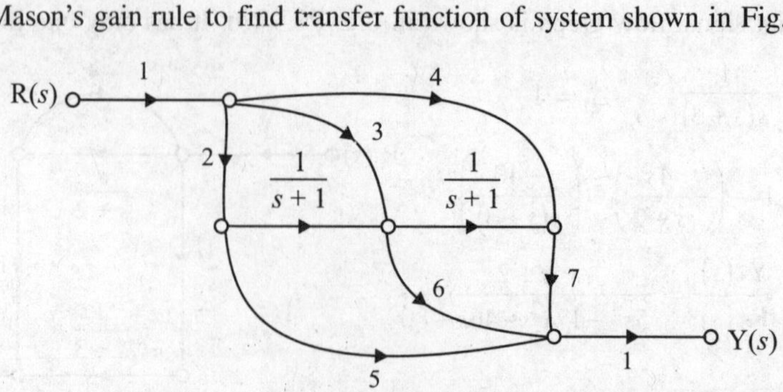

Fig. P3.10

Solution: The signal flow graph has 6 forward paths and no loop.

$$P_1 = (1)\,(4)\,(7)\,(1) = 28, \qquad \Delta_1 = 1$$

$$P_2 = (1)\,(3)\,(6)\,(1) = 18, \qquad \Delta_2 = 1$$

$$P_3 = (1)\,(2)\left(\frac{1}{s+1}\right)\left(\frac{1}{s+1}\right)(7)\,(1) = \frac{14}{(s+1)^2}, \qquad \Delta_3 = 1$$

$$P_4 = (1)\,(2)\,(5)\,(1) = 10, \qquad \Delta_4 = 1$$

$$P_5 = (1)\,(3)\left(\frac{1}{s+1}\right)(7)\,(1) = \frac{21}{s+1}, \qquad \Delta_5 = 1$$

$$P_6 = (1)\,(2)\left(\frac{1}{s+1}\right)(6)\,(1) = \frac{12}{s+1}, \qquad \Delta_6 = 1$$

and $\Delta = 1$

$$\frac{Y(s)}{R(s)} = \sum_{i=1}^{6}\frac{P_i\Delta_i}{\Delta} = 28 + 18 + \frac{14}{(s+1)^2} + 10 + \frac{21}{s+1} + \frac{12}{s+1}$$

$$= \frac{56s^2 + 145s + 103}{(s+1)^2}$$

P 3.11: Convert the block diagram shown in Fig. P3.11 into signal flow graph and obtain following system functions.

(*a*) $T_{11}(s) = \dfrac{Y_1(s)}{R_1(s)}$ (*b*) $T_{22}(s) = \dfrac{Y_2(s)}{R_2(s)}$ (*c*) $T_{12}(s) = \dfrac{Y_1(s)}{R_2(s)}$ (*d*) $T_{21}(s) = \dfrac{Y_2(s)}{R_1(s)}$

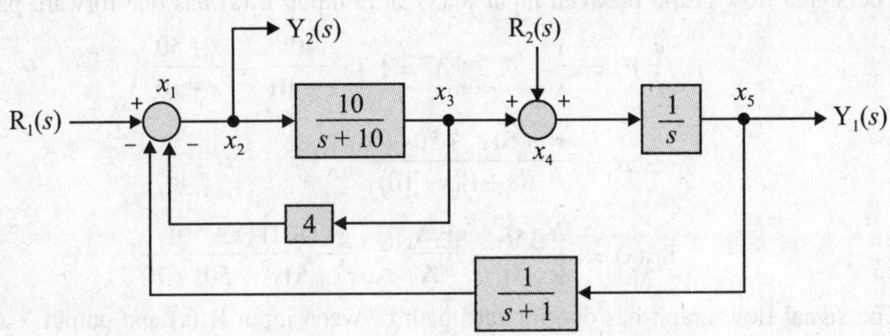

Fig. P3.11

Solution: Choosing the nodes x_1, x_2, x_3, x_4 and x_5 as shown in problem figure, the signal flow graph is constructed as follows:

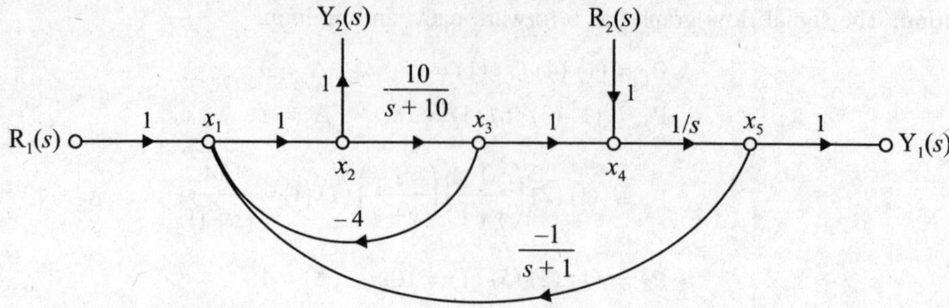

(*a*) The signal flow graph has only one forward path between input $R_1(s)$ and output $Y_1(s)$ with path gain

$$P_1 = \frac{10}{s(s+10)} \qquad \text{Cofactor } \Delta_1 = 1$$

and

$$\text{Determinant } \Delta = 1 - \left[\frac{-40}{s+10} - \frac{10}{s(s+1)(s+10)}\right] = \frac{s^3+51s^2+50s+10}{s(s+1)(s+10)}$$

$$T_{11}(s) = \frac{Y_1(s)}{R_1(s)} = \frac{10\,(s+1)}{s^3+51s^2+50s+10}$$

(*b*) The signal flow graph has one forward path between input $R_2(s)$ and output $Y_2(s)$.

$$P_1 = \frac{-1}{s(s+1)} \qquad \Delta_1 = 1$$

and

$$\Delta = \frac{s^3+51s^2+50s+10}{s(s+1)(s+10)}$$

$$T_{22}(s) = \frac{Y_2(s)}{R_2(s)} = \frac{-(s+10)}{s^3+51s^2+50s+10}$$

(*c*) The signal flow graph between input $R_2(s)$ and output $Y_1(s)$ has one forward path.

$$P_1 = \frac{1}{s} \qquad \Delta_1 = 1 + \frac{40}{s+10} = \frac{s+50}{s+10}$$

and

$$\Delta = \frac{s^3+51s^2+50s+10}{s(s+1)(s+10)}$$

$$T_{12}(s) = \frac{Y_1(s)}{R_2(s)} = \frac{P_1\Delta_1}{\Delta} = \frac{(s+1)\,(s+50)}{s^3+51s^2+50s+10}$$

(*d*) The signal flow graph has one forward path between input $R_1(s)$ and output $Y_2(s)$.

$$P_1 = 1, \qquad \Delta_1 = 1$$

and

$$\Delta = \frac{s^3+51s^2+50s+10}{s(s+1)(s+10)}$$

$$T_{21}(s) = \frac{Y_2(s)}{R_1(s)} = \frac{P_1\,\Delta_1}{\Delta} = \frac{s(s+1)(s+10)}{s^3+51s^2+50s+10}$$

P3.12: Construct signal flow graph and find system function $\frac{X(S)}{U(S)}$ for the system characterised by following equations:

$$x = x_1 + b_3 u$$
$$\dot{x}_1 = -a_1 x_1 + x_2 + b_2 u$$
$$\dot{x}_2 = -a_2 x_1 + b_1 u$$

Solution: The Laplace transformed equations of given system are as follows:

$$X(s) = X_1(s) + b_3 U(s)$$
$$sX_1(s) = -a_1 X_1(s) + X_2(s) + b_2 U(s)$$

or
$$X_1(s) = -\frac{a_1}{s}X_1(s) + \frac{1}{s}X_2(s) + \frac{b_2}{s}U(s)$$

and
$$sX_2(s) = -a_2 X_1(s) + b_1 U(s)$$

or
$$X_2(s) = -\frac{a_2}{s}X_1(s) + \frac{b_1}{s}U(s)$$

The combined signal flow graph with the help of above equations is drawn below.

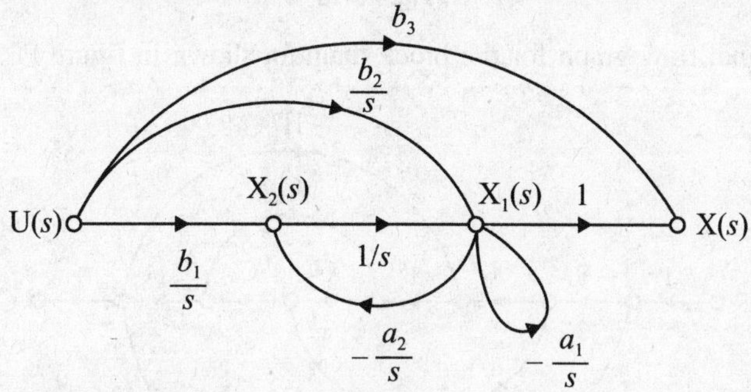

There are 3 forward paths between X(s) and U(s).

$$P_1 = \frac{b_1}{s^2}, \qquad \Delta_1 = 1$$

$$P_2 = \frac{b_2}{s}, \qquad \Delta_2 = 1$$

$$P_3 = b_3, \qquad \Delta_3 = 1 - \left(\frac{-a_1}{s} - \frac{a_2}{s^2}\right) = \frac{s^2 + a_1 s + a_2}{s^2}$$

$$\Delta = 1 - \left(\frac{-a_1}{s} - \frac{a_2}{s^2}\right) = \frac{s^2 + a_1 s + a_2}{s^2}$$

$$\frac{X(s)}{U(s)} = \frac{P_1\Delta_1 + P_2\Delta_2 + P_3\Delta_3}{\Delta} = \frac{\frac{b_1}{s^2} + \frac{b_2}{s} + b_3\left(\frac{s^2 + a_1 s + a_2}{s^2}\right)}{\left(s^2 + a_1 s + a_2\right)/s^2}$$

$$= \frac{b_3\left(s^2 + a_1 s + a_2\right) + b_2 s + b_1}{s^2 + a_1 s + a_2}$$

P 3.13: The block diagram of a Feedback Control System is shown in Fig. P 3.13. Draw the equivalent signal flow graph and determine C/R.

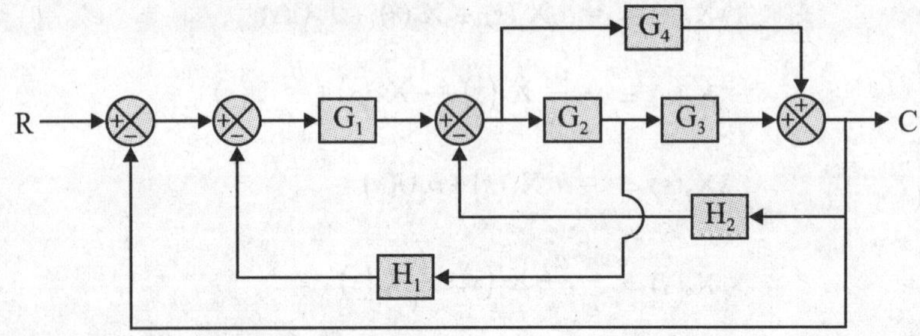

Fig. P3.13

Solution: The signal flow graph for the block diagram shown in figure P3.13 is as follows.

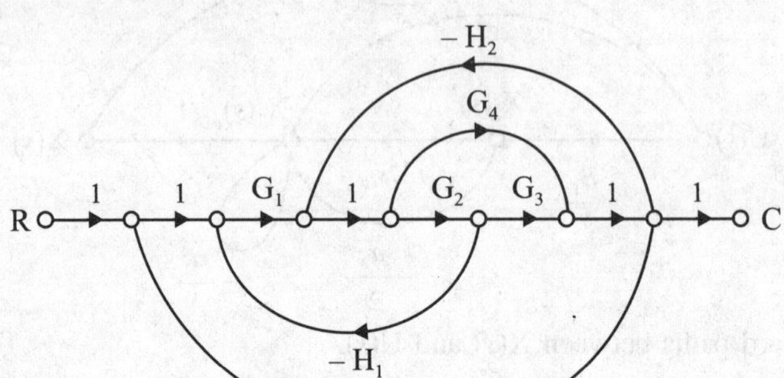

The above signal flow graph has two forward paths.

$$P_1 = G_1G_2G_3 \qquad \Delta_1 = 1$$

$$P_2 = G_1G_4 \qquad \Delta_2 = 1$$

$$\Delta = 1 - [- G_1G_2G_3 - G_1G_4 - G_1G_2H_1 - G_2G_3H_2 - G_4H_2]$$

$$\frac{C}{R} = \frac{G_1G_4 + G_1G_2G_3}{1 + G_1G_2G_3 + G_1G_4 + G_1G_2H_1 + G_2G_3H_2 + G_4H_2}$$

P 3.14: Find the transfer function C(*s*)/R(*s*) for a system whose signal flow graph is shown in Fig. P3.14.

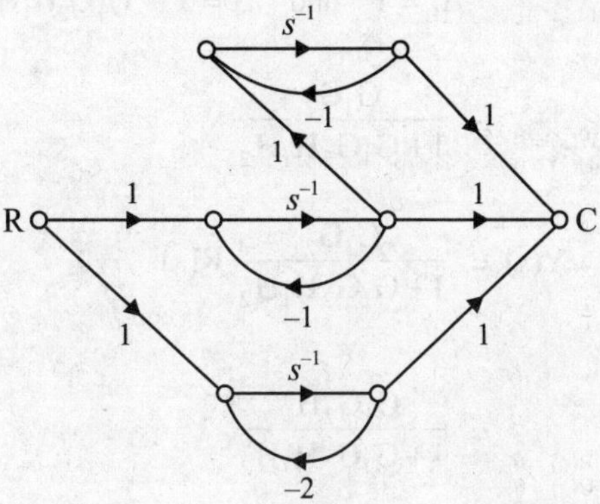

Fig. P3.14

Solution: The signal flow graph has 3 forward paths between input R and output C.

$$P_1 = s^{-1} \qquad \Delta_1 = 1 - (-s^{-1} - 2s^{-1}) + 2s^{-2} = 1 + 3s^{-1} + 2s^{-2}$$

$$P_2 = (1)(s^{-1})(1)(s^{-1})(1) = s^{-2} \qquad \Delta_2 = 1 + 2s^{-1}$$

$$P_3 = (1)(s^{-1})(1) = s^{-1} \qquad \Delta_3 = 1 - (-s^{-1} - s^{-1}) + s^{-2} = 1 + 2s^{-1} + s^{-2}$$

Signal flow graph has 3 sets of two non touching loops and a set of three non touching loops.

$$\Delta = 1 - (s^{-1} - s^{-1} - 2s^{-1}) + (s^{-2} + 2s^{-2} + 2s^{-2}) - (-2s^{-3})$$

$$= 1 + 4s^{-1} + 5s^{-2} + 2s^{-3}$$

$$\frac{C}{R} = \frac{P_1\Delta_1 + P_2\Delta_2 + P_3\Delta_3}{\Delta}$$

$$= \frac{(s^{-1})(1+3s^{-1}+2s^{-2}) + (s^{-2})(1+2s^{-1}) + (s^{-1})(1+2s^{-1}+s^{-2})}{1+4s^{-1}+5s^{-2}+2s^{-3}}$$

$$= \frac{2s^2+6s+5}{s^3+4s^2+5s+2}$$

P3.15: Find Y(*s*) in signal flow graph shown in Fig. P3.15.

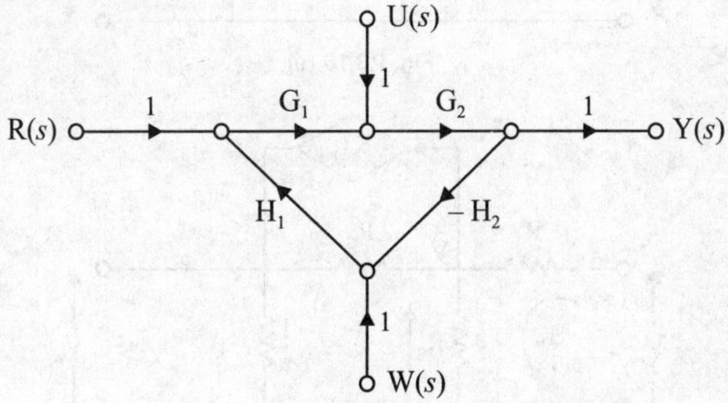

Fig. P3.15

Solution: Signal flow graph has one forward path between R(*s*) and Y(*s*)

$$P_1 = G_1G_2, \qquad \Delta_1 = 1 \quad \text{and} \quad \Delta = 1 + G_1G_2H_1H_2$$

$$\left.\frac{Y(s)}{R(s)}\right|_{\substack{W(s)=0\\U(s)=0}} = \frac{G_1G_2}{1+G_1G_2H_1H_2}$$

or

$$Y(s) = \frac{G_1G_2}{1+G_1G_2H_1H_2}R(s)$$

Similarly

$$\left.\frac{Y(s)}{W(s)}\right|_{\substack{R(s)=0\\U(s)=0}} = \frac{G_1G_2H_1}{1+G_1G_2H_1H_2}$$

or

$$Y(s) = \frac{G_1G_2H_1}{1+G_1G_2H_1H_2}W(s)$$

and

$$\left.\frac{Y(s)}{U(s)}\right|_{\substack{R(s)=0\\W(s)=0}} = \frac{G_2}{1+G_1G_2H_1H_2}$$

or

$$Y(s) = \frac{G_2U(s)}{1+G_1G_2H_1H_2}$$

Applying superposition

$$Y(s) = \frac{G_2}{1+G_1G_2H_1H_2}\left[G_1R(s) + G_1H_1W(s) + U(s)\right]$$

P3.16: Construct signal flow graph for electrical networks shown in Fig. P3.16(*a*) and (*b*).

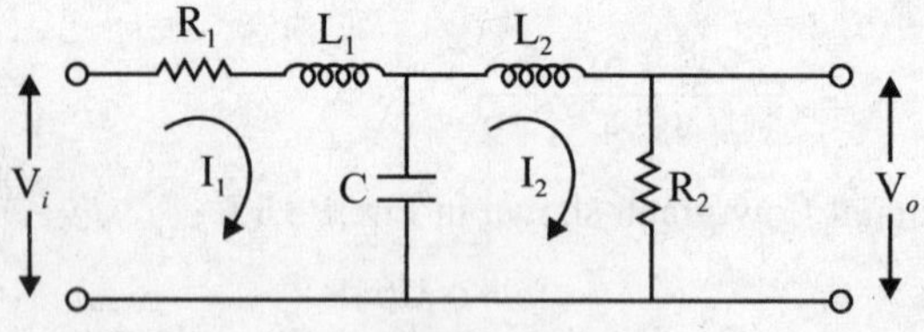

Fig. P3.16 (*a*)

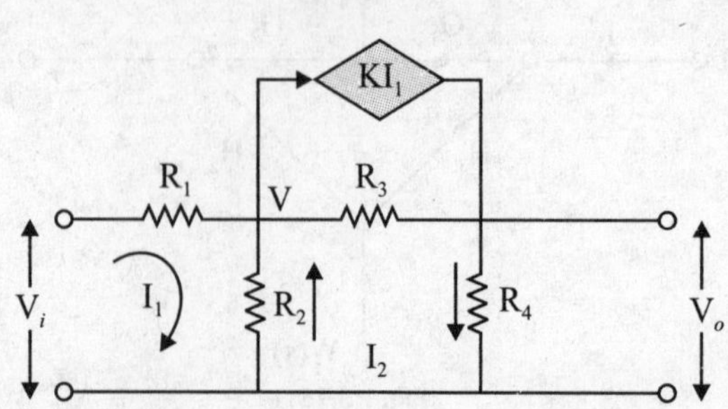

Fig. P3.16 (*b*)

Solution: (*a*) The laplace transformed network of Fig. 3.16(*a*) is as follows:

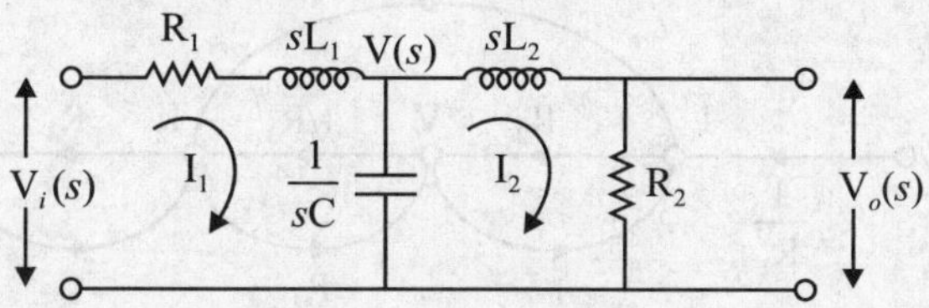

The network equations in terms of branch currents and node voltages are as follows:

$$I_1 = \frac{V_i(s) - V(s)}{R_1 + sL_1} = \frac{1}{R_1 + sL_1} V_i(s) - \frac{1}{R_1 + sL_1} V(s)$$

$$V(s) = \frac{1}{sC} I_1 - \frac{1}{sC} I_2$$

$$I_2 = \frac{1}{sL_2} V(s) - \frac{1}{sL_2} V_o(s)$$

$$V_o(s) = I_2 R_2$$

The signal flow graph for above equations is drawn below.

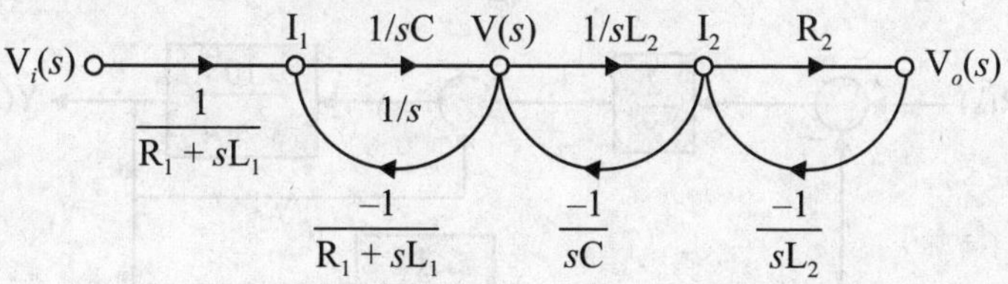

(*b*) The network equations in terms of branch currents and node voltages for Fig. 3.16(*b*) are as follows:

$$I_1 = \frac{V_i - V}{R_1} = \frac{1}{R_1} V_i - \frac{1}{R_1} V$$

$$V = R_2 I_1 - R_2 I_2$$

$$I_2 = \frac{V - V_o}{R_3} + KI_1$$

$$= \left(\frac{1}{R_3}\right) V - \left(\frac{1}{R_3}\right) V_o + KI_1$$

Note that I_2 is the current through R_2 and R_4 but not through R_3

$$V_o = I_2 R_4$$

The signal flow graph for above equations is drawn as follows:

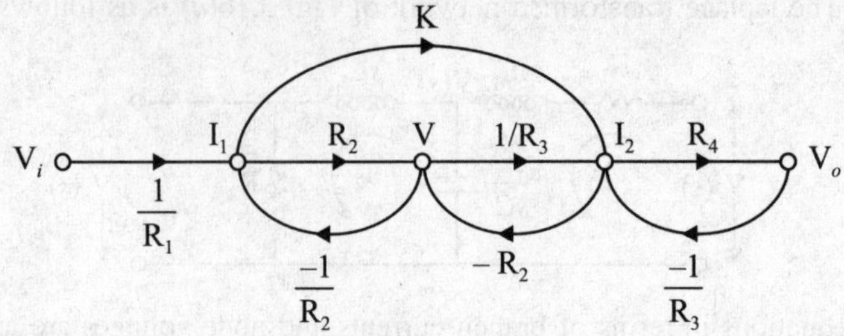

DRILL PROBLEMS

D3.1: Use the rules of block diagram algebra to find the transfer function of the systems shown in Fig. D3.1(*a*), (*b*) and (*c*).

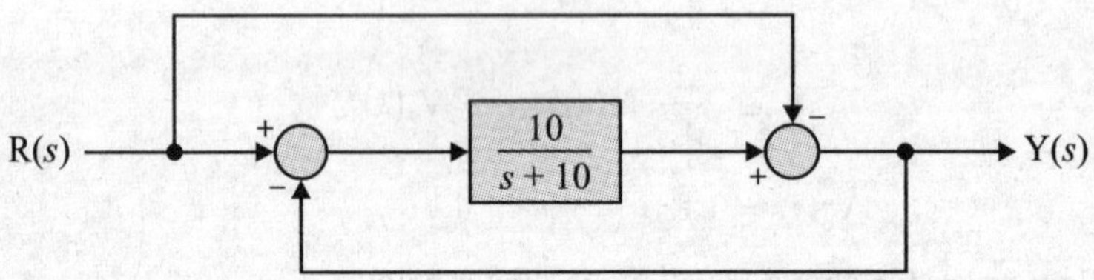

(*a*)

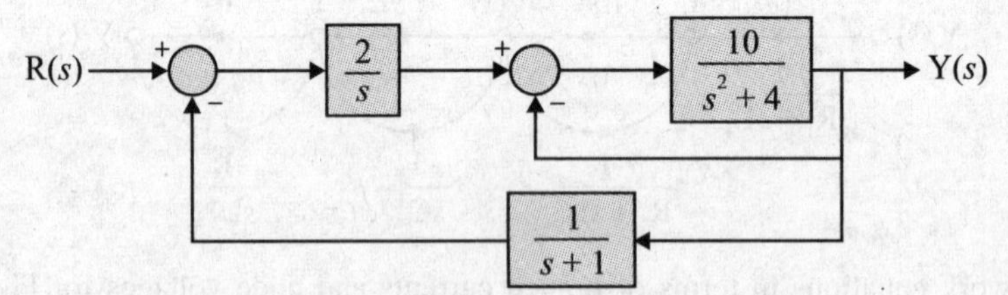

(*b*)

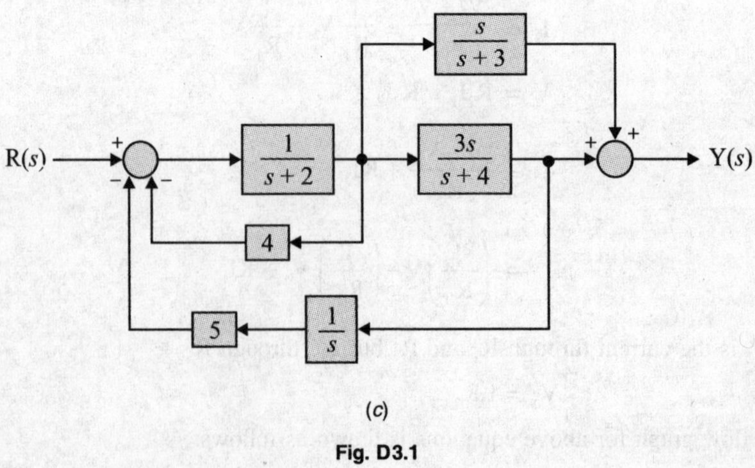

(*c*)

Fig. D3.1

Ans.

(*a*) $-s/(s + 20)$

(*b*) $20(s+1)/(s^4 + s^3 + 14s^2 + 14s + 20)$

(*c*) $\dfrac{s(s+4)(4s+13)}{(s^2+10s+39)(s+3)(s+4)}$

D3.2: Reduce the block diagram shown in Fig. D3.2 to obtain six transfer functions.

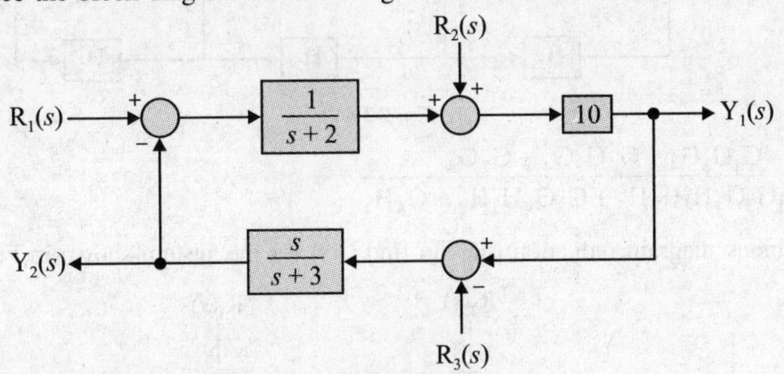

Fig. D3.2

Ans. $10 (s + 3) /(s^2 + 15s + 6)$,

$10 (s + 2) (s + 3) /(s^2 + 15s + 6)$,

$10s/(s^2 + 15s + 6)$,

$10s/(s^2 + 15s + 6)$,

$10s (s + 2)/(s^2 + 15s + 6)$,

$-s (s + 2)/(s^2 + 15s + 6)$

D3.3: Use block diagram reduction technique and obtain the following transfer functions for the system shown in Fig. D3.3:

(*i*) $\dfrac{C(s)}{R(s)}$

(*ii*) $\dfrac{C(s)}{X(s)}$

(*iii*) $\dfrac{C(s)}{Y(s)}$

also find total output C(*s*).

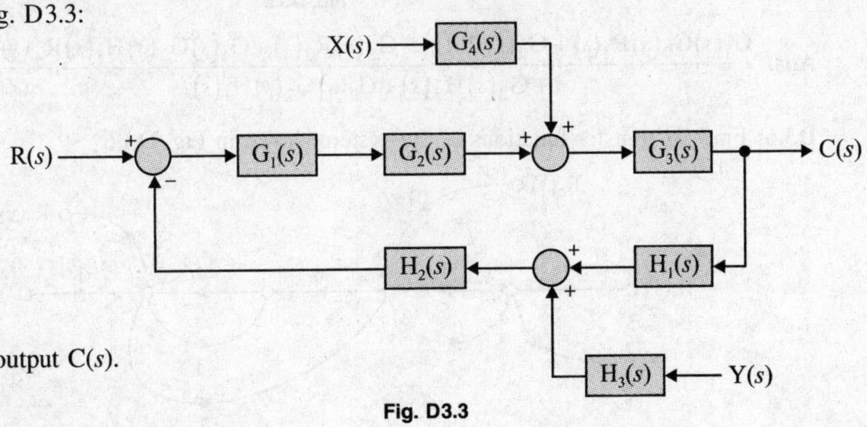

Fig. D3.3

Ans. $C(s) = \dfrac{G_1(s)G_2(s)G_3(s)R(s) + G_3(s)G_4(s)X(s) - G_1(s)G_2(s)G_3(s)H_2(s)H_3(s)Y(s)}{1 + G_1(s)G_2(s)G_3(s)H_1(s)H_2(s)}$

D3.4: A block diagram of a linear feedback system is shown in Fig. D3.4. Obtain a signal flow graph for the system and hence calculate the overall gain C(*s*)/R(*s*) for the system.

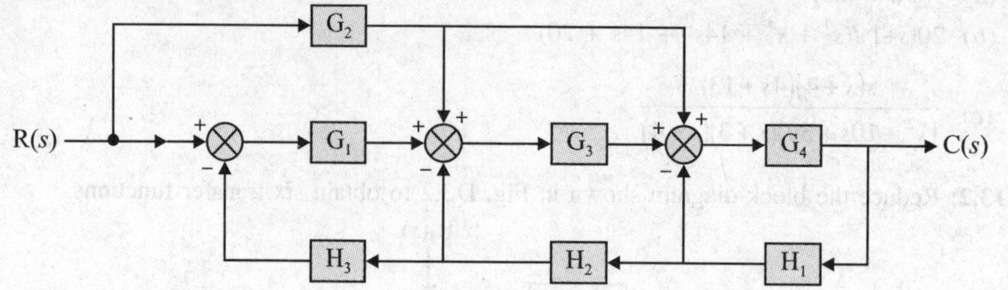

Fig. D3.4

Ans. $\dfrac{G_1G_3G_4 + G_2G_3G_4 + G_2G_4}{1 + G_1G_3G_4H_1H_2H_3 + G_3G_4H_1H_2 + G_4H_1}$

D 3.5: Use block diagram reduction rules to find C(*s*) for the system shown in Fig. D3.5.

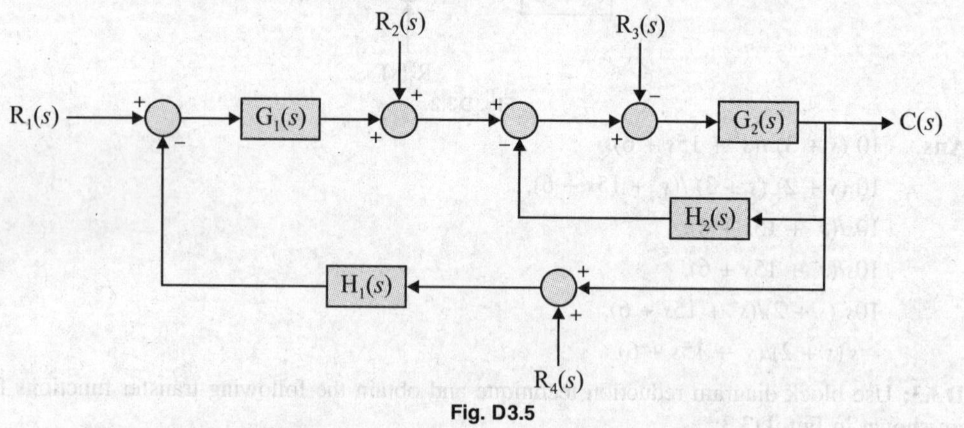

Fig. D3.5

Ans. $\dfrac{G_1(s)G_2(s)R_1(s) + G_2(s)R_2(s) - G_2(s)R_3(s) - G_1(s)G_2(s)H_1(s)R_4(s)}{1 + G_2(s)H_2(s) + G_1(s)G_2(s)H_1(s)}$

D3.6: Find six transfer functions of the system shown in Fig. D3.6.

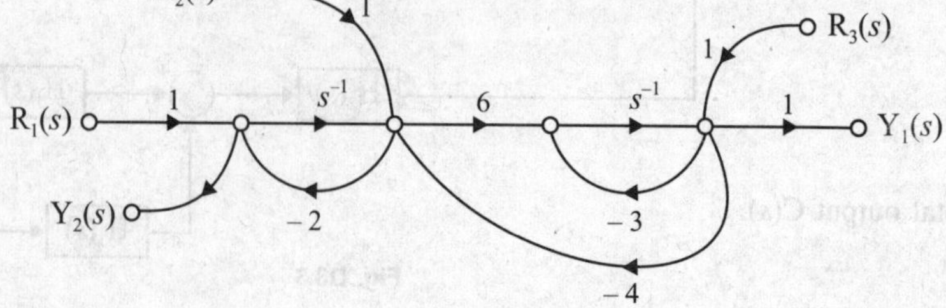

Fig. D3.6

Ans. $6/(s^2 + 29s + 6)$,

$6s/(s^2 + 29s + 6)$,

$s(s + 2)/(s^2 + 29s + 6)$,

$s(s + 27)/(s^2 + 29s + 6)$,
$-2s(s + 3)/(s^2 + 29s + 6)$,
$8s^2/(s^2 + 29s + 6)$

D3.7: Compute Z/Y of system shown in Fig. D3.7.

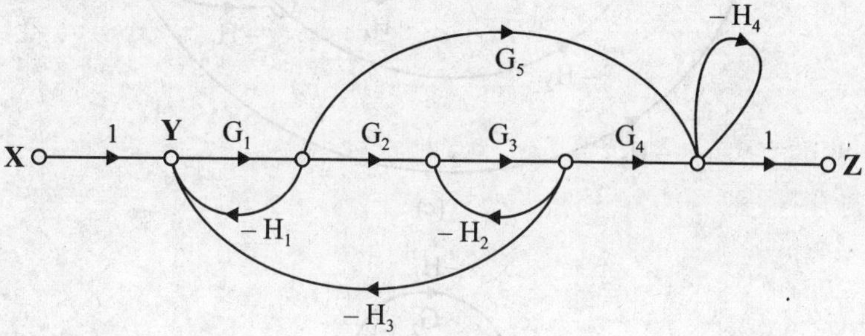

Fig. D3.7

Ans. $\dfrac{Z}{Y} = \dfrac{(Z/X)}{(Y/X)} = \dfrac{G_1G_2G_3G_4 + G_1G_5(1 + G_3H_2)}{1 + H_4 + G_3H_2 + G_3H_2H_4}$

D3.8: Construct signal flow graph for following set of system equations and hence find x_4/x_1.

$$x_2 = k_1x_1 + k_3x_3$$
$$x_3 = k_4x_1 + k_2x_2 + k_5x_3$$
$$x_4 = k_6x_2 + k_7x_3$$

Ans. $\dfrac{k_1k_2k_7 + k_4k_7 + k_1k_6(1 - k_5) + k_3k_4k_6}{1 - k_2k_3 - k_5}$

D3.9: Find C(*s*)/R(*s*) for systems shown in Fig. D3.9 (*a*), (*b*), (*c*) and (*d*).

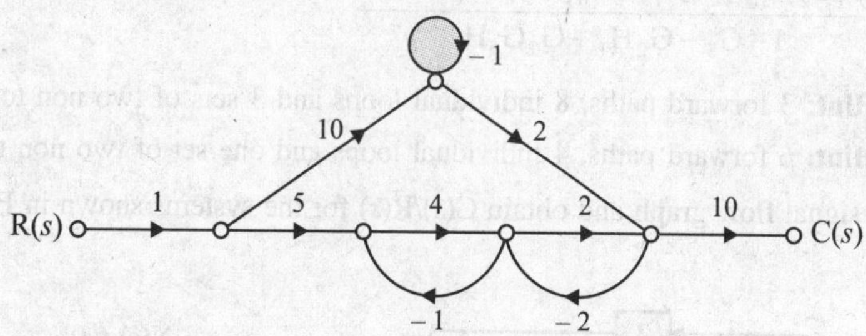

(*a*)

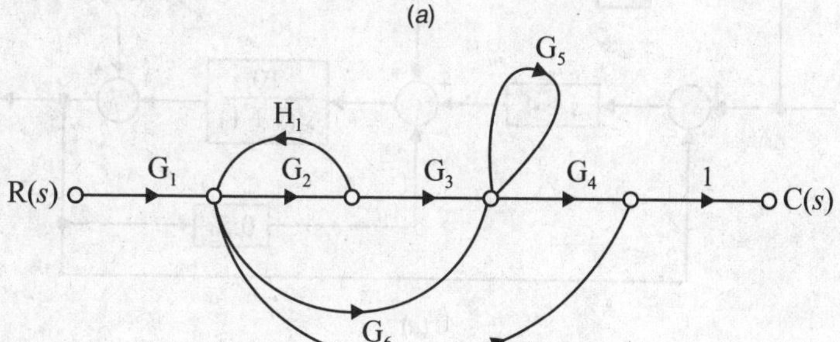

(*b*)

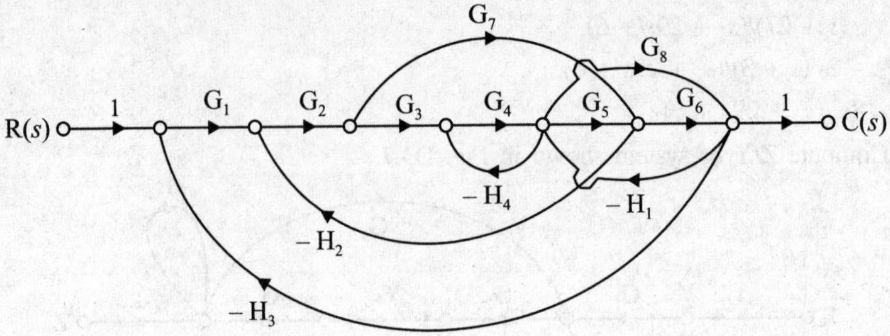

(c)

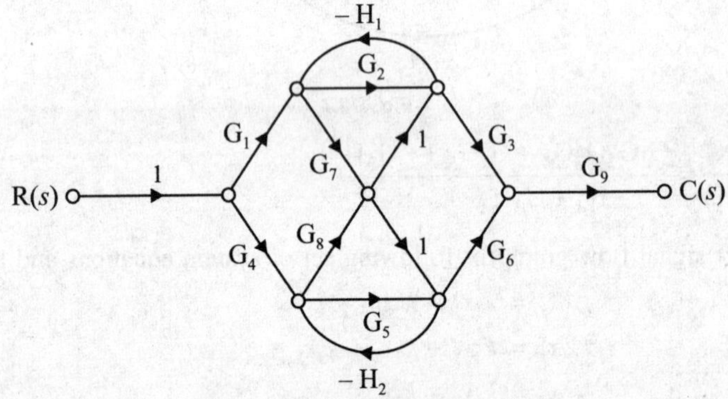

(d)

Fig. D3.9

Ans. (*a*) 100

(*b*) $\dfrac{G_1G_2G_3G_4 + G_1G_4G_6 + G_1G_7(1-G_5)}{1-G_5-G_2H_1+G_2G_5H_1}$

(*c*) **Hint:** 3 forward paths, 8 individual loops and 3 sets of two non touching loops.

(*d*) **Hint:** 6 forward paths, 4 individual loops and one set of two non touching loops.

D3.10: Draw signal flow graph and obtain C(*s*)/R(*s*) for the systems shown in Fig. D3.10 (*a*), (*b*) and (*c*).

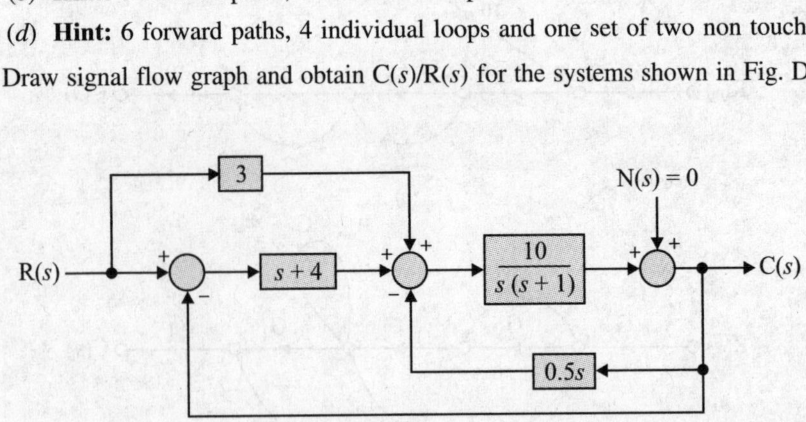

(*a*)

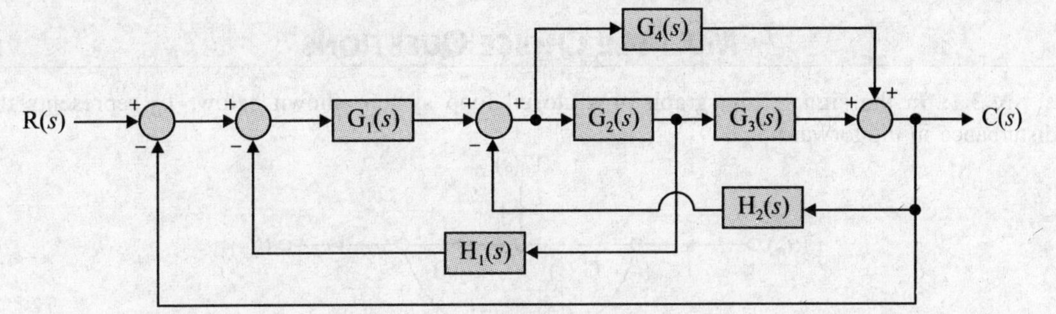

(*b*)

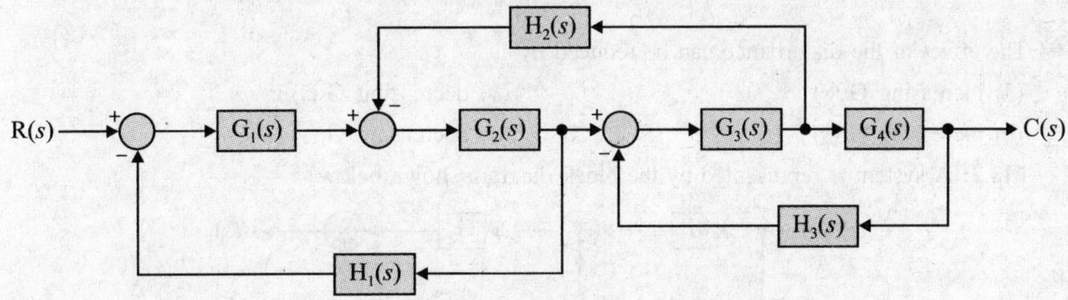

(*c*)

Fig. D3.10

Ans. (*a*) $\dfrac{10(s+7)}{s^2+16s+40}$

(*b*) $\dfrac{G_1(s)\left[G_2(s)G_3(s)+G_4(s)\right]}{1+G_1(s)G_4(s)+G_1(s)G_2(s)G_3(s)+G_1(s)G_2(s)H_1(s)+G_2(s)G_3(s)H_2(s)+G_4(s)H_2(s)}$

(*c*) $\dfrac{G_1(s)G_2(s)G_3(s)G_4(s)}{1+G_1(s)G_2(s)H_1(s)+G_2(s)G_3(s)H_2(s)+G_3(s)G_4(s)H_3(s)+G_1(s)G_2(s)G_3(s)G_4(s)H_1(s)H_3(s)}$

D 3.11: Draw signal flow graph and hence obtain transfer function of network shown in Fig. D3.11.

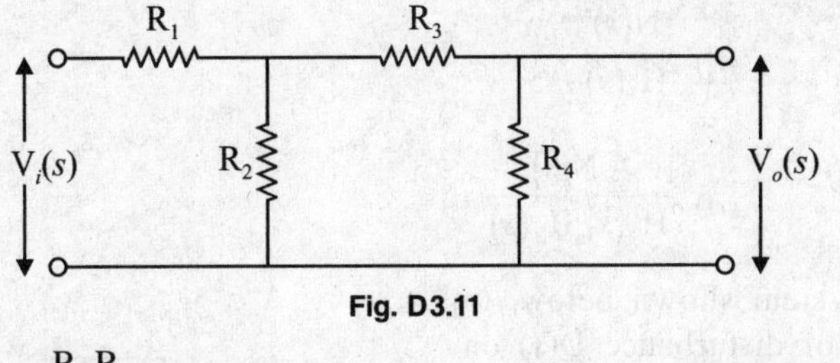

Fig. D3.11

Ans. $\dfrac{R_2R_4}{R_2R_4+R_1R_4+R_1R_2+R_2R_3+R_1R_3}$.

MULTIPLE CHOICE QUESTIONS

M 3.1: In the signal flow graph of a closed loop system shown below, T_D represents the disturbance in the forward path.

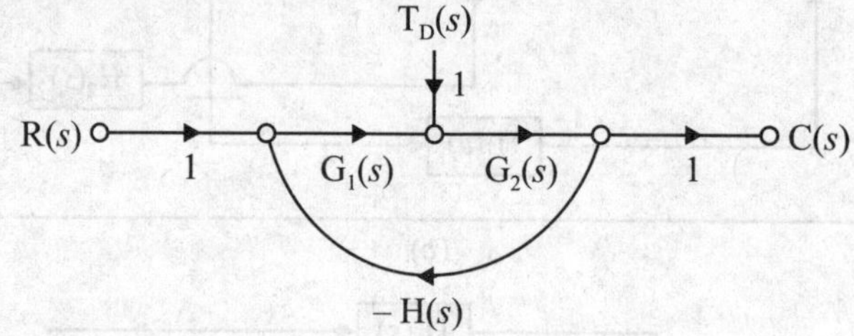

The effect of the disturbance can be reduced by

(*a*) increasing $G_2(s)$ (*b*) decreasing $G_2(s)$

(*c*) increasing $G_1(s)$ (*d*) decreasing $G_1(s)$

M3.2: A system is represented by the block diagram shown below.

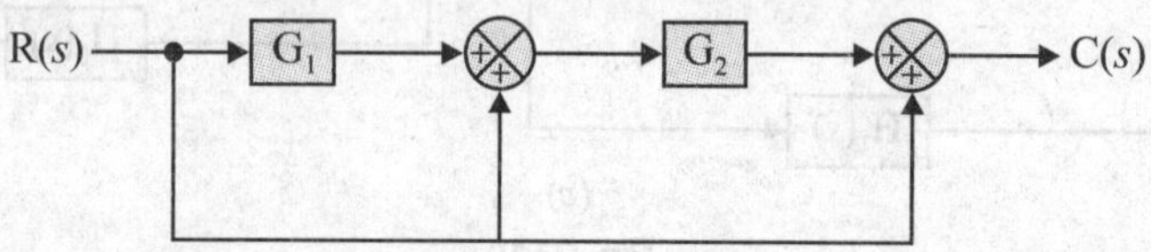

Which one of the following represents the input-output relationship of the above diagram?

(*a*) R(*s*) → $\boxed{G_1G_2}$ → C(*s*) (*b*) R(*s*) → $\boxed{G_1 + G_2}$ → C(*s*)

(*c*) R(*s*) → $\boxed{1 + G_1 + G_1G_2}$ → C(*s*) (*d*) R(*s*) → $\boxed{1 + G_2 + G_1G_2}$ → C(*s*)

M 3.3: In the feedback system shown below, the noise component of output is given by (assume high loop gain at frequencies of interest):

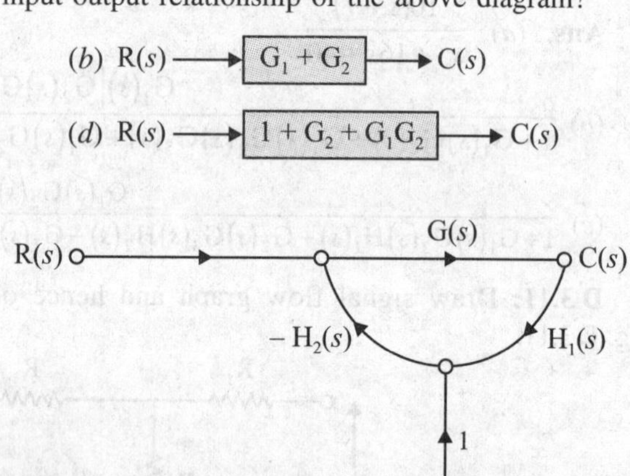

(*a*) $\dfrac{-N(s)}{H_1(s)}$ (*b*) $\dfrac{N(s)}{H_1(s)}$

(*c*) $\dfrac{N(s)}{H_1(s)\,H_2(s)}$ (*d*) $\dfrac{-N(s)}{H_1(s)\,H_2(s)}$

M 3.4: In the system shown below, to eliminate the effect of disturbance D(*s*) on C(*s*), the transfer function $G_d(s)$ should be

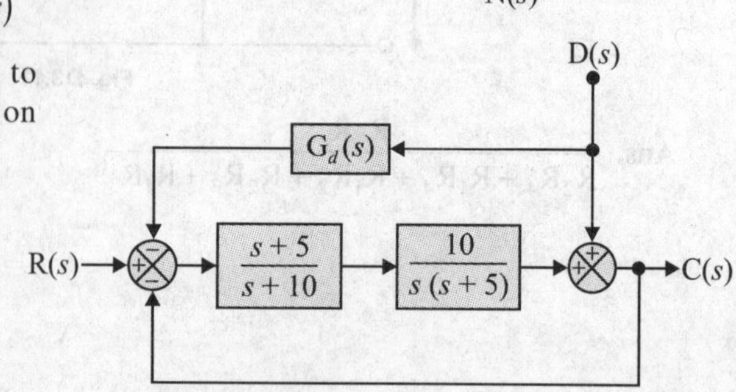

(*a*) $\dfrac{(s+10)}{10}$ (*b*) $\dfrac{s(s+10)}{10}$

(*c*) $\dfrac{10}{(s+10)}$ (*d*) $\dfrac{10}{s(s+10)}$

M3.5: The closed-loop system shown below is subjected to a disturbance N(*s*). The transfer function C(*s*)/N(*s*) is given by

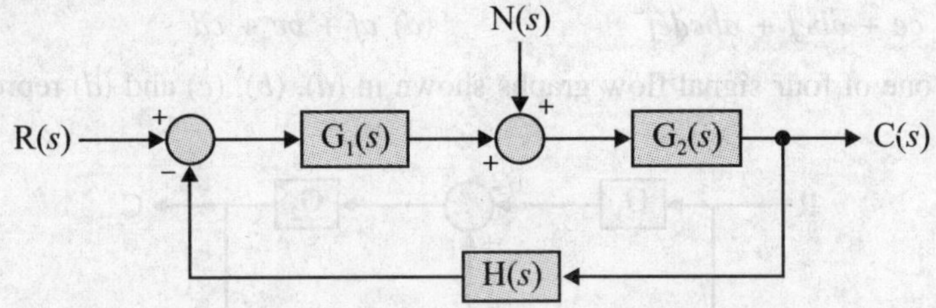

(*a*) $\dfrac{G_1(s)G_2(s)}{1+G_1(s)\,G_2(s)H(s)}$ (*b*) $\dfrac{G_1(s)}{1+G_1(s)H(s)}$

(*c*) $\dfrac{G_2(s)}{1+G_2(s)H(s)}$ (*d*) $\dfrac{G_2(s)}{1+G_1(s)\,G_2(s)H(s)}$

M3.6: The transfer function of the system shown below is

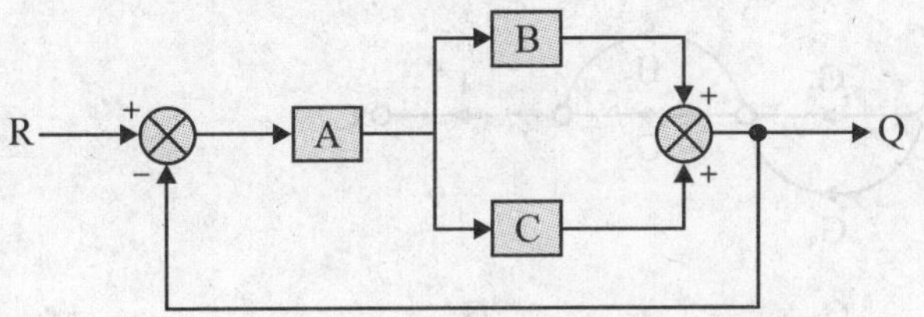

(*a*) $\dfrac{Q}{R}=\dfrac{ABC}{1+ABC}$ (*b*) $\dfrac{Q}{R}=\dfrac{A+B+C}{1+AB+AC}$

(*c*) $\dfrac{Q}{R}=\dfrac{AB+AC}{ABC}$ (*d*) $\dfrac{Q}{R}=\dfrac{AB+AC}{1+AB+AC}$

M3.7: For block diagram shown below, C(*s*)/R(*s*) is given by

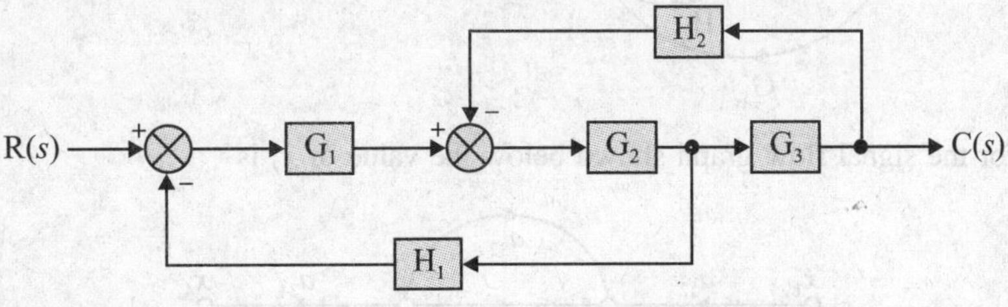

(*a*) $\dfrac{G_1G_2G_3}{1+H_2G_2G_3+H_1G_1G_2}$ (*b*) $\dfrac{G_1G_2G_3}{1+G_1G_2G_3H_1H_2}$

(*c*) $\dfrac{G_1G_2G_3}{1+G_1G_2G_3H_1+G_1G_2G_3H_2}$ (*d*) $\dfrac{G_1G_2G_3}{1+G_1G_2G_3H_1}$

M3.8:

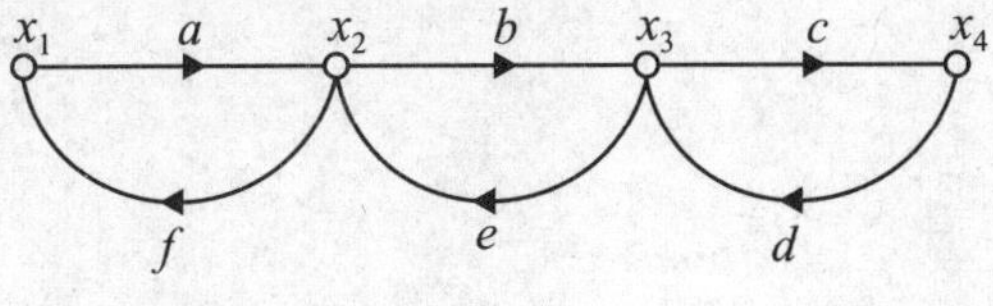

The sum of the gains of the feedback paths in the above signal flow graph is

(*a*) $af + be + cd + abef + bcde + abcdef$ (*b*) $af + be + cd + abef + bcde$

(*c*) $af + be + ce + abef + abcdef$ (*d*) $af + be + cd$

M3.9: Which one of four signal flow graphs shown in (*a*), (*b*), (*c*) and (*d*) represents the diagram shown below?

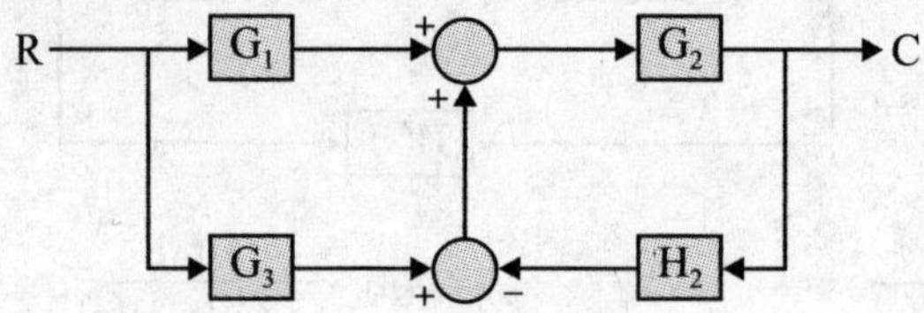

(*a*)

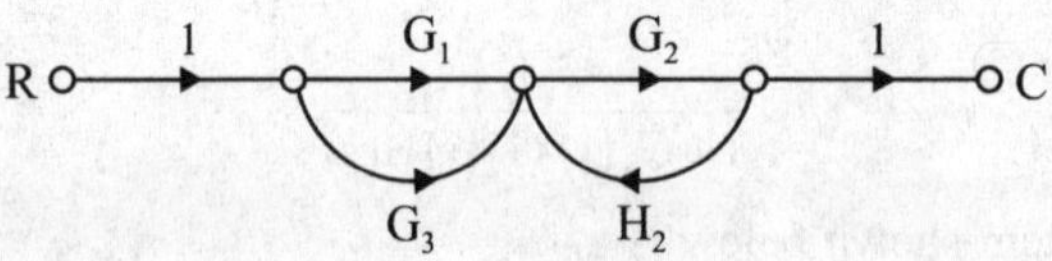

(*b*)

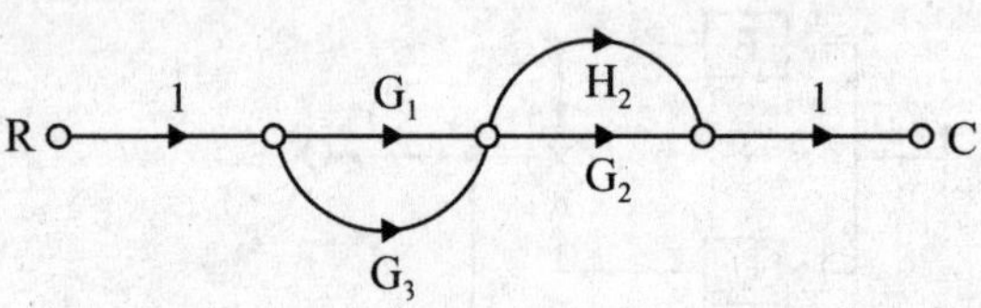

(*c*)

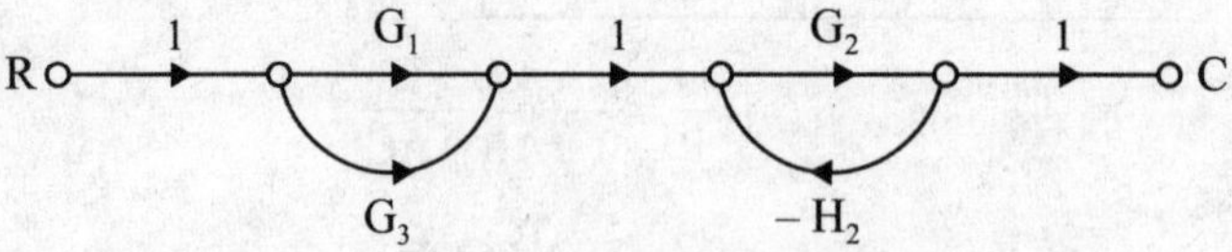

(*d*)

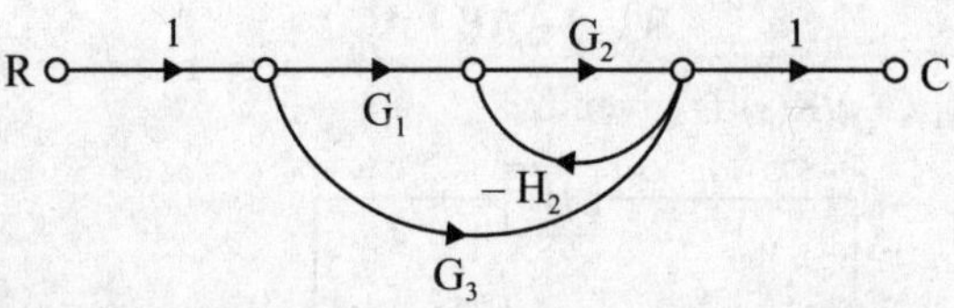

M3.10: For the signal flow graph shown below the value of x_2 is

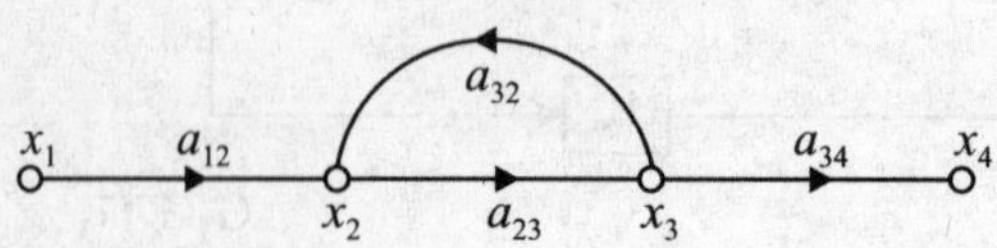

(*a*) $a_{12}\,x_1$ (*b*) $a_{12}\,x_1 + a_{32}\,x_3$

(*c*) $a_{12}\,x_1 - a_{23}\,x_3 + a_{32}\,x_3$ (*d*) $-\,a_{23}\,x_3 - a_{34}\,x_4$

M3.11: Consider the following signal flow graphs.

1.

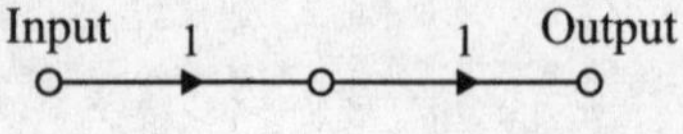

2. 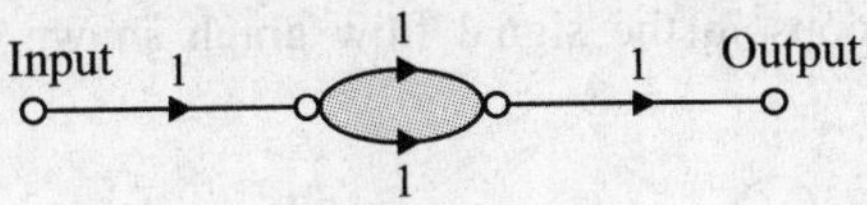

3. 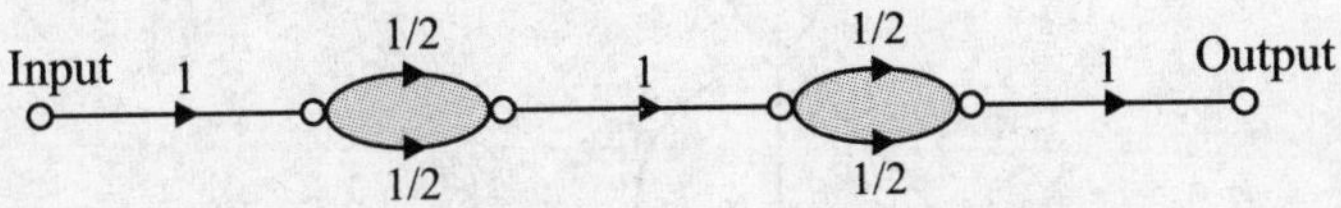

The value of gain is two for

(*a*) 1 (*b*) 2 (*c*) 2 and 3 (*d*) 1 , 2 and 3

M3.12: The signal flow graph of system is shown below.

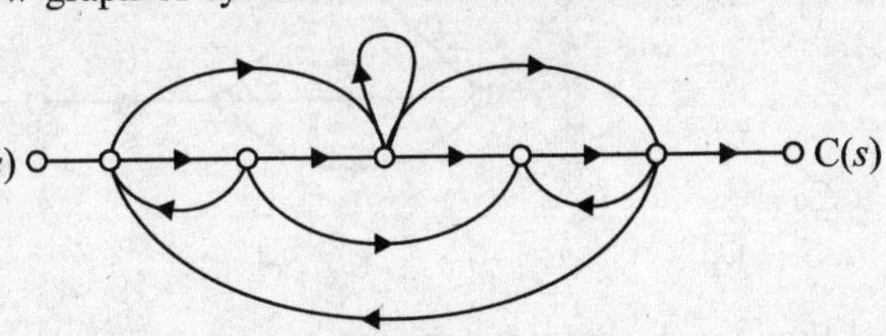

In this graph, the number of three non-touching loops is

(*a*) zero (*b*) 1 (*c*) 2 (*d*) 3.

M 3.13: The sum of gain products of all possible combinations of two non-touching loops in the following signal flow graph is:

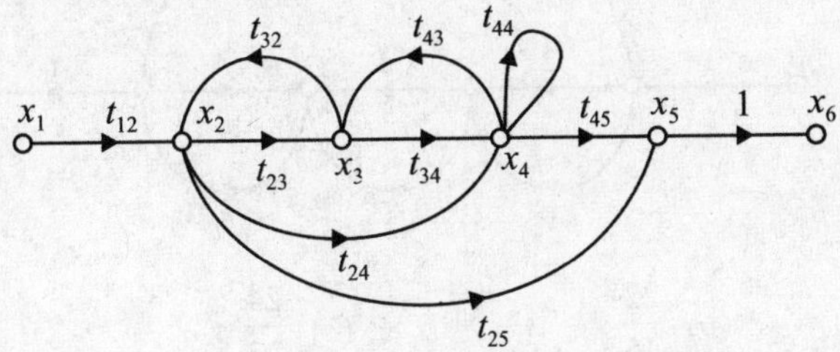

(*a*) $t_{23}\ t_{32}\ t_{44}$ (*b*) $t_{23}\ t_{32} + t_{34}\ t_{43}$

(*c*) $t_{23}\ t_{32} + t_{34}\ t_{43} + t_{44}$ (*d*) $t_{24}\ t_{43}\ t_{32} + t_{44}$

M 3.14: In the signal flow graph shown below, the value of the C/R ratio is

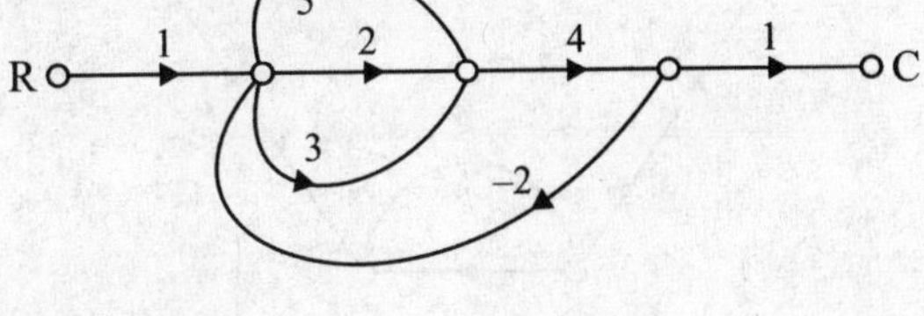

(*a*) $\dfrac{28}{57}$ (*b*) $\dfrac{40}{57}$ (*c*) $\dfrac{40}{81}$ (*d*) $\dfrac{28}{81}$

M 3.15: The forward paths and feedback loops in the signal flow graph shown below are respectively

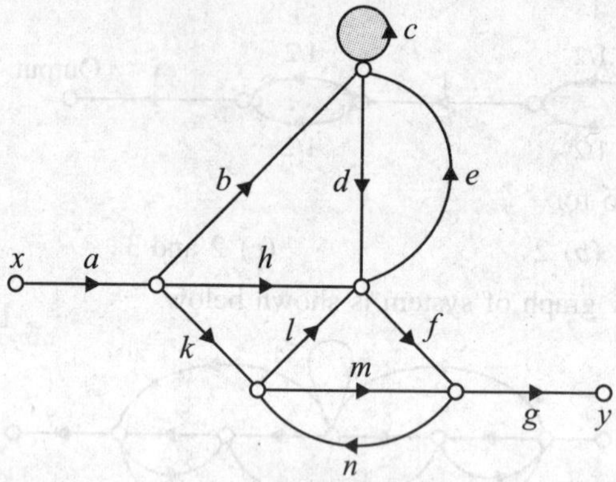

(*a*) 4,4 (*b*) 4,3 (*c*) 3,4 (*d*) 3,3

M 3.16: In the signal flow graph shown below, the gain *c*/*r* will be

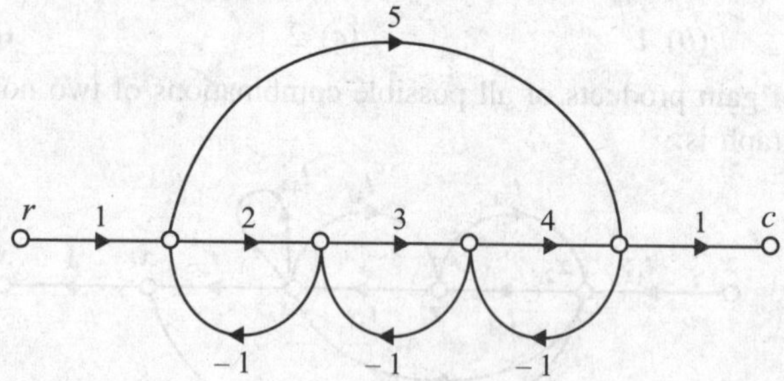

(*a*) $\frac{11}{9}$ (*b*) $\frac{22}{15}$ (*c*) $\frac{24}{23}$ (*d*) $\frac{44}{23}$

M 3.17: For signal flow graph shown below, the transmittance between x_2 and x_1 is

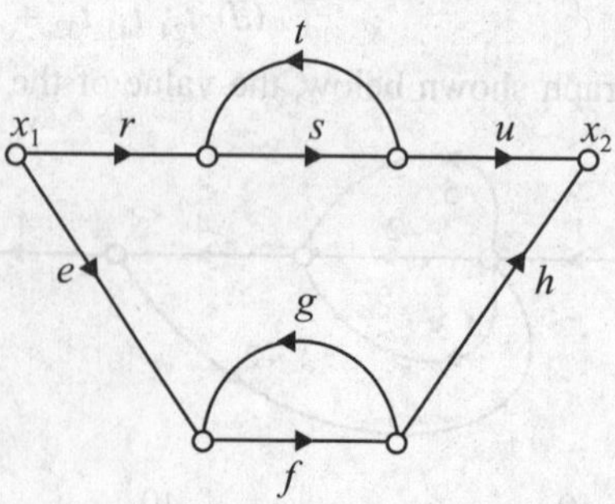

(*a*) $\frac{rsu}{1-st}+\frac{efh}{1-fg}$ (*b*) $\frac{rsu}{1-fg}+\frac{efh}{1-st}$

(*c*) $\frac{efh}{1-ru}+\frac{rsu}{1-eh}$ (*d*) $\frac{rst}{1-eh}+\frac{rsu}{1-st}$

M3.18: Consider the system I and system II shown below. The system I can be reduced to the form as shown in system II with

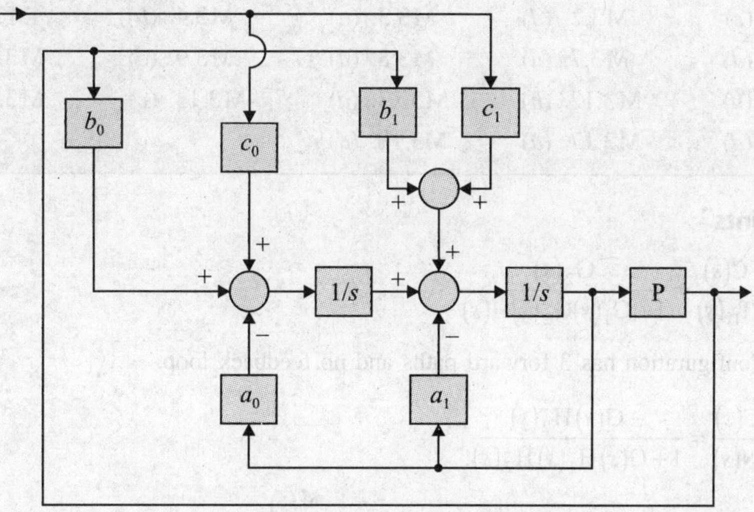

System I

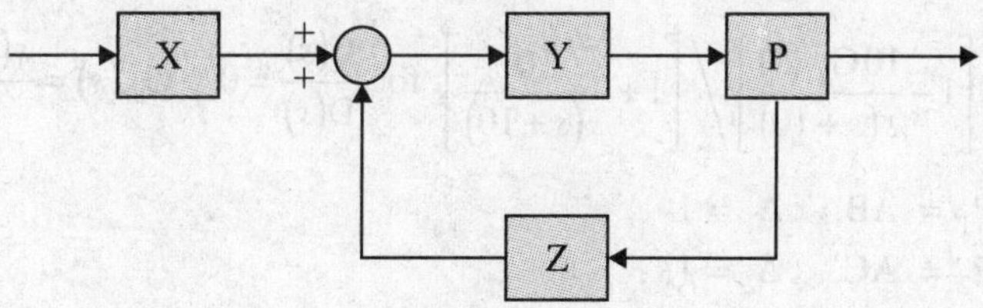

System II

(*a*) $X = c_0s + c_1$, $Y = \dfrac{1}{s^2 + a_0s + a_1}$, $Z = b_0s + b_1$

(*b*) $X = 1$, $Y = \dfrac{c_0s + c_1}{s^2 + a_0s + a_1}$, $Z = b_0s + b_1$

(*c*) $X = c_1s + c_0$, $Y = \dfrac{b_1s + b_0}{s^2 + a_0s + a_1}$, $Z = 1$

(*d*) $X = c_1s + c_0$, $Y = \dfrac{1}{s^2 + a_0s + a_1}$, $Z = b_1s + b_0$

ANSWERS

M3.1. (*c*)	M3.2. (*d*)	M3.3. (*a*)	M3.4. (*b*)	M3.5. (*d*)
M3.6. (*d*)	M3.7. (*a*)	M3.8. (*d*)	M3.9. (*c*)	M3.10. (*b*)
M3.11. (*b*)	M3.12. (*b*)	M3.13. (*a*)	M3.14. (*c*)	M3.15. (*a*)
M3.16. (*d*)	M3.17. (*a*)	M3.18. (*d*)		

Important Hints

M3.1: $$\frac{C(s)}{T_D(s)} = \frac{G_2(s)}{1+G_1(s)G_2(s)H(s)}$$

M3.2: Configuration has 3 forward paths and no feedback loop.

M3.3: $$\frac{C(s)}{N(s)} = \frac{-G(s)H_2(s)}{1+G(s)H_1(s)H_2(s)}$$

Given $G(s)H_1(s)H_2(s) >> 1 \Rightarrow C(s) = -\dfrac{N(s)}{H_1(s)}$

M3.4: $$\frac{C(s)}{D(s)} = \left[1-\frac{10G_d(s)}{s(s+10)}\right]\Big/\left[1+\frac{10}{s(s+10)}\right] \text{ for } \frac{C(s)}{D(s)} = 0\ ,\ G_d(s) = \frac{s(s+10)}{10}$$

M3.6:
$$P_1 = AB, \quad \Delta_1 = 1$$
$$P_2 = AC, \quad \Delta_2 = 1$$
$$\Delta = 1 + AB + AC$$
$$\frac{Q}{R} = \frac{P_1\Delta_1 + P_2\Delta_2}{\Delta}$$

M3.10: Value depends on only incoming paths.

M3.11: Two forward paths of unity gain and no feedback loop.

M3.14:
$$P_1 = 8, \quad \Delta_1 = 1$$
$$P_2 = 20, \quad \Delta_2 = 1$$
$$P_3 = 12, \quad \Delta_3 = 1$$
$$\Delta = 1 - [-16 - 40 - 24] = 81$$
$$\frac{C}{R} = \frac{P_1\Delta_1 + P_2\Delta_2 + P_3\Delta_3}{\Delta}$$

M3.16:
$$P_1 = 24, \quad \Delta_1 = 1$$
$$P_2 = 5, \quad \Delta_2 = (1+3) = 4$$
$$\Delta = 1 - [-4 - 3 - 2 - 5] + 8 = 23$$
$$\frac{c}{r} = \frac{P_1\Delta_1 + P_2\Delta_2}{\Delta} = \frac{44}{23}$$

M3.18:

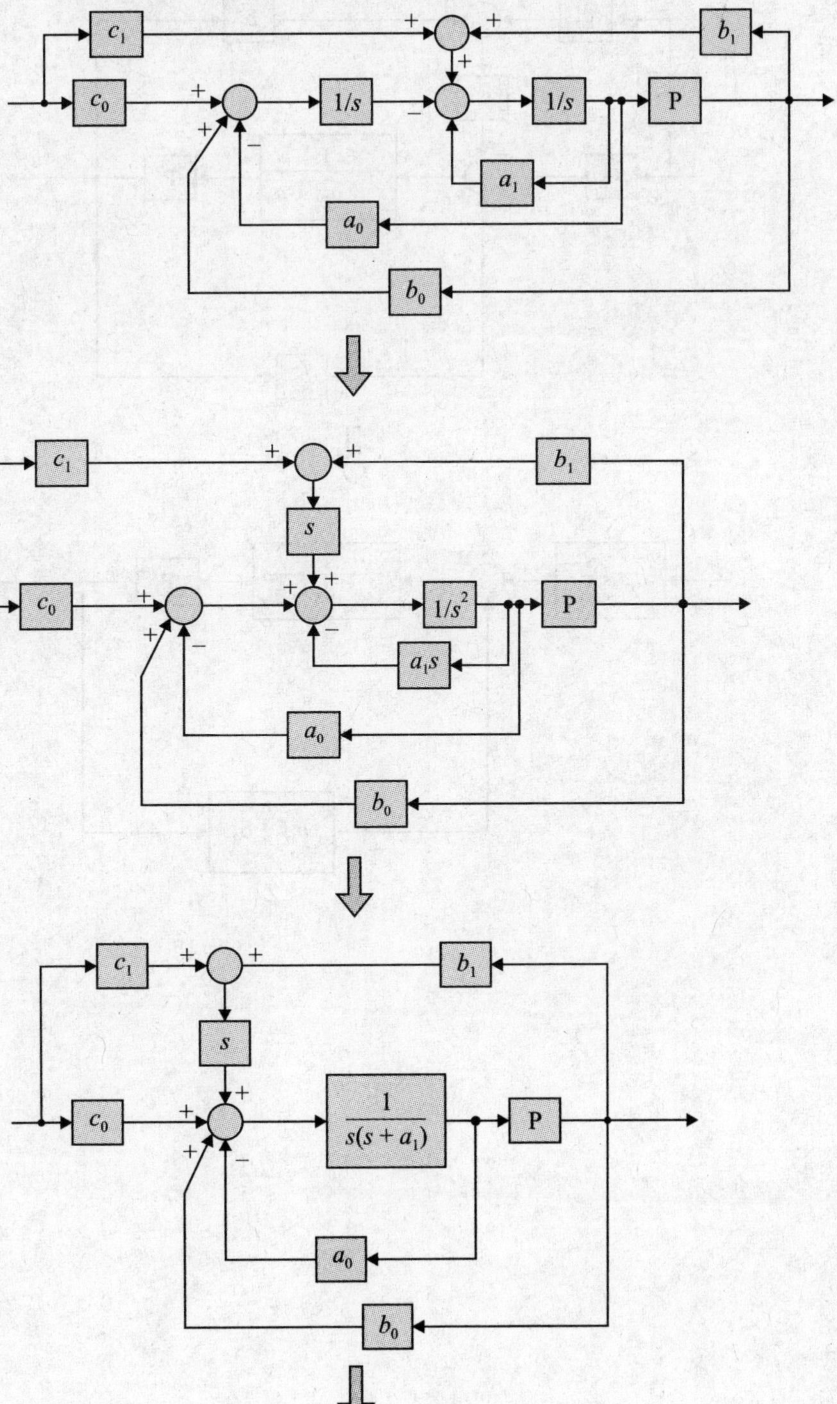

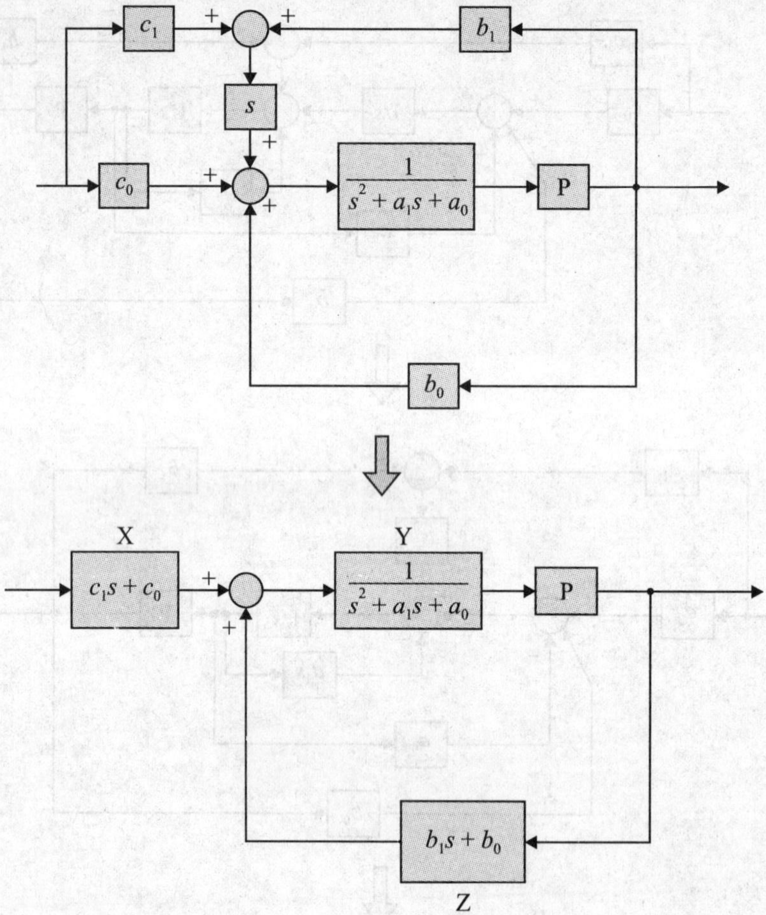
c_1
b_1
s
c_0
$\frac{1}{s^2 + a_1 s + a_0}$
P
b_0
X
Y
$c_1 s + c_0$
$\frac{1}{s^2 + a_1 s + a_0}$
P
$b_1 s + b_0$
Z

4

SYSTEM STABILITY

4.1 INTRODUCTION

The time response of a system has two parts: the transient and the steady state. Every system travels through transients for small amount of time before reaching the steady state. Whether or not transients will die out and the system will reach the finite steady state as desired, is termed **stability analysis**. The stability is an important characterisation of transient part of system behaviour. The transient response is governed by the roots of the characteristic equation. The control system design generally involves a procedure where the roots are so located that system satisfies the prescribed performance specifications. Among various performance specifications, the most important requirement is that system must be stable. An unstable system is, generally useless.

The stability is defined in different ways for different class of systems: linear, non linear, time varying and time invariant. Here we shall discuss only LTI systems for which some common definitions are as follows:

Asymptotic stability

A system is said to be asymptotically stable if for all possible initial conditions, its response decays asymptotically to zero with time. The term asymptotic stability is generally used for autonomous systems that have no external input (s) and are excited by only initial conditions.

Impulse response stability

A system is said to be stable in the sense of impulse response if and only if its response to an impulse input decays asymptotically to zero with time. Recall that the transfer function is the Laplace transform of unit impulse response of initially relaxed system. So, system is stable if and only if all of its characteristic roots (poles) are to the left of the imaginary axis of the complex s plane. For example, a system with transfer function $G(s) = \dfrac{-8s}{(s+3)(s+4)}$ has characteristic roots (poles) at $s = -3$ and $s = -4$, and thus is stable. The impulse response is of form

$$g(t) = a\,e^{-3t} + be^{-4t}$$

which decays with time. *i.e.* $\lim_{t\to\infty} g(t) = 0$.

Now consider a system with transfer function $G(s) = \dfrac{(-s^2+8)}{(s^2+2s+5)(s-4)}$ has poles at $s = -1 \pm j2$ and $s = 4$.

Due to right half plane pole at $s = 4$, the system is unstable. The impulse response is of form

$$g(t) = ae^{-t}[\cos(2t + \theta)] + be^{4t}$$

and term e^{4t} grows with time.

The characteristic roots on imaginary axis, if not repeated, contribute an impulse response that neither grows nor decays with time. For example, the system with characteristic roots at $s = \pm jb$, has impulse response of form

$$g(t) = a\cos(bt + \theta)$$

that exhibits constant amplitude oscillation.

A system is said to be **marginally** stable if it has no right half plane and/or no repeated imaginary axis roots, but there are non repeated imaginary axis roots. If the system has one or more characteristic roots to the right of imaginary axis and/or any repeated imaginary axis roots, its impulse response grows with time and the system is unstable.

Bounded input bounded output (BIBO) stability

A system is said to be BIBO stable if, for every bounded input, its output is bounded.

Let us consider a system whose impulse response is $g(t)$. Let $r(t)$ be input to it and $y(t)$ be its output. The convolution relating $r(t)$, $y(t)$ and $g(t)$ is

$$y(t) = g(t) * r(t) = \int_0^\infty r(t-\tau)\, g(\tau)\, d\tau \qquad \text{...(4.1)}$$

and

$$|y(t)| = \left|\int_0^\infty r(t-\tau)\, g(\tau)\, d\tau\right|$$

Since the absolute value of integral is not greater than the integral of absolute value of integrand

$$|y(t)| \le \int_0^\infty |r(t-\tau)||g(\tau)|d\tau$$

Thus for bounded input $r(t)$, the requirement for stability that $y(t)$ must be bounded, is satisfied if impulse response $g(t)$ is absolutely integrable *i.e.*

$$\int_0^\infty |g(\tau)|\, d\tau \text{ is finite} \qquad \text{...(4.2)}$$

or area under $|g(\tau)|$ vs τ curve is finite.

Further, the stability requirement that $\int_0^\infty |g(\tau)|\, d\tau$ must be finite, can also be related to the location of roots of characteristic equation in s-plane as follows:

The Laplace transform of impulse response $g(t)$ is

$$G(s) = \int_0^\infty g(t)\, e^{-st}\, dt$$

or $$|G(s)| = \left| \int_0^\infty g(t)\, e^{-st}\, dt \right| \le \int_0^\infty |g(t)||e^{-st}|dt \qquad ...(4.3)$$

but $$|e^{-st}| = |e^{-\sigma t}|$$

where $$\sigma = Re[s]$$

When s assumes the values coincident with location of poles of G(s), G(s) = ∞, then (4.3) can be written as

$$\infty \le \int_0^\infty |g(t)||e^{-\sigma t}|dt \qquad ...(4.4)$$

If one or more roots of characteristic equation lie in right half of s plane or on $j\omega$ axis, then

$$\sigma \ge 0 \quad \text{and} \quad |e^{-\sigma t}| \le 1$$

and in such a situation (4.4) becomes

$$\infty \le \int_0^\infty |g(t)|\, dt \qquad ...(4.5)$$

and violates the BIBO stability requirement. Thus BIBO stability requires that all the roots of characteristic equation or poles of G(s) must lie in left half of s plane.

Note the following from foregoing analysis:

(*a*) For linear time invariant system, all stability definitions are equivalent; each implies the others and each holds if and only if all of the system's characteristic roots are in left half of complex plane.

(*b*) It is common practice to characterise systems by mutually exclusive terms: **stable, marginally stable** and **unstable.** These terms can be related to the location of roots of characteristic equation as follows:

(*i*) If all the roots of characteristic equation, lie in left half of s-plane or have –ve real part then ***g(t)*** and $\int_0^\infty g(\tau)\, d\tau$ both are finite and system is BIBO stable or simply **stable.**

(*ii*) If characteristic equation has repeated imaginary axis roots and/or right half plane root(s), then ***g(t)*** and $\int_0^\infty g(\tau)\, d\tau$ both are **infinite** and system is **unstable.**

(*iii*) If characteristic equation has one or more non repeated imaginary axis roots but no right half plane roots, then ***g(t)*is finite** but $\int_0^\infty g(\tau)\, d\tau$ **is infinite** and the system is said to be **marginally stable** or **marginally unstable.** A marginally stable system is neither stable nor unstable. Let us discuss some systems with non repeated imaginary axis roots.

A perfect integrater with G(s) = $1/s$ when excited by unit step signal R(s) = $1/s$, yields continuously growing response as

$$C(s) = G(s) \cdot R(s) = 1/s^2$$

and $$c(t) = t$$

However integrater is an useful system. Similarly a system with roots on $j\omega$ axis at s = $\pm j\omega_0$, when excited by sinusoidal input sin $\omega_0 t$, yields infinite response of form t sin $\omega_0 t$. But the same system with roots at $s = \pm j\omega_0$ when excited by an impulse input, produces response of form sin $\omega_0 t$, which is finite but not asymptotic *i.e.*, $\lim_{t\to\infty} c(t) \neq 0$.

Thus depending upon requirement a system having one or more non repeated roots on $j\omega$ axis but no right half plane roots, may be acceptable with finite response or unacceptable with response of growing nature. Such a situation refers to **marginally** or **limitedly** stable systems.

(*c*) In control system analysis, the stability is also sometimes classified as **absolute stability** and **relative stability**. The absolute stability refers to the condition whether or not a system is stable. It is answer in only yes or no to the stability question. Once the system is stable, it is also of major interest to the designer as to how stable the system is. It refers to relative stability. The relative stability is quantitative measure of how fast the transients die out.

A system with all its characteristic roots in the left half plane but with one or more roots only slightly to the left of imaginary axis, has transient response which decays very slowly. The larger the distance from the imaginary axis to the nearest characteristic root of a stable system, the faster the slowest decaying term in transient response, dies out. The distance on s-plane between the nearest characteristic root and the imaginary axis is termed as relative stability of system. This is further discussed in sec 4.5 of this chapter.

(*d*) A system is said to be **conditionally stable** with respect to a parameter, if it is stable only for a finite range of this parameter. Outside this range system becomes unstable. This is further discussed in section 4.6 of this chapter.

4.2 COEFFICIENT TEST FOR STABILITY

In this section we shall discuss that some information about stability of system can be derived by mere inspection of coefficients of characteristic polynomial. A first or second order polynomial has all roots in left half of s-plane if and only if all polynomial coefficients have same algebraic sign (all positive or all negative) and no one zero. For example

$$s^2 + 2s + 2$$

is characteristic polynomial of a stable system as it satisfies the necessary and sufficient condition that all coefficients bear same sign, and no one missing but

$$s^2 + 2s - 2$$

characterises unstable system as all coefficients do not have same sign. Note that the coefficient test (all of same sign and no one missing) is necessary and sufficient both only for a system of order one and two.

For higher order systems, the condition that characteristic polynomial coefficients have same algebraic sign and none of them is missing, is only necessary but not sufficient. If coefficients are not of same sign and/or at least one coefficient is missing, system is guaranteed to be unstable. If the characteristic polynomial has all its coefficients non zero and of same sign, it is quite possible that it may not have all roots in left half of s-plane.

For example, the charactistic polynomial

$$7s^6 + 5s^4 - 2s^3 - 2s^2 + s + 10$$

definitely has one or more right half plane roots, characterising an unstable system as all coefficients do not have same sign and s^5 term is missing, but the characteristic polynomial

$$4s^4 + 3s^3 + 10s^2 + 8s + 1$$

has no missing coefficient and all have same sign. The characteristic polynomial, although, passes the coefficient test (a necessary condition) but it does not provide definite information about stability or root locations. The system may or may not be stable. Further investigation is necessary.

4.3 ROUTH'S STABILITY TEST

A definitive stability test is the Routh's test which is a numerical procedure for determining whether or not a characteristic polynomial has right half plane (RHP) and imaginary axis (IA) roots without actually solving them. Though it does not give specific root locations as factoring does but performing this test is far easier than factoring. Irrespective of whether or not the polynomial passes the coefficient test, performing Routh's test will provide number of RHP, LHP and IA roots without actually giving their specific location in s-plane.

Routh suggested a method of tabulating the coefficients of characteristic polynomial in a particular way. The tabulation of coefficients gives an array called Routh's array. Consider the 6th order polynomial

$$a_0s^6 + a_1s^5 + a_2s^4 + a_3s^3 + a_4s^2 + a_5s + a_6$$

as an example for Routh array construction. Note that the polynomial must be written in descending powers of s. The initial part of array is formed from polynomial. Write descending powers of s, starting with highest power (s^6) in the polynomial, through s^0, in a column to the left. Enter the coefficients of polynomial in the first two rows with first row consisting of first, third, fifth, .,.,.,., coefficients and second row consisting of second, fourth, sixth, .,.,. coefficients. The construction of rest of the array is as follows:

s^6	a_0	a_2	a_4	a_6
s^5	a_1	a_3	a_5	0
s^4	$\dfrac{a_1a_2 - a_0a_3}{a_1} = p$	$\dfrac{a_1a_4 - a_0a_5}{a_1} = q$	$\dfrac{a_1a_6 - a_0 \times 0}{a_1} = a_6$	0
s^3	$\dfrac{pa_3 - a_1q}{p} = r$	$\dfrac{pa_5 - a_1a_6}{p} = t$	$\dfrac{p \times 0 - a_1 \times 0}{p} = 0$	0
s^2	$\dfrac{rq - pt}{r} = u$	$\dfrac{ra_6 - p \times 0}{r} = a_6$	$\dfrac{r \times 0 - p \times 0}{r} = 0$	0
s^1	$\dfrac{ut - ra_6}{u} = v$	$\dfrac{u \times 0 - r \times 0}{u} = 0$	0	0
s^0	$\dfrac{v \times a_6 - u \times 0}{v} = a_6$	0	0	0

Note the following regarding array construction:

(*a*) The complete array of coefficients is triangular.

(*b*) The missing terms in Routh's array are regarded as zero.

(*c*) An entire row may be divided or multiplied by a positive number to ease out subsequent calculation without altering stability conclusion.

Routh's stability criterion states that the number of right half plane (RHP) roots of the polynomial is equal to the number of algebraic sign changes in the left column of array, going from top to bottom. It should be noted that exact values of the terms in the left column need not be known, instead only signs are needed. The necessary and sufficient condition that all the roots of characteristic equation lie in left half plane (LHP) is that all coefficients of characteristic equation be positive and all terms in the left column of array have positive signs. Consider the following examples for comprehensive exposure to this stability criterion.

Example 1: How many roots of following polynomials are in right half of complex plane?

(*a*) $s^4 + 2s^3 + 3s^2 + 4s + 5$

(*b*) $2s^5 + s^4 + 2s^3 + 4s^2 + s + 6$

Solution: (*a*) Let us follow the procedure just presented and construct the Routh's array. The entries in the first two rows are made directly from polynomial and rest of the array is evaluated as follows:

$$\begin{array}{llll}
s^4 & 1 & 3 & 5 \\
s^3 & \cancel{2}\;\; 1 & \cancel{4}\;\; 2 & 0 \\
 & & & \text{(entire row is divided by 2)} \\
s^2 & \dfrac{(1)(3)-(1)(2)}{1} = 1 & \dfrac{(1)(5)-(1)(0)}{1} = 5 & 0 \\
s^1 & \dfrac{(1)(2)-(1)(5)}{1} = -3 & \dfrac{(1)(0)-(1)(0)}{1} = 0 & 0 \\
s^0 & \dfrac{(-3)(5)-(1)(0)}{(-3)} = 5 & \dfrac{(-3)(0)-(1)(0)}{(-3)} = 0 & 0
\end{array}$$

Note that number of left column sign changes scanning from top to bottom, is two, from 1 to – 3 and from – 3 to 5, therefore the given polynomial has two RHP roots.

(*b*) The Routh array is constructed as follows:

$$\begin{array}{llll}
s^5 & 2 & 2 & 1 \\
s^4 & 1 & 4 & 6 \\
s^3 & \dfrac{(1)(2)-(2)(4)}{1} = -6 & \dfrac{(1)(1)-(2)(6)}{1} = -11 & 0 \\
s^2 & \dfrac{(-6)(4)-(1)(-11)}{-6} = \dfrac{13}{6} & \dfrac{(-6)(6)-(1)(0)}{-6} = 6 & 0 \\
s^1 & \dfrac{\left(\frac{13}{6}\right)(-11)-(-6)(6)}{13/6} = \dfrac{73}{13} & \dfrac{\left(\frac{13}{6}\right)(0)-(-6)(0)}{-6} = 0 & 0 \\
s^0 & \dfrac{\left(\frac{73}{13}\right)(6)-\left(\frac{13}{6}\right)(0)}{73/13} = 6 & 0 & 0
\end{array}$$

There are two changes in sign in the left column, from 1 to – 6 and from – 6 to 13/6 while scanning from top to bottom. So the polynomial has two RHP roots.

Note the following array properties which serve as a partial check on correct array construction:

(*a*) The number of non zero row entries is normally reduced by one after every two rows, with just one non zero element in s^1 row and s^0 row.

(*b*) The last coefficient of polynomial, appears periodically as the last non zero entry in every other row i.e. second, fourth, sixth and so on if order of polynomial is odd (see example 1*b*) and first, third, fifth and so on if the order of polynomial is even (see example 1*a*)

4.4 LEFT COLUMN ZERO OF ARRAY

Consider the system with characteristic polynomial

$$3s^4 + 6s^3 + 2s^2 + 4s + 5$$

for which the Routh array is constructed as follows:

s^4	3	2	5
s^3	~~6~~ 3	~~4~~ 2	0 (entire row is divided by 2)
s^2	$\dfrac{(3)(2)-(3)(2)}{3}=0$	$\dfrac{(3)(5)-(3)(0)}{3}=5$	0
s^1	$\dfrac{(0)(2)-(3)(5)}{0}=\infty$		
s^0			

Note that a snag develops in Routh array construction. This situation where a zero appears in left column of array but the entire row does not consist of zeros, is referred to as a **left column zero**. Because of this zero, all the entries in next row become infinite and Routh construction breaks down. The following three methods are used to resolve this situation.

1. Replace the left column zero by a tiny non zero number $\in$ and continue to construct the rest of the array. $\in$ may be positive or negative but it is usually easier to consider it to be positive. Replacing first entry in s^2 row by $\in$, rest of array is evaluated as follows.

s^4	3	2	5
s^3	3	2	0
s^2	$\in$	5	0
s^1	$\dfrac{(\in)(2)-(3)(5)}{\in}=\dfrac{2\in-15}{\in}$	0	0
s^0	5	0	0

Recall that we are interested in only changes in sign in the left column and not in exact values. The left column entries as $\in \to 0$ are shown below.

s^4	3
s^3	3

$$
\begin{array}{ll}
s^2 & \in = 0+ \\
s^1 & \lim\limits_{\in \to 0}\left(2 - \dfrac{15}{\in}\right) = -\infty \\
s^0 & 3
\end{array}
$$

There are two changes in sign in left column, from 0^+ to $-\infty$ and from $-\infty$ to 3 indicating that polynomial has two RHP roots.

2. The left column zero situation can also be resolved by replacing s by z^{-1} in the polynomial, rearranging the new polynomial in z in descending powers of z and constructing the Routh array. For example the characteristic equation

$$3s^4 + 6s^3 + 2s^2 + 4s + 5 = 0$$

on replacing s by $1/z$ becomes

$$\frac{3}{z^4} + \frac{6}{z^3} + \frac{2}{z^2} + \frac{4}{z} + 5 = 0$$

or
$$5z^4 + 4z^3 + 2z^2 + 6z + 3 = 0$$

for which the Routh array is constructed below.

$$
\begin{array}{llll}
z^4 & 5 & 2 & 3 \\
z^3 & \not{4}\ \ 2 & \not{6}\ \ 3 & 0 \quad \text{(entire row is divided by 2)} \\
z^2 & \dfrac{(2)(2)-(5)(3)}{2} = -\dfrac{11}{2} & \dfrac{(2)(3)-(5)(0)}{2} = 3 & 0 \\
z^1 & \dfrac{\left(-\dfrac{11}{2}\right)(3)-(2)(3)}{\left(-\dfrac{11}{2}\right)} = \dfrac{45}{11} & 0 & 0 \\
z^0 & 3 & 0 & 0
\end{array}
$$

There are two changes in sign in the left column, going from top to bottom. So there are 2 RHP roots.

3. An alternative method to resolve the left column zero situation is to introduce additional known roots to the polynomial, increasing its order and changing the coefficients so that left column zero situation does not occur. For example in given polynomial

$$3s^4 + 6s^3 + 2s^2 + 4s + 5$$

introducing known root ($s = -1$), we have a new polynomial

$$(3s^4 + 6s^3 + 2s^2 + 4s + 5)(s + 1),$$

which when arranged in descending powers of s, is as follows.

$$3s^5 + 9s^4 + 8s^3 + 6s^2 + 9s + 5$$

The Routh array for new polynomial is constructed below.

s^5	3	8	9
s^4	9	6	5
s^3	$\frac{(9)(8)-(3)(6)}{9}=6$	$\frac{(9)(9)-(3)(5)}{9}=7.3$	0
s^2	$\frac{(6)(6)-(9)(7.3)}{6}=-4.95$	$\frac{(6)(5)-(9)(0)}{6}=5$	0
s^1	$\frac{(-4.95)(7.3)-(6)(5)}{-4.95}=13.36$	0	0
s^0	5	0	0

Two algebraic sign changes are again seen in the left column when scanned from top to bottom, indicating two RHP roots. Note that all the three methods presented here have shown same result.

4.5 PREMATURE TERMINATION OF ARRAY

A special circumstance termed **premature termination**, occurs when entire row has zero entries. The left column zero methods presented in Section 4.4 may not provide correct results in such a situation. The premature termination usually results when the **original polynomial contains an even polynomial as a factor** and indicates one or more of the following conditions.

(*i*) The divisor polynomial has at least one pair of symmetrical roots on the real axis one in LHP and one in RHP as shown in Fig. 4.1 (*a*).

(*ii*) The divisor polynomial has one or more pairs of complex conjugate roots on the imaginary axis as shown in Fig. 4.1 (*b*)

(*iii*) The divisor polynomial has pairs of complex conjugate roots forming symmetry about the origin of *s* plane as shown in Fig. 4.1 (*c*).

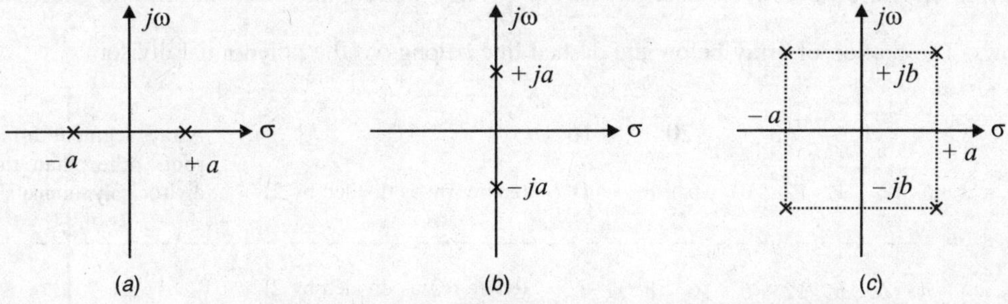

Fig. 4.1: Possible conditions of premature termination of array

It is important to keep a note that roots of divisor polynomial are also roots of original polynomial. The additional symmetry of divisor polynomial roots about imaginary axis eases out the determination of imaginary axis roots. Each RHP root of divisor polynomial must be matched by just one corresponding LHP root. For example, if an eighth order divisor polynomial has three RHP roots, it must have three LHP roots and remaining two on imaginary axis.

Consider the polynomial

$$s^6 + 2s^5 + 8s^4 + 12s^3 + 20s^2 + 16s + 16$$

for which the Routh's array is constructed below.

s^6	1	8	20	16	
s^5	~~2~~ 1	~~12~~ 6	~~16~~ 8	0	(entire row is divided by 2)
s^4	~~2~~ 1	~~12~~ 6	~~16~~ 8	0	(entire row is divided by 2)
s^3	0	0	0	0	(premature termination)
s^2					
s^1					
s^0					

Note the situation of premature termination as entire s^3 row has zero entires. Use the following procedure to circumvent this situation.

(*i*) Form an auxiliary equation A(s) = 0 using the elements of a row just preceding the row containing all zeros.

(*ii*) Find derivative $\frac{d}{ds}\mathrm{A}(s)$

(*iii*) Replace row of all zeros by coefficients of $\frac{d}{ds}\mathrm{A}(s)$ and construct rest of the Routh array.

Using the procedure presented above, we have

$$\mathrm{A}(s) = s^4 + 6s^2 + 8$$

and
$$\frac{d}{ds}\mathrm{A}(s) = 4s^3 + 12s$$

replacing entries of s^3 row by coefficients of $\frac{d}{ds}\mathrm{A}(s)$ the complete Routh array is constructed as follows. The portion of array below the dashed line belongs to the polynomial divisor.

s^6	1	8	20	16		Array segment for testing roots other than those of divisor polynominal
s^5	2′ 1	~~12~~ 6	~~16~~ 8	0	(entire row is divided by 2)	
- - -	- - -	- - -	- - -	- - -	- - -	- - -
s^4	~~2~~ 1	~~12~~ 6	~~16~~ 8		(entire row is divided by 2)	Array segment for testing roots of divisor polynominal
s^3	~~4~~ 1	~~12~~ 3	0		(entire row is divided by 4)	
s^2	3	8				
s^1	1/3	0				
s^0	8	0				

Note that auxiliary polynomial is always even polynomial divisor of the original polynomial. The coefficients of divisor polynomial are those given in the row above the row of zeros. The roots of auxiliary polynomial are also the roots of the original polynomial. The roots of auxiliary polynomial play a significant role in stability investigation. The remaining roots of characteristic polynomial are always in left half and do not play significant role in stability analysis. The roots of polynomial divisor or auxiliary polynomial are tested below the dashed line by looking for any left column sign changes eg. in this case no sign changes in the left column below dashed line and therefore there are no RHP or LHP roots in divisor polynomial. Obviously the fourth order divisor polynomial has four complex conjugate roots on imaginary axis. The location of imaginary axis roots are obtained by solving the auxiliary polynomial whose coefficients are those given in the row above the row of zeros.

Solving the auxiliary/divisor polynomial

$$A(s) = s^4 + 6s^2 + 8 = 0$$

yields

$$s^2 = \frac{-6 \pm \sqrt{36 - 4 \times 8}}{2} = -3 \pm 1 = -2, -4$$

and

$$s = \pm j\sqrt{2}\ ,\ \pm j2$$

Note that these four roots of divisor polynomial are also roots of original polynomial. The remaining two roots of original polynomial are tested using the left column sign changes above the dashed line. There are no sign changes above the dashed line, so the remaining two roots must be in the LHP.

Thus given polynomial has four non repeated roots on imaginary axis and two roots in the left half plane. The system characterised by given polynomial is **marginally stable.**

Example 2: For the following polynomial, how many roots are in LHP, how many are in RHP and how many are on the imaginary axis ?

$$s^5 + s^4 + 6s^3 + 6s^2 + 25s + 25$$

Solution: The Routh array is constructed as

s^5	1	6	25	
s^4	1	6	25	
- - - - - -				
s^3	0	0	0	**(a snag of premature termination develops)**
s^2				
s^1				
s^0				

Noting premature termination at s^3 row, the auxiliary polynomial constructed from entries of s^4 row is

$$A(s) = s^4 + 6s^2 + 25$$

and

$$\frac{d}{ds}A(s) = 4s^3 + 12s$$

Replacing s^3 row by coefficients of $\frac{d}{ds}A(s)$, the rest of the Routh array is constructed as follows:

s^5	1	6	25	
s^4	1	6	25	Array segment for testing roots of divisor polynominal
s^3	4	12		
s^2	3	25		
s^1	– 64/3			
s^0	25			

There are two sign changes in the left column of array, so there are two RHP roots and there must be two LHP roots due to symmetry about imaginary axis. Since the given polynomial has only five roots, there can be no imaginary axis roots as they always appear in conjugate pair. Thus,

$$\text{RHP roots} = 2$$

$$\text{LHP roots} = 3$$

Example 3: Test the polynomial

$$s^6 + s^5 + 5s^4 + s^3 + 2s^2 - 2s - 8$$

For RHP, LHP and imaginary axis roots.

Solution: The polynomial does not satisfy the necessary condition of all coefficients to be of same sign. It indicates the presence of at least one RHP root.

The Routh array is constructed as follows:

s^6	1	5	2	– 8	
s^5	1	1	– 2		
s^4	4	4	– 8		
s^3	0	0			(Premature termination)
s^2					
s^1					
s^0					

All zero entries in s^3 row indicate premature termination

So, $$A(s) = 4s^4 + 4s^2 - 8$$

is auxiliary polynomial or divisor polynomial.

Replacing s^3 row by coefficients of

$$\frac{d}{ds}A(s) = 16s^3 + 8s$$

the array is completed as follows:

$$
\begin{array}{llllll}
s^6 & 1 & 5 & 2 & -8 & \\
s^5 & 1 & 1 & -2 & & \text{Array segment for testing roots other than those of A}(s) \\
\hline
s^4 & 4 & 4 & -8 & & \text{Array segment for testing roots of A}(s) \\
s^3 & 16 & 8 & & & \\
s^2 & 2 & -8 & & & \\
s^1 & 72 & & & & \\
s^0 & -8 & & & &
\end{array}
$$

Below the dashed line, there is one change in sign in the left column. It indicates one RHP root and so only one LHP root. Thus the fourth order A(s) has one RHP, one LHP and two imaginary axis roots. Solving

$$A(s) = 4s^4 + 4s^2 - 8 = 0 \quad \text{or} \quad s^4 + s^2 - 2 = 0$$

yields

$$s^2 = \frac{-1 \pm \sqrt{1+8}}{2} = 1, -2$$

and

$$s = \pm 1, \pm j\sqrt{2}$$

There are no changes in sign in the left column above the dashed line. So, the remaining two roots other than those of A(s) lie in left half of complex plane.

Thus RHP roots = 1

LHP roots = 3

Imaginary axis roots = 2

4.6 RELATIVE STABILITY ANALYSIS

The Routh's stability test, in general, provides information about absolute stability of system *i.e.,* whether or not the system is stable. This is because it only investigates whether or not, all the roots of characteristic polynomial, lie in left half plane. Exact root locations in LHP are not available. In many control situations, the control engineer requires information about relative stability of the system. Normally, the relative stability concept is used only in connection with stable systems. A useful strategy for investigation of relative stability is to shift the imaginary axis of s-plane to the left and apply Routh stability test. This involves the following steps.

(*i*) Shift the imaginary axis σ units to the left as shown below. Substitute $s = z - \sigma$ (σ = constant) in the characteristic polynomial of system to accomplish the axis shift.

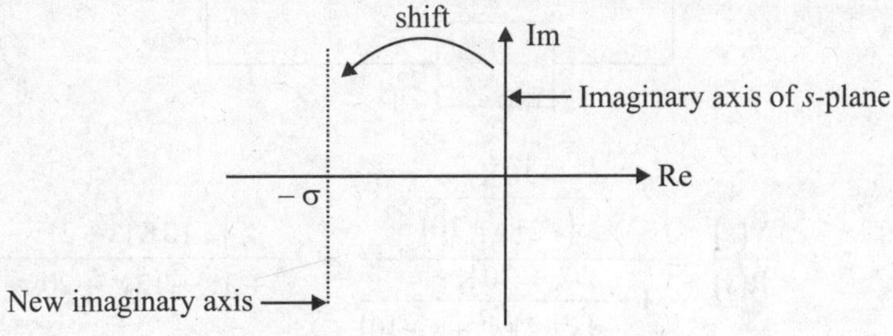

(*ii*) Write the polynomial in z and construct the Routh's array. The number of sign changes in the left column of array, gives the number of roots that are located to the right of vertical line $s = -\sigma$. If there are no sign changes in the left column, it indicates that the original polynomial has all roots to the left of $s = -\sigma$. Then the system characterised by this polynomial is said to have relative stability of at least σ units. The following example demonstrates the implementation of above steps to investigate relative stability.

Example 4: Are all roots of polynomial

$$s^3 + 10s^2 + 30s + 29$$

to the left of $s = -2$ in the complex plane?

Solution: To shift the imaginary axis to the left by 2 unit, substituting $s = (z - 2)$ in given polynomial, a new polynomial in z is as follows:

$$P(z) = (z-2)^3 + 10\,(z-2)^2 + 30\,(z-2) + 29$$

$$= z^3 + 4z^2 + 2z + 1$$

The Routh array is constructed as follows:

z^3	1	2
z^2	4	1
z^1	7/4	0
z^0	1	

There are no sign changes in the left column of array. So all the roots of original polynomial, lie to the left of $s = -2$ in s-plane and the system characterised by given polynomial has relative stability of at least 2 units.

4.7 ROUTH'S STABILITY TEST IN CONTROL SYSTEM ANALYSIS

The Routh's stability test has limited applicability in linear control analysis due to the reason that it does not suggest how to improve relative stability or how to stabilize an unstable system. However, it is possible to investigate the effect of varying one or two system parameters on system stability. Let us consider the following example to investigate for what range or ranges of adjustable parameter, the system is stable.

Example 5: Find range of positive constant K for which the system shown below is stable.

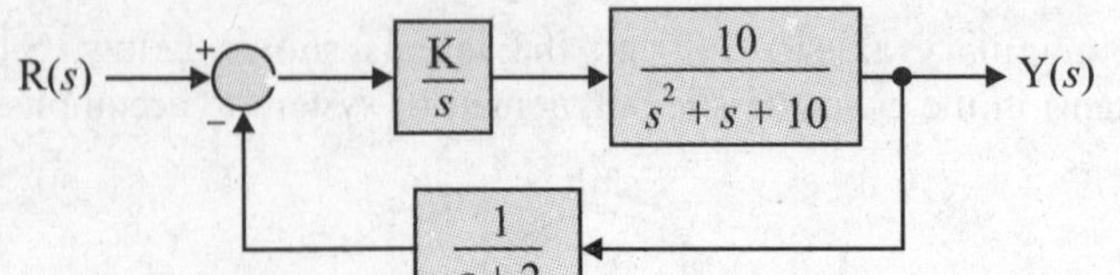

Solution:
$$\frac{Y(s)}{R(s)} = \frac{\dfrac{10K}{s(s^2+s+10)}}{1+\dfrac{10K}{s(s+2)(s^2+s+10)}} = \frac{10K(s+2)}{s^4+3s^3+12s^2+20s+10K}$$

The Routh array for characteristic polynomial

$$P(s) = s^4 + 3s^3 + 12s^2 + 20s + 10K$$

is constructed as follows:

$$\begin{array}{llll} s^4 & 1 & 12 & 10K \\ s^3 & 3 & 20 & \\ s^2 & 16/3 & 10K & \\ s^1 & \dfrac{\frac{320}{3} - 30K}{(16/3)} & 0 & \\ s^0 & 10K & & \end{array}$$

All the left column array entries must be of the same algebraic sign for the polynomial to have no RHP roots. Thus

$$\frac{\frac{320}{3} - 30K}{(16/3)} > 0 \quad \text{and} \quad 10K > 0$$

or

$$0 < K < \frac{32}{9} \quad \text{for system stability.}$$

PROBLEMS AND SOLUTIONS

P 4.1: For the following polynomials, determine how many roots are in RHP, how many are in LHP and how many are on the imaginary axis. Also comment on stability of systems characterised by the polynomials.

(*a*) $2s^4 + 3s^3 + 14s^2 + 2s + 6$

(*b*) $s^5 + 2s^4 + s^3 + 2s^2 + s + 4$

(*c*) $2s^4 + 2s^2 + 3s + 4$

(*d*) $s^6 + 2s^5 + 3s^4 + 4s^3 + 5s^2 + 8s + 1$

(*e*) $s^5 + 3s^4 + 4s^3 + 7s^2 + 4s + 2$

(*f*) $s^5 + s^3 + 2s$

Solution: (*a*) The polynomial

$$2s^4 + 3s^3 + 14s^2 + 2s + 6$$

has a Routh's array that begins as follows:

$$\begin{array}{lllll} s^4 & \not{2}\ 1 & \not{14}\ 7 & \not{6}\ 3 & \text{(entire row is divided by 2)} \\ s^3 & 3 & 2 & 0 & \\ s^2 & 19/3 & 3 & & \\ s^1 & 11/19 & 0 & & \\ s^0 & 3 & 0 & & \end{array}$$

There are no algebraic sign changes in the left column of array going from top to bottom. All the four roots of polynomial are in LHP. The system characterised by the polynomial is **stable**.

(*b*) The Routh array for the polynomial

$$s^5 + 2s^4 + s^3 + 2s^2 + s + 4$$

is constructed as follows:

$$\begin{array}{llll}
s^5 & 1 & 1 & 1 \\
s^4 & \not{2}\ 1 & \not{2}\ 1 & \not{4}\ 2 \\
s^3 & \not{0}\ \in & -1 & \\
s^2 & \dfrac{\in+1}{\in} & 2 & \\
s^1 & \dfrac{-\left(2\in^2+\in+1\right)}{\in+1} & 0 & \\
s^0 & 2 & 0 &
\end{array}$$

(entire row is divided by 2)

(a snag of left column zero develops; zero is replaced by small positive number $\in$ and rest of the array is constructed)

The left column entries as $\in \to 0$, are tabulated as follows:

$$\begin{array}{ll}
s^5 & 1 \\
s^4 & 1 \\
s^3 & \in = 0^+ \\
s^2 & \lim\limits_{\in \to 0}\left(\dfrac{\in+1}{\in}\right) = +\infty \\
s^1 & \lim\limits_{\in \to 0}\left[-\dfrac{\left(2\in^2+\in+1\right)}{(\in+1)}\right] = -1 \\
s^0 & 2
\end{array}$$

There are two changes in sign in the left column going from top to bottom, from + ∞ to – 1 and from – 1 to + 2. So there are 2 roots in RHP and remaining 3 in LHP. The system characterised by the polynomial is **unstable**.

(*c*) The polynomial

$$2s^4 + 2s^2 + 3s + 4$$

has Routh array that begins as follows:

$$\begin{array}{llll}
s^4 & \not{2}\ 1 & \not{2}\ 1 & \not{4}\ 2; \\
s^3 & \in & 3 & 0; \\
s^2 & \dfrac{\in-3}{\in} & 2 &
\end{array}$$

(entire row is divided by 2)

(a snag of left column zero develops; zero is replaced by small positive number $\in$ and rest of the array is constructed)

$$s^1 \quad \frac{3\in -9-2\in^2}{\in -3} \quad 0$$

$$s^0 \quad 2 \quad 0$$

The left column entries as as $\in \to 0$, are tabulated as follows:

$$
\begin{array}{ll}
s^4 & 1 \\
s^3 & \in = 0^+ \\
s^2 & \lim_{\in \to 0}\left(\frac{\in -3}{\in}\right) = -\infty \\
s^1 & \lim_{\in \to 0}\left(\frac{3\in -9-2\in^2}{\in -3}\right) = 3 \\
s^0 & 2
\end{array}
$$

There are two algebraic sign changes in the left column going from top to bottom. So, there are two RHP roots and two LHP roots. The system characterised by the polynomial is **unstable.**

(*d*) The polynomial

$$s^6 + 2s^5 + 3s^4 + 4s^3 + 5s^2 + 8s + 1$$

has Routh array as constructed below:

$$
\begin{array}{lllll}
s^6 & 1 & 3 & 5 & 1 \\
s^5 & \not{2}\ 1 & \not{4}\ 2 & \not{8}\ 4 & 0 \quad \text{(entire row is divided by 2)} \\
s^4 & 1 & 1 & 1 & \\
s^3 & 1 & 3 & 0 & \\
s^2 & -2 & 1 & & \\
s^1 & -\frac{7}{2} & 0 & & \\
s^0 & 1 & 0 & &
\end{array}
$$

There are two algebraic sign changes in the left column, going from top to bottom. So, there are two RHP roots and four LHP roots. The system characterised by the polynomial is **unstable.**

(*e*) The polynomial

$$s^5 + 3s^4 + 4s^3 + 7s^2 + 4s + 2$$

has Routh array that begins as follows:

$$
\begin{array}{llll}
s^5 & 1 & 4 & 4 \\
s^4 & 3 & 7 & 2 \\
s^3 & 5/3 & 10/3 & 0 \\
\hline
s^2 & 1 & 2 & \\
s^1 & 0 & 0 & \quad \textbf{(a snag of premature termination develops)} \\
s^0 & & &
\end{array}
$$

The divisor polynomial whose coefficients are the entries of s^2 row, is

$$A(s) = s^2 + 2$$

and
$$\frac{d}{ds}A(s) = 2s$$

Replacing s^1 row by coefficient of $\frac{d}{ds}A(s)$, the rest of the array is evaluated as

s^5	1	4	4	↑
s^4	3	7	2	Array segment for testing roots other than those of A(*s*)
s^3	5/3	10/3		↓
- - - -	- - - -	- - - -	- - - -	- - - -
s^2	1	2		↑
s^1	2	0		Array segment for testing roots of A(*s*) ↓
s^0	2			

Below the dashed line, there are no changes in the left column going from top to bottom. So, there are no RHP roots. The second order divisor polynomial A(*s*) must have imaginary axis roots whose location obtained by solving A(*s*) = 0, is

$$s = \pm j\sqrt{2}$$

There are no changes in sign above dashed line also. So, there are three LHP roots. The system characterised by the polynomial is **limitedly stable**.

(*f*) The polynomial

$$P(s) = s^5 + s^3 + 2s$$
$$= s(s^4 + s^2 + 2)$$

has one root at origin. To investigate the roots of polynomial

$$s^4 + s^2 + 2,$$

the Routh array is constructed as follows:

s^4	1	1	2	
s^3	0	0	0	(a snag of premature termination develops)
s^2				
s^1				
s^0				

The divisor polynomial is

$$A(s) = s^4 + s^2 + 2$$

Replacing row of zeros by coefficients of the derivative of the divisor polynomial,

$$\frac{d}{ds}\left(s^4 + s^2 + 2\right) = 4s^3 + 2s$$

the completed array is as follows:

s^4	1	1	2
s^3	~~4~~ 2	~~2~~ 1	(entire row is divided by 2)
s^2	1/2	2	
s^1	– 7	0	
s^0	2		

There are two sign changes in the left column, going from top to bottom. So, there are two RHP roots and thus two LHP roots as even divisor polynomial must satisfy the property of symmetry about imaginary axis. The system characterised by the polynomial is **unstable**.

P 4.2: Are the systems shown in Fig P 4.2 (*a*), (*b*) and (*c*) stable or unstable?

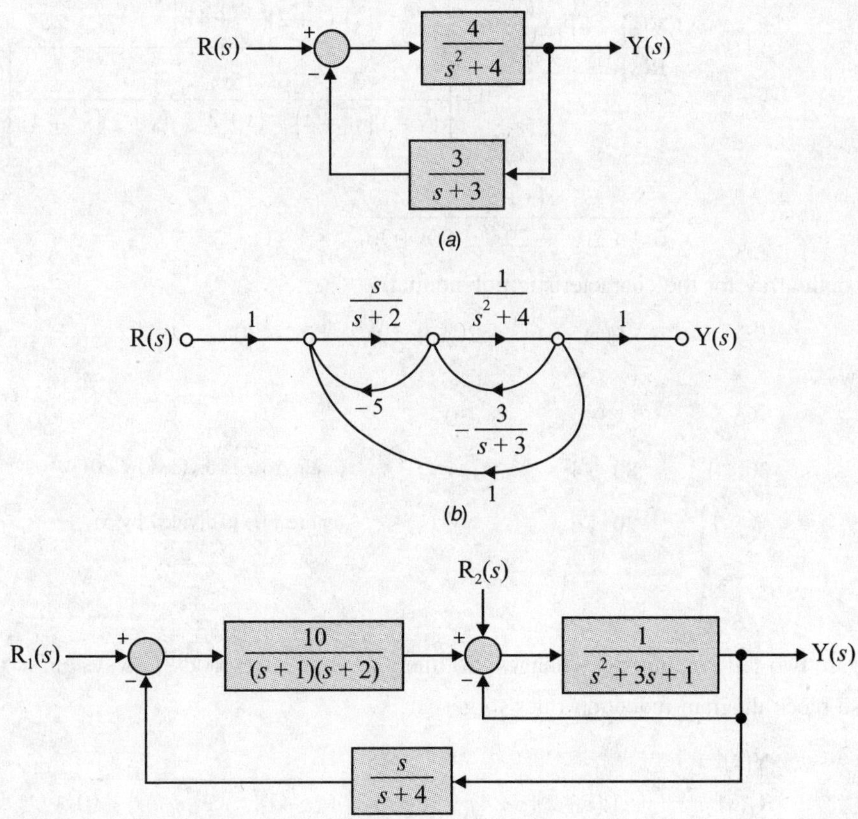

Fig. P 4.2

Solution: (*a*)

$$T(s) = \frac{Y(s)}{R(s)} = \frac{4/(s^2+4)}{1+\dfrac{3}{(s+3)}\cdot\dfrac{4}{(s^2+4)}} = \frac{4(s+3)}{s^3+3s^2+4s+24}$$

The characteristic polynomial

$$P(s) = s^3 + 3s^2 + 4s + 24$$

has Routh array which begins as follows:

$$\begin{array}{llll} s^3 & 1 & 4 & \\ s^2 & \not{3}\ 1 & \not{24}\ 8 & \text{(entire row is divided by 3)} \\ s^1 & -4 & & \\ s^0 & 8 & & \end{array}$$

There are two left column sign changes. So, there are two RHP roots. The system is unstable.

(*b*)
$$T(s) = \frac{Y(s)}{R(s)} = \frac{P_1\Delta_1}{\Delta} = \frac{\left(\dfrac{s}{(s+2)(s^2+4)}\right)(1)}{1-\left[\dfrac{-3}{(s+3)(s^2+4)} - \dfrac{5s}{(s+2)} + \dfrac{s}{(s+2)(s^2+4)}\right]}$$

$$= \frac{s(s+3)}{6s^4+20s^3+29s^2+80s+30}$$

The Routh array for the characteristic polynomial

$$P(s) = 6s^4 + 20s^3 + 29s^2 + 80s + 30$$

is as follows:

$$\begin{array}{lllll} s^4 & 6 & 29 & 30 & \\ s^3 & 20\ 1 & \not{80}\ 4 & & \text{(entire row is divided by 20)} \\ s^2 & \not{5}\ 1 & \not{30}\ 6 & & \text{(entire row is divided by 5)} \\ s^1 & -2 & & & \\ s^0 & 6 & & & \end{array}$$

There are two left column sign changes. So, there are two RHP roots. The system is **unstable**.

(*c*) Use block diagram reduction rules to get

$$\frac{Y(s)}{R_1(s)} = \frac{10(s+4)}{(s+1)(s+2)(s+4)(s^2+3s+1)+(s+1)(s+2)(s+4)+10s}$$

and
$$\frac{Y(s)}{R_2(s)} = \frac{(s+1)(s+2)(s+4)}{(s+1)(s+2)(s+4)(s^2+3s+1)+(s+1)(s+2)(s+4)+10s}$$

The characteristic polynomial

$$P(s) = (s+1)(s+2)(s+4)(s^2+3s+1) + (s+1)(s+2)(s+4) + 10s$$

$$= s^5 + 10s^4 + 37s^3 + 64s^2 + 62s + 16$$

has Routh array that begins as follows:

s^5	1	37	62	
s^4	~~10~~ 5	~~64~~ 32	~~16~~ 8	(entire row is divided by 2)
s^3	30.6	60.4	0	
s^2	22.1	8		
s^1	49.3	0		
s^0	8			

There are no left column sign changes. So, all the characteristic roots lie in left half of s-plane. The system is **stable**.

P 4.3: For what range(s) of the adjustable parameter K, do the following polynomials have all roots in the LHP ?

(*a*) $s^3 + (2 + K)\, s^2 + (8 + K)\, s + 6$

(*b*) $s^4 + (10 + K)\, s^3 + 3s^2 + 9s + 11$

Solution: (*a*) The Routh array in terms of K is as follows:

$$\begin{array}{lll} s^3 & 1 & (8+K) \\ s^2 & (2+K) & 6 \\ s^1 & \dfrac{(2+K)(8+K)-6}{(2+K)} & 0 \\ s^0 & 6 & \end{array}$$

All the left column array entries must be of same algebraic sign for the polynomial to have all roots in the LHP. Thus

$$(2 + K) > 0 \quad \Rightarrow K > -2$$

or $$\frac{(2+K)(8+K)-6}{(2+K)} > 0 \quad \Rightarrow (K + 2)(K + 8) - 6 > 0$$

$$\Rightarrow \quad K^2 + 10\,K + 10 > 0 \quad \Rightarrow (K + 1.12)(K + 8.87) > 0$$

$$\Rightarrow K > -1.12 \quad \text{or} \quad K < -8.87$$

Thus K > – 1.12 or K < – 8.87 will guarantee all the roots in LHP.

(*b*) The Routh array in terms of K is as follows:

$$\begin{array}{llll} s^4 & 1 & 3 & 11 \\ s^3 & 10+K & 9 & 0 \\ s^2 & \dfrac{3K+21}{K+10} & 11 & \\ s^1 & -\dfrac{\left(11K^2+193K+911\right)}{3\,(K+7)} & 0 & \\ s^0 & 11 & & \end{array}$$

All the left column array entries must be of same algebraic sign for the polynomial to have all LHP riooots. Thus

$$K + 10 > 0 \quad \Rightarrow \quad K > -10$$

and
$$-\frac{\left(11K^2 + 193K + 911\right)}{3\left(K+7\right)} > 0 \quad \Rightarrow \quad (11\,K^2 + 193\,K + 911) < 0$$

$$\Rightarrow \qquad (K + 8.77 + j\,2.42)\,(K + 8.77 - j2.42) < 0$$

No value of K satisfies all the above requirements simultaneously. The polynomial has all roots in the LHP for **no real value of K**.

P4.4: Find the range(s) of adjustable parameter K > 0 for which the systems of Fig. P4.4 (*a*) and (*b*) are stable.

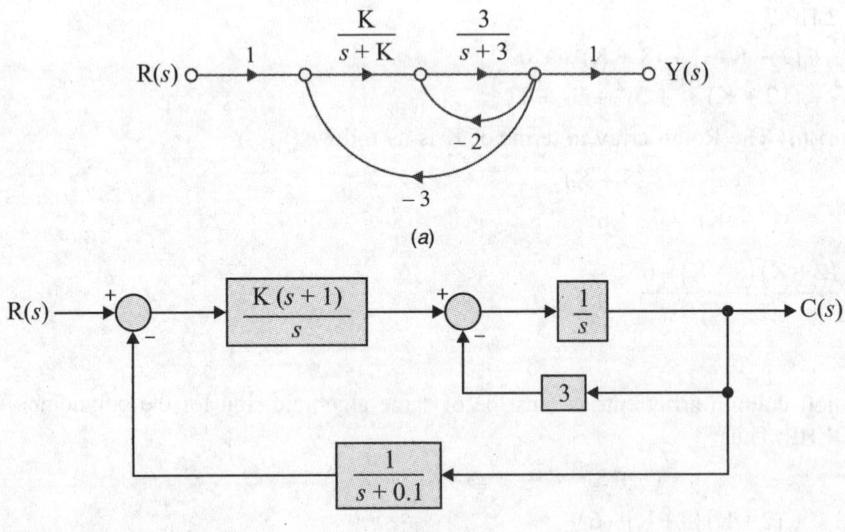

(*a*)

(*b*)

Fig. P4.4

Solution: (*a*)
$$\frac{Y(s)}{R(s)} = \frac{P_1\Delta_1}{\Delta} = \frac{\dfrac{3K}{(s+3)(s+K)}}{1+\left(\dfrac{6}{s+3}\right)+\dfrac{9K}{(s+3)(s+K)}}$$

$$= \frac{3K}{(s+3)(s+K)+6(s+K)+9K}$$

The characteristic polynomial

$$(s + 3)\,(s + K) + 6\,(s + K) + 9K = s^2 + (9 + K)s + 18K$$

has Routh array that begins as follows:

s^2	1	18K
s^1	9 + K	0
s^0	18K	

For the system to be stable, the characteristic polynomial must have all LHP roots for which the left column entries must be of same algebraic sign. Thus

$$9 + K > 0 \quad \Rightarrow \quad K > -9$$

and

$$18\,K > 0 \quad \Rightarrow \quad K > 0$$

so, the system of Fig. P4.4 (*a*) is stable for **all values of K > 0**.

(*b*) Use block diagram reduction rules to get

$$\frac{C(s)}{R(s)} = \frac{K(s+1)(s+0.1)}{s(s+3)(s+0.1)+K(s+1)}$$

and characteristic equation:

$$s^3 + 3.1\, s^2 + (K + 0.3)\, s + K = 0$$

Construct Routh array as follows:

s^3	1	K + 0.3
s^2	3.1	K
s^1	0.68 K + 0.3	0
s^0	K	

For the system to be stable, the characteristic polynomial must have all LHP roots for which all the left column entries must be of same algebraic sign. Thus

$$0.68\,K + 0.3 > 0 \quad \Rightarrow \quad K > -0.44$$

and

$$K > 0$$

So, the system of Fig. 4.4(b) is stable for **all values of K > 0**

P4.5: The open loop transfer function of a unity feedback control system is given by

$$G(s)\,H(s) = \frac{K}{s(s+1)(1+2s)(1+3s)}$$

Determine the value of K

(*i*) for which the system is stable.

(*ii*) which will cause sustained oscillations in the closed-loop system.

Solution: The characteristic equation

$$1 + G(s)\,H(s) = 0$$

or

$$1 + \frac{K}{s(s+1)(1+2s)(1+3s)} = 0$$

or

$$s\,(s + 1)\,(1 + 2s)\,(1 + 3s) + K = 0$$

or

$$6\,s^4 + 11\,s^3 + 6\,s^2 + s + K = 0$$

has Routh array that begins as follows:

$$\begin{array}{llll} s^4 & 6 & 6 & K \\ s^3 & 11 & 1 & 0 \\ s^2 & \dfrac{60}{11} & K & 0 \\ s^1 & \dfrac{60-121K}{60} & 0 & 0 \\ s^0 & K & 0 & 0 \end{array}$$

(*i*) For the system to be stable, the characteristic equation must have all LHP roots for which all the left column entries must be of same algebraic sign. Thus

$$K > 0$$

and $$\frac{60-121K}{60} > 0 \quad \Rightarrow \quad K < \frac{60}{121}$$

So, Range of K for stability is

$$0 < K < \frac{60}{121}$$

(*ii*) $K = \dfrac{60}{121}$ will cause sustained oscillations in system, for this value of K, s^1 row will have all zero entries (a possibility for imaginary axis roots).

The divisor polynomial from s^2 row is

$$A(s) = \frac{60}{11}s^2 + \frac{60}{121} = 0$$

and its solution yields

$$s = \pm j\left(\frac{1}{\sqrt{11}}\right)$$

Thus the system oscillates with frequency $\dfrac{1}{\sqrt{11}}$ rad/sec.

P4.6: A linear feedback control system has an open-loop transfer function

$$A(s) = \frac{\mu(s+\alpha)^2}{\alpha^2(1+s)s^3}$$

where μ and α are adjustable parameters. Find the relation between μ and α so that the system is stable with unity feedback.

Solution: The characteristic equation

$$1 + A(s) = 0$$

or $$1 + \frac{\mu(s+\alpha)^2}{\alpha^2(1+s)s^3} = 0$$

or $$\alpha^2 s^4 + \alpha^2 s^3 + \mu s^2 + 2\alpha\mu s + \mu\alpha^2 = 0$$

has Routh array that begins as follows:

$$\begin{array}{llll} s^4 & \alpha^2 & \mu & \mu\alpha^2 \\ s^3 & \alpha^2 & 2\mu\alpha & 0 \\ s^2 & \mu(1-2\alpha) & \mu\alpha^2 & \\ s^1 & 2\mu\alpha - \dfrac{\alpha^4}{1-2\alpha} & 0 & \\ s^0 & \mu\alpha^2 & & \end{array}$$

For the system to be stable, the characteristic equation must have all LHP roots for which all the left column entries must be of same algebraic sign. Thus

(*i*) $\mu > 0$

(*ii*) $\mu(1-2\alpha) > 0 \Rightarrow 1-2\alpha > 0 \Rightarrow \dfrac{1}{2} > \alpha$

(*iii*) $2\mu\alpha - \dfrac{\alpha^4}{1-2\alpha} > 0 \Rightarrow \mu > \dfrac{\alpha^3}{2(1-2\alpha)}$

P 4.7: The loop transfer function of a feedback control system is given by:

$$G(s)\,H(s) = \frac{K(s+1)}{s(1+sT)(1+2s)},\; K > 0$$

Use Routh Hurwitz criterion to determine the region of K–T plane in which the closed-loop system is stable.

Solution: The characteristic equation

$$1 + G(s)\,H(s) = 0$$

or
$$1 + \frac{K(s+1)}{s(1+sT)(1+2s)} = 0$$

or
$$2T\,s^3 + (2+T)\,s^2 + (1+K)\,s + K = 0$$

has Routh array that begins as follows:

$$\begin{array}{lll} s^3 & 2T & 1+K \\ s^2 & 2+T & K \\ s^1 & \dfrac{(2+T)(1+K)-2KT}{(2+T)} & 0 \\ s^0 & K & \end{array}$$

For the system to be stable, the characteristic equation must have all LHP roots for which all the left column entries must be of same algebraic sign. Thus,

(*i*) $T > 0$

(*ii*) $T + 2 > 0 \Rightarrow T > -2$

(*iii*) $(2 + T)(1 + K) - 2TK > 0$

$\Rightarrow (1 - K)T > -2(1 + K) \quad \Rightarrow (K - 1)T < 2(1 + K)$

$$\Rightarrow T < \frac{2(K+1)}{K-1}$$

(*iv*) $K > 0$

Note from inequality $T < \dfrac{2(K+1)}{K-1}$ that T approaches ∞ as K approaches 1 and T approaches 2 as K approaches ∞. The region for stability is shown in Fig. P4.7.

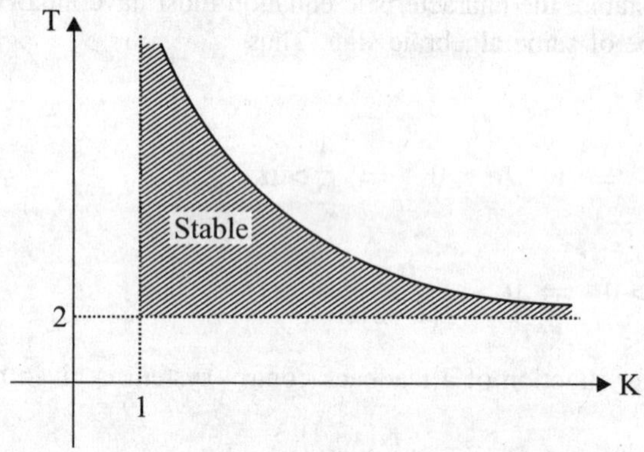

Fig. P4.7: Stable region on K-T plane

P4.8: Consider the closed-loop feedback control system shown in Fig. P4.8

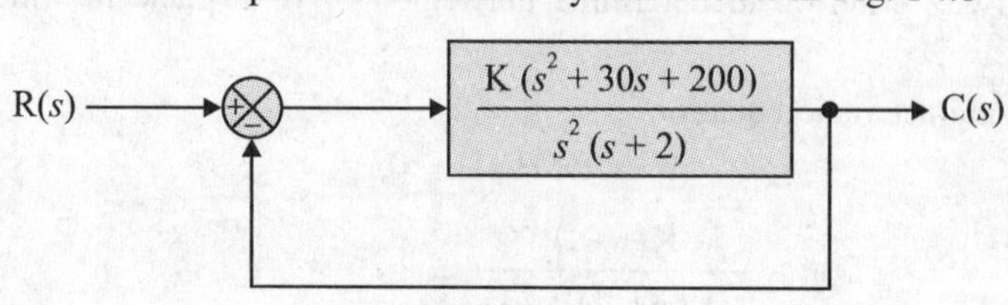

Fig. P4.8

Use Routh-Hurwitz criterion to determine the range of K for which the system is stable. Find also the number of roots of the characteristic equation that are in the right half of *s*-plane for K = 0.5.

Solution: The characteristic equation

$$1 + \frac{K\left(s^2 + 30s + 200\right)}{s^2(s+2)} = 0$$

or

$$s^3 + (2 + K)s^2 + 30Ks + 200K = 0$$

has Routh array that begins as follows:

s^3	1	30K
s^2	2 + K	200K

$$s^1 \qquad \frac{(2+K)\,30K - 200K}{(2+K)} \qquad 0$$

$$s^0 \qquad 200K$$

For the system to be stable, the characteristic equation must have all the LHP roots for which all the left column entries must be of same algebraic sign. Thus,

(*i*) $(2 + K) > 0 \quad \Rightarrow K > -2$

(*ii*) $\dfrac{(2+K)\,30K - 200K}{(2+K)} > 0 \quad \Rightarrow K > \dfrac{14}{3}$

(*iii*) $K > 0$

So, system is stable all values of $K > \dfrac{14}{3}$

For K = 0.5, the characteristic equation

$$s^3 + 2.5\,s^2 + 15\,s + 100 = 0$$

has Routh array that begins as follows:

s^3	1	15
s^2	2.5	100
s^1	– 25	0
s^0	100	

Note two algebraic sign changes in the left column of the array, going from top to bottom. So there are two RHP roots.

P 4.9: For the system shown in Fig. P 4.9, establish relation between '*k*' and '*a*' so that system is stable and show stable region on a plane with '*k*' on *y* axis and '*a*' on *x* axis.

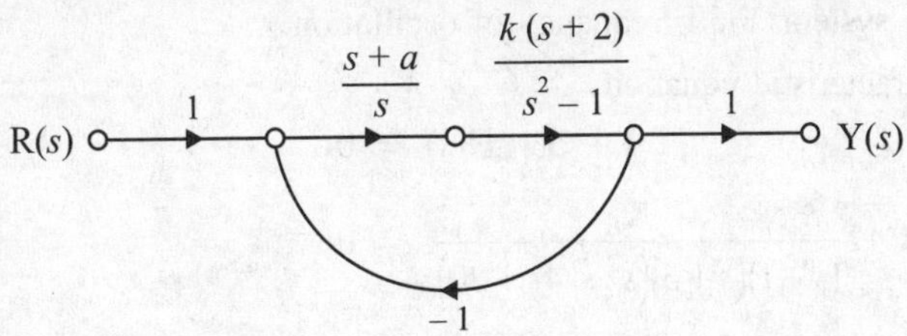

Fig. P 4.9

Solution: From signal flow graph

$$\frac{Y(s)}{R(s)} = \frac{\left(\dfrac{s+a}{s}\right)\dfrac{k(s+2)}{(s^2-1)}}{1+\left(\dfrac{s+a}{s}\right)\dfrac{k(s+2)}{(s^2-1)}} = \frac{k(s+a)(s+2)}{s^3 + ks^2 + (2k + ak - 1)s + 2ak}$$

The Routh array for characteristic polynomial

$$s^3 + ks^2 + (2k + ak - 1)\,s + 2ak$$

is developed below.

s^3	1	$(2k + ak - 1)$
s^2	k	$2ak$
s^1	$k\,(a + 2) - (1 + 2a)$	0
s^0	$2ak$	

For all the left column entries to have same algebraic sign, a requirement for stable system

$$2ak > 0 \quad \Rightarrow \quad ak > 0$$

and

$$k(a + 2) - (1 + 2a) > 0 \quad \Rightarrow \quad k > (1 + 2a)/(a + 2)$$

The region of stability on *a*-*k* plane is shown below.

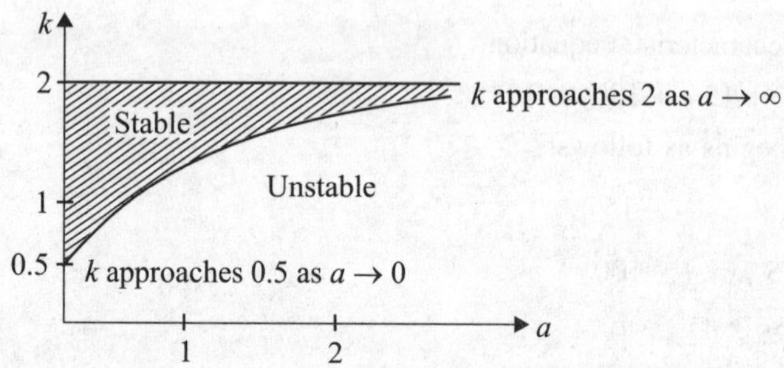

P4.10: A unity feedback system has open loop transfer function

$$G(s)\,H(s) = \frac{K}{(s+1)(s+3)(s^2+4s+20)}$$

Determine stability of closed-loop system as a function of K. Determine value of K that will cause sustained oscillation in system. Find frequency of oscillation.

Solution: The characteristic equation

$$1 + G(s)\,H(s) = 0$$

or

$$1 + \frac{K}{(s+1)(s+3)(s^2+4s+20)} = 0$$

or

$$s^4 + 8s^3 + 39s^2 + 92s + K + 60 = 0$$

has Routh array as follows:

s^4	1	39	K + 60
s^3	8	92	0
s^2	27.5	K + 60	
s^1	75.545 – 0.29 K	0	
s^0	K + 60		

The characteristic polynomial must have all LHP roots for the system to be stable. Thus

$$75.545 - 0.29\,\mathrm{K} > 0 \quad \text{or} \quad \mathrm{K} < 260.5$$

and

$$\mathrm{K} + 60 > 0 \quad \text{or} \quad \mathrm{K} > -60$$

Thus range of K for system stability is

$$-60 < \mathrm{K} < 260.5$$

For K = 260.5, s^1 row will have all zero entries. The auxiliary equation A(s) = 0 from s^2 row is constructed as

$$27.5\,s^2 + \mathrm{K} + 60 = 0$$

or

$$s = \pm j\left(\sqrt{\frac{260.5+60}{27.5}}\right) = \pm j3.41$$

So, system oscillates with frequency 3.41 rad/sec. for K = 260.5

P4.11: Determine the values of K and α, so that system of Fig. P4.11 oscillates at a frequency of 2 rad/sec.

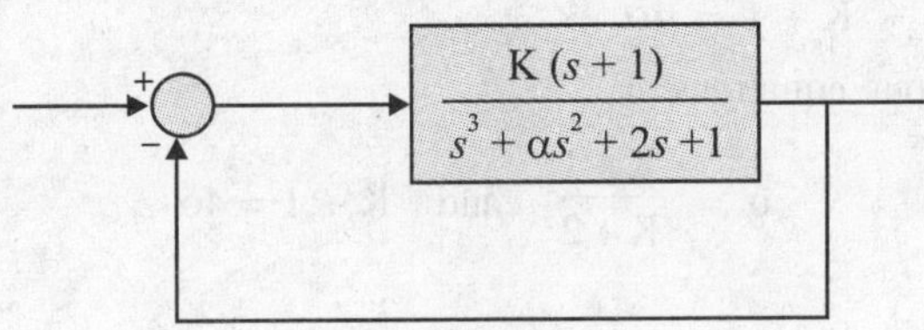

Fig. P4.11

Solution: The closed-loop transfer function is

$$\mathrm{T}(s) = \frac{\mathrm{K}(s+1)/(s^3+\alpha s^2+2s+1)}{1+\mathrm{K}(s+1)/(s^3+\alpha s^2+2s+1)}$$

$$= \frac{\mathrm{K}(s+1)}{s^3+\alpha s^2+(\mathrm{K}+2)s+(1+\mathrm{K})}$$

The Routh array for the characteristic polynomial

$$s^3 + \alpha s^2 + (\mathrm{K}+2)s + (\mathrm{K}+1)$$

is developed below.

$$\begin{array}{lcc} s^3 & 1 & (\mathrm{K}+2) \\ s^2 & \alpha & (\mathrm{K}+1) \\ s^1 & \dfrac{\alpha(\mathrm{K}+2)-(\mathrm{K}+1)}{\alpha} & 0 \\ s^0 & (\mathrm{K}+1) & \end{array}$$

A system with non repeated imaginary axis roots, exhibits sustained oscillation. A polynomial whose Routh array has a row with all zero entries has imaginary axis roots. s^1 row will have all zero entries if

$$\frac{\alpha(K+2)-(K+1)}{\alpha} = 0$$

$$\alpha = \frac{K+1}{K+2}$$

The divisor polynomial constructed from s^2 row

$$A(s) = \alpha s^2 + (K + 1) = 0$$

gives $$s = \pm j\sqrt{\frac{K+1}{\alpha}}$$

where frequency of oscillation

$$\sqrt{\frac{K+1}{\alpha}} = 2$$

gives $$K + 1 = 4\alpha$$

Solving the simultaneous equations

$$\alpha = \frac{K+1}{K+2} \quad \text{and} \quad K + 1 = 4\alpha$$

yields $$K = 2 \quad \text{and} \quad \alpha = \frac{3}{4}$$

P 4.12: A unity feedback system has open-loop transfer function

$$G(s) = \frac{Ke^{-s}}{s(s^2+5s+9)}$$

Determine maximum value of K for closed-loop system to be stable.

Solution: The characteristic equation is

$$1 + G(s)\, H(s) = 0$$

where $$H(s) = 1$$

or $$1 + \frac{Ke^{-s}}{s(s^2+5s+9)} = 0$$

or $$s^3 + 5s^2 + 9s + Ke^{-s} = 0$$

But $$e^{-s} = 1 - s + \frac{s^2}{2!} + \cdots$$

Consider first two terms of series and truncate the rest of series. Then the approximate characteristic equation

$$s^3 + 5s^2 + 9s + K(1 - s) = 0$$

or $$s^3 + 5s^2 + (9 - K)\, s + K = 0$$

has Routh array as follows:

$$\begin{array}{lll} s^3 & 1 & (9-K) \\ s^2 & 5 & K \\ s^1 & 9-\frac{6}{5}K & 0 \\ s^0 & K & 0 \end{array}$$

For system to be stable, all the left column entries must be of same algebraic sign. Thus

$$K > 0$$

and $$9-\frac{6}{5}K > 0 \quad \text{or} \quad K < \frac{45}{6}$$

or K < 7.5, note that system will exhibit sustained oscillation for K = 7.5 with corresponding frequency of oscillation 1.225 rad/sec. Thus, maximum value of K so that closed-loop stability is preserved, is 7.5 – ε (ε is small positive constant).

P 4.13: Show that the system with closed-loop transfer function

$$T(s) = \frac{20}{(s+2)^2(s^2+5s+12)}$$

has relative stability of at least 2 units.

Solution: The characteristic equation is

$$(s+2)^2 \ (s^2+5s+12) = 0$$

or $$s^4+9s^3+36s^2+68s+48 = 0$$

Shifting the imaginary axis 2 units to the left by substituting $(p-2)$ for each s in the original polynomial above, a new polynomial in p results as

$$(p-2)^4+9\,(p-2)^3+36(p-2)^2+68\,(p-2)+48$$

or $$p^4+p^3+6p^2$$

or $$p^2\,(p^2+p+6)$$

whose 4 roots are $p = 0, 0, -0.5 \pm j\,2.398$. The original polynomial has two roots to the left of $s = -2$ and remaining two roots on imaginary axis shifted two units to the left. The system thus has relative stability of at least 2 units.

P 4.14: Determine whether the largest time constant of characteristic equation given below is greater than, less than or equal to 1 sec.

$$s^3+4s^2+6s+4 = 0$$

Solution: The largest time constant of system corresponds to the root closest to the imaginary axis. Thus testing the largest time constant greater than, less than or equal to 1 sec, is equivalent to testing the roots lying to the right of, to the left of or at $s = -1$.

Shifting the imaginary axis 1 unit to the left by substituting $(p-1)$ for each s in the original characteristic polynomial.

$$s^3 + 4s^2 + 6s + 4,$$

the resulting new polynomial in p

$$(p - 1)^3 + 4\,(p - 1)^2 + 6\,(p - 1) + 4$$

or $$p^3 + p^2 + p + 1$$

has Routh array as follows.

p^3	1	1	
p^2	1	1	
p^1	0	0	**(Premature termination)**
p^0			

The divisor polynomial

$$A(p) = p^2 + 1 \text{ has roots } p = \pm\, j1$$

$$\frac{d}{dp}\,A(p) = 2p$$

Replacing p^1 row by coefficient of $\frac{d}{dp}$ A(p) the Routh array is reconstructed as follows:

p^3	1	1
- - - - -	- - -	- - -
p^2	1	1
p^1	2	0
p^0	1	0

Below the dashed line, there are no sign changes in left column. So, there are no roots to the right of imaginary axis shifted by 1 unit *i.e.* $s = -1$. Both the roots must lie on imaginary axis shifted by 1 unit and one to the left of imaginary axis shifted by 1 unit.

Since the root closest to original imaginary axis is 1 unit far to the left, the largest time constant is 1 sec. The roots $p = \pm\, j1$ are actually located at $s = -1 \pm j1$.

DRILL PROBLEMS

D 4.1: For each of following polynomials, how many roots are in the LHP, how many are in the RHP, and how many are on the imaginary axis ?

(*a*) $s^4 + 3s^2 + 4$

(*b*) $s^5 + 2s^4 + 3s^3 + 6s^2 + 2s + 4$

(*c*) $s^6 + 4s^5 + 3s^4 - 16s^2 - 64s - 48$

Ans. (*a*) 2 RHP, 2 LHP

(*b*) 1 LHP, 4 IMAGINARY AXIS

(*c*) 1 RHP, 3 LHP, 2 IMAGINARY AXIS

D 4.2: For what range(s), if any, of the adjustable constant K, are all the roots of following polynomials in the left half of the complex plane ?

(*a*) $s^4 + s^3 + 3s^2 + 2s + 4 + K$

(*b*) $s^3 + Ks^2 + 2Ks + K$

Ans. (*a*) $-4 < K < -2$

(*b*) $0.5 < K < \infty$

D 4.3: Find the range(s) of positive constant K, if any, for which the systems shown in Fig. D4.3 (*a*) and (*b*), are stable.

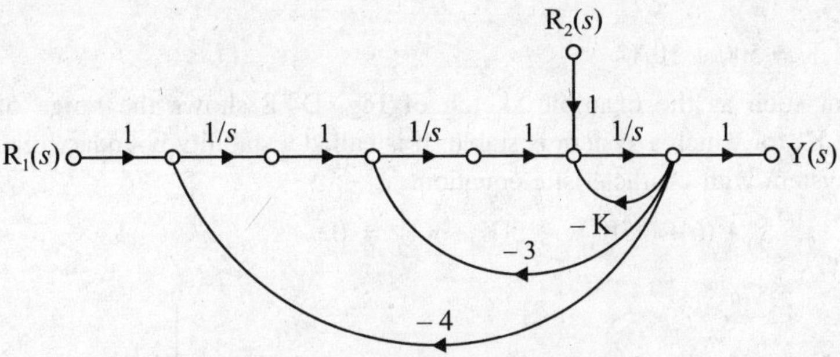

(*a*)

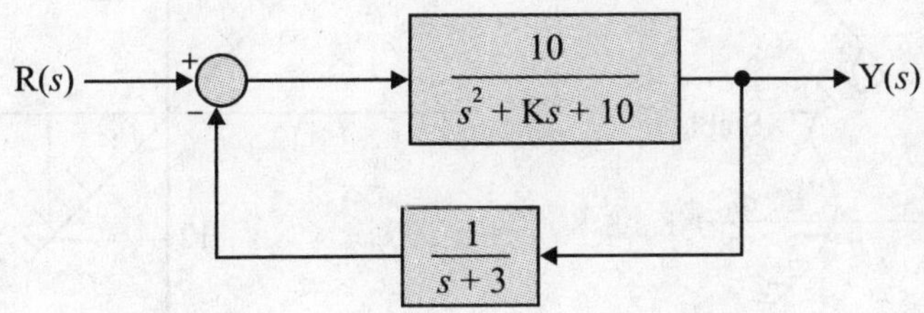

(*b*)

Fig. D4.3

Ans. (*a*) $4/3 < K$

(*b*) $K > 0.49$

D 4.4: Show that the characteristic polynomial

$$s^4 + 14s^3 + 73s^2 + 168s + 144$$

has relative stability of at least 2 units.

D 4.5: The unity feedback system given by

$$G(s) = \frac{4}{s\left(s^2 + as + 2b\right)}; \text{ } a \text{ and } b \text{ are positive constants.}$$

is limitedly stable and oscillates with frequency 4 rad/sec. Find *a* and *b*.

Ans. 0.25, 8

D 4.6: Investigate stability of unity feedback system whose open-loop transfer function is

$$G(s) = \frac{e^{-sT}}{s(s+1)}$$

Ans. Stable for T < 1

D 4.7: The characteristic equation of a system is

$$s^3 + 10s^2 + (50 + A)\, s + K = 0$$

Determine the regions on A-K plane (A on X axis and K on Y axis) in which the closed-loop system is asymptotically stable and unstable. Indicate boundary on which the system is marginally stable.

Ans. K > 0, K = 500 + 10A

D 4.8: A plot such as the example sketch of Fig. D4.8 shows the range of values of two parameters K_1 and K_2 for which a system is stable. It is called a stability boundary diagram. Draw such a diagram for a system with characteristic equation.

$$s^2 + (6 + 0.5K_1)s + 3(K_1 + K_2) = 0$$

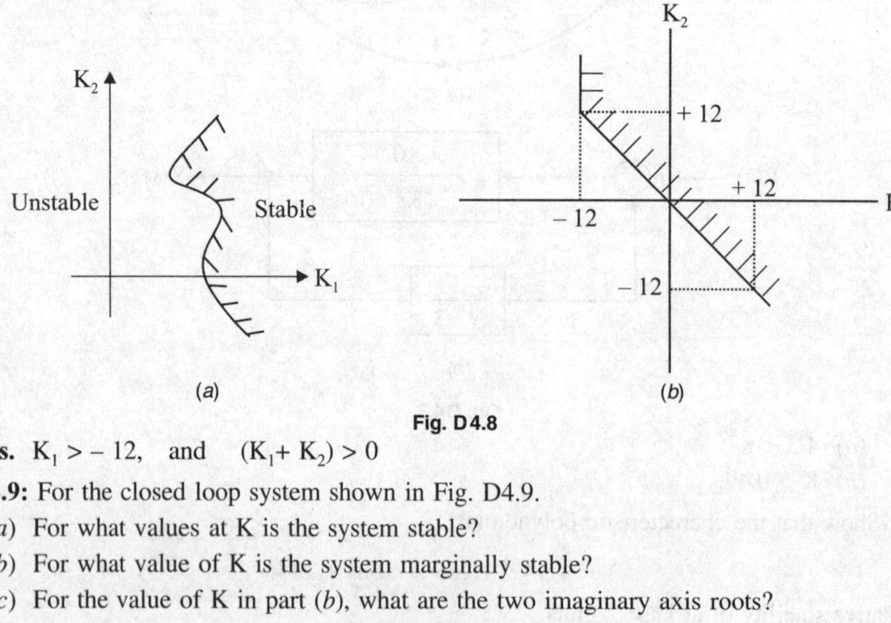

Fig. D4.8

Ans. $K_1 > -12$, and $(K_1 + K_2) > 0$

D 4.9: For the closed loop system shown in Fig. D4.9.

(*a*) For what values at K is the system stable?

(*b*) For what value of K is the system marginally stable?

(*c*) For the value of K in part (*b*), what are the two imaginary axis roots?

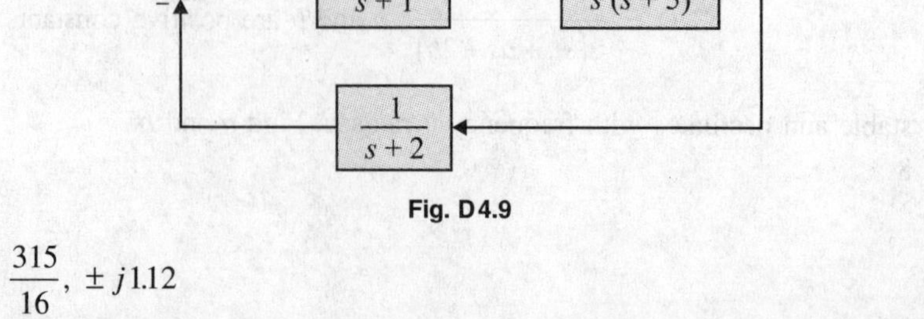

Fig. D4.9

Ans. $K < \frac{315}{16}$, $\pm j1.12$

D 4.10: The closed-loop transfer function of a control system is

$$T(s) = \frac{K}{s^4 + 6s^3 + 30s^2 + 60s + K}$$

What should be the upper limit on K if all the poles of T(*s*) are required to lie to the left of the line σ = – 1?

Ans. 112

MULTIPLE CHOICE QUESTIONS

M 4.1: The open-loop transfer functions with unity feedback are given below for different systems.

1. $G(s) = \frac{2}{s+2}$ 2. $G(s) = \frac{2}{s(s+2)}$ 3. $G(s) = \frac{2}{s^2(s+2)}$ 4. $G(s) = \frac{2(s+1)}{s(s+2)}$

Among these systems, the unstable system is

(*a*) 1 (*b*) 2 (*c*) 3 (*d*) 4

M 4.2: The open-loop transfer function of a control system is given by

$$\frac{K(s+10)}{s(s+2)(s+a)}$$

The smallest possible value of '*a*' for which this system is stable in the closed-loop for positive values of K is

(*a*) 0 (*b*) 8 (*c*) 10 (*d*) 12

M 4.3: The open-loop transfer function of a unity negative feedback control system is given by

$$G(s) = \frac{K(s+2)}{[(s+1)(s-7)]}$$

For K > 6, the stability characteristics of the open-loop configuration and closed-loop configuration of the system are respectively

(*a*) stable and stable (*b*) unstable and stable

(*c*) stable and unstable (*d*) unstable and unstable

M 4.4: For the block diagram shown in the Figure below, the limiting values of K for stability of inner loop is found to be X < K < Y. The overall system will be stable if and only if

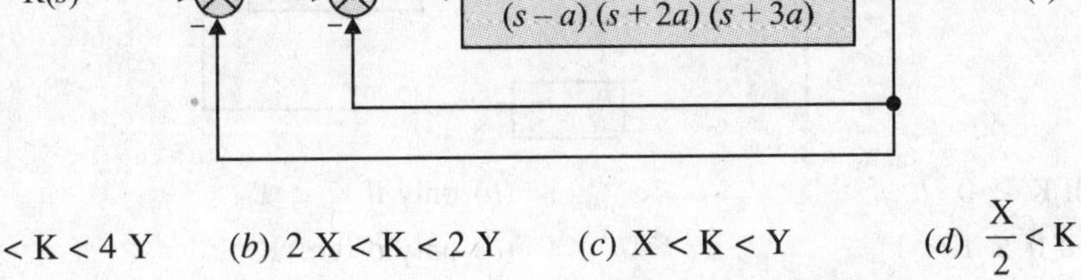

(*a*) 4 X < K < 4 Y (*b*) 2 X < K < 2 Y (*c*) X < K < Y (*d*) $\frac{X}{2} < K < \frac{Y}{2}$

M4.5:. The open-loop transfer function of a unity feedback control system is

$$G(s)\ H(s) = \frac{30}{s(s+1)(s+T)}$$

where T is a variable parameter. The closed-loop system will be stable for all values of

(*a*) T > 0 (*b*) 0 < T < 3 (*c*) T > 5 (*d*) 3 < T < 5

M4.6: While forming Routh's array, the situation of a row of zeros indicates that the system

(*a*) has symmetrically located roots (*b*) is not sensitive to variations in gain

(*c*) is stable (*d*) unstable.

M4.7: None of the poles of a linear control system lie in the right half of *s*-plane. For a bounded input the output of this system

(*a*) is always bounded (*b*) could be unbounded

(*c*) always tends to zero (*d*) none of the above.

M4.8: For what range of K is the following system asymptotically stable ? Assume K ≥ 0

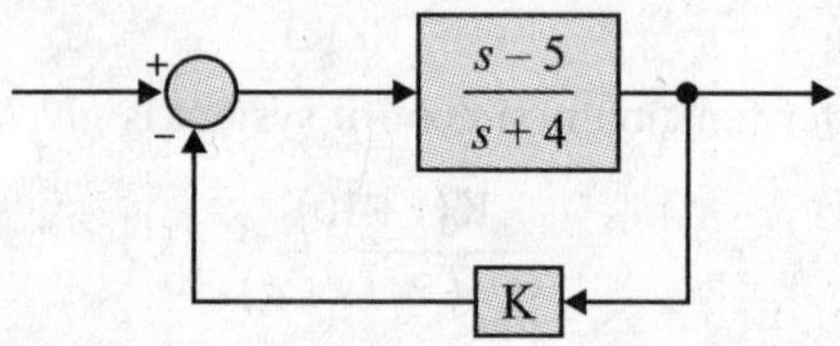

(*a*) 0 < K < 0.8 (*b*) 0 < K < 0.1 (*c*) 0 < K < 8.0 (*d*) 1 < K < 2.0

M4.9: The system represented by the transfer function $G(s) = \dfrac{s^2 + 10s + 24}{s^4 + 6s^3 - 39s^2 + 18s + 84}$ has

(*a*) 2 poles in the right half *s*-plane (*b*) 4 poles in the left half *s*-plane

(*c*) 3 poles in the right half *s*-plane (*d*) 3 poles in the left half *s*-plane.

M 4.10: An electromechanical closed-loop control system has the following characteristic equation

$$s^3 + 6\,K\,s^2 + (K + 2)\,s + 8 = 0$$

where K is the forward gain of the system. The condition for closed-loop stability is

(*a*) K = 0.528 (*b*) K = 2 (*c*) K = 0 (*d*) K = – 2.528

M4.11: The feedback control system shown in Figure below is stable.

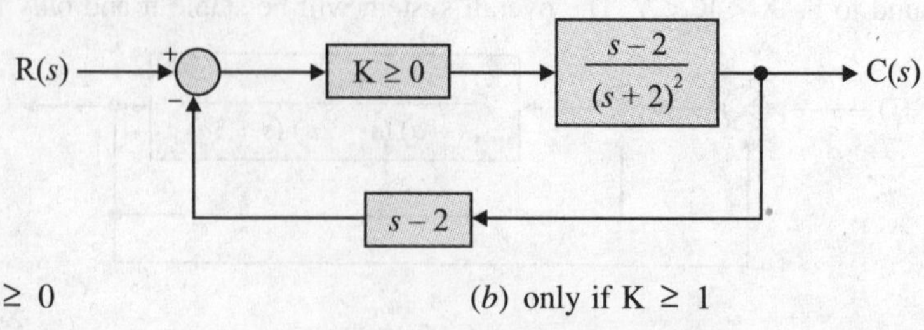

(*a*) for all K ≥ 0 (*b*) only if K ≥ 1

(*c*) only if 0 ≤ K < 1 (*d*) only if 0 ≤ K ≤ 1

M4.12: The first two rows of Routh's array of a fourth-order system are

$$\begin{array}{llll} s^4 & 1 & 10 & 5 \\ s^3 & 2 & 20 & \end{array}$$

The number of roots of the system lying on the right half of s-plane is

(*a*) zero (*b*) 2 (*c*) 3 (*d*) 4

M4.13: The first element of each of the rows of a Routh-Hurwitz stability test showed the sign as follows

Rows	I	II	III	IV	V	VI	VII
Signs	+	−	+	+	+	−	+

The number of roots of the system lying in the right half of s-plane is

(*a*) 2 (*b*) 3 (*c*) 4 (*d*) 5

M4.14: The first two rows of Routh array of a third order system are

$$\begin{array}{lll} s^3 & 2 & 2 \\ s^2 & 4 & 4 \end{array}$$

select the correct answer from the following:

(*a*) system has one root in the RHP.

(*b*) system has two roots on $j\omega$ axis at $s = \pm j$ and third root in the LHP.

(*c*) System has two roots on $j\omega$ axis at $s = \pm j\,2$ and third root in the LHP.

(*d*) System has two roots on $j\omega$ axis at $s = \pm j\,2$ and third root in the RHP.

M4.15: Consider the system of order three with characteristics equation

$$s^3 + Ts^2 + (K + 2)s + (1 + K) = 0$$

The values of K and T such that the system has two roots at $s = \pm j\,2$, are respectively

(*a*) 2, 4/3 (*b*) 2, 3/4 (*c*) 0, 1/2 (*d*) 1/2, 3/4

M4.16: Consider the following statements.

I. A continuous system generates output of form $y = t$ when excited by a step function. This system is unstable.

II. The dynamics of an integrator is given by $dy/dt = u$. The integrator is marginally stable.

Of these statements

(*a*) I and II both are false (*b*) I is true but II is false

(*c*) I is false but II is true (*d*) I and II both are true.

M4.17: Consider the system shown in figure below. Which one of the following statements is true?

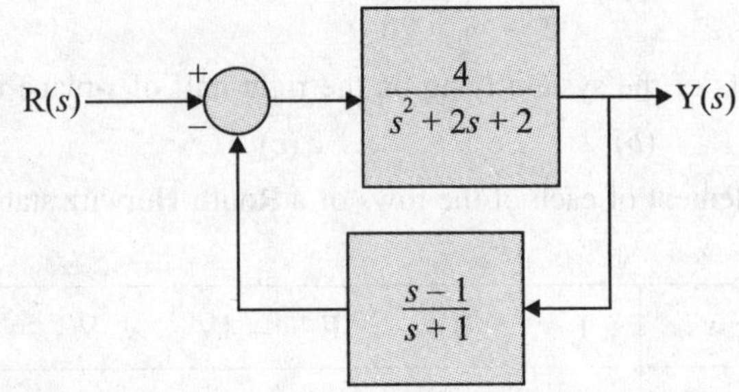

(*a*) Both open loop and closed loop system are stable.

(*b*) Both open loop and closed loop system are unstable.

(*c*) Open loop system is stable but closed loop system is unstable.

(*d*) Open loop system is unstable but closed loop system is stable.

M4.18: A continuous system with time delay, is described by characteristic equation

$$s^2 + s + e^{-sT} = 0$$

Now, consider the following statements.

I. Strictly speaking, this equation has an infinite number of roots.

II. Approximate stability analysis is possible by replacing e^{-sT} in the equation by first two terms of its Taylor series, that is $e^{-sT} = 1 - sT$.

III. T must be less than 1 to preserve the stability.

Of these statements

(*a*) I, II and III all are correct (*b*) only II and III are correct

(*c*) only III is correct (*d*) only I and II are correct.

M4.19: The system shown in figure below is designed to be an oscillator. What is corresponding frequency of oscillation in rad/sec.?

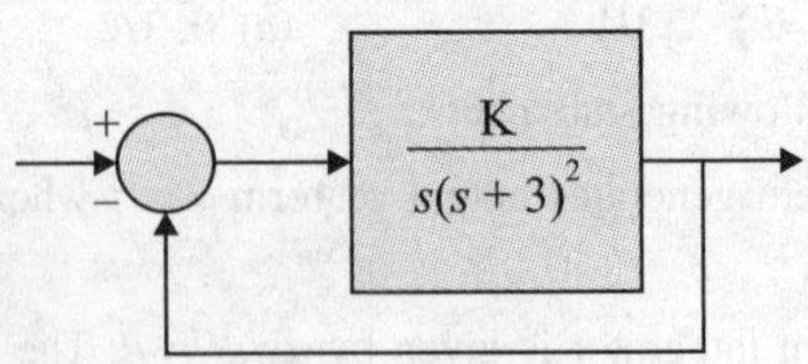

(*a*) 1 (*b*) 2 (*c*) 3 (*d*) 4

M4.20: Consider the following statements in relation with an active network with transfer function model

$$H(s) = \frac{V_o(s)}{V_i(s)} = \frac{K}{\tau^2 s^2 + (3-\alpha)\tau s + 1}$$

where K is gain and τ = RC.

I. The network is unstable for all values of α.

II. The poles of the network function depend on parameter α.

Of these

(*a*) I and II both are false (*b*) I is false but II is true

(*c*) I and II both are true (*d*) I is true but II is false.

M4.21: The design goals for a unity feedback control system having an open loop transfer function

$$G(s) = \frac{K}{s(s+1)(s+2)} \text{ are}$$

1. velocity error coefficient $K_V \geq 10 \text{ sec}^{-1}$.
2. stable open loop operation.

The value of K that should be chosen by designer, is

(*a*) $K < 6$ (*b*) $6 < K < 10$

(*c*) $K > 10$ (*d*) None

M4.22: An open loop transmittance is $G(s) = \frac{K(1-s)}{(1+s)}$. This is unity feedback configuration, will be stable for

(*a*) $|K| > 1$ (*b*) $K > 1$

(*c*) $K < -1$ (*d*) $|K| < 1$.

M4.23: A negative feedback system has loop transfer function $K(s + 3)/(s + 8)^2$ where K is be adjusted that the system exhibits sustained oscillation. The corresponding frequency of oscillation, is

(*a*) $4/\sqrt{3}$ rad/sec. (*b*) 4 rad/sec.

(*c*) $4\sqrt{3}$ rad/sec. (*d*) No such K exists.

M4.24: The system shown in figure below is

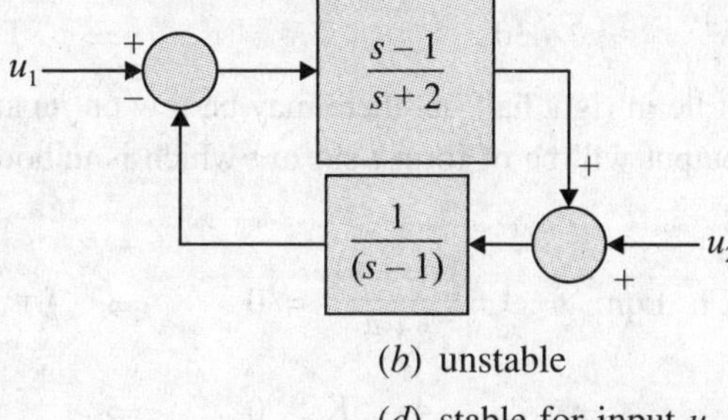

(*a*) stable (*b*) unstable

(*c*) conditionally stable (*d*) stable for input u_1 but unstable for input u_2.

ANSWERS

M4.1. (*c*)	M4.2. (*b*)	M4.3. (*b*)	M4.4. (*d*)	M4.5. (*c*)
M4.6. (*a*)	M4.7. (*b*)	M4.8. (*a*)	M4.9. (*a*)	M4.10. (*b*)
M4.11. (*c*)	M4.12. (*b*)	M4.13. (*c*)	M4.14. (*b*)	M4.15. (*b*)
M4.16. (*d*)	M4.17. (*c*)	M4.18. (*a*)	M4.19. (*c*)	M4.20. (*b*)
M4.21. (*d*)	M4.22. (*d*)	M4.23. (*b*)	M4.24. (*a*)	

Important Hints

M4.1: Ch. Eqn.
1. $s + 4 = 0$ (stable)
2. $s^2 + 2s + 2 = 0$ (stable)
3. $s^3 + 2s^2 + 2 = 0$ (unstable)
4. $s^2 + 4s + 2 = 0$ (stable)

M4.3: Ch. Eqn.: $s^2 - 6s - 7 + Ks + 2K = 0$

s^2	1	2 K − 7		
s^1	K − 6	0	⇒	K > 6
s^0	2 K − 7	0	⇒	K > 3.5

M4.4: Every loop added will add K to ch. polynomial.

M4.5: Ch. Eqn.: $s^3 + (1 + T)\, s^2 + Ts + 30 = 0$

s^3	1	T		
s^2	(1 + T)	30	⇒	T > − 1
s^1	$\dfrac{(1+T)T-30}{1+T}$	0	⇒	$T^2 + T - 30 > 0$
			⇒	(T + 6) (T − 5) > 0
s^0	30	0	⇒	T > 5 , T > − 6.

M4.7: Poles do not lie in right half but there may be few on $j\omega$ axis e.g. ; at $s = \pm j\omega_0$. If input is $\sin \omega_0 t$, output will be of form $t \sin \omega_0 t$ which is unbounded.

M4.8: $0 < K < 0.8$

$$\text{Ch. Eqn:} \quad 1 + \frac{K(s-5)}{s+4} = 0 \quad \Rightarrow \quad s + \frac{4-5K}{1+K} = 0$$

$$\Rightarrow \quad 4 - 5K > 0 \quad \Rightarrow \quad \frac{4}{5} > K \quad \Rightarrow \quad K < 0.8$$

$$\text{also} \quad 1 + K > 0 \quad \Rightarrow \quad K > -1$$

M 4.9: Poles of G(s) are required to be investigated and not the characteristic roots of system. The Routh array for denominator polynomial of G(s) is as follows:

$$\begin{array}{llll}
s^4 & 1 & -39 & 84 \\
s^3 & \cancel{6}\ \ 1 & \cancel{18}\ \ 3 & 0 \\
s^2 & -\cancel{42}\ \ -2 & \cancel{84}\ \ 4 & 0 \\
s^1 & 5 & 0 & 0 \\
s^0 & 4 & 0 & 0
\end{array}$$

Two changes in sign.

M 4.10:

$$\begin{array}{lll}
s^3 & 1 & K+2 \\
s^2 & 6\,K & 8 \\
s^1 & \dfrac{6K(K+2)-8}{6K} & 0 \\
s^0 & 8 & 0
\end{array}$$

$$\Rightarrow K(K+2) > \frac{4}{3} \Rightarrow K > +0.528$$

or $K > -2.528$

M 4.11: Ch. Eqn.: $1+\dfrac{K(s-2)^2}{(s+2)^2}=0 \Rightarrow (1+K)s^2 + 4(1-K)s + 4K = 0$

$$\begin{array}{lll}
s^2 & 1+K & 4\,K \\
s^1 & 4\,(1-K) & 0 \\
s^0 & 4\,K & 0
\end{array}$$

s^1: $\Rightarrow 1 - K > 0 \Rightarrow 1 > K$

s^0: $\Rightarrow K > 0.$

M 4.12:

$$\begin{array}{llll}
s^4 & 1 & 10 & 5 \\
s^3 & \cancel{2}\ \ 1 & \cancel{20}\ \ 10 & \\
s^2 & \cancel{0}\ \ \varepsilon & 5 & \\
s^1 & \dfrac{10\epsilon-5}{\epsilon} & 0 & \\
s^0 & 5 & 0 &
\end{array}$$

$$\lim_{\epsilon\to 0} \frac{10\epsilon-5}{\epsilon} = -\frac{5}{\epsilon} \Rightarrow \text{Two changes in sign.}$$

M4.15:

$$\begin{array}{lll}
s^3 & 1 & K+2 \\
s^2 & T & K+1 \\
s^1 & \dfrac{T(K+2)-(K+1)}{T} & 0 \\
s^0 & K+1 &
\end{array}$$

For two roots at $s = \pm j2$

$$Ts^2 + (K + 1) = 0 \Rightarrow s^2 = -\frac{K+1}{T} = -4 \Rightarrow K+1 = 4T$$

and $$T(K + 2) - (K + 1) = 0 \Rightarrow \frac{K+1}{K+2} = T \Rightarrow \frac{4T}{K+2} = T \Rightarrow K = 2$$

so, $$T = \frac{K+1}{4} = 3/4$$

M4.16: I. Bounded input generates unbounded output. System is unstable.

II. $s = 0$ is characteristic equation of integrator. Integrator is not stable because root does not have negative real part. Also it is not unstable because root does not have positive real part. In fact, integrator is marginally stable.

M4.17: $$\text{OLTF } G(s)\, H(s) = \frac{4(s-1)}{\left(s^2+2s+2\right)(s+1)} = \frac{4(s-1)}{(s+1+j)(s+1-j)(s+1)}$$

is stable. All open loop poles lie in left half of s plane.

$$\text{CLTF } T(s) = \frac{G(s)}{1+G(s)\,H(s)} = \frac{4(s+1)}{s^3+3s^2+8s-2}$$

Since denominator polynomial does not have all coefficients of same sign, this has at least one rhp pole. Closed loop system is unstable.

M4.18: The characteristic equation may be approximated as

$$s^2 + (1 - T)s + 1 = 0$$

To preserve stability $1 - T > 0$ or $T < 1$.

M4.19: For characteristic equation $s^3 + 6s^2 + 9s + K = 0$, Routh array is constructed as

$$\begin{array}{lll} s^3 & 1 & 9 \\ s^2 & 6 & K \quad \Rightarrow 6s^2 + K = 0 \text{ and } s = \pm j3 \text{ for } K = 54 \\ s^1 & \dfrac{54-K}{6} & \quad \Rightarrow \dfrac{54-K}{6} = 0 \Rightarrow K = 54 \\ s^0 & K & \end{array}$$

M4.20: The coefficient test on denominator polynomial, reveals that system will be stable for $\alpha < 3$.

M4.21: $$K_V = \underset{s \to 0}{\text{lt}}\ sG(s) = \frac{K}{2}$$

$$K_V > 10 \text{ requires } \frac{K}{2} > 10 \text{ or } K > 20$$

Characteristic equation is $s^3 + 3s^2 + 2s + K = 0$

s^3	1	2
s^2	3	K
s^1	$\frac{6-K}{3}$	
s^0	K	

Stability requires $\frac{6-K}{3} > 0$ or $K < 6$

The two design requirements cannot be simultaneously met.

M4.22: The characteristic equation: $(1 + s) + K(1 - s) = 0$

or $(1 - K)s + (K + 1) = 0$

Stability requires $1 - K > 0$ or $-K > -1$ or $K < 1$

and $K + 1 > 0$ or $K > -1$

Combine these two conditions to get $-1 < K < 1$ or $|K| < 1$

M4.23: Characteristic equation is $s^2 + (16 + K)s + (64 + 3K) = 0$

For sustained oscillation $16 + K = 0$ or $K = -16$

then

$$s^2 + (64 - 48) = 0 \text{ or } s = \pm j4$$

M4.24: The sfg of system is as shown below.

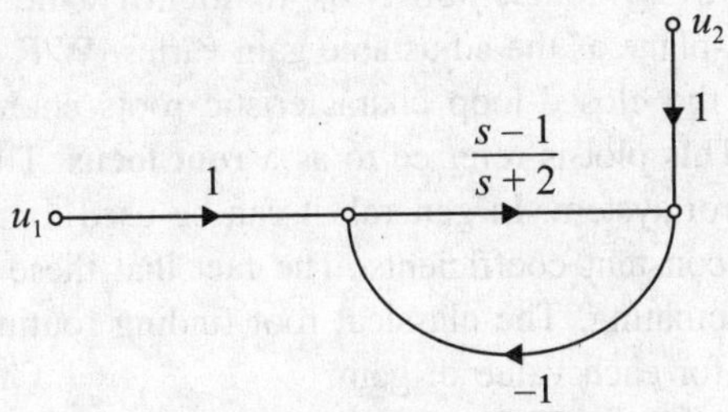

determinant
$$\Delta = 1 - \left[\left(\frac{s-1}{s+2}\right)\cdot\frac{-1}{(s-1)}\right]$$
$$= 1 + \frac{1}{s+2}$$

The characteristic equation describing system dynamics is $1 + \frac{1}{s+2} = 0 \Rightarrow s + 3 = 0$; system is stable as root lies in the lhp. Note that stability of a linear system is independent of input.

5

ROOT LOCUS

5.1 INTRODUCTION

The Routh's stability test provides information about absolute stability of a system. It provides answer in only yes or no to the stability question. Although, determination of relative stability is possible, but it is tedious and requires trial and error procedure. Also a range of values do emerge for a variable gain through this test in which the closed-loop system is stable, but there is no real advice as to which of these gain values are preferable.

In this chapter, a much broader and more useful measure of stability is discussed. A logical approach to determination of system stability is to identify the location of the roots of the characteristic polynomial in s-plane as the adjustable gain varies. **W R Evans** developed a set of rules whereby the path traced by the closed-loop characteristic roots could be sketched to a reasonable accuracy as the gain varies. This plot is referred to as a **root locus**. The root locus is not confined to the study of only linear control system. In general, it can be used to study the behaviour of roots of any algebraic equation with constant coefficients. The fact that these rules do not involve any root finding routine, is indeed fascinating. The classical root finding routines are not convenient because the routine must be repeated for each value of gain.

Gain is usually the variable parameter chosen in the root locus but any other variable of open-loop transfer function may also be used as well. Unless otherwise stated, we shall here after, assume that the gain of open-loop transfer function is the parameter to be varied through all values i.e. from 0 to ∞.

The following example demonstrates the versatility of the root locus. Consider a second order system with unity feedback, whose open-loop transfer function is

$$G(s) = \frac{K}{s(s+2)}$$

and characteristic equation

$$s^2 + 2s + K = 0$$

has roots s_1, s_2 as follows:

$$s_1, s_2 = -1 \pm \sqrt{1-K}$$

As K varies from 0 to ∞, it is easy to observe the following.

(*i*) for K = 0; s_1, s_2 = 0, – 2. Note that root locations for K = 0, coincide with locations of open-loop poles at $s = 0$ and $s = -2$.

(*ii*) for 0 < K < 1, roots are negative, real and distinct. As K varies from 0 to 1, both the roots move towards point (– 1, 0) along the negative real axis from opposite directions.

(*iii*) for K = 1; s_1, s_2 = – 1: the roots are negative, real and equal.

(*iv*) for K > 1; $s_1, s_2 = -1 \pm j\sqrt{K-1}$

The roots are complex conjugate with real part remaining constant equal to – 1. The complete root locus is drawn in Fig. 5.1 using the information furnished just above.

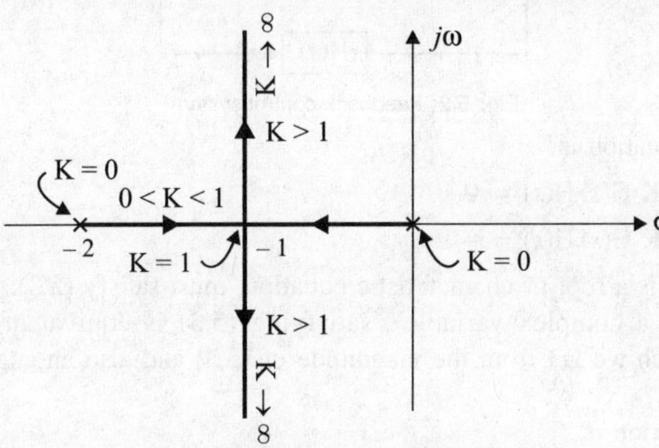

Fig. 5.1: Root locus for characteristic equation $s^2 + 2s + K = 0$

A keen observation of the root locus of Fig. 5.1 reveals the following about system behaviour.

(*a*) The system will exhibit overdamped dynamics for 0 < K < 1 as roots are real and distinct in this range of gain K.

(*b*) For K = 1, system will exhibit critically damped dynamics as both the roots coincide at $s = -1$.

(*c*) For K > 1, system will exhibit underdamped dynamics as the roots become complex conjugate.

(*d*) For K > 1, although roots are complex conjugate, their real part remains constant equal to – 1. The settling time of underdamped dynamics remains constant as it is inversely proportional to only real part of the root.

(*e*) System continues to remain stable for all values of K > 0, as the characteristic equation always has LHP roots. This agrees with the fact that all positive coefficients in a second order system alone sufficiently guarantee system's stability.

From foregoing analysis we conclude that the root locus, an effective graphical procedure for finding characteristic roots, proves quite useful since it indicates the manner in which the characteristic roots should be modified so that response meets prescribed performance specifications. The root locus of Fig. 5.1 for a second order system, is drawn from direct root evaluation which becomes highly tedious for higher order systems in the sense that root finding routine has to be repeated for each value of gain K. An alternative approach is discussed in the following section.

5.2 ROOT LOCUS FOR FEEDBACK SYSTEMS

The root locus for a feedback system is the path traced by the roots of the characteristic polynomial (the poles of closed-loop transfer function) as some system parameter is varied. In general, the control system configuration can be of form as shown in Fig. 5.2, where the closed-loop transfer function T(*s*) is

$$T(s) = \frac{KG(s)}{1+KG(s)\,H(s)} \qquad ...(5.1)$$

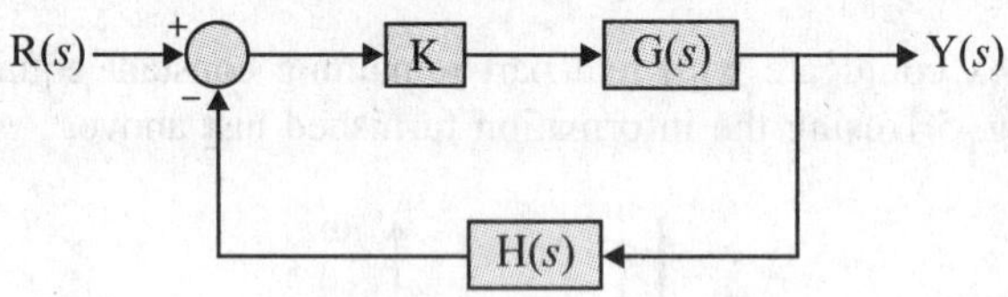

Fig. 5.2: Feedback control system

and the characteristic equation is

$$1 + K\,G(s)\,H(s) = 0 \qquad ...(5.2)$$

or

$$K\,G(s)\,H(s) = -1 \qquad ...(5.3)$$

Any value of *s* that is a root of characteristic equation, must satisfy (5.3). Since KG(*s*) H(*s*) is a complex quantity (*s* is a complex variable), satisfying (5.3) is equivalent to satisfying both a magnitude criterion which we get from the magnitude of (5.3) and also an angle criterion which we get from angle of (5.3).

The magnitude criterion is

$$|K\,G(s)\,H(s)| = 1 \qquad ...(5.4)$$

and the angle criterion is

$$\angle K\,G(s)\,H(s) = \text{odd multiple of } \pm 180° \qquad ...(5.5)$$

$$= \pm\, 180°\,(2q+1)\;;\; q = 0, 1, 2, .., .., ..$$

The angle criterion is more significant than magnitude criterion because some values of *s* that satisfy (5.4) might not satisfy (5.5), while every *s* that satisfies (5.5) can be used to find a value of K. Thus a plot of the points of the complex plane satisfying angle criterion (5.5) is the root locus. The roots of characteristic equation corresponding to a given value of gain K or the value of gain K corresponding to a root (a point on root locus), can be determined from magnitude criterion (5.4). A set of rules (to be discussed later in this chapter) may be found to identify values of *s* satisfying (5.5).

The following example demonstrates that angle criterion will suffice to sketch the root locus.

Consider the feedback control system of Fig. 5.2 where let

$$G(s) = \frac{(s+\beta)}{s(s+\alpha)} \quad \text{and} \quad H(s) = 1;\;\; \beta > \alpha \text{ (both positive)}$$

Apply angle criterion (5.5), to get

$$\angle K\,G(s)\,H(s)\Big|_{s=\sigma+j\omega} = \angle\left[\frac{K(\sigma+j\omega+\beta)}{(\sigma+j\omega)\,(\sigma+j\omega+\alpha)}\right]$$

$$= \pm\, 180°\,(2q+1);\quad q = 0, 1, 2\; ....$$

or $$\tan^{-1}\left(\frac{\omega}{\sigma+\beta}\right) - \tan^{-1}\left(\frac{\omega}{\sigma}\right) - \tan^{-1}\left(\frac{\omega}{\sigma+\alpha}\right) = -\pi \quad \text{(for } q = 0\text{)}$$

or $$\pi + \tan^{-1}\left(\frac{\omega}{\sigma+\beta}\right) = \tan^{-1}\left(\frac{\omega}{\sigma}\right) + \tan^{-1}\left(\frac{\omega}{\sigma+\alpha}\right)$$

Take **tan** of both the sides to get

$$\tan\left[\pi + \tan^{-1}\left(\frac{\omega}{\sigma+\beta}\right)\right] = \tan\left[\tan^{-1}\left(\frac{\omega}{\sigma}\right) + \tan^{-1}\left(\frac{\omega}{\sigma+\alpha}\right)\right]$$

or $$\frac{\tan\pi + \left(\frac{\omega}{\sigma+\beta}\right)}{1 - \tan\pi\left(\frac{\omega}{\sigma+\beta}\right)} = \frac{\frac{\omega}{\sigma} + \frac{\omega}{\sigma+\alpha}}{1 - \left(\frac{\omega}{\sigma}\right)\left(\frac{\omega}{\sigma+\alpha}\right)}$$

or $$\frac{\omega}{\sigma+\beta} = \frac{\omega(2\sigma+\alpha)}{\sigma^2 + \sigma\alpha - \omega^2}$$

or $$\omega\,(\sigma^2 + \omega^2 + 2\beta\sigma + \alpha\beta) = 0$$

either $\omega = 0$ or $\sigma^2 + \omega^2 + 2\beta\sigma = -\alpha\beta$ where

adding β^2 on both sides together with little algebraic manipulation yields

$$\omega = 0 \text{ and } (\sigma + \beta)^2 + \omega^2 = \beta^2 - \alpha\beta \qquad ...(5.6)$$

$\omega = 0$ represents straight line on real axis and $(\sigma + \beta)^2 + \omega^2 = \beta^2 - \alpha\beta$

is equation of a circle of radius $\sqrt{\beta^2 - \alpha\beta}$ with centre at $(-\beta, 0)$. The centre of circle coincides with location of zero of G(*s*) H(*s*). The complete root locus is drawn in Fig. 5.3.

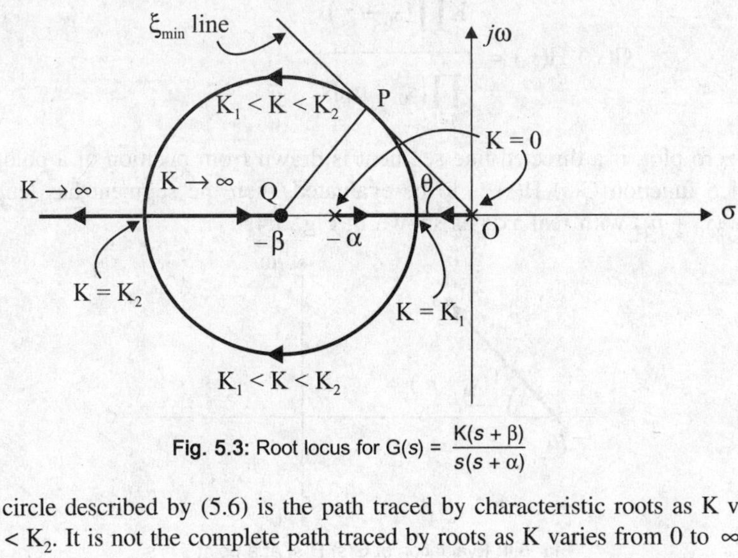

Fig. 5.3: Root locus for $G(s) = \dfrac{K(s+\beta)}{s(s+\alpha)}$

Note that the circle described by (5.6) is the path traced by characteristic roots as K varies in certain range, say $K_1 < K < K_2$. It is not the complete path traced by roots as K varies from 0 to ∞.

For some range of K, the roots trace some segments on real axis also as shown in Fig. 5.3. The construction rules discussed in the subsequent section, will explain how to determine the segments of the real axis lying on root locus.

For range $0 < K < K_1$, and $K > K_2$ the characteristic roots are negative, real and distinct, so the system exhibits overdamped dynamics. For $K = K_1$ and $K = K_2$, the two characteristic roots are negative, real and equal. So, the system exhibits critical damping.

For a range $K_1 < K < K_2$ the complex conjugate roots lie on upper and lower half circles, the system exhibits underdamped dynamics. The location of roots corresponding to minimum damping ratio (ξ_{min}) can be obtained by drawing tangent OP on the circle.

$$OP = \sqrt{OQ^2 - PQ^2} = \sqrt{\beta^2 - (\beta^2 - \alpha\beta)} = \sqrt{\alpha\beta}$$

and minimum damping ratio

$$\xi_{min} = \cos\theta = \frac{OP}{OQ} = \frac{\sqrt{\alpha\beta}}{\beta} = \sqrt{\frac{\alpha}{\beta}}$$

In foregoing analysis, we have used only angle criterion for sketching the root locus. The behaviour of system for all values of K, from 0 to ∞ has also been discussed. The magnitude criterion can be used to determine the value of K for any particular location of roots.

Graphical evaluation of angle and magnitude of G(s) H(s)

Consider the system with

$$G(s)\,H(s) = \frac{K\prod_{i=1}^{m}(s+z_i)}{\prod_{j=1}^{n}(s+p_j)}$$

The system function G(*s*) H(*s*) when evaluated at a specific value of the variable, say $s = s_0$, is

$$G(s_0)\,H(s_0) = \frac{K\prod_{i=1}^{m}(s_0+z_i)}{\prod_{j=1}^{n}(s_0+p_j)}$$

On a pole-zero plot, if a directed line segment is drawn from position of a pole, say $-p_1$, to the value s_0 at which function G(*s*) H(*s*) is to be evaluated, then the segment has length $|s_0 + p_1|$ and makes an angle $(s_0 + p_1)$ with real axis as shown in Fig. 5.4.

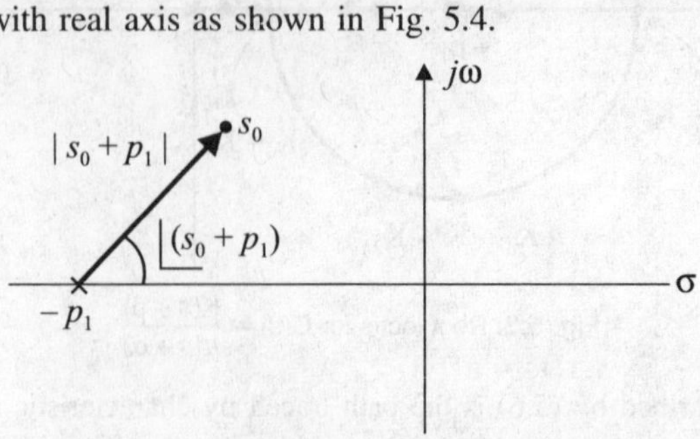

Fig. 5.4: Evaluation of G(s) H(s) at a point $s = s_0$

Thus,

$$|G(s_0)\,H(s_0)| = \frac{|K|\,(\text{Product of lengths of directed line segments from zeros of } G(s)H(s) \text{ to } s_0)}{\text{Product of lengths of directed line segments from poles of } G(s)H(s) \text{ to } s_0}$$

and $\underline{|G(s_0)\,H(s_0)}$ = Σ angles of directed line segments from zeros of G(*s*) H(*s*) to s_0 – Σ angles of directed line segments from poles of G(*s*) H(*s*) to s_0.

For example, for G(*s*) H(*s*) with pole-zero plot of Fig. 5.5, the graphical evaluation at $s_0 = -1 + j3$ gives.

$$|G(s)\,H(s)|_{s=-1+j3} = \frac{(5)}{(3)(3)(5.4)(2.2)} = 0.047$$

and

$$\underline{|G(s)\,H(s)}\Big|_{s=-1+j3} = 143° - 90° - 90° - 68° - 27° = -132°$$

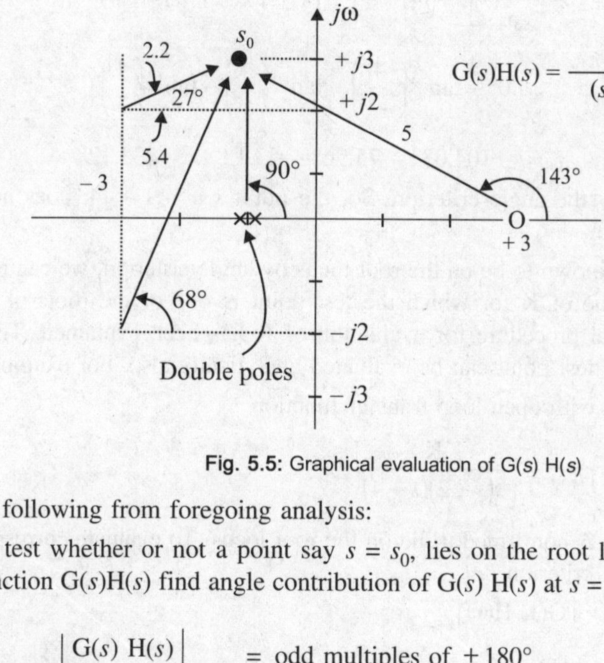

$$G(s)H(s) = \frac{(s-3)}{(s+1)2\,(s+3+j2)\,(s+3-j2)}$$

Fig. 5.5: Graphical evaluation of G(*s*) H(*s*)

Note the following from foregoing analysis:

(*i*) To test whether or not a point say $s = s_0$, lies on the root locus of a system with transfer function G(*s*)H(*s*) find angle contribution of G(*s*) H(*s*) at $s = s_0$. If it satisfies angle criterion *i.e.*

$$\underline{|G(s)\,H(s)}\Big|_{s=s_0} = \text{odd multiples of } \pm 180°$$

then it lies on the root locus otherwise not. Graphical evaluation of angle contribution, has been explained above. However this test can also be performed analytically. For example Consider a system with

$$G(s)\,H(s) = \frac{K}{s(s+4)}$$

To test a point $s = -2 + j5$ for its existence on root locus, we find

$$\underline{|G(s)\,H(s)}\Big|_{s=-2+j5} = \underline{\Bigg|\frac{(K+j0)}{(-2+j5)\,(-2+j5+4)}}$$

$$= 0° - \tan^{-1}\left(\frac{5}{-2}\right) - \tan^{-1}\left(\frac{5}{2}\right)$$

$$= -111.8° - 68.2° = -180°$$

which is odd multiple of $\pm 180°$. The angle criterion is satisfied. So, the point $s = -2 + j5$ lies on root locus of given system.

Consider one more system with

$$G(s)\,H(s) = \frac{K}{s(s+2)(s+4)}$$

To test a point $s = -1 + j4$, whether or not it lies on root locus, we calculate

$$\angle G(s)\,H(s)\Big|_{s=-1+j4} = \angle \frac{(K+j0)}{(-1+j4)\,(-1+j4+2)\,(-1+j4+4)}$$

$$= 0° - \tan^{-1}\left(\frac{4}{-1}\right) - \tan^{-1}\left(\frac{4}{1}\right) - \tan^{-1}\left(\frac{4}{3}\right)$$

$$= -104.03° - 75.96° - 53.13° = -233.12°$$

This does not satisfy the angle criterion. So, the point $s = -1 + j4$, does not lie on root locus of given system.

(*ii*) Once a test point is known to lie on the root locus by angle criterion, we can use magnitude criterion to find value of K for which the test point is one of the roots of characteristic equation. A graphical procedure for evaluation of K has been explained. The value of K corresponding to the test point can be evaluated analytically also. For example

Consider the system with open-loop transfer function

$$G(s)\,H(s) = \frac{K}{s(s+2)(s+4)}$$

The point $s = -0.75$ is confirmed to lie on the root locus. To evaluate corresponding value of K, use magnitude criterion *i.e.*

$$|G(s)\,H(s)|_{s=-0.75} = 1$$

or
$$\frac{|K|}{|-0.75|\,|-0.75+2|\,|-0.75+4|} = 1$$

or
$$K = 3.0468$$

Consider one more system with

$$G(s)\,H(s) = \frac{K}{s(s+4)}$$

The point $s = -2 + j5$ is confirmed to lie on the root locus. To evaluate corresponding value of K, use magnitude criterion *i.e.*

$$|G(s)\,H(s)|_{s=-2+j5} = 1$$

$$\text{or} \qquad \frac{|K|}{|-2+j5||-2+j5+4|} = 1$$

$$\text{or} \qquad K \cong 29$$

So, for K $\cong 29$, the point $s = -2 + j5$ is one of the root of characteristic equation of system

with $G(s)\ H(s) = \dfrac{K}{s(s+4)}$

5.3 ROOT LOCUS CONSTRUCTION

The application of angle criterion (5.5) to the system discussed in Secton 5.2 gave equations (5.6) that were easily identified to be straight line and a circle. But it is not always possible for any general system of any order, to identify the shape of path traced by characteristic roots from the mathematical equation produced by applying the angle criterion. Hence the need to develop a set of rules that can easily identify the values of s satisfying angle criterion (5.5) for a given G(s) H(s) product so that tracing these values, results in the root locus.

In this very discussion, it might seem logical to let today's computers do all the work, and skip the sketching rules altogether. Although a computer could easily plot the roots for us, but we want to predict here the rough shape of the root locus so as to be able to double check the computer results and analyse the systems with some independence from software.

The simple rules that allow approximate root locus sketches to be made easily and rapidly, are as follows:

(a) The root locus is symmetrical about real axis

This follows since the characteristic equation can have complex roots only in conjugate pairs.

(b) Loci branches

The branches of the root locus are continuous curve that start at each of n poles of G(s) H(s), for K = 0. As K $\to \infty$, the locus branches approach m zeros of G(s) H(s). If $n > m$, then ($n - m$) locus branches for excess poles, extend to ∞ and if $n < m$, then ($m - n$) locus branches for excess zeros, extend from ∞.

Note that locus branches never extend from a pole to ∞ and then back from ∞ to a zero.

Consider a feedback control system with configuration as shown in Fig. 5.2. The product G(s) H(s) in general pole–zero form can be written as:

$$G(s)\ H(s) = \frac{\prod_{i=1}^{m}(s+z_i)}{\prod_{j=1}^{n}(s+p_j)} \qquad \text{...(5.7)}$$

The application of magnitude criterion (5.4) gives

$$|G(s)\ H(s)| = \frac{\prod_{i=1}^{m}|(s+z_i)|}{\prod_{j=1}^{n}|(s+p_j)|} = \frac{1}{|K|} \qquad \text{...(5.8)}$$

For K = 0, |G(s) H(s)| = ∞ and (5.8) suggests that it is possible only if $s = -p_j$ or s approaches poles of G(s) H(s). The root locus branches, therefore, start at each of n poles of G(s) H(s). similarly, substituting K = ∞ in (5.8), | G(s) H(s) | = 0 and it is possible only if $s = -z_i$ or s approaches zeros of G(s) H(s). The root locus branches, therefore, approach m zeros of G(s) H(s) as K → ∞.

(c) Real axis segments

A Point on the real axis lies on the locus if and only if it is to the left of an odd number of poles plus zeros of G(*s*) H(*s*) on the real axis.

Any point on root locus must satisfy the angle criterion. The angle contribution of each real axis pole or zero is either 0° or 180°, depending upon whether the pole or zero is to the right or to the left of the real axis point under test, as shown in Fig. 5.6 (*a*) and (*b*).

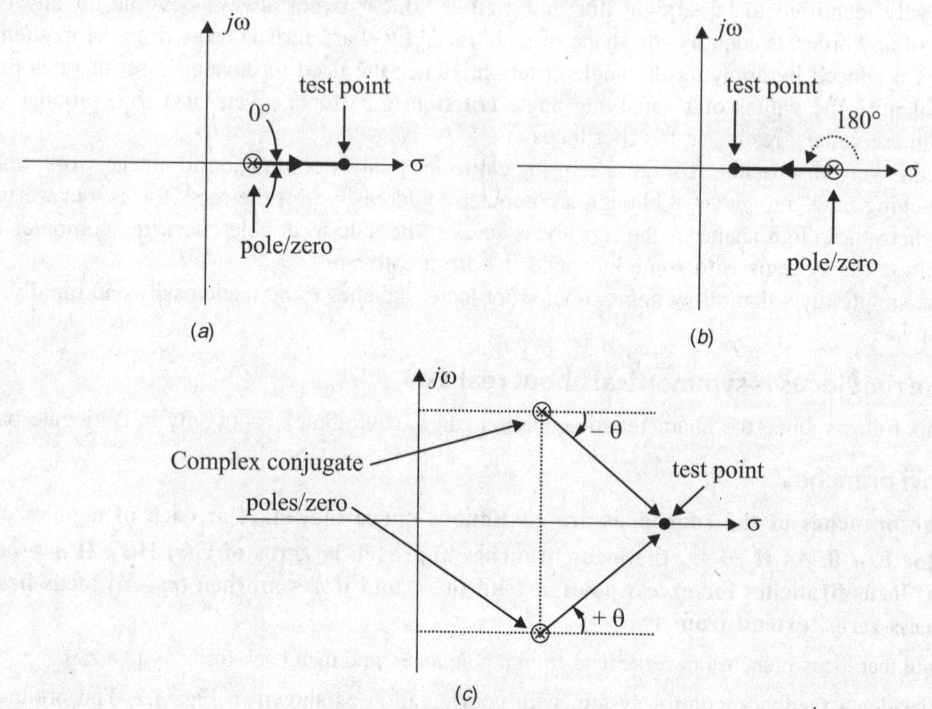

Fig. 5.6: Testing point on real axis

Note that 180° or – 180° are the same angle. A set of complex conjugate poles/zeros, contributes angles to real axis points, that are negatives of one another, so the net angle contribution of a complex set of poles/zeros is zero as shown in Fig. 5.6 (*c*). A point on the real axis is thus on root locus if and only if it is to the left of odd number of poles plus zeros of G(s) H(s), so that the angle of G(s) H(s) at that point is an odd multiple of ±180°.

The real axis root locus segments of several systems are shown by thick lines in Fig. 5.7 (*a*), (*b*), (*c*) and (*d*).

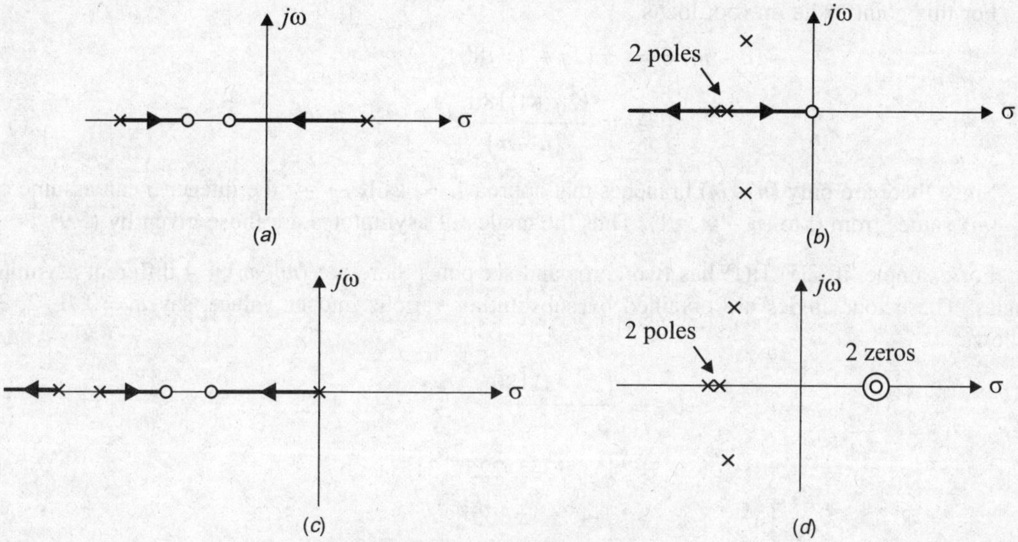

Fig. 5.7: Real axis root locus segments

In Fig. 5.7 (*a*) and (*c*) the loci are entirely along real axis. In Fig. 5.7 (*b*) one segment extends from double pole to zero at $s = 0$ and one extends from double pole to ∞. The two other segments will extend from complex poles, the sketching of which is discussed later. In Fig. 5.7 (*d*), there are no real axis locus segments.

(*d*) Asymptotic angles

If number of finite zeros of G(*s*) H(*s*), *m*, is less than number of finite poles of G(*s*) H(*s*), *n*, then (*n* – *m*) branches of root locus must approach ∞ as $K \rightarrow \infty$. These branches do so along straight line asymptotes whose angles are

$$\theta_A = \frac{(2q+1)180^o}{n-m}; \quad q = 0, 1, 2, \ldots (n-m-1) \qquad ...(5.9)$$

At a point on complex plane very far from all the poles and zeros of G(*s*) H(*s*), $\underline{|G(s)\, H(s)}$ is virtually equal to $-(n-m)\,\theta$ where θ is angle of point itself as shown in Fig. 5.8.

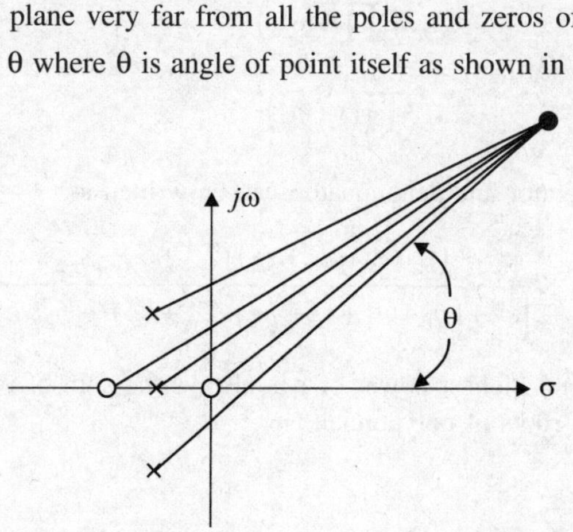

Fig. 5.8: Angle contribution at a point far from poles and zeros of G(*s*) H(*s*)

For this point to lie on root locus

$$-(n-m)\,\theta = \pm(2q+1)\,180°$$

or
$$\theta = \pm\frac{(2q+1)180°}{(n-m)} \qquad \text{...(5.10)}$$

Since there are only $(n - m)$ branches that approach ∞ as $K \to \infty$, the integer q can assume only $(n - m)$ values from 0 to $(n - m - 1)$. Thus the angles of asymptotes are those given by (5.9).

For example, if G(s) H(s) has two zeros and six poles there are $(n - m) = 4$ different asymptotic angles. These four angles are obtained by substituting various integer values, say $q = 0, 1, 2, 3$ as follows.

$$\theta_A = \frac{(2q+1)\,180°}{4}, \qquad q = 0, 1, 2, 3$$

$$= 45°,\ 135°,\ 225°,\ 315°$$

$$= \pm 45°,\ \ \pm 135°$$

Note that substitution of additional integers or choosing negative sign of (5.10), simply gives repetition of same angles and therefore need not be used.

(e) Centroid of the asymptotes

The asymptotes of the root locus branches which approach ∞, intersect each other at a common point on the real axis, called as centroid. The centroid is located at

$$\sigma_A = \frac{\sum \text{poles} - \sum \text{zeros}}{n-m} \qquad \text{...(5.11)}$$

where n = Number of poles of G(s) H(s)

and m = Number of zeros of G(s) H(s)

Consider an open-loop transfer function

$$G(s)\,H(s) = \frac{K\prod_{i=1}^{m}(s+z_i)}{\prod_{j=1}^{n}(s+p_j)}$$

which with expanded numerator and denominator, can be written as

$$G(s)\,H(s) = K\frac{\left[s^m + (z_1 + z_2 + ... + z_m)\,s^{m-1} + ... + z_1 z_2 .... z_m\right]}{\left[s^n + (p_1 + p_2 + ... + p_m)\,s^{n-1} + ... + p_1 p_2 .... p_n\right]} \qquad \text{...(5.12)}$$

because coefficient of second highest power of s is always the sum of roots of that polynomial and constant term is product of roots of polynomial.

Assuming large s and dividing the denominator by numerator in (5.12), we have

$$G(s)\,H(s) = \frac{K}{s^{n-m} + \left(\sum_{j=1}^{n} p_j - \sum_{i=1}^{m} z_i\right) s^{n-m-1} + \ldots}$$

or

$$G(s)\,H(s) \cong \frac{K}{\left[s + \frac{\left(\sum_{j=1}^{n} p_j - \sum_{i=1}^{m} z_i\right)}{n-m}\right]^{n-m}} \quad \ldots(5.13)$$

Then the point of intersection of asymptotes and the real axis can be obtained by equating the denominator of right side of (5.13) to zero and solving for s as

$$s = -\frac{\left(\sum_{j=1}^{n} p_j - \sum_{i=1}^{m} z_i\right)}{n-m} \quad \ldots(5.14)$$

Note the following while computing centroid:

(*i*) The centroid is not used when $(n - m)$ is either zero or one. For these values of $n - m$, the roots do not move off the real axis on the way towards ∞. When $n - m$ is zero, all open-loop poles have an open-loop zero to approach as gain K approaches ∞. When $n - m = 1$, one root moves towards ∞ along the negative real axis.

(*ii*) The imaginary part contributions of conjugate sets of poles and zeros of G(s) H(s) always cancel one another. So, only real parts of complex poles and complex zeros need to be included in the centroid calculations.

(*iii*) The point of intersection of asymptotes is always real and is a sort of **centre of gravity** of root locus.

(*iv*) A root locus branch may lie on one side of the corresponding asymptote or may cross the corresponding asymptote from one side to other side.

Consider a system with pole-zero plot as shown in Fig. 5.9(a)

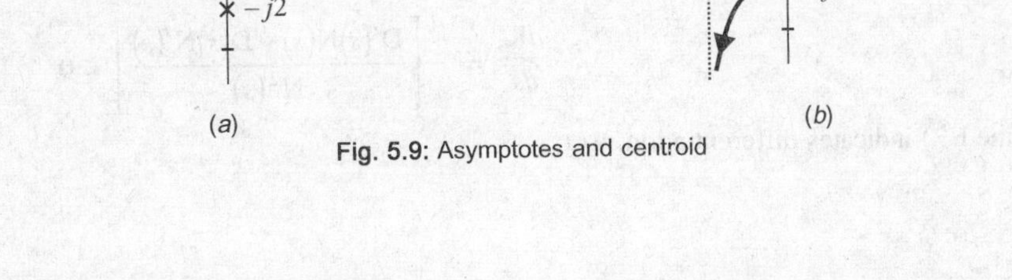

Fig. 5.9: Asymptotes and centroid

Real axis root locus branch begins from pole at $s = -3$ and terminates at zero at origin. The other two branches beginning from imaginary axis poles, terminate at ∞ as shown in Fig. 5.9(*b*).

Asymptotic angles are

$$\theta_A = \frac{(2q+1)\,180^o}{3-1}, \quad q = 0, 1$$

$$= 90°, 270° = \pm 90°$$

and centroid
$$\sigma_A = \frac{-3}{2} = -1.5$$

(*f*) Break away and Break in (entry) points

The break away and entry points on the root locus are determined from roots of $dK/ds = 0$. The root locus branches leave the break away point or approach the entry point also called as break in point, at an angle of $\pm\pi/r$, where r is the number of branches leaving or approaching the point.

Note the following:

(*i*) If the root locus branches move off the real axis at a gain K_1, then K_1 is the maximum value in the range of K for which the branches are on real axis. Similarly if the root locus branches enter the real axis at a gain K_2, then K_2 is minimum value in range of K for which the branches are on real axis. These points corresponding to K_1 and K_2 are called break away and entry points respectively. See Fig. 5.3.

(*ii*) Because of conjugate symmetry of the root loci, the break away and break in (entry) points either lie on real axis or occur in complex conjugate pairs.

(*iii*) If a root locus lies between two adjacent open-loop poles on the real axis, then there exists at least one break away point between the two poles. Similarly, if the root locus lies between two adjacent zeros (one zero may be at $-\infty$) on the real axis, there always exists at least one entry point between the two zeros. If the root locus lies between an open-loop pole and a zero (finite or infinite) on the real axis, then there may exist no break away or entry points or there may exist both break away and entry points.

(*iv*) The characteristic equation of a general system, is given by

$$1 + K\,G(s)\,H(s) = 0$$

or
$$1 + K\frac{N(s)}{D(s)} = 0$$

where $N(s)$ = Numerator polynomial of $G(s)\,H(s)$

and $D(s)$ = Denominator polynomial of $G(s)\,H(s)$.

or
$$K = -\frac{D(s)}{N(s)}$$

The break away/break in points are determined from roots of following equation

$$\frac{dK}{ds} = -\left[\frac{D'(s)N(s) - D(s)N'(s)}{N^2(s)}\right] = 0 \qquad ...(5.15)$$

where the prime (′) indicates differentiation w.r.t. ***s***.

Although the break away/break in points must be the roots of (5.15), but not all the roots of (5.15) are break away or break in points. If a real root of (5.15) lies on root locus portion of the real axis, then it is an actual break away or break in point. If a real root of (5.15) is not on the root locus portion of the real axis, then this root corresponds to neither a break away point nor break in point. If (5.15) gives complex conjugate roots and if it is not certain that the root lies on root locus, then it is necessary to check the value of K corresponding to the complex root of dK/ds = 0. If K is positive, the corresponding complex roots of (5.15) are actual break away or break in points and if K is negative or complex, then the point is neither break away nor break in point.

The following two examples will demonstrate the evaluation of break away and entry points.

Consider a system with open-loop transfer function

$$G(s)\ H(s) = \frac{K(s+10)}{s(s+5)}$$

For which the pole-zero plot is shown in Fig. 5.10 (*a*).

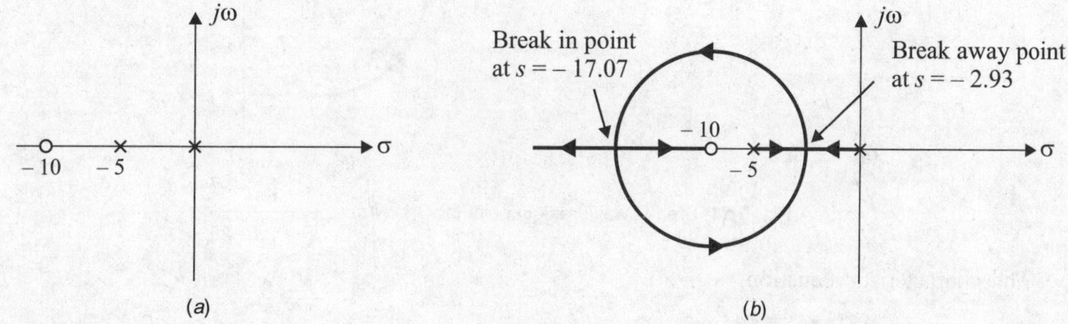

Fig. 5.10: (*a*) Pole-zero plot (*b*) Break away and entry points for $G(s)\ H(s) = \frac{K(s+10)}{s(s+5)}$

The real axis segments of root locus lie between $s = 0$ and $s = -5$ and between $s = -10$ and $s = -\infty$. As K increases from 0 to ∞, the root locus branches begin from two poles of G(*s*) H(*s*) ($s = 0$ and $s = -5$), get closer and closer with increasing K and at some K, there are two repeated roots. For still larger K, the roots break away from real axis. This break away point can be determined as follows;

From the characteristic equation

$$1 + \frac{K(s+10)}{s(s+5)} = 0$$

$$K = -\frac{s(s+5)}{s+10}$$

and

$$-\frac{dK}{ds} = \frac{s^2+20s+50}{(s+10)^2} \qquad ...(5.16)$$

Equating (5.16) to zero and solving, we have

$$s = -10 \pm 7.07 = -17.07 \text{ or } -2.93$$

Note that both the roots (– 17.07 and – 2.93) lie on root locus segment on real axis. So, the roots break away at $s = -2.93$ and break in at $s = -17.07$ as shown in Fig. 5.10(*b*).

Consider yet another system with open-loop transfer function

$$G(s)\,H(s) = \frac{Ks}{s^2 + 2s + 25}$$

for which the pole-zero plot is shown in Fig. 5.11(*a*).

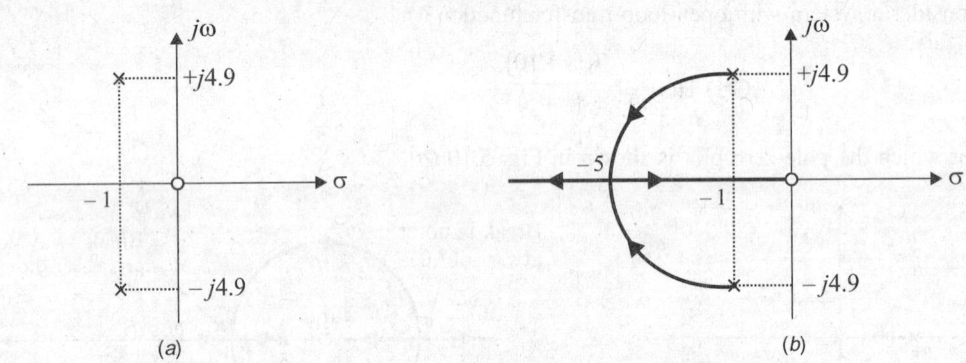

Fig. 5.11: Break away/break in point for $G(s)\,H(s) = \dfrac{Ks}{s^2 + 2s + 25}$

The characteristic equation

$$1 + \frac{Ks}{s^2 + 2s + 25} = 0 \qquad \text{...(5.17)}$$

gives
$$K = -\frac{s^2 + 2s + 25}{s}$$

and
$$-\frac{dK}{ds} = \frac{s^2 - 25}{s^2} \qquad \text{...(5.18)}$$

Equating (5.18) to zero and solving, we have

$$s = \pm 5$$

Note that $s = -5$ lies on real axis root locus segment and therefore, this is an actual break in point as the root locus branches begin from poles, terminate at either zero or ∞ and entire negative real axis is part of root locus. Since $s = +5$, does not lie on root locus segment on real axis, this is neither a break away nor the break in point.

(g) Angles of departure and approach

The angle of departure ϕ_d of the root locus from a complex pole is given by

$$\phi_d = 180° - \Sigma \text{ angles of other poles of } G(s)\,H(s) + \Sigma \text{ angles of zeros of } G(s)\,H(s) \qquad \text{...(5.19)}$$

The angle of approach ϕ_a of the root locus to a complex zero is given by

$$\phi_a = 180° + \Sigma \text{ angles of poles of } G(s)\,H(s) - \Sigma \text{ angles of other zeros of } G(s)\,H(s) \qquad \text{...(5.20)}$$

It is important to note that angle of departure from a real pole of G(s) H(s) or angle of approach to a real zero of G(s) H(s), is always either 0° or 180°. So, the angle of departure/approach calculation is required only for complex poles/complex zeros.

The angles of departure ϕ_{dm} of the root locus from the complex pole of order m, are given by

$$\phi_{dm} = [(2q + 1)180° - \Sigma \text{ angles of other poles of G}(s)\text{ H}(s) + \Sigma \text{ angles of zeros of G}(s)\text{ H}(s)]/m; \quad q = 0, 1, 2, ... (m - 1). \qquad ...(5.21)$$

The angles of approach ϕ_{am} of root locus to complex zero of order m, are given by

$$\phi_{am} = [(2q + 1)180° + \Sigma \text{ angles of poles of G}(s)\text{ H}(s) - \Sigma \text{ angles of other zeros of G}(s)\text{ H}(s)]/m; \quad q = 0, 1, 2, ... (m - 1). \qquad ...(5.22)$$

Note that more than one locus segment begins from repeated complex poles and also more than one locus segment ends at repeated complex zeros, so more than one angle of departure or approach will be found, a different angle for each locus segment.

The following examples demonstrate the angle of departure/approach calculations.

Consider a system with pole-zero plot as shown in Fig. 5.12(a).

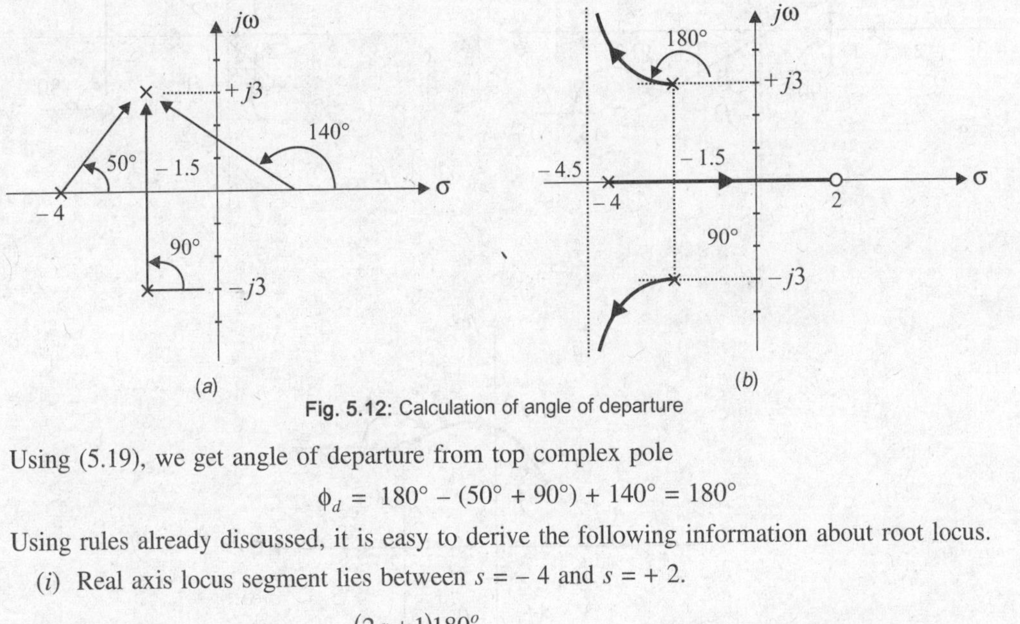

Fig. 5.12: Calculation of angle of departure

Using (5.19), we get angle of departure from top complex pole

$$\phi_d = 180° - (50° + 90°) + 140° = 180°$$

Using rules already discussed, it is easy to derive the following information about root locus.

(i) Real axis locus segment lies between $s = -4$ and $s = +2$.

(ii) Angles of asymptotes

$$= \frac{(2q+1)180^o}{n-m}; \; q = 0,1,........(n-m-1)$$

$$= \frac{(2q+1)\,180^o}{3-1}; \; q = 0, 1$$

$$= 90°, 270°$$

(iii) Centroid

$$\sigma_A = \frac{\sum \text{poles} - \sum \text{zeros}}{n-m} = \frac{-4 - 1.5 - 1.5 - 2}{3-1} = -4.5$$

Using above information, the complete root locus is sketched in Fig. 5.12(b).

Consider another system with pole-zero plot as shown in Fig. 5.13 (*a*). The angle of departure for the top complex pole using (5.19) is

$$\phi_d = 180° - 90° + (135° + 200°) = 425°$$

where 425° is equivalent to 65°.

The angle of approach to top complex zero as demonstrated in Fig. 5.13(*b*) using (5.20) is

$$\phi_a = 180° + (45° + 20°) - 90° = 155°$$

The complete root locus is shown in Fig. 5.13 (*c*). The complete root locus sketch also requires information about imaginary axis crossing point which we shall discuss in subsequent section.

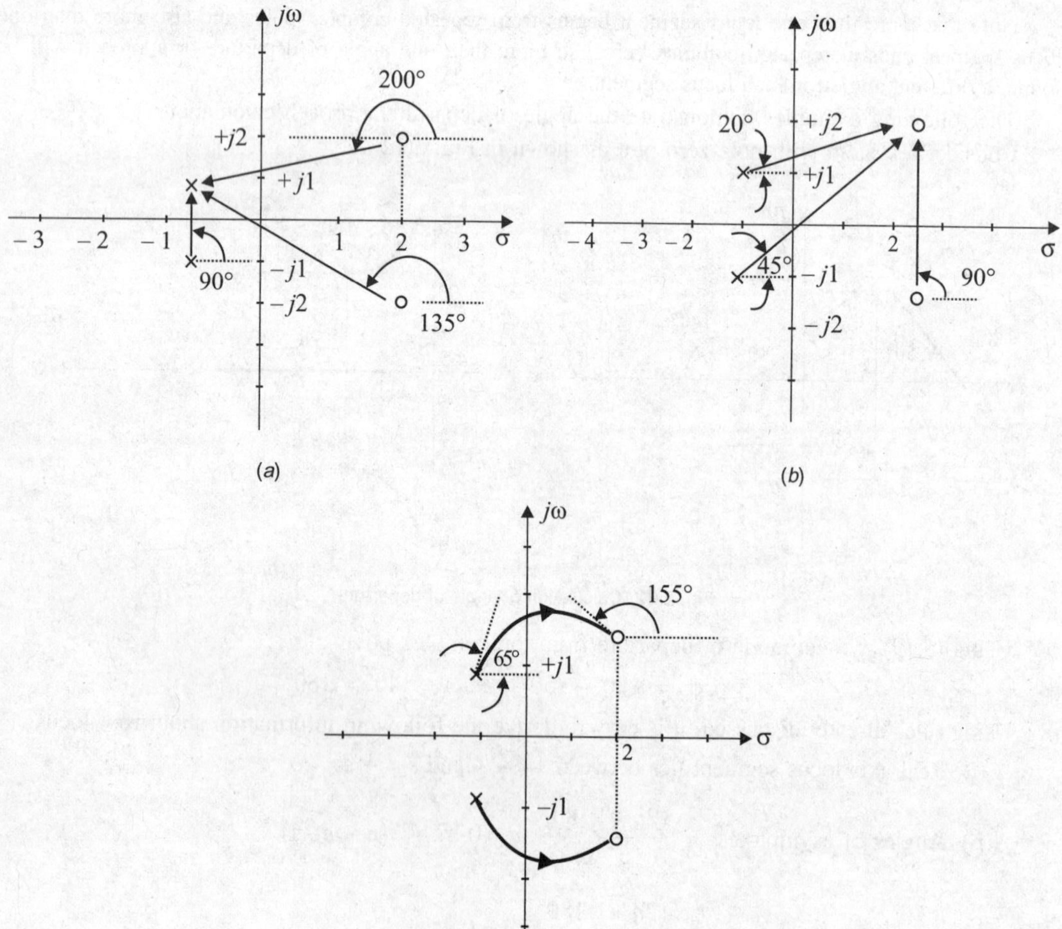

Fig. 5.13: Root locus construction involving angles of departure and angles of approach

Consider yet another system with repeated complex poles as shown in Fig. 5.14 (*a*). The angles of departure from top double complex poles using (5.21) are given by

$$\phi_{dm} = [(2q + 1)180° - (90° + 90° + 108°) + 124°]/2 \; ; q = 0, 1$$

$$= (2q + 1)90° - 82° \; ; q = 0, 1$$
$$= 8° \; (q = 0), \; 188° \; (q = 1)$$

It is easy to see that multiple angles of departure, will be evenly spaced around the multiple complex poles. The complete root locus is shown in Fig. 5.12 (*b*). The root locus segment on the real axis is between $s = 0$ and $s = -10$.

$$\text{Centroid} = \frac{-20 - 20 - 20 - 20 - 10}{5 - 1} = -22.5$$

Angles of asymptotes are given by

$$\theta_A = \frac{(2q+1)180°}{5-1} \; ; \quad q = 0, 1, 2, 3 = 45°, 135°, 225° \text{ and } 315°$$

The evaluation of point at which root locus crosses the imaginary axis, will be discussed in subsequent section.

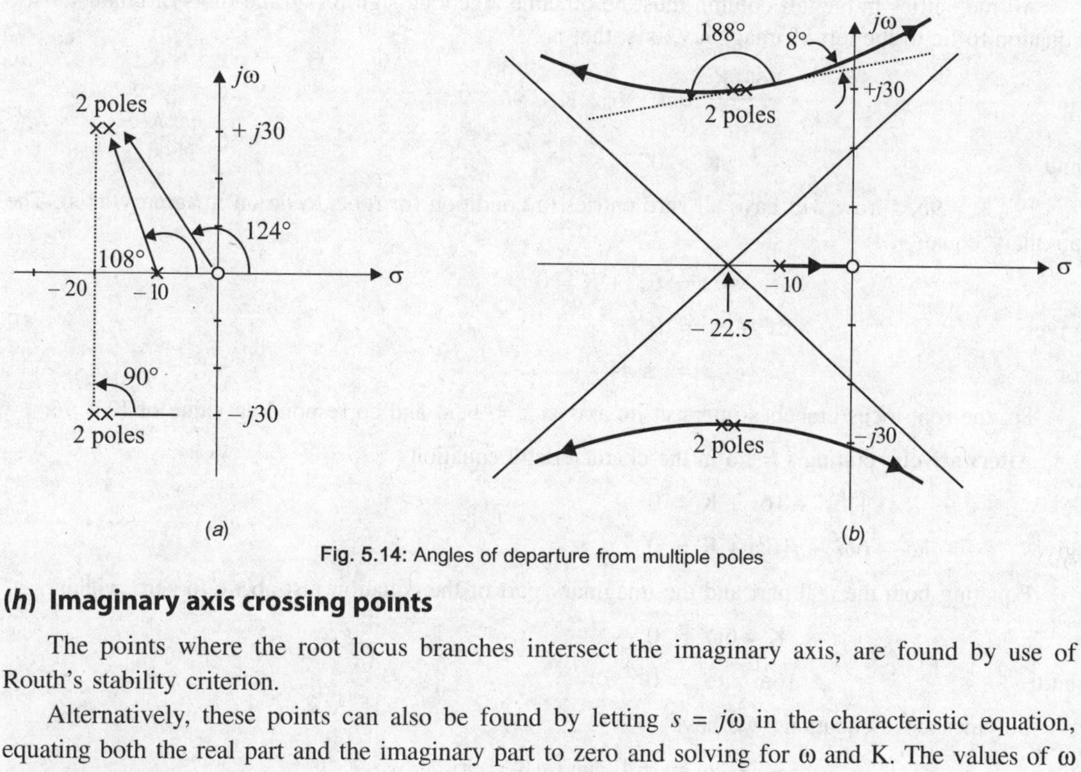

Fig. 5.14: Angles of departure from multiple poles

(*h*) Imaginary axis crossing points

The points where the root locus branches intersect the imaginary axis, are found by use of Routh's stability criterion.

Alternatively, these points can also be found by letting $s = j\omega$ in the characteristic equation, equating both the real part and the imaginary part to zero and solving for ω and K. The values of ω gives the frequencies at which root loci intersect imaginary axis and value of K corresponds to the gain at crossing point.

The following example demonstrates the evaluation of imaginary axis crossing points.

Consider a system with open-loop transfer function

$$G(s)\,H(s) = \frac{K}{s\left(s^2 + 6s + 16\right)}$$

The characteristic equation is

$$1+\frac{K}{s\left(s^2+6s+16\right)}=0$$

or $$s^3 + 6s^2 + 16s + K = 0$$

The Routh array is as follows:

$$\begin{array}{lcc} s^3 & 1 & 16 \\ s^2 & 6 & K \\ s^1 & \dfrac{96-K}{6} & 0 \\ s^0 & K & 0 \end{array}$$

All the entries in the left column must be of same algebraic sign for all the roots of characteristic equation to lie to the left of imaginary axis, that is

$$\frac{96-K}{6} > 0 \quad \text{or} \quad K < 96$$

and $$K > 0$$

For K = 96, s^1 row will have all zero entries (a condition for roots to lie on imaginary axis). The auxiliary equation

$$A(s) = 6s^2 + K = 0$$

gives $$6s^2 + 96 = 0$$

or $$s = \pm j4$$

So, the root locus branches intersect $j\omega$ axis at $s = \pm j4$ and corresponding value of K is 96.

Alternatively, putting $s = j\omega$ in the characteristic equation

$$s^3 + 6s^2 + 16s + K = 0$$

gives $$-j\omega^3 - 6\omega^2 + j16\omega + K = 0$$

Equating both the real part and the imaginary part of the equation just above to zero, yields

$$K - 6\omega^2 = 0$$

and $$16\omega - \omega^3 = 0$$

Solving above equations, we have

$$\omega = \pm 4 \quad \text{and} \quad K = 96.$$

The result obtained is same as given by Routh's stability criterion.

The root locus construction rules, discussed so far in detail, have been summarised in **Table 5.1** for a quick reference.

TABLE 5.1: Root locus construction rules

S. No.	Rule
1.	The root locus is symmetrical about real axis.
2.	The root locus branches are continuous curves that originate from each of n poles of G(s) H(s) for K = 0. As K $\to \infty$, either the locus branches approach m zeros of G(s) H(s) if $n = m$ or the $(n - m)$ branches extend to ∞ from $(n - m)$ excess poles if $n > m$ or the $(m - n)$ branches extend from ∞ to $(m - n)$ excess zeros if $m > n$.
3.	The number of branches of the root locus is equal to either the number of poles or number of zeros, whichever is greater. For a system with n poles and m zeros, $$\text{Number of branches } = n \text{ if } n > m$$ $$= m \text{ if } m > n$$
4.	The locus includes the segment of real axis if and only if there are odd number of poles plus zeros of G(s) H(s) to the right of that segment.
5.	As K $\to \infty$, the ($n - m$ or $m - n$ depending on whether $n > m$ or $m > n$ respectively) branches of the locus become asymptotic to straight lines with angles. $$\theta_A = \pm \frac{(2q+1)180°}{(n-m)}; \qquad q = 0, 1, 2, 3, ..., (n - m - 1)$$
6.	The centroid of asymptotes is always on real axis at $$\sigma_A = \frac{\Sigma \text{poles of G}(s)\text{ H}(s) - \Sigma \text{zeros of G}(s)\text{ H}(s)}{n-m}$$
7.	The loci leave the real axis at maximum value of gain K in that region of the real axis. The loci enter the real axis at the minimum value of K in that region of the real axis. These points are termed break away and entry points respectively and are determined from roots of equation $d\text{K}/ds = 0$. The r segments of root locus leave or enter the real axis at angles of $\pm 180°/r$.
8.	The points of intersection of the root locus branches with imaginary axis are determined either by use of Routh's stability criterion or by putting $s = j\omega$ in the characteristic equation, equating both the real part and the imaginary part to zero and solving for ω and K. ω is frequency at which root loci cross imaginary axis and K is the gain corresponding to the crossing point.
9.	The angle of departure ϕ_d of the root locus from a complex pole is given by $$\phi_d = 180° - \Sigma \text{ angles of other poles of G}(s)\text{ H}(s) + \Sigma \text{ angles of zeros of G}(s)\text{ H}(s)$$ The angle of approach ϕ_a of the root locus to a complex zero is given by $$\phi_a = 180° + \Sigma \text{ angles of poles of G}(s)\text{ H}(s) - \Sigma \text{ angles of other zeros of G}(s)\text{ H}(s)$$ where each angle (pole or zero) is calculated to complex pole for ϕ_d and to complex zero for ϕ_a. If the complex pole or complex zero is of order m, the m angles of departure $\phi_{d,\, m}$ and arrival $\phi_{a,\, m}$ are respectively given by $$\phi_{d,\, m} = [(2q + 1)180° - \Sigma \text{ angles of other poles of G}(s)\text{ H}(s) + \Sigma \text{ angles of zeros of G}(s)\text{ H}(s)]/m$$ $$\phi_{a,\, m} = [(2q + 1)180° - \Sigma \text{ angles of other zeros of G}(s)\text{ H}(s) + \Sigma \text{ angles of poles of G}(s)\text{ H}(s)]/m$$ $$q = 0, 1, 2, ... (m - 1).$$
10.	The open loop gain K in pole zero form at any point s_0 on the root locus is given by $$\text{K}\Big\|_{s = s_0} = \frac{\text{Product of the lengths of directed line segments from poles of G}(s)\text{ H}(s)\text{ to } s_0}{\text{Product of the lengths of directed line segments from zeros of G}(s)\text{ H}(s)\text{ to } s_0}$$

Some root locus plots

The root locus plots for some pole-zero configurations, are put together in Table 5.2 for quick reference.

TABLE 5.2: Root locus plots for typical poles-zeros configurations

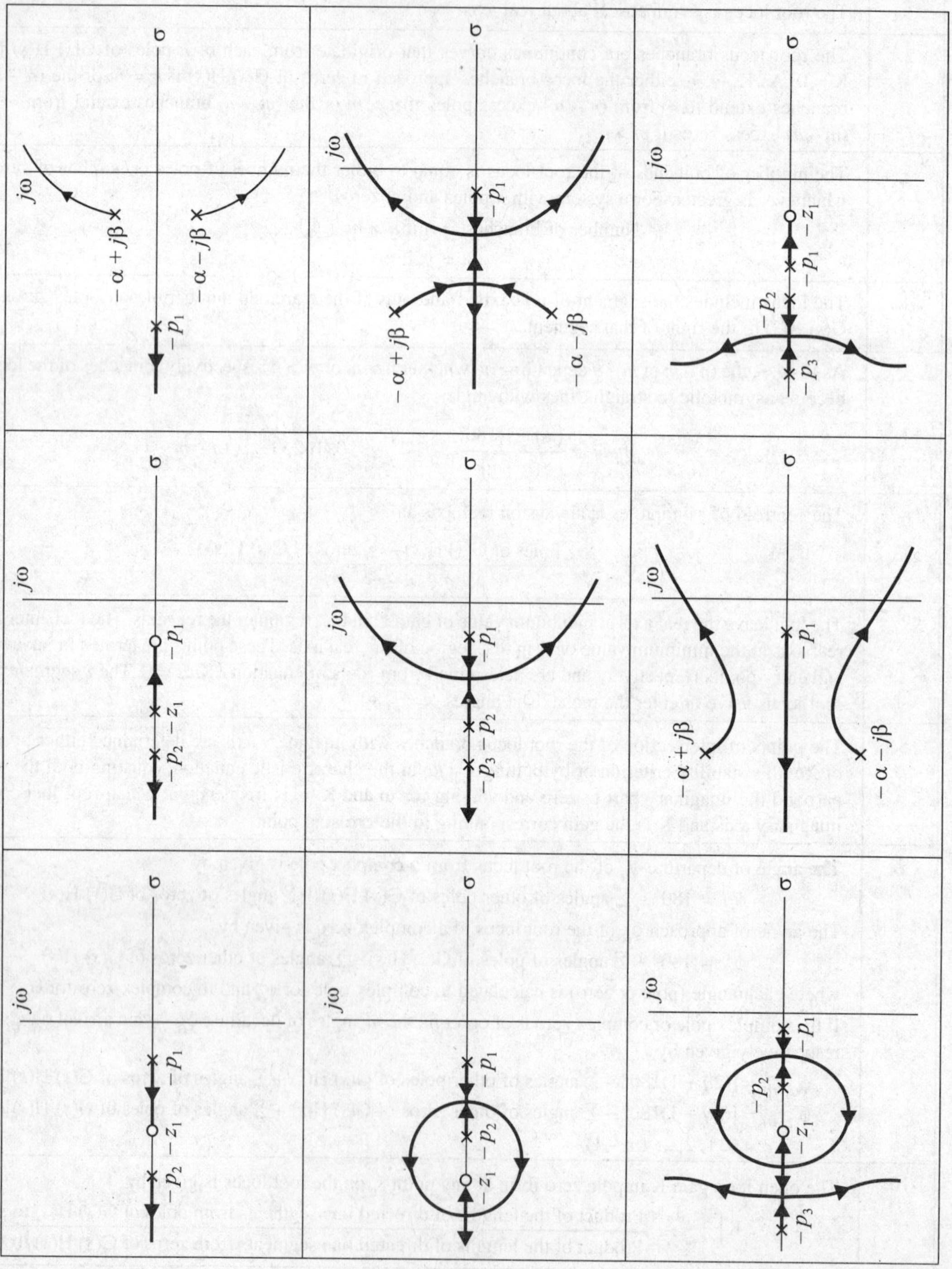

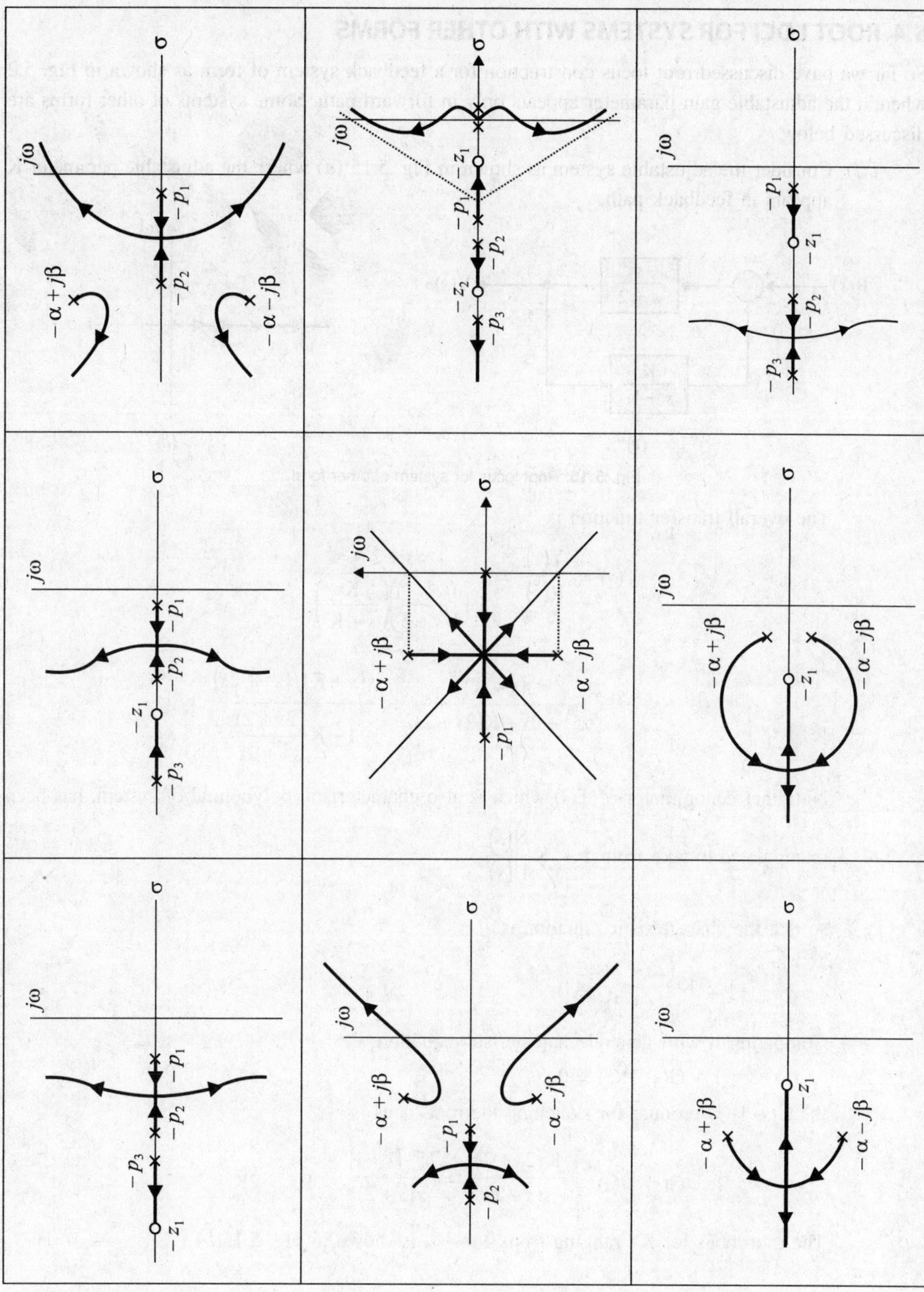
σ
jω
$-p_1$
$-p_2$
$-p_3$
$-z_1$
$-z_2$
$-\alpha + j\beta$
$-\alpha - j\beta$

5.4 ROOT LOCI FOR SYSTEMS WITH OTHER FORMS

So far we have discussed root locus construction for a feedback system of form as shown in Fig. 5.2 wherein the adjustable gain parameter appears only in forward path. Some systems of other forms are discussed below.

(*i*) Consider the adjustable system as shown in Fig. 5.15 (*a*) where the adjustable parameter K appears in feedback path.

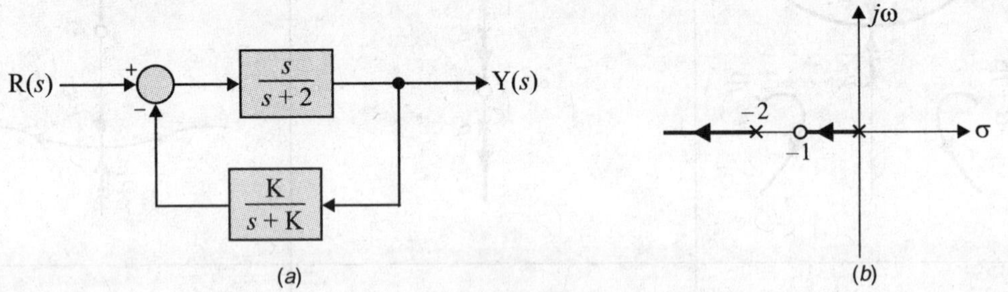

Fig. 5.15: Root locus for system of other form

The overall transfer function is

$$T(s) = \frac{Y(s)}{R(s)} = \frac{s/s+2}{1+\left(\frac{s}{s+2}\right)\left(\frac{K}{s+K}\right)}$$

$$= \frac{s(s+K)}{s^2+2s+K(2s+2)} = \frac{s(s+K)/(s^2+2s)}{1+K\frac{(2s+2)}{s^2+2s}}$$

Note that denominator of T(*s*) which is also characteristic polynomial of system, has been manipulated to take from $1 + K\frac{N(s)}{D(s)}$

so that the characteristic equation is

$$1 + K\frac{(2s+2)}{s(s+2)} = 0$$

Comparing it with general characteristic equation

$$1 + G(s)\,H(s) = 0$$

the G(*s*) H(*s*) product for sketching the root locus is

$$G(s)\,H(s) = \frac{K(2s+2)}{s(s+2)} = \frac{K^*(s+1)}{s(s+2)} \;;\; K^* = 2K$$

The root locus for K* ranging from 0 to ∞, is shown in Fig. 5.15(*b*).

(*ii*) Consider the unity feedback system with adjustable damping ratio ξ as shown in Fig. 5.16 (*a*). The overall transfer function is

$$T(s) = \frac{Y(s)}{R(s)} = \frac{(2s-4)/(s^2+6\xi s+9)}{1+\dfrac{(2s-4)}{s^2+6\xi s+9}}$$

$$= \frac{(2s-4)}{s^2+2s+5+6\xi s} = \frac{(2s-4)/(s^2+2s+5)}{1+\xi\dfrac{6s}{s^2+2s+5}}$$

Note that denominator of T(*s*) which is also characteristic polynomial of system, has been manipulated to take from $1 + K\dfrac{N(s)}{D(s)}$ so that the characteristic equation is

$$1 + \xi\frac{6s}{(s^2+2s+5)} = 0$$

Comparing it with general characteristic equation

$$1 + G(s)\,H(s) = 0$$

the G(*s*) H(*s*) product for sketching the root locus as ξ varies from 0 to ∞, may be taken as

$$G(s)\,H(s) = \frac{Ks}{(s^2+2s+5)}$$

where $K = 6\xi$

The root locus for K in the range from 0 to ∞, is shown in Fig. 5.16(*b*).

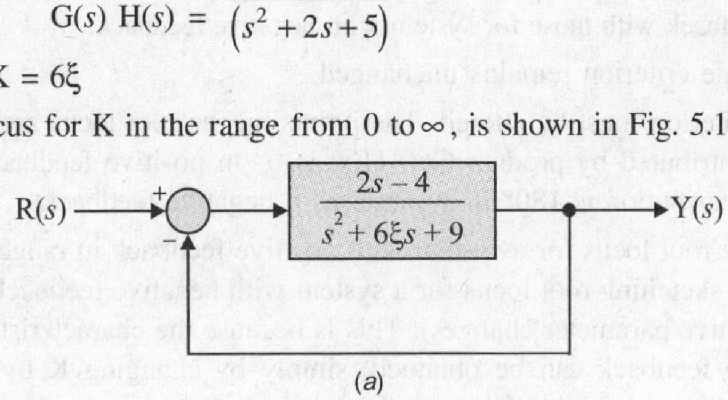

(*a*)

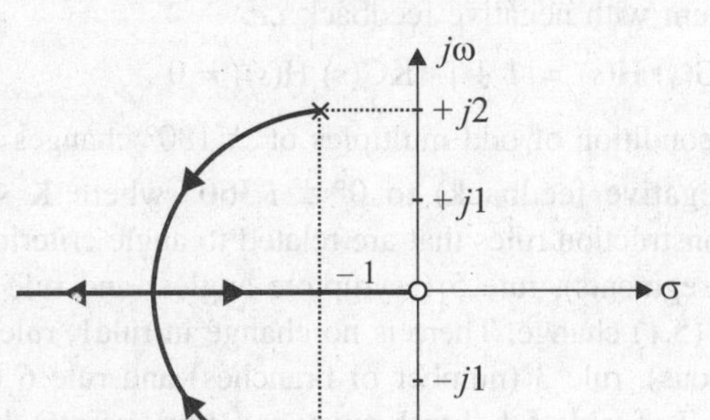

(*b*)

Fig. 5.16: Root locus of a system with damping as adjustable parameter

5.5 ROOT LOCI FOR SYSTEMS WITH POSITIVE FEEDBACK

Consider a system with positive feedback as shown in Fig. 5.17.

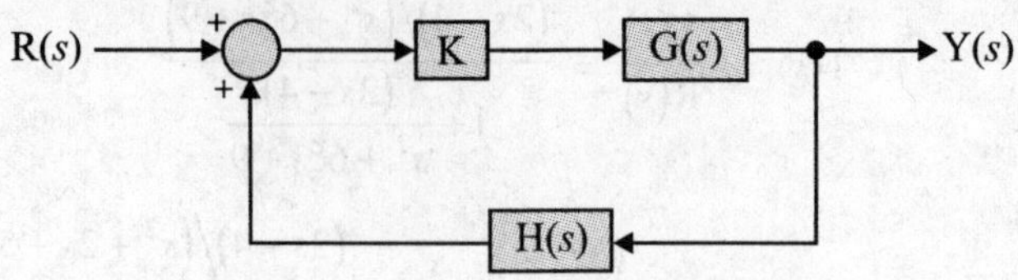

Fig. 5.17: System with positive feedback

The overall transfer function is

$$T(s) = \frac{Y(s)}{R(s)} = \frac{KG(s)}{1-KG(s)\,H(s)}$$

and characteristic equation is

$$1 - KG(s)\,H(s) = 0 \qquad ...(5.23)$$

or

$$KG(s)\,H(s) = 1 \qquad ...(5.24)$$

The angle and magnitude relations of root locus as explained in Section 5.2 can be derived from (5.24) as

$$|K\,G(s)\,H(s)| = 1$$

and

$$\underline{\left|K G(s)\,H(s)\right.} = 0° \pm i360° \qquad (i = 0, 1, 2, ...)$$

Note the following while comparing the angle and the magnitude criterion derived just above for the system with positive feedback with those for system with negative feedback.

(*a*) The magnitude criterion remains unchanged.

(*b*) The angle criterion must be altered. The points on the root locus are values of s for which the angle contributed by product G(*s*) H(*s*) is 0° in positive feedback configuration while this angle contribution is 180° in systems with negative feedback.

(*c*) Sketching the root locus for a system with positive feedback in range of K from 0 to ∞, is equivalent to sketching root locus for a system with negative feedback in range of K from 0 to – ∞, (negative parameter changes). This is because the characteristic equation for system with positive feedback can be obtained, simply by changing K by – K in characteristic equation for system with negative feedback *i.e.*

$$1 - KG(s)\,H(s) = 1 + [-KG(s)\,H(s)] = 0$$

(*d*) Since the angle condition of odd multiples of ± 180° changes from the case where K > 0 (system with negative feedback) to 0° ± *i* 360° where K < 0 (system with positive feedback), the construction rules that are related to angle criterion, also change. Thus rule 4 (real axis locus segments), rule 5 (asymptote angles) and rule 9 (angles of departure and arrival) of Table (5.1) change. There is no change in rule1, rule 2 (starting and terminating points of root locus), rule 3 (number of branches) and rule 6 (centroid) of table 5.1. The statement of rule 7 of table 5.1 (break away and entry points) does not change, but the part of the real axis to be searched for actual break away and entry points, does change to agree with rule 4 (real axis locus segments). The construction rules related to angle criterion of table 5.1 when applied to negative parameter changes are modified as follows:

(*i*) **Rule 4 (real axis locus segment)** is modified as follows:

The root locus includes the segment of real axis if and only if there are even number of poles plus zeros of G(*s*) H(*s*) to the right of that segment.

(*ii*) **Rule 5 (Asymptotic angles)** is modified as follows:

The angles of asymptotes

$$\theta_A = \frac{\pm i\, 360°}{n - m}$$

For $i = 0, 1, 2$ ..... until all $m - n$ or $n - m$ angles are obtained,

where n = number of poles of G(*s*) H(*s*) and m = number of zeros of G(*s*) H(*s*).

(*iii*) **Rule 9 (Angles of departure and arrival)** is modified as follows:

The angle of departure ϕ_d of a locus branch from a complex pole is given by

$\phi_d = 0° - \Sigma$ angles of other poles of G(*s*) H(*s*) + Σ angles of zeros of G(*s*) H(*s*).

The angle of approach ϕ_a of the root locus to a complex zero is given by

$\phi_a = 0° + \Sigma$ angles of poles of G(*s*) H(*s*) – Σ angles of other zeros of G(*s*) H(*s*).

where each angle (pole or zero) is calculated to complex pole for ϕ_d and complex zero for ϕ_a.

If the complex pole or complex zero is of order m, the m angles of departure ϕ_{dm} and arrival ϕ_{am} are respectively given by

$\phi_{dm} = [i360° - \Sigma$ angles of other poles of G(*s*) H(*s*) + Σ angles of zeros of G(*s*) H(*s*)]/m

$\phi_{am} = [i360° + \Sigma$ angles of poles of G(*s*) H(*s*) – Σ angles of other zeros of G(*s*) H(*s*)]/m

for $i = 0, 1, 2, ... (m - 1)$.

Note that other construction rules given in Table 5.1 remain unchanged. The following example demonstrates the application of modified rules.

Consider a system with product G(*s*) H(*s*) = $\frac{K(s+3)}{s(s-2)(s^2+2s+5)}$ and pole-zero plot as shown in Fig. 5.18 (*a*). It is desired to determine the root locus for a parameter K which ranges from 0 to $-\infty$.

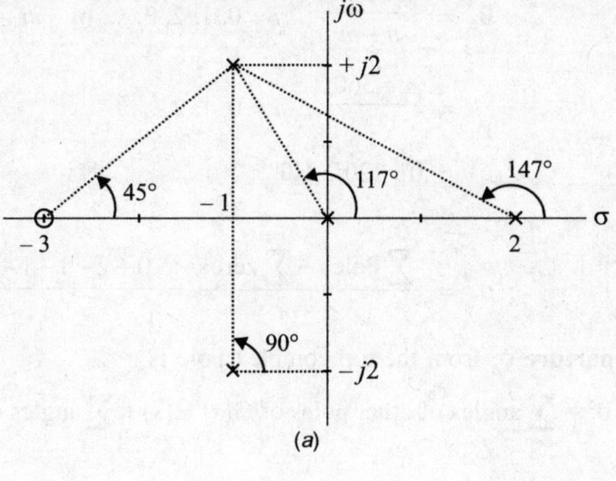

(*a*)

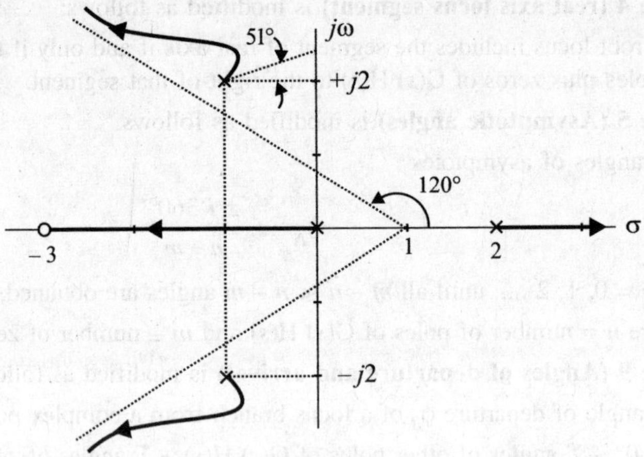

(b)

Fig. 5.18: Root locus construction for negative K

The step by step approach to the construction of root locus is as follows:

(*i*) The root locus is symmetrical about real axis.

(*ii*) Number of poles = $n = 4$

Number of zeros = $m = 1$

Number of separate root locus branches = 4 as $n > m$

The locus segments begin on the poles of G(*s*) H(*s*) (for $K \to 0$) and approach either the zeros of G(*s*) H(*s*) (for $K \to -\infty$) or ∞. Thus starting points are $s = 0$, $s = +2$ and $s = -1 \pm j2$. One locus branch terminates at $s = -3$ and remaining three branches at ∞.

(*iii*) **Real axis locus segment:** The real axis segments between $s = 0$ and $s = -3$ and between $s = +2$ and $s = +\infty$, are part of root locus as there are **even** number of poles plus zeros of G(*s*) H(*s*) on the real axis to the right of these segments.

(*iv*) **Asymptotic angles:**

$$\theta_A = \frac{\pm i\, 360°}{n-m}; \quad i = 0, 1, 2, 3, \ldots (n-m-1)$$

$$= \frac{\pm i\, 360°}{3}; \quad i = 0, 1, 2$$

$$= 0°, 120°, 240°$$

(*v*) **Centroid:**

$$\sigma_A = \frac{\sum \text{Poles} - \sum \text{zeros}}{n-m} = \frac{0+2-1-1-(-3)}{3} = 1$$

(*vi*) **Angle of departure** ϕ_d from the top complex pole is

$$\phi_d = 0° - \sum \text{angles of other poles of } G(s)\,H(s) + \sum \text{angles of zeros of } G(s)\,H(s)$$

$$= 0° - (147° + 117° + 90°) + 45° = -309° = +51°$$

Using the information furnished above, the complete root locus is drawn in Fig. 5.18 (*b*).

For the purpose of quick comparison and better insight into differences in construction rules between root locus of systems with negative feedback (positive parameter changes) and systems with positive feedback (negative parameter changes), a Table 5.3 is given below. It tabulates the root locus for both the positive and negative parameter variation for some typical pole-zero configurations.

TABLE 5.3: Root locus plots of negative feedback and positive feedback systems

Pole zero plot	***Root locus for negative feedback system (Positive parameter changes)***	***Root locus for positive feedback system (Negative parameter changes)***
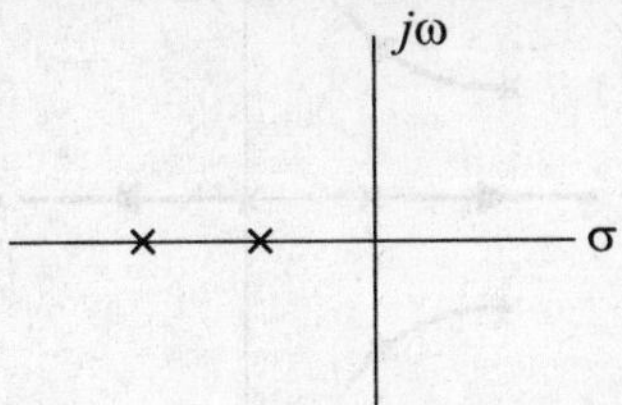	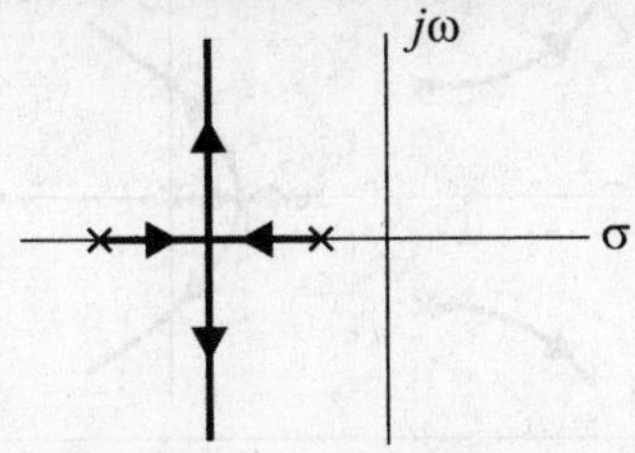	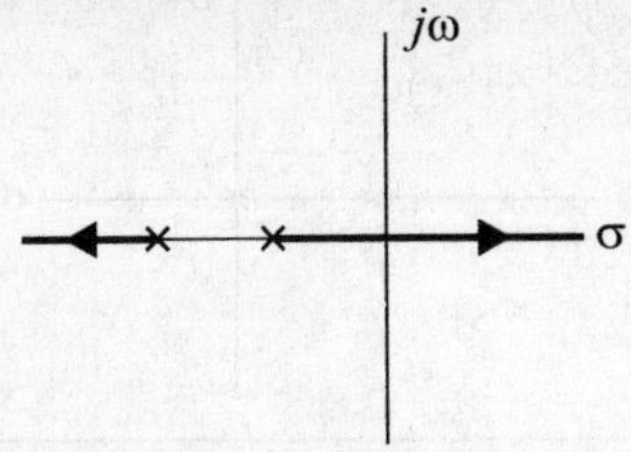
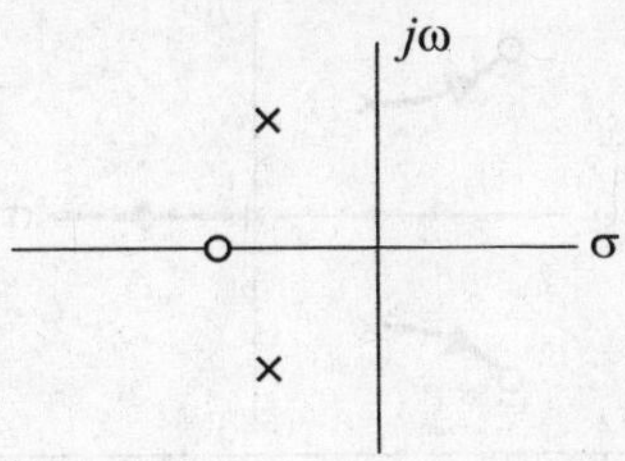	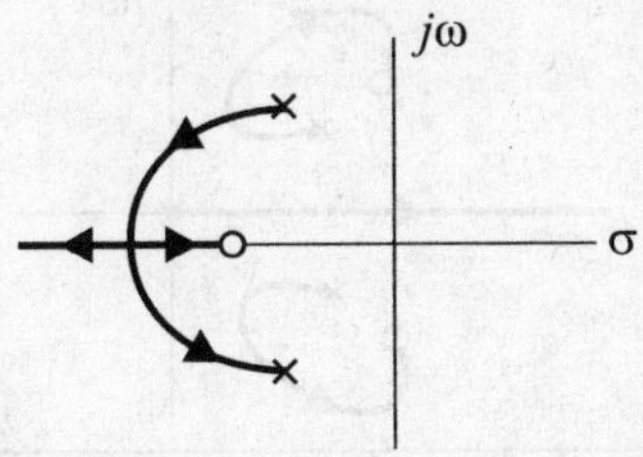	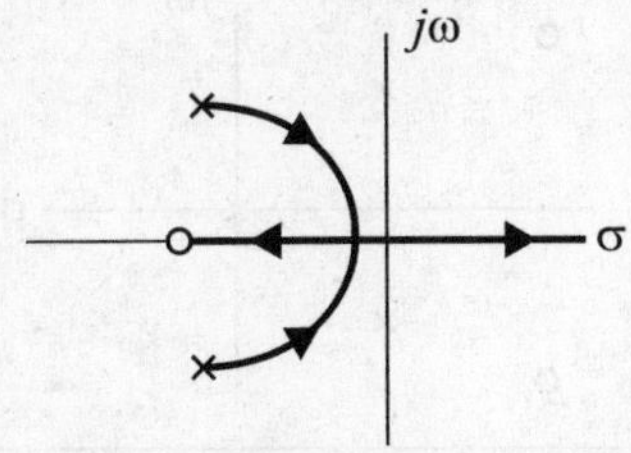
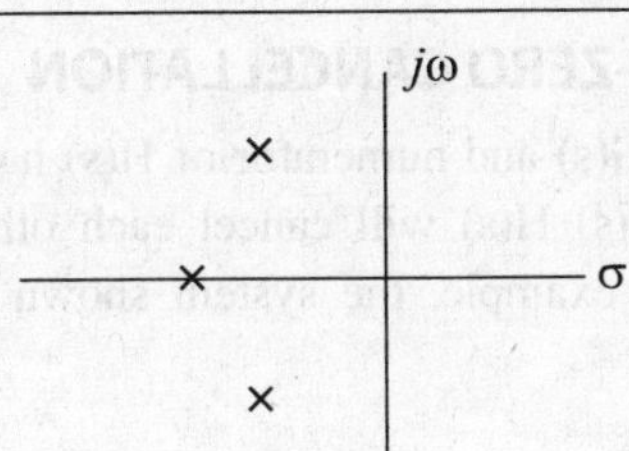	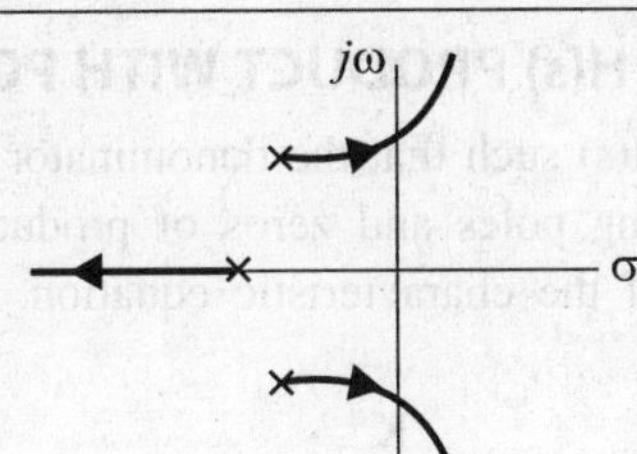	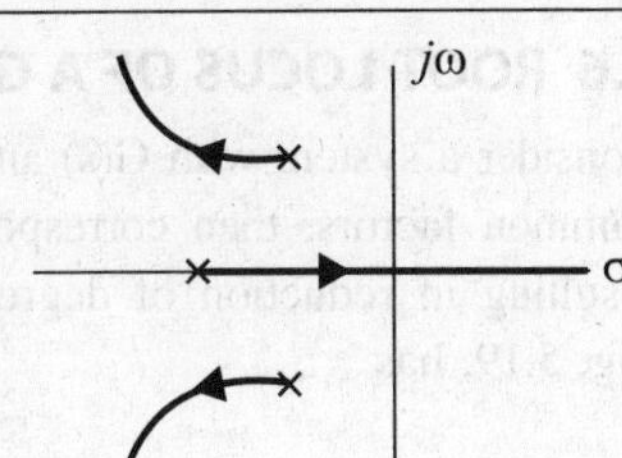
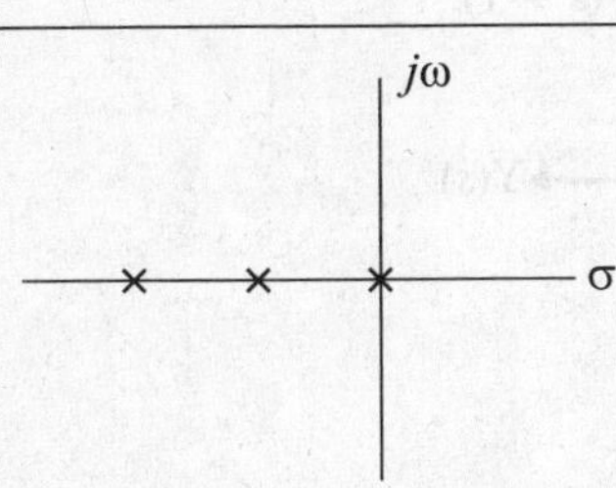	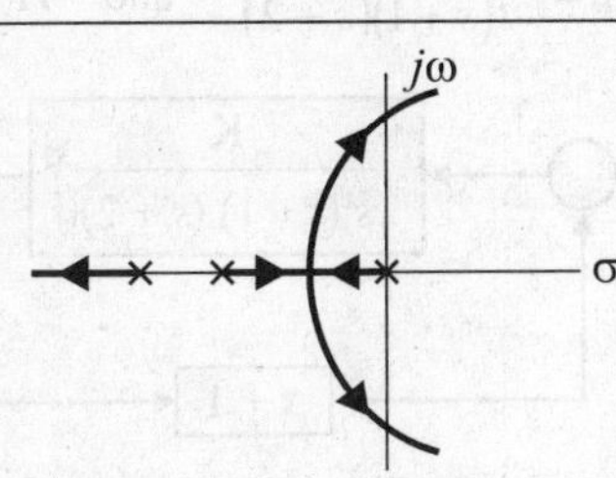	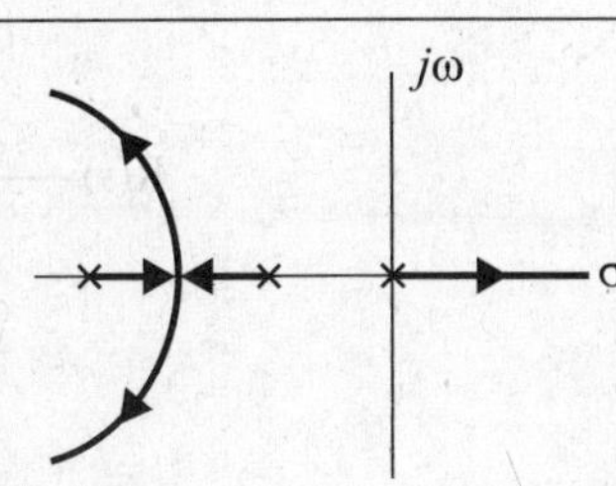

Pole zero plot	***Root locus for negative feedback system (Positive parameter changes)***	***Root locus for positive feedback system (Negative parameter changes)***
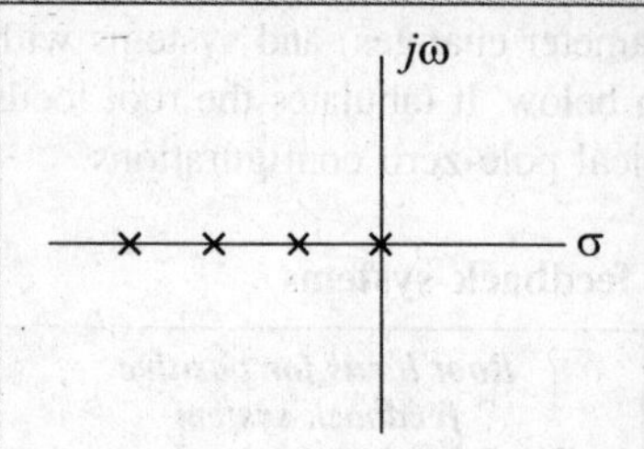	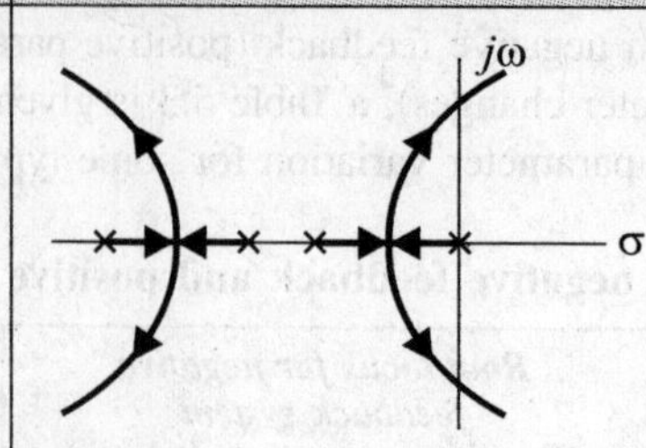	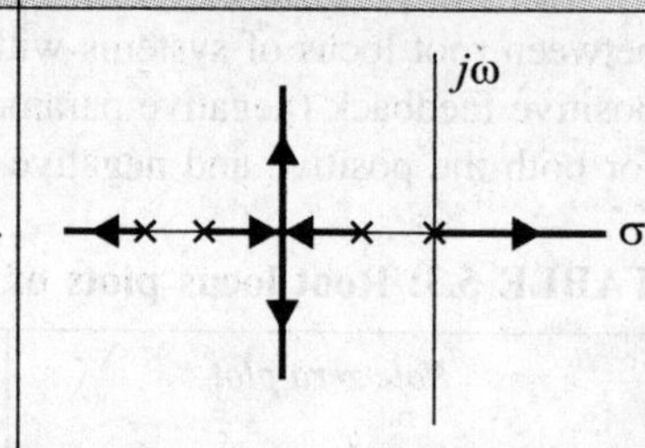
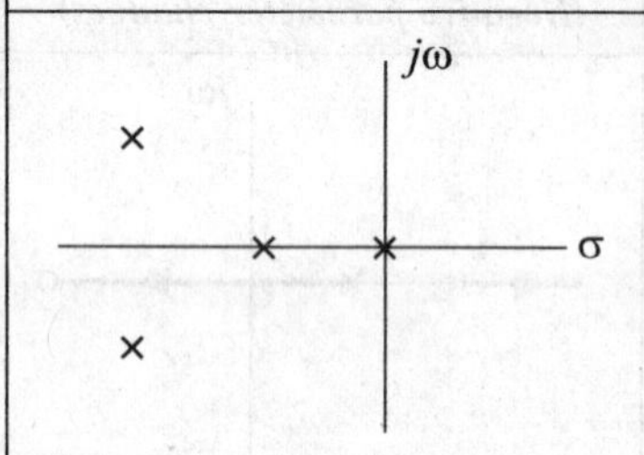	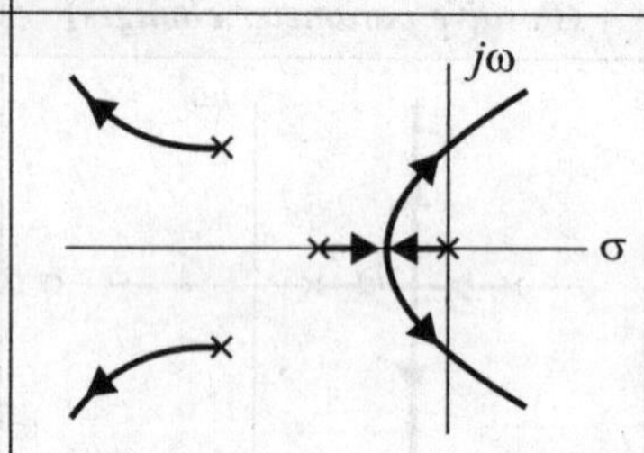	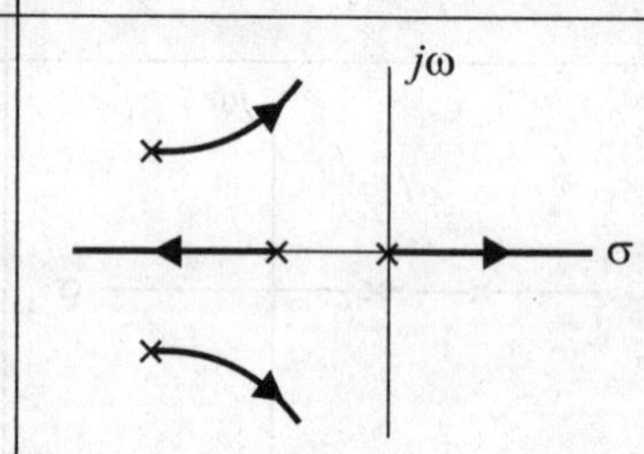
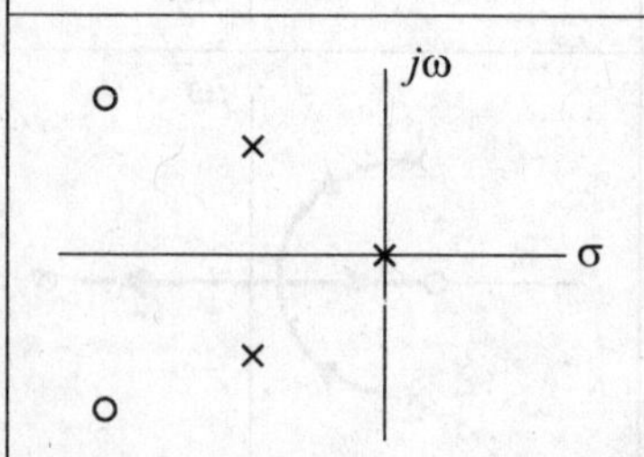	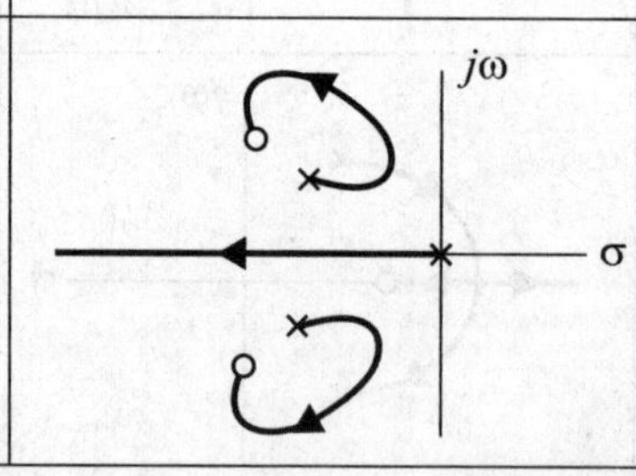	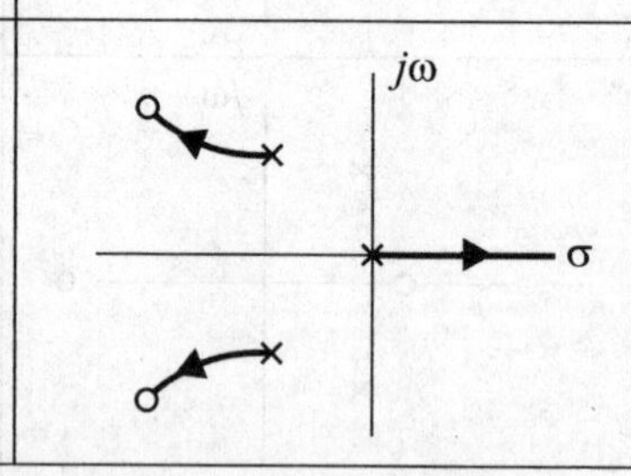

5.6 ROOT LOCUS OF A G(*s*) H(*s*) PRODUCT WITH POLE-ZERO CANCELLATION

Consider a system with G(s) and H(s) such that the denominator of G(s) and numerator of H(s) have common factors, then corresponding poles and zeros of product G(s) H(s) will cancel each other resulting in reduction of degree of the characteristic equation. For example, the system shown in Fig. 5.19, has

$$G(s) = \frac{K}{s(s+1)(s+2)} \quad \text{and} \quad H(s) = (s + 1)$$

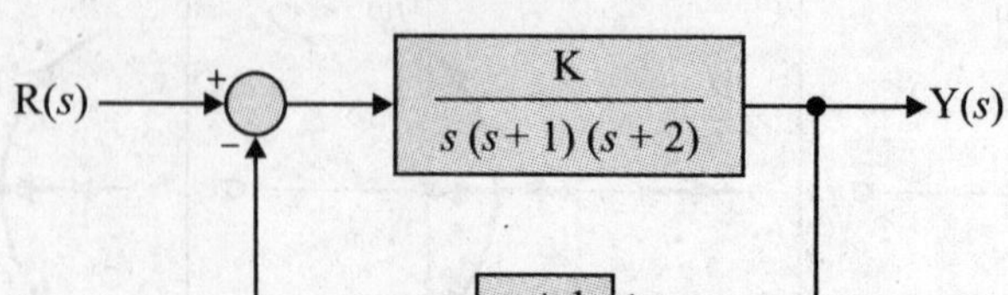

Fig. 5.19: System with G(*s*) and H(*s*) having Common factor

The overall transfer function is

$$T(s) = \frac{Y(s)}{R(s)} = \frac{K}{s(s+1)(s+2)+K(s+1)}$$

and the characteristic equation is

$$(s+1)\,[s\,(s+2)+K]=0$$

Note that one root of characteristic equation at $s=-1$, is independent of K. The product G(s) H(s), after pole-zero cancellation is

$$G(s)\,H(s)=\frac{K}{s\,\cancel{(s+1)}\,(s+2)}\cdot\cancel{(s+1)}$$

and reduced characteristic equation is

$$1+G(s)\,H(s)=0$$

or

$$s^2+2s+K=0$$

The root locus of $G(s)\,H(s)=\dfrac{K}{s(s+2)}$ will obviously show only two roots for any value of K. Thus the complete set of roots for a value of K, will be the two roots obtained from root locus plus the cancelled pole ($s=-1$).

5.7 EFFECTS OF ADDING POLES AND ZEROS TO THE PRODUCT G(*s*) H(*s*) ON SHAPE OF ROOT LOCUS

Addition of poles

Adding a pole to the product G(s) H(s) in the left half of s-plane has an effect of pushing the root locus towards right half of s-plane. This is qualitatively demonstrated with the help of following examples.

Consider a system with product

$$G(s)\,H(s)=\frac{K}{s(s+\alpha)},\quad \alpha>0$$

whose root locus plot is shown in Fig. 5.20 (a). The centroid σ is located at $s=-\alpha/2$ and asymptotic angles are $\pm\,90°$.

Introducing a pole at $s=-\beta$ ($\beta>\alpha$), the product G(s) H(s) becomes

$$G(s)\,H(s)=\frac{K}{s(s+\alpha)(s+\beta)}$$

whose root locus plot is shown in Fig. 5.20 (b). The centroid moves from $-\alpha/2$ to $\dfrac{-(\alpha+\beta)}{3}$ and angles of asymptotes are $\pm\,60°$. Note that addition of pole at $s=-\beta$, has an effect of bending the complex part of root loci towards right half of s-plane.

Introducing yet another pole at $s=-\delta$ ($\delta>\beta$), the product G(s) H(s) is modified to

$$G(s)\,H(s)=\frac{K}{s(s+\alpha)(s+\beta)(s+\delta)}$$

whose root locus plot is shown in Fig. 5.20 (c). The centroid moves to $\dfrac{-(\alpha+\beta+\delta)}{4}$ and asymptotic angles are $\pm 45°$ and $\pm 135°$. The complex part of root loci bends further into right half of s-plane.

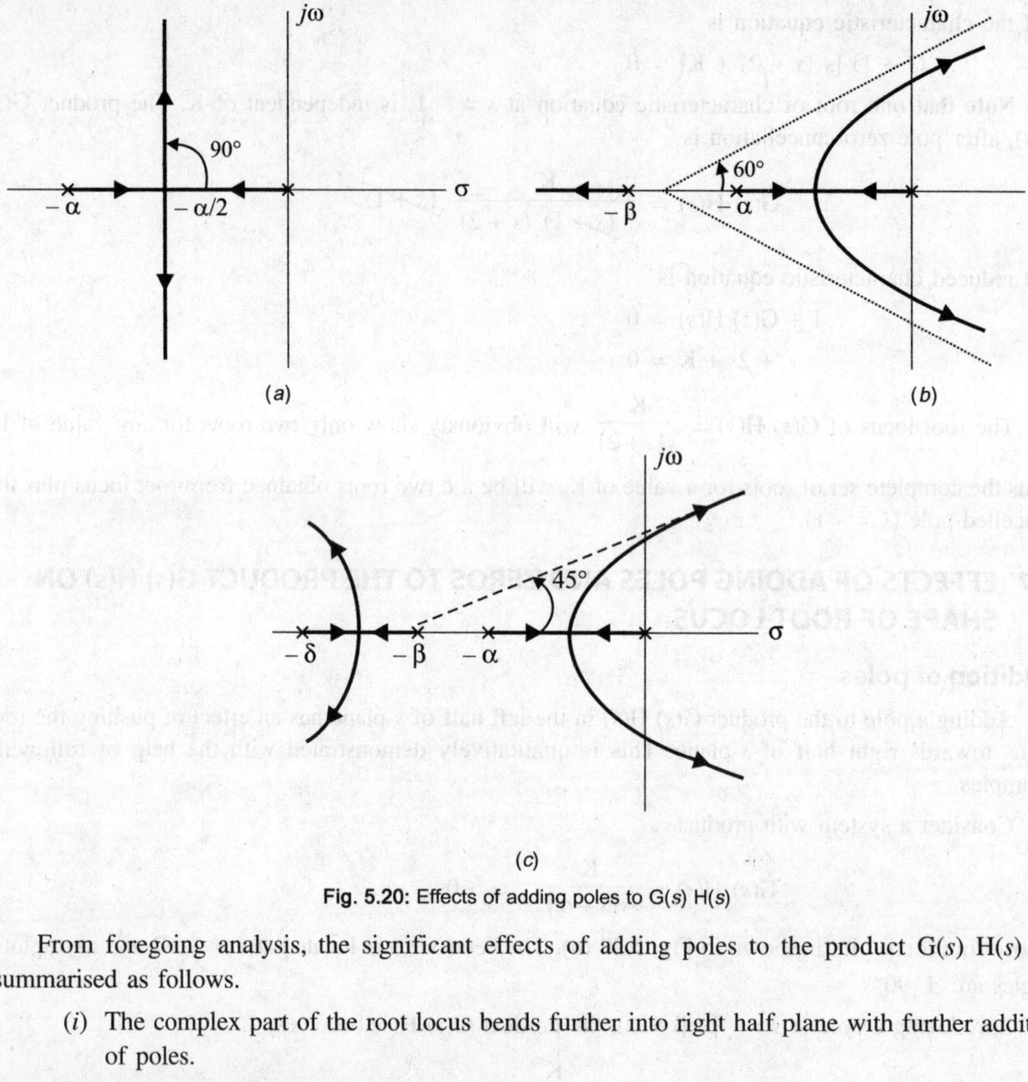

Fig. 5.20: Effects of adding poles to G(*s*) H(*s*)

From foregoing analysis, the significant effects of adding poles to the product G(*s*) H(*s*) are summarised as follows.

(*i*) The complex part of the root locus bends further into right half plane with further addition of poles.

(*ii*) The system stability relatively reduces.

(*iii*) The system becomes more oscillatory.

(*iv*) The range of adjustable parameter K for system stability decreases.

(*v*) The effect of adding pole to G(*s*) H(*s*) is almost equivalent to the effect of introducing **integral control** into the system (to be discussed later in chapter 8).

Addition of zeros

Adding a zero to the product G(*s*) H(*s*), generally has an effect of moving and bending the root locus towards left half of *s*-plane. This is qualitatively demonstrated with the help of following examples.

For system with

$$G(s)\,H(s) = \frac{K}{s(s+\alpha)}; \quad \alpha > 0$$

The root locus has already been shown in Fig. 5.20 (*a*) introducing a zero at $s = -\beta$ ($\beta > \alpha$), the product G(*s*) H(*s*) becomes

$$G(s)\,H(s) = \frac{K(s+\beta)}{s(s+\alpha)}$$

whose root locus also, has already been shown in Fig. 5.3. Note that complex conjugate part of root locus, bends toward left half of *s*-plane forming a circle. A similar effect on the shape of root locus can be seen by further introducing a pair of complex conjugate zeros at

$$s = -\beta \pm j\delta \text{ as shown in Fig. 5.21.}$$

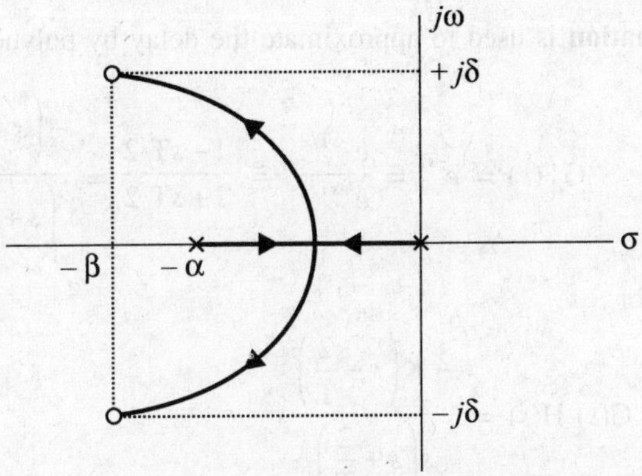

Fig. 5.21: Effect of adding zeros to G(*s*) H(*s*)

From foregoing analysis, the significant effects of adding zeros to G(s) H(s), are summarised as follows:

(*i*) The complex conjugate part of the root locus bends into left half of *s*-plane

(*ii*) The system becomes relatively more stable.

(*iii*) The system becomes less oscillatory.

(*iv*) The range of adjustable parameter K for system stability, increases.

(*v*) The effect of adding zeros to G(*s*) H(*s*) is almost equivalent to the effect of introducing **derivative control** into the system (to be discussed later in chapter 8).

5.8 EFFECTS OF DELAY ON ROOT LOCUS

It might happen in some systems that a delay is intentionally included in the control loop to bring about some improvement in control strategy. For example, a chemical process might require measurements of some physical variable to evolve a better quality control strategy. The measurements take some finite time and thereby implementation of result is delayed. In computer controlled systems also, a delay equal to one or more computational cycle, may some times be required to be introduced.

The introduction of delay in a system, makes the system relatively less stable. This is demonstrated by the following example.

Consider a system as shown in Fig. 5.22.

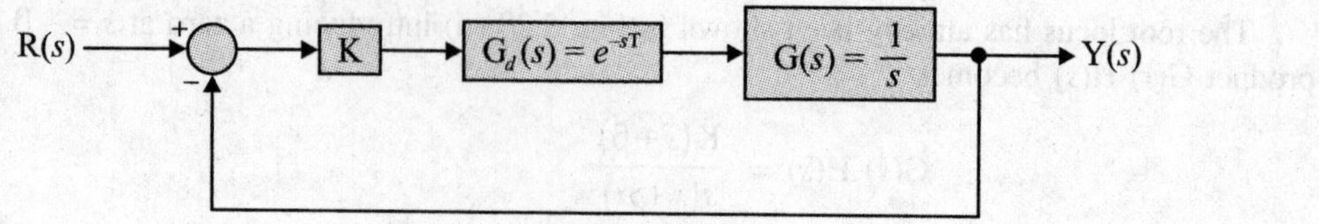

Fig. 5.22: System with delay

Note that a delay $G_d(s) = e^{-sT}$ is included in forward path. $G_d(s)$ contributes true delay of T seconds. Since the construction rules of root locus discussed so far, require the $G(s)\,H(s)$ product to be rational (Polynomial ratios), the system with delay cannot be analysed through root locus technique because delay e^{-sT} is irrational.

The **Padee approximation** is used to approximate the delay by polynomial ratio as follows:

$$G_d(s) = e^{-sT} = \frac{e^{-sT/2}}{e^{+sT/2}} \cong \frac{1 - sT/2}{1 + sT/2} = \frac{-\left(s - \frac{2}{T}\right)}{\left(s + \frac{2}{T}\right)}$$

Then the root locus for

$$G(s)\,H(s) = \frac{-K\left(s - \frac{2}{T}\right)}{s\left(s + \frac{2}{T}\right)}$$

can easily be sketched. The root loci for T = 0 and T = 1 are shown in Fig. 5.23(*a*) and 5.23(*b*) respectively.

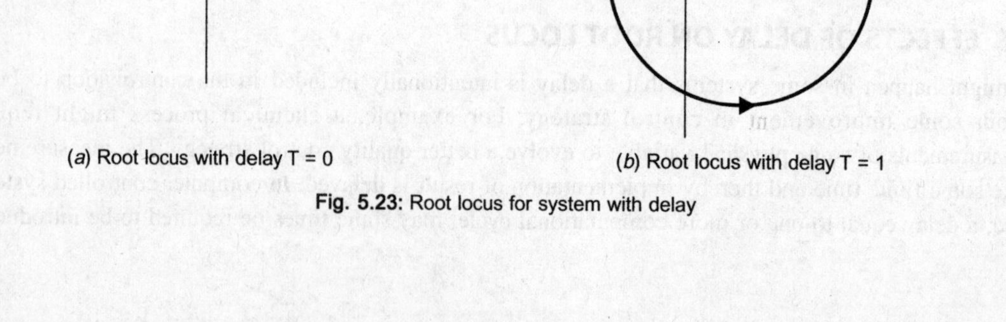

(*a*) Root locus with delay T = 0 (*b*) Root locus with delay T = 1

Fig. 5.23: Root locus for system with delay

It is easy to compute imaginary axis crossing point that occurs for K = 2/T. The inclusion of delay makes the system relatively less stable, with delay T = 1 second, system becomes unstable for K > 2 (see Fig. 5.23 (*b*)). For larger T, the range of K for which the system is stable, further decreases.

5.9 ROOT CONTOURS (MULTIPLE PARAMETER VARIATION)

In so far discussion of construction of root locus, we have used only one variable parameter. In some control situations, the effects of varying more than one parameter, are required to be investigated. When multiple parameters vary continuously from 0 to ∞ in a system, the resulting root loci are referred to as **root contours**. It will be demonstrated little later that root contours still possess the same properties as single parameter root loci, so the construction rules discussed so far, are all applicable.

The overall transfer function of a system involving two variable parameters K_1 and K_2, can be, in general, written in the form

$$T(s) = \frac{N(s)}{D(s)+K_1D_1(s)+K_2D_2(s)} \qquad ...(5.25)$$

The first step in construction of root contour, is to set one of the parameter to zero. For example, setting $K_2 = 0$ in (5.25)

$$T(s) = \frac{N(s)}{D(s)+K_1D_1(s)}$$

and characteristic equation is

$$D(s) + K_1D_1(s) = 0$$

or

$$1 + \frac{K_1D_1(s)}{D(s)} = 0 \qquad ...(5.26)$$

Since the characteristic equation (5.26) is of form $1 + K_1G_1(s)\ H_1(s) = 0$, the root locus can be easily sketched based on pole-zero configuration of $G_1(s)\ H_1(s) = \frac{D_1(s)}{D(s)}$

Next, K_2 is choosen as variable parameter while considering that K_1 is fixed at some value. Then the characteristic equation corresponding to T(*s*) given by (5.25), can be written as

$$1 + \frac{K_2D_2(s)}{D(s)+K_1D_1(s)} = 0 \qquad ...(5.27)$$

Since, (5.27) is of form $1 + K_2G_2(s)\ H_2(s) = 0$, the root contours of (5.27) when K_2 varies from 0 to ∞ (while K_1 is fixed), can be easily constructed based on pole-zero configuration of

$$G_2(s)\ H_2(s) = \frac{D_2(s)}{D(s)+K_1D_1(s)} \qquad ...(5.28)$$

The following example demonstrates the construction of root contours in case of multi parameter variation.

Consider a system with characteristic equation $s^3 + K_2 s^2 + K_1 (s + 1) = 0$

where K_1 and K_2 are variable parameters, varying from 0 to ∞.

Setting $K_2 = 0$, we have

$$s^3 + K_1 (s + 1) = 0$$

or $$1 + \frac{K_1(s+1)}{s^3} = 0$$

The root locus for $G_1(s)\, H_1(s) = \dfrac{(s+1)}{s^3}$ as K_1 varies from 0 to ∞, is shown in Fig. 5.24(*a*) with the help of following information.

(*i*) **Locate poles and zeros of G(*s*) H(*s*) and plot:** Number of poles of $G_1(s)\, H_1(s) = n = 3$, all the three located at $s = 0$

Number of zero of $G_1(s)\, H_1(s) = m = 1$, located at $s = -1$

Number of separate root locus branches = 3

Starting points are $s = 0$

One locus branch terminates at zero ($s = -1$) and remaining two branches at ∞.

(*ii*) **Real axis locus segment:** between $s = 0$ and $s = -1$

(*iii*) **Centroid**

$$\sigma_A = \frac{\sum \text{poles} - \sum \text{zeros}}{n - m} = \frac{0-(-1)}{2} = +0.5$$

(*iv*) **Asymptotic angles**

$$\theta_A = \frac{(2q+1)\,180^o}{2}, \quad q = 0, 1$$

$$= 90°, 270° = \pm 90°$$

(*v*) **Break away points:** from characteristic equation

$$1 + \frac{K_1(s+1)}{s^3} = 0$$

$$-K_1 = \frac{s^3}{s+1}$$

and $$-\left(\frac{dK_1}{ds}\right) = \frac{3s^2(s+1) - s^3}{(s+1)^2} = \frac{2s^3 + 3s^2}{(s+1)^2}$$

The roots of $\dfrac{dK_1}{ds} = 0$ are $s = 0$ and -1.5. Thus break away point is located at $s = 0$. The point $s = -1.5$ does not lie on real axis locus segment.

(vi) **Imaginary axis crossing point** can be easily evaluated to lie at $s = 0$ using Routh's array.

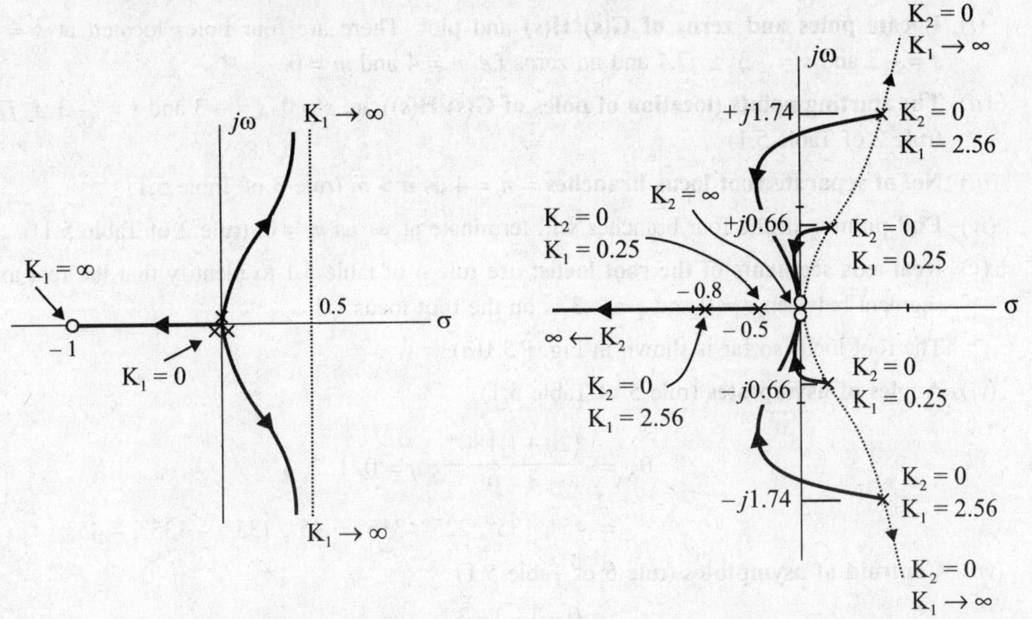

(*a*) $K_2 = 0$ (*b*) K_2 varies and K_1 is constant

Fig. 5.24: Root contours of $s^3 + K_2s^2 + K_1(s + 1) = 0$

Now, choosing K_2 as variable parameter while keeping K_1 constant at some non zero value, the given characteristic equation can be written as

$$1 + \frac{K_2 s^2}{s^3 + K_1(s+1)} = 0$$

The root contours for $G_2(s)\ H_2(s) = \dfrac{s^2}{s^3 + K_1(s+1)}$ as K_2 varies from 0 to ∞, is shown in

PROBLEMS AND SOLUTIONS

Fig. 5.24 (*b*) for $K_1 = 0.25$ and $K_1 = 2.56$. Note that the root contours when K_2 varies from 0 to ∞ while K_1 is fixed, must all emanate from root locus of Fig. 5.24 (*a*). The construction rules discussed earlier, remain unchanged. The information required to sketch the root locus [Fig. 5.24(*b*)] has been left as an exercise to the reader.

P5.1: For a system with open-loop transfer function

$$G(s)\,H(s) = \frac{K}{s(s+3)(s^2+6s+64)}$$

Sketch the root locus showing all details thereon. Comment on the stability of the system.

Solution: The step by step procedure for sketching the root locus is as follows.

(*i*) **Locate poles and zeros of G(*s*) H(*s*)** and plot. There are four poles located at $s = 0$, $s = -3$ and $s = -3 \pm j7.4$ and no zeros *i.e.* $n = 4$ and $m = 0$

(*ii*) **The starting points (location of poles of G(*s*) H(*s*))** are $s = 0$, $s = -3$ and $s = -3 \pm j7.4$ (rule 2 of Table 5.1)

(*iii*) **No. of separate root locus branches** = $n = 4$ as $n > m$ (rule 3 of Table 5.1)

(*iv*) **End points:** all the four branches will terminate at ∞ as $m = 0$. (rule 2 of Table 5.1)

(*v*) **Real axis segments of the root locus:** use rule 4 of table 5.1 to identify that the real axis segment between $s = 0$ and $s = -3$ is on the root locus.

The root locus so far is shown in Fig. P5.1(*a*).

(*vi*) **Angles of asymptotes** (rule 5 of Table 5.1)

$$\theta_A = \frac{(2q+1)180^o}{4-0}\ ;\ q = 0, 1, 2, 3$$

$$= 45°, 135°, 225°, 315° = 45°, 135°, -135°, -45°$$

(*vii*) **Centroid of asymptotes** (rule 6 of Table 5.1)

$$\sigma_A = \frac{0-3-3-3}{4} = -2.25$$

(*viii*) **Break away point** (rule 7 of Table 5.1)

From characteristic equation

$$1+\frac{K}{s(s+3)(s^2+6s+64)} = 0$$

$$K = -s\,(s+3)\,(s^2+6s+64)$$

and
$$-\frac{dK}{ds} = 4s^3 + 27s^2 + 164s + 192$$

The break away point must be the root of equation $dK/ds = 0$ and lie on real axis locus segment between $s = 0$ and $s = -3$. The trial and error solution of $dK/ds = 0$, gives the break away point at $= -1.45$

Alternatively, use magnitude criterion to get

$$K = \left|\frac{1}{G(s)H(s)}\right| = |s\,(s+3)\,(s^2+6s+64)|$$

and choose few points between $s = 0$ and $s = -3$, calculate the corresponding values of K and prepare a table as follows:

Value of s	Value of K
– 1.3	127.93
– 1.4	128.93
– 1.45	129.01
– 1.5	128.81
– 1.6	127.59

From table $s = -1.45$ can be taken as break away point as it corresponds to maximum K in range $-3 < s < 0$.

(*ix*) **Angles of departure** from the top complex pole at $s = -3 + j7.4$ (rule 9 of Table 5.1) from Fig. P 5.1(*a*)

$$\phi_d = 180° - (90° + 90° + 112°) = -112°$$

(*x*) **Imaginary axis crossing point** (rule 8 of Table 5.1)

The characteristic equation

$$1 + \frac{K}{s(s+3)(s^2+6s+64)} = 0$$

so $$s^4 + 9s^3 + 82s^2 + 192s + K = 0$$

has Routh array as follows:

s^4	1	82	K
s^3	9	192	0
s^2	60.67	K	
s^1	$192 - \frac{9K}{60.67}$	0	
s^0	K	0	

For $K = \frac{192 \times 60.67}{9} = 1294.29$, s^1 row will have all zero entries (a condition for roots to lie on imaginary axis). The auxiliary equation constructed from s^2 row for this value of K

$$60.67\, s^2 + K = 0$$

or $$60.67\, s^2 + 1294.29 = 0$$

gives $$s = \pm\, j\, 4.62$$

So, the root locus branches intersect imaginary axis at $s = \pm j\, 4.62$ and corresponding value of K is 1294.29.

The complete root locus is shown in Fig. P5.1(*b*).

(*xi*) **Stability analysis:** For K > 1294.29, the roots move towards right half plane. So, the system is stable for K in range 0 < K < 1294.29. For K = 1294.29, system exhibits sustained oscillation with frequency 4.62 rad/sec. For K > 1294.29 system becomes unstable.

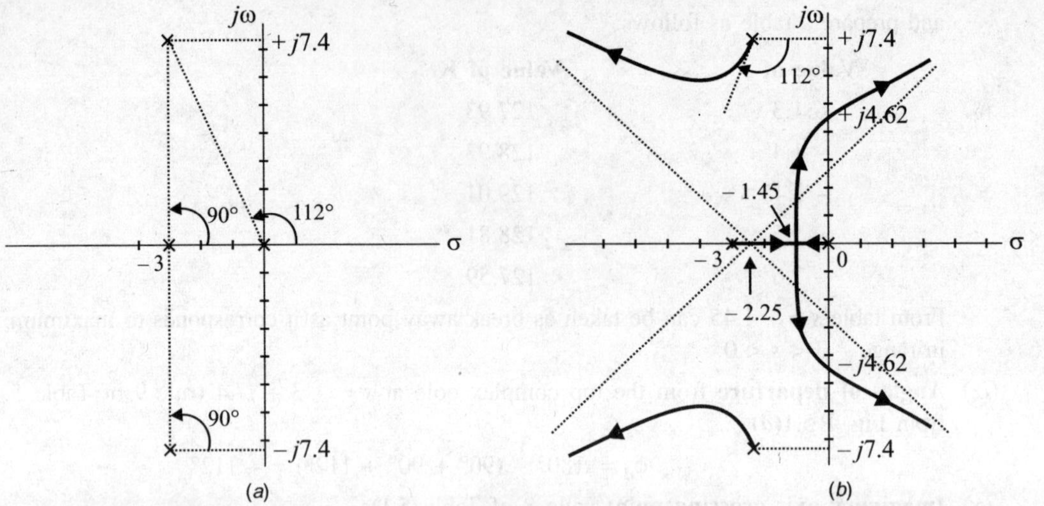

Fig. P5.1: Root locus

P 5.2: Consider the system shown in Fig. P 5.2 (*a*).

(*a*) Sketch root loci as K varies from 0 to ∞.

(*b*) Find the range of K for which system exhibits underdamped and overdamped dynamics.

(*c*) Find value of K for the system to be critically damped.

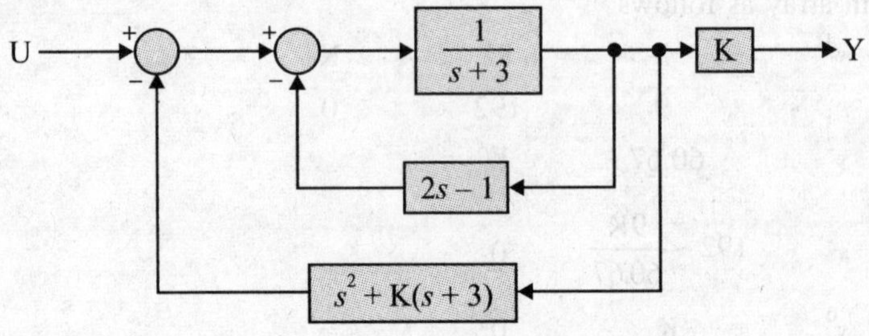

Fig. P5.2 (*a*): System

Solution:

(*a*) Use block diagram reduction rules to get

$$T(s) = \frac{Y(s)}{U(s)} = \frac{K}{s^2 + 3s + 2 + K(s+3)} = \frac{K}{1 + \dfrac{K(s+3)}{(s+1)(s+2)}}$$

and the characteristic equation

$$1 + \frac{K(s+3)}{(s+1)(s+2)} = 0$$

Compare it with general characteristic equation

$$1 + G(s)\,H(s) = 0$$

to get the product G(s) H(s) as

$$G(s)\,H(s) = \frac{K(s+3)}{(s+1)(s+2)}$$

The step by step approach to sketch the root locus for G(s) H(s) product is as follows:

(*i*) **Root locus is symmetrical about real axis**

(*ii*) **Locate poles and zeros of G(s) H(s) and plot**

Number of poles = n = 2

Location of poles: $s = -1$ and $s = -2$

Number of zeros = m = 1

Location of zeros: $s = -3$

Number of separate root locus branches = 2 as $n > m$.

Starting points: $s = -1$ and $s = -2$

Terminating points: one branch terminates at $s = -3$ and another at $s = -\infty$ along negative real axis.

(*iii*) **Root locus segments on the real axis:** between $s = -1$ and $s = -2$ and between $s = -3$ and $s = -\infty$.

Since $n - m = 1$, computation of centroid and asymptotic angles, is irrelevant.

(*iv*) **Break away/entry points:** from characteristic equation

$$1 + \frac{K(s+3)}{(s+1)(s+2)} = 0$$

$$-K = \frac{(s+1)(s+2)}{(s+3)}$$

and $$-\frac{dK}{ds} = \frac{d}{ds}\left[\frac{(s+1)(s+2)}{(s+3)}\right] = \frac{s^2+6s+7}{(s+3)^2}$$

Since, break away/entry points must be the roots of $\frac{dK}{dS} = 0$

$$s^2 + 6s + 7 = 0 \text{ gives}$$

$$s = -4.41,\ -1.59$$

Note that the points $s = -4.41$ and $s = -1.59$, both lie on the real axis locus segment. So, $s = -1.59$ is actual break away point and $s = -4.41$ is actual entry point.

(*v*) **Break away/break in angles**

$$= \pm\frac{\pi}{r};\quad r = 2$$

$$= \pm 90°$$

The complete root locus, using the information furnished above, is shown in Fig. P5.2(*b*).

(*b*) Use rule 10 of table 5.1 to find the value of K at which the loci leave the real axis

$$= \frac{0.59 \times 0.41}{1.41} \cong 0.17$$

and the value of K at which the loci enter the real axis

$$= \frac{3.41 \times 2.41}{1.41} \cong 5.83$$

So the system exhibits underdamped dynamics in the range $0.17 < K < 5.83$. In this range, the roots are complex conjugate.

The system exhibits overdamped dynamics in the range $0 \le K < 0.17$ and $K > 5.83$ as in this range, the roots are negative real and distinct.

(*c*) For K = 0.17 and K = 5.83, the roots are negative real and equal, so, system will be critically damped.

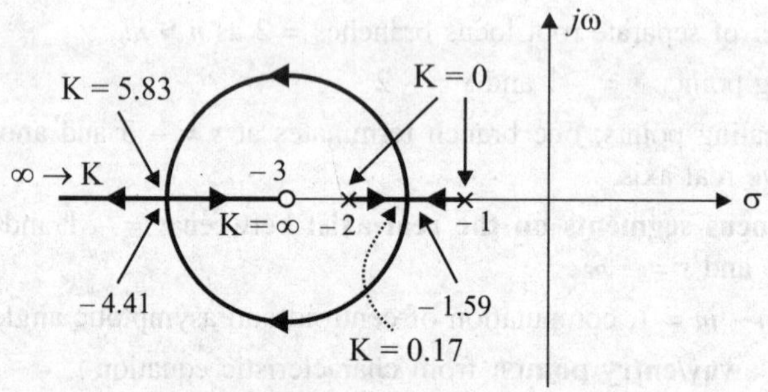

Fig. P5.2 (*b*): Root locus

P5.3: A simplified model of an auto pilot of an aircraft is given by

$$G(s)\ H(s) = \frac{K(s+\alpha)}{s(s-\beta)\left(s^2+2\xi\omega_n s+\omega_n^2\right)}$$

Choose $\alpha = \beta = 1$, $\xi = 0.5$ and $\omega_n = 4$. Sketch the root locus and find the range of gain K for stability.

Solution: For given α, β, ξ and ω_n, the product

$$G(s)\ H(s) = \frac{K(s+1)}{s(s-1)\left(s^2+4s+16\right)}$$

The step by step approach to sketch the root locus is as follows:

(*i*) **Locate poles and zeros of G(*s*) H(*s*) and plot**

Number of poles = $n = 4$

Number of zeros = $m = 1$

Locations of poles: $s = 0$, $s = +1$ and $s = -2 \pm j3.46$

Locations of zeros: $s = -1$

Number of separate root locus branches = 4 as $n > m$.

Starting points: $s = 0$, $s = 1$ and $s = -2 \pm j3.46$

Terminating points: one locus branch terminates at $s = -1$ and remaining three branches do so at ∞.

(*ii*) **Root locus segments on the real axis:** between $s = 0$ and $s = +1$ and between $s = -1$ and $s = -\infty$.

(*iii*) **Centroid**

$$\sigma_A = \frac{\sum \text{poles} - \sum \text{zeros}}{n-m}$$

$$= \frac{0+1-2-2-(-1)}{4-1} = -\frac{2}{3}$$

(*iv*) **Angles of asymptotes**

$$\theta_A = \frac{(2q+1)180^o}{3}; \quad q = 0, 1, 2$$

$$= 60°, 180°, 300°$$

(*v*) **Break away/entry points**

from characteristic equation

$$1 + \frac{K(s+1)}{s(s-1)(s^2+4s+16)} = 0$$

$$-K = \frac{s(s-1)(s^2+4s+16)}{(s+1)}$$

and

$$-\frac{dK}{ds} = \frac{3s^4+10s^3+21s^2+24s-16}{(s+1)^2}$$

The break away/entry point must be the root of equation $dK/ds = 0$

or $$3s^4 + 10s^3 + 21s^2 + 24s - 16 = 0$$

and the real roots must be located at real axis segments of the root locus. The trial and error gives two roots on real axis at $s = 0.45$ and $s = -2.26$. These roots also lie on real axis locus segments. So, these are actual break away and entry points. Now other two roots can easily be evaluated as $s = -0.76 \pm j2.16$, but they are neither break away nor entry points as they do not satisfy the angle criterion.

(*vi*) **Break away angles** $= \pm 180°/r = \pm 90°$ as $r = 2$

(*vii*) **Imaginary axis crossing points:** The characteristic equation

$$1 + \frac{K(s+1)}{s(s-1)(s^2+4s+16)} = 0$$

or $$s^4 + 3s^3 + 12s^2 + (K-16)s + K = 0$$

has Routh array that begins as follows:

s^4	1	12	K
s^3	3	K–16	0

$$
\begin{array}{lcc}
s^2 & \dfrac{52-\mathrm{K}}{3} & \mathrm{K} \\
s^1 & \dfrac{-\mathrm{K}^2+59\mathrm{K}-832}{52-\mathrm{K}} & 0 \\
s^0 & \mathrm{K} & 0
\end{array}
$$

s^1 row will have all zero entries for

$$\frac{-\mathrm{K}^2+59\mathrm{K}-832}{52-\mathrm{K}} = 0$$

or $$\mathrm{K}^2 - 59\mathrm{K} + 832 = 0$$

or $$(\mathrm{K} - 35.7)(\mathrm{K} - 23.3) = 0$$

or $$\mathrm{K} = 35.7 \text{ and } 23.3$$

The auxiliary equation constructed from s^2 row is

$$\frac{52-\mathrm{K}}{3}s^2 + \mathrm{K} = 0$$

Solving it for K = 35.7 and 23.3, the imaginary axis crossing points are

$$s = \pm\, j\,2.56 \quad \text{and} \quad s = \pm j1.56.$$

(*viii*) **Angle of departure from top complex pole**

$$\phi_d = 180° - (120° + 130° + 90°) + 106°$$
$$= -54°$$

The complete root locus using the information furnished above is shown in Fig. P5.3 (*b*). It is easy to see from root locus that the system is stable in range, 23.3 < K < 35.7. Outside this range two roots of characteristic equation lie in right half plane.

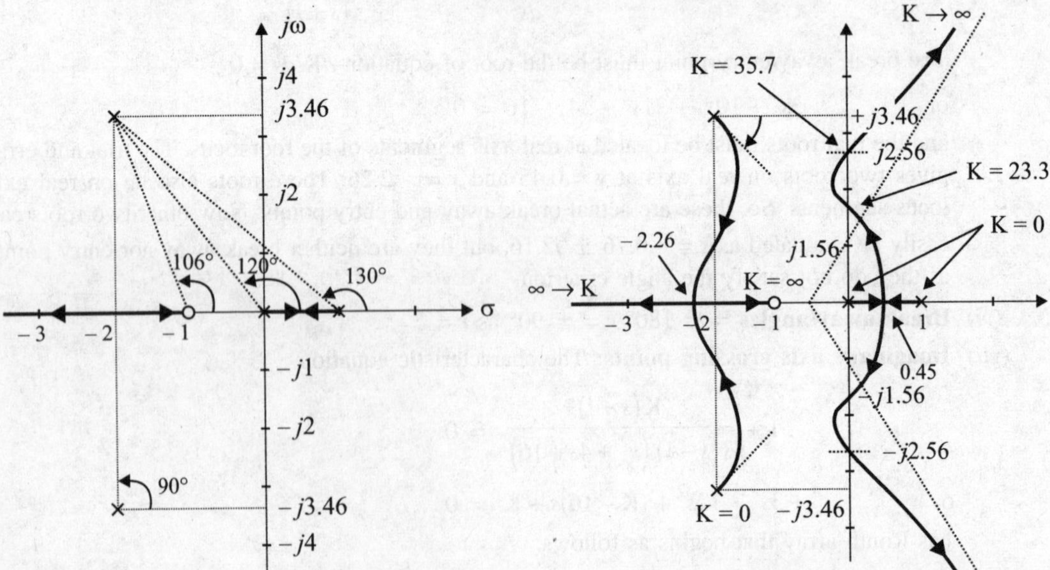

(*a*) Pole-zero plot together with real axis locus segments

(*b*) Complete root locus

Fig. P5.3: Root locus

P 5.4: Sketch the root loci for the system shown in Fig. P5.4 (*a*) and show that system is stable for all K > 0.

Solution: The step by step approach to sketch the root locus is as follows:

(*i*) **Locate poles and zeros of G(*s*) H(*s*) and plot**

$$G(s)\,H(s) = \frac{K(s+0.4)}{s^2(s+3.6)}$$

Number of poles of G(*s*) H(*s*) = n = 3

Locations of poles: at s = 0, s = 0 and s = – 3.6

Number of zeros of G(*s*) H(*s*) = m = 1

Location of zeros: at s = – 0.4

Number of separate root locus branches = 3

Starting points: s = 0, s = 0 and s = – 3.6

End points: One branch terminates at zero located at s = – 0.4 and remaining two branches terminate at ∞.

(*ii*) **Real axis locus segments:** between s = – 0.4 and s = – 3.6.

(*iii*) **Centroid**

$$\sigma_A = \frac{\sum \text{poles} - \sum \text{zeros}}{n-m} = \frac{-0-0-3.6-(-0.4)}{3-1} = -1.6$$

(*iv*) **Angles of asymptotes**

$$\theta_A = \frac{(2q+1)180^o}{3-1} = \pm 90°; \quad q = 0, 1$$

(*v*) **Break away/entry points**

From characteristic equation

$$1 + G(s)\,H(s) = 0$$

or $$1 + \frac{K(s+0.4)}{s^2(s+3.6)} = 0$$

or $$-K = \frac{s^2(s+3.6)}{(s+0.4)}$$

and $$-\frac{dK}{ds} = \frac{(3s^2+7.2s)(s+0.4)-(s^3+3.6s^2)}{(s+0.4)^2}$$

The break away/entry point must be the root of equation dK/ds = 0

or $$s^3 + 2.4s^2 + 1.44s = 0$$

or $s(s + 1.2)^2 = 0$

or $s = 0, -1.2$

Thus $s = 0$ is break away point and $s = -1.2$ is break in point. Note that $s = -1.2$ is repeated root. When repeated roots occur in the equation $\frac{dK}{ds} = 0$, $\frac{d^2K}{ds^2} = 0$ at that point.

The value of gain K at $s = -1.2$ is

$$K = \left[-\frac{s^3 + 3.6s^2}{s + 0.4} \right]\Bigg|_{s=-1.2} = 4.32$$

Hence, three root locus branches meet at point $s = -1.2$ for $K = 4.32$.

(*vi*) **Break away angles** $= \pm 180°/r = \pm 60°$ as $r = 3$.

(*vii*) **Imaginary axis crossing points:** The characteristic equation

$$1 + \frac{K(s+0.4)}{s^2(s+3.6)} = 0$$

or $s^3 + 3.6s^2 + Ks + 0.4K = 0$

has Routh array that begins as follows:

s^3	1	K
s^2	3.6	0.4 K
s^1	$\frac{3.2K}{3.6}$	0
s^0	0.4 K	0

s^1 row will have all zero entries for $K = 0$ and for this value of K, solving the auxiliary equation constructed from s^2 row, gives $s = 0$, so, the root locus is tangent to the imaginary axis due to presence of double pole at origin and there are no other points where the locus branches cross the imaginary axis.

Since, for all values of $K > 0$, the root locus branches remain in left half of s-plane, the system is stable for all values of $K > 0$.

The complete root locus in shown in Fig. P5.4(*b*).

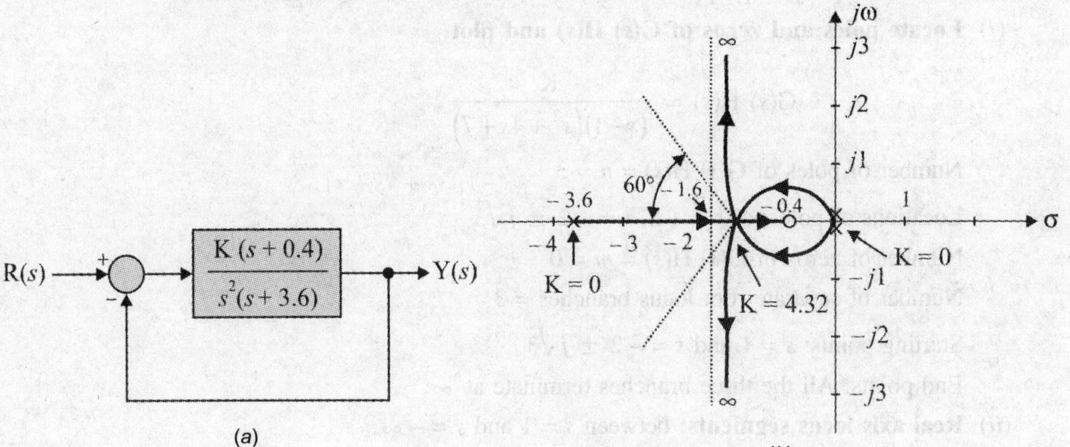

Fig. P 5.4: (*a*) System (*b*) Root locus

P5.5: Sketch the root loci for the control system shown in Fig. P5.5(*a*). Determine the range of parameter K for stability. Apply the angle criterion to show that root locus branches consist of three straight lines.

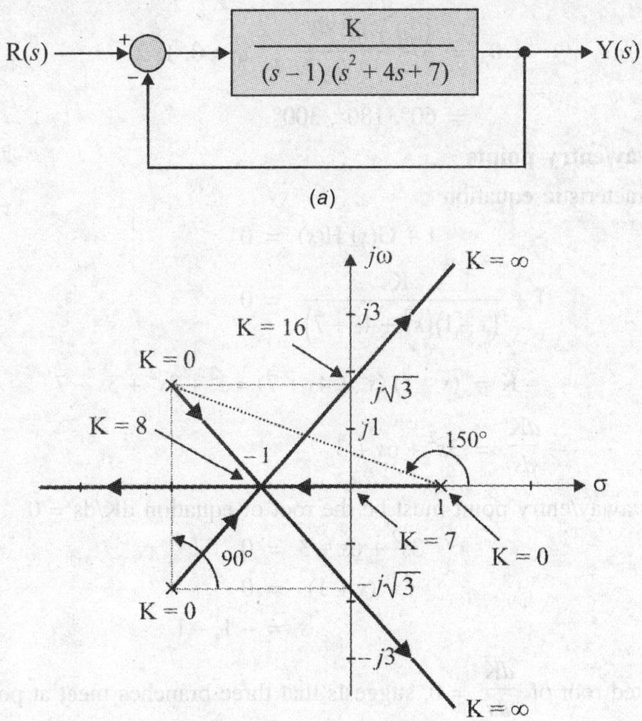

Fig. P 5.5: (*a*) System (*b*) Root locus

Solution: The step by step approach to sketch the root locus is as follows:

(*i*) **Locate poles and zeros of G(*s*) H(*s*) and plot**

$$G(s)\,H(s) = \frac{K}{(s-1)(s^2+4s+7)}$$

Number of poles of G(*s*) H(*s*) = $n = 3$

Locations of poles: at $s = +1$, $s = -2 \pm j\sqrt{3}$

Number of zeros of G(*s*) H(*s*) = $m = 0$

Number of separate root locus branches = 3

Starting points: $s = 1$ and $s = -2 \pm j\sqrt{3}$

End points: All the three branches terminate at ∞.

(*ii*) **Real axis locus segments:** between $s = 1$ and $s = -\infty$.

(*iii*) **Centroid**

$$\sigma_A = \frac{\sum \text{poles} - \sum \text{zeros}}{n-m} = \frac{1-2-2-0}{3} = -1$$

(*iv*) **Asymptotic angles**

$$\theta_A = \frac{(2q+1)180^o}{3}\ ; \quad q = 0, 1, 2$$

$$= 60°, 180°, 300°$$

(*v*) **Break away/entry points**

From characteristic equation

$$1 + G(s)\,H(s) = 0$$

or
$$1 + \frac{K}{(s-1)(s^2+4s+7)} = 0$$

or
$$-K = (s-1)(s^2+4s+7) = s^3 + 3s^2 + 3s - 7$$

and
$$-\frac{dK}{ds} = 3s^2 + 6s + 3$$

The break away/entry point must be the root of equation dK/ds = 0

or
$$3s^2 + 6s + 3 = 0$$

or
$$(s+1)^2 = 0$$

or
$$s = -1, -1$$

The repeated root of $\frac{dK}{ds} = 0$, suggests that three branches meet at point $s = -1$ for

$$K = \left|\left[-(s-1)(s^2+4s+7)\right]\right|_{s=-1} = 8$$

(*vi*) **Break away angles:** $\pm \left.\dfrac{180^o}{r}\right|_{r=3} = \pm\ 60°$

(*vii*) **Imaginary axis crossing points:** The characteristic equation

$$1 + \frac{K}{(s-1)(s^2+4s+7)} = 0$$

or $$s^3 + 3s^2 + 3s - 7 + K = 0$$

has Routh array that begins as follows:

s^3	1	3
s^2	3	K – 7
s^1	$\dfrac{16-K}{3}$	0
s^0	K – 7	0

s^1 row will have all zero entries for

$$\frac{16-K}{3} = 0$$

or $$K = 16$$

For K = 16, the auxiliary equation constructed from s^2 row, is

$$3s^2 + (K-7) = 0$$

or $$3s^2 + 9 = 0$$

or $$s = \pm j\sqrt{3}$$

Thus, the root locus branches cross the imaginary axis at $s = \pm j\sqrt{3}$ for K = 16.

It is also obvious from Routh array that K – 7 > 0 or K > 7 is the requirement for the system to be stable. Thus, one of the roots lies at origin for K = 7.

(*viii*) **Angle of departure from top complex pole:**

$$\phi_d = 180° - (90° + 150°) = -60°$$

The complete root locus in shown in Fig. P5.5 (*b*). Now it is easy to find that system is stable for range 7 > K > 16.

To show that root locus branches consist of straight lines, apply the angle criterion *i.e.*

$$\left.\angle \frac{K}{(s-1)(s^2+4s+7)}\right|_{s=\sigma+j\omega} = \pm\ 180°\ (2q+1)$$

or $$\angle \frac{K}{(\sigma+j\omega-1)\left[(\sigma+j\omega)^2+4(\sigma+j\omega)+7\right]} = \pm\ 180°\ (2q+1)$$

or $$-\tan^{-1}\left(\frac{\omega}{\sigma-1}\right) - \tan^{-1}\left[\frac{(2\omega\sigma+4\omega)}{\sigma^2+4\sigma-\omega^2+7}\right] = \pm\ 180°\ (2q+1)$$

Taking **tan** of both the sides, we have

$$-\frac{\dfrac{\omega}{\sigma-1}+\dfrac{2\omega(\sigma+2)}{\sigma^2+4\sigma-\omega^2+7}}{1-\left(\dfrac{\omega}{\sigma-1}\right)\left[\dfrac{2\omega(\sigma+2)}{\sigma^2+4\sigma-\omega^2+7}\right]} = 0$$

which with little algebraic manipulation, yields

$$\omega\,(3\sigma^2 + 6\sigma - \omega^2 + 3) = 0$$

This equation can further, be factorised to take the form

$$\omega\left(\sigma+1+\frac{\omega}{\sqrt{3}}\right)\left(\sigma+1-\frac{\omega}{\sqrt{3}}\right) = 0$$

which represents the following three straight lines.

$$\omega = 0$$

$$(\sigma+1)+\frac{\omega}{\sqrt{3}} = 0$$

and

$$(\sigma+1)-\frac{\omega}{\sqrt{3}} = 0$$

P5.6: The G(*s*) H(*s*) product of a feedback system is given by

$$G(s)\,H(s) = \frac{K}{s(s+1)(s+9)}$$

Determine if the points given below, lie on the locus of characteristic roots. If yes, find the corresponding value of K.

(*a*) $s = -0.5$ (*b*) $s = -2$ (*c*) $s = j3$ (*d*) $s = -0.5 + j0.5$

Solution:

(*a*)
$$\left.\angle\frac{K}{s\,(s+1)\,(s+9)}\right|_{s=-0.5} = \angle\frac{K}{-0.5\,(-0.5+1)\,(-0.5+9)} = 180°$$

Thus G(*s*)H(*s*) satisfies the angle criterion at $s = -0.5$. The point $s = -0.5$ lies on the root locus.

The characteristic equation

$$1+\frac{K}{s(s+1)(s+9)} = 0$$

$$K|_{s=-0.5} = -s\,(s+1)\,(s+9)|_{s=-0.5}$$

$$K = 2.125$$

(*b*)
$$\left.\angle\frac{K}{s\,(s+1)\,(s+9)}\right|_{s=-2} = \angle\frac{K}{(-2)\,(-2+1)\,(-2+9)} = 0°$$

Angle criterion is not satisfied and therefore point $s = -2$ does not lie on the root locus.

(*c*) $$\angle \frac{K}{s(s+1)(s+9)}\bigg|_{s=j3} = \angle \frac{K}{j3(j3+1)(j3+9)}$$

$$= -90° - \tan^{-1}\left(\frac{3}{1}\right) - \tan^{-1}\left(\frac{3}{9}\right)$$

$$= -90° - 71.57° - 18.43° = -180°$$

Angle criterion is satisfied and therefore, point $s = j3$ lies on the root locus.

The corresponding value of K is obtained by using magnitude criterion as follows:

$$K\Big|_{s=j3} = \left|s(s+1)(s+9)\right|_{s=j3}$$

$$= |j3(1+j3)(9+j3)|$$

$$= 3\cdot\sqrt{10}\cdot\sqrt{90} = 90$$

(*d*) $$= \angle \frac{K}{s(s+1)(s+9)}\bigg|_{s=-0.5+j0.5}$$

$$= \angle \frac{K}{(-0.5+j0.5)(-0.5+j0.5+1)(-0.5+j0.5+9)}$$

$$= 45° - 45° - 3.37°$$

$$= -3.37° \neq \pm 180°$$

Angle criterion is not satisfied and therefore point $s = -0.5 + j\,0.5$ does not lie on the root locus.

P5.7: Sketch the root locus for the product

$$G(s)\,H(s) = \frac{K}{s(s+3)\left(s^2+3s+11.25\right)}$$

and comment on stability.

Solution: The step by step approach to sketch the root locus is as follows:

(*i*) **Locate poles and zeros of G(*s*) H(*s*) and plot**

Number of poles = $n = 4$

Locations of poles: $s = 0$, $s = -3$, $s = -1.5 \pm j3$

Number of zeros = $m = 0$

Number of separate root locus branches = 4

Starting points: $s = 0$, $s = -3$ and $s = -1.5 \pm j3$

Terminating points: all the four locus branches terminate at ∞

(*ii*) **Root locus segments on the real axis:** between $s = 0$ and $s = -3$

The root locus so far is shown in Fig. P5.7(*a*).

(*iii*) **Centroid**

$$\sigma_A = \frac{\sum \text{poles} - \sum \text{zeros}}{n-m} = \frac{-0-3-1.5-1.5-0}{4} = -1.5$$

(*iv*) **Asymptotic angles**

$$\theta_A = \frac{(2q+1)180°}{4}; \quad q = 0, 1, 2, 3$$

$$= 45°, 135°, 225°, 315° = \pm 45°, \pm 135°$$

(*v*) **Break away/entry points**

From characteristic equation

$$1 + G(s)\,H(s) = 0$$

or
$$1 + \frac{K}{s(s+3)(s^2+3s+11.25)} = 0$$

we have

$$-K = s\,(s+3)\,(s^2 + 3s + 11.25)$$

and
$$-\frac{dK}{ds} = \frac{d}{ds}\,[s^4 + 6s^3 + 20.52s^2 + 33.75s]$$

$$= 4s^3 + 18s^2 + 40.5s + 33.75$$

Equating $\frac{dK}{ds}$ to zero and solving, we have

$$s = -1.5 \quad \text{and} \quad -1.5 \pm j1.84.$$

The point $s = -1.5$ lies on real axis locus segment. So, this is an actual break away point.

To test whether or not, the points $s = -1.5 \pm j1.84$, are actual break away/entry points, use angle criterion.

$$\left.\angle \frac{K}{s\,(s+3)\,(s+1.5+j3)\,(s+1.5-j3)}\right|_{s=-1.5+j1.84}$$

$$= -\tan^{-1}\left(\frac{1.84}{-1.5}\right) - \tan^{-1}\left(\frac{1.85}{1.5}\right) - \tan^{-1}(\infty) - \tan^{-1}(-\infty)$$

$$= -180°$$

Thus, angle criterion is satisfied and points $s = -1.5 \pm j1.84$, are actual break away/entry points.

Use magnitude criterion *i.e.* $|G(s)\,H(s)| = 1$, to obtain the corresponding values of K as follows:

$$K\big|_{s=-1.5} = \left|s(s+3)(s^2+3s+11.25)\right|_{s=-1.5} = 20.25$$

and
$$K\big|_{s=-1.5\pm j1.84} = \left|s(s+3)(s^2+3s+11.25)\right|_{s=-1.5\pm j1.84} = 31.64$$

(*vi*) **Intersection of root locus with imaginary axis**

The characteristic equation

$$s(s + 3)(s^2 + 3s + 11.25) + K = 0$$

or $$s^4 + 6s^3 + 20.25s^2 + 33.75s + K = 0$$

has Routh array that begins as follows:

s^4	1	20.25	K
s^3	6	33.75	
s^2	14.625	K	
s^1	$\dfrac{493.59 - 6K}{14.625}$	0	
s^0	K	0	

s^1 row will have all zero entries for

$$\frac{493.59 - 6K}{14.625} = 0$$

or $$K = 82.27$$

The auxiliary equation from s^2 row is

$$14.625s^2 + K = 0$$

or $$s^2 = -\frac{82.27}{14.625}$$

or $$s = \pm\, j2.37$$

So, the locus branches cross the imaginary axis at $s = \pm\, j2.37$.

(*vii*) **Angle of departure from top complex pole**

$$\phi_d = 180° - (90° + 63.4° + 116.6°) = -90°$$

Using the information furnished above together with property of root locus that it is symmetrical about real axis, the complete root locus in drawn in Fig. P5.7 (*b*).

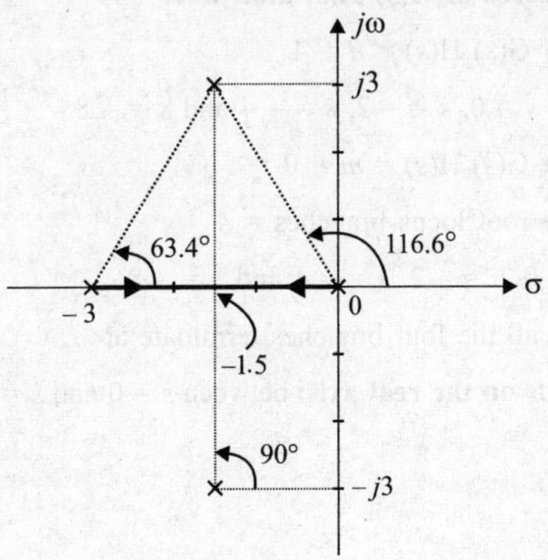

(*a*)

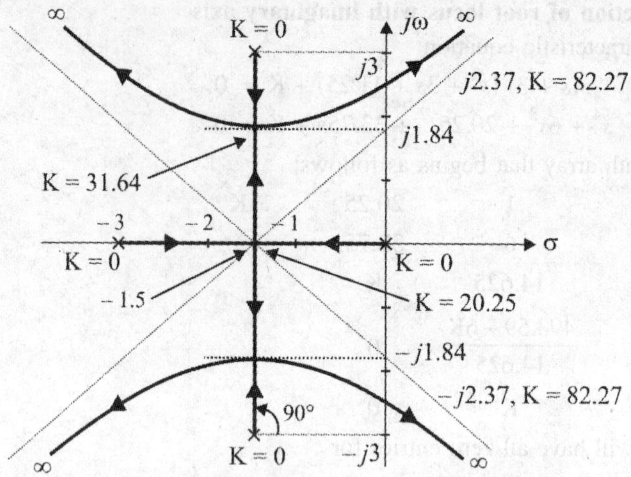

(*b*)

Fig. P 5.7: (*a*) pole-zero plot (*b*) Root locus

(*viii*) **Comment on stability**

For $0 < K < 82.27$ system is stable

For $K = 82.27$ system is marginally stable

For $K > 82.27$ system is unstable.

P 5.8: Sketch the root locus for a system with product

$$G(s)\,H(s) = \frac{K}{s(s+2)(s+4)(s+8)}$$

and determine the location of dominant characteristic roots for damping ratio of 0.707.

Solution: The step by step approach to sketch the root locus is as follows:

(*i*) **Locate poles and zeros of G(*s*) H(*s*) and plot**

Number of poles of $G(s)\,H(s) = n = 4$

Locations of poles: $s = 0$, $s = -2$, $s = -4$ and $s = -8$

Number of zeros of $G(s)\,H(s) = m = 0$

Number of separate root locus branches = 4

Starting points: $s = 0$, $s = -2$, $s = -4$ and $s = -8$

Terminating points: all the four branches terminate at ∞.

(*ii*) **Root locus segments on the real axis:** between $s = 0$ and $s = -2$ and between $s = -4$ and $s = -8$

(*iii*) **Centroid**

$$\sigma_A = \frac{\sum \text{poles} - \sum \text{zeros}}{n-m} = \frac{-0-2-4-8}{4} = -3.5$$

(*iv*) **Asymptotic angles**

$$\theta_A = \frac{(2q+1)180°}{4}; \quad q = 0, 1, 2, 3$$

$$= \pm 45°, \pm 135°$$

(*v*) **Break away/entry points:** The characteristic equation

$$1 + G(s)\,H(s) = 0$$

or
$$1 + \frac{K}{s(s+2)(s+4)(s+8)} = 0$$

gives
$$-K = s(s+2)(s+4)(s+8)$$

and
$$-\frac{dK}{ds} = \frac{d}{ds}\left[s^4 + 14s^3 + 56s^2 + 64s\right]$$

$$= 4s^3 + 42s^2 + 112s + 64$$

equating $\frac{dK}{ds}$ to zero and solving, with trial and error we have approximately

$s = -0.8, -6.66, -3.05$

Since, the points $s = -0.8$ and -6.66 lie on real axis segment of root locus and point $s = -3.05$ does not, the actual break away points are $s = -0.8$ and $s = -6.66$.

(*vi*) **Break away angles** $= \pm \frac{180°}{r}; \quad r = 2$

$= \pm 90°$

(*vii*) **Intersection of root locus with imaginary axis**

The characteristic equation

$$s(s+2)(s+4)(s+8) + K = 0$$

or
$$s^4 + 14s^3 + 56s^2 + 64s + K = 0$$

has Routh array that begins as follows:

s^4	1	56	K
s^3	14	64	0
s^2	51.43	K	
s^1	64 – 0.27 K	0	
s^0	K	0	

s^1 row will have all zero entries for

$$64 - 0.27\,K = 0$$

or
$$K \cong 237$$

The auxiliary equation from s^2 row is

$$51.43s^2 + K = 0$$

or $$51.43s^2 + 237 = 0$$

gives the imaginary axis crossing point at $s = \pm\, j2.15$.

Using the information as furnished above together with property of root locus that it is symmetrical about real axis, the complete root locus is shown in Fig. P5.8.

The characteristic roots for $\xi = 0.707$ is given by the point of intersection of root locus with a line making an angle of $\cos^{-1} \xi = 45°$ with negative real axis. These roots are $s = -0.75 \pm j0.7$ as shown in Fig. P5.8.

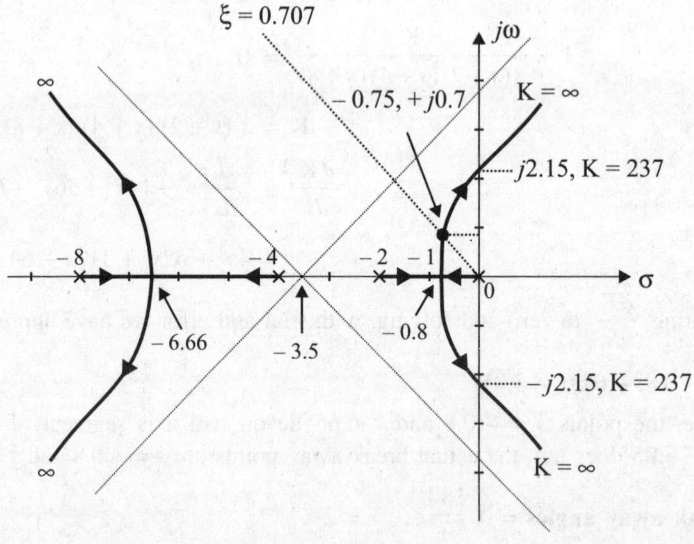

Fig. P5.8: Root locus

P5.9: Consider the system shown in Fig. P5.9(*a*).

(*a*) Sketch the root locus as z varies from 0 to ∞.

(*b*) Determine the value of z so that damping ratio of dominant closed-loop poles is 0.4.

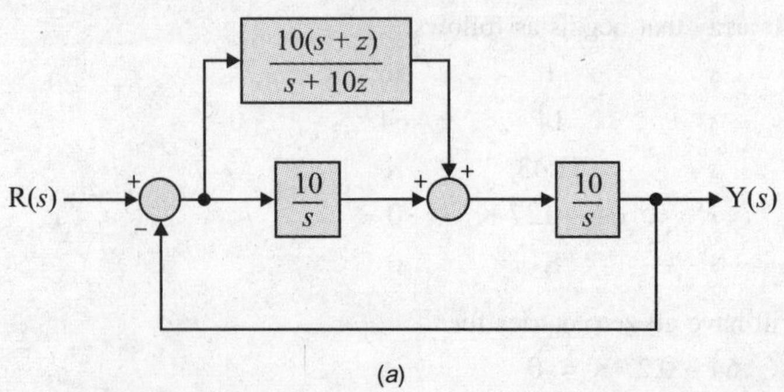

(*a*)

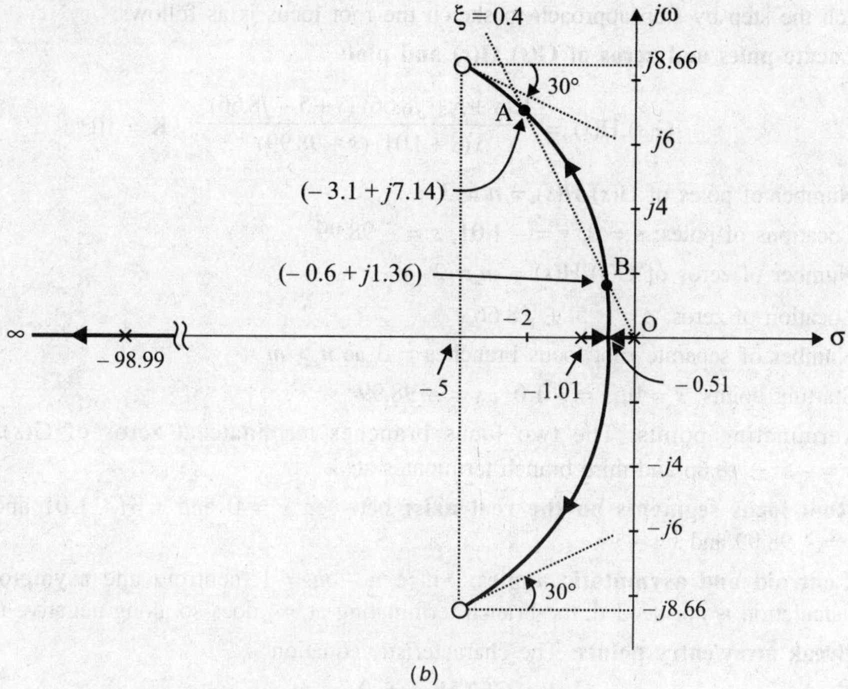

(b)

Fig. P5.9: (a) System (b) Root locus

Solution: (a) Use block diagram reduction technique to get overall transfer function

$$T(s) = \frac{Y(s)}{R(s)} = \frac{100s^2 + s(z+1) + 10z}{s^3 + (10z+100)s^2 + 100(z+1)s + 1000z}$$

and characteristic equation

$$s^3 + (10z + 100)s^2 + 100\,(z + 1)s + 1000z = 0$$

or

$$1 + \frac{10z(s^2 + 10s + 100)}{s(s^2 + 100s + 100)}$$

Compare it with general characteristic equation

$$1 + G(s)\,H(s) = 0$$

to get the product

$$G(s)\,H(s) = \frac{10z(s^2 + 10s + 100)}{s(s^2 + 100s + 100)}$$

$$= \frac{K(s + 5 + j8.66)(s + 5 - j8.66)}{s(s + 1.01)(s + 98.99)}; \quad K = 10z$$

for which the step by step approach to sketch the root locus is as follows:

(*i*) **Locate poles and zeros of G(*s*) H(*s*) and plot**

$$G(s)\,H(s) = \frac{K(s+5+j8.66)\,(s+5-j8.66)}{s(s+1.01)\,(s+98.99)}; \quad K = 10z$$

Number of poles of G(*s*) H(*s*) = n = 3

Locations of poles: $s = 0$, $s = -1.01$, $s = -98.99$

Number of zeros of G(*s*) H(*s*) = m = 2

Location of zeros: $s = -5 \pm j8.66$

Number of separate root locus branches = 3 as $n > m$

Starting points: $s = 0$, $s = -1.01$, $s = -98.99$

Terminating points: The two locus branches terminate at zeros of G(*s*) H(*s*) *i.e.* $s = -5 \pm j8.66$ and third branch terminates at ∞.

(*ii*) **Root locus segments on the real axis:** between $s = 0$ and $s = -1.01$ and between $s = -98.99$ and $s = -\infty$.

(*iii*) **Centroid and asymptotic angles:** Since $n - m = 1$, centroid and asymptotic angles calculation is not needed, the branch terminating at ∞, does so along negative real axis.

(*iv*) **Break away/entry points:** The characteristic equation

$$1 + G(s)\,H(s) = 0$$

or

$$1 + \frac{K\left(s^2+10s+100\right)}{s\left(s^2+100s+100\right)} = 0$$

gives

$$-K = \frac{s^3+100s^2+100s}{s^2+10s+100}$$

and

$$-\frac{dK}{ds} = \frac{\left(3s^2+200s+100\right)\left(s^2+10s+100\right)-(2s+10)\left(s^3+100s^2+100s\right)}{\left(s^2+10s+100\right)^2}$$

equating $-\dfrac{dK}{ds}$ to zero, we have

$$s^4 + 20s^3 + 1200s^2 + 20000s + 10000 = 0$$

Searching the break away point on real axis locus segment between $s = 0$ and $s = -1.01$ through trial and error, we have one approximate root $s = -0.51$ of the equation just above.

(*v*) **Angle of approach to top complex zero**

$$\phi_A = 180° + \tan^{-1}\left(\frac{8.66}{93.99}\right) - \tan^{-1}\left(\frac{8.66}{5}\right) - \tan^{-1}\left(\frac{8.66}{3.99}\right) - 90°$$

$$= 180° + 5.26° - 59.9° - 65.26° - 90°$$

$$= -29.9° \cong -30°$$

(*vi*) **Imaginary axis crossing point**

The characteristic equation

$$1 + \frac{K(s^2 + 10s + 100)}{s(s^2 + 100s + 100)} = 0$$

or $\quad s^3 + (100 + K)s^2 + (10K + 100)s + 100K = 0$

has Routh array that begins as follows:

s^3	1	(10 K + 100)
s^2	(100 + K)	100 K
s^1	$\frac{(100+K)(10K+100)-100K}{100+K}$	0
s^0	100 K	0

In the Routh array above s^1 row will have all zero entries for

$$\frac{(100+K)(10K+100)-100K}{100+K} = 0$$

or $\quad K^2 + 100K + 1000 = 0$

or $\quad K = -50 \pm 38.72$

Since these values of K are negative, the root locus does not cross imaginary axis at any point.

The complete root locus is shown in Fig. P5.9(*b*) using the information as furnished above, together with property of root locus that it is symmetrical about real axis.

(*b*) For $\xi = 0.4$, $\cos^{-1}(0.4) = 66.4°$, a radial line OBA is drawn making an angle of 66.4° with negative real axis as shown in Fig. P5.9(*b*). This intersects the root locus at two points A and B as shown.

For point A, $s = -3.1 \pm j7.14$ and third root is approximately located at $s = -2389$ with corresponding value of $K \cong 145$ and $z \cong 14.5$

For point B, $s = -0.6 \pm j1.4$, $K \cong 2.2$, $z \cong 0.22$ and third root is located approximately at $s = -101$.

P 5.10: Consider the system with positive feedback as shown in Fig. P5.10(*a*). Sketch the root locus for K varying from 0 to ∞ and comment on stability. Show salient points on root locus.

Solution: The product

$$G(s)\,H(s) = \frac{K(s+2)}{(s+3)(s^2+2s+2)}$$

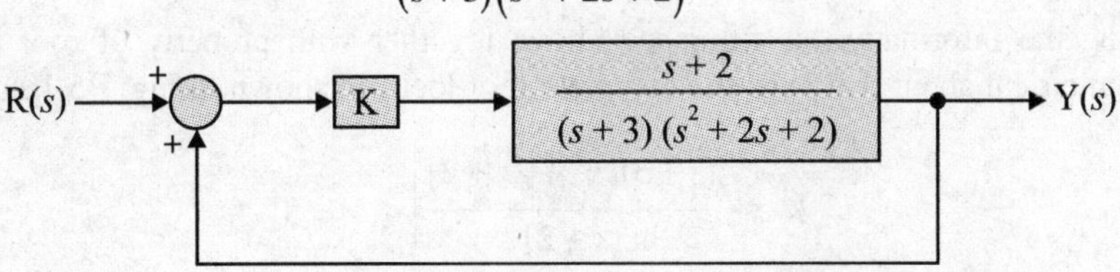

Fig. P5.10: (*a*) Positive feedback system

Note that varying K from 0 to ∞ in a system with positive feedback is equivalent to varying K from 0 to $-\infty$ in a system with negative feedback. Using the modified rules discussed in section 5.5, the step by step approach to sketch the root locus is as follows:

(*i*) **Locate poles and zeros of G(*s*) H(*s*) and plot** as shown in Fig. P5.10(*b*).

Number of poles of G(*s*) H(*s*) = $n = 3$

Locations of poles: $s = -3$, $s = -1 \pm j1$

Number of zeros of G(*s*) H(*s*) = $m = 1$

Location of zeros: $s = -2$

Number of separate root locus branches = $n = 3$ as $n > m$.

Starting points: $s = -3$, and $s = -1 \pm j1$

Terminating points: one root locus branch will terminate at zero located as s = – 2 and other two branches at ∞.

(*ii*) **Root locus segments on the real axis** (Use modified rule 4 discussed in Section 5.5): between $s = -2$ and $s = +\infty$ and between $s = -3$ and $s = -\infty$.

(*iii*) **Asymptotic angles**

$$\theta_A = \frac{\pm i\, 360°}{n-m} = \frac{\pm i\, 360°}{3-1} = \pm 180°$$

So, asymptotes are on real axis

(*iv*) **Break away/entry points:** The characteristic equation

$$(s+3)(s^2+2s+2) - K(s+2) = 0$$

gives
$$K = \frac{(s+3)(s^2+2s+2)}{(s+2)}$$

and
$$\frac{dK}{ds} = \frac{2s^3+11s^2+20s+10}{(s+2)^2}$$

equating $\frac{dK}{ds}$ to zero and solving by trial and error, we have one of root $s = -0.8$ and other two roots $s = -2.35 \pm j0.77$. Since $s = -0.8$ lies on root locus segment on real axis, it is an actual entry point. The points $s = -2.35 \pm j0.77$ are neither break away nor entry points as they do not satisfy angle criterion.

(*v*) **Angle of departure from top complex pole**

$$\phi_d = 0° - 90° - 27° + 45° = -72°$$

Using the information as furnished above together with property of root locus that it is symmetrical about real axis, the complete root locus in shown in Fig. P5.10 (*c*). **Note** that if

$$K > \left.\frac{(s+3)(s^2+2s+2)}{(s+2)}\right|_{s=0} = 3$$

one real root drifts into right half of s-plane. So, for K > 3, the system is unstable.

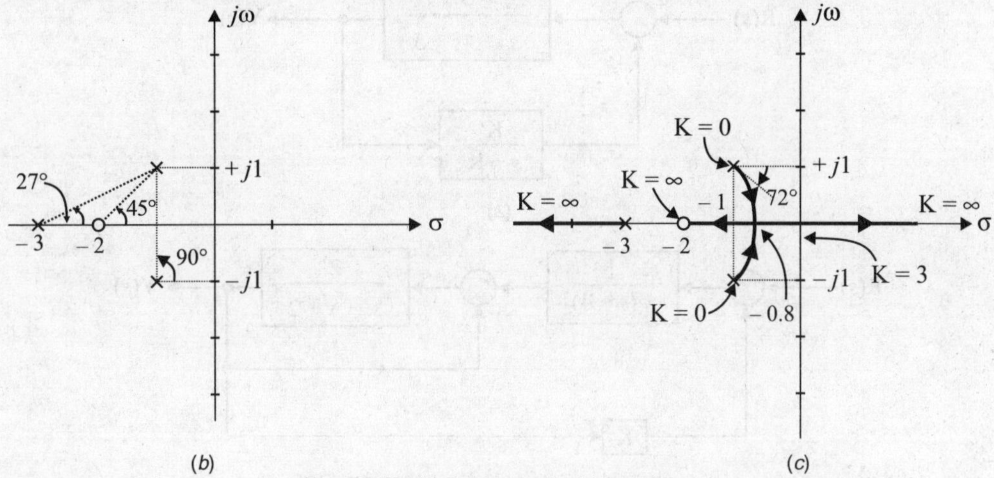

Fig. P5.10: (*b*) Pole-zero plot (*c*) Root locus

DRILL PROBLEMS

D5.1: Sketch root locus for the systems with pole-zero configuration shown in Fig. D5.1(*a*) and (*b*). Find asymptotic angles, centroid, approximate break away/entry points, angles of departure and angles of arrival where applicable.

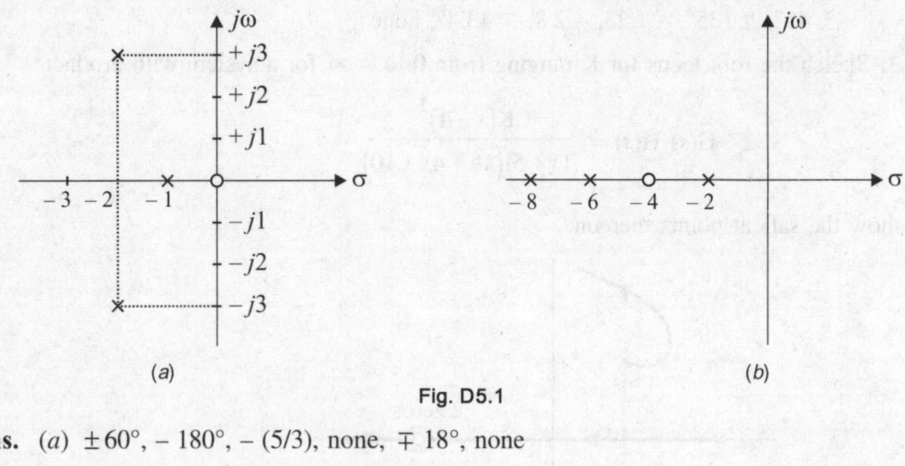

Fig. D5.1

Ans. (*a*) $\pm 60°$, $-180°$, $-(5/3)$, none, $\mp 18°$, none

(*b*) $\pm 90°$, -6, -6.87, none, none

D5.2: Develop root locus plots for the systems shown in Fig. D5.2 (*a*) and (*b*) for K ranging from 0 to ∞. Find asymptotic angles, centroid, approximate break away/entry points, angles of departure and angles of arrival where applicable.

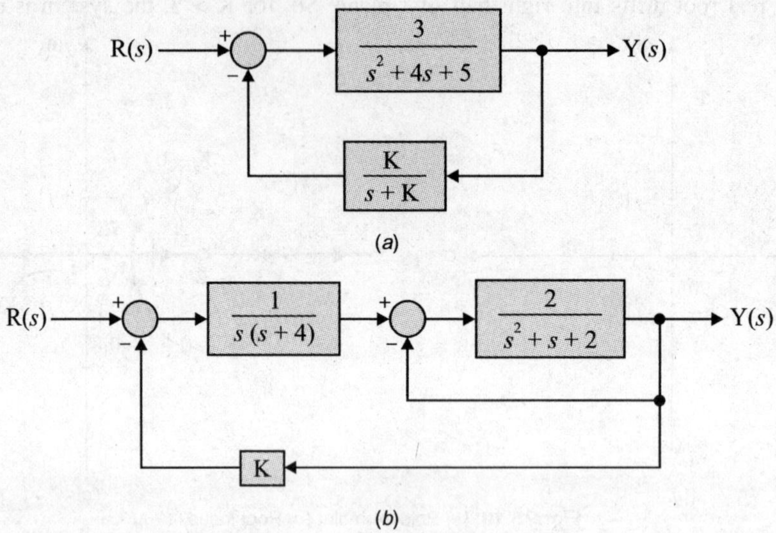

Fig. D5.2

Ans. (*a*) $G(s)\,H(s) = \dfrac{Ks^2 + 4s + 8}{s\left(s^2 + 4s + 5\right)}$, 180°, none, none, ∓ 63.4°, ± 45°

(*b*) $G(s)\,H(s) = \dfrac{2K}{s(s+4)\left(s^2 + s + 4\right)}$

± 45°, ± 135°, – 1.25, – 2.8, ∓ 43.4°, none

D5.3: Sketch the root locus for K ranging from 0 to – ∞ for a system with product

$$G(s)\,H(s) = \frac{K(s-4)^2}{(s+5)\left(s^2 + 4s + 10\right)}$$

and show the salient points thereon.

Ans.

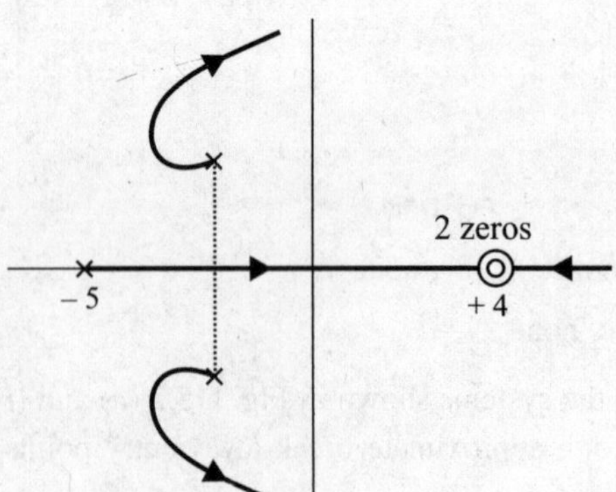

D 5.4: Sketch the root locus for the system with G(s) H(s) product given as

$$G(s)\,H(s) = \frac{K}{s^4 - 16}$$

Ans.

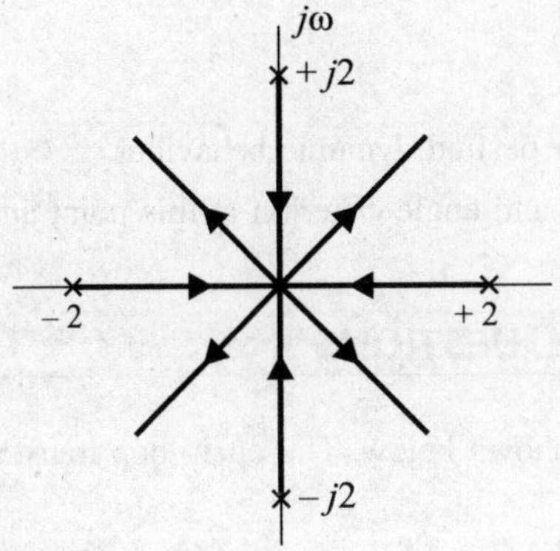

D 5.5: The pitch control system of an aircraft is shown in Fig. D 5.5. Sketch the root locus for K ranging from 0 to ∞. Determine the value of K so that damping ratio of dominant poles is 0.6. What is location of corresponding poles.

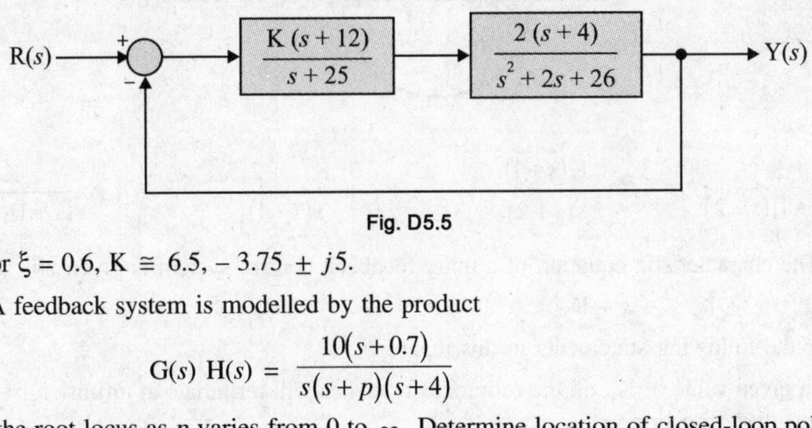

Fig. D5.5

Ans. For $\xi = 0.6$, K ≅ 6.5, $-3.75 \pm j5$.

D 5.6: A feedback system is modelled by the product

$$G(s)\,H(s) = \frac{10(s+0.7)}{s(s+p)(s+4)}$$

Sketch the root locus as p varies from 0 to ∞. Determine location of closed-loop poles for $p = 2$.

Ans. For variable parameter p(=K), $G(s)\,H(s) = \dfrac{K(s+4)}{(s+1)(s^2+3s+7)}$

for $p = 2$ locations: -0.45, $-2.8 \pm j2.8$.

D 5.7: Sketch the root locus for open-loop transfer function

$$G(s)\,H(s) = \frac{K(s^2+4)}{s(s+2)}$$

Calculate the value of K at

(*a*) Break away point (*b*) $s = -0.7 \pm j0.9$.

Ans. (*a*) 0.2 (*b*) 0.46

D 5.8: The characteristic equation for a control system is given by:

$$(s + 2)\ (s + 4)\ (s + \alpha) + K = 0$$

To achieve a good dynamic behaviour, it is desired that damping ratio $\xi = 0.5$ and that the natural frequency $\omega_n = 4$ rad/sec, determine α and K.

Ans. $\alpha = 8$ K = 95.92

Hints: Obtain dominant roots $s = -2 \pm j\,3.464$ for desired dynamic behaviour.

Constitute simultaneous equations using magnitude and angle criterion at this point and solve.

MULTIPLE CHOICE QUESTIONS

M5.1: The root locus of unity feedback system is shown below. The open-loop transfer function is given by

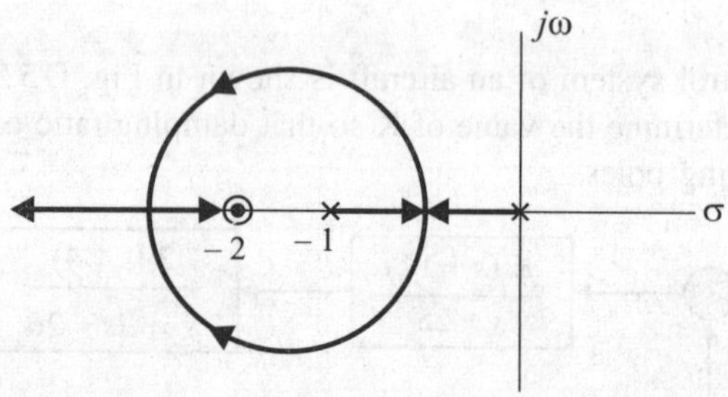

(*a*) $\dfrac{K}{s(s+1)(s+2)}$ (*b*) $\dfrac{K(s+1)}{s(s+2)}$ (*c*) $\dfrac{K(s+2)}{s(s+1)}$ (*d*) $\dfrac{Ks}{(s+1)(s+2)}$

M5.2: The characteristic equation of a unity feedback control system is given by

$$s^3 + K_1 s^2 + s + K_2 = 0$$

Consider the following statements in this regard.

1. For a given value of K_1, all the root locus branches will terminate at infinity for variable K_2 in the positive direction.
2. For a given value of K_2, all the root locus branches will terminate at infinity for variable K_1 in the positive direction.
3. For a given value of K_2, only one root locus branch will terminate at infinity for variable K_1 in the positive direction.

Of these statements:

(*a*) 1 and 2 are correct (*b*) 3 alone is correct

(*c*) 2 alone is correct (*d*) 1 & 3 are correct.

M5.3: The closed-loop transfer function of a feedback control system is given by

$$\frac{C(s)}{R(s)} = \frac{Ks}{s^2 + (3+K)s + 2}$$

Which one of the following diagrams represents the root locus diagram of the system for K > 0?

(*a*)

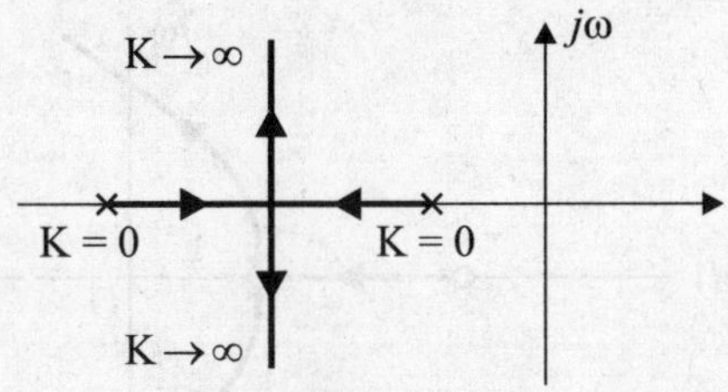

(*b*)

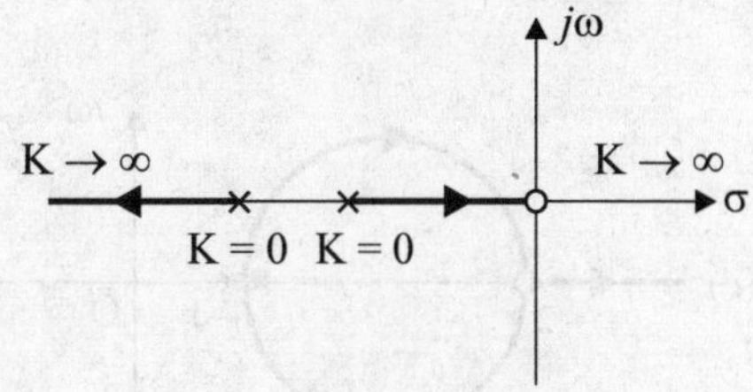

(*c*)

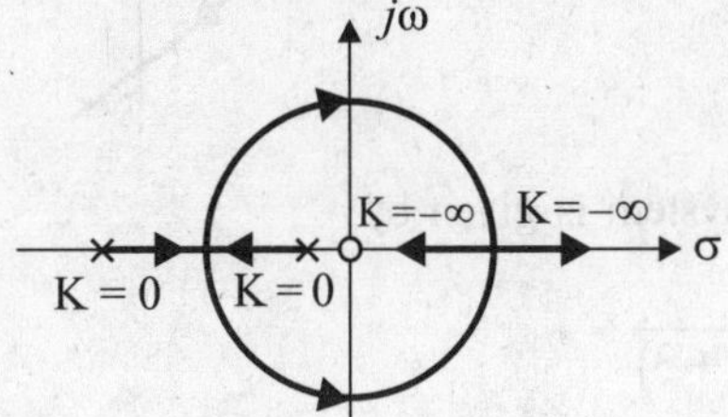

(*d*)

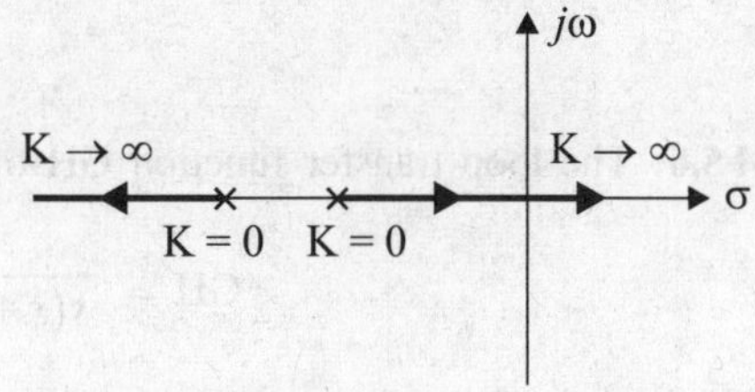

M 5.4: The figure shown below is the root locus of open-loop transfer function of a control system where

× represents pole

● represents zero

O – Origin

PQ = 2.6 = PQ′

PR = 1.4

OR = 2.0

OQ = 1.4 = OQ′

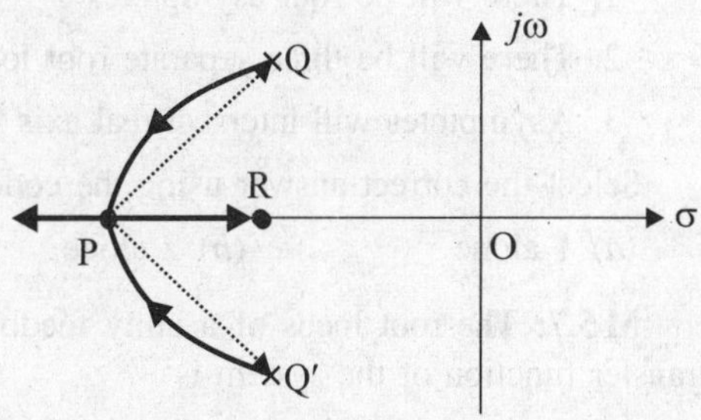

The value of the forward path gain K at the point P is

(*a*) 0.2 (*b*) 1.4 (*c*) 3.4 (*d*) 4.8

M 5.5: Given a unity feedback system with open-loop transfer function:

$$G(s) = \frac{K(s+2)}{(s+1)^2}$$

The correct root locus plot of the system is

(*a*)

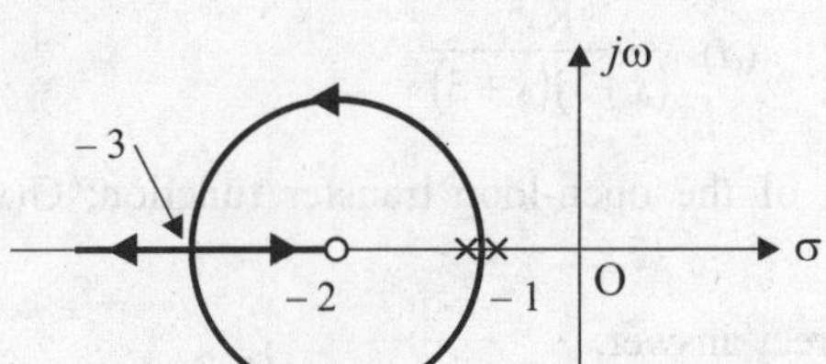

(*b*)

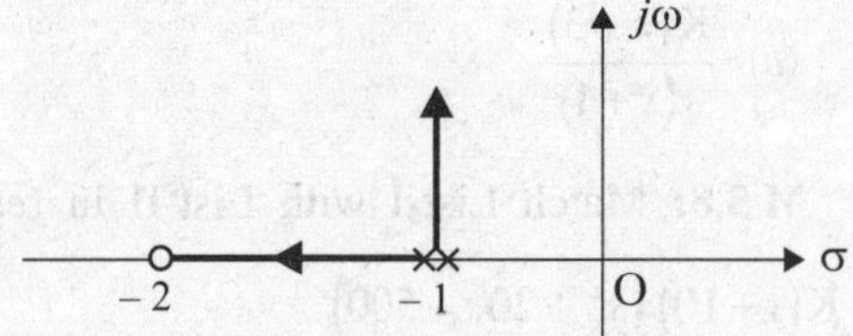

(*c*) 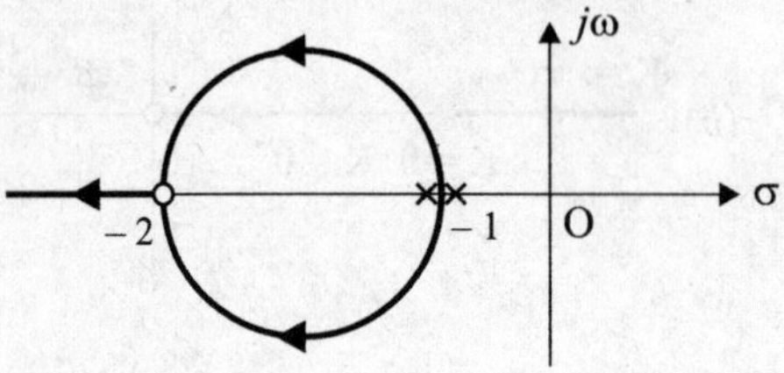

(*d*) 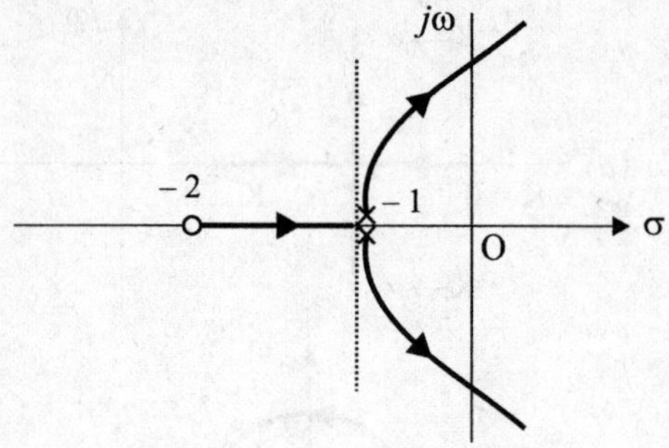

M 5.6: The loop transfer function GH of a control system is given by

$$\text{GH} = \frac{\text{K}}{s(s+1)(s+2)(s+3)}$$

Which of the following statements regarding the conditions of the system root loci diagram is/are correct?

1. There will be four asymptotes
2. There will be three separate root loci
3. Asymptotes will intersect real axis at $\sigma_A = -2/3$.

Select the correct answer using the codes given below:

(*a*) 1 alone (*b*) 2 alone (*c*) 3 alone (*d*) 1, 2 and 3.

M 5.7: The root locus of a unity feedback system is shown in the figure below. The open-loop transfer function of the system is

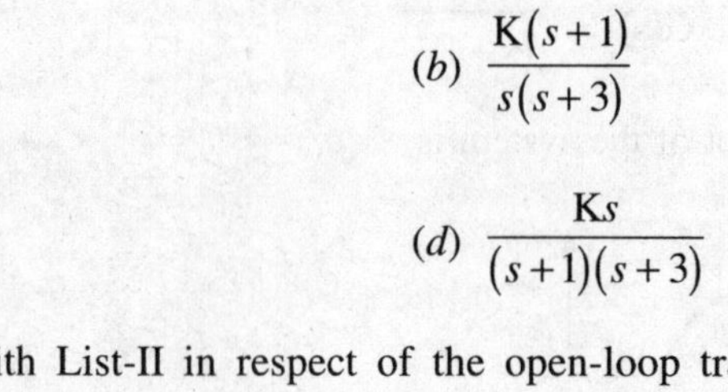

(*a*) $\dfrac{\text{K}}{s(s+1)(s+3)}$ (*b*) $\dfrac{\text{K}(s+1)}{s(s+3)}$

(*c*) $\dfrac{\text{K}(s+3)}{s(s+1)}$ (*d*) $\dfrac{\text{K}s}{(s+1)(s+3)}$

M 5.8: Match List-I with List-II in respect of the open-loop transfer function; G(*s*) H(*s*) = $\dfrac{\text{K}(s+10)\,(s^2+20s+500)}{s(s+20)\,(s+50)\,(s^2+4s+5)}$ and select the correct answer.

List I (Type of loci)	List II (Numbers)
A Separate loci	1. One
B Loci on the real axis	2. Two
C Asymptotes	3. Three
D Break away points	4. Five

Codes:	A	B	C	D
(*a*)	3	4	2	1
(*b*)	3	4	1	2
(*c*)	4	3	1	2
(*d*)	4	2	2	1

M 5.9: Which of the following effects are correct in respect of addition of a pole to the system loop transfer function?

1. The root locus is pulled to the right.
2. The system response becomes slower
3. The steady state error increases

Of these statements:

(*a*) 1 and 2 are correct (*b*) 1, 2 and 3 are correct

(*c*) 2 and 3 are correct (*d*) 1 and 3 are correct.

M 5.10: A unity feedback system has an open-loop transfer function of the form

$$\text{KG}(s) = \frac{\text{K}(s+a)}{s^2(s+b)}; \quad b > a$$

Which of the loci shown below can be valid root loci for the system?

(*a*) 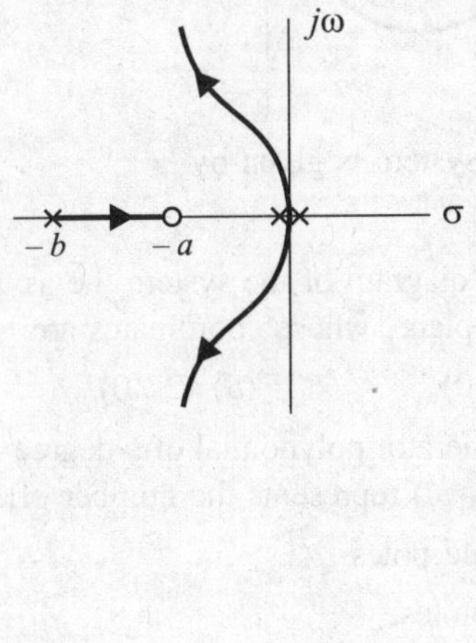

(*b*) 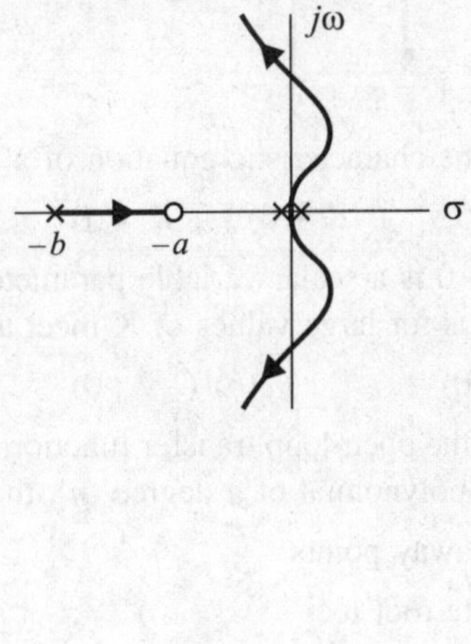

(c) 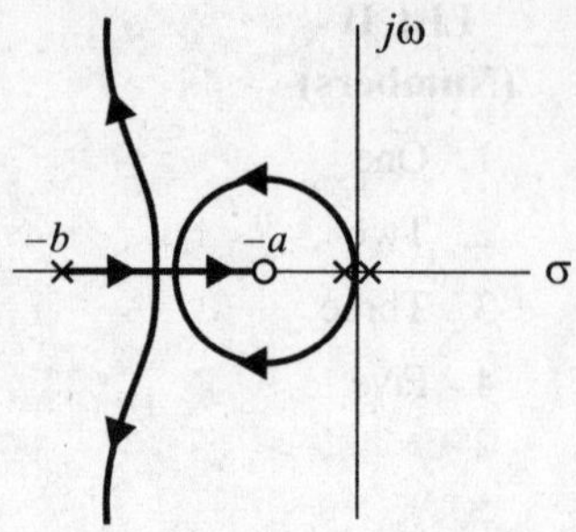

(d) 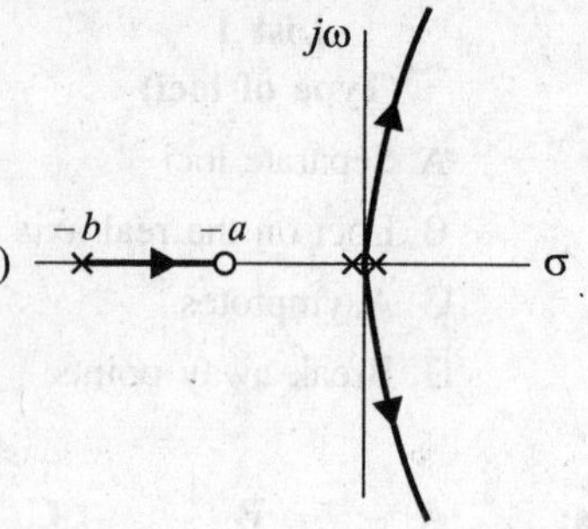

M 5.11: The transfer function of a closed-loop system is

$$T(s) = \frac{K}{s^2 + (3 - K)s + 1}$$

where K is the forward path gain. The root locus plot of the system is

(a) 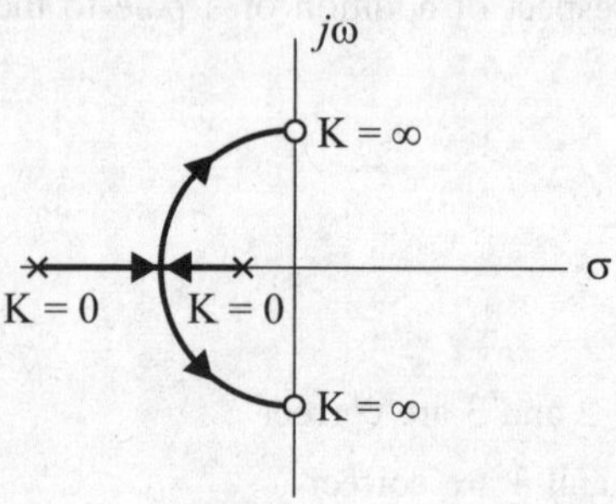

(b) 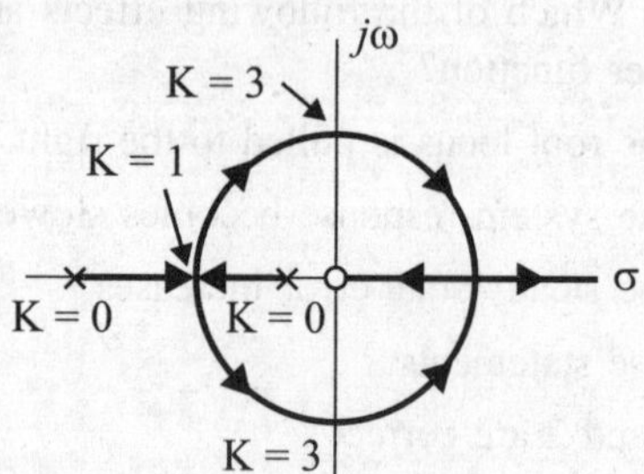

(c) 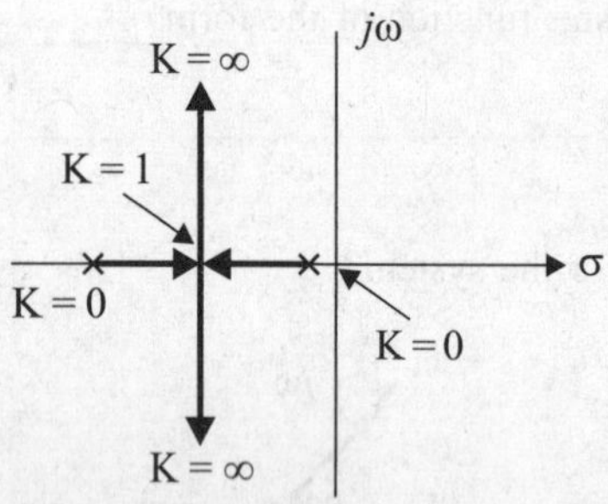

(d) 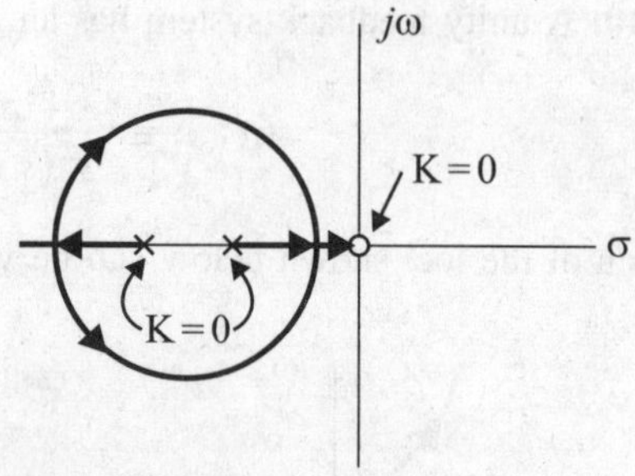

M 5.12: The characteristic equation of a feedback control system is given by

$$s^3 + 5s^2 + (K + 6)s + K = 0$$

where K > 0 is a scalar variable parameter. In the rootloci diagram of the system the asymptotes of the root locus for large values of K meet at a point in the *s*-plane, whose coordinates are:

(*a*) (– 3 , 0) (*b*) (– 2 , 0) (*c*) (– 1 , 0) (*d*) (2 , 0).

M5.13: If the open-loop transfer function is a ratio of a numerator polynomial of a degree '*m*' and a denominator polynomial of a degree '*n*', then the integer (*n* – *m*) represents the number of:

(*a*) break away points

(*b*) unstable poles

(*c*) separate root loci

(*d*) asymptotes.

M 5.14: Consider the points $s_1 = -3 + j4$ and $s_2 = -3 - j2$ in the s-plane. Then, for a system with the open-loop transfer function:

$$G(s)\,H(s) = \frac{K}{(s+1)^4}$$

(*a*) s_1 is on the root locus, but not s_2 (*b*) s_2 is on the root locus, but not s_1

(*c*) both s_1 and s_2 are on the root locus (*d*) neither s_1 nor s_2 is on the root locus.

M 5.15: The root locus diagram for a closed-loop feedback system is shown below. The system is overdamped:

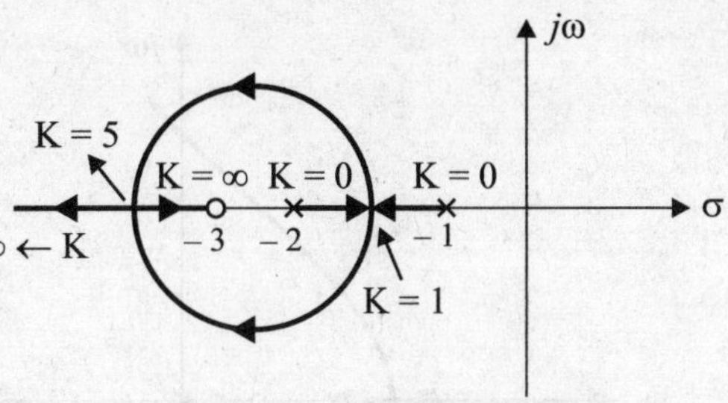

(*a*) only if $0 \le K \le 1$ (*b*) only if $1 < K < 5$

(*c*) only if $K > 5$ (*d*) only if $0 \le K < 1$ or $K > 5$.

M5.16: A closed loop system is described by characteristic equation

$$(s^2 - 4)(s + 1) + K(s - 1) = 0$$

The root locus with K varying from 0 to ∞, is

(*a*)

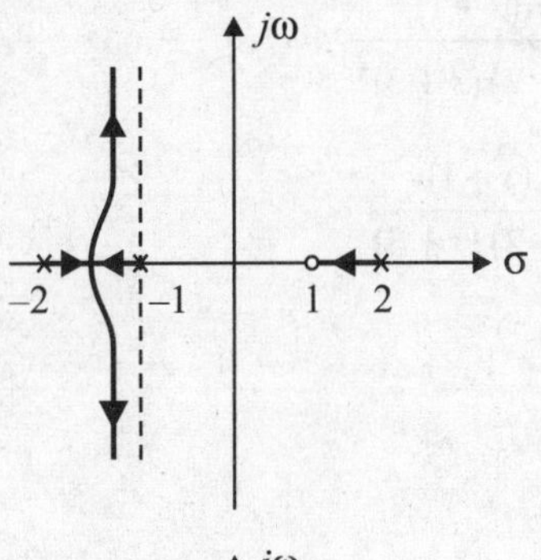

(*b*)

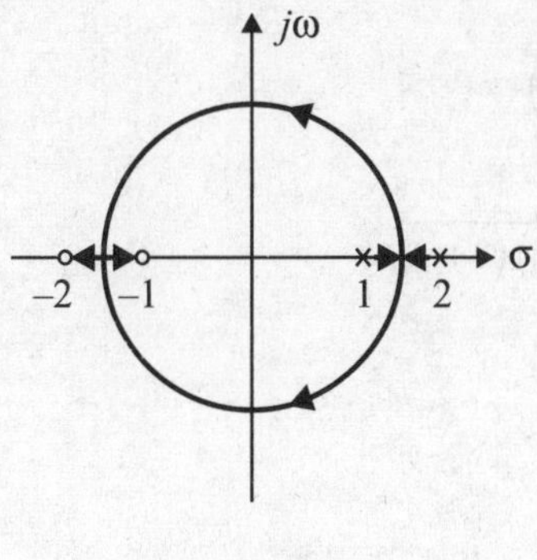

(*c*)

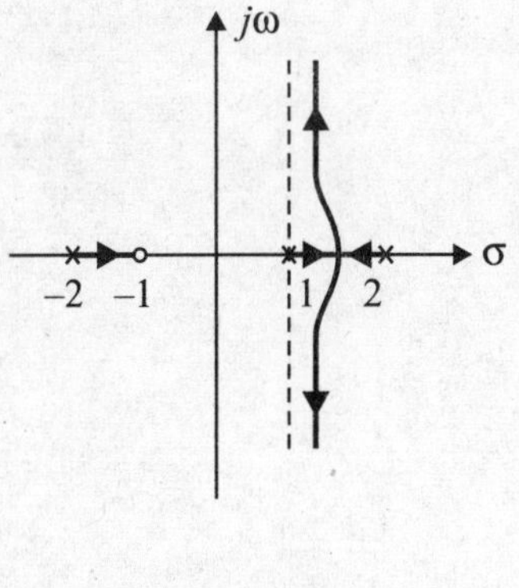

(*d*)

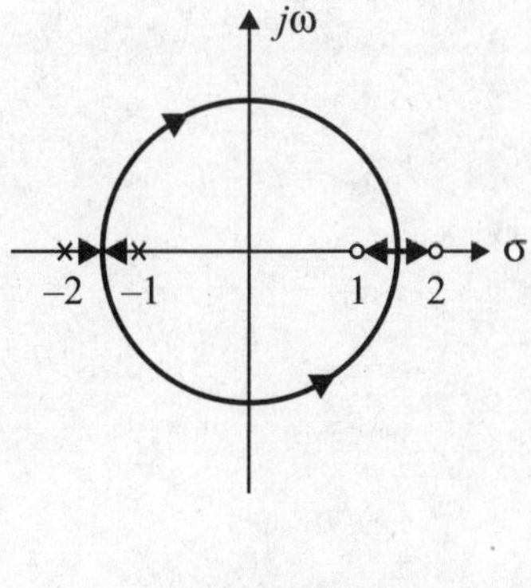

M5.17: The characteristic equation of a closed loop system is $s(s + 1)(s + 3) + K(s + 2) = 0$; $K > 0$. Which of the following statements is true?

(*a*) The roots are always real.

(*b*) The break away point can not lie in the range $-1 < Re[s] < 0$.

(*c*) Two of its roots tend to ∞ along the asymptotes $Re[s] = -1$.

(*d*) It may have complex roots in the rhp.

M5.18: The root locus of a system is shown in figure below. The corresponding open loop transfer function is

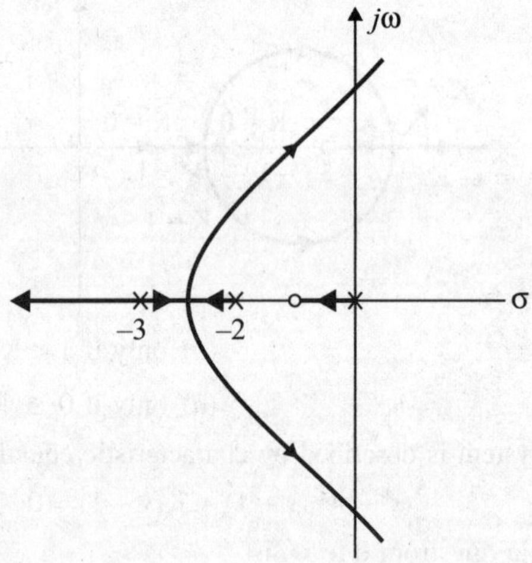

(*a*) $\dfrac{K\,s(s+1)}{(s+2)(s+3)}$ (*b*) $\dfrac{K(s+1)}{s(s+2)(s+3)^2}$

(*c*) $\dfrac{K}{s(s-1)(s+2)(s+3)}$ (*d*) $\dfrac{K(s+1)}{s(s+2)(s+3)}$.

ANSWERS

M5.1. (*c*)	M5.2. (*d*)	M5.3. (*b*)	M5.4. (*d*)	M5.5. (*a*)
M5.6. (*a*)	M5.7. (*c*)	M5.8. (*d*)	M5.9. (*b*)	M5.10. (*a*)
M5.11. (*b*)	M5.12. (*b*)	M5.13. (*d*)	M5.14. (*b*)	M5.15. (*d*)
M5.16. (*a*)	M5.17. (*c*)	M5.18. (*b*)		

Important Hints

M5.2: For K_1 as parameter $G(s) = \dfrac{K_1 s^2}{s^3 + s + K_2}$

for K_2 as parameter $G(s) = \dfrac{K_2}{s^3 + K_1 s^2 + s}$

M5.3: The characteristic equation is

$$s^2 + (3 + K)\, s + 2 = 0$$

or $\quad 1 + \dfrac{Ks}{s^2 + 3s + 2} = 0$ and $G(s)\,H(s) = \dfrac{Ks}{(s+1)(s+2)}$

M5.4:

$$K = \frac{PQ \times PQ'}{PR} = \frac{2.6 \times 2.6}{1.4} = 4.83$$

M5.5:

$$1 + G(s) = 0 \;\Rightarrow\; (s+1)^2 + K(s+2) = 0 \;\Rightarrow\; K = -\frac{(s+1)^2}{s+2}$$

$$\frac{dK}{ds} = 0 \;\Rightarrow\; (s+1)(s+3) = 0$$

break away points are – 1 and – 3.

M5.6: GH product has 4 poles and no zeros.

P = 4, Z = 0

No. of asymptotes = 4

No. of separate loci = 4

$$\sigma = \frac{-3-2-1}{4} = -\frac{6}{4} = -\frac{3}{2}$$

M5.7: Poles at origin and – 1

Zero at – 3.

M5.10: $\sigma = \dfrac{-b+a}{2}$

angles of asymptotes = 90°, 270°

for $b > a$, loci do not intersect imaginary axis except at $s = 0$.

from Routh's array system is stable for all $K > 0$ and $b > a$.

M5.11: Ch. eqn.: $s^2 + 3s + 1 - Ks = 0$

or $1 - \dfrac{Ks}{s^2+3s+1} = 0$ and $G(s)\,H(s) = -\dfrac{Ks}{s^2+3s+1}$

Real axis segment for negative parameter changes: between $s \cong -3.5$ and $s \cong -0.5$ and $s = 0$ and $s = +\infty$.

$$K = \frac{s^2+3s+1}{s}, \quad \frac{dK}{ds} = 0 \quad \Rightarrow s = \pm 1 \quad \textit{(Break away points)}$$

Routh Table

s^2	1	1		
s^1	$3-K$	0	$\Rightarrow$	$K = 3$
s^0	1	0		$A(s) = s^2 + 1 = 0$
			$\Rightarrow$	$s = \pm j1$

Imaginary axis intersection at $s = \pm j1$.

M5.12: Ch. eqn $= 1 + \dfrac{K(s+1)}{s(s+2)(s+3)}$ and $\sigma = \dfrac{-2-3-(-1)}{2} = -2$

M5.14:

$$G(s)\,H(s)\big|_{s=-3-j2} = \frac{K}{(-3-j2+1)^4} = \frac{K}{(-2-j2)^4}$$

$$\angle G(s)\,H(s)\big|_{s=-3-j2} = -180°$$

$$\angle G(s)\,H(s)\big|_{s=-3+j4} = -253.74° \neq \pm 180°$$

M5.15: Over damped system has two real negative but unequal roots.

M5.16: $(s^2 - 4)(s + 1) + K(s - 1) = 0 \Rightarrow 1 + \dfrac{K(s-1)}{(s^2-4)(s+1)} = 0$

Root locus is to be drawn for $G(s)\,H(s) = \dfrac{K(s-1)}{(s+2)(s-2)(s+1)}, \; 0 \le K \le \infty$

$G(s)\,H(s)$ has 2 poles in lhp at $s = -2$ and $s = -1$. $G(s)\,H(s)$ also has a pole at $s = +2$ and a zero at $s = +1$ in the rhp.

$$\sigma_A = \frac{\sum \text{poles} - \sum \text{zeros}}{n-m} = \frac{(-2+2-1)-(1)}{2} = -1$$

M5.17: $$G(s)\,H(s) = \frac{K(s+2)}{s(s+1)(s+3)}$$

$$\sigma_A = -1$$

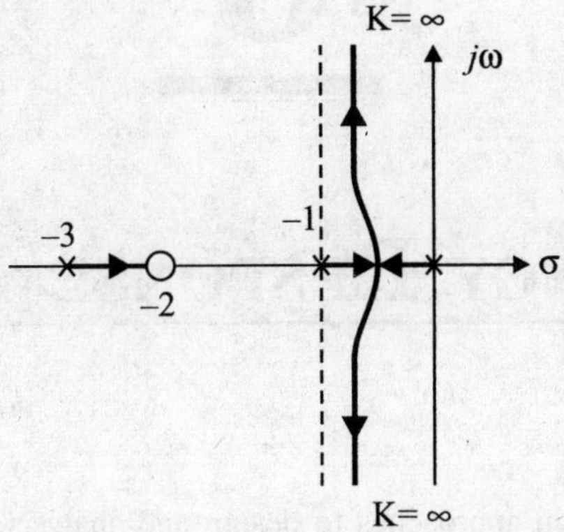

M5.18: The loop transmittance has a zero at $s = -1$ and poles at $s = 0$, $s = -2$, $s = -3$, $s = -3$. There are two poles at $s = -3$; real axis segment between $s = -2$ and $s = -\infty$ lies on root locus.

6

FREQUENCY RESPONSE ANALYSIS

6.1 INTRODUCTION

The two significant time domain approaches to design and analysis of closed loop control systems: the **Routh Stability** and the **Root Locus**, have been rigorously discussed in chapters 4 and 5 respectively. Recall that the Routh criterion provides information about whether or not a closed loop system is stable by working on the characteristic polynomial (the denominator polynomial of the transfer function of the closed loop system) without actually determining the roots thereof and root locus technique provides information about relative stability in terms of damping ratio of dominant roots. The Root Locus shows how the roots of characteristic polynomial move in the complex plane with one or more parameters varying from zero to infinity. These techniques do provide sufficient insight into transient behaviour of closed loop system. However, the applicability of these control tools is based on assumption that the system model, the G(*s*) H(*s*) product, is precisely known and that the G(*s*) H(*s*) product, additionally be a rational function of complex variable *s i.e.,* a ratio of two finite degree polynomials of *s*. These techniques become inapplicable if the system contains a pure delay of form e^{-Ts}, although an approximate analysis is possible by replacing e^{-Ts} by a truncated power series or a rational function such as padee' approximant.

Consider a class of systems such as communication systems, ac control systems, etc. wherein the signals to be processed are either sinusoidal or composed of sinusoidal components of different amplitudes and frequencies. In designing such systems, it is more logical to understand their behaviour as a function of incoming frequencies. The study of system behaviour as a function of frequency requires use of $s = j\omega$. The present chapter is entirely devoted to this approach. The time domain and the frequency domain, both the approaches together provide comprehensive view point of strengths and weaknesses of a system. Both the approaches are, therefore, required to be applied in order to fully understand and improve the system behaviour.

The **strengths** and **weaknesses** of frequency response approach, an interest of current topic, are highlighted as follows:

Strengths of frequency response approach

(*a*) The frequency response computations of a stable open loop system, are fairly easy to perform experimentally for which the prior information about transfer function model of the system is not necessary. This approach, simply, requires the application of a sinusoidal input to the system and measurement of the ratio of the magnitudes of output and input sinusoids as well as phase difference between them after steady state conditions are reached. These measurements can be easily and accurately, made by use of readily available sinusoidal signal generators and precise gain-phase meters. The measurements are repeated over an entire range of frequencies of interest.

(*b*) Whenever it is not possible to obtain the transfer function model of the system through analytical techniques, the necessary information to develop transfer function model, can be extracted from the frequency response.

(*c*) This approach can be applied to the systems that do not have rational transfer function (*e.g.*, system with delay of form e^{-Ts}) without actually using any approximation.

(*d*) There exist the correlating relations between the frequency domain and time domain performances in a linear system, so that the time domain properties of the system can be predicted based on the information gathered from frequency response. For the case of second order system, there is direct correlation between frequency response and transient response (we shall discuss in the following section). For the case of high order systems, there exists indirect correlation and therefore qualitative picture of transient response can always be predicted from its frequency response.

(*e*) The design and parameter adjustment of G(*s*) H(*s*) product for a prescribed closed loop performance, is easier in this approach. In addition, it is also relatively easy to visualize and assess the effects of undesirable noise and parameter variations.

(*f*) While the frequency response is usually found for linear systems, the method can be applied experimentally to certain class of non linear systems as well.

(*g*) The Nyquist criterion (to be discussed later in this chapter) is a frequency response tool which provides information about relative stability of system without actually determining the roots of characteristic polynomial. For difficult cases such as conditionally stable systems, the Nyquist criterion is probably the only tool to analyse stability.

Weaknesses of frequency response approach

(*a*) The experimental evaluation of frequency response, becomes fairly time consuming and cumbersome to perform for the systems with large time constants. This is because the time taken by system response to reach steady state is fairly long. So, the frequency response approach is generally not recommended for systems with large time constant.

(*b*) This approach is inapplicable to the systems of non interruptable nature. For such systems step or impulse response approach is preferred.

(*c*) Although this approach may be considered to be obsolete particularly in view of advent of PC computational packages. However, we again lay stress on the importance of the user backing up and double checking computer results with manual sketches and numerical values. In many cases, very accurate frequency response data and sketches can be arrived at without use of a PC or other computational device.

6.2 FREQUENCY RESPONSE

Steady state sinusoidal response

Consider a stable, linear, time invariant system as shown in Fig. 6.1. The input and response of the system with transfer function G(*s*), are represented by *r*(*t*) and *y*(*t*) respectively.

Fig. 6.1: A stable, linear, time invariant system

The response $y(t)$ to a sinusoidal input

$$r(t) = B\sin(\omega t + \alpha)$$

In general, is

$$y(t) = y_t(t) + y_s(t)$$

where $y_t(t)$ is transient part and $y_s(t)$ is steady state part of response. As far as frequency response approach (an interest of current topic) is concerned, we shall be interested in only steady state response. The steady state part of response, is also sinusoidal with same frequency ω. Generally, the amplitude and phase angle of steady state sinusoidal response, are different from those of input sinusoid and they depend upon input frequency. So, let the steady state part be

$$y_s(t) = C\sin(\omega t + \beta)$$

$$= A(\omega)\, B\sin[\omega t + \alpha + \phi(\omega)]$$

where $A(\omega)$ is ratio of output amplitude to input amplitude and is equal to magnitude of $G(s)$ evaluated at $s = j\omega$ *i.e.*,

$$A(\omega) = |G(s)|_{s=j\omega} = \frac{C}{B}$$

and $\phi(\omega)$ is phase difference between input and output and is equal to the angle of $G(s)$ evaluated at $s = j\omega$ *i.e.*,

$$\phi(\omega) = \angle G(s)\big|_{s=j\omega} = \beta - \alpha$$

Thus $$y_s(t) = |G(j\omega)|\, B\sin\left[\omega t + \alpha + \angle G(j\omega)\right]$$

The following example demonstrates the steady state response evaluation to sinusoidal input.

Example 1: Find the steady state sinusoidal response of a system with $G(s) = s/(s + 3)$ to input signal $r(t) = 7\cos(3t - 40°)$

Solution: Since the frequency of input sinusoid = 3 rad/sec.

$$\text{Amplitude ratio} = |G(s)|_{s=j3} = \left|\frac{j3}{j3+3}\right| = \frac{3}{\sqrt{18}} = \frac{1}{\sqrt{2}}$$

and $$\text{phase difference} = \angle G(s)\big|_{s=j3} = \angle\frac{j3}{j3+3} = 90° - \tan^{-1}(3/3) = 45°$$

Thus $$G(s)\big|_{s=j3} = \frac{1}{\sqrt{2}}\, e^{j45°}$$

and response $$y(t) = \frac{7}{\sqrt{2}}\cos(3t + 45° - 40°) = \frac{7}{\sqrt{2}}\cos(3t + 5°)$$

Frequency response evaluation

When a stable, linear, time invariant system with transfer function G(s), is driven by a sinusoidal input, the ratio of the amplitude of steady state response to the amplitude of input is given by

$$A(\omega) = \frac{\text{Amplitude of sinusoidal output}}{\text{Amplitude of sinusoidal input}} = |G(s)|_{s=j\omega} \qquad ...(6.1)$$

and the difference in phase angles of the output and the input is given by

$$\phi(\omega) = \angle G(s)\big|_{s=j\omega} \qquad ...(6.2)$$

where ω is frequency of input sinusoid in rad/sec. Note that frequency of output sinusoid is same as that of input sinusoid.

The magnitude $|G(j\omega)|$ versus ω together with phase $\angle G(j\omega)$ versus ω under steady state condition, is called **frequency response**. The amplitude ratio and phase difference, both are functions of ω in LTI systems, but the frequency response is independent of the amplitude and phase of the input sinusoid.

To evaluate frequency response of a stable system at some frequency, apply sinusoidal input of that frequency. Choose suitable amplitude of input sinusoid so that it is neither so large to overload the system nor so small to get masked by noise. Wait until transients have died out and then measure the amplitude ratio given by (6.1) and phase difference given by (6.2). With a convenient number of such measurements of amplitude ratio and phase difference at various frequencies, the curves for **A(ω) vs ω** and **ϕ(ω) vs ω** can be sketched.

Graphical evaluation of frequency response

Given the pole-zero plot of a system with transfer function G(s), $|G(s)|$ and $\angle G(s)$ for various values of $s = j\omega$, can be evaluated graphically by drawing set of directed line segments from poles and zeros of G(s) to the points of evaluation on the imaginary axis.

To demonstrate this, consider the pole-zero plot as shown in Fig. 6.2. The transfer function under consideration is

$$G(s) = \frac{10(s+1)}{s^2+4s+13} = \frac{10(s+1)}{(s+2+j3)(s+2-j3)}$$

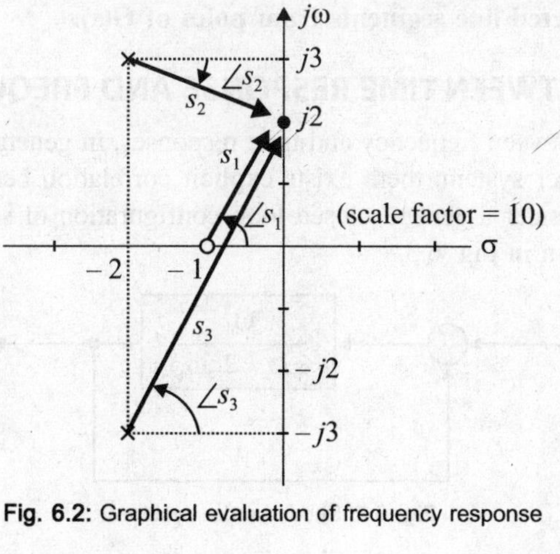

Fig. 6.2: Graphical evaluation of frequency response

The directed line segments are represented by s_1, s_2 and s_3 for s restricted to imaginary axis for example let the present point of evaluation be $s = j2$ as shown in Fig. 6.2. Then

$$G(s)\big|_{s=j2} = \frac{10\,|s_1|\,\left[\angle s_1 - \angle s_2 - \angle s_3\right]}{|s_2|\,|s_3|}$$

$$= \frac{10\times 2.24}{5.39\times 2.24}\,\angle\left\{63.4° - (-26.6°) - 68.2°\right\}$$

$$= 1.85\,\angle 21.8°$$

Also analytically

$$G(j2) = \frac{10(j2+1)}{(j2+2+j3)(j2+2-j3)}$$

$$= \frac{10\sqrt{5}\,\angle 63.4°}{(\sqrt{29}\,\angle 68.2°)\;(\sqrt{5}\,\angle -26.6°)} = 1.857\,\angle 21.8°$$

Thus $\quad |G(j2)| = 1.857$ and $\angle G(j2) = 21.8°$

In general, for

$$G(s) = \frac{K\prod_{i=1}^{m}(s+z_i)}{\prod_{j=1}^{n}(s+p_j)}$$

$$|\mathbf{G}(s)|_{s=j\omega_1} = \frac{\textbf{K [Product of line segments directed from zeros of G(}s\textbf{) to point } s = j\omega_1\textbf{]}}{\textbf{Product of line segments directed from poles of G(}s\textbf{) to point } s = j\omega_1}$$

and

$\angle \mathbf{G}(s)|_{s=j\omega_1}$ = $\sum$ **angles of directed line segments from zeros of G(s) –** $\sum$ **angles of directed line segments from poles of G(s).**

6.3 CORRELATION BETWEEN TIME RESPONSE AND FREQUENCY RESPONSE

Although, the correlation between frequency and time responses, in general, for high order systems, is indirect; but for second order system, there exists explicit correlation between the two. To establish the explicit correlation, consider a general closed loop configuration of second order system without any open loop zero as shown in Fig. 6.3.

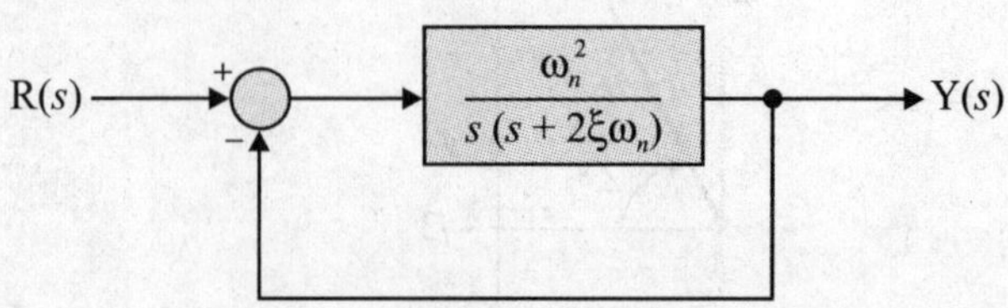

Fig. 6.3: Second order system

$$\frac{Y(s)}{R(s)} = T(s) = \frac{\omega_n^2}{s^2 + 2\xi\,\omega_n s + \omega_n^2}$$

Putting $s = j\omega$, we have

$$T(j\omega) = \frac{\omega_n^2}{-\omega^2 + j2\xi\,\omega\cdot\omega_n + \omega_n^2} = \frac{1}{1 - \dfrac{\omega^2}{\omega_n^2} + j2\xi\dfrac{\omega}{\omega_n}}$$

$$= \frac{1}{(1-u^2) + j2\xi u} \quad \text{...(6.3)}$$

where $u = \dfrac{\omega}{\omega_n}$ is generally termed normalized driving signal frequency. As usual ω_n is natural undamped frequency and ξ is damping ratio. From (6.3), it is easy to find.

$$|T(j\omega)| = M = \frac{1}{\sqrt{(1-u^2)^2 + (2\xi u)^2}} \quad \text{...(6.4)}$$

and

$$\underline{|T(j\omega)} = \phi = -\tan^{-1}\frac{2\xi u}{(1-u^2)} \quad \text{...(6.5)}$$

Using (6.4) and (6.5), the steady state response of system of Fig. 6.3 for a sinusoidal input of unity magnitude and variable ω, *i.e.*, $r(t) = \sin \omega t$ can be written in form as follows:

$$y(t) = M \sin(\omega t + \phi)$$

$$= \frac{1}{\sqrt{(1-u^2)^2 + (2\xi u)^2}} \sin\left[\omega t - \tan^{-1}\left(\frac{2\xi u}{1-u^2}\right)\right] \quad \text{...(6.6)}$$

Frequency response specifications

The following frequency response specifications are often used in practice.

(a) Peak resonance (M_r)

The magnitude $|T(j\omega)|$ attains maximum value greater than 1 in certain range of ξ. This maximum value as shown in Fig. 6.4 (*a*), is called peak resonance M_r.

(b) Resonant frequency (ω_r)

The resonant frequency ω_r is defined as frequency at which the peak resonance (M_r) occurs. The peak resonance (M_r) and resonant frequency (ω_r) can be determined by setting derivative dM/du equal to zero.

Thus,

$$\frac{dM}{du} = 0, \text{ gives}$$

$$-\frac{1}{2}\left[(1-u^2)^2 + (2\xi u)^2\right]^{-3/2}\left[4u^3 - 4u + 8u\xi^2\right] = 0$$

or
$$4u\,(u^2 - 1 + 2\xi^2) = 0$$

or
$$u = 0,\ \sqrt{1-2\xi^2}$$

Since $(dM/du) = 0$ gives maxima of M, the resonant frequency is given by

$$u_r = \sqrt{1-2\xi^2}$$

or
$$\omega_r = \omega_n\sqrt{1-2\xi^2} \qquad ...(6.7)$$

$u = 0$ merely indicates that the slope of M versus ω curve is zero at $\omega = 0$. Since frequency is a real quantity, (6.7) is valid only for $2\xi^2 < 1$ or $\xi < \frac{1}{\sqrt{2}}$. This simply means that for all values of $\xi \geq 0.707$, the resonant frequency ω_r is zero. See Fig. 6.4 (*a*).

The value of M at $u = u_r$ gives resonant peak *i.e.*,

$$M_r = M\big|_{u=u_r} = \frac{1}{\sqrt{\left[1-\left(1-2\xi^2\right)^2+\left(2\xi\sqrt{1-2\xi^2}\right)^2\right]}}$$

or
$$M_r = \frac{1}{2\xi\sqrt{1-\xi^2}} \qquad ...(6.8)$$

(c) Band width (ω_b)

The band width of a closed loop system is defined as the frequency at which $M = |T(j\omega)|$ drops to $\frac{1}{\sqrt{2}}$ of its value at $\omega = 0$.

Thus,

$$M\big|_{u=u_b} = \frac{1}{\sqrt{\left(1-u_b^2\right)^2+(2\xi\,u_b)^2}} = \frac{1}{\sqrt{2}}$$

Simplifying we have

$$u_b^2 = \left(1-2\xi^2\right) \pm \sqrt{4\xi^4-4\xi^2+2}$$

Since u must be a positive real quantity for any value of ξ, choosing only plus sign

$$u_b = \frac{\omega_b}{\omega_n} = \left[\left(1-2\xi^2\right)+\sqrt{4\xi^4-4\xi^2+2}\right]^{1/2}$$

or
$$\omega_b = \omega_n\left[\left(1-2\xi^2\right)+\sqrt{4\xi^4-4\xi^2+2}\right]^{1/2} \qquad ...(6.9)$$

Note the following intricate points in the analysis so far.

(*i*) Using (6.4) and (6.5) the table showing the values of M and ϕ for u = 0, 1 and ∞, is prepared as follows.

u	M	ϕ (rad)
0	1	0
1	$\frac{1}{2\xi}$	$-\frac{\pi}{2}$
∞	0	$-\pi$

Using this table, the magnitude and phase responses are shown in Fig. 6.4 (*a*) and (*b*) respectively.

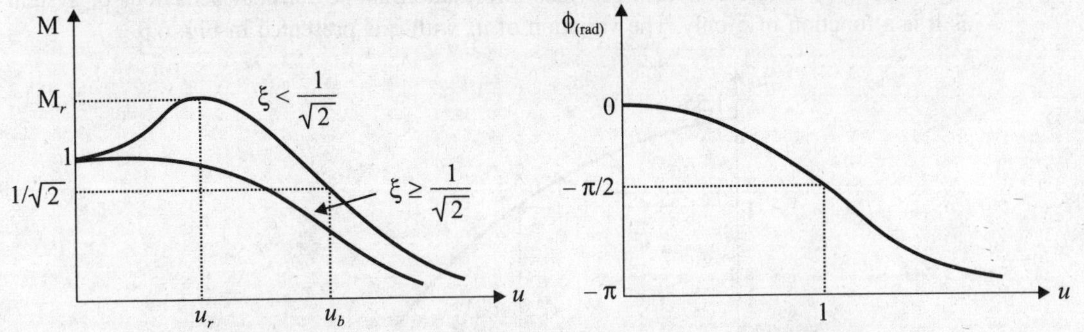

(*a*) Magnitude (*b*) Phase

Fig. 6.4: Typical Frequency Response of second order system

Note from Fig. 6.4 that for $\xi \geq \frac{1}{\sqrt{2}}$ the slope of magnitude curve, does not become zero for any real value of ω and magnitude M decreases monotonically from M = 1 at $u = 0$ with increasing u. So, for $\xi \geq \frac{1}{\sqrt{2}}$, there is no resonance peak and largest value of M equals unity.

(*ii*) As $\xi \to 0$, $\omega_r \to \omega_n$ and $M_r \to \infty$.

For $\xi \geq \frac{1}{\sqrt{2}}$, $M_r = 1$, $\omega_r = 0$

For $0 < \xi < \frac{1}{\sqrt{2}}$, the resonant frequency always has a value less than ω_n and the resonant peak has a value greater than 1. The variation of M_r and ω_r with ξ, is depicted in Fig. 6.5 (*a*) and (*b*) respectively. Note that $M_r = 1.52$ and $\omega_r = 0.82\ \omega_n$ for $\xi = 0.4$.

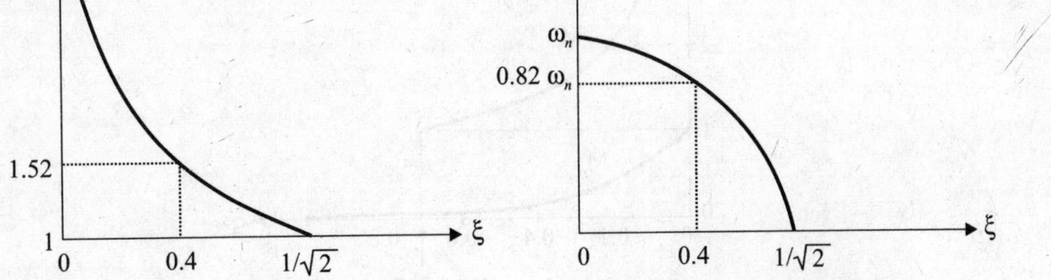

(*a*) (*b*)

Fig. 6.5: (*a*) Variation of M_r with ξ (*b*) Variation of ω_r with ξ

(*iii*) The resonant peak M_r given by (6.8) is function of damping ratio ξ only. Recall that the peak overshoot M_p in step response of second order system, is also purely a function of ξ and indicative of relative stability. Thus value of M_r is indicative of the relative stability. The resonant frequency ω_r of frequency response given by (6.7) is indicative of natural frequency ω_n for a given ξ and hence indicative of speed of response because settling time t_s, is inversely proportional to product $\xi\omega_n$. Thus the larger value of ω_r means faster time response. In other words, the rise time t_r varies inversely with ω_r.

(*iv*) The band width ω_b of a closed loop system, presents filtering property of the system for high frequency noise. This also provides information about transient behaviour of system as it is a function of ξ only. The variation of u_b with ξ is presented in Fig. 6.6.

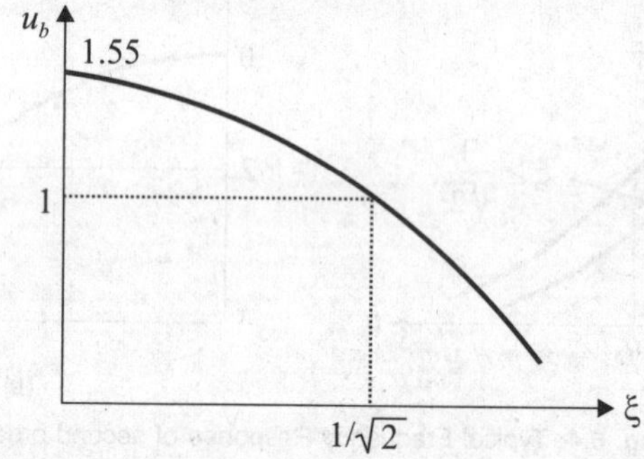

Fig. 6.6: Variation of u_b with ξ

(*v*) For satisfactory operation of a system, the percentage peak overshoot M_p in step response is, in general, required to be about 10 to 15% while the resonant peak M_r in frequency response, is required to be about 1 to 1.4. So, ξ is generally designed to lie in the range $0.4 < \xi < 0.707$. The variation of M_p together with that of M_r with ξ, is graphically demonstrated in Fig. 6.7.

(*vi*) The rise time of step response increases with ξ. The band width of the system decreases with the increase in ξ for given ω_n. Therefore, band width and rise time are inversely proportional to each other.

(*vii*) Bandwidth is directly proportional to ω_n.

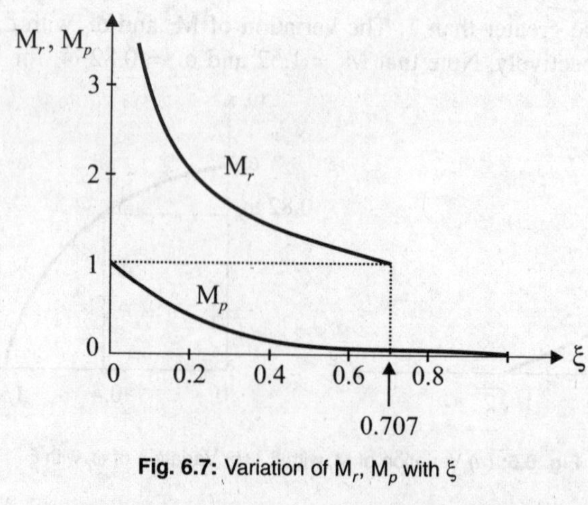

Fig. 6.7: Variation of M_r, M_p with ξ

6.4 GRAPHICAL REPRESENTATION OF FREQUENCY RESPONSE

In so far discussion, it has been shown that the sinusoidal transfer function which is a complex function of frequency ω, is characterised by its magnitude and phase angle with frequency as the variable parameter. The commonly used graphical representations of sinusoidal transfer functions, in control analysis, are as follows:

(*a*) **Polar plot or Nyquist plot:** A plot of the magnitude versus phase in polar coordinates as ω varies from 0 to ∞.

(*b*) **Logarithmic plot or Bode plot:** A plot of the magnitude in dB versus ω (or $\log_{10}$ **ω**) in semilog rectangular co-ordinates.

(*c*) **Log magnitude phase plot (Nichols plots):** A plot of the magnitude in dB versus phase in rectangular coordinates with ω as variable parameter.

Polar plots

The polar plot of a sinusoidal transfer function G(*j*ω) is a plot of the magnitude of G(*j*ω) versus the phase angle of G(*j*ω) on polar co-ordinates as ω is varied from 0 to ∞. The values of ω is usually labelled at significant points on the plot. The following points are worth noting regarding polar plots.

(*a*) It follows from above definition of polar plot that it is the locus of vectors $|G(j\omega)| \angle G(j\omega)$ as ω varies from 0 to ∞.

(*b*) Mathematically, the polar plot may be regarded as the mapping of the positive half of the imaginary axis of the *s*-plane into the G(*j*ω) plane.

(*c*) In polar plots, the phase measured counterclockwise from positive real axis, is referred to as positive and clockwise as negative.

(*d*) The projections of G(*j*ω) on the real and the imaginary axes of G(*j*ω) plane, are its real and imaginary components.

(*e*) The polar plot fascinates in the sense that it depicts the frequency response of the system over the entire frequency range in a single plot while it irks in the sense that the polar plot does not clearly indicate the contributions of each individual factor of G(*j*ω).

(*f*) The polar plot is very useful in stability investigation of a class of closed loop systems. We shall discuss it in detail little later in this chapter.

Polar plot construction

This has been already stated that polar plotting open loop transmittance G(*s*) H(*s*) is a process of mapping of the positive half of the imaginary axis of the *s*-plane into GH plane. The following example demonstrates this mapping or transformation:

Consider

$$G(s)\,H(s) = \frac{s-1}{s+4}$$

Setting $s = j\omega$ and expressing in terms of magnitude and phase, we have

$$G(j\omega)\,H(j\omega) = \frac{j\omega-1}{j\omega+4} = \sqrt{\frac{1+\omega^2}{16+\omega^2}}\left[\angle\tan^{-1}(-\omega)-\tan^{-1}\left(\frac{\omega}{4}\right)\right]$$

so,

$$G(j\omega)\,H(j\omega)\Big|_{\omega=0} = 0.25\,e^{j180°}$$

$$G(j\omega)\,H(j\omega)\Big|_{\omega=1} = 0.34\,e^{j120^\circ}$$

$$G(j\omega)\,H(j\omega)\Big|_{\omega=4} = 0.73\,e^{j60^\circ}$$

and $$G(j\omega)\,H(j\omega)\Big|_{\omega=\infty} = 1\,e^{j0^\circ}$$

Using the information furnished above the polar frequency response plot is shown in Fig. 6.8 (*a*).

The function G(s) H(s) can also be evaluated for various values of $s = j\omega$ on the imaginary axis by drawing directed line segments from poles and zeros of G(s) H(s) to the points of evaluation on the imaginary axis and using the procedure as explained in Sec. 6.2. This process is termed as mapping or transformation of a point on the s-plane into the corresponding G($j\omega$) H($j\omega$) plane. The pole-zero plot and four points of evaluation $s = j0$, $s = j1$, $s = j4$ and $s = j\infty$ are shown in Fig. 6.8 (*b*). This mapping is demonstrated in Fig. 6.8 (*c*), (*d*), (*e*) and (*f*) for four points $s = j0$, $s = j1$, $s = j4$ and $s = j\infty$ respectively.

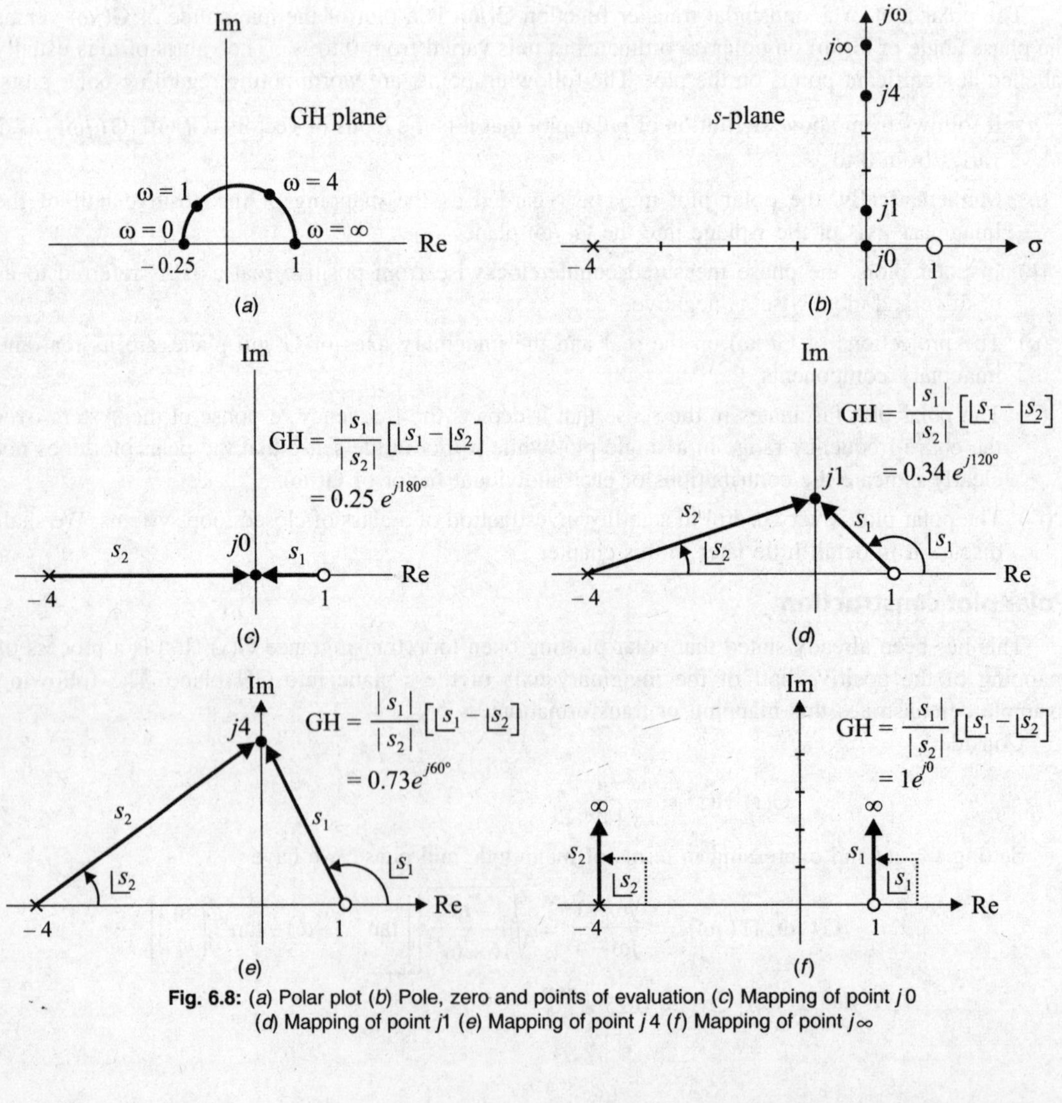

Fig. 6.8: (*a*) Polar plot (*b*) Pole, zero and points of evaluation (*c*) Mapping of point $j0$ (*d*) Mapping of point $j1$ (*e*) Mapping of point $j4$ (*f*) Mapping of point $j\infty$

In many situations of control system analysis, such as Nyquist stability analysis (to be discussed later in this chapter), the exact shape of polar plot is not essential. Very often, only a rough sketch of polar plot, is adequate for the purpose of system analysis through frequency response approach. The rough sketch of polar plot of a transmittance G(*j*ω) H(*j*ω), can be determined from the following set of information.

1. The magnitude and phase of G(*j*ω) H(*j*ω) at ω = 0 and at ω = ∞.
2. The points of intersections of polar plot with the real and imaginary axes together with corresponding values of ω.

Example 2: Sketch rough polar plot for the following transfer functions and label significant points.

(*a*) $G(s) = 1/s$ (*b*) $G(s) = s$

(*c*) $G(s) = 1 + \alpha s$ (*d*) $G(s) = 1/(1 + \alpha s)$

(*e*) $G(s) = \dfrac{\omega_n^2}{s^2 + 2\xi\omega_n s + \omega_n^2}$ (*f*) $G(s) = \dfrac{s^2 + 2\xi\omega_n s + \omega_n^2}{\omega_n^2}$

(*g*) $G(s) = \dfrac{1}{(1+as)(1+bs)}$ (*h*) $G(s) = \dfrac{1}{(1+as)(1+bs)(1+cs)}$

(*i*) $G(s) = \dfrac{1}{s(1+\alpha s)}$ (*j*) $G(s) = \dfrac{1}{s^2(1+\alpha s)}$

(*k*) $G(s) = \dfrac{1}{s^3(1+\alpha s)}$ (*l*) $G(s) = \dfrac{1+as}{1+bs}$

(*m*) $G(s) = \dfrac{10(s-10)}{s(s+2)(s+5)}$ (*n*) $G(s) = \dfrac{e^{-sT}}{1+sT}$

Solution:

(*a*)
$$G(j\omega) = \frac{1}{j\omega} = \frac{1}{\omega}\angle -90^\circ$$

$$\lim_{\omega\to 0} G(j\omega) = \infty\angle -90^\circ$$

$$\lim_{\omega\to\infty} G(j\omega) = 0\angle -90^\circ$$

So, the polar plot is negative imaginary axis as shown in Fig. 6.9(*a*).

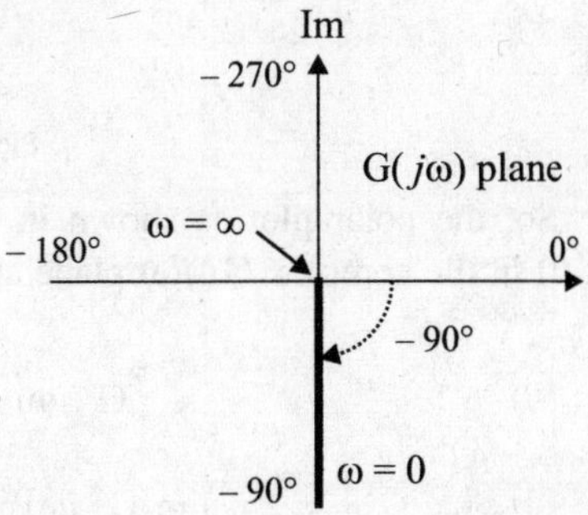

Fig. 6.9: (*a*) Polar plot of G(*s*) = 1/*s*

(*b*)
$$G(j\omega) = j\omega = \omega\angle 90^\circ$$

$$\lim_{\omega\to 0} G(j\omega) = 0\angle 90^\circ$$

$$\lim_{\omega\to\infty} G(j\omega) = \infty\angle 90^\circ$$

So, the polar plot is positive imaginary axis as shown in Fig. 6.9(*b*).

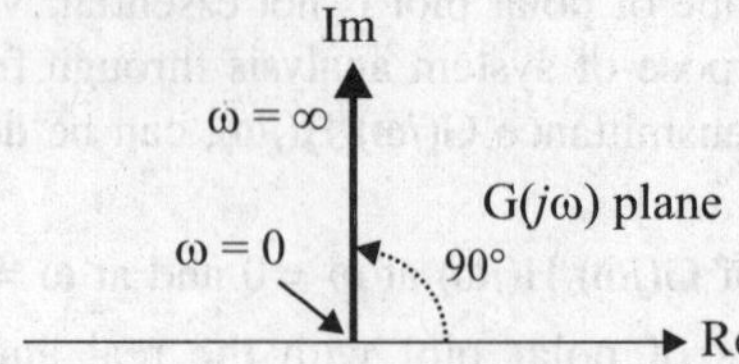

Fig. 6.9: (*b*) Polar plot of G(*s*) = *s*

(*c*)

$$G(j\omega) = (1 + j\alpha\omega)$$

$$\text{Re}\,[G(j\omega)] = 1$$

$$\text{Im}\,[G(j\omega)] = \alpha\omega$$

Note that real part is constant (= 1) and imaginary part is linearly related to ω.

Also,

$$\lim_{\omega\to 0} G(j\omega) = \lim_{\omega\to 0} (1) = 1\angle 0^\circ$$

$$\lim_{\omega\to \infty} G(j\omega) = \lim_{\omega\to \infty} (j\alpha\omega) = \infty\angle 90^\circ$$

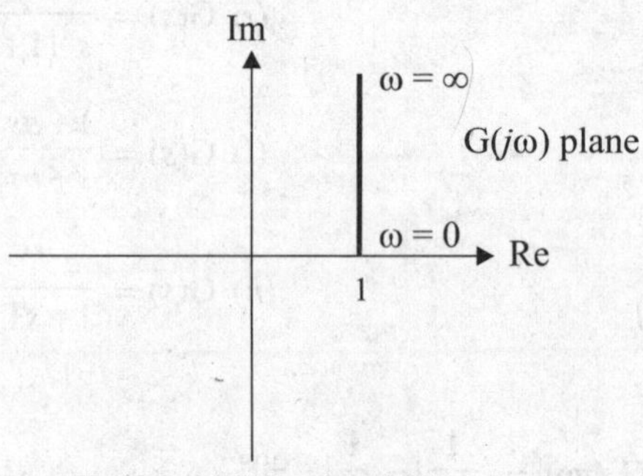

Fig. 6.9: (*c*) Polar plot of G(*s*) = 1 + α*s*

So, the polar plot as shown in Fig. 6.9(*c*), is upper half of straight line passing through point (1, 0) in the complex G(*j*ω) plane and parallel to imaginary axis.

(*d*)

$$G(j\omega) = \frac{1}{1 + j\alpha\omega}$$

$$\lim_{\omega\to 0} G(j\omega) = \lim_{\omega\to 0} (1) = 1\angle 0^\circ$$

$$\lim_{\omega\to \infty} G(j\omega) = \lim_{\omega\to \infty} \left(\frac{1}{j\alpha\omega}\right) = 0\angle -90^\circ$$

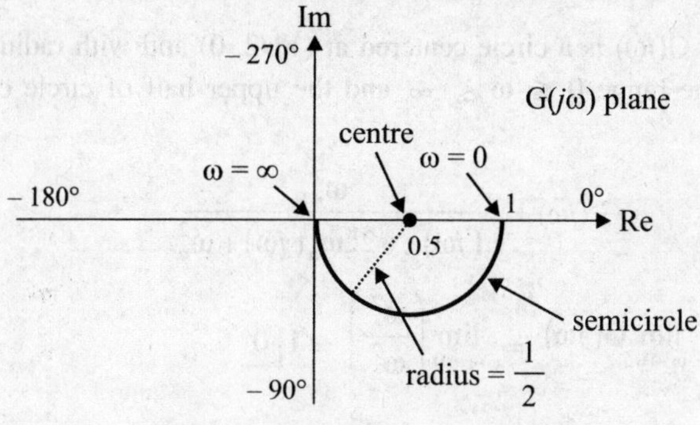

Fig. 6.9: (*d*) Polar plot of G(*s*) $\frac{1}{1+\alpha s}$

Note that G(*j*ω) in terms of magnitude and phase, can be written as

$$G\,(j\omega) = \frac{1}{\sqrt{1+\alpha^2\omega^2}}\ \underline{\big|-\tan^{-1}\alpha\omega}$$

so

$$G(j\omega)\Big|_{\omega=0} = G(j0) = 1\underline{|0^\circ}$$

As ω increases, $|G(j\omega)|$ decreases and $\underline{|G(j\omega)}$ becomes more negative. As $\omega \to \infty$, $|G(j\omega)| \to 0$ and $\underline{|G(j\omega)} \to -90^\circ$. The polar plot is shown in Fig. 6.9 (*d*). The exact polar plot is a semicircle. This can be mathematically demonstrated as follows:

$$G(j\omega) = \frac{1}{1+j\alpha\omega} = \frac{1-j\alpha\omega}{(1+j\alpha\omega)(1-j\alpha\omega)}$$

$$= \left(\frac{1}{1+\alpha^2\omega^2}\right) + j\left(\frac{-\alpha\omega}{1+\alpha^2\omega^2}\right)$$

Let,

$$x = \text{Re}\,[G\,(j\omega)] = \frac{1}{1+\alpha^2\omega^2}$$

and

$$y = \text{Im}\,\{G\,(j\omega)\} = \frac{-\alpha\omega}{1+\alpha^2\omega^2}$$

$$\frac{y}{x} = -\alpha\omega \text{ gives } \omega = -\frac{y}{\alpha x}$$

Putting this value of ω in equation

$$x = \frac{1}{1+\alpha^2\omega^2}, \text{ we have } x = \frac{1}{1+\alpha^2\left(y^2/\alpha^2x^2\right)}$$

which with little algebraic manipulation, gives

$$\left(x-\frac{1}{2}\right)^2 + y^2 = \left(\frac{1}{2}\right)^2$$

Thus, in x–y plane, G($j\omega$) is a circle centered at (+1/2, 0) and with radius 1/2. The lower half of circle corresponds to the range $0 \le \omega \le \infty$ and the upper half of circle corresponds to the range $-\infty \le \omega \le 0$.

(*e*) $$G(j\omega) = \frac{\omega_n^2}{(j\omega)^2 + 2\xi\omega_n(j\omega) + \omega_n^2}$$

$$\lim_{\omega \to 0} G(j\omega) = \lim_{\omega \to 0}\left(\frac{\omega_n^2}{\omega_n^2}\right) = 1\underline{|0^\circ}$$

$$\lim_{\omega \to \infty} G(j\omega) = \lim_{\omega \to \infty}\frac{\omega_n^2}{(j\omega)^2} = 0\underline{|-180^\circ}$$

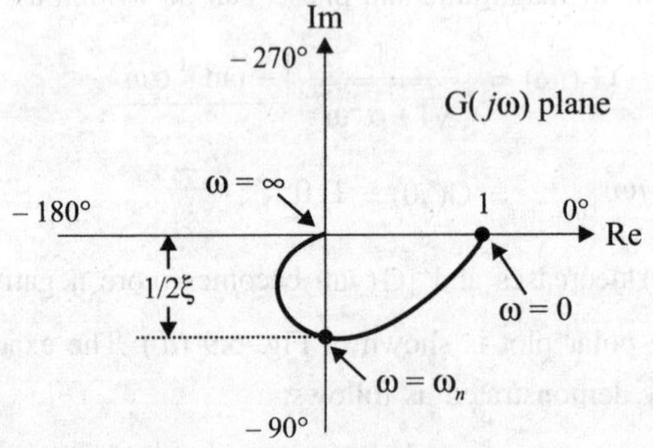

Fig. 6.9: (*e*) Polar plot of G(*s*) = $\omega_n^2/(s^2 + 2\xi\omega_n s + \omega_n^2)$

Note that the term containing highest power of ω, is only retained as $\omega \to \infty$. $\underline{|G(j\omega)}\big|_{\omega=\infty} = -180^\circ$ suggests that G($j\omega$) phasor will be tangent to negative real axis as $\omega \to \infty$. The polar plot is shown in Fig. 6.9 (*e*). The exact shape of polar plot depends on ξ.

To evaluate intersection points let us rewrite G($j\omega$) in the form

$$G(j\omega) = \frac{1}{\left(1 - \frac{\omega^2}{\omega_n{}^2}\right) + j2\xi\left(\frac{\omega}{\omega_n}\right)}$$

It is easy to find that Re [G($j\omega$)] = 0 for $\omega = \omega_n$. So, G($j\omega$) locus will intersect imaginary axis at $\omega = \omega_n$.

Also at $\omega = \omega_n$,

$$\text{Im}\,[G(j\omega)] = -\frac{1}{2\xi} \quad \text{and} \quad \underline{|G(j\omega)}\big|_{\omega=\omega_n} = -90^\circ$$

It is important to note the following from this analysis:

(*i*) The frequency at which G($j\omega$) locus intersects imaginary axis is undamped natural frequency ω_n.

(*ii*) The frequency point whose distance from origin is maximum corresponds to the resonant frequency ω_r.

(*iii*) The resonant peak M_r can be obtained by ratio of magnitude of G(*j*ω) phasor at $\omega = \omega_r$ to the magnitude of G(*j*ω) phasor at ω = 0.

(*iv*) Although the general shape of polar plot is the same for both underdamped ($0 \le \xi < 1$) and overdamped ($\xi > 1$) systems, but as ξ increases beyond unity, the G(*j*ω) locus approaches a semicircle. This is due to the fact that, for $\xi > 1$, both the characteristic roots are real and one is much smaller than the other. The root closer to imaginary axis, plays more significant role in deciding the system dynamics. So, the second order system is closely approximated by a first order system.

(*f*)

$$G(j\omega) = \frac{(j\omega)^2 + j\omega\, 2\xi\, \omega_n + \omega_n^2}{\omega_n^2}$$

$$= 1 + 2\xi\left(j\frac{\omega}{\omega_n}\right) + \left(j\frac{\omega}{\omega_n}\right)^2$$

$$= \left(1 - \frac{\omega^2}{\omega_n^{\ 2}}\right) + j\left(\frac{2\xi\omega}{\omega_n}\right)$$

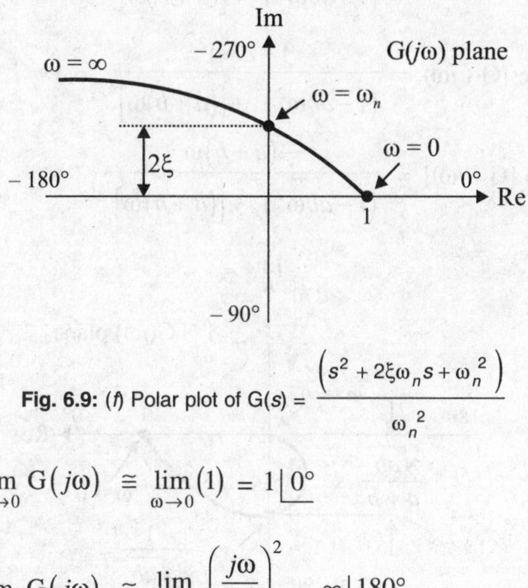

Fig. 6.9: (*f*) Polar plot of $G(s) = \dfrac{\left(s^2 + 2\xi\omega_n s + \omega_n^{\ 2}\right)}{\omega_n^{\ 2}}$

$$\lim_{\omega \to 0} G(j\omega) \cong \lim_{\omega \to 0} (1) = 1\angle 0°$$

$$\lim_{\omega \to \infty} G(j\omega) \cong \lim_{\omega \to \infty} \left(\frac{j\omega}{\omega_n}\right)^2 = \infty \angle 180°$$

The polar plot is shown in Fig. 6.9 (*f*).

Note that

$$\text{Re}\,[G(j\omega)] = 1 - \frac{\omega^2}{\omega_n^{\ 2}}$$

and $$\text{Re}[G(j\omega)]\Big|_{\omega = \omega_n} = 0$$

This corresponds to imaginary axis crossing point.

Also $$\text{Im}\,[G(j\omega)] = \frac{2\xi\omega}{\omega_n}$$

and $$\text{Im}\,[G(j\omega)]\Big|_{\omega=\omega_n} = 2\xi$$

For any value of $\omega > 0$, *Im* [G($j\omega$)] is positive and is monotonically increasing.

(*g*) $$G(j\omega) = \frac{1}{(1+ja\omega)(1+jb\omega)} = \frac{1}{(1-ab\omega^2)+j(a+b)\omega}$$

$$\lim_{\omega\to 0} G(j\omega) \cong \lim_{\omega\to 0}(1) = 1\angle 0^\circ$$

$$\lim_{\omega\to\infty} G(j\omega) \cong \lim_{\omega\to\infty}\left(\frac{1}{j^2ab\omega^2}\right) = 0\angle -180^\circ$$

Rationalizing G($j\omega$), we have

$$G(j\omega) = \frac{(1-ab\omega^2)-j(a+b)\omega}{(1-ab\omega^2)^2+[(a+b)\omega]^2}$$

Thus $$\text{Re}\,[G(j\omega) = \frac{1-ab\omega^2}{(1-ab\omega^2)^2+[(a+b)\omega]^2}$$

and $$\text{Im}\,[G(j\omega)] = \frac{-(a+b)\omega}{(1-ab\omega^2)^2+[(a+b)\omega]^2}$$

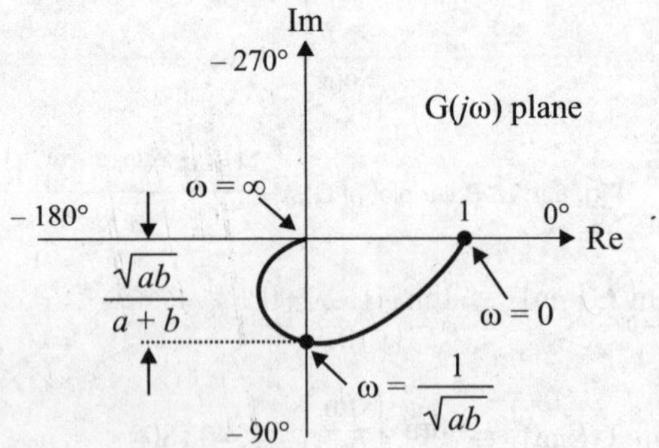

Fig. 6.9: (*g*) Polar plot of $G(s) = \dfrac{1}{(1+as)(1+bs)}$

Im [G($j\omega$)] is always negative for $\omega > 0$ and is zero at $\omega = 0$ and $\omega = \infty$

$$\text{Re}\,[G(j\omega)] = \frac{1-ab\omega^2}{(1-ab\omega^2)^2+[(a+b)\omega]^2}$$

is zero for $\omega = \dfrac{1}{\sqrt{ab}}$ and

$$\text{Im}\left[G(j\omega)\right]\Big|_{\omega=\frac{1}{\sqrt{ab}}} = \frac{\sqrt{ab}}{a+b}$$

The polar plot is shown in Fig. 6.9(*g*) and significant points have been labelled.

(*h*) $$G(j\omega) = \frac{1}{(1+ja\omega)(1+jb\omega)(1+jc\omega)}$$

Rationalizing G(*j*ω), we have

$$G(j\omega) = \frac{1}{\left[1-ab\omega^2-(a+b)c\omega^2\right]+j\left[(a+b+c)\omega-abc\,\omega^3\right]}$$

$$= \frac{1-ab\omega^2-(a+b)c\omega^2}{\left[1-ab\omega^2-(a+b)c\omega^2\right]^2+\left[(a+b+c)\omega-abc\,\omega^3\right]^2}$$

$$-j\frac{(a+b+c)\omega-abc\omega^3}{\left[1-ab\omega^2-(a+b)c\omega^2\right]^2+\left[(a+b+c)\omega-abc\,\omega^3\right]^2}$$

$$\lim_{\omega\to 0} G(j\omega) \cong \lim_{\omega\to 0}(1) = 1\angle 0^\circ$$

$$\lim_{\omega\to\infty} G(j\omega) \cong \lim_{\omega\to\infty}\left(\frac{1}{j^3\,abc\,\omega^3}\right) = 0\angle -270^\circ$$

It is easy to conclude the following :

(*a*) Im [G(*j*ω)] = 0 for ω = 0, $\sqrt{\frac{a+b+c}{abc}}$ and ∞.

(*b*) Re [G(*j*ω)] = 0 for $\omega = \frac{1}{\sqrt{ab+bc+ac}}$

The polar plot together with significant points is shown in Fig. 6.9(*h*).

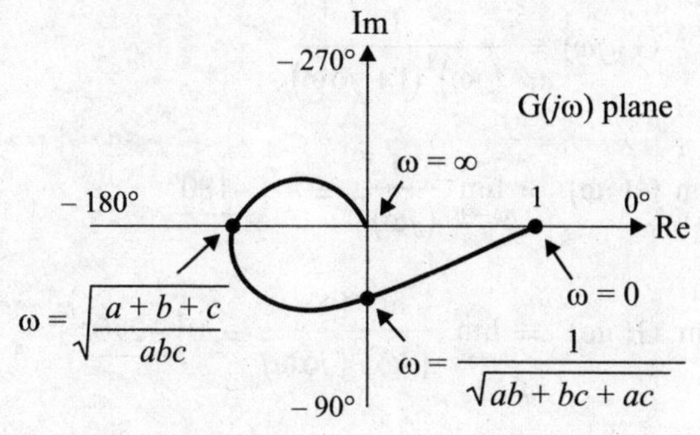

Fig. 6.9: (*h*) Polar plot of $G(s) = \frac{1}{(1+as)(1+bs)(1+cs)}$

(*i*) $$G(j\omega) = \frac{1}{j\omega(1 + j\alpha\omega)}$$

$$\lim_{\omega\to 0} G(j\omega) \cong \lim_{\omega\to 0} \frac{1}{j\omega} = \infty\angle{-90^\circ}$$

$$\lim_{\omega\to \infty} G(j\omega) \cong \lim_{\omega\to \infty} \left(\frac{1}{j\omega \cdot j\alpha\omega}\right) = 0\angle{-180^\circ}$$

G (*j*ω) can also be written as

$$G(j\omega) = \frac{1}{-\alpha\omega^2 + j\omega} = \frac{-\alpha\omega^2}{\alpha^2\omega^4 + \omega^2} - j\frac{\omega}{\alpha^2\omega^4 + \omega^2}$$

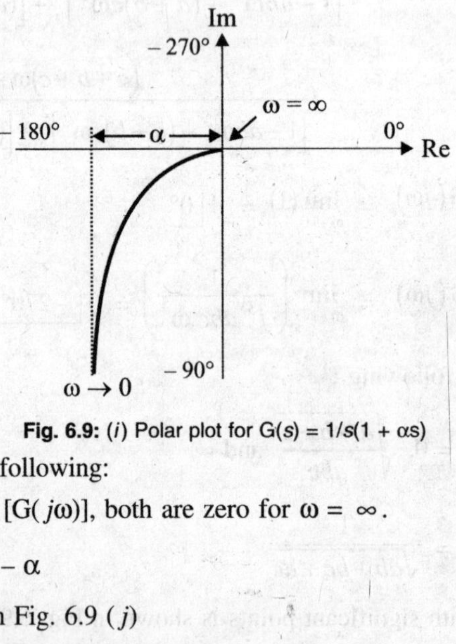

Fig. 6.9: (*i*) Polar plot for G(*s*) = 1/*s*(1 + α*s*)

It is easy to conclude the following:

(*i*) Re [G(*j*ω)] and Im [G(*j*ω)], both are zero for ω = ∞.

(*ii*) $\text{Re}\,[G(j\omega)|_{\omega=0} = -\alpha$

The polar plot is shown in Fig. 6.9 (*j*)

(*j*) $$G(j\omega) = \frac{1}{(j\omega)^2(1 + j\alpha\omega)}$$

$$\lim_{\omega\to 0} G(j\omega) \cong \lim_{\omega\to 0} \frac{1}{(j\omega)^2} = \infty\angle{-180^\circ}$$

$$\lim_{\omega\to \infty} G(j\omega) \cong \lim_{\omega\to \infty} \frac{1}{(j\omega)^2(j\alpha\omega)} = 0\angle{-270^\circ}$$

G ($j\omega$) in terms of real and imaginary part can be written as follows:

$$G(j\omega) = \frac{-\omega^2}{\omega^4+\alpha^2\omega^6} + \frac{j\alpha\omega^3}{\omega^4+\alpha^2\omega^6} = \frac{-1}{\omega^2+\alpha^2\omega^4} + \frac{j\alpha}{\omega+\alpha^2\omega^3}$$

Note the following:

(*a*) Re [G ($j\omega$)] = 0 for $\omega = \infty$

(*b*) Im [G ($j\omega$)] = 0 for $\omega = \infty$

(*c*) The only intersect on the real axis when $\omega = \infty$, is at origin.

The polar plot is shown in Fig. 6.9 (*j*)

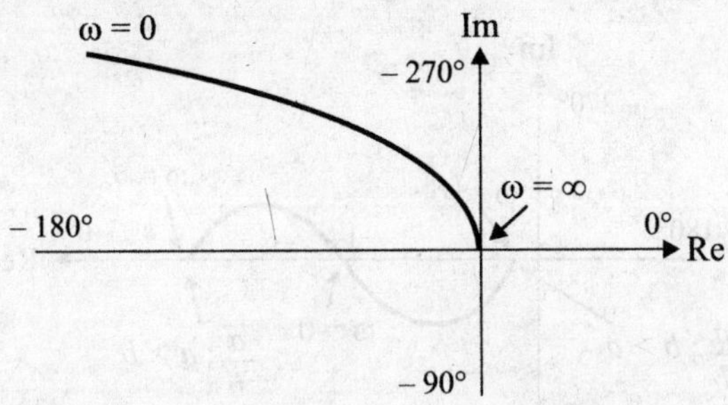

Fig. 6.9: (*j*) Polar plot for G(*s*) H(*s*) = $\frac{1}{s^2(1+\alpha s)}$

(*k*) $G(j\omega) = \frac{j\omega}{1+j\alpha\omega}$

$$\lim_{\omega\to 0} G(j\omega) \cong \lim_{\omega\to 0} (j\omega) = 0\angle 90^\circ$$

$$\lim_{\omega\to\infty} G(j\omega) \cong \lim_{\omega\to\infty}\left(\frac{1}{\alpha}\right) = \frac{1}{\alpha}\angle 0^\circ$$

G ($j\omega$) in terms of real and imaginary part can be written as

$$G(j\omega) = \frac{\alpha\omega^2}{1+\alpha^2\omega^2} + j\frac{\omega}{1+\alpha^2\omega^2}$$

Note the following:

(*a*) $\text{Re}\,[G(j\omega)|_{\omega=0} = 0$

(*b*) $\text{Im}\,[G(j\omega)|_{\omega=0} = 0$

(*c*) $\text{Re}\,[G(j\omega)|_{\omega=\infty} = \frac{1}{\alpha}$

(*d*) $\text{Im}\,[G(j\omega)|_{\omega=0} = 0$

The polar plot is shown in Fig. 6.9 (*k*)

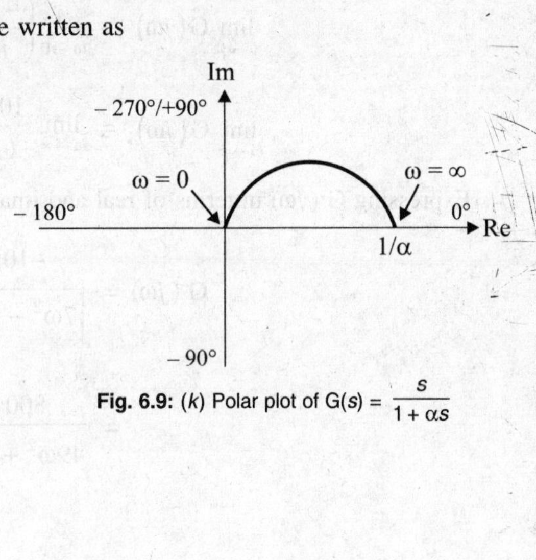

Fig. 6.9: (*k*) Polar plot of G(*s*) = $\frac{s}{1+\alpha s}$

(*l*)
$$G(j\omega) = \frac{1 + ja\omega}{1 + jb\omega}$$

$$\lim_{\omega \to 0} G(j\omega) \cong \lim_{\omega \to 0} (1) = 1\underline{|0^\circ}$$

$$\lim_{\omega \to \infty} G(j\omega) \cong \lim_{\omega \to \infty} \left(\frac{a}{b}\right) = \frac{a}{b}\underline{|0^\circ}$$

$$\underline{|G(j\omega)} = \tan^{-1} a\omega - \tan^{-1} b\omega$$

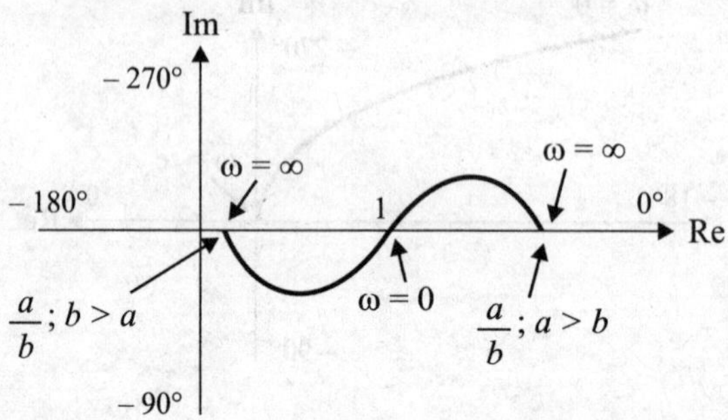

Fig. 6.9: (*l*) Polar plot for $G(s) = \frac{1 + as}{1 + bs}$

For $a > b$, $\underline{|G(j\omega)} \geq 0^\circ$ in range $0 \leq \omega \leq \infty$ and for $b > a$, $G(j\omega) \leq 0^\circ$ in range $0 \leq \omega \leq \infty$

The polar plot is shown in Fig. 6.9(*l*).

(*m*)
$$G(j\omega) = \frac{10(j\omega - 10)}{j\omega(j\omega + 2)(j\omega + 5)}$$

$$\lim_{\omega \to 0} G(j\omega) \cong \lim_{\omega \to 0} \left(\frac{-10}{j\omega}\right) = \infty\underline{|90^\circ}$$

$$\lim_{\omega \to \infty} G(j\omega) \cong \lim_{\omega \to \infty} \frac{10(j\omega)}{(j\omega)^3} = 0\underline{|-180^\circ}$$

Expressing G ($j\omega$) in terms of real and imaginary part, we have

$$G(j\omega) = \frac{-10(j\omega - 10)\left[7\omega^2 + j\omega(10 - \omega^2)\right]}{\left[7\omega^2 - j\omega(10 - \omega^2)\right]\left[7\omega^2 + j\omega(10 - \omega^2)\right]}$$

$$= \frac{800 - 10\omega^2}{49\omega^2 + (10 - \omega^2)^2} + j\frac{100(10 - \omega^2) - 70\omega^2}{49\omega^3 + \omega(10 - \omega^2)^2}$$

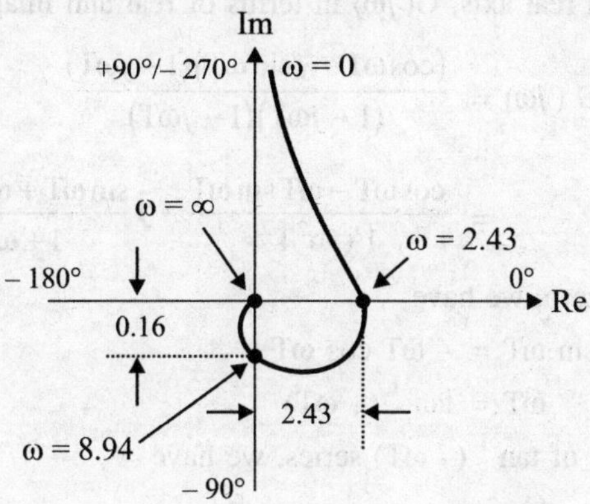

Fig. 6.9: *(m)* Polar plot of G(*s*) = $\frac{10(s-10)}{s(s+2)(s+5)}$

Note the following observations from this expression:

(*i*) Re [G(*j*ω)] = 0 for ω = $\sqrt{80} \cong 8.94$ rad/sec and for ω = 8.94, Im [G(*j*ω)] = – 0.16

(*ii*) Im [G(*j*ω)] = 0 for ω = $\sqrt{\frac{100}{17}} \cong 2.43$ rad/sec and for ω = 2.43, Re [G(*j*ω)] = 2.43

Using the points so far the polar plot is shown in Fig. 6.9 (*m*).

(*n*)
$$G(j\omega) = \frac{e^{-j\omega T}}{1+j\omega T} = \frac{\cos\omega T - j\sin\omega T}{1+j\omega T}$$

The magnitude and phase angle are, respectively,

$$|G(j\omega)| = \frac{1}{\sqrt{1+\omega^2 T^2}}$$

$$\angle G(j\omega) = \tan^{-1}\left[\frac{-\sin\omega T}{\cos\omega T}\right] - \tan^{-1}\omega T$$

$$= -\omega T - \tan^{-1}\omega T$$

$$\left|G(j\omega)\right|_{\omega=0} = 1 \text{ and } \angle G(j\omega)\big|_{\omega=0} = 0°$$

$$\left|G(j\omega)\right|_{\omega=\infty} = 0.$$

So, the magnitude decreases monotonically from 1 at ω = 0 to zero as ω→ ∞ and phase angle also decreases mono-tonically and indefinitely. The polar plot of given transfer function spirals about origin as shown in Fig. 6.9(*n*).

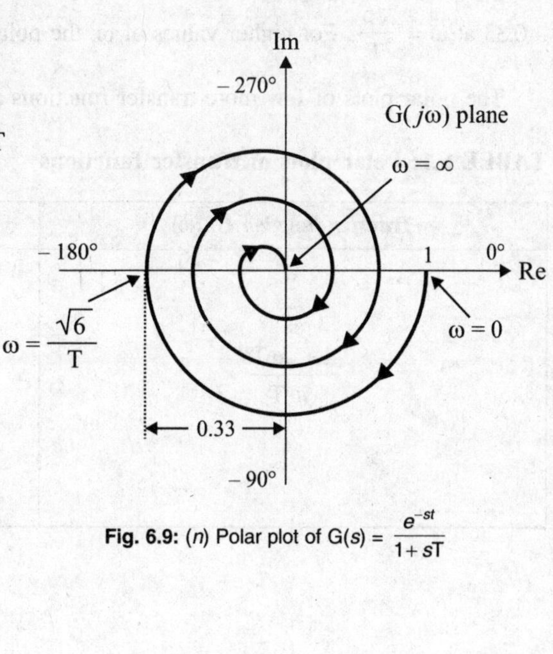

Fig. 6.9: *(n)* Polar plot of G(*s*) = $\frac{e^{-st}}{1+sT}$

To find inter section with real axis, G($j\omega$) in terms of real and imaginary part can be written as

$$G\,(j\omega) = \frac{(\cos\omega T - j\sin\omega T)(1 - j\omega T)}{(1 + j\omega T)(1 - j\omega T)}$$

$$= \frac{\cos\omega T - \omega T\sin\omega T}{1+\omega^2T^2} - j\frac{\sin\omega T + \omega T\cos\omega T}{1+\omega^2T^2}$$

Equating Im [G($j\omega$)] to zero, we have

$$\sin\,\omega T = -\,\omega T\,\cos\,\omega T$$

or

$$\omega T = \tan^{-1}\,(-\,\omega T)$$

Retaining only two terms of $\tan^{-1}$ (– ωT) series, we have

$$\omega T = -\,\omega T - \frac{(-\,\omega T)^3}{3}$$

and solving this, we have

$$\omega = 0 \quad \text{or} \quad \frac{\sqrt{6}}{T}$$

and

$$\text{Re}\left[G(j\omega)\right]\Big|_{\omega=\sqrt{6}/T} = \frac{\cos\sqrt{6} - \sqrt{6}\sin\sqrt{6}}{1+6} = -\,0.33$$

Using the points so far the polar plot is shown in Fig. 6.9 (*p*).

Since $\tan^{-1}$ (– ωT) series consists of infinite number of terms, there will be infinite points of intersections on real axis, the smallest value of ω just greater than zero, the real axis intersection is – 0.33 at $\omega = \frac{\sqrt{6}}{T}$. For higher values of ω, the polar plot spirals around the origin.

The polar plots of few more transfer functions are put together in Table 6.1.

TABLE 6.1: Polar plots of transfer functions

Transfer function G(jω)	*Polar plot*
$\dfrac{1 + j\omega T}{j\omega T}$	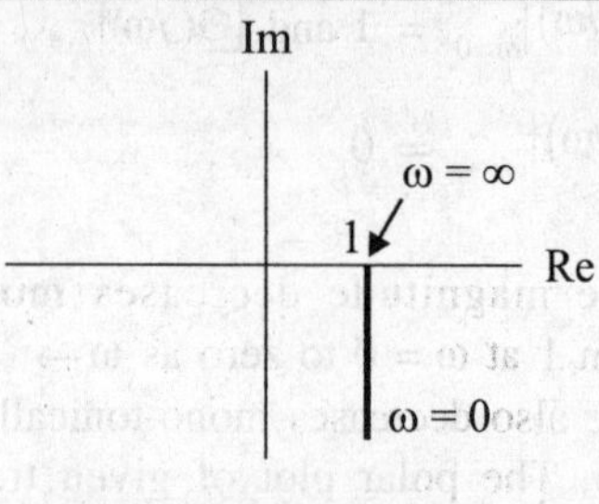

Transfer function G(jω)	*Polar plot*
$\dfrac{1}{(j\omega)^2}$	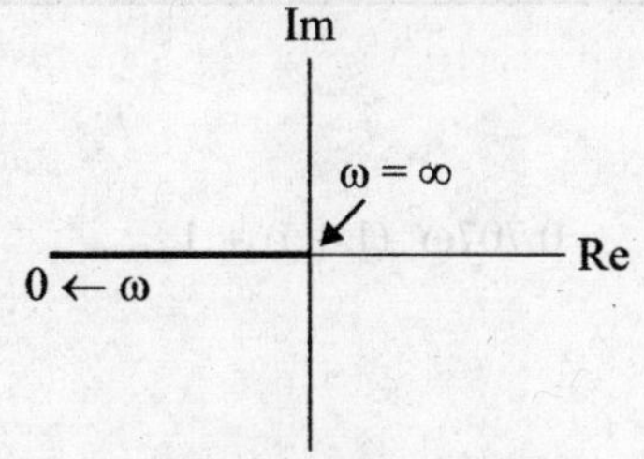
$\dfrac{1}{j\omega(1+ja\omega)(1+jb\omega)}$	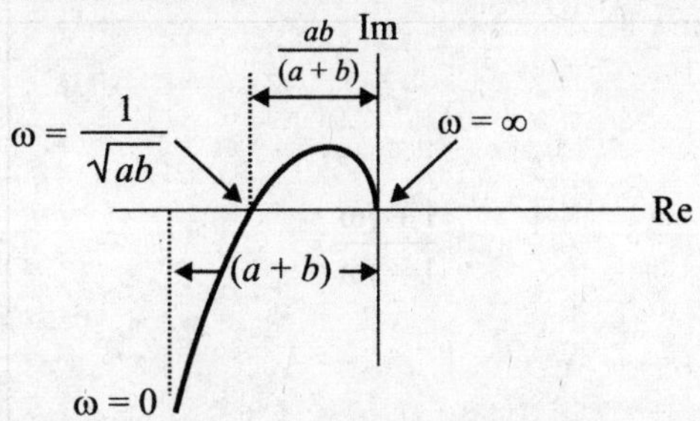
$\dfrac{1}{(j\omega)^2(1+ja\omega)(1+jb\omega)}$	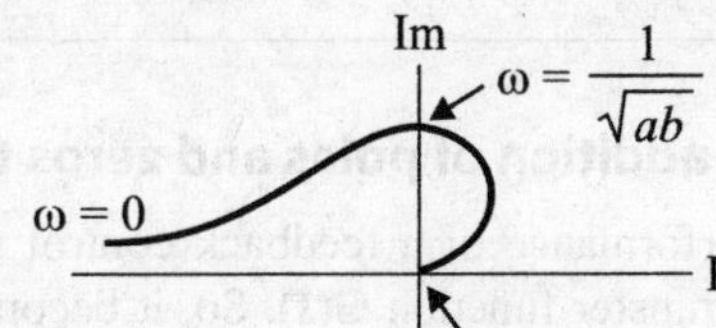
$\dfrac{1}{(j\omega)^2(1+ja\omega)(1+jb\omega)(1+jc\omega)}$	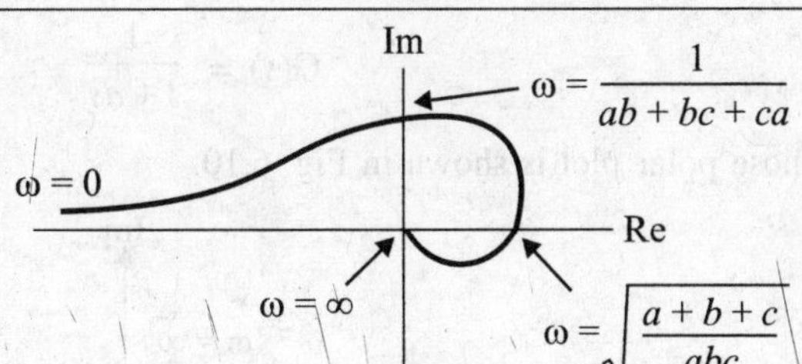
$\dfrac{j\omega-a}{j\omega(j\omega+b)}$	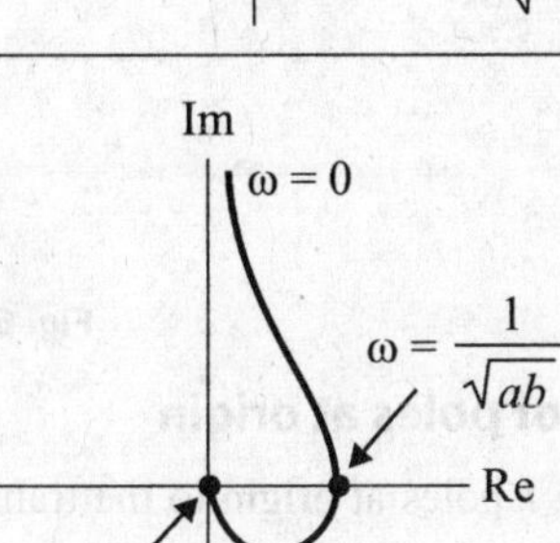

<table>
<tr><th>Transfer function G(jω)</th><th>Polar plot</th></tr>
<tr><td>$0.707\omega^2 (1 + j) + 1$</td><td>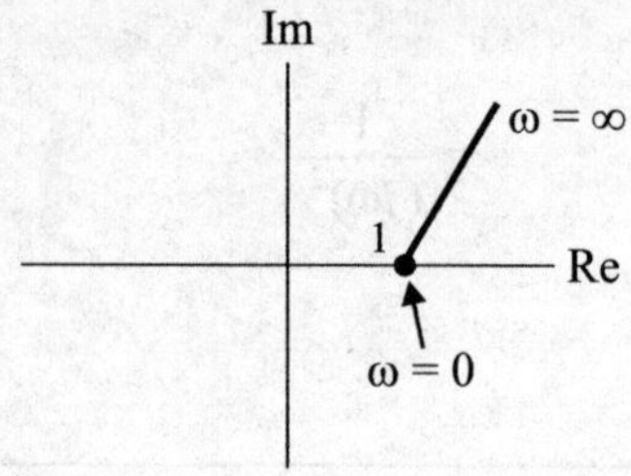
</td></tr>
<tr><td>$\dfrac{1 + j\omega}{1 - j\omega}$</td><td>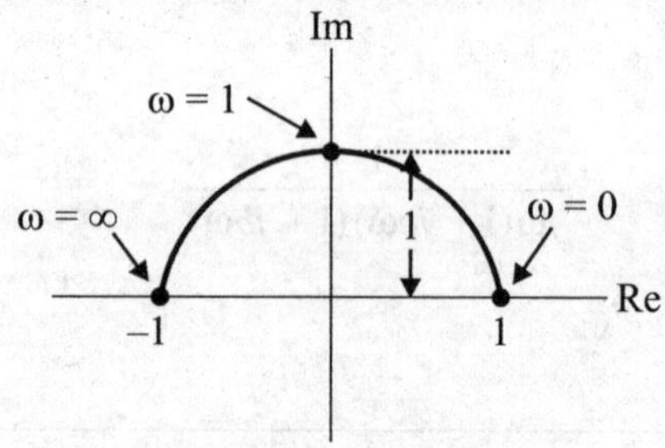
</td></tr>
</table>

Effects of addition of poles and zeros to G(*s*) on the shape of polar plots

The performance of a feedback control system is usually affected by adding poles and zeros to open loop transfer function G(*s*). So, it becomes significant to investigate how the shape of polar plot changes by adding poles/zeros.

To begin with this investigation, consider a system with transfer function

$$G(s) = \frac{1}{1 + as} \quad ...(6.10)$$

whose polar plot is shown in Fig. 6.10.

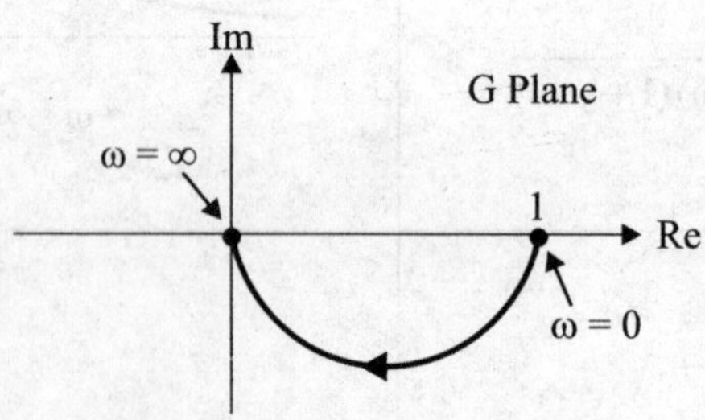

Fig. 6.10: Polar plot of G(*s*) = 1/(1+ *as*)

Addition of poles at origin

Adding *r* poles at origin to the transfer function G(*s*) = 1/(1 + *as*) the modified transfer function is

$$G_m(s) = \frac{1}{s^r(1 + as)}$$

Setting r = 1, 2 and 3, the transfer functions respectively, are

$$G_1(s) = \frac{1}{s(1+as)} \quad ...(6.11)$$

$$G_2(s) = \frac{1}{s^2(1+as)} \quad ...(6.12)$$

$$G_3(s) = \frac{1}{s^3(1+as)} \quad ...(6.13)$$

The polar plots of G(s), $G_1(s)$, $G_2(s)$ and $G_3(s)$ respectively given by equations (6.10), (6.11), (6.12) and (6.13) are shown together in Fig. 6.11.

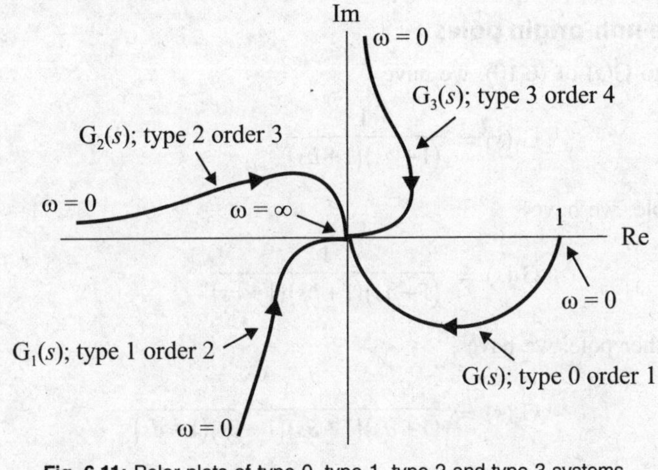

Fig. 6.11: Polar plots of type 0, type 1, type 2 and type 3 systems

Note the following from polar plots of Fig. 6.11:

(*i*) Adding a pole at origin, the phase angle reduces by 90° at $\omega = 0$ and $\omega = \infty$ both *i.e.,* the polar plot is rotated by an angle of – 90° at $\omega = 0$ and $\omega = \infty$ both with the addition of each pole at origin.

(*ii*) At $\omega = 0$

$$\angle G(j\omega) = 0° \text{ for type 0 system}$$

$$\angle G(j\omega) = -90° \text{ for type 1 system}$$

$$\angle G(j\omega) = -180° \text{ for type 2 system}$$

and $$\angle G(j\omega) = -270° \text{ for type 3 system}$$

Thus, the possible type number of system may be identified from polar plot by observing the angle contributed by system at $\omega = 0$.

(*iii*) at $\omega = \infty$, the magnitude of all the systems, becomes zero but phase angle depends on order of system as follows :

$$\angle G(j\omega)\big|_{\omega=\infty} = -90° \text{ for system of order 1}$$

$$\angle G_1(j\omega)\big|_{\omega=\infty} = -180^\circ \text{ for system of order 2}$$

$$\angle G_2(j\omega)\big|_{\omega=\infty} = -270^\circ \text{ for system of order 3}$$

and $$\angle G_3(j\omega)\big|_{\omega=\infty} = -360^\circ \text{ for system of order 4}$$

Thus, the order of system may be identified from polar plot by observing the angle contributed by system at $\omega = \infty$.

(*iv*) The addition of poles at origin to open loop transfer function, adversely affects the stability of closed loop system (stability analysis from polar plot is to be discussed later in this chapter). A system with more than one poles at origin (type 2 or higher) is likely to be unstable or nearly so.

Addition of finite non-origin poles

Adding a pole to G(*s*) of (6.10), we have

$$G_1(s) = \frac{1}{(1+as)(1+bs)} \quad \text{...(6.14)}$$

adding one more pole, we have

$$G_2(s) = \frac{1}{(1+as)(1+bs)(1+cs)} \quad \text{...(6.15)}$$

and adding yet another pole, we have

$$G_3(s) = \frac{1}{(1+as)(1+bs)(1+cs)(1+ds)} \quad \text{...(6.16)}$$

The polar plots of $G_1(s)$, $G_2(s)$, $G_3(s)$ respectively given by equations (6.14), (6.15) and (6.16) together with equation that of G(*s*) (6.10), are shown in Fig. 6.12.

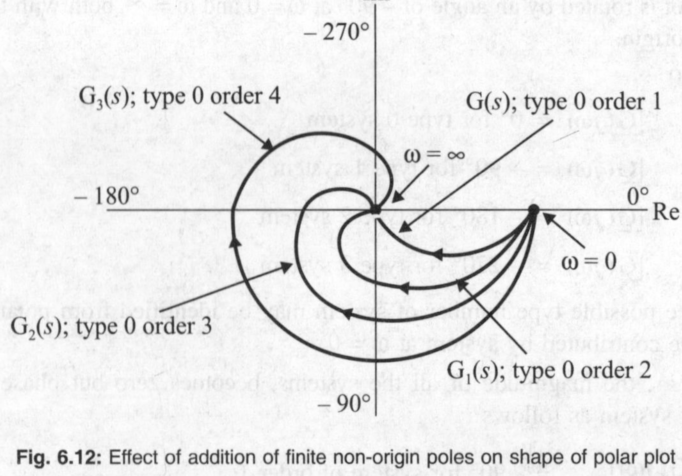

Fig. 6.12: Effect of addition of finite non-origin poles on shape of polar plot

Note the following from polar plots of Fig. 6.12:

(*i*) the polar plot at $\omega = \infty$, is rotated through an angle of $-90°$ with the addition of each non-origin pole.

(*ii*) The addition of poles to open loop transfer function, poses an adverse effect on closed loop stability (to be discussed in following section). It has been already pointed out in chapter 5 that addition of non-origin poles, is some what equivalent to introduction of integral control and tends to make the system relatively less stable.

Addition of zeros at origin

The addition of a zero at origin to G(*s*) of (6.10), gives

$$G_1(s) = \frac{s}{1+as} \quad ...(6.17)$$

Adding one more zero at origin, we have

$$G_2(s) = \frac{s^2}{1+as} \quad ...(6.18)$$

and adding yet another zero at origin, we have

$$G_3(s) = \frac{s^3}{1+as} \quad ...(6.19)$$

The polar plots $G_1(s)$, $G_2(s)$, $G_3(s)$ given by equations (6.17), (6.18) and (6.19) respectively are shown in Fig. 6.13. It is easy to see that addition of a zero to open loop transfer function at origin, rotates the polar plot counterclockwise by 90° at $\omega = 0$ and $\omega = \infty$ both.

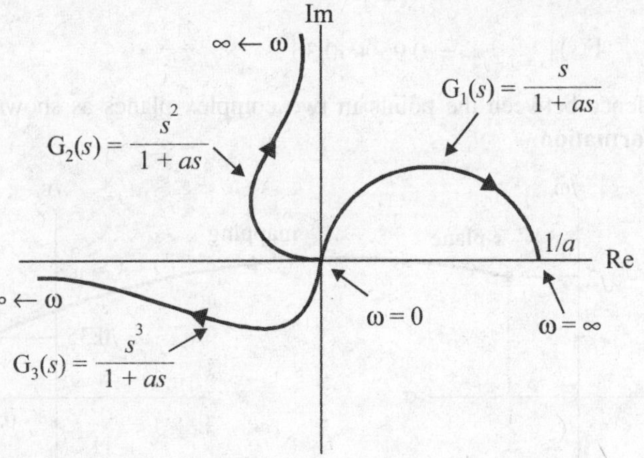

Fig. 6.13: Effect of addition of zeros at origin on shape of polar plots

6.5 THE NYQUIST STABILITY CRITERION

Recall that the stability investigation through **Root locus technique** and **Routh Criterion** necessiate computation of roots of characteristic equation (the closed loop poles). The criterion, derived by H. Nyquist, is a semigraphical procedure that can determine whether or not a system is stable by directly relating the location of roots of characteristic equation to the frequency response of the open loop transmittance G(s) H(s). There is no as such any need to actually determine the closed loop poles. This criterion fascinates in the sense as follows:

(*i*) The Nyquist procedure provides the same information about absolute stability of system as does the Routh stability criterion.

(*ii*) The Nyquist procedure, additionally, provides information about degree of stability (relative stability) of a stable system, the degree of instability of an unstable system and an indication as to how the stability may be improved.

(*iii*) The stability of an irrational closed loop system, for example, a system with a pure delay of form e^{-Ts}, can also be investigated.

We will discuss the Nyquist procedure in two parts. First we will show how to create a Nyquist plot and then stability investigation will be made by interpreting the Nyquist plot.

The entire Nyquist procedure is based on a theorem from theory of complex variables due to **Cauchy**, commonly known as **'principle of arguments'**. To understand this, we shall first discuss **mapping of contours in complex planes.**

Mapping

Let F(s) be a function of complex variable $s = \sigma + j\omega$. Since F(s) is complex, it can always be arranged as a sum of real and imaginary parts as follows

$$F(s) = u(\sigma, \omega) + jv(\sigma, \omega) \qquad ...(6.20)$$

This equation suggests, any point in s-plane at which F(s) is analytic, may be uniquely mapped in F(s) plane by locating values of u and v for given value of s. For example

$$F(s) = \frac{2s+3}{s+5}$$

$$F(s)\big|_{s=1+j2} = 0.95 + j0.35$$

This correspondence between the points in two complex planes as shown in Fig. 6.14 is called **mapping or transformation.**

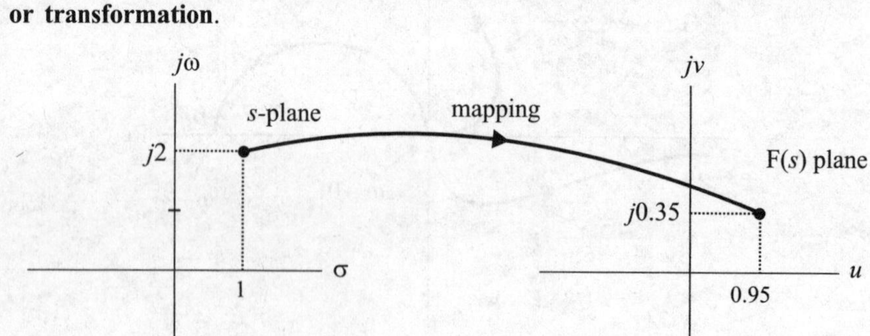

Fig. 6.14: Mapping between two points

It is important to note the following in ongoing discussion:

(*i*) Although the mapping from s plane into F(s) plane is unique (single valued), the reverse process is usually not a single valued mapping. For example, consider the function

$$F(s) = \frac{K}{s(s+1)(s+2)}$$

for each value of s in s-plane, a unique corresponding point is found in F(s) plane. However, for each point in F(s)-plane, the function maps into three corresponding points in s-plane. This can be demonstrated by writing the function under consideration in the form

$$s\,(s+1)\,(s+2) - \frac{K}{F(s)} = 0$$

Supposing F(s) is a constant, the resulting third order equation, provides three roots in s-plane.

(*ii*) A function F(s) is said to be analytic in s-plane if the function and all its derivatives exist. The points in the s-plane where the function (or its derivatives) does not exist, are called **singular points**, for example, poles of a function are singular points. At poles the function becomes infinite.

(*iii*) In the discussion ahead, we shall assume that the G(s) H(s) product is represented as ratio of polynomials in s. For a physically realizable system, the degree of denominator polynomial must be greater than or equal to that of numerator polynomial *i.e.*, $\lim_{s\to\infty} G(s)$ H(s) must be either zero or constant.

(*iv*) Any number of points in s plane, can be uniquely mapped into F(s) plane. It follows that an arbitrarily chosen closed contour in the s-plane which does not pass through any singular points, can also be mapped into an unique closed contour in F(s) plane.

Considering mapping of a closed contour C_s in s-plane into closed contour C_f in F plane, the **principle of argument** is stated as follows.

If F(s) is an analytic function, except for a finite number of poles within C_s, then with complete traversal along C_s which does not pass through any singular points, in clockwise direction, the corresponding contour C_f in F-plane will encircle origin of F-plane N times in the same direction such that

$$\mathbf{N = Z - P}$$

where **Z = number of zeros within C_s**

P = number of poles within C_s

N = number of encirclements of origin of F-plane by the mapped contour

To demonstrate the principle of argument consider the example

$$F(s) = s + 2$$

Let us determine map of closed contours C_{s1} and C_{s2} in the s-plane which are circles with centre of both at origin but with radius 1 unit and 3 units respectively as shown in Fig. 6.15(*a*).

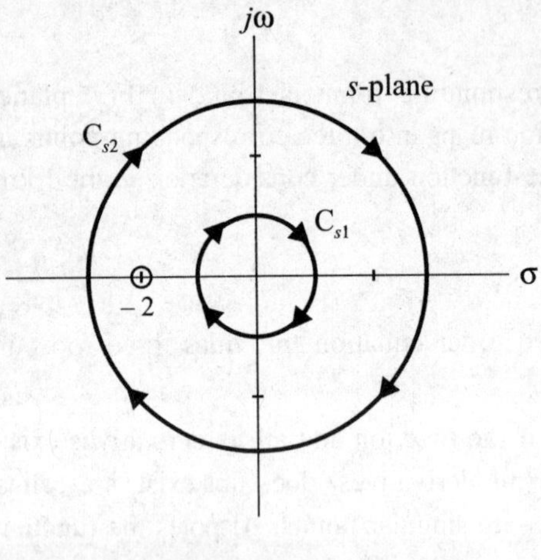

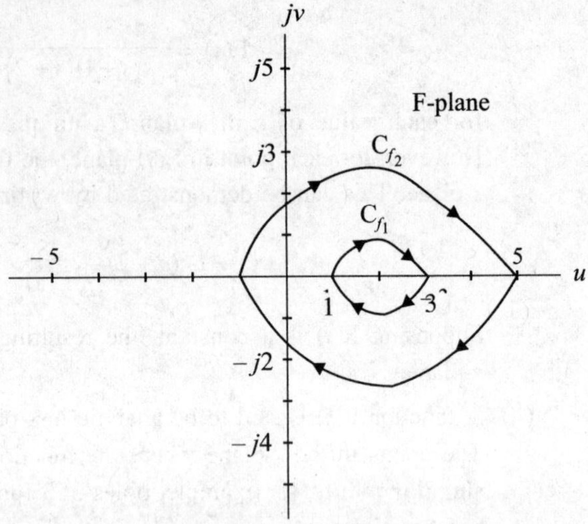

Fig. 6.15: (*a*) Closed contours C_{s1} and C_{s2} in *s*-plane (*b*) Mapped contours C_{f_1} and C_{f_2} in F-plane

using the information

$$F(s)\big|_{s=-1+j0} = -1 + j0 + 2 = 1 + j0$$

$$F(s)\big|_{s=0+j1} = 0 + j1 + 2 = 2 + j1$$

$$F(s)\big|_{s=1+j0} = 1 + j0 + 2 = 3 + j0$$

and $$F(s)\big|_{s=0-j1} = 0 - j1 + 2 = 2 - j1$$

The map of contour C_{s1} in *s*-plane is sketched roughly in Fig. 6.15(*b*) and shown as C_{f_1}. Further using the information

$$F(s)\big|_{s=-3+j0} = -3 + j0 + 2 = -1 + j0$$

$$F(s)\big|_{s=0+j3} = 0 + j3 + 2 = 2 + j3$$

$$F(s)\big|_{s=3+j0} = 3 + j0 + 2 = 5 + j0$$

and $$F(s)\big|_{s=0-j3} = 0 - j3 + 2 = 2 - j3$$

the map of contour C_{s2} in *s*-plane is sketched roughly in Fig. 6.15 (*b*) and shown as C_{f_2}.

Note that C_{s1} does not enclose any pole or zero in *s* plane (Z = 0, P = 0), therefore C_{f_1} does not encircle origin of F-plane *i.e.,* N = 0. C_{s2} encloses one zero and no pole (Z = 1, P = 0), therefore C_{f_2} encloses origin of F-plane once in clockwise direction (N = 1).

Note the following:

(*i*) Considering the clockwise traversal of closed contour C_s chosen in the *s*-plane, the number of encirclements of origin of F-plane by the mapped contour C_f can be positive (N > 0), zero (N = 0), or negative (N < 0). The clockwise encirclements of origin of F-plane are considered to be positive and counter clockwise negative.

If Z > P, contour C_s in *s* plane encloses more zeros than poles. Then the mapped contour C_f in F-plane will encircle the origin of F-plane N-times in the same direction as that of C_s. Thus the encirclement is clockwise and N > 0.

If Z = P, contour C_s in *s*-plane encloses equal number of poles and zeros or no poles and zeros. The mapped contour C_f in F-plane will not encircle the origin of F-plane and N = 0.

If Z < P, contour C_s in *s*-plane encloses more poles than zeros. Then the mapped contour C_f in F-plane will encircle the origin of F-plane N times in opposite direction as that of traversal of C_s. Thus the encirclement is counter clockwise and N < 0.

(*ii*) In order to determine the number of encirclements N w.r.t. origin (or any other point), a line is drawn from the point in any direction to a point as far as necessary, then the number of intersections of this line with the mapped closed contour in F-plane, gives the integer value of N. This is demonstrated in Fig. 6.16(*a*), (*b*) and (*c*).

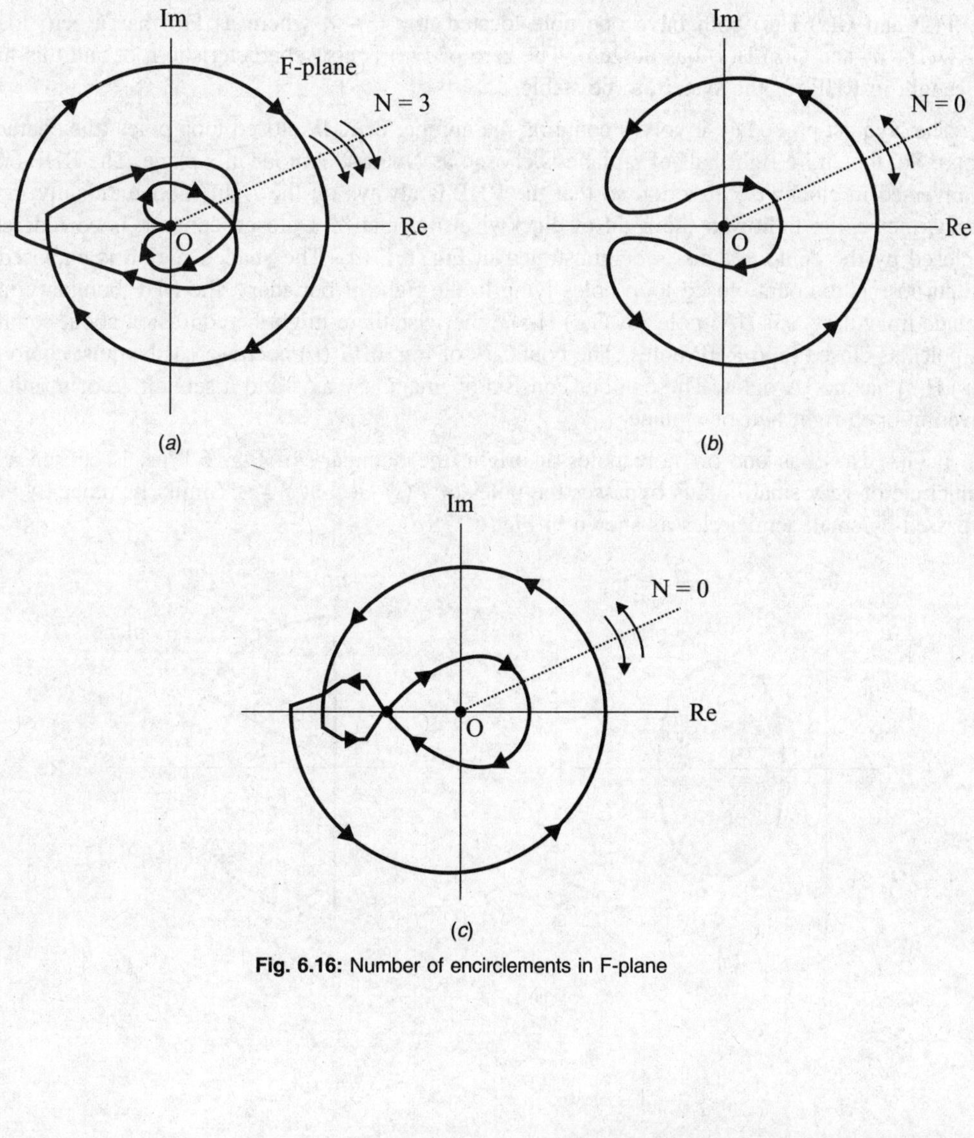

Fig. 6.16: Number of encirclements in F-plane

Nyquist stability criterion

The Nyquist stability criterion relates the roots of characteristic equation to the frequency response of open loop transmittance G(s) H(s). The closed loop stability investigation, in fact, requires determination of presence of any RHP root of characteristic equation of form

$$F(s) = 1 + G(s)\,H(s) = 0.$$

Note that poles of F(s) are also the poles of G(s) H(s) but the zeros are different. The zeros of F(s) are also roots of characteristic equation. To demonstrate

let $$G(s)\,H(s) = \frac{1}{s+a}$$

then $$F(s) = 1 + G(s)\,H(s) = \frac{s+(1+a)}{s+a}$$

F(s) and G(s) H(s) both have one pole located at $s = -a$ where as F(s) has a zero located at $s = -(1 + a)$ and G(s) H(s) has no zero. The zero of F(s) is also characteristic root and this must not be caught in RHP for the system to be stable.

The Nyquist procedure involves counting the number of RHP closed loop poles (the characteristic roots). So, the entire right half of s-plane is chosen as Nyquist contour in s plane. The RHP boundary is traversed in clockwise direction so that the RHP is always on the right. Conventionally in control theory, the region falling to the right of clockwise traversal of a closed contour, is considered to be enclosed by the contour. This is demonstrated in Fig. 6.17(a). The shaded region is enclosed. Since the purpose is to count closed loop poles lying to the right of boundary, the RHP boundary must not include imaginary axis (IA) poles of G(s) H(s), otherwise there might be confusion about counting the IA poles as closed loop RHP poles. The boundary of Fig. 6.17 (b) is chosen as Nyquist contour when G(s) H(s) has no IA poles. The contour consists of imaginary axis and a semicircle of infinite radius covering open right half of s plane.

If G(s) H(s) has one or more poles at origin, the boundary of Fig. 6.17(c) is chosen so that a semicircle of very small radius bypasses the poles of G(s) H(s) at $s = 0$. Similarly, other IA poles are bypassed by small semicircles as shown in Fig. 6.17 (d).

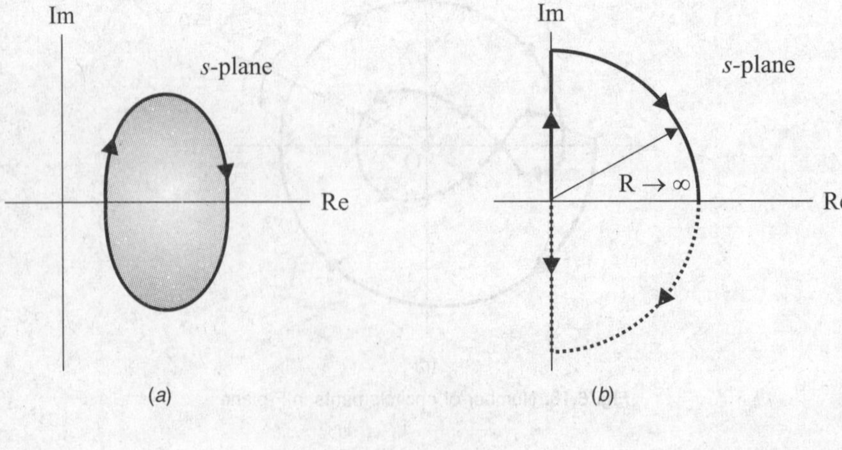

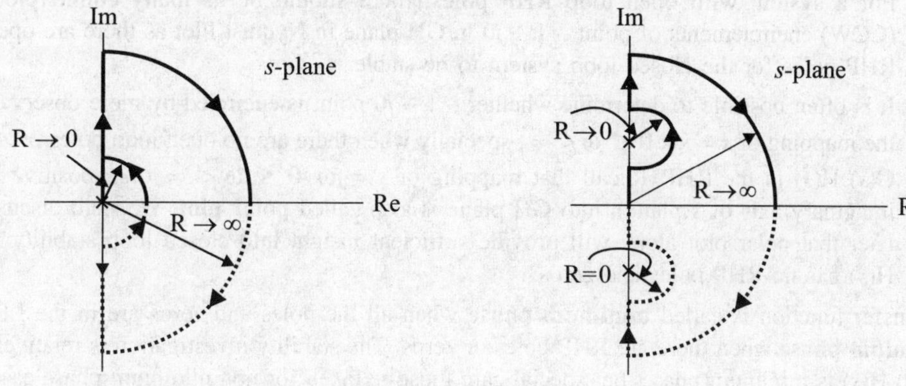

(c) (d)

Fig. 6.17: RHP boundary (*a*) enclosed region of a closed contour (*b*) G(*s*) H(*s*) with no IA poles (*c*) G(*s*) H(*s*) with poles at origin (*d*) G(*s*) H(*s*) with IA poles

Let there be P poles and Z zeros within Nyquist contour chosen in *s* plane as discussed just now. Recall that zeros of F(*s*) = 1 + G(*s*) H(*s*) are also closed loop poles or roots of characteristic equation 1 + G(*s*) H(*s*) = 0.

For the closed loop system to be stable Z must be equal to 0 *i.e.,* Characteristic polynomial must not have any root within the Nyquist contour. From the principle of argument, a map of Nyquist contour in F plane will encircle the origin of F plane N times in clockwise direction where

$$N = Z - P$$

Substituting Z = 0 which is the sufficient condition for the closed loop system to be stable, the principle of argument modifies to N = – P; –ve sign signifies counterclockwise encirclement

So, the **Nyquist stability criterion** may be stated as follows.

A closed loop system will be stable if and only if the number of counterclockwise encirclements (N) of origin of F plane by the map of Nyquist contour in *s* plane, is equal to the number of poles (P) of G(*s*) H(*s*) within the Nyquist contour.

Note that the poles of G(*s*) H(*s*) are same as the poles of F(*s*) and origin of F plane is the point – 1 + *j*0 in G(*s*) H(*s*) plane.

In general, the open loop transmittance (G(*s*) H(*s*) product) of system is known. So, the Nyquist contour choosen in *s* plane is mapped into G(*s*) H(*s*) plane instead of F plane in Nyquist stability investigation procedure. The map of Nyquist contour in GH plane, is called Nyquist plot. Now, the stability criterion may be restated as follows.

"A closed loop system will be stable if and only if the number of counterclockwise encirclements (N) of point – 1 + *j*0 by the map of Nyquist contour in GH plane is equal to number of poles (P) of G(*s*) H(*s*) within the Nyquist contour."

A summary of Nyquist procedure: The salient points of Nyquist stability procedure are summarised as follows.

(*i*) For a system with no open loop RHP poles (P = 0), no encirclements (N = 0) of point – 1 + *j*0 in GH plane by Nyquist Plot is the condition for the closed loop system to be stable.

(*ii*) For a system with open loop RHP poles, there should be as many counterclockwise (CCW) encirclements of point – 1 + *j*0 in GH plane in Nyquist Plot as there are open loop RHP poles for the closed loop system to be stable.

(*iii*) It is often possible to determine whether – 1 + *j*0 point is encircled by mere observation of the mapping of $s = j\omega$; $0 \le \omega < \infty$, specially when there are no open loop poles or zeros of G(*s*) H(*s*) in the RHP. Recall that mapping of $s = j\omega$: $0 \le \omega < \infty$ (*i.e.,* positive half of imaginary axis of *s* plane), into GH plane is also called **polar plot**. We shall discuss little later that polar plot alone will provide sufficient insight into closed loop stability if G(*s*) H(*s*) has no RHP poles and zeros.

A transfer function is called **minimum phase** when all the poles and zeros are in the LHP and **non minimum phase** when there are RHP poles or zeros. The stability investigation is relatively easy when G(*s*) H(*s*) is minimum phase, but special care must be taken for non minimum phase cases. For a minimum phase transfer function, a portion of Nyquist plot corresponding to $s = j0$ to $s = j\infty$ (polar plot) is sufficient for stability analysis.

Nyquist procedure for minimum phase system

Consider that open loop transmittance G(*s*) H(*s*) is of minimum phase type; it does not contain any RHP poles. The principle of agrument N = Z – P, with P = 0, modifies to

$$N = Z$$

But Z must be 0 for closed loop stability. Thus N = 0 will guarantees closed loop stability. This means that the critical point – 1 + *j*0 must not be encircled or simply enclosed by the Nyquist plot in GH plane. So, the simplified Nyquist criterion for minimum phase system is stated as:

"If G(*s*) H(*s*) is of minimum phase type, the corresponding closed loop system will be stable if (– 1 + *j*0) point is not enclosed by map of upper half of imaginary axis of *s* plane (*j*0 to *j*∞) into GH plane; the polar plot."

Note the following points:

(*i*) Imagine a journey from ω = 0 to ω = ∞ on polar plot. All the region to the right of journey is said to be enclosed. This is demonstrated in Fig. 6.18; the polar plots (*a*), (*b*) represent stable systems and (*c*), (*d*) unstable systems. The plot (*e*) represents marginally stable system.

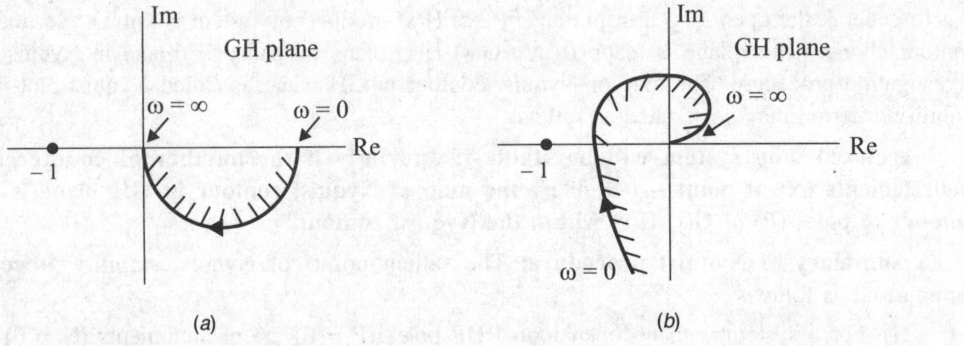

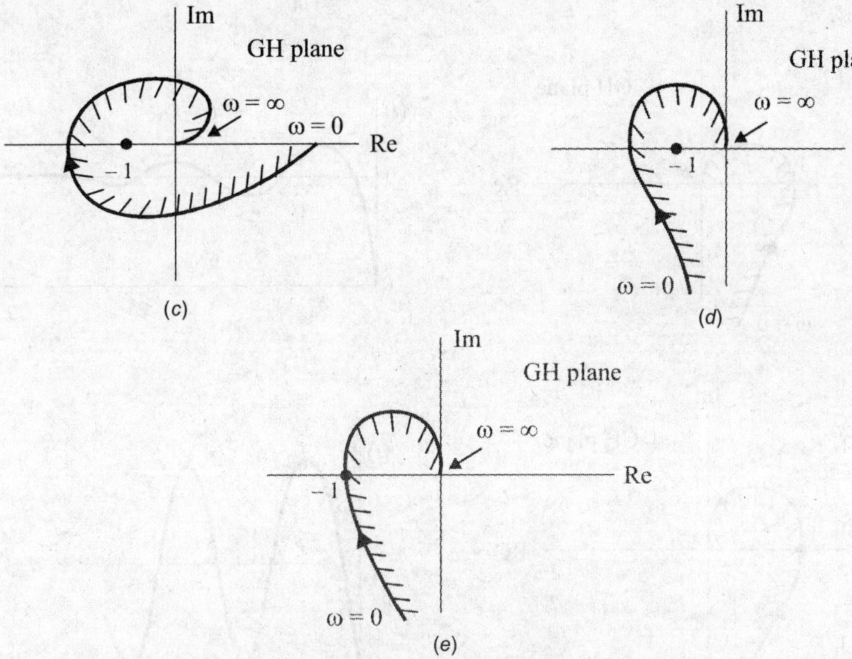

Fig. 6.18: Stability on polar plots

(*ii*) The practical systems are generally minimum phase. So, the simplified Nyquist procedure using enclosure property of critical point ($-1 + j0$) by the polar plot in GH plane, may be conveniently applied to stability investigation. The only drawback is that the enclosure property does not provide information about how many roots of characteristic equation are in the RHP if the system is unstable. The complete Nyquist plot is required to be sketched (to be discussed later) to extract the information about number of RHP roots.

Relative stability using Nyquist procedure

In the current discussion we shall assume that, unless otherwise stated, that the systems are minimum phase. The stability criterion is the non encirclement of the critical point ($-1 + j0$) in GH plane. So, mere inspection of polar plot of G(s) H(s) reveals information about system stability. Intuitively the closeness of the polar plot to ($-1 + j0$) point gives an indication of how stable or unstable the closed loop system is. To demonstrate this the polar plots and corresponding step responses $y(t)$ of a typical third order system are shown in Fig. 6.19 (*a*), (*b*), (*c*) and (*d*).

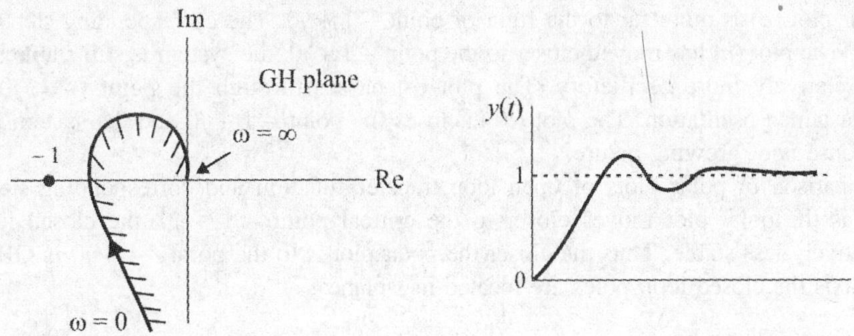

(*a*)

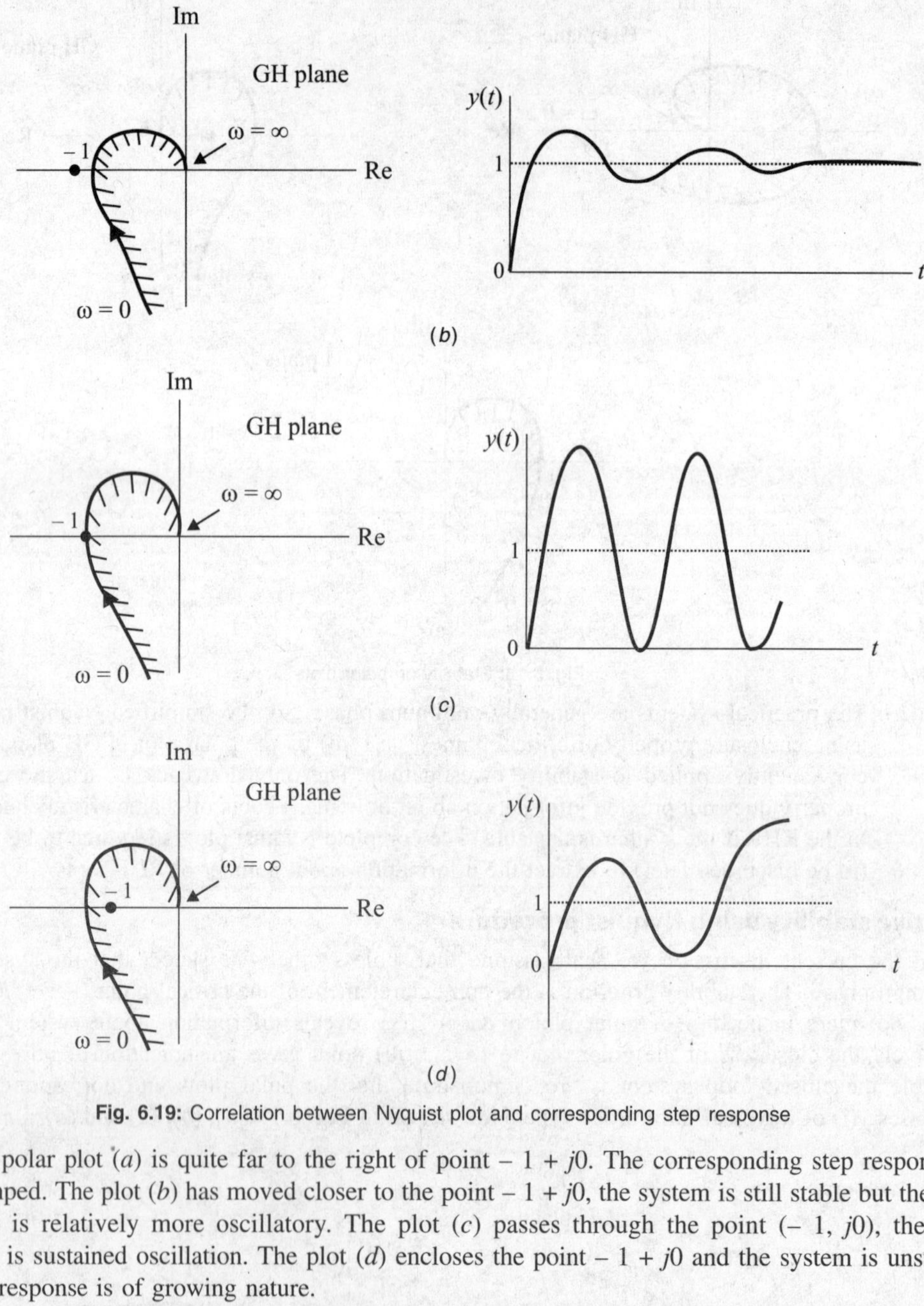

Fig. 6.19: Correlation between Nyquist plot and corresponding step response

The polar plot (*a*) is quite far to the right of point – 1 + *j*0. The corresponding step response is well damped. The plot (*b*) has moved closer to the point – 1 + *j*0, the system is still stable but the step response is relatively more oscillatory. The plot (*c*) passes through the point (– 1, *j*0), the step response is sustained oscillation. The plot (*d*) encloses the point – 1 + *j*0 and the system is unstable, the step response is of growing nature.

The comparison of polar plots of open loop transfer function and corresponding step response reveals that as the polar plot moves closer to the critical point – 1 + *j*0, the closed loop system becomes relatively less stable. Thus the closer the polar plot is to the point – 1 + *j*0 in GH plane, the closer to *j*ω axis the closed loop poles are located in *s*-plane.

The measures of relative stability: Gain margin and phase margin

Consider the polar plot of typical third order minimum phase open loop system as shown in Fig. 6.20. The polar plot intersects negative real axis at a frequency $\omega = \omega_{pc}$ with an intercept of A and unit circle centered at origin at $\omega = \omega_{gc}$. The line joining the origin and the point corresponding to $\omega = \omega_{gc}$ makes an angle ϕ with negative real axis.

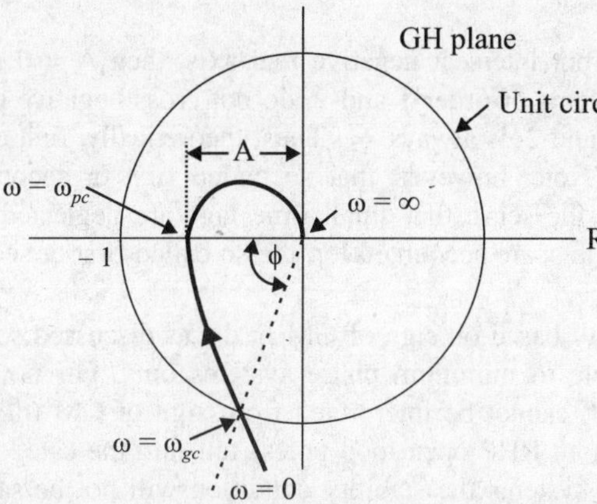

Fig. 6.20: Polar plot of a typical third order system

It can be observed that the polar plot gets closer to $(-1 + j0)$ point as A approaches unity and /or ϕ approaches zero. This has been just discussed that the closeness of polar plot to the point $(-1 + j0)$ is related to the degree of stability. Thus A and ϕ can be used as measure of relative stability. These view points are used to bring out the following definitions.

(*i*) **Phase cross over point and phase cross over frequency:** The phase cross over point in GH plane is a point at which G(*j*ω) H(*j*ω) locus (polar plot) intersects the negative real axis and frequency corresponding to this point is called phase cross over frequency (ω_{pc}). ω_{pc} can also be determined by setting.

$$\angle G(j\omega)\, H(j\omega) \Big|_{\omega = \omega_{pc}} = 180°$$

or

$$\text{Im}\,[G(j\omega)\, H(j\omega)] = 0$$

(*ii*) **Gain margin:** Having defined phase cross over frequency ω_{pc}, it is easy to see from Fig. 6.20 that

$$|G(j\omega)\, H(j\omega)|_{\omega = \omega_{pc}} = A$$

Then the gain Margin of closed loop system is defined as

$$GM = \frac{1}{A}$$

and in terms of *d*B

$$GM_{dB} = 20 \log \frac{1}{A} = -20 \log A$$

Note the following in current discussion:

(*a*) If the intercept A < 1, the polar plot will not enclose (– 1 + *j*0) point, GM > 0 dB (+ ve GM) and system will be stable.

(*b*) If the intercept A > 1, the polar plot will enclose (– 1 + *j*0) point, GM < 0 dB (– ve GM) and system will be unstable.

(*c*) If polar plot passes through – 1 + *j*0 point, A = 1 and GM = 0 dB. This situation indicates presence of roots on *j*ω axis and sustained oscillations in the system. The system is marginally stable.

(*d*) It polar plot does not intersect negative real axis, then A = 0 and GM = ∞. Recall that polar plots for systems of order 1 and 2, do not cross negative real axis. Therefore GM of system of order 1 and 2, is always ∞. Thus, theoretically, first or second order system can not be unstable. (Note, however, that so called first or second order systems are only approximations in the sense that small time lags are neglected in developing the system model. If the time lags are accounted for, the so called first or second order systems may be unstable.)

(*e*) Interpreting stability based on sign of GM in dB as discussed so far in (*a*), (*b*), (*c*) and (*d*) above, is applicable to minimum phase systems only. For non minimum phase systems stability in general, cannot be interpreted from sign of GM (dB). The unstable open loop systems (one or more RHP open loop poles) fall into the category of non minimum phase systems. For such systems the stability condition will not be satisfied unless G(*j*ω) H(*j*ω) plot encircles the (– 1 + *j*0) point. Hence, such **stable non minimum phase system will have negative phase and gain margins**. So in case of non minimum phase systems, it is safer to examine the complete Nyquist plot (to be discussed little later) rather than to rely only on sign of GM (dB).

(*f*) It is still more important to point out that conditionally stable systems exhibit two or more phase cross over frequencies as shown in Fig. 6.21. The system will be stable only if (– 1 + *j*0) point lies between P and Q or further to the left of R; the polar plot will not enclose the point (– 1 + *j*0). The system will be unstable if (– 1 + *j*0) point lies between Q and R or O and P; the point (– 1 + *j*0) will be enclosed by the polar plot. In case of multiple phase cross over frequencies, there results a set of gain margins. Then the **smallest in the set, is considered to be the gain margin** of stable system. If there is no phase cross over frequency, the gain margin is said to be infinite.

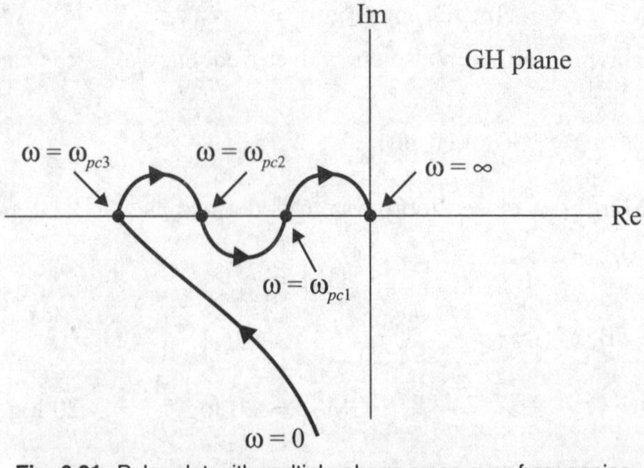

Fig. 6.21: Polar plot with multiple phase cross over frequencies

(*g*) It has been seen that

$$\left| G(j\omega)\, H(j\omega) \right|_{\omega = \omega_{pc}} = A$$

If the system gain is increased by a factor of 1/A, the new intercept on negative real axis by polar plot will be A. (1/A) = 1 and GM = 0 dB. The system is driven to the verge of instability. In this sense the gain margin is the amount of gain in dB that is allowed to be increased in the loop before the closed loop system reaches instability.

(*h*) The gain margin is only one-dimensional specification of relative stability. In fact gain margin provides insight into system stability with respect to the loop gain. Occasionally the system may be subjected to variations in parameters other than loop gain. These variations might affect $\angle G(j\omega)\, H(j\omega)$ and result in polar plots as shown in Fig. 6.22. Note that both the polar plots S_1 and S_2 have same gain margins. However S_1 is relatively more stable than S_2 due to simple observation that polar plot of S_2 passes closer to $(-1 + j0)$ point. Thus gain margin fails to be the measure of relative stability. To tackle such a situation, the term phase margin (PM) (to be discussed little later) is introduced. The gain margin together with phase margin always provide sufficient insight into relative stability.

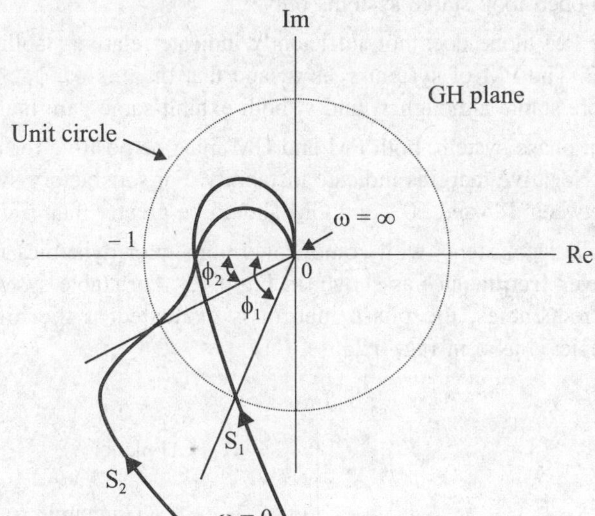

Fig. 6.22: Systems with same GM but different degrees of relative stability

(*iii*) **Gain cross over point and gain cross over frequency:**
The gain cross over point is the point at which the polar plot intersects the unit circle centered at origin and the corresponding frequency is called gain cross over frequency ω_{gc} as shown in Fig. 6.20 place.

$$\left| G(j\omega)\, H(j\omega) \right|_{\omega = \omega_{gc}} = 1$$

(*iv*) **Phase margin (PM): The phase margin is that amount of additional phase lag at the gain cross over frequency required to bring the system to the verge of instability. Alternatively PM is defined as the angle in degrees through which the polar plot must be rotated about the origin in order that gain cross over point passes through – 1 + *j*0 point.**

The PM is depicted by an angle ϕ in Fig. 6.20 where it is easy to see that

$$\angle G(j\omega)\, H(j\omega)\Big|_{\omega = \omega_{gc}} = -180^\circ + \phi$$

If an additional phase lag ϕ is introduced at ω_{gc}, then

$$\angle G(j\omega)\, H(j\omega)\Big|_{\omega = \omega_{gc}} = -180^\circ$$

while $$|G(j\omega)\, H(j\omega)|_{\omega = \omega_{gc}} = 1$$

The polar plot will pass through $(-1 + j0)$ point, driving the system into verge of instability. This additional phase lag ϕ is called PM.

Thus $$\text{PM} = \angle G(j\omega)\, H(j\omega)\Big|_{\omega = \omega_{gc}} + 180^\circ \qquad ...(6.21)$$

Note the following:

(*a*) Let it be pointed out once again that GM and PM as measures of relative stability are applicable to open loop stable systems only.

(*b*) GM alone or PM alone does not sufficiently indicate relative stability, both must be used. See Fig. 6.22. The PM of system s_1 is ϕ_1 and that of s_2 is ϕ_2. $\phi_1 > \phi_2$ indicates that s_1 is relatively more stable although s_1 and s_2 both exhibit same gain margin.

(*c*) For minimum phase system, both PM and GM must be positive for the closed loop system to be stable. Negative margins indicate instability. For satisfactory system performance PM should lie between 30° and 60°, and GM should be greater than 6 dB.

(*d*) Some higher order systems with complicated numerator dynamics might exhibit **multiple gain cross over frequencies** as shown in Fig. 6.23. For stable systems with multiple gain cross over frequencies, the phase margin is evaluated at the highest gain cross over frequency depicted as ϕ in Fig. 6.23.

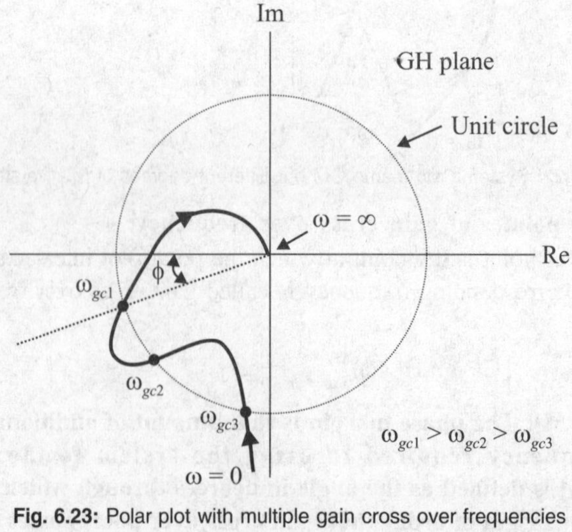

Fig. 6.23: Polar plot with multiple gain cross over frequencies

The analytical evaluation of gain margin (GM), phase margin (PM), gain cross over frequency (ω_{gc}) and phase cross over frequency (ω_{gc}) is demonstrated by the examples as follows.

Example 3: Determine phase cross over frequency, GM and stability for G(*s*) H(*s*) = $\dfrac{100}{(s+1)^3}$

Solution: $$G(j\omega)H(j\omega) = \frac{100}{(j\omega+1)^3}$$

The phase cross over frequency can be determined from

$$\angle G(j\omega)H(j\omega)\Big|_{\omega=\omega_{pc}} = -3\tan^{-1}\omega_{pc} = -180°$$

or $$\omega_{pc} = \tan 60° = 1.73 \text{ rad/sec}$$

The intercept A on negative real axis is determined as:

$$A = |G(j\omega)H(j\omega)|_{\omega=\omega_{pc}}$$

$$= \frac{100}{\left[1+\omega_{pc}^2\right]^{3/2}} = \frac{100}{\left[1+1.73^2\right]^{3/2}} = 12.5$$

so $$GM = \frac{1}{12.5} = 0.08$$

and $$GM_{dB} = 20\log 0.08 = -21.95 \text{ dB}$$

The system has GM with negative dB; hence the system is unstable.

Note that G(*s*) H(*s*) has no RHP poles or zeros, it characterizes minimum phase system and therefore system stability can be interpreted from sign of dB (GM).

Alternatively, the polar plot using the information below is shown in Fig. 6.24.

(*i*) $\lim_{\omega\to 0} G(j\omega)H(j\omega) = 100\angle 0°$

(*ii*) $\lim_{\omega\to\infty} G(j\omega)H(j\omega) \cong \lim_{\omega\to\infty}\dfrac{100}{(j\omega)^3} = 0\angle -270°$

(*iii*) To determine the intersection points of polar plot with real axis and corresponding values of ω, G(*j*ω) H(*j*ω) is put in the form

$$G(j\omega)H(j\omega) = \frac{100}{(j\omega+1)^3}$$

$$= \frac{100\left(1-3\omega^2\right)-j\left(3\omega-\omega^3\right)}{\left[\left(1-3\omega^2\right)+j\left(3\omega-\omega^3\right)\right]\left[\left(1-3\omega^2\right)-j\left(3\omega-\omega^3\right)\right]}$$

$$= \frac{100(1-3\omega^2)}{(1-3\omega^2)^2+(3\omega-\omega^3)^2} - j\frac{(3\omega-\omega^3)}{(1-3\omega^2)^2+(3\omega-\omega^3)^2}$$

It is easy to see that

$$\text{Im}\,[G(j\omega)] = 0 \text{ for } \omega = 0, \sqrt{3}$$

$$\text{Re}\left[G(j\omega)\right]\Big|_{\omega=\sqrt{3}} = -\frac{100}{8} = -12.5$$

So, phase cross over frequency $= \sqrt{3}$ rad/sec.

$$\text{GM} = \frac{1}{12.5} = 0.08$$

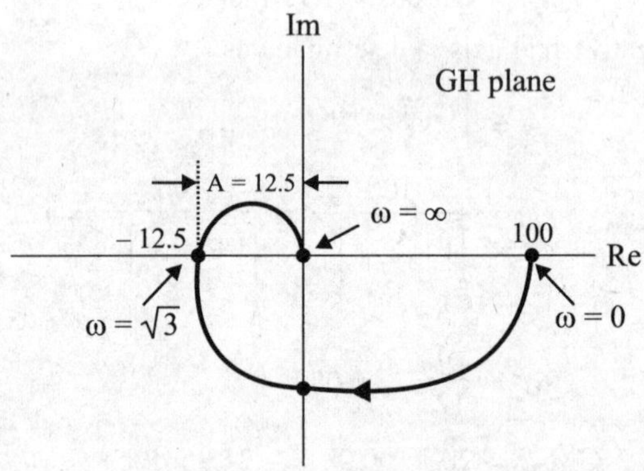

Fig. 6.24: Phase cross over frequency and GM for $G(s)\,H(s) = \frac{100}{(s+1)^3}$

Example 4: Determine gain cross over frequency and phase margin for $G(s)\,H(s) = \frac{10s}{(s+1)^2}$

Solution: To determine gain cross over frequency ω_{gc}, the equation

$$\left|G(j\omega)\,H(j\omega)\right|_{\omega=\omega_{gc}} = 1$$

gives

$$\frac{10\omega_{gc}}{1+\omega_{gc}^2} = 1$$

or

$$\omega_{gc}^2 - 10\,\omega_{gc} + 1 = 0$$

or

$$\omega_{gc} = 5 \pm 4.89 = 9.89 \text{ or } 0.1 \text{ rad/sec}$$

These two gain cross over frequencies give two phase margins as follows:

$$\text{PM} = \underline{\left|G(j\omega)\,H(j\omega)\right.}\Big|_{\omega=\omega_{gc}} + 180°$$

$$= 90° - 2\tan^{-1}\omega_{gc} + 180° = 270° - 2\tan^{-1}\omega_{gc}$$

So, $$\text{PM}\big|_{\omega_{gc}=9.89} = 101.5°$$

and $$\text{PM}\big|_{\omega_{gc}=0.1} = 258.5°$$

To depict this, the polar plot is drawn in Fig. 6.25.

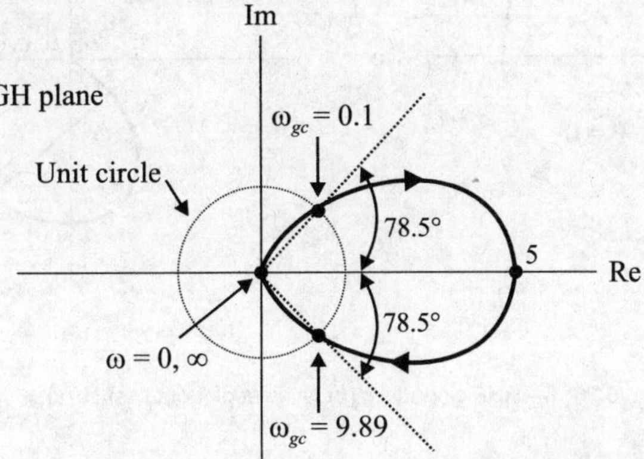

Fig. 6.25: Two gain cross over frequencies for G(*s*) H(*s*) = $\dfrac{10s}{(1+s)^2}$

Generating complete Nyquist plot and interpretting stability

It has been discussed in the preceding section that only a segment of complete Nyquist plot corresponding to $s = j\omega$ where $0 \leq \omega \leq \infty$ (polar plot), is sufficient for stability analysis if G(*s*) H(*s*) is minimum phase transfer function. But this procedure based on enclosure property of polar plot of point $(-1 + j0)$ provides information about whether or not, closed loop system is stable. If closed loop system is stable, the procedure further provides information about relative stability in terms of GM and PM. In case the closed loop system turns to be unstable, it does not furnish information on number of RHP roots due to which it is unstable. Then number of RHP roots can be determined by generating complete Nyquist plot. For non minimum phase G(*s*) H(*s*), the stability analysis is not at all possible through polar plot procedure. Again the complete Nyquist plot is required to be generated to predict closed loop stability. The procedure for generating the complete Nyquist plot is demonstrated with the examples as follows.

Example 5: Sketch Nyquist plot and interpret stability for open loop transmittance

$$G_1(s)\ H_1(s) = \frac{6}{(s+1)(s+2)}$$

Note that $G_1(s)\ H_1(s)$ has no open loop IA (Imaginary axis) pole, the RHP boundary in *s* plane is choosen as shown in Fig. 6.26 (*a*) where in poles of $G_1(s)\ H_1(s)$ are also depicted. The RHP boundary does not enclose any pole of $G_1(s)\ H_1(s)$; P = 0.

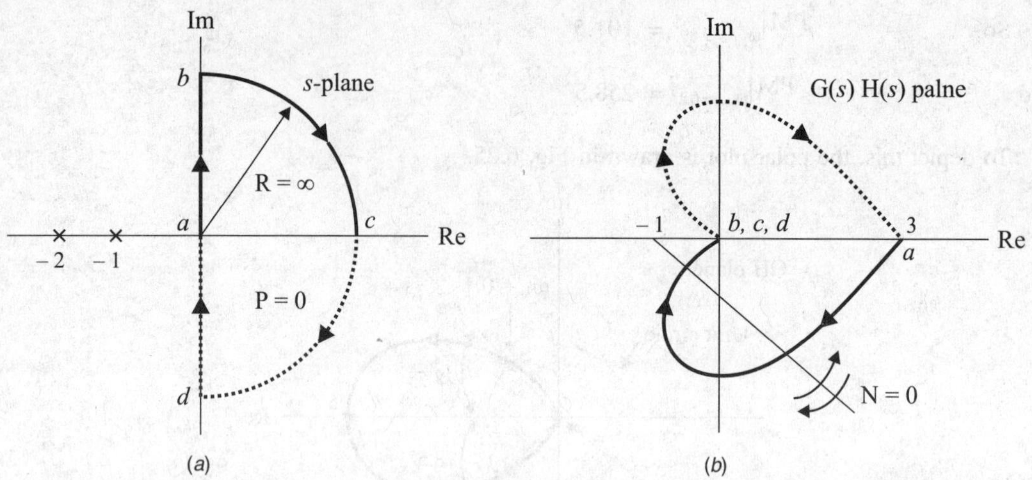

Fig. 6.26: (*a*) RHP boundary (*b*) Nyquist plot of $G_1(s)\ H_1(s) = \dfrac{6}{(s+1)\ (s+2)}$

The mapping of RHP boundary *abcda* into GH plane is demonstrated in the steps as follows:

(*i*) **Map of upper half of IA (path *ab*)**: Put $s = j\omega$ for $0 \le \omega \le \infty$ (polar plot)

$$G_1(j\omega)\ H_1(j\omega) = \frac{6}{(j\omega+1)(j\omega+2)}$$

$$\lim_{\omega\to 0} G_1(j\omega)\ H_1(j\omega) = 3\underline{|0^\circ} \quad \text{(map of point } a)$$

$$\lim_{\omega\to\infty} G_1(j\omega)\ H_1(j\omega) \cong \lim_{\omega\to\infty} \frac{6}{(j\omega)^2} = 0\underline{|-180^\circ} \quad \text{(map of point } b)$$

These points are plotted in Fig. 6.26 (*b*) and interconnected by an arrow. This is, in fact polar plot. Sketching polar plot has been already discussed in detail.

(*ii*) **Map of path *bcd* (semicircle of ∞ radius):**

Put $$s = \lim_{R\to\infty} Re^{j\theta}$$

so that $$G_1(s)\ H_1(s) = \lim_{R\to\infty} \frac{6}{(Re^{j\theta}+1)(Re^{j\theta}+2)} = 0$$

and mapped values *b*, *c* and *d* lie at origin of GH plane.

(*iii*) **Map of path *da*:** It is not necessary to evaluate the mapping of lower half of imaginary axis of Fig. 6.26 (*a*). The Nyquist plot corresponding to mapping of lower half of IA is just the mirror image of mapping of upper half of IA. This is reflected about the real axis and depicted by dashed line in Fig. 6.26 (*b*).

Note that it is common practice to draw arrows to connect the points mapped in GH plane in the same order as the RHP boundary in *s* plane is traversed so that the mapped locations *a b c d a* are oriented in that order.

Interpreting Nyquist plot for stability: Recall the principle of argument (6.20) as

$$N = Z - P$$

Since G(*s*) H(*s*) has no pole within RHP boundary in *s* plane (Fig. 6.26 (*a*)), P = 0

The determination of number of encirclements (N) is demonstrated by drawing a vector outward from point $(-1 + j0)$. The vector is crossed once in each direction (CW and CCW). The conclusion is that there are no encirclements.

So, $$N = 0$$

and $$N = Z - P \text{ gives } Z = 0$$

There is no RHP root and the system is **stable**.

Example 6: Sketch Nyquist plot and interpret stability for open loop transmittance

$$G_2(S)\, H_2(s) = \frac{6}{s^2(s+2)}$$

Solution: Note that $G_2(s)\, H_2(s)$ contains two open loop poles at origin and therefore the RHP boundary in *s* plane, is chosen as shown in Fig. 6.27 (*a*). The poles of $G_2(s)\, H_2(s)$ are also depicted and a semicircle of very small radius ($\rho \to 0$) by passes the poles at origin.

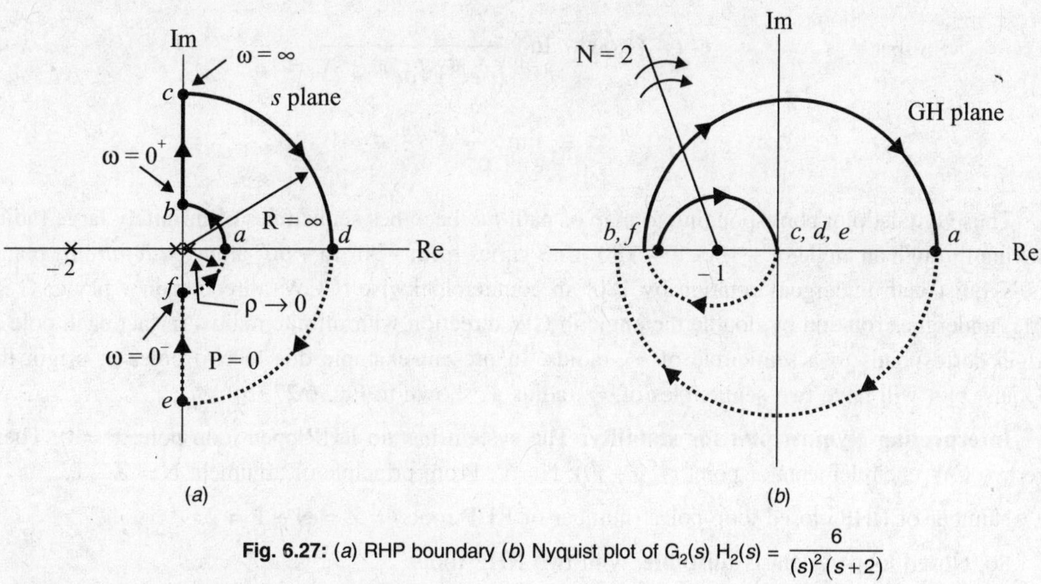

Fig. 6.27: (*a*) RHP boundary (*b*) Nyquist plot of $G_2(s)\, H_2(s) = \dfrac{6}{(s)^2(s+2)}$

The step by step mapping of RHP boundary of Fig. 6.27 (*a*) is demonstrated as follows.

(*i*) **Map of upper half of IA (Path *bc*):** Put $s = j\omega$ for $0 \le \omega \le \infty$

$$G_2(j\omega)\, H_2(j\omega) = \frac{6}{(j\omega)^2 (j\omega + 2)}$$

$$\lim_{\omega \to 0} G_2(j\omega) H_2(j\omega) \cong \lim_{\omega \to 0} \frac{6}{(j\omega)^2} = \infty \angle -180^\circ \quad \text{(map of point } b\text{)}$$

$$\lim_{\omega\to\infty} G_2(j\omega)H_2(j\omega) \cong \lim_{\omega\to\infty} \frac{6}{(j\omega)^3} = 0\angle -270^\circ \quad \text{(map of point } c\text{)}$$

It is easy to verify that there is no intermediate intersection on negative real axis. The above two mapped points are plotted and interconnected by an arrow as shown in Fig. 6.27 (*b*).

(*ii*) **Map of path *cde* (semicircle of ∞ radius):**

Put $$s = \lim_{R\to\infty} Re^{j\theta}$$

so that $$G_2(s)\,H_2(s) = \lim_{R\to\infty} \frac{6}{\left(Re^{j\theta}\right)^2\left(Re^{j\theta}+2\right)} = 0$$

and mapped values *c*, *d* and *e* lie at origin of GH plane.

(*iii*) **Map of path *ef*:** There is no need to evaluate this. It is simply mirror image of map of path *bc* as shown by dashed line in Fig. 6.27 (*b*).

(*iv*) **Map of path *fab* (semicircle of small radius):**

Put $$s = \lim_{\rho\to 0} \rho e^{j\theta}$$

So that $$G_2(s)\,H_2(s) = \lim_{\rho\to 0} \frac{6}{\left(\rho e^{j\theta}\right)^2\left(\rho e^{j\theta}+2\right)}$$

$$\cong \lim_{\rho\to 0} \frac{3}{\rho^2} e^{-j2\theta} = \infty\angle -2\theta$$

Thus Nyquist plot corresponding to map of path *fab* becomes semicircles of infinitely large radius moving through an angle of + 180° to – 180° as θ varies from – 90° to + 90° along path *fab* in *s* plane. As Nyquist path undergoes rotation by 180° in counterclockwise (CCW) direction in *s* plane, $G_2(s)$ $H_2(s)$ undergoes rotation by double the angle in CW direction with infinite radius. In fact each pole at origin corresponds to a semicircle of ∞ radius. In present example due to two poles at origin the Nyquist plot will have two semicircles of ∞ radius as shown in Fig. 6.27 (*b*).

Interpreting Nyquist plot for stability: The system has no RHP open loop pole; P = 0. There are two CW encirclements of point (– 1 + *j*0); N = 2. From principle of argument N = Z – P,

Number of RHP closed loop poles (number of RHP roots) = Z = N + P = 2.

So, closed loop system is **unstable**, with two RHP roots.

Example 7: Sketch Nyquist plot and interpret stability for open loop transmittance

$$G_3(s)\,H_3(s) = \frac{K(s+3)}{s(s-1)} \quad (K > 1)$$

Solution: Since $G_3(s)$ $H_3(s)$ has one pole at origin, the RHP boundary in *s* plane is chosen as shown in Fig. 6.28(*a*). The Poles and zeros of $G_3(s)$ $H_3(s)$ are also depicted and semicircle of very small radius bypasses the pole at origin. Note that there lies one open loop pole within chosen RHP boundary: P =1.

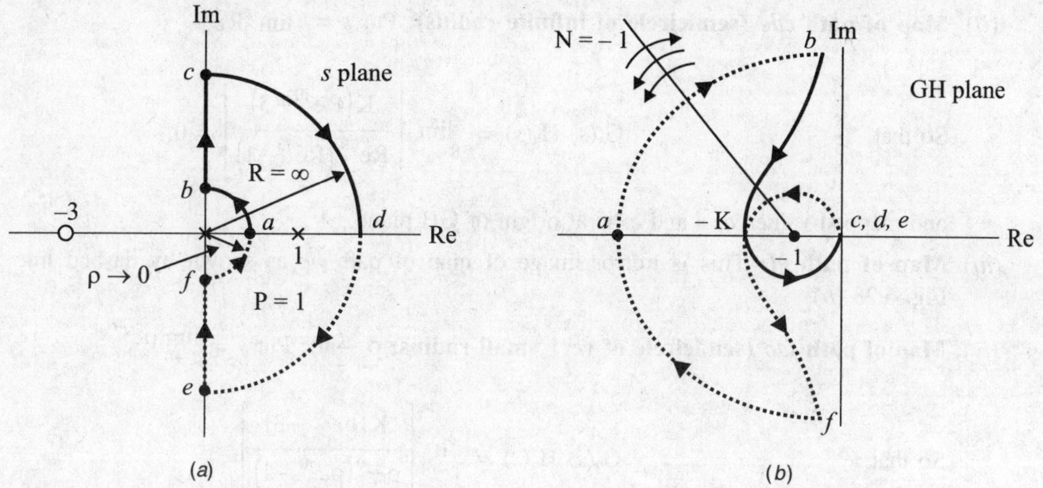

Fig. 6.28: (*a*) RHP boundary (*b*) Nyquist plot for $G_3(s)\,H_3(s) = \dfrac{K(s+3)}{s(s-1)}$; K > 1

The step by step approach to sketch the map of RHP boundary of Fig. 6.28 (*a*) is as follows:

(*i*) **Map of upper half of IA (path *bc*):** Put $s = j\omega$ for $0 \le \omega \le \infty$

$$G_3(j\omega)\,H_3(j\omega) = \frac{K(j\omega+3)}{j\omega(j\omega-1)}$$

$$\lim_{\omega\to 0} G_3(j\omega)\,H_3(j\omega) \cong \lim_{\omega\to 0}\left[\frac{-3K}{j\omega}\right] = \infty\angle -270^\circ \quad \text{(map of point } b)$$

$$\lim_{\omega\to\infty} G_3(j\omega)\,H_3(j\omega) \cong \lim_{\omega\to\infty}\left[\frac{K}{j\omega}\right] = 0\angle -90^\circ \quad \text{(map of point } b)$$

To search the intersection points on negative real axis, we transform $G_3(j\omega)\,H_3(j\omega)$ as a sum of real and imaginary parts as follows.

$$\frac{K(j\omega+3)}{j\omega(j\omega-1)} = \frac{-K}{\omega}\left[\frac{(j\omega+3)(\omega-j)}{(\omega+j)(\omega-j)}\right]$$

$$= \frac{-K}{\omega(1+\omega^2)}\left[4\omega + j(\omega^2-3)\right]$$

It is easy to determine that

$$\text{Im}\,[G_3(j\omega)\,H_3(j\omega)] = 0 \text{ for } \omega = \sqrt{3}$$

and $$\text{Re}\left[G_3(j\omega)H_3(j\omega)\right]\Big|_{\omega=\sqrt{3}} = -K;\quad K > 1$$

(*ii*) **Map of path *cde* (semicircle of infinite radius):** Put $s = \lim_{R \to \infty} Re^{j\theta}$

So that $$G_3(s)\, H_3(s) = \lim_{R \to \infty} \left[\frac{K(Re^{j\theta} + 3)}{Re^{j\theta}(Re^{j\theta} - 1)} \right] = 0$$

and mapped values *c*, *d* and *e* lie at origin of GH plane.

(*iii*) **Map of path *ef*:** This is mirror image of map of path *bc* as shown by dashed line in Fig. 6.28 (*b*).

(*iv*) **Map of path *fab* (semicircle of very small radius; $\rho \to 0$):** Put $s = \lim_{\rho \to 0} \rho e^{j\theta}$

So that $$G_3(s)\, H_3(s) = \underset{\rho \to 0}{\text{lt}} \left[\frac{K(\rho e^{j\theta} + 3)}{\rho e^{j\theta}(\rho e^{j\theta} - 1)} \right]$$

$$\cong \lim_{\rho \to 0} \left(\frac{-3K}{\rho} \right) e^{-j\theta} = -\infty \angle -\theta$$

thus the Nyquist plot corresponding to map of path *f a b* becomes a semicircle of infinitely large radius moving through the angles from – 90° to + 90° as shown in Fig. 6.28 (*b*).

Interpreting Nyquist plot for stability: There is one open loop RHP pole; P = 1. For K > 1 there is one CCW encirclement of point – 1 + *j*0 as shown by drawing outward vector, N = – 1. So, from principle of argument.

Z = N + P = – 1 + 1 = 0, there is no RHP root and closed loop system is **stable**. For K < 1, the intersection point on negative real axis will lie to the right of point (– 1 + *j*0). This is shown in Fig. 6.29.

$$N = 1 \quad \text{and} \quad Z = N + P = 1 + 1 = 2.$$

There will be two RHP closed loop poles and system will turn to become unstable.

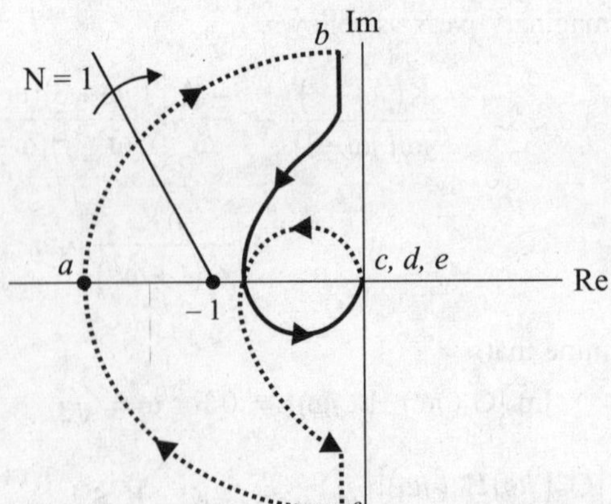

Fig. 6.29: Nyquist plot of $= \dfrac{K(s+3)}{s(s-1)}$; K < 1

6.6 BODE PLOT

Yet another very useful graphical representation of sinusoidal transfer function, termed as **Bode Plot** after **Hendrik W Bode** is presented in this section. In fact the Bode plot is composed of two plots: one, a plot of magnitude $|G(j\omega)\ H(j\omega)|$ in decibel (dB) two, a plot of phase angle $\underline{|G(j\omega)\ H(j\omega)}$ in degrees; both plotted against logarithmic scale for ω. The Bode plot is also called as the **corner plot** or **asymptotic plot.** These names originate from the fact that the **Bode plot** can be sketched by using the straight line approximations that are asymptotic to actual plot.

Note the following special features of Bode plot:

(*i*) In general, the Bode plot is convenient to apply to minimum phase systems, so that the enclosure criterion as discussed in Nyquist procedure can be used for closed loop stability investigation. The utmost care has to be taken while interpreting closed loop stability from Bode plot in case of non minimum phase transmittance G(*s*) H(*s*).

(*ii*) The Bode plot corresponds to positive half of IA of *s* plane; $s = j\omega,\ 0 \le \omega \le \infty$. So, its applicability to stability investigation, will remain confined to determination of gain and phase cross over frequencies and corresponding gain and phase margins.

(*iii*) A transmittance G(*s*) H(*s*) which is, generally, a product of several simple terms, has magnitude that is product of magnitudes of individual terms. Since the magnitude $|G(j\omega)\ H(j\omega)|$ in Bode plot is dealt in dB, the multiplication of individual magnitudes, becomes addition. The overall phase angle $\underline{|G(j\omega)\ H(j\omega)}$ is also algebraic sum of individual phase angles.

(*iv*) The Bode plot displays information much more clearly than the corresponding linear plots. The magnitude and frequency both vary over many powers of ten so that most of the information would be compressed near the origin by a linear plot, while the dB scale expands magnitude in a Bode plot and the log frequency scale expands frequency. Since the vertical axes of Bode plots are linear in dB and linear in phase, the Bode plots are termed semilog plots.

(*v*) Expanding low frequency range by use of logarithmic scale for ω, is highly advantageous in the sense that low frequencies are of greater importance in practical systems. Although it is not possible to plot right down to zero frequency due to log ω ($\log 0 = -\infty$), this does not pose a serious problem.

(*vi*) The magnitude (dB) plot can be approximated by straight line segments (to be discussed little later). So, it becomes quite easy to sketch the magnitude plot without too much computation.

The product terms of G(*j*ω) H(*j*ω)

For a rational transmittance G(*j*ω) H(*j*ω), it is only necessary to be able to plot magnitude and phase for following type of terms.

(*i*) constant K

(*ii*) Poles and zeros at origin of *s* plane

(*iii*) Real axis poles and zeros

(*iv*) Complex conjugate pairs of poles and zeros

A rational G($j\omega$) H($j\omega$) can always be factored in terms of these type. Once we become familiar with plots of terms of these type, we can use them to sketch the composite plot for any G($j\omega$) H($j\omega$). The composite magnitude (dB) plot is then the sum of individual dB plots and composite phase plot is sum of individual phase plots.

Constant K: A positive constant K has magnitude in dB

$$20 \log_{10} K$$

and phase angle 0°. A negative constant K has dB magnitude

$$20 \log |K|$$

and phase angle 180°. For example G($j\omega$) H($j\omega$) = 10 has

$$|GH|_{dB} = 20 \log_{10} 10 = 20 \text{ and}$$

$$\underline{|GH} = 0° \text{ (independent of } \omega\text{) as plotted in Fig. 6.30}(a).$$

Similarly G($j\omega$) H($j\omega$) = – 1/10 has

$$|GH|_{dB} = 20 \log_{10} (1/10) = -20 \text{ and}$$

$$\underline{|GH} = -180° \text{ (independent of } \omega\text{) as plotted in Fig. 6.30}(b)$$

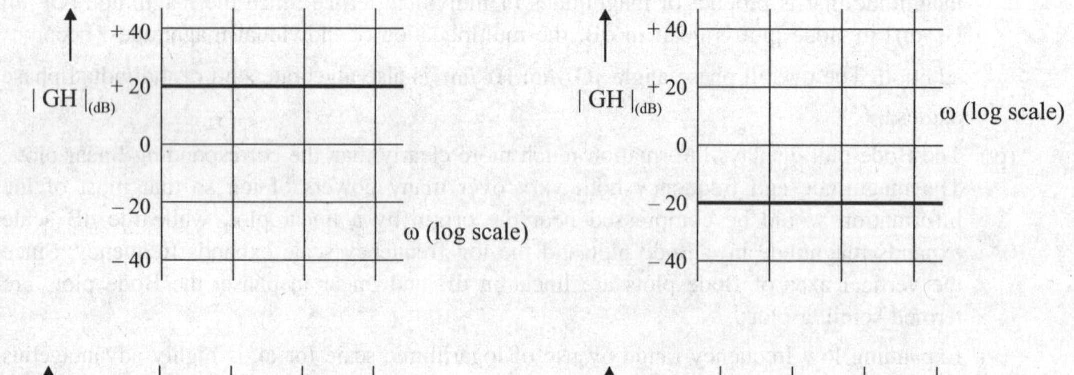

Fig. 6.30: Bode plot for (*a*) K = 10 (*b*) K = – 0.1

Poles and zeros at origin of *s* plane

Consider the transmittance with a single pole at origin;

$$G(s)\ H(s) = 1/s$$

$$G(j\omega)\ H(j\omega) = 1/j\omega$$

$$|G(j\omega)\,H(j\omega)| = 1/\omega$$

$$|G(j\omega)\,H(j\omega)|_{dB} = -\,20\log_{10}\omega \qquad ...(6.22)$$

Since magnitude in dB (y-axis) is plotted versus $\log_{10}\omega$ (x-axis), the equation (6.22) is of form $y = mx$ and represents a straight line passing through 0 dB at $\omega = 1$ with the slope of – 20 dB per unit change in $\log_{10}\omega$ as shown in Fig. 6.31.

Note that unit change in $\log_{10}\omega$ for change in ω from ω_1 to ω_2, means

$$\log_{10}\left(\frac{\omega_2}{\omega_1}\right) = 1$$

or

$$\omega_2 = 10\,\omega_1$$

This range of frequencies from ω_1 to ω_2 such that $\boldsymbol{\omega_2 = 10\,\omega_1}$, is called as a **decade**. Similarly, the range of frequencies from ω_1 to ω_2: $\boldsymbol{\omega_2 = 2\,\omega_1}$, is called as **octave**. Thus the slope of line represented by eqn. (6.22) is – 20 dB/dec. Since $-\,20\log_{10}2 \cong -\,6$, the slope of line of eqn. (6.22) can also be expressed as – 6 dB/octave.

Consider the transmittance with two poles at origin;

$$G(s)\,H(s) = 1/s^2$$

$$G(j\omega)\,H(j\omega) = 1/(j\omega)^2$$

$$|G(j\omega)\,H(j\omega)| = 1/\omega^2$$

$$|G(j\omega)\,H(j\omega)|_{dB} = -\,20\log_{10}\omega^2 = -\,40\log_{10}\omega \qquad ...(6.23)$$

The magnitude (dB) versus $\log_{10}\omega$ plot of (6.23) is also a straight line passing through **0 dB at $\boldsymbol{\omega = 1}$** but with the slope of **– 40 dB/dec or – 12 dB/oct** as shown in Fig. (6.30). The plot of **G(s) H(s) = $1/s^2$** is sum of two **G(s) H(s) = $1/s$** plots, that is, double the plots for $1/s$. In fact, the nth power of transmittance **[G(s) H(s) = $1/s^n$]** has magnitude plot which is n times the magnitude plot of original transmittance **[G(s) H(s) = $1/s$]**. Thus the plot for $1/s^n$ is also a straight line passing through **0 db at $\boldsymbol{\omega = 1}$** but with the slope **– 20 n dB/dec** or **– $6n$ dB/oct**. This is demonstrated in Fig. (6.31).

Consider the transmittance with a zero at origin ;

$$G(s)\,H(s) = s$$

$$G(j\omega)\,H(j\omega) = j\omega$$

$$|G(j\omega)\,H(j\omega)|_{dB} = 20\log_{10}\omega \qquad ...(6.24)$$

The magnitude (dB) versus $\log_{10}\omega$ plot of (6.24) is again a straight line passing through **0 dB at $\boldsymbol{\omega = 1}$** but with the slope of **+ 20 dB/dec** or **+ 6 dB/oct** as shown in Fig. (6.30).

In general, the transmittance with n zeros at origin;

$$G(s)\,H(s) = s^n$$

$$|G(j\omega)\,H(j\omega)| = \omega^n$$

$$|G(j\omega)\,H(j\omega)|_{dB} = 20\,n\log_{10}\omega \qquad ...(6.25)$$

The magnitude (dB) versus $\log_{10}\omega$ plot of (6.25) is also a straight line passing through **0 dB at $\boldsymbol{\omega = 1}$** but with the slope of **+ 20 n dB/dec or + $6n$ dB/oct**. Thus the plot of transmittance s^n is $\boldsymbol{n}$ times the plot of original transmittance **G(s) H(s) = s**. The plot sketching is demostrated in Fig. 6.31.

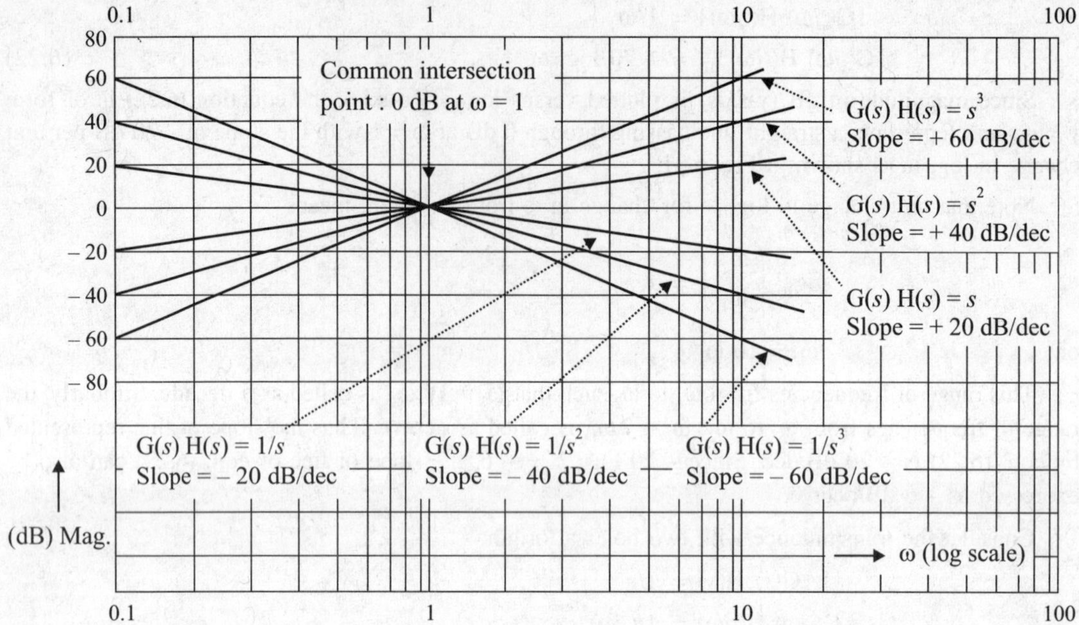

Fig. 6.31: Bode magnitude plot for transmittances having poles and zeros at origin

The sketching of **Bode phase plot** for transmittances with poles/zeros at origin, is demostrated as follows:

For $\quad G(s)\,H(s) = s; \quad G(j\omega)\,H(j\omega) = j\omega$

and $\quad \angle G(j\omega)\,H(j\omega) = +\,90°$

For $\quad G(s)\,H(s) = s^2; \quad G(j\omega)\,H(j\omega) = (j\omega)^2$

and $\quad \angle G(j\omega)\,H(j\omega) = +\,180°$

In general, for $\quad G(s)\,H(s) = s^n$

$$\angle G(j\omega)\,H(j\omega) = +\,(90\,n)°$$

For $\quad G(s)\,H(s) = 1/s; \quad G(j\omega)\,H(j\omega) = 1/j\omega$

and $\quad \angle G(j\omega)\,H(j\omega) = -\,90°$

For $\quad G(s)\,H(s) = 1/s^2; \quad G(j\omega)\,H(j\omega) = 1/(j\omega)^2$

and $\quad \angle G(j\omega)\,H(j\omega) = -\,180°$

In general, for $\quad G(s)\,H(s) = 1/s^n$

$$\angle G(j\omega)\,H(j\omega) = -\,(90\,n)°$$

The phase plots of G(s) H(s) = s, s^2, 1/s and $1/s^2$ are depicted in Fig. 6.32.

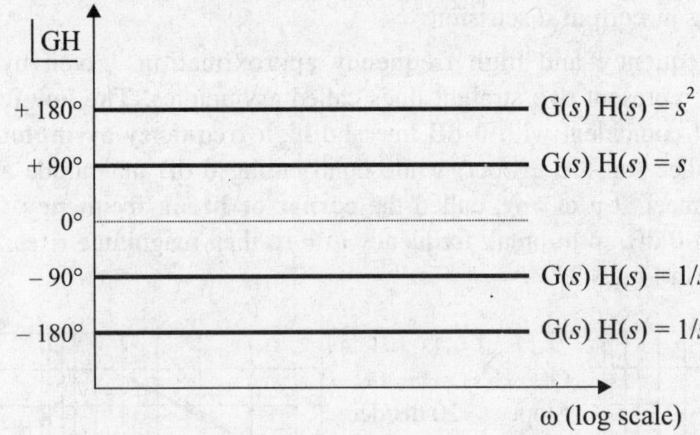

Fig. 6.32: Bode phase plot for transmittances having poles/zeros at origin

Real axis poles or zeros

Consider the transmittance G(s) H(s) having a left half plane (LHP) zero;

$$G(s)\,H(s) = 1 + \frac{s}{\alpha}$$

$$G(j\omega)\,H(j\omega) = 1 + j\frac{\omega}{\alpha}$$

$$|G(j\omega)\,H(j\omega)| = \sqrt{1+\frac{\omega^2}{\alpha^2}}$$

$$|G(j\omega)\,H(j\omega)|_{dB} = 20\log\left(1+\frac{\omega^2}{\alpha^2}\right)^{1/2} = 10\log\left(1+\frac{\omega^2}{\alpha^2}\right) \quad \text{...(6.26)}$$

$$\underline{|G(j\omega)\,H(j\omega)} = \tan^{-1}\left(\frac{\omega}{\alpha}\right) \quad \text{...(6.27)}$$

For ease in sketching **magnitude plot**, (6.26) is approximated as follows.

For $\omega << \alpha$ or $\frac{\omega^2}{\alpha^2} << 1$ (low frequency approximation)

$$|G(j\omega)\,H(j\omega)|_{dB} = 10\log_{10}1 = 0 \quad \text{...(6.28)}$$

and $$\underline{|G(j\omega)\,H(j\omega)} = \tan^{-1}(0) = 0°; \quad \text{Im}\,[G(j\omega)\,H(j\omega)] \to 0 \quad \text{...(6.29)}$$

For $\omega >> \alpha$ or $\frac{\omega^2}{\alpha^2} >> 1$ (high frequency approximation)

$$|G(j\omega)\,H(j\omega)|_{dB} = 10\log\frac{\omega^2}{\alpha^2}$$

or $$|G(j\omega)\,H(j\omega)|_{dB} = 20\log\omega - 20\log\alpha \quad \text{...(6.30)}$$

and $$\underline{|G(j\omega)\,H(j\omega)} = 90° \quad \text{...(6.31)}$$

Note the following in current discussion:

(*i*) The low frequency and high frequency approximations given by (6.28) and (6.30) respectively represent two straight lines called asymptotes. **The low frequency asymptote** (6.28) is flat coincident with **0 dB** line and **high frequency asymptote** (6.30) has a slope of + 20 dB/dec (or + 6 dB/oct) while contributing 0 dB magnitude at $\omega = \alpha$. Thus both asymptotes meet at a $\omega = \alpha$, called the **corner or break frequency**. The magnitude (dB) plot is along 0 dB up to break frequency $\omega = \alpha$, then magnitude rises 20 dB per decade of ω as shown in Fig. 6.33.

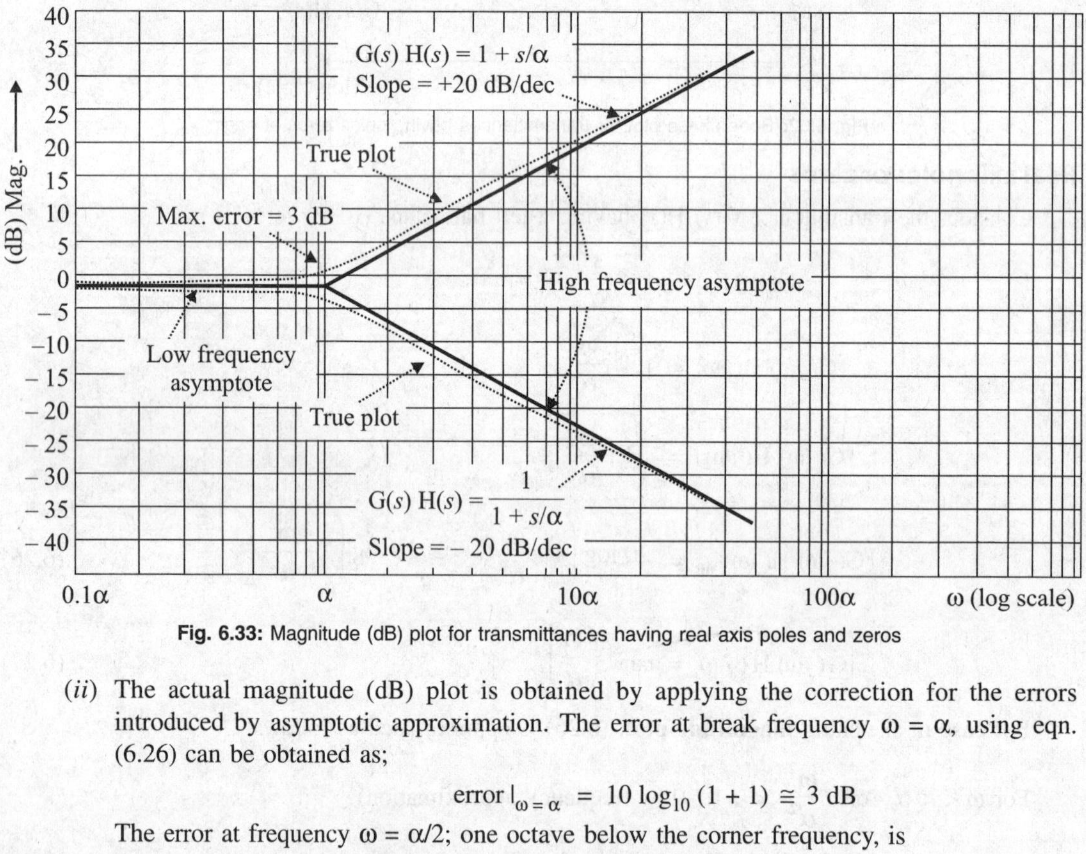

Fig. 6.33: Magnitude (dB) plot for transmittances having real axis poles and zeros

(*ii*) The actual magnitude (dB) plot is obtained by applying the correction for the errors introduced by asymptotic approximation. The error at break frequency $\omega = \alpha$, using eqn. (6.26) can be obtained as;

$$\text{error}\Big|_{\omega = \alpha} = 10 \log_{10} (1 + 1) \cong 3 \text{ dB}$$

The error at frequency $\omega = \alpha/2$; one octave below the corner frequency, is

$$\text{error}\Big|_{\omega = \frac{\alpha}{2}} = 10 \log_{10} \left(1 + \frac{1}{4}\right) \cong 1\text{dB}$$

Similarly, the error at frequency $\omega = 2\alpha$, one octave above the corner frequency, can be obtained using (6.26) and (6.30)

$$\text{error}\Big|_{\omega = 2\alpha} = 10 \log_{10} (1 + 4) - 10 \log_{10} (4) \cong 1\text{dB}$$

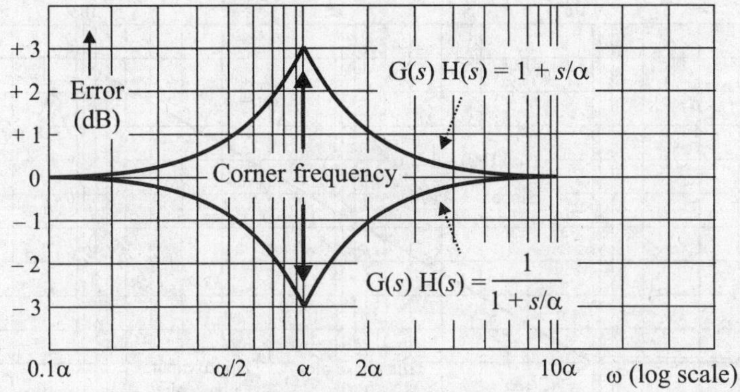

Fig. 6.34: Errors due to approximation for transmittances having real axis poles and zeros

These errors at three significant frequencies; the corner frequency, one octave below and above the corner frequency, are depicted in Fig. 6.34.

(*iii*) The true magnitude (dB) plot can be sketched with reasonably good accuracy from the approximate plot by correcting the asymptotic plot by + 3 dB at corner frequency and by + 1 dB, one octave below and above the corner frequency and then by drawing a smooth plot through these three points approaching the low and high frequency asymptotes as demonstrated in Fig. 6.33.

(*iv*) The **phase vs log ω** plot can also be approximated by three segment straight lines plot. Since $\underline{|G(j\omega)\, H(j\omega)}$ approaches 0° as $\omega \to 0°$ (eqn. 6.29), $\underline{|G(j\omega)\, H(j\omega)}$ approaches 90° as $\omega \to \infty$ (eqn. 6.31), and $\underline{|G(j\omega)\, H(j\omega)}\Big|_{\omega=\alpha} = \tan^{-1}(1) = 45°$, the phase plot changes slope by + 45°/dec at one tenth of break frequency and by – 45°/dec at ten times the break frequency.

Thus, in the approximation, the phase angle is 0° up to one tenth of the break frequency, rises 45° per decade through + 45° at the break frequency and continues at + 45° per decade slope up to ten times the break frequency. Beyond ten times the break frequency, the approximate angle is 90°. At the break frequency ω = α, the actual and approximate plots are equal. The approximate phase plot has maximum error of less than 6°. This is demonstrated in Fig. 6.35.

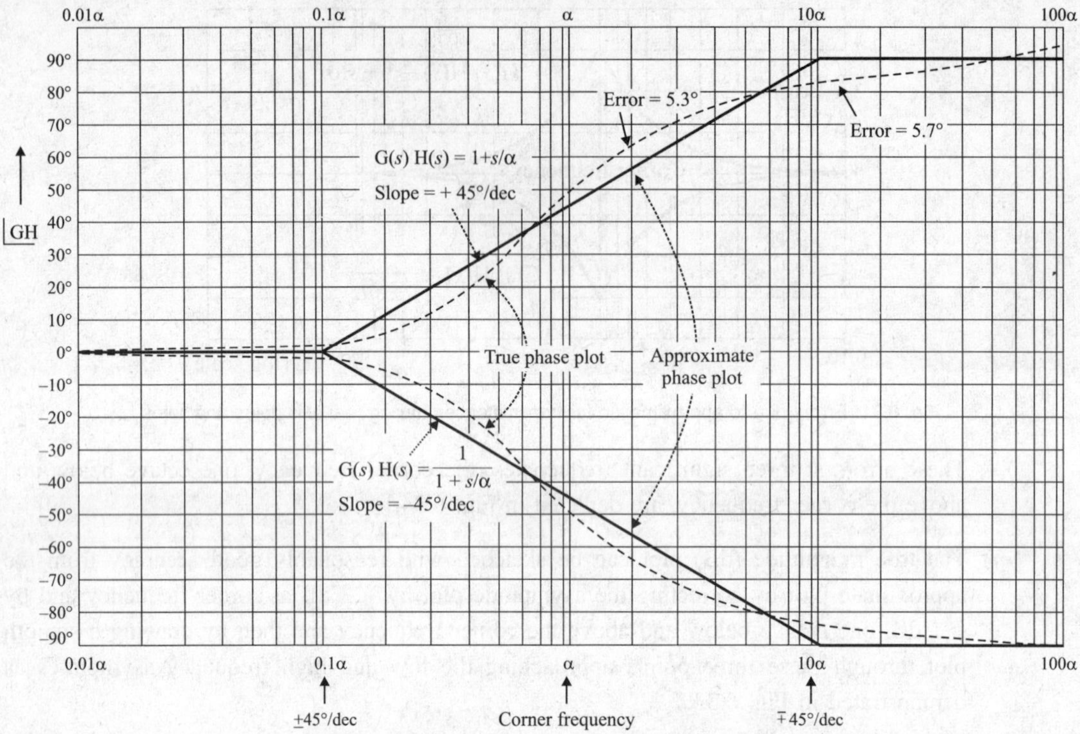

Fig. 6.35: Bode phase plot for real axis LHP pole and zero

Now consider a transmittance G(*s*) H(*s*) having a left half plane (LHP) pole;

$$G(s)\,H(s) = \frac{1}{1+\dfrac{s}{\alpha}}$$

$$G(j\omega)\,H(j\omega) = \frac{1}{1+\dfrac{j\omega}{\alpha}}$$

$$|G(j\omega)\,H(j\omega)| = \frac{1}{\left(1+\dfrac{\omega^2}{\alpha^2}\right)^{\frac{1}{2}}}$$

$$|G(j\omega)\,H(j\omega)|_{dB} = 20\log\left(1+\frac{\omega^2}{\alpha^2}\right)^{-\frac{1}{2}} = -10\log\left(1+\frac{\omega^2}{\alpha^2}\right) \quad ...(6.32)$$

$$\underline{|G(j\omega)\,H(j\omega)} = -\tan^{-1}\left(\frac{\omega}{\alpha}\right) \quad ...(6.33)$$

Comparing magnitude (dB) plot for LHP zero (6.26) with that of LHP pole (6.32) and phase plot for LHP zero (6.27) with that of LHP pole (6.33), it is easy to see that LHP pole has magnitude and phase plots that are negative of plots for a corresponding LHP zero. Sketching magnitude (dB) plot is shown in Fig 6.33. The error between asymptotic dB plot and actual plot is depicted in Fig. 6.34 the phase plot is shown in Fig. 6.35. For still more insight into sketching the Bode plot for transmittance having a real axis LHP pole note the following.

(*i*) the asymptotic magnitude (dB) plot is along 0 dB line up to break frequency $\omega = \alpha$, then magnitude falls 20 dB/dec of ω.

(*ii*) The true magnitude (dB) plot is sketched from asymptotic plot by correcting it by – 3 dB at break frequency and by – 1 dB, one octane below and above break frequency and then by drawing a smooth plot through these three points approaching the low and high frequency asymptotes.

(*iii*) The approximate phase plot is 0° up to one tenth of break frequency, falls 45° per decade through – 45° at the break frequency and continues at – 45° per decade slope up to ten times the break frequency. Beyond ten times the break frequency, the approximate angle is – 90°.

Consider a transmittance with a right half plane (RHP) zero:

$$G(s)\,H(s) = 1 - \frac{s}{\alpha}$$

$$G\,(j\omega)\,H(j\omega) = 1 - \frac{j\omega}{\alpha}$$

$$|G\,(j\omega)\,H(j\omega)| = \sqrt{1 + \frac{\omega^2}{\alpha^2}}$$

$$|G\,(j\omega)\,H\,(j\omega)|_{dB} = 10\log\left(1 + \frac{\omega^2}{\alpha^2}\right)$$

and $$\underline{|G(j\omega)\,H(j\omega)} = -\tan^{-1}\left(\frac{\omega}{\alpha}\right)$$

Similarly, consider a transmittance with RHP pole;

$$G(s)\,H(s) = \frac{1}{1 - \dfrac{s}{\alpha}}$$

$$G\,(s)\,H(s) = \frac{1}{1 - \dfrac{j\omega}{\alpha}}$$

$$|G\,(j\omega)\,H(j\omega)|_{dB} = -10\log\left(1 + \frac{\omega^2}{\alpha^2}\right)$$

and $$\underline{|G(j\omega)\,H(j\omega)} = -\tan^{-1}\left(-\frac{\omega}{\alpha}\right) = \tan^{-1}\left(\frac{\omega}{\alpha}\right)$$

Comparing the magnitudes and phase angles of RHP pole and RHP zero with their LHP counter parts, it is obvious that RHP zeros and RHP poles differ from their LHP counter parts by algebraic sign of the phase angle; magnitudes are same for LHP roots of the same type. Thus the magnitude (dB) plots for transmittance $1 - (s/\alpha)$ and $1 + (s/\alpha)$ are same. Similarly, the dB plot for $1/(1 - s/\alpha)$ and $1/(1 + s/\alpha)$ are also same. But the phase plot of RHP zero; $1 - (s/\alpha)$ is same as that of LHP pole; $1/(1 + s/\alpha)$ and phase plot of RHP pole; $1/(1 - s/\alpha)$ is same as that of LHP zero; $1 + (s/\alpha)$.

Complex conjugate poles or zeros

Consider a transmittance with complex conjugate LHP poles;

$$G(s)\,H(s) = \frac{\omega_n^2}{s^2 + 2\xi\omega_n s + \omega_n^2}$$

$$G\,(j\omega)\,H(j\omega) = \frac{1}{\left(1 - \dfrac{\omega^2}{\omega_n^2}\right) + j2\xi\dfrac{\omega}{\omega_n}}$$

$$|G\,(j\omega)\,H(j\omega)| = \frac{1}{\sqrt{\left(1 - \dfrac{\omega^2}{\omega_n^2}\right)^2 + \left(2\xi\dfrac{\omega}{\omega_n}\right)^2}}$$

$$|G\,(j\omega)\,H(j\omega)|_{dB} = -10\log\left[\left(1 - \frac{\omega^2}{\omega_n^2}\right)^2 + \left(2\xi\frac{\omega}{\omega_n}\right)^2\right] \quad \text{...(6.34)}$$

$$\underline{|G(j\omega)\,H(j\omega)} = -\tan^{-1}\left[\frac{2\xi\dfrac{\omega}{\omega_n}}{1 - \left(\dfrac{\omega^2}{\omega_n^2}\right)}\right] \quad \text{...(6.35)}$$

Sketching magnitude (dB) plot

The straight line asymptotes similar to simple real axis poles and zeros can be obtained as under.

For $\omega << \omega_n$ or $\omega/\omega_n << 1$ (low frequency approximation)

$$|G(j\omega)\,H(j\omega)| \cong 1$$

and

$$|G(j\omega)\,H(j\omega)|_{dB} \cong 0 \quad \text{...(6.36)}$$

For $\omega >> \omega_n$ or $\omega/\omega_n >> 1$ (high frequency approximation)

$$|G(j\omega)\,H(j\omega)| = 1/(\omega/\omega_n)^2$$

and

$$|G(j\omega)\,H(j\omega)|_{dB} = -40\log(\omega/\omega_n) \quad \text{...(6.37)}$$

Note that the low frequency asymptote (6.36), is a straight line coincident with 0 dB and the high frequency asymptote (6.37), is a straight line with slope of – 40 dB/dec (or – 12 dB/oct) while contributing 0 dB at $\omega = \omega_n$. Thus both the asymptotes meet at a frequency $\omega = \omega_n$, called the **corner or break frequency.**

The two asymptotes just derived are independent of ξ. However, near break frequency $\omega = \omega_n$, a resonant peak is usually expected and the magnitude of resonant peak depends on ξ. The errors introduced by asymptotic approximation, is evaluated as under.

$$\text{error}\Big|_{\omega = \omega_n} = -10 \log (2\xi)^2 = -20 \log 2\xi \qquad \text{(from eqn. 6.34)}$$

$$\text{error}\Big|_{\omega = \frac{\omega_n}{2}} = -10 \log \left(\frac{9}{16} + \xi^2\right) \qquad \text{(from eqn. 6.34)}$$

and
$$\text{error}\Big|_{\omega = 2\omega_n} = -10 \log (9 + 16\xi^2) + 40 \log (2) \qquad \text{(from eqns. 6.34 and 6.37)}$$

Note that error at break frequency $\omega = \omega_n$, is zero for $\xi = 0.5$. For $0.5 < \xi < 1$, the error is negative, the true dB plot lies generally below the asymptotic plot. For $0 < \xi < 0.5$, the error is positive, the true dB plot lies generally above the asymptotic plot. The plots of magnitude for different values of ξ are depicted in Fig. 6.36(*a*). For $\xi \geq 1$, the poles are, not complex, they are real, and therefore can be handled as real axis pole methods.

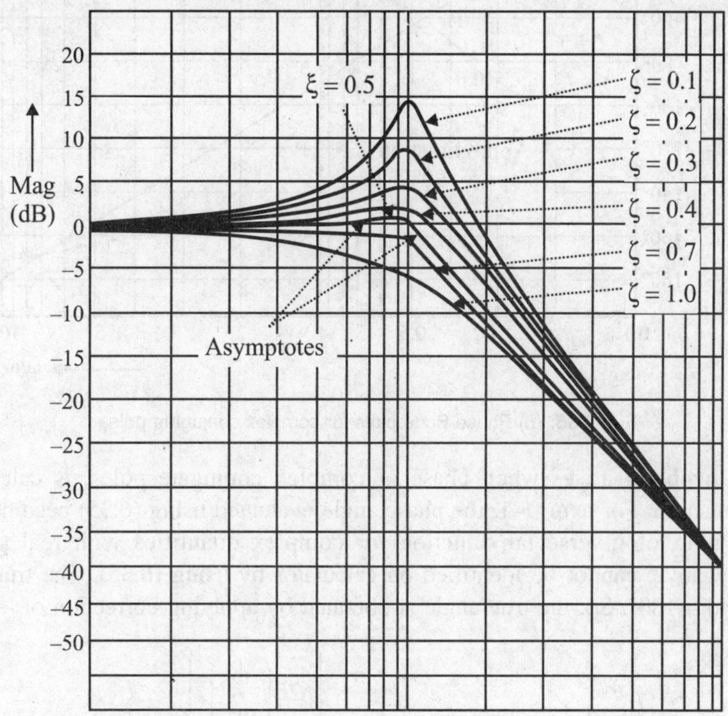

Fig. 6.36: (*a*) Magnitude Bode plots for complex conjugate pole

Sketching phase plot

The phase angle given by (6.35) is function of both ω and ξ.

$$\angle G(j\omega) H(j\omega)\Big|_{\omega = 0} = 0°$$

$$\angle G(j\omega) H(j\omega)\Big|_{\omega = \omega_n} = -\tan^{-1}(\infty) = -90° \quad \text{(independent of } \xi)$$

$$\angle G(j\omega) H(j\omega)\Big|_{\omega = \infty} = -180°$$

The phase angle approximation is 0° to one tenth of break frequency, then – 90°/decade slope, passing through – 90° at the break frequency and continues until ten times the break frequency. Beyond ten times the break frequency, the approximate angle is – 180°. The phase plots for various value of ξ are shown in Fig. 6.35(*b*). It can be observed that the phase plots become sharper while moving from low frequency to high frequency as ξ decreases, until for $\xi = 0$, the plot exhibits jump discontinuity from 0° down to – 180° at $\omega = \omega_n$.

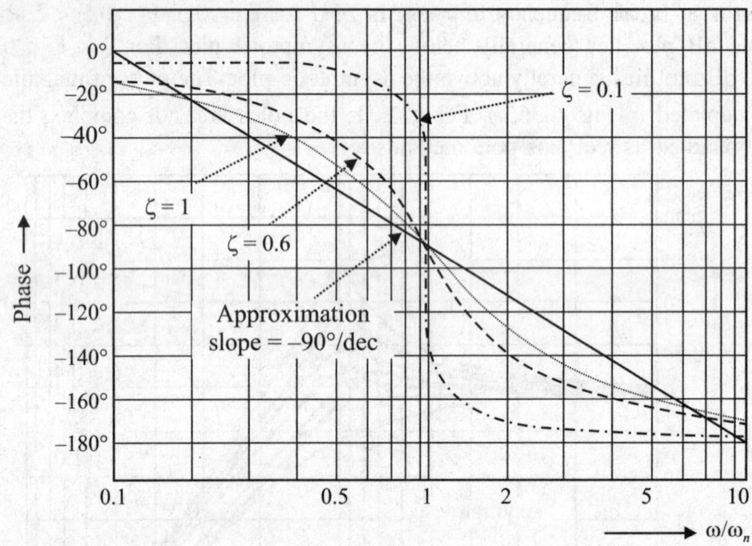

Fig. 6.36: (*b*) Phase Bode plots for complex conjugate poles

A potential problem arises when phase of complex conjugate poles, is calculated using an inverse tangent function. For $\omega/\omega_n > 1$, the phase angle evaluated using (6.35) becomes positive. This is because behaviour of inverse tan function for complex quantities with real part negative or imaginary part negative, cannot be identified on calculator by using (6.35). The true angle must lie between – 90° and – 180°. So, the true angle is obtained by applying correction of – 180°, that is

$$\underline{|G(j\omega)\,H(j\omega)}\Big|_{\frac{\omega}{\omega_n}>1} = -\,180° + \tan^{-1}\left[\frac{2\xi\left(\dfrac{\omega}{\omega_n}\right)}{\left(\dfrac{\omega^2}{\omega_n^{\,2}}\right)-1}\right] \quad ...(6.38)$$

Now consider a transmittance with complex conjugate LHP zeros of form

$$G(s)\ H(s) = \frac{s^2 + 2\xi\omega_n s + \omega_n^{\,2}}{\omega_n^{\,2}}$$

$$G\,(j\omega)\ H(j\omega) = \left(1 - \frac{\omega^2}{\omega_n^{\,2}}\right) + j2\xi\left(\frac{\omega}{\omega_n}\right)$$

$$|G(j\omega)\,H(j\omega)| = \sqrt{\left(1-\frac{\omega^2}{\omega_n^2}\right)^2+\left(2\xi\frac{\omega}{\omega_n}\right)^2}$$

$$|G(j\omega)\,H(j\omega)|_{dB} = 10\log\left[\left(1-\frac{\omega^2}{\omega_n^2}\right)^2+\left(2\xi\frac{\omega}{\omega_n}\right)^2\right] \quad \text{...(6.39)}$$

$$\angle G(j\omega)\,H(j\omega) = \tan^{-1}\left[\frac{2\xi\left(\frac{\omega}{\omega_n}\right)}{1-\frac{\omega^2}{\omega_n^2}}\right] \quad \text{...(6.40)}$$

Comparing magnitude (dB) plot (6.34) and phase plot (6.35) for complex conjugate LHP poles with magnitude (dB) plot (6.39), and phase plot (6.40) for complex conjugate LHP zeros, it is easy to observe that complex conjugate zeros have Bode plots that are negative of the plots for corresponding complex conjugate poles. Note the following for still more insight into Bode plots of complex poles/ zeros.

(*i*) The asymptotic magnitude (dB) plot is along 0 dB line up to break frequency $\omega = \omega_n$ for both complex conjugate poles and zeros. Then magnitude rises + 40 dB/dec of ω for complex conjugate LHP zeros and falls –40 dB/dec of ω for complex conjugate LHP poles. The error due to asymptotic approximation can again be evaluated as discussed for complex poles with only exception that algebraic sign will change.

(*ii*) The phase approximation is 0° to one tenth of break frequency, then + 90°/dec slope, passing through + 90° at break frequency and continues until ten times the break frequency. Beyond ten times the break frequency, the approximate angle is + 180° for complex conjugate LHP zeros. The complex conjugate poles have phase approximation of 0° up to one tenth of break frequency $\omega = \omega_n$, then –90°/dec slope passing through –90° at break frequency and continuing until ten times the break frequency. Beyond ten times the break frequency, the approximate phase is –180°.

The error fitting for complex conjugate zero is same as discussed for complex conjugate poles except that change in algebraic sign must be made.

Gain margin and phase margin from Bode plot

Recall that gain margin (GM) in dB is given as

$$GM_{dB} = -20\log\,|G(j\omega)\,H(j\omega)|_{\omega=\omega_{pc}}$$

where ω_{pc} is phase cross over frequency. $|G(j\omega)\,H(j\omega)|_{\omega=\omega_{pc}}$ in dB and hence the gain margin can be directly read out from Bode magnitude (dB) plot.

Consider the Bode plots; the dB plot and the phase plot both together, the phase plot just below the dB plot as shown in Fig. 6.36. Identify the phase cross over point; the point of intersection of phase plot and – 180° line and corresponding phase cross over frequency ω_{pc}. Extend $\omega = \omega_{pc}$ line vertically upward. Identify the point at which this line intersects magnitude plot and note the magnitude corresponding to this point. The corresponding dB value is $|G(j\omega_{pc})\,H(j\omega_{pc})|$ and

GM = $-\,|G(j\omega_{pc})\,H(j\omega_{pc})|_{dB}$. As shown in Fig. 6.37, GM is positive and the system is stable when the intersect on magnitude plot at phase cross over frequency is below 0 dB line. If the intersect is above the 0 dB line, the GM is negative and the system is unstable as shown in Fig. 6.38.

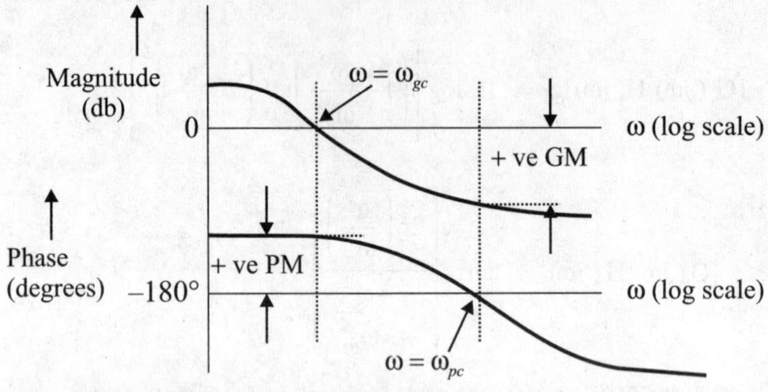

Fig. 6.36: Positive GM, PM for stable minimum phase system

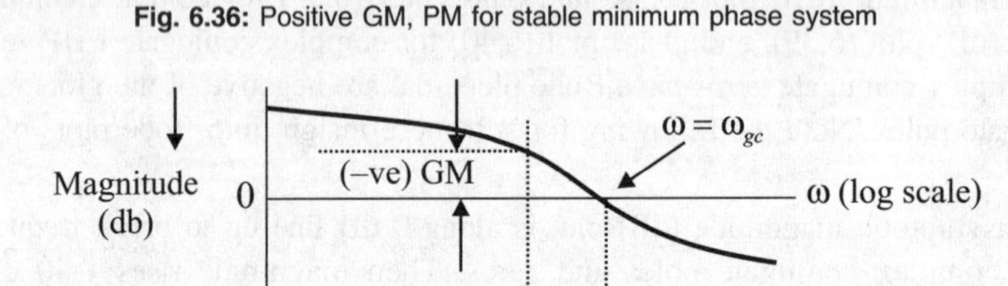

Fig. 6.37: Negative GM and PM for unstable minimum phase system

Similarly, PM can also be read out from Bode phase plot. Identify the gain cross over point; the point of intersection of gain plot and 0 dB line and corresponding gain cross over frequency ω_{gc} Extend $\omega = \omega_{gc}$ line vertically downward. Identify the point at which this line intersects phase plot and note the phase angle $\underline{|G(j\omega_{gc})\,H(j\omega_{gc})}$ corresponding to this point. Then

$$PM = 180^\circ + \underline{|G(j\omega_{gc})\,H(j\omega_{gc})}$$

As shown in Fig. 6.36, PM is positive and system is stable when intersect on phase plot at the gain cross over frequency is above – 180° line. If the intersect is below – 180° line, the PM is negative and system is unstable as shown in Fig. 6.37. It should be emphasized again that the analysis is valid only for minimum phase transfer functions. It is important to note the following.

(*i*) GM and PM both must be positive for the system to be stable. It is easy to see from Fig. 6.36 that $\omega_{gc} < \omega_{pc}$ in a situation where GM and PM both are positive.

(*ii*) If GM and PM both are negative, system will be unstable and $\omega_{gc} > \omega_{pc}$. The system will also be unstable in case either GM or PM is negative.

(*iii*) For a marginally stable system, GM and PM both are zero and $\omega_{gc} = \omega_{pc}$.

Relationship between Bode magnitude (dB) plot and number type of a system

Recall that the position, velocity and acceleration error constants are given as:

$$K_p = \lim_{s\to 0} G(s)\, H(s)$$

$$K_v = \lim_{s\to 0} sG(s)\, H(s)$$

$$K_a = \lim_{s\to 0} s^2\, G(s)\, H(s)$$

For a given number type of system only one of K_p, K_v and K_a, is finite and significant. Recall that K_p for type 0 system, K_v for type 1 system and K_a for type 2 system, assume a finite value. K_p is infinitely large for a system of type number greater then 0. K_v is zero for system of type 0 and infinitely large for systems of type number greater than 1. Similarly K_a is zero for a system of type number less than 2 and infinitely large for a system of type number greater than 2.

The error constants K_p, K_v and K_a describe low frequency behaviour of type 0, type 1 and type 2 systems respectively. The number type of the system determines the initial slope of Bode magnitude (dB) plot. Thus information about existence and magnitude of steady state error of a system to a given input can be determined by simply inspecting the dB plot in low frequency range (initial slope). This is demonstrated as follows:

Evaluation of K_p from Bode plot

The transmittance G(*s*) H(*s*) of type 0 system, is generally given as

$$G(s)\, H(s) = \frac{K\prod_{i=1}^{m}(1 + sz_i)}{\prod_{j=1}^{n}(1 + sp_j)}$$

and $$K_p = \lim_{s\to 0} G(s)\, H(s) = K$$

Note that G(*s*) H(*s*) does not contain any pole/zero at origin. The magnitude (dB) plot at low frequencies, is 20 log K_p, that is, the low frequency asymptote is flat coincident with dB line equal to 20 log K_p as depicted in Fig. 6.38.

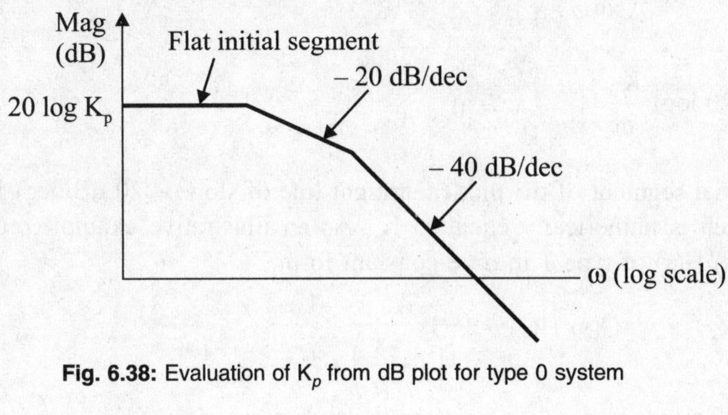

Fig. 6.38: Evaluation of K_p from dB plot for type 0 system

Evaluation of K_v from Bode plot

The transmittance G(*s*) H(*s*) of type 1 system, is generally given as

$$G(s)\,H(s) = \frac{K\prod_{i=1}^{m}(1+sz_i)}{s\prod_{j=1}^{n}(1+sp_j)}$$

and $$K_v = \lim_{s\to 0} sG(s)\,H(s) = K$$

Note that G(*s*) H(*s*) has one pole at origin. So, the magnitude (dB) plot at low frequencies will be asymptotic to straight line having slope of – 20 dB/dec as shown in Fig. 6.39.

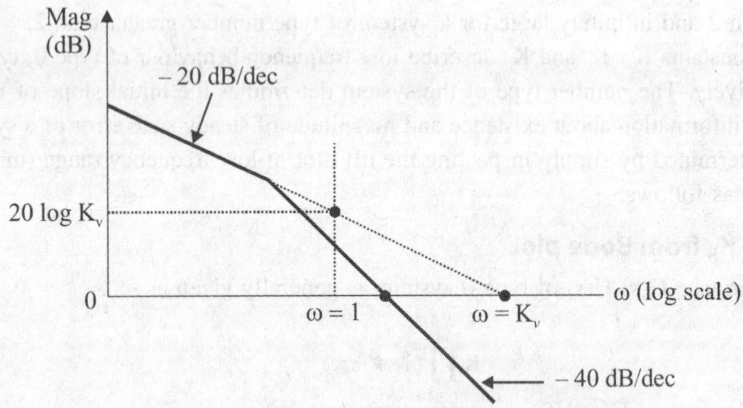

Fig. 6.39: Evaluation of K_v from dB plot for type 1 system

$$G(j\omega)\,H(j\omega)\big|_{\omega<<1} \cong \frac{K}{j\omega} = \frac{K_v}{j\omega}$$

$$\big|G(j\omega)\,H(j\omega)\big|_{\omega<<1} = \frac{K_v}{\omega}$$

$$20\log\left(\frac{K_v}{\omega}\right)\Bigg|_{\omega=1} = 20\log K_v$$

and $$20\log\left(\frac{K_v}{\omega}\right)\Bigg|_{\omega=K_v} = 0$$

Thus, the initial segment of dB plot (a straight line of slope – 20 dB/dec) intersects 0 dB line at frequency ω which is numerically equal to K_v. As an illustrative example, consider the open loop transmittance G(*s*) H(*s*) of type 1 in time constant form

$$G(s)\,H(s) = \frac{K}{s\left(1+\dfrac{s}{\alpha}\right)}$$

$$K_v = \lim_{s \to 0} s\, G(s)\, H(s) = K$$

The asymptotic dB plot is shown in Fig. 6.40. Let the corner frequency for pole be identified as $\omega_2 = \alpha$. Initial segment of slope – 20 dB/dec, intersects 0 dB line at frequency $\omega_1 = K_v = K$. Let the segment of slope – 40 dB/dec, intersect 0 dB line at frequency ω_3. This frequency ω_3, in approximation, can be evaluated as follows:

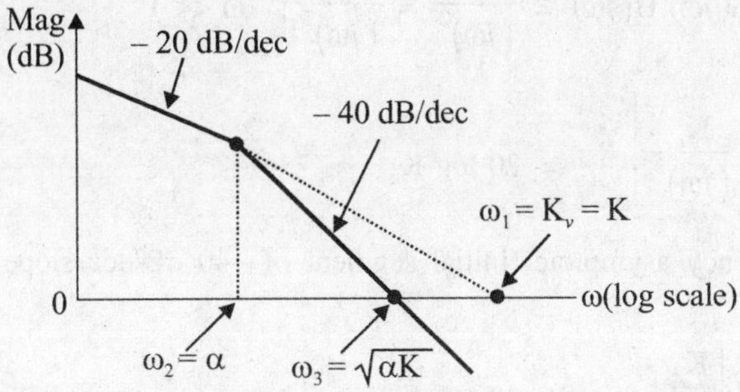

Fig. 6.40: K_v together with significant frequencies from dB plot of $\dfrac{K}{s\left(1 + \dfrac{s}{\alpha}\right)}$

$$\left| G(j\omega)\, H(j\omega) \right|_{\omega >> 1} \cong \frac{\alpha K}{\omega^2}$$

and
$$\left. \frac{\alpha K}{\omega^2} \right|_{\omega = \omega_3} = 1$$

so
$$\omega_3^{\,2} = \alpha K$$

inter relating $\omega_1 = K$, $\omega_2 = \alpha$ and $\omega_3 = \sqrt{\alpha K}$, we have

$$\omega_3^{\,2} = \omega_1\, \omega_2$$

or
$$\frac{\omega_3}{\omega_2} = \frac{\omega_1}{\omega_3}$$

or
$$\log \omega_3 - \log \omega_2 = \log \omega_1 - \log \omega_3$$

Thus, the frequency point ω_3 lies midway between frequency points ω_1 and ω_2 on log scale of ω. The damping ratio is given as:

$$\xi = \frac{1}{2}\sqrt{\frac{\alpha}{K}} = \frac{\omega_2}{2\omega_3}$$

Evaluation of K_a from Bode plot

The general from of the transmittance G(*s*) H(*s*) of type 2 is

$$G(s)\, H(s) = \frac{K \prod_{i=1}^{m} (1 + s z_i)}{s^2 \prod_{j=1}^{n} (1 + s p_j)}$$

and $$K_a = \lim_{s \to 0} s^2\, G(s)\, H(s) = K$$

The initial segment of Bode magnitude (dB) plot, will have slope – 40 dB/dec. The magnitude in dB corresponding to the intersection point of this segment with ω = 1 line, is 20 log K = 20 log K_v. For low frequencies.

$$G(j\omega)\,H(j\omega) \cong \frac{K}{(j\omega)^2} = \frac{K_a}{(j\omega)^2}; \quad \omega << 1$$

and $$20 \log \left| \frac{K_a}{(j\omega)^2} \right|_{\omega = 1} = 20 \log K_a$$

Let this low frequency asymptote (initial segment of –40 dB/dec slope) intersect 0 dB line at $\omega = \omega_a$. Then

$$20 \log \left| \frac{K_a}{\omega^2} \right|_{\omega = \omega_a} = 20 \log 1$$

or $$\omega_a = \sqrt{K_a}$$

So, K_a can be evaluated by simply inspecting the frequency point at which the initial segment of slope – 40 dB/dec, intersects 0 dB line, when extended as depicted in Fig. 6.41.

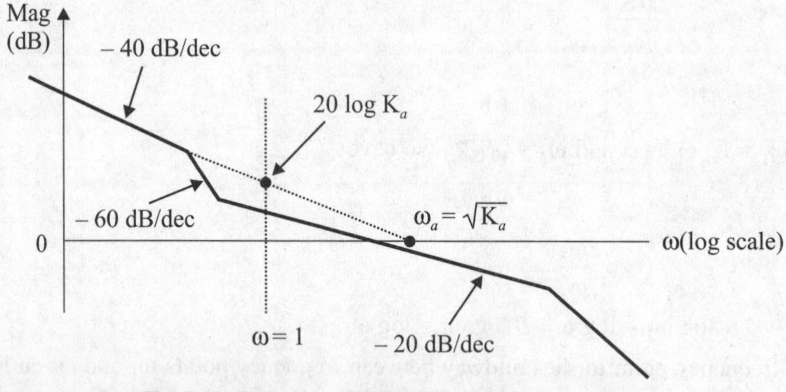

Fig. 6.41: Evaluation of K_a from dB plot of type 2 system

Sketching Bode plot: The steps involved in sketching the complete Bode plot for a given transmittance G(jω) H(jω) are as follows:

(*i*) Express G(jω) H(jω) in time constant form.

(*ii*) Identify the corner frequencies associated with each factor of G(jω) H(jω). Arrange them in increasing order and keep a note about each corner frequency whether it corresponds to pole or zero.

(*iii*) Draw the asymptotic magnitude (dB) plot. The information that follows will be useful in sketching the dB plot. Initial segment will be flat, coincident with 20 log K dB line if the transmittance KG(*s*) H(*s*) contains no poles or zeros at origin. If KG(*s*) H(*s*) has poles of multiplicity *n*, the initial segment will be a straight line with slope of – 20 *n* dB/dec and passing through intersection point of ω = 1 and 20 log K dB. If KG(*s*) H(*s*) has zeros of multiplicity *n*, the initial segment will be a straight line with slope of + 20 *n* dB/dec and again passing through intersection point of ω = 1 and 20 log K dB. The entire asymptotic magnitude (dB) plot consists of straight line segments with slope changing at each corner frequency by + 20 *n* dB/dec if corner frequency corresponds to zeros of multiplicity *n* and – 20 *n* dB/dec if the corner frequency corresponds to poles of multiplicity *n*. At a corner frequency corresponding to a pair of complex conjugate zero the slope changes by + 40 dB/dec and at a corner frequency corresponding to a pair of complex conjugate poles, the slope changes by – 40 dB/dec.

(*iv*) Prepare a table of corrections to be applied to the asymptotic plot at significant frequencies which usually are, the corner frequencies and one octave below and above the corner frequencies.

(*v*) Sketch a smooth curve through the corrected points such that it is asymptotic to the line segments. This curve is true dB plot.

(*vi*) Prepare a table of phase angles by actual calculation at few suitably chosen values of ω. Plot these points and sketch a smooth curve through them. This is Bode phase plot.

To demonstrate the **step by step approach** to entire Bode plot sketching, consider the following transmittance for an example:

$$G(s)\,H(s) = \frac{242(s+5)}{s(s+1)(s^2+5s+121)}$$

(*i*) **Arrange the transmittance in time constant form**

$$G(s)\,H(s) = \frac{5\times 242\left(1+\frac{s}{5}\right)}{121s(1+s)\left(1+\frac{5s}{121}+\frac{s^2}{121}\right)}$$

and

$$G(j\omega)\,H(j\omega) = \frac{10\left(1+\frac{j\omega}{5}\right)}{j\omega(1+j\omega)\left[\left(1-\frac{\omega^2}{121}\right)+j0.04\omega\right]}$$

(*ii*) **Identify the corner frequencies**

ω_{c_1} = 1 corresponding to a pole

ω_{c_2} = 5 corresponding to a zero

ω_{c_3} = 11 corresponding to a pair of complex conjugate poles.

(*iii*) **dB plot:** Since G(*s*) H(*s*) has a pole at origin, the initial segment is a straight line of slope – 20 dB/dec passing through point B at (ω = 1, dB = 20 log K$|_{K=10}$ = + 20). As demonstrated in fig. 6.42 (*a*) locate one more point A, one decade below point B at (ω = 0.1, dB = + 40) and draw initial segment as shown in fig. 6.42 (*a*) terminating at corner frequency ω_{c_1} = 1.

The corner frequency $\omega_{c_1} = 1$ corresponds to a simple pole. The slope changes by – 20 dB/dec at this frequency. The resultant slope of – 40 dB/dec from $\omega_{c_1} = 1$ (point B) onwards, continues till the next corner frequency $\omega_{c_2} = 5$. Between $\omega_{c_1} = 1$ and $\omega_{c_2} = 5$, the straight line has a slope of – 40 dB/dec and one point B at (ω = 1, dB = 20) is already known. Locate one more point at (ω = 10, dB = – 20) as demonstrated in Fig. 6.42 (*b*).

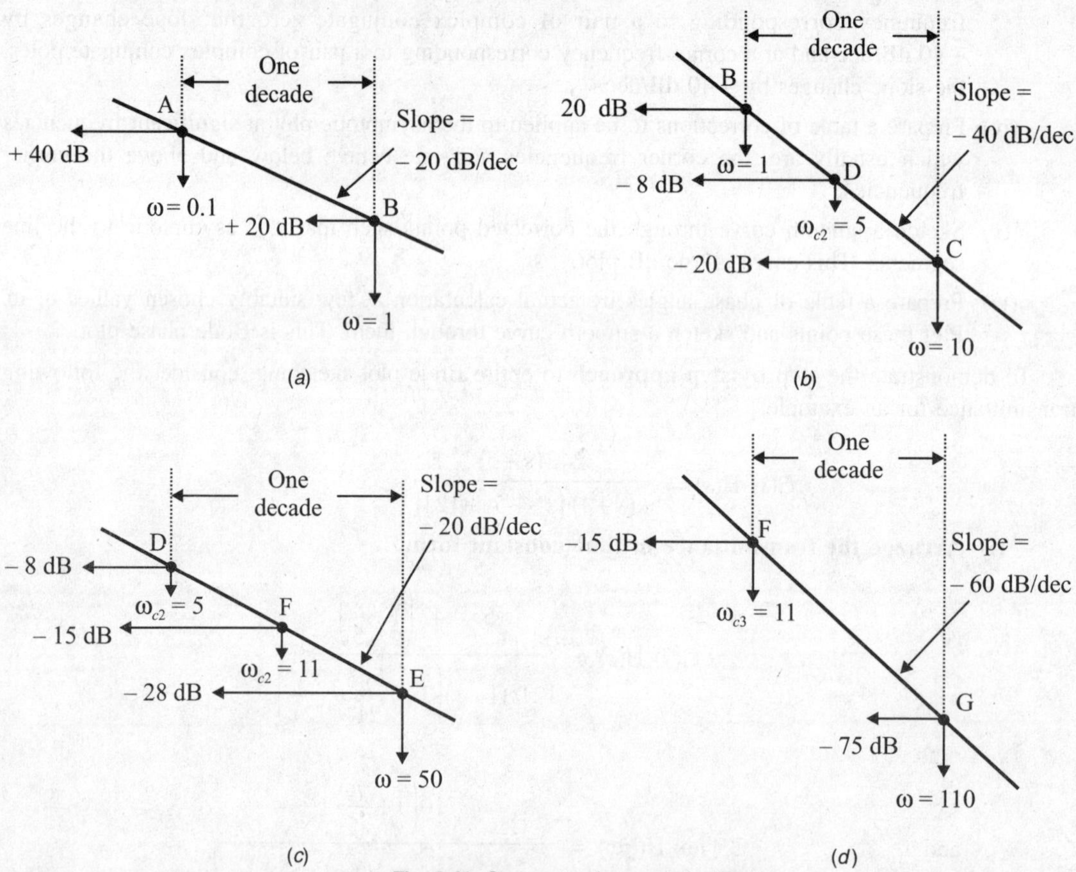

Fig. 6.42: Segments of asymptotic dB plot

This segment of slope – 40 dB/dec passing through the two points B and C, is drawn while terminating it at next corner frequency $\omega_{c_2} = 5$ (point D) as shown in Fig. 6.43 (*a*). Note dB = – 8 at $\omega_{c_2} = 5$ from Fig. 6.43(*a*).

The corner frequency $\omega_{c_2} = 5$ belongs to a simple zero. The slope changes by + 20 dB/dec. So, the line between $\omega_{c_2} = 5$ and next higher corner frequency $\omega_{c_3} = 11$ will have slope of – 20 dB/dec (= – 40 dB/dec + 20 dB/dec). Out of the two points required to draw this line, one point D is already known to be (ω = 5, dB = – 8) and other point E can be identified to be (ω = 50, dB = – 28) as demonstrated in Fig. 6.42 (*c*). A line of slope – 20 dB/dec, passing through these two points, is drawn as shown in Fig. 6.43 (*a*) while terminating it again at next corner frequency $\omega_{c_3} = 11$ (point F). Note dB = – 15 from Fig. 6.43 (*a*).

The corner frequency $\omega_{c_3} = 11$ belongs to a pair of complex conjugate poles. The slope changes by – 40 dB/dec. So, the line beyond $\omega_{c_3} = 11$, will have slope of – 60 dB/dec

(= – 20 dB/dec – 40 dB/dec). Out of the two points required to draw this line, one point F is already known to be (ω = 11, dB = – 15) and the other point G can be easily identified to be (ω = 110, dB = – 75) as demonstrated in Fig. 6.42 (*d*). A line joining points F and G is drawn without terminating at any higher frequency as these is no corner frequency beyond ω_{c_3}. The entire asymptotic dB plot is shown in Fig. 6.43 (*a*).

(*iv*) The table of correction is constructed as follows:

	$\frac{1}{2}\omega_{c_1}$	Pole ω_{c_1}	$2\,\omega_{c_1}$	$\frac{1}{2}\omega_{c_2}$	Zero ω_{c_2}	$\frac{1}{2}\omega_{c_3}$	$2\,\omega_{c_2}$	Complex pole ω_{c_3}	$2\,\omega_{c_3}$
ω	0.5	1	2	2.5	5	5.5	10	11	22
Error	– 1	– 3	– 1	+ 1	+ 3	+ 2.12	+ 1	+ 6.8	+ 2.12

Recall that error at corner frequency corresponding to a real axis zero is + 3 dB and that at one octave above and below the corner frequency is + 1 dB. The error at corner frequency and that at one octave below and above the corner frequency corresponding to a real axis pole is same as that of real axis zero except for change in algebraic sign.

For quadratic factor $\dfrac{1}{1+5\dfrac{j\omega}{121}+\dfrac{(j\omega)^2}{121}}$

Undamped natural frequency and damping ratio are

$$\omega_n = 11$$

$$\xi = 5/22 = 0.227$$

$$\text{Error}\Big|_{\omega=\omega_{c_3}} = -20\log 2\xi = +6.8\text{ dB}$$

$$\text{Error}\Big|_{\omega=\frac{1}{2}\omega_{c_3}} = -10\log\left[\left(1-\frac{\omega^2}{\omega_n^2}\right)^2+4\xi^2\frac{\omega^2}{\omega_n^2}\right];\ \omega_{c_3} = \omega_n$$

$$= -10\log\left[\left(1-\frac{1}{4}\right)^2+4(0.227)^2\cdot\frac{1}{4}\right]$$

$$= +2.12\text{ dB}$$

$$\text{Error}\Big|_{\omega=2\omega_{c_3}} = -10\log\left[\left(1-\frac{\omega^2}{\omega_n^2}\right)^2+4\xi^2\frac{\omega^2}{\omega_n^2}\right]+40\log\frac{\omega}{\omega_n}$$

$$= -10\log[(1-4)^2+4(0.227)^2\cdot 4]+40\log 2$$

$$= +2.12\text{ dB}$$

Applying the correction as shown in table just above, the true dB plot is shown in Fig. 6.43(*a*). The error table shows error of + 1 dB at ω = 10 because error due to quadratic pole corner frequency $\omega_{c_3} = 11$ has not been added to it. So, the true error will be little larger than + 1 dB as shown in Fig. 6.43(*a*).

(*v*) **Phase plot:** A table of phase angle for some arbitrarily chosen values of ω is constructed below by actually calculating the phase angle as follows. Note that – 180° has to be added for ω > 11 due to quadratic factor.

$$\underline{|G(j\omega)\,H(j\omega)} = -90° - \tan^{-1}\omega + \tan^{-1}\frac{\omega}{5} - \tan^{-1}\left[\frac{0.04\,\omega}{1-\dfrac{\omega^2}{121}}\right]$$

ω	0.1	1	5	8	10	20	∞
$\underline{\|GH}$	– 94.7°	– 126°	– 138°	– 149.8°	– 177.8°	– 261.6°	– 270°

Locate these points to sketch a smooth phase plot as shown in Fig. 6.43 (*b*).

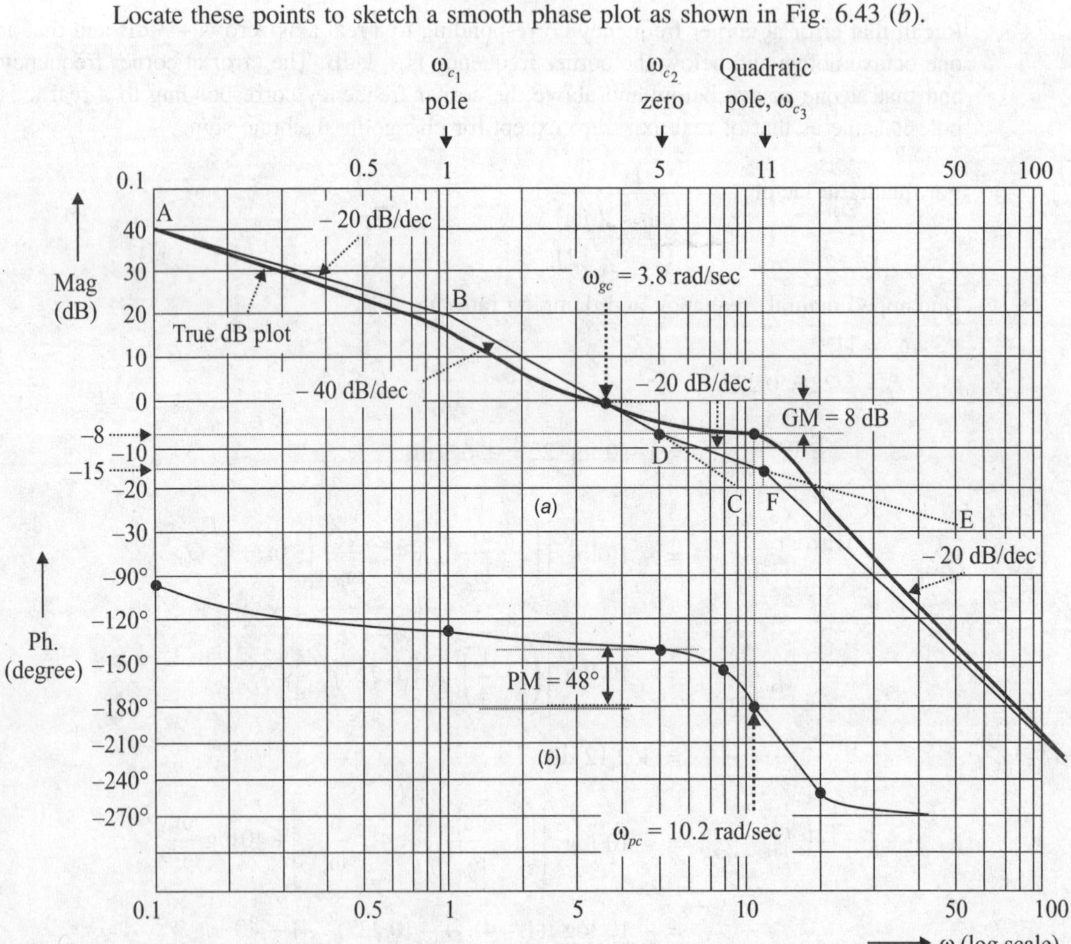

Fig. 6.43: Bode plots for G(*s*) H(*s*) = $\dfrac{242(s+5)}{s(s+1)(s^2+5s+121)}$ (*a*) Magnitude (dB) plot (*b*) Phase plot

(*vi*) **Interpreting stability from Bode plot:** It is easy to read the following:

(*i*) gain cross over frequency ω_{gc} = 3.8 rad/sec (frequency corresponding to 0 dB gain)

(*ii*) phase cross over frequency ω_{pc} = 10.2 rad/sec (frequency corresponding to – 180° phase)

(*iii*) gain margin GM = + 8 dB

(*iv*) Phase margin PM = + 48°

The system is closed loop stable as GM and PM both are positive.

Irrational transmittances

An advantage of frequency response tool is that it is not restricted to the transmittances of rational polynomials. An irrational transfer function such as transportation lag of form $e^{-\tau s}$ where τ is a positive constant, can also be handled at ease. The transportation lag represents the time delay of incoming signal by τ sec. In some practical systems, the time delay is deliberately incorporated. For example, in micro controlled systems the difficulty in matching the speed of micro controller with that of supporting peripherals, is over come by intentionally providing a suitable time delay in micro controller.

A steady state sinusoidal signal which is only delayed in time emerges with no change in amplitude, but it does undergo a phase change. The higher the frequency of sinusoid, the greater the phase shift for same time delay.

Consider
$$G(j\omega) = e^{-j\omega\tau} = \cos\omega\tau - j\sin\omega\tau$$
$$|G(j\omega)| = \sqrt{\sin^2\omega\tau + \cos^2\omega\tau} = 1$$
$$|G(j\omega)|_{dB} = 0$$
$$\angle G(j\omega) = \tan^{-1}\left[-\frac{\sin\omega\tau}{\cos\omega\tau}\right] = -\omega\tau \text{ radians} = -57.3\ \omega\tau \text{ degrees}$$

The phase angle varies linearly with frequency ω. On a logarithmic frequency scale, the phase angle is more and more compressed for larger values of ω as depicted in Fig. 6.44.

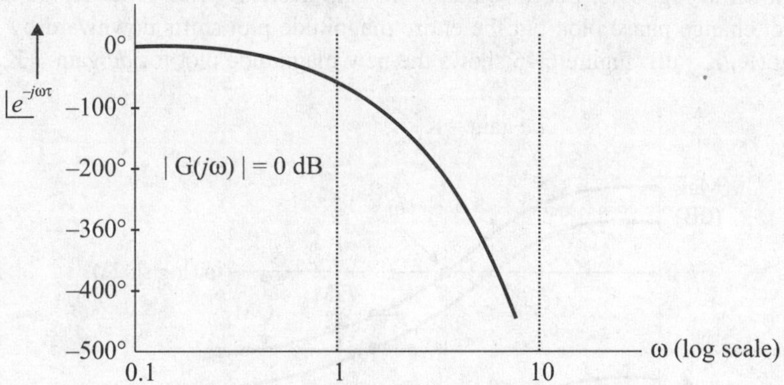

Fig. 6.44: Typical Bode phase plot for $G(j\omega) = e^{-j\omega\tau}$

All pass systems

Consider a transmittance containing a zero in mirror image position in right half of *s* plane corresponding to every pole in left half of *s* plane, for example

$$G(j\omega) = \frac{1 - j\omega\tau}{1 + j\omega\tau}$$

$$| G(j\omega) | = 1, \quad | G(j\omega) |_{dB} = 0$$

$$\underline{|G(j\omega)} = \tan^{-1}(-\omega\tau) - \tan^{-1}(\omega\tau) = -2\tan^{-1}\omega\tau$$

Note that the magnitude contributed by the system is unity (0 dB) for all frequencies and phase varies from 0° to – 180° as ω varies from 0 to ∞. Systems with such transmittance, are called all pass systems in the sense that magnitude is unity and independent of ω. The poles are not permitted to lie in the right half plane because such a system would be unstable. The all pass systems fall into the category of non-minimum phase system, one or more zeros may lie in right half plane. The addition of an all pass function to a transmittance will bring about change in phase plot without affecting the magnitude plot. The magnitude (dB) plot of an all pass function is a line coincident with 0 dB. The phase plot is depicted in Fig. 6.45.

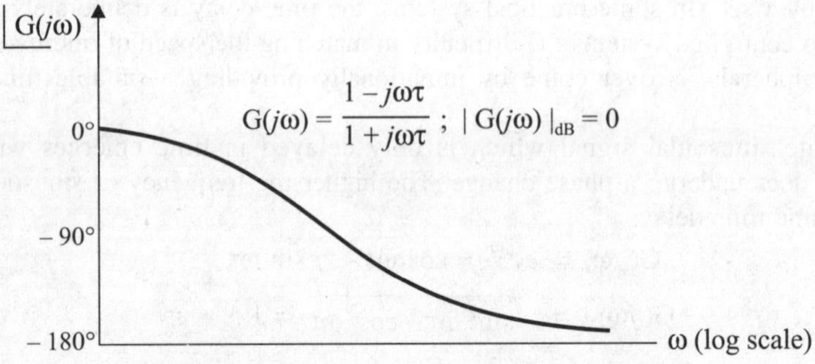

Fig. 6.45: Bode phase plot for $G(j\omega) = \frac{1 - j\omega\tau}{1 + j\omega\tau}$

Effect of variation in gain K on Bode plot

Consider the Bode plot of a typical transmittance $G(j\omega)\, H(j\omega)$ in time constant from with *dc* gain = K_1 as shown in Fig. 6.46. Let the value of *dc* gain decrease from K_1 to K_2. Note that change in *dc* gain does not change phase plot; but the entire magnitude plot shifts downward by 20 log K_1 – 20 log K_2 = 20 log (K_1/K_2) dB. Figure 6.46 shows the new magnitude plot for *dc* gain = K_2 also; $K_2 < K_1$.

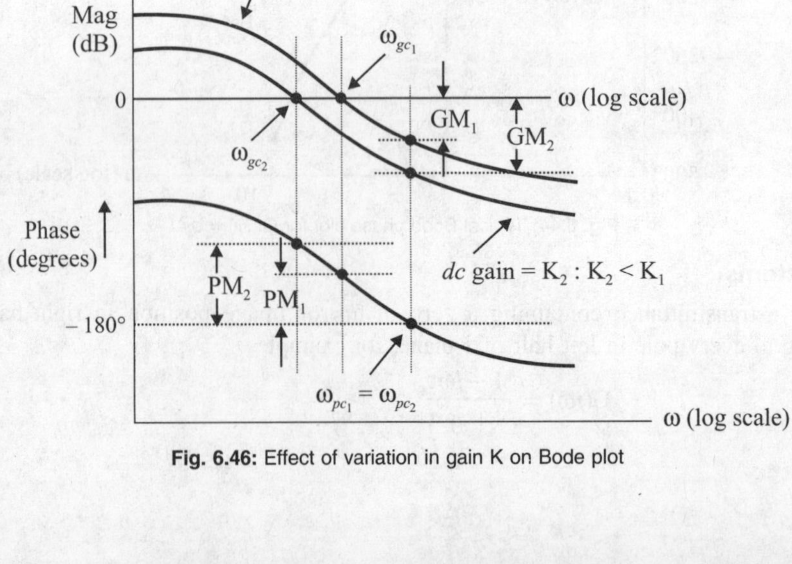

Fig. 6.46: Effect of variation in gain K on Bode plot

The gain cross over frequency, phase cross over frequency, gain margin and phase margin are depicted in Fig. 6.46 as ω_{gc1}, ω_{pc1}, GM_1 and PM_1 for dc gain = K_1 and ω_{gc2}, ω_{pc2}, GM_2 and PM_2 for *dc* gain = K_2 respectively. As *dc* gain decreases, it is easy to observe the following:

(*i*) phase cross over frequency remains unchanged

(*ii*) Gain margin and phase margin both increase ($GM_2 > GM_1$ and $PM_2 > PM_1$). System becomes relatively more stable.

(*iii*) Gain cross over frequency decreases ($\omega_{gc_2} < \omega_{gc_1}$)

(*iv*) Should dc gain increase, the effects just listed above are reverse.

Effect of presence of delay in system on Bode plot

Consider the Bode plot of a typical rational transmittance $G(j\omega)$ as shown in Fig. 6.47. Let a delay of form $e^{-j\omega\tau}$ be introduced in system. Note that delay of τ sec., does not change the magnitude plot. The phase plot changes as $e^{-j\omega\tau}$ additionally contributes negative phase angle. The new phase plot is also shown in Fig. 6.47.

The gain cross over frequency, phase cross over frequency, gain margin and phase margin are depicted in Fig. 6.47 as ω_{gc_1}, ω_{pc_1}, GM_1 and PM_1 for a typical system without delay and ω_{gc_2}, ω_{pc_2}, GM_2 and PM_2 for the same system with delay respectively. The presence of delay in a system, brings about the following effects;

(*i*) gain plot remains unchanged and phase plot changes.

(*ii*) gain cross over frequency remains unchanged ($\omega_{gc_1} = \omega_{gc_2}$).

(*iii*) gain margin decreases ($GM_2 < GM_1$)

(*iv*) phase margin also decreases ($PM_2 < PM_1$)

(*v*) system becomes relatively less stable.

(*vi*) phase cross over frequency decreases ($\omega_{pc_2} < \omega_{pc_1}$).

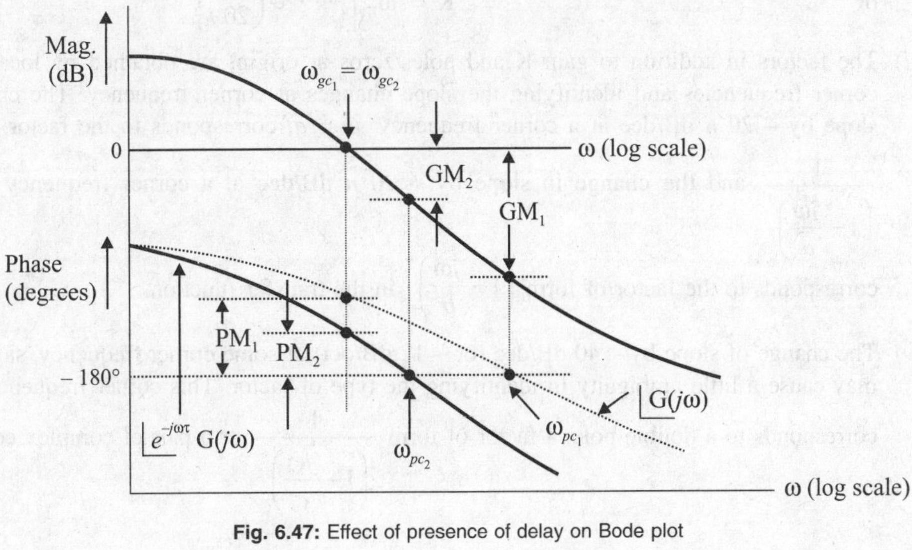

Fig. 6.47: Effect of presence of delay on Bode plot

Finding transfer function models

There exist many practical system components, such as pneumatic valves and air frames, for which it is difficult to obtain analytic expressions for transmittances. If it is possible to carry out frequency response test, however, the transmittance can be experimentally determined. The dB gain and phase shift versus logarithmic frequency are plotted. The straight line asymptotes are fitted to the experimental plots for both gain and phase. While fitting the asymptotes, keep in view the significant point that the slopes of asymptotes must be an integral multiples of ± 20 dB/dec (±6 dB/oct) and that the dB corrections for the type of factor revealed, are in close agreement. Then using the slopes and corresponding corner frequencies, the transfer function is determined.

Note the following while finding the transfer function model from asymptotic plot fitted from experimental data.

(*i*) The flat initial segment of asymptotic dB plot, indicates that there are no poles/zeros at origin. Let this flat initial segment be coincident with G dB line, the value of gain K is given by

$$20 \log K = G$$

or

$$K = \text{antilog}\left(\frac{G}{20}\right)$$

(*ii*) The initial segment of dB plot with slope of – 20 *n* dB/dec, indicates presence of *n* poles at origin and that with slope of + 20 *n* dB/dec, indicates *n* zeros at origin. In both the cases, the dB value corresponding to one frequency point, if known, will suffice to get value of gain K. To demonstrate let this point be G dB at $\omega = \omega_1$ on initial segment of slope – 40 dB/dec (or – 12 dB/oct). Then

$$20 \log \left.\frac{K}{\omega^2}\right|_{\omega = \omega_1} = G$$

or

$$K = \omega_1^2 \left[\text{antilog}\left(\frac{G}{20}\right)\right]$$

(*iii*) The factors in addition to gain K and poles/zeros at origin, are obtained by locating the corner frequencies and identifying the slope changes at corner frequency. The change in slope by – 20 *n* dB/dec at a corner frequency $\omega = a$, corresponds to the factor of form $\dfrac{1}{\left(1+\dfrac{j\omega}{a}\right)^n}$ and the change in slope by + 20 *n* dB/dec at a corner frequency $\omega = b$, corresponds to the factor of form $\left(1+\dfrac{j\omega}{b}\right)^n$ in the transfer function.

(*iv*) The change of slope by – 40 dB/dec (or – 12 dB/oct) at some corner frequency, say $\omega = c$, may cause a little ambiguity in identifying the type of factor. This corner frequency either corresponds to a double pole, a factor of form $\dfrac{1}{\left(1+\dfrac{j\omega}{c}\right)^2}$ or a pair of complex conjugate poles, a factor of form $\dfrac{1}{1+\left(\dfrac{j\omega}{c}\right)^2+j\left(2\xi\dfrac{\omega}{c}\right)}$. In such a situation, the error between the asymptotic dB plot and the true plot, helps in identifying the type of factor. If error is about – 6 dB at corner frequency $\omega = c$, it indicates presence of a double pole. If the error is positive, it indicates presence of pair of complex conjugate poles.

Finding transfer function model from given dB plot, is demonstrated as follows. Consider the dB plot shown in Fig. 6.48.

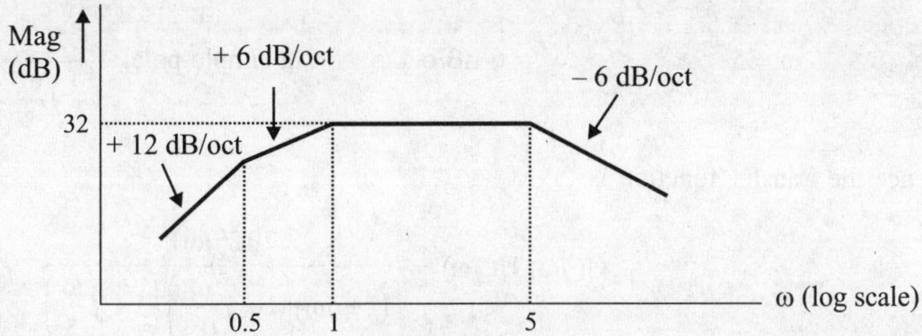

Fig. 6.48: dB plot

The initial segment of dB plot has slope of +12 dB/oct. So, there are two zeros at origin in the transfer function model. To find gain K, it is required to locate a point on initial segment. Since $dB|_{\omega=1} = 32$, $\omega = 0.5$ is one octave below $\omega = 1$ and the corresponding segment has slope of + 6 dB/oct., dB at $\omega = 0.5$ will be 6 dB down from 32 dB; that is $dB|_{\omega=0.5} = 32 - 6 = 26$. This is demonstrated as follows:

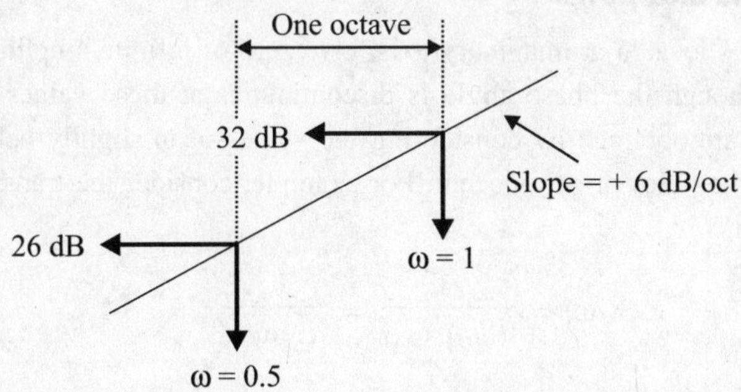

Having located the point ($\omega = 0.5$, dB = 26) on initial segment, the gain K can be evaluated as

$$20\log K\omega^2\Big|_{\omega=0.5} = 26$$

or
$$K = \frac{1}{(0.5)^2}\left[\text{anti log}\,\frac{26}{20}\right]$$

$$= 79.8$$

In addition to gain K and two zeros at origin, the other factors in the transfer function are identified by locating the corner frequencies and slope changes there at as follows:

Corner frequency	Slope change	Type of factor
$\omega = 0.5$	– 6 dB/oct.	a simple pole; $\dfrac{1}{1+\dfrac{j\omega}{0.5}}$
$\omega = 1$	– 6 dB/oct.	a simple pole; $\dfrac{1}{1+j\omega}$
$\omega = 5$	– 6 dB/oct.	a simple pole; $\dfrac{1}{1+j\dfrac{\omega}{5}}$

Hence the transfer function is

$$G(j\omega)\,H(j\omega) = \frac{79.8(j\omega)^2}{(1+j\omega)\left(1+j\dfrac{\omega}{0.5}\right)\left(1+j\dfrac{\omega}{5}\right)}$$

and

$$G(s)\,H(s) = \frac{79.8s^2}{(1+s)(1+2s)(1+0.2s)}$$

$$= \frac{199.5s^2}{(s+1)(s+0.5)(s+5)}$$

Imaginary axis zeros and poles

The zeros and poles located at imaginary axis, give zero or infinite amplitude for the values of ω where they occur. Although the phase angle is discontinuous at these values of ω, but the limiting values of phase angle are obtained by considering the values of ω slightly below and slightly above the value of ω where the poles or zeros occur. For example, consider the transmittance with a pair of imaginary axis poles.

$$G(j\omega) = \frac{1}{(j\omega)^2 + a} = \frac{1}{a-\omega^2}$$

Noting that $\left|G(j\omega)\right|_{\omega=\sqrt{a}} = \infty$

The magnitude and phase plots are sketched in Fig. 6.49.

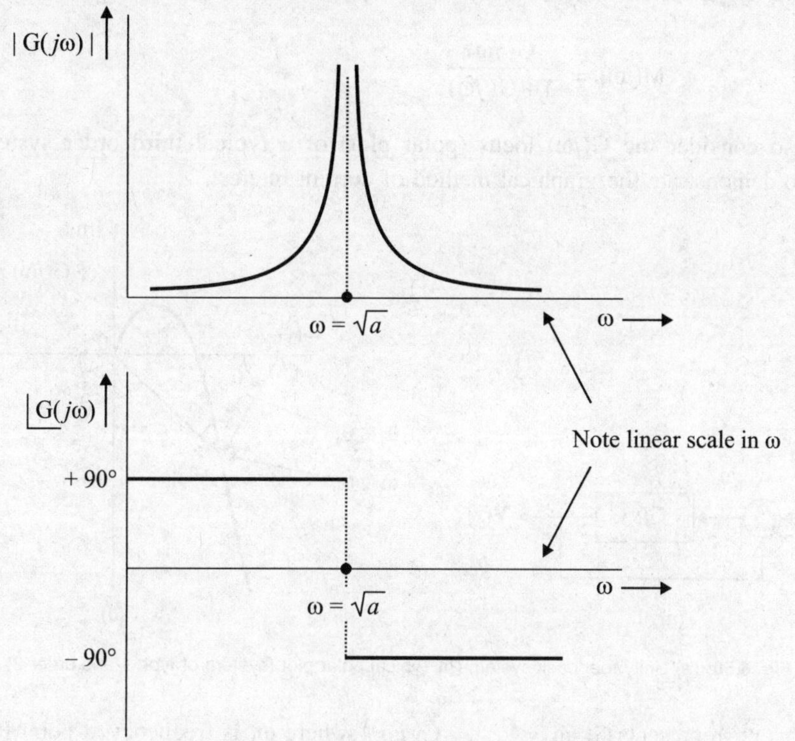

Fig. 6.49: Approximate frequency response for $G(s) = 1/(s^2 + a)$

The natural behaviour of a system with complex conjugate poles on imaginary axis, is of sinusoidal nature, exhibiting neither growing nor decaying response with time. The magnitude plot shown in Fig. 6.49, obviously suggests that the amplitude response grows larger and larger if such a system is excited by sinusoidal signal of frequency equal to that of natural behaviour. This theoretical perception has evolved because the nonlinearities involved in the system, have been ignored. The evaluation routine has used a linear, time invariant system model. In practical systems the response does not become infinite; the non linearties eventually, limit the response to a finite value.

6.7 CLOSED LOOP FREQUENCY RESPONSE OF UNITY FEEDBACK SYSTEM

The focus of current discussion is to develop a graphical method for determination of closed loop frequency response using open loop frequency response. In the discussion that follows, we shall see that the method is directly applicable to systems with unity feedback. However, with little modification it can also be applied to non unity feedback systems as well.

Let us consider unity feedback system shown in Fig. 6.50(a). The closed loop transfer function is

$$M(s) = \frac{Y(s)}{R(s)} = \frac{G(s)}{1+G(s)}$$

Replacing s by $j\omega$ for sinusoidal steady state response, we have

$$M(j\omega) = \frac{G(j\omega)}{1+G(j\omega)} \quad ...(6.41)$$

Let us also consider the $G(j\omega)$ locus (polar plot) of a typical third order system shown in Fig. 6.50(b) to demonstrate the graphical method of current interest.

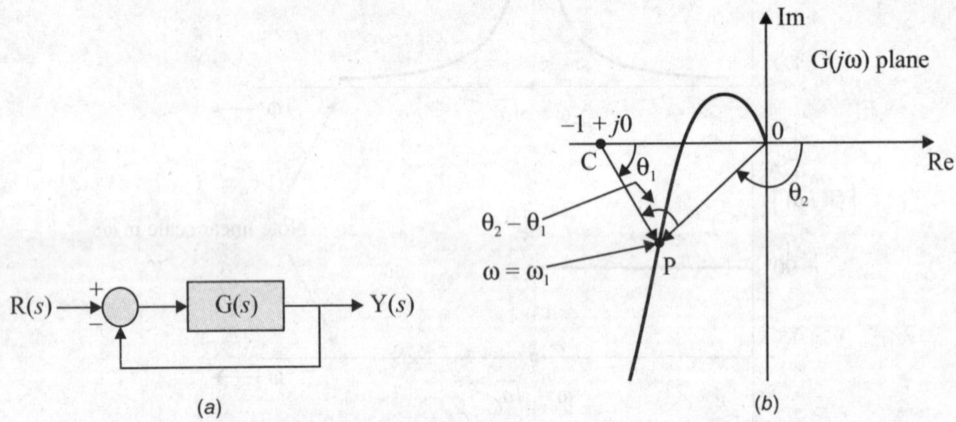

Fig. 6.50: (a) Unity feedback system (b) Typical polar plot (system of type 1 and order 3)

The vector $\overrightarrow{OP}$ represents $G(j\omega)|_{\omega=\omega_1} = G(j\omega_1)$ where ω_1 is frequency at point P on the polar plot. The length OP is equal to $|G(j\omega_1)|$ and angle θ_2 is equal to $\angle G(j\omega_1)$. With similar reasoning vector $\overrightarrow{CP}$ represents $1 + G(j\omega_1)$. The point C corresponds to $-1 + j0$ point on $G(j\omega)$ plane. The length CP is equal to $|1+G(j\omega_1)|$ and angle θ_1 is equal to $\angle 1+G(j\omega_1)$. Thus, according to (6.41) the ratio of $\overrightarrow{OP}$ to $\overrightarrow{CP}$ represents the closed loop frequency response, that is

$$M(j\omega_1) = \frac{\overrightarrow{OP}}{\overrightarrow{CP}} = \frac{G(j\omega_1)}{1+G(j\omega_1)}$$

The magnitude of closed loop frequency response at $\omega = \omega_1$ is $|M(j\omega)|_{\omega=\omega_1} = |M(j\omega_1)| = \left|\frac{\overrightarrow{OP}}{\overrightarrow{CP}}\right|$ and the phase angle thereof at $\omega = \omega_1$ is $\angle M(j\omega)|_{\omega=\omega_1} = \angle M(j\omega_1) = \theta_2 - \theta_1$. Similarly, $|M(j\omega)|$ and $\angle M(j\omega)$ can be obtained at various frequency points in the range of practical interest.

The discussion just above, reveals that the entire closed loop frequency response can be determined by measuring the magnitudes and phase angles at different frequency points on $G(j\omega)$ locus. Now, we shall introduce **constant magnitude loci (M circles)** and **constant phase angle loci (N circles)** that are particularly convenient in determining the closed loop frequency response from polar plot (open loop frequency response curve).

Constant magnitude loci (M circles)

$G(j\omega)$ is a complex quantity and can be written as

$$G(j\omega) = \text{Re}\,[G(j\omega)] + j\,\text{Im}\,[G(j\omega)] = x + jy \qquad ...(6.42)$$

where, $x = \text{Re}\,[G(j\omega)]$ and $y = \text{Im}\,[G(j\omega)]$

Now,
$$|M(j\omega)| = \left|\frac{G(j\omega)}{1+G(j\omega)}\right| = \left|\frac{x+jy}{1+x+jy}\right| = \sqrt{\frac{x^2+y^2}{(1+x)^2+y^2}}$$

For simplicity, let us denote $|M(j\omega)|$ by M, so that

$$M^2 = \frac{x^2+y^2}{(1+x)^2+y^2} = \frac{x^2+y^2}{1+2x+x^2+y^2}$$

and little algebraic manipulation, gives

$$(1 - M^2)\,x^2 - 2M^2\,x - M^2 + (1 - M^2)\,y^2 = 0 \qquad ...(6.43)$$

For M = 1, (6.43) takes the form

$$x = -\frac{1}{2} \qquad ...(6.44)$$

This is the equation of a straight line parallel to Y-axis and passing through the point (–1/2, 0) in $G(j\omega)$ plane.

For M ≠ 1, (6.43) can be written as

$$x^2 + y^2 + \frac{2M^2}{M^2-1}x + \frac{M^2}{M^2-1} = 0$$

and adding $\frac{M^2}{(M^2-1)^2}$ on both sides of this equation, we have

$$x^2 + \frac{2M^2}{M^2-1}x + \left[\frac{M^2}{M^2-1}\right]^2 + y^2 = \frac{M^2}{(M^2-1)^2}$$

or
$$\left(x + \frac{M^2}{M^2-1}\right)^2 + y^2 = \frac{M^2}{(M^2-1)^2} \qquad ...(6.45)$$

For a given value of M, (6.45) represents a circle with the centre at $x = \text{Re}\,[G(j\omega)] = -M^2/(M^2-1)$, $y = \text{Im}\,G(j\omega) = 0$ and radius $r = \left|\frac{M}{M^2-1}\right|$. When M takes on different values, (6.45) describes in $G(j\omega)$ plane a family of circles that are called the **constant M loci** or the **constant M circles.** For

M = 1.3, 1.5 and 2 (values in increasing order and greater than 1), the values of ***x*** (centre) and ***r*** (radius) are computed and put together in **Table 6.2(*a*).** The corresponding circles are shown in Fig. 6.51. For M = 0.4, 0.6 and 0.8 (values in decreasing order and less than 1), the values of ***x*** and ***r*** are again computed and put together in **Table 6.2(*b*).** The corresponding circles are also shown in Fig. 6.51. The circles are not drawn to the scale. The key idea of showing this figure is to only have an integrated view of constant M circles for few values of M equal to 1, greater than 1 and less than 1.

TABLE 6.2(*a*)

M	1.3	1.5	2
Radius *r*	1.88	1.2	0.67
x	–2.44	–1.8	–1.33

TABLE 6.2(*b*)

M	0.4	0.6	0.8
r	0.48	0.94	2.2
x	0.19	0.56	1.78

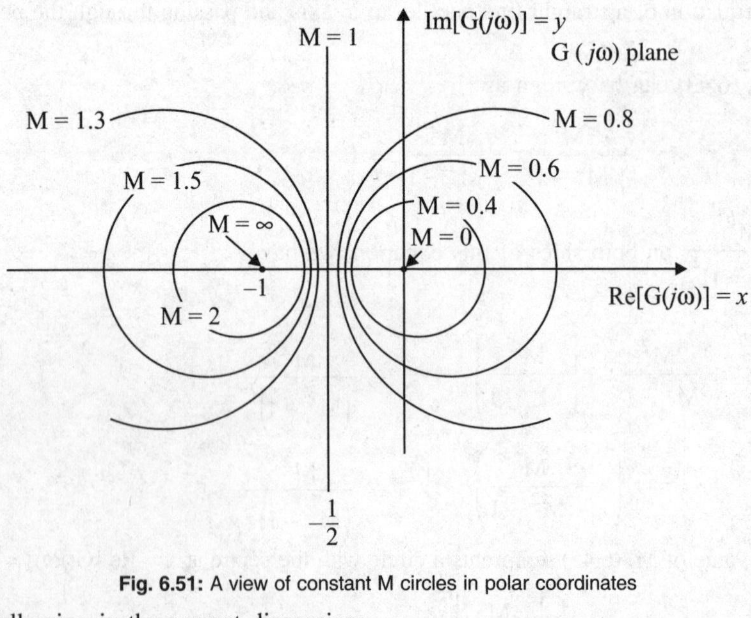

Fig. 6.51: A view of constant M circles in polar coordinates

Note the following in the current discussion:

(*i*) M = 1 locus is a straight line passing through (–1/2, 0) and parallel to the imaginary axis. In fact M = 1 is the locus of points equidistant from origin and –1 + *j*0 point.

(*ii*) The circles to the left of M = 1 line correspond to the values of M greater than 1 and those to the right of M = 1 line correspond to values of M less than 1. The M circles are symmetrical with respect to M = 1 line and the real axis.

(*iii*) As M becomes larger compared with 1, the M circles become smaller and their centre converges to $-1 + j0$ point. Similarly, as M becomes smaller compared with 1, the M circles become smaller and their centre converges to the origin. In fact, the centre of M circles lie to the right of the origin for $0 < M < 1$.

Interpreting magnitude response of closed loop system from constant M circles and polar plot (open loop frequency response):

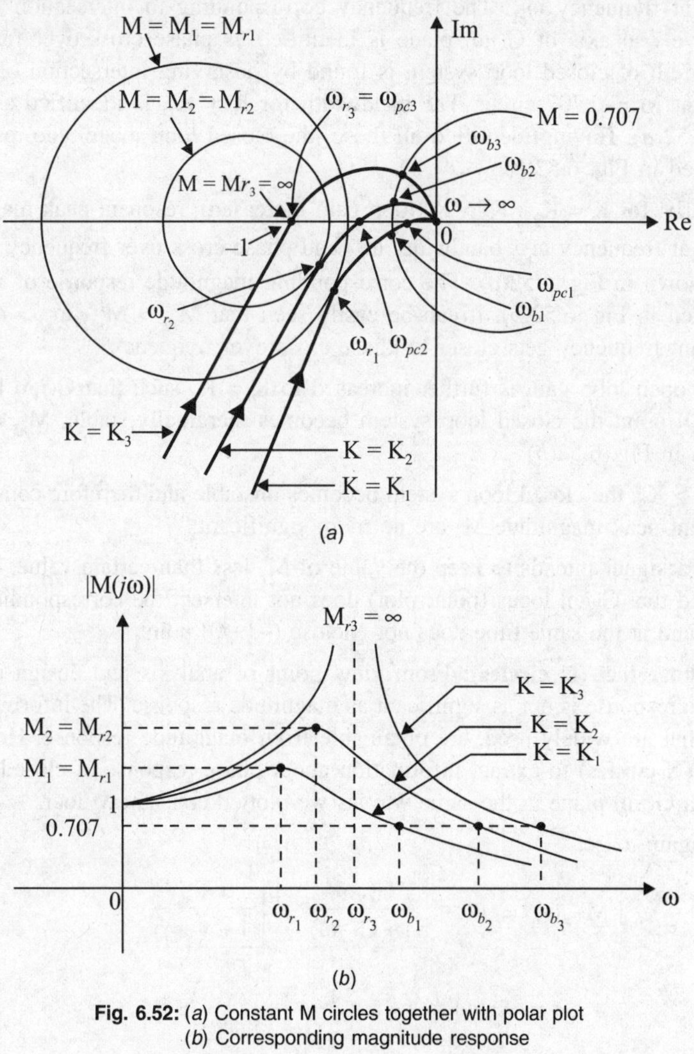

Fig. 6.52: (*a*) Constant M circles together with polar plot
(*b*) Corresponding magnitude response

The magnitude response of closed loop system, can be directly determined from polar plot of $G(j\omega)$. This is accomplished by first drawing constant M circles as demonstrated in Fig. 6.52(*a*). Graphically, the intersection of $G(j\omega)$ curve (polar plot) and the constant M circle gives the value of magnitude (M) at the corresponding frequency. Several M circles are superimposed on the polar plot in G(*j*ω) plane and the curve **M** versus **ω** is easily plotted from these intersection points. This is demonstrated by drawing few M circles together with polar plot in Fig. 6.52(*a*) and corresponding magnitude response is shown in Fig. 6.52(*b*). Let us **note** the following points that are of still greater interest in the current reference.

(*i*) The constant M circle with smallest radius that is tangent to G(*j*ω) locus (polar plot) gives the value of the resonant peak magnitude M*r* and the resonant frequency ω_r is read off at the point of tangency. As demonstrated in Fig. 6.52(*a*), the polar plot of G(*j*ω) for $K = K_1$ is tangent to the smallest circle corresponding to $M = M_1$ and this is identified as peak resonance M_{r1}, the corresponding frequency at the tangent point is identified as resonant frequency ω_{r1}. The frequency corresponding to intersection of polar plot with negative real axis of G(*j*ω) plane is identified as phase cross over frequency ω_{pc1}. The bandwidth of closed loop system is found by observing intersection of G(*j*ω) locus with constant M = 0.707 circle. The bandwidth for $K = K_1$, is identified as ω_{b1} as shown in Fig. 6.52(*a*). Having identified all these, the closed loop magnitude response is typically sketched in Fig. 6.52(*b*).

(*ii*) Similarly, for $K = K_2 : K_2 > K_1$(loop gain increased), resonant peak magnitude $M_2 = M_{r2}$, resonant frequency ω_{r2}, bandwidth ω_{b2} and phase cross over frequency ω_{pc2} are identified and shown in Fig. 6.52(*a*). The corresponding magnitude response of closed loop is also sketched in Fig. 6.52(*b*). It can be easily seen that $M_{r2} > M_{r1}$, $\omega_{r2} > \omega_{r1}$, $\omega_{b2} > \omega_{b1}$ and resonant frequency gets closer to phase cross over frequency.

(*iii*) When open loop gain is further increased to $K = K_3$ such that G(*j*ω) locus pass through (–1, *j*0) point the closed loop system becomes marginally stable; $M_{r3} = \infty$, $\omega_{r3} = \omega_{pc3}$ as shown in Fig. 6.52(*b*).

(*iv*) For $K > K_3$, the closed loop system becomes unstable and therefore constant M circles and resonant peak magnitude M_r are no more significant.

(*v*) If the designer intends to keep the value of M_r less than certain value, it is required to be ensured that G(*j*ω) locus (polar plot) does not intersect the corresponding M circle at any point and at the same time does not enclose (–1, *j*0) point.

Constant phase loci (N circles): From view point of analysis and design of the closed loop system, the phase response is not as significant as magnitude response. The information about M*r*, ω_r and bandwidth that are widely used, are obtainable from magnitude response. However, the **loci of constant phase (N circles)** to extract information about phase response of closed loop system, may also be plotted in G(*j*ω) plane in the same way as we plotted constant M loci.

Use (6.42) again to get

$$\angle M(j\omega) = \alpha = \angle \frac{G(j\omega)}{1+G(j\omega)} = \angle \frac{x+jy}{1+x+jy}$$

or $$\alpha = \tan^{-1}\frac{y}{x} - \tan^{-1}\frac{y}{(1+x)}$$

let us define $$\tan \alpha = N$$

then $$N = \tan\left[\tan^{-1}\left(\frac{y}{x}\right) - \tan^{-1}\left(\frac{y}{(1+x)}\right)\right]$$

or $$N = \frac{\dfrac{y}{x} - \dfrac{y}{1+x}}{1 + \dfrac{y}{x}\left(\dfrac{y}{1+x}\right)} = \frac{y}{x^2 + x + y^2}$$

or $$x^2 + x + y^2 - \frac{y}{N} = 0$$

Adding $\left[\dfrac{1}{4} + \dfrac{1}{(2N)^2}\right]$ on both sides of this equation, we have

$$x^2 + x + \frac{1}{4} + y^2 - \frac{y}{N} + \frac{1}{(2N)^2} = \frac{1}{4} + \frac{1}{(2N)^2}$$

or $$\left(x + \frac{1}{2}\right)^2 + \left(y - \frac{1}{2N}\right)^2 = \frac{1}{4} + \frac{1}{(2N)^2} \qquad ...(6.46)$$

For a given value of N, (6.46) represents a circle with centre at $x = -\dfrac{1}{2}$, $y = \dfrac{1}{2}N$ and radius $\sqrt{\dfrac{1}{4} + \dfrac{1}{4N^2}}$.

For example, if $\alpha = 30^o$, then $N = \tan\alpha = 0.577$, (6.46) describes a circle with centre at (–0.5, 0.866) and radius $r = 1$. A family of constant N circles, is shown in Fig. 6.53 with α as parameter.

Note the following in current discussion:

(*i*) The two points ($x = 0$, $y = 0$) and ($x = -1$, $y = 0$) satisfy (6.46) regardless of the value of N. So, each circle passes through the origin and $-1 + j0$ point in G($j\omega$) plane.

(*ii*) The constant N locus for a given value of α, is actually not the entire circle, but only an **arc**. For example, $\alpha = 30^o$ and $\alpha = -150^o$ arcs are part of same circle. This is so because the **tan** of an angle remains same if $\pm 180^o$ or multiples thereof, is added to the angle (circle for $\alpha = \alpha_1$ and that for $\alpha = \alpha_1 \pm 180^o\, i$; $i = 1, 2, \ldots\ldots$ are same).

(*iii*) The intersections of G($j\omega$) locus and N circles, give the values of N at frequency points of G($j\omega$) locus. While using N circles for determination of phase angle, care must be taken to interpret proper values of α. To avoid any error, it is better to start at zero frequency ($\alpha = 0^o$) and proceed to higher frequencies while bearing in mind that phase curve must be continuous.

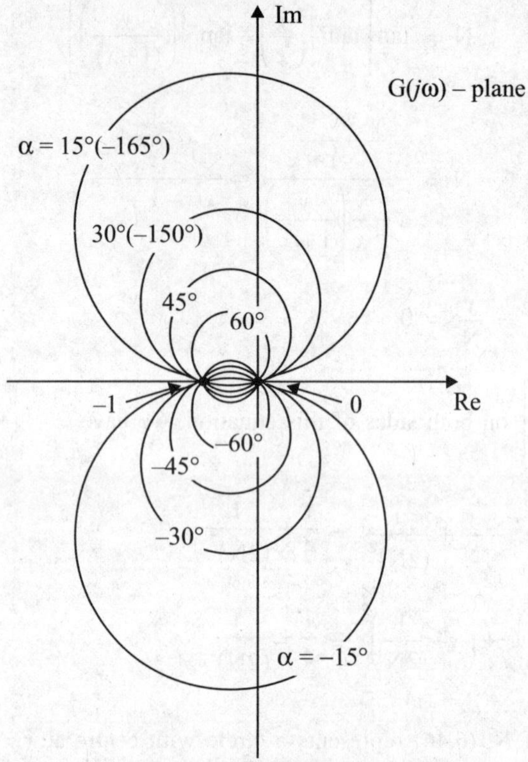

Fig. 6.53: A family of constant N circles

Nichols chart: The design of the control system, often involves change in open loop gain or addition of series controller to the system or both. Such a design effort becomes easy if one works with the magnitude – phase plot of G(*j*ω). Note that the entire G(*j*ω) locus shifts upward or downward while retaining its shape when open loop gain is increased or decreased respectively. Further, the magnitude – phase plot changes only in the horizontal direction if the design effort is made to change the phase of G(*j*ω) while preserving open loop gain. These predictable changes in G(*j*ω) locus, help in arriving at final successful design.

Thus, for design convenience, it makes sense to construct the **M** and **N** loci in log magnitude versus phase plane. The chart consisting of the **M** and **N** loci in **log magnitude versus phase plane is called Nichols chart** (named after NB Nichols). Although these charts are no longer commercially available, but they can be easily generated with the help of computer. The current discussion will include the brief description of Nichols chart and interpreting it to extract information about frequency response specifications like M_r, ω_r and bandwidth.

Note the following significant points about Nichols chart:

(*i*) The constant M circles in the polar co-ordinates, transform to a family of ellipses in the logarithmic gain-phase plane.

(*ii*) The Nichols chart is symmetrical about the –180° axis.

(*iii*) The critical point (–1 + *j*0) of G(*j*ω) plane is mapped to Nichols chart as point (0 dB, –180°).

(*iv*) The M and N loci repeat for every 360° and there is symmetry at every 180° interval. The M loci are centered about the critical point (0 dB, –180°).

(*v*) The frequency response of closed loop system can be determined from that of open loop system by sketching G(*j*ω) locus on the Nichols Chart. The intersections of G(*j*ω) locus (open loop frequency response curve) and **M** and **N** loci give the values of the magnitude **M** and phase **α** of closed loop frequency response at the corresponding frequencies of G(*j*ω). The resonant peak magnitude M*r*, is determined by identifying the smallest of the constant M locus (M ≥ 1) to which the G(*j*ω) locus is tangent. The resonant frequency ω_r is given by the frequency at the point of tangency. The bandwidth of closed loop system, is the frequency at which G(*j*ω) locus intersects M = –3 dB (0.707) locus. For the sake of demonstration Fig. 6.54(*a*) shows a typical G(*j*ω) locus together with the **M** and **N** loci. Having noted the magnitude and the phase at intersections of G(*j*ω) locus and the **M** and **N** loci together with corresponding frequencies, the closed loop frequency response curves are drawn in Fig. 6.54(*b*). It is easy to note from Fig. 6.54(*a*) that M*r* = 5dB, ω_r = 0.8 rad/sec, bandwidth ω_b = 1.3 rad/sec, phase cross over frequency ω_{pc} = 1.24 rad/sec (recall that a phase cross over point is one where G(*j*ω) locus intersects –180° axis) and gain cross over frequency ω_{gc} = 0.76 rad/sec (recall that gain cross over point is one where G(*j*ω) locus intersects 0 dB axis).

The phase margin (PM) is the horizontal distance in degrees between gain cross over point and the critical point (0 dB, –180°). It is easy to read from Fig. 6.54(*a*) that PM ≅ 32°. The gain margin (GM) is the vertical distance in dB between phase cross over point and the critical point (0 dB, –180°). It can easily be read from Fig. 6.53(*a*) that GM ≅ 6.5 dB.

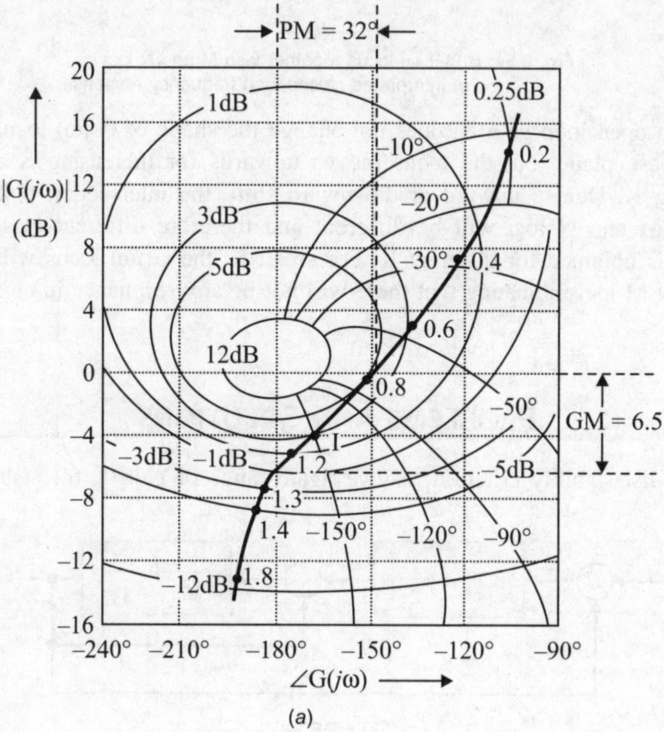

(*a*)

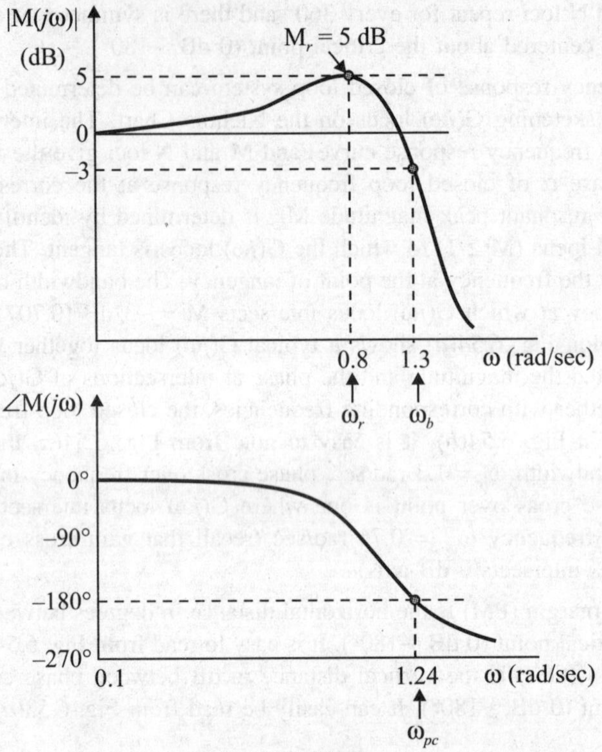

(b)

Fig. 6.54: (a) G(jω) locus together with M and N loci
(b) Interpreted closed loop frequency response.

(*vi*) Change in open loop gain K, does not change the shape of G(*j*ω) locus in log magnitude versus phase plane, but the locus moves upwards for increasing K and downward for decreasing K. Due to upward or downward shift, the intersection points of G(*j*ω) locus with the **M** and **N** loci will be different and therefore different closed loop frequency response is obtained for different K. For small K, the G(*j*ω) locus will not be tangent to any of the M loci indicating that there will not be any resonance in closed loop frequency response.

PROBLEMS AND SOLUTIONS

P 6.1: Use Nyquist stability criterion to investigate range of gain K for stability of the system shown in Fig. P6.1.

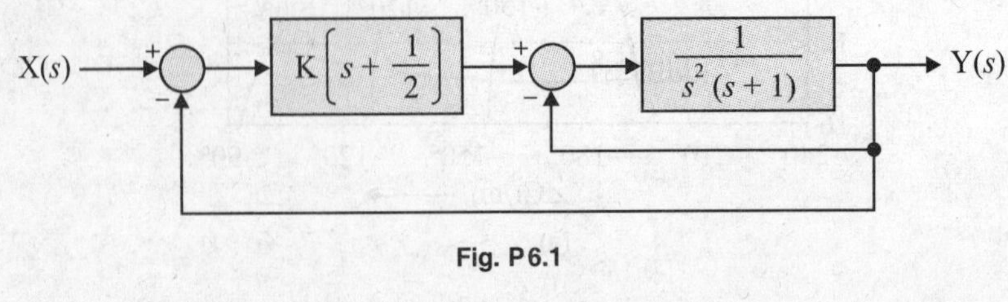

Fig. P6.1

Solution: Let $$G_1(s) = K\left(s+\frac{1}{2}\right)$$

and $$G_2(s) = \frac{1/\left[s^2(s+1)\right]}{1+1/\left[s^2(s+1)\right]} = \frac{1}{s^3+s^2+1}$$

Note that the stability investigation of given system requires sketching Nyquist plot of $G(s) = G_1(s)\,G_2(s)$. But poles of $G(s)$ are not known. Let us use Routh array constructed below to identify RHP poles of $G(s)$.

s^3	1	0
s^2	1	1
s^1	−1	0
s^0	1	

The left column exhibits two sign changes. So, $G(s)$ has two RHP poles.

For sketching Nyquist plot, let us chose RHP boundary in s-plane as shown in Fig. P6.1 (*a*).

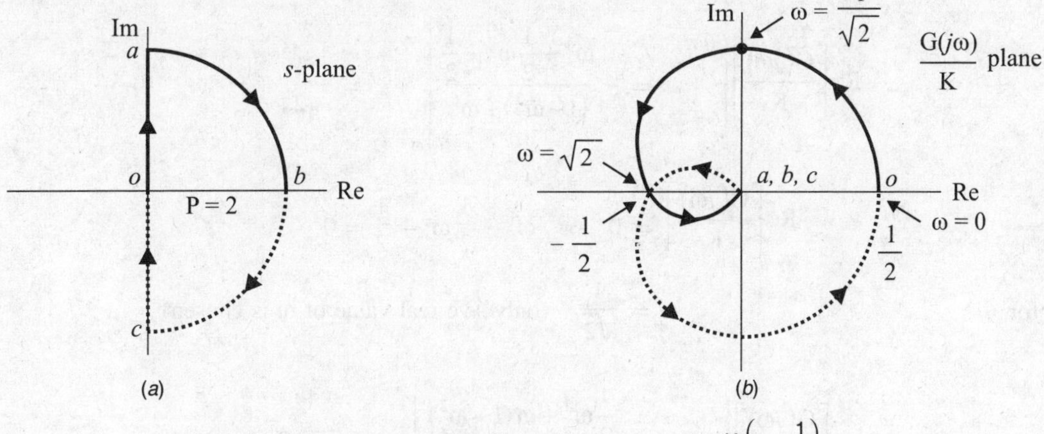

Fig. P6.1: (*a*) RHP boundary; (*b*) Nyquist plot for $G(s) = \dfrac{K\left(s+\frac{1}{2}\right)}{(s^3+s^2+1)}$

The mapping of RHP boundary of Fig. P6.1(*a*) into $\dfrac{G(j\omega)}{K}$ plane is demonstrated as follows.

(*i*) **Map of path *oa*:** Put $s = j\omega$ for $0 \le \omega \le \infty$

$$G(j\omega) = \frac{K\left(j\omega+\frac{1}{2}\right)}{(j\omega)^3+(j\omega)^2+1} = \frac{K\left(j\omega+\frac{1}{2}\right)}{\left[(1-\omega^2)-j\omega^3\right]}\frac{\left[(1-\omega^2)+j\omega^3\right]}{\left[(1-\omega^2)+j\omega^3\right]}$$

or $$\frac{G(j\omega)}{K} = -\frac{\omega^4+\frac{1}{2}\omega^2-\frac{1}{2}}{(1-\omega^2)^2+\omega^6} + j\,\frac{\frac{1}{2}\omega^3+\omega(1-\omega^2)}{(1-\omega^2)^2+\omega^6}$$

where $$\text{Re}[G(j\omega)/K] = -\frac{\omega^4 + \frac{1}{2}\omega^2 - \frac{1}{2}}{(1-\omega^2)^2 + \omega^6}$$

and $$\text{Im}[G(j\omega)/K] = \frac{\frac{1}{2}\omega^3 + \omega(1-\omega^2)}{(1-\omega^2)^2 + \omega^6}$$

Note the following significant points

$$\lim_{\omega \to 0} G(j\omega)/K = 0.5\angle 0^\circ \quad \text{(map of point } o\text{)}$$

$$\lim_{\omega \to \infty} G(j\omega)/K \cong \lim_{\omega \to \infty} \frac{1}{(j\omega)^2} = 0\angle -180^\circ \quad \text{(map of point } a\text{)}$$

$$\text{Im}\left[\frac{G(j\omega)}{K}\right] = 0 \quad \text{or} \quad \frac{1}{2}\omega^3 + \omega(1-\omega^2) = 0$$

for $$\omega = 0 \quad \text{and} \quad \omega = \sqrt{2}$$

$$\text{Re}\left[\frac{G(j\omega)}{K}\right]\Bigg|_{\omega=\sqrt{2}} = -\frac{\omega^4 + \frac{1}{2}\omega^2 - \frac{1}{2}}{(1-\omega^2)+\omega^6}\Bigg|_{\omega=\sqrt{2}} = -\frac{1}{2}$$

$$\text{Re}\left[\frac{G(j\omega)}{K}\right] = 0 \quad \text{or} \quad \omega^4 + \frac{1}{2}\omega^2 - \frac{1}{2} = 0$$

for $$\omega = \frac{1}{\sqrt{2}} \quad \text{(only +ve real value of } \omega \text{ is chosen)}$$

$$\text{Im}\left[\frac{G(j\omega)}{K}\right]\Bigg|_{\omega=\frac{1}{\sqrt{2}}} = \frac{\frac{1}{2}\omega^3 + \omega(1-\omega^2)}{(1-\omega^2)^2 + \omega^6}\Bigg|_{\omega=\frac{1}{\sqrt{2}}} = 0.94$$

The two mapped points o for $\omega = 0$ and a for $\omega = \infty$ together with intersection points on real and imaginary axes are plotted and interconnected by an arrow as shown in Fig. P6.1 (*b*).

(*ii*) **Map of path *abc* (semicircle of ∞ radius)**

Put $$s = \lim_{R \to \infty} Re^{j\theta}$$

so that $$\frac{G(s)}{K} = \lim_{R \to \infty} \frac{\left(Re^{j\theta} + \frac{1}{2}\right)}{\left(Re^{j\theta}\right)^3 + \left(Re^{j\theta}\right)^2 + 1} = 0$$

and mapped points a, b, c lie at origin of $G(j\omega)/K$ plane as shown in Fig. P6.1(*b*).

(*iii*) **Map of path *co*:** This is mirror image of map of path *oa* as shown by dashed line in Fig. P6.1 (*b*).

Interpreting stability: G(*s*) has two RHP poles; P = 2. For the system to be closed loop stable (Z = 0) requires N = Z – P = – 2 meaning there must be two CCW encirclements of point (– 1 + *j*0). This would happen if

$$-\frac{1}{2}K < -1 \quad \text{or} \quad K > 2$$

Thus G(*s*) will be stable in closed loop for +ve values of K larger than 2.

P 6.2: Sketch the complete Nyquist plot and interpret stability therefrom for the systems with following loop transmittances.

(*a*) $G_1(s) = \dfrac{4s+1}{s^2(s+1)(2s+1)}$

(*b*) $G_2(s) = \dfrac{1}{s^4(s+\alpha)}$

Solution: (*a*) $G_1(s)$ has two open loop poles at origin. The RHP boundary in *s*-plane is chosen as shown in Fig. P6.2 (*a*). The poles of $G_1(s)$ are also depicted and a semicircle of arbitrarily small radius ($\rho \to 0$) bypasses the poles at origin. $G_1(s)$ has no RHP poles (P = 0).

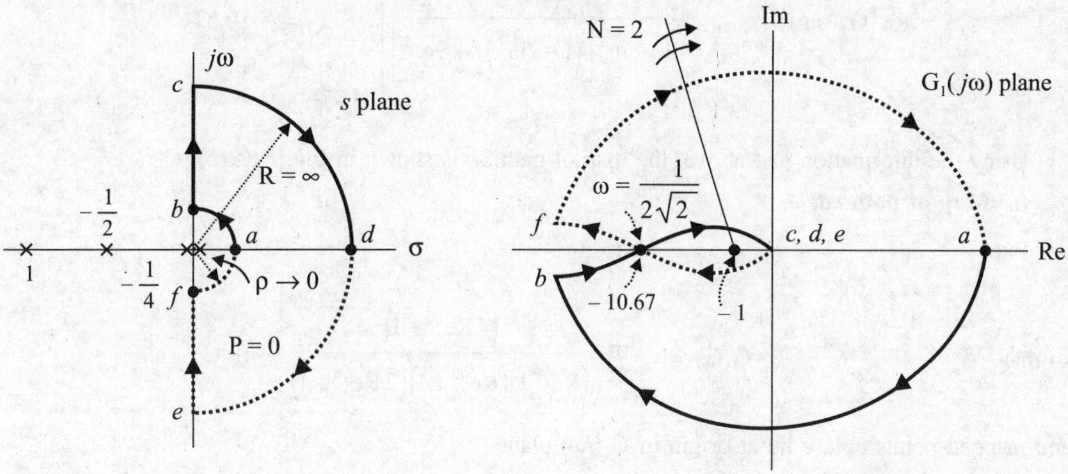

(*a*) RHP boundary

(*b*) Nyquist plot for $G_1(s) = \dfrac{4s+1}{s^2(s+1)(2s+1)}$

Fig. P6.2

The step by step approach to the mapping of RHP boundary of Fig. P6.2 (*a*) is demonstrated as follows:

(*i*) **Map of path *bc*:** Put $s = j\omega$ for $0 \le \omega \le \infty$

$$G_1(j\omega) = \frac{j4\omega+1}{(j\omega)^2(1+j\omega)(1+j2\omega)}$$

$$\lim_{\omega \to 0} G_1(j\omega) \cong \lim_{\omega \to 0} \frac{1}{(j\omega)^2} = \infty \angle -180° \quad \text{(map of point } b\text{)}$$

$$\lim_{\omega \to \infty} G_1(j\omega) \cong \lim_{\omega \to \infty} \frac{j4\omega}{(j\omega)^2 (j\omega)(j2\omega)} = 0 \angle -270° \quad \text{(map of point } c\text{)}$$

To investigate any real axis intersections, transform $G_1(j\omega)$ in following form:

$$G_1(j\omega) = \frac{(j4\omega + 1)}{-\omega^2 \left[\left(1 - 2\omega^2\right) + j3\omega\right]} \frac{\left[\left(1 - 2\omega^2\right) - j3\omega\right]}{\left[\left(1 - 2\omega^2\right) - j3\omega\right]}$$

$$= \frac{1 + 10\omega^2}{-\omega^2 \left[\left(1 - 2\omega^2\right)^2 + 9\omega^2\right]} + j \frac{4\omega\left(1 - 2\omega^2\right) - 3\omega}{-\omega^2 \left[\left(1 - 2\omega^2\right)^2 + 9\omega^2\right]}$$

$$\text{Im}\,[G_1(j\omega)] = 0 \quad \text{for } \omega = 0 \text{ and } \omega = \frac{1}{2\sqrt{2}}$$

and

$$\text{Re}\left[G_1(j\omega)\right]\Big|_{\omega = \frac{1}{2\sqrt{2}}} = \frac{\left(1 + 10\omega^2\right)}{-\omega^2 \left[\left(1 - 2\omega^2\right)^2 + 9\omega^2\right]}\Bigg|_{\omega = \frac{1}{2\sqrt{2}}} = -10.67$$

Using the information just above, the map of path *bc* is shown in Fig. P6.2(*b*).

(*ii*) **Map of path *cde***

Put

$$s = \lim_{R \to \infty} Re^{j\theta}$$

so that

$$G_1(s) = \lim_{R \to \infty} \frac{\left(4\,Re^{j\theta} + 1\right)}{\left(Re^{j\theta}\right)^2 \left(Re^{j\theta} + 1\right)\left(2\,Re^{j\theta} + 1\right)} = 0$$

and mapped points *c*, *d*, *e* lie at origin of $G_1(j\omega)$ plane.

(*iii*) **Map of path *ef***: This is mirror image of map of path *bc* as shown by dashed line in Fig. P 6.2(*b*).

(*iv*) **Map of path *fab***

Put

$$s = \lim_{\rho \to 0} \rho e^{j\theta}$$

so that

$$G_1(s) = \lim_{\rho \to 0} \frac{(4\rho e^{j\theta} + 1)}{\left(\rho e^{j\theta}\right)^2 \left(\rho e^{j\theta} + 1\right)\left(2\rho e^{j\theta} + 1\right)}$$

$$\cong \lim_{\rho \to 0} \left[\frac{1}{\rho^2} e^{-j2\theta}\right] = \infty \angle -2\theta$$

The Nyquist plot corresponding to map of path *fab* becomes semicircles of ∞ radius moving through an angle of + 180° to – 180° as θ varies from – 90° to + 90° along path *fab* in *s*-plane. As segment *fab* of RHP boundary undergoes rotation by 180° in CCW direction in *s* plane, $G_1(s)$ undergoes rotation by double the angle (360°) in CW direction with ∞ radius as shown in Fig. P6.2(*b*).

Interpreting stability: $G_1(s)$ contains no RHP pole; P = 0. There are two CW encirclements of point (– 1 + *j*0); N = 2. From principle of argument N = Z – P,

$$\text{number of RHP roots} = Z = N + P = 2.$$

So, closed loop system is **unstable** with two **RHP roots**.

(*b*) $G_2(s)$ has four open loop poles at origin. The RHP boundary in *s*-plane is chosen as shown in Fig. P6.2(*c*). The poles of $G_2(s)$ are also depicted and a semicircle of very small radius ($\rho \to 0$) bypasses the poles at origin. $G_2(s)$ has no RHP poles; P = 0.

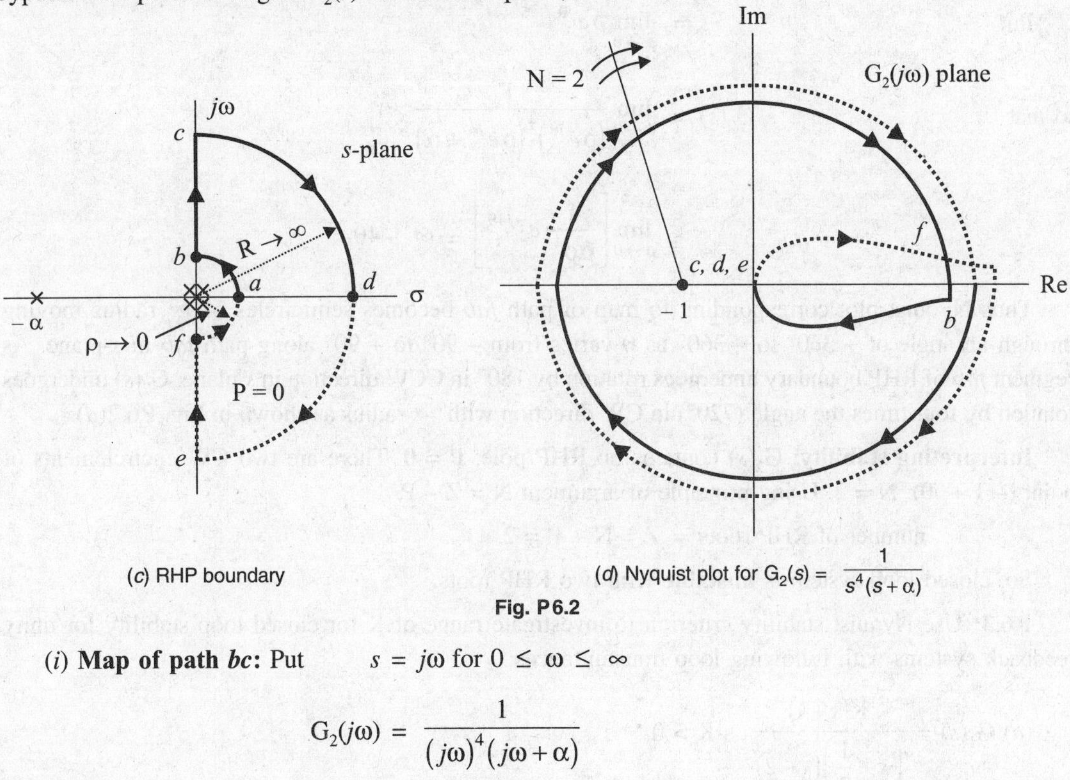

(*c*) RHP boundary (*d*) Nyquist plot for $G_2(s) = \dfrac{1}{s^4(s+\alpha)}$

Fig. P6.2

(*i*) **Map of path *bc*:** Put $s = j\omega$ for $0 \le \omega \le \infty$

$$G_2(j\omega) = \frac{1}{(j\omega)^4(j\omega+\alpha)}$$

$$\lim_{\omega \to 0} G_2(j\omega) \cong \lim_{\omega \to 0} \frac{1}{(j\omega)^4} = \infty \angle -360° \quad \text{(map of point } b)$$

$$\lim_{\omega \to \infty} G_2(j\omega) \cong \lim_{\omega \to \infty} \frac{1}{(j\omega)^5} = 0 \angle -450° = 0 \angle -90° \quad \text{(map of point } c)$$

Note that there is no possibility of any intermediate intersection on real axis. The two mapped points are plotted and interconnected by an arrow as shown in Fig. P6.2(*d*).

(*ii*) **Map of path *cde***

Put $$s = \lim_{R \to \infty} Re^{j\theta}$$

so that $$G_2(s) = \lim_{R \to \infty} \frac{1}{\left(Re^{j\theta}\right)^4 \left(Re^{j\theta} + \alpha\right)} = 0$$

and mapped points *c*, *d*, *e* lie at origin of $G_2(j\omega)$ plane.

(*iii*) **Map of path *ef*:** This is mirror image of map of path *bc* as shown by dashed line in Fig. P6.2(*d*).

(*iv*) **Map of path *fab***

Put $$s = \lim_{\rho \to 0} \rho e^{j\theta}$$

so that $$G_2(s) = \lim_{\rho \to 0} \frac{1}{\left(\rho e^{j\theta}\right)^4 \left(\rho e^{j\theta} + \alpha\right)}$$

$$\cong \lim_{\rho \to 0} \left[\frac{1}{\alpha \rho^4} e^{-j4\theta} \right] = \infty \ \angle -4\theta$$

Thus Nyquist plot corresponding to map of path *fab* becomes semicircles of ∞ radius moving through an angle of + 360° to – 360° as θ varies from – 90° to + 90° along path *fab* in *s*-plane. As segment *fab* of RHP boundary undergoes rotation by 180° in CCW direction in *s* plane, $G_2(s)$ undergoes rotation by four times the angle (720°) in CW direction with ∞ radius as shown in Fig. P6.2(*d*).

Interpreting stability: $G_2(s)$ contains no RHP pole; P = 0. There are two CW encirclements of point (– 1 + *j*0); N = 2. Using principle of argument N = Z – P,

number of RHP roots = Z = N + P = 2.

So, closed loop system is **unstable** with two RHP roots.

P 6.3: Use Nyquist stability criterion to investigate range of K for closed loop stability for unity feedback systems with following loop transmittances.

(*a*) $G_1(s) = \dfrac{K(s+1)}{\left(s+\frac{1}{2}\right)(s-2)}$; K > 0

(*b*) $G_2(s) = \dfrac{K}{s(\tau s - 1)}$; τ > 0

(*c*) $G_3(s) = \dfrac{K}{s(s+1)(s^2+2s+2)}$

Solution: (*a*) Note that $G_1(s)$ has no IA pole, the RHP boundary in *s* plane in chosen as shown in Fig. P 6.3(*a*). The zeros and poles of $G_1(s)$ are also depicted. The RHP boundary encloses one pole; P = 1.

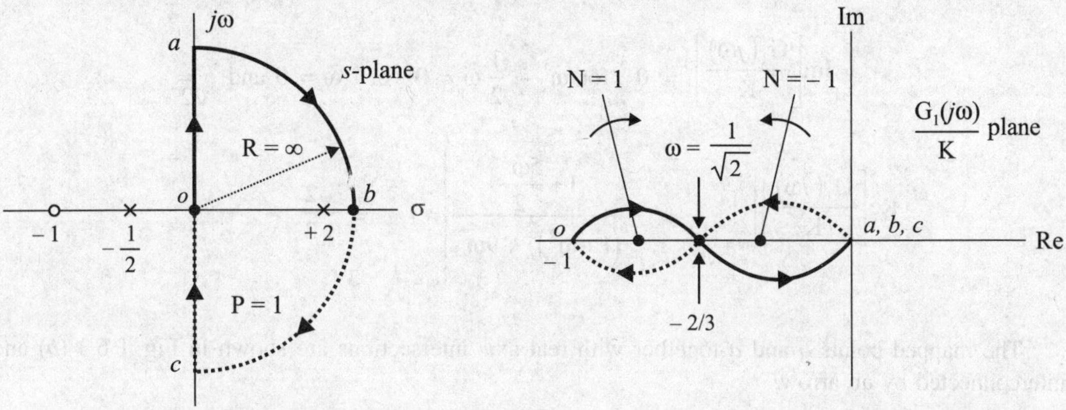

(*a*) RHP boundary (*b*) Nyquist plot for $G_1(s) = \dfrac{K(s+1)}{\left(s+\frac{1}{2}\right)(s-2)}$

Fig. P6.3

The mapping of RHP boundary *oabco* into $\dfrac{G_1(j\omega)}{K}$ plane is demonstrated in the steps as follows:

(*i*) **Map of path *oa*:** Put $s = j\omega$ for $0 \le \omega \le \infty$

$$G_1(j\omega) = \frac{K(j\omega+1)}{\left(j\omega+\frac{1}{2}\right)(j\omega-2)}$$

$$\lim_{\omega\to 0}\frac{G_1(j\omega)}{K} = \lim_{\omega\to 0}\frac{(j\omega+1)}{\left(j\omega+\frac{1}{2}\right)(j\omega-2)} = 1\ -180^\circ \quad \text{(map of point } o\text{)}$$

$$\lim_{\omega\to\infty}\frac{G_1(j\omega)}{K} = \lim_{\omega\to\infty}\frac{(j\omega)}{(j\omega)\cdot(j\omega)} = 0\ -90^\circ \quad \text{(map of point } a\text{)}$$

To investigate any real axis intersections, transform $\dfrac{G_1(j\omega)}{K}$ in following form.

$$\frac{G_1(j\omega)}{K} = -\frac{j\omega+1}{(1+\omega^2)+j\frac{3}{2}\omega}$$

$$= -\left[\frac{(j\omega+1)\left[(1+\omega^2)-j\frac{3}{2}\omega\right]}{\left[(1+\omega^2)+j\frac{3}{2}\omega\right]\left[(1+\omega^2)-j\frac{3}{2}\omega\right]}\right]$$

$$= -\left[\frac{1+\frac{5}{2}\omega^2}{(1+\omega^2)^2+9\omega^2}+j\frac{\omega^3-\frac{1}{2}\omega}{(1+\omega^2)^2+9\omega^2}\right]$$

$$\text{Im}\left[\frac{G_1(j\omega)}{K}\right] = 0 \quad \text{for } \omega^3 - \frac{1}{2}\omega = 0 \quad \text{or} \quad \omega = 0 \text{ and } \frac{1}{\sqrt{2}}$$

$$\text{Re}\left[\frac{G_1(j\omega)}{K}\right]\Bigg|_{\omega=\frac{1}{\sqrt{2}}} = -\frac{1+\dfrac{5\omega^2}{2}}{(1+\omega^2)^2+9\omega^2}\Bigg|_{\omega=\frac{1}{\sqrt{2}}} = -\frac{2}{3}$$

The mapped points *o* and *a* together with real axis intersections are shown in Fig. P6.3 (*b*) and interconnected by an arrow.

(*ii*) **Map of path *abc*:**

Put
$$s = \lim_{R\to\infty} Re^{j\theta}$$

so that
$$\frac{G_1(s)}{K} = \lim_{R\to\infty} \frac{(Re^{j\theta}+1)}{\left(Re^{j\theta}+\frac{1}{2}\right)(Re^{j\theta}-2)} = 0$$

and mapped points *a*, *b*, *c* lie at origin of $\frac{G_1(j\omega)}{K}$ plane.

(*iii*) **Map of path *co*:** This is mirror image of map of path *oa* as shown by dashed line in Fig. P6.3 (*b*).

Interpreting stability: $G_1(s)$ has one RHP pole; P = 1. From principle of argument N = Z – P, Z must be zero for closed loop stability. So N = Z – P = – 1 meaning there must be one CCW encirclements of point (– 1 + *j*0). For

$$-\frac{2K}{3} < -1 \quad \text{or} \quad K > \frac{3}{2}$$

the point (– 1 + *j*0) lies between origin of $G_1(j\omega)$ plane and real axis intersection at $\omega = \frac{1}{\sqrt{2}}$ of Nyquist plot; a requirement for the closed loop system to be stable. For $-\frac{2K}{3} > -1$ or $K < \frac{3}{2}$ the point (– 1 + *j*0) will lie to the left of real axis intersection at $\omega = \frac{1}{\sqrt{2}}$; N = 1 and Z = N + P = 2.

So, the closed loop system will be unstable with two RHP roots for $K < \frac{3}{2}$ and stable for $K > \frac{3}{2}$.

(*b*) $G_2(s)$ has one RHP pole and one pole at origin, the RHP boundary in *s*-plane is chosen as shown in Fig. P6.3 (*c*). The poles of $G_2(s)$ are also depicted and a semicircle of very small radius ($\rho \to 0$) bypasses the pole at origin.

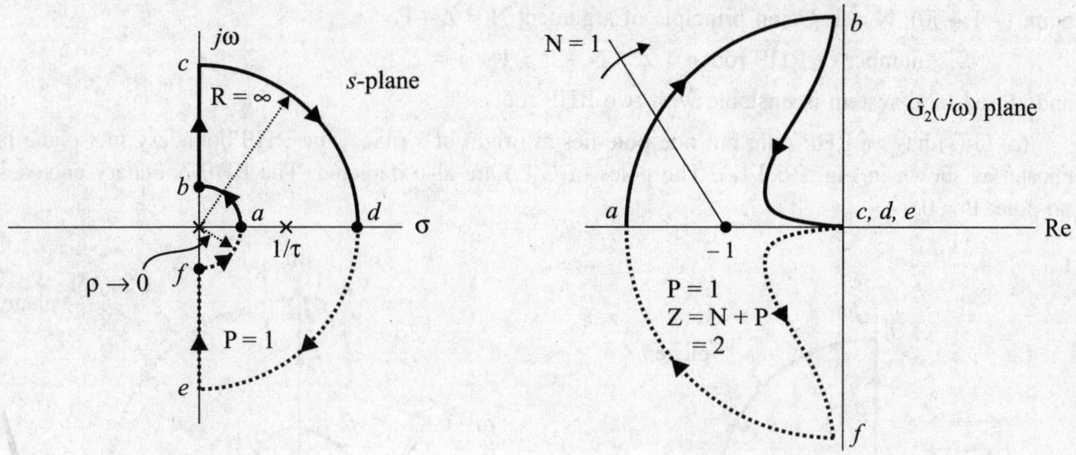

(c) RHP boundary

(d) Nyquist plot for $G_2(s) = \dfrac{K}{s(\tau s - 1)}$

Fig. P6.3

(*i*) **Map of path *bc*:** Put $s = j\omega$ for $0 \le \omega \le \infty$

$$G_2(j\omega) = \frac{K}{j\omega(j\omega\tau - 1)}$$

$$\lim_{\omega \to 0} G_2(j\omega) = \lim_{\omega \to 0} \frac{K}{j\omega(j\omega\tau - 1)} \cong \lim_{\omega \to 0} \frac{-K}{j\omega} = \infty \angle -270^\circ \quad \text{(map of point } b\text{)}$$

$$\lim_{\omega \to \infty} G_2(j\omega) = \lim_{\omega \to \infty} \frac{K}{j\omega(j\omega\tau - 1)} \cong \lim_{\omega \to \infty} \frac{K}{j\omega(j\omega\tau)} = 0 \angle -180^\circ \quad \text{(map of point } c\text{)}$$

These two mapped points are plotted and interconnected by an arrow as shown in Fig. P6.3(*d*).

(*ii*) **Map of path *cde*:** Put $\quad s = \lim_{R \to \infty} Re^{j\theta}$

so that
$$G_2(s) = \lim_{R \to \infty} \frac{K}{(Re^{j\theta})(\tau Re^{j\theta} - 1)} = 0$$

and mapped points *c*, *d*, *e* lie at origin of $G_2(j\omega)$ plane.

(*iii*) **Map of path *ef*:** This is mirror image of map of path *bc* as shown by dashed line in Fig. P 6.3(*d*).

(*iv*) **Map of path *fab*:** Put $\quad s = \lim_{\rho \to 0} \rho e^{j\theta}$

so that
$$G_2(s) = \lim_{\rho \to 0} \frac{1}{\rho e^{j\theta}(\rho\tau e^{j\theta} - 1)} \cong \lim_{\rho \to 0} \frac{-K}{\rho} e^{-j\theta} = -\infty \angle -\theta$$

Thus Nyquist plot corresponding to map of path *fab* becomes semicircle of ∞ radius moving through an angle of – 90° to + 90° as θ varies from – 90° to + 90° along path *fab* in *s*-plane as shown in Fig. P6.3(*d*).

Interpreting stability: $G_2(s)$ contains one RHP pole; P = 1. There is one CW encirclements of point $(-1 + j0)$; N = 1. Using principle of argument N = Z – P,

$$\text{number of RHP roots} = Z = N + P = 1 + 1 = 2$$

and closed loop system is unstable with two RHP roots.

(*c*) $G_3(s)$ has no RHP pole but one pole lies at origin of *s* plane; the RHP boundary in *s*-plane is chosen as shown in Fig. P6.3 (*e*). The poles of $G_3(s)$ are also depicted. The RHP boundary encloses no pole; P = 0.

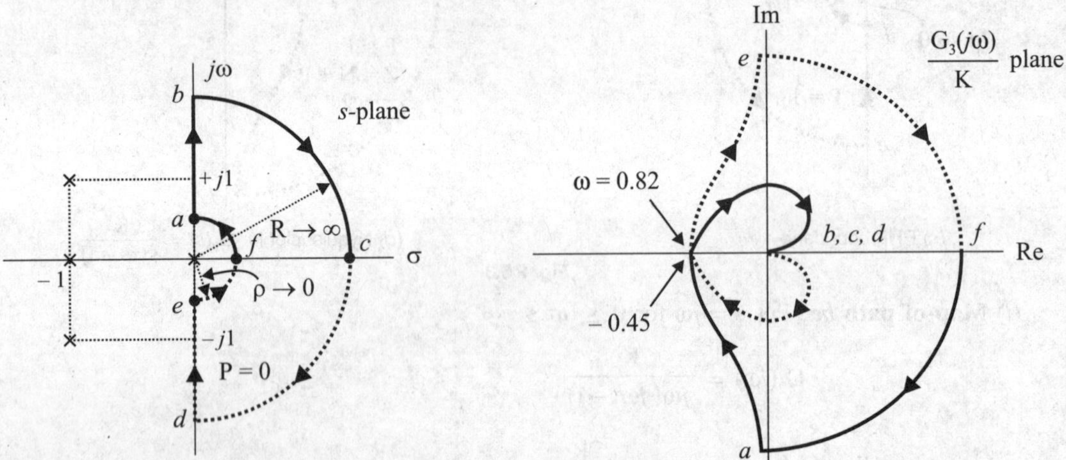

(*e*) RHP boundary (*f*) Nyquist plot for $G_3(s) = \dfrac{K}{s(s+1)(s^2+2s+2)}$

Fig. P6.3

The mapping of RHP boundary *abcdefa* into G(*j*ω)/K plane is demonstrated in the steps as follows:

(*i*) **Map of path *ab*:** Put $s = j\omega$ for $0 \le \omega \le \infty$

$$\frac{G_3(j\omega)}{K} = \frac{1}{j\omega(j\omega+1)\left[(j\omega)^2 + j2\omega + 2\right]}$$

$$\lim_{\omega\to 0}\frac{G_3(j\omega)}{K} = \lim_{\omega\to 0}\frac{1}{j2\omega} = \infty\angle{-90^\circ} \quad \text{(map of point } a)$$

$$\lim_{\omega\to \infty}\frac{G_3(j\omega)}{K} = \lim_{\omega\to \infty}\frac{1}{(j\omega)(j\omega)(j\omega)^2} = 0\angle{-360^\circ} \quad \text{(map of point } b)$$

To investigate any real axis intersections, transform $\dfrac{G_3(j\omega)}{K}$ in the form as follows:

$$\frac{G_3(j\omega)}{K} = \frac{1}{j\omega(j\omega+1)\left[(j\omega)^2 + j2\omega + 2\right]}$$

$$= -\frac{(\omega + j1)\left[(2-\omega^2) - j2\omega\right]}{\omega(\omega - j1)(\omega + j1)\left[(2-\omega^2) + j2\omega\right]\left[(2-\omega^2) - j2\omega\right]}$$

$$= -\left[\frac{(4-\omega^2)}{(1+\omega^2)\left[(2-\omega^2)^2 + 4\omega^2\right]} + j\frac{2-3\omega^2}{\omega(1+\omega^2)\left[(2-\omega^2)^2 + 4\omega^2\right]}\right]$$

$$\text{Im}\left[\frac{G_3(j\omega)}{K}\right] = 0 \quad \text{for } \omega = \sqrt{\frac{2}{3}} = 0.82$$

$$\text{Re}\left[\frac{G_3(j\omega)}{K}\right]\Bigg|_{\omega = \sqrt{\frac{2}{3}}} = -\frac{9}{20} = -0.45$$

The mapped points *a* and *b* together with real axis intersections are shown in Fig. P6.3(f) and interconnected by an arrow.

(*ii*) **Map of path *bcd*:** Put $s = \lim_{R \to \infty} Re^{j\theta}$

so that
$$\frac{G_3(s)}{K} = \lim_{R \to \infty} \frac{K}{Re^{j\theta}(Re^{j\theta}+1)\left[(Re^{j\theta})^2 + 2Re^{j\theta} + 2\right]} = 0$$

and mapped points *b*, *c*, *d* lie at origin of $G_3(j\omega)/K$ plane.

(*iii*) **Map of path *de*:** This is mirror image of map of path *ab* as shown by dashed line in Fig. P6.3 (f).

(*iv*) **Map of path *efa*:** Put $s = \lim_{\rho \to 0} \rho e^{j\theta}$

so that
$$\frac{G_3(s)}{K} = \lim_{\rho \to 0} \frac{1}{\rho e^{j\theta}(\rho e^{j\theta}+1)\left[(\rho e^{j\theta})^2 + 2\rho e^{j\theta} + 2\right]}$$

$$\cong \lim_{\rho \to 0} \frac{1}{2\rho} e^{-j\theta} = \infty \angle -\theta$$

and Nyquist plot corresponding to map of path *efa* is a semicircle of ∞ radius moving through an angle of + 90° to – 90° as θ varies from – 90° to + 90° along path *efa* in *s*-plane as depicted in Fig. P6.3(f).

Interpreting stability: For – 0.45 K > – 1 or 0.45 K < 1 or K < 2.22, the Nyquist plot does not encircle point (– 1 + *j*0); N = 0. $G_3(s)$ does not have any RHP pole; P = 0. From principle of argument N = Z – P, number of RHP roots = Z = N + P = 0. So, closed loop system is stable for K < 2.22.

For – 0.45 K < – 1 or K > 2.22, there are two CW encirclements of point (– 1 + *j*0); N = 2, number of RHP roots = Z = N + P = 2. So, closed loop system will be **unstable** with two RHP roots for K > 2.22.

P 6.4: Sketch Nyquist plot for each of the following GH functions. Determine whether or not each system is stable using Nyquist plot.

$$(a)\ G_1(s)\ H_1(s) = \frac{(s^2+5)}{s^2(s^2+4s+8)}$$

$$(b)\ G_2(s)\ H_2(s) = \frac{1}{(s+2)(s^2+4)}$$

Solution: (*a*) $G_1(s)\ H_1(s)$ has two poles at origin. The RHP boundary in *s*-plane is chosen as shown in Fig. P 6.4 (*a*). $G_1(s)\ H_1(s)$ has no RHP pole; P = 0.

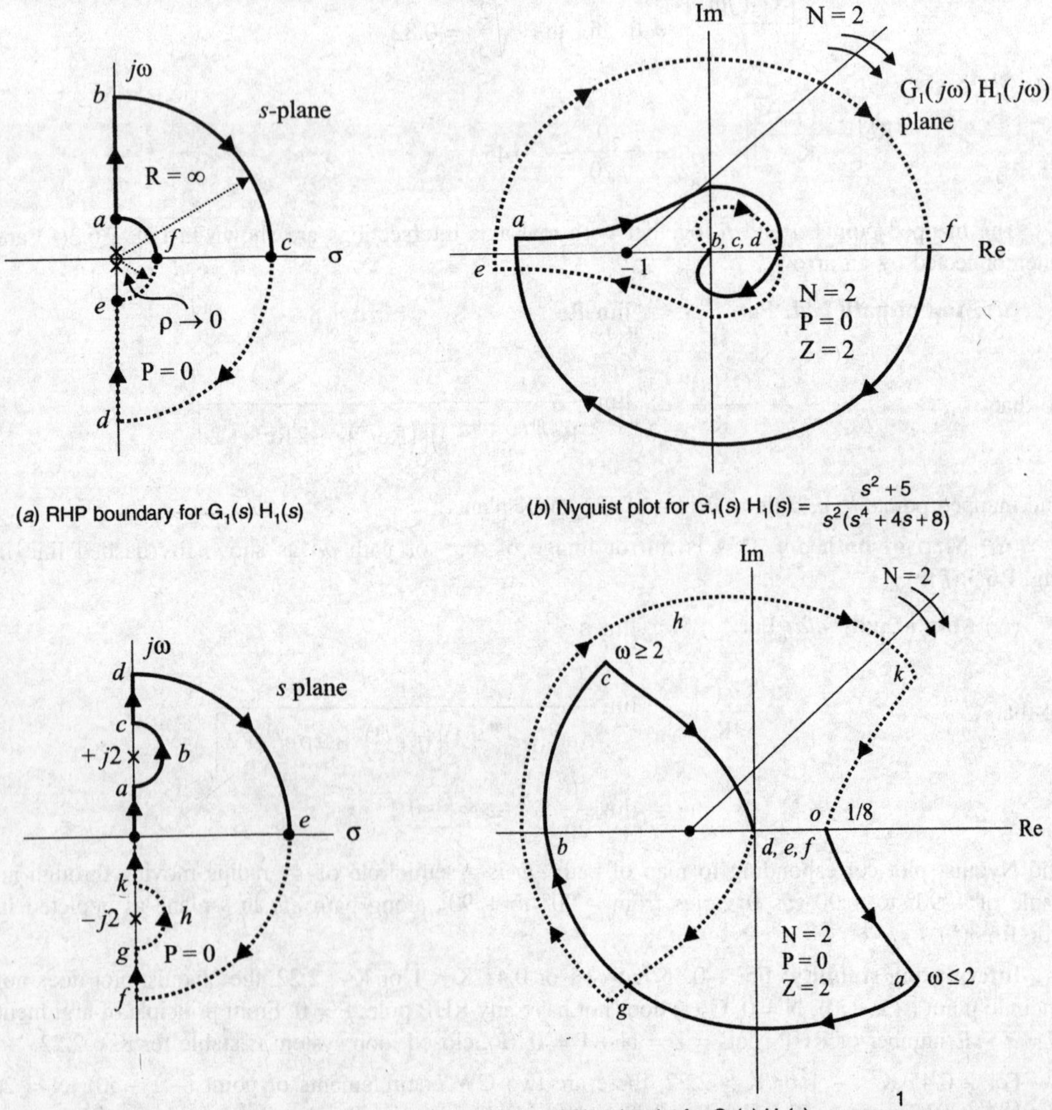

(*a*) RHP boundary for $G_1(s)\ H_1(s)$

(*b*) Nyquist plot for $G_1(s)\ H_1(s) = \dfrac{s^2+5}{s^2(s^4+4s+8)}$

(*c*) RHP boundary for $G_2(s)\ H_2(s)$

(*d*) Nyquist plot for $G_2(s)\ H_2(s) = \dfrac{1}{(s+2)(s^2+4)}$

Fig. P6.4

Mapping of RHP boundary of Fig. P6.4(*a*) is demonstrated in following steps.

(*i*) Map of path *ab*: Put $s = j\omega$ for $0 \le \omega \le \infty$

$$G_1(j\omega)\, H_1(j\omega) = \frac{(j\omega)^2 + 5}{(j\omega)^2\left[(j\omega)^2 + j4\omega + 8\right]}$$

$$\lim_{\omega \to 0} G_1(j\omega)\, H_1(j\omega) = \infty \angle -180^\circ \quad \text{(map of point } a)$$

$$\lim_{\omega \to \infty} G_1(j\omega)\, H_1(j\omega) = 0 \angle -180^\circ \quad \text{(map of point } b)$$

The two mapped points are plotted and interconnected by an arrow as shown in Fig. P6.4(*b*). The real and imaginary axis intersections are left for the reader to investigate.

(*ii*) Map of path *bcd*: Put $s = \lim_{R \to \infty} Re^{j\theta}$

As discussed on multiple occasions before the mapped points *b*, *c*, *d* will lie at origin of $G_1(s)$ $H_1(s)$ plane.

(*iii*) Map of path *de*: This is mirror image of map of path ab as shown by dashed line in Fig. P6.4(*b*).

(*iv*) Map of path *efa*: Put $s = \lim_{\rho \to 0} \rho e^{j\theta}$

so that $$G_1(s)\, H_1(s) = \infty \angle -2\theta$$

and map represents two semicircles of ∞ radius moving through an angle of + 180° to – 180° (CW) as θ varies from – 90° to + 90° (CCW) in *s* plane as shown in Fig. P6.4(*b*).

Interpreting stability: $G_1(s)$ $H_1(s)$ contains no RHP pole; P = 0. There are two CW encirclements of point (– 1 + *j*0); N = 2. Using principle of argument N = Z – P, number of RHP roots = Z = N + P = 2. So, the system is **unstable with two RHP roots**.

(*b*) $G_2(s)$ $H_2(s)$ has two IA poles at $s = \pm j2$. The RHP boundary in *s* plane is chosen as shown in Fig. P6.4(*c*). Two semicircles *a b c* and *g h k* of very small radius (ρ → 0) bypass IA poles at $s = \pm j2$.

Mapping of RHP boundary is demonstrated in the steps as follows:

(*i*) **Map of path *oa*:** Put $s = j\omega$ for $0 \le \omega \le 2$

so that $$G_2(j\omega)\, H_2(j\omega) = \frac{1}{(j\omega + 2)(4 - \omega^2)}$$

$$\lim_{\omega \to 0} G_2(j\omega)\, H_2(j\omega) = \frac{1}{8} \angle 0^\circ \quad \text{(map of point } o)$$

$$\lim_{\omega \to 2} G_2(j\omega)\, H_2(j\omega) = \infty \angle -45^\circ \quad \text{(map of point } a)$$

Note that the term (*j*ω +2) in denominator contributes – 45° angle and the term $(4 - \omega^2)$ in denominator contributes ∞ magnitude. These two mapped points are plotted and interconnected by an arrow as shown in Fig. P6.4(*d*).

(*ii*) **Map of path *abc*:** Put $s = \lim_{\rho \to 0} \left(j2 + \rho e^{j\theta} \right)$; $-90° \le \theta \le +90°$

so that
$$G_2(s)\, H_2(s) = \lim_{\rho \to 0} \frac{1}{\left(j2 + \rho e^{j\theta} + 2\right)\left[\left(j2 + \rho e^{j\theta}\right)^2 + 4\right]}$$

$$= \lim_{\rho \to 0} \frac{1}{\rho e^{j\theta}\left(\rho e^{j\theta} + j2 + 2\right)\left(\rho e^{j\theta} + j4\right)}$$

$$= \infty \lfloor -\theta - 45° - 90° = \infty \lfloor -\theta - 135°$$

and map represents a semicircle moving through an angle of – 45° to – 225° as θ varies from – 90° to + 90° along *abc* in *s*-plane as shown in Fig. P 6.4(*d*). Note that denominator term $\rho e^{j\theta} + j2 + 2$ contributes – 45° angle for $\rho \to 0$ and denominator term $\rho e^{j\theta} + j4$ contributes – 90° angle for $\rho \to 0$.

(*iii*) **Map of path *cd*:** Put $s = j\omega$; $2 \le \omega \le \infty$

For $\omega \ge 2$, $|G_2(j\omega)\, H_2(j\omega)| = \infty$ and $\lfloor G_2(j\omega)\, H_2(j\omega) = -225°$ (map of point *c*)

and
$$\lim_{\omega \to \infty} G_2(j\omega)\, H_2(j\omega) = 0 \lfloor -270° \quad \text{(map of point } d\text{)}$$

The mapped points *c* and *d* are interconnected by an arrow as shown in Fig. P6.4(*d*).

(*iv*) **Map of path *def***

The path *def* maps at origin of $G_2(j\omega)\, H_2(j\omega)$ plane as discussed on multiple occasions before.

(*v*) **Map of path *fg*:**

This is mirror image of map of path *cd* as shown by dashed line in Fig. P6.4(*d*).

(*vi*) **Map of path *ghk*:** Put $s = \lim_{\rho \to 0} \left(-j2 + \rho e^{j\theta} \right)$; $-90° \le \theta \le +90°$

so that
$$G_2(s)\, H_2(s) = \lim_{\rho \to 0} \frac{1}{\left(-j2 + \rho e^{j\theta} + 2\right)\left[\left(-j2 + \rho e^{j\theta}\right)^2 + 4\right]}$$

$$= \lim_{\rho \to 0} \frac{1}{\rho e^{j\theta}\left(2 - j2 + \rho e^{j\theta}\right)\left(\rho e^{j\theta} - j4\right)}$$

$$= \infty \lfloor -\theta - \tan^{-1}(-1) - \tan^{-1}(-\infty) = \infty \lfloor -\theta - 225°$$

and map represents a semicircle moving through an angle of – 135° to – 315° as θ varies from – 90° to + 90° along *ghk* in *s* plane as shown in Fig. P6.4 (*d*).

(*vii*) **Map of path *ko***

This is mirror image of map of path oa as shown in Fig. P6.4 (*d*).

Interpreting stability: $G_2(s)\, H_2(s)$ contains no RHP pole; P = 0. There are two CW encirclements of point (– 1 + *j*0) ; N = 2. using principle of argument N = Z – P, number of RHP roots = Z = N + P = 1 + 1 = 2. So closed loop system is **unstable with two RHP roots.**

P 6.5: Use Nyquist criterion to determine range of K for closed loop stability of the open loop function with unity feedback given by:

$$G(s) = \frac{Ke^{-0.8s}}{(s+1)}$$

Solution: $$G(s)\big|_{s=j\omega} = G(j\omega) = \frac{Ke^{-j0.8\omega}}{(j\omega+1)}$$

$$\lim_{\omega \to 0} G(j\omega) = K\angle 0°$$

$$\lim_{\omega \to \infty} G(j\omega) = 0; \text{ since magnitude is zero, angle information is irrelevant}$$

To investigate any real axis intersections, transform $G(j\omega)$ in the form as follows.

$$G(j\omega) = \frac{K(1-j\omega)(\cos 0.8\omega - j\sin 0.8\omega)}{(1+j\omega)(1-j\omega)}$$

$$\frac{G(j\omega)}{K} = \frac{\cos 0.8\omega - \omega\sin 0.8\omega}{1+\omega^2} - j\frac{\sin 0.8\omega + \omega\cos 0.8\omega}{1+\omega^2}$$

Equating $\text{Im}\left[\frac{G(j\omega)}{K}\right]$ to zero

$$\sin 0.8\,\omega + \omega \cos 0.8\,\omega = 0$$

or $$\tan 0.8\,\omega = -\omega$$

or $$0.8\,\omega = \tan^{-1}(-\omega)$$

where $$\tan^{-1}(-\omega) = -\omega - (-\omega)^3/3;$$

only first two terms are retained from series of $\tan^{-1}(-\omega)$

so $$0.8\,\omega = -\omega + \omega^3/3$$

or $$\omega = 0,\ 2.32$$

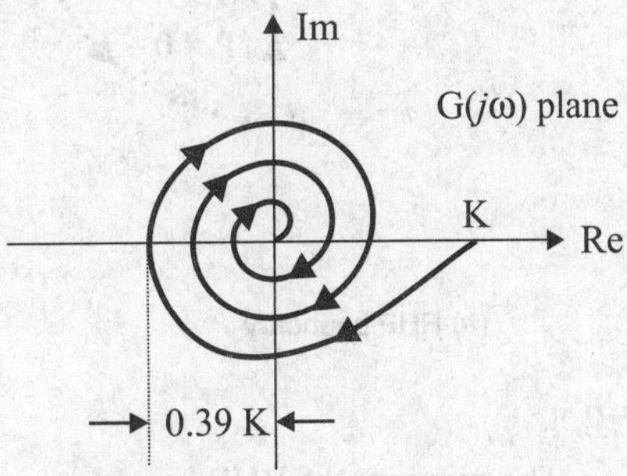

Fig. P6.5: Polar plot for $G(s) = \frac{Ke^{-0.8s}}{(s+1)}$

$$\text{Re}\left[\frac{G(j\omega)}{K}\right]\Bigg|_{\omega=2.32} = \frac{\cos(0.8\times 2.32) - 2.32\sin(0.8\times 2.32)}{1+2.32^2} = -0.39$$

and $$\text{Re}\,[G(j\omega)]\,|_{\omega=2.32} = -0.39\text{ K}$$

Since $\tan^{-1}(-\omega)$ series consists of infinite terms, there will be infinite points of intersections on real axis. For the smallest value of ω just greater than zero, the real axis intersection is – 0.39 K. For higher values of ω, the polar plot spirals around the origin as shown in Fig. P6.5. According to enclosure property of polar plot, closed loop stability is guaranteed for 0.39 K < 1 or K < 2.56.

P 6.6: Sketch Nyquist plot for unity feedback system with loop transmittance

$$G(s) = \frac{K(s+10)^2}{s^3}; \quad K > 0$$

and investigate range of gain K for stability.

Solution: G(s) has three poles at origin, the RHP boundary in s plane is chosen as shown in Fig. P6.6(a).

G(s) has no RHP pole ; P = 0.

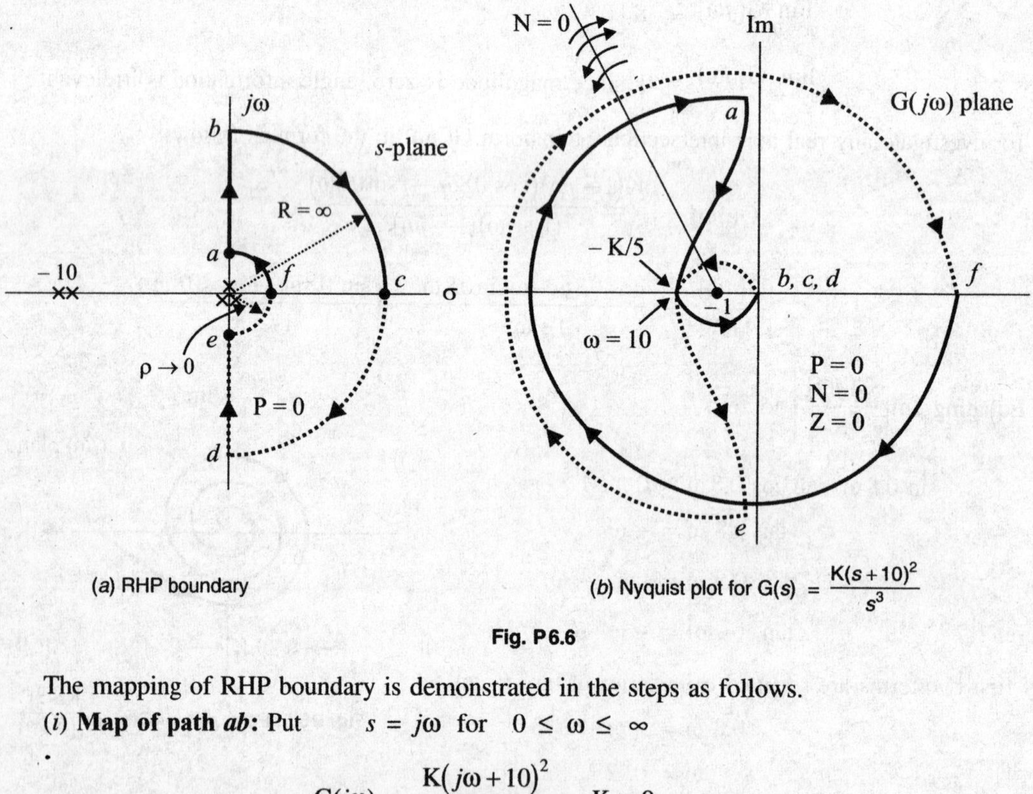

(a) RHP boundary (b) Nyquist plot for $G(s) = \frac{K(s+10)^2}{s^3}$

Fig. P6.6

The mapping of RHP boundary is demonstrated in the steps as follows.

(i) **Map of path *ab*:** Put $s = j\omega$ for $0 \le \omega \le \infty$

$$G(j\omega) = \frac{K(j\omega+10)^2}{(j\omega)^3}; \quad K > 0$$

$$\lim_{\omega \to 0} G(j\omega) = \infty\angle{-270^\circ} \quad \text{(map of point } a\text{)}$$

$$\lim_{\omega \to \infty} G(j\omega) = \lim_{\omega \to \infty} \frac{K(j\omega)^2}{(j\omega)^3} = 0\angle{-90^\circ} \quad \text{(map of point } b\text{)}$$

To investigate any real axis intersections, G($j\omega$) is rationalised as follows:

$$\frac{G(j\omega)}{K} = \frac{(100-\omega^2)+j20\omega}{-j\omega^3} = -\frac{20}{\omega^2} + j\frac{100-\omega^2}{\omega^3}$$

$$\text{Im}\left[\frac{G(j\omega)}{K}\right] = 0 \quad \text{for } \omega = 10$$

and
$$\text{Re}\left[G(j\omega)\right]\Big|_{\omega=10} = -\frac{K}{5}$$

Using the information just above, the map of path *ab* is shown in Fig. P6.6(*b*).

(*ii*) **Map of path *bcd*:** Put $\quad s = \lim_{R\to\infty} Re^{j\theta}$,

the mapped points *b, c, d* will lie at origin of G(*j*ω) plane.

(*iii*) **Map of path *de*:** This is mirror image of map of path *ab* as shown by dashed line in Fig. P6.6(*b*).

(*iv*) **Map of path *efa*:** Put $\quad s = \lim_{\rho\to 0} \rho e^{j\theta}$

so that
$$G(s) = \lim_{\rho\to 0} \frac{\left(\rho e^{j\theta}+10\right)^2}{\left(\rho e^{j\theta}\right)^3}$$

$$= \infty\angle{-3\theta}; \quad -90° \le \theta \le +90°$$

and map of path *efa* represents three semicircles moving clockwise through an angle of + 270° to – 270° as θ varies from – 90° to + 90° (CCW) in *s* plane as shown in Fig. P6.6(*b*).

Interpreting stability: G(*s*) has no RHP pole; P = 0. From principle of argument, N = Z – P, Z must be zero for the system to be closed loop stable. Then the number of encirclement of point (– 1 + *j*0), N = 0 if

$$-\frac{K}{5} < -1 \quad \text{or} \quad K > 5$$

So, closed loop system is stable for K > 5.

P 6.7: Draw a Nyquist plot for a system with open loop transmittance

$$G(s)\,H(s) = \frac{s^2+4s+6}{s^2+5s+4}$$

Solution: G(*s*) H(*s*) contains neither RHP poles nor IA poles; the RHP boundary is chosen as shown in Fig. P6.7(*a*).

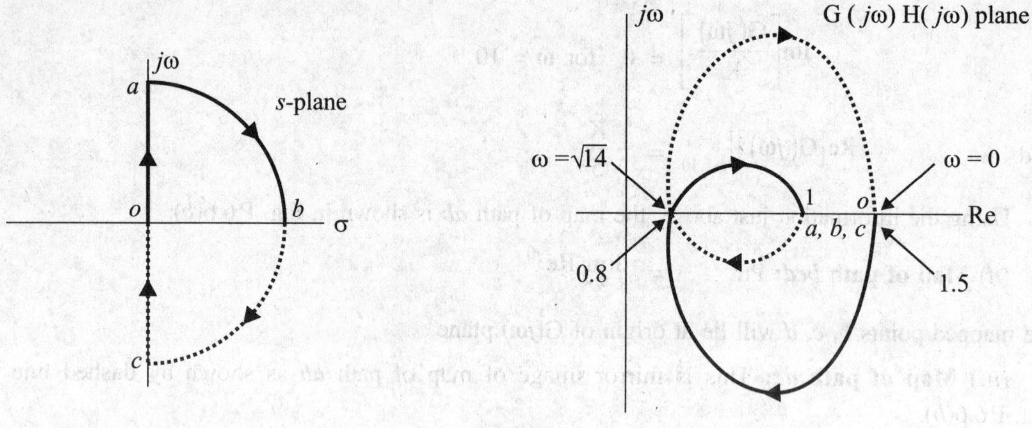

(a) RHP boundary (b) Nyquist plot for G(s) H(s) = $\frac{s^2+4s+6}{s^2+5s+4}$

Fig. P6.7

(*i*) **Map of path *oa*:** Put $s = j\omega \quad \text{for } 0 \le \omega \le \infty$

so that
$$G(j\omega)\,H(j\omega) = \frac{(j\omega)^2 + j4\omega + 6}{(j\omega)^2 + j5\omega + 4}$$

$$\lim_{\omega \to 0} G(j\omega)\,H(j\omega) = 1.5\angle 0^\circ \quad \text{(map of point } o)$$

$$\lim_{\omega \to \infty} G(j\omega)\,H(j\omega) = 1\angle 0^\circ \quad \text{(map of point } a)$$

To investigate any real axis intersection, express G(jω) H(jω) in the form as follows:

$$G(j\omega)\,H(j\omega) = \frac{\left[(6-\omega^2) + j4\omega\right]\left[(4-\omega^2) - j5\omega\right]}{\left[(4-\omega^2) + j5\omega\right]\left[(4-\omega^2) - j5\omega\right]}$$

$$= \frac{20\omega^2 + (6-\omega^2)(4-\omega^2)}{(4-\omega^2)^2 + 25\omega^2} + j\,\frac{4\omega(4-\omega^2) - 5\omega(6-\omega^2)}{(4-\omega^2)^2 + 25\omega^2}$$

$$\text{Im}\left[G(j\omega)H(j\omega)\right] = 0 \quad \text{for } \omega = 0 \text{ and } \omega = \sqrt{14}$$

and
$$\text{Re}\left[G(j\omega)H(j\omega)\right]\Big|_{\omega=\sqrt{14}} = 0.8$$

The two mapped points *o* and *a* together with real axis intersections are plotted and interconnected by an arrow as shown in Fig. P6.7(*b*).

(*ii*) **Map of path *abc*:** Put $\quad s = \lim\limits_{R \to \infty} Re^{j\theta}$

the mapped points *a*, *b*, *c*, will lie at the same point on real axis of G(*j*ω) plane as shown in Fig. P6.7(*b*).

(*iii*) **Map of path *co*:** This is mirror image of map of path *oa* as shown by dashed segment in Fig. P6.7(*b*).

Interpreting stability: The Nyquist plot does not encircle (– 1 + *j*0) point; N = 0. G(*s*) H(*s*) has no RHP pole: P = 0. From principle of argument N = Z – P, number of RHP roots Z = N + P = 0. So, closed loop system is **stable**.

P6.8: Sketch Bode plot for a feedback system with loop transmittance

$$G(s)\,H(s) = \frac{100(s+4)}{s(s+0.5)(s+10)}$$

and find (*a*) GM (*b*) PM (*c*) gain cross over frequency (*d*) phase cross over frequency. Comment on closed loop stability.

Solution: Sketching Bode plot is demonstrated in the steps as follows:

(*i*) Write the loop transmittance in time constant form

$$G(j\omega)\,H(j\omega) = \frac{80\left(1+\dfrac{j\omega}{4}\right)}{j\omega\left(1+\dfrac{j\omega}{0.5}\right)\left(1+\dfrac{j\omega}{10}\right)}$$

(*ii*) **Identify the corner frequencies**

$\omega_{c_1} = 0.5$ (a real axis pole)

$\omega_{c_2} = 4$ (a real axis zero)

$\omega_{c_3} = 10$ (a real axis pole)

(*iii*) **dB plot:** The initial segment is a straight line of slope – 20 dB/dec passing through the point B at ($\omega = 1$, dB = $20 \log K|_{K=80} = +38$) on account of a pole at origin. In order to draw this initial segment of dB plot, locate one more point A at ($\omega = 0.1$, dB = + 38 + 20 = + 58), one decade below point B as demonstrated in Fig. P6.8 (*a*).

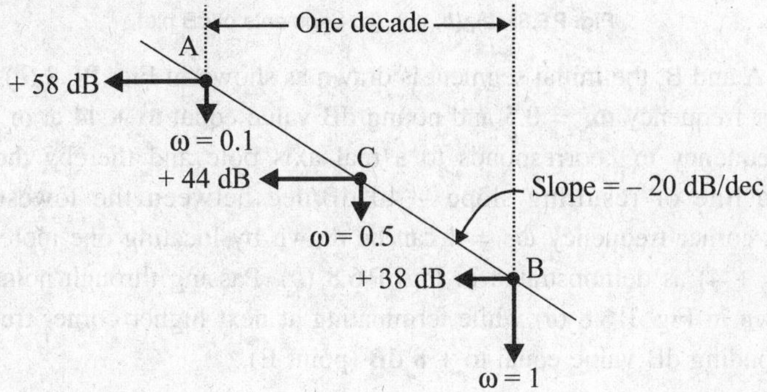

(*a*)

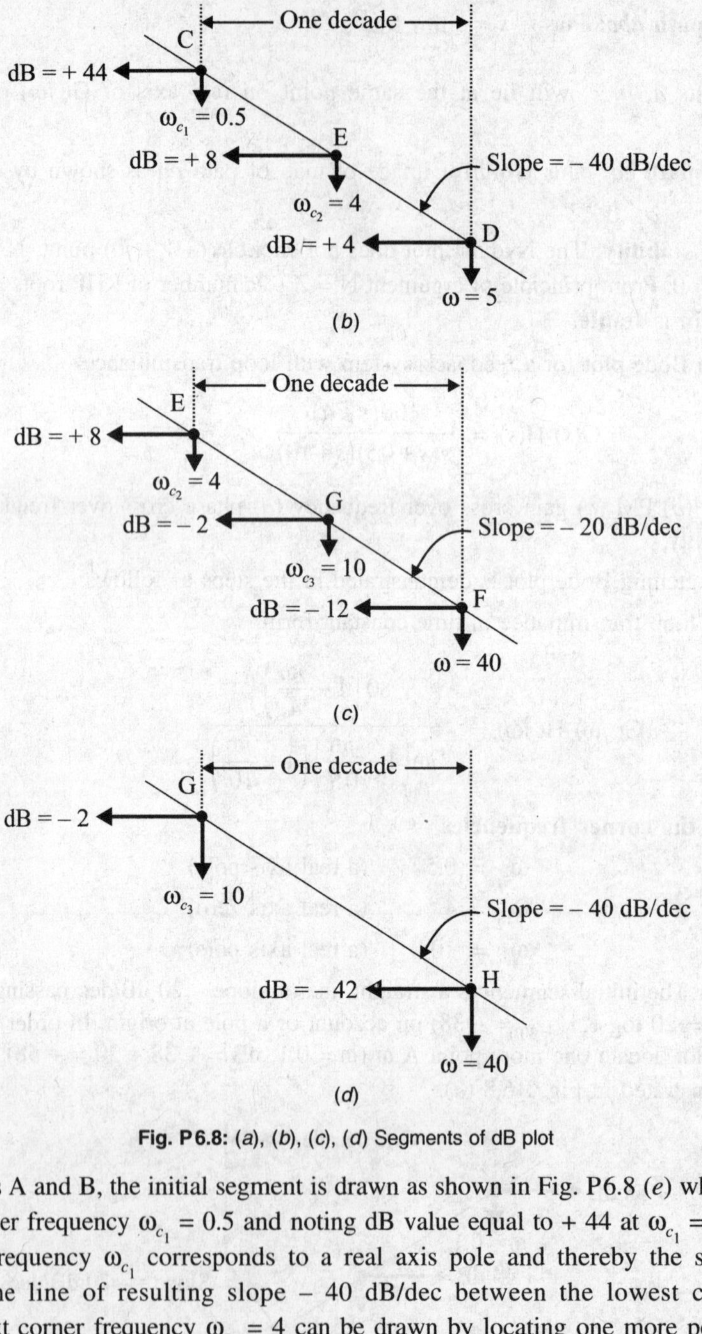

Fig. P6.8: (*a*), (*b*), (*c*), (*d*) Segments of dB plot

Joining points A and B, the initial segment is drawn as shown in Fig. P6.8 (*e*) while terminating it at the lowest corner frequency $\omega_{c_1} = 0.5$ and noting dB value equal to + 44 at $\omega_{c_1} = 0.5$ (point C).

The corner frequency ω_{c_1} corresponds to a real axis pole and thereby the slope changes by – 20 dB/dec. The line of resulting slope – 40 dB/dec between the lowest corner frequency $\omega_{c_1} = 0.5$ and next corner frequency $\omega_{c_2} = 4$ can be drawn by locating one more point D at ($\omega = 5$, dB = + 44 – 40 = + 4) as demonstrated in Fig. P6.8 (*b*). Passing through points C and D, the dB segment is as shown in Fig. P6.8 (*e*) while terminating at next higher corner frequency $\omega_{c_2} = 4$ and noting the corresponding dB value equal to + 8 dB (point E).

The corner frequency $\omega_{c_2} = 4$, corresponds to a real axis zero and thereby slope changes by + 20 dB/dec. The line of resulting slope – 20 dB/dec between $\omega_{c_2} = 4$ and $\omega_{c_3} = 10$, can be drawn by locating one more point F at (ω = 40, dB = + 8 – 20 = – 12) as demonstrated in Fig. P6.8(*c*). This segment is depicted in P6.8 (*e*) while terminating at next higher corner frequency $\omega_{c_3} = 10$ (point G). The dB value at point G is noted to be equal to – 2.

Since the corner frequency $\omega_{c_3} = 10$ corresponds to a real axis pole, the slope again changes by – 20 dB/dec. The straight line segment of resulting slope – 40 dB/dec beyond $\omega_{c3} = 10$ can be drawn by locating one more point H at (ω = 100, dB = – 2 – 40 = – 42). This segment is depicted in Fig. P 6.8 (*e*) without any termination as there is no corner frequency beyond ω_{c_3}. The entire asymptotic dB plot is shown in Fig. P6.8 (*e*).

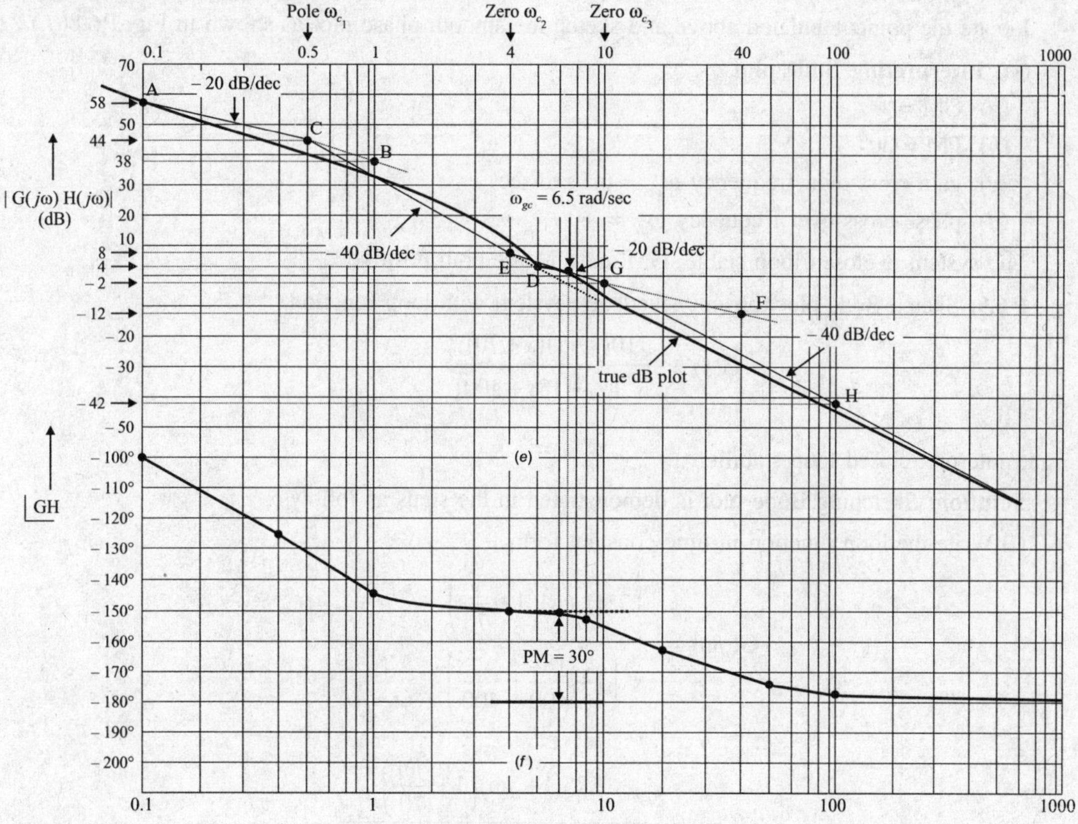

Fig. P6.8: (*e*) Complete dB plot (*f*) Phase plot

(*iv*) The table of correction to straight line approximated dB plot is constructed as follows:

	$\frac{\omega_{c_1}}{2}$ ↓	ω_{c_1} ↓	$2\,\omega_{c_1}$ ↓	$\frac{1}{2}\omega_{c_2}$ ↓	ω_{c_2} ↓	$\frac{1}{2}\omega_{c_3}$ ↓	$2\,\omega_{c_2}$ ↓	ω_{c_3} ↓	$2\,\omega_{c_3}$ ↓
ω	0.25	0.5	1	2	4	5	8	10	20
Error	– 1	– 3	– 1	+ 1	+ 3	– 1	+ 1	– 3	– 1

Fitting the correction as per the table above, the true dB plot is shown in Fig. P 6.8 (*e*) by the bold smooth curve.

(*v*) **Phase plot:** A table of phase angle for few arbitrarily chosen values of ω is constructed below by actually calculating the phase angle as follows:

$$\angle G(j\omega)\,H(j\omega) = -90° - \tan^{-1}\left(\frac{\omega}{0.5}\right) + \tan^{-1}\left(\frac{\omega}{4}\right) - \tan^{-1}\left(\frac{\omega}{10}\right)$$

ω	0.1	0.4	1	4	8	20	50	100
∠GH	– 100°	– 125°	– 145°	– 150°	– 152°	– 163°	– 172°	– 176°

Locate the points tabulated above and sketch the smooth phase plot as shown in Fig. P 6.8(*f*).

(*vi*) **Interpreting Bode plot**

(*a*) GM = ∞

(*b*) PM = 30°

(*c*) gain cross over frequency ω_{gc} = 6.5 rad/sec

(*d*) phase cross over frequency ω_{pc} = ∞

The system is closed loop stable; GM and PM are both positive.

P 6.9: Sketch Bode plot for unity feedback system with loop function

$$G(s) = \frac{10(s+1)(s+70)}{s^2(s^2+18s+400)}$$

and interpret closed loop stability.

Solution: Sketching Bode plot is demonstrated in the steps as follows:

(*i*) Write the loop function in time constant form;

$$G(j\omega) = \frac{1.75(1+s)\left(1+\dfrac{s}{70}\right)}{s^2\left(1+\dfrac{18s}{400}+\dfrac{s^2}{400}\right)}$$

and

$$G(j\omega) = \frac{1.75(1+j\omega)\left(1+\dfrac{j\omega}{70}\right)}{(j\omega)^2\left[\left(1-\dfrac{\omega^2}{400}\right)+j0.045\omega\right]}$$

(*ii*) **Identify the corner frequencies**

ω_{c_1} = 1 (a real axis zero)

ω_{c_2} = 20 (a pair of complex conjugate poles, ω_n = 20, ξ = 0.45)

ω_{c_3} = 70 (a real axis zero)

(*iii*) **dB plot:** Choose an appropriate scale on semilog graph paper and mark the corner frequencies as shown in Fig. P6.9(*e*). Due to presence of two poles at origin, the initial segment of dB plot is a straight line of slope – 40 dB/dec passing through point B ($\omega = 1$, dB = 20 log $K|_{K = 1.75}$ = + 4.8 ≅ + 5) locate one more point A, one decade below point B at ($\omega = 0.1$, dB = + 40 + 5 = + 45) on this segment as demonstrated in P 6.9 (*a*). The straight line segment joining points A and B is shown in Fig. P 6.9(*e*) while terminating it at corner frequency $\omega_{c_1} = 1$ (point B).

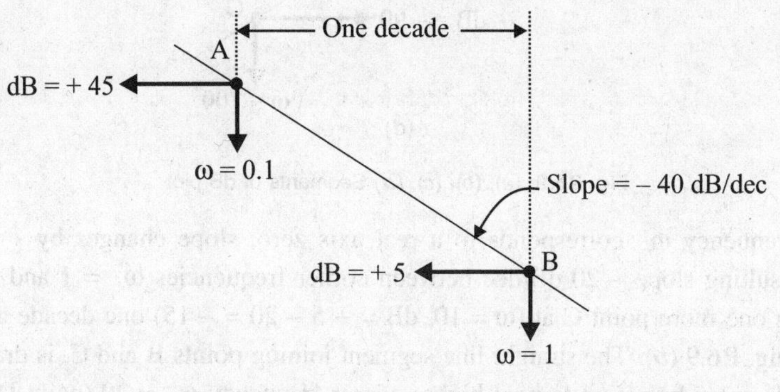

(*a*)

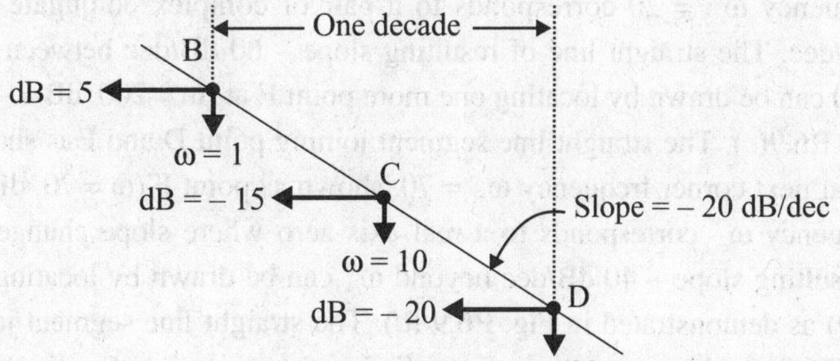

(*b*)

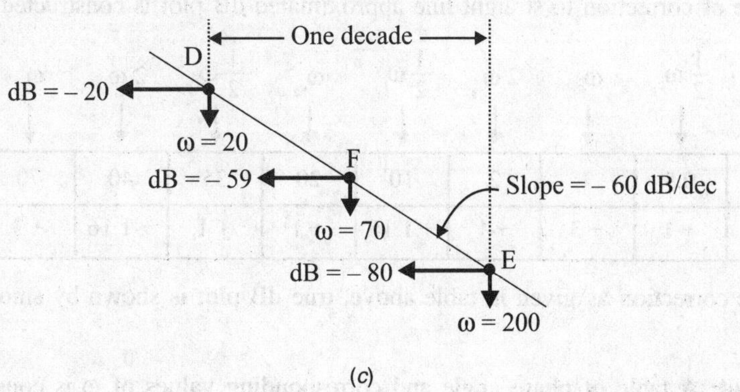

(*c*)

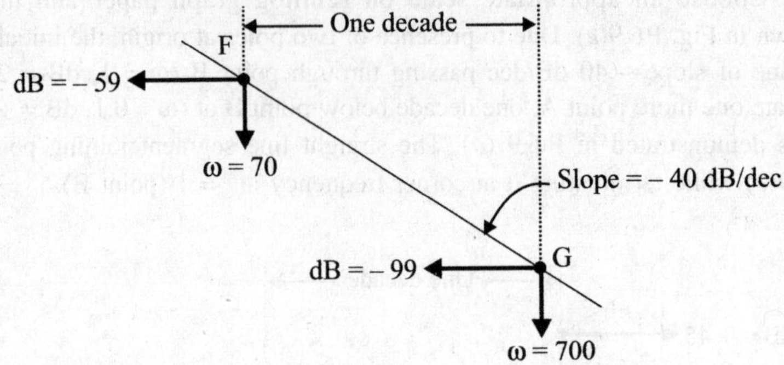

(*d*)

Fig. P6.9: (*a*), (*b*), (*c*), (*d*) Segments of dB plot

The corner frequency ω_{c_1} corresponds to a real axis zero, slope changes by + 20 dB/dec. The straight line of resulting slope – 20 dB/dec between corner frequencies $\omega_{c_1} = 1$ and $\omega_{c_2} = 20$ can be drawn by locating one more point C at ($\omega = 10$, dB = + 5 – 20 = – 15) one decade above point B as demonstrated in Fig. P6.9 (*b*). The straight line segment joining points B and C, is drawn as shown in Fig. P6.9 (*e*) while extending it up to next higher corner frequency $\omega_{c_2} = 20$ (point D) and noting the corresponding dB value equal to – 20 dB.

The corner frequency $\omega_{c_2} = 20$ corresponds to a pair of complex conjugate poles where slope changes by – 40 dB/dec. The straight line of resulting slope – 60 dB/dec between corner frequencies $\omega_{c_2} = 20$ and $\omega_{c_3} = 70$ can be drawn by locating one more point E at ($\omega = 200$, dB = – 20 – 60 = – 80) as demonstrated in Fig. P6.9(*c*). The straight line segment joining point D and E is shown in Fig. P6.9(*e*) while terminating it at next corner frequency $\omega_{c_3} = 70$, shown as point F ($\omega = 70$, dB = – 59).

The corner frequency ω_{c_3} corresponds to a real axis zero where slope changes by + 20 dB/dec. The asymptote of resulting slope – 40 dB/dec beyond ω_{c_3} can be drawn by locating one more point G ($\omega = 700$, dB = – 99) as demonstrated in Fig. P6.9 (*d*). The straight line segment joining points F and G is shown in Fig. P6.9(*e*). this segment does not find any termination due to absence of any other higher corner frequency.

(*iv*) The table of correction to straight line approximated dB plot is constructed as follows:

	$\frac{1}{2}\omega_{c_1}$	ω_{c_1}	$2\omega_{c_1}$	$\frac{1}{2}\omega_{c_2}$	ω_{c_2}	$\frac{1}{2}\omega_{c_3}$	$2\omega_{c_2}$	ω_{c_3}	$2\omega_{c_3}$
ω	0.5	1	2	10	20	35	40	70	140
Error	+ 1	+ 3	+ 1	+ 1.16	+ 1	+ 1	+ 1.16	+ 3	+ 1

Applying the correction as given in table above, true dB plot is shown by smooth dotted line in Fig. P6.9(*e*).

(*v*) **Phase plot:** A table of phase angle and corresponding values of ω is constructed below by actually calculating the phase angle using the relationship as follows: (values of phase angles are expressed in nearest integer).

$$\underline{|G(j\omega)} = -180° + \tan^{-1}(\omega) + \tan^{-1}\left(\frac{\omega}{70}\right) - \tan^{-1}\left[\frac{0.045\,\omega}{1-\frac{\omega^2}{400}}\right]$$

ω	0.1	0.2	1	4	10	50	100	200	500
$\underline{|GH}$	– 175°	– 169°	– 137°	– 112°	– 119°	– 212°	– 205°	– 194°	– 188°

The points [ω, $\underline{|G(j\omega)}$] tabulated above are plotted and smooth phase curve is drawn as shown in P6.9 (*f*).

(*vi*) **Interpreting Bode plot:** It is easy to identify the following from Bode plot.

(*a*) GM = + 26 dB

(*b*) PM = + 62°

(*c*) gain cross over frequency ω_{gc} = 2.4 rad/sec

(*d*) phase cross over frequency ω_{pc} = 26 rad/sec

The system is closed loop stable; GM and PM are both positive.

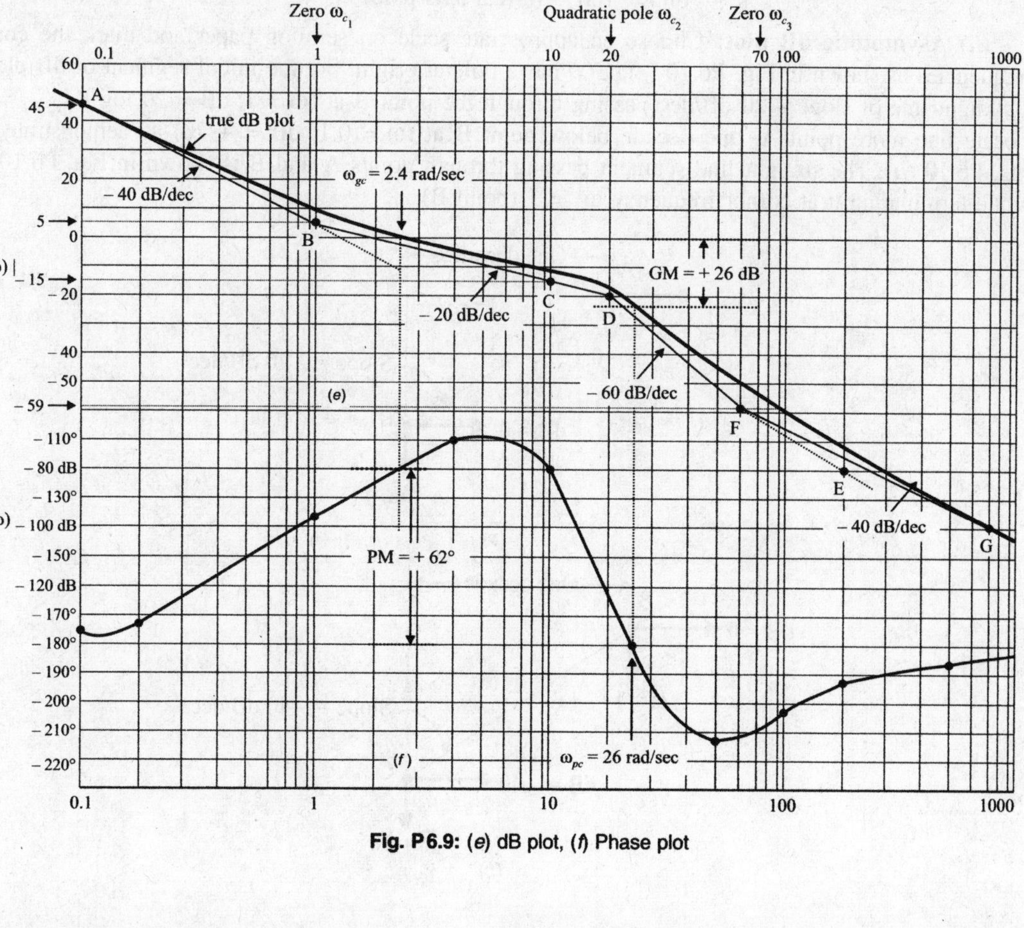

Fig. P6.9: (*e*) dB plot, (*f*) Phase plot

P 6.10: The loop transmittance of unity feedback system is

$$G(s) = \frac{K}{s(1+s)(1+0.1s)(1+0.01s)}$$

(*a*) Sketch asymptotic dB plot and phase plot for K = 1 to find gain cross over frequency, phase cross over frequency, GM, PM and whether closed loop system is stable?

(*b*) Determine valve of K so that GM = + 10 dB

(*c*) Determine value of K so that PM = + 25°

(*d*) Determine value of K so that gain cross over frequency is 0.3 rad/sec. Is the system closed loop stable for this value of K?

Solution: (*a*) Sketching Bode plot is demonstrated in the steps as follows:

(*i*) Note that the loop function is already in time constant form;

$$G(j\omega)\big|_{K=1} = \frac{1}{(j\omega)(1+j\omega)(1+j0.1\omega)(1+j0.01\omega)}$$

Identify the corner frequencies as follows:

$\omega_{c_1} = 1$ (a real axis pole)

$\omega_{c_2} = 10$ (a real axis pole)

$\omega_{c_3} = 100$ (a real axis pole)

(*ii*) **Asymptotic dB plot:** Choose an appropriate scale on semilog paper and mark the corner frequencies as shown in Fig. P6.10 (*e*). G(*s*) has a pole at origin. So, the initial segment of dB plot is a straight line of slope – 20 dB/dec passing through the point B at (ω = 1, dB = 20 log $K|_{K=1}$ = 0). Locate one more point A, one decade below point B at (ω = 0.1, dB = + 20) as demonstrated in Fig. P6.10 (*a*). The straight line segment passing through points A and B is shown in Fig. P6.10 (*e*) while terminating it at corner frequency ω_{c_1} = 1 (point B).

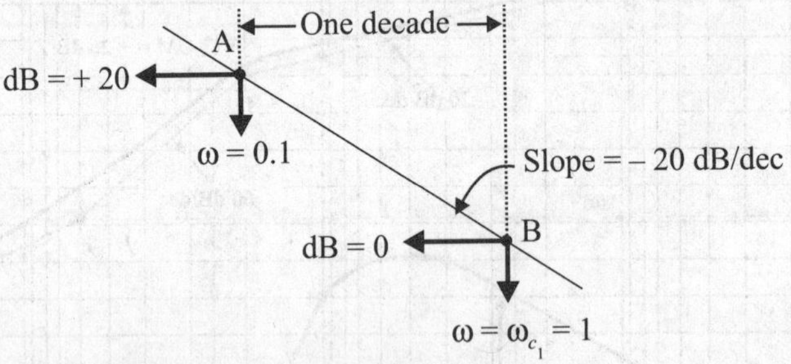

(*a*)

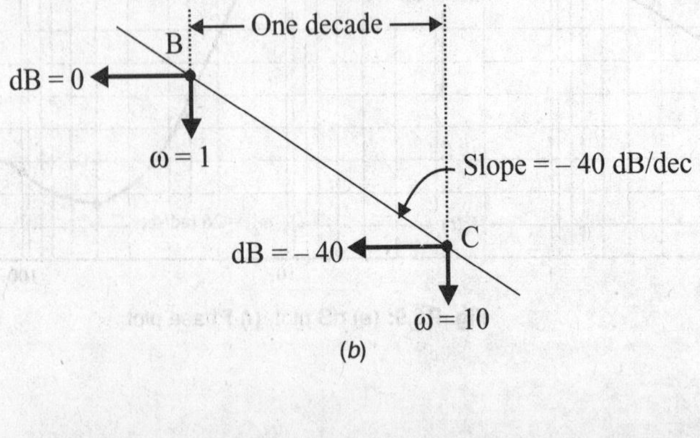

(*b*)

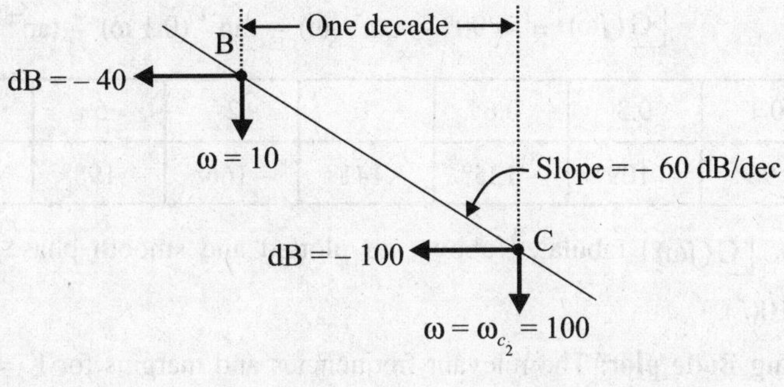

(c)

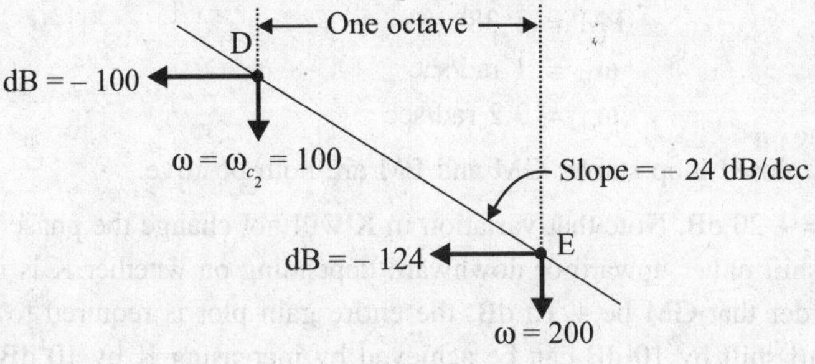

(d)

Fig. P6.10: (a), (b), (c), (d) Segments of dB plot (e) Complete dB plot (f) Phase plot.

The corner frequency ω_{c_1} corresponds to a real axis pole, slope changes by – 20 dB/dec. The straight line of resulting slope – 40 dB/dec between corner frequencies $\omega_{c_1} = 1$ and $\omega_{c_2} = 10$ can be drawn by locating one more point C at (ω = 10, dB = – 40) as demonstrated in Fig. P6.10(*b*). The straight line segment passing through points B and C is shown in Fig. P6.10(*e*) while terminating it at next corner frequency $\omega_{c_2} = 10$.

The corner frequency ω_{c_2} corresponds to a real axis pole; slope changes by – 20 dB/dec. The straight line of resulting slope – 60 dB/dec can be drawn by locating one more point D at (ω = 100, dB = – 100) as demonstrated in Fig. P6.10(*c*). The straight line segment passing through points C and D is shown in Fig. P6.10(*e*) while terminating it at next higher corner frequency $\omega_{c_3} = 100$.

The corner frequency ω_{c_3} corresponds to a real axis pole; slope again changes by – 20 dB/dec. The straight line of resulting slope – 80 dB/dec or – 24 dB/octave can be drawn by locating one more point E, one octave above point D (ω = 200, dB = – 124) as demonstrated in Fig. P6.10(*d*). This segment is also shown in Fig. P6.10(*e*) without any termination due to absence of any other higher corner frequency.

(*iii*) **Phase plot:** A table of phase angle (expressed by nearest integer) together with corresponding values of ω arbitrarily chosen is constructed below by actually calculating the phase angles using the relationship.

$$\angle G(j\omega) = -90° - \tan^{-1}(\omega) - \tan^{-1}(0.1\,\omega) - \tan^{-1}(0.01\,\omega)$$

ω	0.1	0.3	0.6	1	2	5	10	∞
$\angle G(j\omega)$	– 96°	– 109°	– 125°	– 141°	– 166°	– 198°	– 225°	– 360°

The points (ω, $\angle G(j\omega)$) tabulated above are plotted and smooth phase curve is sketched as shown in Fig. P6.10(*f*).

(*iv*) **Interpreting Bode plot:** The relevant frequencies and margins for K = 1 are identified from Fig. P6.10(*e*) and (*f*) as follows:

$$GM = +20 \text{ dB}$$
$$PM = +38°$$
$$\omega_{gc} = 1 \text{ rad/sec}$$
$$\omega_{pc} = 3.2 \text{ rad/sec}$$

The system is closed loop stable; GM and PM are both positive.

(*b*) $GM|_{K=1} = +20$ dB. Note that variation in K will not change the phase plot and ω_{pc}, it causes only gain plot to shift either upward or downward depending on whether K is increased or decreased respectively. In order that GM be + 10 dB, the entire gain plot is required to be shifted upward by 10 dB. The upward shift by 10 dB can be achieved by increasing K by 10 dB or multiplying initial value of K = 1 by antilog (10/20) = 3.16 *i.e.,*

$$K|_{GM=+10 \text{ dB}} = 3.16$$

(*c*) **Note** that PM = $\angle G(j\omega)\big|_{\omega=\omega_{gc}} + 180°$. For PM = 25°, $\angle G(j\omega)\big|_{\omega=\omega_{gc}} = -155°$. Identity point P on phase plot where $\angle G(j\omega) = -155°$ as shown in Fig. P 6.10 (*f*). Draw a vertical line passing through point P and identify the corresponding ω_{gc} = 1.7 rad/sec. For ω_{gc} to be 1.7 rad/sec, the entire gain plot is required to be shifted upward by 8 dB. The upward shift by 8 dB can be achieved by increasing K by 8 dB or multiplying initial value of K = 1 by antilog (8/20) = 2.51. So, $K|_{PM=+25°} = 2.51$.

(*d*) Observe $|G(j\omega)|$ = 12 dB at ω = 0.3 rad/sec from dB plot for K = 1. For ω = 0.3 to be the gain crossover frequency (ω_{gc}), entire gain plot has to be shifted downward by 12 dB as demonstrated in Fig. P6.10(*f*). This can be achieved by decreasing K by 12 dB or by dividing initial value of K = 1 by antilog (12/20) = 3.98. So $K|_{\omega_{gc}=0.3 \text{ rad/sec}} = 0.25$, $GM|_{K=0.25} = 32$ dB and $PM|_{K=0.25} = 71°$. The closed loop system remains stable.

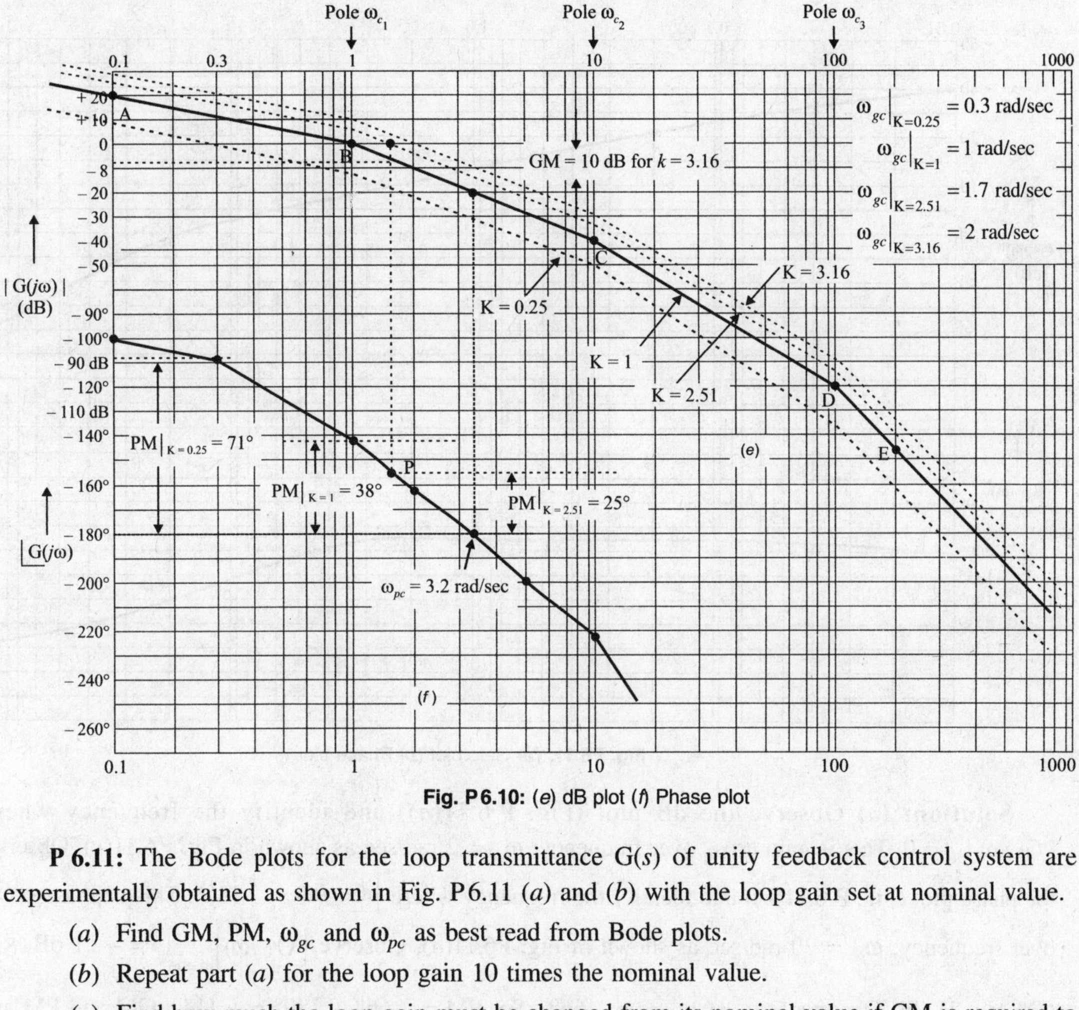

Fig. P6.10: (*e*) dB plot (*f*) Phase plot

P 6.11: The Bode plots for the loop transmittance G(*s*) of unity feedback control system are experimentally obtained as shown in Fig. P6.11 (*a*) and (*b*) with the loop gain set at nominal value.

(*a*) Find GM, PM, ω_{gc} and ω_{pc} as best read from Bode plots.

(*b*) Repeat part (*a*) for the loop gain 10 times the nominal value.

(*c*) Find how much the loop gain must be changed from its nominal value if GM is required to be + 40 dB.

(*d*) Find how much the loop gain must be changed from its nominal value if PM is required to be + 45°.

(*e*) Find the steady state error of the system if the reference input to it is a unit step function.

(*f*) Find modified values of GM and PM if the system has delay t_d = 0.05 seconds and loop gain at nominal.

(*g*) With the gain at nominal, find maximum time delay t_d the system can tolerate without going into instability.

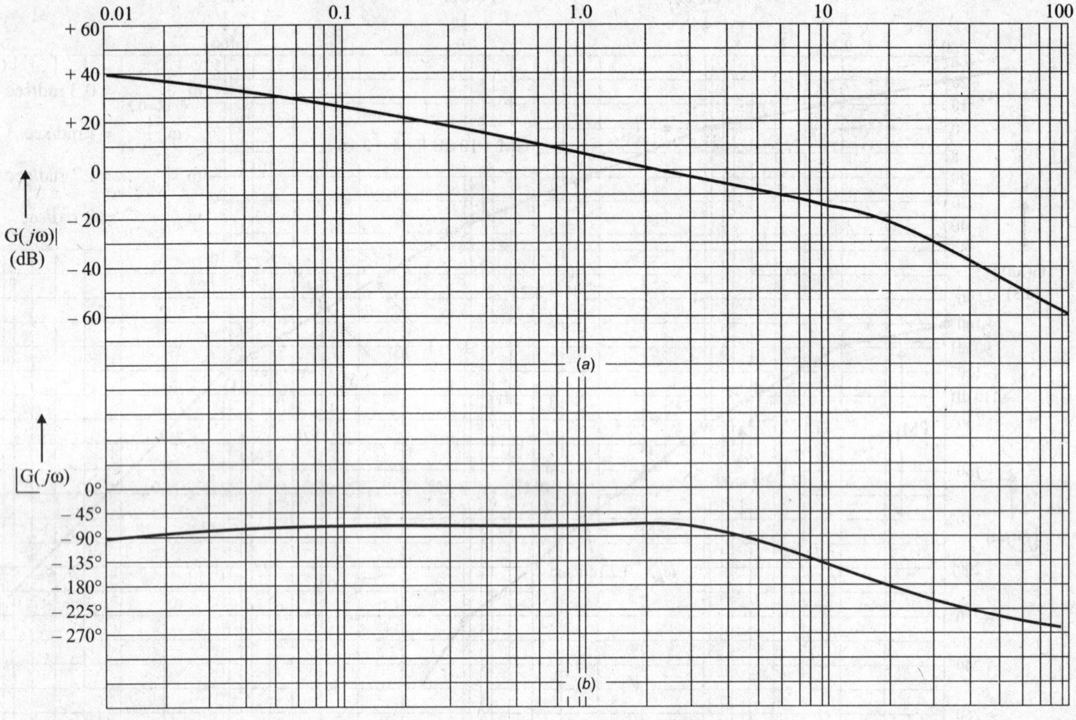

Fig. P6.11: *(a)* Gain plot *(b)* Phase plot

Solution: (*a*) Observe the dB plot (Fig. P 6.11(*a*)) and identify the frequency where $|G(j\omega)|_{dB} = 0$. This is gain cross over frequency; $\omega_{gc} = 2$ rad/sec as shown in Fig. P6.11(*c*). Observe the phase plot (Fig. P 6.11(*b*)) and identify the frequency where $\angle G(j\omega) = -180°$. This is phase cross over frequency; $\omega_{pc} = 20$ rad/sec as shown in Fig. P6.11(*d*). Observe $\left. |G(j\omega)| \right|_{\omega = \omega_{pc}} = -21$ dB. So, GM = + 21 dB. Observe $\left. \angle G(j\omega) \right|_{\omega = \omega_{gc}} = -66°$. So, PM = – 66° + 180° = + 114°. GM and PM are also shown in Fig. P6.11(*c*) and (*d*) respectively.

(*b*) **Note** that ten fold increment in loop gain from its nominal value is equivalent to increment by 20 log 10 = 20 dB. The dB plot will shift upward by 20 dB as demonstrated by dotted line in Fig. P6.11(*c*) together with phase plot remaining unaltered. Following the procedure as explained in part (*a*), ω_{gc}* (gain cross over frequency), ω_{pc}* (phase cross over frequency), GM* (gain margin), PM* (phase margin) can be approximately read as demonstrated in Fig. P6.11(*c*) and (*d*).

$$\omega_{gc}^* = 18 \text{ rad/sec}$$

$$\omega_{pc}^* = 20 \text{ rad/sec} = \omega_{pc}$$

$$GM^* = +1 \text{ dB}$$

$$PM^* = +4°$$

(*c*) The GM for nominal loop gain has been read to be + 21 dB in part (*a*). For requirement of + 40 dB GM, the gain plot must be shifted downward by (40 – 21) = 19 dB. Since antilog (19/20) ≅ 8.9, required loop gain = (nominal loop gain/8.9) ≅ 0.112 × nominal gain; decreasing the loop gain causes the dB plot to shift downward.

(*d*) PM = $\left.\angle G(j\omega)\right|_{\omega=\omega_{gc}}$ + 180°. In order that PM = + 45°, $\left.\angle G(j\omega)\right|_{\omega=\omega_{gc}}$ = – 135° and ω_{gc} = 12 rad/sec. Observe $|G(j\omega)|_{dB}$ at $\omega = \omega_{gc}$ = 12 rad/sec from dB plot. It is – 11 dB. So, loop gain must be increased by 11 dB so that gain cross over frequency becomes 12 rad/sec. Since antilog (11/20) = 3.55, required loop gain = 3.55 × nominal loop gain.

(*e*) Since the initial segment of dB plot has slope of – 20 dB/dec, G(*s*) is of type 1. The position error constant $K_P = \lim_{s\to 0} G(s) = \infty$ and steady state error $e(\infty) = 1/(1 + K_P) = 0$.

(*f*) Delay t_d = 0.05 (transmittance $e^{-0.05s}$) alters the phase plot and dB plot remains unaltered. The phase angle gets augmented by – 0.05ω radians or $\frac{-0.05\omega\times 180°}{\pi}$ degrees = – 2.86 ω degrees. The augmentation of phase angle as per table below has been demonstrated by dotted phase plot. PM = + 100° and GM = + 12 dB can be easily read as demonstrated in Fig. P 6.11(*c*) and (*d*) respectively.

ω	0.1	1	2	4	8	10	20
Phase augmented	– 0.29°	– 2.86°	– 5.7°	– 11.5°	– 23°	– 28.6°	– 57.2°

(*g*) $PM|_{t_d=0}$ = + 114°. For $\left.\frac{-\omega t_d\times 180°}{\pi}\right|_{\omega=\omega_{gc}=2} \leq -114°$ or $t_d \geq \frac{114°\times\pi}{2\times 180}$ or $t_d \geq 0.995$, the phase margin. PM will be negative. So, the system can tolerate delay of 0.995 sec without going into instability.

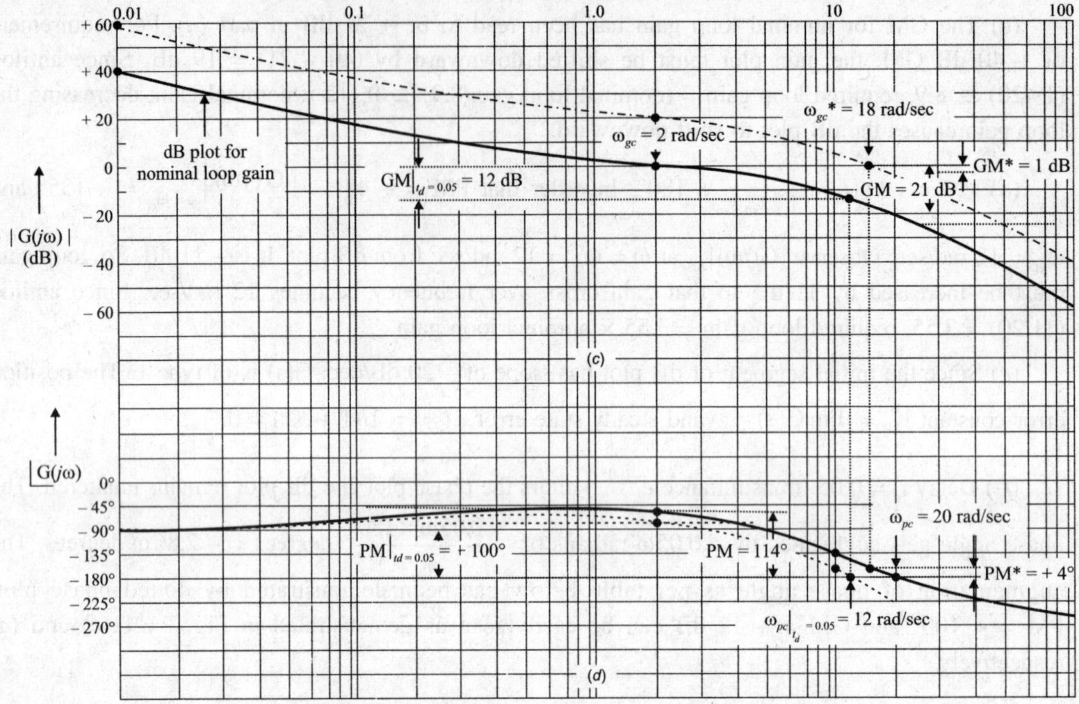

Fig. P6.11: Relevant changes in (c) Gain plot (d) Phase plot

P6.12: The open loop transfer function of a unity feedback system is $G(s) = \dfrac{K}{s(1+0.1s)}$. If the steady state error of the system is specified to be limited to 1% on unit ramp excitation, determine the values of gain margin (GM), phase margin (PM), gain cross over frequency (ω_{gc}), phase cross over frequency (ω_{pc}), resonant frequency (ω_r) and resonant peak (M_r).

Solution: Recall that $e(\infty)$ for unit ramp excitation $= \dfrac{1}{K_v}$, K_v = velocity error constant

where
$$K_v = \lim_{s \to 0} sG(s) = \lim_{s \to 0} \frac{K}{(1+0.1s)} = K$$

Since steady state error is required to be limited to 1%, $\dfrac{1}{K} \leq 0.01$ or $K \geq 100$

$$G(s)\big|_{K=100} = \frac{100}{s(1+0.1s)}$$

and
$$G(j\omega) = \frac{100}{j\omega(1+j0.1\omega)} = \frac{100}{j\omega - 0.1\omega^2}$$

The polar plot is shown in Fig. P6.12. It is easy to determine that $\omega_{pc} = \infty$ and GM = ∞ .

To investigate gain cross over frequency, use the relationship

$$|G(j\omega)|\Big|_{\omega=\omega_{gc}} = 1$$

$$\frac{100}{\omega_{gc}\sqrt{1+(0.1\omega_{gc})^2}} = 1$$

$$\omega_{gc}^2(1+0.01\omega_{gc}^2) = 10^4$$

$$0.01\omega_{gc}^4+\omega_{gc}^2-10^4 = 0$$

$$\omega_{gc}^2 = \frac{-1\pm\sqrt{1+4\times10^4\times10^{-2}}}{2\times10^{-2}}$$

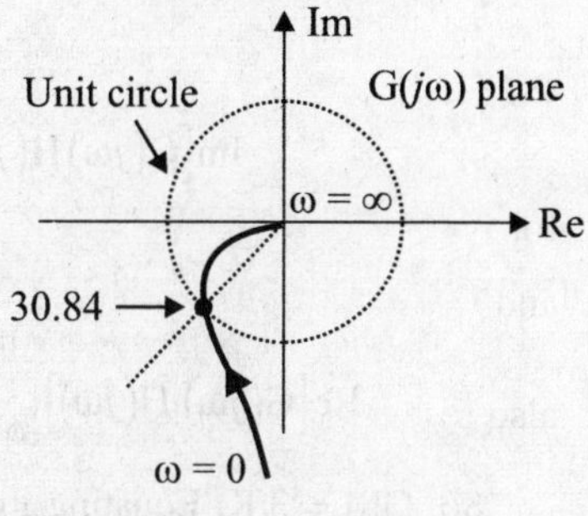

Fig. P6.12: Polar plot

So, $\omega_{gc} = 30.84$ rad/sec

$$PM = \angle G(j\omega)\Big|_{\omega=\omega_{gc}} + 180^\circ$$

$$= -90^\circ - \tan^{-1}(0.1\times30.84)+180^\circ$$

$$= +17.96^\circ$$

To determine M_r and ω_r, compare the characteristic equation

$$1+G(s) = 0$$

or
$$1+\frac{100}{s(1+0.1s)} = 0$$

or
$$s^2+10s+1000 = 0$$

with
$$s^2+2\xi\omega_n s+\omega_n^2 = 0$$

so that
$$\omega_n = \sqrt{1000}$$

$$\xi = \frac{5}{\sqrt{1000}}$$

$$M_r = \frac{1}{2\xi\sqrt{1-\xi^2}} = 3.24 = 10.2 \text{ dB}$$

$$\omega_r = \omega_n\sqrt{1-2\xi^2} = 30.82 \text{ rad/sec}$$

P6.13: A feedback system has loop function $G(s)\,H(s) = \dfrac{K}{s(1+0.5s)(1+s)}$, determine analytically

(*a*) the value of K so that system exhibits GM of + 6 dB.

(*b*) The value of K so that system exhibits PM of + 30°.

Solution: (*a*) $G(j\omega)\,H(j\omega) = \dfrac{K}{j\omega(1+j0.5\omega)(1+j\omega)} = \dfrac{K}{-1.5\omega^2+j\omega(1-0.5\omega^2)}$

$$= \frac{-K\left[1.5\omega^2 + j\omega\left(1 - 0.5\omega^2\right)\right]}{\left(1.5\omega^2\right)^2 + \omega^2\left(1 - 0.5\omega^2\right)^2}$$

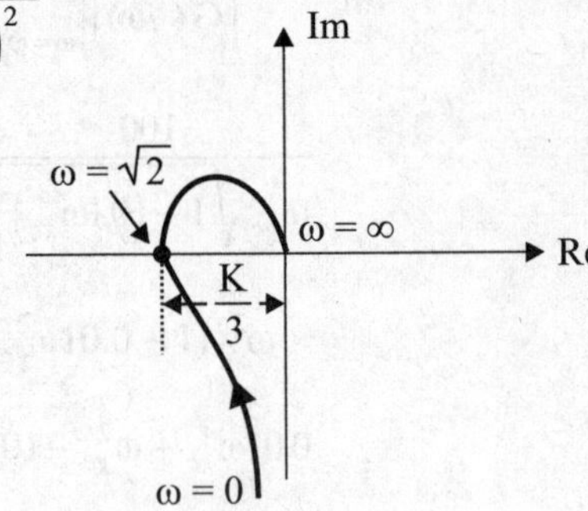

Fig. P6.13: Polar plot

$$\text{Im}\left[G(j\omega)H(j\omega)\right] = 0$$

for $\omega = 0$

and $\omega = \sqrt{2}$

also $\text{Re}\left[G(j\omega)\,H(j\omega)\right]\Big|_{\omega=\sqrt{2}} = -\frac{K}{3}$

So, GM = 3/K. Equating it to required GM = antilog (6/20) = 1.995, we have

$$\frac{3}{K} = 1.995 \quad \text{or} \quad K = \frac{3}{1.995} = 1.5$$

Polar plot along with significant points is shown in Fig. P6.13.

(*b*) The relationship PM = $\angle G(j\omega)\,H(j\omega)\Big|_{\omega=\omega_{gc}} + 180°$ gives

$$\angle G(j\omega)\,H(j\omega)\Big|_{\omega=\omega_{gc}} = 30° - 180°$$

or $-90° - \tan^{-1}\omega_{gc} - \tan^{-1}0.5\,\omega_{gc} = -150°$

or $\tan^{-1}\omega_{gc} + \tan^{-1}0.5\,\omega_{gc} = 60°$

Take tan of both sides to get

$$\frac{\omega_{gc} + 0.5\omega_{gc}}{1 - 0.5\omega_{gc}^2} = \tan 60° = 1.732$$

and solve it to get

$$0.866\,\omega_{gc}^2 + 1.5\,\omega_{gc} - 1.732 = 0$$

or $\omega_{gc} = 0.79$ (Choosing only +ve ω_{gc})

use the relation

$$\left|G(j\omega)\,G(j\omega)\right|_{\omega=\omega_{gc}} = 1$$

to get

$$\frac{K}{\omega_{gc}\sqrt{1+\omega_{gc}^{2}}\sqrt{1+\left(0.5\omega_{gc}\right)^2}} = 1$$

Put value of ω_{gc} in the equation just above and solve it to get K = 1.07.

P6.14: Sketch the Bode plots for following loop function and find the loop gain K for gain cross over frequency ω_{gc} to be 3 rad/sec.

$$G(s) = \frac{Ke^{-0.1s}}{s(1+s)(1+0.1s)}$$

Solution: Note that $G(j\omega) = \dfrac{Ke^{-j0.1\omega}}{j\omega(1+j\omega)(1+j0.1\omega)}$

$$= \frac{K(\cos 0.1\omega - j\sin 0.1\omega)}{j\omega(1+j\omega)(1+j0.1\omega)}$$

is in time constant form. Identify the corner frequencies as follows:

$$\omega_{c_1} = 1 \text{ rad/sec} \qquad \text{(a real axis pole)}$$

$$\omega_{c_2} = 10 \text{ rad/sec} \qquad \text{(a real axis pole)}$$

Sketching dB and phase plot is demonstrated as follows:

dB Plot: Choose an appropriate scale on semilog graph paper and mark the corner frequencies as shown in Fig. P6.14 (*d*). Note that the factor (cos 0.1 ω – *j* sin 0.1 ω) contributes 0 dB magnitude and therefore need not be considered for sketching dB plot.

Assuming K = 1, the initial segment of dB plot is straight line of slope – 20 dB/dec passing through point B at (ω = 1, dB = 0). Locate one more point A, one decade below point B at (ω = 0.1, dB = + 20) on this segment as demonstrated in Fig. P6.14 (*a*). The straight line segment joining points A and B is shown in Fig. P6.14 (*d*) while terminating at corner frequency $\omega_{c_1} = 1$ (point B).

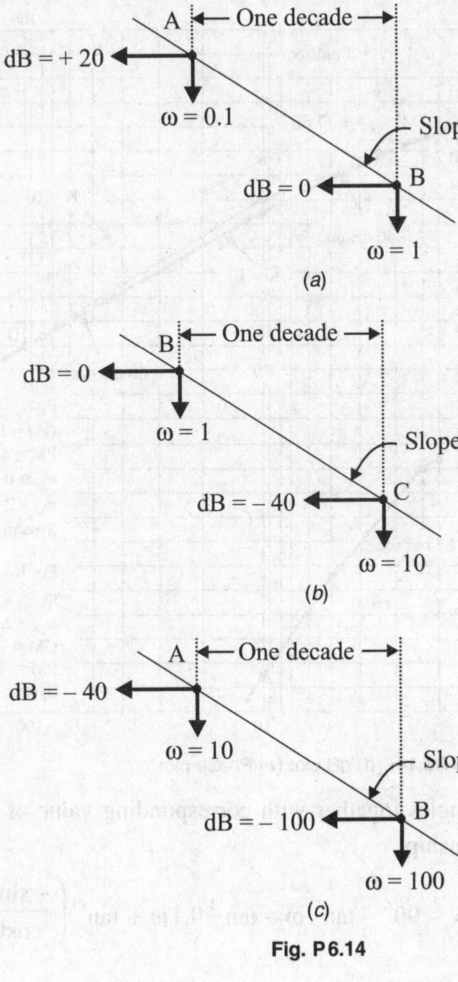

Fig. P6.14

The corner frequency $\omega_{c_1} = 1$ corresponds to a real axis pole; slope changes by – 20 dB/dec. The straight line of resulting slope – 40 dB/dec between corner frequencies $\omega_{c_1} = 1$ and $\omega_{c_2} = 10$ can be drawn by locating one more point C at (ω = 10, dB = – 40) as demonstrated in Fig. P6.14(*b*). The straight line segment passing through points B and C is shown in Fig. P6.14(*d*) while terminating it at next higher corner frequency $\omega_{c_2} = 10$.

The corner frequency $\omega_{c_2} = 10$ corresponds to another real axis pole; slope changes by another – 20 dB/dec. The straight line of resulting slope – 60 dB/dec can be drawn by locating one more point, D at (ω = 100, dB = – 100) as demonstrated in Fig. P6.14(*c*). The straight line segment passing through points C and D is shown in Fig. P6.14(*d*) without any further termination. Smooth true dB plot is sketched by applying the correction to the straight line approximated plot as per table below.

ω	0.5	1	2	5	10	20
Correction (dB)	– 1	– 3	– 1	– 1	– 3	– 1

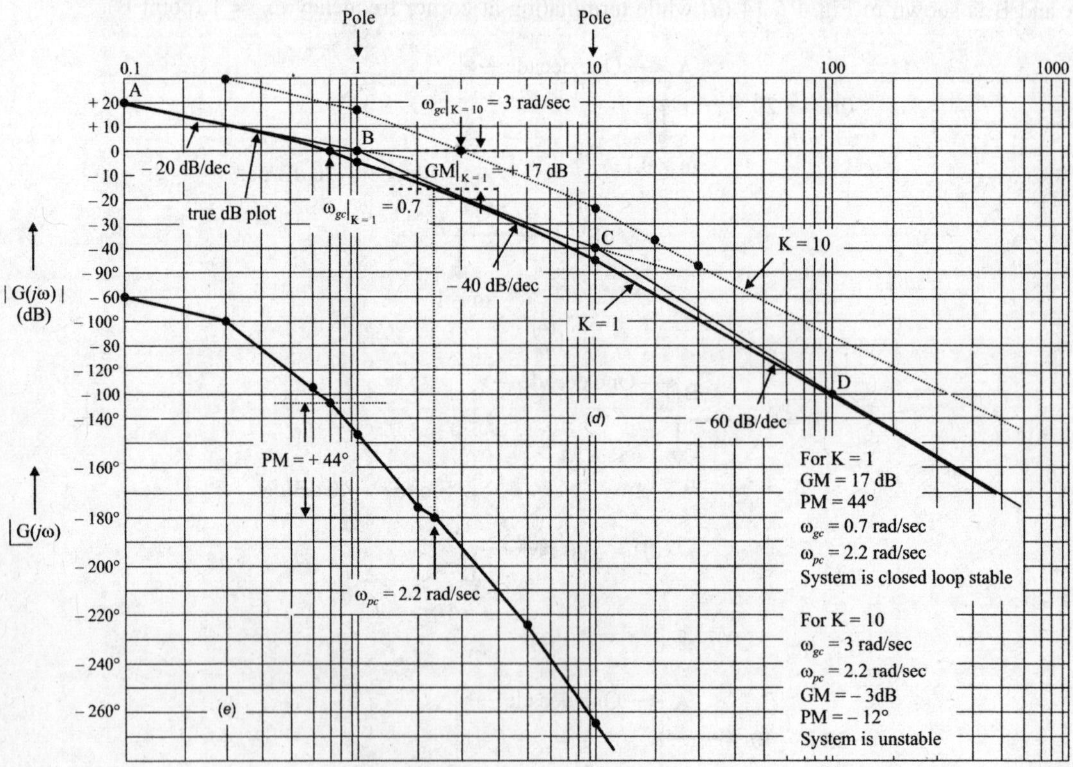

Fig. P6.14: (*d*) dB plot (*e*) Phase plot

Phase plot: The table of phase angles together with corresponding value of ω arbitrarily chosen is constructed below using the relationship.

$$\angle G(j\omega) = -90^\circ - \tan^{-1}\omega - \tan^{-1}0.1\omega + \tan^{-1}\left(\frac{-\sin 0.1\omega}{\cos 0.1\omega}\right)$$

$$= -90° - \tan^{-1}\omega - \tan^{-1} 0.1\,\omega - (0.1\,\omega) \times \frac{180°}{\pi}$$

$$= -90° - \tan^{-1}\omega - \tan^{-1} 0.1\,\omega - 5.73\,\omega$$

ω	0.1	0.3	0.6	1	2	5	10
$\angle G(j\omega)$	– 96.9°	– 110°	– 127.8°	– 146.4°	– 176.2°	– 224°	– 276.6°

The points (ω, $\angle G(j\omega)$) tabulated above are plotted and smooth phase curve is sketched as shown in Fig. P6.14 (*e*).

Interpreting Bode plot: For K = 1, ω_{gc} = 0.7 rad/sec. To determine the loop gain so that ω_{gc} = 3 rad/sec, read dB contribution at ω = 3 from dB plot for K = 1. It is – 20 dB. So, the dB plot must be raised upward by 20 dB as demonstrated in Fig. P6.14 (*d*). This can be accomplished by increasing K by 20 dB or multiplying initial value of K = 1 by antilog (20/20) = 10. So, loop gain K for ω_{gc} = 3 rad/sec is 10.

P 6.15: Determine the transfer function model of systems with asymptotic dB plots shown in Fig. P6.15(*a*), (*b*), (*c*), (*d*) and (*e*). Assume minimum phase characteristics possessed by each system.

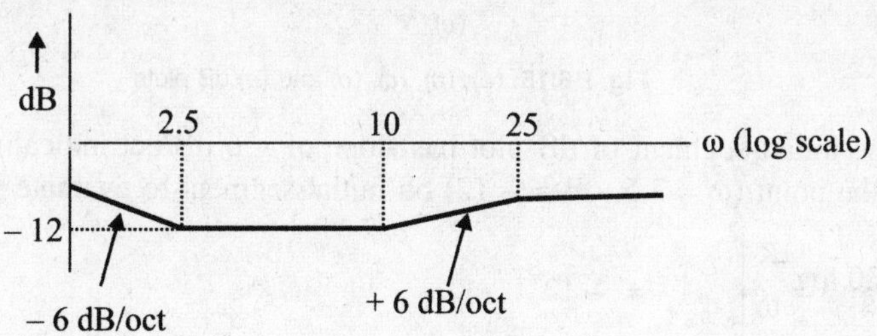

(*a*)

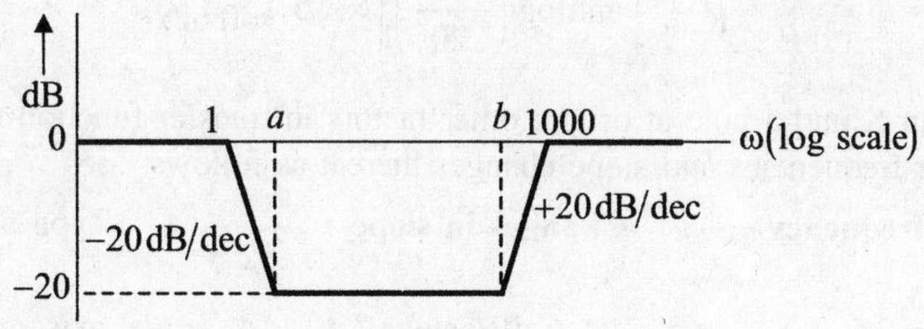

(*b*)

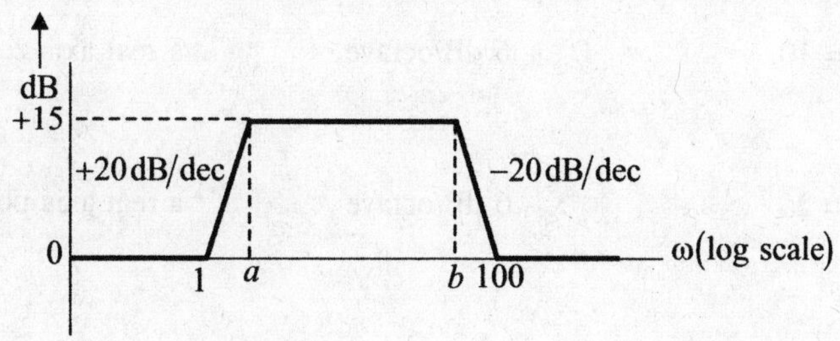

(*c*)

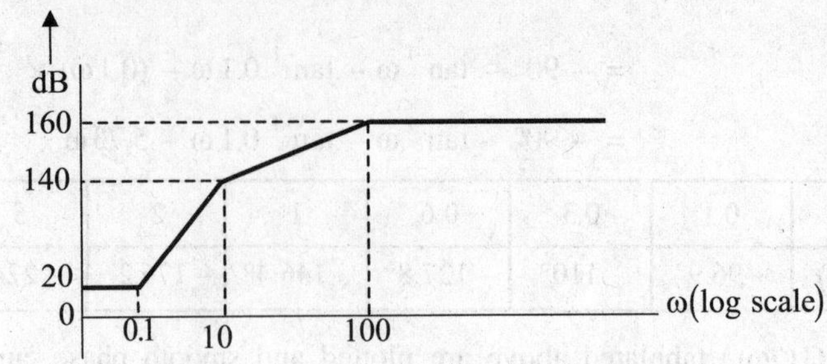

(d)

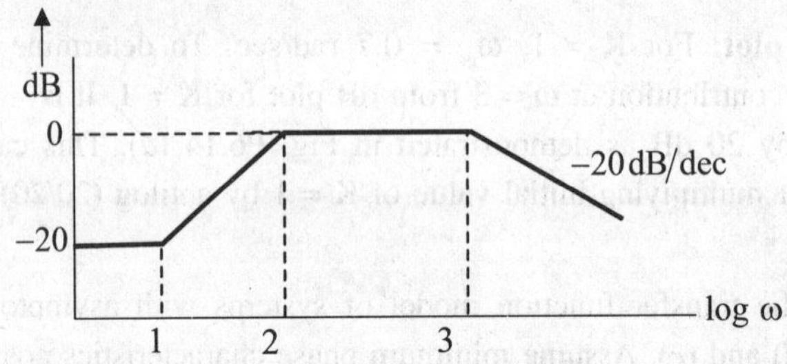

(e)

Fig. P6.15: (a), (b), (c), (d) and (e) dB plots

Solution: (a) The initial segment of dB plot has slope of – 6 dB/oct indicating presence of one pole at origin. Note the point (ω = 2.5, dB = – 12) on initial segment to evaluate gain K as follows:

$$20 \log \left.\frac{K}{\omega}\right|_{\omega = 2.5} = -12$$

or

$$K = \left[\text{antilog}\left(\frac{-12}{20}\right)\right] \times 2.5 = 0.628$$

In addition to gain K and a pole at origin, other factors in transfer function model are identified by locating the corner frequencies and slope changes thereat as follows:

Corner frequency	Changes in slope	Type of factor
ω = 2.5	+ 6 dB/octave	a real axis zero; $\left(1 + \frac{j\omega}{2.5}\right)$
ω = 10	+ 6 dB/octave	a real axis zero; $\left(1 + \frac{j\omega}{10}\right)$
ω = 25	– 6 dB/octave	a real axis pole; $\left(\frac{1}{1 + \frac{j\omega}{25}}\right)$

So, the transfer function model in time constant form is

$$G(j\omega) = \frac{0.628\left(1+\dfrac{j\omega}{2.5}\right)\left(1+\dfrac{j\omega}{10}\right)}{j\omega\left(1+\dfrac{j\omega}{25}\right)}$$

and in pole-zero form

$$G(s) = \frac{0.628(s+2.5)(s+10)}{s(s+25)}.$$

(*b*) The initial segment of dB plot is flat indicating absence of any pole or zero at origin. Since this segment coincides with 0 dB, 20 log K = 0 or K = 1. The other factors are identified as follows:

Corner frequency	Changes in slope	Type of factor
$\omega = 1$	– 20 dB/dec	$\left(\dfrac{1}{1+j\omega}\right)$; a real axis pole
$\omega = a$	+ 20 dB/dec	$\left(1+\dfrac{j\omega}{a}\right)$; a real axis zero
$\omega = b$	+ 20 dB/dec	$\left(1+\dfrac{j\omega}{b}\right)$; a real axis zero
$\omega = 1000$	– 20 dB/dec	$\left(\dfrac{1}{1+j\dfrac{\omega}{1000}}\right)$; a real axis pole

Note that the corner frequency $\omega = a$ lies one decade above the frequency $\omega = 1$ on the straight line with slope – 20 dB/dec. Since $dB|_{\omega=1} = 0$, $dB|_{\omega=a=10} = -20$ as demonstrated below:

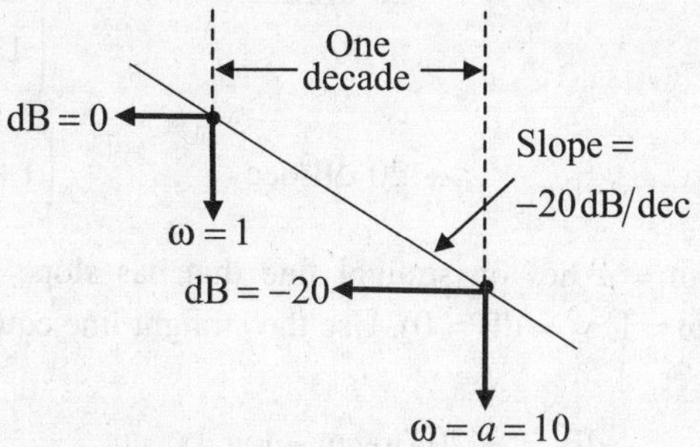

The corner frequency $\omega = b$ lies one decade below the corner frequency $\omega = 1000$ on straight line with slope + 20 dB/dec. Since $dB|_{\omega=1000} = 0$, $dB|_{\omega=b=100} = -20$ as demonstrated on the following page:

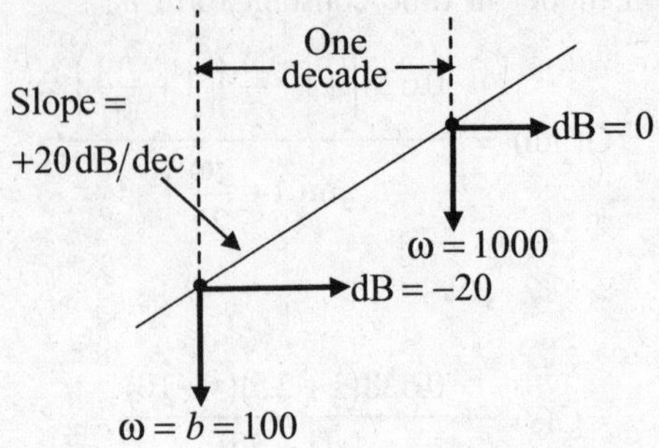

Collecting the factors identified above, the transfer function model in time constant form is obtained as

$$G(j\omega) = \frac{\left(1+j\dfrac{\omega}{a}\right)\left(1+j\dfrac{\omega}{b}\right)}{(1+j\omega)\left(1+j\dfrac{\omega}{1000}\right)}; \quad a = 10, \quad b = 100$$

and in pole-zero form

$$G(s) = \frac{(s+10)(s+100)}{(s+1)(s+1000)}$$

(*c*) The initial segment is flat coinciding with 0 dB line; there is no pole or zero at origin and 20 log K = 0 gives K = 1. The other factors are identified as follows:

Corner frequency	Changes in slope	Type of factor
$\omega = 1$	+ 20 dB/dec	$(1 + j\omega)$; a real axis zero
$\omega = a$	– 20 dB/dec	$\left(\dfrac{1}{1+j\dfrac{\omega}{a}}\right)$; a real axis pole
$\omega = b$	– 20 dB/dec	$\left(\dfrac{1}{1+j\dfrac{\omega}{b}}\right)$; a real axis pole
$\omega = 100$	+ 20 dB/dec	$\left(1+j\dfrac{\omega}{100}\right)$; a real axis zero

The corner frequency $\omega = a$ lies on straight line that has slope m = + 20 dB/dec and passes through the point ($x_1 = \log\omega = 1$, y_1 = dB = 0). Use the straight line equation $y - y_1 = m(x - x_1)$ where $x \equiv \log\omega$ and $y \equiv$ dB to get

$$\text{dB} = +\,20\,(\log\omega - \log 1)$$

$$y = \text{dB}|_{\omega = a} = 15 \text{ gives}$$

$$\log(a) - \log(1) = (15/20)$$

or

$$a = \text{antilog}\,(15/20) = 5.62.$$

Note that ω is on logarithmic scale.

Similarly, the corner frequency $\omega = b$ lies on straight line that has slope $m = -20$ dB/dec and passes through the point (ω = 100, dB = 0). Using the straight line equation $y - y_1 = m(x - x_1)$, we have dB = – 20 (log ω – log 100); ($x \equiv \log \omega$ and $y \equiv$ dB)

and $$\text{dB}\,|_{\omega=b} = +15 \text{ gives}$$

$$+15 = -20\,[\log(b) - \log 100]$$

or $$b = 100 \text{ antilog } (-15/20) = 17.78.$$

Collecting the factors identified just above together with values of a and b, the transfer function model in time constant form is

$$G(j\omega) = \frac{(1+j\omega)\left(1+j\dfrac{\omega}{100}\right)}{\left(1+j\dfrac{\omega}{5.62}\right)\left(1+j\dfrac{\omega}{17.78}\right)}$$

$$G(s) = \frac{(1+s)(1+0.01s)}{(1+0.178s)(1+0.056s)}$$

and in pole-zero form

$$G(s) = \frac{(s+1)(s+100)}{(s+5.62)(s+17.78)}$$

(*d*) The initial segment of dB plot is flat at 20 dB. So, 20 log K = 20 gives K = 10. Moving along dB plot from ω = 0.1 to ω = 10, dB change is 140 – 20 = + 120 dB for change in ω equal to 2 decades. The frequency ω = 10 is two decades above the frequency ω = 0.1. So, the slope change at corner frequency ω = 0.1 is + 120 dB/two decades or + 60 dB/dec indicating presence of three real axis zeros. It is easy to reason out that slope change at corner frequency ω = 10 is – 40 dB/dec. The slope of dB segment between ω = 10 and ω = 100 is + 20 dB/dec. With discussion so far the factors constituting the entire transfer function model can be identified as follows:

Corner frequency	**Changes in slope**	**Type of factor**
$\omega = 0.1$	+ 60 dB/dec	$\left(1+\dfrac{j\omega}{0.1}\right)^3$; three real axis zeros
$\omega = 10$	– 40 dB/dec	$\dfrac{1}{\left(1+\dfrac{j\omega}{10}\right)^2}$; two real axis poles
$\omega = 100$	– 20 dB/dec	$\dfrac{1}{1+\dfrac{j\omega}{100}}$; a real axis pole

Putting together the factors identified above, the transfer function model in time constant form is

$$G(j\omega) = \frac{10\left(1+\dfrac{j\omega}{0.1}\right)^3}{\left(1+\dfrac{j\omega}{10}\right)^2\left(1+\dfrac{j\omega}{100}\right)}$$

and in pole-zero form

$$G(s) = \frac{10^8(s+0.1)^3}{(s+10)^2(s+100)}$$

(*e*) The initial segment of dB plot is flat at – 20 dB. So, 20 log K = – 20 gives K = 0.1. The segment between log ω = 1 (ω = 10) and log ω = 2 (ω = 100) has slope of + 20 dB/dec. The factors of which the transfer function model is composed of, are identified as follows:

Corner frequency	Changes in slope	Type of factor
ω = 10 (log ω = 1)	+ 20 dB/dec	$\left(1+j\frac{\omega}{10}\right)$ a real axis zero
ω = 100 (log ω = 2)	– 20 dB/dec	$\left(\frac{1}{1+\frac{j\omega}{100}}\right)$ a real axis pole
ω = 1000 (log ω = 3)	– 20 dB/dec	$\left(\frac{1}{1+\frac{j\omega}{1000}}\right)$ a real axis pole

Transfer function in time constant form is

$$G(j\omega) = \frac{0.1\left(1+\frac{j\omega}{10}\right)}{\left(1+\frac{j\omega}{100}\right)\left(1+\frac{j\omega}{1000}\right)};$$

$$G(s) = \frac{0.1(1+0.1s)}{(1+0.01s)(1+0.001s)}$$

and in pole-zero form

$$G(s) = \frac{10^3(s+10)}{(s+100)(s+1000)}$$

DRILL PROBLEMS

D 6.1: Sketch frequency response curves (both dB plot and phase plot) for the transmittance

$$G(s) = \frac{s^2+2s+100}{s^2+10s+100}$$

Ans.

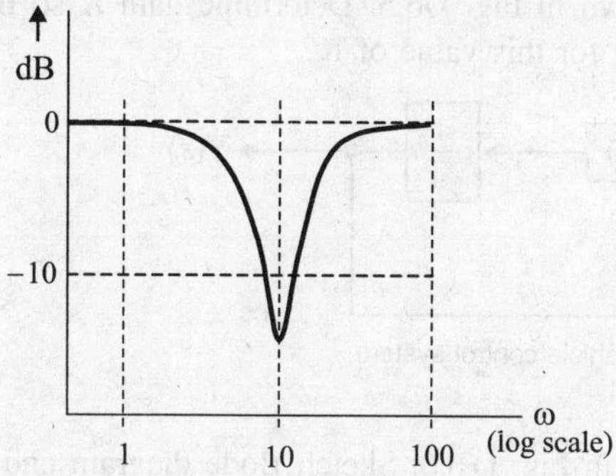

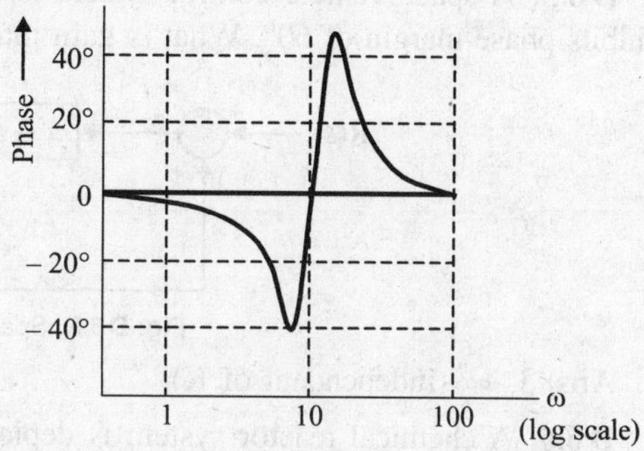

D 6.2: Draw Bode plot for the system with following loop transfer functions:

(*a*) $G_1(s)\ H_1(s) = \dfrac{3(s+1)(s+6)}{s^2(s^2+18s+400)}$ (*b*) $G_2(s)\ H_2(s) = \dfrac{20(s+1)}{s(s+5)(s^2+2s+100)}$

and determine each of the following therefrom (as best read)

(*i*) gain cross over frequency (ω_{gc})

(*ii*) phase cross over frequency (ω_{pc})

(*iii*) Gain margin (GM)

(*iv*) Phase margin (PM)

Comment on stability

Ans. (*a*) ω_{gc} = 0.2 rad/sec, ω_{pc} = ∞ rad/sec, GM = ∞, PM = 15°, stable

(*b*) ω_{gc} = 0.45 rad/sec, ω_{pc} = 4 rad/sec, GM = 10 dB, PM = 104°, stable

D 6.3: Draw Bode diagram for unity feedback system

$$G(s) = \frac{K}{s(s+2)(s+10)}$$

Read as best you can, the value of ω_{pc}. For what value of K will ω_{pc} be equal to ω_{gc}. Comment on stability for this value of K.

Ans. ω_{pc} = 4.7 rad/sec, 240, marginally stable.

D 6.4: The open loop transfer function of a closed loop system is

$$G(s)\ H(s) = \frac{K}{s(s+1)(2s+1)}$$

Determine phase cross over frequency ω_{pc} and gain margin in terms of K. Is the closed loop system stable for K = 2 ? What is critical value of K for stability ?

Ans. $\omega_{pc} = 1/\sqrt{2}$ rad/sec, GM = 1.5/K, unstable, $K_{critical}$ = 1.5.

D 6.5: A space vehicle control system is shown in Fig. D6.5. Determine gain K so that system exhibits phase margin of 60°. What is gain margin for this value of K?

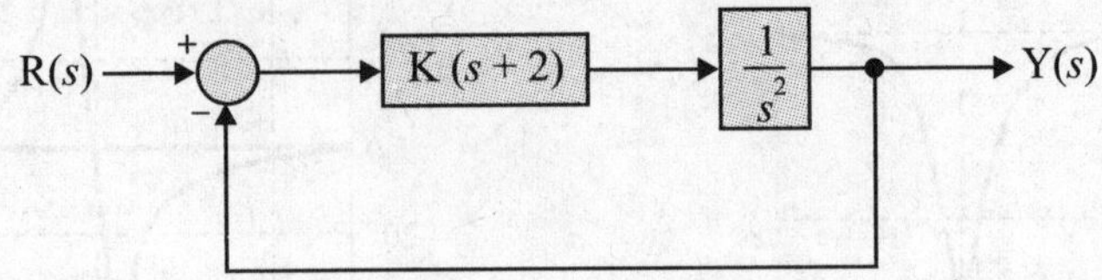

Fig. D 6.5: Space vehicle control system

Ans. 3, ∞ (independent of K).

D 6.6: A chemical reactor system is depicted in Fig. D 6.6. Sketch Bode diagram and determine therefrom as best read the following:

(*i*) gain cross over frequency (ω_{gc})

(*ii*) phase cross over frequency (ω_{pc})

(*iii*) Gain margin (GM)

(*iv*) Phase margin (PM)

Is the system stable?

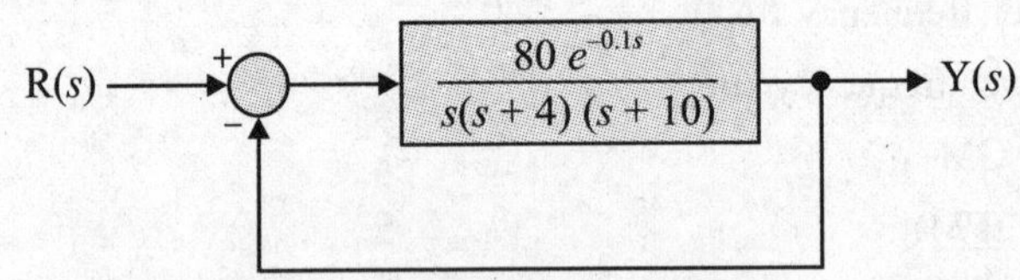

Fig. D 6.6: Chemical reactor system

Ans. (*i*) 2 rad/sec (*ii*) 4 rad/sec (*iii*) 8 dB (*iv*) 45°, stable.

D 6.7: Find transfer function model for each of the following dB plots (assume minimum phase characteristics).

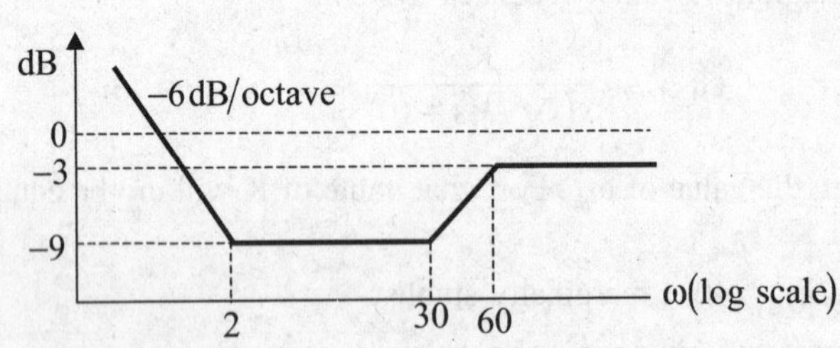

(*a*)

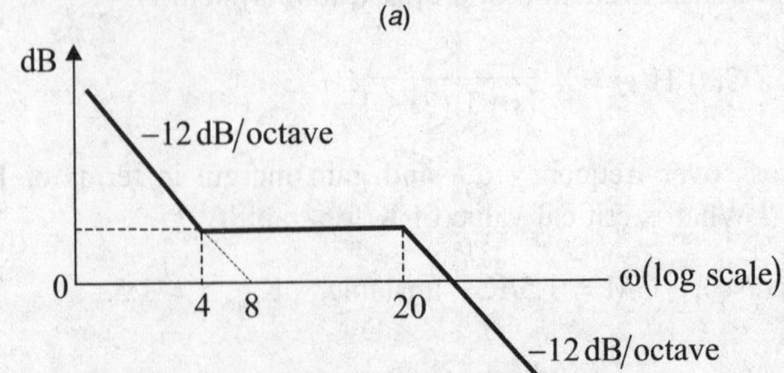

(*b*)

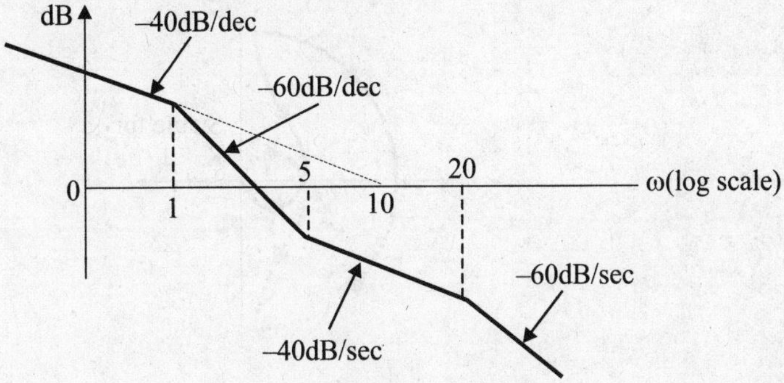

(c)

Ans. (*a*) $\dfrac{0.7(s+2)(s+30)}{s(s+60)}$ (*b*) $\dfrac{16\times10^2(s+4)^2}{s^2(s+20)^2}$ (*c*) $\dfrac{400(s+5)}{s^2(s+1)(s+20)}$

D 6.8: Determine value of α such that a closed loop system with loop transmittance

$$G(s) = \frac{\left(1+\dfrac{s}{\alpha}\right)^2}{s^3}$$

exhibits phase margin (PM) of 30°.

Ans. α = 0.917.

D 6.9: Sketch Nyquist plot for each of the following loop transmittances and investigate range of K for closed loop stability:

(*a*) $G(s)\,H(s) = \dfrac{K(s+1)}{s^2(s+2)(s+4)}$; K > 0

(*b*) $G(s)\,H(s) = \dfrac{K(s-1)}{s(s+5)(s+6)(s+7)}$; K > 0

(*c*) $G(s)\,H(s) = \dfrac{K(s+3)}{s(s-1)}$; K > 0

Ans. (*a*)

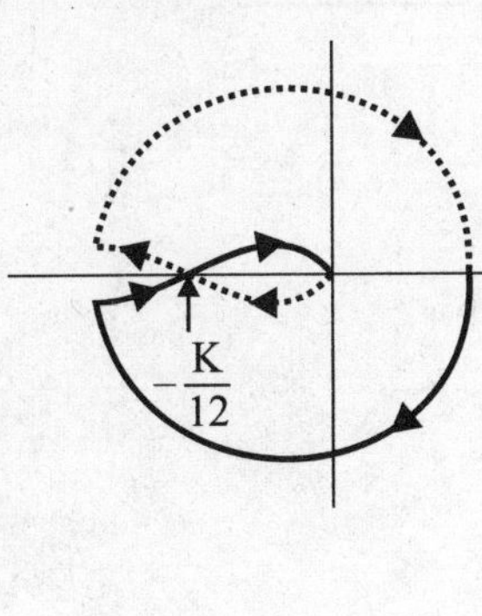

(*b*)

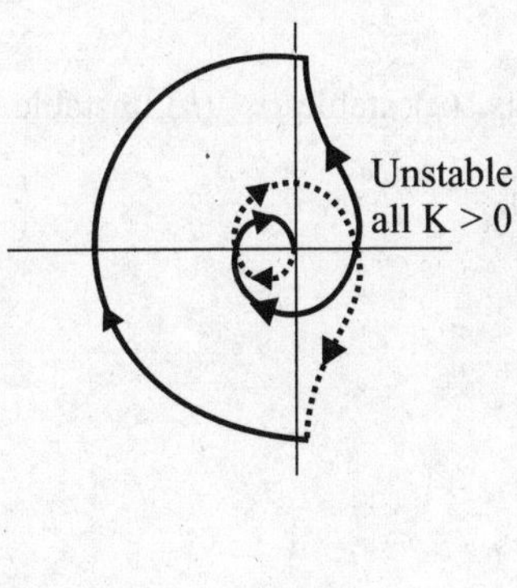

(*c*)

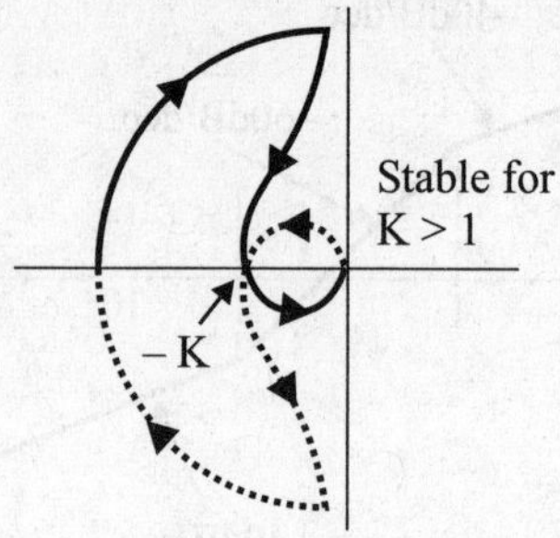

D 6.10: Sketch Nyquist plots for the following feedback systems and determine therefrom whether or not the system is stable.

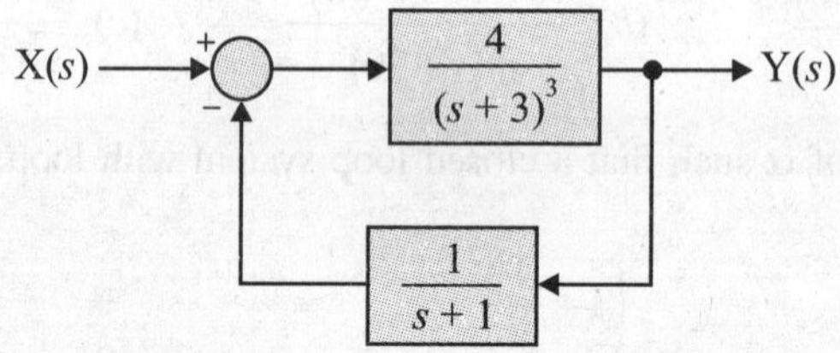

(*a*)

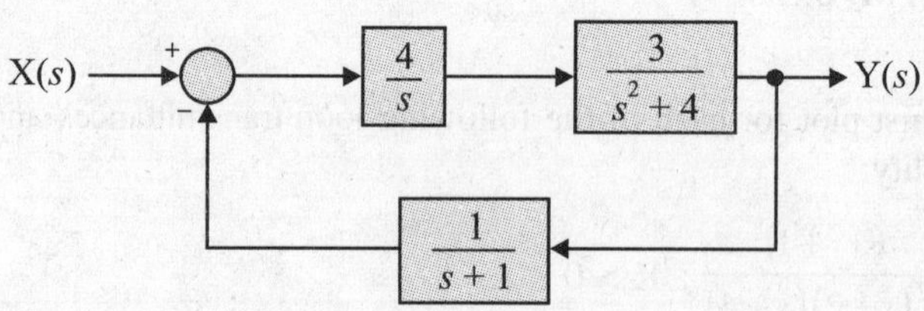

(*b*)

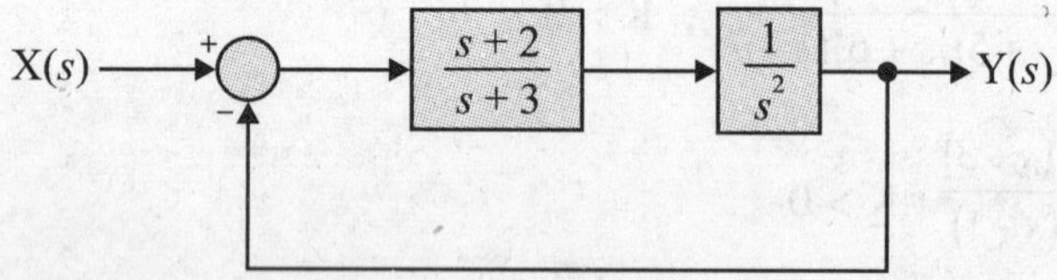

(*c*)

Ans. (*a*) stable (*b*) unstable (*c*) stable.

MULTIPLE CHOICE QUESTIONS

M 6.1: Consider Nyquist plots (*i*) and (*ii*) shown below. Assume loop transmittance having no RHP poles. Of the following the correct statement is

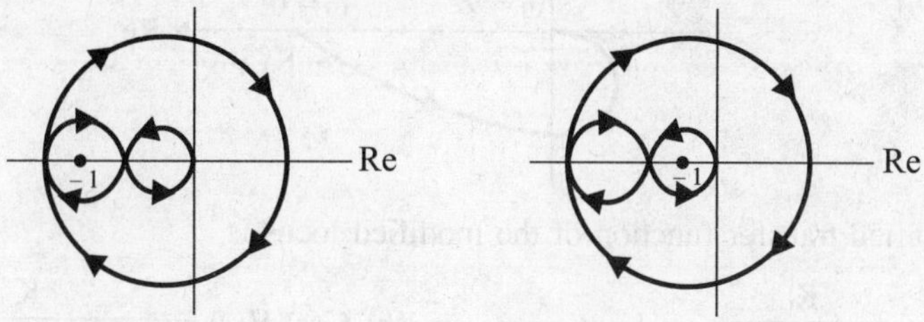

(*i*) (*ii*)

(*a*) (*i*) and (*ii*) both are stable (*b*) (*i*) and (*ii*) both are unstable

(*c*) (*i*) is stable but (*ii*) is unstable (*d*) (*i*) is unstable but (*ii*) is stable

M 6.2: The amplitude ratio and phase shift function for a transmittance are

$$A(\omega) = 4 \qquad \phi(\omega) = -3\omega \text{ (rad)}$$

The forced sinusoidal response to the input signal $r(t) = 10 \cos (5t - 30°)$ is

(*a*) $40 \cos (5t - 170°)$ (*b*) $30 \cos (5t - 170°)$

(*c*) $40 \cos (5t - 45°)$ (*d*) $30 \cos (5t - 45°)$.

M 6.3: The amplitude ratio A (ω) and phase shift function ϕ (ω) for irrational transfer function $G(s) = \dfrac{e^{-4s}}{s}$ respectively are

(*a*) $A(\omega) = -4\omega$, $\phi(\omega) = -\pi/2$ (*b*) $A(\omega) = 1/\omega$, $\phi(\omega) = -4\omega - \pi/2$

(*c*) $A(\omega) = 1/\omega$, $\phi(\omega) = -4\omega$ (*d*) $A(\omega) = -4\omega$, $\phi(\omega) = -4\omega$.

M 6.4: For $G(s) = \sqrt{s}$ consider the following statements

(*i*) $|G(j\omega)| = \sqrt{\omega}$ (*ii*) $\angle G(j\omega) = 45°$.

Select the correct answer

(*a*) (*i*) and (*ii*) both are true (*b*) (*i*) is true and (*ii*) is false

(*c*) (*i*) and (*ii*) both are false (*d*) (*i*) is false and (*ii*) is true.

M 6.5: The approximate GM and PM for unity feedback system with loop transmittance $\dfrac{e^{-0.1s}}{s}$ are respectively

(*a*) 24 dB, 84° (*b*) ∞, ∞ (*c*) 0, ∞ (*d*) ∞, 0.

M 6.6: The Nyquist locus of a transfer function is given below:

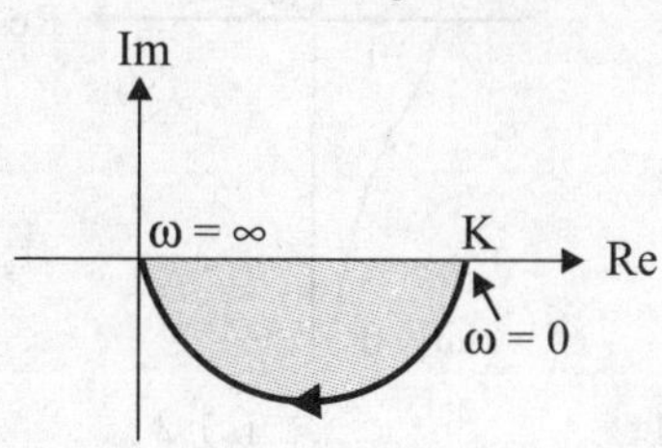

The locus is modified as shown below on addition of pole or poles to the original G(*s*) H(*s*).

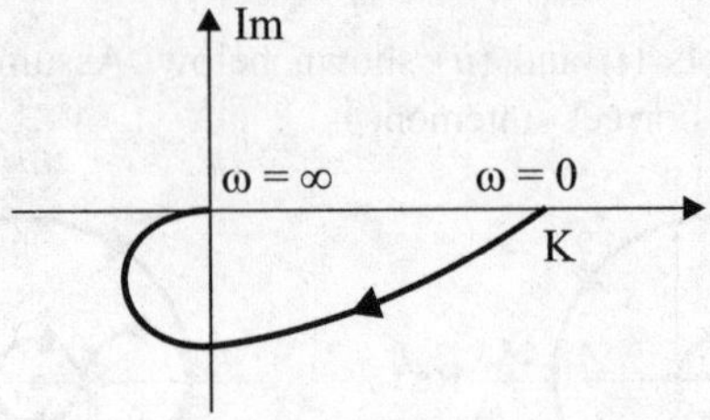

Then, the modified transfer function of the modified locus is

(*a*) $G(s)\,H(s) = \dfrac{K}{s(1+sT_1)}$ (*b*) $G(s)\,H(s) = \dfrac{K}{(1+sT_1)\,(1+sT_2)}$

(*c*) $G(s)\,H(s) = \dfrac{K}{s(1+sT_1)\,(1+sT_2)}$ (*d*) $G(s)\,H(s) = \dfrac{K}{s(1+sT_1)(1+sT_2)(1+sT_3)}$

M 6.7: Which one of the following statements is true for gain margin and phase margin of two closed loop systems having loop transfer functions G(*s*) H(*s*) and exp (– *s*) G(*s*) H(*s*)?

(*a*) Both gain and phase margins of the two systems will be identical.

(*b*) Both gain and phase margins of G(*s*) H(*s*) will be more.

(*c*) Gain margins of the two systems are the same but phase margin of G(*s*) H(*s*) will be more.

(*d*) Phase margins of the two systems are the same but gain margin of G(*s*) H(*s*) will be less.

M 6.8: The gain margin for feedback system with $G(s)\,H(s) = \dfrac{10^7(s+1)^2(s+100)^2}{(s+10)^4}$ is

(*a*) 10 dB (*b*) 20 dB (*c*) 30 dB (*d*) ∞

M 6.9: The open loop transfer function of a unity feedback control system is $\dfrac{10}{(s+5)^3}$. The gain margin of the system will be

(*a*) 20 dB (*b*) 40 dB (*c*) 60 dB (*d*) 80 dB

M 6.10: The Nyquist plot of the open loop transfer function of a feedback control system is shown in the figure below. If the open loop poles and zeros are all located in the left half of the *s* plane, then the number of closed loop poles in the right half of the *s* plane will be

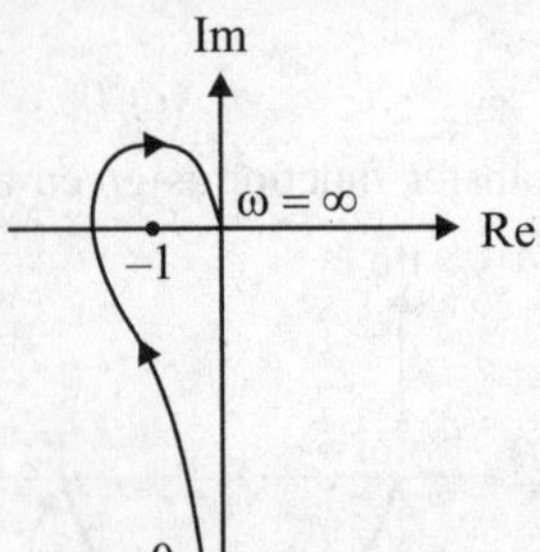

(*a*) zero (*b*) 1 (*c*) 2 (*d*) 3

M 6.11: A system with transfer function $G(s) = \frac{s}{(1+s)}$ is subjected to a sinusoidal input $r(t) = \sin \omega t$ in steady state, the phase angle of the output relative to the input at $\omega = 0$ and $\omega = \infty$ will be respectively

(*a*) 0° and – 90° (*b*) 0° and 0° (*c*) 90° and 0° (*d*) 90° and – 90°

M 6.12: Consider the following Nyquist plots of different control systems. The plot of unstable system is:

(*a*)

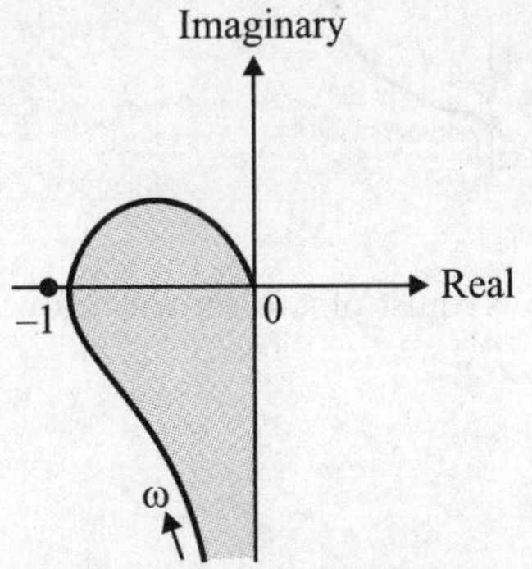

(*b*)

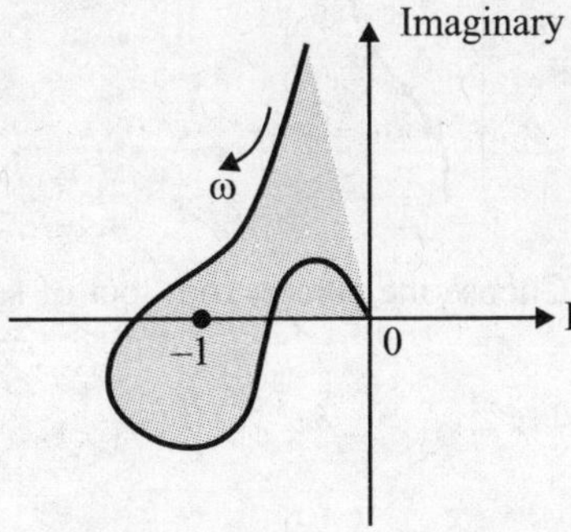

(*c*)

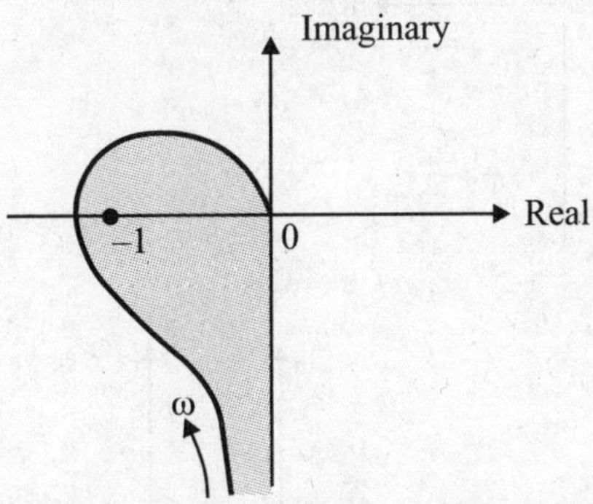

(*d*)

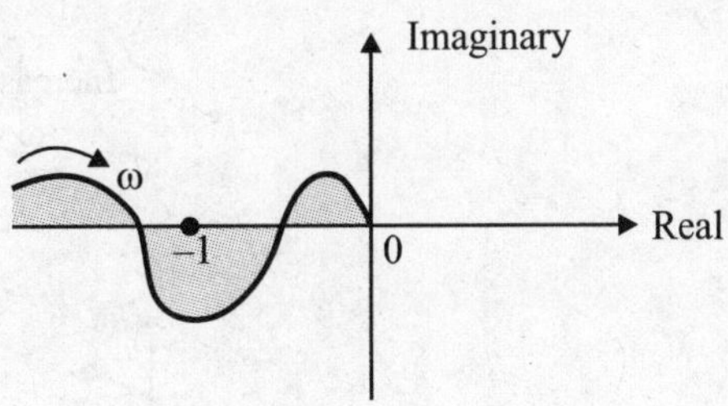

M 6.13: The phase angle of the system $G(s) = \frac{s+5}{s^2+4s+9}$ varies between

(*a*) 0° and 90° (*b*) 0° and – 90° (*c*) 0° and – 180° (*d*) – 90° and – 180°

M 6.14: The transfer function of a certain system is given by $G(s) = \frac{s}{(1+s)}$. The Nyquist plot of the system is

(*a*)

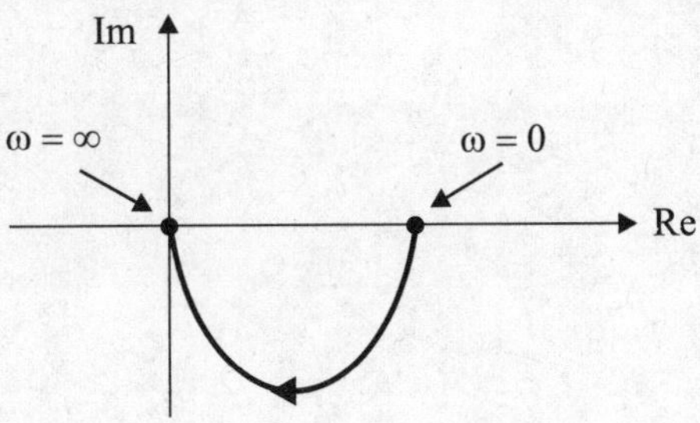

(*b*)

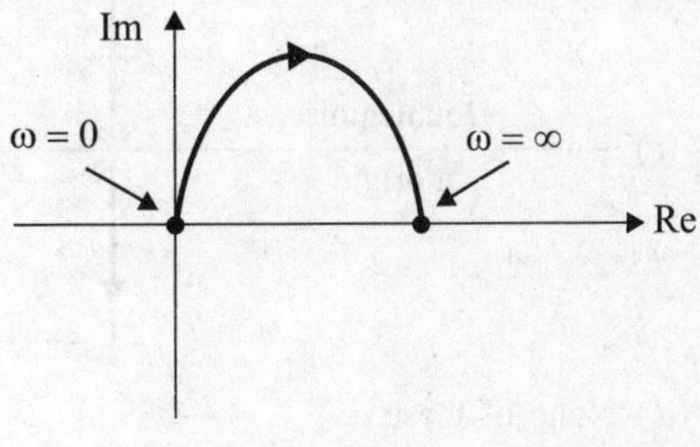

(*c*)

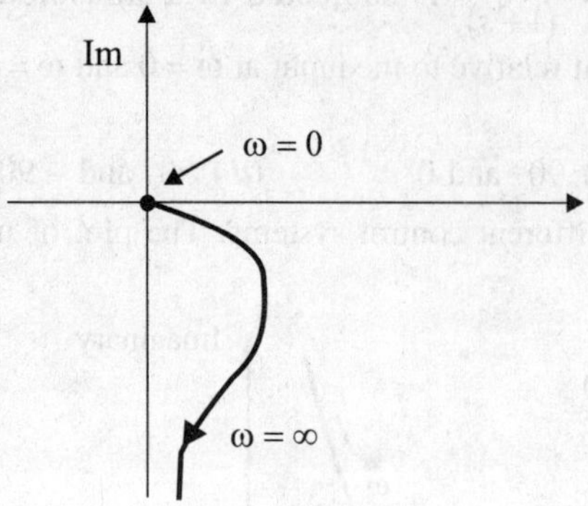

(*d*)

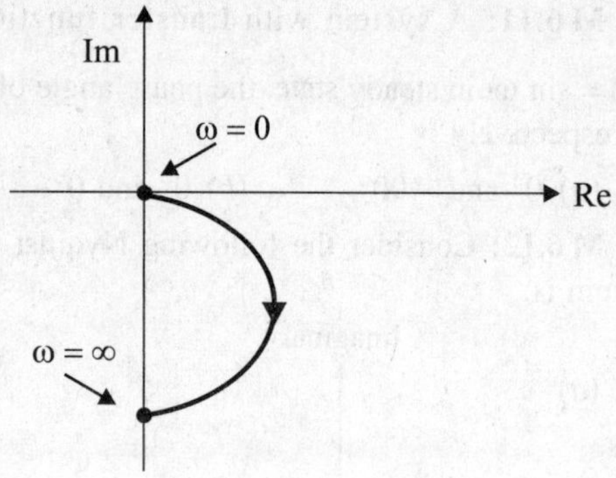

M 6.15: Choose the correct root loci of servo system whose Nyquist plot is shown below.

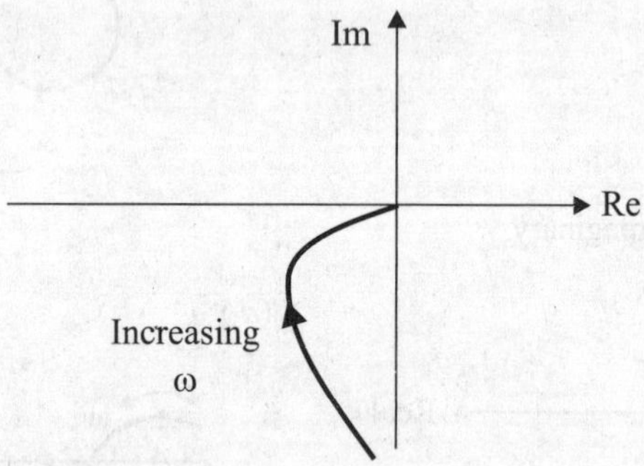

(*a*)

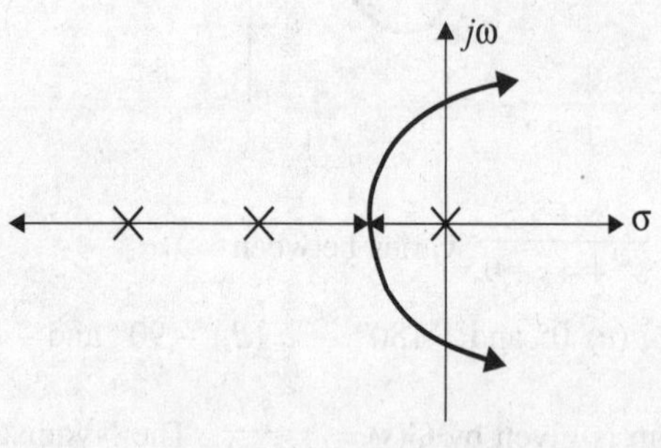

(*b*)

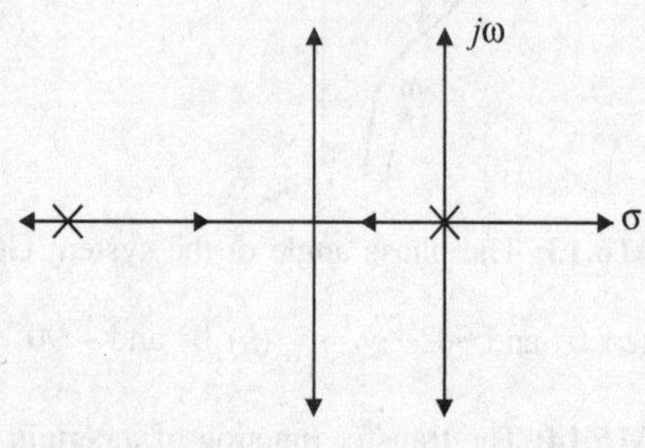

(*c*)

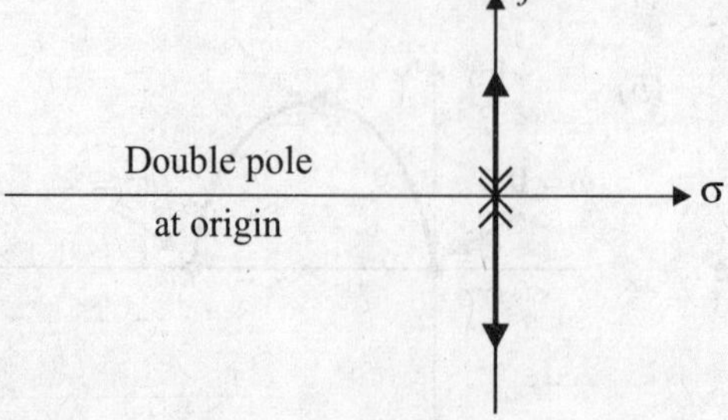

(*d*) None of these.

M6.16: If the Nyquist plot cuts the negative real axis at a distance of 0.4, then the gain margin of the system is

(*a*) 0.4 (*b*) – 0.4 (*c*) 4 (*d*) 2.5

M6.17: The Nyquist plot of G(*s*) H(*s*) = $\dfrac{10}{s^2(1+0.5s)(1+s)}$

(*a*) will start (ω = ∞) in the first quadrant and will terminate (ω = 0) in the second quadrant

(*b*) will start (ω = ∞) in the fourth quadrant and will terminate (ω = 0) in the second quadrant

(*c*) will start (ω = ∞) in the second quadrant and will terminate (ω = 0) in the third quadrant

(*d*) will start (ω = ∞) in the first quadrant and will terminate (ω = 0) in the fourth quadrant.

M6.18: Consider the following statements associated with phase and gain margins.

1. They are a measure of closeness of the polar plot to the – 1 + *j*0 point.
2. For a non-minimum phase to be stable it must have positive phase and gain margins.
3. For a minimum phase system to be stable, both the margins must be positive.

Which of the above statements is/are correct?

(*a*) 2 and 3 (*b*) 1 and 3 (*c*) 1 and 2 (*d*) 1 alone

M6.19: For loop transmittance G(*s*) = $\dfrac{\omega_n{}^2}{s^2+2\xi\omega_n s+\omega_n{}^2}$, the value of $|G(j\omega_n)|$ is

(*a*) $\dfrac{1}{2\xi}$ (*b*) $\dfrac{1}{4\xi}$ (*c*) $\dfrac{\xi}{2}$ (*d*) $\dfrac{2}{\xi}$

M6.20: The Nyquist plot shown below, matches with the transfer function

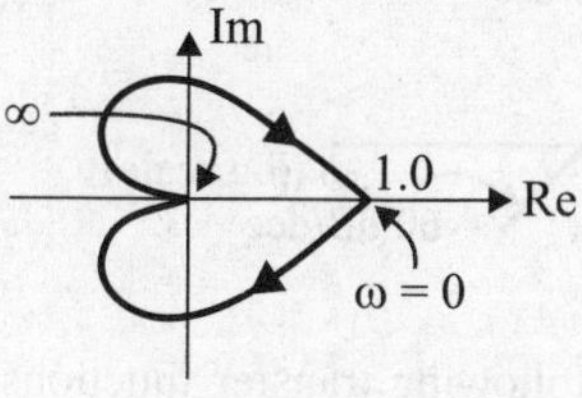

(*a*) $\dfrac{1}{(s+1)^3}$ (*b*) $\dfrac{1}{(s+1)^2}$ (*c*) $\dfrac{1}{(s^2+2s+2)}$ (*d*) $\dfrac{1}{(s+1)}$

M6.21: Which one of the following statements is correct in respect of the theory of stability?

(*a*) phase margin is the phase angle lagging, in short of 180°, at the frequency corresponding to a gain of 10.

(*b*) gain margin is the value by which the gain falls short of unity, at a frequency corresponding to 90° phase lag.

(*c*) Routh-Hurwitz criterion can determine the degree of stability.

(*d*) gain margin and phase margin are the measure of the degree of stability.

M6.22: The Nyquist plot for a control system is shown below. The Bode plot for the same system will be

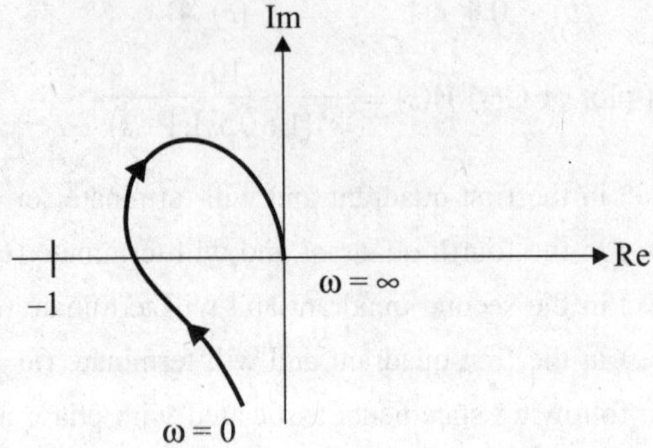

(*a*)

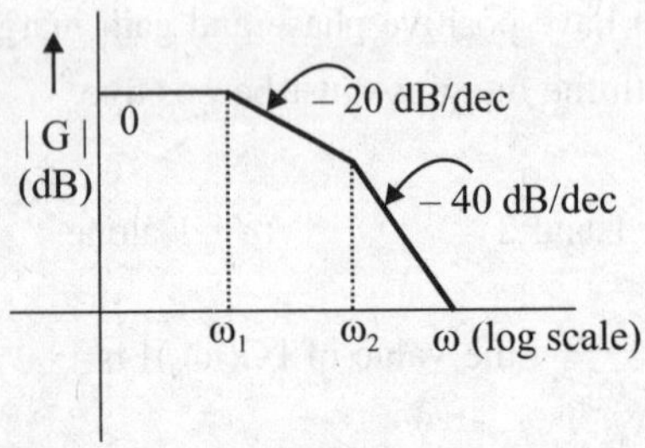

(*b*)

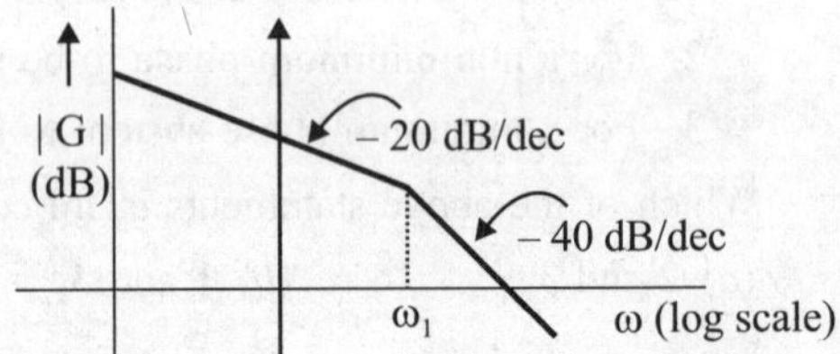

(*c*)

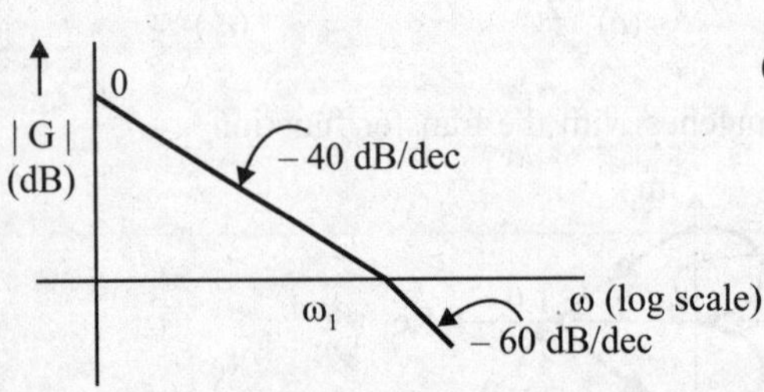

(*d*)

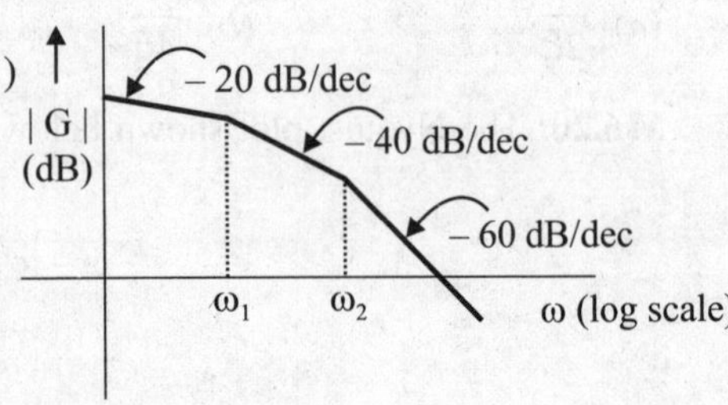

M6.23: Which one of the following transfer functions represents the Bode plot shown below?

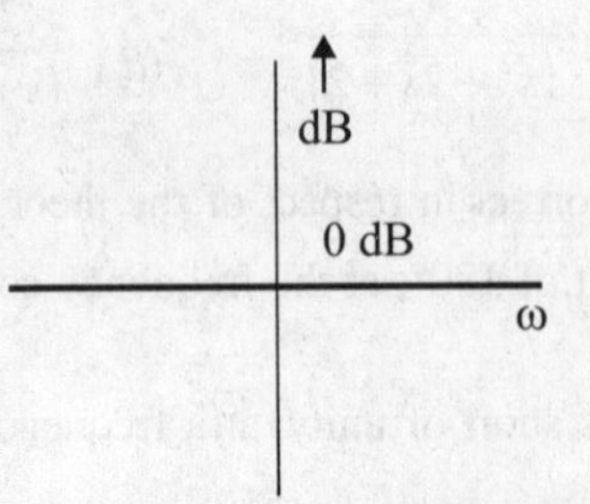

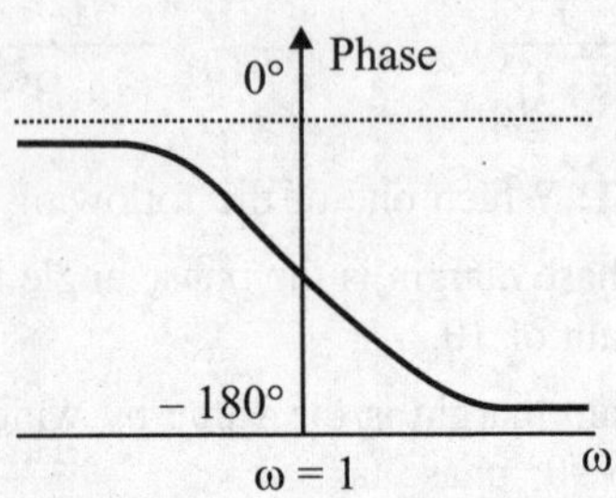

(*a*) $G = \dfrac{1-s}{1+s}$ (*b*) $G = \dfrac{1}{(1+s)^2}$ (*c*) $G = \dfrac{1}{s^2}$ (*d*) $G = \dfrac{1}{s(1+s)}$.

M6.24: The magnitude frequency response of a control system is shown below. The value of ω_1 and ω_2 are respectively

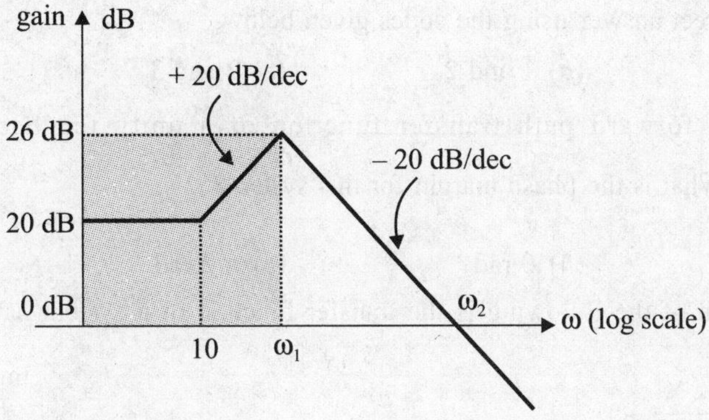

(*a*) 10 and 200 (*b*) 20 and 200 (*c*) 20 and 400 (*d*) 100 and 400

M6.25: The Bode plot shown below has G (*j*ω) as

(*a*) $\dfrac{100}{j\omega(1+j0.5\omega)(1+j\,0.1\omega)}$

(*b*) $\dfrac{100}{j\omega(2+j\omega)(10+j\omega)}$

(*c*) $\dfrac{10}{j\omega(1+2j\omega)(1+10j\omega)}$

(*d*) $\dfrac{10}{j\omega(1+0.5j\omega)(1+0.1j\omega)}$

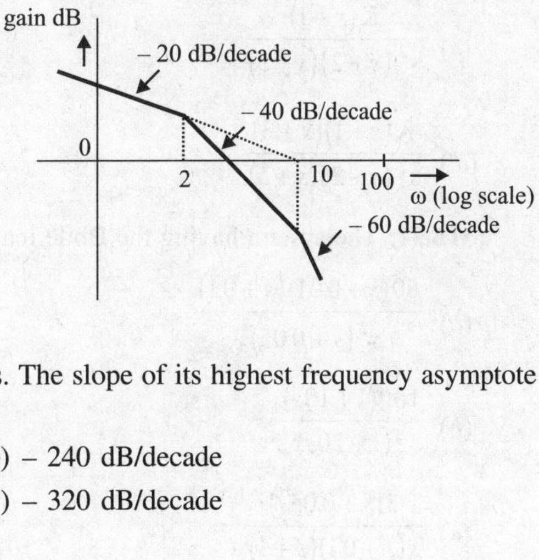

M6.26: A system has fourteen poles and two zeros. The slope of its highest frequency asymptote in its magnitude plot is

(*a*) – 40 dB/decade (*b*) – 240 dB/decade

(*c*) – 280 dB/decade (*d*) – 320 dB/decade

M6.27: Consider the following techniques.

1. Bode plot
2. Nyquist plot
3. Nichol's chart
4. Routh-Hurwitz criterion

Which of these techniques are used to determine relative stability of a closed loop linear system?

(*a*) 1 and 2 (*b*) 1 and 4 (*c*) 1, 2 and 3 (*d*) 2, 3 and 4

M6.28: Consider the following statements regarding the frequency response of a system as shown.

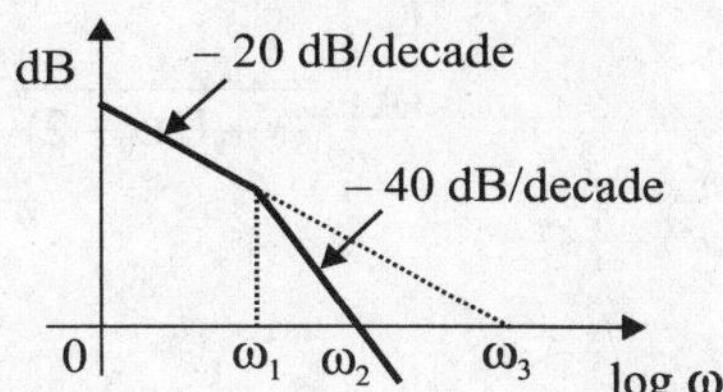

1. The type of the system is one
2. ω_3 = static error coefficient (K_v)

3. $\omega_2 = \dfrac{\omega_1 + \omega_3}{2}$

Select the correct answer using the codes given below.

(*a*) 1, 2 and 3 (*b*) 1 and 2 (*c*) 2 and 3 (*d*) 1 and 3

M 6.29: The forward path transfer function of a unity feedback system is given by $G(s) = \dfrac{1}{(1+s)^2}$. What is the phase margin for this system?

(*a*) $-\pi$ rad (*b*) 0 rad (*c*) $\pi/2$ rad (*d*) π rad

M 6.30: Which of the following is the transfer function of a system having the Nyquist plot as shown below?

(*a*) $\dfrac{K}{s(s+2)^2(s+5)}$

(*b*) $\dfrac{K}{s^2(s+2)(s+5)}$

(*c*) $\dfrac{K(s+1)}{s^2(s+2)(s+5)}$

(*d*) $\dfrac{K(s+1)(s+3)}{s^2(s+2)(s+5)}$

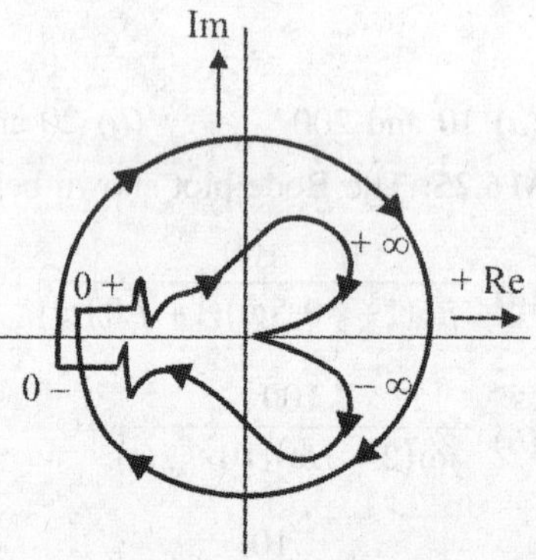

M 6.31: The system having the Bode magnitude plot as shown below has the transfer function

(*a*) $\dfrac{60(s+0.01)(s+0.1)}{s^2(s+0.05)^2}$

(*b*) $\dfrac{100(1+10s)}{s(1+20s)}$

(*c*) $\dfrac{3(s+0.05s)}{s(s+0.1)(s+1)}$

(*d*) $\dfrac{5(s+0.1)}{s(s+0.05)}$

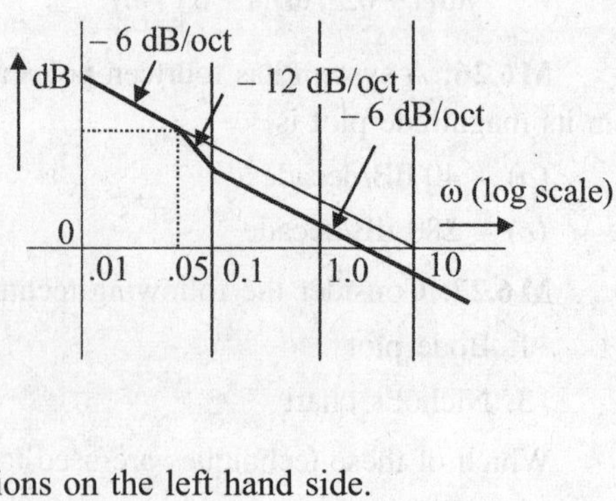

M 6.32: Match the polar plots for the following functions on the left hand side.

(A) $\dfrac{s}{(s+1)(s+2)}$

(P)

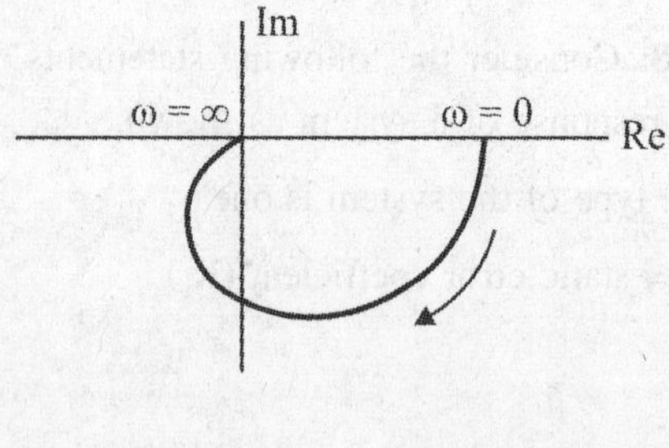

(B) $\dfrac{s^2+1}{s^3}$ (Q) 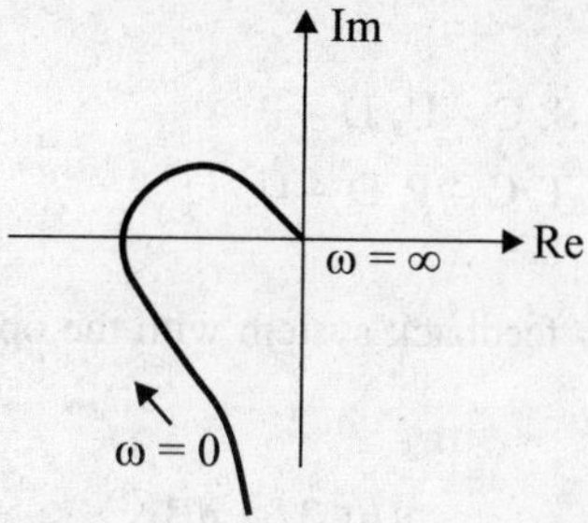

(C) $\dfrac{s^2-1}{s^2+1}$ (R) 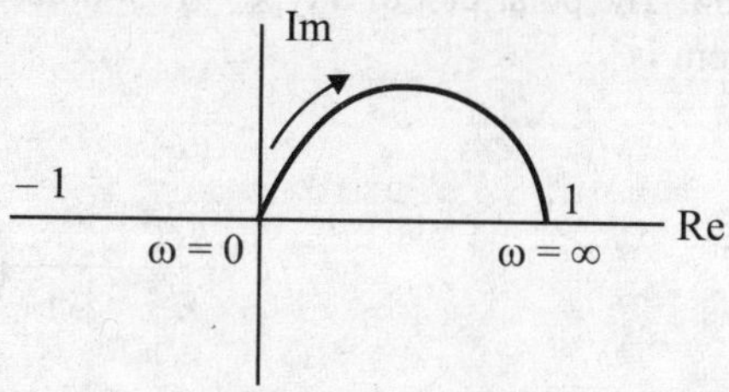

(D) $\dfrac{1}{s^2+10}$ (S) 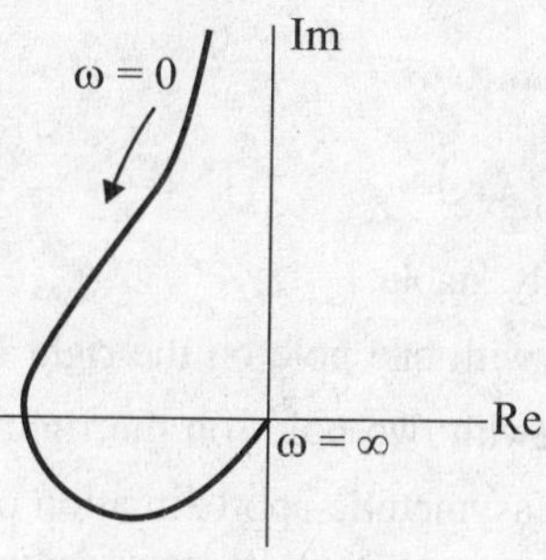

(T) 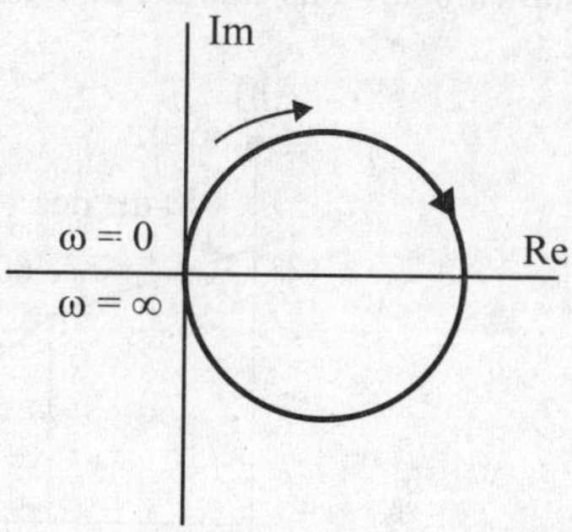

(U) 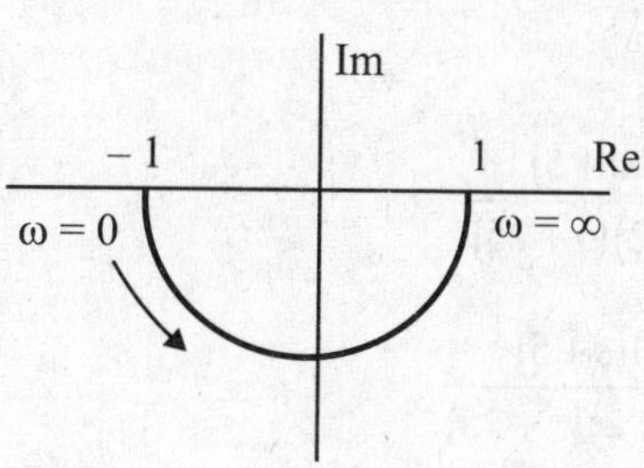

Codes:

(*a*) A – T, B – S, C – U, D – P (*b*) A – R, B – T, C – U, D – P

(*c*) A – R, B – T, C – P, D – U (*d*) A – U, B – R, C – U, D – P

M6.33: A unity feedback system with the open loop transfer function $G(s) = \dfrac{1}{s(s+2)(s+4)}$ has gain margin of

(*a*) 33.6 dB (*b*) 32.4 dB (*c*) 32.6 dB (*d*) 30.6 dB

M6.34: The polar plot of a type – 1, 3–pole, stable open-loop system is shown below. The closed-loop system is

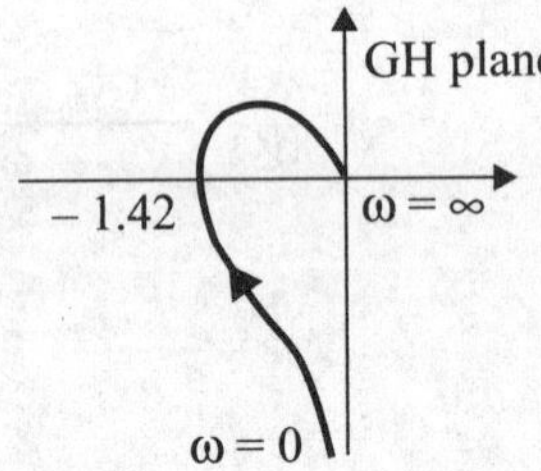

(*a*) always stable

(*b*) marginally stable

(*c*) unstable with one pole on the right half *s*-plane

(*d*) unstable with two poles on the right half *s*-plane.

M 6.35: The asymptotic approximation of the log-magnitude versus frequency plot of a minimum phase system is shown below. Its transfer function is

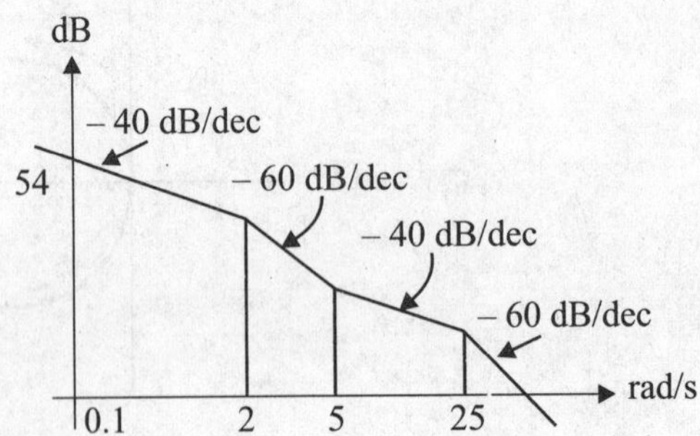

(*a*) $\dfrac{20(s+5)}{s(s+2)(s+25)}$ (*b*) $\dfrac{10(s+5)}{(s+2)^2(s+25)}$

(*c*) $\dfrac{20(s+5)}{s^2(s+2)(s+25)}$ (*d*) $\dfrac{50(s+5)}{s^2(s+2)(s+25)}$

M 6.36: The pole-zero pattern of a certain filter is shown below. The filter must be of the following type (× depicts pole and o depicts zero):

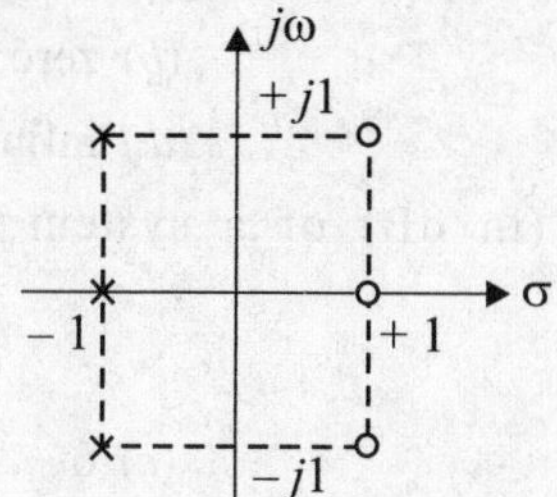

(*a*) low-pass (*b*) high-pass (*c*) all-pass (*d*) band-pass

M 6.37: The open-loop transfer function of a feedback control system is

$$G(s)\ H(s) = \frac{1}{(s+1)^3}$$

The gain margin of the system is

(*a*) 2 (*b*) 4 (*c*) 8 (*d*) 16

M 6.38: Bode plot of a stable system is shown below. The transfer function of the system is

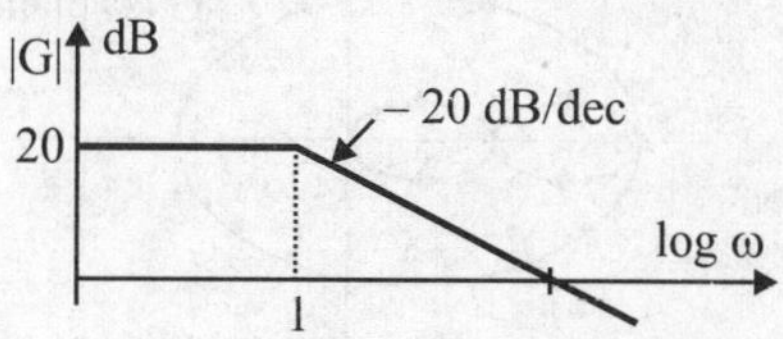

(*a*) $G(s) = \dfrac{100}{s+10}$ (*b*) $G(s) = \dfrac{10}{s+10}$

(*c*) $G(s) = \dfrac{10}{s+1}$ (*d*) $G(s) = \dfrac{10}{s+100}$

M 6.39: Non-minimum phase transfer function is defined as the transfer function

1. which has zeros in the right-half of *s* plane.
2. which has zeros only in the left-half of *s* plane.
3. which has poles in the right-half of *s* plane.
4. which has poles in the left-half of *s* plane.

Of these

(*a*) 1 and 3 are correct (*b*) 1 and 4 are correct

(*c*) 2 alone is correct (*d*) 2 and 3 are correct.

M 6.40: In the Bode plot of a unity feedback control system, the value of phase of G (*j*ω) at the gain cross over frequency is – 125°. The phase margin of the system is

(*a*) – 125° (*b*) – 55° (*c*) 55° (*d*) 125°

M 6.41: The Nyquist plot of a loop transfer function G(*j*ω) H (*j*ω) of a system encloses the (– 1, *j*0) point. The gain margin of the system is

(*a*) less than zero (*b*) zero

(*c*) greater than zero (*d*) infinity

M 6.42: The gain margin (in dB) of a system having the loop transfer function $G(s)\,H(s) = \frac{\sqrt{2}}{s(s+1)}$ is

(*a*) 0 (*b*) 3 (*c*) 6 (*d*) ∞

M 6.43: The phase margin (in degrees) of a system having the loop transfer function $G(s)\,H(s) = \frac{2\sqrt{3}}{s(s+1)}$ is

(*a*) 45 (*b*) – 30 (*c*) 60 (*d*) 30

M 6.44: The Nyquist plot for the open-loop transfer function G(*s*) of a unity negative feedback system is shown below. If G(*s*) has no pole in the right-half of *s* plane, the number of roots of the system characteristic equation in the right-half of *s*-plane is

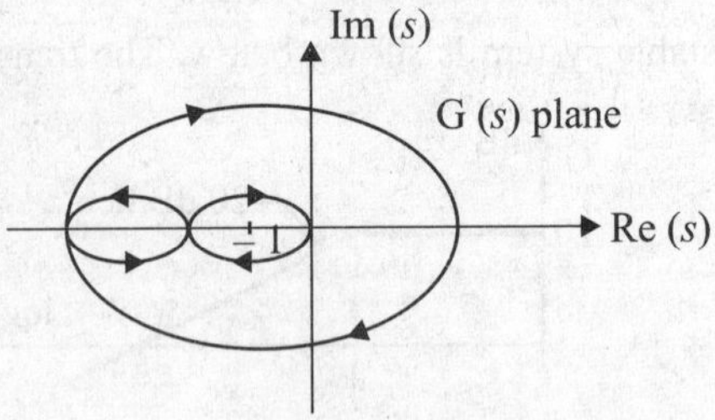

(*a*) 0 (*b*) 1 (*c*) 2 (*d*) 3

M6.45: Figure shows the Nyquist plot of the open-loop transfer function G(*s*) H(*s*) of a system. If G(*s*) H(*s*) has one right-hand pole, the closed-loop system is

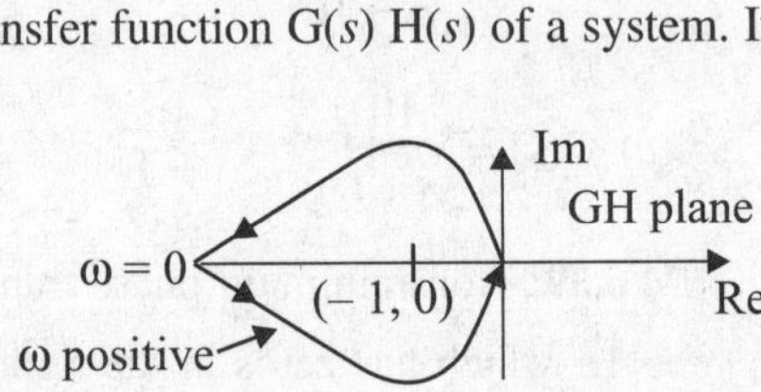

(*a*) always stable.

(*b*) unstable with one closed-loop right hand pole.

(*c*) unstable with two closed-loop right hand poles.

(*d*) unstable with three closed-loop right hand poles.

M6.46: The gain-phase plot of a linear control system is shown below.

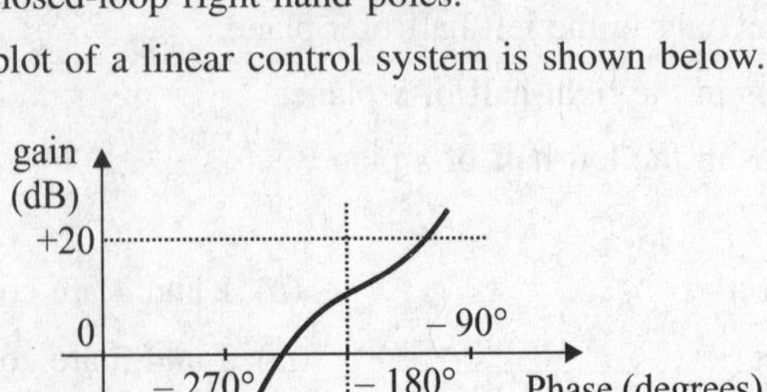

What are gain margin (GM) and phase margin (PM) of the system?

(*a*) GM > 0 dB and PM > 0 degree (*b*) GM < 0 dB and PM < 0 degree

(*c*) GM > 0 dB and PM < 0 degree (*d*) GM < 0 dB and PM > 0 degree

M6.47: The asymptotic Bode plot of the transfer function $\frac{K}{1+\frac{s}{a}}$ is given in figure. The error in phase angle and dB gain at a frequency of $\omega = 0.5\,a$ are respectively

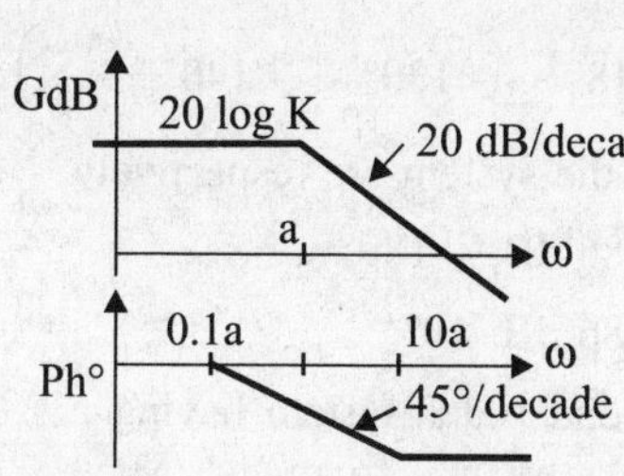

(*a*) 4.9°, 0.97 dB (*b*) 5.7°, 3 dB (*c*) 4.9°, 3 dB (*d*) 5.7°, 0.97 dB

M6.48: Consider the Nyquist diagram for given KG(*s*)H(*s*). The transfer function KG(*s*)H(*s*) has no poles and zeros in the right half of *s* plane. If the (– 1, *j*0) point is located first in region I and then in region II, the change in stability of the system will be from

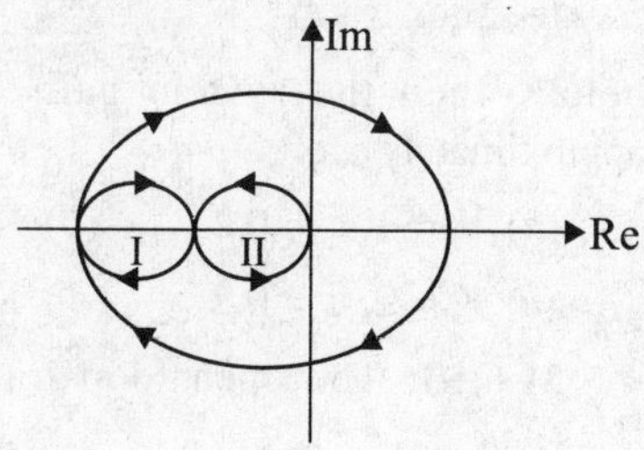

(*a*) unstable to stable (*b*) stable to stable

(*c*) unstable to unstable (*d*) stable to unstable

M6.49: The Nyquist plot of a unity feedback system having open loop transfer function $G(s) = \frac{K(s+3)(s+5)}{(s-2)(s-4)}$ for K = 1 is as shown. For the system to be stable, the range of values of K is

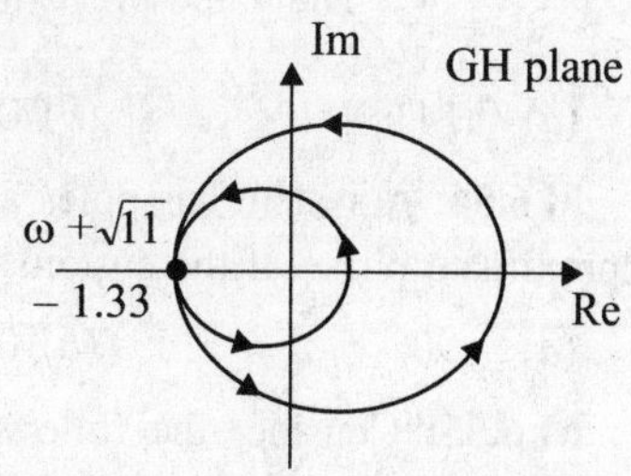

(*a*) 0 < K < 1.33 (*b*) 0 < K < 1/1.33

(*c*) K > 1.33 (*d*) K > 1/1.33

M6.50: The pole-zero map and the Nyquist plot of the loop transfer function GH(*s*) of a feedback system are shown below. For this

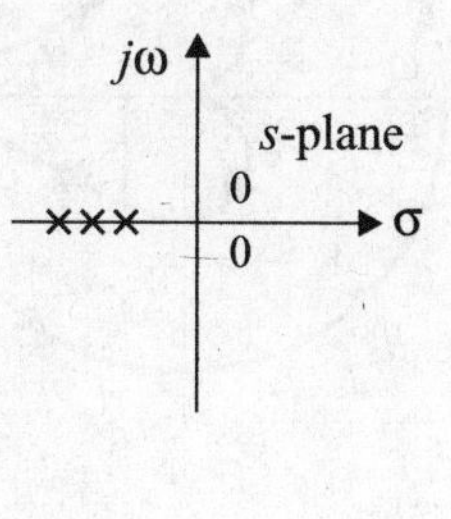

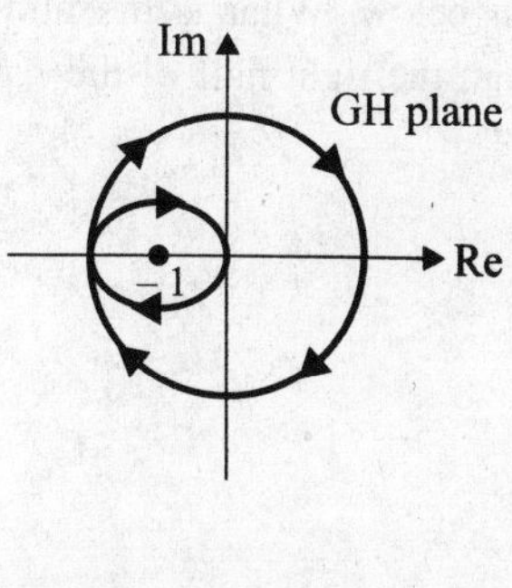

(*a*) both open loop and closed loop systems are stable.

(*b*) open loop system is stable but closed loop system is unstable.

(*c*) open loop system is unstable but closed loop system is stable.

(*d*) both open loop and closed loop systems are unstable.

M 6.51: Consider the following open loop frequency response of a unity feedback system.

ω, rad/s →	2	3	4	5	6	8	10
$\lvert G(j\omega)\rvert$ →	7.5	4.8	3.15	2.25	1.70	1.00	0.64
$\angle G(j\omega)$ →	– 118°	– 130°	– 140°	– 150°	– 157°	– 170°	– 180°

The gain and phase margin of the system are respectively

(*a*) 0.00 dB, – 180° (*b*) 3.86 dB, – 180°

(*c*) 0.00 dB, – 10° (*d*) 3.86 dB, 10°

M 6.52: An underdamped second order system having a transfer function of the form $M(s) = \dfrac{K\omega_n^2}{s^2 + 2\xi\omega_n s + \omega_n^2}$ has frequency response plot as shown below, then the system gain K and the damping ratio approximately are

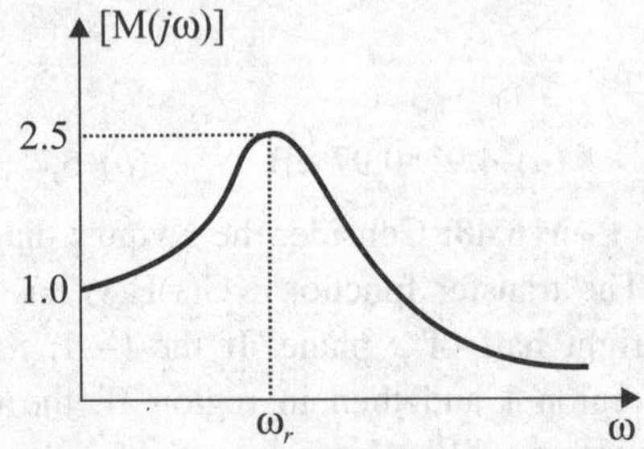

(*a*) K = 1, ξ = 0.2 (*b*) K = 1, ξ = 0.3

(*c*) K = 2, ξ = 0.2 (*d*) K = 3, ξ = 0.3.

M 6.53: The open loop transfer function of a unity feedback control system is given as $G(s) = \dfrac{as+1}{s^2}$. The value of '*a*' to give a phase margin of 45° is equal to

(*a*) 0.141 (*b*) 0.441 (*c*) 0.841 (*d*) 1.141

M 6.54: A system has poles at 0.01 Hz, 1 Hz and 80 Hz; zeros at 5 Hz, 100 Hz and 200 Hz. The approximate phase of the system response at 20 Hz is

(*a*) – 90° (*b*) 0° (*c*) 90° (*d*) – 180°

M 6.55: Consider the following Nyquist plot of a feedback system having open loop transfer function $GH(s) = (s + 1)/[s^2(s - 2)]$ as shown in the diagram given below. What is the number of closed loop poles in the right half of the *s* plane?

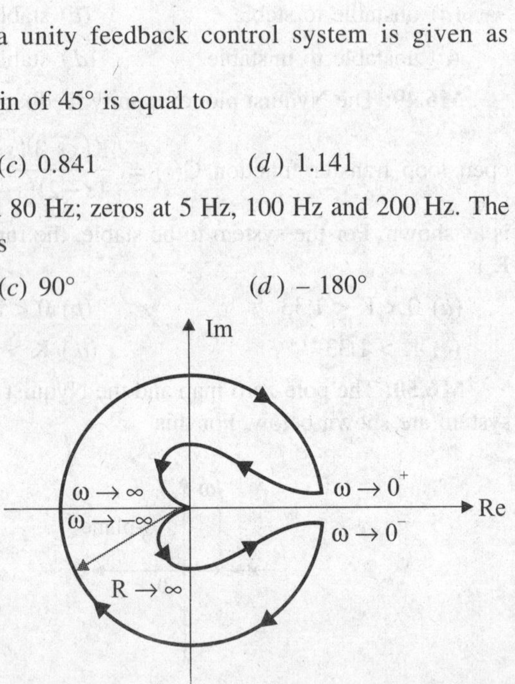

(*a*) 0

(*b*) 1

(*c*) 2

(*d*) 3

M 6.56: Consider the following statements for a counter clockwise Nyquist path.

1. For a stable closed loop system, the Nyquist plot of G(*s*) H(*s*) should encircle (– 1, *j*0) point as many times as there are poles of G(*s*) H(*s*) in the right half of the *s* plane, the encirclements, if there are any must be made in the counter clockwise direction.
2. If the loop gain function G(*s*) H(*s*) is a stable function, the closed loop system is always stable.
3. If the loop gain function G(*s*) H(*s*) is a stable function, for a stable closed-loop system, the Nyquist plot of G(*s*) H(*s*) must not enclose the critical point (– 1, *j*0).

Which of the statements given above is/are correct?

(*a*) Only 1 (*b*) 1 and 2 (*c*) 1 and 3 (*d*) Only 3

M 6.57: The gain margin of a unity feedback control system with the open loop transfer function

$G(s) = \dfrac{s+1}{s^2}$ is

(*a*) 0 (*b*) $1/\sqrt{2}$ (c) $\sqrt{2}$ (*d*) ∞

M 6.58: In the GH(*s*) plane, the Nyquist plot of the loop transfer function $G(s)\,H(s) = \dfrac{\pi e^{-0.25s}}{s}$ passes through the negative real axis at the point

(*a*) (– 0.25, *j*0) (*b*) (– 0.5, *j*0) (*c*) (– 1, *j*0) (*d*) (– 2, *j*0)

M 6.59: If the compensated system shown in figure has a phase margin of 60° at the crossover frequency of 1 rad/sec, the value of the gain K is

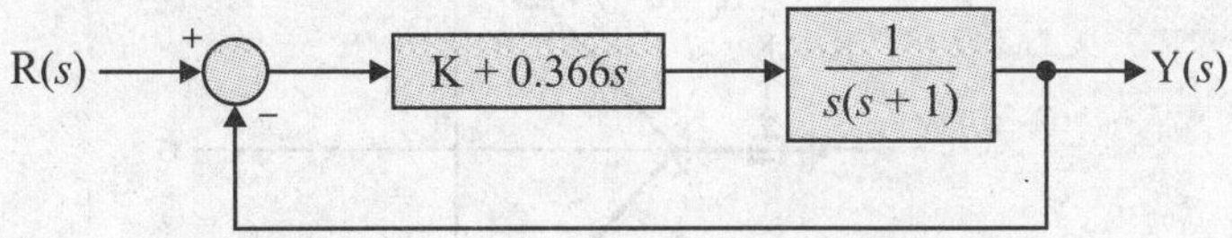

(*a*) 0.366 (*b*) 0.732 (*c*) 1.366 (*d*) 2.738

M 6.60: The polar diagram of a conditionally stable system for open loop gain K = 1 is shown in figure. The open loop transfer function of the system is known to be stable. The closed loop system is stable for

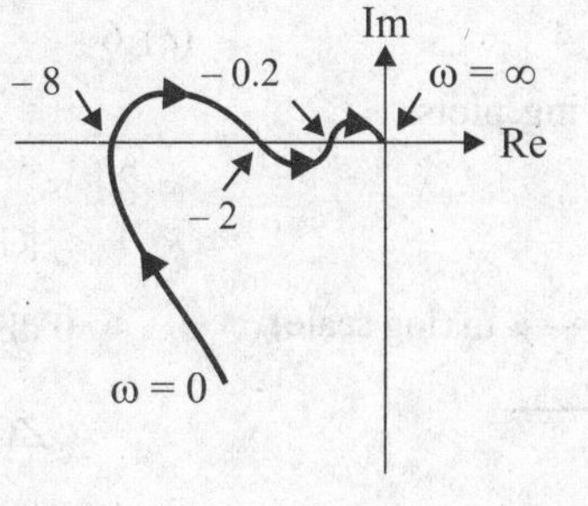

(*a*) K < 5 and 1/2 < K < 1/8 (*b*) K < 1/8 and 1/2 < K < 5

(*c*) K < 1/8 and 5 < K (*d*) K > 1/8 and K < 5

M6.61: The radius of constant N circle of N = 1, is

(*a*) 2 (*b*) $\sqrt{2}$ (*c*) 1 (*d*) $1/\sqrt{2}$.

M6.62: A constant M circle is described by equation

$$x^2 + 2.25x + y^2 = -11.25$$

where x = Re[G($j\omega$)] and y = Im[G($j\omega$)]. The value of M, is

(*a*) 1 (*b*) 2 (*c*) 3 (*d*) 4.

M6.63: A constant N circle has centre at $-\frac{1}{2} + j0$ in G($j\omega$) plane. It represents phase angle equal to

(*a*) 180° (*b*) 90° (*c*) 45° (*d*) 0°.

M6.64: The N loci is described by equation

$$x^2 + x + y^2 = 0 \text{ where } x = \text{Re}[G(j\omega)] \text{ and } y = \text{Im}[G(j\omega)].$$

The value of phase angle, is

(*a*) –45° (*b*) 0° (*c*) 45° (*d*) 90°.

M6.65: The root locus of a unity feedback system is shown in figure below. For design value of gain K = 8, the root locations are shown by small square. The gain margin of system is

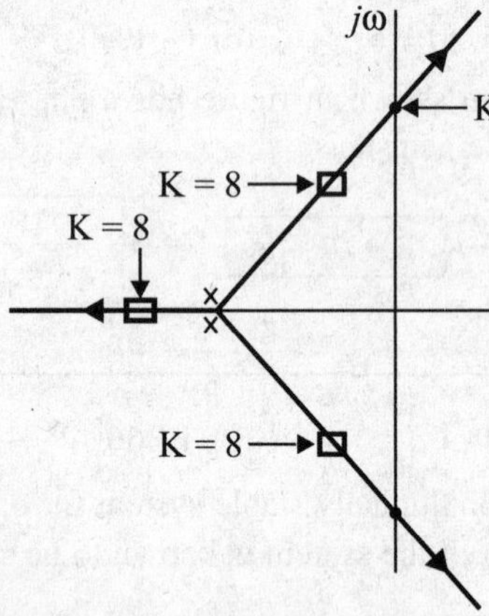

(*a*) 2 (*b*) 4 (*c*) 6 (*d*) 8.

M6.66: Consider the following plots.

1.

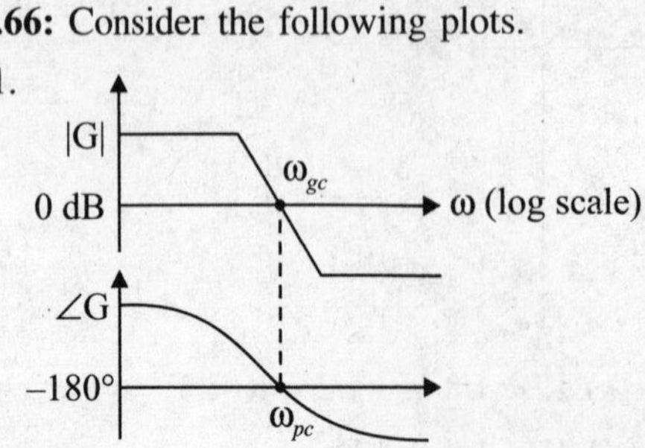

2.

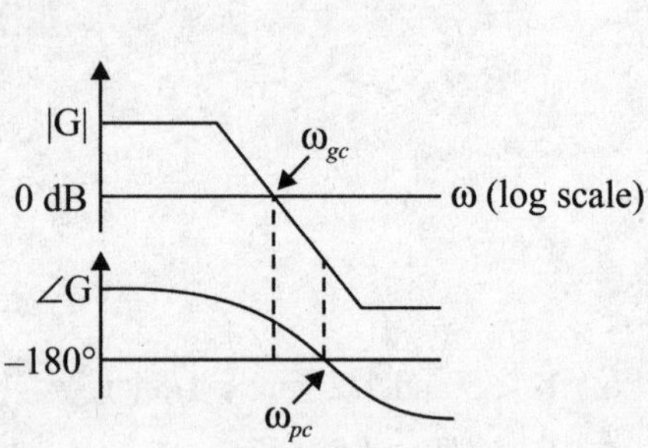

3.

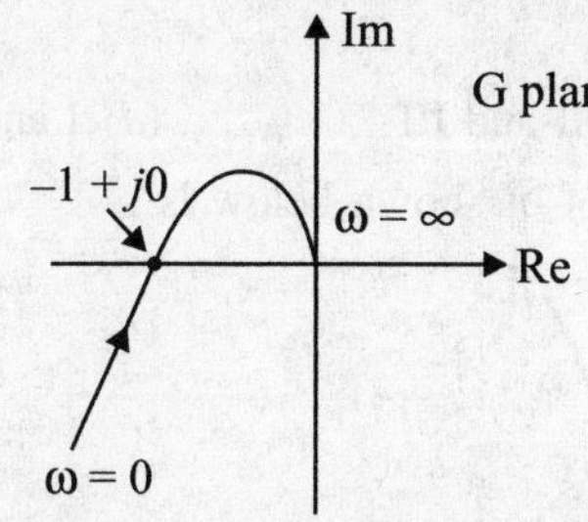

4.

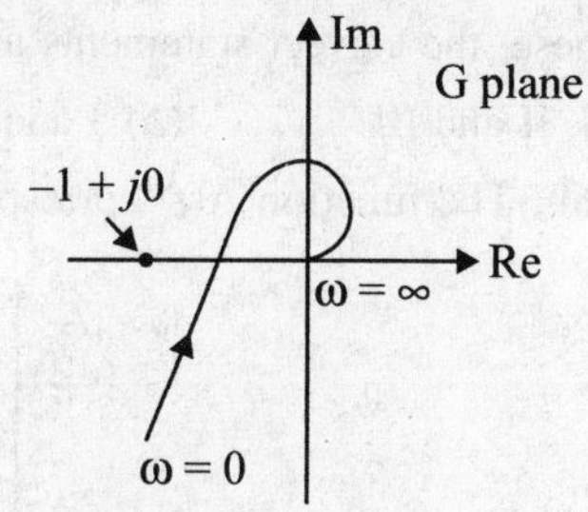

The plots which represent marginally stable systems, would include

(*a*) 1 and 2 (*b*) 3 and 4 (*c*) 1 and 3 (*d*) 2 and 4.

M6.67: The Bode plots of an open loop transfer function of a control system are shown in figure below. The gain margin of the system is

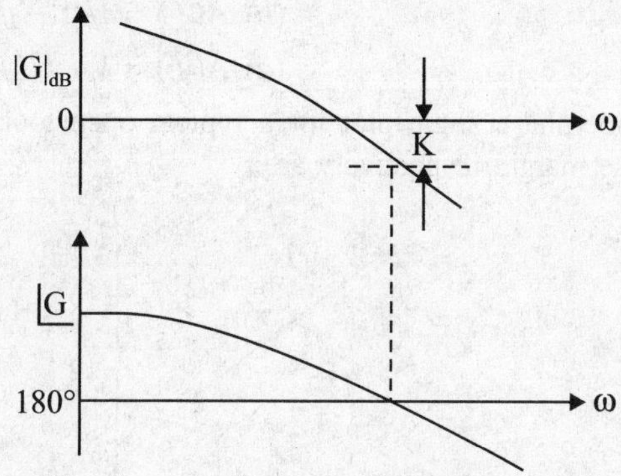

(*a*) K (*b*) –K (*c*) 1/K (*d*) –1/K

M6.68: The dB (Bode plot) of transfer function G(*s*) is shown in figure below.

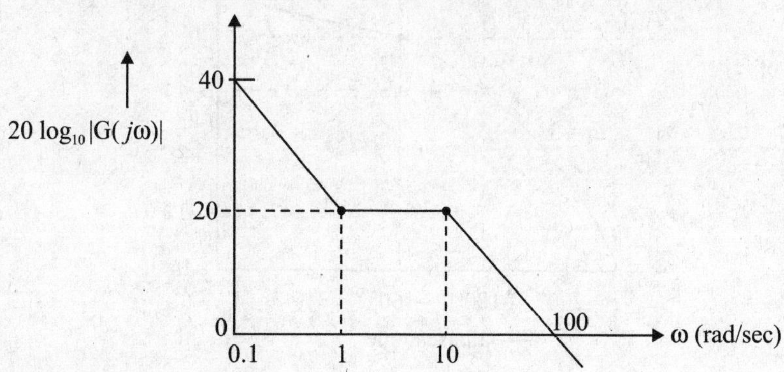

Now, consider the following statements.

I. G(*s*) has corner frequencies at ω = 0.1, 1 and 10.

II. $G(s) = \dfrac{100(s+1)}{s(s+10)}$.

III. The magnitude, $20 \log_{10} |G(j\omega)|$ at ω = 1000, is –20 dB.

Of these, the correct statements are

(*a*) I, II and III (*b*) I and II (*c*) II and III (*d*) I and III.

M6.69: The function A(*f*) corresponding to Bode plot of shown below is

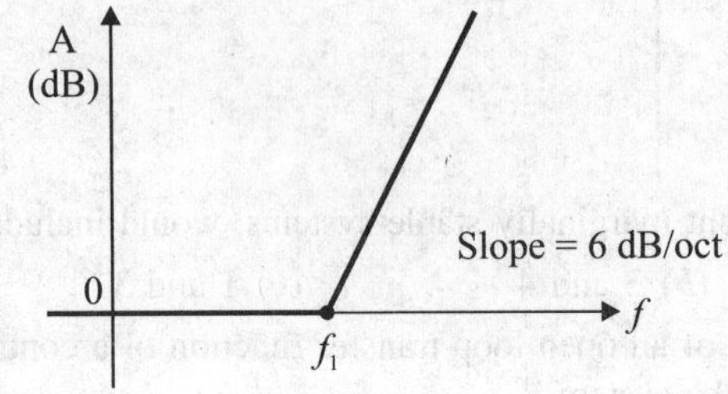

(*a*) $A(f) = jf/f_1$ (*b*) $A(f) = 1/(1 - jf_1/f)$

(*c*) $A(f) = 1/(1 + jf/f_1)$ (*d*) $A(f) = 1 + jf/f_1$.

M6.70: The dB magnitude-phase angle plot for a typical open loop transfer function is shown below. Gain margin and phase margin respectively are

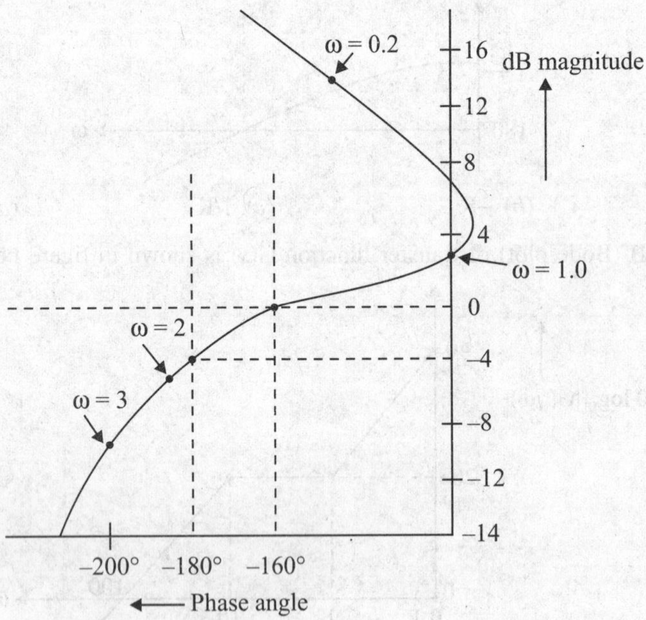

(*a*) 4 dB, 20° (*b*) – 4 dB, –20° (*c*) – 4 dB, 20° (*d*) 4 dB, –20°

M6.71: The Bode dB plot is shown below. The corresponding transfer function model is

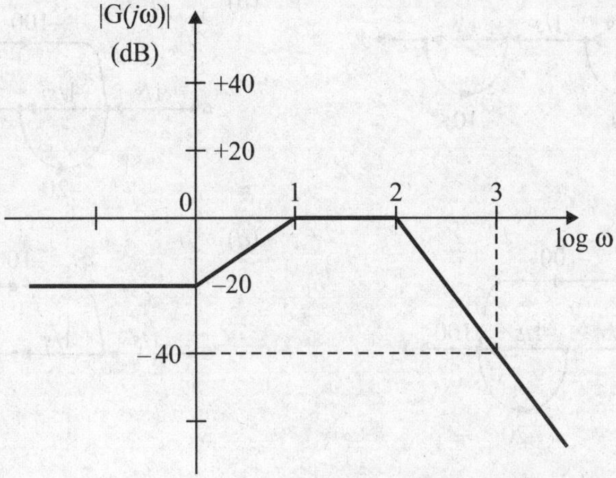

(*a*) $\dfrac{10^4(1+j\omega)}{(10+j\omega)(100+j\omega)^2}$ (*b*) $\dfrac{10^{-1}(1+j\omega)}{(10+j\omega)(100+j\omega)^2}$

(*c*) $\dfrac{10^{-4}(1+j\omega)}{j\omega(j\omega+100)^2}$ (*d*) $\dfrac{10^4(1+j\omega)}{\left(1+j\dfrac{\omega}{10}\right)\left(1+j\dfrac{\omega}{100}\right)^2}$

M6.72: If x = Re[G($j\omega$)] and y = Im[G($j\omega$)], then for $\omega \to 0^+$, the Nyquist plot for G(s) = $1/s(s+1)(s+2)$, is

(*a*) $x = 0$ (*b*) $x = -3/4$

(*c*) $x = y - (1/6)$ (*d*) $x = y/\sqrt{3}$

Common Data for questions M6.73 and M6.74

The input-output transfer function of a plant H(s) = $\dfrac{100}{s(s+10)^2}$. The plant is placed in a unity negative feedback configuration as shown in the figure below.

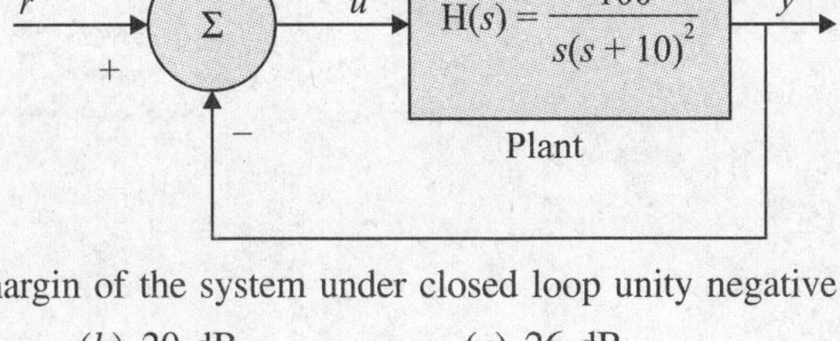

M6.73: The gain margin of the system under closed loop unity negative feedback is

(*a*) 0 dB (*b*) 20 dB (*c*) 26 dB (*d*) 46 dB.

M6.74: The signal flow graph that **DOES NOT** model the plant transfer function H(s) is

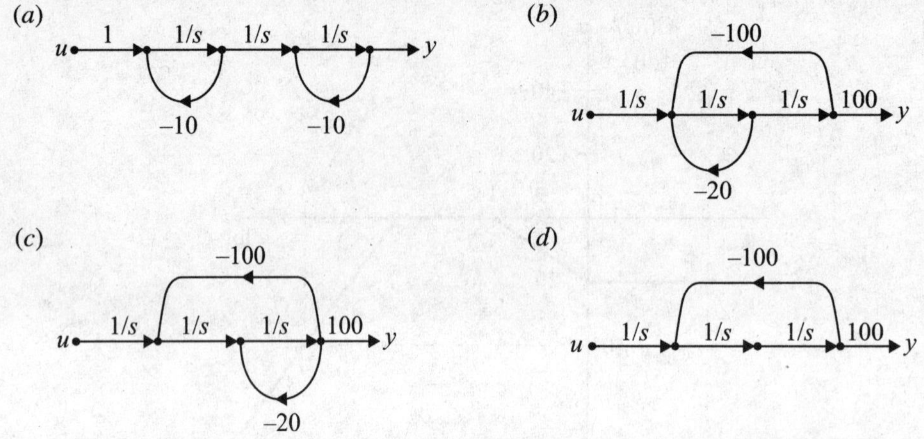

ANSWERS

M6.1. (*d*)	M6.2. (*a*)	M6.3. (*b*)	M6.4. (*a*)	M6.5. (*a*)
M6.6. (*b*)	M6.7. (*b*)	M6.8. (*d*)	M6.9. (*b*)	M6.10. (*c*)
M6.11. (*c*)	M6.12. (*c*)	M6.13. (*b*)	M6.14. (*b*)	M6.15. (*b*)
M6.16. (*d*)	M6.17. (*a*)	M6.18. (*b*)	M6.19. (*a*)	M6.20. (*b*)
M6.21. (*d*)	M6.22. (*d*)	M6.23. (*a*)	M6.24. (*c*)	M6.25. (*d*)
M6.26. (*b*)	M6.27. (*c*)	M6.28. (*a*)	M6.29. (*d*)	M6.30. (*b*)
M6.31. (*d*)	M6.32. (*a*)	M6.33. (*a*)	M6.34. (*d*)	M6.35. (*d*)
M6.36. (*c*)	M6.37. (*c*)	M6.38. (*a*)	M6.39. (*a*)	M6.40. (*c*)
M6.41. (*a*)	M6.42. (*d*)	M6.43. (*d*)	M6.44. (*c*)	M6.45. (*a*)
M6.46. (*b*)	M6.47. (*a*)	M6.48. (*a*)	M6.49. (*d*)	M6.50. (*b*)
M6.51. (*d*)	M6.52. (*a*)	M6.53. (*c*)	M6.54. (*a*)	M6.55. (*c*)
M6.56. (*d*)	M6.57. (*d*)	M6.58. (*b*)	M6.59. (*c*)	M6.60. (*b*)
M6.61. (*d*)	M6.62. (*c*)	M6.63. (*b*)	M6.64. (*d*)	M6.65. (*d*)
M6.66. (*c*)	M6.67. (*a*)	M6.68. (*c*)	M6.69. (*d*)	M6.70. (*a*)
M6.71. (*a*)	M6.72. (*b*)	M6.73. (*c*)	M6.74. (*d*)	

Important Hints:

M 6.1:

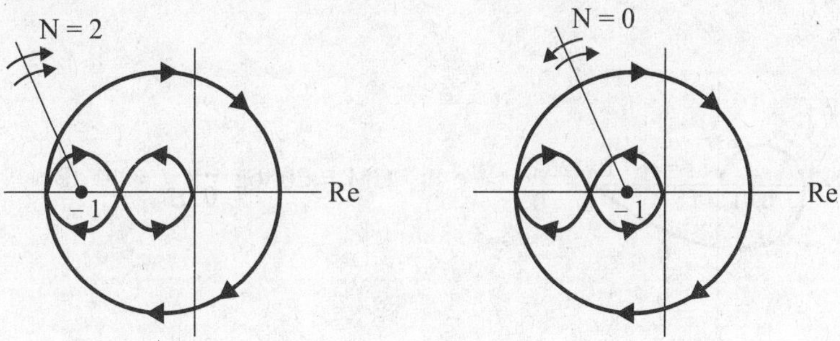

(*i*) (*ii*)

M 6.2:

$$A(\omega)\big|_{\omega=5} = 4 \quad \text{and} \quad \phi(\omega)\big|_{\omega=5} = -15 \text{ rad} = \frac{-15\times180°}{\pi}$$

$$= -859.5° \text{ or } -139.5° \cong -140°$$

$$\text{Forced response } = 10 \times 4 \cos(5t - 30° - 140°)$$

M 6.3:

$$G(j\omega) = \frac{e^{-j4\omega}}{j\omega} = \frac{\cos 4\omega - j\sin 4\omega}{j\omega}$$

$$A(\omega) = |G(j\omega)| = \frac{1}{\omega}, \qquad \phi(\omega) = \angle G(j\omega) = \frac{-\pi}{2} - 4\omega$$

M 6.4: $G(j\omega) = (j\omega)^{1/2}, \quad |G(j\omega)| = \omega^{1/2}$ and $\angle G(j\omega) = \dfrac{1}{2} \times 90° = 45°$

M 6.5: $G(j\omega) = \dfrac{\cos 0.1\omega - j\sin 0.1\omega}{j\omega}, \quad |G(j\omega)|_{\omega = \omega_{gc}} = 1$ gives

$$\omega_{gc} = 1, \; PM = 180° - 90° - 0.1\,\omega_{gc} \times \frac{180°}{\pi} = 84.27°$$

Im [G ($j\omega$)] = 0 gives cos 0.1 ω = 0 or $0.1\,\omega = \dfrac{\pi}{2}$ or $\omega = \dfrac{\pi}{0.2} = 15.7$ rad/sec

So, $$\omega_{pc} = 15.7 \text{ rad/sec}$$

$$\text{Re}\,[G(j\omega)]\big|_{\omega = 15.7} = \frac{-\sin(0.1 \times 15.7)}{15.7} = -6.37 \times 10^{-2}$$

So, $$GM_{dB} = 20 \log \left[\frac{1}{(6.37 \times 10^{-2})}\right] = 23.9$$

Note that all calculations are worked out in radians.

M 6.6: $\angle G(j\omega)\big|_{\omega = 0} = 0$ gives number type of G(s) H(s) = 0. Real axis pole is added.

$\angle G(j\omega)\big|_{\omega = \infty} = -180°$ gives order of G(s) H(s) = 2

M 6.7: Transportation lag in system decreases both GM and PM.

M 6.8: $\omega_{pc} = \infty$

M 6.9:

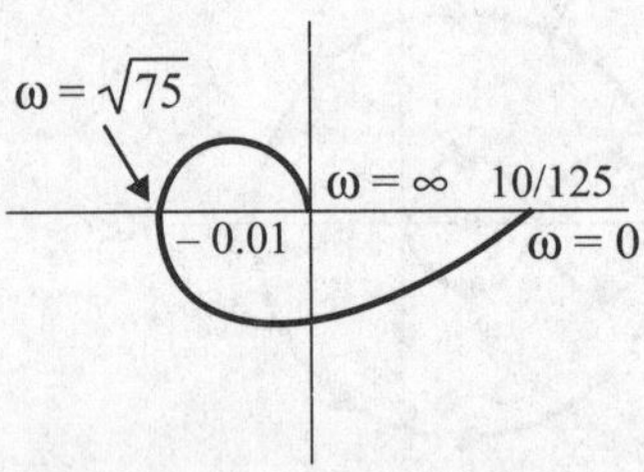

$$GM = 20 \log \frac{1}{0.01} = 40 \text{ dB}$$

M 6.10:

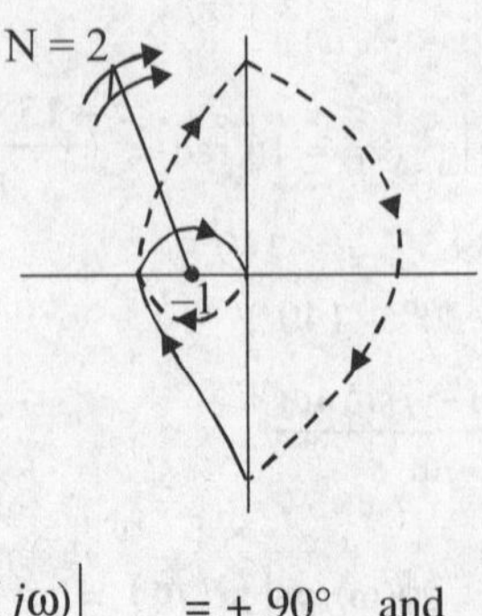

N = Z – P gives Z = 2 while P = 0.

M 6.11: $\angle G(j\omega)\big|_{\omega = 0} = +90°$ and $\angle G(j\omega)\big|_{\omega = \infty} = 0°$

M 6.12: (– 1 + *j*0) point is not enclosed by Nyquist plots (*a*), (*b*) and (*d*).

M 6.13: $\angle G(j\omega)\big|_{\omega=0} = 0°$ and $\angle G(j\omega)\big|_{\omega=\infty} = -90°$

M 6.14: $\angle G(j\omega)$ monotonically decreases from + 90° at ω = 0 to 0° at ω = ∞.

M 6.15: $\angle G(j\omega)\big|_{\omega=0} = -90°$, system is of type 1 and order 2; system has one pole at origin and another real axis LHP.

M 6.16: $$\text{GM} = \left|\frac{1}{\text{real axis intercept}}\right| = \frac{1}{0.4} = 2.5$$

M 6.17:

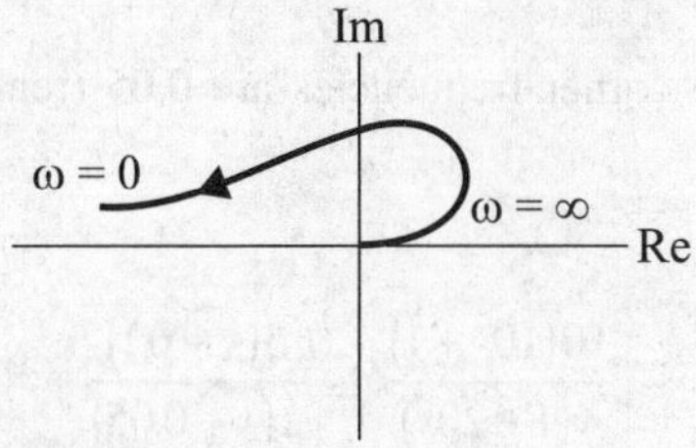

M 6.19: $$|G(j\omega)| = \frac{1}{\left[\left(1-\frac{\omega^2}{\omega_n^2}\right)^2 + \left(2\xi\frac{\omega}{\omega_n}\right)^2\right]^{1/2}} \quad \text{and} \quad |G(j\omega)|_{\omega=\omega_n} = \frac{1}{2\xi}$$

M 6.20: Function is of type 0 and order 2, note $G(j\omega)\big|_{\omega=0} = 1\angle 0°$ while $\left.\frac{1}{s^2+2s+2}\right|_{s=0}$

$= 0.5\angle 0°$.

M 6.22: Type = 1, order = 3. System has one pole at origin and two LHP poles.

M 6.23: Gain is unity or 0 dB and independent of frequency, all pass function.

M 6.24: ± 20 dB/dec = ± 6 dB/octave, $\omega_1 = 10 \times 2 = 20$ and $\omega_2 = 20 \times 2 \times 10 = 400$

M 6.25: $20\log\left.\frac{K}{\omega}\right|_{\omega=10} = 0$ gives K = 10; $G(s) = \frac{10}{s\left(1+\frac{s}{2}\right)\left(1+\frac{s}{10}\right)} = \frac{200}{s(s+2)(s+10)}$

M 6.26: Slope = (– 14 × 20 + 2 × 20) dB/dec = – 240 dB/dec.

M 6.27: Routh criterion gives absolute stability.

M 6.28: ω_2 is just midway between ω_1 and ω_3, $\omega_3 - \omega_2 = \omega_2 - \omega_1$ or $\omega_2 = 1/2\,(\omega_1 + \omega_3)$. Note log ω scale. Had the scale been ω on log scale, relation would have been $\log\omega_3 - \log\omega_2 = \log\omega_2 - \log\omega_1$ or $\omega_2^2 = \omega_1\omega_3$. Refer section 6.5 (relation between dB plot and number type).

M 6.29: $\left|G(j\omega)\right|_{\omega=\omega_{gc}} = 1$ gives $\dfrac{1}{1+\omega_{gc}^2} = 1$, or $\omega_{gc} = 0$

and $PM = 180° + \angle G(j\omega)\big|_{\omega=\omega_{gc}} = 180° = \pi$ rad.

M 6.30: $\angle G(j\omega)$ varies from – 180° at ω = 0 to – 360° at ω = ∞ monotonically. Two semicircles indicate presence of two poles at origin. G(*s*) also has two LHP poles.

M 6.31: $20\log\dfrac{K}{\omega}\bigg|_{\omega=10} = 0 = 20\log 1$ gives K = 10. Initial segment of slope – 6 dB/oct indicates presence of one pole at origin. The corner frequencies are 0.05 (real axis pole) and 0.1 real axis zero.

$$G(j\omega) = \frac{10\left(1+\dfrac{s}{0.1}\right)}{s\left(1+\dfrac{s}{0.05}\right)} = \frac{10(10s+1)}{s(1+20s)} = \frac{5(s+0.1)}{s(s+0.05)}$$

M 6.33: $$G(j\omega) = \frac{-1}{6\omega^2 + j\omega(\omega^2-8)} = \frac{-6\omega^2 + j\omega(\omega^2-8)}{36\omega^4 + \omega^2(\omega^2-8)^2}$$

$$\text{Im}[G(j\omega)] = 0 \text{ for } \omega = 0 \text{ or } \omega = 2\sqrt{2} \text{ and } \text{Re}[G(j\omega)]\big|_{\omega=2\sqrt{2}} = -\frac{1}{48}$$

$$GM = 20\log_{10}(48) = 33.6 \text{ dB}$$

M 6.34:

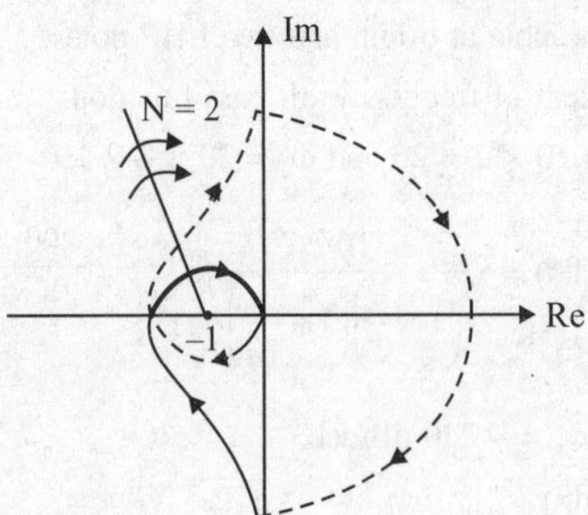

$$N = 2, \quad P = 0 \text{ and } Z = N + P = 2$$

M 6.35: $20\log\dfrac{K}{\omega^2}\bigg|_{\omega=0.1} = 54$ gives K = 5, $G(s) = \dfrac{5\left(1+\dfrac{s}{5}\right)}{s^2\left(1+\dfrac{s}{2}\right)\left(1+\dfrac{s}{25}\right)} = \dfrac{50(s+5)}{s^2(s+2)(s+25)}$

M 6.36: Mirror image RHP zeros for all LHP poles.

M 6.37: $$G(j\omega)\,H(j\omega) = \frac{-1}{(1-3\omega)^2 + j\omega(3-\omega^2)} = \frac{(1-3\omega^2) - j\omega(3-\omega^2)}{(1-3\omega^2)^2 + \omega^2(3-\omega^2)^2}$$

$\text{Im}\,[G(j\omega)\,H(j\omega)] = 0$ for $\omega = 0$ and $\omega = \sqrt{3}$

$$\text{Re}\big[G(j\omega)H(j\omega)\big]\Big|_{\omega=\sqrt{3}} = -\frac{1}{8}, \quad GM = 8$$

M 6.38: $20 \log K = 20$ gives $K = 10$, $G(s) = \left(\dfrac{10}{1+\dfrac{s}{10}}\right)$; pole is located at $\log \omega = 1$ or $\omega = 10$.

M 6.40: $\underline{|PM} = 180° + G(j\omega)\,|_{\omega = \omega_{gc}}$

M 6.42: Second order system has $\omega_{pc} = \infty$ and $GM = \infty$

M 6.43: $|G(j\omega)|\Big|_{\omega=\omega_{gc}} = 1$ gives $\dfrac{2\sqrt{3}}{\omega_{gc}\sqrt{1+\omega_{gc}^{\;2}}} = 1$

or $\omega_{gc} = \sqrt{3}$ and $PM = 180° + \underline{|G(j\omega)}\Big|_{\omega=\sqrt{3}}$

M 6.44: From $N = Z - P$, $P = 0$, $N = 2$ and $Z = 2$

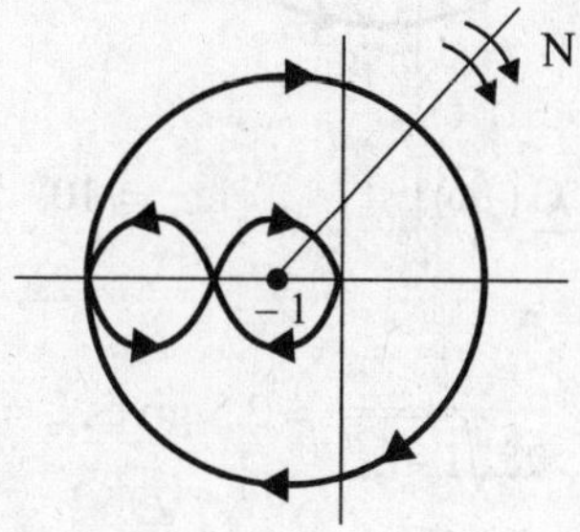

M 6.45: $N = -1$, $P = 1$, $Z = N + P = 0$

M 6.48:

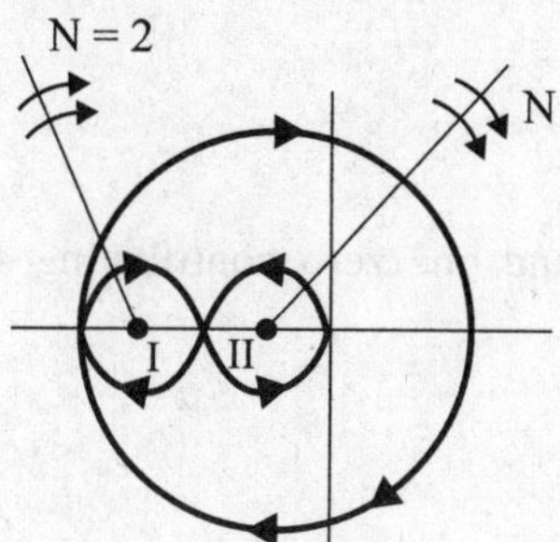

M 6.49: For 1.33 K > 1, N = – 2 and Z = N + P = – 2 + 2 = 0, note two RHP poles.

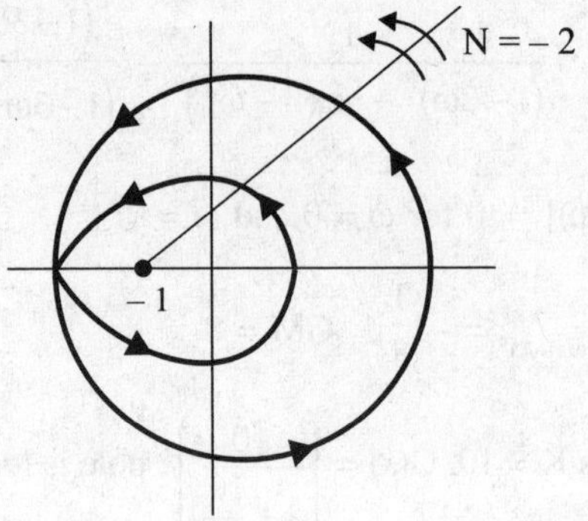

M 6.50: No RHP poles, P = 0 (stable open loop); N = 2, Z = 2 (unstable closed loop with 2 RHP roots)

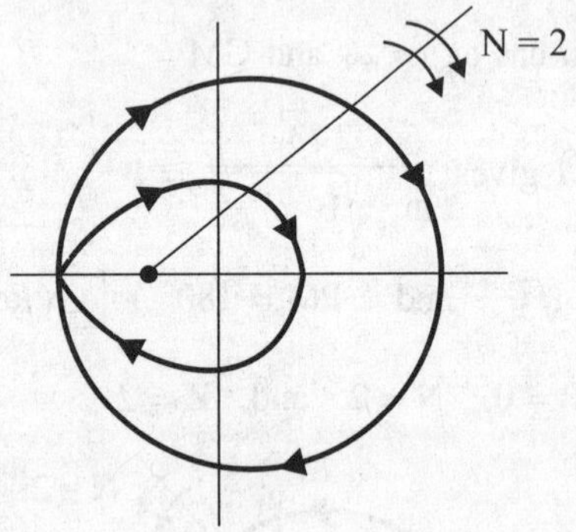

M 6.51: $\omega_{gc} = 8$, $\text{PM} = 180° + \underline{\lfloor G(j\omega)}\big|_{\omega=\omega_{gc}}$, $\omega_{pc} = 10$, $\text{GM} = 20\log\dfrac{1}{|G(j\omega)|_{\omega=\omega_{pc}}}$

M 6.52: $|M(j\omega)|_{\omega=0} = K$, $M_r = \dfrac{1}{2\xi\sqrt{1-\xi^2}} = 2.5$ gives $\xi = 0.2$

M 6.53: $\text{PM} = 180° + \underline{\lfloor G(j\omega)}\big|_{\omega=\omega_{gc}}$ gives $\omega_{gc} = \dfrac{1}{a}$ but $|G(j\omega)|_{\omega=\omega_{gc}} = 1$

gives $\dfrac{\sqrt{1+a^2\omega_{gc}^2}}{\omega_{gc}^2} = 1$ or $a^2\sqrt{2} = 1$

M 6.54: for $\omega < 20$ Hz there are two poles and one zero contributing approximately phase $= -90° \times 2 + 90°$

M 6.55: (*c*) N = 1, P = 1, Z = N + P = 2

M 6.57: Order of G(*s*) = 2

M 6.58: $G(j\omega) = \frac{\pi}{j\omega}[\cos 0.25\,\omega - j \sin 0.25\,\omega] = -\frac{\pi}{\omega}\sin 0.25\,\omega - j\frac{\pi}{\omega}\cos 0.25\,\omega$

$\text{Im}\,[G(j\omega)] = 0$ for $\omega = \infty$ and $\omega = 2\pi$ and $\text{Re}\left[G(j\omega)\right]\Big|_{\omega = 2\pi} = -0.5$

M 6.59: $\text{PM} = \angle G(j\omega)\Big|_{\omega = \omega_{gc}} + 180°$ gives

$$180° - 90° - \tan^{-1}\omega_{gc} + \tan^{-1}\left(\frac{0.366\,\omega_{gc}}{K}\right) = 60° \text{ where } \omega_{gc} = 1.$$

M 6.60: For $8K < 1$ or $K < 1/8$, $2K > 1$ or $K > 1/2$ and $0.2\,K < 1$ or $K < 5$ the point $(-1 + j0)$ will not be enclosed by polar plot. So, closed loop system will be stable for $K < 1/8$ and $1/2 < K < 5$.

M6.61: radius $= \sqrt{\frac{1}{4} + \left(\frac{1}{2N}\right)^2} = 1/\sqrt{2}$

M6.62: Compare with general equation of M circle ($M \neq 1$)

$$x^2 + y^2 + \frac{2M^2}{M^2 - 1}x + \frac{M^2}{M^2 - 1} = 0$$

to get $$\frac{M^2}{M^2 - 1} = 1.125 \Rightarrow M = 3$$

M6.63: Compare with general expression for centre of N circles, $\left(-\frac{1}{2}, \frac{1}{2N}\right)$ to get $N = \infty$ or $\tan\alpha = \infty$ or $\alpha = 90°$

M6.64: Compare with $x^2 + x + y^2 - \frac{y}{N} = 0$ to get $N = \infty$ or $\alpha = 90°$

M6.65: $\text{GM} = \frac{\text{gain K at stability boundary}}{\text{design value of K}} = \frac{64}{8} = 8$

M6.66: Plots 2 and 4 represent stable systems.

M6.68: $20\log\frac{K}{\omega}\Big|_{\omega = 1} = 20 \Rightarrow K = 10$ and $G(s) = \frac{10(1+s)}{s\left(1 + \frac{s}{10}\right)} = \frac{100(s+1)}{s(s+10)}$

$G(j\omega)$ has only two corner frequencies at $\omega = 1$ and $\omega = 10$.

M6.70: Curve crosses 0 dB line at a phase angle of $-160° \Rightarrow \text{PM} = 180° - 160° = 20°$

Curve crosses $-180°$ line at $\text{dB} = -4 \Rightarrow \text{GM} = 4$ dB

M6.71: 20 log K = −20 ⇒ K = 10^{-1}

The corner frequencies are log ω = 0 or ω = 1 (a real zero), log ω = 1 or ω = 10(a real pole) and log ω = 2 or ω = 100 (two real poles)

$$G(j\omega) = \frac{10^{-1}(1+j\omega)}{\left(1+\dfrac{j\omega}{10}\right)\left(1+\dfrac{j\omega}{100}\right)^2} = \frac{10^{-1}\times 10^5(1+j\omega)}{(10+j\omega)(100+j\omega)^2}$$

M6.72:

$$G(j\omega) = \frac{1}{j\omega(j\omega+1)(j\omega+2)}$$

$$= \frac{(1-j\omega)(2-j\omega)}{j\omega(1+\omega^2)(4+\omega^2)}$$

$$= \frac{-3}{(1+\omega^2)(4+\omega^2)} - j\frac{2\omega}{(1+\omega^2)(4+\omega^2)}$$

and $$G(j\omega)\Big|_{\omega=0} = -\frac{3}{4} - j0 \Rightarrow x = -3/4$$

M6.73:

$$H(j\omega) = \frac{100}{-20\omega^2 + j(100\omega - \omega^3)}$$

$$\text{Im}[H(j\omega)] = 0 \text{ for } 100\omega - \omega^3 = 0 \text{ or } \omega = 10$$

$$\text{Re}[H(j\omega)]_{\omega=10} = \frac{-100}{20\times 10^2} = -\frac{1}{20}$$

$$GM = 20 \log 20 = 26 \text{ dB.}$$

M6.74: sfg (*d*) has only one feedback loop.

7

STATE SPACE ANALYSIS AND DESIGN

7.1 INTRODUCTION

The classical control tools Routh, Root locus, Bode and Nyquist already discussed in depth in chapter four, five and six respectively belong to pre 1950, first era of development of control system analysis and design. Those days the computers were not available. The emphasis was laid on computational and graphical routines. The techniques of classical control theory despite being powerful, are conceptually simple but need a reasonable amount of computation. Their unique characteristic is that all are based on transfer function model. The major limitations of classical transfer function model approach are as follows.

1. It is generally applicable to single input single output (SISO), linear time invariant (LTI) systems. The design and analysis of multiple input multiple output (MIMO) systems, is approached while taking one loop at a time and therefore it becomes difficult and time consuming. It is powerless for time varying and non linear systems except for a class of simple non linear systems.
2. The transfer function model approach does not incorporate initial conditions in itself. Transfer function is defined under zero initial conditions.
3. For a given input, the transfer function model provides information only about response of system, revealing no information about internal behaviour thereof. There might arise a control situation where it becomes necessary and advantageous to provide feedback proportional to some internal variables of the system rather than the output alone, in order to bring about improvement in system response.
4. The classical techniques do not apply to the design of optimal and adaptive control systems which are often time varying and/or non linear.
5. The classical control approach is essentially based on trial and error routine which, in general, may not lead to control system that is optimal with respect to given performance index.

The second era of development of control system analysis and design is known in literature as **modern control approach** widely popular as **state space approach**, of course not so modern as on today. This allows the system to be modelled directly in time domain and in turn the analysis and design are also tried in time domain. The analysis and design techniques that are strictly manual, have lost their importance due to advent of digital computers. The computers can be used to numerically solve or simulate even large system that are non linear and time varying. Thus the state space approach which is easily amenable to solution through digital computers, alleviates the set of limitations listed just above. State space model provides standard equation arrangement which offers economy of notation and ease of data entry into digital computer processing. The state space approach although, enjoys great success in terms of providing adequate insight into system structure and properties, but some how fails to perform so well in evolving control strategy in real applications, specially where the system models are prone to uncertainty.

The third era, 1970s and 1980s of development of control system analysis and design goes around evolving control strategy that combines the features of classical approach and modern approach. It is commonly popular as **robust control**. It can tackle the problem of uncertainty in system model.

In order to have complete insight into the modern approach, let us first understand the fundamentals of state space modelling. Note that most of the mathematics involved in state space modelling and analysis, relies heavily on matrix algebra. In further discussion it is deemed that reader is fully acquainted with fundamentals of matrix algebra.

7.2 STATE SPACE REPRESENTATION

Before we begin with the routine of state space modelling, let us define the following terms:

State

The state of a dynamical system is the tiniest set of variables called as **state variables**. These variables are so chosen that information about them at $t = 0$ along with information about inputs for $t \geq 0$, completely specifies the system response for any time $t \geq 0$.

If chosen state variables are $x_1(t)$, $x_2(t)$, ..., ...,.....$x_n(t)$, then, in general, they need not be real world recognizable quantities for the state space modelling routine to follow. However, practically, it is better to choose easily recognizable quantities as state variables. Note that evolving optimal control strategy will require the feedback of all such state variables with suitable weights.

State vector

A vector $x(t)$ whose components are n state variables chosen in a dynamical system, is called state vector.

State space

The n-dimensional space with n coordinate axes: x_1, x_2......x_n is called a state space. The state of system at any time $t \geq 0$ can be depicted by a point in the state space. For $n = 2$, the two dimensional space is called **state plane** or **phase plane**.

In general, the state space model for a dynamical system characterised by n^{th} order linear differential equation, can be developed by choosing n state variables and creating n first order differential equations. Recall that Laplace transformed differential equation yields transfer function relating input and output of system. System order is determined from denominator polynomial of transfer function and the order identifies the number of state variables that are needed. The only problem lying here is to determine a set of n state variables such that they are independent of each other. The following three methods are usually employed for proper selection of n independent state variables.

(*i*) If transfer function of dynamical system is known without any information about true system structure, phase or dual phase variables are used to create linkage, called

simulation diagram, between Transfer function model and state space model. The output of each integrator (as explained little later) is defined as a state variable.

(*ii*) If structure is known in the form of block diagram, the simulation diagram follows the structure. The integrator outputs are again defined as state variables but they now have physical significance. These physical variables might include voltage, current, velocity, position, etc depending on system structure.

(*iii*) Either of the two methods provide a set of matrices as a terminal result. Third method involves certain operations performed on set of matrices obtained, such that a new set of matrices are obtained, particularly in some more convenient form, known as **canonical variable** form. The sense in which canonical variable form is more convenient, will be explained later.

With either of three methods as listed above applied, the result is a state space model consisting of a set of matrices. In general, the state space model of a MIMO system may be depicted by a block diagram as shown in Fig. 7.1. The set of n variables x_1, x_2 ...., x_n are state variables.

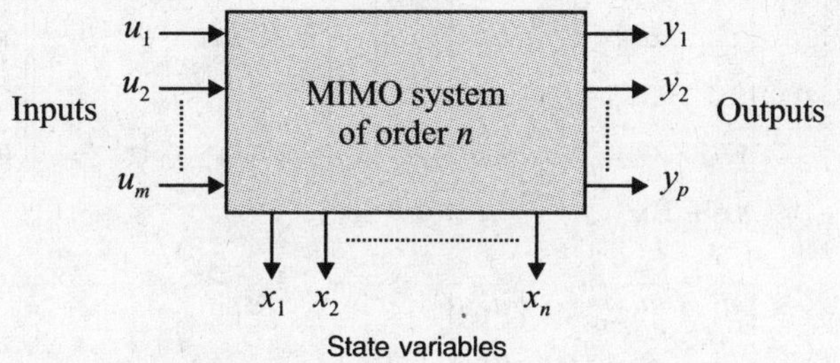

Fig. 7.1: State Space Model of MIMO system

The column matrix of state variables

$$x = \begin{bmatrix} x_1 \\ x_2 \\ \vdots \\ x_n \end{bmatrix}$$

is called $n \times 1$ **state vector**. The system has m inputs u_1, u_2, u_3, ... u_m and the column matrix

$$u = \begin{bmatrix} u_1 \\ u_2 \\ \vdots \\ u_m \end{bmatrix}$$

is called $m \times 1$ **input vector**. The system has p outputs y_1, y_2 .... y_p and column matrix

$$y = \begin{bmatrix} y_1 \\ y_2 \\ \vdots \\ y_p \end{bmatrix}$$

is called $p \times 1$ **output vector**.

A general state space model of nth order MIMO system involves *n* integrators, the outputs of which are *n* state variables. The inputs of each of the integrators are driven with linear combination of state variables and inputs. Note that input of each integrator is first derivative of its own output. Thus a set of *n* first derivatives are represented as linear combination of *n* state variable and *m* inputs as follows:

$$\begin{aligned}
\dot{x}_1 &= a_{11}x_1 + a_{12}x_2 + ... + a_{1n}x_n + b_{11}u_1 + b_{12}u_2 + ... + b_{1m}u_m \\
\dot{x}_2 &= a_{21}x_1 + a_{22}x_2 + ... + a_{2n}x_n + b_{21}u_1 + b_{22}u_2 + ... + b_{2m}u_m \\
\vdots \quad & \quad \vdots \qquad\qquad\qquad\qquad \vdots \qquad\qquad\qquad\qquad \vdots \\
\dot{x}_n &= a_{n1}x_1 + a_{n2}x_2 + ... + a_{nn}x_n + b_{n1}u_1 + b_{n2}u_2 + ... + b_{nm}u_m
\end{aligned} \qquad ...(7.1)$$

These first order differential equations are called **state equations** and are usually written compactly in matrix form as:

$$\begin{bmatrix} \dot{x}_1 \\ \dot{x}_2 \\ \vdots \\ \dot{x}_n \end{bmatrix} = \begin{bmatrix} a_{11} & a_{12} & \cdots & a_{1n} \\ a_{21} & a_{22} & \cdots & a_{2n} \\ \vdots & & & \vdots \\ a_{n1} & a_{n2} & \cdots & a_{nn} \end{bmatrix} \begin{bmatrix} x_1 \\ x_2 \\ \vdots \\ x_n \end{bmatrix} + \begin{bmatrix} b_{11} & b_{12} & \cdots & b_{1m} \\ b_{21} & b_{22} & \cdots & b_{2m} \\ \vdots & & & \vdots \\ b_{n1} & b_{n2} & \cdots & b_{nm} \end{bmatrix} \begin{bmatrix} u_1 \\ u_2 \\ \vdots \\ u_m \end{bmatrix} \qquad ...(7.2)$$

or

$$\dot{x} = \mathrm{A}x + \mathrm{B}u \qquad ...(7.3)$$

where

$$\mathrm{A} = \begin{bmatrix} a_{11} & a_{12} & \cdots & a_{1n} \\ a_{21} & a_{22} & \cdots & a_{2n} \\ \vdots & & & \vdots \\ a_{n1} & a_{n2} & \cdots & a_{nn} \end{bmatrix}$$

is called $n \times n$ **system matrix** and

$$\mathrm{B} = \begin{bmatrix} b_{11} & b_{12} & \cdots & b_{1m} \\ b_{21} & b_{22} & \cdots & b_{2m} \\ \vdots & & & \vdots \\ b_{n1} & b_{n2} & \cdots & b_{nm} \end{bmatrix}$$

is called $n \times m$ **input matrix**.

The system outputs are similarly related linearly to state variables and inputs as follows:

$$\begin{aligned}
y_1 &= c_{11}x_1 + c_{12}x_2 + ... + c_{1n}x_n + d_{11}u_1 + d_{12}u_2 + ... + d_{1m}u_m \\
y_2 &= c_{21}x_1 + c_{22}x_2 + ... + c_{2n}x_n + d_{21}u_1 + d_{22}u_2 + ... + d_{2m}u_m \\
\vdots \quad & \quad \vdots \qquad\qquad\qquad\qquad \vdots \qquad\qquad\qquad\qquad \vdots \\
y_p &= c_{p1}x_1 + c_{p2}x_2 + ... + c_{pn}x_n + d_{p1}u_1 + d_{p2}u_2 + ... + d_{pm}u_m
\end{aligned} \qquad ...(7.4)$$

These equations are called **output equations** and are usually written compactly in matrix form as follows:

$$\begin{bmatrix} y_1 \\ y_2 \\ \vdots \\ y_p \end{bmatrix} = \begin{bmatrix} c_{11} & c_{12} & \cdots & c_{1n} \\ c_{21} & c_{22} & \cdots & c_{2n} \\ \vdots & & & \vdots \\ c_{p1} & c_{p2} & \cdots & c_{pn} \end{bmatrix} \begin{bmatrix} x_1 \\ x_2 \\ \vdots \\ x_n \end{bmatrix} + \begin{bmatrix} d_{11} & d_{12} & \cdots & d_{1m} \\ d_{21} & d_{22} & \cdots & d_{2m} \\ \vdots & & & \vdots \\ d_{p1} & d_{p2} & \cdots & d_{pm} \end{bmatrix} \begin{bmatrix} u_1 \\ u_2 \\ \vdots \\ u_m \end{bmatrix} \quad ...(7.5)$$

or

$$y = \mathrm{C}x + \mathrm{D}u \quad ...(7.6)$$

where

$$\mathrm{C} = \begin{bmatrix} c_{11} & c_{12} & \cdots & c_{1n} \\ c_{21} & c_{22} & \cdots & c_{2n} \\ \vdots & & & \vdots \\ c_{p1} & c_{p2} & \cdots & c_{pn} \end{bmatrix}$$

is called $p \times n$ **output matrix** and

$$\mathrm{D} = \begin{bmatrix} d_{11} & d_{12} & \cdots & d_{1m} \\ d_{21} & d_{22} & \cdots & d_{2m} \\ \vdots & & & \vdots \\ d_{p1} & d_{p2} & \cdots & d_{pm} \end{bmatrix}$$

is called $p \times m$ **transmission matrix**.

In order to predict system behaviour from the model governed by state equation (7.3) and output equation (7.6), we need to simulate the system where synthesis of transfer function through the interconnection of simple components becomes necessary. The synthesis leads to the methods of system description, analysis and design. These methods are systematic, compact and suitable for computer analysis. The basic component of synthesis is the integrator, which is a block with gain **1/*s*** in block diagram or a branch with transmittance **1/*s*** in signal flow graph. A block diagram or signal flow graph composed of constant transmittances and integrators is called **simulation diagram**. Signal flow graph is particularly convenient in the sense that transfer function is evident by mere inspection using Mason's gain rule.

The state equation (7.3) and output equation (7.6) are used to develop block diagram and signal flow graph as shown in Fig. 7.2(*a*) and 7.2(*b*) respectively. Double arrows have been used to indicate vector quantities.

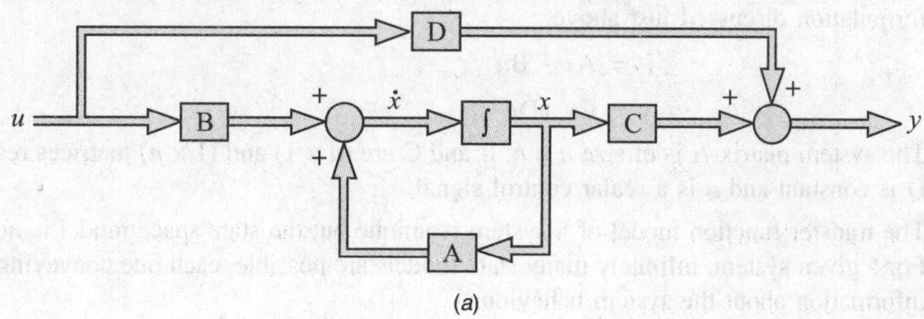

(*a*)

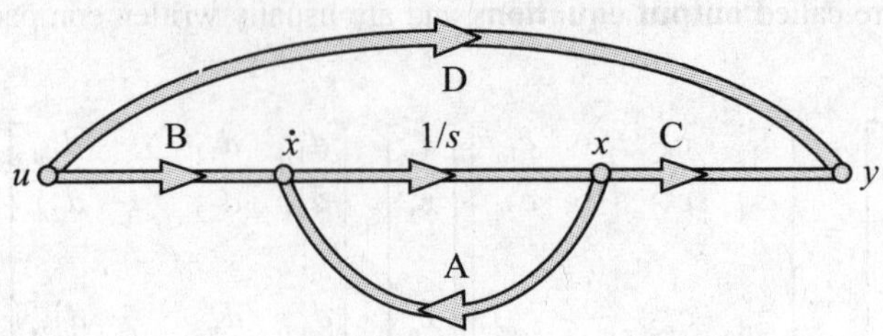

(b)

Fig. 7.2: Simulation diagram (a) Block diagram (b) Signal flow graph

Note the following in foregoing discussion:

(i) The set of four matrices obtained from state equation (7.3) and output equation (7.6) is called a **quadruplet** (A, B, C, D). The quadruplet completely characterises the system dynamics. The state and output equations together constitute state space model or simply state model.

(ii) The state equations describe how the system state vector evolves in time. The tip of state vector determines a point which is called state point in n dimensional space (state space). The curve traced by tip of vector from time $t = 0$ to any time $t > 0$ (say t_1) is called state trajectory in state space. The two dimensional state space is called state plane or phase plane. The output equations describe how the system outputs are related to system states.

(iii) In state model, only state equations are dynamical and required to be solved. The output equation is non dynamical. The solution of state equations provide information about system states and mere substitution of solution of state equations in output equations, provides information about system outputs. In this sense, the output equation is also called as **read out function**.

(iv) The presence of transmission matrix D in state model identifies direct coupling between input and output. It rarely occurs in control systems as power amplification of control signal is very often required. If we have transfer function where order of numerator polynomial is equal to that of denominator polynomial or number of poles is equal to that of zeros, the corresponding state model will have non zero transmission matrix D.

(v) The state model for a linear time varying system has the form same as given by (7.3) and (7.6) with the only change that the elements of quadruplet (A, B, C, D) are no more constants but are functions of time. However we shall constrict our discussion to only LTI systems.

(vi) The state model for SISO LTI system can be obtained by simply assuming $p = m = 1$ in the formulation discussed just above.

$$\dot{x} = Ax + Bu \qquad ...(7.7)$$

$$y = Cx + Du \qquad ...(7.8)$$

The system matrix A is of size $n \times n$. B and C are $(n \times 1)$ and $(1 \times n)$ matrices respectively. D is constant and u is a scalar control signal.

(vii) The transfer function model of a system is unique but the state space model is non unique. For a given system, infinitely many state models are possible, each one conveying the same information about the system behaviour.

Phase variable form of state model

A transfer function T(s) that is expressed as a ratio of numerator polynomial N(s) to denominator polynomial D(s) such that T(s) = $\dfrac{N(s)}{D(s)}$ is said to be **rational**. If degree of N(s) is less than or equal to that of D(s), then T(s) is said to be **proper** as well. T(s) is said to be **strictly proper** if order of N(s) is strictly less than that of D(s). The simulation diagram may be developed for any proper, rational transfer function. The simulation diagram is composed of integrator, constant multiplier and summer. The phase variable form, an useful realization, is the focus of current discussion.

Consider the transfer function

$$T(s) = \frac{Y(s)}{U(s)} = \frac{0.4s^2 + 1.4s + 0.8}{s^3 + 0.3s^2 + 1.7s + 0.2}$$

Dividing numerator and denominator both by s^3 (highest power of s in denominator) in order that the terms in both numerator and denominator have inverse powers of s (s^{-r} for $r > 1$ represents multiple integrations) we have

$$T(s) = \frac{Y(s)}{U(s)} = \frac{\dfrac{0.4}{s} + \dfrac{1.4}{s^2} + \dfrac{0.8}{s^3}}{1 + \dfrac{0.3}{s} + \dfrac{1.7}{s^2} + \dfrac{0.2}{s^3}} = \frac{\dfrac{0.4}{s} + \dfrac{1.4}{s^2} + \dfrac{0.8}{s^3}}{1 - \left(-\dfrac{0.3}{s} - \dfrac{1.7}{s^2} - \dfrac{0.2}{s^3}\right)}$$

$$\equiv \frac{p_1 + p_2 + p_3}{1 - (L_1 + L_2 + L_3)}$$

Comparing it with Mason's gain formula T(s) = $\dfrac{\Sigma p_i \Delta_i}{\Delta}$, the following observations are made.

(*i*) The numerator terms identify the three forwards paths with path gains $p_1 = \dfrac{0.4}{s}$, $p_2 = \dfrac{1.4}{s^2}$ and $p_3 = \dfrac{0.8}{s^3}$ corresponding cofactors $\Delta_1 = \Delta_2 = \Delta_3 = 1$. Note that signal flow graph is to be so constructed that forward paths are intermingled through minimal number of integrators as shown in Fig. 7.3(*a*).

(*ii*) The denominator terms identify the three individual loops with loops gains $L_1 = -0.3/s$, $L_2 = -1.7/s^2$ and $L_3 = -0.2/s^3$. Note that signal flow graph is to be so constructed that all the loops have at least one common node so that no product of loop gain terms evolve. All the loops must also touch each of the forward paths so that each path cofactor is unity. For this purpose each of the loops are placed through the node to which input U(s) is coupled as shown in Fig. 7.3 (*b*). In order to develop state space model, the integrator outputs from right to left, are chosen as state variables x_1, x_2 and x_3 as shown in Fig. 7.3(*b*).

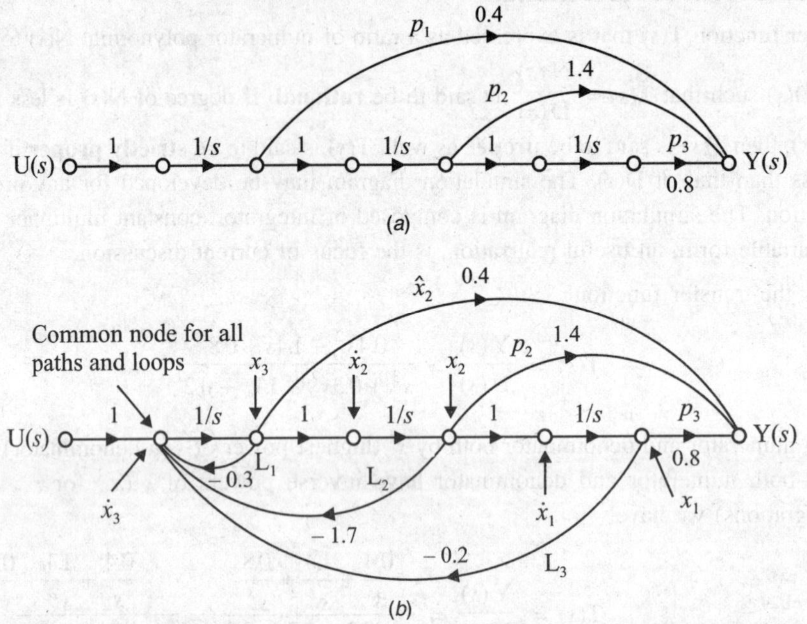

Fig. 7.3: Phase Variable form (*a*) Forward paths in simulation diagram (*b*) Complete simulation diagram

The corresponding input nodes of each integrator is labelled $\dot{x}_1, \dot{x}_2$ and $\dot{x}_3$ as $\dot{x}_1$ integrated yields x_1 itself and so on. From simulation diagram, the state and output equations in time domain are developed as follows:

$$\dot{x}_1 = x_2$$

$$\dot{x}_2 = x_3$$

$$\dot{x}_3 = -0.2x_1 - 1.7x_2 - 0.3x_3 + u$$

$$y = 0.8x_1 + 1.4x_2 + 0.4x_3$$

This is a set of coupled first order differential equations. Defining the vectors x and $\dot{x}$ as

$$x = \begin{bmatrix} x_1 \\ x_2 \\ x_3 \end{bmatrix} \quad \text{and} \quad \dot{x} = \begin{bmatrix} \dot{x}_1 \\ \dot{x}_2 \\ \dot{x}_3 \end{bmatrix}$$

above equations can be written in matrix-vector form as

$$\begin{bmatrix} \dot{x}_1 \\ \dot{x}_2 \\ \dot{x}_3 \end{bmatrix} = \begin{bmatrix} 0 & 1 & 0 \\ 0 & 0 & 1 \\ -0.2 & -1.7 & -0.3 \end{bmatrix} \begin{bmatrix} x_1 \\ x_2 \\ x_3 \end{bmatrix} + \begin{bmatrix} 0 \\ 0 \\ 1 \end{bmatrix} u$$

$$y = \begin{bmatrix} 0.8 & 1.4 & 0.4 \end{bmatrix} \begin{bmatrix} x_1 \\ x_2 \\ x_3 \end{bmatrix}$$

Dual phase variable form of state model

Recall that signal flow graph (simulation diagram) in phase variable format of state model was so constructed that all of the forward paths and all of the loops did touch a node to which the input signal was coupled. If the signal flow graph is so constructed that **all of the forward paths and all of the loops touch the output node**, the result is the state model in dual phase variable format. The idea behind doing so is again to ensure that each forward path cofactor is unity and no product of loop gain term evolves in Mason's gain rule. Consider again the transfer function discussed just before

$$T(s) = \frac{Y(s)}{U(s)} = \frac{0.4s^2 + 1.4s + 0.8}{s^3 + 0.3s^2 + 1.7s + 0.2} = \frac{\frac{0.4}{s} + \frac{1.4}{s^2} + \frac{0.8}{s^3}}{1 + \frac{0.3}{s} + \frac{1.7}{s^2} + \frac{0.2}{s^3}}$$

$$\equiv \frac{p_1 + p_2 + p_3}{1 - (L_1 + L_2 + L_3)}$$

In order to develop state model in dual phase variable form, the signal flow graph with three forward paths and three individual loops is constructed as shown in Fig. 7.4. Note that each of three paths and each of the three loops are intermingled and touch a common node, that is output node.

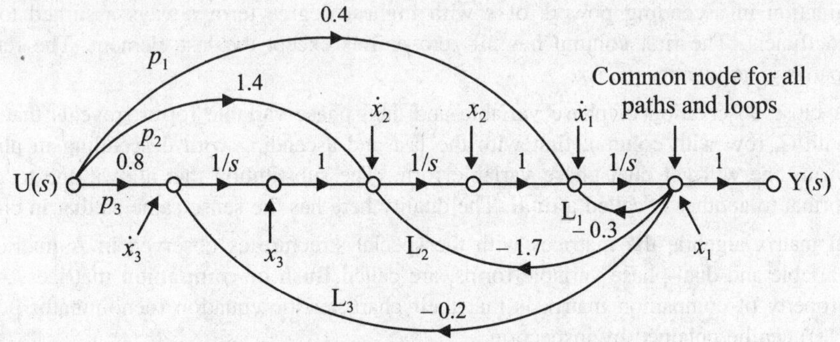

Fig. 7.4: Simulation diagram (Dual phase variable format)

The integrator outputs from right to left are chosen as state variables x_1, x_2 and x_3 as shown in Fig. 7.4 and the corresponding input node of each integrator is labelled $\dot{x}_1$, $\dot{x}_2$ and $\dot{x}_3$.From simulation diagram, the state and output equations are developed in time domain as follows:

$$\dot{x}_1 = -0.3\,x_1 + x_2 + 0.4\,u$$

$$\dot{x}_2 = -1.7\,x_1 + x_3 + 1.4\,u$$

$$\dot{x}_3 = -0.2\,x_1 + 0.8\,u$$

$$y = x_1$$

This is again a set of coupled first order differential equations. With vectors x and $\dot{x}$ as defined previously, these equations can be written in matrix vector form as:

$$\begin{bmatrix} \dot{x}_1 \\ \dot{x}_2 \\ \dot{x}_3 \end{bmatrix} = \begin{bmatrix} -0.3 & 1 & 0 \\ -1.7 & 0 & 1 \\ -0.2 & 0 & 0 \end{bmatrix} \begin{bmatrix} x_1 \\ x_2 \\ x_3 \end{bmatrix} + \begin{bmatrix} 0.4 \\ 1.4 \\ 0.8 \end{bmatrix} u$$

$$y = \begin{bmatrix} 1 & 0 & 0 \end{bmatrix} \begin{bmatrix} x_1 \\ x_2 \\ x_3 \end{bmatrix}$$

Note the following in respect of phase variable and dual phase variable formats of state space model.

(*i*) The phase variable and dual phase variable forms are called **canonical forms** of state model. Both the forms always provide a specific quadruplet (A, B, C, D).

(*ii*) If state variables x_1, x_2 and x_3 are chosen from right to left in phase variable form, then B is a column vector with all elements zero except that the last element is unity. C is a row vector whose elements are coefficients of numerator polynomial of transfer function in ascending powers of s. D is a scalar and is zero if transfer function is strictly proper, that is degree of numerator is strictly less than degree of denominator. The system matrix A when partitioned as

$$A = \left[\begin{array}{c|cc} 0 & 1 & 0 \\ 0 & 0 & 1 \\ \hline -0.2 & -1.7 & -0.3 \end{array}\right]$$

reveals that the elements of the last row are negative of coefficients of denominator of transfer function in ascending powers of s with highest degree term always assumed to have unity coefficient. The first column has all zero entries except the last element. The remaining sub matrix is an identity matrix.

(*iii*) A close observation of phase variable and dual phase variable forms, reveals that replacing B with C, row with column, first with the last and ascending with descending in phase variable form, one will get dual phase variable form. The substitution that allows one to go from one format to another is called a **dual**. The duality here has the sense same as that in circuit theory.

(*iv*) In matrix algebra, the matrices with the special structure as observed in A matrices of phase variable and dual phase variable forms, are called **Bush or companion** matrices. An important property of companion matrix is that their characteristic equation (denominator polynomial of T(s)) can be obtained by inspection.

(*v*) With the observations made just above, it is obvious that phase variable and dual phase variable formats can be obtained directly by mere inspection of transfer function and vice versa. In this sense, the phase variable formulations are the powerful techniques to establish link between classical transfer function approach and modern time domain **design** approach.

(*vi*) Although phase variable formats are simple to realize but in general, the phase variables are not real world recognizable quantities and therefore not available for the purpose of measurement and control. So, phase variable models are not advantageous from view point of measurement and control.

State model for systems with single input and multiple outputs

The state model for a system with single input and multiple outputs can be easily developed using phase variable format in particular. For illustration, consider the transfer functions, describing a system with one input and two outputs.

$$T_1(s) = \frac{Y_1(s)}{U(s)} = \frac{0.4s^2 + 1.4s + 0.8}{s^3 + 0.3s^2 + 1.7s + 0.2}$$

$$T_2(s) = \frac{Y_2(s)}{U(s)} = \frac{-0.5s^2 + 0.7s - 1.9}{s^3 + 0.3s^2 + 1.7s + 0.2}$$

Note that both the transfer functions $T_1(s)$ and $T_2(s)$ share the same denominator polynomial and input. So, system matrix A and input vector B are same as discussed before, that is

$$A = \begin{bmatrix} 0 & 1 & 0 \\ 0 & 0 & 1 \\ -0.2 & -1.7 & -0.3 \end{bmatrix}, \quad B = \begin{bmatrix} 0 \\ 0 \\ 1 \end{bmatrix}$$

and the state equation is

$$\dot{x} = \begin{bmatrix} 0 & 1 & 0 \\ 0 & 0 & 1 \\ -0.2 & -1.7 & -0.3 \end{bmatrix}\begin{bmatrix} x_1 \\ x_2 \\ x_3 \end{bmatrix} + \begin{bmatrix} 0 \\ 0 \\ 1 \end{bmatrix} u$$

Since the system has two outputs y_1 and y_2, the output vector y will be two dimensional and output matrix C together with transmission matrix D will have two rows. Using the property of phase variable format that the row elements of C are coefficients of numerator polynomial of transfer function in ascending powers of s, two dimensional output vector y can be written directly by inspecting $T_1(s)$ and $T_2(s)$ as

$$\begin{bmatrix} y_1 \\ y_2 \end{bmatrix} = \begin{bmatrix} 0.8 & 1.4 & 0.4 \\ -1.9 & 0.7 & -0.5 \end{bmatrix}\begin{bmatrix} x_1 \\ x_2 \\ x_3 \end{bmatrix} + \begin{bmatrix} 0 \\ 0 \end{bmatrix} u$$

where D has all zero entries because $T_1(s)$ and $T_2(s)$ are strictly proper. The simulation diagram (signal flow graph) for state model in phase variable format as shown above, is drawn in Fig. 7.5.

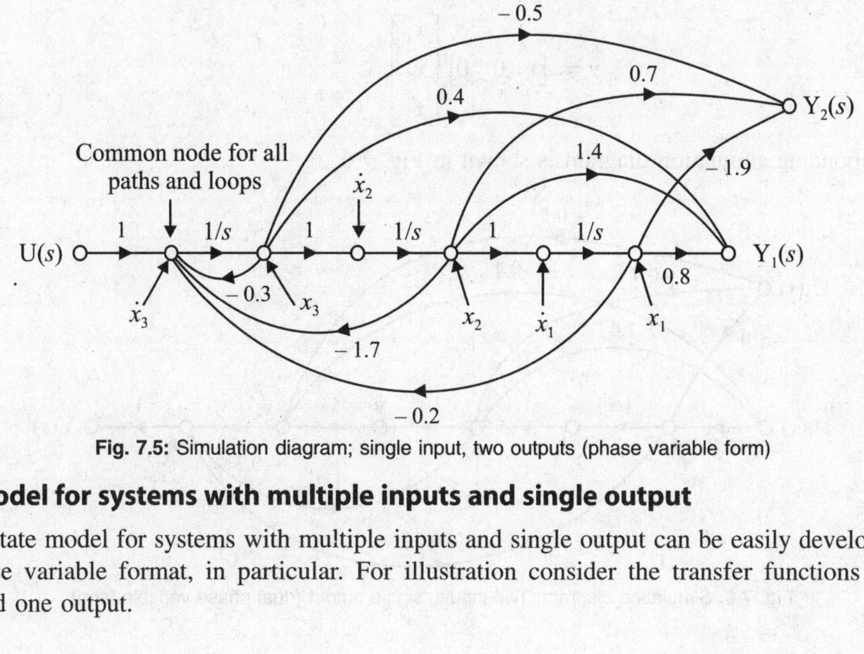

Fig. 7.5: Simulation diagram; single input, two outputs (phase variable form)

State model for systems with multiple inputs and single output

The state model for systems with multiple inputs and single output can be easily developed using dual phase variable format, in particular. For illustration consider the transfer functions with two inputs and one output.

$$T_1(s) = \frac{Y(s)}{U_1(s)} = \frac{0.4s^2 + 1.4s + 0.8}{s^3 + 0.3s^2 + 1.7s + 0.2}$$

$$T_2(s) = \frac{Y(s)}{U_2(s)} = \frac{-0.5s^2 + 0.7s - 1.9}{s^3 + 0.3s^2 + 1.7s + 0.2}$$

Note that both the transfer functions $T_1(s)$ and $T_2(s)$ share the same denominator polynomial and output. So, system matrix A and output matrix C are same as discussed before, that is

$$A = \begin{bmatrix} -0.3 & 1 & 0 \\ -1.7 & 0 & 1 \\ -0.2 & 0 & 0 \end{bmatrix} \quad \text{and} \quad C = \begin{bmatrix} 1 & 0 & 0 \end{bmatrix}$$

Since the system has two inputs u_1 and u_2, the input vector u will be two dimensional and input matrix B will have two columns. Using the property of dual phase variable format that the column elements of matrix B are coefficients of numerator polynomial of transfer function in descending powers of s, from top to bottom, the matrix B can written directly by inspecting $T_1(s)$ and $T_2(s)$ as:

$$B = \begin{bmatrix} 0.4 & -0.5 \\ 1.4 & 0.7 \\ 0.8 & -1.9 \end{bmatrix}$$

So, the state and output equations are:

$$\begin{bmatrix} \dot{x}_1 \\ \dot{x}_2 \\ \dot{x}_3 \end{bmatrix} = \begin{bmatrix} -0.3 & 1 & 0 \\ -1.7 & 0 & 1 \\ -0.2 & 0 & 0 \end{bmatrix} \begin{bmatrix} x_1 \\ x_2 \\ x_3 \end{bmatrix} + \begin{bmatrix} 0.4 & -0.5 \\ 1.4 & 0.7 \\ 0.8 & -1.9 \end{bmatrix} \begin{bmatrix} u_1 \\ u_2 \end{bmatrix}$$

$$y = \begin{bmatrix} 1 & 0 & 0 \end{bmatrix} \begin{bmatrix} x_1 \\ x_2 \\ x_3 \end{bmatrix}$$

and corresponding simulation diagram is shown in Fig. 7.6.

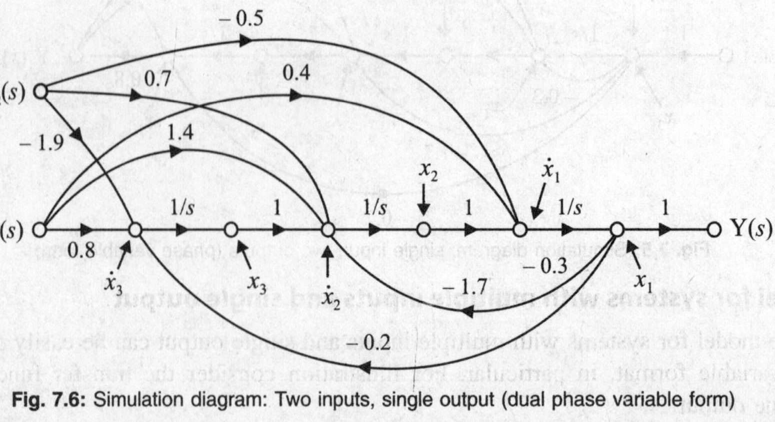

Fig. 7.6: Simulation diagram: Two inputs, single output (dual phase variable form)

State model for systems with multiple inputs and multiple outputs

Consider the simulation diagram for a system with two inputs and two outputs as shown in Fig. 7.7.

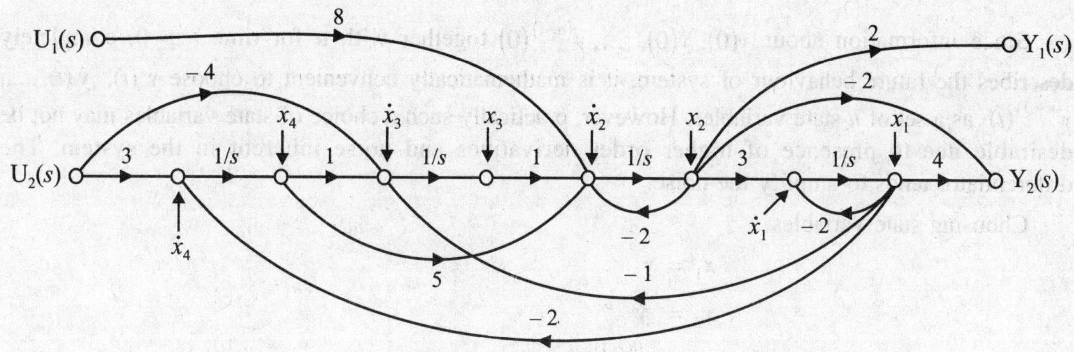

Fig. 7.7: Simulation diagram: System with two inputs and two outputs

The integrator outputs from right to left are chosen as state variables x_1, x_2, x_3 and x_4 as shown in Fig. 7.7 and corresponding input node of each integrator is labelled $\dot{x}_1, \dot{x}_2, \dot{x}_3$ and $\dot{x}_4$. The state and output equations in time domain, are developed as follows:

$$\begin{aligned}
\dot{x}_1 &= -2x_1 + 3x_2 \\
\dot{x}_2 &= -2x_2 + x_3 + 5x_4 + 8u_1 \\
\dot{x}_3 &= -x_1 + x_4 - 4u_2 \\
\dot{x}_4 &= -2x_1 + 3u_2 \\
y_1 &= 2x_2 \\
y_2 &= 4x_1 - 2x_2
\end{aligned}$$

The matrix form of above equation is as follows:

$$\begin{bmatrix} \dot{x}_1 \\ \dot{x}_2 \\ \dot{x}_3 \\ \dot{x}_4 \end{bmatrix} = \begin{bmatrix} -2 & 3 & 0 & 0 \\ 0 & -2 & 1 & 5 \\ -1 & 0 & 0 & 1 \\ -2 & 0 & 0 & 0 \end{bmatrix} \begin{bmatrix} x_1 \\ x_2 \\ x_3 \\ x_4 \end{bmatrix} + \begin{bmatrix} 0 & 0 \\ 8 & 0 \\ 0 & -4 \\ 0 & 3 \end{bmatrix} \begin{bmatrix} u_1 \\ u_2 \end{bmatrix}$$

$$\begin{bmatrix} y_1 \\ y_2 \end{bmatrix} = \begin{bmatrix} 0 & 2 & 0 & 0 \\ 4 & -2 & 0 & 0 \end{bmatrix} \begin{bmatrix} x_1 \\ x_2 \\ x_3 \\ x_4 \end{bmatrix} + \begin{bmatrix} 0 & 0 \\ 0 & 0 \end{bmatrix} u$$

Note that matrices A, B, C and D are not in any particular form.

Other ways of modelling

In the discussion so far it has been observed that the integrators are useful for describing the systems of all kinds. In general, the state space model of nth order system involves n integrators and the inputs of each of integrators are driven with a linear combination of state variables and inputs.

There are other ways of developing state space model without aid of simulation diagram. For example consider the nth order system described by differential equation.

$$y^{(n)} + \alpha_1 y^{(n-1)} + \ldots + \alpha_{n-1}\dot{y} + \alpha_n y = u \qquad \ldots(7.9)$$

Since information about $y(0)$, $\dot{y}(0)$, ..., $y^{(n-1)}(0)$ together with u for time $t \geq 0$, completely describes the future behaviour of system, it is mathematically convenient to choose $y(t)$, $\dot{y}(t)$, ..., $y^{(n-1)}(t)$ as a set of n state variables. However, practically such a choice of state variables may not be desirable due to presence of higher order derivatives and noise inherent in the system. The differentiator tends to amplify the noise.

Choosing state variables

$$x_1 = y$$
$$x_2 = \dot{y}$$
$$x_n = y^{(n-1)}$$

The state equations from (7.9) can be written as

$$\dot{x}_1 = x_2$$
$$\dot{x}_2 = x_3$$
$$\dot{x}_n = u - \alpha_n x_1 - \alpha_{n-1} x_2 \ldots - \alpha_1 x_n$$

or in matrix form

$$\begin{bmatrix} \dot{x}_1 \\ \dot{x}_2 \\ \vdots \\ \dot{x}_n \end{bmatrix} = \begin{bmatrix} 0 & 1 & 0 & \cdots & 0 \\ 0 & 0 & 1 & \cdots & 0 \\ \vdots & \vdots & \vdots & & \vdots \\ 0 & 0 & 0 & \cdots & 1 \\ -\alpha_n & -\alpha_{n-1} & -\alpha_{n-2} & \cdots & -\alpha_1 \end{bmatrix} \begin{bmatrix} x_1 \\ x_2 \\ \vdots \\ x_n \end{bmatrix} + \begin{bmatrix} 0 \\ 0 \\ \vdots \\ 0 \\ 1 \end{bmatrix} u \qquad \ldots(7.10)$$

and output equation as

$$y = \begin{bmatrix} 1 & 0 & \cdots & 0 \end{bmatrix} \begin{bmatrix} x_1 \\ x_2 \\ \vdots \\ x_n \end{bmatrix} \qquad \ldots(7.11)$$

for the sake of more insight into the approach discussed just above, consider the system described by

$$\dddot{y} + 6\ddot{y} + 7\dot{y} + 12y = 2u$$

choose state variables

$$x_1 = y$$
$$x_2 = \dot{y}$$
$$x_3 = \ddot{y}$$

so that state equations are

$$\dot{x}_1 = x_2$$

$$\dot{x}_2 = x_3$$

or

$$\dot{x}_3 = -12x_1 - 7x_2 - 6x_3 + 2u$$

$$\begin{bmatrix} \dot{x}_1 \\ \dot{x}_2 \\ \dot{x}_3 \end{bmatrix} = \begin{bmatrix} 0 & 1 & 0 \\ 0 & 0 & 1 \\ -12 & -7 & -6 \end{bmatrix} \begin{bmatrix} x_1 \\ x_2 \\ x_3 \end{bmatrix} + \begin{bmatrix} 0 \\ 0 \\ 2 \end{bmatrix} u$$

and output equation is

$$y = \begin{bmatrix} 1 & 0 & 0 \end{bmatrix} \begin{bmatrix} x_1 \\ x_2 \\ x_3 \end{bmatrix}$$

Note that the differential equation describing the system just above does not involve derivatives of input and the corresponding transfer function does not have zeros. Consider a system with transfer function having a zero as

$$\frac{Y(s)}{U(s)} = \frac{10(s+4)}{(s+1)(s+2)(s+3)} = \frac{10s+40}{s^3+6s^2+11s+6}$$

The cross multiplication together with the transform inverted, gives differential equation model as

$$\dddot{y} + 6\ddot{y} + 11\dot{y} + 6y = 10\dot{u} + 40u$$

Choosing the state variables such that right hand side of state equation does not have derivatives of u

$$x_1 = y$$

$$x_2 = \dot{y}$$

$$x_3 = \ddot{y} - 10u$$

so that

$$\dot{x}_1 = x_2$$

$$\dot{x}_2 = x_3 + 10u$$

$$\dot{x}_3 = -6x_1 - 11x_2 - 6(x_3 + 10u) + 40u$$

$$= -6x_1 - 11x_2 - 6x_3 - 20u$$

and state equation in matrix form is

$$\begin{bmatrix} \dot{x}_1 \\ \dot{x}_2 \\ \dot{x}_3 \end{bmatrix} = \begin{bmatrix} 0 & 1 & 0 \\ 0 & 0 & 1 \\ -6 & -11 & -6 \end{bmatrix} + \begin{bmatrix} 0 \\ 10 \\ -20 \end{bmatrix} u$$

The output equation is

$$y = \begin{bmatrix} 1 & 0 & 0 \end{bmatrix} \begin{bmatrix} x_1 \\ x_2 \\ x_3 \end{bmatrix}$$

Consider one more transfer function involving more number of zeros

$$\frac{Y(s)}{U(s)} = \frac{s^3 + 8s^2 + 17s + 8}{s^3 + 6s^2 + 11s + 6}$$

This transfer function may be broken into two parts involving a fictitious variable W(s) as

$$\frac{Y(s)}{U(s)} = \frac{Y(s)}{W(s)} \times \frac{W(s)}{U(s)} = \left[s^3 + 8s^2 + 17s + 8\right]\left[\frac{1}{s^3 + 6s^2 + 11s + 6}\right]$$

where $$\frac{W(s)}{U(s)} = \left[\frac{1}{s^3 + 6s^2 + 11s + 6}\right]$$

gives differential equation (cross multiply and invert the transform)

$$\dddot{w} + 6\ddot{w} + 11\dot{w} + 6w = u$$

choose state variables

$$x_1 = w$$
$$x_2 = \dot{w}$$
$$x_3 = \ddot{w}$$

so that state equations are

$$\dot{x}_1 = x_2$$
$$\dot{x}_2 = x_3$$
$$\dot{x}_3 = u - 6x_1 - 11x_2 - 6x_3$$

and in matrix from

$$\begin{bmatrix} \dot{x}_1 \\ \dot{x}_2 \\ \dot{x}_3 \end{bmatrix} = \begin{bmatrix} 0 & 1 & 0 \\ 0 & 0 & 1 \\ -6 & -11 & -6 \end{bmatrix} \begin{bmatrix} x_1 \\ x_2 \\ x_3 \end{bmatrix} + \begin{bmatrix} 0 \\ 0 \\ 1 \end{bmatrix} u$$

$$\frac{Y(s)}{W(s)} = s^3 + 8s^2 + 17s + 8$$

gives differential equation

$$y = \dddot{w} + 8\ddot{w} + 17\dot{w} + 8w$$

where w and its derivatives can be replaced by state variables x_1, x_2, x_3 so that

$$y = (u - 6x_1 - 11x_2 - 6x_3) + 8x_3 + 17x_2 + 8x_1$$
$$= 2x_1 + 6x_2 + 2x_3 + u$$

which in matrix form is

$$y = \begin{bmatrix} 2 & 6 & 2 \end{bmatrix} \begin{bmatrix} x_1 \\ x_2 \\ x_3 \end{bmatrix} + u$$

State space model using canonical variables

In the discussion so far it has been observed that state space representation is not unique. In fact, there are infinitely many representations depending on how the state variables are chosen. If the state variables can be so chosen that the resulting system matrix A in the state model, has a simple diagonal form, then the state variables are often called **canonical variables**. Developing state space model with canonical variables, involves partial fraction expansion. For example consider the transfer function

$$T(s) = \frac{Y(s)}{U(s)} = \frac{3s^2 - 2}{s^3 + 15s^2 + 54s + 40}$$

$$= \frac{1/27}{s+1} + \frac{-23/9}{s+4} + \frac{149/27}{s+10} = T_1(s) + T_2(s) + T_3(s)$$

where $T_1(s) = \dfrac{1/27}{s+1}$

$$T_2(s) = \frac{-23/9}{s+4}$$

and $T_3(s) = \dfrac{149/27}{s+10}$

T(*s*) decomposed in terms of $T_1(s)$, $T_2(s)$ and $T_3(s)$ can be put in the form of parallel (or tandem) connection of first order sub systems as shown by signal flow graph in Fig. 7.8(*a*). The simulation diagram connecting together each of these sub systems for canonical variable representation of system is shown in Fig. 7.8(*b*).

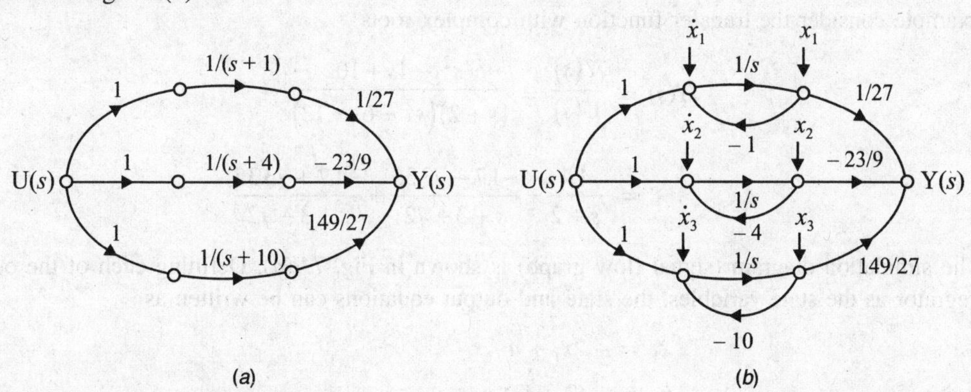

Fig. 7.8: (*a*) Parallel first order sub systems (*b*) Simulation diagram for canonical variable representation

Choosing as before the integrator outputs as state variables x_1, x_2 and x_3, the state equations from Fig. 7.8(*b*), can be written as

$$\dot{x}_1 = -x_1 + u$$

$$\dot{x}_2 = -4x_2 + u$$

$$\dot{x}_3 = -10x_3 + u$$

and the output equation as

$$y = \frac{1}{27}x_1 + \frac{-23}{9}x_2 + \frac{149}{27}x_3$$

The state and output equations in matrix form are

$$\begin{bmatrix} \dot{x}_1 \\ \dot{x}_2 \\ \dot{x}_3 \end{bmatrix} = \begin{bmatrix} -1 & 0 & 0 \\ 0 & -4 & 0 \\ 0 & 0 & -10 \end{bmatrix} \begin{bmatrix} x_1 \\ x_2 \\ x_3 \end{bmatrix} + \begin{bmatrix} 1 \\ 1 \\ 1 \end{bmatrix} u$$

$$y = \begin{bmatrix} \frac{1}{27} & \frac{-23}{9} & \frac{149}{27} \end{bmatrix} \begin{bmatrix} x_1 \\ x_2 \\ x_3 \end{bmatrix}$$

Note the following in current discussion of canonical variable representation:

(*i*) The system matrix A is diagonal and the state equations are **decoupled** from each other in the sense that the first order differential equations are independent of each other.

(*ii*) The diagonal elements of matrix A are the poles (or characteristic roots) of T(s). Little later in subsequent sections we shall see that the information about system stability, controllability and observability can be extracted directly by mere inspection of state model in canonical variable form if there is no pole zero cancellation in the transfer function. No mathematical routine will be required for this purpose.

The transfer function discussed just above has all the poles (or characteristic roots) real and distinct. However, the complex characteristic roots or repeated characteristic roots might also appear. For example consider the transfer function with complex roots

$$T(s) = \frac{Y(s)}{U(s)} = \frac{s^2 - 4s + 10}{(s+2)(s^2 + 6s + 13)}$$

$$= \frac{4.4}{s+2} + \frac{-1.7 - j3.35}{s+3+j2} + \frac{-1.7 + j3.35}{s+3-j2}$$

The simulation diagram (signal flow graph) is shown in Fig. 7.9(a). Defining each of the output of integrator as the state variables, the state and output equations can be written as

$$\dot{x}_1 = -2x_1 + u$$

$$\dot{x}_2 = -(3 + j2)\, x_2 + u$$

$$\dot{x}_3 = -(3 - j2)\, x_3 + u$$

$$y = 4.4x_1 - (1.7 + j3.35)\, x_2 - (1.7 - j3.35)\, x_3$$

which in matrix form are

$$\begin{bmatrix} \dot{x}_1 \\ \dot{x}_2 \\ \dot{x}_3 \end{bmatrix} = \begin{bmatrix} -2 & 0 & 0 \\ 0 & -(3+j2) & 0 \\ 0 & 0 & -(3-j2) \end{bmatrix} \begin{bmatrix} x_1 \\ x_2 \\ x_3 \end{bmatrix} + \begin{bmatrix} 1 \\ 1 \\ 1 \end{bmatrix} u$$

$$y = \begin{bmatrix} 4.4 & -(1.7 + j3.55) & -(1.7 - j3.35) \end{bmatrix} \begin{bmatrix} x_1 \\ x_2 \\ x_3 \end{bmatrix}$$

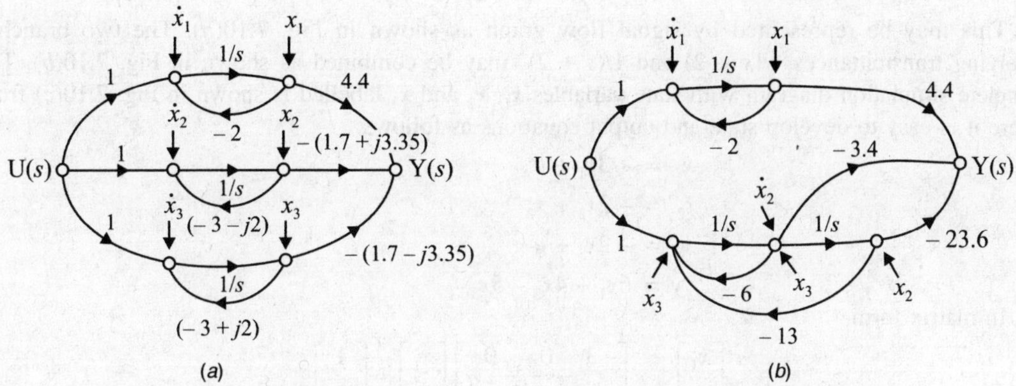

Fig. 7.9: (*a*) State model in canonical variable form (*b*) State model involving real numbers: canonical cum phase variable form

Note that state space model involves complex numbers. Although mathematical form is valid, but individual physical components cannot be assembled. Another convenient form of modelling without involving complex numbers, can be realized by combining the corresponding complex conjugate partial fractions, that is

$$\frac{-(1.7 + j3.55)}{s + 3 + j2} + \frac{-(1.7 - j3.55)}{s + 3 - j2} = \frac{-3.4s - 23.6}{s^2 + 6s + 13}$$

This part of system may be modelled in phase variable form involving real numbers. The simulation diagram is shown in Fig. 7.9(*b*) where the state variables and their derivatives are also labelled. The state and output equations can be written as

$$\dot{x}_1 = -2x_1 + u$$

$$\dot{x}_2 = x_3$$

$$\dot{x}_3 = -13x_2 - 6x_3 + u$$

$$y = 4.4x_1 - 23.6x_2 - 3.4x_3$$

and in matrix form

$$\begin{bmatrix} \dot{x}_1 \\ \dot{x}_2 \\ \dot{x}_3 \end{bmatrix} = \begin{bmatrix} -2 & 0 & 0 \\ 0 & 0 & 1 \\ 0 & -13 & -6 \end{bmatrix} \begin{bmatrix} x_1 \\ x_2 \\ x_3 \end{bmatrix} + \begin{bmatrix} 1 \\ 0 \\ 1 \end{bmatrix} u$$

$$y = \begin{bmatrix} 4.4 & -23.6 & -3.4 \end{bmatrix} \begin{bmatrix} x_1 \\ x_2 \\ x_3 \end{bmatrix}$$

The portion within dotted rectangle is in phase variable form. Thus the systems with one or more pairs of complex conjugate characteristic roots, may be modelled either by diagonalised state equations involving complex numbers or in block diagonal form involving real numbers.

Next consider a system with repeated characteristic roots as

$$T(s) = \frac{Y(s)}{U(s)} = \frac{s^2 + 3s + 6}{(s+2)^2 (s+3)} = \frac{6}{s+3} + \frac{-5}{s+2} + \frac{4}{(s+2)^2}$$

This may be represented by signal flow graph as shown in Fig. 7.10(*a*). The two branches involving transmittances $1/(s + 2)$ and $1/(s + 2)^2$ may be combined as shown in Fig. 7.10(*b*). The complete simulation diagram with state variables x_1, x_2 and x_3 labelled is shown in Fig. 7.10(*c*) from where it is easy to develop state and output equations as follows:

$$\dot{x}_1 = -3x_1 + u$$

$$\dot{x}_2 = -2x_2 + x_3$$

$$\dot{x}_3 = -2x_3 + u$$

$$y = 6x_1 + 4x_2 - 5x_3$$

In matrix form,

$$\begin{bmatrix} \dot{x}_1 \\ \dot{x}_2 \\ \dot{x}_3 \end{bmatrix} = \begin{bmatrix} -3 & 0 & 0 \\ 0 & -2 & 1 \\ 0 & 0 & -2 \end{bmatrix} \begin{bmatrix} x_1 \\ x_2 \\ x_3 \end{bmatrix} + \begin{bmatrix} 1 \\ 0 \\ 1 \end{bmatrix} u$$

$$y = \begin{bmatrix} 6 & 4 & -5 \end{bmatrix} \begin{bmatrix} x_1 \\ x_2 \\ x_3 \end{bmatrix}$$

Note that the state model shown just above, is in block diagonal form which is often termed as **Jordan Canonical** form.

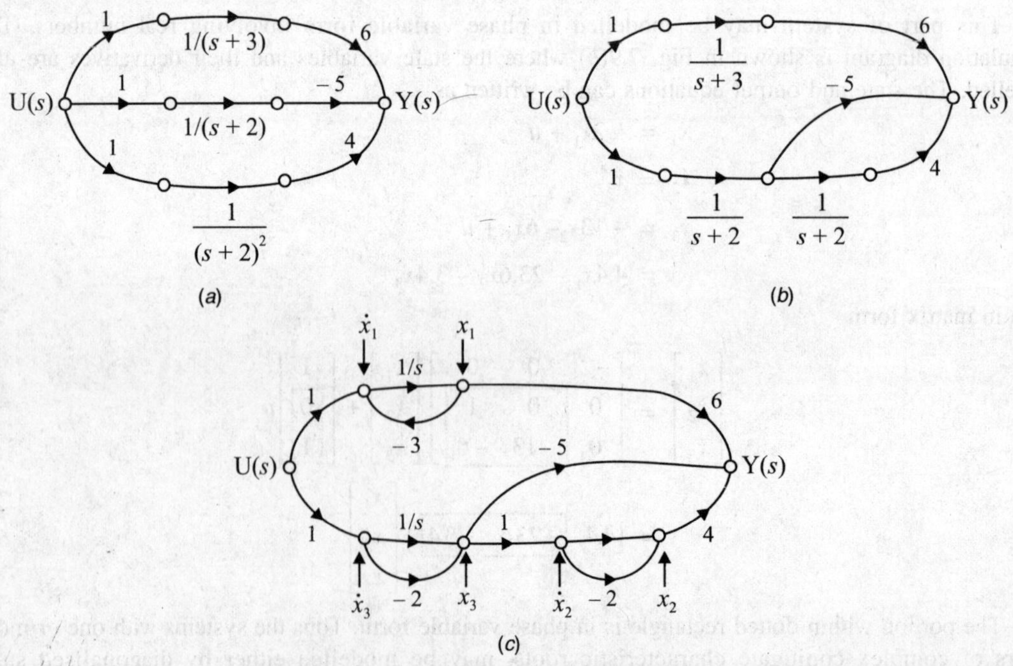

Fig. 7.10: (*a*) Signal flow graph depicting partial fractions (*b*) Signal flow graph combining repeated transmittances (*c*) Signal flow graph with labelled state variables

7.3 MODELLING ELECTRICAL AND MECHANICAL SYSTEMS

While developing state space model for electrical system, it is a common practice to choose voltage across capacitors and current through inductors as physical state variables, although the choice of state variables for a given system is not unique. For example, consider an electrical network as shown in Fig. 7.11.

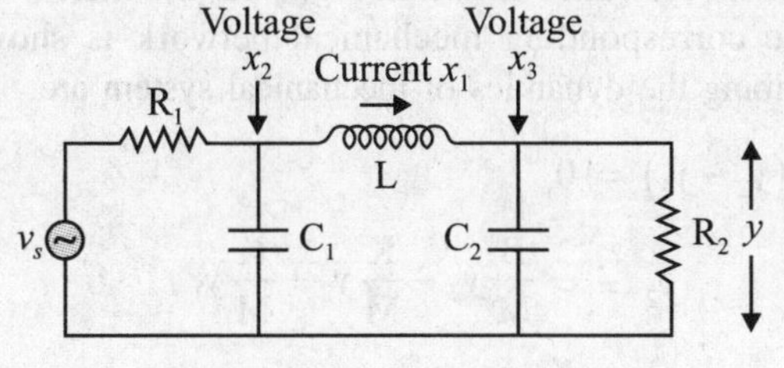

Fig. 7.11: An electrical network

Assign current x_1 through inductor L and voltages x_2 and x_3 across capacitors C_1 and C_2 respectively, as state variables as shown in Fig. 7.11. Current x_1 through inductor L is

$$x_1 = \frac{1}{L}\int (x_2 - x_3)\, dt$$

so that
$$\dot{x}_1 = \frac{1}{L}x_2 - \frac{1}{L}x_3$$

KCL at node x_2 yields

$$C_1\dot{x}_2 + \frac{x_2 - v_s}{R_1} + x_1 = 0$$

so that
$$\dot{x}_2 = -\frac{1}{C_1}x_1 - \frac{1}{R_1C_1}x_2 + \frac{v_s}{R_1C_1}$$

KCL at node x_3 yields

$$x_1 = C_2\dot{x}_3 + \frac{x_3}{R_2}$$

so that
$$\dot{x}_3 = \frac{1}{C_2}x_1 - \frac{1}{R_2C_2}x_3$$

The state equations in vector matrix form are

$$\begin{bmatrix}\dot{x}_1\\ \dot{x}_2\\ \dot{x}_3\end{bmatrix} = \begin{bmatrix} 0 & \frac{1}{L} & -\frac{1}{L} \\ -\frac{1}{C_1} & -\frac{1}{R_1C_1} & 0 \\ \frac{1}{C_2} & 0 & -\frac{1}{R_2C_2}\end{bmatrix}\begin{bmatrix}x_1\\ x_2\\ x_3\end{bmatrix} + \begin{bmatrix}0\\ \frac{1}{R_1C_1}\\ 0\end{bmatrix}u; \qquad (u \text{ stands for } v_s)$$

The output equation depends on what is output quantity of interest. Suppose voltage across capacitor C_2 (state variable x_3) is output quantity of interest, then

$$y = x_3$$

or
$$y = \begin{bmatrix} 0 & 0 & 1 \end{bmatrix} \begin{bmatrix} x_1 \\ x_2 \\ x_3 \end{bmatrix}$$

Next consider a mechanical system as shown in Fig. 7.12 (*a*) where F is input force and y_1, y_2 are output displacements. The corresponding mechanical network is shown in Fig. 7.12 (*b*). The differential equations describing the dynamics of mechanical system are

$$\mathrm{M}\ddot{y}_2 + \mathrm{B}\dot{y}_2 + \mathrm{K}(y_2 - y_1) = 0$$

or
$$\ddot{y}_2 = -\frac{\mathrm{B}}{\mathrm{M}}\dot{y}_2 - \frac{\mathrm{K}}{\mathrm{M}}y_2 + \frac{\mathrm{K}}{\mathrm{M}}y_1$$

and
$$\mathrm{F} = \mathrm{K}\,(y_1 - y_2)$$

or
$$y_1 = y_2 + \frac{1}{\mathrm{K}}\mathrm{F}$$

The block diagram representation of above equations is shown in Fig. 7.12 (*c*). Choosing state variables x_1 and x_2 at output of each integrator from right to left as shown in Fig. 7.12 (*c*), the state and output equation can be written as

$$\dot{x}_1 = x_2$$

$$\dot{x}_2 = -\frac{\mathrm{B}}{\mathrm{M}}x_2 - \frac{\mathrm{K}}{\mathrm{M}}x_1 + \frac{\mathrm{K}}{\mathrm{M}}y_1$$

$$\dot{x}_2 = -\frac{\mathrm{B}}{\mathrm{M}}x_2 - \frac{\mathrm{K}}{\mathrm{M}}x_1 + \frac{\mathrm{K}}{\mathrm{M}}\left[\frac{\mathrm{F}}{\mathrm{K}} + x_1\right]$$

$$\dot{x}_2 = -\frac{\mathrm{B}}{\mathrm{M}}x_2 + \frac{1}{\mathrm{M}}\mathrm{F}$$

$$y_1 = x_1 + \frac{1}{\mathrm{K}}\mathrm{F}$$

$$y_2 = x_1$$

These in matrix form are

$$\begin{bmatrix} \dot{x}_1 \\ \dot{x}_2 \end{bmatrix} = \begin{bmatrix} 0 & 1 \\ 0 & -\frac{\mathrm{B}}{\mathrm{M}} \end{bmatrix} \begin{bmatrix} x_1 \\ x_2 \end{bmatrix} + \begin{bmatrix} 0 \\ \frac{1}{\mathrm{M}} \end{bmatrix} u; \quad (u \text{ stands for force F})$$

$$\begin{bmatrix} y_1 \\ y_2 \end{bmatrix} = \begin{bmatrix} 1 & 0 \\ 1 & 0 \end{bmatrix} \begin{bmatrix} x_1 \\ x_2 \end{bmatrix} + \begin{bmatrix} \frac{1}{\mathrm{K}} \\ 0 \end{bmatrix} u$$

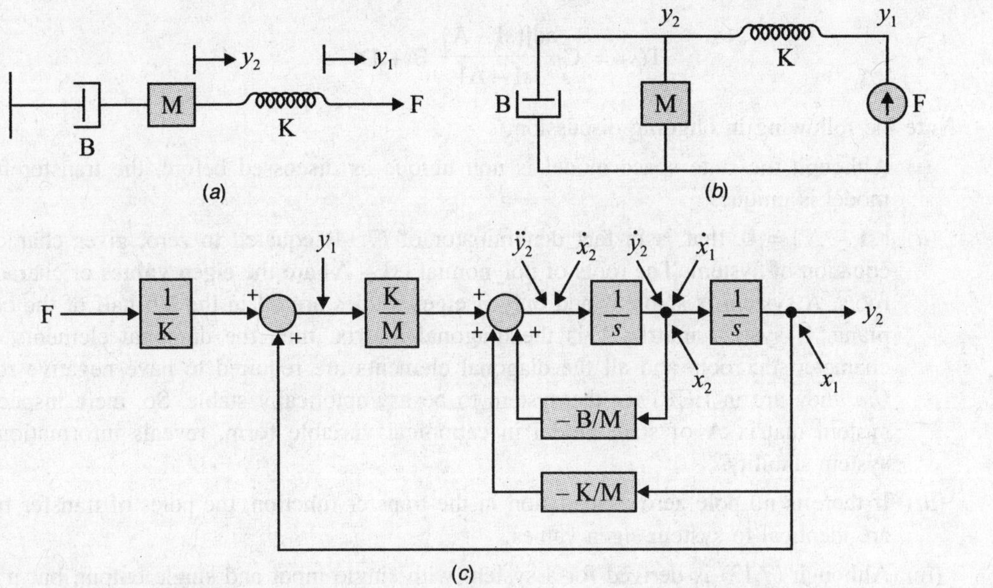

Fig. 7.12: (*a*) Mechanical System (*b*) Mechanical network (*c*) Simulation diagram with labelled state variables

7.4 FINDING TRANSFER FUNCTION FROM STATE SPACE MODEL

It has been already learnt that certain formats of state models for example, phase variable, dual phase variable and canonical variable models have direct link with transfer function. So, the transfer function model of the system can be determined by mere inspection of state space model without going through any computational routine. The current interest is finding transfer function model from a general state space model. Consider a single input single output system given by (equations 7.3 and 7.6 are rewritten below for convenience).

$$\dot{x} = \mathrm{A}x + \mathrm{B}u$$

$$y = \mathrm{C}x + \mathrm{D}u$$

Recall that transfer function model of a system is defined with zero initial conditions. Laplace transforming state equation with zero initial conditions yields

$$s\mathrm{X}(s) = \mathrm{AX}(s) + \mathrm{BU}(s)$$

$$\mathrm{X}(s) = (s\mathrm{I} - \mathrm{A})^{-1}\,\mathrm{BU}(s) \qquad \text{...(7.12)}$$

where I is an identity matrix.

Laplace transforming output equation gives

$$\mathrm{Y}(s) = \mathrm{CX}(s) + \mathrm{DU}(s)$$

And substituting X(*s*) from 7.12, we have

$$\mathrm{Y}(s) = \mathrm{C}(s\mathrm{I} - \mathrm{A})^{-1}\,\mathrm{BU}(s) + \mathrm{DU}(s)$$

$$\mathrm{T}(s) = \frac{\mathrm{Y}(s)}{\mathrm{U}(s)} = \mathrm{C}(s\mathrm{I} - \mathrm{A})^{-1}\,\mathrm{B} + \mathrm{D} \qquad \text{...(7.13)}$$

or $$T(s) = C\frac{\text{adj}(sI - A)}{|sI - A|}B + D \qquad ...(7.14)$$

Note the following in ongoing discussion:

(*i*) Although the state space model is non unique as discussed before, the transfer function model is unique.

(*ii*) $|sI - A| = 0$, that is in fact denominator of (7.14) equated to zero, gives characteristic equation of system. The roots of polynomial $|sI - A|$ are the **eigen values** or characteristic roots. A system is stable if and only if eigen values are all in the left half of the complex plane. If system matrix A is the diagonal matrix, then the diagonal elements are the characteristic roots and all the diagonal elements are required to have negative real part (*i.e.* they are in LHP) for the system to be asymptotically stable. So, mere inspection of system matrix A of state model in canonical variable form, reveals information about system stability.

(*iii*) If there is no pole zero cancellation in the transfer function, the poles of transfer function are identical to system eigen values.

(*iv*) Although (7.13) is derived for a system with single input and single output, but it can be extended to a system with multiple inputs and multiple outputs as well. For example, if there are m inputs and p outputs, the matrix $C(sI - A)^{-1}B + D$ of size $p \times m$ is arranged as

$$C(sI - A)^{-1}B + D = \begin{bmatrix} T_{11}(s) & T_{12}(s) & \cdots & T_{1m}(s) \\ T_{21}(s) & T_{22}(s) & \cdots & T_{2m}(s) \\ \vdots & \vdots & \cdots & \vdots \\ T_{p1}(s) & T_{p2}(s) & \cdots & T_{pm}(s) \end{bmatrix} \qquad ...(7.15)$$

where $$T_{11}(s) = \left.\frac{Y_1(s)}{U_1(s)}\right|_{U_2(s) = U_3(s) = ... = U_m(s) = 0}$$

$$T_{21}(s) = \left.\frac{Y_1(s)}{U_2(s)}\right|_{U_1(s) = U_3(s) = ... = U_m(s) = 0}$$

and so on.

In order to have more insight into finding transfer function from state space model, consider two input, single output system with equations

$$\begin{bmatrix} \dot{x}_1 \\ \dot{x}_2 \end{bmatrix} = \begin{bmatrix} -2 & 3 \\ -1 & -1 \end{bmatrix}\begin{bmatrix} x_1 \\ x_2 \end{bmatrix} + \begin{bmatrix} 4 & 0 \\ -5 & 6 \end{bmatrix}\begin{bmatrix} u_1 \\ u_2 \end{bmatrix}$$

$$y = [7 \quad 8]\begin{bmatrix} x_1 \\ x_2 \end{bmatrix}$$

Note system matrix $$A = \begin{bmatrix} -2 & 3 \\ -1 & -1 \end{bmatrix}$$

$$[s\mathrm{I} - \mathrm{A}] = s\begin{bmatrix}1 & 0\\0 & 1\end{bmatrix} - \begin{bmatrix}-2 & 3\\-1 & -1\end{bmatrix} = \begin{bmatrix}s+2 & -3\\1 & s+1\end{bmatrix}$$

$$[s\mathrm{I} - \mathrm{A}]^{-1} = \begin{bmatrix}\dfrac{s+1}{s^2+3s+5} & \dfrac{3}{s^2+3s+5}\\ \dfrac{-1}{s^2+3s+5} & \dfrac{s+2}{s^2+3s+5}\end{bmatrix}$$

and
$$\mathrm{T}(s) = \mathrm{C}[s\mathrm{I} - \mathrm{A}]^{-1}\mathrm{B}$$

$$= \left[\begin{bmatrix}7 & 8\end{bmatrix}\begin{bmatrix}\dfrac{s+1}{s^2+3s+5} & \dfrac{3}{s^2+3s+5}\\ \dfrac{-1}{s^2+3s+5} & \dfrac{s+2}{s^2+3s+5}\end{bmatrix}\begin{bmatrix}4 & 0\\-5 & 6\end{bmatrix}\right]$$

$$= \begin{bmatrix}7 & 8\end{bmatrix}\begin{bmatrix}\dfrac{4s-11}{s^2+3s+5} & \dfrac{18}{s^2+3s+5}\\ \dfrac{-5s-14}{s^2+3s+5} & \dfrac{6s+12}{s^2+3s+5}\end{bmatrix}$$

$$= \begin{bmatrix}\dfrac{-12s-189}{s^2+3s+5} & \dfrac{48s+222}{s^2+3s+5}\end{bmatrix}$$

so
$$\frac{\mathrm{Y}(s)}{\mathrm{U}_1(s)} = \frac{-12s-189}{s^2+3s+5}$$

and
$$\frac{\mathrm{Y}(s)}{\mathrm{U}_2(s)} = \frac{48s+222}{s^2+3s+5}$$

Next, consider a system with single input and two outputs described by equations.

$$\begin{bmatrix}\dot{x}_1\\ \dot{x}_2\end{bmatrix} = \begin{bmatrix}-3 & 4\\-2 & 0\end{bmatrix}\begin{bmatrix}x_1\\x_2\end{bmatrix} + \begin{bmatrix}2\\1\end{bmatrix}u$$

$$\begin{bmatrix}y_1\\y_2\end{bmatrix} = \begin{bmatrix}-4 & 6\\5 & -1\end{bmatrix}\begin{bmatrix}x_1\\x_2\end{bmatrix}$$

Note system matrix
$$\mathrm{A} = \begin{bmatrix}-3 & 4\\-2 & 0\end{bmatrix}$$

$$[s\mathrm{I} - \mathrm{A}] = s\begin{bmatrix}1 & 0\\0 & 1\end{bmatrix} - \begin{bmatrix}-3 & 4\\-2 & 0\end{bmatrix} = \begin{bmatrix}s+3 & -4\\2 & s\end{bmatrix}$$

$$[s\mathrm{I} - \mathrm{A}]^{-1} = \frac{1}{|s\mathrm{I} - \mathrm{A}|}\begin{bmatrix}s & 4\\-2 & s+3\end{bmatrix} = \begin{bmatrix}\dfrac{s}{s^2+3s+8} & \dfrac{4}{s^2+3s+8}\\ \dfrac{-2}{s^2+3s+8} & \dfrac{s+3}{s^2+3s+8}\end{bmatrix}$$

and $$T(s) = C\,[sI - A]^{-1}\,B$$

$$= \begin{bmatrix} -4 & 6 \\ 5 & -1 \end{bmatrix} \begin{bmatrix} \dfrac{s}{s^2+3s+8} & \dfrac{4}{s^2+3s+8} \\ \dfrac{-2}{s^2+3s+8} & \dfrac{s+3}{s^2+3s+8} \end{bmatrix} \begin{bmatrix} 2 \\ 1 \end{bmatrix}$$

$$= \begin{bmatrix} -4 & 6 \\ 5 & -1 \end{bmatrix} \begin{bmatrix} \dfrac{2s+4}{s^2+3s+8} \\ \dfrac{s-1}{s^2+3s+8} \end{bmatrix} = \begin{bmatrix} \dfrac{-2s-22}{s^2+3s+8} \\ \dfrac{9s+21}{s^2+3s+8} \end{bmatrix}$$

so, $$\frac{Y_1(s)}{U(s)} = \frac{-2s-22}{s^2+3s+8} \quad \text{and} \quad \frac{Y_2(s)}{U(s)} = \frac{9s+21}{s^2+3s+8}$$

7.5 FINDING TIME RESPONSE FROM STATE MODEL

The current discussion is about finding general solution of linear time invariant state equation. We shall first consider first order system and then proceed to higher order system.

First order systems

The Laplace transform approach is commonly used to find time response of system. The state equations are Laplace transformed and solved for state signals. The transform solution of state equations is inverted and then substituted in output equation. The system outputs, being linear combinations of state signals, are then easily found. For example, consider a first order system in state variable form

$$\dot{x} = ax + bu \qquad \text{...(7.16)}$$

$$y = cx \qquad \text{...(7.17)}$$

Laplace transform of state equation (7.16) gives

$$sX(s) - x(0) = aX(s) + bU(s)$$

$$X(s) = \frac{x(0)}{s-a} + \frac{bU(s)}{s-a}$$

The inverted transform gives

$$x(t) = x(0)\,e^{at} + bu(t) * e^{at}; \quad * \text{ denotes convolution}$$

$$= x(0)\,e^{at} + b\int_0^t e^{a(t-\tau)}\,u(\tau)\,d\tau \qquad \text{...(7.18)}$$

Note that the inverse transform of product of Laplace transforms is convolution of pertinent time functions.

Then

$$y = cx(0)\,e^{at} + c\cdot bu(t) * e^{at}$$

$$= cx(0)\,e^{at} + cb\int_0^t e^{a(t-\tau)}\,u(\tau)\,d\tau \qquad \text{...(7.19)}$$

As a numerical example, consider a system described by

$$\dot{x} = -2x + u(t)$$
$$y = 10x$$
$$x(0) = 3$$
$$u(t) = 4e^{5t}$$

Use (7.18) to obtain

$$x(t) = 3e^{-2t} + [4e^{5t} * e^{-2t}]$$

$$= 3e^{-2t} + \int_0^t e^{-2(t-\tau)} \cdot 4e^{5t}\, d\tau$$

$$= 3e^{-2t} + 4e^{-2t}\int_0^t e^{7\tau}\, d\tau = \frac{17}{7}e^{-2t} + \frac{4}{7}e^{5t}$$

and $$y(t) = 10x(t) = \frac{170}{7}e^{-2t} + \frac{40}{7}e^{5t}$$

State transition matrix

The state transition matrix (STM) plays a significant role in finding time response of higher order systems. For initial state vector $x(0)$, STM $\phi(t)$ is defined by the matrix equation

$$x(t) = \phi(t)\, x(0) \quad \text{...(7.20)}$$

Consider a linear homogeneous state equation

$$\dot{x}(t) = \mathrm{A}x(t) \quad \text{...(7.21)}$$

whose Laplace transform is

$$s\mathrm{X}(s) - \mathrm{X}(0) = \mathrm{AX}(s)$$

or

$$\mathrm{X}(s) = (s\mathrm{I} - \mathrm{A})^{-1}\, x(0) \quad \text{...(7.22)}$$

$$x(t) = \mathcal{L}^{-1}[(s\mathrm{I} - \mathrm{A})^{-1}]\, x(0) \quad \text{...(7.23)}$$

Comparison of (7.20) with (7.23) reveals that STM $\phi(t)$ is inverse Laplace transform of $\phi(s) = (s\mathrm{I} - \mathrm{A})^{-1}$. $\phi(s)$ is often called **resolvent matrix**. I is an identity matrix.

Using the classical approach to solve linear differential equation, the solution to equation (7.21) may also be written as

$$x(t) = e^{\mathrm{A}t}\, \mathrm{X}(0);\ t \geq 0 \quad \text{...(7.24)}$$

where $e^{\mathrm{A}t}$ in power series of matrix At is

$$e^{\mathrm{A}t} = \mathrm{I} + \mathrm{A}t + \frac{\mathrm{A}^2t^2}{2!} + \frac{\mathrm{A}^3t^3}{3!} + ... \quad \text{...(7.25)}$$

Thus STM

$$\phi(t) = \mathcal{L}^{-1}\,\phi(s) = \mathcal{L}^{-1}\,[(s\mathrm{I} - \mathrm{A})^{-1}]$$

$$= 1 + \mathrm{A}t + \frac{\mathrm{A}^2t^2}{2!} + \frac{\mathrm{A}^3t^3}{3!} + ... \quad \text{...(7.26)}$$

Note the following in current discussion:

(*i*) From equation (7.20) it is seen that solution of equation (7.21) is simply a transformation of the initial condition. This justifies the name state transition matrix to the unique matrix $\phi(t)$. The transition of states from the initial time $t = 0$ to any time $t \geq 0$ for zero inputs, is governed by the matrix exponential e^{At}.

(*ii*) STM describes free response of system as it satisfies the homogeneous state equation (7.21). STM depends on only system matrix A. So, it is sometimes referred to as STM of A. If matrix A is diagonal, then

$$\phi(t) = e^{At} = \begin{bmatrix} e^{\lambda_1 t} & 0 & \cdots & 0 \\ 0 & e^{\lambda_2 t} & \cdots & 0 \\ \vdots & \vdots & \vdots & \vdots \\ 0 & 0 & \cdots & e^{\lambda_n t} \end{bmatrix}$$

where $\lambda_1, \lambda_2, ... \lambda_n$, are distinct eigen values of matrix A. If there is multiplicity in eigen values, for example, if eigen values of A are

$$\lambda_1, \lambda_1, \lambda_1, \lambda_2, ... \lambda_n$$

then $\phi(t)$ will contain, in addition to the exponentials $e^{\lambda_1 t}, e^{\lambda_2 t}, ... e^{\lambda_n t}$, terms like $te^{\lambda_1 t}$ and $t^2 e^{\lambda_1 t}$.

Properties of STM

STM $\phi(t) = e^{At}$ has following properties

1. $\phi(0) = e^{At}\Big|_{t=0} = I$; (I is identity matrix)
2. $\phi^{-1}(t) = \phi(-t)$

   $\phi(t) = e^{At} = [e^{-At}]^{-1} = [\phi(-t)]^{-1}$ or $\phi^{-1}(t) = \phi(-t)$
3. $[\phi(t)]^n = \phi(nt)$
4. $\phi(t_1 + t_2) = \phi(t_1)\,\phi(t_2) = \phi(t_2) \cdot \phi(t_1)$

   $\phi(t_1 + t_2) = e^{A(t_1 + t_2)} = e^{At_1} \cdot e^{At_2} = \phi(t_1) \cdot \phi(t_2)$
5. $\phi(t_2 - t_1)\,\phi(t_1 - t_0) = \phi(t_2 - t_0)$

   $\phi(t_2 - t_1)\,\phi(t_1 - t_0) = e^{A(t_2 - t_1)} \cdot e^{A(t_1 - t_0)} = e^{(t_2 - t_0)} = \phi(t_2 - t_0)$

This property implies that entire state transition process can be divided into multiple small steps. As illustrated in Fig. 7.13 the transition from $t = t_0$ to $t = t_2$ is equal to the transition from $t = t_0$ to $t = t_1$ and then from $t = t_1$ to $t = t_2$.

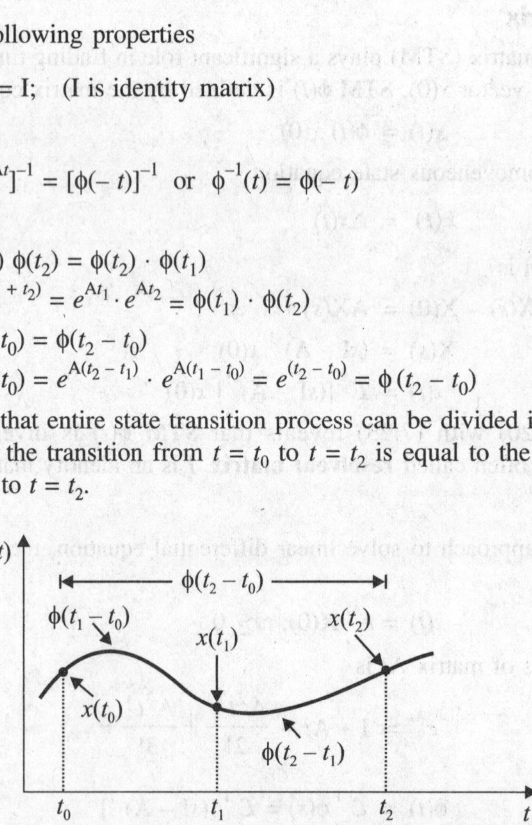

Fig. 7.13: State transition in steps

Higher order systems

In general, an higher order system in state variable format is given by

$$\dot{x}(t) = Ax(t) + Bu(t)$$

$$y(t) = Cx(t) + Du(t)$$

Laplace transform of state equation gives

$$sX(s) - x(0) = AX(s) + BU(s)$$

$$(sI - A)\,X(s) = x(0) + BU(s)$$

$$X(s) = (sI - A)^{-1}\,x(0) + (sI - A)^{-1}\,BU(s)$$

where $(sI - A)^{-1}$ denoted by $\phi(s)$ gives

$$X(s) = \phi\,(s)\,X(0) + \phi\,(s)\,BU(s)$$

whose inverse Laplace transform gives state response

$$x(t) = \underbrace{\phi(t)\,x(0)}_{\text{Zero input response}} + \underbrace{\int_0^t \phi(t-\tau)\,Bu(\tau)\,d\tau}_{\text{Zero state response}} \qquad ...(7.27)$$

The state response $x(t)$ is clearly the sum of two terms, one due to initial condition $x(0)$ referred to as **zero input response** and the other arising from input vector $u(t)$ referred to as **zero state response** or **forced response**. The system output vector $y(t)$ is easily found by mere substitution of solution (7.27) in output equation.

As a numerical example, consider the following second order system.

$$\begin{bmatrix}\dot{x}_1\\ \dot{x}_2\end{bmatrix} = \begin{bmatrix}0 & 1\\ -12 & -7\end{bmatrix}\begin{bmatrix}x_1\\ x_2\end{bmatrix} + \begin{bmatrix}1\\ 1\end{bmatrix}u; \quad \text{input } u(t) \text{ is unit step function}$$

$$y = \begin{bmatrix}1 & -1\end{bmatrix}\begin{bmatrix}x_1\\ x_2\end{bmatrix}$$

$$[x_1(0) \quad x_2(0)]^t = [10 \quad 0]; \quad t \text{ stands for transpose}$$

The resolvent matrix $\phi(s)$ is given by

$$\phi(s) = [sI - A]^{-1}$$

$$= \left[s\begin{bmatrix}1 & 0\\ 0 & 1\end{bmatrix} - \begin{bmatrix}0 & 1\\ -12 & -7\end{bmatrix}\right]^{-1} = \begin{bmatrix}s & -1\\ 12 & s+7\end{bmatrix}^{-1}$$

$$= \begin{bmatrix}\dfrac{s+7}{(s+3)(s+4)} & \dfrac{1}{(s+3)(s+4)}\\[2ex] \dfrac{-12}{(s+3)(s+4)} & \dfrac{s}{(s+3)(s+4)}\end{bmatrix}$$

The STM is

$$\phi(t) = \mathcal{L}^{-1}\,[\phi(s)] = \begin{bmatrix}4e^{-3t} - 3e^{-4t} & e^{-3t} - e^{-4t}\\ -12e^{-3t} + 12e^{-4t} & -3e^{-3t} + 4e^{-4t}\end{bmatrix}$$

The state response $x(t)$ is

$$x(t) = \phi(t)\,x(0) + \int_0^t \phi(t-\tau)\,\mathrm{B}u(\tau)\,d\tau$$

$$= \begin{bmatrix} 4e^{-3t} - 3e^{-4t} & e^{-3t} - e^{-4t} \\ -12e^{-3t} + 12e^{-4t} & -3e^{-3t} + 4e^{-4t} \end{bmatrix} \begin{bmatrix} 10 \\ 0 \end{bmatrix}$$

$$+ \int_0^t \begin{bmatrix} 4e^{-3(t-\tau)} - 3e^{-4(t-\tau)} & e^{-3(t-\tau)} - e^{-4(t-\tau)} \\ -12e^{-3(t-\tau)} + 12e^{-4(t-\tau)} & -3e^{-3(t-\tau)} + 4e^{-4(t-\tau)} \end{bmatrix} \begin{bmatrix} 1 \\ 1 \end{bmatrix} u(t)d\tau$$

where $u(\tau) = 1$ for $t \geq 0$

$$x(t) = \begin{bmatrix} 40e^{-3t} - 30e^{-4t} \\ -120e^{-3t} + 120e^{-4t} \end{bmatrix} + \int_0^t \begin{bmatrix} 5e^{-3(t-\tau)} - 4e^{-4(t-\tau)} \\ -15e^{-3(t-\tau)} + 16e^{-4(t-\tau)} \end{bmatrix} d\tau$$

$$= \begin{bmatrix} 40e^{-3t} - 30e^{-4t} \\ -120e^{-3t} + 120e^{-4t} \end{bmatrix} + \begin{bmatrix} \frac{2}{3} - \frac{5}{3}e^{-3t} + e^{-4t} \\ -1 + 5e^{-3t} - 4e^{-4t} \end{bmatrix}$$

$$= \begin{bmatrix} \frac{2}{3} + \frac{115}{3}e^{-3t} - 29e^{-4t} \\ -1 - 115e^{-3t} + 116e^{-4t} \end{bmatrix}$$

The unit step response $y(t)$ is

$$y(t) = \begin{bmatrix} 1 & -1 \end{bmatrix} \begin{bmatrix} \frac{2}{3} + \frac{115}{3}e^{-3t} - 29e^{-4t} \\ -1 - 115e^{-3t} + 116e^{-4t} \end{bmatrix}$$

$$= \frac{5}{3} + \frac{460}{3}e^{-3t} - 145e^{-4t}\ ;\ t \geq 0$$

Note that initial response $y(t)\big|_{t=0} = 10$ and steady state response $y(t)\big|_{t=\infty} = y(\infty) = 5/3$. In between for $0 < t < \infty$, the system response has exponential dynamics.

7.6 CONTROLLABILITY AND OBSERVABILITY

In control system analysis and design, it is often required to bring about an improvement in system performance. In an attempt to do so, the two basic questions straightway etch the designer's mind: one, regarding ability to bring about desirable change in system performance and two, regarding realizability of an appropriate measurement needed therein. The two terms controllability and observability introduced here take care of these two requirements, respectively.

A system is said to be completely state controllable if every state variable describing the system dynamics can be driven to any final desired value, say $x(t_f)$ from its initial state $x(t_0)$ by applying unconstrained control input $u(t)$ in finite time.

A system is said to be completely observable if every initial state $x(t_0)$ is determinable by observing the output over a finite time.

Note the following in current discussion:

1. The test for controllability is carried out assuming zero initial state. The initial conditions do not necessarily help in evolving appropriate control strategy. The test of observability is carried out assuming zero input. The input does not necessarily help in determination of earlier state. The test of controllability and observability is discussed in the section to just follow.
2. The definition of controllability does not pose any restriction on selection of control input $u(t)$ and definition of observability does not pose any restriction on output $y(t)$. If a system turns out to be uncontrollable it probably means that the system has not been so constructed that allows the appropriate control strategy to evolve. If the system turns out to be unobservable, it means that earlier value of all the state variables is not determinable by watching output.
3. Practically all-physical systems (provided there is no pole-zero cancellation in transfer function) are controllable and observable. However a designer might sometimes get a model of the system that might not be completely controllable and/or observable. In such a situation, the model, perhaps, does not accurately represent the system and designer may search another completely controllable and observable model. The facts that state space model is not unique, has already been established. For a completely controllable system, it is always possible to evolve an optimal control. An uncontrollable model either does not provide an optimal control or if it does so, the obtained optimal control may turn to be unimplementable. Similarly, for a completely observable system, it is always possible to design an observer to perform the task of state reconstruction. The optimal control is not discussed in this book. The term optimal in this perspective fundamentally refers to performing the prescribed task with minimal effort. The need and design of observer is discussed later in this chapter.

Test of state controllability for diagonal systems

For a system model with the system matrix A in diagonal form, the test of controllability is easy and straightforward. For example, consider the following third order single input single output (SISO) system.

$$\begin{bmatrix} \dot{x}_1 \\ \dot{x}_2 \\ \dot{x}_3 \end{bmatrix} = \begin{bmatrix} \lambda_1 & 0 & 0 \\ 0 & \lambda_2 & 0 \\ 0 & 0 & \lambda_3 \end{bmatrix} \begin{bmatrix} x_1 \\ x_2 \\ x_3 \end{bmatrix} + \begin{bmatrix} b_1 \\ b_2 \\ b_3 \end{bmatrix} u \quad \text{...(7.28)}$$

$$y = \begin{bmatrix} c_1 & c_2 & c_3 \end{bmatrix} \begin{bmatrix} x_1 \\ x_2 \\ x_3 \end{bmatrix} \quad \text{...(7.29)}$$

The signal flow graph describing the state equation (7.28) and output equation (7.29) is shown in Fig. 7.14.

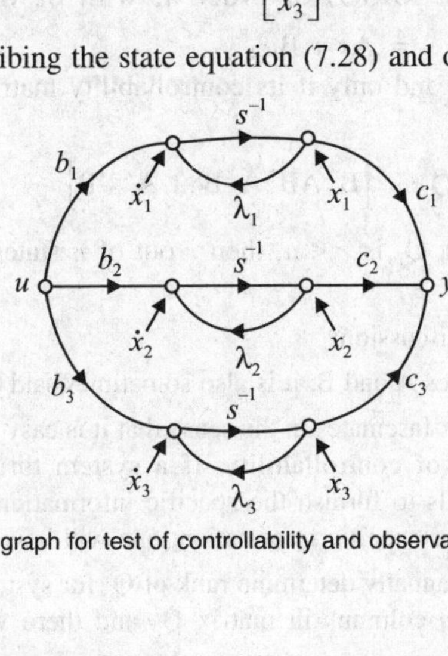

Fig. 7.14: Signal flow graph for test of controllability and observability of diagonal system

We present here (without proof) the condition which is necessary and sufficient both for complete controllability of a system in diagonal form. The control vector $B = \begin{bmatrix} b_1 \\ b_2 \\ b_3 \end{bmatrix}$ is examined and complete controllability requires that all the elements of vector B should be non zero. If any element is zero, then the corresponding state variable will be uncontrollable. This can be seen from signal flow graph, Fig. (7.14) that any zero element of B will mean the corresponding state variable being decoupled from control u and then it is no longer possible to influence that particular state variable by u. For example, say $b_2 = 0$, the control u is delinked with dynamics of state variable x_2 and then x_2 becomes uncontrollable.

The necessary and sufficient condition for a diagonal system with single input, discussed just above can be extended to a diagonal system with multiple inputs as well. For example, the $n \times m$ control matrix B for a system of order n and with m inputs is given by

$$B = \begin{bmatrix} b_{11} & b_{12} & \cdots & b_{1m} \\ b_{21} & b_{22} & \cdots & b_{2m} \\ \vdots & \vdots & \vdots & \vdots \\ b_{n1} & b_{n2} & \cdots & b_{nm} \end{bmatrix}$$

and the necessary and sufficient condition for controllability is that the matrix B must not have any row with all zero entries. A row with all zero entries will mean that the corresponding state variable is not coupled to any of m inputs and therefore remains uncontrollable.

Kalman's controllability test: For a system in non diagonal form neither the signal flow graph nor the zero entries in control vector B, reveals any information about controllability in general. One way to handle this situation is to diagonalise and then test controllability as discussed just above (diagonalisation is discussed later in this chapter). Another easier method that just follows, termed as **Kalman's test** is also fortunately available.

A general system (SISO or MISO) of order n, with or without repeated eigen values,

$$\dot{x} = Ax + Bu$$

is completely state controllable if and only if its controllability matrix of size $n \times n$ for single input and $n \times nm$ for m inputs

$$Q_c = \left[B \vdots AB \vdots A^2B \ldots. A^{n-1}B\right] \qquad \ldots(7.30)$$

is of full rank n. If rank of matrix Q_c is $r < n$, then r out of n states are controllable and remaining $(n - r)$ states are uncontrollable.

Note the following in current discussion:

(*i*) Since Q_c involves matrices A and B, it is also sometimes said that the pair (A, B) is controllable.

(*ii*) The controllability matrix fascinates in the sense that it is easy to apply. It only provides a 'yes or no' answer to the test of controllability. If a system turns out to be uncontrollable, the controllability matrix fails to furnish the specific information as to which particular states are uncontrollable. To get this specific information, one will have to diagonalise the system.

(*iii*) It becomes difficult to manually determine rank of Q_c for systems with multiple inputs. Even for $m = 2$, there will be $2n$ columns in matrix Q_c and there will be large number of possible combinations of $n \times n$ matrices. An easier method here is to construct matrix QQ^t (Q^t stands for transpose of Q) of size $n \times n$, then if QQ^t is non-singular, that is $|QQ^t| \neq 0$, Q_c has rank n.

(*iv*) The state controllability of a system depends on how the state variables are assigned. For example, consider the system with differential equation

$$\ddot{y}(t)+2\dot{y}(t)+y(t) = \dot{u}(t)+u(t) \quad ...(7.31)$$

Let the state variable assignment be

$$x_1(t) = y(t)$$

$$x_2(t) = \dot{y}(t) - u(t)$$

then the state equations

$$\dot{x}_1(t) = x_2(t) + u(t)$$

$$\dot{x}_2(t) = -2\,[x_2(t) + u(t)] - x_1(t) + u(t)$$

$$= -x_1(t) - 2x_2(t) - u(t)$$

in matrix from is

$$\left.\begin{aligned} \begin{bmatrix} \dot{x}_1(t) \\ \dot{x}_2(t) \end{bmatrix} &= \begin{bmatrix} 0 & 1 \\ -1 & -2 \end{bmatrix}\begin{bmatrix} x_1(t) \\ x_2(t) \end{bmatrix} + \begin{bmatrix} 1 \\ -1 \end{bmatrix} u(t) \\ \text{and} \quad y(t) &= \begin{bmatrix} 1 & 0 \end{bmatrix}\begin{bmatrix} x_1(t) \\ x_2(t) \end{bmatrix} \end{aligned}\right\} \quad ...(7.32)$$

The controllability matrix

$$Q_c = \begin{bmatrix} B & AB \end{bmatrix} = \begin{bmatrix} 1 & -1 \\ -1 & 1 \end{bmatrix}$$

is singular, that is $|Q_c| = 0$. The system is **not completely state controllable**.

Now let us reassign the state variables so as to develop state space model in phase variable form. Laplace transforming the differential equation (7.31) with zero initial conditions to get transfer function as

$$\frac{Y(s)}{U(s)} = \frac{s+1}{s^2+2s+1}$$

The state and output equation in phase variable form, can be written by mere inspection of this transfer function as

$$\left.\begin{aligned} \begin{bmatrix} \dot{x}_1 \\ \dot{x}_2 \end{bmatrix} &= \begin{bmatrix} 0 & 1 \\ -1 & -2 \end{bmatrix}\begin{bmatrix} x_1 \\ x_2 \end{bmatrix} + \begin{bmatrix} 0 \\ 1 \end{bmatrix} u \\ \text{and} \quad y &= \begin{bmatrix} 1 & 1 \end{bmatrix}\begin{bmatrix} x_1 \\ x_2 \end{bmatrix} \end{aligned}\right\} \quad ...(7.33)$$

The controllability matrix

$$Q_c = \begin{bmatrix} B & AB \end{bmatrix} = \begin{bmatrix} 0 & 1 \\ 1 & -2 \end{bmatrix}$$

is non-singular as $|Q_c| \neq 0$. Q_c is of full rank 2 and system is **completely state controllable**.

Test of output controllability

Sometimes it might be typically required to control the output of the system rather than states. The state controllability discussed just above, becomes less significant and there stands a need for defining output controllability.

Consider an nth order MIMO system with m inputs and p outputs described by

$$\dot{x} = Ax + Bu$$

$$y = Cx + Du$$

The system is said to be **completely output controllable if it is possible to drive initial output, say $y(t_0)$ to any desired final output $y(t_f)$ in a finite time interval $t_0 \leq t \leq t_f$ by applying an unconstrained control vector u.**

It can be shown that the system as described above is **completely output controllable if and only if the matrix Q_{oc} of size $p \times (n + 1)m$, given by**

$$Q_{oc} = \left[CB \vdots CAB \vdots CA^2B \vdots ..... CA^{n-1}B \vdots D\right]$$

is of rank p.

Test of observability for diagonal systems

The test of observability is again easy when the system model is in diagonal form. The necessary and sufficient condition for complete state observability of system described by (7.28) and (7.29), is that all the elements of $C = [c_1 \quad c_2 \quad c_3]$ should be non zero. If any element is zero, then the corresponding state variable will be unobservable. It is apparent from signal flow graph shown in Fig. (7.5) that any zero element of C will mean the corresponding state variable being decoupled from output y and then the particular state variable may assume any value without its effects being shown in output y. For example, say $c_1 = 0$, output y is decoupled from state variable x_1. Now changes in x_1 will no more be reflected in output y, and therefore x_1 becomes unobservable.

The test of observability just discussed for a diagonal system with single output can be extended for multiple outputs as well. For example the output matrix C for an nth order diagonal system with p outputs, is given by

$$C = \begin{bmatrix} c_{11} & c_{12} & \cdots & c_{1n} \\ c_{21} & c_{22} & \cdots & c_{2n} \\ \vdots & \vdots & \vdots & \vdots \\ c_{p1} & c_{p2} & \cdots & c_{pn} \end{bmatrix}$$

and the **necessary and sufficient condition for complete state observability is that the matrix C must not have any column with all zero entries.** A column with all zero entries will mean that the corresponding state variable is not coupled to any of p outputs and therefore remains unobservable.

Kalman's observability test: For non diagonal system, one may proceed with test of observability by first diagonalising and then examining entries in matrix C. Fortunately, the Kalman's test given just below, is much simpler method of determining system observability. Kalman's test requires that the observability matrix $Q_o = \begin{bmatrix} C \\ CA \\ \vdots \\ CA^{n-1} \end{bmatrix}$ be formed. The system is completely observable if and only if the matrix Q_o is of full rank n, that is, if and only if Q_o has linearly independent rows. Note that the condition is also referred to as the pair [A, C] being observable.

To illustrate Kalman's test of observability consider the system described by (7.31). Recall that controllability property of system model depends on state variable assignment. The observability property also depends on how the state variables are assigned. For example, the state space model (7.32) of differential equation (7.31) gives observability matrix

$$Q_o = \begin{bmatrix} C \\ \cdots \\ CA \end{bmatrix} = \begin{bmatrix} 1 & 0 \\ 0 & 1 \end{bmatrix}$$

which is non singular, and system if modelled as (7.32) is completely observable, while the model given by (7.33) gives observability matrix

$$Q_o = \begin{bmatrix} C \\ \cdots \\ CA \end{bmatrix} = \begin{bmatrix} 1 & 1 \\ -1 & -1 \end{bmatrix}$$

which is singular, and the system if modelled as (7.33) is not completely observable.

Causes of uncontrollability and/or unobservability

Some of the causes of uncontrollability and/or unobservability exhibited by some of the system models are as follows.

(*i*) Pole-zero cancellation

If the transfer function of the system does not have pole-zero cancellation, there does exist a state space model such that it is completely controllable and observable. Conversely, the pole-zero cancellation in a transfer function either leads to uncontrollability or unobservability or both, depending on how the state variables are assigned.

For example, consider the system once again described by differential equation (7.31). The two different models given by (7.32) and (7.33) have already been developed and their controllability and observability properties have also been investigated. The model given by (7.32) is not completely state controllable but completely observable. The model given by (7.33) is completely controllable but not completely observable. Now let us find out the unique transfer function for the state models (7.32) and (7.33)

$$T(s) = \frac{Y(s)}{U(s)} = C(sI - A)^{-1}B$$

$$= \begin{bmatrix} 1 & 0 \end{bmatrix} \begin{bmatrix} s & -1 \\ 1 & s+2 \end{bmatrix}^{-1} \begin{bmatrix} 1 \\ -1 \end{bmatrix} = \frac{s+1}{(s+1)^2} = \frac{1}{s+1}$$

There is pole-zero cancellation and this is the reason why the property of uncontrollability or unobservability is being exhibited by the developed models.

(*ii*) System symmetry

Yet another reason for lack of controllability and/or observability in a system model is symmetry of form

$$\left.\begin{aligned}\begin{bmatrix}\dot{x}_1\\ \dot{x}_2\end{bmatrix} &= \begin{bmatrix}1 & 1\\ 1 & 1\end{bmatrix}\begin{bmatrix}x_1\\ x_2\end{bmatrix}+\begin{bmatrix}1\\ 1\end{bmatrix}u\\ \text{and}\qquad y &= \begin{bmatrix}1 & 1\end{bmatrix}\begin{bmatrix}x_1\\ x_2\end{bmatrix}\end{aligned}\right\} \quad ...(7.34)$$

where controllability matrix

$$Q_c = \begin{bmatrix}1 & 2\\ 1 & 2\end{bmatrix}$$

and observability matrix

$$Q_o = \begin{bmatrix}1 & 1\\ 2 & 2\end{bmatrix}$$

are both singular, $|Q_c| = |Q_o| = 0$. The system is neither controllable nor observable. In a physical system such a symmetry is rare. It is also important to note that the transfer function of second order system described by model (7.34)

$$T(s) = \frac{Y(s)}{U(s)} = \begin{bmatrix}1 & 1\end{bmatrix}\begin{bmatrix}s-1 & -1\\ -1 & s-1\end{bmatrix}^{-1}\begin{bmatrix}1\\ 1\end{bmatrix}$$

$$= \frac{2s}{s(s-2)} = \frac{2}{s-2}$$

exhibits pole-zero cancellation, the pole at origin is cancelled out.

(*iii*) Redundancy in state variable assignment

While developing state space model for a complex system, if redundant or unnecessary state variables are assigned, the system model will again lack controllability and/or observability property. Let us demonstrate it with the help of simple RL circuit shown in Fig. 7.15 where voltage $u(t)$ is input and current $x(t)$, flowing through inductor L is output, $y(t)$. The dynamics of this first order circuit can be described by assigning a single state variable $x(t)$, the current through inductor as usual.

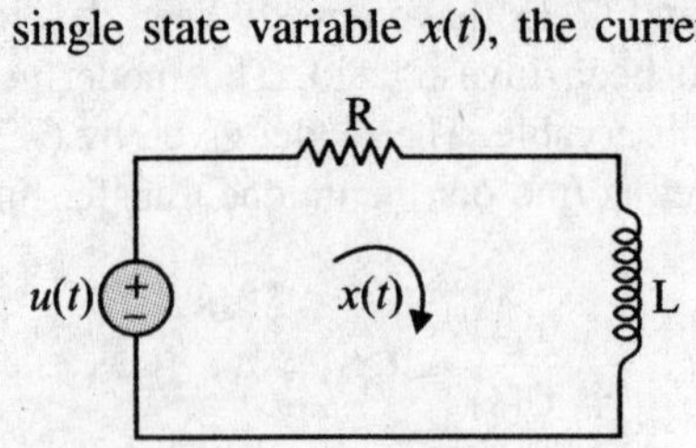

Fig. 7.15: RL circuit

Using KVL, the state and output equations are

$$u(t) = Rx(t) + L\frac{d}{dt}x(t)$$

$$\text{or}\qquad \left.\begin{aligned}\dot{x}(t) &= -\frac{R}{L}x(t) + \frac{1}{L}u(t)\\ y(t) &= x(t)\end{aligned}\right\} \quad ...(7.35)$$

Now let us introduce a redundant state variable, the charge q such that the two state variables $x_1 = q$ and $x_2 = x$, the current in circuit, describe the system dynamics. Then the state model comprising of state equations and output equation, is developed as:

$$\dot{x}_1 = \dot{q} = x = x_2$$

$$\dot{x}_2 = -\frac{R}{L}x_2 + \frac{1}{L}u$$

$$y = x_2$$

which in matrix form is

$$\begin{bmatrix}\dot{x}_1\\ \dot{x}_2\end{bmatrix} = \begin{bmatrix}0 & 1\\ 0 & -\frac{R}{L}\end{bmatrix} + \begin{bmatrix}0\\ \frac{1}{L}\end{bmatrix}u$$

$$y = \begin{bmatrix}0 & 1\end{bmatrix}\begin{bmatrix}x_1\\ x_2\end{bmatrix} \qquad \text{...(7.36)}$$

The controllability matrix

$$Q_c = \begin{bmatrix}0 & \frac{1}{L}\\ \frac{1}{L} & -\frac{R}{L^2}\end{bmatrix}$$

is non singular while the observability matrix

$$Q_o = \begin{bmatrix}0 & 1\\ 0 & -\frac{R}{L}\end{bmatrix}$$

is singular. The model is controllable but unobservable.

Principle of duality

The principle of duality establishes the relation between the two concepts: controllability and observability. A general n^{th} order system with m inputs and p outputs, is modelled as

$$\left.\begin{aligned}\dot{x}_{(n\times 1)} &= A_{(n\times n)}\, x_{(n\times 1)} + B_{(n\times m)}\, u_{(m\times 1)}\\ y_{(p\times 1)} &= C_{(p\times n)}\, X_{(n\times 1)}\end{aligned}\right\} \qquad \text{...(7.37)}$$

where each subscript denotes size of matrix and its controllability and observability matrices respectively are

$$Q_c = \left[B \,\vdots\, AB \,\vdots\, ... \; A^{n-1}B\right]$$

$$Q_o = \begin{bmatrix}C\\ CA\\ \vdots\\ CA^{n-1}\end{bmatrix}$$

The dual of system model (7.37) is defined as

$$\left.\begin{aligned} \dot{z}_{(n\times 1)} &= A^t_{(n\times n)} z_{(n\times 1)} + C^t_{(n\times p)} v_{(p\times 1)} \\ n_{(m\times 1)} &= B^t_{(m\times n)} z_{(n\times 1)} \end{aligned}\right\} \qquad \text{...(7.38)}$$

where t denotes transpose. The controllability and observability matrices of dual system are respectively

$$Q_{cd} = \left[C^t \vdots A^t C^t \vdots \ldots. (A^t)^{n-1} C^t\right] = \begin{bmatrix} C \\ CA \\ \vdots \\ CA^{n-1} \end{bmatrix}$$

$$Q_{od} = \begin{bmatrix} B^t \\ B^t A^t \\ \vdots \\ B^t (A^t)^{n-1} \end{bmatrix} = \left[B \vdots AB \vdots \ldots. A^{n-1} B\right]$$

The comparison of controllability and observability matrices Q_c and Q_o respectively of a system (7.37) with controllability and observability matrices Q_{cd} and Q_{od} respectively of dual system (7.38) reveals the following:

(*i*) The controllability matrix Q_c of a system is also the observability matrix Q_{od} of its dual.

(*ii*) The observability matrix Q_o of a system is also the controllability matrix Q_{cd} of its dual.

With this observation the principle of duality is stated as:

"A system is completely state controllable (observable) if and only if its dual system is completely observable (state controllable)."

7.7 FINDING DECOUPLED STATE EQUATIONS (DIAGONALISATION)

It is well understood by now that the investigation of system properties such as stability, controllability, observability and evaluation of time response, is easy and straight forward if we have the state model in canonical variable form where system matrix A is diagonal and state equations are in decoupled form. There stands no need for any computational routine, mere inspection of state model reveals information about all of them. For systems in general (nondiagonalised) form, these properties cannot be determined by inspection. In order to make these properties determinable by inspection, it becomes imperative to transform a general state model into a canonical variable model. Such a transformation is referred to as **diagonalisation**.

Consider, as usual, a general state model

$$\left.\begin{aligned} \dot{x} &= Ax + Bu \\ y &= Cx + Du \end{aligned}\right\} \qquad \text{...(7.39)}$$

where A is nondiagonal. Let us assign a new state vector z such that

$$x = Pz$$

where P is non singular constant matrix. This modifies the original state model to

$$P\dot{z} = APz + Bu$$

$$\left.\begin{aligned} \dot{z} &= P^{-1}APz + P^{-1}Bu \\ n &= CPz + Du \end{aligned}\right\} \qquad ...(7.40)$$

which may be written in form

$$\dot{z} = \overline{A}z + \overline{B}u$$

$$n = \overline{C}z + Du$$

where $\overline{A} = P^{-1}AP$, $\overline{B} = P^{-1}B$ and $\overline{C} = CP$. Such an operation on matrices is called **similarity transformation**. From knowledge of matrix algebra a similarity transformation **P** may be found provided A is square matrix with distinct eigen values such that

$$\overline{A} = P^{-1}AP$$

is a diagonal matrix with eigen values appearing as diagonal elements. Then the state model (7.40) will be in canonical variable form. The matrix P is also referred to as **diagonalising or modal matrix.** Note that A and $\overline{A}$ both have same characteristic equation and eigen values remain unchanged with this transformation. The diagonalising matrix P can be determined through the use of a set of eigen vectors, one for each eigen value. Note that eigen vectors are not unique, any non zero multiple of any eigen vector also works. The steps to follow to determine P are as follows:

(*i*) Find eigen values λ_i from the characteristic equation

$$|\lambda I - A| = 0 \qquad ...(7.41)$$

(*ii*) Find an eigen vector x_i for each λ_i. Let $x = x_i$ be the solution of equation

$$(\lambda_i I - A)\, x = 0 \qquad ...(7.42)$$

This solution x_i is called **eigen vector of A** for eigen values λ_i.

(*iii*) Construct matrix P by collecting the eigen vectors as

$$P = [x_1 \quad x_2 \; .... \; x_n]$$

For example consider the system in non diagonal form

$$\begin{bmatrix} \dot{x}_1 \\ \dot{x}_2 \end{bmatrix} = \begin{bmatrix} -6 & 1 \\ -5 & 0 \end{bmatrix} \begin{bmatrix} x_1 \\ x_2 \end{bmatrix} + \begin{bmatrix} 0 \\ 1 \end{bmatrix} u$$

$$y = \begin{bmatrix} 1 & -2 \end{bmatrix} \begin{bmatrix} x_1 \\ x_2 \end{bmatrix}$$

use step (*i*) to get eigen values as follows:

$$|\lambda I - A| = 0$$

$$\begin{bmatrix} \lambda + 6 & -1 \\ 5 & \lambda \end{bmatrix} = 0$$

$$(\lambda + 1)(\lambda + 5) = 0$$

$\lambda_1 = -1$, $\lambda_2 = -5$ (eigen values can be selected in any order)

Use step (*ii*) to find eigen vectors. The first eigen vector for $\lambda_1 = -1$ is found by solving

$$[\lambda_1 I - A] \begin{bmatrix} x_{11} \\ x_{12} \end{bmatrix} = \begin{bmatrix} 0 \\ 0 \end{bmatrix}$$

$$\begin{bmatrix} 5 & -1 \\ 5 & -1 \end{bmatrix} \begin{bmatrix} x_{11} \\ x_{12} \end{bmatrix} = \begin{bmatrix} 0 \\ 0 \end{bmatrix}$$

$$5x_{11} - x_{12} = 0$$

$$5x_{11} - x_{12} = 0$$

Both the equations are equivalent. An infinite number of solutions exist. Choose arbitrarily $x_{11} = 1$. This gives $x_{12} = 5$. So, the first eigen vector for $\lambda_1 = -1$ is

$$x_1 = \begin{bmatrix} x_{11} \\ x_{12} \end{bmatrix} = \begin{bmatrix} 1 \\ 5 \end{bmatrix}$$

Similarly second eigen vector for $\lambda_2 = -5$ is found by solving

$$[\lambda_2 I - A] \begin{bmatrix} x_{21} \\ x_{22} \end{bmatrix} = \begin{bmatrix} 0 \\ 0 \end{bmatrix}$$

$$\begin{bmatrix} 1 & -1 \\ 5 & -5 \end{bmatrix} \begin{bmatrix} x_{21} \\ x_{22} \end{bmatrix} = \begin{bmatrix} 0 \\ 0 \end{bmatrix}$$

$$x_{21} - x_{22} = 0$$

$$5x_{21} - 5x_{22} = 0$$

These two equations are also equivalent yielding infinite number of solutions. Again choosing arbitrarily $x_{21} = 1$ gives $x_{22} = 1$ and second eigen vector for $\lambda_2 = -5$ is

$$x_2 = \begin{bmatrix} x_{21} \\ x_{22} \end{bmatrix} = \begin{bmatrix} 1 \\ 1 \end{bmatrix}$$

Collect these eigen vectors to construct matrix

$$P = \begin{bmatrix} 1 & 1 \\ 5 & 1 \end{bmatrix}$$

$$P^{-1} = \begin{bmatrix} -\frac{1}{4} & \frac{1}{4} \\ \frac{5}{4} & -\frac{1}{4} \end{bmatrix}$$

$$\overline{A} = P^{-1}AP = \begin{bmatrix} -\frac{1}{4} & \frac{1}{4} \\ \frac{5}{4} & -\frac{1}{4} \end{bmatrix} \begin{bmatrix} -6 & 1 \\ -5 & 0 \end{bmatrix} \begin{bmatrix} 1 & 1 \\ 5 & 1 \end{bmatrix} = \begin{bmatrix} -1 & 0 \\ 0 & -5 \end{bmatrix}$$

$$\overline{B} = P^{-1}B = \begin{bmatrix} -\frac{1}{4} & \frac{1}{4} \\ \frac{5}{4} & -\frac{1}{4} \end{bmatrix} \begin{bmatrix} 0 \\ 1 \end{bmatrix} = \begin{bmatrix} \frac{1}{4} \\ -\frac{1}{4} \end{bmatrix}$$

$$\bar{C} = CP = \begin{bmatrix} 1 & -2 \end{bmatrix} \begin{bmatrix} 1 & 1 \\ 5 & 1 \end{bmatrix} = \begin{bmatrix} -9 & -1 \end{bmatrix}$$

The new system model (diagonal) with new state variables assignment is

$$\begin{bmatrix} \dot{z}_1 \\ \dot{z}_2 \end{bmatrix} = \begin{bmatrix} -1 & 0 \\ 0 & -5 \end{bmatrix} \begin{bmatrix} z_1 \\ z_2 \end{bmatrix} + \begin{bmatrix} \frac{1}{4} \\ -\frac{1}{4} \end{bmatrix} u$$

$$n = \begin{bmatrix} -9 & -1 \end{bmatrix} \begin{bmatrix} z_1 \\ z_2 \end{bmatrix}$$

Note that both the old and new models represent the same characteristic equation. So, the similarity transformation keeps the eigen values preserved.

If the system model is available in phase variable form where the quadruplet {A, B, C, D} for system function

$$G(s) = \frac{b_1 s^{n-1} + \ldots. + b_{n-1} s + b_n}{s^n + a_1 s^{n-1} + \ldots. + a_n}$$

is given by

$$A = \begin{bmatrix} 0 & 1 & 0 & \ldots. & 0 \\ 0 & 0 & 1 & \ldots. & 0 \\ \vdots & \vdots & \vdots & & \vdots \\ -a_n & -a_{n-1} & \ldots. & & -a_1 \end{bmatrix} \qquad B = \begin{bmatrix} 0 \\ 0 \\ \vdots \\ 1 \end{bmatrix}$$

$$C = [b_n \quad b_{n-1} \quad \ldots. \quad b_1], \qquad D = [0]$$

then, the matrix p can be selected as

$$p = \begin{bmatrix} 1 & 1 & \cdots & 1 \\ \lambda_1 & \lambda_2 & \cdots & \lambda_n \\ \vdots & \vdots & \cdots & \vdots \\ \lambda_1^{n-1} & \lambda_2^{n-1} & \cdots & \lambda_n^{n-1} \end{bmatrix}$$

which is called **Vander Monde** Matrix. λ_i for i = 1, 2, ..., n are the distinct eigen values of matrix A. To demonstrate the transformation with Vander Monde Matrix, consider a non diagonal system.

$$\begin{bmatrix} \dot{x}_1 \\ \dot{x}_2 \\ \dot{x}_3 \end{bmatrix} = \begin{bmatrix} 0 & 1 & 0 \\ 0 & 0 & 1 \\ -40 & -54 & -15 \end{bmatrix} \begin{bmatrix} x_1 \\ x_2 \\ x_3 \end{bmatrix} + \begin{bmatrix} 0 \\ 0 \\ 1 \end{bmatrix} u$$

$$y = \begin{bmatrix} -2 & 0 & 3 \end{bmatrix} \begin{bmatrix} x_1 \\ x_2 \\ x_3 \end{bmatrix}$$

Compute eigen values from

$$|\lambda I - A| = 0$$

$$\begin{vmatrix} \lambda & -1 & 1 \\ 0 & \lambda & -1 \\ 40 & 54 & \lambda+15 \end{vmatrix} = 0$$

to get $\lambda_1 = -1$, $\lambda_2 = -4$, $\lambda_3 = -10$ and Vander Monde Matrix

$$P = \begin{bmatrix} 1 & 1 & 1 \\ -1 & -4 & -10 \\ 1 & 16 & 100 \end{bmatrix}$$

$$P^{-1} = \begin{bmatrix} \frac{40}{27} & \frac{14}{27} & \frac{1}{27} \\ -\frac{15}{27} & -\frac{11}{18} & -\frac{1}{18} \\ \frac{2}{27} & \frac{5}{54} & \frac{1}{54} \end{bmatrix}$$

$$\overline{A} = P^{-1}AP = \begin{bmatrix} -1 & 0 & 0 \\ 0 & -4 & 0 \\ 0 & 0 & -10 \end{bmatrix}$$

$$\overline{B} = P^{-1}B = \begin{bmatrix} \frac{1}{27} \\ -\frac{1}{18} \\ \frac{1}{54} \end{bmatrix}$$

$$\overline{C} = CP = [1 \quad 46 \quad 298]$$

The new system model (diagonal) with new state variables assignment is

$$\begin{bmatrix} \dot{z}_1 \\ \dot{z}_2 \\ \dot{z}_3 \end{bmatrix} = \begin{bmatrix} -1 & 0 & 0 \\ 0 & -4 & 0 \\ 0 & 0 & -10 \end{bmatrix} \begin{bmatrix} z_1 \\ z_2 \\ z_3 \end{bmatrix} + \begin{bmatrix} \frac{1}{27} \\ -\frac{1}{18} \\ \frac{1}{54} \end{bmatrix} u$$

$$n = \begin{bmatrix} 1 & 46 & 298 \end{bmatrix} \begin{bmatrix} z_1 \\ z_2 \\ z_3 \end{bmatrix}$$

We terminate the discussion on finding transformation matrix P that diagonalises a non diagonal square matrix A. Truly speaking, this is the fundamental technique of linear algebra. For more details refer to the text on linear algebra.

7.8 STATE FEEDBACK AND POLE PLACEMENT

The classical approach to evolve a typical control strategy, involves use of standard compensators such as lead lag, PID (to be discussed in detail in chapter 8). The system output is fedback and processed by these compensators to meet the system stability and performance requirements. In doing so, the designer in fact attempts to bring about meaningful change in the shape of root locus or Bode Plot. The imperfection that lies in this approach is that the compensators have only few free tuning parameters to control the system dynamics. The practical systems seeking modification in their dynamical behaviour, are usually of order larger than the number of free parameters, the compensators have. So, the compensators are incapable of exercising independent control over all the closed loop poles that are also the same as roots of characteristic equation.

In state space design, the entire state vector is fedback in order to bring about desired change in dynamical behaviour of system and designer enjoys the total control over location of all the closed loop poles of sytem. This is in contrast with classical design where designer can enjoy relocating at will only a pair of complex conjugate poles that are dominant. However, the arbitrary pole placement at will as claimed in state space design, can be apparently achieved only under certain condition.

If the system is completely state controllable, it can always be modelled in phase variable form. Conversely, if the system is described in phase variable form, it is always completely state controllable. The state space design strategy involves the sensing of the state variables and feeding them back to the input through appropriate gain. Provided the system is completely state controllable, the state feedback in this way, always succeeds in placing the system's characteristic roots at any desired location.

Consider the system of order n, described by the state model

$$\dot{x} = \mathrm{A}x + \mathrm{B}u \qquad \ldots(7.43)$$
$$y = \mathrm{C}x$$

The state feedback control is

$$u = -\mathrm{K}x + r \qquad \ldots(7.44)$$

where K is $1 \times n$ feedback matrix with constant gain elements. The state feedback system is depicted in Fig. 7.16 (*a*). The substitution of (7.44) into (7.43) modifies the state equation to

$$\dot{x} = (\mathrm{A} - \mathrm{BK})x + \mathrm{B}r \qquad \ldots(7.45)$$

The closed loop system matrix has been modified from A to (A – BK) as illustrated in Fig. 7.16 (*b*) and (*c*) through rearrangement of Fig. 7.16 (*a*).

The characteristic equation of closed loop system is

$$|\lambda\mathrm{I} - \mathrm{A} + \mathrm{BK}| = 0 \qquad \ldots(7.46)$$

The characteristic equation (7.46) is an nth order polynomial in s consisting of gains K_1, K_2, ..., K_n. The state feedback design involves picking up of these gains so that the roots of characteristic equation (7.46) are placed at desirable locations. Let the desired locations be

$$s = \lambda_1, \lambda_2, \ldots, \lambda_n$$

Then the desired characteristic equation is

$$(s - \lambda_1)(s - \lambda_2) \ldots (s - \lambda_n) = 0 \qquad \ldots(7.47)$$

The required n elements of feedback matrix K are obtained by matching coefficients in (7.46) and (7.47).

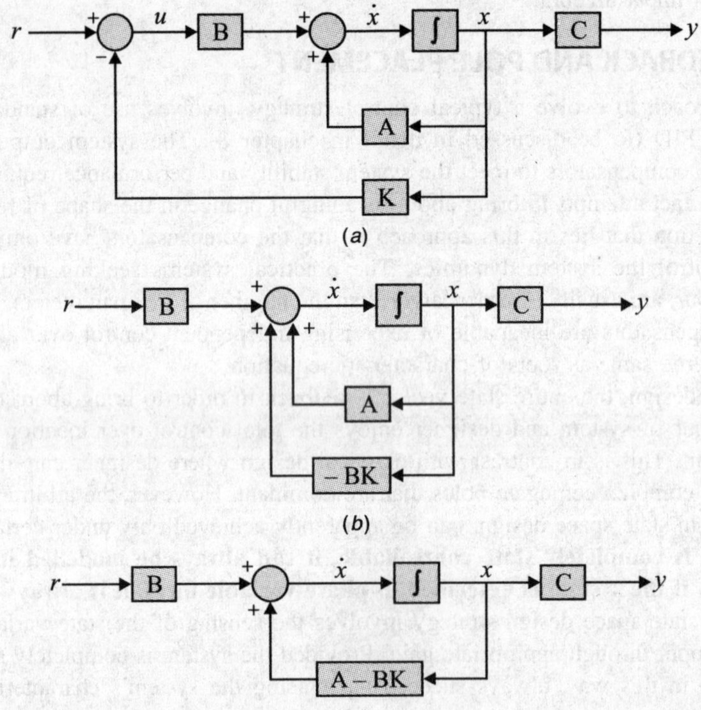

Fig. 7.16: (*a*) State feedback (*b*), (*c*) Rearranged diagram to show system matrix modified as (A – BK)

If the original system (7.43) is available in phase variable form, state feedback design is especially convenient in the sense that matrix A will be in companion form and the characteristic polynomial can be written by inspection. In fact the coefficients of characteristic polynomial can be read off the last row of the matrix A. In phase variable form of original system (7.43), A and B are of forms

$$A = \begin{bmatrix} 0 & 1 & 0 & .... & 0 \\ 0 & 0 & 1 & .... & 0 \\ \vdots & \vdots & \vdots & & \vdots \\ 0 & 0 & 0 & & 1 \\ -a_n & -a_{n-1} & -a_{n-2} & & -a_1 \end{bmatrix}$$

$$B = \begin{bmatrix} 0 \\ 0 \\ \vdots \\ 1 \end{bmatrix}$$

and characteristic equation is of form

$$s^n + a_1 s^{n-1} + ... + a_{n-1} s + a^n = 0 \qquad ...(7.48)$$

The modified closed loop system matrix is of form

$$A - BK = \begin{bmatrix} 0 & 1 & 0 & \dots & 0 \\ 0 & 0 & 1 & \dots & 0 \\ \vdots & \vdots & \vdots & & \vdots \\ \vdots & \vdots & \vdots & & \vdots \\ -a_n - K_1 & -a_{n-1} - K_2 & \cdots & & -a_1 - K_n \end{bmatrix}$$

and modified characteristic equation (7.46) is of form

$$|\lambda I - A + BK| = \lambda^n + (a_1 + K_n)\lambda^{n-1} + \dots + (a_{n-1} + K_2)\lambda + (a_n + K_1) = 0 \quad \dots(7.49)$$

Note that the feedback gains K_1, K_2, ..., K_n are isolated in each coefficient of characteristic equation (7.49). So, any desired polynomial can be achieved by selecting the feedback gains. The discussion so far on state feedback/pole placement design, can be generalised as follows:

The characteristic roots (closed loop poles) of a system can be arbitrarily placed anywhere in complex plane if and only if the system is completely state controllable.

As a numerical example of placing the system poles as desired with state feedback, consider the following single input, single output system described in phase variable form:

$$\begin{bmatrix} \dot{x}_1 \\ \dot{x}_2 \\ \dot{x}_3 \end{bmatrix} = \begin{bmatrix} 0 & 1 & 0 \\ 0 & 0 & 1 \\ -3 & -6 & -7 \end{bmatrix} \begin{bmatrix} x_1 \\ x_2 \\ x_3 \end{bmatrix} + \begin{bmatrix} 0 \\ 0 \\ 1 \end{bmatrix} u$$

$$y = \begin{bmatrix} 2 & 0 & -1 \end{bmatrix} \begin{bmatrix} x_1 \\ x_2 \\ x_3 \end{bmatrix}$$

It is desired that the system poles be located at $s = -3$, -4 and -5.

With state feedback

$$u = -\begin{bmatrix} K_1 & K_2 & K_3 \end{bmatrix} \begin{bmatrix} x_1 \\ x_2 \\ x_3 \end{bmatrix} + r$$

the modified closed loop system matrix is of form

$$[A - BK] = \begin{bmatrix} 0 & 1 & 0 \\ 0 & 0 & 1 \\ -3 & -6 & -7 \end{bmatrix} - \begin{bmatrix} 0 \\ 0 \\ 1 \end{bmatrix} \begin{bmatrix} K_1 & K_2 & K_3 \end{bmatrix}$$

$$= \begin{bmatrix} 0 & 1 & 0 \\ 0 & 0 & 1 \\ -K_1 - 3 & -K_2 - 6 & -K_3 - 7 \end{bmatrix}$$

and the characteristic equation is

$$|\lambda I - A + BK| = s^3 + (K_3 + 7)s^2 + (K_2 + 6)s + (K_1 + 3) = 0$$

while the desired characteristic equation is

$$(s + 3)(s + 4)(s + 5) = 0$$

or $$s^3 + 12\,s^2 + 47\,s + 60 = 0$$

Match the coefficients of characteristic equations just above to get the values of gain elements K_1, K_2 and K_3 of feedback matrix as

$$K_1 = 57$$
$$K_2 = 41$$
$$K_3 = 5$$

It is true in general that appropriate state feedback will place the poles of any controllable system arbitrarily. However, in doing so, the designer should also be **cautious** about the following:

(*i*) The state feedback helps out arbitrary pole placement, it does not affect the system zeros. If zeros are found to be at undesired locations, the state feedback cannot do anything about. Note that the steady state behaviours of a system depends on poles and zeros both. So, the state feedback alone may fail to meet steady state requirement.

(*ii*) If a system is not completely state controllable, the state feedback will be incapable of moving all the poles. Only few which are controllable, can be moved.

(*iii*) If the designer tempts to move the poles too far away from imaginary axis into LHP, it results in smaller time constant and faster dynamics alright but the difficulty then emerging is that the faster system requires more accurate measuring device and larger actuator (motor). This adds to the cost of overall control configuration. Pushing the poles too far into LHP also results in larger system bandwidth whereby the system becomes more sensitive to noise.

(*iv*) An useful suggestion emerges from optical control theory that the RHP poles if any must be reflected about imaginary axis, for example a pole at $s = +\,2.5$ must be relocated at $s = -\,2.5$. In doing so, the designer minimizes the energy required to implement the control.

(*v*) The state feedback is not practically realizable in general. It is either infeasible or impractical to sense all the state variables directly. The states that are not directly measurable, will require state estimators also called as **observers**. The observers if used, will increase the system order and overall cost.

(*vi*) The control through state feedback is some what equivalent to PD compensator that has infinite bandwidth while the real world components have only finite bandwidth.

7.9 OBSERVER DESIGN

The arbitrary pole placement as discussed in preceding section, involves generating a control signal by feeding back the entire state vector through constant real gains. In high order systems, the large number of states involved would require large number of transducers to sense the state variables for the purpose of feedback. The cost of sensing may turn to be prohibitive. Even for low order systems, often not all the state variables are directly accessible for measurement. However, in reality, the output variables are guaranteed to be always accessible for measurement with unlimited precision. Hereby one conceives an idea of estimating the state variables from measurement of output variables.

An observer is a device that uses the inputs and outputs of a system to produce estimates of its states. The term device here refers to either hardware or software implementation. If all the state variables of a system are estimated irrespective of whether few of them can be directly measured, the observer is said to be **full order state observer or identity observer**. If only those state variables which cannot be directly measured, are estimated and, the remaining states being accurately measured and feedback, the observer is said to be **reduced order state observer**. Our discussion here shall remain restricted to only identity observer. If the system is of order n, the identity observer will also be of same order. When the observer estimates are fedback into the system, the order of closed loop system will be $2n$. Note that it is always possible to place the system poles arbitrarily or stabilise the system using observer state estimates in lieu of actual states.

The structure of full order state observer for the system with single input and single output modelled as usual as

$$\dot{x} = Ax + Bu; \qquad x(0) = x_0$$
$$y = Cx$$

is shown in Fig. 7.17. The observer estimates the state vector of an observable linear system while using the input and output thereof. The dynamics of observer can be described as

$$\dot{\hat{x}} = A\hat{x} + Bu + M(y - C\hat{x}); \qquad \hat{x}(0) = \hat{x}_0 \qquad ...(7.50)$$

where $\hat{x}$ is the estimate of state vector x and M is $n \times 1$ real constant gain matrix. The matrix M which is in fact a column vector for single input single output system, is called the **observer gain**. The design of observer begins with defining the state error vector

$$\tilde{x}(t) = x(t) - \hat{x}(t) \qquad ...(7.51)$$

Differentiating (7.51) gives the error dynamics as

$$\dot{\tilde{x}} = \dot{x} - \dot{\hat{x}}; \qquad \tilde{x}(0) = \tilde{x}_0$$

Substituting $\dot{x}$ and $\dot{\hat{x}}$ from equations just above, we have

$$\begin{aligned}\dot{\tilde{x}} &= (Ax + Bu) - [(A\hat{x} + Bu) + M(Cx - C\hat{x})] \\ &= A(x - \hat{x}) - MC(x - \hat{x}) \\ &= (A - MC)\,\tilde{x} \qquad ...(7.52)\end{aligned}$$

It is apparent from observer error dynamics (7.52) that the error will asymptotically decay to zero inspective of $\tilde{x}(0)$ if and only if the eigen values of matrix (A – MC) or the roots of characteristic equation

$$|\, sI - (A - MC)\,| = 0 \qquad ...(7.53)$$

are in the LHP. So, the observer design now concentrates on choosing matrix M such that $\tilde{x}(t)$ converges to zero or $\hat{x}(t)$ converges to $x(t)$ regardless of the value of $\hat{x}(0)$ and additionally the convergence be reasonably fast.

It turns out that **if the system is completely state observable, the matrix M can be chosen such that the eigen values of (A – MC) are arbitrarily placed**. The eigen values of (A – MC) are the observer poles. Thus the arbitrary observer pole assignment requires that the system be completely observable. Recall that complete state controllability is the requirement for arbitrary placement of system poles. More categorically, the condition of state controllability of system is closely related to existence of solution of state feedback for the purpose of placing the eigen values arbitrarily while the concept of observability relates to the condition of observing or estimating the state variables from output variable which are guaranteed to be measurable.

The selection of matrix M so that error system is stable and the error is acceptably small, is tried in exactly the same manner as we did for selection of matrix K in state feedback design. First we specify the location of roots of characteristic equation (7.53) keeping in view the two requirements: one, roots must be in the LHP and two, the roots must be reasonably fast. Let the desired roots be

$$s = \lambda_1, \lambda_2, ..., \lambda_n$$

The observer characteristic equation with this set of roots is constructed as

$$(s - \lambda_1)(s - \lambda_2) .... (s - \lambda_n) = 0 \qquad ...(7.54)$$

Then we determine M by comparing the coefficients of (7.54) with those of (7.53). While solving for K in state feedback design, we had seen that phase variable form of system model was particularly convenient. The dual phase variable form of system model is particularly convenient for finding solution for M. In this sense the phase variable form is also called as **controllable canonical form** and dual phase variable form is also called as **observable canonical form**. In dual phase variable form, the matrices A and C are of form

$$A = \begin{bmatrix} 0 & 0 & \cdots & 0 & -a_n \\ 1 & 0 & \cdots & 0 & -a_{n-1} \\ 0 & 1 & \cdots & 0 & -a_{n-2} \\ \vdots & \vdots & & & \vdots \\ 0 & 0 & \cdots & 1 & -a_1 \end{bmatrix}, \qquad C = [0 \;\; 0 \;\; ... \;\; 1]$$

and characteristic equation is of form same as (7.48)

The observer error matrix is of form

$$(A - MC) = \begin{bmatrix} 0 & 0 & \cdots & 0 & -a_n - m_1 \\ 1 & 0 & \cdots & 0 & -a_{n-1} - m_2 \\ 0 & 1 & \cdots & 0 & -a_{n-2} - m_2 \\ \vdots & \vdots & \vdots & & \vdots \\ 0 & 0 & \cdots & 1 & -a_1 - m_n \end{bmatrix}$$

which gives the characteristic equation of form

$$s^n + (a_1 + m_n) s^{n-1} + ... + (a_{n-1} + m_2) s + (a_n + m_1) = 0 \qquad ...(7.55)$$

Now the observer gain matrix $M = \begin{bmatrix} m_1 \\ m_2 \\ \vdots \\ m_n \end{bmatrix}$ can be determined by comparing the coefficients of (7.55) with those of (7.54).

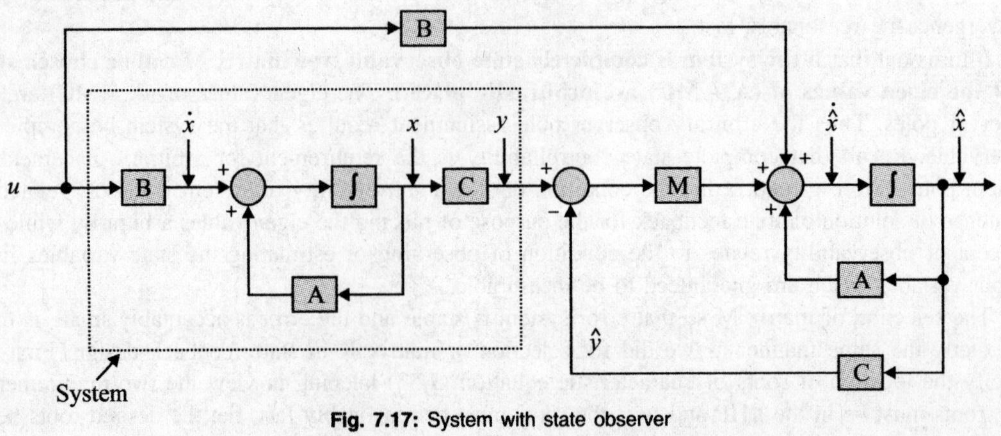

Fig. 7.17: System with state observer

As a numerical example for observer design consider the following system.

$$\begin{bmatrix} \dot{x}_1 \\ \dot{x}_2 \end{bmatrix} = \begin{bmatrix} -2 & -4 \\ 1 & -4 \end{bmatrix} \begin{bmatrix} x_1 \\ x_2 \end{bmatrix} + \begin{bmatrix} 2 \\ 0 \end{bmatrix} u$$

$$y = \begin{bmatrix} 1 & 0 \end{bmatrix} \begin{bmatrix} x_1 \\ x_2 \end{bmatrix}$$

It is required to find observer gain M so that the observer eigen values are placed at (– 50, – 50). Note that the system is completely observable. The observability matrix

$$Q_o = \begin{bmatrix} C \\ \cdots \\ CA \end{bmatrix} = \begin{bmatrix} 1 & 0 \\ -2 & -4 \end{bmatrix}$$

is non singular. The observer characteristic polynomial is

$$| sI - (A - MC) | = \left| \begin{bmatrix} s & 0 \\ 0 & s \end{bmatrix} - \begin{bmatrix} -2 & -4 \\ 1 & -4 \end{bmatrix} + \begin{bmatrix} m_1 \\ m_2 \end{bmatrix} \begin{bmatrix} 1 & 0 \end{bmatrix} \right|$$

$$= s^2 + (6 + m_1)\, s + (12 + 4m_1 - 4m_2)$$

while the desired characteristic polynomial is

$$(s + 50)\,(s + 50) = s^2 + 100s + 2500$$

Matching the coefficients of observer characteristic polynomial with those of desired polynomial, we have

$$m_1 = 94, \qquad m_2 = -528$$

and
$$M = \begin{bmatrix} 94 \\ -528 \end{bmatrix}$$

Use (7.50) to obtain the equations describing the observer dynamics as

$$\begin{bmatrix} \dot{\hat{x}}_1 \\ \dot{\hat{x}}_2 \end{bmatrix} = \begin{bmatrix} -2 & -4 \\ 1 & -4 \end{bmatrix} \begin{bmatrix} \hat{x}_1 \\ \hat{x}_2 \end{bmatrix} + \begin{bmatrix} 2 \\ 0 \end{bmatrix} u + \begin{bmatrix} 94 \\ -528 \end{bmatrix} \begin{bmatrix} 1 & 0 \end{bmatrix} \begin{bmatrix} x_1 - \hat{x}_1 \\ x_2 - \hat{x}_2 \end{bmatrix}$$

which gives

$$\dot{\hat{x}}_1 = -2\hat{x}_1 - 4\hat{x}_2 + 2u + 94(x_1 - \hat{x}_1)$$

$$\dot{\hat{x}}_2 = \hat{x}_1 - 4\hat{x}_2 - 528(x_1 - \hat{x}_1)$$

Using the dynamical equations just above a block diagram for the interconnected system and observer, is shown in Fig. 7.18.

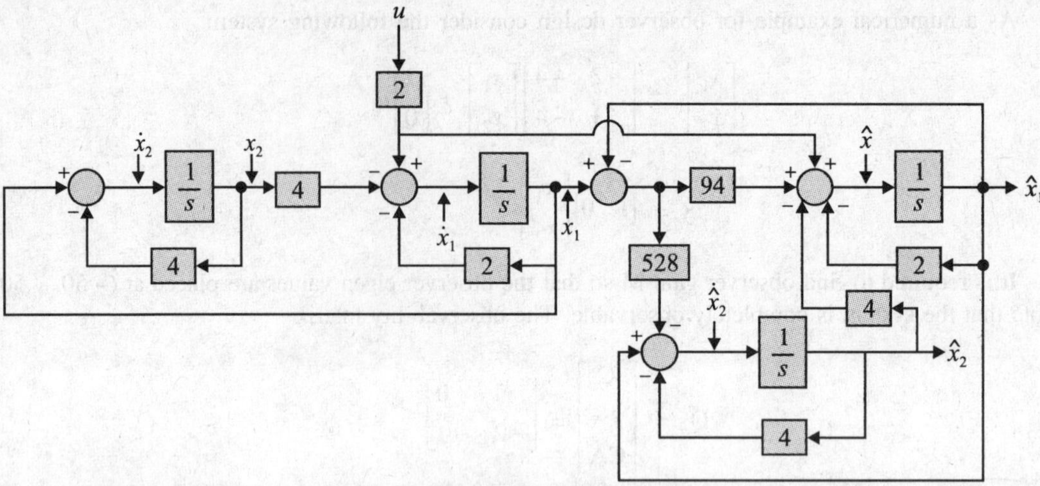

Fig. 7.18: Interconnected system and observer

PROBLEMS AND SOLUTIONS

P 7.1: Construct simulation diagram in phase variable form for systems with following transfer functions and develop state space model in matrix form.

$$(a)\ T(s) = \frac{Y(s)}{U(s)} = \frac{11s}{s^3+4s^2+3s+2}$$

$$(b)\ T_{11}(s) = \frac{Y_1(s)}{U(s)} = \frac{-s^2+4}{s^3+3s^2+s+2}$$

$$T_{21}(s) = \frac{Y_2(s)}{U(s)} = \frac{s^2+s+5}{s^3+3s^2+s+2}$$

Solution: *(a)*

$$T(s) = \frac{Y(s)}{U(s)} = \frac{11/s^2}{1+\dfrac{4}{s}+\dfrac{3}{s^2}+\dfrac{2}{s^3}} = \frac{11/s^2}{1-\left[-\dfrac{4}{s}-\dfrac{3}{s^2}-\dfrac{2}{s^3}\right]}$$

To construct simulation diagram in phase variable form, compare the transfer function with Mason's gain formula. The numerator identifies one forward path with path gain $P_1 = 11/s^2$ and the denominator identifies three individual loops with loop gains $L_1 = -4/s$, $L_2 = -3/s^2$ and $L_3 = -2/s^3$. The simulation diagram is shown in Fig. P 7.1(*a*). It has been ensured that the forward path and all of the loops touch a node to which input is coupled so that forward path cofactor is unity and no product of loop gain term evolves in Mason's gain formula.

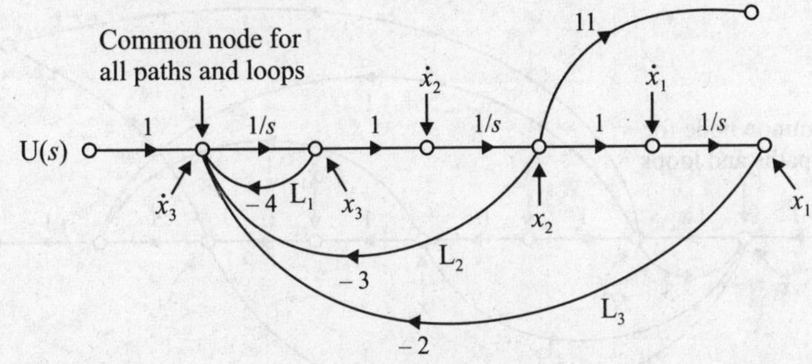

Fig. P7.1: (*a*) Simulation diagram

With state variable assignment x_1, x_2, x_3 as shown in simulation diagram Fig. P 7.1(*a*), the state and output equations are

$$\dot{x}_1 = x_2$$

$$\dot{x}_2 = x_3$$

$$\dot{x}_3 = -2x_1 - 3x_2 - 4x_3 + u$$

$$y = 11x_2$$

and in matrix form

$$\begin{bmatrix} \dot{x}_1 \\ \dot{x}_2 \\ \dot{x}_3 \end{bmatrix} = \begin{bmatrix} 0 & 1 & 0 \\ 0 & 0 & 1 \\ 2 & 3 & 4 \end{bmatrix} \begin{bmatrix} x_1 \\ x_2 \\ x_3 \end{bmatrix} + \begin{bmatrix} 0 \\ 0 \\ 1 \end{bmatrix} u$$

$$y = [0 \;\; 11 \;\; 0] \begin{bmatrix} x_1 \\ x_2 \\ x_3 \end{bmatrix}$$

(*b*) $$\frac{Y_1(s)}{U(s)} = \frac{-\dfrac{1}{s} + \dfrac{4}{s^3}}{1 - \left[-\dfrac{3}{s} - \dfrac{1}{s^2} - \dfrac{2}{s^3}\right]}$$

$$\frac{Y_2(s)}{U(s)} = \frac{\dfrac{1}{s} + \dfrac{1}{s^2} + \dfrac{5}{s^3}}{1 - \left[-\dfrac{3}{s} - \dfrac{1}{s^2} - \dfrac{2}{s^3}\right]}$$

Note that both the transfer functions share the same denominator and input. The phase variable simulation diagram is shown in Fig. P7.1(*b*) keeping in view the points discussed in the solution just above.

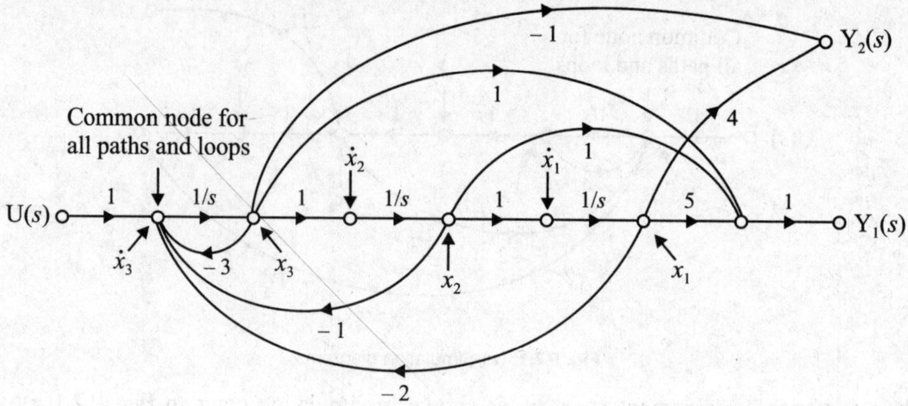

Fig. P 7.1: (*b*) Simulation diagram

Choosing the integrator outputs from right to left as state variables x_1, x_2 and x_3 as shown in Fig. P 7.1(*b*), the state and output equations in time domain are as follows:

$$\dot{x}_1 = x_2$$

$$\dot{x}_2 = x_3$$

$$\dot{x}_3 = -2x_1 - x_2 - 3x_3 + u$$

$$y_1 = 4x_1 - x_3$$

$$y_2 = 5x_1 + x_2 + x_3$$

These equations in matrix form are

$$\begin{bmatrix} \dot{x}_1 \\ \dot{x}_2 \\ \dot{x}_3 \end{bmatrix} = \begin{bmatrix} 0 & 1 & 0 \\ 0 & 0 & 1 \\ -2 & -1 & -3 \end{bmatrix} \begin{bmatrix} x_1 \\ x_2 \\ x_3 \end{bmatrix} + \begin{bmatrix} 0 \\ 0 \\ 1 \end{bmatrix} u$$

$$\begin{bmatrix} y_1 \\ y_2 \end{bmatrix} = \begin{bmatrix} 4 & 0 & -1 \\ 5 & 1 & 1 \end{bmatrix} \begin{bmatrix} x_1 \\ x_2 \\ x_3 \end{bmatrix}$$

P 7.2: Construct simulation diagram in dual phase variable form for systems with following transfer functions and develop state space model in matrix form.

(*a*) $T(s) = \dfrac{Y(s)}{U(s)} = \dfrac{2s+6}{3s^3 + 2s^2 + 8s + 10}$

(*b*) $T_{11}(s) = \dfrac{Y_1(s)}{U_1(s)} = \dfrac{3s^2 + 2}{s^3 + 3s^2 + s + 2}$

$T_{12}(s) = \dfrac{Y_1(s)}{U_2(s)} = \dfrac{s - 10}{s^3 + 3s^2 + s + 2}$

Solution: (*a*)

$$T(s) = \frac{Y(s)}{U(s)} = \frac{2s+6}{3s^3+2s^2+8s+10} = \frac{\frac{2s}{3}+2}{s^3+\frac{2}{3}s^2+\frac{8}{3}s+\frac{10}{3}}$$

$$= \frac{\frac{2/3}{s^2}+\frac{2}{s^3}}{1 \left[\frac{2/3}{s} \;\; \frac{8/3}{s^2} \;\; \frac{10/3}{s^3} \right]} \equiv \frac{P_1+P_2}{1 \; [L_1+L_2+L_3]}$$

Dual phase variable simulation diagram is shown in Fig. P 7.2(*a*). It has two forward paths $P_1 = (2/3)/s^2$ and $P_2 = 2/s^3$ together with three individual loops $L_1 = -(2/3)/s$, $L_2 = -(8/3)/s^2$ and $L_3 = -(10/3)/s^3$. The forward paths and individual loops have been intermingled such that all of them touch a common node, that is coupled to output node.

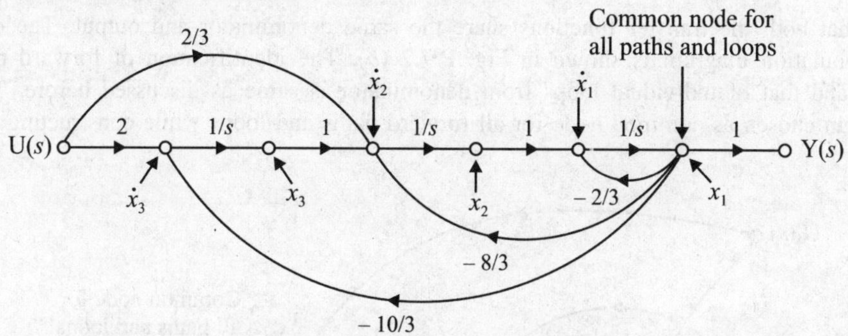

Fig. P 7.2: (*a*) Simulation diagram

The integrator outputs from right to left are chosen as state variables x_1, x_2 and x_3 as shown in Fig. P 7.2(*a*). The state and output equations are

$$\dot{x}_1 = \frac{2}{3}x_1 + x_2$$

$$\dot{x}_2 = \frac{8}{3}x_1 + x_3 + \frac{2}{3}u$$

$$\dot{x}_3 = \frac{10}{3}x_1 + 2u$$

$$y = x_1$$

These equations in matrix form are

$$\begin{bmatrix} \dot{x}_1 \\ \dot{x}_2 \\ \dot{x}_3 \end{bmatrix} = \begin{bmatrix} \frac{2}{3} & 1 & 0 \\ \frac{8}{3} & 0 & 1 \\ \frac{10}{3} & 0 & 0 \end{bmatrix} \begin{bmatrix} x_1 \\ x_2 \\ x_3 \end{bmatrix} + \begin{bmatrix} 0 \\ \frac{2}{3} \\ 2 \end{bmatrix} u$$

$$y = \begin{bmatrix} 1 & 0 & 0 \end{bmatrix} \begin{bmatrix} x_1 \\ x_2 \\ x_3 \end{bmatrix}$$

(*b*) $$\frac{Y_1(s)}{U_1(s)} = \frac{3s^2+2}{s^3+3s^2+s+2} = \frac{\dfrac{3}{s}+\dfrac{2}{s^3}}{1-\left[-\dfrac{3}{s}-\dfrac{1}{s^2}-\dfrac{2}{s^3}\right]}$$

$$\frac{Y_1(s)}{U_2(s)} = \frac{s-10}{s^3+3s^2+s+2} = \frac{\dfrac{1}{s^2}+\dfrac{10}{s^3}}{1-\left[-\dfrac{3}{s}-\dfrac{1}{s^2}-\dfrac{2}{s^3}\right]}$$

Note that both the transfer functions share the same denominator and output. The dual phase variable simulation diagram is shown in Fig. P 7.2 (*b*). The identification of forward paths from numerator and that of individual loops from denominator is same as discussed before. The output node is again chosen as common node for all forward paths and loops while constructing simulation diagram.

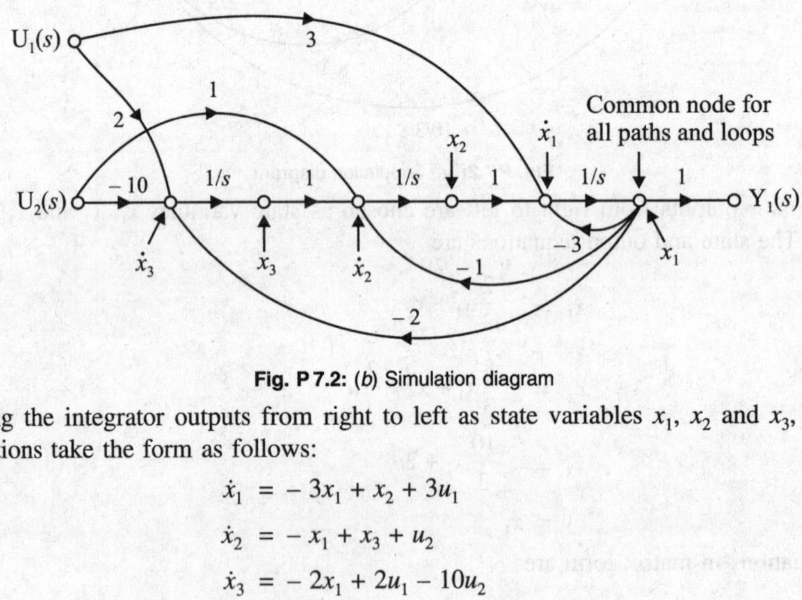

Fig. P 7.2: (*b*) Simulation diagram

Choosing the integrator outputs from right to left as state variables x_1, x_2 and x_3, the state and output equations take the form as follows:

$$\dot{x}_1 = -3x_1 + x_2 + 3u_1$$

$$\dot{x}_2 = -x_1 + x_3 + u_2$$

$$\dot{x}_3 = -2x_1 + 2u_1 - 10u_2$$

$$y_1 = x_1$$

These equations in matrix form are

$$\begin{bmatrix} \dot{x}_1 \\ \dot{x}_2 \\ \dot{x}_3 \end{bmatrix} = \begin{bmatrix} -3 & 1 & 0 \\ -1 & 0 & 1 \\ -2 & 0 & 0 \end{bmatrix} \begin{bmatrix} x_1 \\ x_2 \\ x_3 \end{bmatrix} + \begin{bmatrix} 3 & 0 \\ 0 & 1 \\ 2 & -10 \end{bmatrix} \begin{bmatrix} u_1 \\ u_2 \end{bmatrix}$$

$$y_1 = \begin{bmatrix} 1 & 0 & 0 \end{bmatrix} \begin{bmatrix} x_1 \\ x_2 \\ x_3 \end{bmatrix}$$

P 7.3: Develop state space model for the system with signal flow graph shown in Fig. P7.3.

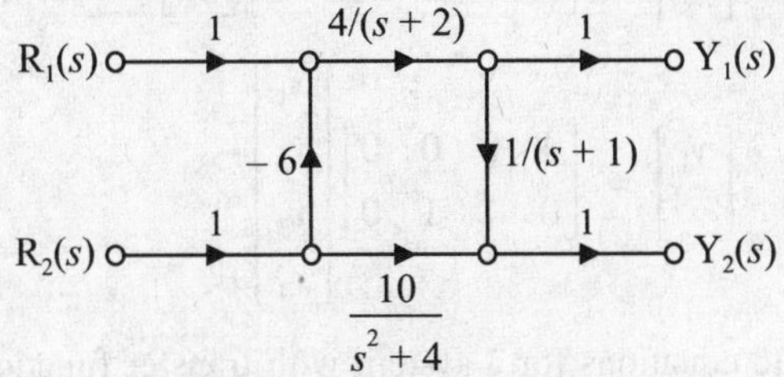

Fig. P 7.3: Signal flow graph

Solution: Expand the signal flow graph of Fig. P7.3 into a simulation diagram as shown in Fig. P7.3(a) by replacing individual transmittances. Use phase variable form in each of these transmittances for ease. While doing so care must be taken to preserve the signal relationship

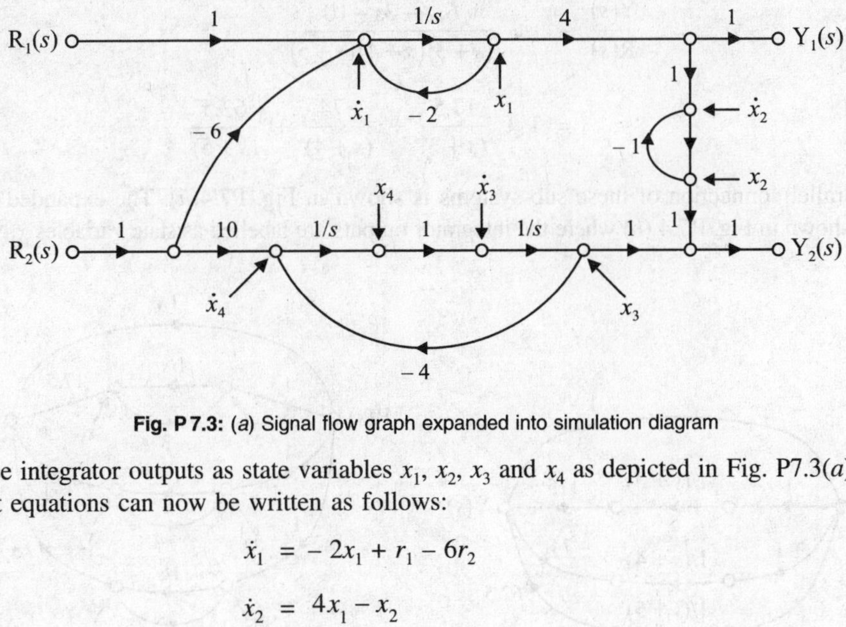

Fig. P 7.3: (a) Signal flow graph expanded into simulation diagram

Choose integrator outputs as state variables x_1, x_2, x_3 and x_4 as depicted in Fig. P7.3(a). The state and output equations can now be written as follows:

$$\dot{x}_1 = -2x_1 + r_1 - 6r_2$$

$$\dot{x}_2 = 4x_1 - x_2$$

$$\dot{x}_3 = x_4$$

$$\dot{x}_4 = -4x_3 + 10r_2$$

$$y_1 = 4x_1$$

$$y_2 = x_2 + x_3$$

These in matrix form are

$$\begin{bmatrix} \dot{x}_1 \\ \dot{x}_2 \\ \dot{x}_3 \\ \dot{x}_4 \end{bmatrix} = \begin{bmatrix} -2 & 0 & 0 & 0 \\ 4 & -1 & 0 & 0 \\ 0 & 0 & 0 & 1 \\ 0 & 0 & -4 & 0 \end{bmatrix} \begin{bmatrix} x_1 \\ x_2 \\ x_3 \\ x_4 \end{bmatrix} + \begin{bmatrix} 1 & -6 \\ 0 & 0 \\ 0 & 0 \\ 0 & 10 \end{bmatrix} \begin{bmatrix} r_1 \\ r_2 \end{bmatrix}$$

$$\begin{bmatrix} y_1 \\ y_2 \end{bmatrix} = \begin{bmatrix} 4 & 0 & 0 & 0 \\ 0 & 1 & 1 & 0 \end{bmatrix} \begin{bmatrix} x_1 \\ x_2 \\ x_3 \\ x_4 \end{bmatrix}$$

P 7.4: Find diagonal state equations for a system with transfer function

$$T(s) = \frac{Y(s)}{R(s)} = \frac{s^3 + 18s^2 + 50s + 50}{(s+3)(s+4)(s+5)}$$

Solution: Expanding transfer function in partial fractions, we have

$$\frac{Y(s)}{R(s)} = 1 + \frac{6s^2 + 3s - 10}{(s+3)(s+4)(s+5)}$$

$$= 1 + \frac{17.5}{(s+3)} + \frac{-74}{(s+4)} + \frac{62.5}{(s+5)}$$

The parallel connection of these sub-systems is shown in Fig. P7.4(a). The expanded simulation diagram is shown in Fig. P7.4 (b) where the integrator outputs are labelled as state variables x_1, x_2 and x_3.

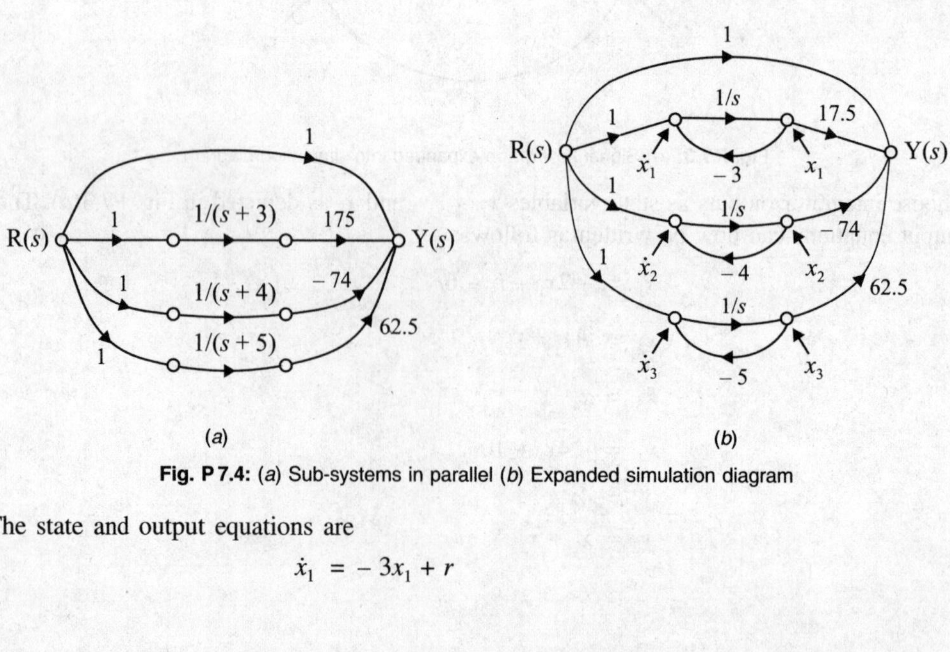

(a) (b)

Fig. P 7.4: (a) Sub-systems in parallel (b) Expanded simulation diagram

The state and output equations are

$$\dot{x}_1 = -3x_1 + r$$

$$\dot{x}_2 = -4x_2 + r$$

$$\dot{x}_3 = -5x_3 + r$$

$$y = 17.5x_1 - 74x_2 + 62.5x_3 + r$$

and in matrix form

$$\begin{bmatrix} \dot{x}_1 \\ \dot{x}_2 \\ \dot{x}_3 \end{bmatrix} = \begin{bmatrix} -3 & 0 & 0 \\ 0 & -4 & 0 \\ 0 & 0 & -5 \end{bmatrix} \begin{bmatrix} x_1 \\ x_2 \\ x_3 \end{bmatrix} + \begin{bmatrix} 1 \\ 1 \\ 1 \end{bmatrix} r$$

$$y = \begin{bmatrix} 17.5 & -74 & 62.5 \end{bmatrix} \begin{bmatrix} x_1 \\ x_2 \\ x_3 \end{bmatrix} + (1)r$$

P 7.5: The SISO system with transfer function

$$\mathrm{T}(s) = \frac{\mathrm{Y}(s)}{\mathrm{U}(s)} = \frac{9}{(s+2)(s^2+4s+13)}$$

involves complex characteristic roots. Obtain the diagonal form of state space model. Also obtain an alternative block diagonal model which does not involve complex numbers.

Solution: Expand the transfer function in partial fractions to get

$$\frac{\mathrm{Y}(s)}{\mathrm{U}(s)} = \frac{1}{s+2} + \frac{-0.5}{s+2+j3} + \frac{-0.5}{s+2-j3}$$

The simulation diagram (parallel connection of these sub-systems) is shown in Fig. P7.5(*a*) where the integrator outputs are labelled as state variables x_1, x_2 and x_3. The state and output equations are

$$\dot{x}_1 = -2x_1 + r$$

$$\dot{x}_2 = -(2 + j3)\, x_2 + r$$

$$\dot{x}_3 = -(2 - j3)\, x_3 + r$$

$$y = x_1 - 0.5\, x_2 - 0.5x_3$$

which in matrix form

$$\begin{bmatrix} \dot{x}_1 \\ \dot{x}_2 \\ \dot{x}_3 \end{bmatrix} = \begin{bmatrix} -2 & 0 & 0 \\ 0 & -(2+j3) & 0 \\ 0 & 0 & -(2-j3) \end{bmatrix} \begin{bmatrix} x_1 \\ x_2 \\ x_3 \end{bmatrix} + \begin{bmatrix} 1 \\ 1 \\ 1 \end{bmatrix} r$$

$$y = \begin{bmatrix} 1 & -0.5 & -0.5 \end{bmatrix} \begin{bmatrix} x_1 \\ x_2 \\ x_3 \end{bmatrix}$$

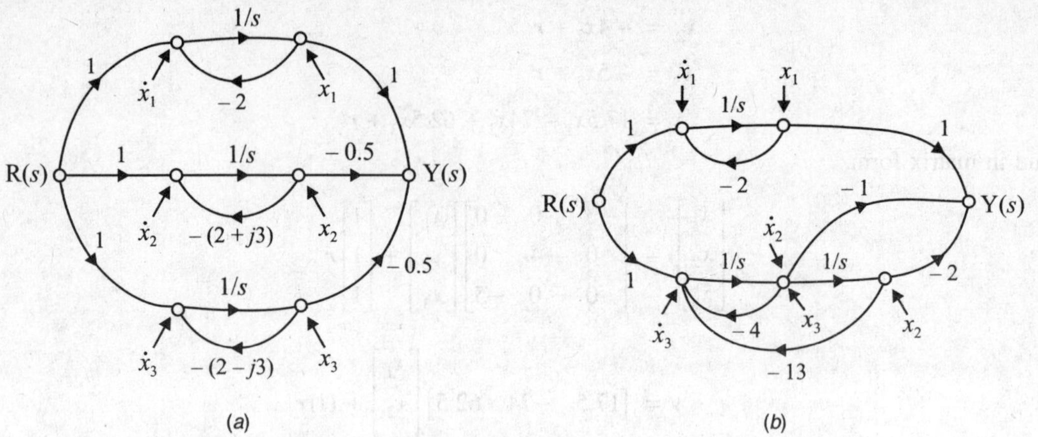

Fig. P 7.5: Simulation diagram (*a*) Diagonal form with complex numbers (*b*) Diagonal cum phase variable form with real numbers

In order to obtain a model that does not involve the complex numbers, the complex terms of partial fraction are combined as

$$\frac{-0.5}{s+2+j3}+\frac{-0.5}{s+2-j3}=\frac{s+2}{s^2+4s+13}=\frac{-\frac{1}{s}-\frac{2}{s^2}}{1+\frac{4}{s}+\frac{13}{s^2}}$$

and then placed in phase variable form as demonstrated in simulation diagram shown in Fig. P7.5(*b*) where the sate variables and their derivatives are also labelled. The state and output equations are

$$\dot{x}_1 = -2x_1 + r$$

$$\dot{x}_2 = x_3$$

$$\dot{x}_3 = -13x_2 - 4x_3 + r$$

$$y = x_1 - 2x_2 - x_3$$

which in matrix form are

$$\begin{bmatrix}\dot{x}_1\\ \dot{x}_2\\ \dot{x}_3\end{bmatrix}=\begin{bmatrix}-2 & 0 & 0\\ 0 & 0 & 1\\ 0 & -13 & -4\end{bmatrix}\begin{bmatrix}x_1\\ x_2\\ x_3\end{bmatrix}+\begin{bmatrix}1\\ 0\\ 1\end{bmatrix}r$$

$$y=\begin{bmatrix}1 & -2 & -1\end{bmatrix}\begin{bmatrix}x_1\\ x_2\\ x_3\end{bmatrix}$$

P 7.6: The SISO system with repeated characteristic roots is described by transfer function

$$T(s)=\frac{Y(s)}{U(s)}=\frac{7s^3}{(s+2)^2(s+6)^2}$$

Find state space model in block diagonal Jordan canonical form.

Solution: Expanding in partial fractions we have

$$\frac{Y(s)}{U(s)} = \frac{7s^3}{(s+2)^2(s+6)^2}$$

$$= \frac{-3.5}{(s+2)^2} + \frac{7}{(s+2)} + \frac{-94.5}{(s+6)^2} + \frac{0}{(s+6)}$$

The signal flow graph showing each of these partial fraction terms is shown in Fig. P7.6(*a*). An alternative signal flow graph combining two transmittances $(-3.5)/(s+2)^2$ and $7/(s+2)$, is shown in Fig. P7.6(*b*). The simulation diagram with state variables x_1, x_2, x_3 and x_4 labelled, is shown in Fig. P7.6(*c*).

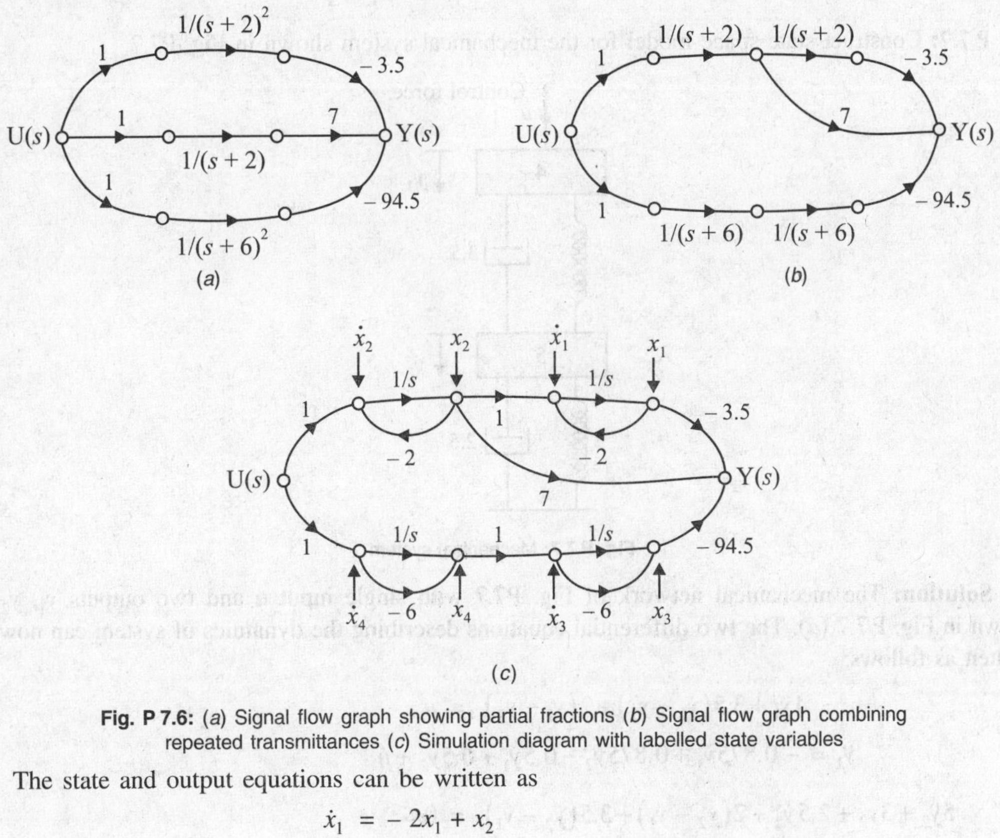

Fig. P 7.6: (*a*) Signal flow graph showing partial fractions (*b*) Signal flow graph combining repeated transmittances (*c*) Simulation diagram with labelled state variables

The state and output equations can be written as

$$\dot{x}_1 = -2x_1 + x_2$$

$$\dot{x}_2 = -2x_2 + u$$

$$\dot{x}_3 = -6x_3 + x_4$$

$$\dot{x}_4 = -6x_4 + u$$

$$y = -3.5x_1 + 7x_2 - 94.5x_3$$

which in matrix form

$$\begin{bmatrix} \dot{x}_1 \\ \dot{x}_2 \\ \dot{x}_3 \\ \dot{x}_4 \end{bmatrix} = \begin{bmatrix} -2 & 1 & 0 & 0 \\ 0 & -2 & 0 & 0 \\ 0 & 0 & -6 & 1 \\ 0 & 0 & 0 & -6 \end{bmatrix} \begin{bmatrix} x_1 \\ x_2 \\ x_3 \\ x_4 \end{bmatrix} + \begin{bmatrix} 0 \\ 1 \\ 0 \\ 1 \end{bmatrix} u$$

$$y = \begin{bmatrix} -3.5 & 7 & -94.5 & 0 \end{bmatrix} \begin{bmatrix} x_1 \\ x_2 \\ x_3 \\ x_4 \end{bmatrix}$$

P 7.7: Construct state space model for the mechanical system shown in Fig. P7.7.

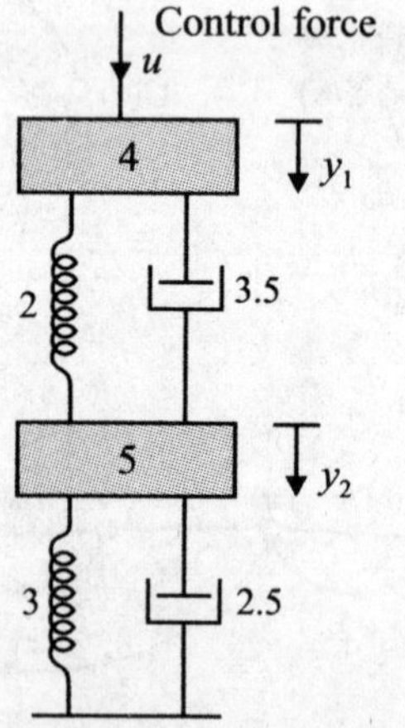

Fig. P 7.7: Mechanical system

Solution: The mechanical network of Fig. P7.7 with single input u and two outputs y_1, y_2, is shown in Fig. P 7.7 (*a*). The two differential equations describing the dynamics of system can now be written as follows:

$$4\ddot{y}_1 + 3.5(\dot{y}_1 - \dot{y}_2) + 2(y_1 - y_2) = u$$

or $$\ddot{y}_1 = -0.875\dot{y}_1 + 0.875\dot{y}_2 - 0.5y_1 + 0.5y_2 + u$$

and $$5\ddot{y}_2 + 3y_2 + 2.5\dot{y}_2 + 2(y_2 - y_1) + 3.5(\dot{y}_2 - \dot{y}_1) = 0$$

or $$\ddot{y}_2 = -1.2\dot{y}_2 + 0.7\dot{y}_1 - y_2 + 0.4y_1$$

These equations with state variable assignment

$$x_1 = y_1$$
$$x_2 = \dot{y}_1$$
$$x_3 = y_2$$
$$x_4 = \dot{y}_2$$

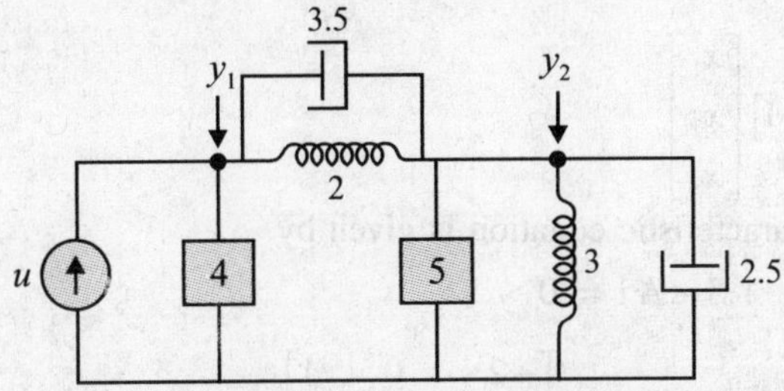

Fig. P 7.7: (*a*) Mechanical network

and little mathematical manipulation can be put in form of state and output equations as follows:

$$\dot{x}_1 = x_2$$
$$\dot{x}_2 = -0.5x_1 - 0.875x_2 + 0.5x_3 + 0.875x_4 + u$$
$$\dot{x}_3 = x_4$$
$$\dot{x}_4 = 0.4x_1 + 0.7x_2 - x_3 - 1.2x_4$$
$$y_1 = x_1$$
$$y_2 = x_3$$

These equations in matrix form are written as follows:

$$\begin{bmatrix} \dot{x}_1 \\ \dot{x}_2 \\ \dot{x}_3 \\ \dot{x}_4 \end{bmatrix} = \begin{bmatrix} 0 & 1 & 0 & 0 \\ -0.5 & -0.875 & 0.5 & 0.875 \\ 0 & 0 & 0 & 1 \\ 0.4 & 0.7 & -1 & -1.2 \end{bmatrix} \begin{bmatrix} x_1 \\ x_2 \\ x_3 \\ x_4 \end{bmatrix} + \begin{bmatrix} 0 \\ 1 \\ 0 \\ 0 \end{bmatrix} u$$

$$\begin{bmatrix} y_1 \\ y_2 \end{bmatrix} = \begin{bmatrix} 1 & 0 & 0 & 0 \\ 0 & 0 & 1 & 0 \end{bmatrix} \begin{bmatrix} x_1 \\ x_2 \\ x_3 \\ x_4 \end{bmatrix}$$

P 7.8: Find characteristic equation for each of the following systems. Then for each, determine if they are stable.

(*a*) $$\begin{bmatrix} \dot{x}_1 \\ \dot{x}_2 \\ \dot{x}_3 \end{bmatrix} = \begin{bmatrix} -2 & 0 & 1 \\ 0 & 0 & 1 \\ 0 & -10 & -6 \end{bmatrix} \begin{bmatrix} x_1 \\ x_2 \\ x_3 \end{bmatrix} + \begin{bmatrix} 1 \\ 0 \\ 1 \end{bmatrix} u$$

$$y = \begin{bmatrix} -1 & 3 & 3 \end{bmatrix} \begin{bmatrix} x_1 \\ x_2 \\ x_3 \end{bmatrix}$$

(*b*) $$\begin{bmatrix} \dot{x}_1 \\ \dot{x}_2 \\ \dot{x}_3 \end{bmatrix} = \begin{bmatrix} 1 & -2 & 3 \\ 4 & 0 & 6 \\ -1 & 2 & 1 \end{bmatrix} \begin{bmatrix} x_1 \\ x_2 \\ x_3 \end{bmatrix} + \begin{bmatrix} 2 & -1 \\ 0 & 0 \\ 3 & 6 \end{bmatrix} \begin{bmatrix} u_1 \\ u_2 \end{bmatrix}$$

$$y = \begin{bmatrix} 1 & 0 & -1 \end{bmatrix} \begin{bmatrix} x_1 \\ x_2 \\ x_3 \end{bmatrix}$$

Solution: (*a*) The characteristic equation is given by

$$| sI - A | = 0$$

where
$$A = \begin{bmatrix} -2 & 0 & 1 \\ 0 & 0 & 1 \\ 0 & -10 & -6 \end{bmatrix}$$

$$| sI - A | = \begin{vmatrix} s+2 & 0 & -1 \\ 0 & s & -1 \\ 0 & 10 & s+6 \end{vmatrix} = s^3 + 8s^2 + 22s + 20$$

Now stability can be determined by constructing Routh array as follows:

s^3	1	22
s^2	8	20
s^1	19.5	
s^0	20	

There is no algebraic sign changes in the left column. So, the characteristic polynomial has all LHP roots and system is stable.

(*b*) $| sI - A | = 0$ gives

$$\begin{vmatrix} s-1 & 2 & -3 \\ -4 & s & -6 \\ 1 & -2 & s-1 \end{vmatrix} = 0$$

or
$$s^3 - 2s^2 - 32 = 0$$

The characteristic polynomial fails the coefficient test. So, the system is unstable. However, the Routh test can be performed as follows:

s^3	1	0
s^2	– 2	– 32
s^1	– 16	0
s^0	– 32	0

There is one sign change in the left column. This is indicative of one RHP root, system is unstable.

P 7.9: The following transfer functions do not share a common denominator polynomial, but they may be made to do so by multiplying their numerators and denominators by appropriate factors. Obtain simulation diagram involving only three integrators. Then obtain state space model in matrix form.

$$T_{11}(s) = \frac{Y_1(s)}{U_1(s)} = \frac{3s+1}{(s+2)(s+3)}$$

$$T_{21}(s) = \frac{Y_2(s)}{U_1(s)} = \frac{16s}{(s+1)(s+3)}$$

Solution: In order to make both transfer functions share a common denominator polynomial, let us multiply numerator and denominator of $T_{11}(s)$ by $(s + 1)$ and those of $T_{21}(s)$ by $(s + 2)$ to get

$$T_{11}(s) = \frac{Y_1(s)}{U_1(s)} = \frac{(3s+1)(s+1)}{(s+1)(s+2)(s+3)}$$

$$= \frac{3s^2+4s+1}{s^3+6s^2+11s+6} = \frac{\frac{3}{s}+\frac{4}{s^2}+\frac{1}{s^3}}{1-\left[-\frac{6}{s}-\frac{11}{s^2}-\frac{6}{s^3}\right]}$$

and
$$T_{21}(s) = \frac{Y_2(s)}{U_1(s)} = \frac{16s(s+2)}{(s+1)(s+2)(s+3)}$$

$$= \frac{16s^2+32s}{s^3+6s^2+11s+6} = \frac{\frac{16}{s}+\frac{32}{s^2}}{1-\left[-\frac{6}{s}-\frac{11}{s^2}-\frac{6}{s^3}\right]}$$

The form of transfer functions shown above can be translated into simulation diagram using three integrators as shown in Fig. P 7.9.

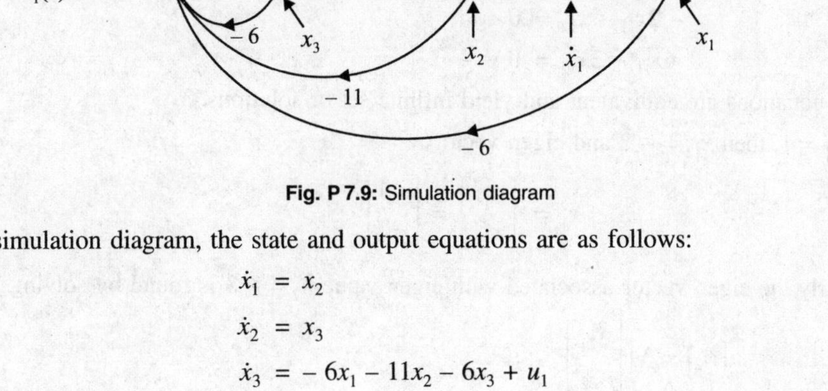

Fig. P 7.9: Simulation diagram

From simulation diagram, the state and output equations are as follows:

$$\dot{x}_1 = x_2$$

$$\dot{x}_2 = x_3$$

$$\dot{x}_3 = -6x_1 - 11x_2 - 6x_3 + u_1$$

$$y_1 = x_1 + 4x_2 + 3x_3$$
$$y_2 = 32x_2 + 16x_3$$

These in matrix form are

$$\begin{bmatrix} \dot{x}_1 \\ \dot{x}_2 \\ \dot{x}_3 \end{bmatrix} = \begin{bmatrix} 0 & 1 & 0 \\ 0 & 0 & 1 \\ -6 & -11 & -6 \end{bmatrix} \begin{bmatrix} x_1 \\ x_2 \\ x_3 \end{bmatrix} + \begin{bmatrix} 0 \\ 0 \\ 1 \end{bmatrix} u_1$$

$$\begin{bmatrix} y_1 \\ y_2 \end{bmatrix} = \begin{bmatrix} 1 & 4 & 3 \\ 0 & 32 & 16 \end{bmatrix} \begin{bmatrix} x_1 \\ x_2 \\ x_3 \end{bmatrix}$$

P 7.10: Find decoupled state equations for the system described as

$$\begin{bmatrix} \dot{x}_1 \\ \dot{x}_2 \end{bmatrix} = \begin{bmatrix} 0 & 1 \\ -6 & -5 \end{bmatrix} \begin{bmatrix} x_1 \\ x_2 \end{bmatrix} + \begin{bmatrix} 1 & 0 \\ 0 & 1 \end{bmatrix} \begin{bmatrix} u_1 \\ u_2 \end{bmatrix}$$

$$y = \begin{bmatrix} 0 & 1 \end{bmatrix} \begin{bmatrix} x_1 \\ x_2 \end{bmatrix}$$

Solution: The eigen values are found as follows:

$$| \lambda I - A | = 0$$

$$\begin{vmatrix} \lambda & -1 \\ 6 & \lambda + 5 \end{vmatrix} = 0$$

$$\lambda_1 = -2, \qquad \lambda_2 = -3$$

The eigen vector associated with eigen value $\lambda_1 = -2$ is found by solving

$$[\lambda_1 I - A] \begin{bmatrix} x_{11} \\ x_{12} \end{bmatrix} = 0$$

$$\begin{bmatrix} -2 & -1 \\ 6 & -3 \end{bmatrix} \begin{bmatrix} x_{11} \\ x_{12} \end{bmatrix} = 0$$

$$-2x_{11} - x_{12} = 0$$
$$6x_{11} + 3x_{12} = 0$$

These equations are equivalent and yield infinite set of solutions.

Let $x_{11} = 1$, then $x_{12} = -2$ and eigen vector

$$x_1 = \begin{bmatrix} x_{11} \\ x_{12} \end{bmatrix} = \begin{bmatrix} 1 \\ -2 \end{bmatrix}$$

Similarly the eigen vector associated with eigen value $\lambda_2 = -3$ is found by solving

$$[\lambda_2 I - A] \begin{bmatrix} x_{21} \\ x_{22} \end{bmatrix} = 0$$

$$\begin{bmatrix} -3 & -1 \\ 6 & 2 \end{bmatrix} \begin{bmatrix} x_{21} \\ x_{22} \end{bmatrix} = 0$$

$$-3x_{21} - x_{22} = 0$$

$$6x_{21} + 2x_{22} = 0$$

Let $x_{21} = 1$, then $x_{22} = -3$ and eigen vector

$$x_2 = \begin{bmatrix} x_{21} \\ x_{22} \end{bmatrix} = \begin{bmatrix} 1 \\ -3 \end{bmatrix}$$

Collect eigen vectors to get

$$P = \begin{bmatrix} 1 & 1 \\ -2 & -3 \end{bmatrix}$$

Then

$$P^{-1} = \begin{bmatrix} 3 & 1 \\ -2 & -1 \end{bmatrix}$$

$$\bar{A} = P^{-1}AP = \begin{bmatrix} 3 & 1 \\ -2 & -1 \end{bmatrix} \begin{bmatrix} 0 & 1 \\ -6 & -5 \end{bmatrix} \begin{bmatrix} 1 & 1 \\ -2 & -3 \end{bmatrix} = \begin{bmatrix} -2 & 0 \\ 0 & -3 \end{bmatrix}$$

$$\bar{B} = P^{-1}B = \begin{bmatrix} 3 & 1 \\ -2 & -1 \end{bmatrix} \begin{bmatrix} 1 & 0 \\ 0 & 1 \end{bmatrix} = \begin{bmatrix} 3 & 1 \\ -2 & -1 \end{bmatrix}$$

$$\bar{C} = CP = \begin{bmatrix} 0 & 1 \end{bmatrix} \begin{bmatrix} 1 & 1 \\ -2 & -3 \end{bmatrix} = \begin{bmatrix} -2 & -3 \end{bmatrix}$$

and diagonalised state model in matrix form is

$$\begin{bmatrix} \dot{z}_1 \\ \dot{z}_2 \end{bmatrix} = \begin{bmatrix} -2 & 0 \\ 0 & -3 \end{bmatrix} \begin{bmatrix} z_1 \\ z_2 \end{bmatrix} + \begin{bmatrix} 3 & 1 \\ -2 & -1 \end{bmatrix} \begin{bmatrix} u_1 \\ u_2 \end{bmatrix}$$

$$n = \begin{bmatrix} -2 & -3 \end{bmatrix} \begin{bmatrix} z_1 \\ z_2 \end{bmatrix}$$

P 7.11: Obtain controllability and observability matrices and investigate whether or not the following systems are completely controllable and/or completely observable.

$$(a) \begin{bmatrix} \dot{x}_1 \\ \dot{x}_2 \\ \dot{x}_3 \end{bmatrix} = \begin{bmatrix} 1 & 0 & -2 \\ 3 & -3 & 0 \\ 0 & 0 & 1 \end{bmatrix} \begin{bmatrix} x_1 \\ x_2 \\ x_3 \end{bmatrix} + \begin{bmatrix} 1 & -1 \\ 2 & 0 \\ 0 & 0 \end{bmatrix} \begin{bmatrix} u_1 \\ u_2 \end{bmatrix}$$

$$\begin{bmatrix} y_1 \\ y_2 \end{bmatrix} = \begin{bmatrix} 0 & 4 & 1 \\ 0 & -2 & 3 \end{bmatrix} \begin{bmatrix} x_1 \\ x_2 \\ x_3 \end{bmatrix}$$

(*b*)

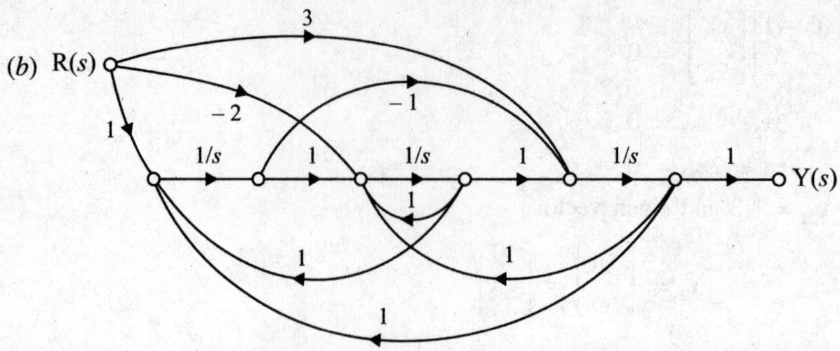

Solution: (*a*) The controllability matrix Q_c and observability matrix Q_o for given third order system are

$$Q_c = \begin{bmatrix} B \vdots AB \vdots A^2B \end{bmatrix}, \qquad Q_o = \begin{bmatrix} C \\ CA \\ CA^2 \end{bmatrix}$$

where
$$A = \begin{bmatrix} 1 & 0 & -2 \\ 3 & -3 & 0 \\ 0 & 0 & 1 \end{bmatrix}, \quad B = \begin{bmatrix} 1 & -1 \\ 2 & 0 \\ 0 & 0 \end{bmatrix}, \quad C = \begin{bmatrix} 0 & 4 & 1 \\ 0 & -2 & 3 \end{bmatrix}$$

$$AB = \begin{bmatrix} 1 & -1 \\ -3 & -3 \\ 0 & 0 \end{bmatrix}, \quad A^2B = \begin{bmatrix} 1 & -1 \\ 12 & 6 \\ 0 & 0 \end{bmatrix}$$

$$Q_c = \begin{bmatrix} 1 & -1 & 1 & -1 & 1 & -1 \\ 2 & 0 & -3 & -3 & 12 & 6 \\ 0 & 0 & 0 & 0 & 0 & 0 \end{bmatrix}$$

The controllability matrix Q_c is not of full rank, rather it is of rank 2. This system is not completely controllable.

$$CA = \begin{bmatrix} 12 & -12 & 1 \\ -6 & 6 & 3 \end{bmatrix}$$

$$CA^2 = \begin{bmatrix} -24 & 36 & -23 \\ 12 & -18 & 15 \end{bmatrix}$$

$$Q_o = \begin{bmatrix} 0 & 4 & 1 \\ 0 & -2 & 3 \\ 12 & -12 & 1 \\ -6 & 6 & 3 \\ -24 & 36 & -23 \\ 12 & -18 & 15 \end{bmatrix}$$

The absorbability matrix Q_o is of full rank and the system is completely observable.

(*b*) Choosing integrator outputs from right to left as state variables x_1, x_2 and x_3 as labelled in Fig. P 7.11 (*a*), the state and output equations can be written as follows:

$$\dot{x}_1 = x_2 - x_3 + 3r$$
$$\dot{x}_2 = x_1 + x_2 + x_3 - 2r$$
$$\dot{x}_3 = x_1 + x_2 + r$$
$$y = x_1$$

$$\begin{bmatrix} \dot{x}_1 \\ \dot{x}_2 \\ \dot{x}_3 \end{bmatrix} = \begin{bmatrix} 0 & 1 & -1 \\ 1 & 1 & 1 \\ 1 & 1 & 0 \end{bmatrix} + \begin{bmatrix} 3 \\ -2 \\ 1 \end{bmatrix} r$$

$$y = \begin{bmatrix} 1 & 0 & 0 \end{bmatrix} \begin{bmatrix} x_1 \\ x_2 \\ x_3 \end{bmatrix}$$

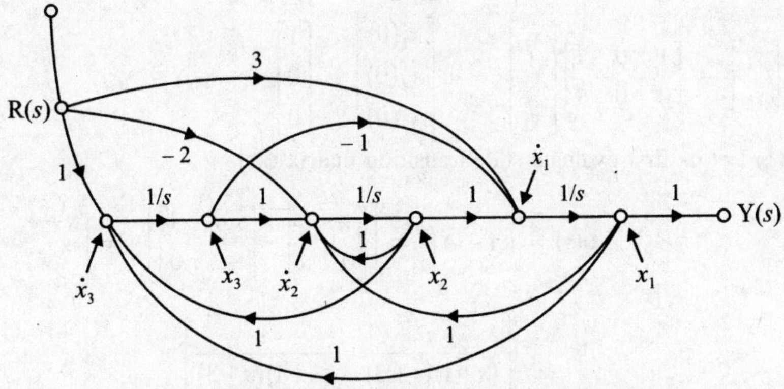

Fig. P 7.11: (*a*) Simulation diagram with labelled state variables

$$A = \begin{bmatrix} 0 & 1 & -1 \\ 1 & 1 & 1 \\ 1 & 1 & 0 \end{bmatrix}, \quad B = \begin{bmatrix} 3 \\ -2 \\ 1 \end{bmatrix}, \quad C = \begin{bmatrix} 1 & 0 & 0 \end{bmatrix}$$

$$AB = \begin{bmatrix} -3 \\ 2 \\ 1 \end{bmatrix}, \quad A^2B = \begin{bmatrix} 1 \\ 0 \\ -1 \end{bmatrix}, \quad CA = \begin{bmatrix} 0 & 1 & -1 \end{bmatrix}$$

$$CA^2 = [0 \quad 0 \quad 1]$$

$$Q_c = \begin{bmatrix} B \vdots AB \vdots A^2B \end{bmatrix} = \begin{bmatrix} 3 & -3 & 1 \\ 2 & 2 & 0 \\ -1 & 1 & -1 \end{bmatrix}$$

The controllability matrix Q_c is of full rank. The system is completely controllable.

$$Q_o = \begin{bmatrix} C \\ CA \\ CA^2 \end{bmatrix} = \begin{bmatrix} 1 & 0 & 0 \\ 0 & 1 & -1 \\ 0 & 0 & 1 \end{bmatrix}$$

The observability matrix Q_o is also of full rank. So, the system is completely observable as well.

P 7.12: Find the state response and system response for the systems described as follows:

(*a*)
$$\begin{bmatrix} \dot{x}_1 \\ \dot{x}_2 \end{bmatrix} = \begin{bmatrix} -4 & 1 \\ -3 & 0 \end{bmatrix}\begin{bmatrix} x_1 \\ x_2 \end{bmatrix} + \begin{bmatrix} 1 \\ 0 \end{bmatrix} r$$

$$y = \begin{bmatrix} 1 & 0 \end{bmatrix}\begin{bmatrix} x_1 \\ x_2 \end{bmatrix}, \quad \begin{bmatrix} x_1(0) \\ x_2(0) \end{bmatrix} = \begin{bmatrix} 0 \\ 0 \end{bmatrix}, \quad r(t) = \delta(t), \text{ unit impulse}$$

(*b*)
$$\begin{bmatrix} \dot{x}_1 \\ \dot{x}_2 \\ \dot{x}_3 \end{bmatrix} = \begin{bmatrix} -3 & 1 & 0 \\ -2 & 0 & 1 \\ 0 & 0 & 0 \end{bmatrix}\begin{bmatrix} x_1 \\ x_2 \\ x_3 \end{bmatrix} + \begin{bmatrix} 0 \\ 1 \\ 0 \end{bmatrix} r$$

$$\begin{bmatrix} y_1 \\ y_2 \end{bmatrix} = \begin{bmatrix} 1 & 0 & 1 \\ 0 & 0 & 1 \end{bmatrix}\begin{bmatrix} x_1 \\ x_2 \\ x_3 \end{bmatrix}, \quad \begin{bmatrix} x_1(0) \\ x_2(0) \\ x_3(0) \end{bmatrix} = \begin{bmatrix} 1 \\ 0 \\ 0 \end{bmatrix}, \quad r(t) = 2u(t)$$

Solution: (*a*) Let us first evaluate state transition matrix $\phi(t)$

$$\phi(s) = [sI - A]^{-1} = \left[\begin{bmatrix} s & 0 \\ 0 & s \end{bmatrix} - \begin{bmatrix} -4 & 1 \\ -3 & 0 \end{bmatrix}\right]^{-1} = \begin{bmatrix} s+4 & -1 \\ 3 & s \end{bmatrix}^{-1}$$

$$= \begin{bmatrix} \dfrac{s}{(s+1)(s+3)} & \dfrac{1}{(s+1)(s+3)} \\ \dfrac{-3}{(s+1)(s+3)} & \dfrac{s+4}{(s+1)(s+3)} \end{bmatrix}$$

$$\phi(t) = \mathcal{L}^{-1}[\phi(s)] = \begin{bmatrix} -0.5e^{-t} + 1.5e^{-3t} & 0.5e^{-t} - 0.5e^{-3t} \\ -1.5e^{-t} + 1.5e^{-3t} & 1.5e^{-t} - 0.5e^{-3t} \end{bmatrix}$$

The state response

$$x(t) = \phi(t)\, x(0) + \int_0^t \phi(t-\tau)\, Br(\tau)\, d\tau$$

$$= \begin{bmatrix} 0 \\ 0 \end{bmatrix} + \int_0^t \begin{bmatrix} -0.5e^{-(t-\tau)} + 1.5e^{-3(t-\tau)} & 0.5e^{-(t-\tau)} - 0.5e^{-3(t-\tau)} \\ -1.5e^{-(t-\tau)} + 1.5e^{-3(t-\tau)} & 1.5e^{-(t-\tau)} - 0.5e^{-3(t-\tau)} \end{bmatrix}\begin{bmatrix} 1 \\ 0 \end{bmatrix}\delta(\tau)\, d\tau$$

gives
$$\begin{bmatrix} x_1 \\ x_2 \end{bmatrix} = \begin{bmatrix} -0.5e^{-t} + 1.5e^{-3t} \\ -1.5e^{-t} + 1.5e^{-3t} \end{bmatrix}, \quad \delta(\tau) = 0 \text{ for } \tau \neq 0$$

and system response

$$y(t) = \begin{bmatrix}1 & 0\end{bmatrix}\begin{bmatrix}x_1\\ x_2\end{bmatrix} = -0.5e^{-t} + 1.5e^{-3t}$$

(*b*) The state transition matrix $\phi(t)$ is computed as follows:

$$\phi(s) = [s\mathrm{I} - \mathrm{A}]^{-1} = \left[\begin{bmatrix}s & 0 & 0\\ 0 & s & 0\\ 0 & 0 & s\end{bmatrix} - \begin{bmatrix}-3 & 1 & 0\\ -2 & 0 & 1\\ 0 & 0 & 0\end{bmatrix}\right]^{-1}$$

$$= \begin{bmatrix}s+3 & -1 & 0\\ 2 & s & -1\\ 0 & 0 & s\end{bmatrix}^{-1} = \begin{bmatrix}\dfrac{s}{(s+1)(s+2)} & \dfrac{1}{(s+1)(s+2)} & \dfrac{1}{s(s+1)(s+2)}\\ \dfrac{-2}{(s+1)(s+2)} & \dfrac{s+3}{(s+1)(s+2)} & \dfrac{s+3}{s(s+1)(s+2)}\\ 0 & 0 & 1/s\end{bmatrix}$$

$$\phi(t) = \mathcal{L}^{-1}[\phi(s)] = \begin{bmatrix}-e^{-t}+2e^{-2t} & e^{-t}-e^{-2t} & \dfrac{1}{2}-e^{-t}+\dfrac{1}{2}e^{-2t}\\ -2e^{-t}+2e^{-2t} & 2e^{-t}-e^{-2t} & \dfrac{3}{2}-2e^{-t}+\dfrac{1}{2}e^{-2t}\\ 0 & 0 & 1\end{bmatrix}$$

The state response

$$x(t) = \phi(t)\,x(0) + \int_0^t \phi(t-\tau)\,\mathrm{B}\,r(\tau)\,d\tau$$

$$= \begin{bmatrix}-e^{-t}+2e^{-2t} & e^{-t}-e^{-2t} & \dfrac{1}{2}-e^{-t}+\dfrac{1}{2}e^{-2t}\\ -2e^{-t}+2e^{-2t} & 2e^{-t}-e^{-2t} & \dfrac{3}{2}-2e^{-t}+\dfrac{1}{2}e^{-2t}\\ 0 & 0 & 1\end{bmatrix}\begin{bmatrix}1\\ 0\\ 0\end{bmatrix}$$

$$+ \int_0^t \begin{bmatrix}-e^{-(t-\tau)}+2e^{-2(t-\tau)} & e^{-(t-\tau)}-e^{-2(t-\tau)} & \dfrac{1}{2}-e^{-(t-\tau)}+\dfrac{1}{2}e^{-2(t-\tau)}\\ -2e^{-(t-\tau)}+2e^{-2(t-\tau)} & 2e^{-(t-\tau)}-e^{-2(t-\tau)} & \dfrac{3}{2}-2e^{-(t-\tau)}+\dfrac{1}{2}e^{-2(t-\tau)}\\ 0 & 0 & 1\end{bmatrix}\begin{bmatrix}0\\ 1\\ 0\end{bmatrix} 2u(\tau)\,d(\tau)$$

$$= \begin{bmatrix} -e^{-t}+2e^{-2t} \\ -2e^{-t}+2e^{-2t} \\ 0 \end{bmatrix} + 2\int_0^t \begin{bmatrix} e^{-(t-\tau)}-e^{-2(t-\tau)} \\ 2e^{-(t-\tau)}-e^{-2(t-\tau)} \\ 0 \end{bmatrix} d\tau$$

$$= \begin{bmatrix} -e^{-t}+2e^{-2t} \\ -2e^{-t}+2e^{-2t} \\ 0 \end{bmatrix} + \begin{bmatrix} 1-2e^{-t}+e^{-2t} \\ 3-4e^{-t}+e^{-2t} \\ 0 \end{bmatrix} = \begin{bmatrix} 1-3e^{-t}+3e^{-2t} \\ 3-6e^{-t}+3e^{-2t} \\ 0 \end{bmatrix}$$

So, state response

$$\begin{bmatrix} x_1(t) \\ x_2(t) \\ x_3(t) \end{bmatrix} = \begin{bmatrix} 1-3e^{-t}+3e^{-2t} \\ 3-6e^{-t}+3e^{-2t} \\ 0 \end{bmatrix}$$

and system response

$$\begin{bmatrix} y_1(t) \\ y_2(t) \end{bmatrix} = \begin{bmatrix} 1 & 0 & 1 \\ 0 & 0 & 1 \end{bmatrix} \begin{bmatrix} 1-3e^{-t}+3e^{-2t} \\ 3-6e^{-t}+3e^{-2t} \\ 0 \end{bmatrix} = \begin{bmatrix} 1-3e^{-t}+3e^{-2t} \\ 0 \end{bmatrix}$$

P 7.13: A system modelled as

$$\dot{x}(t) = \mathrm{A}x(t)$$

generates state response $x(t) = \begin{bmatrix} e^{-2t} \\ -2e^{-2t} \end{bmatrix}$ for initial vector $x(0) = \begin{bmatrix} 1 \\ -2 \end{bmatrix}$ and $x(t) = \begin{bmatrix} e^{-t} \\ -e^{-t} \end{bmatrix}$ for $x(0) = \begin{bmatrix} 1 \\ -1 \end{bmatrix}$. Find the system matrix A and state transition matrix (STM).

Solution: Let STM $\phi(t) = \begin{bmatrix} \phi_{11} & \phi_{12} \\ \phi_{21} & \phi_{22} \end{bmatrix}$

The state response due to initial conditions given by

$$x(t) = \phi(t)\, x(0)$$

gives the following set of simultaneous equations

$$\begin{bmatrix} \phi_{11} & \phi_{12} \\ \phi_{21} & \phi_{22} \end{bmatrix} \begin{bmatrix} 1 \\ -2 \end{bmatrix} = \begin{bmatrix} e^{-2t} \\ -2e^{-2t} \end{bmatrix}$$

or

$$\phi_{11} - 2\phi_{12} = e^{-2t}$$

$$\phi_{21} - 2\phi_{22} = -\,2e^{-2t}$$

and

$$\begin{bmatrix} \phi_{11} & \phi_{12} \\ \phi_{21} & \phi_{22} \end{bmatrix} \begin{bmatrix} 1 \\ -1 \end{bmatrix} = \begin{bmatrix} e^{-t} \\ -e^{-t} \end{bmatrix}$$

or

$$\phi_{11} - \phi_{12} = e^{-t}$$

$$\phi_{21} - \phi_{22} = -e^{-t}$$

Solving the equations just above, gives

or $$\phi_{11} = 2e^{-t} - e^{-2t}, \qquad \phi_{12} = e^{-t} - e^{-2t}$$

$$\phi_{21} = -2e^{-t} + 2e^{-2t}, \qquad \phi_{22} = 2e^{-2t} - e^{-t}$$

and $$\phi(t) = \begin{bmatrix} 2e^{-t} - e^{-2t} & e^{-t} - e^{-2t} \\ -2e^{-t} + 2e^{-2t} & 2e^{-2t} - e^{-t} \end{bmatrix}$$

$$\phi(s) = \begin{bmatrix} \dfrac{s+3}{(s+1)(s+2)} & \dfrac{1}{(s+1)(s+2)} \\ -\dfrac{2}{(s+1)(s+2)} & \dfrac{s}{(s+1)(s+2)} \end{bmatrix}$$

but $\phi(s) = [sI - A]^{-1}$.

Let system matrix $$A = \begin{bmatrix} a_{11} & a_{12} \\ a_{21} & a_{22} \end{bmatrix}$$

then $$[sI - A]^{-1} = \frac{1}{(s-a_{11})(s-a_{22}) - a_{12}a_{21}} \begin{bmatrix} s-a_{22} & a_{12} \\ a_{21} & s-a_{11} \end{bmatrix}$$

Equating $\phi(s)$ to $[sI - A]^{-1}$, we have

$$\frac{\begin{bmatrix} s-a_{22} & a_{12} \\ a_{21} & s-a_{11} \end{bmatrix}}{(s-a_{11})(s-a_{22}) - a_{12}a_{21}} = \frac{\begin{bmatrix} s+3 & 1 \\ -2 & s \end{bmatrix}}{(s+1)(s+2)}$$

Matching the elements of matrices on either side we have

$$a_{11} = 0, \qquad a_{12} = 1$$

$$a_{21} = -2, \qquad a_{22} = -3$$

and $$A = \begin{bmatrix} 0 & 1 \\ -2 & -3 \end{bmatrix}$$

P 7.14: Develop state space model for each of the electrical networks shown below. Investigate if each one of them is completely controllable and/or completely observable. Substantiate the result with suitable comments if any.

(*a*)

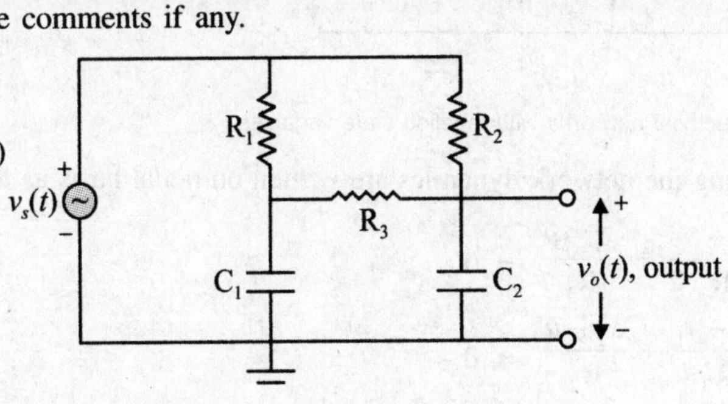

$R_1 = R_2 = R_3 = 1\ M\Omega$

$C_1 = C_2 = 4.7\ \mu F$

(*b*)

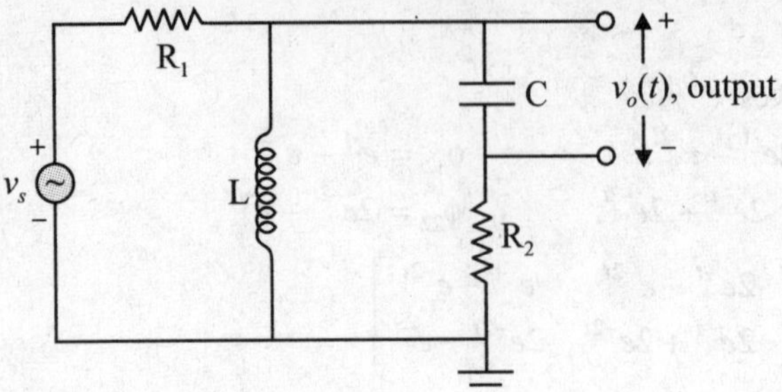

$R_1 = R_2 = 1\ K\Omega$

$L = 10$ mH

$C = 470\ \mu F$

(*c*)

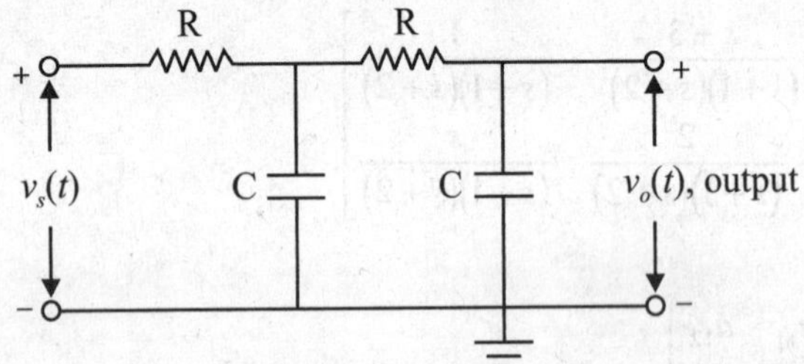

$R = 1\ M\Omega$

$C = 4.7\ \mu F$

Solution: (*a*) The electrical network is redrawn in Fig. P7.14 (*a*) with capacitor voltages labelled as state variables x_1, x_2 together with input *u* and output *y* (*u* stands for $v_s(t)$ and *y* stands for $v_o(t)$).

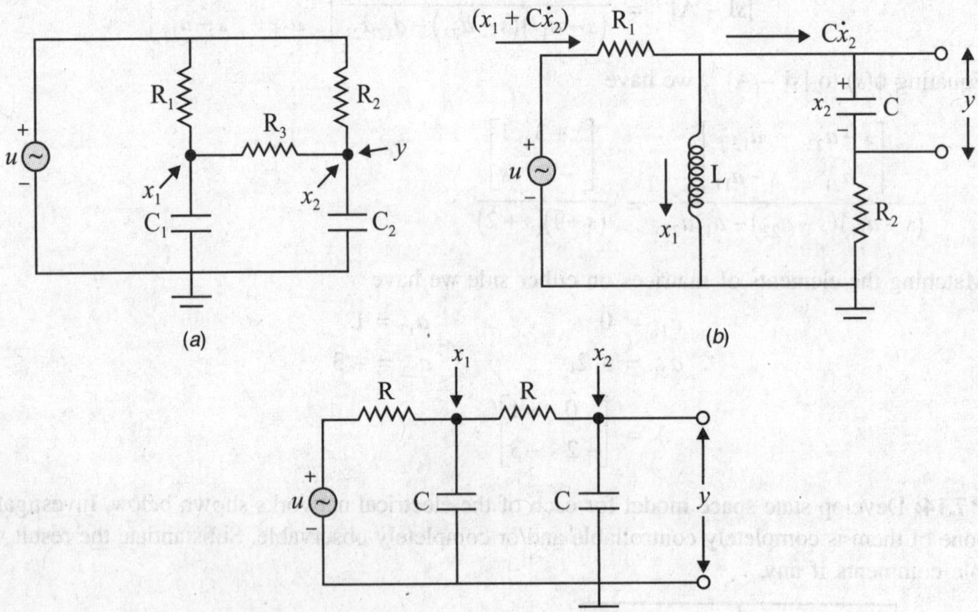

Fig. P 7.14: Electrical networks with labelled state variables

The differential equations describing the network dynamics are written on nodal basis as follows:

$$C_1 \frac{d}{dt}(x_1 - 0) + \frac{x_1 - x_2}{R_3} + \frac{x_1 - u}{R_1} = 0$$

$$C_2 \frac{d}{dt}(x_2 - 0) + \frac{x_2 - x_1}{R_3} + \frac{x_2 - u}{R_2} = 0$$

The state and output equations with little algebraic manipulation in the differential equations just above, take the form as follows:

$$\dot{x}_1 = -\frac{R_1+R_3}{R_1R_3C_1}x_1 + \frac{1}{R_3C_1}x_2 + \frac{1}{R_1C_1}u$$

$$\dot{x}_2 = \frac{1}{R_3C_2}x_1 - \frac{R_2+R_3}{R_2R_3C_2}x_2 + \frac{1}{R_2C_2}u$$

$$y = x_2$$

Having fitted the component values, these equations in matrix form are

$$\begin{bmatrix}\dot{x}_1\\ \dot{x}_2\end{bmatrix} = \begin{bmatrix}-\frac{20}{47} & \frac{10}{47}\\ \frac{10}{47} & -\frac{20}{47}\end{bmatrix}\begin{bmatrix}x_1\\ x_2\end{bmatrix} + \begin{bmatrix}\frac{10}{47}\\ \frac{10}{47}\end{bmatrix}u$$

$$y = \begin{bmatrix}0 & 1\end{bmatrix}\begin{bmatrix}x_1\\ x_2\end{bmatrix}$$

The controllability matrix

$$Q_c = [B \vdots AB] = \begin{bmatrix}\frac{10}{47} & -\frac{100}{2209}\\ \frac{10}{47} & -\frac{100}{2209}\end{bmatrix}$$

is not of full rank. The system is not completely controllable. This is due to the fact that the bridge is balanced with the component values given. The state variables x_1, x_2 (voltage across R_3) cannot be influenced by input u.

The observability matrix

$$Q_o = \begin{bmatrix}C\\ \cdots\\ CA\end{bmatrix} = \begin{bmatrix}0 & 1\\ \frac{10}{47} & -\frac{20}{47}\end{bmatrix}$$

is of full rank and system is completely observable.

(*b*) The electrical network with labelled state variables x_1, x_2 , the output y [y stands for $v_o(t)$] and input u [u stands for $v_s(t)$] is shown in Fig. P 7.14(*b*). The current through inductor L is chosen as state variable x_1 and voltage across capacitor C as state variable x_2. KVL applied to the two loops involving input u, yields the following set of differential equations governing the dynamics of the network.

$$u - (x_1 + C\dot{x}_2)R_1 - L\dot{x}_1 = 0$$

or

$$\dot{x}_1 = -\frac{R_1}{L}x_1 - \frac{CR_1}{L}\dot{x}_2 + \frac{1}{L}u$$

and $$u-(x_1+C\dot{x}_2)R_1-x_2-C\dot{x}_2R_2 = 0$$

or $$\dot{x}_2 = -\frac{R_1}{C(R_1+R_2)}x_1-\frac{1}{C(R_1+R_2)}x_2+\frac{u}{C(R_1+R_2)}$$

Substituting this value of $\dot{x}_2$ in the equation above involving $\dot{x}_1$ together with little algebraic manipulation, we have

$$\dot{x}_1 = -\frac{R_1R_2}{L(R_1+R_2)}x_1+\frac{R_1}{L(R_1+R_2)}x_2+\frac{R_2}{L(R_1+R_2)}u$$

Fitting the component values in the equations with L.H.S. having first derivative of state variables, the state and output equations can be developed as follows:

$$\dot{x}_1 = -5\times 10^4\,x_1 + 50x_2 + 50u$$

$$\dot{x}_2 = -1063.8x_1 - 1.06x_2 + 1.06u$$

$$y = x_2$$

These in matrix form, are

$$\begin{bmatrix}\dot{x}_1\\ \dot{x}_2\end{bmatrix} = \begin{bmatrix}-5\times10^4 & 50\\ -1063.8 & -1.06\end{bmatrix}\begin{bmatrix}x_1\\ x_2\end{bmatrix}+\begin{bmatrix}50\\ 1.06\end{bmatrix}u$$

$$y = [0 \;\; 1]\begin{bmatrix}x_1\\ x_2\end{bmatrix}$$

(*c*) The given electrical network is redrawn in Fig. P7.14 (*c*) with capacitor voltages labelled as state variables x_1, x_2 together with input u [u stands for $v_s(t)$] and output y [y stands for $v_o(t)$]. The differential equations describing the network dynamics are written on nodal basis as follows:

$$\frac{x_1-u}{R}+C\frac{d}{dt}(x_1-0)+\frac{x_1-x_2}{R} = 0$$

$$C\frac{d}{dt}(x_2-0)+\frac{x_2-x_1}{R} = 0$$

The state and output equations with some algebraic manipulation in the equations just above, can be written as

$$\dot{x}_1 = -\frac{2}{RC}x_1+\frac{1}{RC}x_2+\frac{1}{RC}u$$

$$\dot{x}_2 = \frac{1}{RC}x_1-\frac{1}{RC}x_2$$

$$y = x_2$$

Having fitted the component values, these equations in matrix form are

$$\begin{bmatrix}\dot{x}_1\\ \dot{x}_2\end{bmatrix} = \begin{bmatrix}-\frac{20}{47} & \frac{10}{47}\\ \frac{10}{47} & -\frac{10}{47}\end{bmatrix}\begin{bmatrix}x_1\\ x_2\end{bmatrix}+\begin{bmatrix}\frac{10}{47}\\ 0\end{bmatrix}u$$

$$y = \begin{bmatrix} 0 & 1 \end{bmatrix} \begin{bmatrix} x_1 \\ x_2 \end{bmatrix}$$

The controllability and observability matrices Q_c and Q_o respectively are:

$$Q_c = \begin{bmatrix} \frac{10}{47} & -\frac{200}{2209} \\ 0 & \frac{100}{2209} \end{bmatrix}, \quad Q_o = \begin{bmatrix} 0 & 1 \\ \frac{10}{47} & -\frac{10}{47} \end{bmatrix}$$

Q_c and Q_o both are of full rank, the system is completely controllable and observable.

P 7.15: A system using state feedback is governed by the following set of equations:

$$\begin{bmatrix} \dot{x}_1 \\ \dot{x}_2 \\ \dot{x}_3 \end{bmatrix} = \begin{bmatrix} 0 & 1 & 0 \\ 0 & 0 & 1 \\ -10 & -5 & -2 \end{bmatrix} \begin{bmatrix} x_1 \\ x_2 \\ x_3 \end{bmatrix} + \begin{bmatrix} 0 & -1 \\ 0 & 0 \\ 1 & 7 \end{bmatrix} \begin{bmatrix} u \\ r \end{bmatrix}$$

$$u = -\begin{bmatrix} K_1 & K_2 & K_3 \end{bmatrix} \begin{bmatrix} x_1 \\ x_2 \\ x_3 \end{bmatrix}$$

$$y = \begin{bmatrix} 1 & 0 & 0 \end{bmatrix} \begin{bmatrix} x_1 \\ x_2 \\ x_3 \end{bmatrix}$$

Determine feedback gain constants K_i so as to place closed loop poles at $s = -4$ and $-4 \pm j2$.

Solution: With state feedback $u = -\begin{bmatrix} K_1 & K_2 & K_3 \end{bmatrix} \begin{bmatrix} x_1 \\ x_2 \\ x_3 \end{bmatrix}$ the state equations get modified as follows:

$$\begin{bmatrix} \dot{x}_1 \\ \dot{x}_2 \\ \dot{x}_3 \end{bmatrix} = \begin{bmatrix} 0 & 1 & 0 \\ 0 & 0 & 1 \\ -10 & -5 & -2 \end{bmatrix} \begin{bmatrix} x_1 \\ x_2 \\ x_3 \end{bmatrix} + \begin{bmatrix} 0 \\ 0 \\ 1 \end{bmatrix} \begin{bmatrix} -K_1 & -K_2 & -K_3 \end{bmatrix} \begin{bmatrix} x_1 \\ x_2 \\ x_3 \end{bmatrix} + \begin{bmatrix} -1 \\ 0 \\ 7 \end{bmatrix} r$$

$$= \begin{bmatrix} 0 & 1 & 0 \\ 0 & 0 & 1 \\ -10 - K_1 & -5 - K_2 & -2 - K_3 \end{bmatrix} \begin{bmatrix} x_1 \\ x_2 \\ x_3 \end{bmatrix} + \begin{bmatrix} 0 \\ 0 \\ 7 \end{bmatrix} r$$

The modified characteristic equation is of form

$$|\lambda I - A| = \begin{vmatrix} \lambda & -1 & 0 \\ 0 & \lambda & -1 \\ 10 + K_1 & 5 + K_2 & \lambda + 2 + K_3 \end{vmatrix} = 0$$

or $$\lambda^3 + (2 + K_3)\lambda^2 + (5 + K_2)\lambda + (10 + K_1) = 0$$

while the desired characteristic equation is

$$(s + 4)\,(s + 4 + j2)\,(s + 4 - j2) = 0$$

or
$$s^3 + 12s^2 + 52s + 80 = 0$$

Matching the coefficients of modified characteristic equation with those of desired one, the feedback gain constants are

$$K_1 = 70, \quad K_2 = 47, \quad K_3 = 10$$

P 7.16: A system with state feedback is depicted below in Fig. P7.16. Find the values of K_1, K_2 and K_3 so that the system satisfies the following performance requirements

Peak overshoot $\leq 15\%$

Settling time ≤ 4 sec

Velocity error constant ≥ 1.5

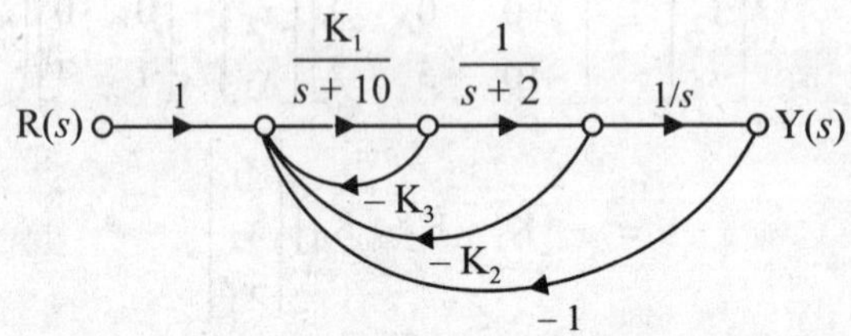

Fig. P 7.16: System with state feedback

Solution: The specifications

$$\text{Peak overshoot } e^{-\pi\xi/\sqrt{1-\xi^2}} \leq 0.15$$

and
$$\text{settling time } \frac{4}{\xi\omega_n} \leq 4$$

dictate $\xi \cong 0.52$, $\omega_n \cong 1.94$ and the dominant closed loop poles be located at $s = -\left(-\xi\omega_n \pm j\omega_n\sqrt{1-\xi^2}\right)$ or $s \cong -1 \pm j1.66$. In order to preserve the dominance of these poles, let us select the third pole at $s = -10\xi\omega_n$ or $s = -10$. So, the desired characteristic equation is

$$(s + 10)\,(s + 1 + j\,1.66)\,(s + 1 - j\,1.66) = 0$$

or
$$s^3 + 12s^2 + 23.8s + 37.6 = 0$$

Let us also test whether the specification $K_v \geq 1.5$ is met. Putting the characteristic equation in the form of

$$1 + G(s)\,H(s) = 0, \quad \text{we have}$$

$$G(s)\,H(s) = \frac{37.6}{s\left(s^2 + 12s + 23.8\right)}$$

and
$$K_v = \lim_{s\to 0} sG(s)H(s) = 1.58 \text{ meets the prescribed specification.}$$

The simulation diagram showing state variables is shown in Fig. P7.16(a). The state equations can be developed as follows:

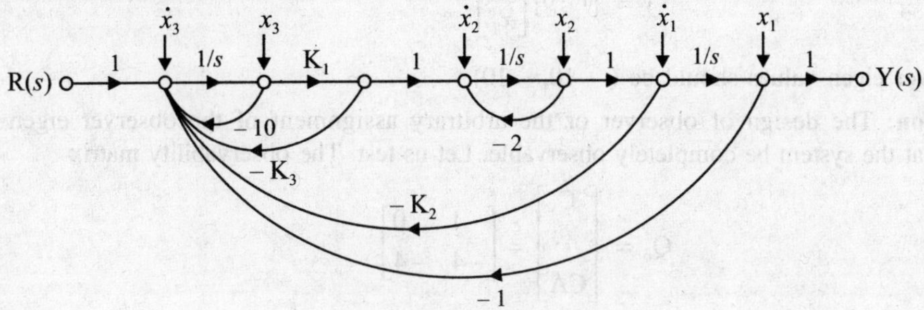

Fig. P 7.16: (a) Simulation diagram

$$\dot{x}_1 = x_2$$

$$\dot{x}_2 = -2x_2 + K_1 x_3$$

$$\dot{x}_3 = -x_1 - K_2 x_2 - K_1 K_3 x_3 - 10x_3 + r$$

or
$$\begin{bmatrix} \dot{x}_1 \\ \dot{x}_2 \\ \dot{x}_3 \end{bmatrix} = \begin{bmatrix} 0 & 1 & 0 \\ 0 & -2 & K_1 \\ -1 & -K_2 & -K_1K_3 - 10 \end{bmatrix} \begin{bmatrix} x_1 \\ x_2 \\ x_3 \end{bmatrix} + \begin{bmatrix} 0 \\ 0 \\ 1 \end{bmatrix} r$$

This gives the characteristic equation is

$$|sI - A| = 0$$

$$\begin{vmatrix} s & -1 & 0 \\ 0 & s+2 & -K_1 \\ 1 & K_2 & s + K_1K_3 + 10 \end{vmatrix} = 0$$

or $$s^3 + (K_1K_3 + 12)s^2 + (2K_1K_3 + 20 + K_1K_2)s + K_1 = 0$$

Comparing which with desired characteristic equation, we have

$$K_1 = 37.6$$

$$2K_1K_3 + K_1K_2 + 20 = 23.8$$

$$K_1K_3 + 12 = 12$$

or $$K_1 = 37.6, \qquad K_2 = 0.1, \qquad K_3 = 0$$

P 7.17: For the system given below, an observer is to be designed to estimate the state variables. Select the observer gain and write the equations describing the observer dynamics. Also develop the block diagram for the interconnected system and observer.

$$\begin{bmatrix} \dot{x}_1 \\ \dot{x}_2 \end{bmatrix} = \begin{bmatrix} -4 & -4 \\ 1 & -2 \end{bmatrix} \begin{bmatrix} x_1 \\ x_2 \end{bmatrix} + \begin{bmatrix} 0 \\ 2 \end{bmatrix} u$$

$$y = \begin{bmatrix} 1 & 0 \end{bmatrix} \begin{bmatrix} x_1 \\ x_2 \end{bmatrix}$$

Observer eigen values should be (– 10, – 10).

Solution: The design of observer or the arbitrary assignment of the observer eigen values, requires that the system be completely observable. Let us test. The observability matrix

$$Q_o = \begin{bmatrix} C \\ \cdots \\ CA \end{bmatrix} = \begin{bmatrix} 1 & 0 \\ -4 & -4 \end{bmatrix}$$

is non singular, $|Q_o| \neq 0$. System is completely observable. In order to determine observer gain matrix $M = \begin{bmatrix} m_1 \\ m_2 \end{bmatrix}$ let us first determine the observer characteristic polynomial as follows:

$$|sI - (A - MC)| = \left| \begin{bmatrix} s & 0 \\ 0 & s \end{bmatrix} - \begin{bmatrix} -4 & -4 \\ 1 & -2 \end{bmatrix} + \begin{bmatrix} m_1 \\ m_2 \end{bmatrix} \begin{bmatrix} 1 & 0 \end{bmatrix} \right|$$

$$= s^2 + (6 + m_1)\, s + (12 + 2m_1 - 4m_2)$$

The desired characteristic polynomial is

$$(s + 10)\,(s + 10) = s^2 + 20s + 100$$

Matching the coefficients of observer characteristic polynomial with those of desired polynomial, we have

$$m_1 = 14, \qquad m_2 = -15$$

and

$$M = \begin{bmatrix} 14 \\ -15 \end{bmatrix}$$

The equations describing the observer dynamics are

$$\dot{\hat{x}} = A\hat{x} + Bu + MC(x - \hat{x})$$

$$\begin{bmatrix} \dot{\hat{x}}_1 \\ \dot{\hat{x}}_2 \end{bmatrix} = \begin{bmatrix} -4 & -4 \\ 1 & -2 \end{bmatrix} \begin{bmatrix} \hat{x}_1 \\ \hat{x}_2 \end{bmatrix} + \begin{bmatrix} 0 \\ 2 \end{bmatrix} u + \begin{bmatrix} 14 \\ -15 \end{bmatrix} \begin{bmatrix} 1 & 0 \end{bmatrix} \begin{bmatrix} x_1 - \hat{x}_1 \\ x_2 - \hat{x}_2 \end{bmatrix}$$

which gives

$$\dot{\hat{x}}_1 = -4\hat{x}_1 - 4\hat{x}_2 + 14(x_1 - \hat{x}_1)$$

$$\dot{\hat{x}}_2 = \hat{x}_1 - 2\hat{x}_2 + 2u - 15(x_1 - \hat{x}_1)$$

Using these equations, the block diagram for the interconnected system and observer, is shown in Fig. P7.17.

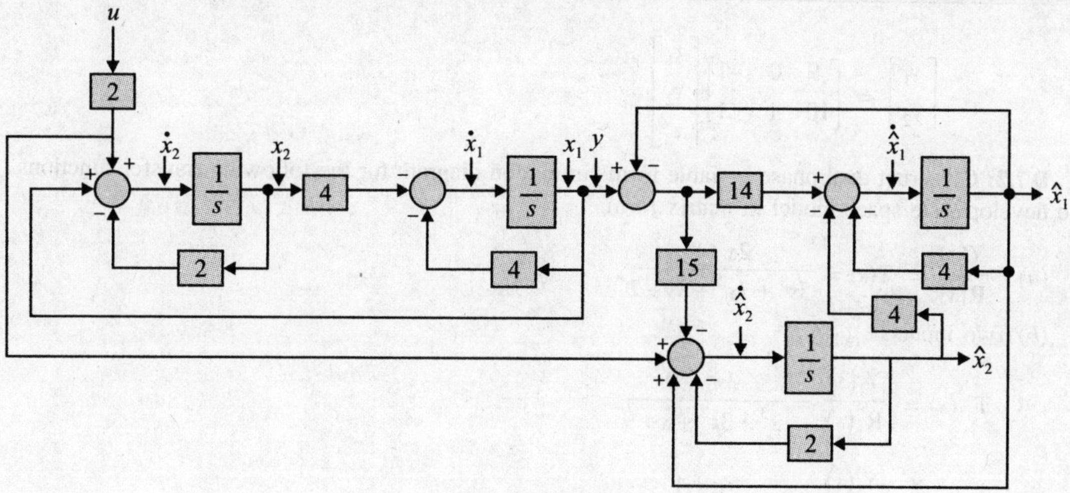

Fig. P 7.17: Block diagram

DRILL PROBLEMS

D 7.1: Construct phase variable form simulation diagram for the following transfer functions and develop state space model in matrix form

(*a*) $\quad T(s) = \dfrac{Y(s)}{U(s)} = \dfrac{10s}{s^3 + 12s^2 + 7s + 2}$

(*b*) Two outputs:

$$T_{11}(s) = \frac{Y_1(s)}{U_1(s)} = \frac{-s^2 + 9}{s^3 + 3s^2 + s + 4}$$

$$T_{21}(s) = \frac{Y_2(s)}{U_1(s)} = \frac{s^2 + s + 10}{s^3 + 3s^2 + s + 4}$$

Ans. (*a*) $\begin{bmatrix} \dot{x}_1 \\ \dot{x}_2 \\ \dot{x}_3 \end{bmatrix} = \begin{bmatrix} 0 & 1 & 0 \\ 0 & 0 & 1 \\ -2 & -7 & -12 \end{bmatrix} \begin{bmatrix} x_1 \\ x_2 \\ x_3 \end{bmatrix} + \begin{bmatrix} 0 \\ 0 \\ 1 \end{bmatrix} u$

$$y = \begin{bmatrix} 0 & 10 & 0 \end{bmatrix} \begin{bmatrix} x_1 \\ x_2 \\ x_3 \end{bmatrix}$$

(*b*) $\begin{bmatrix} \dot{x}_1 \\ \dot{x}_2 \\ \dot{x}_3 \end{bmatrix} = \begin{bmatrix} 0 & 1 & 0 \\ 0 & 0 & 1 \\ -4 & -1 & -3 \end{bmatrix} \begin{bmatrix} x_1 \\ x_2 \\ x_3 \end{bmatrix} + \begin{bmatrix} 0 \\ 0 \\ 1 \end{bmatrix} u$

$$\begin{bmatrix} y_1 \\ y_2 \end{bmatrix} = \begin{bmatrix} 9 & 0 & -1 \\ 10 & 1 & 1 \end{bmatrix} \begin{bmatrix} x_1 \\ x_2 \\ x_3 \end{bmatrix}$$

D 7.2: Construct dual phase variable form simulation diagram for the following transfer functions and develop state space model in matrix form.

(*a*) $\dfrac{Y(s)}{R(s)} = T(s) = \dfrac{2s+8}{3s^3+7s^2+8s+2}$

(*b*) Two inputs:

$$T_{11}(s) = \frac{Y_1(s)}{R_1(s)} = \frac{3s^2+9}{s^3+3s^2+s+9}$$

$$T_{12}(s) = \frac{Y_1(s)}{R_2(s)} = \frac{s-4}{s^3+3s^2+s+9}$$

Ans. (*a*) $\begin{bmatrix} \dot{x}_1 \\ \dot{x}_2 \\ \dot{x}_3 \end{bmatrix} = \begin{bmatrix} -7/3 & 1 & 0 \\ -8/3 & 0 & 1 \\ -2/3 & 0 & 0 \end{bmatrix} \begin{bmatrix} x_1 \\ x_2 \\ x_3 \end{bmatrix} + \begin{bmatrix} 0 \\ 2/3 \\ 8/3 \end{bmatrix} r$

$$y = \begin{bmatrix} 1 & 0 & 0 \end{bmatrix} \begin{bmatrix} x_1 \\ x_2 \\ x_3 \end{bmatrix}$$

(*b*) $\begin{bmatrix} \dot{x}_1 \\ \dot{x}_2 \\ \dot{x}_3 \end{bmatrix} = \begin{bmatrix} -3 & 1 & 0 \\ -1 & 0 & 1 \\ -9 & 0 & 0 \end{bmatrix} \begin{bmatrix} x_1 \\ x_2 \\ x_3 \end{bmatrix} + \begin{bmatrix} 3 & 0 \\ 9 & 1 \\ 0 & -4 \end{bmatrix} \begin{bmatrix} r_1 \\ r_2 \end{bmatrix}$

$$y = \begin{bmatrix} 1 & 0 & 0 \end{bmatrix} \begin{bmatrix} x_1 \\ x_2 \\ x_3 \end{bmatrix}$$

D 7.3: Find diagonal state equations for a system with transfer function

$$T(s) = \frac{Y(s)}{R(s)} = \frac{2s^2+3s-7}{(s+2)(s^2+13s+40)}$$

Ans. $\begin{bmatrix} \dot{x}_1 \\ \dot{x}_2 \\ \dot{x}_3 \end{bmatrix} = \begin{bmatrix} -2 & 0 & 0 \\ 0 & -8 & 0 \\ 0 & 0 & -5 \end{bmatrix} \begin{bmatrix} x_1 \\ x_2 \\ x_3 \end{bmatrix} + \begin{bmatrix} 1 \\ 1 \\ 1 \end{bmatrix} r$

$$Y = \begin{bmatrix} -\dfrac{5}{18} & \dfrac{97}{18} & -\dfrac{28}{9} \end{bmatrix} \begin{bmatrix} x_1 \\ x_2 \\ x_3 \end{bmatrix}$$

D 7.4: Determine the transfer functions for the system modelled as

$$\begin{bmatrix}\dot{x}_1\\ \dot{x}_2\end{bmatrix} = \begin{bmatrix}0 & 1\\ -3 & -7\end{bmatrix}\begin{bmatrix}x_1\\ x_2\end{bmatrix} + \begin{bmatrix}0 & 2\\ 4 & -8\end{bmatrix}\begin{bmatrix}u_1\\ u_2\end{bmatrix}$$

$$y = \begin{bmatrix}1 & -4\end{bmatrix}\begin{bmatrix}x_1\\ x_2\end{bmatrix}$$

Ans. $\begin{bmatrix}\dfrac{-16s+4}{s^2+7s+3} & \dfrac{34s+30}{s^2+7s+3}\end{bmatrix}$

D 7.5: Given a system

$$\begin{bmatrix}\dot{x}_1\\ \dot{x}_2\\ \dot{x}_3\end{bmatrix} = \begin{bmatrix}0 & -2 & 3\\ 0 & -4 & -1\\ 0 & 1 & -8\end{bmatrix}\begin{bmatrix}x_1\\ x_2\\ x_3\end{bmatrix} + \begin{bmatrix}2\\ -8\\ 4\end{bmatrix}u$$

$$y = \begin{bmatrix}4 & 1 & 6\end{bmatrix}\begin{bmatrix}x_1\\ x_2\\ x_3\end{bmatrix}$$

Find characteristic equation and determine if the system is stable.

Ans. $s^3 + 12s^2 + 34s$, marginally stable.

D 7.6: Consider the following SISO system with complex characteristic roots. Develop state space model in matrix form with diagonal state equations. Also find block diagonal model involving real numbers.

$$T(s) = \frac{Y(s)}{U(s)} = \frac{2s^2+3s-4}{(s+2)(s^2+6s+10)}$$

Ans.

$$\begin{bmatrix}\dot{x}_1\\ \dot{x}_2\\ \dot{x}_3\end{bmatrix} = \begin{bmatrix}-2 & 0 & 0\\ 0 & -3-j & 0\\ 0 & 0 & -3+j\end{bmatrix}\begin{bmatrix}x_1\\ x_2\\ x_3\end{bmatrix} + \begin{bmatrix}1\\ 1\\ 1\end{bmatrix}u$$

$$y = \begin{bmatrix}-1 & 1.5-j3 & 1.5+j3\end{bmatrix}\begin{bmatrix}x_1\\ x_2\\ x_3\end{bmatrix}$$

$$\begin{bmatrix}\dot{z}_1\\ \dot{z}_2\\ \dot{z}_3\end{bmatrix} = \begin{bmatrix}-2 & 0 & 0\\ 0 & 0 & 1\\ 0 & -10 & -6\end{bmatrix}\begin{bmatrix}z_1\\ z_2\\ z_3\end{bmatrix} + \begin{bmatrix}1\\ 0\\ 1\end{bmatrix}u$$

$$y = \begin{bmatrix}-1 & 3 & 3\end{bmatrix}\begin{bmatrix}z_1\\ z_2\\ z_3\end{bmatrix}$$

D 7.7: Obtain controllability and observability matrices and investigate whether or not the following system is completely controllable and/or completely observable.

$$\begin{bmatrix} \dot{x}_1 \\ \dot{x}_2 \\ \dot{x}_3 \end{bmatrix} = \begin{bmatrix} 3 & 0 & -5 \\ -2 & 1 & 5 \\ 0 & 0 & -2 \end{bmatrix} \begin{bmatrix} x_1 \\ x_2 \\ x_3 \end{bmatrix} + \begin{bmatrix} 1 & 0 \\ 2 & 0 \\ 0 & -1 \end{bmatrix} \begin{bmatrix} u_1 \\ u_2 \end{bmatrix}$$

$$\begin{bmatrix} y_1 \\ y_2 \end{bmatrix} = \begin{bmatrix} 4 & 1 & -3 \\ 3 & 2 & -1 \end{bmatrix} \begin{bmatrix} x_1 \\ x_2 \\ x_3 \end{bmatrix}$$

Ans. Completely controllable but not completely observable.

D 7.8: The systems together with inputs and initial conditions are given below. Determine system response in each.

(*a*)
$$\dot{x} = -2x + 3r(t)$$
$$y = 4x$$
$$x(0) = 10$$
$$r(t) = 5u(t)$$

(*b*)
$$\begin{bmatrix} \dot{x}_1 \\ \dot{x}_2 \end{bmatrix} = \begin{bmatrix} 0 & 1 \\ -6 & -8 \end{bmatrix} \begin{bmatrix} x_1 \\ x_2 \end{bmatrix} + \begin{bmatrix} 1 \\ 1 \end{bmatrix} r$$

$$y = \begin{bmatrix} 1 & -1 \end{bmatrix} \begin{bmatrix} x_1 \\ x_2 \end{bmatrix}$$

$$\begin{bmatrix} x_1(0) \\ x_2(0) \end{bmatrix} = \begin{bmatrix} 7 \\ -3 \end{bmatrix}$$

$r(t) = \delta(t)$; the unit impulse function.

Ans. (*a*) $y(t) = 10\, e^{-2t} + 30; \quad t \geq 0$

(*b*) $y(t) = 51\, e^{-2t} - 47\, e^{-4t}; \quad t \geq 0$

D 7.9: A system is described by the following signal flow graph. Write state and output equations in matrix form:

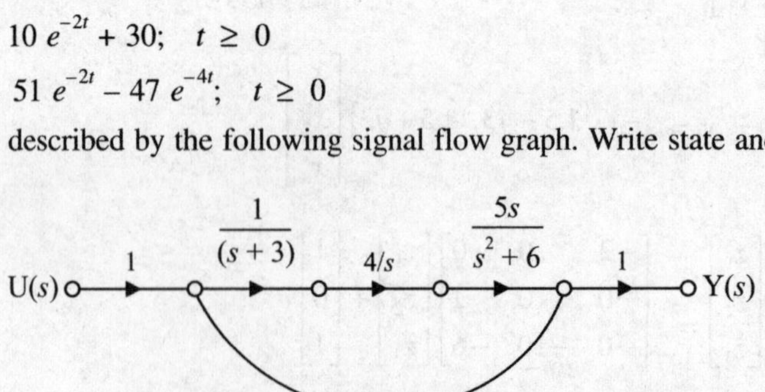

Ans. One possible solution is

$$\begin{bmatrix} \dot{x}_1 \\ \dot{x}_2 \\ \dot{x}_3 \\ \dot{x}_4 \end{bmatrix} = \begin{bmatrix} 0 & 1 & 0 & 0 \\ -6 & 0 & 1 & 0 \\ 0 & 0 & 0 & 4 \\ 0 & -10 & 0 & -3 \end{bmatrix} \begin{bmatrix} x_1 \\ x_2 \\ x_3 \\ x_4 \end{bmatrix} + \begin{bmatrix} 0 \\ 0 \\ 0 \\ 1 \end{bmatrix} u$$

$$y = \begin{bmatrix} 0 & 5 & 0 & 0 \end{bmatrix} \begin{bmatrix} x_1 \\ x_2 \\ x_3 \\ x_4 \end{bmatrix}$$

D 7.10: A system using state feedback control is governed by the following set of equations. Determine feedback gains so as to place the closed loop system poles at $s = -4$ and $-4 \pm j2$.

$$\begin{bmatrix} \dot{x}_1 \\ \dot{x}_2 \\ \dot{x}_3 \end{bmatrix} = \begin{bmatrix} 0 & 0 & 1 \\ 3 & -2 & -1 \\ 0 & 1 & 0 \end{bmatrix} \begin{bmatrix} x_1 \\ x_2 \\ x_3 \end{bmatrix} + \begin{bmatrix} 0 & -1 \\ 1 & 0 \\ 0 & 1 \end{bmatrix} \begin{bmatrix} r \\ u \end{bmatrix}$$

$$y = -\begin{bmatrix} K_1 & K_2 & K_3 \end{bmatrix} \begin{bmatrix} x_1 \\ x_2 \\ x_3 \end{bmatrix}$$

$$y = \begin{bmatrix} 1 & 0 & 0 \end{bmatrix} \begin{bmatrix} x_1 \\ x_2 \\ x_3 \end{bmatrix}$$

Ans. $K_1 = 83$, $K_2 = 10$, $K_3 = 51$

D 7.11: A system is described as follows:

$$\dot{x} = Ax + Bu$$

$$y = Cx$$

where $$A = \begin{bmatrix} 1 & 1 \\ 0 & 1 \end{bmatrix}, \quad B = \begin{bmatrix} \alpha_1 \\ \alpha_2 \end{bmatrix}, \quad C = \begin{bmatrix} \beta_1 & \beta_2 \end{bmatrix}$$

What restrictions should be imposed on α_1, α_2, β_1 and β_2 so that the system is completely controllable and observable.

Ans. $\alpha_2 \neq 0$ for controllability and $\beta_1 \neq 0$ for observability.

D 7.12: The block diagram of a system together with state variable assignment as labelled therein, is shown in Fig. P7.12. Test controllability and observability. Comment on test result.

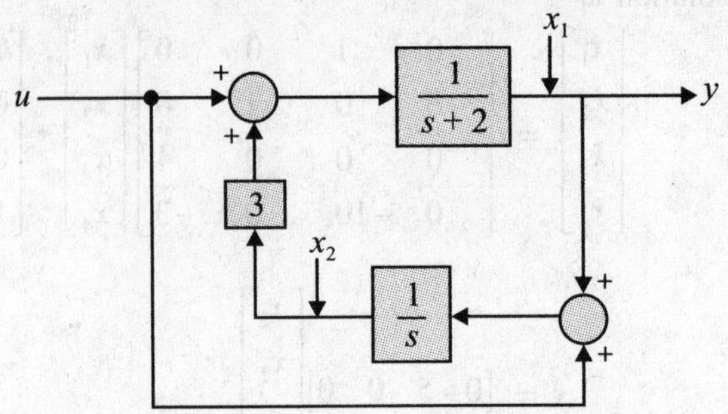

Fig. D7.12: Block diagram

Ans. Completely observable but not controllable. Transfer function has pole-zero cancellation

D 7.13: For the system with matrices

$$A = \begin{bmatrix} 0 & 1 \\ -10 & -3 \end{bmatrix}, \quad B = \begin{bmatrix} 0 \\ 10 \end{bmatrix} \quad \text{and} \quad C = \begin{bmatrix} 1 & 0 \end{bmatrix}$$

Design an observer such that the observer eigen values are placed at (– 20, – 20). Develop observer equations and signal flow graph showing interconnection of system and observer.

Ans. $M = \begin{bmatrix} 37 \\ 279 \end{bmatrix}$

$$\dot{\hat{x}}_1 = \hat{x}_2 + 37(x_1 - \hat{x}_1)$$

$$\dot{\hat{x}}_2 = -10\hat{x}_1 - 3\hat{x}_2 + 10u + 279(x_1 - \hat{x}_1)$$

MULTIPLE CHOICE QUESTIONS

M 7.1: The state variable description of a single input single output linear system is given by

$$\dot{x}(t) = Ax(t) + Bu(t)$$

$$y(t) = Cx(t)$$

where $$A = \begin{bmatrix} 1 & 1 \\ 2 & 0 \end{bmatrix}, \quad B = \begin{bmatrix} 0 \\ 1 \end{bmatrix} \quad \text{and} \quad C = \begin{bmatrix} 1 & -1 \end{bmatrix}$$

The system is

(*a*) controllable and observable (*b*) controllable but unobservable

(*c*) uncontrollable but observable (*d*) uncontrollable and unobservable

M 7.2: Which of the following properties are associated with the state transition matrix $\phi(t)$?

1. $\phi(t_1/t_2) = \phi(t_1) \cdot \phi^{-1}(t_2)$
2. $\phi(-t) = \phi^{-1}(t)$
3. $\phi(t_1 - t_2) = \phi(-t_2) \cdot \phi(t_1)$

Select the correct answer using the codes given below:

(*a*) 1, 2 and 3 (*b*) 1 and 2 (*c*) 2 and 3 (*d*) 1 and 3

M 7.3: A linear system is described by the state equations

$$\begin{bmatrix}\dot{x}_1\\ \dot{x}_2\end{bmatrix} = \begin{bmatrix}1 & 0\\ 1 & 1\end{bmatrix}\begin{bmatrix}x_1\\ x_2\end{bmatrix} + \begin{bmatrix}0\\ 1\end{bmatrix} r$$

$$y = x_2$$

where r and y are the input and output respectively. The transfer function is

(*a*) $\dfrac{1}{(s+1)}$ (*b*) $\dfrac{1}{(s+1)^2}$ (*c*) $\dfrac{1}{(s-1)}$ (*d*) $\dfrac{1}{(s-1)^2}$

M 7.4: Consider the following properties attributed to state model of a system.

1. State model is unique.
2. State model can be derived from the system transfer function.
3. State model can be derived for time variant systems.

Of these statements

(*a*) 1, 2 and 3 are correct (*b*) 1 and 2 are correct

(*c*) 2 and 3 are correct (*d*) 1 and 3 are correct.

M 7.5: A system is described by the state equation

$$\begin{bmatrix}\dot{x}_1\\ \dot{x}_2\end{bmatrix} = \begin{bmatrix}2 & 0\\ 0 & 2\end{bmatrix}\begin{bmatrix}x_1\\ x_2\end{bmatrix} + \begin{bmatrix}1\\ 1\end{bmatrix} u$$

The state transition matrix of the system is

(*a*) $\begin{bmatrix}e^{2t} & 0\\ 0 & e^{2t}\end{bmatrix}$ (*b*) $\begin{bmatrix}e^{-2t} & 0\\ 0 & e^{-2t}\end{bmatrix}$ (*c*) $\begin{bmatrix}e^{2t} & 1\\ 1 & e^{2t}\end{bmatrix}$ (*d*) $\begin{bmatrix}e^{-2t} & 1\\ 1 & e^{-2t}\end{bmatrix}$.

M 7.6: The state and output equations of a system are

$$\begin{bmatrix}\dot{x}_1\\ \dot{x}_2\end{bmatrix} = \begin{bmatrix}0 & 1\\ -1 & -2\end{bmatrix}\begin{bmatrix}x_1\\ x_2\end{bmatrix} + \begin{bmatrix}0\\ 1\end{bmatrix} u(t)$$

$$y(t) = \begin{bmatrix}1 & 1\end{bmatrix}\begin{bmatrix}x_1\\ x_2\end{bmatrix}$$

The systems is

(*a*) neither state controllable nor output controllable

(*b*) state controllable but not output controllable

(*c*) output controllable but not state controllable

(*d*) both state controllable and output controllable.

M 7.7: The state equation of a linear system is given by $\dot{x} = \mathrm{A}x + \mathrm{B}u$, where;

$$\mathrm{A} = \begin{bmatrix}0 & 2\\ -2 & 0\end{bmatrix} \text{ and } \mathrm{B} = \begin{bmatrix}0\\ -1\end{bmatrix}$$

The state transition matrix of the system is

(*a*) $\begin{bmatrix} e^{2t} & 0 \\ 0 & e^{2t} \end{bmatrix}$ (*b*) $\begin{bmatrix} e^{-2t} & 0 \\ 0 & e^{2t} \end{bmatrix}$

(*c*) $\begin{bmatrix} \sin 2t & \cos 2t \\ -\cos 2t & \sin 2t \end{bmatrix}$ (*d*) $\begin{bmatrix} \cos 2t & \sin 2t \\ -\sin 2t & \cos 2t \end{bmatrix}$

M 7.8: The state variable description of a linear autonomous system is $\dot{x} = \text{A}x$ where x is a state vector and

$$\text{A} = \begin{bmatrix} 0 & 2 \\ -2 & 0 \end{bmatrix}$$

The poles of the system are located at

(*a*) – 2 and + 2 (*b*) – 2 and – 2 (*c*) – 2 *j* and + 2 *j* (*d*) + 2 and + 2

M 7.9: Consider the Laplace transform of state transition matrix;

$$\phi(s) = \begin{bmatrix} \dfrac{s+6}{s^2+6s+5} & \dfrac{1}{s^2+6s+5} \\ \dfrac{-5}{s^2+6s+5} & \dfrac{s}{s^2+6s+5} \end{bmatrix}$$

The eigen values of the system are

(*a*) 0 and – 6 (*b*) 1 and – 5 (*c*) 0 and + 6 (*d*) – 1 and – 5

M 7.10: Which one of the following is NOT a correct statement about the state-space model of a physical system ?

(*a*) State-space model can be obtained only for a linear system

(*b*) Eigen values of the system represent the roots of the characteristic equation

(*c*) $\dot{x} = \text{A}x + \text{B}u$ represents linear state-space model of a physical system

(*d*) $x(t)$ represents the state vector of the system.

M 7.11: A linear second-order continuous time system is described by the following set of differential equations.

$$\dot{x}_1(t) = -2x_1(t) + 4x_2(t)$$
$$\dot{x}_2(t) = -2x_1(t) - x_2(t) + u(t)$$

where $x_1(t)$ and $x_2(t)$ are the state variables and $u(t)$ is the control variable. The system is

(*a*) controllable and stable (*b*) controllable and unstable

(*c*) uncontrollable and unstable (*d*) uncontrollable and stable

M 7.12: For the system described by the state equation

$$\dot{x} = \begin{bmatrix} 0 & 1 & 0 \\ 0 & 0 & 1 \\ 0.5 & 1 & 2 \end{bmatrix} x + \begin{bmatrix} 0 \\ 0 \\ 1 \end{bmatrix} u$$

if the control signal u is given by $u = [-0.5 \quad -3 \quad -5]\ x + v$, then the eigen values of the closed-loop system will be

(*a*) 0, – 1, – 2 (*b*) 0, – 1, – 3 (*c*) – 1, – 1, – 2 (*d*) 0, – 1, – 1

M 7.13: The matrix of any state-space equations for the transfer function C(*s*)/R(*s*) of the system, shown below in figure is

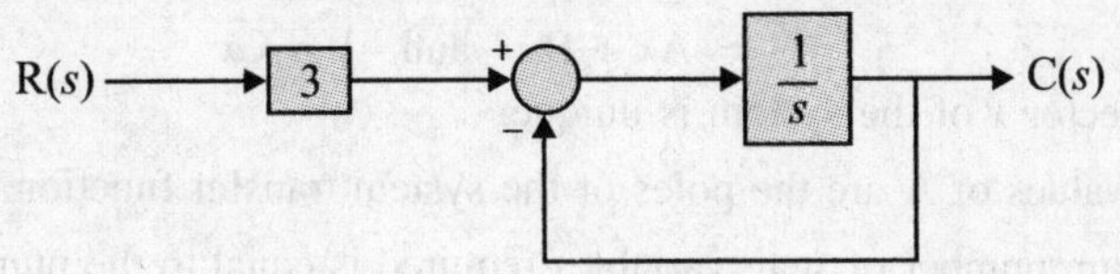

(*a*) $\begin{bmatrix} -1 & 0 \\ 0 & -1 \end{bmatrix}$ (*b*) $\begin{bmatrix} 0 & 1 \\ 0 & -1 \end{bmatrix}$ (*c*) $[-1]$ (*d*) $[3]$

M7.14: Given the homogeneous state-space equation $\dot{x} = \begin{bmatrix} -3 & 1 \\ 0 & -2 \end{bmatrix} x$. The steady state value of

$x_{ss} = \lim_{t \to \infty} x(t)$, given the initial state value of $x(0) = [10 \ -10]^t$, (t stands for transpose) is

(*a*) $x_{ss} = \begin{bmatrix} 0 \\ 0 \end{bmatrix}$ (*b*) $x_{ss} = \begin{bmatrix} -3 \\ -2 \end{bmatrix}$ (*c*) $x_{ss} = \begin{bmatrix} -10 \\ 10 \end{bmatrix}$ (*d*) $x_{ss} = \begin{bmatrix} \infty \\ \infty \end{bmatrix}$

M 7.15: The zero-input response of a system given by

$$\begin{bmatrix} \dot{x}_1 \\ \dot{x}_2 \end{bmatrix} = \begin{bmatrix} 1 & 0 \\ 1 & 1 \end{bmatrix} \begin{bmatrix} x_1 \\ x_2 \end{bmatrix} \quad \text{and} \quad \begin{bmatrix} x_1(0) \\ x_2(0) \end{bmatrix} = \begin{bmatrix} 1 \\ 0 \end{bmatrix} \quad \text{is:}$$

(*a*) $\begin{bmatrix} te^t \\ t \end{bmatrix}$ (*b*) $\begin{bmatrix} e^t \\ t \end{bmatrix}$ (*c*) $\begin{bmatrix} e^t \\ te^t \end{bmatrix}$ (*d*) $\begin{bmatrix} t \\ te^t \end{bmatrix}$

M7.16: The state-space representation in phase-variable form for the transfer function

$$G(s) = \frac{2s+1}{s^2+7s+9} \quad \text{is}$$

(*a*) $\dot{x} = \begin{bmatrix} 0 & 1 \\ -9 & -7 \end{bmatrix} x + \begin{bmatrix} 0 \\ 1 \end{bmatrix} u : y = [1 \ \ 2]x$ (*b*) $\dot{x} = \begin{bmatrix} 0 & 1 \\ -9 & -7 \end{bmatrix} x + \begin{bmatrix} 0 \\ 1 \end{bmatrix} u : y = [0 \ \ 1]x$

(*c*) $\dot{x} = \begin{bmatrix} -9 & 0 \\ 0 & -7 \end{bmatrix} x + \begin{bmatrix} 0 \\ 1 \end{bmatrix} u : y = [2 \ \ 0]x$ (*d*) $\dot{x} = \begin{bmatrix} 9 & -7 \\ 1 & 0 \end{bmatrix} x + \begin{bmatrix} 0 \\ 1 \end{bmatrix} u : y = [1 \ \ 2]x$

M7.17: Let $$\dot{x} = \begin{bmatrix} 1 & 2 \\ 0 & 1 \end{bmatrix} x + \begin{bmatrix} 0 \\ 1 \end{bmatrix} u$$

$$y = [b \ \ 0]x$$

where b is an unknown constant.

This system is

(*a*) observable for all values of *b* (*b*) unobservable for all values of *b*

(*c*) observable for all non-zero values of *b* (*d*) unobservable for all non-zero values of *b*

M 7.18: Consider the following statements with respect to a system represented by its state-space model

$$\dot{x} = Ax + Bu \quad \text{and} \quad y = Cx$$

1. The state vector x of the system is unique.

2. The eigen values of A are the poles of the system transfer function.

3. The minimum number of state variables required is equal to the number of independent energy storage elements in the system.

Which of these statements are correct ?

(*a*) 1 and 2 (*b*) 2 and 3 (*c*) 1 and 3 (*d*) 1, 2 and 3

Statement for linked answer Questions M 7.19 and M 7.20.

A state variable system $\dot{x}(t) = \begin{bmatrix} 0 & 1 \\ 0 & -3 \end{bmatrix} x(t) + \begin{bmatrix} 1 \\ 0 \end{bmatrix} u(t)$, with the initial condition $x(0) = [-1 \;\; 3]^T$ and the unit step input $u(t)$ has.

M 7.19: The state transition matrix

(*a*) $\begin{bmatrix} 1 & \frac{1}{3}\left(1-e^{-3t}\right) \\ 0 & e^{-3t} \end{bmatrix}$ (*b*) $\begin{bmatrix} 1 & \frac{1}{3}\left(e^{-t}-e^{-3t}\right) \\ 0 & e^{-t} \end{bmatrix}$

(*c*) $\begin{bmatrix} 1 & \frac{1}{3}\left(e^{-t}-e^{-3t}\right) \\ 0 & e^{-3t} \end{bmatrix}$ (*d*) $\begin{bmatrix} 1 & \left(1-e^{-t}\right) \\ 0 & e^{-t} \end{bmatrix}$.

M 7.20: The state transition equation

(*a*) $x(t) = \begin{bmatrix} t-e^{-t} \\ e^{-t} \end{bmatrix}$ (*b*) $x(t) = \begin{bmatrix} t-e^{-t} \\ 3e^{-3t} \end{bmatrix}$

(*c*) $x(t) = \begin{bmatrix} t-e^{-3t} \\ 3e^{-3t} \end{bmatrix}$ (*d*) $x(t) = \begin{bmatrix} t-e^{-3t} \\ e^{-t} \end{bmatrix}$

M 7.21: Consider the system s_1 modelled as

$$\dot{x} = \begin{bmatrix} 2 & 0 \\ 0 & -1 \end{bmatrix} x + \begin{bmatrix} 1 \\ 0 \end{bmatrix} u$$

and system s_2 modelled as

$$\dot{z} = \begin{bmatrix} 2 & 0 \\ 0 & 1 \end{bmatrix} z + \begin{bmatrix} 1 \\ 0 \end{bmatrix} u$$

What can be said about stabilizability of these systems?

(*a*) s_1 and s_2 both are stabilizable.
(*b*) only s_1 is stabilizable.
(*c*) only s_2 is stabilizable.
(*d*) neither s_1 nor s_2 is stabilizable.

M 7.22: Consider system s_1 modelled as

$$\dot{x} = \begin{bmatrix} 2 & 0 \\ 0 & -1 \end{bmatrix} x + \begin{bmatrix} 1 \\ 0 \end{bmatrix} u$$

$$y = [1 \quad 0] x$$

and system s_2 modelled as

$$\dot{z} = \begin{bmatrix} 2 & 0 \\ 0 & 1 \end{bmatrix} z + \begin{bmatrix} 1 \\ 0 \end{bmatrix} u$$

$$w = [1 \quad 0] z$$

What can be predicted about detectability of these systems?

(*a*) s_1 and s_2 both are detectable
(*b*) only s_1 is detectable
(*c*) only s_2 is detectable
(*d*) neither s_1 nor s_2 is detectable

M 7.23: The signal flow graph together with state variable assignment is shown below:

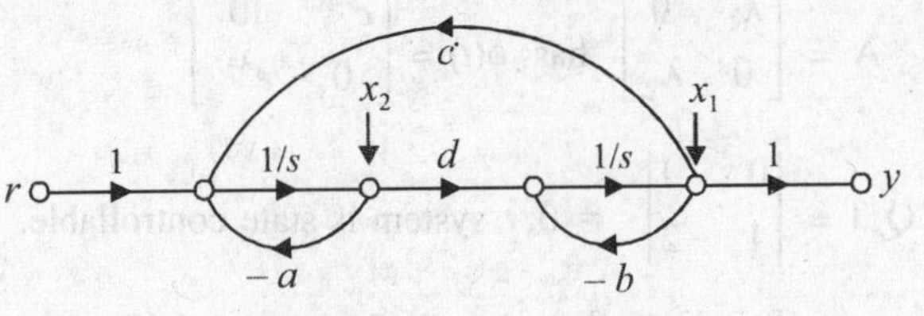

The condition for complete state controllability and complete observability is:

(*a*) $d \neq 0$ and a, b, c can be anything.
(*b*) $a \neq 0$ and b, c, d can be anything.
(*c*) $b \neq 0$ and a, c, d can be anything.
(*d*) $c \neq 0$ and a, b, d can be anything.

M7.24: A system is described by state transition matrix $\phi(t) = \begin{bmatrix} e^{-t} & 0 \\ 0 & e^{-2t} \end{bmatrix}$ and has initial conditions $\begin{bmatrix} x_1(0) \\ x_2(0) \end{bmatrix} = \begin{bmatrix} 1 \\ 2 \end{bmatrix}$. The state of system after 0.5 seconds will be

(*a*) $\begin{bmatrix} 0.6 \\ 0.74 \end{bmatrix}$
(*b*) $\begin{bmatrix} 0.74 \\ 0.6 \end{bmatrix}$
(*c*) $\begin{bmatrix} 0.6 \\ 1.2 \end{bmatrix}$
(*d*) $\begin{bmatrix} 1.2 \\ 0.37 \end{bmatrix}$

ANSWERS

M7.1. (*a*)	M7.2. (*c*)	M7.3. (*a*)	M7.4. (*c*)	M7.5. (*a*)
M7.6. (*d*)	M7.7. (*d*)	M7.8. (*c*)	M7.9. (*d*)	M7.10. (*a*)
M7.11. (*a*)	M7.12. (*a*)	M7.13. (*c*)	M7.14. (*a*)	M7.15. (*c*)
M7.16. (*a*)	M7.17. (*c*)	M7.18. (*b*)	M7.19. (*a*)	M7.20. (*c*)
M7.21. (*b*)	M7.22. (*b*)	M7.23. (*a*)	M7.24. (*a*).	

Important Hints:

M 7.1: $|Q_c| = \begin{vmatrix} 0 & 1 \\ 1 & 0 \end{vmatrix} \neq 0,$ controllable

$|Q_o| = \begin{vmatrix} 0 & -1 \\ -1 & 0 \end{vmatrix} \neq 0,$ observable

M 7.2: $\phi(t_1/t_2) = e^{A(t_1/t_2)} \neq \phi(t_1)\,\phi^{-1}(t_2) = e^{At_1}\, e^{-At_2}$

M 7.3: $\dfrac{Y(s)}{R(s)} = C\,(sI - A)^{-1}\, B = \begin{bmatrix} 0 & 1 \end{bmatrix}\left[\begin{bmatrix} s & 0 \\ 0 & s \end{bmatrix} - \begin{bmatrix} 1 & 0 \\ 1 & 1 \end{bmatrix}\right]^{-1}\begin{bmatrix} 0 \\ 1 \end{bmatrix} = \dfrac{s-1}{(s-1)^2} = \dfrac{1}{s+1}$

M 7.5: System is diagonal. $\phi(t)$ can be written by just inspection. For example, a second order diagonal system matrix

$$A = \begin{bmatrix} \lambda_1 & 0 \\ 0 & \lambda_2 \end{bmatrix} \text{ has } \phi(t) = \begin{bmatrix} e^{\lambda_1 t} & 0 \\ 0 & e^{\lambda_2 t} \end{bmatrix}$$

M 7.6: $|Q_c| = \begin{vmatrix} 0 & 1 \\ 1 & -2 \end{vmatrix} \neq 0;$ system is state controllable.

$|Q_{oc}| = \begin{bmatrix} CB \,\vdots\, CAB \end{bmatrix} = [1 \quad -1]$ has rank 1; system is output controllable.

M 7.7: A is non-diagonal, $\phi(s) = (sI - A)^{-1} = \begin{bmatrix} s & -2 \\ 2 & s \end{bmatrix}^{-1} = \dfrac{1}{s^2+4}\begin{bmatrix} s & 2 \\ -2 & s \end{bmatrix}$

$\phi(t) = \mathcal{L}^{-1}\phi(s)$; $\mathcal{L}^{-1}$ denotes inverse Laplace.

M 7.8: $|sI - A| = s^2 + 4 = 0$ gives $s = \pm\, j2$

M 7.9: $\phi(s) = (sI - A)^{-1} = \dfrac{\text{Adj}(sI - A)}{|sI - A|} = \dfrac{\begin{bmatrix} s+6 & 1 \\ -5 & s \end{bmatrix}}{s^2 + 6s + 5}$

M 7.11: $$A = \begin{bmatrix} -2 & 4 \\ -2 & -1 \end{bmatrix}, \quad B = \begin{bmatrix} 0 \\ 1 \end{bmatrix}$$

$$|Q_c| = \begin{vmatrix} 0 & 4 \\ 1 & -1 \end{vmatrix} \neq 0; \text{ system is controllable.}$$

$$|\lambda I - A| = \begin{vmatrix} \lambda+2 & -4 \\ 2 & \lambda+1 \end{vmatrix} = \lambda^2 + 3\lambda + 10 = 0$$

gives eigen values at $-1.5 \pm j2.78$. System is stable.

M 7.12: $$\begin{bmatrix} \dot{x}_1 \\ \dot{x}_2 \\ \dot{x}_3 \end{bmatrix} = \begin{bmatrix} 0 & 1 & 0 \\ 0 & 0 & 1 \\ 0.5 & 1 & 2 \end{bmatrix} \begin{bmatrix} x_1 \\ x_2 \\ x_3 \end{bmatrix} + \begin{bmatrix} 0 \\ 0 \\ 1 \end{bmatrix} \begin{bmatrix} -0.5 & -3 & -5 \end{bmatrix} \begin{bmatrix} x_1 \\ x_2 \\ x_3 \end{bmatrix} + v$$

gives system matrix $A = \begin{bmatrix} 0 & 1 & 0 \\ 0 & 0 & 1 \\ 0 & -2 & -3 \end{bmatrix}$

$|\lambda I - A| = 0$ gives $\lambda (\lambda^2 + 3\lambda + 2) = 0$ and eigen value are $0, -1, -2$.

M 7.13: First order system.

M 7.14: $\lambda I - A| = 0$ gives $(\lambda + 2)(\lambda + 3) = 0$ and eigen values are $-2, -3$. System is stable and states will approach zero at $t \to \infty$.

M 7.15: $$\phi(s) = (sI - A)^{-1} = \begin{bmatrix} \dfrac{1}{s-1} & 0 \\ \dfrac{1}{(s-1)^2} & \dfrac{1}{s-1} \end{bmatrix} \text{ gives } \phi(t) = \begin{bmatrix} e^t & 0 \\ te^t & e^t \end{bmatrix}$$

zero input response $= \phi(t)\, x(0) = \begin{bmatrix} e^t \\ te^t \end{bmatrix}$.

M 7.16: Matrices A and C can be written by just inspection. The entries in last row of A are coefficients of denominator polynomial of G(s) in ascending powers of s and row vector C is constituted of coefficients of numerator polynomial again in ascending powers of s.

M 7.17: $$|Q_o| = \begin{vmatrix} b & 0 \\ b & 2b \end{vmatrix} = 2b^2; \text{ observable for } b \neq 0.$$

M 7.19: $\phi(s) = (sI - A)^{-1} = \begin{bmatrix} s & -1 \\ 0 & s+3 \end{bmatrix}^{-1} = \begin{bmatrix} \frac{1}{s} & \frac{1}{s(s+3)} \\ 0 & \frac{1}{s+3} \end{bmatrix}$

$$\text{STM } \phi(t) = \begin{bmatrix} 1 & \frac{1}{3} - \frac{1}{3}e^{-3t} \\ 0 & e^{-3t} \end{bmatrix}$$

M 7.20: State transition equation

$$x(t) = \phi(t)\, x(0) + \int_0^t \phi(t-\tau)\, Bu(\tau)\, d\tau = \begin{bmatrix} -e^{-3t} \\ 3e^{-3t} \end{bmatrix} + \int_0^t 1 \cdot d\tau = \begin{bmatrix} t - e^{-3t} \\ 3e^{-3t} \end{bmatrix}$$

M 7.21: Systems s_1 and s_2 both are diagonal. Mere inspection reveals that neither of the systems s_1 and s_2 are controllable. s_1 has uncontrollable but stable mode at – 1 while unstable mode at 2 is controllable. So, s_1 is stabilizable. s_2 has unstable and uncontrollable mode at 1 while mode at 2 is unstable but controllable. s_2 is not stabilizable.

M 7.22: Systems s_1 and s_2 both are diagonal. Just inspection reveals that s_1 and s_2 both are unobservable. s_1 has two modes: one at 2 which is unstable but observable and other at – 1 which is stable but unobservable. So. s_1 is detectable. s_2 has two unstable modes out of which one at 1 is unobservable as well. So, s_2 is not detectable.

M 7.23: $A = \begin{bmatrix} -b & d \\ c & -a \end{bmatrix}$, $B = \begin{bmatrix} 0 \\ 1 \end{bmatrix}$, $C = \begin{bmatrix} 1 & 0 \end{bmatrix}$

$$|Q_c| = \begin{vmatrix} 0 & d \\ 1 & -a \end{vmatrix} = -d; \quad |Q_o| = \begin{vmatrix} 1 & 0 \\ -b & d \end{vmatrix} = d$$

System is completely controllable and observable for $d \neq 0$.

Key note to M7.21 and M7.22

The canonical variable form of system model has state equations which are decoupled from each other. The natural response terms in each of individual equations are called **modes**. The overall response of system is linear combination of these modes. For example, a second order diagonal system

$$\begin{bmatrix} \dot{x}_1 \\ \dot{x}_2 \end{bmatrix} = \begin{bmatrix} 8 & 0 \\ 0 & -4 \end{bmatrix} \begin{bmatrix} x_1 \\ x_2 \end{bmatrix} + \begin{bmatrix} 1 \\ -2 \end{bmatrix} u$$

$$y = \begin{bmatrix} 4 & -5 \end{bmatrix} \begin{bmatrix} x_1 \\ x_2 \end{bmatrix}$$

has two modes one at 8 and other at – 4. **Note** that both the modes are controllable and observable. Matrices B and C do not have any zero entry. Mode at 8 is unstable but controllable and observable while mode at – 4 is stable, controllable and observable as well. The state variables have natural responses of form:

$$x_{1_{\text{natural}}} = K_1 e^{8t}$$

$$x_{2_{\text{natural}}} = K_2 e^{-4t}$$

* The term **stabilizability** refers to the ability to move only unstable modes of system. **A system is stabilizable if unstable modes are controllable** or equivalently, if the **uncontrollable modes are stable.**
* The term **detectability** is dual of stabilizability. A system is **detectable if unstable modes are observable** or equivalently, the **unobservable modes are stable.**

M7.24:

$$x(t) = \phi(t)\, x(0) = \begin{bmatrix} e^{-t} & 0 \\ 0 & e^{-2t} \end{bmatrix} \begin{bmatrix} 1 \\ 2 \end{bmatrix} = \begin{bmatrix} e^{-t} \\ 2e^{-2t} \end{bmatrix}$$

$$x(t)\Big|_{t=0.5} = \begin{bmatrix} e^{-0.5} \\ 2e^{-1} \end{bmatrix} = \begin{bmatrix} 0.6 \\ 0.74 \end{bmatrix}$$

8

CONTROL SYSTEM DESIGN

8.1 INTRODUCTION

An engineering work in general and the control system in particular has two interrelated facets: analysis and design. The analysis, in fact, refers to collecting the information as to how the system works. The design aims at making the system work in a prescribed manner. In the chapters so far we have concentrated on control system analysis. In the current chapter we shall particularly focus on design aspects of control systems. It has been well understood by now that a control system, in general, is expected to possess reasonably good steady state performance (where does the system go?) and reasonably good stability (how does the system reach the steady state?). The steady state performance and stability are specified in two domains; the time and the frequency, in terms of peak overshoot, settling time, rise time, steady state error, gain margin, phase margin, bandwidth, peak resonance etc. A typical control system design includes the constraints related to size, weight, power input and cost of components constituting the control system also in addition to primarily meeting the performance specifications as prescribed.

Every control system designed may not be able to meet the design goals. A common approach to force the system to meet the design goals or more categorically meet the prescribed design specifications, is to use a **controller** or **compensator**.

The widely pervasive computers today do substantially help in any design process but they can neither replace the intuition of designer nor reduce the expected skill of an efficient designer. The computers can only play a role of active design partner. Throughout this chapter the stress is laid on inculcating the design skills and providing insight into design strategies by means of adequate graphical illustrations.

8.2 CONTROLLER CONFIGURATIONS

A design process generally begins with deciding the position of controller relative to the system being controlled and then purposefully selecting the elements of controller so as to meet the design specifications. The commonly used controller configuration for SISO system are shown in Fig. 8.1. In Fig. 8.1(*a*) the controller $G_c(s)$ is inserted into forward path in series with the controlled system $G(s)$ and the configuration is referred to as **series or cascade** compensation. In Fig. 8.1(*b*) the controller $G_c(s)$ is placed in feedback path and the configuration is referred to as **feedback compensation**. Each of the controller configuration of Fig. 8.1 (*a*) and (*b*) is said to have only one degree of freedom in the sense that each one has only one controller. Although each of the controller may have more than one parameter that can be manipulated so as to meet design goals. Yet the number of performance specifications that can be simultaneously met, remains restricted. For example, a system together with controller with one degree of freedom, having been designed to meet reasonably good stability requirement, might still have poor sensitivity to parameter variation. Or a system despite the characteristic roots having been selected to meet certain damping requirement, might still exhibit unacceptably large peak overshoot in the step response owing to zeros in closed loop transfer function. The controller with one degree of freedom, may be incapable of meeting the additional requirement of poor sensitivity and large peak overshoot in the cases cited just above.

The controller configurations of Fig. 8.1 (*c*), (*d*) and (*e*) have two degrees of freedom and so, they are capable of relaxing the restriction on design goals. These have some additional parameters that may be used to perform additional task. Fig. 8.1(*c*) shows a system configuration involving both cascade and feedback compensation, $G_{c1}(s)$ is placed in forward path and $G_{c2}(s)$ in feedback path. Such a configuration is called **series feedback compensation**. Fig. 8.1(*d*) also shows series feedback compensator with the only difference that $G_{c2}(s)$ is now placed in minor feedback loop. In Fig. 8.1(*e*), the controller $G_{c1}(s)$ is placed in series with closed loop system which has controller $G_{c2}(s)$ in forward path. Such a configuration is referred to as **feed forward compensation**. Note that the controller $G_{c1}(s)$ does not fall in the loop of system and therefore, the characteristic equation $1 + G_{c2}(s)\, G(s) = 0$ of original system with cascade controller, remains preserved. $G_{c1}(s)$ having been suitably selected, additionally may serve the purpose of cancelling undesired poles and zeros of closed loop transfer function of original system.

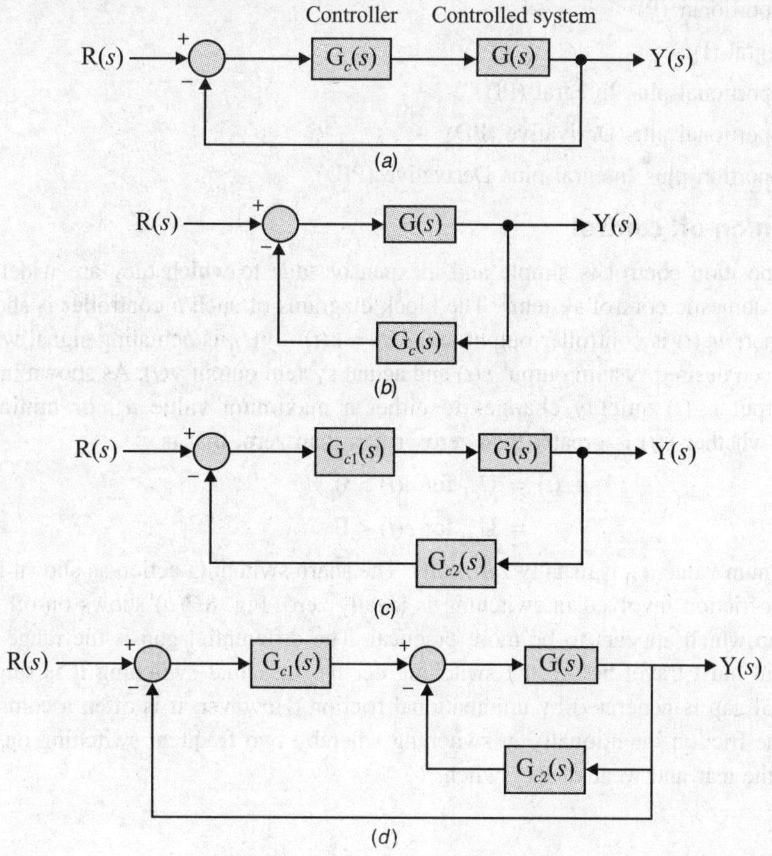

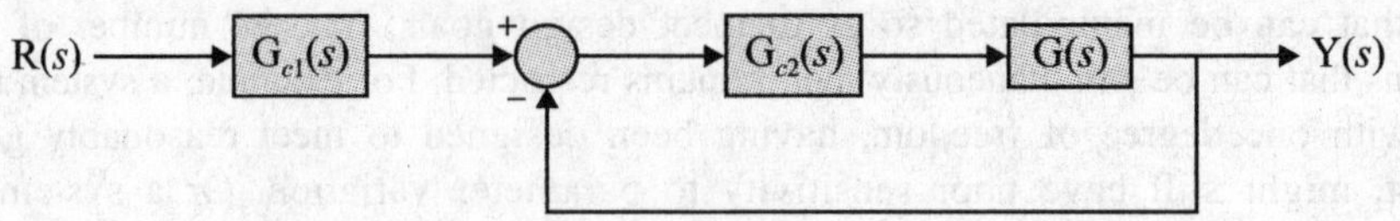

(e)

Fig. 8.1: Controller configurations (*a*) Cascade compensated system, (*b*) Feedback compensated system, (*c*) System with cascade and feedback compensator, (*d*) System with cascade and feedback compensator in minor loop, (*e*) System with forward and cascade compensator

8.3 INDUSTRIAL AUTOMATIC CONTROLLERS

An automatic controller maintains the desired value of system output by measuring the existing system output, comparing it with the desired value and then employing the deviation between these two to initiate a control action so as to reduce the deviation to zero or to an acceptably small value. The automatic controllers, in fact, function in a closed loop without human aid. The power source required in operation of these controllers may be derived from electricity, pressurized fluid such as oil or air, based on which they may be classified as electronic, hydraulic or pneumatic controller respectively. However, the manner in which the control action is initiated, remains common in all of them. Based on the control action initiated, the automatic controllers are classified as:

1. Two position/on-off
2. Proportional (P)
3. Integral (I)
4. Proportional plus Integral (PI)
5. Proportional plus Derivative (PD)
6. Proportion plus Integral plus Derivative (PID).

Two position/on-off control

The two position control is simple and inexpensive due to which they are widely used in both industrial and domestic control systems. The block diagrams of such a controller is shown in Fig. 8.2 (*a*) and (*b*) where $u_c(t)$ is controller output and $e(t) = r(t) - y(t)$, is actualing signal which is, in fact, the error between desired system output $r(t)$ and actual system output $y(t)$. As shown in Fig. 8.2(*a*) the controller output $u_c(t)$ quickly changes to either a maximum value u_{c1} or minimum value u_{c0} depending on whether $e(t)$ is greater than zero or less than zero, that is

$$u_c(t) = U_{c1} \text{ for } e(t) > 0$$
$$= U_{c0} \text{ for } e(t) < 0$$

The minimum value u_{c0} is usually zero (off). The sharp switching action as shown in Fig. 8.1(*a*) is possible if the friction involved in switching is ideally zero. Fig. 8.2(*b*) shows on-off controller with differential gap which appears to be more practical. The differential gap is the range through which the error signal must travel before the switching occurs. In actual switching it is same as hystersis. The differential gap is generated by unintentional friction. However, it is often a common practice to introduce some friction intentionally in switching whereby two frequent switching on/off is avoided. This reduces the tear and wear of the switch.

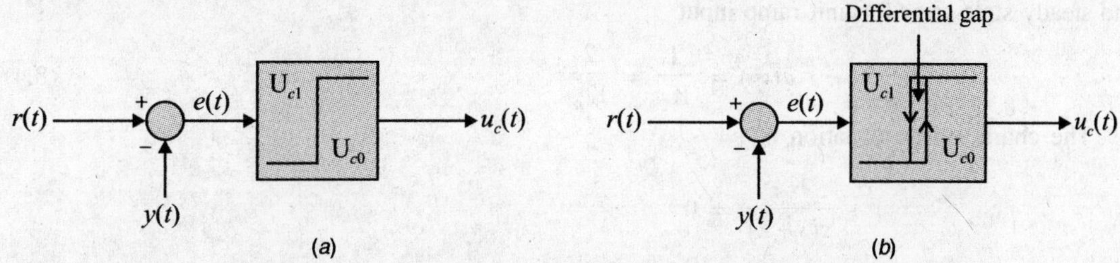

Fig. 8.2: (*a*) On-off controller, (*b*) On-off controller with differential gap

Proportional control

The proportional controller as shown in Fig. 8.3(*a*) has the output $u_c(t)$ that is linearly related to its input error signal $e(t)$, that is

$$u_c(t) \propto e(t)$$

or

$$u_c(t) = K_p e(t) \quad ...(8.1)$$

where K_p is referred to as **proportional gain or sensitivity** and inverse of proportional gain is termed as **proportional band**. The proportional controller is simply an amplifier with adjustable gain. It has the constant transmittance

$$G_c(s) = \frac{U_c(s)}{E(s)} = K_p \quad ...(8.2)$$

Note that a control designer is typically interested in relative stability and steady state error performance of closed loop system. In order to have insight into comparative benefits derived from each of the controllers listed above, let us consider one example system of type 1 with transmittance

$$G(s) = \frac{1}{s(s+2)} \quad ...(8.3)$$

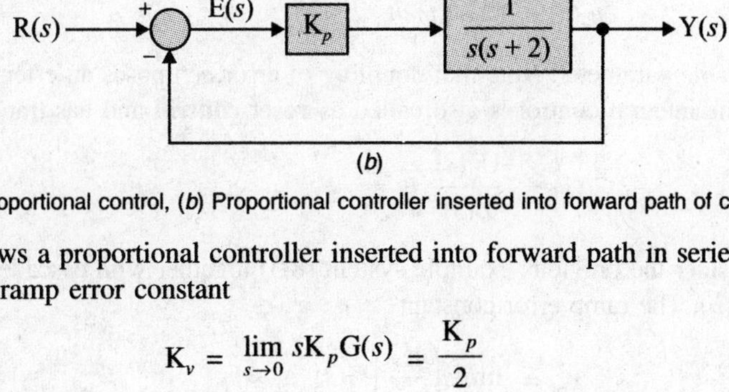

Fig. 8.3: (*a*) Proportional control, (*b*) Proportional controller inserted into forward path of controlled system

Fig. 8.3(*b*) shows a proportional controller inserted into forward path in series with system under consideration. The ramp error constant

$$K_v = \lim_{s\to 0} sK_pG(s) = \frac{K_p}{2}$$

and steady state error for unit ramp input

$$e(\infty) = \frac{1}{K_v} = \frac{2}{K_p} \qquad ...(8.4)$$

The characteristic equation is

$$1 + \frac{K_p}{s(s+2)} = 0$$

or

$$s^2 + 2s + K_p = 0 \qquad ...(8.5)$$

Comparing this with general characteristic equation

$$s^2 + 2\xi\omega_n s + \omega_n^2 = 0$$

we have

$$\omega_n = \sqrt{K_p}$$

and

$$\xi = \frac{1}{\sqrt{K_p}}$$

Note the following in current discussion:

1. It is obvious from (8.4) that the proportional gain K_p should be increased in order to reduce the steady state error. The steady state error can be completely removed by allowing K_p to become infinitely large. But with increase in K_p, damping ratio ξ reduces while undamped natural frequency ω_n increases. Reduced ξ causes the system to be more oscillatory. If designer tries to improve the relative stability by reducing K_p, the steady state error would increase. So, the simultaneous improvement in both the transient and steady state response of system, is not possible with P type of control action.
2. Larger K_p causes the controller output to be larger. The larger controller output demands larger actuator movement, possibly reaching physical limits.

Integral control

Figure 8.4(*a*) shows an integral controller where the controller output $u_c(t)$ changes at the rate proportional to input error signal $e(t)$, that is

$$\frac{d}{dt} u_c(t) \propto e(t)$$

or

$$u_c(t) = K_i \int e(t)\, dt \qquad ...(8.6)$$

where K_i is an adjustable parameter. Note that doubling of error $e(t)$ poses an effect of moving output $u_c(t)$ twice as fast. The integral control is also called as **reset control** and has transmittance

$$G_c(s) = \frac{U_c(s)}{E(s)} = \frac{K_i}{s} \qquad ...(8.7)$$

Let us again consider the previous example system (8.3) together with cascade integral controller as shown in Fig. 8.4(*b*). The ramp error constant

$$K_v = \lim_{s \to 0} s\left(\frac{K_i}{s}\right) G(s) = \infty$$

and steady state error for unit ramp input

$$e(\infty) = \frac{1}{K_v} = 0$$

The characteristic equation is

$$1 + \frac{K_i}{s^2(s+2)} = 0$$

or
$$s^3 + 2s^2 + K_i = 0 \qquad ...(8.8)$$

Note the following from current discussion:

(*i*) The integral controller raises the number type of system. Note that the original system of type one has become of type 2. This is the reason why the same system which for unit ramp input exhibited finite steady state error $e(\infty) = 2/K_p$ with proportional controller, now exhibits zero steady state with integral controller.

(*ii*) The integral controller also increases the order of system. Note that originally a second order system has been converted into third order system. The failure of coefficient test reveals that the characteristic equation (8.8) has RHP roots and the closed loop system is always unstable. An unstable system is of no practical interest. This is the reason why the proportional controller is always added with integral control. The proportional part of control action tries to stabilise the system while the integral part tries to eliminate or mitigate the steady state error.

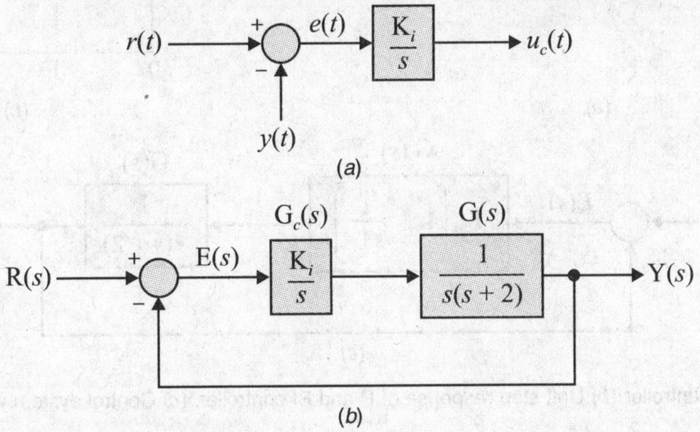

Fig. 8.4: (*a*) Integral controller and (*b*) Control system with integral control

Proportional *plus* Integral control (PI)

In order to derive advantages of both proportional and integral controller, they are additively combined to form a composite configuration called as PI controller as shown in Fig. 8.5(*a*). The following equation describes the dynamics of PI controller.

$$u_c(t) = \underbrace{K_p e(t)}_{\text{Proportional}} + \underbrace{\frac{K_p}{T_i}\int_0^t e(t)\,dt}_{\text{Integral}} \qquad ...(8.9)$$

where K_p is proportional gain and T_i is integral time. K_p and T_i are tuneable parameters. T_i affects only integral part while K_p affects both proportional and integral part of controller. The transmittance of PI controller is

$$G_c(s) = \frac{U_c(s)}{E(s)} = K_p\left[1+\frac{1}{sT_i}\right] \quad \text{...(8.10)}$$

The unit step response of PI controller

$$u_c(t)\Big|_{e(t)=u(t)} = K_p\,u(t) + \frac{K_p}{T_i}\int_0^t u(t)\,dt; \quad t \geq 0$$

$$= K_p + \frac{K_p}{T_i}\,t$$

is shown in Fig. 8.5(*b*). Note that $u_c(t)\big|_{t=T_i} = 2K_p$, $u_c(t)\big|_{t=2T_i} = 3K_p$ and so on. Thus proportional part of control action goes on repeating every after one integral time.

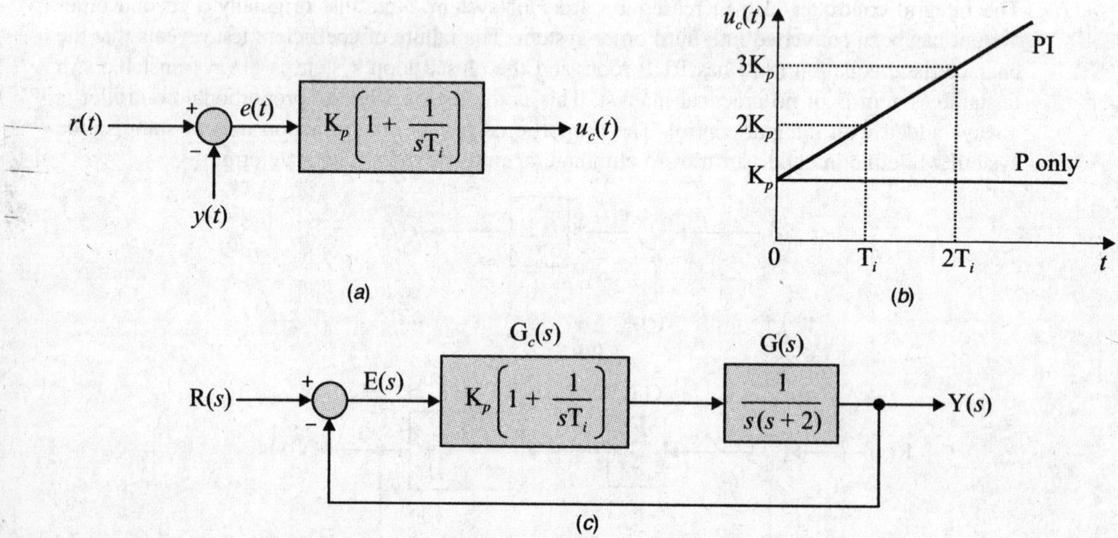

Fig. 8.5: (*a*) PI controller (*b*) Unit step response of P and PI controller, (*c*) Control system with PI controller

Let us again consider example system (8.3) together with cascade PI controller as shown in Fig. 8.5(*c*). The ramp error constant

$$K_v = \lim_{s\to 0} sK_p\left(1+\frac{1}{sT_i}\right)\left(\frac{1}{s(s+2)}\right) = \infty$$

and steady state error for unit ramp input

$$e(\infty) = \frac{1}{K_v} = 0$$

The characteristic equation is

$$1 + K_p\left(1+\frac{1}{sT_i}\right)\left(\frac{1}{s(s+2)}\right) = 0$$

or $$s^3 + 2s^2 + K_p s + \frac{K_p}{T_i} = 0 \qquad ...(8.11)$$

Note the following in ongoing discussion:

(*i*) PI compensator by virtue that it has a pole at origin, raises the system type number, thus eliminating step error for a type 0 system, ramp error for type 1 system and so on.

(*ii*) The PI compensator also raises the order of system by unity. Note that the system has third order characteristic equation (8.11). For $K_p > 0$, it passes the coefficient test. But the closed loop system may be unstable for large value of K_p.

(*iii*) The steady state accuracy is not a problem with PI controller. However K_p has to be suitably chosen so that system exhibits reasonable transient response (stability).

Proportional *plus* Derivative control (PD)

The derivative control is defined as one in which the controller output is proportional to the rate of change of input error signal. The derivative control is also referred to as **rate control**. A derivative control when added to the proportional controller, the composite configuration is referred to as PD control as shown in Fig. 8.6(*a*). The following equation describes the dynamics of PD control.

$$u_c(t) = \underbrace{K_p\, e(t)}_{\text{Proportional}} + \underbrace{K_p T_d \frac{d}{dt} e(t)}_{\text{Derivative}} \qquad ...(8.12)$$

where K_p is proportional gain and T_d is derivative time. K_p and T_d both are tuneable parameters. The derivative time T_d may be defined as the time interval by which the derivative part of control advances the effect of proportional part of control.

The unit step response of PD controller does not reveal any meaningful information. The derivative part produces an infinite response at $t = 0$ for $e(t) = u(t)$ and zero response for $t > 0$. Therefore, the input error signal is chosen to be unit ramp, that is $e(t) = t;\ t \geq 0$. Then from (8.12)

$$u_c(t) = K_p t + K_p T_d = K_p\,(t + T_d)$$

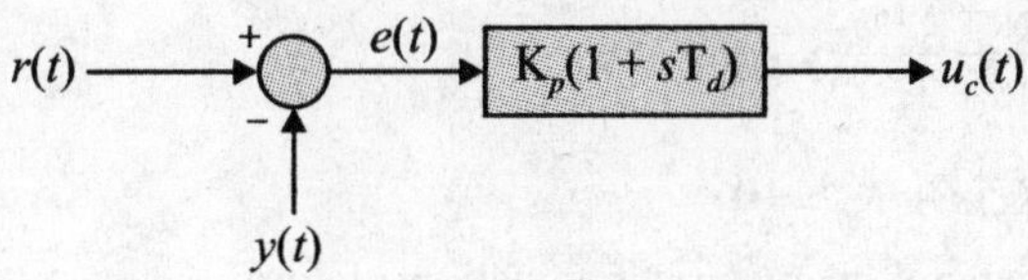

(*a*)

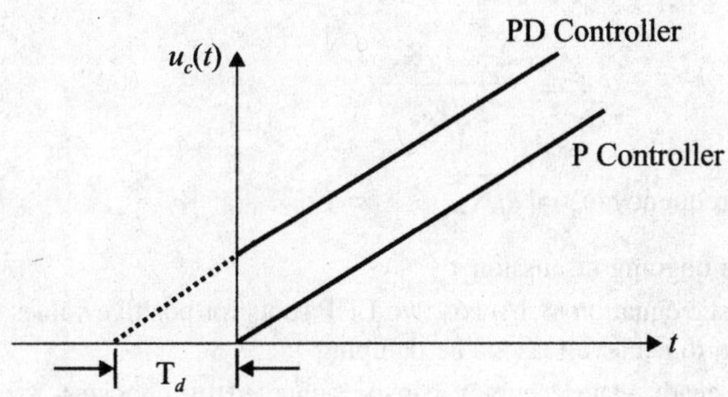

(*b*)

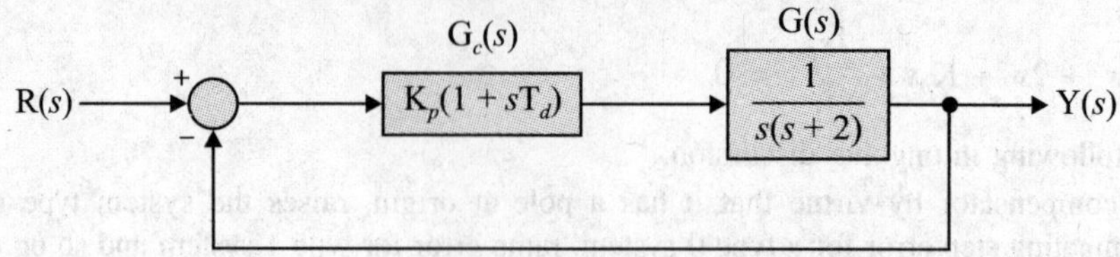

(c)

Fig. 8.6: (a) PD controller (b) Unit ramp response of PD controller, (c) Control system with PD controller

This unit ramp response is shown in Fig. 8.6(b) where it can be seen that the controller output leads the time change of input error signal by an amount equal to T_d, the derivative time. Thus the derivative time may be defined as the time interval by which the derivative part advances the effect of proportional part of controller. In this sense the derivative control is said to possess the property of anticipating the future direction of error signals. However, truly speaking in reality, the derivative control can never anticipate an action that is yet to take place.

PD controller has the transmittance

$$G_c(s) = \frac{U_c(s)}{E(s)} = K_p(1 + sT_d) \qquad ...(8.13)$$

Let us again consider example system (8.3) together with PD controller in series therewith as shown in Fig. 8.6(c). The ramp error constant

$$K_v = \lim_{s \to 0} sG_c(s)\,G(s) = \frac{K_p}{2}$$

and steady state error for unit ramp input

$$e(\infty) = \frac{2}{K_p}$$

The characteristic equation is

$$1 + \frac{K_p(1 + sT_d)}{s(s+2)} = 0$$

or

$$s^2 + (2 + T_dK_p)s + K_p = 0 \qquad ...(8.14)$$

Compare (8.14) with general characteristic equation of form

$$s^2 + 2\xi\omega_n s + \omega_n^2 = 0$$

to get damping ratio

$$\xi = \frac{2 + T_dK_p}{2\sqrt{K_p}}$$

and undamped natural frequency $\omega_n = \sqrt{K_p}$

Note the following in ongoing discussion:

(i) The characteristic equation (8.14) has two LHP roots for positive values of K_p and T_d. T_d may be suitably chosen to achieve reasonable damping.

(ii) A reasonable steady state accuracy can be achieved by choosing K_p. The steady state error, however, cannot be made zero.

(*iii*) PD controller is advantageous in the sense that it is capable of bringing about simultaneous improvement in both transient and steady state response of system. The derivative part of controller has an additional advantage of being anticipatory in character. The drawback of derivative control is that it tends to amplify the noise signals whereby the actuator may get saturated.

(*iv*) The derivative controller is generally, not recommended to be used alone. It produces zero output for any error signal which is constant. It is effective only during transients.

Proportional *plus* Integral *plus* Derivative control (PID)

The proportional, integral and derivative controllers are additively combined to form a composite controller configuration referred to as PID controller as shown in Fig. 8.7(*a*). The PID controller is described by the integro differential equation

$$u_c(t) = \underbrace{K_p e(t)}_{\text{Proportional}} + \underbrace{\frac{K_p}{T_i}\int e(t)\,dt}_{\text{Integral}} + \underbrace{K_p T_d \frac{d}{dt}e(t)}_{\text{Derivative}} \qquad ...(8.15)$$

where K_p, T_i and T_d are tuneable parameters. Let us look into the controller dynamics while assuming the error signal to be a unit ramp, that is

$$e(t) = t; \qquad t \geq 0$$

$$u_c(t)\big|_{e(t)=t} = K_p t + \frac{K_p}{T_i}\int t\,dt + K_p T_d \frac{d}{dt}(t)$$

$$= K_p\left[t + \frac{t^2}{2T_i} + T_d\right] \qquad ...(8.16)$$

The unit ramp response of PID controller is shown in Fig. 8.7(*b*). The proportional part of controller repeats the input error signal. The derivative part together with proportional part plays the role of shifting the controller response ahead in time. The integral part further contributes to the controller output proportional to the area under error line. Thus proportional, integral and derivative all three parts make additive contribution to total response as shown in Fig. 8.7(*b*).

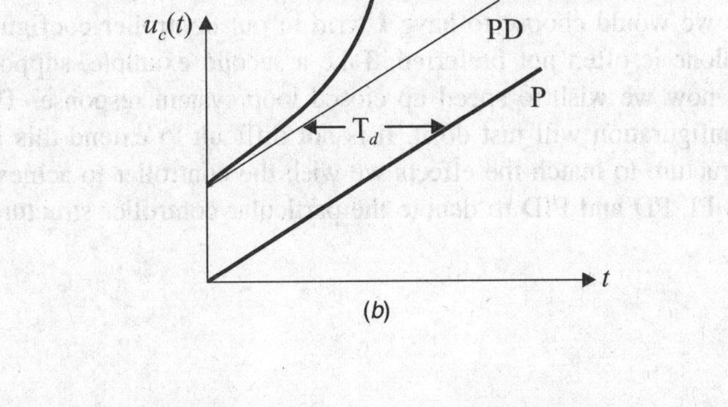

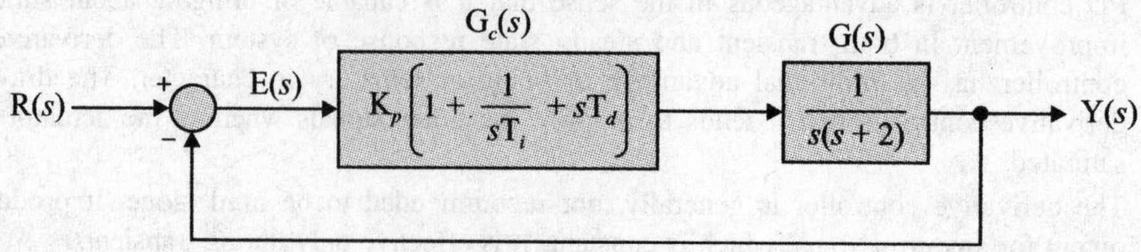

(c)

Fig. 8.7: (*a*) PID controller (*b*) Unit ramp response of P, PD and PID controller, (*c*) Control system with PID controller

PID controller has the transmittance of form

$$G_s(s) = \frac{U_c(s)}{E(s)} = K_p\left(1 + \frac{1}{sT_i} + sT_d\right)$$

Consider again the example system (8.3) together with PID controller as shown in Fig. 8.7(*c*). The ramp error constant

$$K_v = \lim_{s \to 0} sG_c(s)\,G(s) = \infty$$

and steady state error $e(\infty) = 0$

The characteristic equation is

$$1 + G_c(s)\,G(s) = 0$$

or
$$s^3 + (2 + K_pT_d)\,s^2 + K_p s + \frac{K_p}{T_i} = 0 \qquad \text{...(8.17)}$$

Note the following in ongoing discussion:

(*i*) PID controller combines steady state accuracy improvement property of PI controller (with some loss of stability) and stability improvement property of PD controller.

(*ii*) Integral part of controller increments the system type number and system order both by unity. Steady state error is zero for ramp input due to raising of system type number from 1 to 2. The system stability worsens due to raising of system order from 2 to 3. However, reasonably acceptable stability can be achieved by derivative and proportional part of controller.

The effects of different terms of three term control (PID) together with combination of two or all three of them, as addressed in the discussion so far, are contained in **Table 8.1** for an easy reference. The table is similar to a toolbox where each term or their combination does a particular job. For example, if we need to remove steady state error, the table says that **D** term will not do this, that a **P** term will reduce it on increasing proportional gain K_P but that an **I** term will eliminate the error completely. Therefore, we would choose to have **I** term in our controller configuration. Keep a note that **integral** control alone is often not preferred. Take a second example, suppose having chosen **I** term in our controller, now we wish to speed up closed loop system response. The table shows that increasing K_P in PI configuration will just do it. It is not difficult to extend this idea so that we can select the controller structure to match the effects we wish the controller to achieve. The table shows the shorthand tables P, PI, PD and PID to denote the particular controller structures.

TABLE 8.1: Effects of different terms of three term control (PID)

Controller	*Transmittance (a general form)*	*Typical effect on steady state errors*	*Typical effect on relative stability*	*Change in type number*	*Change in system order*
PD	$K(1 + bs)$	Somewhat improves	Increases	0	0
PID	$K\left(1+\frac{a}{s}+bs\right)$	Greatly improves	Increases	+ 1	+ 1
P	K	Improves with larger K	Increases with smaller K	0	0
PI	$K\left(\frac{s+a}{s}\right)$	Greatly improves	Reduces	+ 1	+ 1
I	K/s	Greatly improves	Greatly reduces	+ 1	+ 1

Note: In the general form of transmittance shown in table ***a*** stands for $1/T_i$ and ***b*** stands for T_d. T_i and T_d are integral and derivative times in discussion so far. K stands for K_p.

8.4 GENERATING HARDWARE FOR INDUSTRIAL CONTROLLERS

In the discussion so far, we have been concentrating on mathematical modelling and theoretical aspects of P, PI, PD and PID controllers. In this section, we shall focus on generating hardware.

PI controller

Assuming ideal behaviour of operational amplifier the circuit shown in Fig. 8.8(*a*) is an example of proportional (P) controller whose transmittance

$$|G_c(s)| = \left|\frac{u_c(t)}{e(t)}\right| = \frac{R_2}{R_1}, \quad \text{a constant}$$

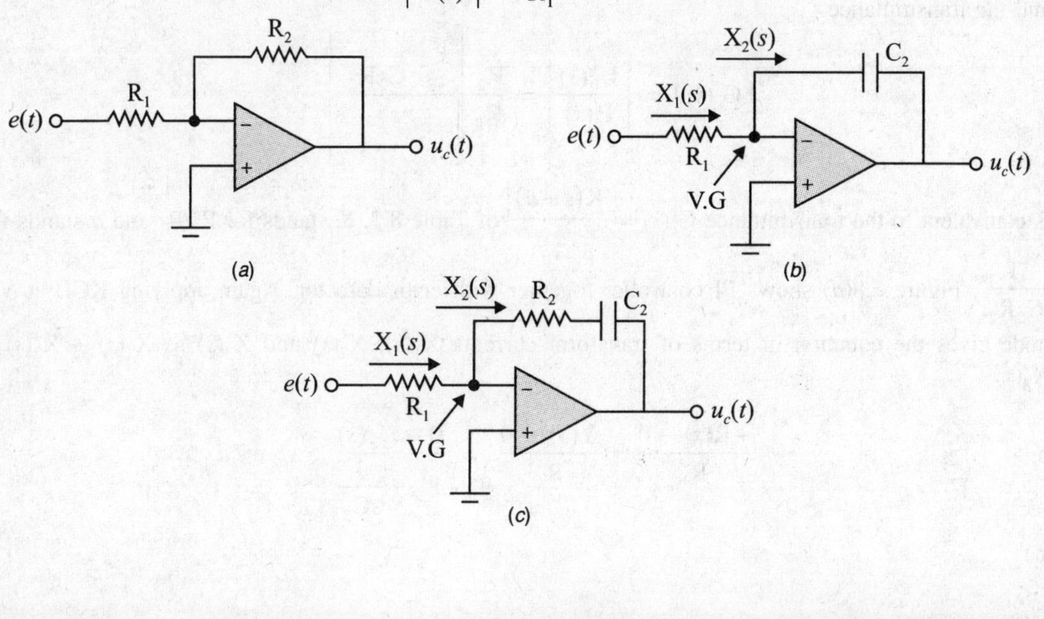

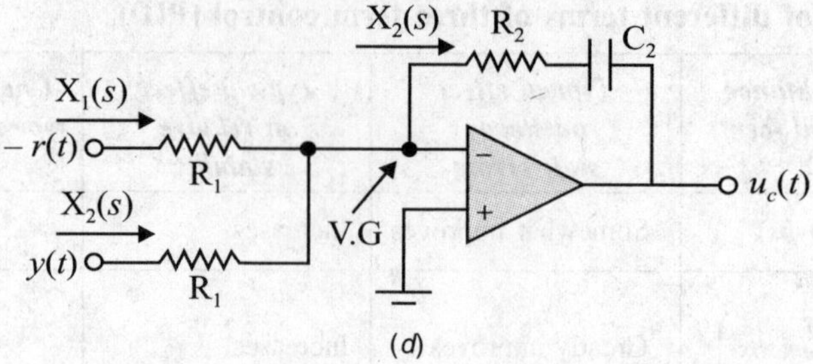

Fig. 8.8: (*a*) P controller (*b*) I controller, (*c*) PI controller, (*d*) PI controller with error detector

Note that circuit of Fig. 8.8(*a*) exhibits very fast dynamics. It does not involve any time constant. The circuit shown in Fig. 8.8(*b*) is an example of integral (I) controller. Ideally inverting input terminal of operational amplifier is virtual ground (VG). $X_1(s)$ and $X_2(s)$ are the transform currents. KCL at node VG in the circuit gives the transform equation $X_1(s) = X_2(s)$

or $$\frac{E(s)-0}{R_1} = \frac{0-U_c(s)}{(1/sC_2)}$$

and the transmittance

$$|G_c(s)| = \left|\frac{U_c(s)}{E(s)}\right| = \frac{1}{sR_1C_2} \equiv \frac{K_i}{s}$$

The circuit shown in Fig. 8.8(*c*) is an example of PI controller. $X_1(s)$ and $X_2(s)$ are transform currents. KCL applied at VG node gives

$$X_1(s) = X_2(s)$$

or $$\frac{E(s)-0}{R_1} = \frac{0-U_c(s)}{\left(R_2+\frac{1}{sC_2}\right)}$$

and the transmittance

$$|G_c(s)| = \left|\frac{U_c(s)}{E(s)}\right| = \frac{R_2}{R_1}\left[\frac{s+\frac{1}{C_2R_2}}{s}\right]$$

is equivalent to the transmittance $G_c(s) = \frac{K(s+a)}{s}$ of Table 8.1. K stands for R_2/R_1 and ***a*** stands for $\frac{1}{C_2R_2}$. Figure 8.8(*d*) shows PI controller together with error detector. Again applying KCL at VG node gives the equation in terms of transform currents $X_1(s)$, $X_2(s)$ and $X_3(s)$ as $X_1(s) + X_2(s) = X_3(s)$

or $$\frac{-R(s)-0}{R_1} + \frac{Y(s)-0}{R_1} = \frac{0-U_c(s)}{R_2+\frac{1}{sC_2}}$$

where R(*s*) is reference signal, Y(*s*) is output and transform error E(*s*) = R(*s*) – Y(*s*). The controller transmittance takes the form as

$$G_c(s) = \frac{U_c(s)}{E(s)} = \frac{R_2}{R_1}\left[\frac{s + \frac{1}{C_2R_2}}{s}\right]$$

which is same as that of circuit of Fig. 8.8 (*c*) but this circuit takes the reference and output signals as inputs.

PD controller

The circuit as shown in Fig. 8.9 is a typical example of PD Controller. KCL at VG node gives the equation in terms of transform current as

$$X_1(s) + X_2(s) = X_3(s)$$

or
$$\frac{E(s) - 0}{R_1} + \frac{E(s) - 0}{(1/sC)} = \frac{0 - U_c(s)}{R_2}$$

which in turn gives the controller transmittance as

$$|G_c(s)| = \left|\frac{U_c(s)}{E(s)}\right| = \frac{R_2}{R_1}(1 + sCR_1) \equiv K(1 + bs);$$

K stands for $\frac{R_2}{R_1}$ and ***b*** stands for CR_1.

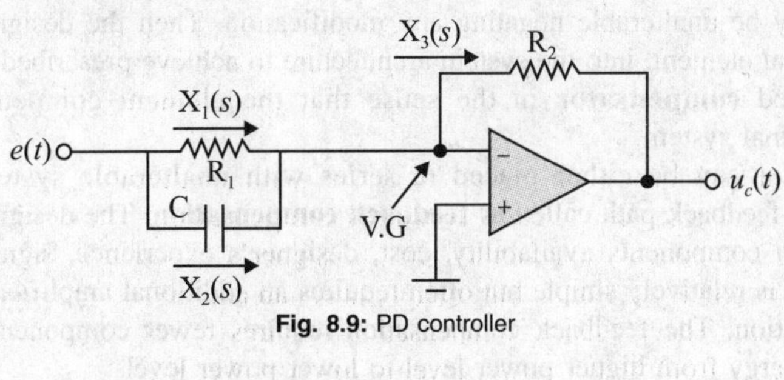

Fig. 8.9: PD controller

PID controller

The circuit shown in Fig. 8.10 is a typical example of PID controller. KCL at node VG gives the equation in terms of transforms currents shown in Fig. 8.10 as

$$X_1(s) = X_2(s)$$

or
$$\frac{E(s) - 0}{Z_1(s)} = \frac{0 - U_c(s)}{Z_2(s)}$$

where $Z_1(s) = \frac{R_1}{1 + sCR_1}$ and $Z_2(s) = R_2 + \frac{1}{sC_2}$

Substituting these values of $Z_1(s)$ and $Z_2(s)$ in the equation just above, the controller transmittance is of form:

$$|G_c(s)| = \left|\frac{U_c(s)}{E(s)}\right| = \frac{C_1R_1 + C_2R_2}{C_2R_1}\left[1 + \frac{1}{(R_1C_1 + R_2C_2)s} + s \cdot \frac{R_1R_2C_1C_2}{(C_1R_1 + C_2R_2)}\right]$$

which is equivalent to transmittance $K\left(1+\frac{a}{s}+bs\right)$ of Table 8.1. K stands for $\frac{C_1R_1+C_2R_2}{C_2R_1}$,

a stands for $\frac{1}{R_1C_1+R_2C_2}$ and *b* stands for $\frac{R_1R_2C_1C_2}{C_1R_1+C_2R_2}$.

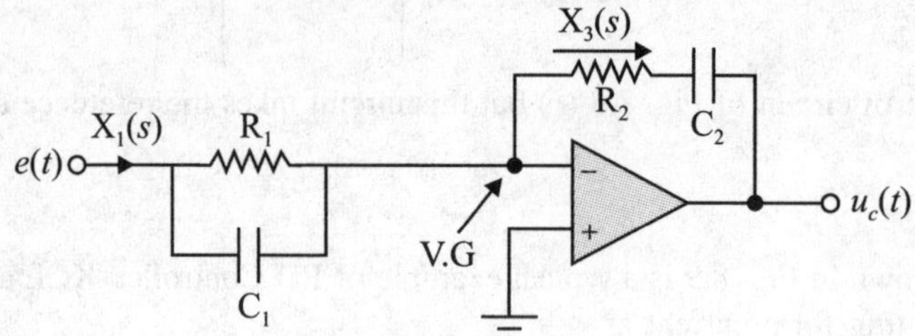

Fig. 8.10: PID controller

8.5 THE COMPENSATOR ELEMENTS

Every control system is designed to do a particular job. What is expected of a control system, is generally spelled out in terms of performance specifications related to steady state accuracy, relative stability and speed of response. Having designed a control system, the designer verifies whether the system meets the prescribed performance specifications. If it does not, the designer may try to repeat the design by adjusting the system parameters or may change system configuration until the prescribed specifications are achieved. The success of such repetitive trial, heavily depends on designer's skill. There are two problems faced by designer: one the repetitive design may not succeed; two, the system may be unalterable negating any modification. Then the designer has an option of inserting an additional element, into the system architecture to achieve prescribed specifications. Such an element is called **compensator** in the sense that the element compensates for deficient performance of original system.

The compensator can be either placed in series with unalterable system called as **series compensation** or in feedback path called as **feedback compensation**. The designer's choice between the two, depends on components availability, cost, designer's experience, signal power levels etc. Series compensation is relatively simple but often requires an additional amplifier to enhance the gain and/or provide isolation. The feedback compensation requires fewer components. This is probably due to transfer of energy from higher power level to lower power level.

Although a family of such series compensators, is found in literature, we shall restrict our discussion to only phase lead, phase lag and phase lag-lead compensators. In the current section, we shall discuss the compensator networks and their properties. In subsequent sections we shall focus on their design aspects.

Phase lead compensator

The phase lead compensator has a pole and a zero on negative real axis with zero being closer to origin than the pole. As shown in Fig. 8.11(*a*) the compensator zero located at $s = -1/\tau$, is to the right of compensator pole located at $s = -(1/\alpha\tau)$; $\alpha < 1$. The phase lead compensator has the transmittance, $G_{lead}(s)$ of form

$$G_{lead}(s) = \frac{s+\frac{1}{\tau}}{s+\frac{1}{\alpha\tau}} = \alpha\frac{\tau s+1}{\alpha\tau s+1}; \qquad \alpha < 1 \qquad ...(8.18)$$

and can be synthesized with electrical network as shown in Fig. 8.11(*b*) Assuming that the network is perfectly non-interactive on source (input) and load (output) both sides, the transmittance can be obtained as:

$$\frac{U_c(s)}{E(s)} = \frac{R_2}{\left(R_1 \| \frac{1}{sC}\right) + R_2} = \left(\frac{R_2}{R_1 + R_2}\right)\left[\frac{sCR_1 + 1}{sC\frac{R_1R_2}{R_1 + R_2} + 1}\right] \quad ...(8.19)$$

It is easy to identify that (8.18) and (8.19) are of same form with

$$\tau = R_1C$$

and
$$\alpha = \frac{R_2}{R_1 + R_2} < 1$$

Figure 8.11(*c*) shows a mechanical lead network whose dynamics is described by following differential equations:

$$B_2(\dot{x}_o - \dot{x}_i) + B_1(\dot{x}_o - \dot{y}) = 0; \quad \text{(node } x_0)$$

$$B_1(\dot{y} - \dot{x}_o) + K(y - 0) = 0; \quad \text{(node } y)$$

Laplace transforming these differential equations with zero initial conditions, we have

$$B_2(s)\,[X_o(s) - X_i(s)] + B_1(s)\,[X_o(s) - Y(s)] = 0$$

and
$$B_1s\,[Y(s) - X_o(s)] + KY(s) = 0$$

Little algebraic manipulation to eliminate $Y(s)$, yields

$$\frac{X_o(s)}{X_i(s)} = \left(\frac{B_2}{B_1 + B_2}\right)\frac{\left(\frac{B_1}{K}s + 1\right)}{\left(s\frac{B_2}{B_1 + B_2}\frac{B_1}{K} + 1\right)} \quad ...(8.20)$$

It is again easy to identify that (8.18) is of form same as (8.20) with

$$\tau = \frac{B_1}{K} \quad \text{and} \quad \alpha = \frac{B_2}{B_1 + B_2} < 1$$

The sinusoidal transmittance $G_{lead}(j\omega)$ is of form

$$G_{lead}(j\omega) = \alpha\frac{j\omega\tau + 1}{j\omega\alpha\tau + 1}; \quad 0 < \alpha < 1$$

which has the polar plot as shown in Fig. 8.11(*d*). Note $\lim_{\omega\to 0} G_{lead}(j\omega) = \alpha\angle 0°$ and $\lim_{\omega\to\infty} G_{lead}(j\omega) = 1\angle 0°$. Since $\alpha < 1$, $\angle G(j\omega)$ is always positive. The polar plot is a semicircle. The angle between the tangent drawn from origin to semicircle and positive real axis, gives peak phase lead ϕ_m as shown in Fig. 8.11(*d*). The frequency corresponding to the tangent point, P or the frequency at which the lead compensator contributes maximum phase lead, is designated as ω_m in Fig. 8.11(*d*). Notice that radius of semicircle (polar plot) is 1/2 $(1 - \alpha)$ and centre, O of semicircle is located on positive real axis at a distance of α + 1/2 $(1 - \alpha)$ = 1/2 $(1 + \alpha)$ from origin. So, the peak phase lead is related to the value of α as:

$$\sin \phi_m = \frac{PQ}{OQ} = \frac{\frac{1}{2}(1-\alpha)}{\frac{1}{2}(1+\alpha)} = \frac{1-\alpha}{1+\alpha} \qquad ...(8.21)$$

The variation of ϕ_m against value of $1/\alpha$ is shown in Fig. 8.11 (*f*). The frequency ω_m which corresponds to peak phase lead ϕ_m, can be obtained in terms of α and τ by equating derivative

$\frac{d}{d\omega}\phi(\omega) = \frac{d}{d\omega}[\angle G_{lead}(j\omega)]$ to zero, that is

$$\frac{d}{d\omega}\left[\tan^{-1}\omega\tau - \tan^{-1}\omega\alpha\tau\right]\Big|_{\omega=\omega_m} = 0$$

$$\frac{\tau}{1+\omega_m^2\tau^2} - \frac{\alpha\tau}{1+\omega_m^2\alpha^2\tau^2} = 0$$

$$\omega_m = \sqrt{\left(\frac{1}{\tau}\right)\left(\frac{1}{\alpha\tau}\right)} \qquad ...(8.22)$$

It is obvious from (8.22) that the frequency ω_m at which the lead compensator contributes maximum phase lead is, in fact, the **geometric mean of the two corner frequencies $1/\tau$ and $1/\alpha\tau$** of the compensator.

Note that *dc* gain $|G_{lead}(j\omega)|_{\omega=0} = \alpha < 1$. So, the lead network causes an attenuation. In order to neutralize this *dc* attenuation, the lead compensator of Fig. 8.11(*b*) is very often combined with an amplifier of gain $1/\alpha$ as shown in Fig. 8.11(*g*). For further discussion in frequency domain, we shall then consider the sinusoidal transmittance $G_{lead}(j\omega)$ of form

$$G_{lead}(j\omega) = \frac{j\omega\tau + 1}{j\alpha\omega\tau + 1}$$

The Bode plots (both dB and phase) for this $G_{lead}(j\omega)$ are shown in Fig. 8.11(*e*). The corner frequencies can easily be identified as $\omega_L = 1/\tau$ and $\omega_H = 1/\alpha\tau$, ω_L and ω_H are low and high corner frequencies respectively. The dB plot shows unity gain (0 dB) at low frequencies. At high frequencies the gain (dB) tends to approach $20 \log_{10} (1/\alpha)$. The gain increases at the rate of + 20 dB/dec from 0 dB at $\omega_L = 1/\tau$ and attains the peak at $\omega_H = 1/\alpha\tau$ where after the gain remains constant at $20 \log (1/\alpha)$. At $\omega = \omega_m$, the gain is half the peak dB value, that is equal to $20 \log (1/\sqrt{\alpha})$. The phase increases from 0° at $\omega = 0$ towards + 90° but the compensator pole affects the phase change and reduces the final phase back to 0° as $\omega \to \infty$.

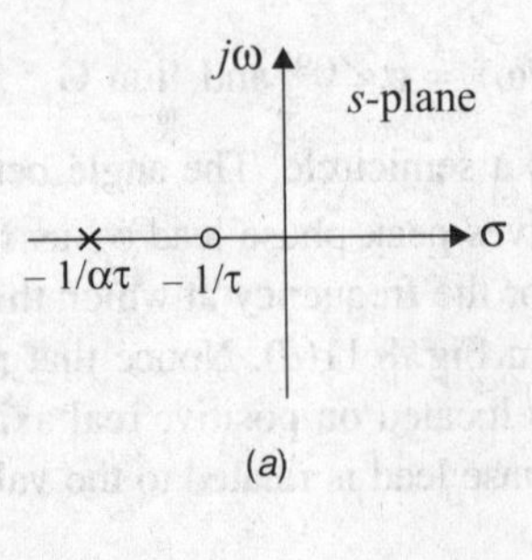

(*a*)

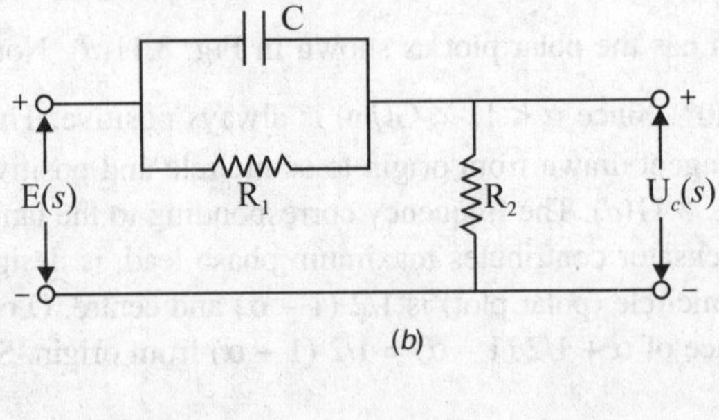

(*b*)

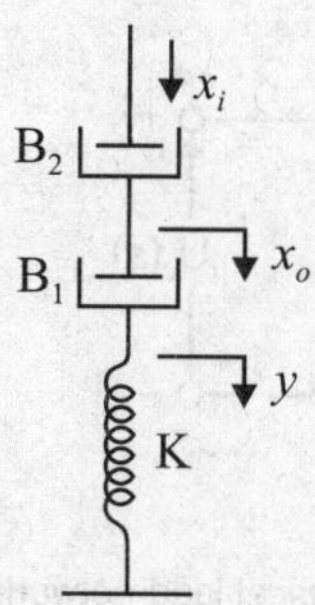
x_i
B_2
B_1
x_o
y
K

(c)

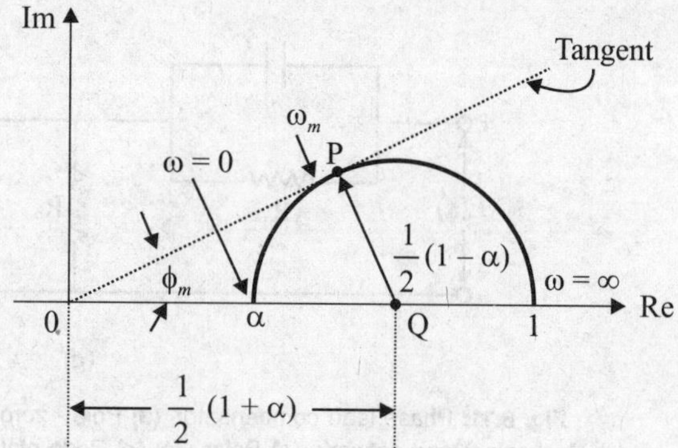
Im
Tangent
ω_m
P
ω = 0
1/2 (1 − α)
ϕ_m
ω = ∞
Re
0
α
Q
1
1/2 (1 + α)

(d)

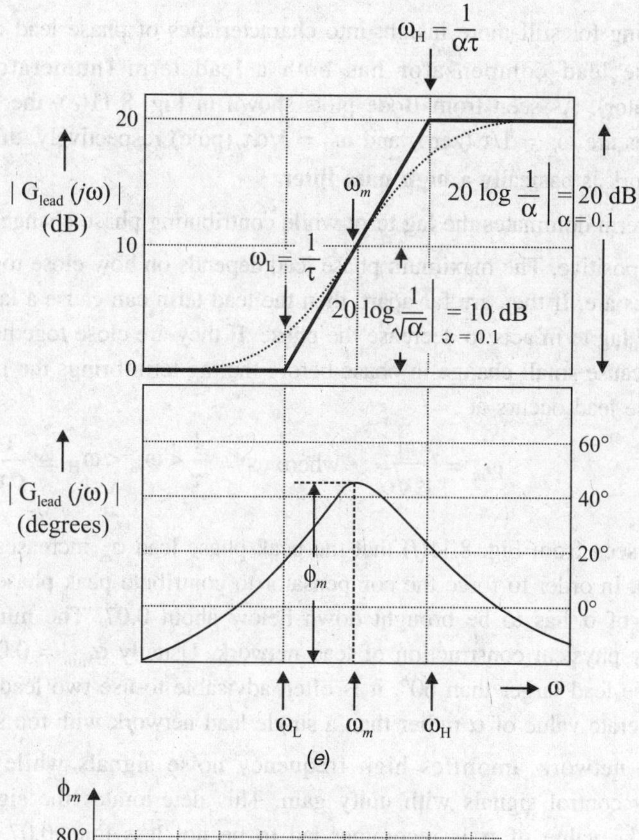
ω_H = 1/ατ
20
| G_lead (jω) | (dB)
ω_m
20 log 1/α |_{α = 0.1} = 20 dB
10
ω_L = 1/τ
20 log 1/√α = 10 dB
α = 0.1
0
60°
| G_lead (jω) | (degrees)
40°
20°
ϕ_m
0°
ω
ω_L
ω_m
ω_H

(e)

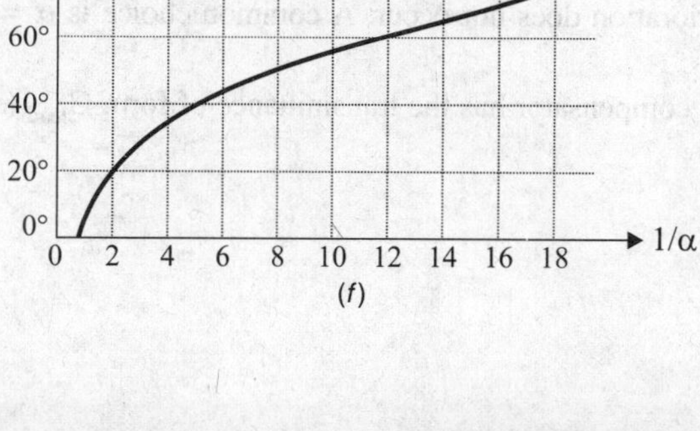
ϕ_m
80°
60°
40°
20°
0°
0
2
4
6
8
10
12
14
16
18
1/α

(f)

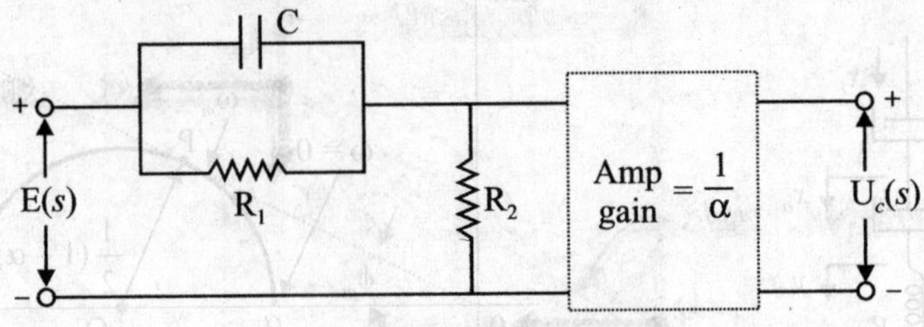

(g)

Fig. 8.11: Phase lead compensator, (*a*) Pole - zero diagram, (*b*) Electrical lead network, (*c*) Mechanical lead network, (*d*) Polar plot (*e*) Bode plot, (*f*) Peak change in phase versus 1/α plot, (*g*) Phase lead compensator combined with an amplifier

Note the following for still more insight into characteristics of phase lead compensator.

(*i*) The phase lead compensator has both a lead term (numerator) and a lag term (denominator). As seen from Bode plots shown in Fig. 8.11(*e*), the low and high corner frequencies are $\omega_L = 1/\tau$ (zero) and $\omega_H = 1/\alpha\tau$ (pole) respectively. dB plot shows that the lead network is basically a **high pass filter.**

(*ii*) The lead term dominates the lag term while contributing phase change and so only $\underline{|G(j\omega)}$ is always positive. The maximum phase lead depends on how close together the two corner frequencies are. If they are far apart, then the lead term can cause a large increase in phase before the lag term acts to decrease the phase. If they are close together, then the lead term can only cause small change in phase before the lag term brings the phase back to 0°. The peak phase lead occurs at

$$\omega_m = \frac{1}{\tau\sqrt{\alpha}} \quad \text{where } \omega_L = \frac{1}{\tau} < \omega_m < \omega_H = \frac{1}{\alpha\tau}$$

(*iii*) It can be seen from Fig. 8.11(*f*) that the peak phase lead ϕ_m increases as α decreases (1/α increases). In order to force the compensator to contribute peak phase lead more than 60°, the value of α has to be brought down below about 0.07. The minimum value of α is limited by physical construction of lead network. Usually $\alpha_{min} = 0.07$. In order to obtain peak phase lead larger than 60°, it is often advisable to use two lead networks in cascade with moderate value of α rather than a single lead network with too small α.

(*iv*) The lead network amplifies high frequency noise signals while processing the low frequency control signals with unity gain. This deteriorates the signal to noise ratio at output. The value of α is recommended to be not less than 0.07 so as to ensure that excessive deterioration does not occur. A common choice is α = 0.1.

(*v*) The phase lead compensator has the transmittance of form $G_{lead}(s) = K\frac{(s+a)}{(s+b)}$ with $b > a$.

Comparing it with (8.18), we have $a = 1/\tau$, $b = 1/\alpha\tau$ and K = 1. If lead compensator is inserted in unity feedback system configuration in series with uncompensated system G(s), then the root locus of $G_{lead}(s)$ G(s) with n poles and m zeros, will move to the left by $-\dfrac{(b-a)}{n-m}$ while preserving the original asymptote angles because $n - m$ remains unchanged. The lead compensator primarily influences system stability.

Phase lag-compensator

The phase lag-compensator also has a pole and a zero on negative real axis but pole remains closer to origin than the zero. As shown in Fig. 8.12(a), the compensator zero located at $s = -1/\tau$, is to the left of pole located at $s = -1/\beta\tau$; $\beta > 1$. The phase lag-compensator has the transmittance $G_{lag}(s)$ of form

$$G_{lag}(s) = \frac{s+\dfrac{1}{\tau}}{s+\dfrac{1}{\beta\tau}} = \beta\frac{\tau s+1}{\beta\tau s+1}; \qquad \beta > 1 \qquad \text{...(8.23)}$$

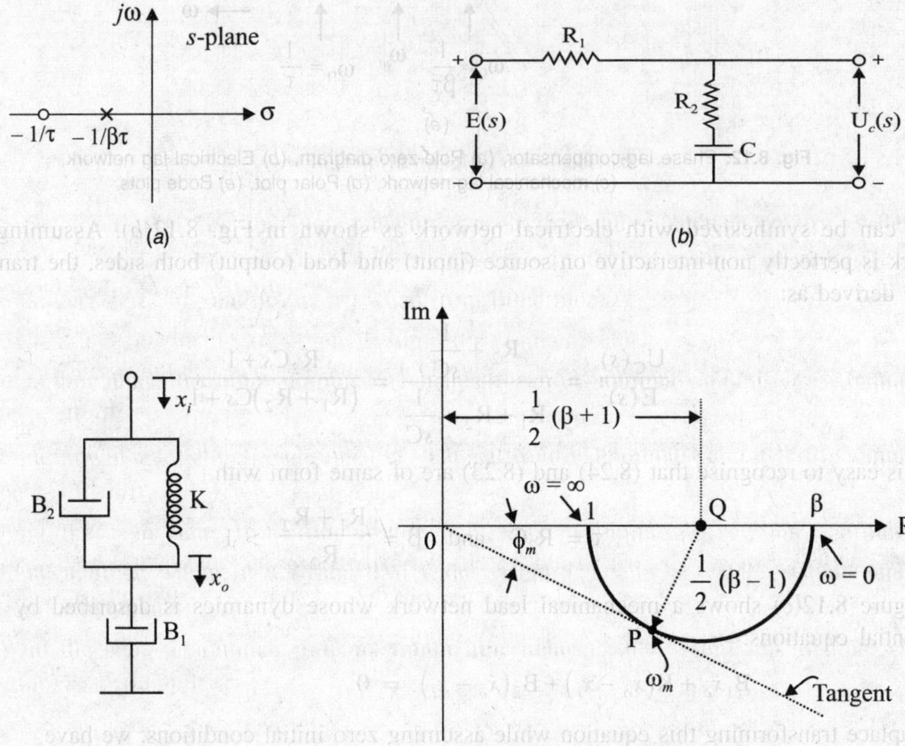

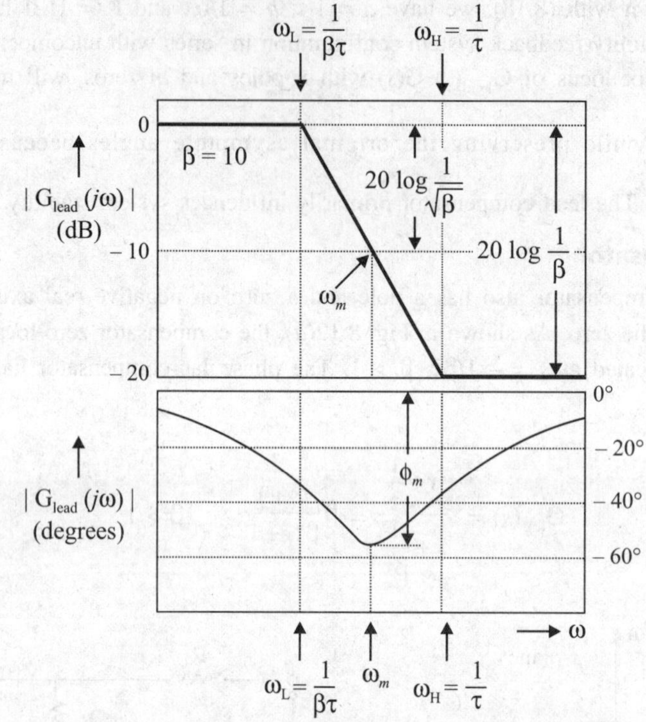

(e)

Fig. 8.12: Phase lag-compensator, (*a*) Pole-zero diagram, (*b*) Electrical lag network, (*c*) mechanical lag network, (*d*) Polar plot, (*e*) Bode plots

which can be synthesized with electrical network as shown in Fig. 8.12(*b*). Assuming that the network is perfectly non-interactive on source (input) and load (output) both sides, the transmittance can be derived as:

$$\frac{U_C(s)}{E(s)} = \frac{R_2 + \dfrac{1}{sC}}{R_1 + R_2 + \dfrac{1}{sC}} = \frac{R_2Cs + 1}{(R_1 + R_2)Cs + 1} \qquad ...(8.24)$$

It is easy to recognise that (8.24) and (8.23) are of same form with

$$\tau = R_2C \quad \text{and} \quad \beta = \frac{R_1 + R_2}{R_2} > 1$$

Figure 8.12(*c*) shows a mechanical lead network whose dynamics is described by following differential equations:

$$B_1\dot{x}_o + K(x_o - x_i) + B_2(\dot{x}_o - \dot{x}_i) = 0$$

Laplace transforming this equation while assuming zero initial conditions, we have

$$B_1s\, X_o(s) + K[X_o(s) - X_i(s)] + B_2s[X_o(s) - X_i(s)] = 0$$

or
$$\frac{X_o(s)}{X_i(s)} = \frac{\frac{B_2}{K}s+1}{\frac{(B_1+B_2)}{K}s+1} \qquad ...(8.25)$$

It is again easy to recognize that (8.25) is of form same as (8.23) with $\tau = \frac{B_2}{K}$ and $\frac{B_1+B_2}{K} = \beta > 1$. The polar plot for

$$G_{lag}(j\omega) = \beta\frac{j\omega\tau+1}{j\beta\omega\tau+1}; \quad \beta > 1$$

is shown in Fig. 8.12(*d*). Note $\lim_{\omega\to 0} G_{lag}(j\omega) = \beta\angle 0°$ and $\lim_{\omega\to\infty} G_{lag}(j\omega) = 1\angle 0°$. Since $\beta > 1$, $\angle G_{lag}(j\omega)$ is always negative. The polar plot of lag network is some what similar to that of lead network with the only differences that phase lead becomes phase lag and $\alpha < 1$ becomes $\beta > 1$. So, we shall only rewrite the significant relationships without detailed discussion. With the same reasoning as we did in case of lead network, the peak phase lag is related to the value of β as

$$\sin\phi_m = \frac{\beta-1}{\beta+1} \qquad ...(8.26)$$

Note that phase lag can also be expressed as negative degrees, for example 30° lag is also – 30° angle. If one wants to include minus sign in (8.26),

$$\phi_m = \sin^{-1}\frac{1-\beta}{1+\beta}$$

The frequency ω_m at which the lead compensator contributes maximum (peak) phase lag, is given as

$$\omega_m = \sqrt{\frac{1}{\tau}\times\frac{1}{\beta\tau}} \qquad ...(8.27)$$

which is again the **geometric mean of two corner frequencies** $1/\tau$ and $1/\beta\tau$.

The Bode plots (dB plot and phase plot) for $G_{lag}(j\omega) = \frac{j\omega\tau+1}{j\omega\beta\tau+1}$ is shown in Fig. 8.23(*e*). The corner frequencies can be identified as $\omega_L = \frac{1}{\beta\tau}$ and $\omega_H = \frac{1}{\tau}$, ω_L and ω_H are low and high corner frequencies respectively. The dB plot shows unity gain (0 dB) at low frequencies and the gain at high frequencies approaches 20 $\log_{10}$ $(1/\beta)$. We can verify this from dB plot where $\beta = 10$. The gain decreases at the constant rate of – 20 dB/dec from 0 dB at $\omega_L = \frac{1}{\beta\tau}$ and attains gain of 20 log $(1/\beta)$ at $\omega_H = \frac{1}{\tau}$ whereafter the gain remains constant. The phase begins to change from 0° at $\omega = 0$ towards – 90° due to compensator pole but the compensator zero contributes the positive phase change bringing back the final phase to 0° as $\omega \to \infty$.

Note the following significant points:

(*i*) dB plot of Fig. 8.12(*e*) shows that the lag-compensator is basically a **low pass filter**.

(*ii*) The dominant lag term forces the phase to decrease from 0° at $\omega = 0$ to $-90°$ as ω approaches ∞, but before it reaches $-90°$ the lead term acts to increase the phase. The final phase change at high frequencies is therefore again zero. The maximum change in phase is determined by how close the two corner frequencies are. If they are far apart, then the lag term causes large reduction in phase before lead term begins to increase the phase. If they are close together, then the lag term can cause only a small change in phase before the lead term brings the phase back to 0°. The peak phase lag occurs at

$$\omega_m = \frac{1}{\tau\sqrt{\beta}} \text{ where } \omega_L = \frac{1}{\beta\tau} < \omega_m < \omega_H = \frac{1}{\tau}.$$

(*iii*) The typical value of β in design is 10.

(*iv*) The lag-compensator processes the low frequency control signals with unity gain while attenuating high frequency noise signals. So, the signal to noise ratio is improved.

(*v*) The phase lag-compensator has the transmittance of form $G_{lag}(s) = K\dfrac{s+a}{s+b}$ with $a > b$. Comparing it with (8.23), we have $a = 1/\tau$, $b = 1/\beta\tau$ and $K = 1$. Note that the PI compensator is a special case of lag-compensator for which the constant b is zero. The lag-compensator can be fabricated using a passive RC circuit as shown in Fig. 8.12(*b*), while the PI compensator requires an active device, perhaps including operational amplifier as shown in Fig. 8.8(*c*). For non-zero b, lag-compensator does not increase the system type number. However, steady state performance can be improved over that of the uncompensated feedback system.

(*vi*) In contrast to the lead compensator, the lag-compensator causes the root locus of lag compensated system $G_{lag}(s)$ $G(s)$ with n poles and m zeros to move to the right by $(a - b)/(n - m)$ while preserving the original asymptote angles because $n - m$ remains unchanged. An usual design approach is to select the ratio a/b to equal the factor by which the error coefficient is to be increased in order to improve steady state error performance. As the ratio a/b approaches infinity, the cascade lag-compensator tends to become cascade PI compensator.

A summary of gain and phase characteristics of phase lead and phase lag-compensators is given in Table 8.2 for easy and quick reference.

TABLE 8.2: Characteristics of phase lead and phase lag elements

Type of compensator		*Phase lead*	*Phase lag*
Transmittance		$G_{lead}(j\omega) = \dfrac{j\omega\tau + 1}{j\omega\alpha\tau + 1}$; $\alpha < 1$	$G_{lag}(j\omega) = \dfrac{j\omega\tau + 1}{j\omega\beta\tau + 1}$; $\beta > 1$
Low (ω_L) and high (ω_H) corner frequencies		$\omega_L = \dfrac{1}{\tau}$, $\omega_H = \dfrac{1}{\alpha\tau}$	$\omega_L = \dfrac{1}{\beta\tau}$; $\omega_H = \dfrac{1}{\tau}$
Gain (dB)	Low frequency	0	0
	High frequency	20 log $(1/\alpha)$; $\alpha < 1$	20 log $(1/\beta)$; $\beta > 1$
Phase (degrees)	Low frequency	0°	0°
	High frequency	0°	0°
Maximum phase change (ϕ_m)		$\phi_m = \sin^{-1}\dfrac{1-\alpha}{1+\alpha}$	$\phi_m = \sin^{-1}\dfrac{1-\beta}{1+\beta}$
Frequency at maxium phase change (ω_m)		$\omega_m = \dfrac{1}{\tau\sqrt{\alpha}}$	$\omega_m = \dfrac{1}{\tau\sqrt{\beta}}$
Magnitude at $\omega = \omega_m$		$20 \log\left(\dfrac{1}{\sqrt{\alpha}}\right)$	$20 \log\left(\dfrac{1}{\sqrt{\beta}}\right)$
Signal to Noise ratio		Worsens	Improves
Effect on steady state accuracy and relative stability		Relative stability improves	Steady state accuracy improves

Lag-lead compensator

A cascade lead compensator improves relative stability while a cascade lag-compensator improves steady state accuracy. The lag-lead compensator combines the best attributes of both the compensators. The lag-lead compensator has two poles and two zeros. Such a compensator increases the order of original system by two, provided pole-zero cancellation does not occur in the compensated system. The pole-zero diagram is shown in Fig. 8.13(*a*).

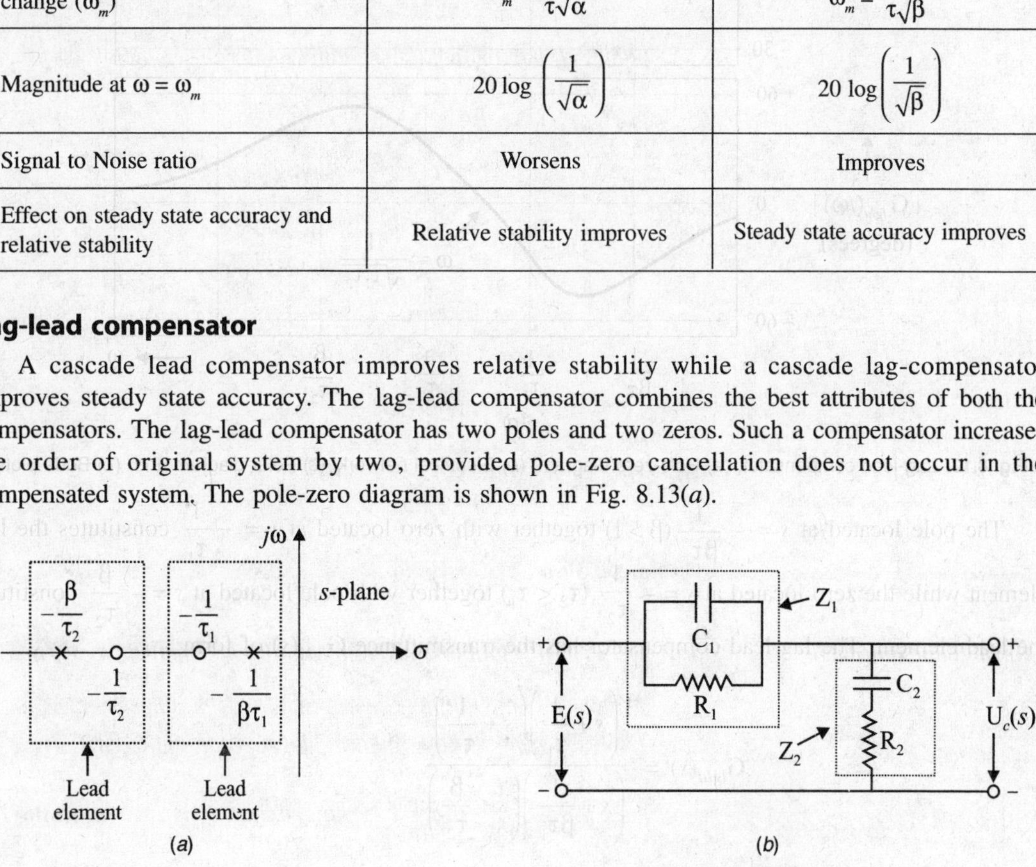

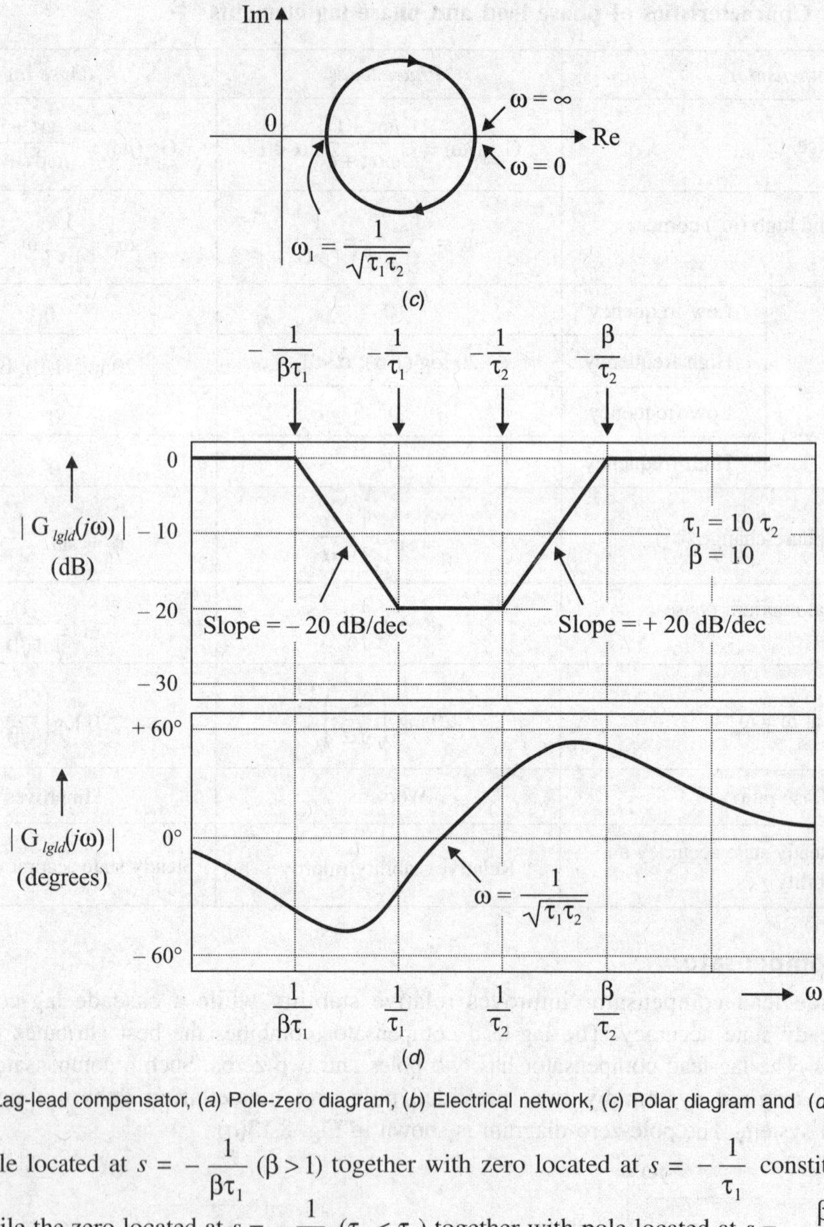

Fig. 8.13: Lag-lead compensator, (*a*) Pole-zero diagram, (*b*) Electrical network, (*c*) Polar diagram and (*d*) Bode plots

The pole located at $s = -\dfrac{1}{\beta\tau_1}$ $(\beta > 1)$ together with zero located at $s = -\dfrac{1}{\tau_1}$ constitutes the lag element while the zero located at $s = -\dfrac{1}{\tau_2}$ $(\tau_2 < \tau_1)$ together with pole located at $s = -\dfrac{\beta}{\tau_2}$ constitute the lead element. The lag-lead compensator has the transmittance $G_{lgld}(s)$ of form

$$G_{lgld}(s) = \frac{\left(s+\dfrac{1}{\tau_1}\right)\left(s+\dfrac{1}{\tau_2}\right)}{\left(s+\dfrac{1}{\beta\tau_1}\right)\left(s+\dfrac{\beta}{\tau_2}\right)}$$

$$= \frac{(\tau_1 s+1)(\tau_2 s+1)}{(\beta\tau_1 s+1)\left(\dfrac{\tau_2}{\beta}s+1\right)}; \quad \beta > 1 \text{ and } \tau_1 > \tau_2 \qquad \text{...(8.28)}$$

An electrical lag-lead network is shown in Fig. 8.13(*b*) for which the transmittance can be derived as:

$$\frac{U_c(s)}{E(s)} = G_{lgld}(s) = \frac{z_2}{z_1+z_2} = \frac{(sC_1R_1+1)(sC_2R_2+1)}{(sC_1R_1+1)(sC_2R_2+1)+sR_1C_2} \qquad \text{...(8.29)}$$

Notice that (8.29) and (8.28) are of same form. Comparing these we have

$$\tau_1 = R_1C_1; \quad \tau_2 = R_2C_2 \quad \text{and} \quad R_1C_1 + R_2C_2 + R_1C_2 = \frac{\tau_2}{\beta} + \beta\tau_1$$

The sinusoidal transmittance of lag-lead compensator is

$$G_{lgld}(j\omega) = \frac{(j\omega\tau_1+1)(j\omega\tau_2+1)}{(j\omega\beta\tau_1+1)\left(j\omega\dfrac{\tau_2}{\beta}+1\right)} \qquad \text{...(8.30)}$$

and the polar plot of $G_{lgld}(j\omega)$ is shown in Fig. 8.13(*c*). For $0 < \omega < \omega_1$, the network acts as a lag element contributing negative phase angle and for $\omega_1 < \omega < \infty$, the network acts as a lead element contributing positive phase angle. The frequency ω_1 at which $\angle G_{lgld}(j\omega) = 0$, is given as:

$$\omega_1 = \frac{1}{\sqrt{\tau_1\tau_2}} \qquad \text{...(8.31)}$$

The lag-lead compensator with transmittance (8.30) has Bode plots (dB plot and phase plot) as shown in Fig. 8.13(*d*) for $\tau_1 = 10\tau_2$ and $\beta = 10$. The compensator exhibits unity gain (0 dB) at low frequencies and also at high frequencies.

8.6 ROOT LOCUS DESIGN

In chapter 5, we saw that the root locus graphically displays the information about location of closed loop poles from the knowledge of the locations of open loop poles and zeros as some parameter, usually gain is varied from zero to infinity. The location of closed loop poles, in turn, reveals both the transient response and stability information. It has been learnt that the root locus can be quickly sketched to get a general idea of the changes in transient response generated by changes in gain. A typical control system design begins by simply adjusting the gain to determine closed loop poles to achieve the transient response as specified. Keep a note that gain adjustment design is limited to only those responses that exist along the root locus. It is also possible in some cases that the system may not be stable for the values of the gain essential for desired performance. If the gain adjustment fails to meet the design objectives, the compensators must be added in order to modify the original root locus.

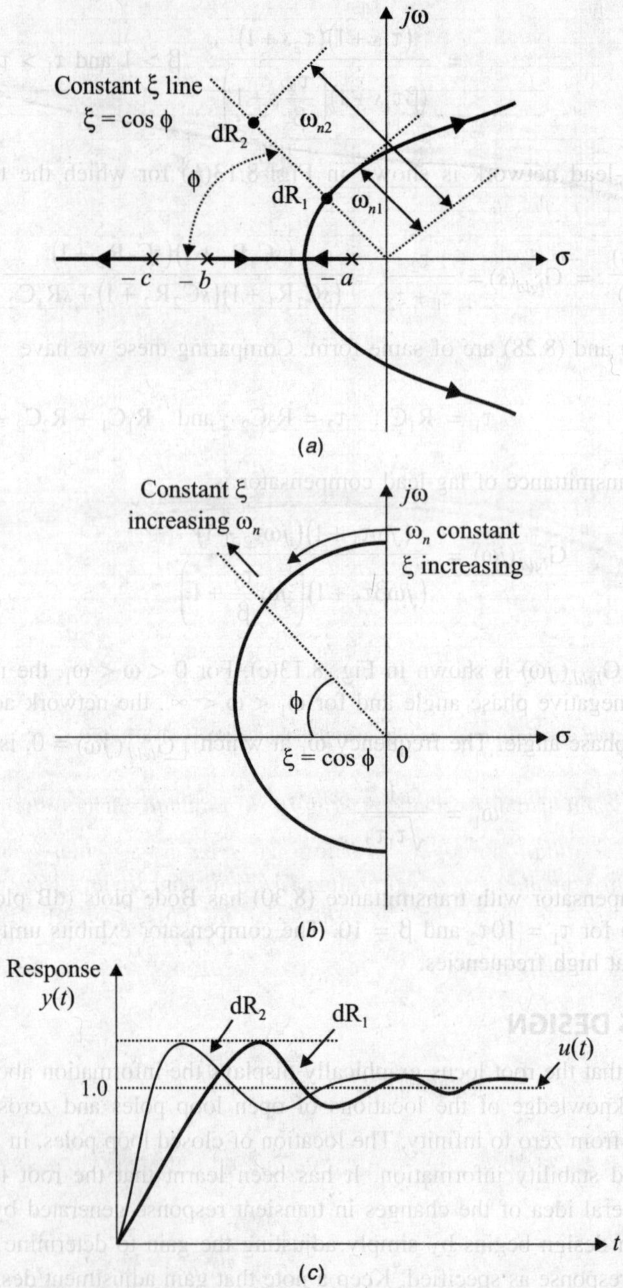

Fig. 8.14: Gain adjustment design with root locus, (*a*) Showing requirement of reshaping the root locus, (*b*) Changes in ξ and ω_n and (*c*) Typical unit step responses for dominant roots dR_1 and dR_2

The effect of compensator on reshaping the root locus, is learnt. Then poles and zeros of the compensator are chosen so as to force the root locus to pass through the desired location of dominant closed loop poles. Each compensator (lead, lag, lag-lead) is inserted in series with uncompensated system to produce the dominant roots. It is possible to evolve the general design rules for each compensator assuming that the dominant roots exist. The idea is to select the compensator poles and zeros such that the reasonable design emerges. Fig. 8.14(*a*) shows a typical root locus for a third order system with poles typically located at say $s = -a$, $s = -b$ **and** $s = -c$. Let the design requirement be specified in terms of peak overshoot and settling time. If the designer tries to fix ξ for specified peak overshoot, it can be done so by drawing a constant ξ line radialy outward from origin making an angle of $\phi = \cos^{-1}\xi$. Let this line intersect the original root locus at a point shown as dR_1 (dominant root) in Fig. 8.14(*a*). The gain value can also be determined for the dominant root dR_1, but now ω_n gets fixed and so only the settling time which is inversely proportional to the product $\xi\omega_n$, also gets fixed. Having achieved peak overshoot as specified, if the designer further intends to speed up the response (reduce settling time) while preserving the peak overshoot, then it is not possible by mere gain adjustment. Note that the solution lies in increasing undamped natural frequency from ω_{n1} corresponding to dominant root dR_1 to ω_{n2} corresponding to dominant root dR_2. The dominant root dR_2 required to achieve specified settling time does not lie on root locus. The constant ξ with increasing ω_n (straight lines radially outward from origin) and constant ω_n with increasing ξ (semi-circle) loci are shown in Fig. 8.14(*b*) for a quick reminder in the current perspective, else it has been discussed in detail in chapter three. The current problem of making the response faster while preserving the peak overshoot is illustrated in terms of unit step response corresponding to dominant roots dR_1 and dR_2 in Fig. 8.14(*c*). It is easy to predict that for constant ξ, the rise time and settling time are inversely proportional to ω_n.

In order to solve this problem, one may conceive an idea of replacing the existing system by yet another that has the root locus passing through dominant root dR_2. But this will, perhaps, be expensive and counter productive. So, the designer prefers to compensate the system with additional poles and zeros so as to force the root locus of compensated system to pass through the desired dominant root dR_2 for some value of gain.

Cascade compensator design for improving steady state performance

Consider a compensator design problem for a system exhibiting satisfactory transient response while seeking improvement in steady state response. Such a design requirement of improving steady state error while almost preserving the transient behaviour can be met in two ways: One, using PI compensator with transmittance of form $G_{PI}(s) = \dfrac{K(s+a)}{s}$ thus adding a pole at $s = 0$ and a zero at $s = -a$. The added open loop pole at origin, increases the type number of system by 1 and therefore, for a stable design, the steady state error reduces to zero. Two, using lag-compensator with transmittance of form $G_{lag}(s) = \dfrac{K(s+a)}{(s+b)}$ which does not involve a pure integrator thereby it cannot as such force the steady error to be zero but yields a measurable reduction in steady state error.

The PI Compensator, although reduces the steady state error to zero but it must be fabricated with active network, such as operational amplifier. The lag-compensator can reduce the steady state error to an acceptably small value but not zero. Yet, lag-compensator is advantageous in the sense that it can be fabricated with less expensive passive network requiring no additional power supply.

PI compensator design

The PI compensator improves steady state performance while preserving the transient response of original uncompensated system. To illustrate this, let us consider an example of uncompensated system $G(s) = \dfrac{1}{(s+1)(s+2)(s+6)}$ together with compensator $G_{PI}(s) = \dfrac{K(s+a)}{s}$ as shown in

Fig. 8.15(*a*). The root locus for uncompensated system is shown in Fig. 8.15(*b*). Let us assume that the desired transient response specification ($\xi = 0.5$, $\omega_n = 2$) is generated by dominant root, dR for $K \cong K_{dR}$ as shown in Fig. 8.15(*b*). The current design objective is to improve steady state error with transient specifications being preserved. One way to reduce steady state error to zero, is to add a pole at origin ($s = 0$) whereby system type number increases. This is demonstrated in Fig. 8.15(*c*) that the root locus no longer passes through the dominant root, dR as angle contribution of open loop poles at dR is no longer 180°.

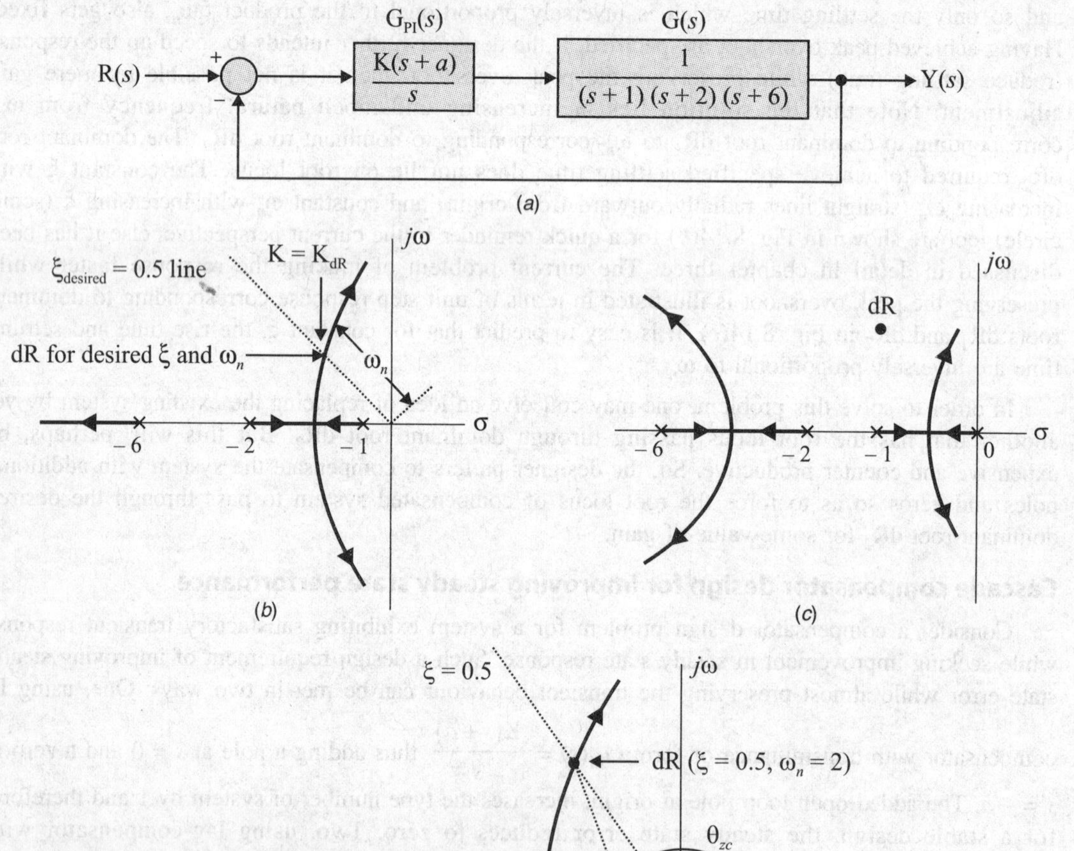

Fig. 8.15: PI compensator design, (*a*) System G(*s*) with cascade PI compensator $G_{PI}(s)$, (*b*) Root locus of uncompensated system G(*s*), (*c*) Root locus of system with a pole at origin (Pure integral compensation) (*d*) Root locus of PI compensated system

In order to ensure that the root locus continues to pass through dR, let us add a zero close to origin at $s = -a$ in addition to the pole at $s = 0$. For small a, (say typically 0.1) $\theta_{pc} \cong \theta_{zc}$, dR is preserved while system type number increases. This is demonstrated in Fig. 8.15(*d*). Notice that the value of K at dR also remains approximately same as before compensation because the ratio of lengths from compensator pole and compensator zero to dR, is almost unity.

Note that the compensator zero will be placed very close to compensator pole at origin so that original root locus remains almost same as that after compensation. The PI compensator added to system will create one more closed loop pole alright but will not pose too much problem as the new closed loop pole being close enough to the zero, will cause pole-zero cancellation.

Lag-compensator design

PI compensator design involves placing a pole at origin in order to improve steady state performance. Then a zero is placed very close to the pole at origin on negative real axis, to ensure that the satisfactory transient response of uncompensated system, remains unaffected. Such a design strategy requires an active integrator. An advantage of using passive network, can be derived by moving the compensator pole and zero to the left. The result is lag compensation. The lag-compensator does not increase the system type number. However, the steady state error performance can be considerably improved. Let us demonstrate this improvement with an example of type 1 system.

$$G(s) = \frac{K\prod^{i}(s+z_i)}{s\prod^{j}(s+p_j)}$$

The error coefficient of this uncompensated system is

$$K_{v_{uc}} = \lim_{s\to 0} sG(s) = \frac{K\prod^{i}(z_i)}{\prod^{j}(p_j)} \quad ...(8.32)$$

A cascade lag-compensator has the transmittance of form

$$G_{\text{lag}}(s) = \frac{s+1/\tau}{s+1/\beta\tau}\ ;\ \beta > 1$$

For a lag-compensated system, the error coefficient is

$$K_{v_c} = \lim_{s\to 0} sG_{lag}(s)\,G(s) = \beta K\frac{\prod^{i}(z_i)}{\prod^{j}(p_j)} \quad ...(8.33)$$

We can substitute (8.32) into (8.33) to get

$$K_{v_c} = \beta K_{v_{uc}} \quad ...(8.34)$$

The error coefficient increases by the factor β. One design approach is to select ratio β to equal the factor by which the error coefficient is to be increased. If β approaches infinity, the cascade lag- compensator becomes cascade PI. However, large β poses the problem of fabricating a reliable RC circuit with a pole and a zero widely separated. The effect of lag-compensator on transient response of system, is demonstrated while taking an example of uncompensated system

$G(s) = \dfrac{K}{(s+\sigma_1)(s+\sigma_2)(s+\sigma_3)}$; $\sigma_1 < \sigma_2 < \sigma_3$ whose typical root locus in shown in Fig. 8.16(*a*). The point dR shown on root locus, is assumed to be dominant root (a requirement to satisfy transient response).

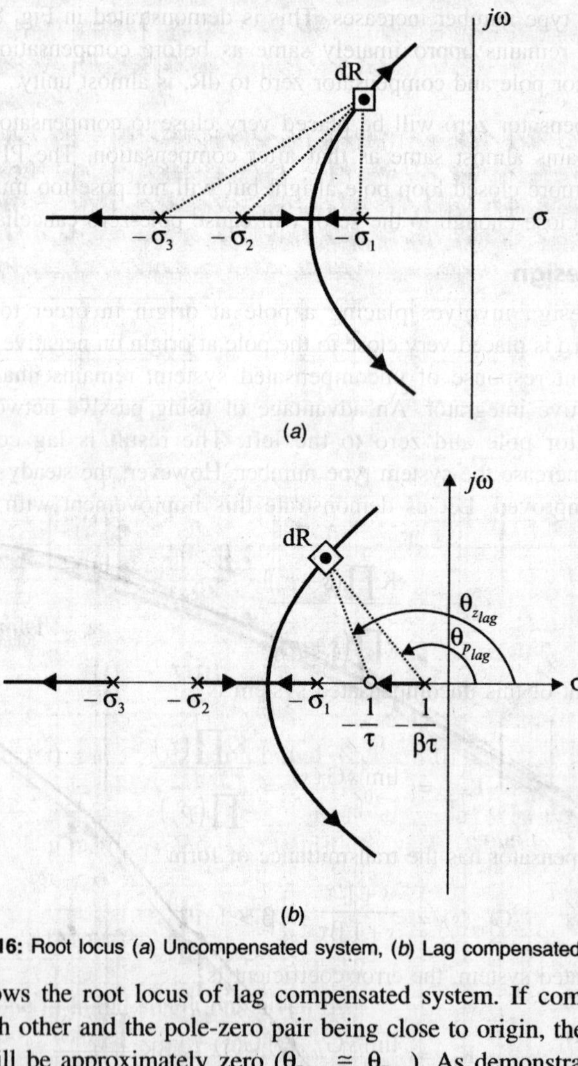

Fig. 8.16: Root locus (*a*) Uncompensated system, (*b*) Lag compensated system

Figure 8.16(*b*) shows the root locus of lag compensated system. If compensator pole and zero are placed close to each other and the pole-zero pair being close to origin, then the angle contributed by the compensator will be approximately zero ($\theta_{z_{lag}} \cong \theta_{p_{lag}}$). As demonstrated in Fig. 8.16(*b*), the dominant root dR remains almost unchanged. The gain K at dR, also remains almost same before and after compensation; the phasor lengths drown from compensator pole to dR and compensator zero to dR are almost same. Thus lag-compensator contributes steady state performance improvement as dictated by (8.34) while keeping the transient response almost unchanged.

Assuming that the simple gain adjustment can meet the transient response specifications, the design procedure for lag-compensator to improve steady state response, can be put in the form of simple steps as follows:

(*i*) Sketch root locus for uncompensated system. Calculate ξ and ω_n for location of dominant poles (dR) as dictated by specified transient response.

(*ii*) Evaluate gain K at dominant root, dR and corresponding error coefficient.

(*iii*) Determine the factor by which the error coefficient is required to be increased. Choose β slightly larger than this ratio.

(*iv*) Locate compensator zero very close to origin and then compensator pole at $s = -1/\beta\tau$ with β known in step (*iii*).

The very idea behind choosing compensator pole-zero pair close to origin is to ensure that $|\theta_{z_{lag}} - \theta_{p_{lag}}| < 5°$ so that there results a negligibly small angular change at dR and satisfactory transient response remains preserved.

As an illustrative example of designing the lag-compensator, consider a unity feedback system with forward transmittance

$$G(s) = \frac{K}{s(s+7)}$$

The design objectives are as follows:

(*i*) peak overshoot ≤ 15%

(*ii*) the steady state error for unit ramp input is to improve by a factor of 20.

Step (*i*): The root locus for G(*s*) is shown in Fig. 8.17.

$$e^{-\pi\xi/\sqrt{1-\xi^2}} = 0.15 \text{ gives } \xi = 0.517 \text{ and } \cos^{-1} 0.517 = 58.9°$$

A line radially outward from origin, making an angle of 58.9° with negative real axis, is also shown in Fig. 8.17.The point of intersection of this line with root locus is dominant root dR as shown in Fig. 8.17.

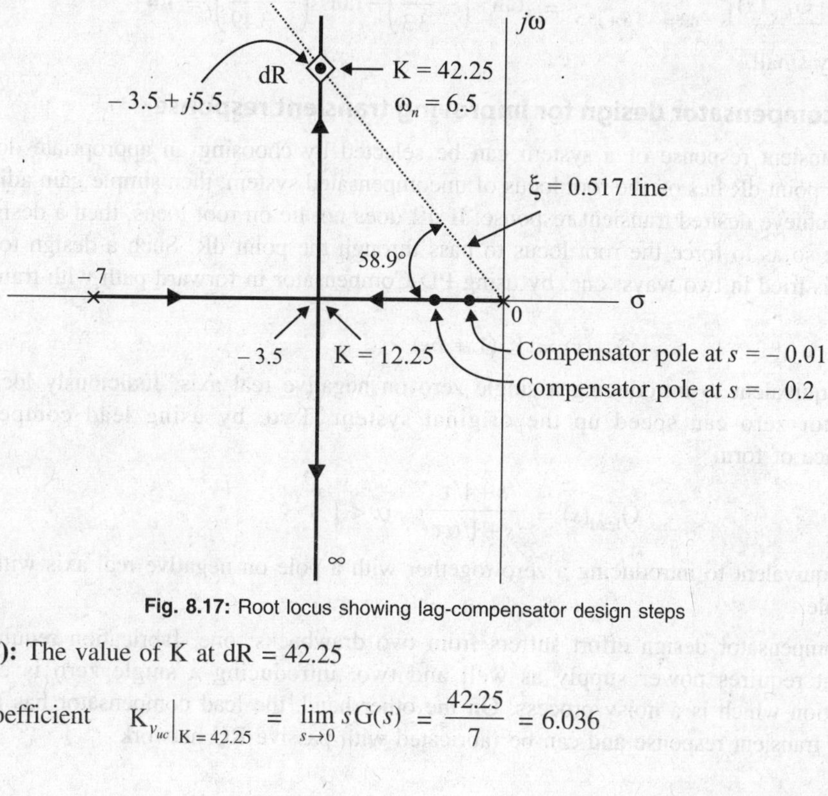

Fig. 8.17: Root locus showing lag-compensator design steps

Step (*ii*): The value of K at dR = 42.25

error coefficient $K_{v_{uc}}\big|_{K=42.25} = \lim_{s\to 0} sG(s) = \frac{42.25}{7} = 6.036$

and steady state error, $$e(\infty)_{uc} = \frac{1}{K_v} = 0.1657$$

Step (*iii*): Choose $\beta = 20$. **Note** that steady state error is required to improve by a factor of 20.

Step (*iv*): Now the compensator pole-zero pair is to be so chosen that the compensated value of error coefficient be

$$K_{v_c} = 20\ K_{v_{uc}} = 120.7$$

If we choose compensator zero at $s = -0.2$, then compensator pole must be located at $s = -(0.2/20) = -0.01$ in compliance with (8.34).

So, $G_{lag}(s) = \dfrac{s+0.2}{s+0.01}$ and lag compensated forward transmittance

$$G_{lag}(s)\ G(s) = \frac{42.25(s+0.2)}{s(s+0.01)(s+7)}$$

Comment: The lag compensated system yields error coefficient

$$K_{v_c} = \underset{s \to 0}{\text{lt}}\ sG_{lag}(s)\ G(s) = \frac{42.25 \times 0.2}{0.01 \times 7} = 120.7$$

shows 20 fold improvement while almost preserving the location of dominant root dR. The angle contributed by $G_{lag}(s)$ at dR

$$\angle G_{lag}(s)\Big|_{s\,=\,dR\,=\,-3.5+j5.5} = \tan^{-1}\left(-\frac{5.5}{3.3}\right) - \tan^{-1}\left(-\frac{5.5}{3.49}\right) = 1.4^\circ$$

is negligibly small.

Cascade compensator design for improving transient response

The transient response of a system can be selected by choosing an appropriate dominant root (dR). If the point dR lies on the root locus of uncompensated system, then simple gain adjustment will suffice to achieve desired transient response. If dR does not lie on root locus, then a design effort has to be made so as to force the root locus to pass through the point dR. Such a design to reshape the root locus is tried in two ways: one, by using PD Compensator in forward path with transmittance of form

$$G_{PD}(s) = K\ (1 + bs)$$

which is equivalent to introducing a single zero on negative real axis. Judiciously locating such a compensator zero can speed up the original system. Two, by using lead compensator with transmittance of form

$$G_{lead}(s) = \frac{s+1/\tau}{s+1/\alpha\tau}\ ; \quad \alpha < 1$$

which is equivalent to introducing a zero together with a pole on negative real axis with zero to the right of pole.

PD compensator design effort suffers from two drawbacks: one, fabrication requires an active circuit that requires power supply as well and two, introducing a single zero is equivalent to differentiation which is a noisy process. On the other hand, the lead compensator has the ability of improving transient response and can be fabricated with passive RC network.

PD compensator design

If the design effort involving only gain adjustment fails to achieve the desired transient response then adding a single zero to the forward path (PD compensator) can help in reshaping the root locus so as to meet the design goals. The design idea is to place a zero on negative real axis at a point such that the modified root locus passes through the design point (dominant root dR) dictated by desired transient response specifications. Adding a zero reduces the number asymptotes in root locus plot by unity, changes the centroid and asymptotic angles, consequently the entire root locus gets reshaped. In order to demonstrate this consider the uncompensated system.

$$G(s) = \frac{K}{s(s+2)(s+4)}$$

whose root locus is shown in Fig. 8.18(*a*). The centroid $\sigma_{uc} = -2$, the asymptotic angles = $\pm 60°$, $180°$, breakaway point at $s = -0.85$ and imaginary axis crossing points at $s = \pm j2.83$ are also shown in Fig. 8.18(*a*).

The system is assumed to be operating with damping ratio $\xi = 0.707$. The dominant roots dR as shown in Fig. 8.18(*a*) lies at $s = -0.8 \pm j0.8$ for gain K = 5.5. Figure 8.18(*b*) shows the root locus of PD compensated system. A compensator zero is added at $s = -1$. The centroid changes from $\sigma_{uc} = -2$ to $\sigma_c = -2.5$. The asymptotic angles are changed to $\pm 90°$. The breakaway point shifts from $s = -0.85$ to $s = -2.9$ and the compensated root locus exhibits no imaginary axis crossing point. The uncompensated system is unstable for $K \geq 48$ while the compensated system is stable for all K > 0. The relocated dominant root shown as dR* in Fig. 8.18(*b*) lies at $s = -2.75 \pm j2.75$ for K = 10.95.

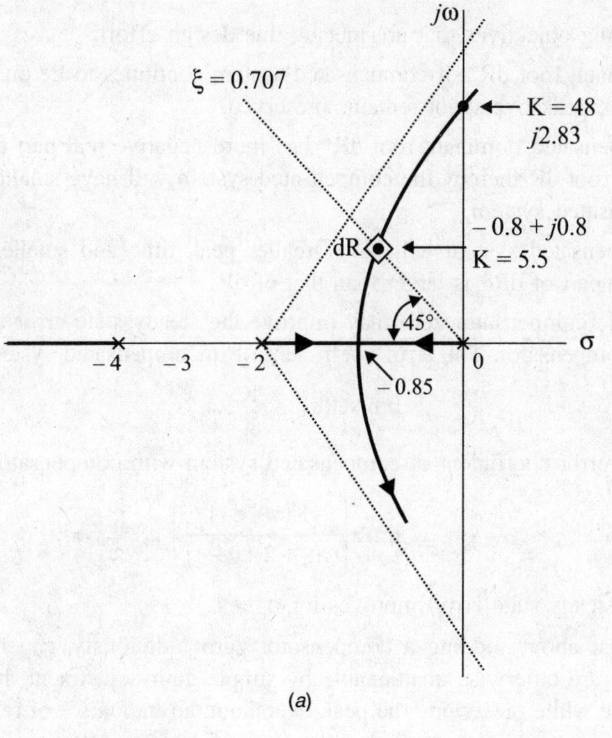

(*a*)

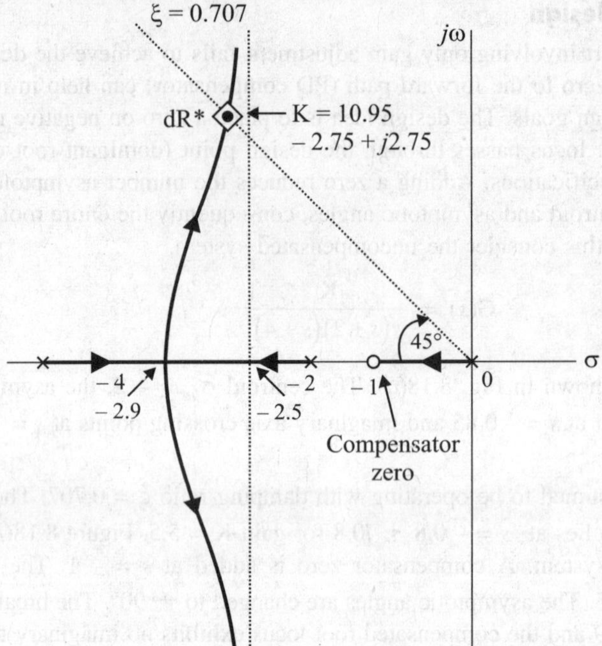

Fig. 8.18: Root locus (*a*) Uncompensated system (*b*) PD compensated system

Note the following objectives that are met by this design effort.

(*i*) The dominant root dR* of compensated system continues to lie on constant $\xi = 0.707$ line thereby the peak overshoot remains preserved.

(*ii*) The compensated dominant root dR* has more negative real part than the uncompensated dominant root dR thereby the compensated system will have smaller settling time than the uncompensated system.

(*iii*) The compensated system will have smaller peak time and smaller rise time because the imaginary part of dR* is larger than that of dR.

(*iv*) Adding the compensator zero may improve the steady state error also even without using **PI/Lag**-compensator. The error coefficient of uncompensated system is

$$K_{v_{uc}} = \lim_{s\to 0} sG(s) = \frac{K}{8}$$

while the error coefficient of compensated system with compensator zero at $s = -\sigma_c$ is

$$K_{v_c} = \lim_{s\to 0} s\left[\frac{K(s+\sigma_c)}{s(s+2)(s+4)}\right] = \frac{K}{8}\cdot\sigma_c$$

Thus the steady state error improves for $\sigma_c > 1$.

As discussed just above, adding a compensator zero judiciously, can meet transient response specifications which are otherwise unattainable by simple gain adjustment. It is possible to quicken the transient response while preserving the peak overshoot, an indicative of relative stability. Now we present the PD compensator design routine in the form of steps as follows:

(*i*) Sketch root locus for uncompensated system and translate the transient response specifications into location of dominant root.

(*ii*) Evaluate the sum of angles from open loop poles and zeros of uncompensated system to the dominant root found in step (*i*).

(*iii*) The difference between 180° and the angle found in step (*ii*) must be the angle contribution of compensator zero. Trigonometry is then used to find the location of compensator zero.

To demonstrate numerically, consider the system with transmittance

$$G(s) = \frac{K}{s(s+4)(s+6)}$$

for which the PD compensator is to be designed such that the compensated system exhibits 12% peak overshoot and has settling time equal to 1 second.

Let us proceed with the design using the steps listed above as follows:

Step (*i*): The root locus for uncompensated system is shown in Fig. 8.19(*a*). Translating the peak overshoot of 12% into damping ratio ξ, we have

$$e^{-\pi\xi/\sqrt{1-\xi^2}} = 0.12$$

or

$$\xi = 0.56$$

The constant $\xi = 0.56$ loci is shown in Fig. 8.19(*a*) making an angle of $\cos^{-1}(0.56) \cong 56°$ with negative real axis and the point at which this line intersects the root locus is marked as dominant root dR (– 1.8 + *j*1.2). The settling time of uncompensated system at *d*R is $t_s = \frac{4}{\xi\omega_n} = \frac{4}{1.8} = 2.22$ for K = 36.04. The calculation based on second order approximation appears to be valid because the third root for K = 36.04 will lie well-beyond $s = -6$ to its left. The specified settling time $t_s = 1$ sec. requires that the real part of compensated dominant root *d*R* be – 4 and imaginary part $\omega_d^* = 4 \tan 56° \cong 5.93$. So, the compensated dominant root must be located at $s = -4 \pm j5.93$ as shown in Fig. 8.19(*a*).

Step (*ii*): The sum of angles from open loop poles and zeros of uncompensated system to *d*R* = – [124° + 90° + 70°] = – 284° and angle contribution required from compensator zero = 284° – 180° = 104°.

Step (*iii*) (locating compensator zero): The angle contribution of 104° by compensator zero dictates that the location of compensator zero ($s = -\sigma_c$), will be somewhere to the right of point $s = -4$. Using trigonometry, σ_c can be calculated as

$$\frac{5.93}{4-\sigma_c} = \tan(180° - 104°)$$

or

$$\sigma_c = 2.52$$

So, adding a compensator zero at $s = -2.52$ meets the design goals. Compensated system will exhibit peak overshoot of 12% because dR* lies on constant $\xi = 0.56$ loci and settling time of 1 second because real part of dR* is – 4. The PD compensated system has a new pole-zero function

$$G_{PD}(s)\,G(s) = \frac{K(s+2.52)}{s(s+4)(s+6)}$$

whose root locus is shown in Fig 8.9(*b*). The value of gain K is 44.45 at dominant root dR*. The centroid of asymptotes is at $s = -3.75$, the asymptotic angles are $\pm 90°$ and the break away point approximately lies at $s = -4.85$.

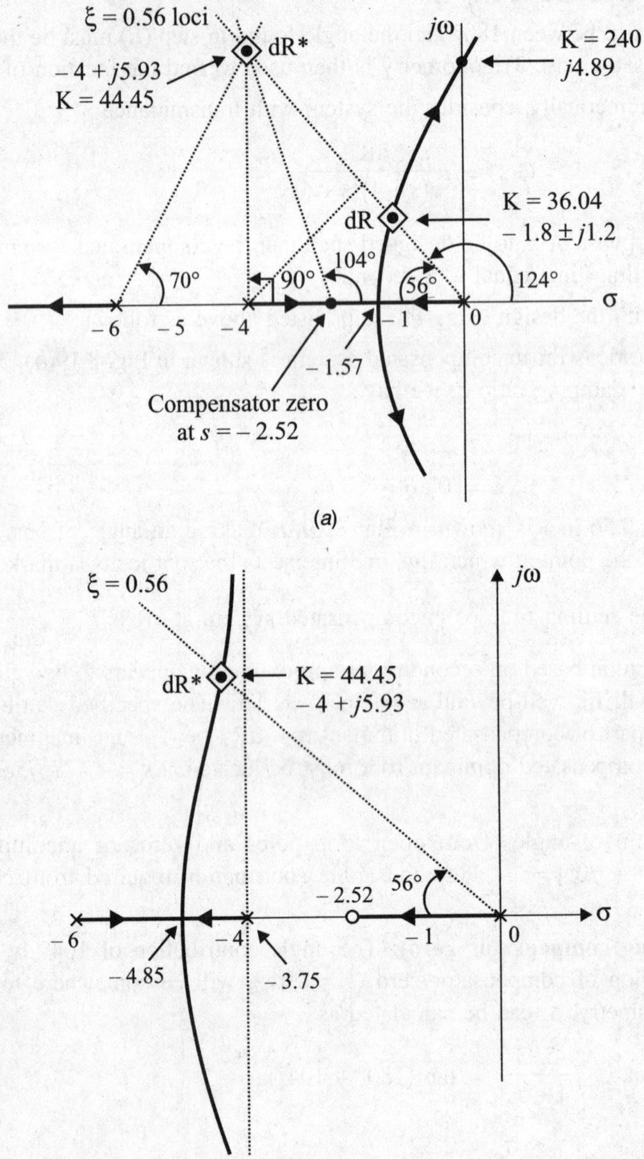

Fig. 8.19: Root locus (*a*) Uncompensated system $G(s) = \dfrac{K}{s(s+4)(s+6)}$ (*b*) PD compensated system

Lead compensator design

We have already seen that PI compensator can be approximated by passive lag network. Similarly, PD compensator can be approximated by passive lead network. The passive lead network cannot generate a single zero, instead results in a pole-zero pair. However, if pole is placed far away from origin than the zero, the net angle contributed by pole-zero pair remains still positive and thus approximates an equivalent single zero of PD compensator. The passive lead network is advantageous over an active PD compensator in the sense that additional power supply is not needed and the noise generated by differentiation is reduced. The disadvantage of passive lead network is that it does not reduce the number of root locus brances tending to cross imaginary axis and moving into RHP. The PD compensator while adding a single zero and no pole, tends to reduce the number of root locus branches tending to move into RHP.

Now we shall present the entire design procedure of lead compensator in the form of small steps for ease in design and then demonstrate the implementation of these steps with an example. The steps are as follows:

Step (*i*): Sketch the root locus of uncompensated system and **translate the performance specifications into the location of dominant root**.

Step (*ii*): If the dominant root lies on root locus then the gain adjustment alone can achieve it. If dominant root does not lie on root locus, then lead compensator is required to be designed. Find the **sum of angles from uncompensated system's poles and zeros to the dominant root**. The difference between 180° and sum of angles must be the angular contribution required of compensator. If the angle is quite large, it is advisable to use two or more lead networks rather than a single one.

Step (*iii*): Determine the location of pole-zero pair of lead compensator. Note that there is no unique location of such a pole-zero pair. As a guide line, the pole-zero pair should be so selected that it yields largest possible α so that the additional gain required of amplifier is as small as possible. Recall that

$$G_{lead}(s) = \frac{s + 1/\tau}{s + \dfrac{1}{\alpha\tau}}; \quad \alpha < 1$$

One possible procedure to select pole-zero pair of lead compensator is demonstrated in Fig 8.20. The point dR shown in Fig. 8.20 is dominant root. Join two points dominant root dR and origin. Also draw a horizontal line dR — A. Bisect the angle formed by two lines dR — A and dR — O. The bisector is dR — C as shown in Fig. 8.20. Draw two lines dR — B and dR — D making an angle of $\pm\phi/2$ with bisector dR — C. ϕ is the angle, the lead compensator is expected to contribute. The intersections of dR — B and dR — D with negative real axis provide the location of pole-zero pair of lead compensator. Note that we have presented only an idea of selecting compensator pole and zero, these could be arbitrarily selected keeping only one thing in mind that the dominance of dominant root found in step (*i*) must be approximately preserved or second order approximation must remain justified.

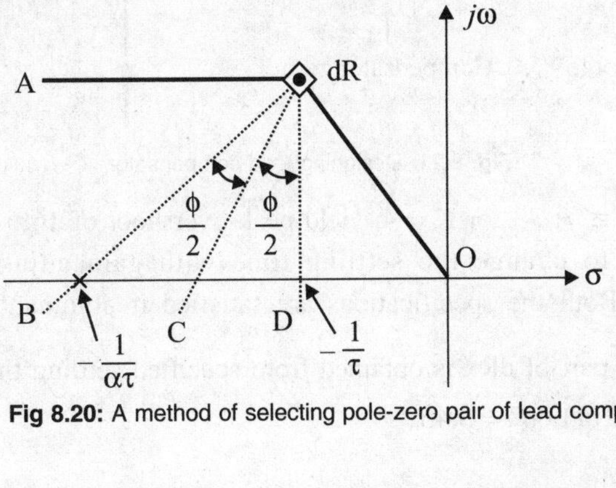

Fig 8.20: A method of selecting pole-zero pair of lead compensator

Step (*iv*): Determine the gain value of lead compensated system at dominant root dR. Having designed the lead compensator, check whether the design objectives have been achieved. If not, repeat the design by adjusting the compensator pole-zero pair until all design specifications are met. If the dominant root found in step (*i*) does not indeed remain dominant, then the effort by trial and error, has to be made to relocate the pole-zero pair of compensator.

As a numerical example, consider the system $G(s) = \dfrac{K}{s(s+2)}$ for which a lead compensator is to be designed to meet the specifications:

peak over shoot $\leq 16\ \%$

settling time ≤ 2 sec.

The step by step procedure to design the lead compensator is as follows:

Step (*i*): The root locus for uncompensated system $G(s) = \dfrac{K}{s(s+2)}$ is shown in Fig. 8.21. The centroid lies at $s = -1$, asymptotic angles are $\pm\ 90°$ and break away point lies at $s = -1$ translating peak over shoot of 16% into damping ratio ξ, we have $e^{-\pi\xi/\sqrt{(1-\xi^2)}} = 0.16$ or $\xi \cong 0.5$. A constant $\xi = 0.5$ line making an angle of $\cos^{-1}(\xi) = \cos^{-1}(0.5) = 60°$ with negative real axis, is also shown.

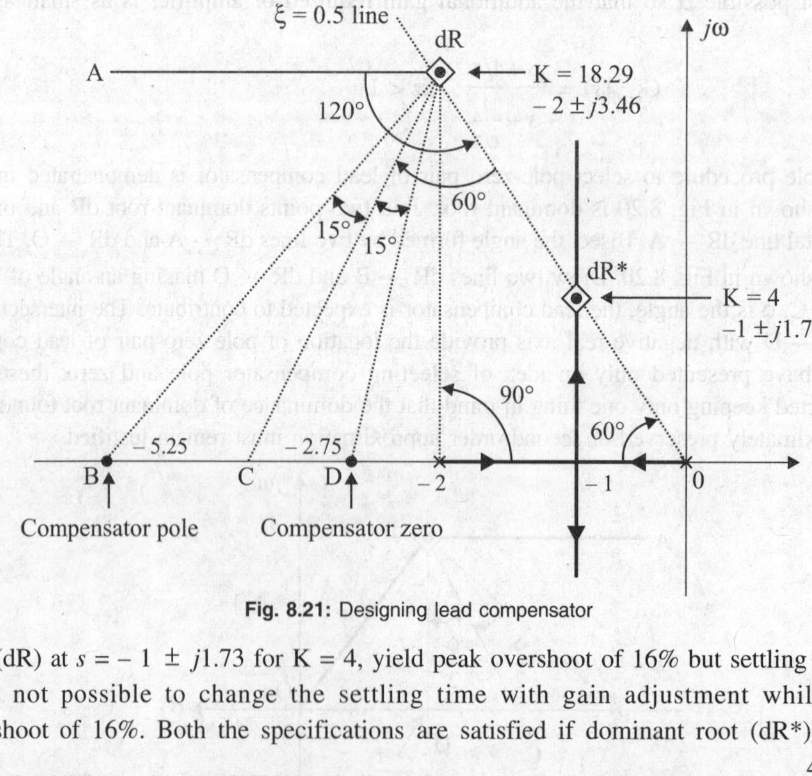

Fig. 8.21: Designing lead compensator

The roots (dR) at $s = -1 \pm j1.73$ for K = 4, yield peak overshoot of 16% but settling time is 4 sec. Now it is not possible to change the settling time with gain adjustment while preserving peak overshoot of 16%. Both the specifications are satisfied if dominant root (dR*) is located at $s = -2 \pm j3.46$. The real part of dR* is obtained from specified settling time as $\dfrac{4}{\text{real part}} = 2$ and imaginary part of dR* is 2 tan 60° = 3.46.

Step (*ii*): Sum of angles from poles at $s = 0$ and $s = -2$ to dR* is – [120° + 90°] = – 210°. Angle contribution required of compensator, $\phi = 210° - 180° = 30°$.

Step (*iii*): Using the procedure as explained the compensator zero at located at $s = -2.75$ and pole at $s = -5.25$

Step (*iv*): The value of gain is K = 18.29.

The new pole-zero function of lead compensated system is

$$G_{lead}(s)\ G(s) = \frac{18.29(s+2.75)}{s(s+2)(s+5.25)}$$

PID controller design

PID compensator design effort is made in case of requirement of simultaneous improvement in transient as well as steady state performance. It is combination of PD and PI compensator design and involves placing two zeros on negative real axis and a pole at origin. One LHP zero together with a pole at origin, designs PI compensator while the other zero corresponds to the design of PD compensator. The entire PID design is tried in following steps:

(*i*) PD compensator is designed to meet the transient response specification. This involves locating a zero on negative real axis and adjusting the loop gain K. The sub-steps of this design has already been explained and demonstrated with an example.

(*ii*) Then PI compensator is designed to achieve required steady state performance. This involves placing a pole at origin and a zero close to origin. The sub-steps of this design has also been explained and demonstrated with an example.

To demonstrate the design steps of PID compensator, let us consider the unity feedback system with transmittance

$$G(s) = \frac{K}{(s+1)(s+4)}$$

The desired performance indices are

$$\text{Peak time} = 1.047 \text{ sec}$$
$$\text{Damping ratio} = 0.8$$
$$\text{Steady state error} = 0 \text{ for step input.}$$

Step (*i*) Designing PD compensator: The root locus for uncompensated system is shown in Fig. 8.22(*a*).

The constant $\xi = 0.8$ line making an angle of $\cos^{-1} 0.8 = 36.87°$ with negative real axis is also shown. The dominant root satisfying specified damping ratio is marked as dR. dR is found at $s = -2.5 \pm j1.875$ with gain K = 5.76. Translating peak time $t_p = 1.047$ into imaginary part of a new dominant root, we have

$$\text{Im[dominant root]} = \omega_d = \frac{\pi}{t_p} \cong 3$$

and real part of dominant root is

$$\text{Re[dominant root]} = \sigma = \frac{3}{\tan(36.87°)} = 4$$

Thus, the new dominant root marked dR* in Fig. 8.22(*a*) at $s = -4 \pm j3$, satisfies specified $t_P = 1.047$ and also $\xi = 0.8$ since dR* lies on constant $\xi = 0.8$ loci.

Sum of angles of open loop poles of uncompensated system to dR* = – 90° – 143.13° = – 233.13°. Angle contribution required of compensator zero = 233.13° – 180° = 53.13°. Let the compensator zero be located at $s = -\sigma_c$, then using trigonometry σ_c can be determined as

$$\frac{3}{\sigma_c - 4} = \tan 53.13° \quad \text{or} \quad \sigma_c = 6.25$$

This is demonstrated in Fig. 8.22(*b*). The transmittance of PD compensator is $G_{PD}(s) = (s + 6.25)$ and that of PD compensated system is

$$G_{PD}(s)\ G(s) = \frac{3.32(s+6.25)}{(s+1)(s+4)}$$

Note that the dominant root dR* satisfies transient specifications for gain K = 3.32.

Step (*ii*) (Designing PI compensator): Having designed the PD compensator and having satisfied the transient specification, we now design PI compensator to additionally meet steady state error requirement while preserving the transient response achieved as specified. In order to reduce the steady state error to zero for step input, we place a pole at origin and a zero close to origin. Let us choose the PI compensator

$$G_{PI}(s) = \frac{s+0.1}{s}$$

The PID compensator has transmittance

$$G_{PID}(s) = \frac{(s+0.1)(s+6.25)}{s}$$

and the new pole-zero function of PID compensated system is

$$G_{PID}(s)\ G(s) = \frac{3.32(s+0.1)(s+6.25)}{s(s+1)(s+4)}$$

The change in gain value due PI compensation, is assumed to be negligibly small. The rough shape of root locus for PD compensated system is shown in Fig. 8.22(*c*) and that for PID compensated system is shown in Fig. 8.22(*d*).

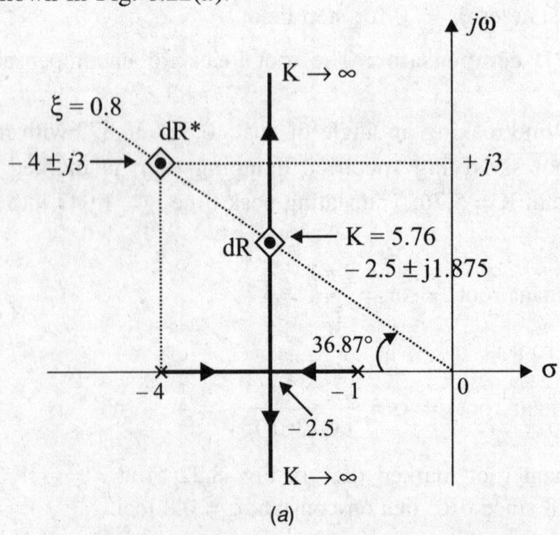

(*a*)

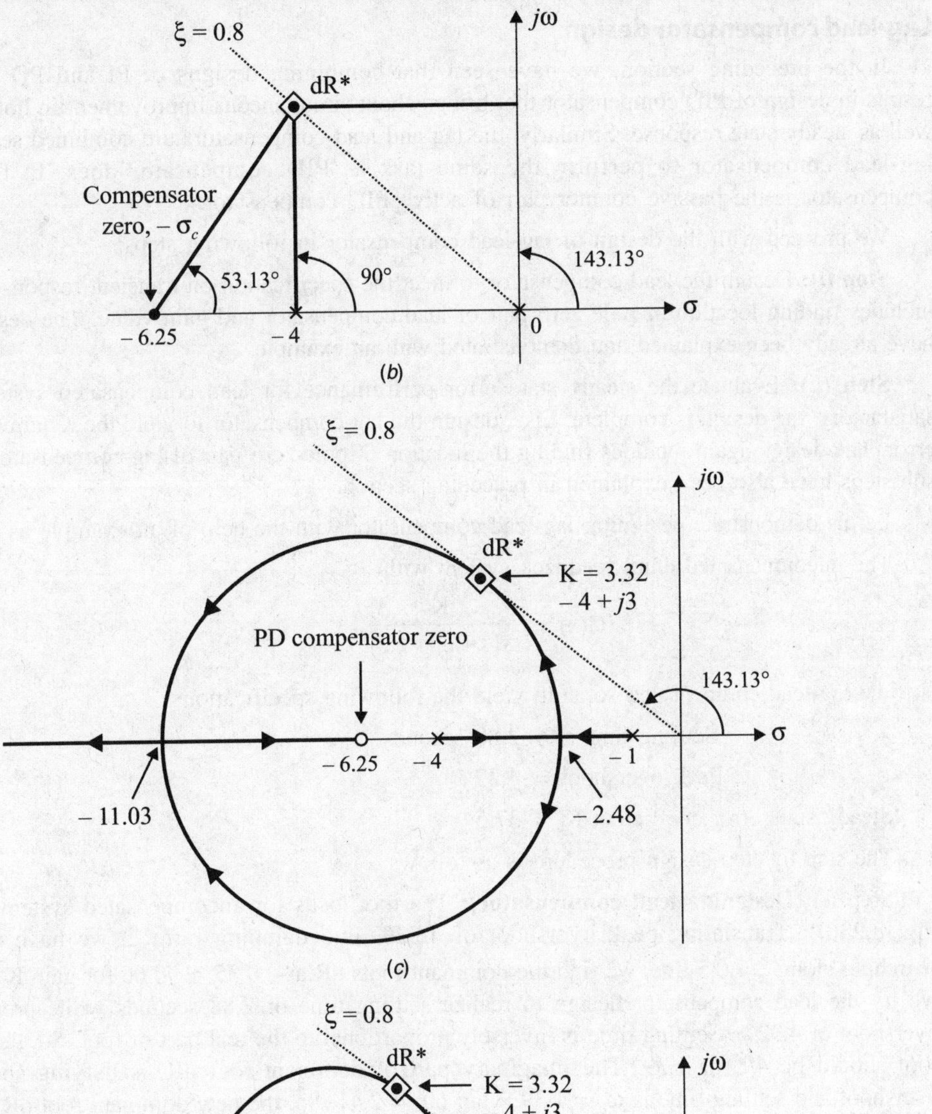

Fig. 8.22: Designing PD compensator (*a*) Root locus for uncompensated system, (*b*) Finding location of PD compensator zero, (*c*) Root locus for PD compensated system and (*d*) Rough root locus (not to the scale) for PID compensated system

Lag-lead compensator design

In the preceding section, we have seen that combining designs of PI and PD compensator, results in design of PID compensator that brings about simultaneous improvement in both transient as well as steady state response. Similarly, the lag and lead compensators are combined serially to form lag-lead compensator to perform the same task as PID compensator does. In fact, lag-lead compensator is the passive counter part of active PID compensator.

We proceed with the design of lag-lead compensator in following steps:

Step (*i*): Design the lead compensator to meet the specification on transient response. The design includes finding location of pole-zero pair of lead compensator and gain value. The design sub-steps have already been explained and demonstrated with an example.

Step (*ii*): Evaluate the steady state error performance for lead compensated system. If this is satisfactory, the design is complete. Else, design the lag-compensator to yield the required steady state error. The design again includes finding the location of pole-zero pair of lag-compensator. The design sub-steps have also been explained in preceding section.

Let us demonstrate designing lag-lead compensator with the help of an example as follows:

The uncompensated unity feedback system with

$$G(s) = \frac{K}{s(s+1)(s+3)}$$

is to be lag-lead compensated so as to yield the following specifications:

Settling time t_s = 2.88 seconds

Peak overshoot = 4.32%

Steady state error coefficient K_v = 37.5

The step by step design procedure is as follows:

Step (*i*) (Designing lead compensator): The root locus for uncompensated system is shown in Fig. 8.23(*a*). Translating peak overshoot of 4.32% into damping ratio ξ, we have $\xi = 0.5$ and searching along $\xi = 0.5$ line, we find the dominant roots dR at $-0.35 \pm j0.66$ for gain K = 1.11. Next we try the lead compensator design to realize settling time of 2.88 seconds while preserving peak overshoot of 4.32%. Settling time is inversely proportional to the real part of root. So, the real part of root should be $4/t_s$ or 1.39. The imaginary part of dominant root dR* satisfying specified peak overshoot and settling time is $\omega_d = 1.39 \times \tan 60° = 2.41$. So, the new dominant root dR* is found at $s = -1.39 \pm j\,2.41$. Now we design the lead compensator. Sum of angles from uncompensated poles to dR* = – 120° – 80° – 55° = – 255°. The angular contribution required from the lead compensator = 255° – 180° = 75°. We arbitrarily select the lead compensator zero at $s = -1.5$. To locate the lead compensator pole, join the compensator zero at $s = -1.5$ to dR* and draw a line making an angle of 75° as demonstrated in Fig 8.23(*a*). This line intersects negative real axis at $s = -12.95$. This is location of lead compensator pole.

The new pole-zero function of lead compensated system is $G_{lead}(s)\ G(s)$

$$= \frac{97.47(s+1.5)}{s(s+1)(s+3)(s+12.95)} \text{ and error coefficient } K_v = \lim_{s\to 0} sG_{lead}(s)\,G(s) = 3.76.$$

The specified K_v = 37.5. So, lag-compensator must be designed to improve K_v by almost ten fold.

Step (*ii*) (Designing lag-compensator): We arbitrarily choose lag-compensator pole at $s = -0.01$, which then dictates the lag-compensator zero to be located at $s = -0.1$ so as to improve K_v from 3.76 to 37.5 (almost ten fold). This gives

$$G_{lag}(s) = \frac{(s+0.1)}{(s+0.01)}$$

and lag-lead compensated system has open loop transfer function

$$G_{lead}(s)\ G_{lag}(s)\ G(s) = \frac{97.47(s+0.1)(s+1.5)}{s(s+0.01)(s+1)(s+3)(s+12.95)}$$

with the assumption that lag-compensator when serially added to lead compensated system, brings about negligibly small changes in location of dominant root dR* and corresponding gain value. The root locus of lead compensated system is shown in Fig. 8.23(*b*) and that of lag-lead compensated system in Fig. 8.23(*c*). These are not drawn to the scale. These will only provide an insight into what has been done to reshape the root locus and how the root locus of compensated system is forced to pass through the dominant roots.

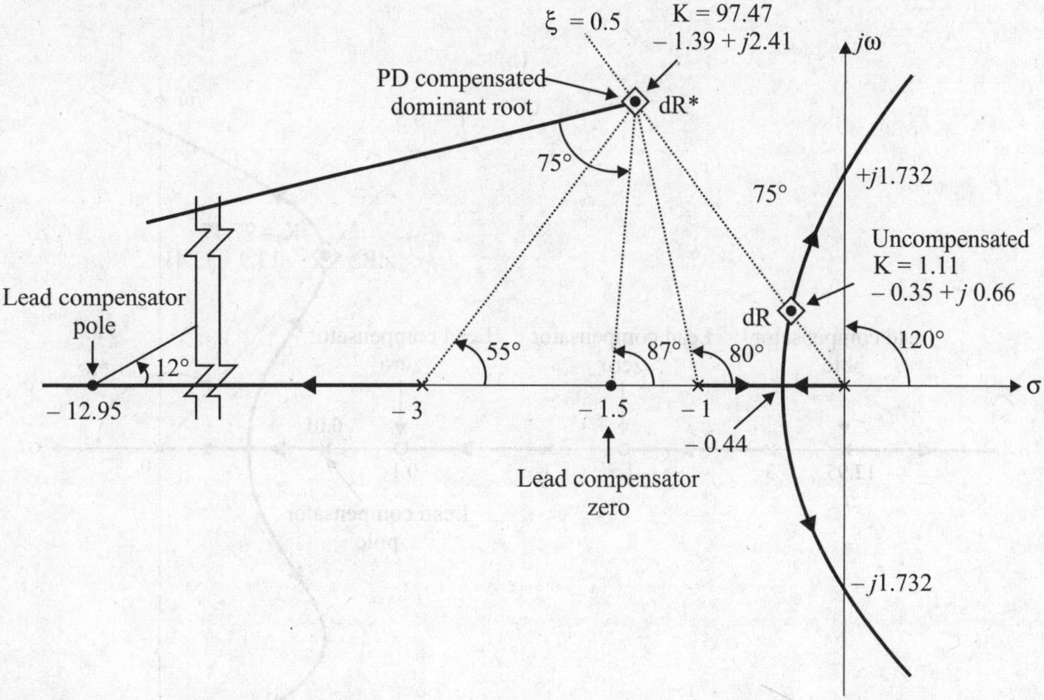

(*a*)

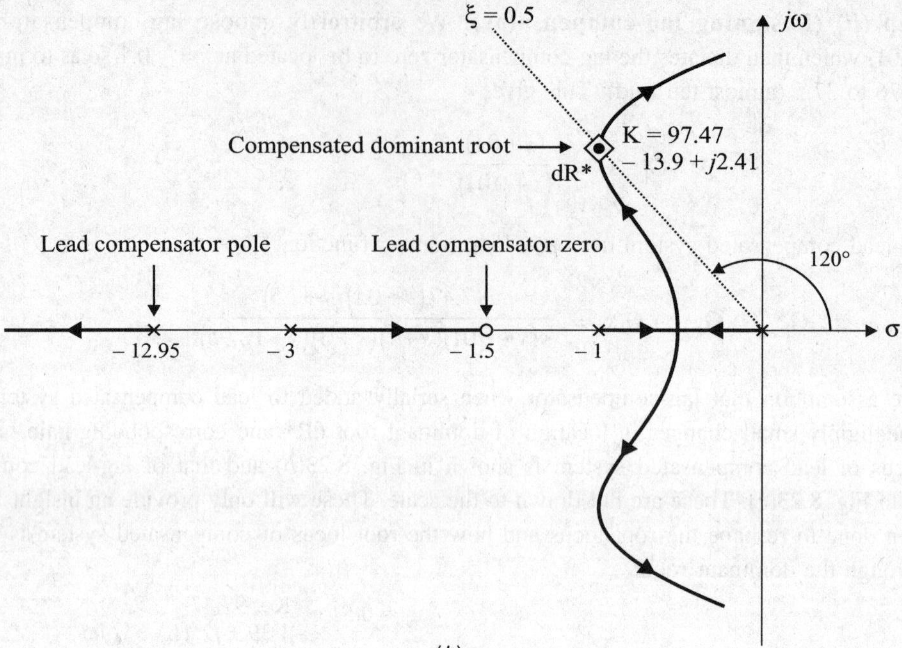

(b)

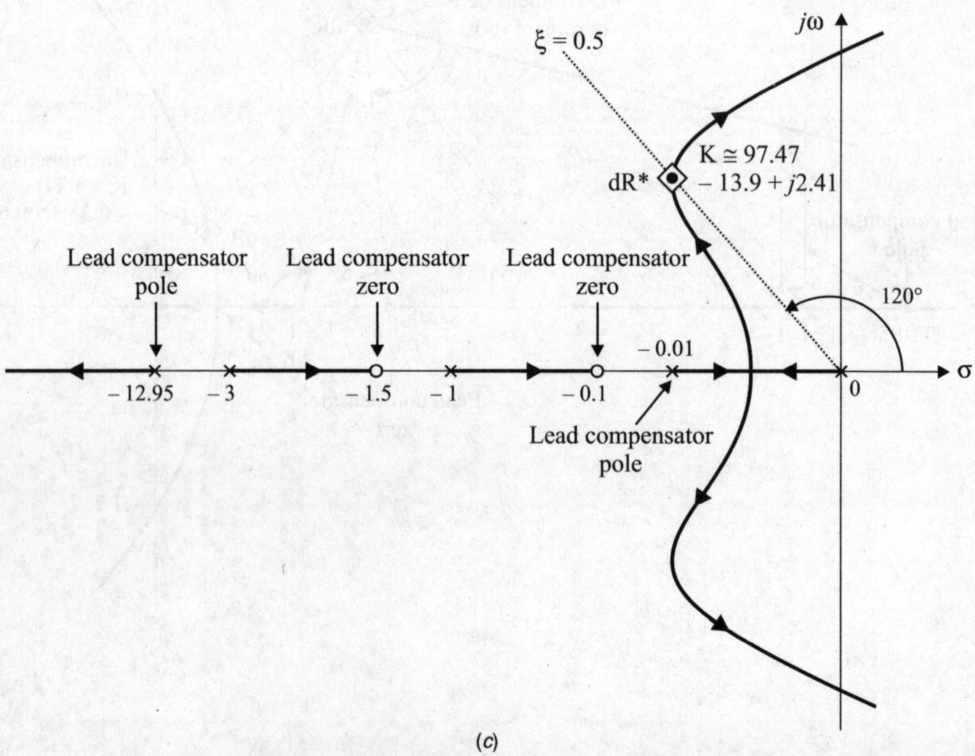

(c)

Fig. 8.23: Designing lag-lead compensator (*a*) Root locus (not to the scale) of uncompensated system and design of lead compensator, (*b*) Root locus (not to the scale) for lead compensated system and (*c*) Root locus (not to the scale) for lag-lead compensated system

8.7 FREQUENCY RESPONSE DESIGN

In preceding section, we concentrated on compensator design to improve the system performance as viewed from the perspective of the root locus. In this section, we use Bode plots (a frequency response technique) to parallel the root locus design. The compensator design via frequency response, is not as intuitive as the root locus, the direct control over time behaviour of system is lost. We have already learned in chapter 6 that phase margin and gain crossover frequency are two significant measures of stability from frequency response view point. The phase margin is related to damping ratio ξ of dominant roots while gain crossover frequency is related to undamped natural frequency, ω_n of dominant roots. Thus, it is evident that the time response, root locus and frequency response are interrelated, although each one follows entirely different computational routine. For an ease and quick reference, the formulas based on second order approximation interrelating the design specifications of time domain and frequency domain, are put together as follows:

(*i*) Phase margin $\phi_m = \tan^{-1}\left[\dfrac{2\xi}{\left(\sqrt{(4\xi^4+1)}-2\xi^2\right)^{1/2}}\right] \cong 100\xi$ ...(8.35)

(*ii*) Gain crossover frequency $\omega_{gc} = \omega_n\sqrt{\sqrt{(4\xi^4+1)}-2\xi^2}$ ...(8.36)

(*iii*) Bandwidth $\omega_b = \omega_n\left[(1-2\xi^2)+\sqrt{4\xi^4-4\xi^2+2}\right]^{1/2}$ ...(8.37(*a*))

$$= \frac{4}{t_s\xi}\left[(1-2\xi^2)+\sqrt{4\xi^4-4\xi^2+2}\right]^{1/2}\ ; \quad t_s = \text{Settling time} \qquad ...(8.37(b))$$

$$= \frac{\pi}{t_p\sqrt{1-\xi^2}}\left[(1-2\xi^2)+\sqrt{4\xi^4-4\xi^2+2}\right]^{1/2}\ ; \quad t_p = \text{Peak time} \qquad ...(8.37(c))$$

(*iv*) Resonant Peak $M_r = \dfrac{1}{2\xi\sqrt{1-\xi^2}}$ ...(8.38)

(*v*) Resonant frequency $\omega_r = \omega_n\sqrt{1-2\xi^2}$ ...(8.39)

(*vi*) For most values of damping ratio ξ, the gain crossover frequency ω_{gc} and undamped natural frequency ω_n are nearly equal.

(*vii*) The phase margin ϕ_m is directly and almost linearly, related to the damping ratio ξ as shown in Fig. 8.24. So, increasing gain crossover frequency in frequency domain while keeping phase margin constant, is equivalent to reducing settling time and rise time in time domain and forcing the root locus to move outward along constant ξ loci. For $\xi \leq 0.5$, $\phi_m \cong 100\,\xi$. As ξ varies from 0.5 to 1, the slope of ϕ_m *vs* ξ curve decreases.

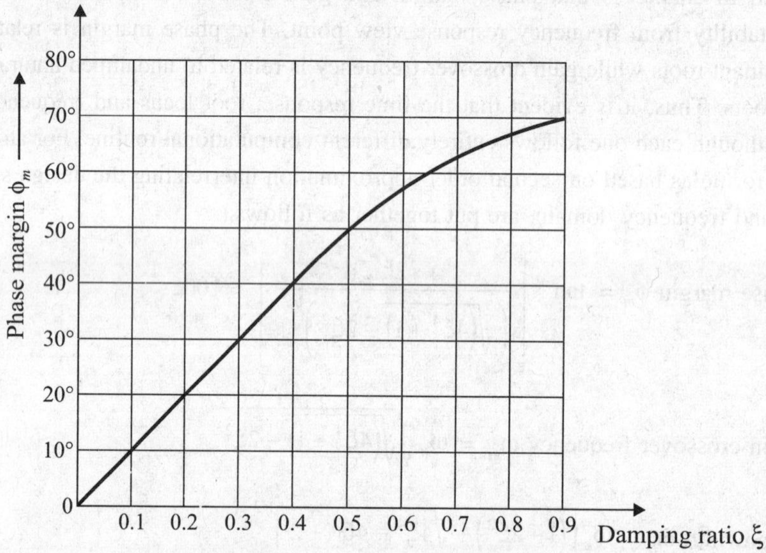

Fig. 8.24: Phase margin vs damping ratio curve

Lag-compensator design

The philosophy behind phase lag-compensator design, is to shift the gain crossover frequency to a lower frequency whereby the phase margin improves. The change in phase occurs at low frequency and has little effect near gain crossover where phase margin is calculated. In fact, improvement in phase margin is brought about by attenuation in high frequency range contributed by lag-compensator and the phase contribution of lag-compensator is of no consequence. As demonstrated in Fig. 8.25, the lag-compensator reduces high frequency gain while preserving low frequency gain. The low frequency gain can be made high to yield a large K_v without creating instability. Thus, the lag-compensator improves steady state error by increasing low frequency gain and improves phase margin to yield desired transient response. GM_{uc} and GM_c as depicted in Fig. 8.25 are gain margins of uncompensated and compensated system respectively while ϕ_{muc} and ϕ_{mc} are uncompensated and compensated phase margins. Increased phase margin reduces peak overshoot.

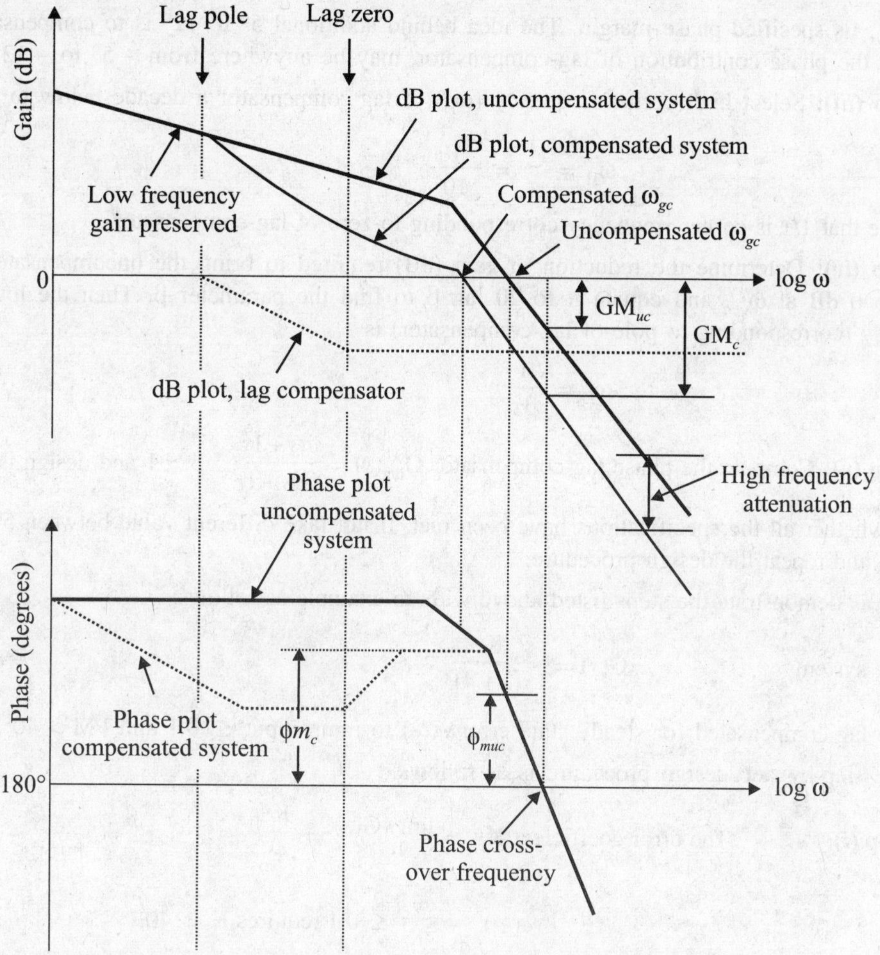

Fig. 8.25: Conceiving idea of reshaping frequency response by lag-compensator

Having understood that the dB plot can be reshaped by judiciously designing the lag-compensator, let us recall that the transmittance of lag-compensator is

$$G_{lag}(j\omega) = \frac{\tau s+1}{\beta\tau s+1}; \quad \beta > 1$$

The design involves selecting the parameters τ and β together with gain K to meet design objectives. The design procedure with design specification on steady state error and phase margin can be put in the form of steps as follows:

Step (*i*): Find gain K to satisfy specified steady state error. Sketch Bode plot for this value of gain K and find phase margin, ϕ_{muc} .

Step (*ii*): If design specification on phase margin is not met, we need to move gain crossover frequency so as to obtain the required phase margin. We therefore find a new gain crossover frequency ω_{gcn} where uncompensated system contributes phase margin

$$\phi_m = \phi_{ms} + 5° \text{ to } 12°$$

where ϕ_{ms} is specified phase margin. The idea behind additional 5° to 12° is to compensate for the fact that the phase contribution of lag-compensator, may be anywhere from – 5° to – 12° at ω_{gcn}.

Step (*iii*): Select high corner frequency (ω_H) of lag-compensator a decade below ω_{gcn} *i.e.*

$$\omega_H = \frac{1}{\tau} = \frac{\omega_{gcn}}{10}$$

Note that $1/\tau$ is corner frequency corresponding to zero of lag-compensator.

Step (*iv*): Determine the reduction in gain (dB) required to bring the uncompensated dB plot down to 0 dB at ω_{gcn} and equate it to 20 lag β to find the parameter β. Then the lower corner frequency (corresponding to pole of lag-compensator) is

$$\omega_L = \frac{1}{\beta\tau}$$

Step (*v*): Compute the phase lag-compensator $G_{lag}(s) = \frac{\tau s+1}{\beta\tau s+1}$; $\beta > 1$ and design is complete.

Check whether all the specifications have been met. If not take different value between 5° to 12° in step (*ii*) and repeat the design procedure.

Let us demonstrate the steps listed above with an example as follows:

The system $$G(s) = \frac{K}{s(s+4)}$$

is to be lag-compensated for steady state error $e(\infty)$ to ramp input ≤ 0.1 and PM > 40°.

The step-by-step design procedure is as follows:

Step (*i*): The error coefficient $K_v = \lim_{s\to 0} sG(s) = \frac{K}{4}$

and $$e(\infty) = \frac{4}{K} \leq 0.1 \text{ requires } K \geq 40.$$

The Bode plot for $\frac{40}{s(s+4)} = \frac{10}{s(1+0.25s)}$ is shown in Fig. 8.26. The phase margin is approximately 32°, which does not meet the design specification.

Step (*ii*): The new gain cross-over frequency ω_{gcn} where compensated system is required to contribute phase margin = ϕ_{ms} + 5° = 45°, is 4 rad/sec as depicted in Fig. 8.26. The phase margin has been raised by 5° from specified value ϕ_{ms} in order to compensate for the phase angle contribution of the lag-compensator.

Step (*iii*): The high corner frequency of lag-compensator is

$$\omega_H = \frac{1}{\tau} = \frac{\omega_{gcn}}{10} = 0.4$$

Step (*iv*): The required reduction in gain (dB) so as to force the dB plot to pass through 0 dB at ω_{gcn} = 4, is approximately 8 dB. Equating it to 20 lag β, we have

$$20 \text{ lag } \beta = 8 \text{ or } \beta = 2.51$$

Then, the lower corner frequency of lag-compensator is

$$\omega_L = \frac{1}{\beta\tau} = \frac{1}{2.51 \times 2.5} = 0.159$$

Step (*v*): The phase lag-compensator is

$$G_{lag}(s) = \frac{2.5s+1}{6.289s+1}$$

and the new pole-zero function of lag-compensated system is

$$G(s)\, G_{lag}(s) = \frac{10(2.5s+1)}{s(0.25s+1)(6.289s+1)}$$

whose bode plot is also shown in Fig. 8.26. Note that the design has been tried with Bode plots using

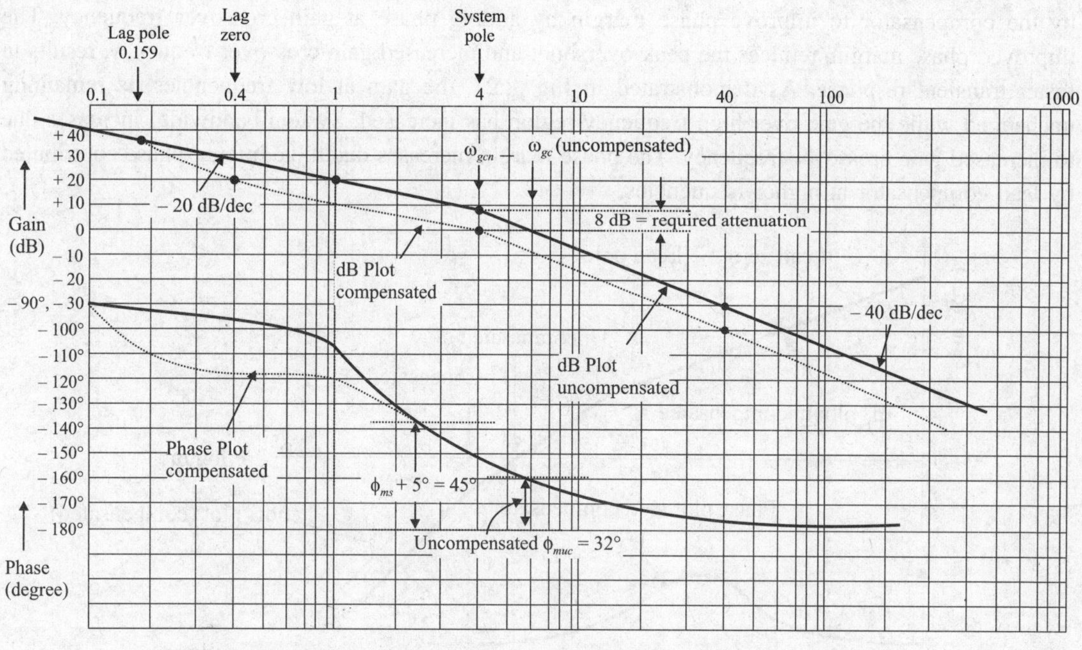

Fig. 8.26: Bode plot of uncompensated system and lag compensated systems

asymptotic approximation. More accurate design could be found with true dB plot.

Note the following in lag-compensator design perspective.

(*i*) The lag-compensator improves steady state error while keeping the transient response relatively unaffected. Improvement in steady state error is brought about by high gain at low frequency. The gain at low frequencies, remains unchanged with lag-compensation while the gain over high frequency region is reduced by 20 lag β so as to avoid system instability.

(*ii*) The gain crossover frequency is reduced. The system bandwidth also reduces. The gain crossover frequency is rough measure of bandwidth. The reduced bandwidth results in slower transient response (longer settling time and rise time).

(*iii*) The phase margin is increased. The damping ratio ξ increases, the system exhibits less overshoot and becomes less oscillatary.

(*iv*) The phase characteristics of lag-compensator, is of no use for compensation purpose.

(*v*) The lag-compensator is essentially a low pass filter and tends to integrate the input signal. In this sense, the lag-compensator acts some what like PI compensator. Note that $\tau s + 1 \cong 1$ at low frequencies and $\beta\tau s + 1 \cong \beta\tau s$ for large β. So, the transmittance of lag-compensator $\frac{\tau s+1}{\beta\tau s+1}$ can be approximated as $\frac{1}{\beta\tau s} \cong \frac{K_i}{s}$. With these assumptions, phase lag-compensator appears to be similar to an integral compensator in low frequency region with large β.

Lead compensator design

The philosophy behind lead compensator design, is to use the additional phase lead contributed by the compensator to improve phase margin by adding phase at gain crossover frequency. The improved phase margin, reduces the peak overshoot and increased gain crossover frequency, results in faster transient response. As demonstrated in Fig. 8.27, the gain at low frequencies, is remaining unchanged while the gain over high frequency region has increased. System bandwidth increases due to increased gain crossover frequency. The phase margin increases due to additional phase contributed by lead compensator at higher frequencies.

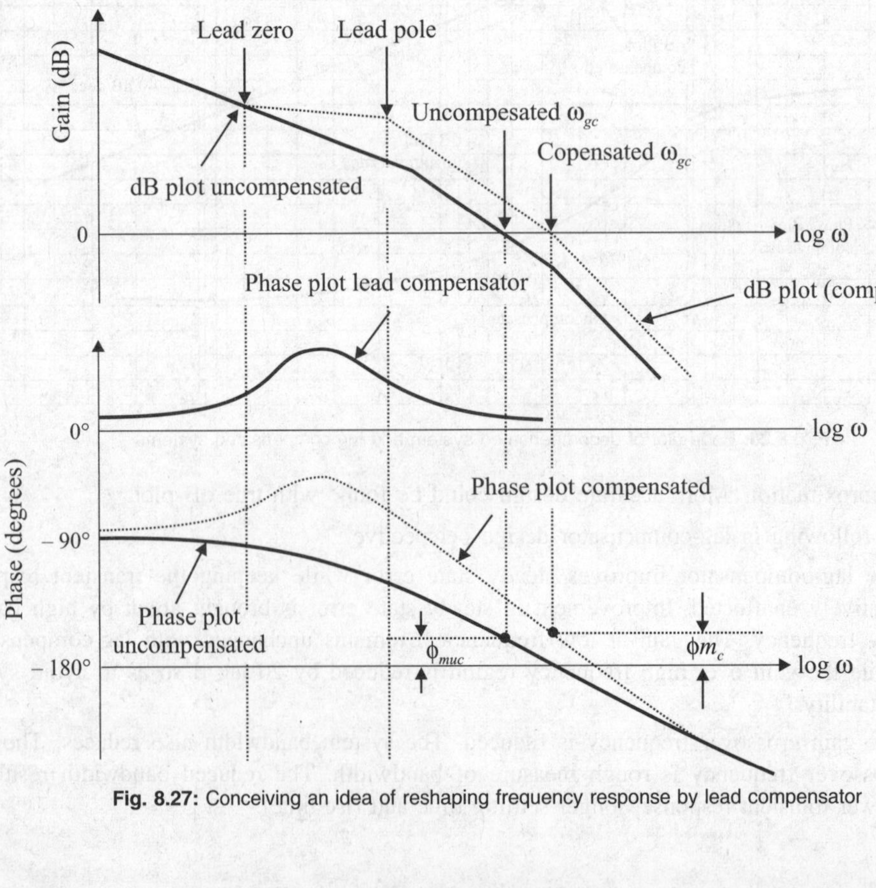

Fig. 8.27: Conceiving an idea of reshaping frequency response by lead compensator

Having understood as to how the Bode plots can be reshaped by judiciously designing the lead compensator, let us recall the transmittance of lead compensator

$$G_{lead}(s) = \frac{\tau s+1}{\alpha\tau s+1}; \quad \alpha < 1$$

with the low (zero) and high (pole) corner frequencies given by:

$$\omega_L = \frac{1}{\tau} \quad \text{and} \quad \omega_H = \frac{1}{\alpha\tau}$$

The design involves finding parameters τ and α together with gain K to meet design objectives. The design procedure can be put in the form of steps as follows:

Step (*i*): Find gain K to satisfy specified steady error. Sketch the Bode plot for this value of gain and find phase margin, ϕ_{muc}.

Step (*ii*): Find additional phase contribution, ϕ_m required from lead compensator using the relation

$$\phi_m = \phi_{ms} + \phi_c - \phi_{muc}$$

where ϕ_{ms} = specified phase margin and $\phi_c = 5°$ to $20°$, the extra phase ϕ_c is added to compensate for shift in gain crossover frequency. If specification is available on peak overshoot or damping ratio, we shall require to translate these into phase margin.

Step (*iii*): Compute α using relation

$$\sin\phi_m = \frac{1-\alpha}{1+\alpha}$$

Step (*iv*): Determine the frequency ω_m where gain (dB) contribution of uncompensated system equals $-$ 10 log (1/α). Select this frequency as the new gain crossover frequency. This frequency corresponds to ω_m at which maximum phase lead occurs.

Step (*v*): Calculate $\tau = \dfrac{1}{\omega_m\sqrt{\alpha}}$ and phase lead compensator $G_{lead}(s) = \dfrac{s+\frac{1}{\tau}}{s+\frac{1}{\alpha\tau}} = \alpha\dfrac{\tau s+1}{\alpha\tau s+1}$.

Finally, insert an amplifier with gain = 1/α in order to neutralize the lead compensator attenuation and keep the *dc* gain of compensated system at unity so that value of K set in step (*i*) continues to yield steady state error as specified.

Now, consider the following example to demonstrate the steps listed above.

The system $G(s) = \dfrac{K}{s(s+8)}$ with unity feedback is to be lead compensated to meet the Design specifications.

Steady state error to unity rate ramp $\leq$ 0.1.

Damping ratio $\xi \geq 0.5$.

The step by step design procedure is as follows:

Step (*i*): Transforming the uncompensated system in time constant form, we have

$$G(j\omega) = \frac{K/8}{j\omega(j0.125\omega+1)}$$

The error coefficient $$K_v = \lim_{s\to 0} sG(s) = \frac{K}{8}$$

and specified $e(\infty) \le 0.1$ calls for $\frac{1}{K_v} = \frac{8}{K} \le 0.1$ or $K \ge 80$.

For this value of K, $G(j\omega) = \frac{10}{j\omega(j0.125\omega+1)}$, the Bode plot for which, is shown in Fig. 8.28.

The uncompensated phase margin $\phi_{muc} = 42°$ and gain crossover frequency $\omega_{gc} = 9$ rad/sec.

Step (*ii*): Translating specified $\xi \ge 0.5$ into phase margin, we have

$$\phi_{ms} = \tan^{-1}\left[\frac{2\xi}{\left(\sqrt{(4\xi^4+1)}-2\xi^2\right)^{1/2}}\right] = 52°$$

The additional phase contribution required from lead compensator is

$$\phi_m = \phi_{ms} + \phi_c - \phi_{muc} = 52° + 5° - 42° = 15°$$

We have added the correction factor $\phi_c = 5°$ to compensate for the shift in gain crossover frequency.

Step (*iii*): $\alpha = \frac{1-\sin\phi_m}{1+\sin\phi_m} = 0.59$

Step (*iv*): The frequency where the uncompensated system contributes dB gain of $-10\log(1/\alpha) = -2.3$ dB, is $\omega_m = 12$ rad/sec.

Step (*v*): $\tau = \frac{1}{\omega_m\sqrt{\alpha}} = \frac{1}{12\sqrt{0.59}} = 0.108$

and lead compensator $G_{lead}(s) = 0.59\,\frac{(0.108s+1)}{(0.064s+1)}$

Finally, we insert an amplifier with gain 0.108/0.064 = 1.69 in order to undo the attenuation caused by lead network.The Bode plots for lead compensated system, are also shown in Fig. 8.28.

Note the following in lead compensator design perspective:

(*i*) Designing lead compensator, using frequency response technique is advantages in the sense that specifications on transient response can be met together with meeting the steady state error requirement. The initial slope of dB plot that determines the steady state error, is not affected by the design for the transient response. The root locus technique provides an infinite number of possible solutions to the lead compensator design, each differing in steady state error.

(*ii*) The lead compensator like lag-compensator, keeps the gain at low frequency unchanged but the gain at high frequencies, increases due to lead compensation. So, the lead compensated system is more sensitive to noise and high frequency disturbances.

(*iii*) Phase margin increases whereby peak overshoot decreases.

(*iv*) The gain crossover frequency and so only the bandwidth increases whereby faster transient response is realized.

(*v*) In mid to high frequency range, the phase lead compensator can be approximated as PD compensator for $\alpha < < 1$.

$$G_{lead}(s) = \left.\frac{\tau s+1}{\alpha\tau s+1}\right|_{\alpha<<1} \cong \tau s + 1$$

and for values of ω in the mid to high frequency range $\tau s + 1 \cong \tau s$. Thus, the phase lead compensator will approach a pure derivative controller as $\alpha \to 0$. The phase lead controller has the advantage that phase lead at low frequencies is zero while a pure derivative controller has 90° phase lead over the entire frequency range.

(*vi*) The maximum phase lead is practically 60° to 70° for a single stage. If we require higher phase lead, it is advised to use two phase lead compensators in cascade.

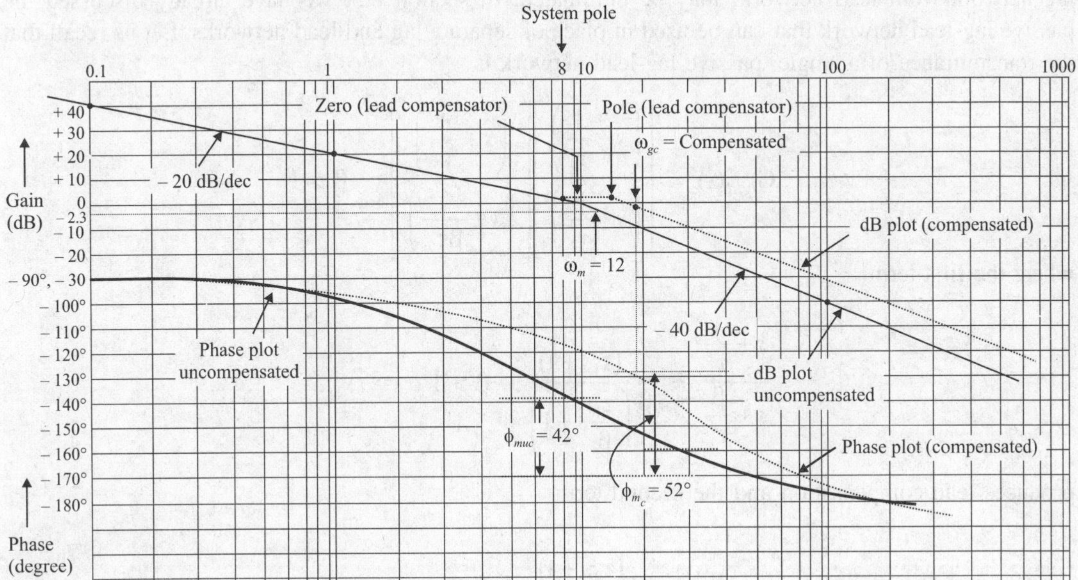

Fig. 8.28: Bode plot for original and lead compensated systems

Phase lag-lead compensator design

We have learnt that the lag-compensator reduces the gain near and above the gain crossover frequency while allowing the increase in gain at low frequency region to improve the steady state error. So, the lag-compensator is used to adjust steady state error while reducing the bandwidth and slowing down the transient response. On the other hand, the lead compensator increases the bandwidth and makes the system faster.

If both fast transient response and good steady state accuracy, are desired, then we combine the lag design and lead design resulting in design of lag-lead compensator that combines the best attributes of each.

Although, it has been seen that good steady state accuracy and faster dynamics can be built in lead compensator design. However, if the system is of high order or the system has large steady state error, the lead compensator design may result in excessively large bandwidth which may be unacceptable in practice due to enhanced susceptibility of system to the noise signals. The lag-lead compensator is again used to tackle such a problem where the lead compensator adjusts the bandwidth and lag-compensator provides additional phase margin.

One possibility of lag-lead compensator design is to use both a lead and a lag as separate elements. However, it may be more economical to use a single, passive lag-lead network that combines both lead and lag elements. In doing so, the need for the buffer amplifier that separates the lag network from lead network, may be eliminated. In section 8.5, we have already discussed the passive lag-lead network that can be used in place of separate lag and lead networks. Let us recall that the transmittance of a single, passive lag-lead network is

$$G_{lgld}(s) = \left[\frac{\left(s+\frac{1}{\tau_1}\right)}{\left(s+\frac{\beta}{\tau_1}\right)}\right]\left[\frac{\left(s+\frac{1}{\tau_2}\right)}{\left(s+\frac{1}{\beta\tau_2}\right)}\right]; \quad \beta > 1$$

where the first term

$$\frac{s+\frac{1}{\tau_1}}{s+\frac{\beta}{\tau_1}} = \frac{1}{\beta}\frac{(\tau_1 s+1)}{\left(\frac{\tau_1}{\beta}s+1\right)}; \quad \beta > 1$$

produces lead compensation and the second term

$$\frac{s+\frac{1}{\tau_2}}{s+\frac{1}{\beta\tau_2}} = \beta\frac{(\tau_2 s+1)}{(\beta\tau_2 s+1)}; \quad \beta > 1$$

produces lag-compensation. Notice that the parameter α used for lead compensator in the discussion so far, is constrained by the relation $\beta = 1/\alpha$. The parameter β has been used for lag-compensator in the discussion so far. The constrained relation $\alpha\beta = 1$ does not permit independent choice of α and β. But it eases out the design of lag-lead compensator. The design procedure for a lag-lead compensator follows the procedure of lag-compensator first. Then having designed the lag-compensator (τ_2 and β determined). The only variable parameter required τ_1 will be found for lead compensator design.

Now, we present the step by step procedure for designing lag-lead compensator as follows:

Step (*i*): Use second order approximation to translate the damping ratio or peak overshoot requirement into phase margin and rise time, peak time or settling time requirement into closed loop bandwidth. Such translations are not required if phase margin and bandwidth are directly specified.

Step (*ii*): Find gain K to satisfy specified steady state error and sketch Bode plot for this value of K.

Step (*iii*): Select a new gain crossover frequency ω_{gcn} tentatively near specified bandwidth. The phase crossover frequency of uncompensated system may also be a choice for the new gain crossover frequency, ω_{gcn}.

Step (*iv*): Design lag-compensator by choosing the higher corner frequency (zero) to be one decade below new gain crossover frequency, that is

$$\omega_H = \frac{1}{\tau_2} = \frac{\omega_{gcn}}{10}$$

Then the lower corner frequency (pole) is $\omega_L = \dfrac{1}{\beta\tau_2}$

where value of β is found from maximum phase lead required to be contributed by lead compensator. In fact, the lead compensator will be required to contribute the additional phase lead to meet phase margin requirement at ω_{gcn}. This additional phase ϕ_m is sum of specified phase margin ϕ_{ms} plus ϕ_c (= 5° to 12°) to compensate for phase lag contributed by lag-compensator at ω_{gcn}, that is

$$\phi_m = \phi_{ms} + \phi_c$$

Recall that maximum phase lead ϕ_m, contributed by lead compensator is related to β as

$$\phi_m = \sin^{-1}\left[\frac{1-\frac{1}{\beta}}{1+\frac{1}{\beta}}\right]$$

Step (*v*): Design the lead compensator. With the values of new gain crossover frequency ω_{gcn} and β already known from the Steps (*iii*) and (*iv*) respectively, the value of lower corner frequency, $\omega_L = 1/\tau_1$ may be found using the relation

$$\omega_{gcn} = \frac{\sqrt{\beta}}{\tau_1}$$

Then upper corner frequency $\omega_H = \beta/\tau_1$.

Let us demonstrate the procedure with an example. Design a lag-lead compensator for a unity feedback system with transfer function

$$G(s) = \frac{K}{s(s+8)(s+30)}$$

to meet the specifications:

$$\text{peak overshoot} = 13.5\%$$
$$\text{peak time} = 0.6 \text{ sec}$$
$$\text{error coefficient } K_v = 10$$

The step by step design procedure for lag-lead compensator, is numerically demonstrated as follows:

Step (*i*): Translating specified peak overshoot of 13.5% into damping ratio ξ, we have

$$e^{-\pi\xi}\Big/\sqrt{1-\xi^2} = 0.135 \quad \text{or} \quad \xi = 0.537$$

and then translating $\xi = 0.537$ into phase margin, we have

$$\phi_{ms} = \tan^{-1}\left[\frac{2\xi}{\left(\sqrt{4\xi^4+1}-2\xi^2\right)^{1/2}}\right] \cong 55°$$

Translating peak time $t_p = 0.6$ sec., into bandwidth, we have

$$\omega_b = \frac{\pi}{t_p\sqrt{1-\xi^2}}\left(1-2\xi^2+\sqrt{4\xi^4-4\xi^2+2}\right)^{1/2} = 9 \text{ rad/sec}$$

Step (*ii*): The error coefficient

$$K_v = \lim_{s\to 0} sG(s) = \frac{K}{240}$$

Equating it to specified $K_v = 10$, we have K = 2400. Notice that K = 2400 is evaluated from pole-zero form of transfer function. Now, for sketching Bode plot let us transform the uncompensated transfer function into time constant form to get

$$G(j\omega) = \frac{2400}{j\omega(j\omega+8)+(j\omega+30)} = \frac{10}{j\omega\left(1+\frac{j\omega}{8}\right)\left(1+\frac{j\omega}{30}\right)}$$

for which the Bode plot is shown in Fig. 8.29.

Step (*iii*): Let us select the new gain crossover frequency $\omega_{gcn} = 6$ near specified bandwidth.

Step (*iv*): The phase of uncompensated system at $\omega = \omega_{gcn} = 6$ is $-140°$ as best read from phase plot of Fig. 8.29. The specified phase margin $\phi_{ms} = 55°$. Let us raise it by $\phi_c = 5°$ assuming that lag-compensator will contribute phase lag of 5° at $\omega = \omega_{gcn} = 6$ rad/sec. Then the phase angle of compensated system at $\omega = \omega_{gcn} = 6$ will be $-180° + 55° + 5° = -120°$ to meet the specified phase margin. Since the uncompensated system already, has phase equal to $-140°$, the lead compensator is required to generate additional phase lead of 20°.

Let us choose $\beta = 3$, since the maximum phase lead that is contributed by lead compensator for $\beta = 3$ is

$$\sin^{-1}\left(\frac{1-\frac{1}{\beta}}{1+\frac{1}{\beta}}\right) = \sin^{-1}\left(\frac{1-\frac{1}{3}}{1+\frac{1}{3}}\right) \cong 30°$$

while the requirement is only 20°, which is quite possible by of single lag-lead network. Now we are ready to design lag-compensator. The upper corner frequency of lag-compensator is

$$\omega_H = \frac{1}{\tau_2} = \frac{\omega_{gcn}}{10} = 0.6$$

and the lower corner frequency is

$$\omega_L = \frac{1}{\beta\tau_2} = \frac{0.6}{3} = 0.2$$

the lag-compensator has the transmittance

$$G_{lag}(s) = \frac{s+0.6}{s+0.2} = 3\left(\frac{1.67s+1}{5s+1}\right)$$

Step (*v*): Next, we design the lead compensator. The value of β is known from step (*iv*). The lower corner frequency of lead compensator is found using the relation

$$\omega_{gcn} = \frac{\sqrt{\beta}}{\tau_1}$$

as $\omega_L = 1/\tau_1 = 3.46$. Then the upper corner frequency is

$$\omega_H = \frac{\beta}{\tau_1} = 10.39$$

and the transmittance of lead compensator is

$$G_{lead}(s) = \frac{s+3.46}{s+10.39} = \frac{1}{3}\left[\frac{0.289s+1}{0.096s+1}\right]$$

The lag-lead compensator has the transmittance

$$G_{lgld}(s) = \left(\frac{1.67s+1}{5s+1}\right)\left(\frac{0.289s+1}{0.096s+1}\right)$$

The Bode plot of lag-lead compensated system with sinusoidal transfer function

$$G_{lgld}(j\omega)\,G(j\omega) = \frac{10(j1.67\omega+1)(j0.289\omega+1)}{j\omega\left(1+\frac{j\omega}{8}\right)\left(1+\frac{j\omega}{30}\right)(j0.096\omega+1)(j5\omega+1)}$$

is also shown in Fig. 8.29. The compensated system meets the specifications on peak overshoot (equivalent phase margin of 55°) and error coefficient (low frequency dB plot remaining unchanged). Now check bandwidth. The closed loop bandwidth is equal to that frequency where dB plot contributes approximately – 7 dB.

Notice that only asymptotic dB plot has been shown in Fig. 8.29. The true dB plot will be approximately 3 dB below that asymptotic plot in neighbourhood of ω = 10 rad/sec due to two pole corner frequencies, one at ω = 8 and another at ω = 10.39. More accurately – 7 dB gain will be found at approximately ω = 10 rad/sec., which is close to the specified bandwidth.

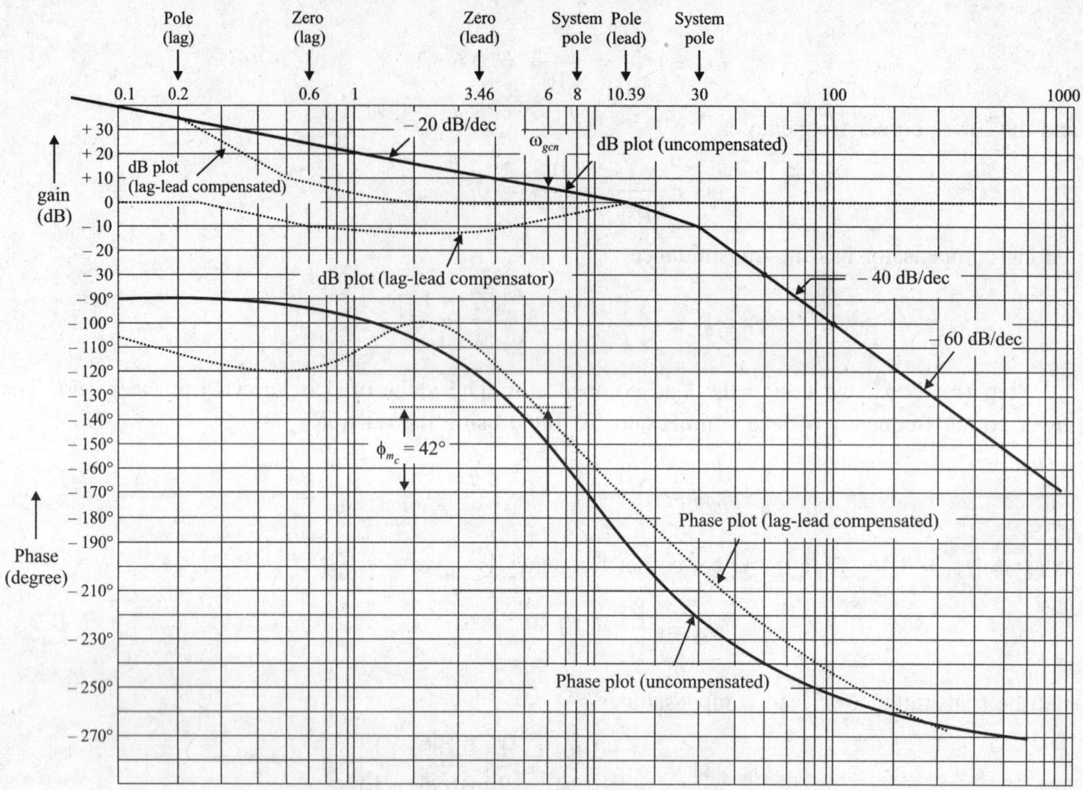

Fig. 8.29: Bode plot for uncompensated and lag-lead compensated systems

8.8 RATE FEEDBACK COMPENSATOR DESIGN

In the discussion so far, we have concentrated on placing the compensator in series with uncompensated system (called as cascade compensator) to meet the design objectives. The compensator can be placed in feedback path (called as feedback compensator) also to perform the same task. Although, the design procedures for feedback compensators might turn to be more difficult, but it may yield faster response. When the design of cascade compensator becomes unsuitable due to noise etc., the feedback compensators are designed. The feedback compensation is also advantageous in the sense that it may not require any additional amplifier. The signal that propagates through the feedback compensator, originates at high level output of forward path and terminates at low level input in the forward path.

A typical rate feedback system as shown in Fig. 8.30(*a*) contains two feedback signals. The outer loop (also called as major loop) contains unity feedback while the inner loop (also called as minor loop) provides rate of change of output. A popular feedback compensator is a rate sensor that acts as a differentiator. Usually, this rate sensor is tachometer. In aircrafts and ships, the rate gyro is also widely used as rate sensor. The rate gyro and tachometer both generate voltage output proportional to input angular velocity.

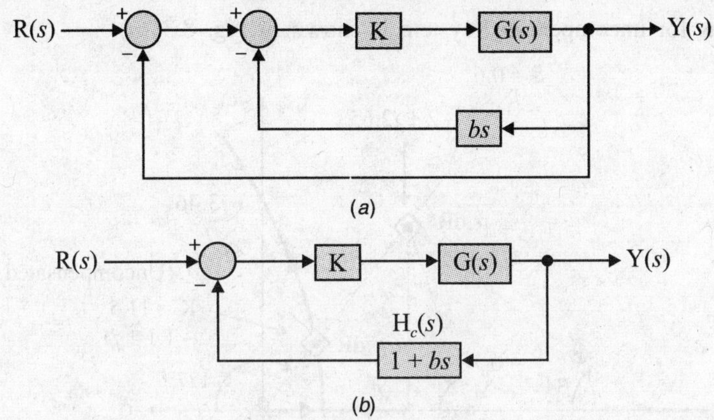

Fig. 8.30: (*a*) Rate feedback compensated system (*b*) Equivalent feedback compensated system

It is easy to visualise that the design procedure of feedback compensator involves finding the values of **K** and ***b*** so as to meet the design goals. An equivalent configuration of feedback compensation is shown in Fig. 8.30(b), which is obtained by combining the two feedback paths major and minor into a single feedback path. The loop gain from Fig. 8.30(*b*) can be obtained as

$$KG(s)\ H_c(s) = KG(s)\ (1 + bs) = Kb(s + 1/b)\ G(s) \qquad ...(8.40)$$

So, placing the rate sensor with transmittance $H(s) = bs$ in the minor loop of Fig. 8.30(*a*), is equivalent to placing a compensator with transmittance

$$H_c(s) = (1 + bs) = b(s + 1/b) \qquad ...(8.41)$$

in the feedback path obtained by combining the major and the minor loop. It is obvious from (8.41) that the feedback compensator $H_c(s)$ has the output which is sum of two components, one is proportional to its input and another is the derivative of its input. This type of compensator, as previously discussed, is referred to as PD compensator. The design of this PD compensator, in fact, involves adding a zero at $s = -1/b$ to the existing open loop poles and zeros of uncompensated system G(*s*). The additional zero at $s = -1/b$ reshapes the root locus in a way to force the root locus to pass through the dominant root (design point). The judicious selection of the location of this zero together with value of K **yields** the desired response.

The addition of one open loop zero, decreases the number of asymptotes by 1. The asymptote angles increase, generally improving stability. Thus, the rate feedback compensator is primarily intended to improve the relative stability, that is, to move the closed loop poles (characteristic roots) to the left of their locations for the uncompensated case. The steady state error may be larger or smaller than that for uncompensated system, depending on the value of K resulting after the feedback compensator is inserted.

The step by step procedure for PD compensator design has been discussed in preceding section. Let us demonstrate it again in the perspective of rate feedback compensator with the help of following numerical example:

Let the uncompensated system

$$G(s) = \frac{K}{s(s+2)(s+6)}$$

be rate feedback compensated so as to meet the design specifications:

Peak overshoot = 9.5%

Settling time = 2 seconds

The root locus for uncompensated system is shown in Fig. 8.31(*a*).

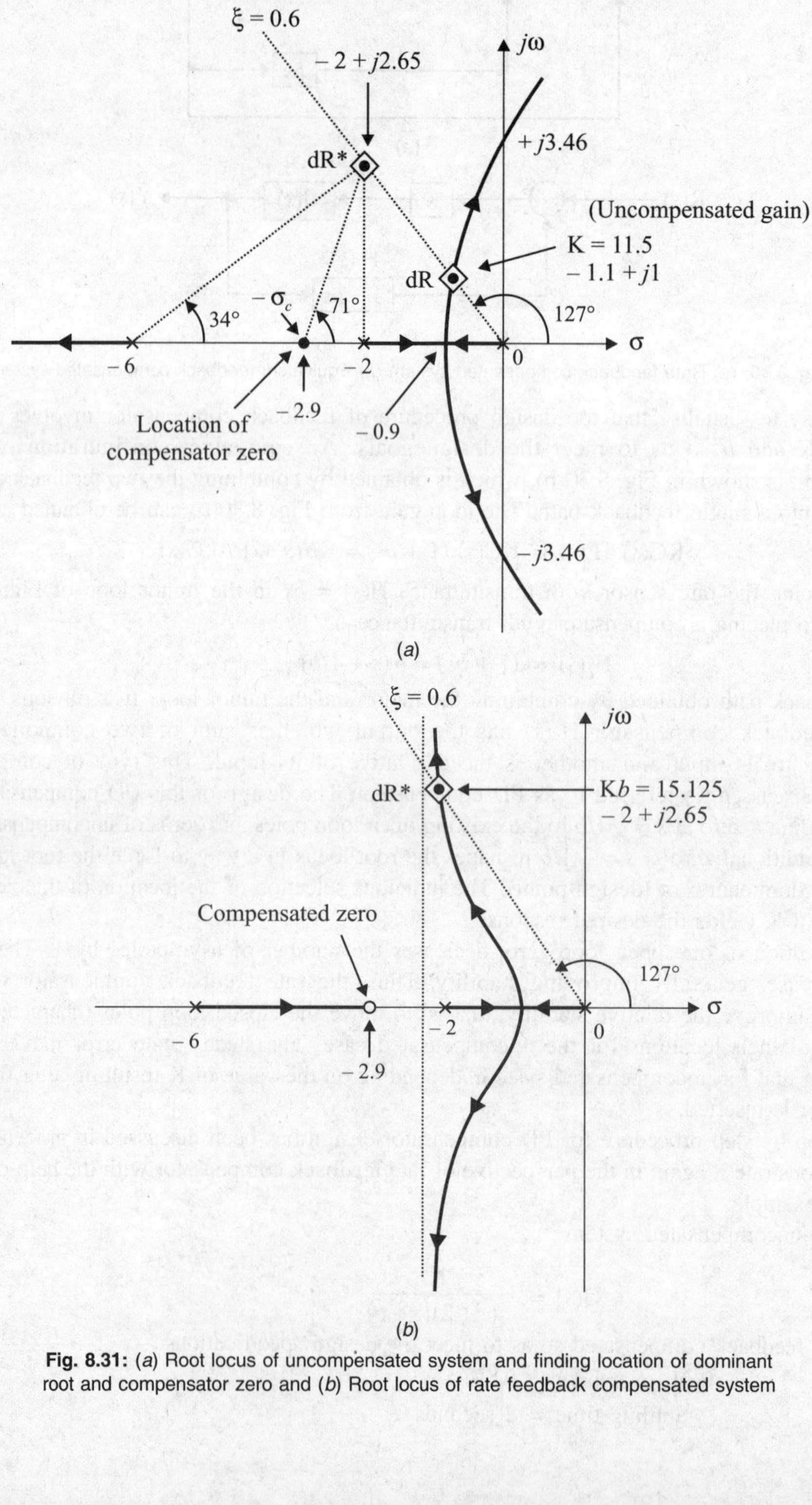

Fig. 8.31: (*a*) Root locus of uncompensated system and finding location of dominant root and compensator zero and (*b*) Root locus of rate feedback compensated system

The dominant root is found at point dR (– 1.1 + j1 for K = 11.5) along the line of 9.5% peak overshoot or ξ = 0.6 as shown in Fig. 8.31(a). Note

$$e^{-\pi\xi/\sqrt{1-\xi^2}}\Big|_{\xi=0.6} = 0.09478 \cong 0.095$$

Using the relation $(4/\sigma) = t_s$, it is easy to find that real part of dominant root must be 2 to satisfy the settling time requirement of 2 seconds. The imaginary part of dominant root satisfying peak overshoot and settling time simultaneously is 2 × tan 53° = 2.65. Note that $\cos^{-1} 0.6 \cong 53°$.

The dominant root dR* that meets the design specifications is also shown in Fig. 8.31(a). The open loop poles of uncompensated system contribute the angle at dR* equal to – (127° + 90° + 34°) or – 251°. The compensator will be required to contribute + 71° (= 251° – 180°) so as to ensure that the sum of angles contributed by poles of uncompensated system together with compensator zero is – 180° at dR*.

Using trigonometry we can locate the compensator zero. As shown in Fig. 8.31(a), the relation

$$\frac{2.65}{\sigma_c - 2} = \tan 71°$$

gives $\sigma_c = 2.9$.

The root locus for the equivalent compensated system is shown in Fig. 8.31(b). The gain (Kb) at dR* is found to be 15.12. So, K ≅ 44.5 and b = 0.34 in the rate feedback configuration as shown in Fig. 8.30(a), meets the design specifications.

Minor loop feedback compensation

Some of the reasons for the feedback compensators to be attractive, have been identified above. Yet another attractive feature of feedback compensation is the possibility that open loop system may continue to work of course with some deteriorated performance even if the feedback path is broken. For example, it is very important in aircrafts and space crafts that the system as a whole be able to sustain the operation even in case of control system damage and component failure.

Sometimes, in a rate feedback compensation architecture as shown in Fig. 8.32(a), it may be advantageous to design the minor loop's transient response separately from closed loop system response. For example, in an aircraft the minor loop may be designed to control the position of aerodynamic surfaces and the overall closed loop system may be designed to control the pitch angle.

The minor loop of Fig. 8.32(a), in fact, models the forward path transfer function, $G_m(s)$ of unity feedback system as shown in Fig. 8.32(b).

$$G_m(s) = \frac{G(s)}{1 + b\, sG(s)} \qquad \text{...(8.42)}$$

The poles of $G_m(s)$ can be adjusted by sketching root locus with **b** as variable parameter. This is how the minor loop's transient response is designed. Now these adjusted poles are, in fact, the open loop poles of the system as a whole. Thus, the minor loop design, actually changes the system poles and reshapes the root locus. Recall that the cascade compensators reshape the root locus by adding poles and zeros. Having designed the minor loop, finally the closed loop poles are adjusted by loop gain K to meet the design specifications by sketching the root locus of open loop system function $KG_m(s)$ with gain K as variable parameter.

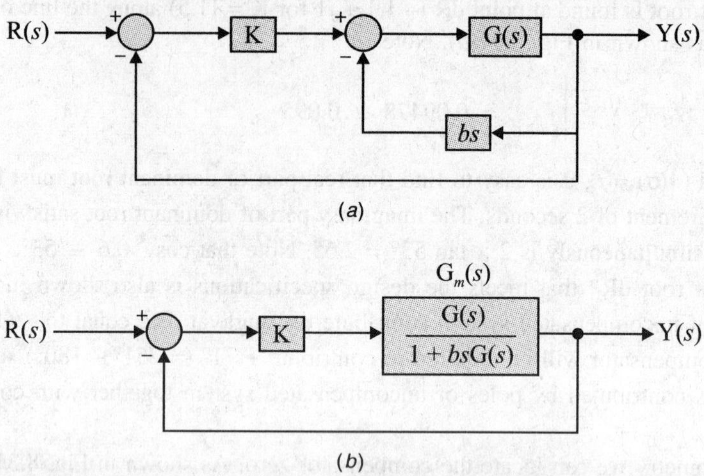

Fig. 8.32: (*a*) An architecture of rate feedback compensation, (*b*) Equivalent feedback compensator configuration with minor loop shown as forward path transfer function

Let us demonstrate the design procedure for minor loop and major loop separately with an example of uncompensated system

$$G(s) = \frac{K}{s(s+2)(s+6)}$$

with the minor and major loop architecture as shown in Fig. 8.33(*a*), the minor loop design is to aim at damping ratio of 0.8 and the closed loop system design is to yield damping ratio of 0.6.

This design problem is solved in two steps. The first step involves drawing the root locus for open loop transfer function of minor loop

$$bs\ G(s) = \frac{bs}{s(s+2)(s+6)} \qquad ...(8.43)$$

with **b** as variable parameter and selecting the value of b so that minor loop yields damping ratio, ξ of 0.8. The second step involves drawing the root locus for open loop transfer function of major loop,

$$G_m(s) = \frac{KG(s)}{1+bsG(s)} = \frac{K}{s\left[s^2+8s+(12+b)\right]} \qquad ...(8.44)$$

with K as variable parameter and selecting the value of K such that the major loop yields damping ratio of 0.6.

The root locus for **bs** $G(s) = \dfrac{bs}{s(s+2)(s+6)}$ is shown in Fig. 8.33(*b*). It is important to note that a zero appearing at origin, comes from feedback transfer function of minor loop, this is not the zero of closed loop transfer function of minor loop. In fact, the closed loop transfer function of minor loop is

$$G_{ml}(s) = \frac{1}{s\left[s^2+8s+(12+b)\right]}$$

So, the pole at origin appears to remain stationary and there is no pole-zero cancellation at origin. The root locus of Fig. 8.33(*b*) shows the two complex conjugate roots (closed loop poles) changing with gain *b*. As demonstrated in Fig. 8.33(*b*), the dominant root, *d*R when searched on $\xi = 0.8$ line, is found at $s = -4 \pm j3$ for $b = 12.98 \cong 13$. Thus, $b \cong 13$ places the minor loop poles at $s = -4 \pm j3$ and meets the minor loop design specification ($\xi = 0.8$).

Now, poles found at $s = -4 \pm j3$ together with one pole at origin, act as open loop poles for the major loop. The major loop design is tried by sketching the root locus for

$$G_m(s) = \frac{K}{s(s+4+j3)(s+4-j3)} = \frac{K}{s(s^2+8s+25)}$$

with K as variable parameter as shown in Fig. 8.33(*c*). Searching the dominant root dR* along $\xi = 0.6$ line, the complex conjugate roots (closed loop poles) are found at $s = -2.1 \pm j2.9$ for K = 21.8. This meets the major loop design specification ($\xi = 0.6$).

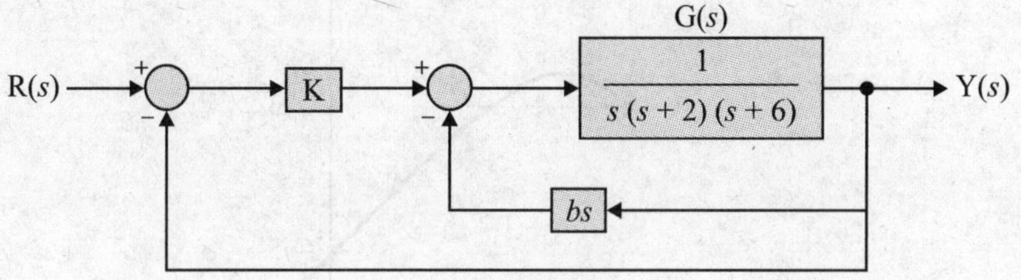

(*a*)

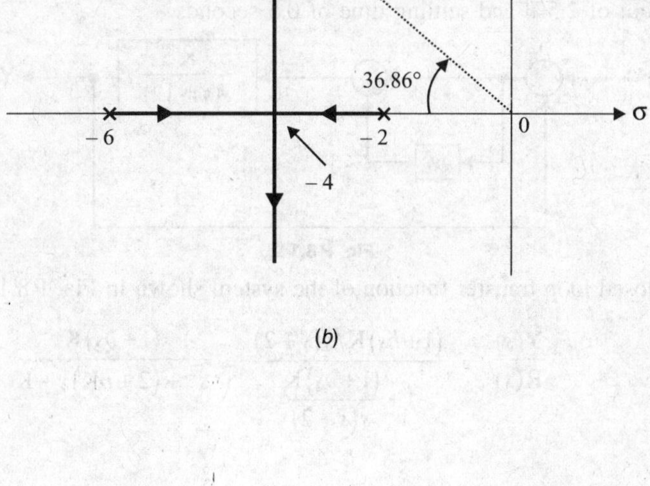

(*b*)

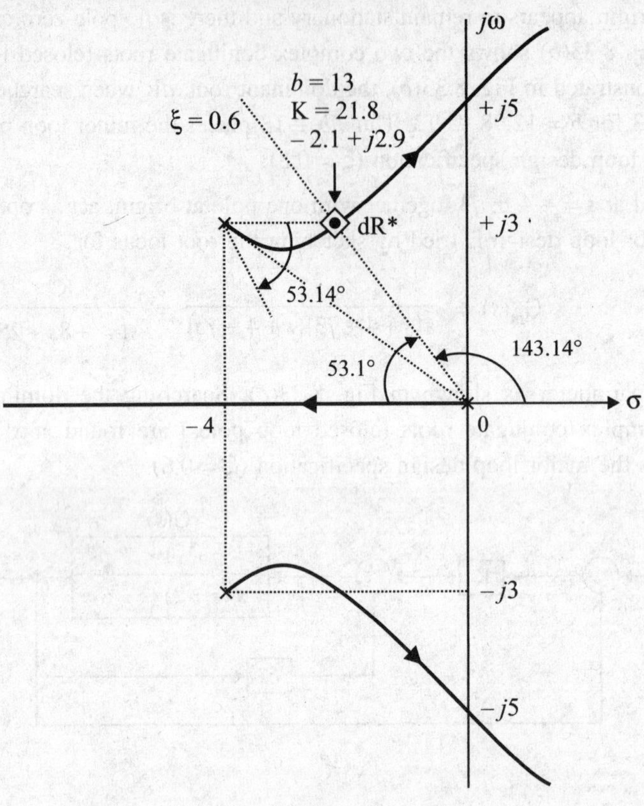

(c)

Fig. 8.33: (*a*) An example for design of minor and major loop, (*b*) Root locus for demonstrating minor loop design and (*c*) Root locus for demonstrating major loop design

PROBLEMS AND SOLUTIONS

P 8.1: For the PD compensated system shown in Fig. P8.1, compute K and **b** such that the system exhibits peak overshoot of 2.5% and settling time of 0.5 seconds.

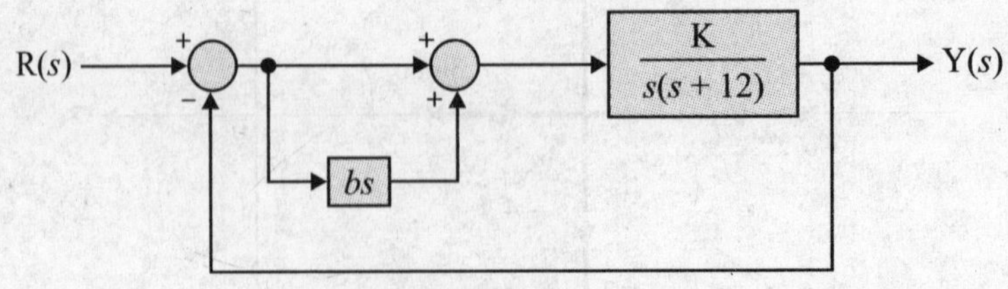

Fig. P 8.1

Solution: The closed loop transfer function of the system shown in Fig. P 8.1 is

$$\frac{Y(s)}{R(s)} = \frac{(1+bs)K/s(s+2)}{1+\dfrac{(1+bs)K}{s(s+2)}} = \frac{(1+bs)K}{s^2+(2+bK)s+K}$$

and the characteristic equation is

$$s^2 + (2 + bK)\, s + K = 0$$

Comparing it with general second order characteristic equation

$$s^2 + 2\xi\omega_n\, s + \omega_n^2 = 0$$

we have $$\omega_n = \sqrt{K} \quad \text{and} \quad \xi = \frac{2 + bK}{2\sqrt{K}}$$

Translating the specified peak overshoot and settling time into damping ratio, ξ and undamped natural frequency, ω_n we have

$$\text{Peak overshoot } M_p = e^{-\pi\xi/\sqrt{1-\xi^2}} = 0.025 \quad \text{or} \quad \xi = 0.58$$

$$\text{Settling time } t_s = \frac{4}{\xi\omega_n} = 0.5 \quad \text{or} \quad \omega_n = 13.79$$

So, $$\sqrt{K} = 13.79 \quad \text{or} \quad K = 190.16$$

and $$\frac{2 + bK}{2\sqrt{K}} = 0.58 \quad \text{or} \quad b = 0.0736$$

P 8.2: A unity feedback system has open loop transfer function

$$G(s) = \frac{K}{(s+3)(s+6)}$$

and operates with peak overshoot of 4.32%. Design a PI compensator via root locus such that the system is forced to track the step input with zero error in steady state.

Solution: The root locus for uncompensated system is shown in Fig. P8.2(*a*). The design specification on peak overshoot of 0.0432 when translated through $e^{-\pi\xi/\sqrt{1-\xi^2}} = 0.0432$ into damping ratio ξ, gives $\xi = 0.707$.The line corresponding to constant $\xi = 0.707$, making an angle of 45° with negative real axis, is also shown in Fig. P8.2(*a*). This line intersects the root locus at point dR located at $s = -4.5 \pm j4.5$ for K = 22.48. The point dR determines the location of dominant, second order poles.

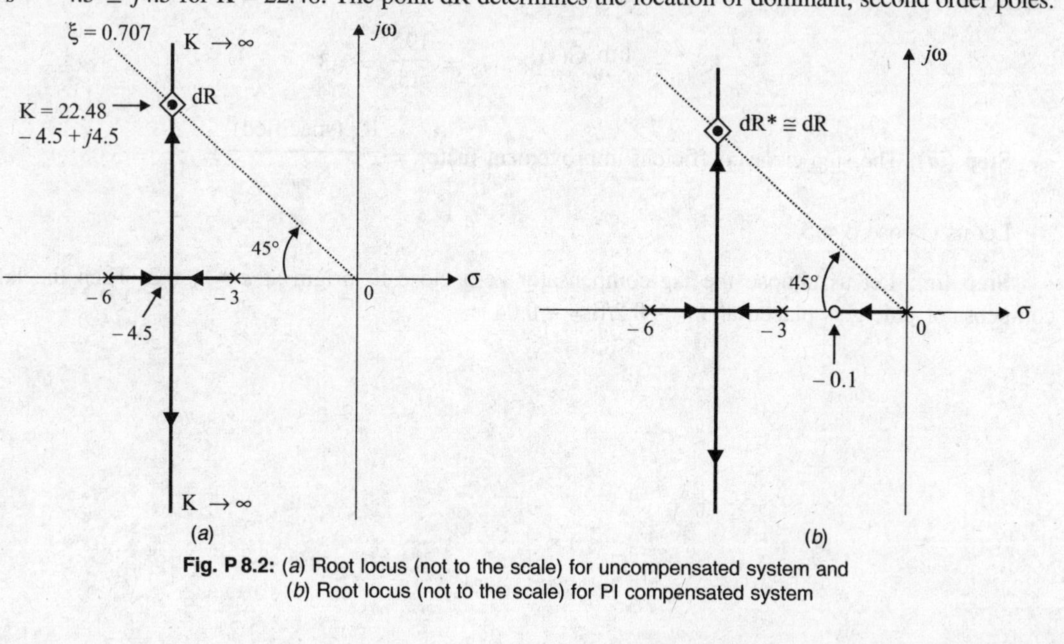

Fig. P 8.2: (*a*) Root locus (not to the scale) for uncompensated system and (*b*) Root locus (not to the scale) for PI compensated system

In order to satisfy the specification on steady state error equal to zero we design PI compensator by placing a pole at origin together with a zero at $s = -0.1$, very close to origin. This almost preserves the location of dominant root dR and so only the transient response while making the system respond with zero steady state error. In fact, the pole added at origin increases the number type of system by 1.

The root locus for PI compensated system $G(s)\,G_{PI}(s) = \dfrac{K(s+0.1)}{s(s+3)(s+6)}$ is shown in Fig. P 8.2(*b*).

P 8.3: Design a lag-compensator for a unity feedback system with forward transmittance

$$G(s) = \frac{K}{(s+1)(s+3)(s+5)}$$

to yield the following specifications:

(*i*) peak overshoot $\leq$ 10%

(*ii*) step error coefficient $\geq$ 4

Solution: The step by step procedure for intended design is as follows:

Step (*i*): The root locus for uncompensated system is shown in Fig. P8.3. Translating specification on peak overshoot of 10% into damping ratio ξ, we have

$$e^{-\pi\xi/\sqrt{1-\xi^2}} = 0.1 \quad \text{or} \quad \xi = 0.59$$

and $$\cos^{-1}\xi\big|_{\xi=0.59} = 53.8°.$$

A line radially outward from origin making an angle of 53.8° with negative real axis, is also shown in Fig. P8.3. The point of intersection of this line with root locus of uncompensated system, is identified as the location of dominant root, dR as depicted in Fig. P 8.3.

Step (*ii*): The dominant root, dR is found at $s = -1.4 \pm j1.9$ for gain K = 19.5. The step error coefficient

$$K_{puc} = \lim_{s\to 0} G(s)\big|_{K=19.5} = \frac{19.5}{15} = 1.3$$

Step (*iii*): The step error coefficient improvement factor $= \dfrac{K_p(\text{specified})}{K_{puc}} = \dfrac{4}{1.3} = 3.08$

Let us choose $\beta = 5$.

Step (*iv*): Let us choose the lag-compensator zero close to origin at $s = -0.2$. Then the lag-compensator pole gets placed at $s = -0.2/\beta = -0.04$.

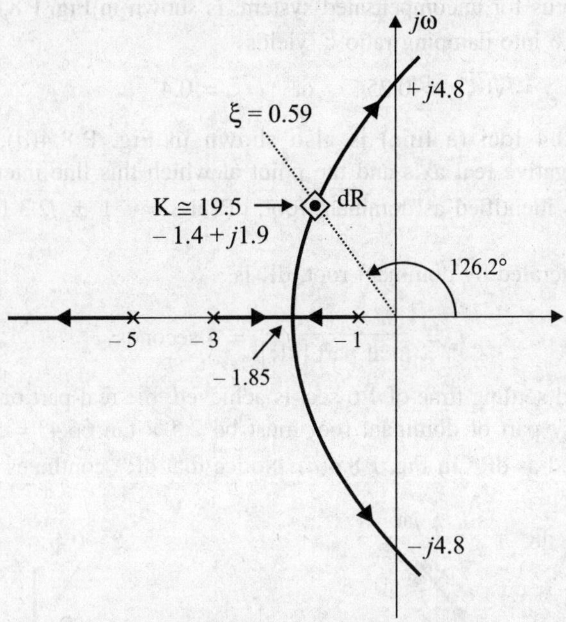

Fig. P 8.3: Locus for uncompensated system

So, $$G_{lag}(s) = \frac{s+0.2}{s+0.04}$$

and lag-compensated forward transmittance is

$$G_{lag}(s)\ G(s) = \frac{19.5(s+0.2)}{(s+0.04)(s+1)(s+3)(s+5)}$$

This lag-compensated system yields the step error coefficient

$$K_{pc} = \lim_{s\to 0} G_{lag}(s)\ G(s) = \frac{19.5\times 0.2}{0.04\times 15} = 6.5$$

which is larger than the specified value while almost, preserving the location of dominant closed loop poles found in step (*ii*) and so only the desired transient response.

P 8.4: Design a cascade PD compensator for the unity feedback system with transmittance

$$G(s) = \frac{K}{(s+2)^2(s+3)}$$

so that the following design objectives are met:

(*i*) % peak overshoot = 25%

(*ii*) settling time, t_s = 1.6 seconds

Solution: The intended design involves placing a zero on negative real axis so as to force the modified root locus to pass through the dominant root satisfying the prescribed design specification. The design procedure is demonstrated through the following steps:

Step (*i*): The root locus for uncompensated system, is shown in Fig. P 8.4(*a*). The translation of per cent overshoot of 25% into damping ratio ξ, yields

$$e^{-\pi\xi/\sqrt{1-\xi^2}} = 0.25 \qquad \text{or} \qquad \xi = 0.4$$

The constant $\xi = 0.4$ loci (a line) is also shown in Fig. P 8.4(*a*), making an angle of $\cos^{-1}0.4 = 66.2°$ with negative real axis and the point at which this line intersects the root locus of uncompensated system is identified as dominant root, dR at $s = -1 \pm j2.3$ for uncompensated gain K = 19.4.

The settling time generated by dominant root, dR is

$$t_s = \frac{4}{\text{real part [dR]}} = 4 \text{ seconds}$$

In order that specified settling time of 1.6 sec. is achieved, the real part of dominant root must be 4/1.6 = 2.5 and imaginary part of dominant root must be 2.5 × tan 66.4° = 5.7. The dominant root ($-2.5 \pm j5.7$) is identified as dR* in Fig. P 8.4(*a*). Notice that dR* continues to lie on $\xi = 0.4$ line.

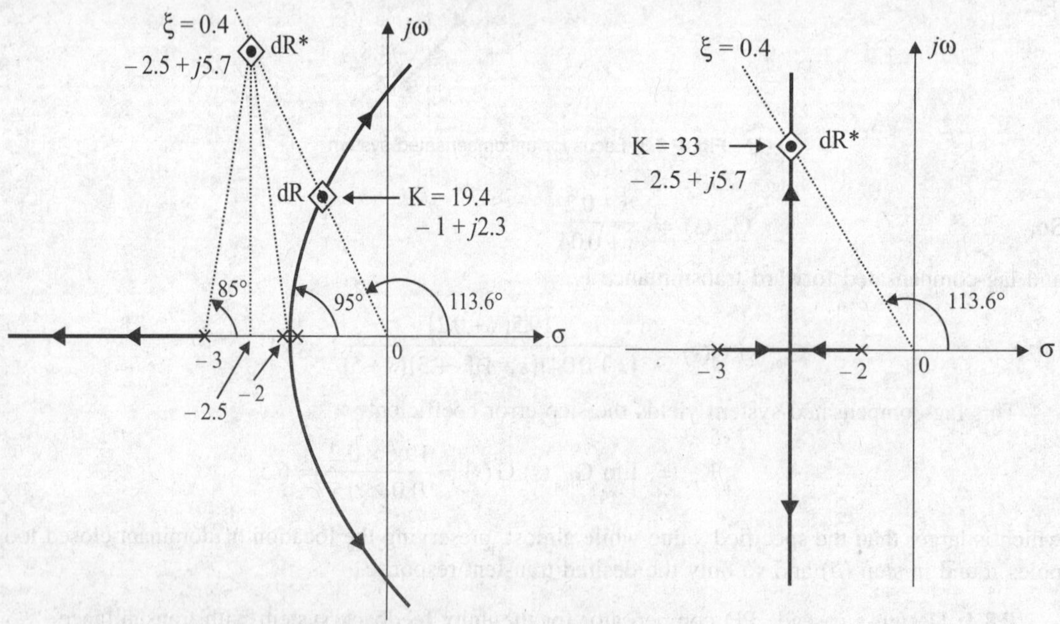

Fig. P 8.4: (*a*) Root locus of uncompensated system and finding the dominant roots that meet the design specifications and (*b*) Root locus of PD compensated system

Step (*ii*): The sum of angles from open loop poles of uncompensated system to dR* = – 95° – 95° – 85° = – 275°.

The angle contribution required from PD compensator zero = 275° – 180° = 95°.

Step (*iii*) (Locating compensator zero): It is apparent from Fig. P 8.4(*a*) that the locating compensator zero at $s = -2$ will contribute + 95° angle so as to make the total sum of angles from open loop poles of uncompensated system together with compensator zero equal to – 180°. This will force the reshaped root locus to pass through the point dR* as shown in Fig. P8.4 (b). The PD compensated transmittance becomes

$$G_{PD}(s)\ G(s) = \frac{33(s+2)}{(s+2)^2(s+3)} = \frac{33}{(s+2)(s+3)}$$

P 8.5: For a system in unity feedback configuration with forward transmittance

$$G(s) = \frac{K}{s^2},$$

a cascade compensator is to be designed to achieve settling time of 1.6 seconds and peak overshoot of 16%. If the compensator zero is located at $s = -1$, do the following:

(a) Locate the dominant roots

(b) Locate the compensator pole

(c) Find the system gain required to meet design specifications.

(d) Find the appropriate error coefficient.

Solution: Before we begin to work out the problem, let us understand the following:

(i) The number type of system is 2. $K_p = K_v = \infty$ and steady state error due to step input and ramp input both will be zero.

(ii) The root locus of uncompensated system, is the entire imaginary axis. The closed loop poles (characteristic roots) will always lie on $j\omega$ axis.

(iii) The peak overshoot of 16% is equivalent to damping ratio $\xi = 0.5$

$$e^{-\pi\xi/\sqrt{1-\xi^2}} = 0.16 \quad \text{gives } \xi = 0.5$$

The dominant roots are to be searched on line $\xi = 0.5$. **Note** that $\cos^{-1} \xi = \cos^{-1} (0.5) = 60°$.

(iv) The settling time of 1.6 seconds is equivalent to real part of dominant root = 4/1.6 = 2.5. The imaginary part of dominant root = 2.5 tan 60° = 4.3 as demonstrated in Fig. P 8.5(a). The dominant root is identified as dR.

(a) The dominant roots are located at $s = -2.5 \pm j4.3$.

(b) Having placed the lead compensator zero at $s = -1$, let us find the sum of angles contributed by system poles together with lead compensator zero as demonstrated in Fig. P 8.5(b). This is equal to $-120° - 120° + 109° = -131°$. So, the lead compensator pole must contribute $-49°$ so that the sum of angles contributed by poles of uncompensated system together with pole-zero pair of lead compensator equals $-180°$ $(= -120° - 120° + 109° - 49°)$ at dominant root, dR. This will ensure that the root of compensated system will pass through dR as shown in Fig. P 8.5(b). Now, using trigonometry, the location of lead compensator pole ($-P_c$) can be found as follows:

$$\frac{4.3}{P_c - 2.5} = \tan 49°$$

or
$$P_c = 6.2$$

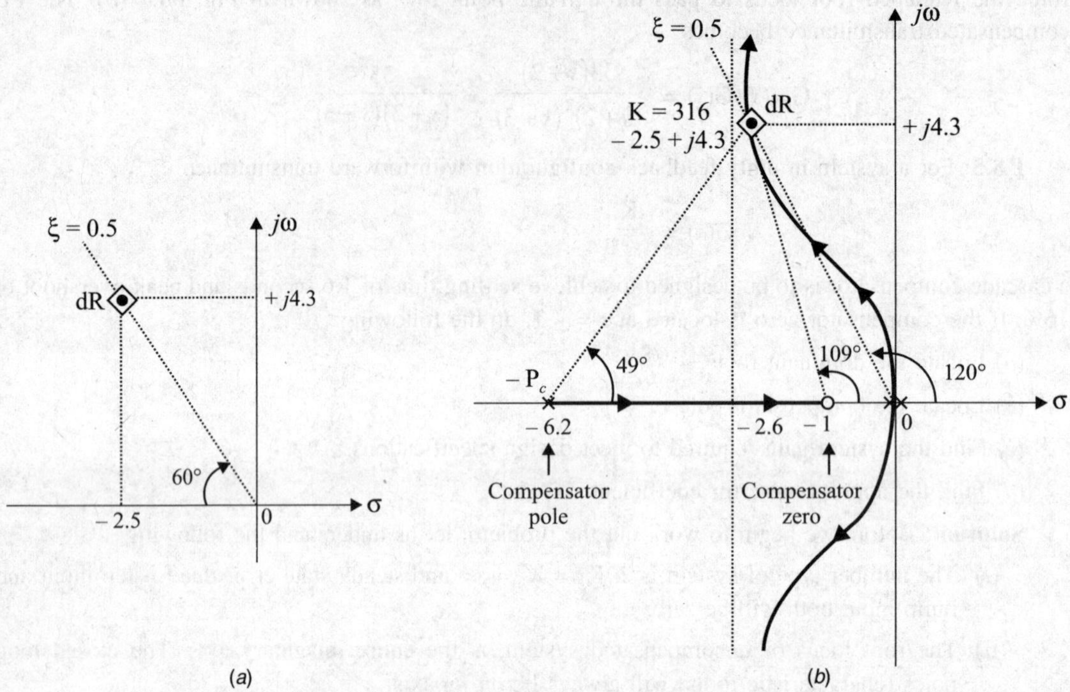

Fig. P 8.5: (*a*) Locating dominant roots and (*b*) Locating compensator pole and root locus of lead of lead compensated system

Thus, the pole of lead compensator is placed at $s = -6.2$ as shown in Fig. P 8.5(*b*).

(*c*) The gain K at dominant root, *d*R is found to be 31.6 using magnitude criterion of root locus.

(*d*) The transmittance of lead compensated system is

$$G_{lead}(s)\ G(s) = \frac{31.6(s+1)}{s^2(s+6.2)}$$

$$K_a = \lim_{s \to 0} s^2\ G_{lead}\ (s)\ G\ (s) = \frac{31.6}{6.2} = 5.096$$

The root locus of compensated system, is also shown in Fig. P8.5(*b*). Notice that centroid $\sigma_A = -2.6$, asymptotic angles = ± 90°, break away point lies at $s = 0$ and system is stable for all $K > 0$.

P 8.6: Design PID compensator for the system with pole-zero function

$$G(s) = \frac{K}{(s+3)(s+6)}$$

in unity feedback configuration so as to meet the following specifications:

Peak overshoot ≤ 1.18%

Settling time ≤ 0.7 seconds

Steady state error for step input = 0

Solution: This design problem is worked out in following two steps:

Step (*i*) (Designing PD compensator): The root locus of uncompensated system is shown in Fig. P 8.6(*a*). From specification on percent overshoot, we have

$$e^{-\pi\xi/\sqrt{1-\xi^2}} = 0.0118 \quad \text{or} \quad \xi = 0.816$$

The constant $\xi = 0.816$ line, making on angle of $\cos^{-1} 0.816 = 35.3°$ with negative real axis, is also shown in Fig. P 8.6(*a*). Searching the dominant root along this line, it is found at point marked dR, at $s = -4.5 \pm j3.2$ for $K_{uc} = 12.96$. K_{uc} is the value of K for uncompensated system. This dominant root, dR generates settling time of 4/4.5 = 0.89 seconds while the specified value is 0.7 seconds.

The specification on settling time of 0.7 seconds, requires that the real part of dominant root be 4/0.7 = 5.7. The imaginary part of dominant root = 5.7 × tan 35.3° ≅ 4. The dominant root dR*, at $s = -5.7 \pm j4$ as depicted in Fig. P 8.6(*a*) satisfies both the specifications on peak overshoot and settling time.

In order to ensure that the root locus of PD compensated system passes through dR*. The location of PD compensator zero is found as follows.

Sum of angles of open loop poles of uncompensated system at dR* = – 124° – 85° = – 209°

The angle required from PD compensator zero = 209° – 180° = 29°. Let the PD compensator zero be located at $s = -\sigma_c$, then using trigonometry σ_c can be determined as:

$$\frac{4}{\sigma_c - 5.7} = \tan 29°$$

or

$$\sigma_c = 12.9$$

This is demonstrated in Fig. P 8.6(*a*). The transmittance of PD compensator is

$$G_{PD}(s) = (s + 12.9)$$

and that of PD compensated system is

$$G_{PD}(s)\, G(s) = \frac{K(s+12.9)}{(s+3)(s+6)}$$

The root locus of PD compensated system with K as variable parameter, is shown in Fig. P 8.6(*b*). The dominant root dR* is found to lie at $s = -5.7 \pm j4$ for K = 2.4.

Step (*ii*) (Designing PI compensator): In order to force the steady state error for step input to be zero, the PI compensator is designed by placing a pole at origin and then PI compensator zero at $s = -0.1$, close to origin.

So, the transmittance of PI compensator is

$$G_{PI}(s) = \frac{s+0.1}{s}$$

and that of PID compensator is

$$G_{PID}(s) = \frac{(s+0.1)(s+12.9)}{s}$$

New pole-zero function of PID compensated system, is

$$G_{PID}(s)\, G(s) = \frac{2.4\,(s+0.1)(s+12.9)}{s(s+3)(s+6)}$$

The root locus of PID compensated system is shown in Fig. P8.6(*c*). It is assumed that the addition of PI compensator introduces negligibly small perturbation in location of dominant root dR* and gain value of K = 2.4. Thus, PI compensator forces the steady state error to zero while preserving the transient response generated by PD compensator.

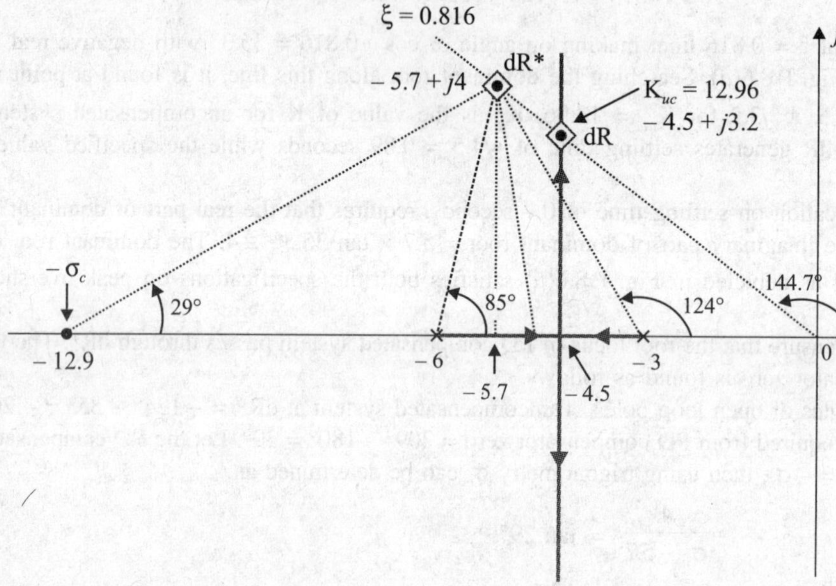

(*a*)

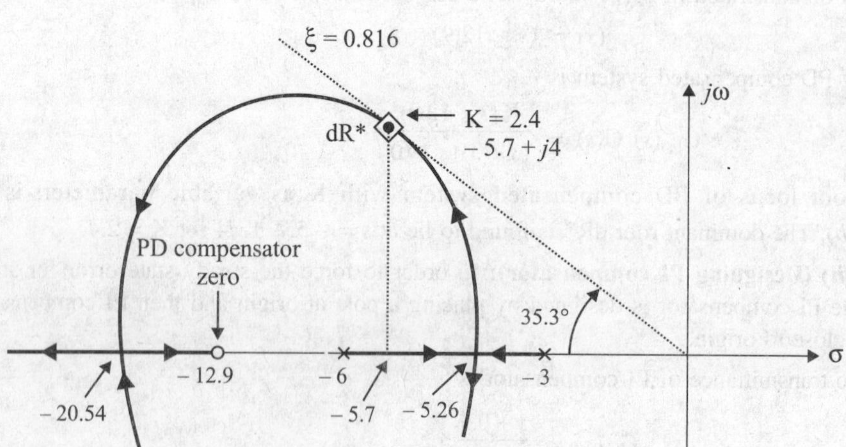

(*b*)

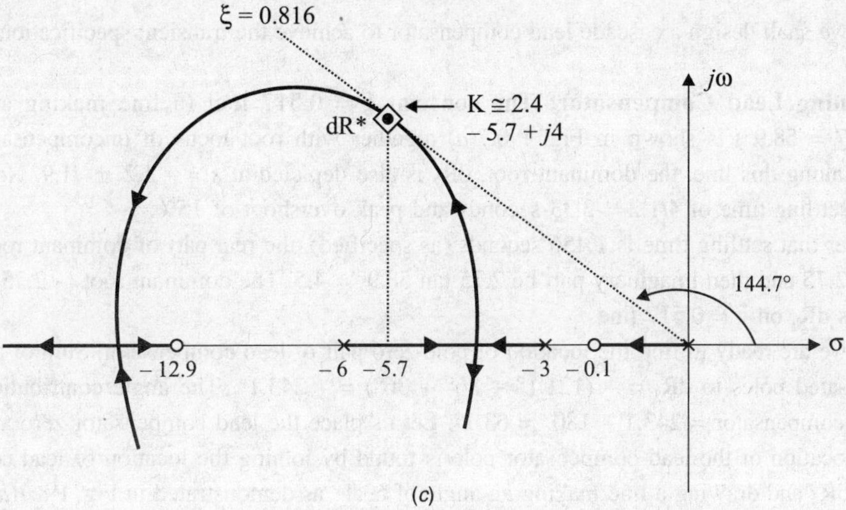

Fig. P 8.6: (*a*) Root locus of uncompensated system and locating zero of PD compensator, (*b*) Root locus (not drawn to the scale) of PD compensated system and (*c*) Root locus (not to the scale) of PID compensated system

P 8.7: A system in unity feedback configuration, has the transmittance

$$G(s) = \frac{K}{s(s+3)(s+9)}.$$

(*a*) What value of K will force the system to exhibit peak overshoot of 20% to a step input?

(*b*) For the value of K found in (*a*), find the settling time, t_s and velocity error coefficient, K_v.

(*c*) Design a cascade compensator such that the following specifications are achieved.

(*i*) Peak overshoot ≤ 15%

(*ii*) Settling time ≤ 2/5 of that found in part (*b*)

(*iii*) $K_v \geq 20$

Solution: (*a*) Translating the peak overshoot of 20% into damping ratio ξ, we have

$$e^{-\pi\xi/\sqrt{1-\xi^2}} = 0.2$$

or

$$\xi = 0.456$$

The root locus for given system, is shown in Fig. P8.7(*a*) constant ξ = 0.456 line making an angle of $\cos^{-1} \xi = 62.9°$, is also shown. The intersection of this line with root locus, shown as dR_1, generates the dominant roots for 20% overshoot. The value of K is found to be 54.9.

(*b*) The real part of the dominant root, *d*R for K = 54.9, is 1.1. This gives settling time $t_s = 4/1.1 = 3.64$ seconds.

Velocity error coefficient $\quad K_v = \lim_{s\to 0} sG(s)\big|_{K=54.9} = \frac{54.9}{27} = 2.03$

(*c*) Before we begin with compensator design, let us find the following:

(*i*) 15% overshoot is equivalent to ξ = 0.517

(*ii*) Specified settling time = 2/5 × 3.64 = 1.456

Now, we shall design a cascade lead compensator to achieve the transient specifications listed just above.

Designing Lead Compensator: The constant ξ = 0.517 loci (a line making an angle of $\cos^{-1}$ 0.517 = 58.9°) is shown in Fig. P 8.7(*a*) together with root locus of uncompensated system. Searching along this line, the dominant root, dR_2 is also depicted at $s = -1.2 \pm j1.9$. Note that dR_2 generates settling time of 4/1.2 = 3.33 seconds and peak overshoot of 15%.

In order that settling time is 1.456 seconds (as specified), the real part of dominant roots must be 4/1.456 = 2.75 and then imaginary part be 2.75 tan 58.9° = 4.5. The dominant root ($-2.75 \pm j\,4.5$) is depicted as dR_3 on ξ = 0.517 line.

Now, we are ready to find the location of pole-zero pair of lead compensator. Sum of angles from uncompensated poles to dR_3 = – (121.1° + 86° + 36°) = – 243.1°. The angle contribution required from lead compensator = 243.1 – 180° = 63.1°. Let us place the lead compensator zero at $s = -3.5$. Then the location of the lead compensator pole is found by joining the location of lead compensator zero with dR_3 and drawing a line making an angle of 63.1° as demonstrated in Fig. P8.7(*a*). This line intersects negative real axis at $s = -18.4$. This is the location of lead compensator pole. So, transmittance of lead compensator is

$$G_{lead}(s) = \frac{s+3.5}{s+18.4}$$

and the pole-zero function of lead compensated system is

$$G_{lead}(s)\ G(s) = \frac{658.6(s+3.5)}{s(s+3)(s+9)(s+18.4)}$$

The root locus of lead compensated system with K as variable parameter, is shown in Fig. P8.7(*b*).

Next, the velocity error coefficient, K_v of lead compensated system is

$$K_v = \lim_{s\to 0} s\, G_{lead}(s)\, G(s) = \frac{658.6\times 3.5}{3\times 9\times 18.4} = 4.64$$

while specified K_v = 20. So, lag-compensator must be designed to improve K_v by almost (20/4.64 = 4.3 ≅ 5) five fold.

Designing Lag-compensator: Let us arbitrarily choose the lag-compensator pole at $s = -0.1$. Then the lag-compensator zero must be located at $s = -0.5$ so as to improve velocity error coefficient K_v from 4.64 to a value ≥ 20. This gives the transmittance of lag-compensator

$$G_{lag}(s) = \frac{s+0.5}{s+0.1}$$

and that of lag-lead compensated system

$$G_{lead}(s)\ G_{lag}(s)\ G(s) = \frac{658.6(s+0.5)(s+3.5)}{s(s+0.1)(s+3)(s+9)(s+18.4)}$$

with the assumption that lag-compensator when serially added to lead compensated system, brings about negligibly small changes in location of dominant root dR_3 and the corresponding gain

K = 658.6. The root locus of lag-lead compensated system is shown in Fig. P 8.7 (*c*). Notice that the root locus is not drawn to the scale. This is merely meant for providing an insight into how the lag-lead compensator reshapes the root locus of uncompensated system to drive it to pass through the dominant root dR_3. Thus, the compensated system meets both the steady state and the relative stability requirements.

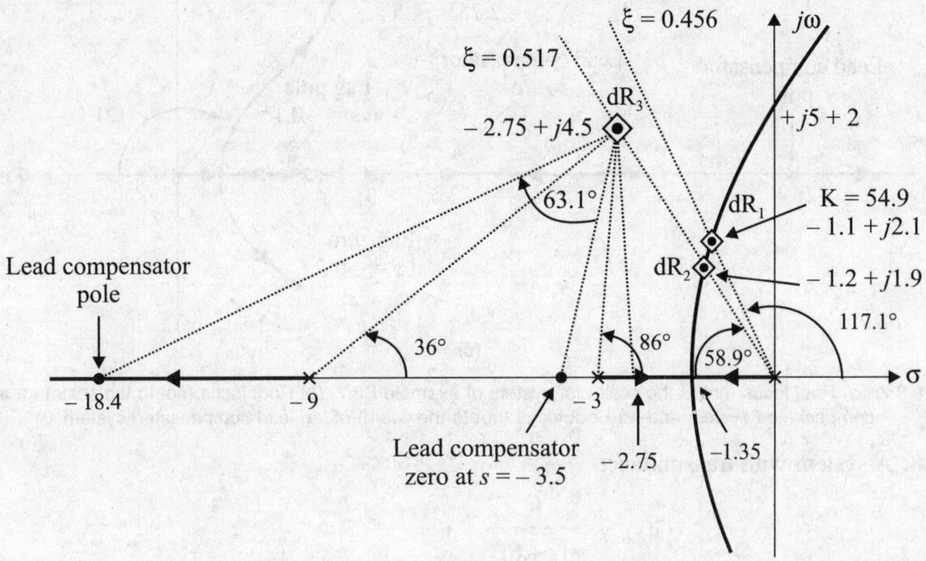

(*a*)

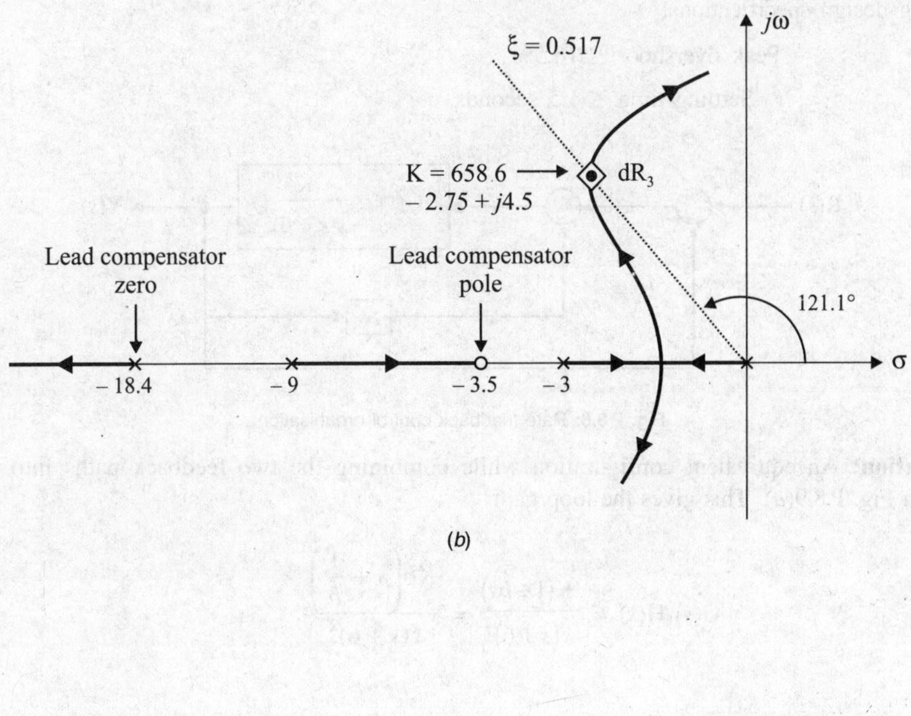

(*b*)

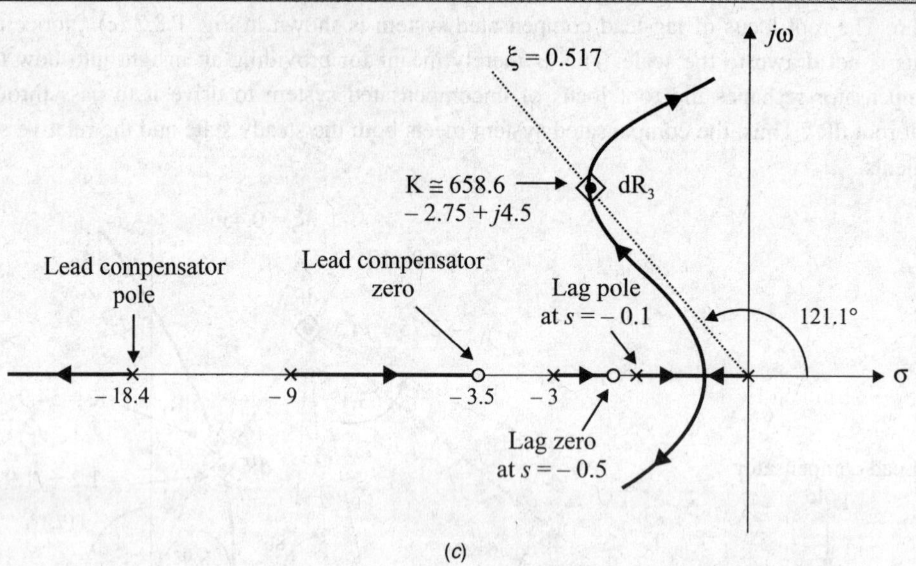

Fig. P 8.7: (*a*) Root locus (not to the scale) for system of example P8.7, (*b*) Root locus (not to the scale) for lead compensated system and (*c*) Root locus (not to the scale) of lag-lead compensated system

P 8.8: A system with transmittance

$$G(s) = \frac{K}{s(s+6)^2}$$

is placed in rate feedback control organisation as shown in Fig. P 8.8. Find **K** and **b** so as to meet the following design specifications:

Peak overshoot $\leq$ 16.5%

Settling time $\leq$ 1.5 seconds

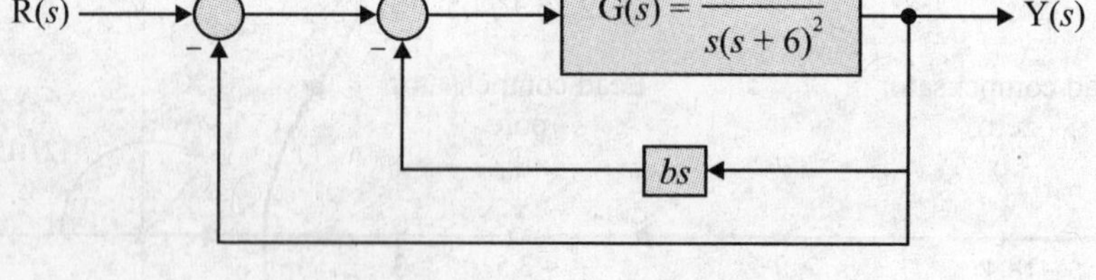

Fig. P 8.8: Rate feedback control organisation

Solution: An equivalent configuration while combining the two feedback paths into one, is shown in Fig. P 8.9(*a*). This gives the loop gain

$$G(s)\ H(s) = \frac{K(1+bs)}{s(s+6)^2} = \frac{Kb\left(s + \dfrac{1}{b}\right)}{s\,(s+6)^2}$$

Thus, the design of rate compensator involves adding a zero at $s = -1/b$ to the existing poles of G(s) at $s = 0$, $s = -6$ and $s = -6$. A final adjustment of gain K, meets the design goals.

The root locus for uncompensated system function G(s) with K as variable parameter, is shown in Fig. P 8.9(b).

The specification on peak overshoot of 16.3% is equivalent to damping ratio $\xi = 0.5$. The dominant root, searched along the line making an angle of $\cos^{-1} 0.5 = 60°$ with negative real axis, is found at $s = -1.4 \pm j2.4$ for $K_{uc} = 75.7$. This is marked as dR in Fig. P 8.9(b). Note that $K_{uc} = 75.7$ is value of K of uncompensated system satisfying the specified value of peak overshoot. This generates settling time of 4/1.4 = 2.857 seconds while the specified value is 1.5 seconds.

In order to achieve settling time of 1.5 seconds, the real part of new dominant root must be 4/1.5 = 2.66. Then, the imaginary part of new dominant root, satisfying both the transient response requirements (peak overshoot of 16.3%. and settling time of 1.5 seconds) will be 2.66 × tan 60° = 4.6. This dominant root is depicted as dR* at $s = -2.66 \pm j4.6$ in Fig. P 8.9(b).

Locating compensator zero as demonstrated in Fig. P8.9(b) sum of angles of poles of uncompensated system at dR* = – (120° + 53° + 53°) = – 226° and angle required to be contributed by compensator zero = 226° – 180° = 46°.

Then, to drive the root locus of compensated system to pass through point dR*, the compensator zero must be so located that sum of angles contributed by poles of existing system together with compensator zero, must be – 180°. The location of compensator zero ($s = -\sigma_c$), as demonstrated in Fig. P 8.9(b), is found using trigonometry as follows:

$$\frac{4.6}{\sigma_c - 2.66} = \tan 46°$$

or $$\sigma_c = 7.1$$

since, $$\sigma_c = \frac{1}{b}, \quad b = 0.1408$$

The root locus for rate feedback compensated system

$$G(s)\,H(s) = \frac{Kb\left(s + \frac{1}{b}\right)}{s\,(s + b)^2}$$

is shown in Fig. P 8.9(c) with Kb as variable parameter. In fact K varies from 0 to ∞ and b remains fixed at 0.1408. **Notice** that the desired dominant roots (dR*) are found at $s = -2.6 \pm j4.6$ for $Kb = 26.9$ where K = 191 and $b = 0.1408$. The design point dR* lies on $\xi = 0.5$ line to satisfy peak overshoot requirements and its real part being equal to 2.66, satisfies settling time requirement.

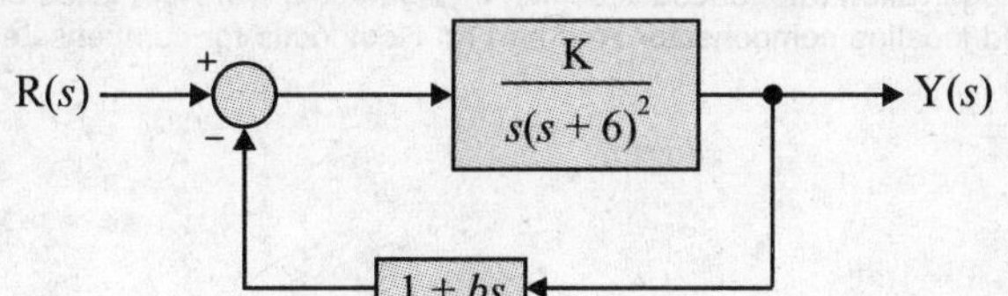

(a)

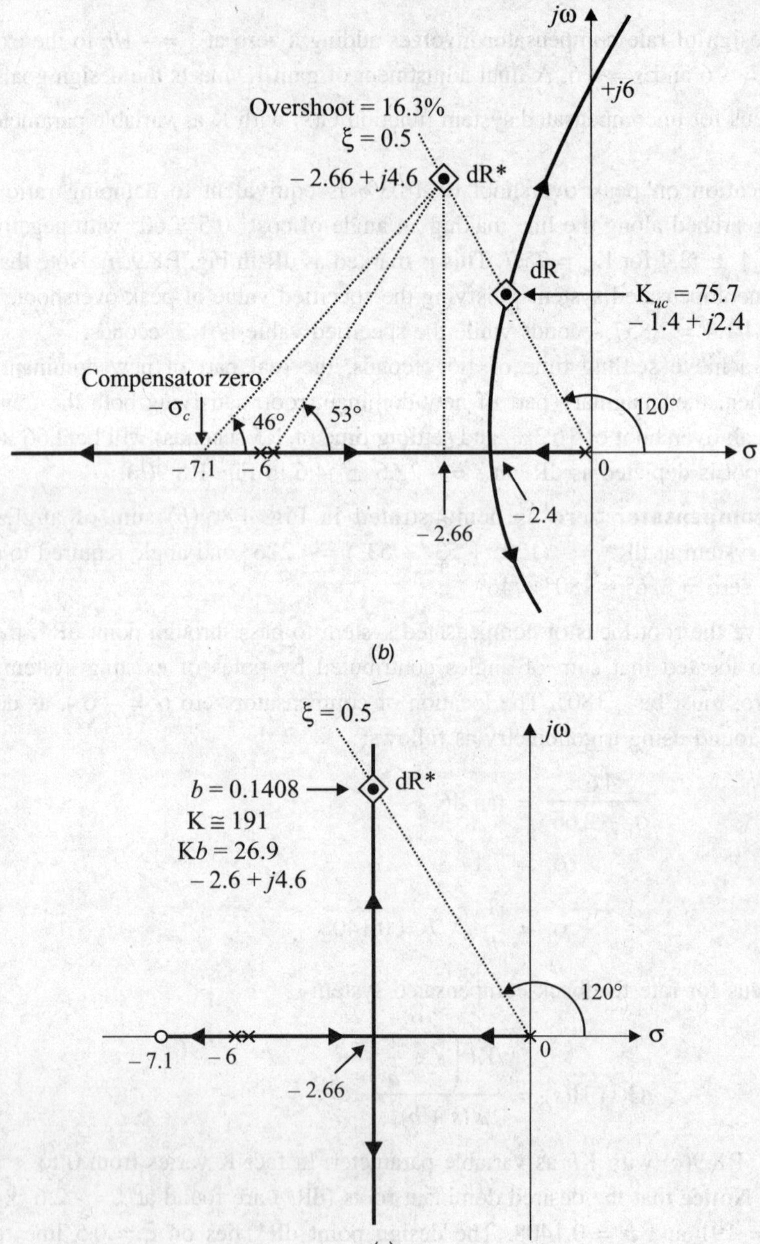

Fig. P 8.9: (*a*) An equivalent rate feedback control configuration, (*b*) Root locus of uncompensated system and locating compensator zero and (*c*) Root locus for compensated system

P 8.9: A feedback compensated system is shown below. Do the following:

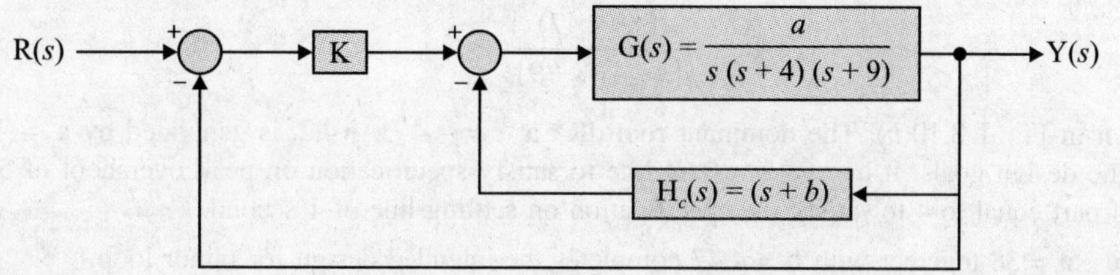

(*a*) Find the value of **a** and **b** in minor feedback loop so as to achieve the settling time of 1 second with 5% peak overshoot for the step response.

(*b*) Find the value of K so as to force the major loop response to exhibit peak overshoot of 10% to a step input.

Solution: (*a*) The open loop transfer function of minor loop is

$$G(s)\,H_c(s) = \frac{a(s+b)}{s(s+4)(s+9)}$$

The intended design, in fact, involves drawing root locus for $G(s) = \dfrac{a}{s(s+4)(s+9)}$ with **a** as variable parameter and finding **a** for 5% peak overshoot (equivalent to damping ratio $\xi = 0.69$). Finally the value of **b** which is location of compensator zero, is found to satisfy the specification on settling time of 1 second.

The root locus for $G(s) = \dfrac{a}{s(s+4)(s+9)}$ with **a** as variable parameter is shown in Fig. P 8.10(*a*).

The point of intersection of this root locus with $\xi = 0.69$ line, is depicted as dR and found to be located at $s = -1.6 + j1.68$. The dominant roots at $s = -1.6 \pm j1.68$, generates 5% peak overshoot but the settling time (4/1.6 = 2.5 seconds) remains larger than the specified value of 1 second.

In order to adjust the settling time equal to 1 second while preserving the specified peak overshoot (5%), the real part of dominant root must be 4 and then imaginary part be $4 \times \tan 46.4° = 4.2$. The location of such a dominant root is depicted as point dR* in Fig. P8.10(*a*) at $s = -4 \pm j4.2$.

Sum of angles contributed by poles of $G(s)$ at dR* $= -133.6° - 90° - 40° = -263.6°$

Angle required to be contributed by compensator zero $= 263.6° - 180° = 83.6°$

As demonstrated in Fig. P 8.10(*a*), the location of compensator zero ($s = -b$), using trigonometry, is found as follows:

$$\frac{4.2}{b-4} = \tan 83.6°$$

or

$$b = 4.47$$

Now, the root locus with **a** as variable parameter for

$$G(s)\ H_c(s) = \frac{a(s+4.47)}{s(s+4)(s+9)}$$

is shown in Fig. P 8.10(*b*). The dominant root dR* at $s = -4 \pm j\,4.2$, is generated by **a** ≅ **38** and meets the design goals. It lies on ξ = 0.69 line to satisfy specification on peak overshoot of 5% and has real part equal to 4 to satisfy the specification on settling line of 1 second.

Thus, **a** = **38** together with **b** = **4.47** completes the intended design for minor loop.

(*b*) The open loop transfer function $G_m(s)$ for major loop is

$$G_m(s) = \frac{KG(s)}{1+G(s)H_c(s)} = \frac{aK}{s(s+4)(s+9)+a(s+b)}$$

which on substituting values of **a** and **b** found above, becomes.

$$G_m(s) = \frac{38K}{s^3+13s^2+74s+169.86}$$

$$\cong \frac{38K}{(s+5)(s+4+j4.2)+(s+4-j4.2)}$$

This is demonstrated in Fig. P 8.10(*c*) by eleminating the minor loop. The root locus for $G_m(s)$ with K* = 38 K as variable parameter, is shown in Fig. P 8.10(*d*). The peak overshoot of 10% is equivalent to damping ratio ξ = 0.59. The constant ξ = 0.59 line is also shown in Fig. P 8.10(*d*). This line intersects the root locus at $s = -3.1 \pm j4.35$ for K* = 37 and K = 37/38. So, K ≅ 1 forces the major loop response to exhibit peak overshoot of 10% for a step input.

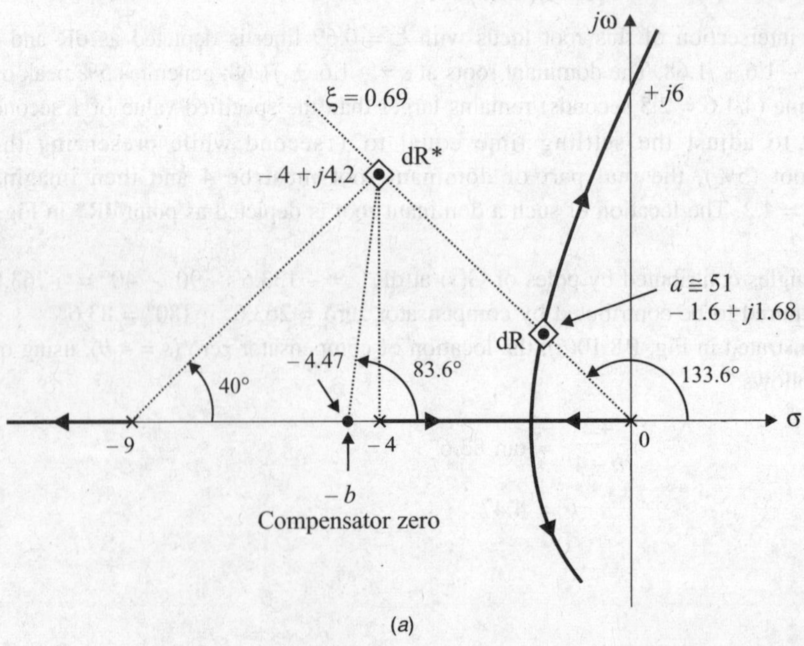

(*a*)

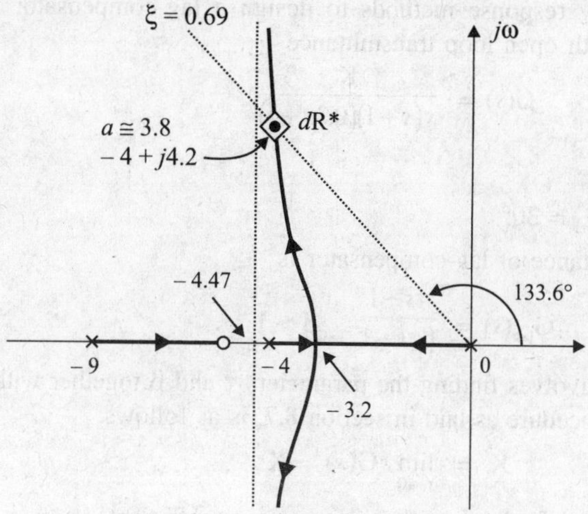

(b)

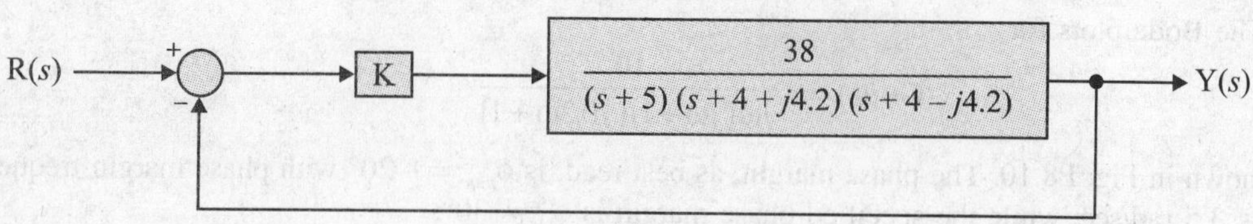

(c)

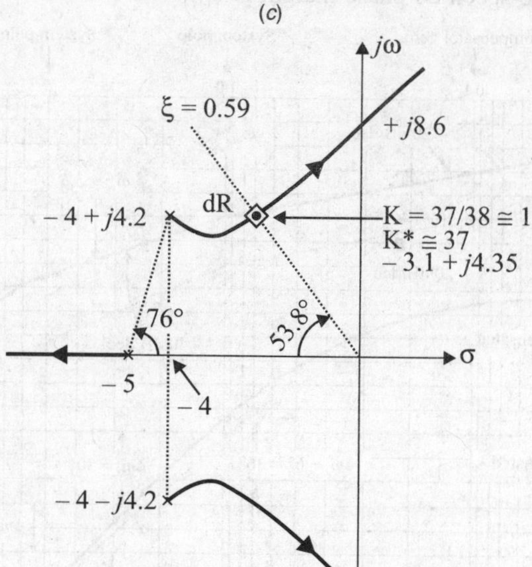

(d)

Fig. P 8.10: (*a*) Root locus for $G(s) = \dfrac{a}{s(s+4)(s+9)}$ and locating compensator zero, (*b*) Root locus compensated minor loop. Fig. P 8.10, (*c*) Equivalent system with unity feedback, minor loop eliminated and (*d*) Root locus for open loop transfer function $G_m(s)$ of major loop

P 8.10: Use frequency response methods to design a lag-compensator for a system in unity feedback configuration, with open loop transmittance

$$G(s) = \frac{K}{s(s+1)(0.2s+1)}$$

The design goals are

(*i*) $K_v = 10$

(*ii*) Phase margin $\phi_{ms} = 30°$

Solution: The transmittance of lag-compensator is

$$G_{lag}(s) = \frac{\tau s+1}{\beta\tau s+1}; \quad \beta > 1$$

The intended design involves finding the parameters τ and β together with K to meet the design goals. The step by step procedure as laid in section 8.7, is as follows:

Step (*i*): $$K_v = \lim_{s\to 0} sG(s) = K$$

So, K must be 10 to satisfy the specification on error coefficient

The Bode plots for

$$G(j\omega) = \frac{10}{j\omega(j\omega+1)(j0.2\omega+1)}$$

are shown in Fig. P8.10. The phase margin, as best read, is $\phi_{muc} = -20°$ with phase margin frequency $\omega_{gc} \cong 3.5$ rad/sec. while the specified phase margin is $\phi_{ms} = 30°$.

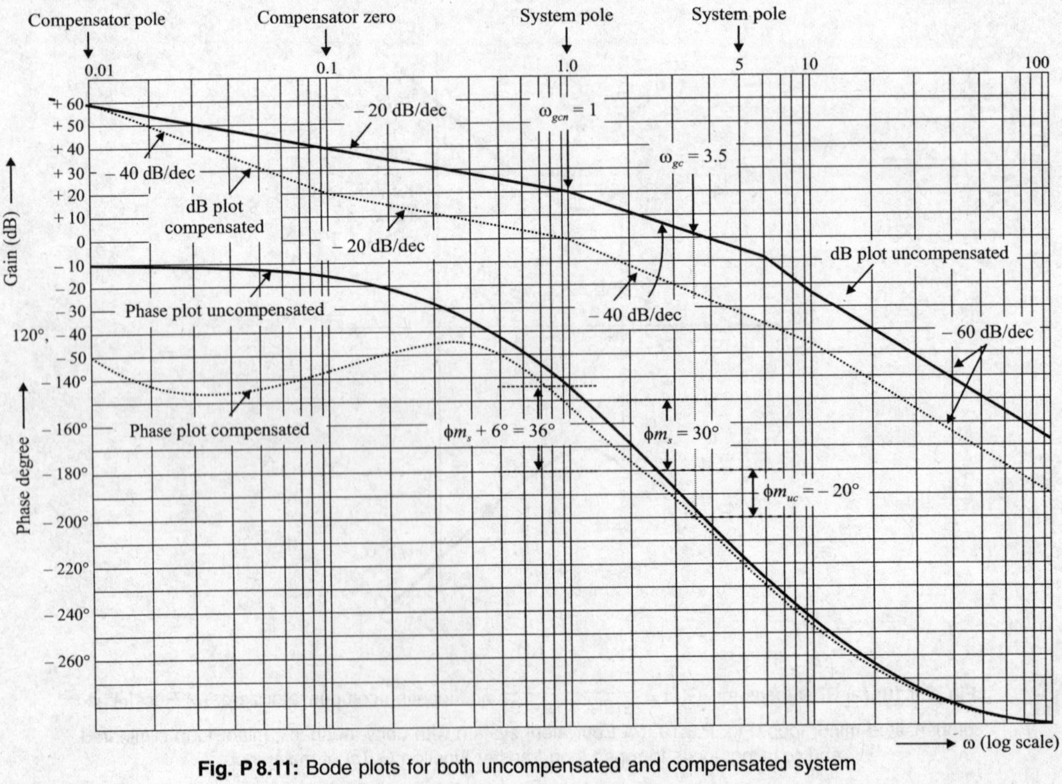

Fig. P 8.11: Bode plots for both uncompensated and compensated system

Step (*ii*): The new gain crossover frequency, ω_{gcn} where the compensated system is required to yield phase margin = ϕ_{ms} + 6° = 36°, is 1 rad/sec. as depicted in Fig. P 8.11. Notice that specified phase margin has been raised by 6° in order to compensate for phase angle contribution of lag-compensator at ω_{gcn}.

Step (*iii*): Selecting upper corner frequency, ω_H of lag-compensator, one decade below ω_{gn} = 1, we have

$$\omega_H = 1/\tau = 0.1$$

Step (*iv*): The required reduction in gain (dB) to bring down the uncompensated dB plot to 0 dB at ω_{gcn} = 1 rad/sec., is 20 dB.

Equating it to 20 lag β, we have

$$20 \text{ lag } \beta = 20 \quad \text{or} \quad \beta = 10$$

Then, the lower corner frequency of lag-compensator is

$$\omega_L = \frac{1}{\beta\tau} = 0.01$$

This gives the transmittance of lag-compensator

$$G_{lag}(s) = \frac{10s+1}{100s+1}$$

and thus, the new pole-zero function of lag compensated system, is

$$G_{lag}(s)\, G(s) = \frac{10\,(10s+1)}{s(s+1)\,(0.2s+1)(100s+1)}$$

The Bode plots (dB and phase both) for lag-compensated system, is also shown in Fig. P 8.11. Specified phase margin ϕ_{ms} = 30° is achieved. The lag-compensator does not alter the slope of initial segment of dB plot. So, the error coefficient adjusted to K_v = 10 by virtue of selecting K = 10 before inserting the lag-compensator, remains unchanged.

P 8.11: A unity feedback system has the loop transmittance

$$G(s) = \frac{1000K}{s(s+40)(s+100)}$$

Design a lead compensator so as to achieve the following specifications:

(*i*) Peak overshoot = 15%

(*ii*) Error coefficient K_v = 40

(*iii*) Peak time = 0.1 second.

Use frequency response approach.

Solution: Before we begin with design, let us translate the time domain specifications into frequency domain specifications.

The relation $$e^{-\pi\xi/\sqrt{1-\xi^2}} = 0.15$$

gives $\xi = 0.517$ and the relation

$$\text{Phase margin} = \tan^{-1}\left[\frac{2\xi}{\left(\sqrt{4\xi^4+1}-2\xi^2\right)^{1/2}}\right]$$

gives phase margin $\phi_{ms} \cong 53°$

The specific bandwidth, $$\omega_b = \frac{\pi}{t_p\sqrt{1-\xi^2}}\left[\left(1-2\xi^2\right)+\sqrt{4\xi^4-4\xi^2+2}\right]^{1/2}$$

$$\cong 46 \text{ rad/sec.}$$

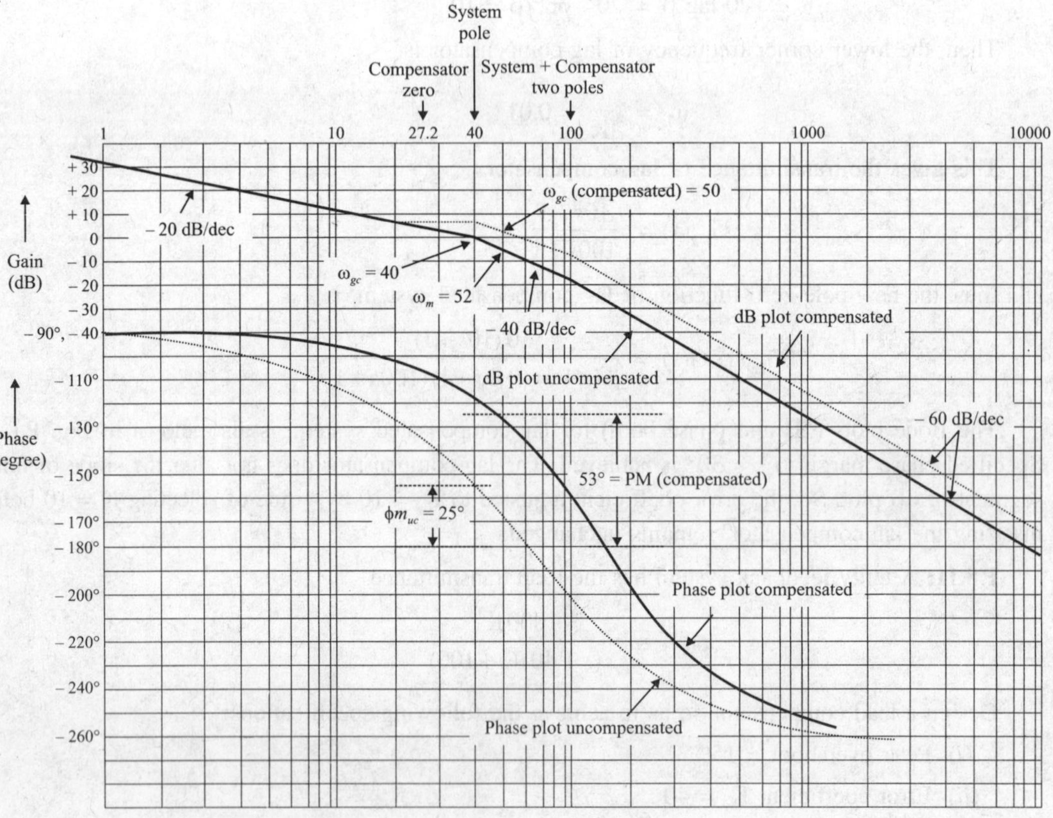

Fig. P8.12: Bode plots for both compensated and uncompensated systems

The uncompensated system in time constant form is

$$G(s) = \frac{K/4}{s\left(1+\dfrac{s}{40}\right)\left(1+\dfrac{s}{100}\right)}$$

The error coefficient $\quad K_v = \lim_{s\to 0} sG(s) = \frac{K}{4}$

Equating it to specified $K_v = 40$, we have K = 160

Design steps: Recall that the transmittance of lead compensator is $G_{lead}(s) = \alpha\frac{\tau s+1}{\alpha\tau s+1}$. So, the design involves finding the parameters τ and α. This is demonstrated in following steps:

Step (*i*): The Bode plot for

$$G(j\omega) = \frac{40}{j\omega\left(1+\frac{j\omega}{40}\right)\left(1+\frac{j\omega}{100}\right)}$$

is shown in Fig. P 8.12. The phase margin, as best read, is $\phi_{muc} = 25°$.

Step (*ii*): The additional phase contribution required from lead compensator is

$$\phi_m = \phi_{ms} + \phi_c - \phi_{muc} = 53° + 7° - 25° = 35°$$

Step (*iii*): $\quad \alpha = \frac{1-\sin\phi_m}{1+\sin\phi_m} = \frac{1-\sin 35°}{1+\sin 35°} = 0.27$

Step (*iv*): $\quad -10\log(1/\alpha) = -5.69$

Searching the frequency where uncompensated system contributes gain equal to – 5.69 dB, we have $\omega_m \cong 52$ rad/sec.

Step (*v*): $\quad \tau = \frac{1}{\omega_m\sqrt{\alpha}} = 0.037 \quad$ and $\quad \alpha\tau = 0.27 \times 0.037 = 9.99 \times 10^{-3}$

and the lead compensator has the transmittance

$$G_{lead}(s) = 0.27\frac{0.037s+1}{9.99\times 10^{-3}s+1} \cong 0.27\left[\frac{0.037s+1}{0.01s+1}\right]$$

Finally, we insert an amplifier with gain = 3.7 in order to offset the attenuation caused by lead network. Note $1/\alpha = 3.7$.

The Bode plot for lead compensated system

$$G_{lead}(j\omega)\,G(j\omega) = \frac{40(j0.037\omega+1)}{j\omega\left(1+\frac{j\omega}{40}\right)\left(1+\frac{j\omega}{100}\right)^2}$$

is also shown in Fig. P 8.12. Notice that the lead compensated system exhibits phase margin of 53°. The frequency corresponding to –7 dB gain, as best read from dB plot (compensated) is approximately 90 rad/sec. This gives the estimate of closed loop bandwidth to be 90 rad/sec which is larger than the requirement of 46 rad/sec. Thus, the peak time specification is also met.

P 8.12: Use frequency response approach to design a lag-lead compensator for a unity feedback system where the loop transmittance is

$$G(s) = \frac{K(s+8)}{s(s+4)(s+20)}$$

The design goals are

(*i*) peak overshoot = 15%

(*ii*) settling time = 0.1 sec

(*iii*) velocity error coefficient K_v = 1000.

Solution: Let us do the following time domain to frequency domain translations before the design is tried.

(*i*) Peak overshoot of 15% is equivalent to damping ratio $\xi = 0.517$ and $\xi = 0.517$ is equivalent to phase margin $\phi_{ms} \cong 53°$ [use equation (8.35)].

(*ii*) The settling time of 0.1 sec together with $\xi = 0.517$, is equivalent to bandwidth $\omega_b = 97.3 \cong 97$ rad/sec. [use equation (8.37(*b*))].

(*iii*) $K_v = \lim_{s\to 0} sG(s) = \frac{K}{10}$. Equating it to the specification on K_v, we have K/10 = 1000 or $K = 10^4$.

(*iv*) Translating the pole-zero format of

$$G(s) = \frac{10^4(s+8)}{s(s+4)(s+20)}$$

into sinusoidal transfer function $G(j\omega)$ in time constant form, we have

$$G(j\omega) = \frac{10^3\left(1+\frac{j\omega}{8}\right)}{j\omega\left(1+\frac{j\omega}{4}\right)\left(1+\frac{j\omega}{20}\right)}$$

whose Bode plot is shown in Fig. P8.13 where on P_s and Z_s denote system poles and zeros respectively. Recall that the transmittance of a single, passive lag-lead compensator is

$$G_{lgld}(s) = \frac{\left(s+\frac{1}{\tau_1}\right)}{\left(s+\frac{\beta}{\tau_1}\right)} \cdot \frac{\left(s+\frac{1}{\tau_2}\right)}{\left(s+\frac{1}{\beta\tau_2}\right)}; \quad \beta > 1$$

So, the design procedure involves choosing the parameters τ_1, τ_2 and β to meet the design goals. This is demonstrated in the following steps:

Step (*i*): Let us arbitrarily select the new gain crossover frequency ω_{gcn} = 60 rad/sec. Phase angle contributed by uncompensated system at $\omega = \omega_{gcn}$ = 60 rad/sec, is – 166°. The specified phase margin ϕ_{ms} = 53°. Let us raise it by 7° to compensate for phase contributed by lag-compensator at $\omega = \omega_{gcn}$. Then, the phase angle of compensated system, is required to be – 180° + 53° + 7° = – 120° at $\omega = \omega_{gcn}$ to achieve specified phase margin.

So, the additional phase lead required to be contributed by lead compensator, is 166° – 120° = 46°. Recall that maximum phase lead contributed by lead compensator in terms of parameter β, is given by

$$\phi_{max} = \sin^{-1}\left[\frac{1-\frac{1}{\beta}}{1+\frac{1}{\beta}}\right]$$

and $$\phi_{max}\big|_{\beta=10} \cong 55°$$

So, selecting $\beta = 10$ will suffice in the present design problem which requires phase lead of only about 46°.

Step (*ii*) (Designing lag-compensator)

The upper corner frequency of lag-compensator $= \omega_H = \dfrac{1}{\tau_2} = \dfrac{\omega_{gcn}}{10} = 6$

The lower corner frequency of lag-compensator $= \omega_L = \dfrac{1}{\beta\tau_2} = \dfrac{6}{10} = 0.6$

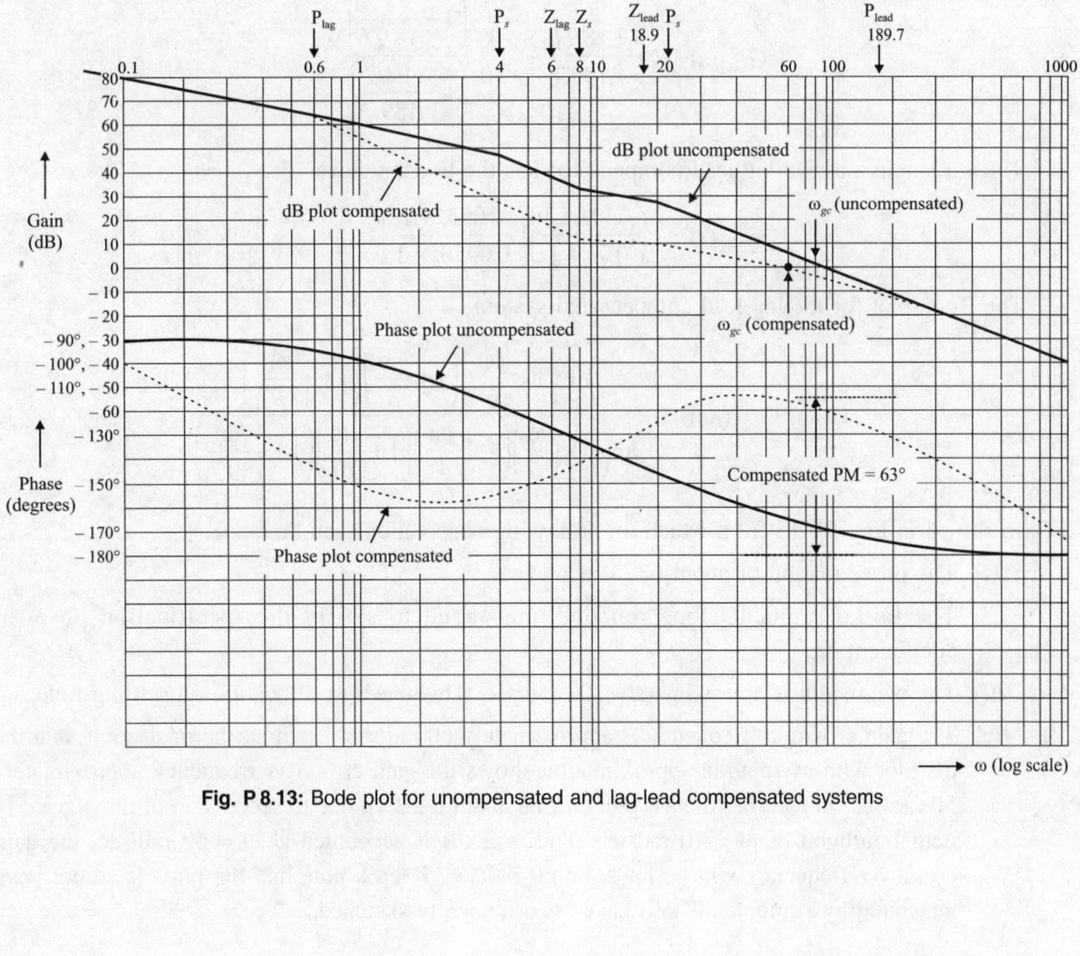

Fig. P 8.13: Bode plot for unompensated and lag-lead compensated systems

The transmittance of lag-compensator is

$$G_{lag}(s) = \frac{s+6}{s+0.6} = 10\frac{\left(1+\dfrac{s}{6}\right)}{\left(1+\dfrac{s}{0.6}\right)}$$

Step (*iii*) (Designing lead compensator)

Use the relation $\omega_{gcn} = \dfrac{\sqrt{\beta}}{\tau_1}$ to get the lower corner frequency of lead compensator,

$$\omega_L = \frac{1}{\tau_1} = \frac{\omega_{gcn}}{\sqrt{\beta}} = \frac{60}{\sqrt{10}} = 18.97$$

Then, upper corner frequency of lead compensator = $\omega_H = \dfrac{\beta}{\tau_1} = 189.7$

and the transmittance of lead compensator is

$$G_{lead}(s) = \frac{s+18.97}{s+189.7} = \frac{1}{10}\frac{\left(1+\dfrac{s}{18.97}\right)}{\left(1+\dfrac{s}{189.7}\right)}$$

Thus, a single, passive lag-lead compensator has the transmittance

$$G_{lgld}(s) = \left(\frac{0.17s+1}{1.7s+1}\right)\left(\frac{0.053s+1}{0.0053s+1}\right)$$

The Bode plot for the lag-lead compensated system

$$G(j\omega)\, G_{lgld}(j\omega) = \frac{10^3\left(1+\dfrac{j\omega}{6}\right)\left(1+\dfrac{j\omega}{8}\right)\left(1+\dfrac{j\omega}{18.97}\right)}{j\omega\left(1+\dfrac{j\omega}{0.6}\right)\left(1+\dfrac{j\omega}{4}\right)\left(1+\dfrac{j\omega}{20}\right)\left(1+\dfrac{j\omega}{189.7}\right)}$$

is also shown in Fig. P 8.13 from which the following observations can be made:

(*i*) The phase margin of about 63° is achieved.

(*ii*) The initial segment slope remains unchanged to satisfy the specification on error coefficient K_v.

(*iii*) The bandwidth is approximately 100 rad/sec. This meets specification on settling time.

(*iv*) The gain crossover frequency is approximately 60 rad/sec. Let it not be misleading that the dB plot with asymptotic approximation shows the gain crossover frequency approximately 50 rad/sec. In fact the true dB plot will be about 3 dB up due to succession of three zeros in neighbourhood of ω = 10 rad/sec. This + 3 dB if accounted at ω = 60 rad/sec, the gain crossover frequency will perhaps, be 60 rad/sec. Keep a note that the plots have not been generated by computer. These have been manually sketched.

P 8.13: A system in unity feedback configuration, has the pole-zero function $\dfrac{K}{s(s+50)(s+100)}$

Design the value of gain K for 15% peak overshoot in closed loop step response using frequency response techniques.

Solution: Transforming the system into time constant form, we have

$$\frac{K/5000}{j\omega\left(1+\dfrac{j\omega}{50}\right)\left(1+\dfrac{j\omega}{100}\right)}$$

Identifying the corner frequencies as 50 and 100 rad/sec., the Bode plot for K = 50,000, is shown in Fig. P 8.14. Notice that 20 log (50,000/5,000) = 20 dB will be the gain at ω = 1 rad/sec; on dB plot. The Bode plot, as best read, reveals the following:

(*i*) The gain crossover frequency ω_{gc} for K = 50,000 is approximately 14 rad/sec.

(*ii*) The phase crossover frequency ω_{pc} for K = 50,000 is approximately 65 rad/sec.

(*iii*) Phase margin for K = 50,000 is approximately 70°.

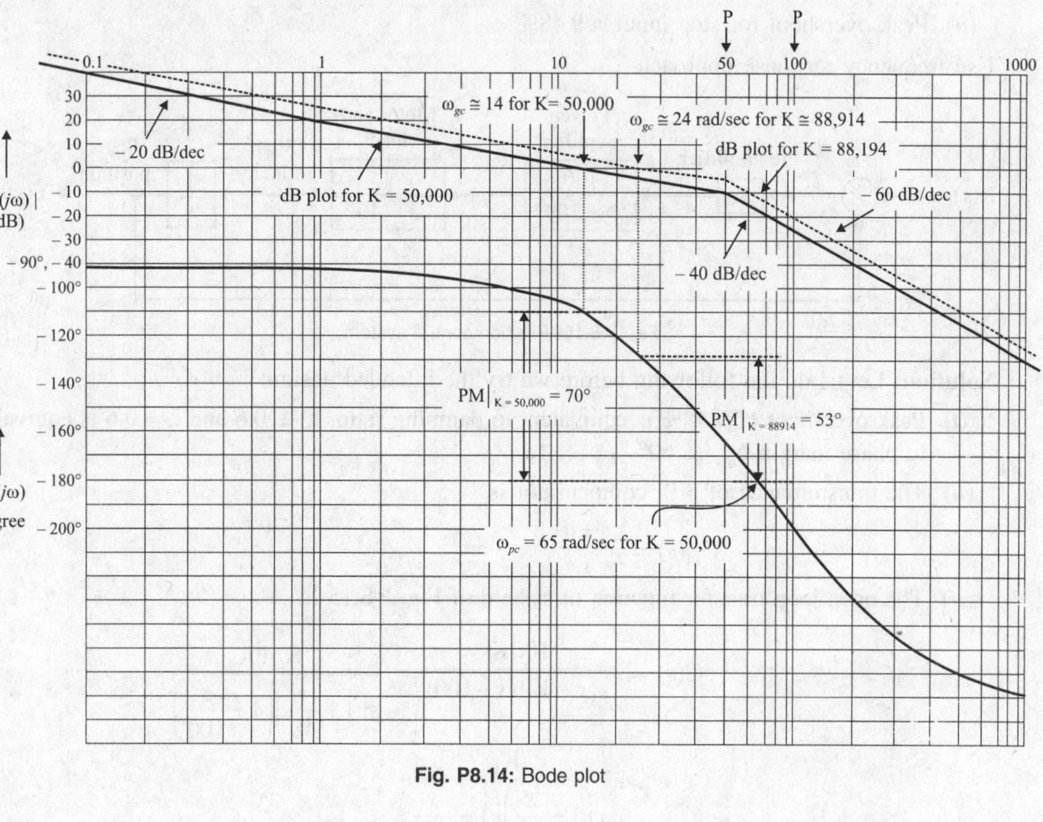

Fig. P8.14: Bode plot

The peak overshoot of 15% when translated into damping ratio ξ using the relationship

$$e^{-\pi\xi/\sqrt{1-\xi^2}} = 0.15$$

gives $\xi = 0.517$ and ξ translated into phase margin (PM) using relationship

$$\text{PM} = \tan^{-1}\left[\frac{2\xi}{\left(\sqrt{4\xi^4+1}-2\xi^2\right)^{1/2}}\right]$$

gives PM $\cong$ 53°

As demonstrated in Fig. P8.14, the gain crossover frequency must be 24 rad/sec for the system to exhibit phase margin of 53°. This requires the dB plot to be shifted upward by approximately 5 dB. Then, the value of gain K to generate phase margin of 53° is

$$K\big|_{PM=53°} = 50{,}000 \times \text{antilog}(5/20) \cong 88914.$$

P 8.14: Design a PI compensator for the position control system shown in Fig. P8.15, to achieve the following design goals:

(*i*) Steady state error for ramp input = 0

(*ii*) Peak overshoot for step input = 9.48%

Use frequency response approach.

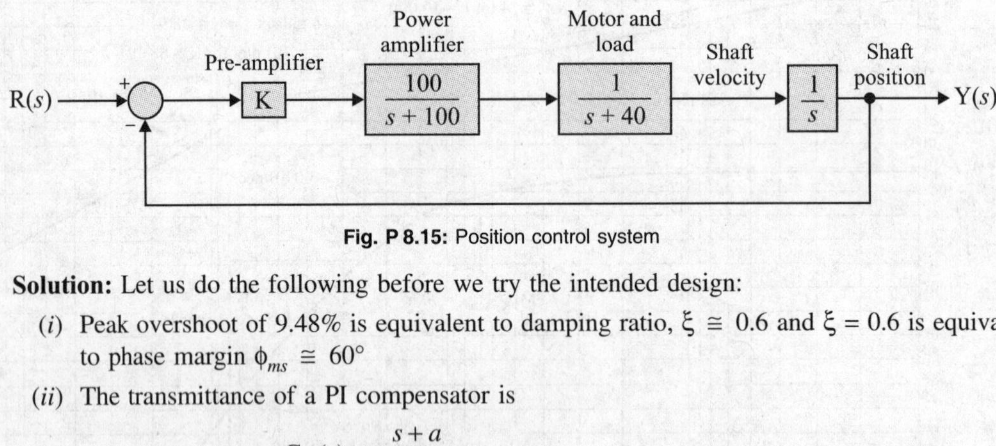

Fig. P 8.15: Position control system

Solution: Let us do the following before we try the intended design:

(*i*) Peak overshoot of 9.48% is equivalent to damping ratio, $\xi \cong 0.6$ and $\xi = 0.6$ is equivalent to phase margin $\phi_{ms} \cong 60°$

(*ii*) The transmittance of a PI compensator is

$$G_{PI}(s) = \frac{s+a}{s}$$

(*iii*) The open loop transfer function of system of Fig. P8.15 is

$$G(s) = \frac{100K}{s(s+40)(s+100)} = \frac{K/40}{s\left(1+\frac{s}{40}\right)\left(1+\frac{s}{100}\right)}$$

$$= \frac{K_1}{s\left(1+\frac{s}{40}\right)\left(1+\frac{s}{100}\right)}$$

where $K_1 = K/40$.

Now, we are ready to attempt the design with following steps:

Step (*i*) The Bode plot for uncompensated system of Fig. P 8.15 for $K_1 = 1$, that is

$$G(j\omega)\big|_{K_1=1} = \frac{1}{j\omega\left(1+\frac{j\omega}{40}\right)\left(1+\frac{j\omega}{100}\right)}$$

is shown in Fig. P8.15(*a*). The specified phase margin is $\phi_{ms} \cong 60°$. Let us raise it by 5° in order to compensate for phase lag contributed by PI compensator at required gain crossover frequency or phase margin frequency. Now, to generate phase margin of 60° + 5° = 65°, the phase angle must be – 180° + 65° = – 115°. As demonstrated in Fig. P 8.15(*a*), the phase angle of – 115°, results at $\omega \cong 14$ rad/sec. as best read from phase plot of uncompensated system.

Step (*ii*): In order that the specification of 60° on phase margin is met, let us choose a new gain crossover frequency $\omega_{gcn} = 14$ rad/sec. We select the corner frequency (zero) of PI compensator, one decade below $\omega = \omega_{gcn} = 14$ rad/sec such that $a = 1.4$. The Bode plots (dB plot and phase plot both) for the PI compensated system for $K_1 a = 1$; that is

$$G_{PI}(j\omega)\, G(j\omega)\big|_{K_1 a = 1} = \frac{\left(1+\frac{j\omega}{1.4}\right)}{(j\omega)^2\left(1+\frac{j\omega}{40}\right)\left(1+\frac{j\omega}{100}\right)}$$

is also shown in Fig. P 8.15(*a*). Notice from phase plot that specified phase margin (60°) is met for gain crossover frequency of $\omega_{gcn} = 14$ rad/sec.

Step (*iii*): Let us bear in mind that variation in $K_1 a$, does not alter the phase plot. Next step is to adjust the gain $K_1 a$ such that $\omega = \omega_{gcn} = 14$ rad/sec. is indeed the gain crossover frequency. The dB value at $\omega = \omega_{gcn} = 14$ rad/sec, as best read from dB plot (compensated) for $K_1 a = 1$, is – 24 dB. Thus, the dB plot is required to be shifted upward by 24 dB which is possible by increasing gain $K_1 a$ from 1 to antilog (24/20) ≅ 15.85. So, the PI compensated system has the transmittance

$$G_{PI}(s)\, G(s) = \frac{K_1 a\left(1+\frac{s}{a}\right)}{s^2\left(1+\frac{s}{40}\right)\left(1+\frac{s}{100}\right)} = \frac{15.85\left(1+\frac{s}{1.4}\right)}{s^2\left(1+\frac{s}{40}\right)\left(1+\frac{s}{100}\right)}$$

where $15.85 = K_1 a$ or $K_1 = 11.32$ and $11.32 = K/40$ or $K = 452.85$. K is gain of pre amplifier. The Bode plot of the PI compensated system for $K_1 a = 15.85$ is shown in Fig. P 8.15(*a*). The design goals are achieved by introducing a pole at origin together with a zero at $s = -1.4$ (PI compensator) and adjusting gain of pre-amplifier, K = 452.85.

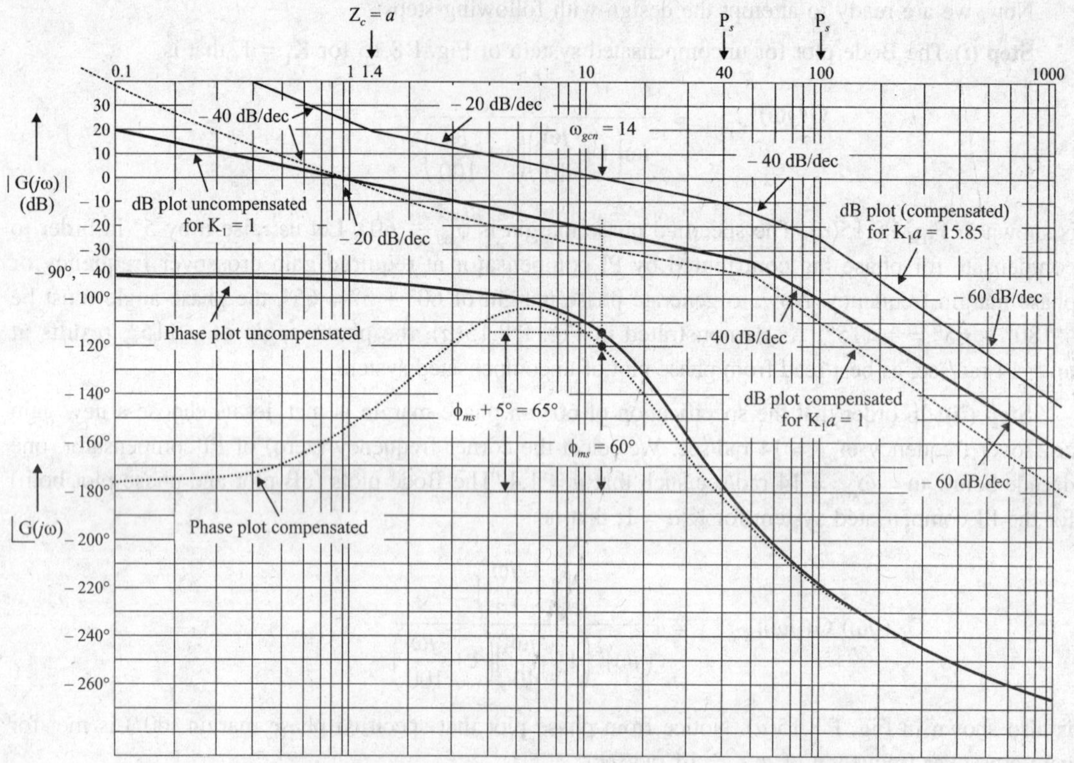

Fig. P 8.15 (*a*): Bode plots uncompensated and compensated system

P 8.15: A unity feedback system with loop transmittance

$$G(s) = \frac{K}{s(s+5)(s+20)}$$

operates with approximately 55% peak overshoot and 0.5 seconds peak time when the gain K is adjusted to yield the velocity error coefficient $K_v = 10$.

Use frequency response approach to design a PD compensator to reduce the peak overshoot to 10% while keeping the peak time and the error coefficient about the same or less.

Solution: The error coefficient

$$K_v = \lim_{s \to 0} sG(s) = \frac{K}{100}$$

Equating it to adjusted value of $K_v = 10$, we have K = 1000 changing the transmittance in pole-zero form for this value of K into time constant form, we have

$$G(j\omega) = \frac{10}{j\omega\left(1+\dfrac{j\omega}{5}\right)\left(1+\dfrac{j\omega}{20}\right)}$$

for which the Bode plot is shown in Fig. P 8.16. It is easy to read the following from this plot:

(*i*) gain crossover frequency (ω_{gc}) ≅ 6 rad/sec.

(*ii*) Phase margin ϕ_{muc} = 21° which is equivalent to damping ratio ξ = 0.187 and peak overshoot of about 55%.

(*iii*) Bandwidth ≅ 10 rad/sec, which is equivalent to peak time t_p = 0.5 seconds.

Recall that the transmittance of a PD compensator, is

$$G_{PD}(s) = (1 + bs)$$

Thus, the intended design involves locating a compensator zero such that the phase margin improves from ϕ_{muc} = 21° to ϕ_{mc} ≅ 64° (equivalent to ξ = 0.6 and peak overshoot ≅ 10%) while preserving the peak time and error coefficient about the same or less.

Note that peak time may be preserved by keeping gain crossover frequency, ω_{gc} = 6 rad/sec., almost same or slightly larger and K_v may be preserved by keeping the slope of initial segment of dB plot, same as before inserting the compensator.

Required enhancement in phase margin = $\phi_{mc} - \phi_{muc}$ = 43°. In order to contribute additional phase angle of 43° at ω = 6 rad/sec., let us introduce a zero with corner frequency 6 rad/sec such that

$$G_{PD}(s) = \left(1+\frac{s}{6}\right)$$

or

$$G_{PD}(s)\ G(s) = \frac{10\left(1+\frac{s}{6}\right)}{s\left(1+\frac{s}{5}\right)\left(1+\frac{s}{20}\right)}$$

It may be noted that

$$\angle G_{PD}(j\omega)\big|_{\omega=6} = +45°$$

The Bode plot for the PD compensated system is also shown in Fig. P 8.16, as demonstrated in Fig. P 8.16, observe the following:

(*i*) Phase margin improves from 21° to 63°. So, the corresponding peak overshoot reduces from 55% to 10%.

(*ii*) The slope of initial segment of dB plot, continues to remain – 20 dB/dec. So, K_v = 10 remains same.

(*iii*) The new gain crossover frequency ω_{gcn} ≅ 8 rad/sec and the new bandwidth ≅ 20 rad/sec. So, t_p = 0.5 or less, is also preserved.

Thus introducing a zero with corner frequency equal to 6 rad/sec, meets the design objectives.

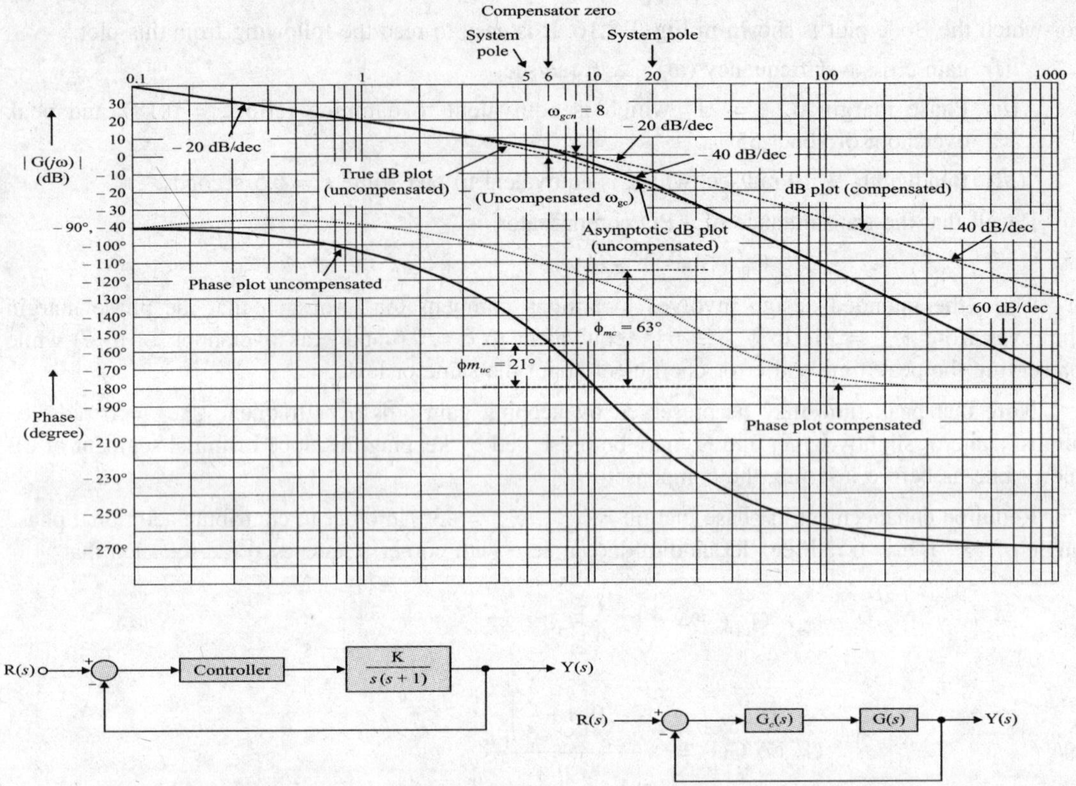

Fig. P 8.16: Bode plot for uncompensated and compensated system of example problem P8.15

DRILL PROBLEMS

D 8.1: A system in unity feedback configuration, has the transmittance

$$G(s) = \frac{K}{(s+3)(s+9)(s+15)}$$

Use frequency response methods to determine the value of K. Such that system is forced to exhibit

(*a*) Gain margin of 10 dB.

(*b*) Percent overshoot in step response of 28.5%.

Ans. (*a*) K ≅ 1640 (*b*) K ≅ 1780

D 8.2: A system is placed in a feedback organisation as shown in Fig. D8.2. Design a lead compensator so as to meet the following design goals:

(*i*) Error coefficient $K_v = 2$

(*ii*) Phase margin $\phi_m = 30°$

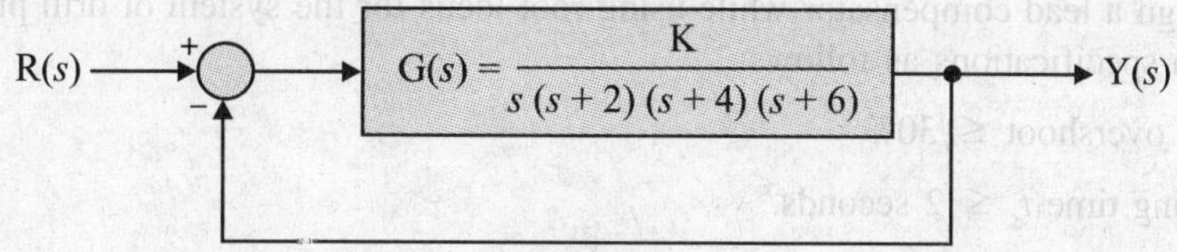

Fig. D8.2

Ans. One possible solution is

$$G_{lead}(s) \cong \frac{s+1.12}{s+4.15}, \qquad K \cong 355$$

D 8.3: The open loop transfer function of a system in unity feedback configuration, is

$$G(s) = \frac{K}{(s+1)(s+2)(s+10)}$$

Use the root locus approach to do the following:

(*a*) Determine the value of K such that system exhibits peak overshoot of 57.2%.

(*b*) Find the steady state error when system is forced to track a step input.

(*c*) Insert a PI compensator in series with system G(*s*) in the forward path with transmittance

$$G_{PI}(s) = \frac{s+0.15}{s}$$

and show that steady state error improves while almost preserving the transient response adjusted in part (*a*)

Ans. (*a*) K ≅ 164 (*b*) 0.109

D 8.4: Use the root locus approach to design a lag compensator for the system of drill problem D8.3 to achieve the following design goals:

(*i*) Peak overshoot ≤ 57.5%

(*ii*) The steady state error for a step input = 0.019

Ans. One possible solution is $G_{lag}(s) = \frac{s+0.11}{s+0.01}$

D 8.5: A system in unity feedback configuration, has the loop transmittance

$$G(s) = \frac{K}{s(s+4)(s+6)}$$

Use the root locus methods and do the following:

(*a*) Sketch root locus and find the value of gain K such that damping ratio, ξ = 0.5.

(*b*) What is settling time of the system for the value of K found in part (*a*)?

(*c*) Design a PD compensator such that the settling time reduces from the value found in part (*b*) to 1.1 seconds. Find the corresponding gain K.

Ans. (*a*) K ≅ 44 (*b*) $t_s \cong 3.3$ (*c*) $G_{PD}(s) = s + 3$, K ≅ 47.5 (a possibility)

D 8.6: Design a lead compensator while using root locus for the system of drill problem D8.5, to meet the design specifications as follows:

(*i*) Peak overshoot $\leq$ 30%

(*ii*) Settling time $t_s \leq$ 2 seconds.

Ans. One possible solution is $G_{lead}(s) = \dfrac{s+4}{s+20}$

D 8.7: Use frequency response approach to design a cascade lag-compensator for the system in unity feedback with loop transmittance

$$G(s) = \frac{K}{s(0.25s+1)}$$

so as to achieve the following design specifications:

(*i*) The steady state error to a ramp input $\leq$ 0.01

(*ii*) Phase margin $\phi_{ms} \geq 40°$

Ans. One solution is $G_{lag}(s) \cong \dfrac{2.85s+1}{63.8s+1}$

D 8.8: A system in unity feedback configuration has the loop transmittance

$$G(s) = \frac{K}{s(2s+1)}$$

Using frequency response techniques, do the following:

(*a*) Find K to meet the specification on steady state error to a unit ramp input $\leq$ 0.5.

(*b*) Sketch Bode plots using the gain K found in part (*a*) and read the phase margin there from.

(*c*) Design a cascade lead compensator so as to improve the phase margin found in part (*b*) to a value $\geq 45°$.

Ans. (*a*) K $\geq$ 2 (*b*) PM $\cong$ 28° (*c*) One possible solution $G_{lead}(s) = \dfrac{1.36s+1}{0.5s+1}$

D 8.9: Use frequency response methods to design a cascade, single, passive lag-lead compensator for a unity feedback system with open loop transmittance

$$G(s) = \frac{K}{s(s+1)(s+2)}$$

to meet the following design goals:

(*i*) the ramp error coefficient $K_v = 10\ \text{sec}^{-1}$

(*ii*) phase margin $\phi_{ms} = 50°$

(*iii*) gain margin $\geq$ 10 dB

Ans. One possible solutions is K = 20, $G_{lgld}(s) \cong \left(\dfrac{1.4s+1}{0.14s+1}\right)\left(\dfrac{6.7s+1}{67s+1}\right)$

D 8.10: An architecture of a feedback compensator is shown in Fig. D 8.10. Design the parameters K and b to achieve the following specifications:

(*i*) Peak overshoot $\leq$ 10%

(*ii*) Settling time $\leq$ 4 seconds.

Use root locus techniques.

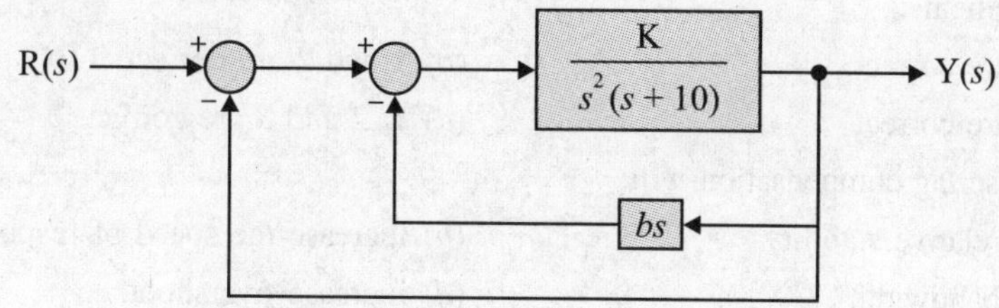

Fig. P 8.10: An architecture of feedback compensator

Ans. K $\cong$ 19 $\alpha \cong 0.9$

MULTIPLE CHOICE QUESTIONS

M 8.1: In the control system shown in the Fig. M8.1, the controller which can give zero steady state error to a ramp input, with K = 9 is

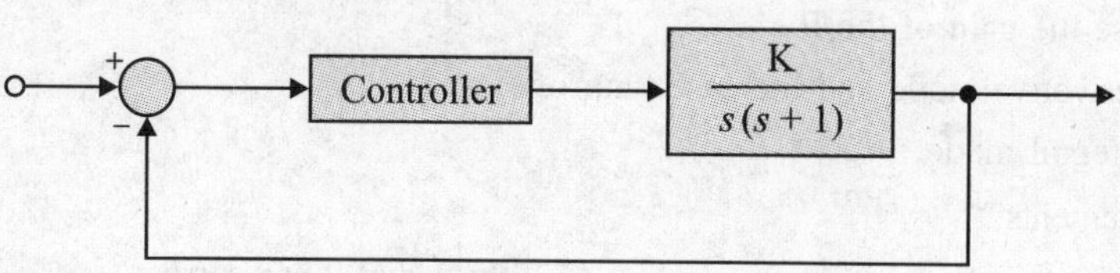

Fig. M 8.1

(*a*) proportional type (*b*) integral type

(*c*) derivative type (*d*) proportional plus derivative type.

M 8.2: The gain crossover frequency and bandwidth of a control system are ω_{cu} and ω_{bu} respectively. A phase lag network is employed for compensating the system. If the gain crossover frequency and bandwidth of the compensated system are ω_{cc} and ω_{bc} respectively, then

(*a*) $\omega_{cc} < \omega_{cu}$; $\omega_{bc} < \omega_{bu}$ (*b*) $\omega_{cc} > \omega_{cu}$; $\omega_{bc} < \omega_{bu}$

(*c*) $\omega_{cc} < \omega_{cu}$; $\omega_{bc} > \omega_{bu}$ (*d*) $\omega_{cc} > \omega_{cu}$; $\omega_{bc} > \omega_{bu}$

M 8.3: Consider the following statements regarding time domain analysis of control systems.

1. Derivative control improves system's transient performance.
2. Integral control does not improve system's steady state performance.
3. Integral control can convert a second order system into a third order system.

Of these statements:

(*a*) 1 and 2 are correct (*b*) 1 and 3 are correct

(*c*) 2 and 3 are correct (*d*) 1, 2 and 3 are correct.

M 8.4: Consider the following statements.

In a feedback control system, lead compensator

1. increases the margin of stability.
2. speeds up transient response.
3. does not affect the system error constant.

Of these statements:

(*a*) 2 and 3 are correct (*b*) 1 and 2 are correct

(*c*) 1 and 3 are correct (*d*) 1, 2 and 3 are correct.

M 8.5: A phase lag compensation will:

(*a*) improve relative stability (*b*) increase the speed of response

(*c*) increase bandwidth (*d*) increase overshoot.

M 8.6: The maximum phase shift that can be obtained by using a lead compensator with transfer function $G(s) = \dfrac{4(1+0.15\,s)}{(1+0.05\,s)}$ is equal to

(*a*) 15° (*b*) 30° (*c*) 45° (*d*) 60°.

M 8.7: Consider the following statements regarding a first order system with a proportional (P) controller which exhibits an offset to a step input.

In order to reduce the offset, it is necessary to

1. increase the gain of the P.
2. add derivative mode and increase gain of P.
3. add integral mode.

Of these statements:

(*a*) 1, 2 and 3 are correct (*b*) 1 and 2 are correct

(*c*) 2 and 3 are correct (*d*) 1 and 3 are correct.

M 8.8: An effect of phase lag-compensation on servo system performance is that

(*a*) for a given relative stability, the velocity constant increases

(*b*) for a given relative stability, the velocity constant decreases

(*c*) the bandwidth of the system is increased

(*d*) the time response is made faster.

M 8.9: Match List I (System) with List II (Transfer function) and select the correct answer using the codes given below the Lists.

List I	List II
(*a*) AC servomotor	1. $\dfrac{s+z}{s+p}$ $(z < p)$
(*b*) DC amplifier	2. $\dfrac{1+T_1 s}{1+T_2 s}$ $(T_1 < T_2)$

(*c*) Lead network		3.	$\frac{K}{1+Ts}$
(*d*) Lag network		4.	$\frac{K}{s(1+Ts)}$

Codes:	**A**	**B**	**C**	**D**
(*a*)	3	4	1	2
(*b*)	4	3	1	2
(*c*)	3	4	2	1
(*d*)	4	3	2	1.

M 8.10: A phase lead compensator has the transfer function

$$G_c(s) = \frac{10(1+0.04s)}{(1+0.01s)}$$

The maximum phase angle lead provided by this compensator will occur at a frequency ω_m equal to

(*a*) 50 rad/sec (*b*) 25 rad/sec (*c*) 10 rad/sec (*d*) 4 rad/sec

M 8.11: The transfer function of a phase lead compensator is given by $\frac{1+aTs}{1+Ts}$ where $a > 1$ and T > 0. The maximum phase shift provided by such a compensator is

(*a*) $\tan^{-1}\left(\frac{a+1}{a-1}\right)$ (*b*) $\tan^{-1}\left(\frac{a-1}{a+1}\right)$ (*c*) $\sin^{-1}\left(\frac{a+1}{a-1}\right)$ (*d*) $\sin^{-1}\left(\frac{a-1}{a+1}\right)$

M 8.12: Consider the following statements regarding a phase-lead compensator.

1. It increases the bandwidth of the system
2. It helps in reducing the steady state error due to ramp input
3. It reduces the overshoot due to step input.

Which of the above statements is/are correct?

(*a*) 1 and 2 (*b*) 1 and 3 (*c*) 2 and 3 (*d*) 1 alone.

M 8.13: Consider the following performance characteristics.

1. Reduced velocity constant for a given relative stability.
2. Reduced gain crossover frequency.
3. Reduced bandwidth.
4. Reduced resonance peak of the system.

Which of these performance characteristics are achieved with the phase-lag compensation?

(*a*) 1 and 2 (*b*) 1 and 3 (*c*) 2, 3 and 4 (*d*) 1, 2, 3 and 4.

M 8.14: A closed loop system, employing lag-lead compensator $G_c(s)$ is shown in the Fig. M8.14:

If G(*s*) has 3 poles in the left half of *s*-plane, then the slope of the Bode plot for $|G(s)G_c(s)|$ in the highest frequency range will be

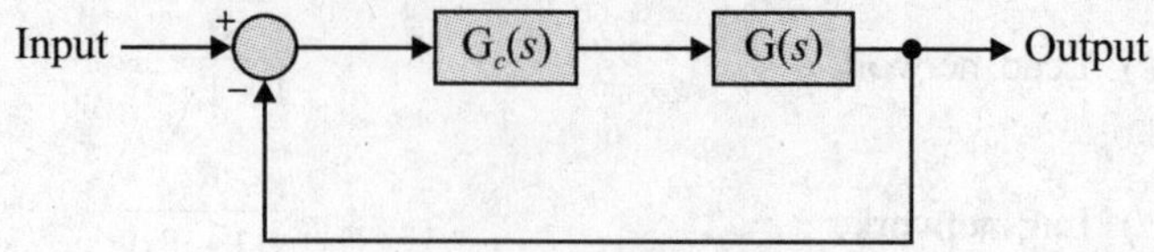

Fig. M 8.14

(*a*) – 20 dB/decade
(*b*) – 40 dB/decade
(*c*) – 60 dB/decade
(*d*) – 80 dB/decade.

M 8.15: The transfer function of a phase lead network, as shown in the Fig. M8.14 is $\frac{K(1+0.3s)}{(1+0.17s)}$

The values of R_1 and R_2 are respectively:

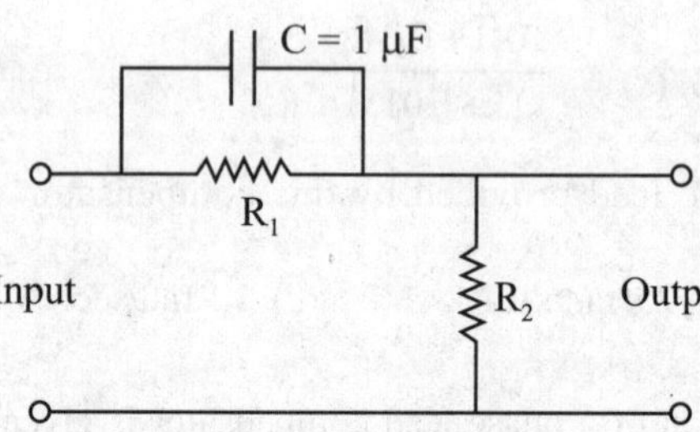

Fig. M 8.15

(*a*) 300 kΩ and 300 kΩ
(*b*) 300 kΩ and 400 kΩ
(*c*) 400 kΩ and 300 kΩ
(*d*) 400 kΩ and 400 kΩ

M 8.16: Which one of the following statements is correct?

A plant is controlled by a proportional controller. If a time delay element is introduced in the loop, its

(*a*) phase margin remains the same
(*b*) phase margin increases
(*c*) phase margin decreases
(*d*) gain margin increases.

M 8.17: Which one of the following statements is correct?

The effects of phase lead compensator on gain crossover frequency (ω_{gc}) and the bandwidth (BW) are

(*a*) that both are decreased
(*b*) that ω_{gc} is decreased but BW is increased
(*c*) that ω_{gc} is increased but BW is decreased
(*d*) that both are increased.

M 8.18: How does cascading an integral controller in the forward path of a control system affect the relative stability (RS) and the steady-state error (SSE) of that system?

(*a*) Both are increased
(*b*) RS is reduced but SSE is increased
(*c*) RS is increased but SSE is reduced
(*d*) Both are reduced.

M 8.19: Consider the following statements for phase-lead compensation:

1. Phase-lead compensation shifts the gain crossover frequency to the right.
2. The maximum phase-lead angle occurs at the arithmetic mean of the corner-frequencies of the phase-lead network.
3. Phase-lead compensation is effective when the slope of the uncompensated system near the gain crossover is low.

Which of the statements given above are correct?

(*a*) 1, 2 and 3 (*b*) 1 and 2 (*c*) 2 and 3 (*d*) 1 and 3.

M 8.20: The phase lead compensation is used to

(*a*) increase rise time and decrease overshoot

(*b*) decrease both rise time and overshoot

(*c*) increase both rise time and overshoot

(*d*) decrease rise time and increase overshoot.

M 8.21 Maximum phase-lead of the compensator $D(s) = (0.5s + 1)/(0.05s + 1)$, is

(*a*) 52 deg at 4 rad/sec.

(*b*) 52 deg at 10 rad/sec.

(*c*) 55 deg at 12 rad/sec.

(*d*) None of the answers in (*a*), (*b*) and (*c*) is correct.

M 8.22: A PD controller is used to compensate a system. Compared to the uncompensated system, the compensated system has

(*a*) a higher type number (*b*) reduced damping

(*c*) higher noise amplification (*d*) larger transient overshoot.

M8.23: The system with $G(s) = \dfrac{900}{s(s+1)(s+9)}$ is to be compensated such that its gain cross over frequency becomes same as its uncompensated phase cross over frequency and provides phase margin of 45°. To achieve this goal, one may use

(*a*) a lag compensator that contributes attenuation of 20 dB and phase lag of 45° at frequency of $3\sqrt{3}$ rad/sec.

(*b*) a lead compensator that contributes amplification of 20dB and phase lead of 45° at frequency of 3 rad/sec.

(*c*) a lag-lead compensator that contributes amplification of 20dB and a phase lag of 45° at frequency of $\sqrt{3}$ rad/sec.

(*d*) a lag-lead compensator that contributes an attenuation of 20dB and phase lead of 45° at frequency of 3 rad/sec.

ANSWERS

M 8.1. (*b*)	M 8.2. (*a*)	M 8.3. (*b*)	M 8.4. (*d*)	M 8.5. (*a*)
M 8.6. (*b*)	M 8.7. (*a*)	M 8.8. (*a*)	M 8.9. (*b*)	M 8.10. (*a*)
M 8.11. (*d*)	M 8.12. (*b*)	M 8.13. (*c*)	M 8.14. (*c*)	M 8.15. (*b*)
M 8.16. (*c*)	M 8.17. (*d*)	M 8.18. (*d*)	M 8.19. (*d*)	M 8.20. (*b*)
M 8.21. (*d*)	M 8.22. (*c*)	M8.23. (*d*)		

Important Hints:

M 8.2: Lag-compensation reduces gain crossover frequency and bandwidth both.

M 8.4: Lead compensation increases phase margin and system bandwidth. Increased bandwidth results in faster transient response. The initial slope of dB plot that determines the steady state error is not affected by design of lead compensator.

M 8.5: Lag-compensator improves phase margin by shifting gain crossover to a lower frequency

M 8.6: $\sin \phi_m = \dfrac{1-\alpha}{1+\alpha}$ gives $\phi_m = \sin^{-1} \dfrac{1-1/3}{1+1/3} = 30°$; $\alpha = \dfrac{0.05}{0.15} = \dfrac{1}{3}$

M 8.8: Lag-compensator primarily improves steady state performance while improving or at least preserving relative stability.

M 8.10: $\omega_m = \sqrt{\dfrac{1}{\tau} \times \dfrac{1}{\alpha\tau}} = \sqrt{\dfrac{1}{0.01} \times \dfrac{1}{0.04}} = 50$ rad/sec

M 8.11: Comparing with $G_{lead}(s) = \dfrac{1+\tau s}{1+\alpha\tau s}$

$$\tau = a\text{T}, \quad \alpha\tau = \text{T} \quad \text{and} \quad \alpha = \frac{1}{a}$$

$$\phi_m = \sin^{-1}\left[\frac{1-\frac{1}{a}}{1+\frac{1}{a}}\right] = \sin^{-1}\left(\frac{a-1}{a+1}\right)$$

M 8.12: Lead compensator primarily improves relative stability and makes time response faster by increasing bandwidth. Some lead compensator design may improve steady state performance also, but it does not always do so.

M 8.14: Lag-lead compensator will introduce two poles and two zeros, in addition to three system poles in the left half plane. Slope of dB plot in the highest frequency range remains unchanged.

M 8.15: $\tau = 0.3, \quad \alpha = \dfrac{1.7}{3}, \quad \tau = R_1C$ gives $R_1 = 300$ K

and $\alpha = \dfrac{R_2}{R_1 + R_2}$ gives $R_2 \cong 400$ K

M 8.16: Time delay element modifies phase plot while keeping the dB plot unaltered. Time delay element in the loop will decrease gain margin and phase margin both.

M 8.21: $\tau = 0.5, \quad \alpha\tau = 0.05 \quad \text{and} \quad \alpha = 0.1$

$$\sin \phi_m = \frac{1-\alpha}{1+\alpha} \text{ gives } \phi_m \cong 52° \text{ and } \omega_m = \sqrt{\frac{1}{\tau} \times \frac{1}{\alpha\tau}} = 6.32 \text{ rad/sec.}$$

M8.23:

$$G(j\omega) = \frac{900}{-10\omega^2 + j\left(9\omega - \omega^3\right)}$$

and $\text{Im}[G(j\omega)] = 0 \text{ for } 9\omega - \omega^3 = 0$

or $\omega = 3 \text{ rad/sec} \Rightarrow \omega_{pc} = 3 \text{ rad/sec}$

$$\left|G(j\omega)\right|_{\omega = \omega_{pc} = 3} = \frac{900}{10 \times 3^2} = 10 = 20 \text{ dB}$$

Uncompensated system has gain of 20 dB at required ω_{gc} = uncompensated ω_{pc} = 3 rad/sec. So, lag compensator is used to contribute attenuation of 20 dB at ω = 3 rad/sec. It is known that the phase characteristic of lag compensator is of no consequence in compensation.

Now, $\angle G(j\omega)|_{\omega = 3} = -180°$. In order to achieve phase margin of 45° at ω = 3 rad/sec, additionally lead compensator is to be used to contribute phase lead of 45° at ω = 3 rad/sec.

9

CONTROL SYSTEM COMPONENTS

9.1 INTRODUCTION

The control strategy appropriately evolved in a system, generally intends to force a parameter which is otherwise undetermined or does not have desired value, to track a reference (input) to a specified degree of accuracy. The parameter to be manipulated is often the system output of particular interest. In fact, the overall response of the system is the pure response due to real input plus the response due to disturbances. The disturbances include both internal as well as external. Since the disturbances have fluctuating nature, the overall response becomes free running. The controllers are then included in system structure to ensure that response continues to stay within prescribed limits despite presence of inevitable disturbances. The variants of controller and various aspects of controller design, have been extensively dealt in chapter **eight**. To vivify further the current chapter will focus on important components associated with the structure of a typical control system. The components of current interest are also often, referred to as servo components.

The term **Servo** in current use, refers to a class of feedback control system wherein the variable to be controlled is either mechanical position or its time derivatives significantly velocity and acceleration. The servo components, in general are **error detectors, servo amplifiers** and **servo motors.** A typical interconnection of servo components briefly described below, is demonstrated in the form of block diagram in Fig. 9.1.

The **error detector** generates output proportional to the difference between true (real) output and reference input (which is also desired output). Note that both polarity and magnitude of error signal (deviation signal) are of interest to evolve control strategy. The error signal so generated is often weak and requires amplification. The demand of amplification is met by another component named as **servo amplifier**. The amplified output emerging from servo amplifier, often drives yet another component called as **servo motor. Gear train** is also usually included between servo motor and mechanical load. A situation generally, arises in mechanical system where **servo motor operating at high speed but low torque,** drives a load that demands **high torque and low speed.** The gear trains meet the demand of **torque magnification** and **speed regulation.** The gear trains have already been discussed in Section 1.6. Let us focus on the remaining components in the current discussion.

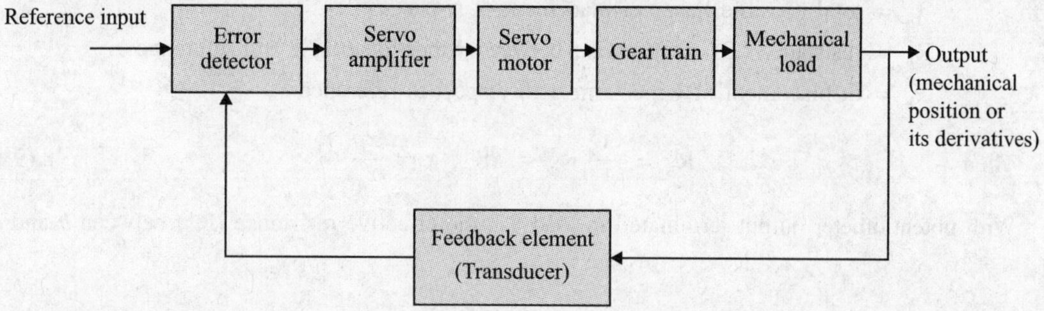

Fig. 9.1: Interconnection of servo components

9.2 POTENTIOMETER

The potentiometer is an **electro mechanical transducer** that transforms mechanical energy into electrical energy. The mechanical energy input is the displacement which can be linear and angular both. In fact, the potentiometer is a voltage divider having three terminals out of which two terminals (*a*, *c*) are fixed and the third terminal (*b*) is movable as shown in Fig. 9.2. The movable terminal is also referred to as wiper arm which can move on a linear (straight line) path along ac as shown in Fig. 9.2(*a*) or on a circular path along ac as shown in Fig. 9.2(*b*). A fixed reference voltage V_R is applied across fixed terminals ***a*** and ***c***.

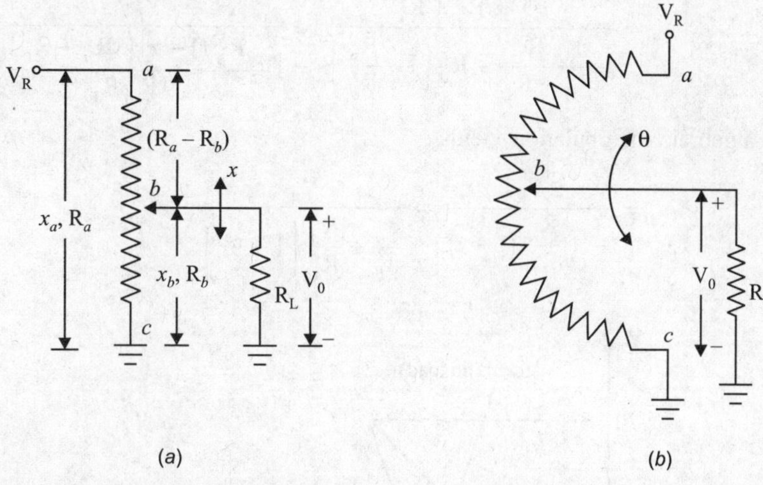

Fig. 9.2: Potentiometer together with load R_L (*a*) Translatory (linear) (*b*) Rotary (angular)

The voltage output developing across moving terminal ***b*** and ground ***c*** is linearly related to displacement in no load condition ($R_L = \infty$). However, it is often seen that the voltage output generated from potentiometer, is applied to an amplifier whose input resistance constitutes load (R_L) on potentiometer. The loading effect forces the voltage-displacement relationship of potentiometer to deviate from linear property. Let us mathematically demonstrate it. Refer to Fig. 9.2(*a*) where

R_a = total resistance of potentiometer between terminals ***a*** and ***c***;

(R_a is assumed to be uniformly distributed over entire length *ac*)

x_a = total possible linear displacement of wiper arm.

R_b = resistance of potentiometer between wiper location ***b*** and reference ***c***.

x_b = displacement of wiper arm with respect to reference ***c***.

Then
$$R_b = \frac{x_b}{x_a} R_a = xR_a; \; x = \frac{x_b}{x_a} \quad ...(9.1)$$

With potentiometer output terminated in load R_L, the effective resistance (R*) between ***b*** and ***c***, is

$$R^* = R_b \parallel R_L = \frac{xR_a R_L}{xR_a + R_L} \quad ...(9.2)$$

and output voltage
$$V_0 = \frac{V_R R^*}{R^* + (R_a - R_b)} \quad ...(9.3)$$

where $R_a - R_b$ = resistance of potentiometer between terminals ***a*** and ***b***.

Substituting value of R* from (9.2) and $x = \frac{R_b}{R_a}$ from (9.1) in (9.3), we get

$$V_0 = \frac{V_R \cdot \dfrac{xR_a R_L}{xR_a + R_L}}{\dfrac{xR_a R_L}{xR_a + R_L} + R_a\left(1 - \dfrac{R_b}{R_a}\right)} = \frac{V_R}{1 + \dfrac{R_a(1-x)(xR_a + R_L)}{xR_a R_L}}.$$

Little more algebraic manipulation yields

$$V_0 = \frac{V_R}{\dfrac{1}{x} + \dfrac{R_a}{R_L}(1-x)} = \frac{V_R}{\dfrac{x_a}{x_b} + \dfrac{R_a}{R_L}\left(1 - \dfrac{x_b}{x_a}\right)} \quad ...(9.4)$$

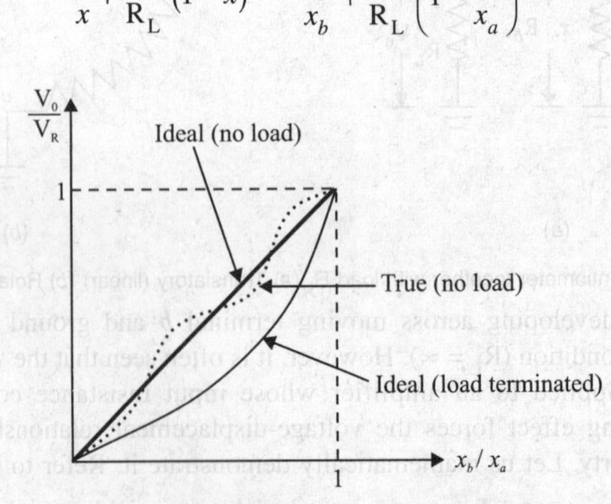

Fig. 9.3: Voltage-displacement relation with and without load

Note that no load ($R_L = \infty$) condition exhibits linear relation between output voltage V_0 and displacement of wiper arm x_b as

$$V_0|_{R_L=\infty} = V_R \cdot \frac{x_b}{x_a} \quad ...(9.5)$$

This is depicted in Fig. 9.3. However, even in no load condition, the true relationship between voltage and displacement may be in some deviation from ideal (straight line) as shown in Fig. 9.3 by virtue of manufacturing imperfection contributing to non-uniform distribution of resistance over entire length. The non-linear relation due to load termination as dictated by (9.4) is also depicted typically in Fig. 9.3.

The non-linearity due to direct load termination can be removed by inserting a voltage follower using operation amplifier as shown in Fig. 9.4. Recall that voltage follower, ideally, has infinite input resistance ($R_i = \infty$) and zero output resistance ($R_0 = 0$). It has the ability to completely remove interactive (loading) error. The circuit of Fig. 9.4 generates output voltage equal to open circuit voltage $V_0 = \frac{V_R}{x_a} \cdot x_b$ regardless of what R_L is

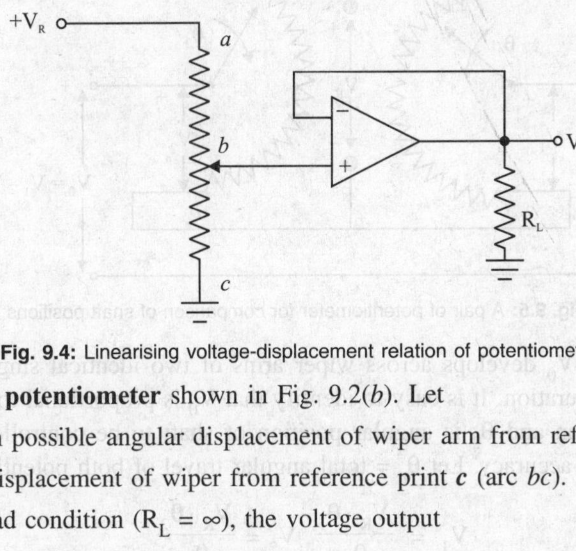

Fig. 9.4: Linearising voltage-displacement relation of potentiometer

Consider **angular potentiometer** shown in Fig. 9.2(*b*). Let

θ_a= maximum possible angular displacement of wiper arm from reference point ***c*** (arc *ac*)

θ_b= angular displacement of wiper from reference print ***c*** (arc *bc*).

Then under no load condition ($R_L = \infty$), the voltage output

$$V_0|_{R_L=\infty} = \frac{V_R \cdot \theta_b}{\theta_a} \quad ...(9.6)$$

is linearly related to wiper angle θ_b. As discussed above in case of translatory potentiometer, the output voltage-wiper angle relation in rotary potentiometer will also turn out to be non-linear when potentiometer is directly terminated in load R_L. It is easy to demonstrate that the direct load terminated rotary potentiometer with uniform resistance distribution, will generate

$$V_0 = \frac{V_R}{\frac{\theta_a}{\theta_b} + \frac{R_a}{R_L}\left(1 - \frac{\theta_b}{\theta_a}\right)} \quad ...(9.7)$$

Note that similar mathematical routine follows for both (9.4) and (9.7).

In the discussion just above, we have considered only single turn rotary potentiometer where maximum possible angular displacement of wiper arm from reference point *c* (arc *ac*), is only 340° typically. If control design demands greater accuracy, finer resolution, higher resistance values and larger total angular travel, we use **multi turn potentiometer**. For example, typical **ten turn** potentiometer provides total angular travel of 3600° (360° × 10) together with better resolution and accuracy. The relation (9.6) for general **N turn rotary potentiometer** gets modified to

$$V_0|_{R_{L=\infty}} = \frac{V_R \cdot \theta_b}{2\pi N}; \ \theta_a = 2\pi N$$

If the control situation demands the comparison of positions of two shafts that are remotely located, a pair of potentiometer can be used as shown in Fig. 9.5. In fact, this configuration acts as error detector. This is also useful in design of a typical position control system where the position of output shaft is required to follow that of reference (input) shaft.

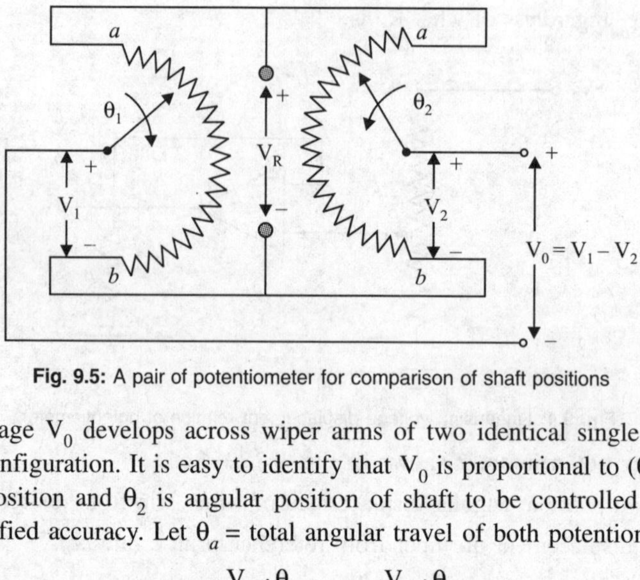

Fig. 9.5: A pair of potentiometer for comparison of shaft positions

The output voltage V_0 develops across wiper arms of two identical single turn potentiometers placed in parallel configuration. It is easy to identify that V_0 is proportional to $(\theta_1 - \theta_2)$ where θ_1 is reference angular position and θ_2 is angular position of shaft to be controlled so as to follow the reference with specified accuracy. Let θ_a = total angular travel of both potentiometers, then

$$V_1 = \frac{V_R \cdot \theta_1}{\theta_a}, V_2 = \frac{V_R \cdot \theta_2}{\theta_a}$$

and
$$V_0 = V_1 - V_2 = \frac{V_R}{\theta_a}(\theta_1 - \theta_2) \quad \text{...[9.8(a)]}$$

Similarly, for multi turn potentiometer (9.8(*a*)) takes the form

$$V_0 = \frac{V_R}{2\pi N}(\theta_1 - \theta_2) \quad \text{...[9.8(b)]}$$

Potentiometer performance indices

The performance of a potentiometer is generally assessed in terms of following specifications:

(*a*) **Accuracy/Precision/Linearity:** The accuracy of a potentiometer is primarily indicative of its linearity. Linearity is specified by the maximum deviation of output voltage-displacement relation from ideal linear (straight line) property. It has been already demonstrated in Fig. 9.3 and it is expressed as a percentage of applied voltage V_R, that is

$$\% \text{ linearity } = \frac{\Delta V_{0_{max}}}{V_R} \times 100$$

Its range typically lies in between 0.5 to 0.02. The following are responsible for deviation from linearity:

(*i*) The turns may not be identical in size.

(*ii*) The turns may not be spaced exactly equally.

(*iii*) The resistance wire may not have uniform diameter over the entire length.

(*iv*) The film deposition may not be accurate in film potentiometers.

(*v*) Some mechanical properties like wear out etc., also contribute to deviation from linearity.

(*b*) **Resolution:** Resolution is the smallest possible change in output voltage generated by potentiometer. It is also expressed in percentage of applied voltage V_R as

$$\% \text{ resolution } = \frac{\Delta V_{0_{min}}}{V_R} \times 100$$

The resolution is, in fact, the output voltage generated per turn in wire wound potentiometer and it is infinite in potentiometers fabricated by uniform spreading of resistance material (deposit film potentiometer). The resolution typically ranges between 0.5% to 0.002%.

(*c*) **Power handling capability:** Power handling capacity (in watts) is the maximum power that a potentiometer can continuously dissipate. It poses limit on maximum fixed reference voltage that can be safely applied, that is

$$V_{R_{max}} = \sqrt{P_r \cdot R_a}$$

where P_r = power rating of potentiometer and R_a = total potentiometer resistance between fixed terminals.

(*d*) **Noise:** The noise is defined as any distortion appearing in output voltage of potentiometer, in addition to those contributed by deviation form linearity. The noise, significantly, includes vibrational noise, high velocity noise, residual noise and noise due to inherent design.

The **vibrational noise** is generated by contact jumping away from winding (thereby opening the contact). As shown in Fig. 9.6, the contact bouncing results in staircase output voltage. The vibrational noise is purely mechanical but generates effect of electrical nature. **High**

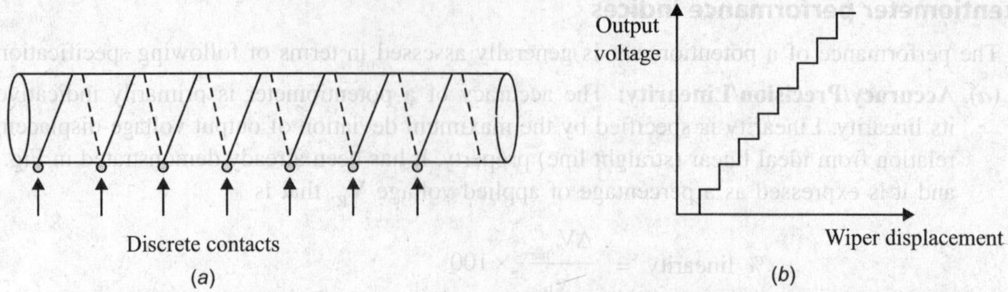

Fig. 9.6: (*a*) Bouncing results in discrete contacts (b) Stair case voltage output due to discrete contacts

velocity noise is caused by slider bouncing along the coil due to very high rotational speed resulting in a series of momentary open circuits. The potentiometer design has some critical speed beyond which high velocity noise will result. The **residual noise** is contributed by foreign objects such as dirty winding or wear products between contact surface and winding. The noise due to **inherent design** includes **loading noise, shorting noise and resolution noise**. Resolution noise is step variation in output voltage. **Shorting noise** is present even in no load condition. As wiper travels over the coil, it may short the adjacent turns thereby a portion of current flows through the slider from one turn to the next and the output voltage is distorted.

(*e*) **Frequency characteristics:** At high frequency, the distributed inductance and capacitance together with winding resistance, force the potentiometer response to become complex, that is, the output voltage V_0 becomes frequency dependent.

Merits and Demerits of potentiometer

The potentiometers are very commonly used as error detector as well as displacement transducer because they

(*i*) are economical with high accuracy and compact size.

(*ii*) demand easy maintenance.

(*iii*) can be easily interfaced with other elements of servo system.

(*iv*) are describable by fixed parameters and possess predictable characteristics.

(*v*) have high reliability and capability to withstand the most severe environmental conditions.

However, the potentiometers have the following **demerits** as well:

(*i*) Low potentiometer sensitivity demands amplifier with high gain in servo system design.

(*ii*) Unsuitable for small change in displacement.

(*iii*) Wiper movement demands larger force.

(*iv*) Tear and wear of the wiper contributes to discontinuous output voltage posing serious problem in servo system design.

Types of potentiometer

The variants of potentiometer commonly used in practice include **wire wound** and **deposit film** potentiometer. The **wire wound** potentiometers are fabricated by wrapping resistance wire around a core and then applying coat of an insulating material (except on contact surface). The insulating material also provides mechanical strength. The cores used in fabrication are **copper mandrels** (insulated copper magnet wire), **cards** (insulated metal or plastic strips with cylindrical shape) and **torroids** (short length of plastic or insulated metal). The **mandrel** type is advantageous in the sense of ease in manufacture and low cost while **card** type provides higher resistance with same diameter. The **torroids** type poses difficulty in accurate winding. The wire wound potentiometers are available in both modes of operation: **linear and rotary.** The rotary type potentiometer again may have **single** turn or **multi** turn. The multi turn potentiometers are inherently capable of providing better accuracy and resolution as compared to single turn.

Deposit film potentiometers are developed using deposited metal, carbon or conductive plastic films. The resolution obtainable is 10 to 100 times better than that of the best wire wound potentiometer. In fact, the resolution is limited by only granularity of the deposited film. These potentiometers are very useful if situation demands **good resolution**, **high operating temperature** and **high accuracy** in small size. Carbon film precision potentiometers generate continuously varying stepless voltage output. The smooth unbroken surface permits high operating speed and long life. The failure of these potentiometers is gradual and is detectable at any point of failure cycle. The plastic film conducting potentiometers have infinite theoretical resolution.

9.3 SYNCHROS

A synchro is an electromagnetic transducer that transforms angular position of a shaft into an electrical signal. It is also commercially referred to as **telsyn, autosyn** or **selsyn.** Synchros are very widely used for remote transmission of shaft position in servo system design. In fact transmission of shaft position through synchro is superior to potentiometric transmission in the sense of following:

(*i*) Irregularities (steps) are not exhibited in synchro signal transmission. Synchros do not exhibit resolution error.

(*ii*) Except very small noncritical wear at slip rings, no other wear is experienced.

(*iii*) Even ordinary synchro provides accuracy higher than the best potentiometer.

(*iv*) Synchros can be operated at much higher speeds together with being adaptable to multispeed operation.

(*v*) Synchros exhibit high reliability. They have useful operating angle of 360° and capable of continuous rotation.

Synchro construction and operation

The design of a typical servo system, often involves controlling of load shaft position in accordance with reference shaft position. In such a design effort, a synchro pair is used as error detector. The synchro pair consists of **synchro transmitter** (or synchro generator) and **synchro receiver** (also refereed to as control transformer or synchro meter). Both of these units are discussed below.

Synchro transmitter (ST)

The synchro transmitter has the construction similar to that of a three phase alternator. The rotor (moving sub-unit) is a **salient pole dumbbell shaped** magnetic structure housing the primary winding of the transmitter. The voltage (ac) is applied to this winding through the slip rings and brushes. The stator (stationary sub-unit) has three balanced, Y (Star) connected windings with their axes 120° apart as shown in Fig. 9.7. Despite the fact that the **windings (being 120° apart) resemble three phase machine,** only **single phase voltage appears across any of these windings.** The flux linking each of these coils, depends on angular position of rotor. Here we shall restrict to only schematic structure which will suffice to understand theoretical aspects. It is not of particular interest in current perspective to discuss too much of constructional details.

Let an *ac* voltage applied to rotor terminals R_1, R_2 of synchro transmitter be

$$V_r(t) = V_r \sin \omega_a t$$

This applied voltage causes magnetizing current in the rotor coil whereby sinusoidally time varying flux is set up along the axis of the rotor. The flux is almost sinusoidally distributed in the air gap along the stator periphery. The voltages are induced in each of the stator winding due to transformer action. The voltage induced in any stator winding is proportional to cosine of angle between rotor axis and corresponding stator axis. The entire action appears to be that of a single phase transformer whose primary winding is the rotor winding and three secondary windings are the three stator windings of synchro transmitter.

As shown in Fig. 9.7, let angle between rotor axis and stator S_2 be θ. The voltages v_{s_1n}, v_{s_2n} and v_{s_3n} induced in stator windings S_1, S_2 and S_3 respectively with respect to neutral, can be written as

$$v_{s_1n}(t) = KV_r \sin \omega_a t \cos (\theta + 120°) \quad ...[9.9(a)]$$

$$v_{s_2n}(t) = KV_r \sin \omega_a t \cos \theta \quad ...[9.9(b)]$$

$$v_{s_3n}(t) = KV_r \sin \omega_a t \cos (\theta + 240°) \quad ...[9.9(c)]$$

where K is the coefficient of coupling. With little mathematical manipulation it can be shown that the stator terminal voltages take the form as

$$v_{s_1s_2} = v_{s_1n} - v_{s_2n} = \sqrt{3}\, KV_r \sin(\theta + 240°) \sin \omega_a t \quad ...[9.10(a)]$$

$$v_{s_2s_3} = v_{s_2n} - v_{s_3n} = \sqrt{3}\, KV_r \sin(\theta + 120°) \sin \omega_a t \quad ...[9.10(b)]$$

$$v_{s_3s_1} = v_{s_3n} - v_{s_1n} = \sqrt{3}\, KV_r \sin\theta \sin \omega_a t \quad ...[9.10(c)]$$

Note that θ = 0 aligns the rotor axis with that of stator S_2 and voltage induced in S_2 is maximum. It is easy to conclude from [9.10(*c*)] that $v_{s_3s_1} = 0$ in this position. This rotor position is said to be **electrical zero** and is used as reference to specify the angular position of rotor. In general, the set of three single voltages (output of synchro transmitter) given by (9.10) provide information about angular position (θ) of rotor shaft (input to synchro transmitter).

Synchro control transformer (CT)

In the current discussion let us keep in view the fact that a typical servo design goal is to use synchro as error detector that converts the difference of two shaft positions into an electrical signal. In order to serve this purpose, the output of synchro transmitter, is applied to stator windings of synchro control transformer. The composite configuration of synchro transmitter (ST) and control transformer (CT) performing role of error detector is shown in Fig. 9.8. The construction of CT is similar to that of ST. The only difference is that **rotor of CT is cylindrical** in shape so that air gap flux is uniformly distributed. This minimises the change in rotor impedance with rotor position.

The stator windings of ST and CT are interconnected and output voltages of ST are applied to corresponding stator coils S_1, S_2 and S_3 of CT. This generates in its air gap exactly same flux distribution as that in the air gap of ST. The voltage drops in the resistances and leakage reactances are assumed to be negligibly small in stator coils of both units ST and CT. The flux distribution in air gap of CT, induces the voltage in its rotor. The voltage induced in rotor of CT, is proportional to cosine of angle between rotor of CT and that of ST. Let us identify it as error voltage $e(t)$:

$$e(t) = KV_r \cos\phi \sin\omega_a t \quad ...(9.11)$$

where ϕ is angular difference between two rotors. Note that $e(t)|_{\phi = 90^\circ} = 0$. This demonstrates that voltage induced in rotor of CT is zero when rotors of CT and ST are at right angles to each other. This position corresponding to zero voltage in rotor of CT, is referred to **electrical zero** of CT. Fig. 9.8 shows both ST and CT in position of electrical zero.

Now let the rotor of ST travel through an angle α and that of CT through an angle β in the direction as shown in Fig. 9.8. The angular difference between two rotors, $\phi = 90^\circ - (\alpha + \beta)$. The voltage induced in rotor of CT becomes

$$e(t) = KV_r \cos[90^\circ - (\alpha - \beta)] \sin\omega_a t = KV_r \sin(\alpha - \beta) \sin\omega_a t \quad ...(9.12)$$

Since $\sin\theta \cong \theta$ for small θ, (9.12) takes the form

$$e(t) = KV_r (\alpha - \beta) \sin\omega_a t = KV_r \delta \sin\omega_a t;\ \delta = \alpha - \beta \quad ...(9.13)$$

and it categorically demonstrates that S T – C T pair performs the role of error detector generating voltage at rotor terminals of CT, proportional to angular separation between ST and CT shaft positions. Note that (9.13) continues to be valid so long as rate of angle change is small. Slow time varying angular difference $\delta(t)$ is typically shown in Fig. 9.9(a). Figure 9.9(b) shows the constant amplitude sinusoidal signal $v_a(t) = KV_r \sin\omega_a t$. It is obvious that the voltage $e(t)$ emerging from Rotor of CT, is

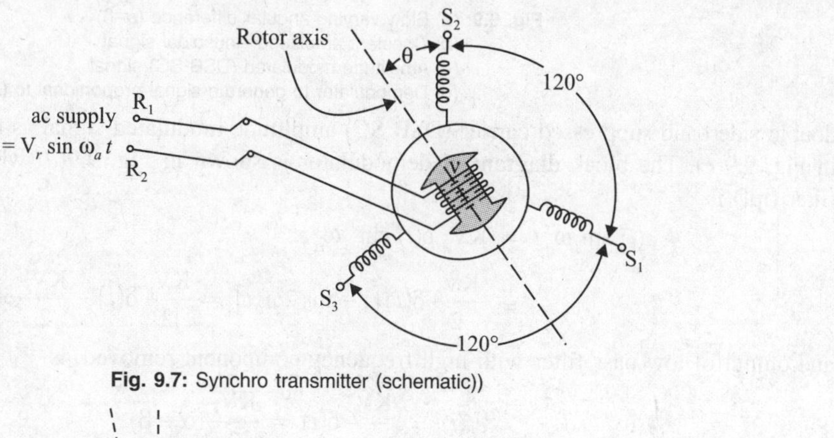

Fig. 9.7: Synchro transmitter (schematic))

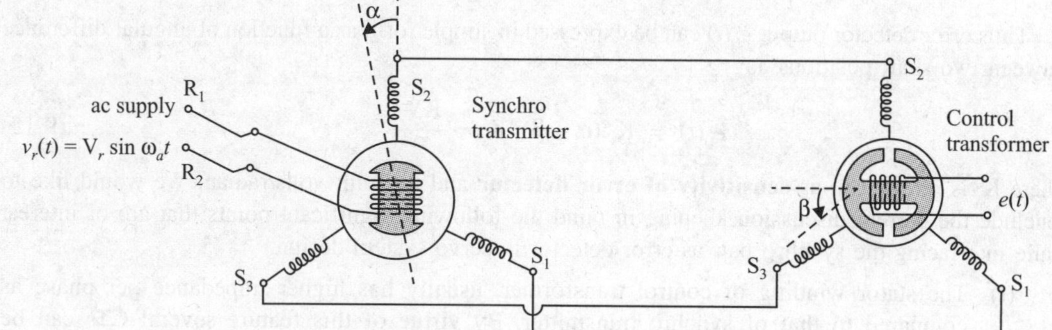

Fig. 9.8: Interconnected ST and CT (Error detector)

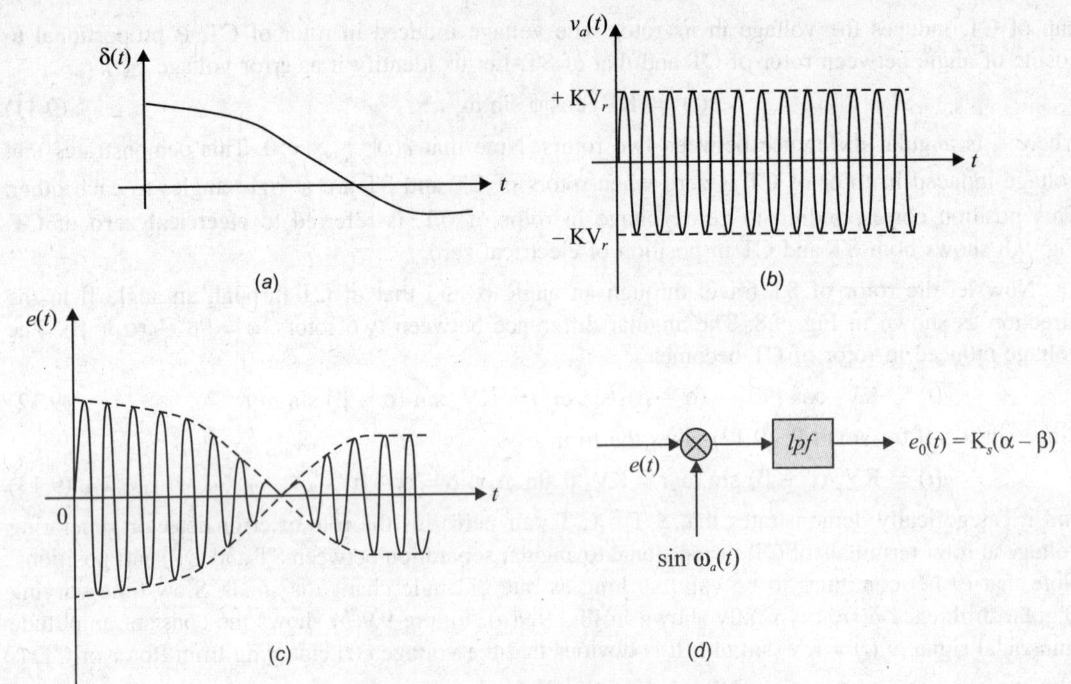

Fig. 9.9: (*a*) Slow varying angular difference (α–β)
(*b*) Constant amplitude sinusoidal signal
(*c*) Amplitude modulated (DSB SC) signal
(*d*) Demodulator to generate signal proportional to (α–β)

double side band suppressed carrier (DSB SC) amplitude modulated signal as typically demonstrated in Fig. 9.9(*c*). The block diagram of demodulator is shown in Fig. 9.9(*d*) where input to low pass filter (lpf) is

$$e(t) \sin \omega_a t = KV_r \, \delta(t) \sin^2 \omega_a t$$

$$= \frac{KV_r}{2} \delta(t) [1 - \cos 2\omega_a t] = \frac{KV_r}{2} \delta(t) - \frac{KV_r}{2} \delta(t) \cos 2\omega_a t$$

and output of low pass filter with high frequency component removed, is

$$e_0(t) = \frac{KV_r}{2} \delta(t) = \frac{KV_r}{2} (\alpha - \beta) \quad ...(9.14)$$

Thus error detector output $e_0(t)$ can be expressed in simple form as a function of angular difference between two shaft positions as

$$e_0(t) = K_s \, (\alpha - \beta); \; K_s = \frac{KV_r}{2} \quad ...(9.15)$$

where K_s is referred to as **sensitivity of error detector** and has unit volts/radian. We would like to conclude the current discussion keeping in mind the following significant points that are of interest while interfacing the synchro pair as error detector in servo system design.

(*i*) The stator winding of control transformer, usually has higher impedance per phase as compared to that of synchro transmitter. By virtue of this feature several CTs can be connected to the same transmitter with little loss of accuracy.

(*ii*) The rotor of CT is mechanically designed to have cylindrical shape so as to ensure that air gap is uniform. This feature minimises the variation in rotor output impedance with rotor shaft position and therefore rotor terminals can be connected to an amplifier without posing any interfacing difficulty. Note that the integral design of servo system generally demands amplification of error signal.

9.4 SERVO MOTOR

The torque required to drive the load in servo system, is supplied by servo motor. The servo motor is driven by amplified error signal. The variants of servo motors include DC servo motor (field controlled/armature controlled), AC servo motor and special servo motor like stepper motor.

DC servo motor

DC servo motor is essentially an ordinary DC motor, of course, with some minor changes as demanded by servo system design. The servo application requirements and changes brought about to meet these requirements, are as follows:

(*i*) On removal of control signal (error signal), the motor should be designed to stop instantly without irksome delay. Motor should also have fast dynamic response, that is, it should generate rapid accelerations from standstill and quickly respond to rapid changes in error signal. These requirements are met by designing motor with **low inertia** and **high starting torque.** For low inertia it is designed with **large length to diameter ratio.** Faster dynamic response is achieved by keeping **torque to weight ratio high.**

(*ii*) Servo motor should be designed to exhibit linear relationship between error signal and rotor speed over a wide range. Linear torque – speed characteristics are also desired and therefore all DC servo motors are essentially separately excited type.

(*iii*) Servo motor should be easily reversible and their operation should be stable without exhibiting oscillations/overshoots.

The DC servo motors are operated in two control modes: one, **armature control** where armature winding voltage is supplied by error amplifier and field winding is driven by a constant current and two, field control where field winding voltage is supplied by error amplifier and armature winding is driven by constant current.

Armature controlled DC servo motor

The schematic diagram of an armature controlled DC motor is shown in Fig. 9.10. The components used in this diagram are as follows:

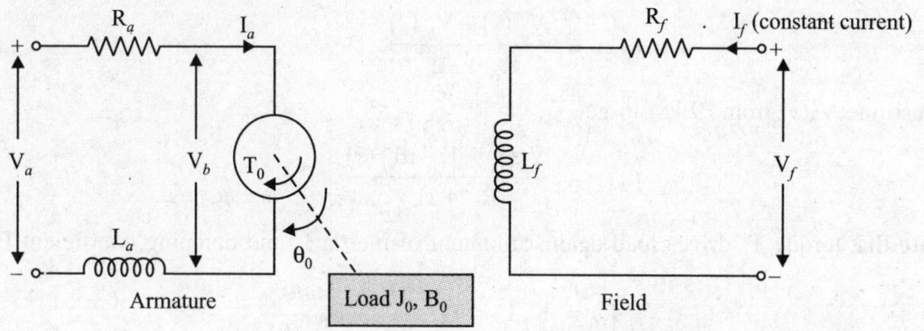

Fig. 9.10: Armature controlled DC motor (schematic)

R_a = armature resistance

L_a = armature winding inductance

I_a = armature current

I_f = field current

V_a = amplified error signal driving armature of motor

V_b = back emf

T_0 = torque developed by motor

θ_0 = angular displacement of motor shaft (in radian)

J_0 = equivalent moment of inertia of motor plus that of load referred to motor shaft

B_0 = equivalent damping coefficient of motor plus that of load referred to motor shaft.

In order to develop **transfer function model**, let us assume that

(*i*) air gap flux ϕ is proportional to the field current. This is fairly valid as DC motors are usually operated in linear region of magnetisation curve. So,

$$\phi = K_f I_f \text{ ; } K_f = \text{proportionality constant} \quad ...(9.16)$$

(*ii*) losses and armature reaction are negligible. The torque T_0 developed is proportional to product of armature current and air gap flux.

$$T_0 = K_m K_f I_f I_a \text{ ; } K_m = \text{proportionality constant}$$

Since I_f is constant

$$T_0 = K_T I_a \quad ...(9.17)$$

where constant $K_T = K_m K_f I_f$ is referred to as **motor – torque constant.**

The motor back emf, V_b is directly proportional to shaft velocity $\omega_0 = d\theta_0/dt$. So,

$$V_b = K_b \frac{d\theta_0}{dt} \text{ ; } K_b = \text{back emf constant}$$

and its transform version $V_b(s) = K_b\, s\, \theta_0(s)$ ...(9.18)

Kirchhoff's law to armature circuit, gives

$$V_a = V_b + I_a R_a + L_a \frac{d}{dt}(I_a)$$

and its Laplace transformation with zero initial conditions, gives

$$V_a(s) = V_b(s) + R_a I_a(s) + sL_a I_a(s)$$

or

$$I_a(s) = \frac{V_a(s) - V_b(s)}{R_a + sL_a} \quad ...[9.19(a)]$$

Substitute $V_b(s)$ from (9.18) to get

$$I_a(s) = \frac{V_a(s) - K_b\, s\theta_0(s)}{R_a + sL_a} \quad ...[9.19(b)]$$

Note that torque T_0 drives load against moment of inertia J_0 and damping coefficient B_0. So,

$$T_0 = K_T I_a = J_0 \frac{d^2\theta_0}{dt^2} + B_0 \frac{d\theta_0}{dt}$$

and its transform version, is

$$T_0(s) = K_T I_a(s) = (J_0 s^2 + B_0 s)\,\theta_0(s) \qquad \text{...[9.20(a)]}$$

or

$$I_a(s) = \frac{1}{K_T}\left[J_0 s^2 + B_0 s\right]\theta_0(s) \qquad \text{...[9.20(b)]}$$

Equating expressions of $I_a(s)$ from [9.19(*b*)] and [9.20(*b*)], we have

$$\frac{\left(J_0 s^2 + B_0 s\right)\theta_0(s)}{K_T} = \frac{V_a(s) - K_b s\theta_0(s)}{\left(R_a + sL_a\right)}$$

and little more algebraic manipulation gives the transfer function

$$G(s) = \frac{\theta_0(s)}{V_a(s)} = \frac{K_T}{s\left[\left(R_a + sL_a\right)\left(B_0 + sJ_0\right) + K_T K_b\right]} \qquad \text{...[9.21(a)]}$$

or

$$G(s) = \frac{K_T}{sR_a B_0\left(1 + s\tau_a\right)\left(1 + s\tau_0\right) + sK_T K_b} \qquad \text{...[9.21(b)]}$$

where $\tau_a = \dfrac{L_a}{R_a}$ is time constant of armature circuit (**electrical time constant**) and

$\tau_0 = \dfrac{J_0}{B_0}$ is **mechanical time constant** (or simply motor time constant).

For still better insight into the overall and interior dynamics of armature controlled DC motor, let us arrange the interrelationship more transparently as follows:

(*a*) The armature current $I_a(s)$ is generated by applying error voltage $V_a(s)$ minus back emf $V_b(s)$ to armature circuit with resistance R_a and inductance L_a in series. This has been established in (9.19(*a*)). This is rewritten below only for easy reference.

$$I_a(s) = \frac{V_a(s)}{R_a + sL_a} - \frac{V_b(s)}{R_a + sL_a}$$

This is depicted in the form of signal flow graph segment in Fig. 9.11(*a*).

(*b*) With constant flux ϕ generated by field circuit supplied with constant current I_f, the motor develops torque $T_0(s)$ that equals $K_T I_a(s)$. This is depicted as signal flow graph segment in Fig. 9.11(*b*).

(*c*) The torque $T_0(s)$ rotates load with angular speed $\omega_0(s) = s\theta_0(s)$ against moment of inertial J_0 and damping coefficient B_0. This action has been established in (9.20(*a*)) and is rewritten below for quick reference.

$$T_0(s) = (B_0 + sJ_0)\, s\theta_0(s) = (B_0 + sJ_0)\,\omega_0(s)$$

so that

$$\omega_0(s) = \frac{1}{B_0 + sJ_0}\cdot T_0(s)$$

and according to (9.18) back emf $V_b(s) = K_b\,\omega_0(s)$. These relations are demonstrated in the form of signal flow graph segment in Fig. 9.11(*c*).

(*d*) The angular displacement of motor shaft θ_0 is obtained by integrating angular velocity ω_0 or multiplying $\omega_0(s)$ by $1/s$. This is demonstrated as signal flow graph segment shown in Fig 9.11(*d*).The integrated signal flow graph is shown in Fig 9.11(*e*) by interconnecting the segments of Fig. 9.11(*a*), (*b*), (*c*) and (*d*). The transfer function already derived as (9.21), is also alternatively determinable by applying Mason's gain rule to signal flow graph shown in Fig. 9.11(*e*).

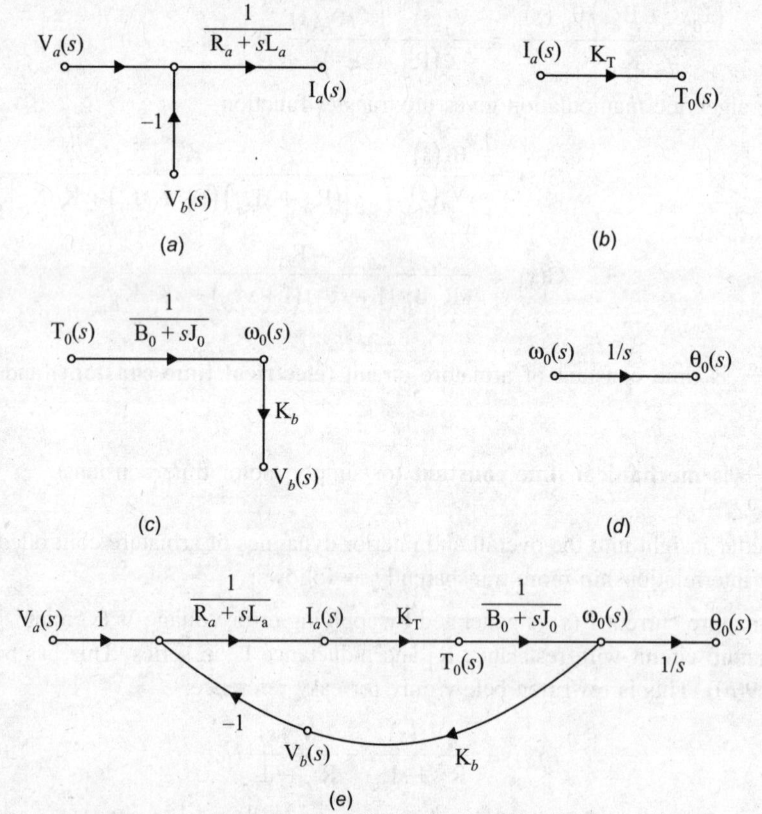

Fig 9.11: Developing signal flow graph for armature controlled DC motor

Note: the following in current discussion:

(*i*) Figure 9.10 of armature controlled DC motor, represents a closed loop system where the feedback signal is back emf which is proportional to angular speed of motor. In this sense the back emf effect resembles the tachometer feedback of a general control system.

(*ii*) It is practically experienced that L_a (armature inductance) is negligibly small. Having ignored L_a, the transfer function derived as 9.21(*a*) is simplified as

$$\frac{\theta_0(s)}{V_a(s)} = \frac{K_T}{s\left[R_a\left(B_0 + sJ_0\right) + K_T K_b\right]}$$

$$= \frac{K_T / R_a}{s^2 J_0 + s\left(B_0 + \dfrac{K_T K_b}{R_a}\right)} = \frac{K_T / R_a}{s^2 J_0 + sB_0^*}$$

where $B_0^* = B_0 + \frac{K_T K_b}{R_a}$ simply demonstrates that back emf has an effect of increasing the damping coefficient of system.

Field controlled DC servo motor

A schematic diagram of field controlled DC motor is shown in Fig. 9.12. The components shown in this diagram are as follows:

R_f = field winding resistance

L_f = field winding inductance

V_f = error voltage applied to field circuit (field control voltage)

I_f = field circuit current

T_0 = Torque developed by motor

J_0 = Moment of inertia of motor plus that of load referred to motor shaft

B_0 = damping coefficient of motor plus that of load referred to motor shaft

θ_0 = angular displacement of motor shaft.

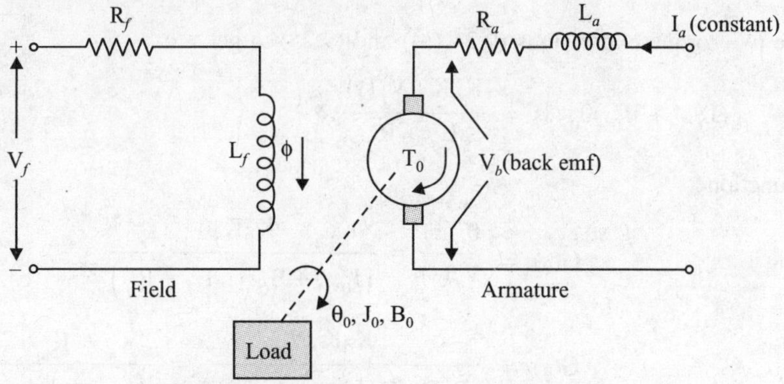

Fig. 9.12: Field controlled DC motor (schematic)

In order to develop **transfer function model**, let us assume that

(*a*) the armature winding is driven by constant current source I_a.

(*b*) the air gap flux ϕ is proportional to field current I_f, that is

$$\phi = K_f I_f \,;\; K_f = \text{proportionality constant}$$

(*c*) the torque (T_0) developed by motor, is proportional to product of air gap flux (ϕ) and armature current (I_a), that is

$$T_0 \propto \phi\, I_a$$

or $$T_0 = K'\phi I_a \qquad K' = \text{proportionality constant}$$

$$T_0 = K' K_f I_f I_a$$

since I_a is constant

$$T_0 = K_0 K_f I_f; \qquad K_0 = K' I_a \qquad ...[9.22(a)]$$

and its Laplace transform is

$$T_0(s) = K_0 K_f I_f(s) \qquad ...[9.22(b)]$$

KVL applied to field circuit, gives

$$L_f \frac{dI_f}{dt} + R_f I_f = V_f$$

and its Laplace transform with zero initial conditions, is

$$(sL_f + R_f)\, I_f(s) = V_f(s) \qquad ...(9.23)$$

Since the developed torque drives load against inertia J_0 and damping coefficient B_0, the torque equation is

$$J_0 \frac{d^2\theta_0}{dt^2} + B_0 \frac{d\theta_0}{dt} = T_0 \qquad ...[9.24(a)]$$

and its Laplace transform with zero initial condition, is

$$(J_0 s^2 + B_0 s)\theta_0(s) = T_0(s) \qquad ...[9.24(b)]$$

Substitute $I_f(s)$ from (9.23) in (9.22 (*b*)), to get

$$T_0(s) = K_0 K_f \left[\frac{V_f(s)}{sL_f + R_f}\right] \qquad ...(9.25)$$

Equate the two torque expressions (9.24 (*b*)) and (9.25) to get

$$(J_0 s^2 + B_0 s)\theta_0(s) = \frac{K_0 K_f V_f(s)}{sL_f + R_f}$$

and transfer function

$$G(s) = \frac{\theta_0(s)}{V_f(s)} = \frac{K_0 K_f}{(J_0 s^2 + B_0 s)(sL_f + R_f)} \qquad ...(9.26)$$

or

$$G(s) = \frac{K_0 K_f}{sB_0 R_f (1 + s\tau_f)(1 + s\tau_m)} = \frac{K_m}{s(1 + s\tau_f)(1 + s\tau_m)} \qquad ...(9.27)$$

where $\tau_f = \frac{L_f}{R_f}$ is **electrical time constant** of field circuit and $\tau_m = \frac{J_0}{B_0}$ is **mechanical time constant.**

The constant $K_m = \frac{K_0 K_f}{B_0 R_f}$ is referred to as **motor time constant**.

For better insight into internal dynamics of field controlled DC motor, the signal flow graph as shown in Fig. 9.13 can be sketched using the equations derived just above. The field current I_f is generated by applying error voltage V_f to field circuit comprising of R_f and L_f in series. Flux ϕ is produced by I_f and equals the product $K_f I_f$. The product $K_f I_f$ further multiplied by K_0 gives torque T_0 (9.22 (*a*)). Refer (9.24 (*a*)) to get $\frac{T_0(s)}{sJ_0 + B_0} = s\theta_0(s)$. Multiplying $\omega_0(s)$ by $1/s$ (ω_0 integrated) gives $\theta_0(s)$. Now, it is obvious that **field controlled DC motor** represents an **open loop system.**

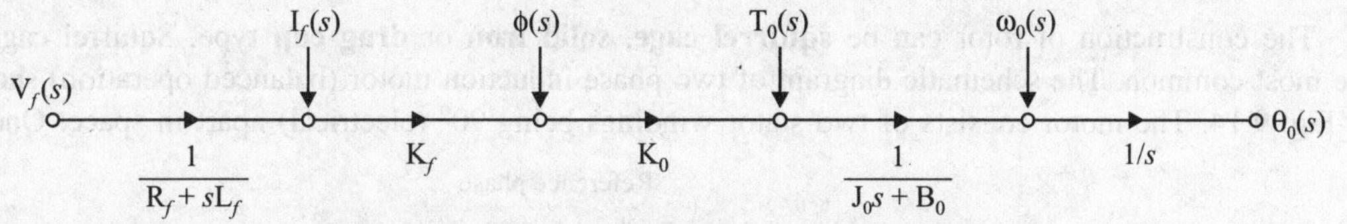

Fig. 9.13: Signal flowgraph for field controlled DC motor

Comparing armature controlled and field controlled DC servo motors

Having discussed the dynamics of armature controlled and field controlled DC motors in depth, it would be of practical interest to bring out the comparison of these two, in the aspects as follows:

(*i*) **Field controlled** DC motor is suitable for **low rating** while **armature controlled** DC motor involves **low cost** if servo design demands motor of **large rating**. In fact, field power requirement in field controlled motor, is less and it is often not necessary to amplify the control (error) signal. On the other hand, in armature controlled motor, the power demand in armature is high. The demand of high power necessitates an additional power amplifier which adds to the cost.

(*ii*) In armature controlled DC motor, the **back emf** developed by armature contributes to **additional damping** and motor operation becomes more stable. The damping in field controlled DC motor, is relatively less. In fact, only motor and load friction contributes entire damping in field controlled DC motor.

(*iii*) The efficiency of field controlled DC motor is less than that of armature controlled DC motor.

(*iv*) The time constant of armature controlled DC motor, is generally small as compared to that of field controlled DC motor. Therefore, armature controlled DC motor exhibits faster dynamical response.

(*v*) The armature controlled DC motor represents a **closed loop system** while field controlled DC motor represents an **open loop system**. The advantage of closed loop structure over open loop is well known.

(*vi*) Permanent magnets may be used instead of field coils in armature controlled DC motor. This makes it less expensive. Field controlled DC motors do require field coils.

(*vii*) Field controlled mode of operation demands constant current source for armature circuit while armature controlled mode of operation demands constant voltage source for field circuit. It is easy to get constant voltage source than a constant current source.

AC servo motor

Most of the small size motors used in servo system design are AC motors. For low power applications, AC servo motor is preferred because they are **light in weight**, **rugged** and have **no brush contacts**. AC servo motor is primarily a two-phase induction motor with following special design features making it particularly useful in servo mechanisms:

(*i*) The rotor of AC servo motor is designed to have high resistance so as to keep X/R ratio (ratio of rotor reactance X to rotor resistance R) small. Small X/R ratio linearises the speed-torque characteristic to meet stability requirement (explained later in this section). In conventional induction motors, X/R ratio is often kept high so as to achieve the maximum torque close to operating region (within 5% slip).

(*ii*) The diameter of rotor is kept small so as to reduce inertia and have fast acceleration.

The construction of rotor can be **squirrel cage**, **solid iron** or **drag cup** type. Squirrel cage is the most common. The schematic diagram of two phase induction motor (balanced operation) shown in Fig. 9.14. The motor consists of two stator windings being 90° (electrical) apart in space. One of

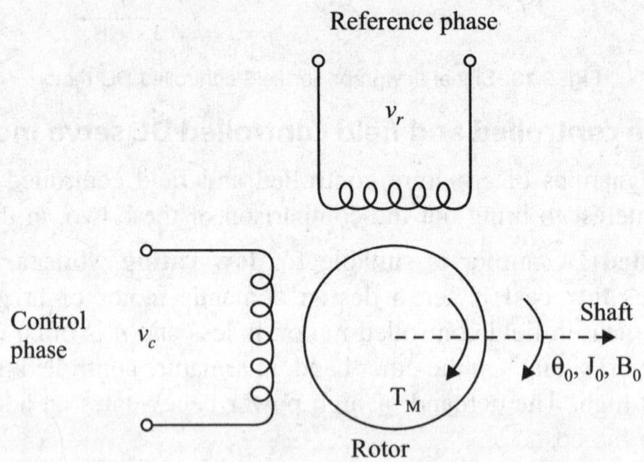

Fig. 9.14: Two-phase induction motor (Schematic)

them is called **control winding** and another **reference winding**. Both windings are excited with AC voltages of equal rms value but with 90° phase difference. The respective magnetic fields of the two windings are 90° apart in both time and space. This results in two magnetic stator field vectors of constant magnitude rotating at synchronous speed. The rotating flux vector sweeps past the conductors in the rotor and induces voltage. As the rotor is a closed circuit, the induced voltage causes the current to flow. With the rotor stationary, the rotor currents are of same frequency as that of stator frequency. The two magnetic stator field vectors which rotate together at synchronous speed, interact with rotor currents and produce a starting torque on rotor in the direction of field rotation. This torque accelerates the rotor until it reaches its operating speed which is determined by friction and load torque. The rotor, however, cannot reach synchronous speed because at this speed rotor conductors become stationary with respect to field and therefore voltage induced becomes zero.

The shape of torque-speed characteristic curve, largely depends on ratio of rotor reactance X to rotor resistance R (X/R ratio). As demonstrated in Fig. 9.15, the characteristic curves are almost

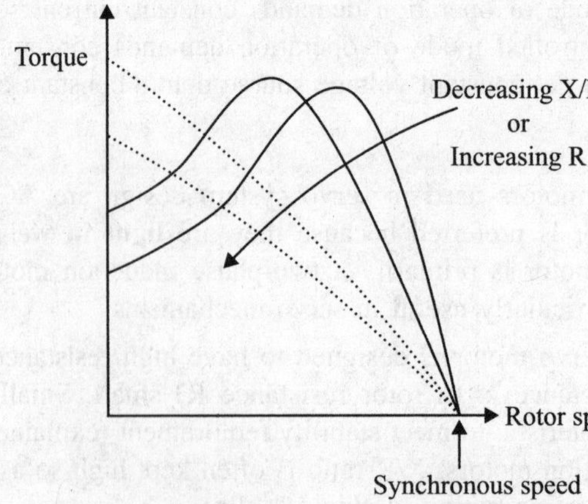

Fig. 9.15: Torque-speed characteristics

linear for smaller X/R ratio. So, servo motors are designed with small X/R ratio to achieve linearity. If the servo system includes a motor operating in the region where the characteristic curve has positive slope (showing increase in torque with increase in speed), it tends to be unstable. The positive slope is exhibited in characteristic curve for large X/R ratio. Small X/R ratio ensures that torque always decreases as speed increases, that is, positive damping prevails and system remains stable.

The ordinary two-phase induction motors with low rotor resistance, are unsuitable for servo applications. It should also be noted that the voltages applied to two stator windings of servo ac motor, are seldom balanced. The reference winding has constant voltage excitation but the control winding is excited by amplified error signal that varies in magnitude and polarity (± 90° w.r.t. reference phase). The motor reverses its direction if error signal reverses its polarity. The output torque is roughly proportional to magnitude of control voltage and direction of torque is determined by the polarity of error signal.

The torque generated by motor, depends on rms control voltage V_c and speed ω (= $d\theta/dt$). A family of torque – speed characteristic curves for different control voltages varied in equal steps are sketched in Fig. 9.16(*a*). Strictly speaking, even after linearising by increasing rotor resistance (or decreasing X/R ratio), the curves are still non-linear to some extent as it is evident from Fig. 9.16(*a*). However, in the region of low speed, the curves are almost linear and equidistant for varied control voltage in equal steps. In servo applications, the motor is generally operated at low speeds. Therefore, approximation of torque-speed curves by straight lines as shown in Fig. 9.16(*b*) holds good particularly in servo system analysis that follows:

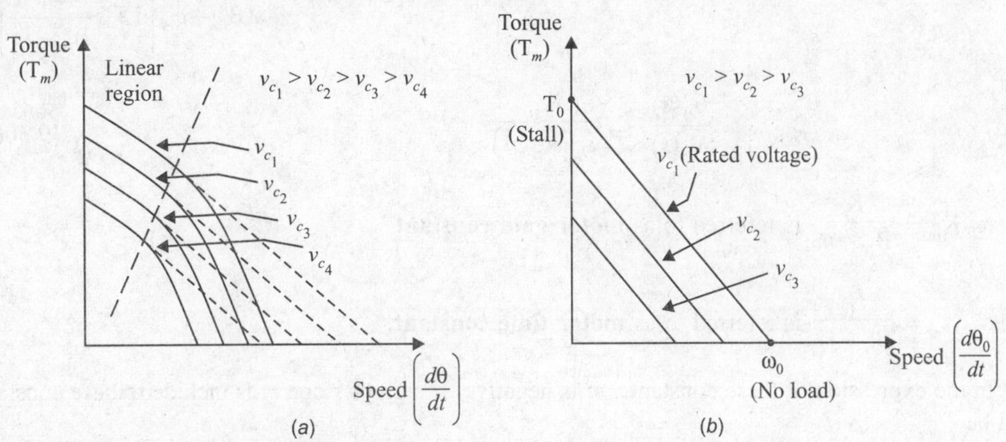

Fig. 9.16: (*a*) Torque-speed curves (linearisation) (*b*) Ideal torque-speed curves (straight line approximated)

The stalling torque T_0 (corresponding to zero rotor speed or rotor held stationary) of the servo motor, is proportional to the voltage applied to control phase. Since the output torque decreases linearly with speed, it may be considered to be equal to stalled torque minus the torque absorbed in damping. The damping coefficient is proportional to slope of torque-speed curve.

For rated voltage V_{c_1} slope $\boldsymbol{m} = -\dfrac{T_0}{\omega_0}$ (minus sign due to negative slope). T_0 and ω_0 are stall torque and no load speed at rated voltage respectively. Let us define a constant K as **'blocked rotor torque (stall torque) per unit rated control voltage'** *i.e.,*

$$K = \frac{T_0}{V_{c_l}}$$

With motor constants **K** and ***m*** as defined just above, the family of straight lines of Fig. 9.16(*b*) can be represented by the equation,

$$T_m = KV_c + m\frac{d\theta_0}{dt} \quad ...(9.28)$$

where V_c is the voltage applied to control winding.

Let J_0 = moment of inertia of motor plus that of load referred to motor and B_0 = viscous friction of motor plus that of load referred to motor. Then torque T_m can also be expressed as

$$T_m = J_0\frac{d^2\theta_0}{dt^2} + B_0\frac{d\theta_0}{dt} \quad ...(9.29)$$

Equating the torque expressions (9.28) and (9.29) and taking Laplace transform with zero initial conditions, we have

$$KV_c(s) + ms\theta_0(s) = J_0 s^2\theta_0(s) + B_0 s\theta_0(s) \quad ...[9.30(a)]$$

and transfer function of two phase AC servo motor, is

$$G(s) = \frac{\theta_0(s)}{V_c(s)} = \frac{K}{J_0 s^2 + s(B_0 - m)} = \frac{K}{s(B_0 - m)\left[1 + s\dfrac{J_0}{B_0 - m}\right]}$$

or

$$\frac{\theta_0(s)}{V_c(s)} = \frac{K_m}{s(1+\tau_m s)} \quad ...[9.30(b)]$$

where $K_m = \dfrac{K}{B_0 - m}$ is referred to as **motor gain constant**

and $\tau_m = \dfrac{J_0}{B_0 - m}$ is referred to as **motor time constant.**

In the expression of these constants, ***m*** is negative. If sign of slope *m* is included, these constants will be

$$K_m = \frac{K}{B_0 + m} \quad \text{and} \quad \tau_m = \frac{J_0}{B_0 + m}$$

Note the following in current discussion:

(*i*) The negative slope (*m*) of torque-speed curve, contributes to additional positive damping. In this sense it is also referred to as **internal electric damping.** If the slope is positive, then effective damping dictated by ($B_0 - m$), might be negative forcing the overall system to exhibit negative damping and become unstable. The conventional two-phase induction motors do exhibit positive slope in the low speed region of torque-speed curve as shown in Fig. 9.15 and this is the reason why the **conventional induction motors are not suitable for servo application.**

(*ii*) The signal flow graph for AC servo motor, revealing information about interior dynamics, is shown in Fig. 9.17. The acceleration (Laplace transformed) $\alpha(s)$ can be expressed using (9.30(*a*)) as

$$\alpha(s) = s^2\theta_0(s) = \frac{K}{J_0}V_c(s) - \frac{(B_0 + m)}{J_0} \cdot s\theta_0(s)$$

$$= \frac{K}{J_0}V_c(s) - \frac{(B_0 + m)}{J_0}\omega_0(s)$$

Note that (–) sign of ***m*** is included in the expression of $\alpha(s)$ just above and three variables $V_c(s)$, $\alpha(s)$ and $\omega_0(s)$ are represented by three nodes in the signal flow graph.

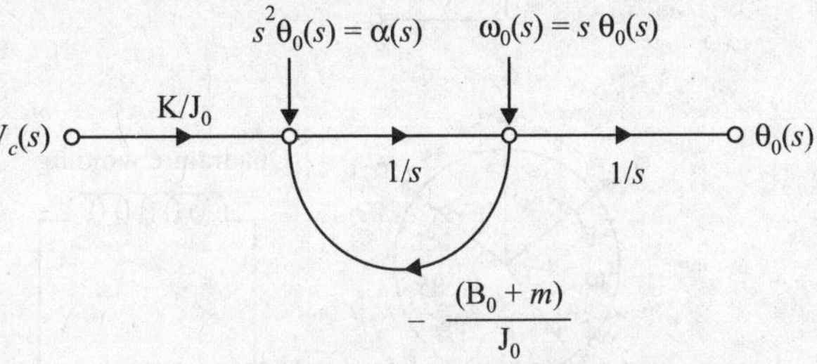

Fig. 9.17: Signal flow graph (AC servo motor)

9.5 AC TACHOMETER

The tachometer is an electromechanical device which generates electrical output proportional to shaft speed (ω). The variants of tachometer include both DC tachometer and AC tachometer.

DC tachometer

The schematic diagram of DC tachometer is shown in Fig 9.18. This consists of a small armature coupled to output shaft. The armature revolves in the field of permanent magnet.

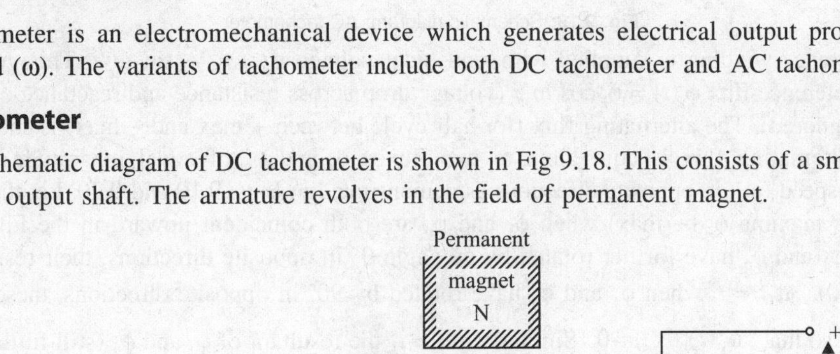

Fig. 9.18: DC tachometer

The voltage (V_0) generated is tapped via commutator and brush arrangement. The voltage generated is proportional to product of flux and speed. Since the flux is constant, output voltage is proportional to speed (ω). The polarity of output voltage is indicative of direction of rotation. The permanent magnet tachometers fascinate in the sense that they are **compact, efficient and reliable** but irk in the sense that they have **high inertia** and **brushes pose maintenance troubles.**

AC tachometer

The schematic diagram of AC tachometer or drag cup generator is shown in Fig. 9.19. This also has two stator windings (reference and quadrature) mounted at right angles to each other (quadrature in space) like AC servo motor. The low inertia rotor is a **thin aluminum cup** rotating in the air gap. The rotor is short circuited.

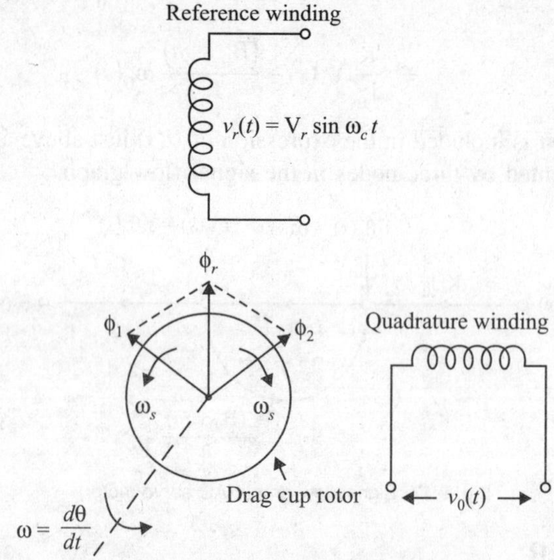

Fig. 9.19: Schematic diagram: AC tachometer

The reference winding is supplied with sinusoidal voltage $v_r(t) = V_r \sin \omega_c t$. This generates an alternating reference flux $\phi_r(t) = \phi_r \cos \omega_c t$ (voltage drop across resistance and reactance of reference winding, is ignored). The alternating flux (for half cycle between + max and – max) is shown in Fig. 9.20. This alternating flux is equivalent to two fluxes ϕ_1 and ϕ_2 of equal magnitude rotating at synchronous speed (ω_s) in opposite directions as demonstrated in Figs. 9.19 and 9.20. Let the reference flux be at its maxima ϕ_r (+ max) when ϕ_1 and ϕ_2 are both coincident upward in the Fig. 9.20. At $t = t_a$ when ϕ_1 and ϕ_2 have further rotated by an angle θ_a in opposite directions, their resultant is ϕ_a (see Fig. 9.20). At $t = t^*$ when ϕ_1 and ϕ_2 have rotated by 90° in opposite directions, these are equal and opposite so that $\phi_r(t)\big|_{t=t^*} = 0$. Similarly at $t = t_b$ the resultant of ϕ_1 and ϕ_2 (still further rotated) is ϕ_b. Direction of ϕ_b is opposite to that of ϕ_a. In fact, $\phi_b = -\phi_a$. Negative peak, ϕ_r (–max) occurs when ϕ_1 and ϕ_2 are both coincident downward. Thus, it is clear that sinusoidal (alternating) flux is set up. With stationary rotor, ϕ_1 and ϕ_2 (components of ϕ_r) induce equal and opposite voltages in the quadrature coil so that open circuit voltage $v_0(t) = 0$.

Now let the rotor rotate at speed $\omega = d\theta/dt$ in the direction ϕ_1 is rotating. The relative speed of rotor with respect to ϕ_1, decreases and that with respect to ϕ_2 increases. So, the current induced in rotor due to ϕ_1 tends to decrease and that due to ϕ_2 tends to increase. As a result, the net ϕ_1 strengthens and net ϕ_2 weakens. The unbalance in ϕ_1 and ϕ_2 causes the voltage to be induced in quadrature winding. The voltage $v_0(t)$ of reverse polarity is generated if the rotor rotates in opposite direction.

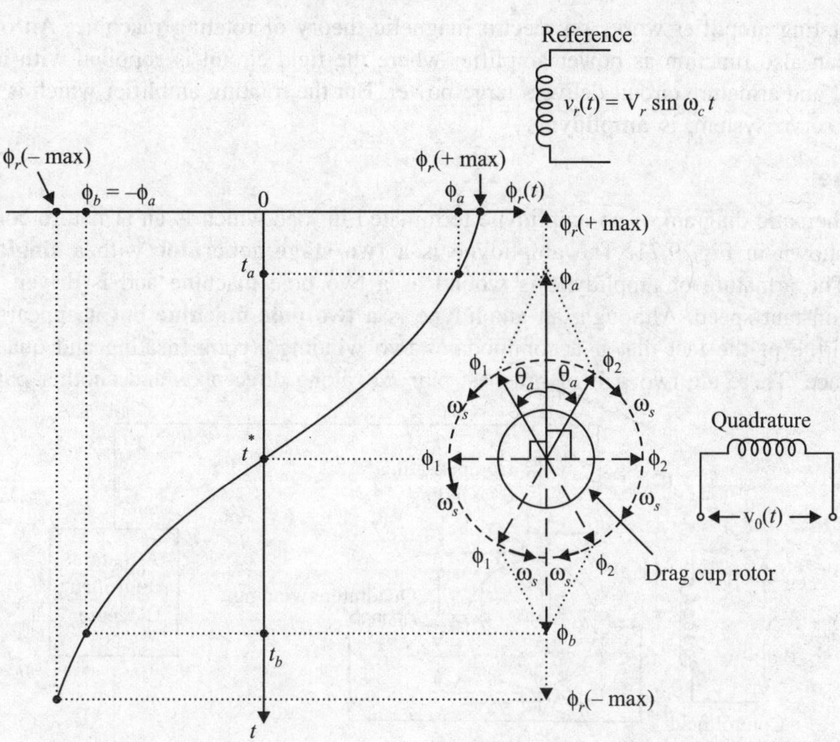

Fig. 9.20: Generation of reference flux (alternating) ϕ_r as phasor sum of ϕ_1 and ϕ_2.

The open circuit voltage $v_0(t)$ across quadrature winding is 90° phase displaced with respect to reference voltage $v_r(t) = V_r \sin \omega_c t$ and its magnitude is proportional to rotor speed ($\omega = d\theta/dt$).

$$v_0(t) = K\omega \cos \omega_c t \qquad ...(9.31)$$

Note that (9.31) also holds good for slow time varying ω ($\omega << \omega_c$). In fact, the output of AC tachometer is an AC voltage. This tachometer can be used in DC servo system by passing this output through demodulator (peak detector) to get corresponding DC output as

$$V_0 = K\omega = K\frac{d\theta}{dt} \qquad ...(9.32)$$

The transfer function model is

$$G(s) = \frac{V_0(s)}{\theta(s)} = Ks \qquad ...(9.33)$$

where K is tachometer constant (volts/rad/sec).This has been already discussed in Section 8.8 that tachometer with transfer function model (9.33) is very popular in rate feedback compensator design.

9.6 SERVO AMPLIFIER

The error signal emerging from error detector is often weak and requires amplification. The electronic amplifiers are known to serve this purpose. If the servo system design demands the power level higher than that achievable from electronic amplifiers, the **rotating amplifier** becomes an obvious choice.

The rotating amplifier works on electro magnetic theory of rotating machine. An ordinary DC generator can also function as power amplifier where the field circuit is supplied with input at low power level and armature circuit delivers large power. But the rotating amplifier which is particularly suitable in servo system, is **amplidyne.**

Amplidyne

The schematic diagram of an amplidyne terminated in load which is an armature controlled DC motor, is shown in Fig. 9.21. The amplidyne is a **two stage generator** with a **single armature winding.** The armature of amplidyne is wound as a two pole machine and is driven by a prime mover at constant speed. Although, an amplidyne is a **two pole machine** but it appears to be four poles by virtue of the fact that it accommodates two windings (compensating and quadrature) 90° apart in space. There are two pair of brushes; pair *dd′* along direct axis and another pair *qq′* along

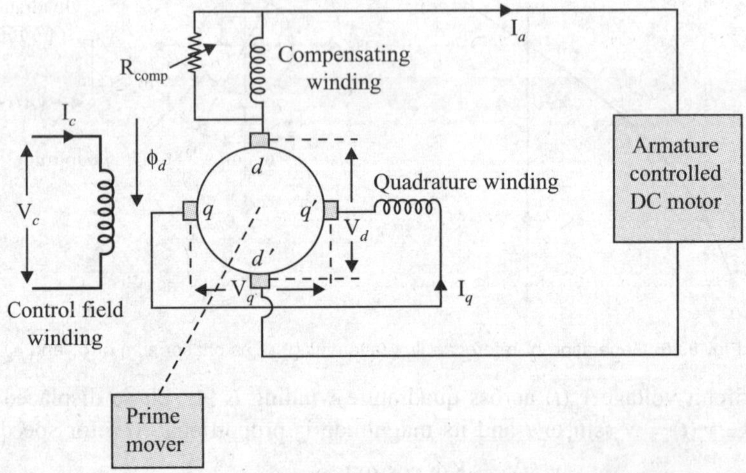

Fig. 9.21: Schematic diagram of an amplidyne terminated in load, an armature controlled DC motor

quadrature axis which is at right angle to direct axis. **The flux established along direct axis, induces the voltage in the armature appearing across brushes *qq′* in quadrature axis. Similarly quadrature axis flux, induces voltage in the armature across the brushes *dd′* in direct axis. Also the current flowing though any pair of brushes (*dd′* or *qq′*) establishes flux along the same brush axis due to armature reaction.**

The input voltage V_c is applied to control field winding which has large number of turns and high impedance so as to keep input power small. The current I_c in control field winding, sets up a small flux ϕ_d along the direct axis. Due to this flux ϕ_d and armature driven by prime mover, voltage V_q is induced across brushes *qq′*. The brushes *qq′* are short circuited through series quadrature winding to allow the current I_q to flow through the armature. Due to current I_q, an armature reaction flux ϕ_q appears at right angle to the direction of flux ϕ_d. In conventional generators **an attempt is made to suppress this armature reaction, but in an amplidyne, a low reluctance magnetic path is provided to strengthen it.** The brushes *qq′* are short circuited so that very strong flux ϕ_q is produced for a given control field current I_c. The key idea of using quadrature winding is to increase effective turns so that desired flux ϕ_q can be established while keeping I_q small. Small I_q eases out commutation which is, indeed, difficult in such type of machine.

Due to strong quadrature flux and rotation of amplidyne armature, a high voltage V_d is generated across brushes *dd′* (perpendicular to flux ϕ_q). The voltage V_d drives the current I_a through the load (armature controlled DC motor in current discussion). The current I_a again sets up armature reaction flux ϕ_a along the direct axis but in the direction to oppose control flux ϕ_d. Due to this degenerative action of I_a, the voltage V_d tends to decrease undesirably.

A **compensating winding** is provided in the direct axis and is so connected as to undo this degenerative action. A potentiometer R_{comp} is also connected across compensating winding to provide almost 100% compensation. In the sense of the addition of this compensating winding, the machine is called amplidyne, a machine that delivers voltage which is only marginally affected by load current I_a. Note that direct axis voltage V_d is much larger than quadrature axis voltage V_q because quadrature axis flux ϕ_q is much larger than direct axis (control) flux ϕ_d.

Transfer function model of amplidyne with open circuited output is developed below while ignoring the magnetic saturation and assuming 100% compensation.

KVL to control field winding, gives

$$V_c(t) = I_c(t)\,R_f + L_f \cdot \frac{d}{dt} I_c(t)$$

and its Laplace transform with zero initial conditions, gives

$$V_c(s) = R_f I_c(s) + L_f \cdot sI_c(s) \qquad \text{...[9.34(a)]}$$

or

$$V_c(s) = R_f I_c(s)\,[1 + s\tau_f] \qquad \text{...[9.34(b)]}$$

where $\tau_f = \dfrac{L_f}{R_f}$ is time constant of control field winding.

The field current I_c sets up direct axis flux and voltage V_q induced along quadrature axis is proportional to I_c.

$$V_q(t) = K_q I_c(t) \qquad \text{...[9.35(a)]}$$

where K_q is quadrature axis emf constant (volts induced per unit control field current)

and

$$V_q(s) = K_q I_c(s) = \frac{K_q V_c(s)}{R_f(1 + s\tau_f)} \qquad \text{...[9.35(b)]}$$

For closed quadrature circuit, KVL gives

$$R_q I_q(t) + L_q \frac{d I_q(t)}{dt} = V_q(t) \qquad \text{...[9.36(a)]}$$

and taking Laplace transform with zero initial conditions

$$I_q(s) = \frac{V_q(s)}{R_q(1 + s\tau_q)} \qquad \text{...[9.36(b)]}$$

where $\tau_q = \dfrac{L_q}{R_q}$ is **time constant of quadrature circuit.**

Since open circuit voltage $V_d(t)$ is proportional to $I_q(t)$,

$$V_d(t) = K_d I_q(t) \qquad \text{...(9.37)}$$

where K_d is **direct axis speed emf constant**.

Thus
$$V_d(s) = K_d I_q(s) = \frac{K_d V_q(s)}{R_q(1+s\tau_q)} = \frac{K_d K_q V_c(s)}{R_f R_q (1+s\tau_f)(1+s\tau_q)}$$

or
$$\frac{V_d(s)}{V_c(s)} = \frac{K_d K_q}{R_f R_q (1+s\tau_f)(1+s\tau_q)} \quad ...(9.38)$$

The generation of open circuit output voltage $V_d(s)$ beginning from input (control) voltage $V_c(s)$, is demonstrated in signal flow graph as shown in Fig. 9.22.

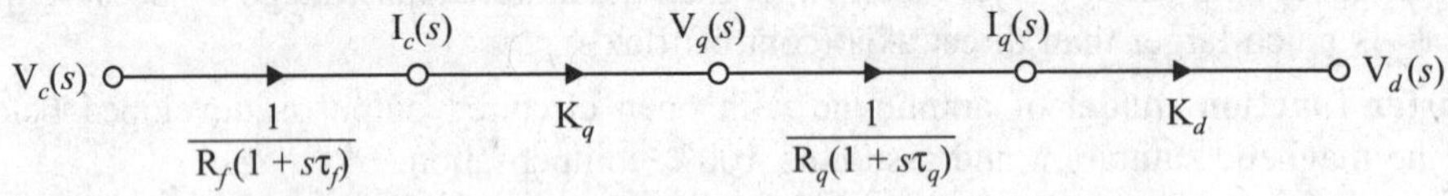

Fig. 9.22: Signal flow graph for amplidyne (open circuit output)

Now, let the amplidyne be terminated in load (armature controlled DC motor). The overall transfer function can be obtained by cascading transfer function model of amplidyne (9.38) and that of armature controlled DC motor (9.21(*a*)). The input $V_a(s)$ in transfer function model (9.21(*a*)), is in fact, the output $V_d(s)$ in transfer function model (9.38). The overall transfer function relates output $\theta_0(s)$ of armature controlled DC motor and input $V_c(s)$ to the amplidyne. In derivation of overall transfer function $G^*(s) = \frac{\theta_0(s)}{V_d(s)}$, 100% compensation via compensating winding together with R_{comp}, is assumed so that the motor armature current does not affect the control flux ϕ_d. Thus, ignoring armature inductance L_a of DC motor, which is usually negligibly small, the overall transfer function given below, has four poles including one at origin:

$$G^*(s) = \frac{\theta_0(s)}{V_d(s)} = \frac{K_d K_q}{R_f R_q (1+s\tau_f)(1+\tau_q)} \times \frac{K_T}{s[R_a(B_0+sJ_0)+K_T K_b]} \quad ...(9.39)$$

Note the following in the current discussion:

(*i*) In order to provide low reluctance on quadrature axis, two more poles called as **interpoles** are placed in quadrature axis. This is to strengthen quadrature axis flux ϕ_q and contribute large amplification.

(*ii*) The amplidyne is a **two stage generator** with a **single armature** winding. It behaves as a **two stage power amplifier**. Stage one lies between control field winding and quadrature winding and stage two lies between quadrature winding and direct axis circuit (compensating winding).

$$\text{Power gain of stage I} = Ap_1 = \frac{V_q I_q}{V_c I_c}$$

and
$$\text{power gain of stage II} = Ap_2 = \frac{V_d I_a}{V_q I_q}$$

In general, Ap_1 is larger than Ap_2. Typically Ap_1 is 200 and Ap_2 is 50. The overall power gain is of the order of 10^4.

(*iii*) In order to accommodate additional control facilities, the amplidynes are also occasionally built with number of different control field windings. In fact, the control field is wound in several sections so that many control signals can be applied simultaneously.

(*iv*) Ideally, an amplidyne appears to respond instantaneously to the changes in control voltage. But in practice, time delay is contributed by winding inductances and eddy currents. Consequently, the time constant of an amplidyne, typically lies in the ranges 0.02 to 0.25 seconds.

9.7 STEPPER MOTOR

So far we have discussed AC and DC servo motors which are driven by analog control voltage. The design of control system today, often includes digital computer. If the digital signals emerging from digital computer or digital storage devices, are used to directly drive the motor, the usual choice is stepper motor. The stepper motor responds to train of input voltage pulses. This is, in fact, special class of synchronous motor designed to rotate through a particular angle (referred to as step) for each voltage pulse inputed to it as control voltage. The steps through which it rotates, are typically 7.5°, 15°, 30° or larger. Smaller steps (half stepping and micro stepping discussed little later in this section) are also achievable.

The advantages of using stepper motor, are listed below:

(*i*) The stepper motor is a digital device. It is easy to interface it with PC, microprocessors and PLC_s.

(*ii*) The accessories like position and speed sensors, amplifiers etc. are not needed when stepper motor is used. The information about position and speed can be directly obtained by counting the input pulses. (discussed later in this section)

(*iii*) These motors can operate at low load speed without using gear train.

(*iv*) Brushes and commutators are not necessary in these motors. The brushless construction potentially eases out maintenance problems. **Tear and wear** in these motors is almost nil.

(*v*) These motors are capable of driving larger load while maintaining desired precision in position and speed.

These motors also have certain **disadvantages** as given below:

(*i*) Stepper motors are less efficient and have low output per unit weight.

(*ii*) The stepping rate is governed by the number of pulses being applied per second (PPS). As stepping rate increases, the rotor has less time to drive the load from one position to the next. If PPS exceeds certain value, the rotor does not strictly follow the command pulses and begins to miss the steps. Refer to torque – PPS curve shown in Fig. (9.23). For a given load torque (T_L), fr_1 is maximum PPS (pulses per second) at which the motor can start and respond to pulses without missing any step. The start range as shown in Fig. 9.23, is one in which the motor can start and synchronise to the input pulses without losing steps for any given load torque (T_L). The point A on PPS axis corresponds to maximum starting frequency. The slew range occurs beyond point A till B where motor synchronises with pulse rate, but cannot start, stop or reverse on command. For given load torque (T_L), fr_2 is maximum

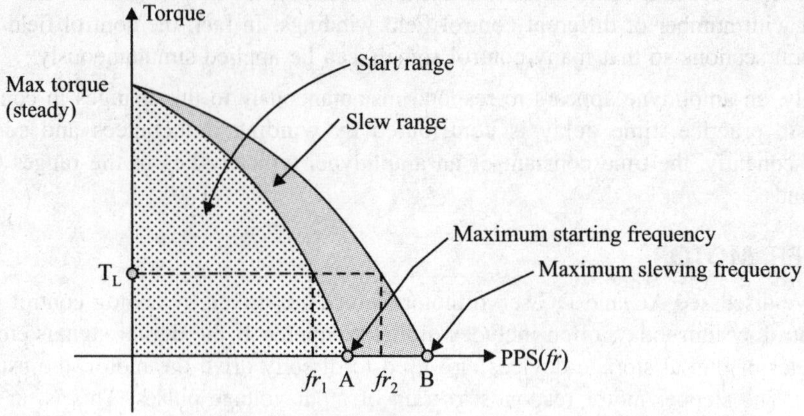

Fig. 9.23: Typical torque–PPS curve (stepper motor)

frequency ($fr_2 > fr_1$) till which motor follows the pulse rate without missing steps but cannot start, stop or reverse. If the pulse rate exceeds fr_2 (for given T_L), motor begins to miss the steps. Missing of steps, may also occur if amplitude of motor oscillation about locking position is too large. The points A and B are also referred to as **maximum pull in** and **maximum pull out** respectively.

The application of stepper motor, extends over a wide range. The significant areas are identified below:

(*i*) They are almost invariably used in **robotics.**

(*ii*) These are extensively used as paper feed motors in type writers and printers, XY plotters, CNC machine tools, recording heads in computer disk drives, positioning of work tables etc. All these demand incremental motion.

(*iii*) Process control systems, IC fabrication etc. use these motors very often.

(*iv*) They are also used in metering, mixing of chemicals, cutting, blending and packing food stuffs.

The application areas continue to grow fast as attempt is being made to further increase the power rating together with reducing the cost.

The **types of stepper motor** in general use, are

(*a*) Variable Reluctance motor

(*b*) Permanent Magnet motor

(*c*) Hybrid motor.

(*a*) **Variable reluctance stepper motor:** The variable reluctance stepper motor consists of a stator and a rotor. The motor may also have multiple stacks of stators and rotors to achieve smaller step angle. We shall discuss this little later in this section. Currently we shall concentrate on a **single stator-rotor structure.** The stator is usually wound for three phases. As shown in Fig. 9.24(*a*) the stator has six poles (teeth) with concentrated exciting windings around each one of them. The coils wound around diametrically opposite poles are connected in series (phase opposition) and the three phases are energised from a DC source with the help of switches. The rotor is made up of slotted steel laminations (ferromagnetic material) and has two poles (teeth) without any exciting winding.

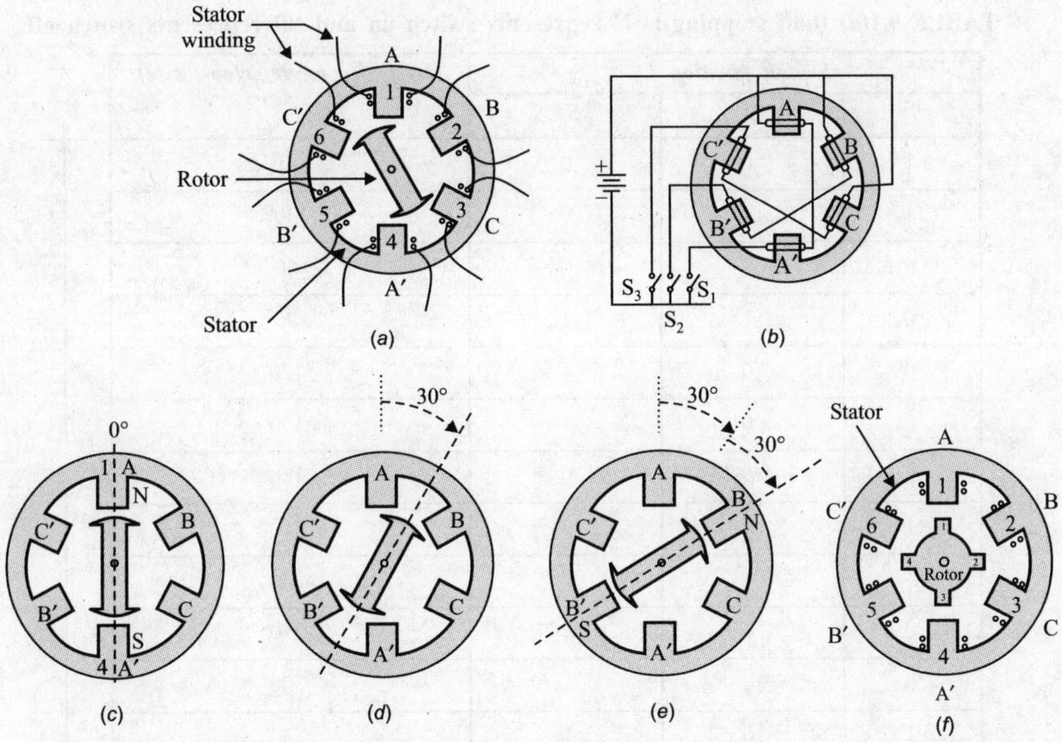

Fig. 9.24: (*a*) Variable reluctance stepper Motor (schematic) (*b*) Drive circuit (*c*) Rotor position with AA′ energised (*d*) Rotor position with AA′ and BB′ simultaneously energised (*e*) Rotor position with BB′ energised (*f*) Motor with three-phase stator and four teeth rotor.

The stator teeth are energised by a drive circuit as shown in Fig. 9.24(*b*). As shown in Fig. 9.24(*a*). The stator teeth 1 and 4 are wound by coils AA′, teeth 2 and 5 by coils BB′ and teeth 3 and 6 by coils CC′. The coils are then connected to DC source through switches S_1, S_2 and S_3. Switching is controlled by solid state devices. Let us understand the motor operation with the following points.

(*i*) When the coils AA′ are energised with switch S_1 closed together with S_2 and S_3 open, teeth 1 and 4 become stator poles (1 forming N pole and 4 forming S pole). Now, ferromagnetic rotor seeks a position where it presents minimum reluctance to stator field *i.e.*, rotor aligns itself to stator field axis (stator teeth 1 and 4). This position is demonstrated in Fig. 9.24(*c*).

(*ii*) Let the phase BB′ also be energised by closing switch S_2 while keeping AA′ energised (S_1, S_2 closed and S_3 open). Stator field axis rotates 30° clockwise and so only the rotor also rotates 30° clockwise to attain new minimum reluctance position. This position is demonstrated in Fig. 9.24(*d*).

(*iii*) If switch S_1 is opened while keeping S_2 closed, only phase BB′ remains energised and rotor further rotates 30° clockwise to adjust itself in still newer minimum reluctance position. This position is demonstrated in Fig. 9.24(*e*).

With the explanation of motor operation just above, it is obvious that the motor can be forced to rotate through **twelve** steps (each of 30°) to perform one complete revolution by successively energising the three phases (AA′, BB′ and CC′) in a particular sequence. Table 9.1(*a*) below demonstrates all possible 12 steps:

TABLE 9.1(*a*) (half stepping): '1' represents switch on and '0' represents switch off

Switch position			*Half step angle (clock wise)*	
S_1	S_2	S_3		
1	0	0	0°	One revolution
1	1	0	30°	
0	1	0	60°	
0	1	1	90°	
0	0	1	120°	
1	0	1	150°	
1	0	0	180°	
1	1	0	210°	
0	1	0	240°	
0	1	1	270°	
0	0	1	300°	
1	0	1	330°	
1	0	0	360°/0°	

TABLE 9.1(*b*) (stepping): '1' represents switch on and '0' represents switch off

Switch position			*Step angle (clockwise)*	
S_1	S_2	S_3		
1	0	0	0°	One revolution
0	1	0	60°	
0	0	1	120°	
1	0	0	180°	
0	1	0	240°	
0	0	1	300°	
1	0	0	360°/0°	

Note also the following in the perspective of current discussion:

(*i*) The **step angle** is defined as the **angular displacement of rotor in response to each distinct pulse.** In general, the step angle α_s is given by

$$\alpha_s = \frac{360°}{mN_r} \qquad ...(9.40)$$

where m = number of stator phases

N_r = number of rotor teeth.

For example, refer to Fig. 9.24(*a*), $\alpha_s = \frac{360°}{3 \times 2} = 60°$ *i.e.,* rotor rotates through an angle of 60°, if excitation changes from phase AA′ to BB′ and so on. In fact, the rotor will rotate through six steps each of 60° to complete one revolution as demonstrated in table 9.1(*b*). Nevertheless, rotor can also be rotated through steps of an angle 30° if two neighbouring phases are simultaneously excited in a particular sequence as shown in Table 9.1(*a*). This is called **half stepping** in the sense that rotor will now rotate through twelve steps each of 30° to complete one revolution.

Let us consider another example, if there is a design demand to reduce the step angle from 60° to 30°, it is possible by constructing rotor with four salient projections (poles or teeth) instead of two while retaining the stator with six poles (three phase) as shown in Fig. 9.24(*f*).

The **step angle** $\alpha_s = \frac{360°}{3 \times 4} = 30°$. Nevertheless, rotor can also be rotated through half steps of an angle of 15° if two neighbouring phases are simultaneously energised in a particular sequence thereby the number of steps would increase to twenty-four.

(*ii*) Still further demand of reducing the step angle, can be met by increasing the number of poles of stator and rotor.

The schematic diagram of a typical **stepper motor** to achieve step angle of 5°, is shown in Fig. 9.25. The stator is wound for **four phases** and has **eight salient poles.** Each stator pole has two teeth. The rotor has 18 teeth (and 18 slots) which are uniformly distributed over angular periphery of 360°. The intervening slot between two teeth of a stator pole has angular periphery equal to that of each rotor teeth or slot. When phase AA′ is energised, the rotor is positioned as shown in Fig. 9.25. By successively energising four phases AA′, BB′, CC′ and DD′ in a particular sequence, the rotor can be rotated through 72 steps to complete one revolution. The step angle using (9.40), is $\alpha_s = \frac{360°}{mN_r} = \frac{360°}{4 \times 18} = 5°$.

Likewise the stepper motor can be designed to achieve any desired step angle with suitable choice of stator phases and rotor teeth.

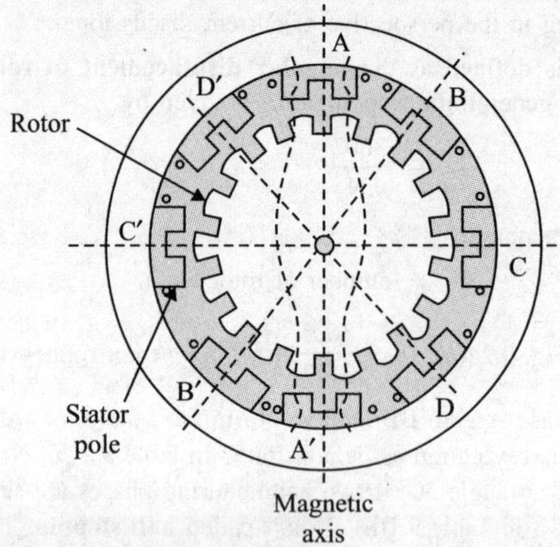

Fig. 9.25: Stator-rotor structure to achieve step angle of 5°

In so far, we have discussed only single stack stepper motor where all phase windings were in the same plane. The demand of reducing step angle, can also be met by arranging the windings in multiple stacks. The motor, so constructed, is referred to as **multi stack stepper motor.** The stators have a common frame and rotors have a common shaft. For illustration, longitudinal cross-section of **three stack motor** is shown in Fig. 9.26(*a*). Teeth structure of stator and rotor is demonstrated in Fig. 9.26(*b*). It is easy to visualise that stator and rotor teeth can be aligned by virtue that they are of same size. Note that stators are pulse energised while rotors remain unexcited. When the stator is energised, the rotor is forced to move into nearest minimum reluctance position that is the position where stator and rotor teeth are aligned. Further note that the teeth on all the rotors, are in perfect alignment while stator teeth of different stacks, have angular separation α given by:

$$\alpha = \frac{360°}{q \cdot N_r} \qquad ...(9.41)$$

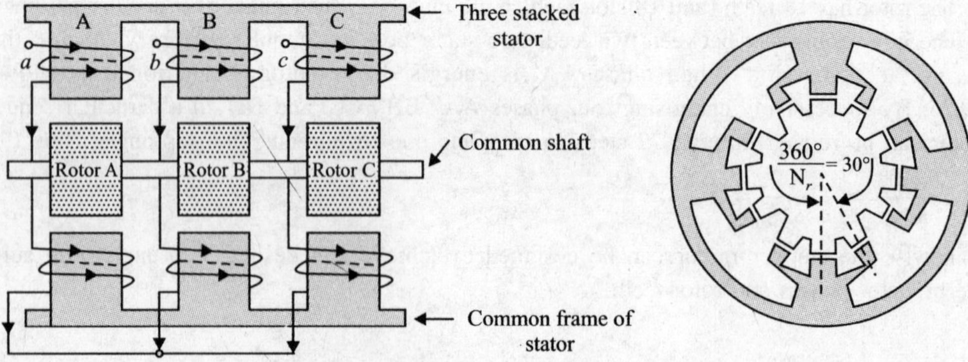

Fig 9.26: Three stack variable reluctance stepper motor (*a*) Longitudinal cross-section (*b*) Teeth structure of stator and rotor

where q = number stacks and N_r = number rotor teeth. In the example under discussion q = 3, N_r = 12 and $\alpha = 10^\circ$. The twelve teeth (N_r = 12) on rotor, have angular separation of 30° as shown in Fig. 9.26(*b*). Let us consider a pair of stator-rotor. Energising the stator, rotor aligns with stator in angular position $\theta = 0^\circ$ or multiples of $360^\circ/N_r = 30^\circ$ (stable positions) as demonstrated in Fig. 9.27(*a*).

However, if we consider rotors of stack A, stack B and stack C together with stack A, stack B and stack C of stator, the rotational dynamics (the dynamics as a whole) of rotor is as demonstrated

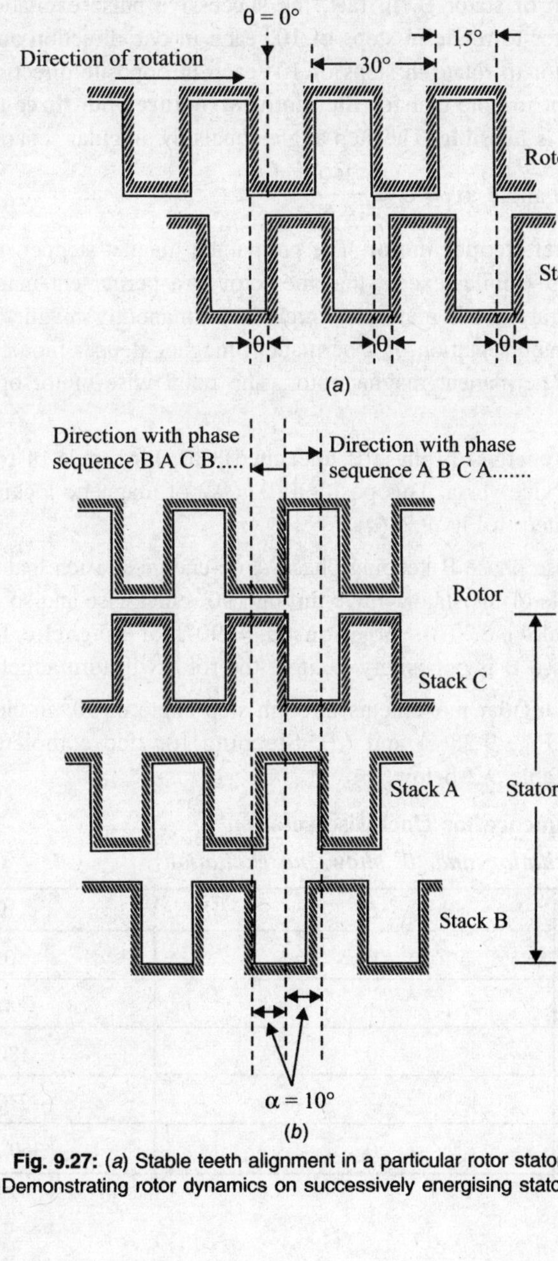

Fig. 9.27: (*a*) Stable teeth alignment in a particular rotor stator set. (*b*) Demonstrating rotor dynamics on successively energising stator phases.

in Fig. 9.27(*b*) where it is initially assumed that stack C rotor teeth are aligned with stack C stator. At this point, let us recall that the teeth of all the rotors of stack A, stack B and stack C, are perfectly aligned. This is the reason why Fig 9.27(*b*) shows only one rotor. Before going into insight of rotational dynamics of rotor, let us keep a note **that the number of stator phases, is equal to the number of stacks.** If phase A (stator A) is energised by a pulse, rotor will rotate by 10° from initial position in the direction shown so that rotor teeth are aligned with teeth of stator A. Instead of phase A, if phase B is pulse energised, rotor will rotate by 10° in opposite direction so that rotor teeth are aligned with teeth of stator B. In fact, the successive pulse excitation in phase sequence A B C A B...causes the rotor to rotate in steps of 10° each in one direction and the phase sequence B A C B A...causes the rotor to rotate in steps of 10° each in opposite direction as demonstrated in Fig. 9.27(*b*). It is significant to note that for **the motor structure** with **three or more phases only, the control over direction is possible.** The step angle equals the angular separation of stator teeth of different stacks, that is step angle $\alpha_s = \alpha = \dfrac{360°}{q \cdot N_r}$.

(*b*) **Permanent magnet stepper motor.** The permanent magnet stepper motor and the variable reluctance stepper motor are similar except that the rotor is a permanent magnet in the former. In fact, rotor is a ferrite or rare earth material which is permanently magnetised. For illustration, Fig. 9.28 gives schematic representation of a permanent magnet stepper motor with two phase (four poles) stator and two pole permanent magnet rotor. The point wise motor operation can be better understood as follows:

(*i*) When phase A is energised (phase B unexcited), the north pole of rotor aligns with south pole of stator and vice versa. This position ($\theta = 0°$) of magnetic locking between stator and rotor is demonstrated in Fig. 9.28(*a*).

(*ii*) Let us now energise phase B keeping phase A de-energised such that the locations of north pole and south pole of the stator, move through 90° clockwise and so only rotor also moves through 90° clockwise. This position ($\theta = 90°$) of magnetic locking is shown in Fig. 9.28(*b*).If phase B is oppositely excited, the rotor will turn anticlockwise by 90°.

(*iii*) Similarly, further angular movements through step angle of 90° in clockwise direction, are demonstrated in Fig. 9.28(*c*) and (*d*). Stepping for one complete revolution, is also demonstrated in Table 9.2 below.

TABLE 9.2 AB phase sequence for clockwise rotation
(+, – show polarity of excitation and '0' shows no excitation)

A	*B*	*θ*
+	0	0°
0	+	90°
–	0	180°
0	–	270°
+	0	360°/0°

(*iv*) The step angle θ_s is given by $\theta_s = \dfrac{N_s - N_r}{N_s \times N_r} \times 360°$ ...(9.42)

where N_s = number of poles (teeth) in stator and N_r = number of poles (teeth) in rotor. In the current discussion, $N_s = 4$, $N_r = 2$ and $\theta_s = \dfrac{4-2}{4 \times 2} \times 360° = 90°$. Note that (9.40) also gives $\theta = 90°$.

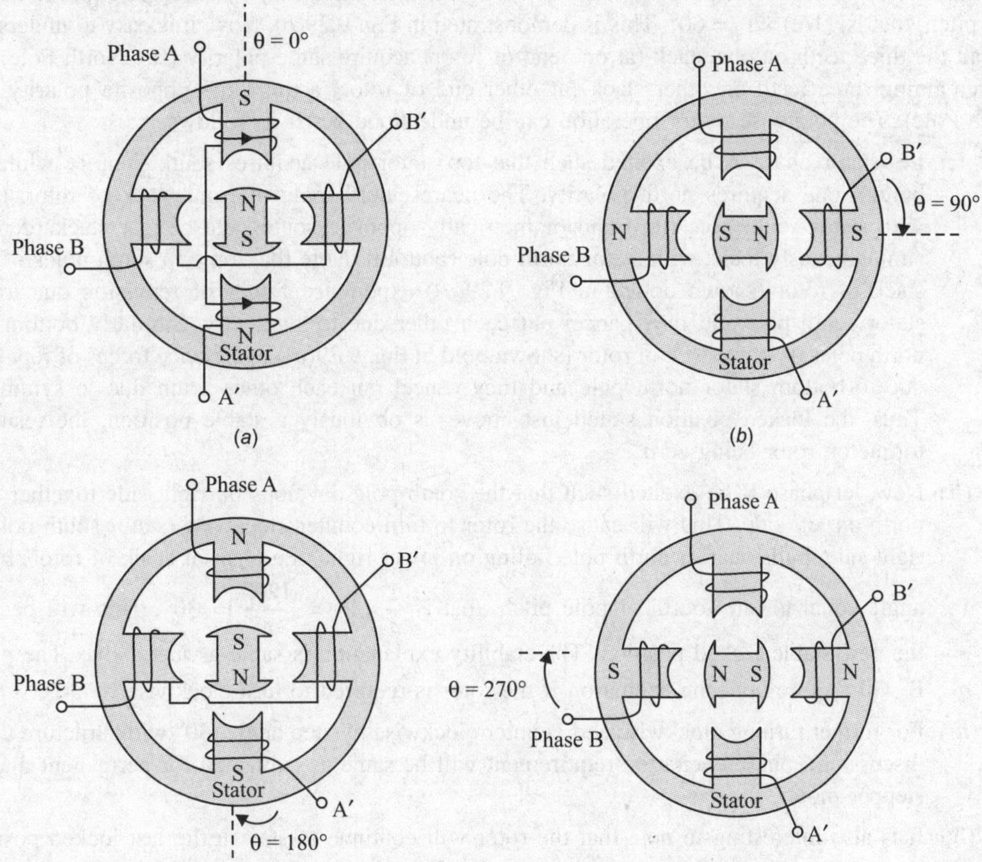

Fig. 9.28: Permanent magnet stepper motor (schematic) (interconnection of AA′ and BB′ in (*b*), (*c*), (*d*) omitted for clarity)

(*v*) It is difficult to manufacture the permanent magnet stepper motor with large number poles. Therefore, small steps in the permanent magnet motor, are not possible. These motors operate at larger steps up to 90° and at maximum response rate of about 300 pps. **The variable reluctance type of stepper motors are used to meet the design demand of smaller step angles.**

(*c*) **Hybrid stepper motor:** The hybrid stepper motor is basically a permanent magnet stepper motor but this has multistack, toothed rotor like multistack variable reluctance motor. In this sense, it is referred to as hybrid motor. The stator structure is similar to that of permanent magnet motor. For illustration, Fig. 9.29(*a*) shows schematic view of hybrid motor with two phase (four poles) stator and two stack rotor. The permanent magnet is placed axially along the rotor in the form of ring shaped cylinder over the motor shaft. Each of two stacks at each end, have angular separation equal to pole pitch or tooth pitch P_t given by

$$P_t = \frac{360°}{\text{Number of teeth}} = \frac{360°}{3} = 120°$$

But the three teeth at both ends (both stacks) are not aligned, rather three teeth on one stack (at one end) are displaced from the three teeth on other stack (at other end) by an angle equal to half of pole pitch, that is, $(1/2) \times P_t = 60°$. This is demonstrated in Fig. 9.29(*b*). Now, it is easy to understand that all the three teeth on one stack (at one end of rotor) acquire same polarity (say North Pole) and the remaining three teeth on other stack (at other end of rotor) acquire the opposite polarity (say South Pole). The pointwise motor operation can be understood better as follows:

(*i*) Let stator phase A be excited such that top stator pole acquires south polarity while the bottom one acquires north polarity. The nearest north pole of front stack of rotor, locks with stator south pole (top) and diametrically opposite south pole of rear stack of rotor, simultaneously locks with stator north pole (bottom). Note that top two south poles of rear stack of rotor (shown dotted in Fig. 9.29(*a*)) experience forces of repulsion due to top stator south pole and they cancel out each other due to symmetry. Similarly bottom two north poles of front stack of rotor (shown bold in Fig. 9.29(*a*)) experience forces of repulsion due to bottom stator north pole and they cancel out each other again due to symmetry. Thus, the locked position stated just above, is obviously a stable position, the resultant torque on rotor being zero.

(*ii*) Now, let phase B be excited such that the south pole develops on right side together with north on left side. This will cause the rotor to turn counter clockwise (stator south pole on right side, pulls nearest north pole falling on lower right side of front stack of rotor) by an angle equal to one fourth of pole pitch, that is $\frac{1}{4} \times P_t = \frac{120°}{4} = 30°$. This will become the new stable locked position. The stability explanation is same as just above. The phase B will require opposite excitation if the rotor is required to turn clockwise by 30°.

(*iii*) For further turning clockwise and counter clockwise by step angle (30° with structure under discussion), phase excitation requirement will be same as explained for permanent magnet stepper motor.

(*iv*) It is also interesting to note that the rotor will continue to stay in the last locked position even if stator excitation were removed. Further movement of rotor is prevented in either direction (clockwise/counter clockwise) due to rotor structure together with permanent magnet.

(*v*) Hybrid motor fascinates in the sense that smaller steps for better resolution are easily obtainable as compared to those of permanent magnet motor. By increasing the number of stack teeth and adding more stack pairs on the rotor, the step angle can be made smaller. For example, if number of teeth, is increased from 3 to 5 in two stack rotor the step angle

$$\theta_s = \frac{1}{4} \times P_t = \frac{1}{4} \times \frac{360°}{5} = \frac{360°}{5 \times 4} = 18°$$

The **resolution** of the stepper motor is defined as number of steps, the motor turns to complete one revolution, that is resolution = $\frac{360°}{\text{Number of steps}}$.

(*vi*) Comparing hybrid stepper motor with variable reluctance stepper motor, the former requires less excitation because rotor is excited by permanent magnet.

(*vii*) If phase A and phase B are simultaneously excited, the rotor will turn by only half the step angle. Not only half stepping, even stepping of any other fraction is also achievable by suitable proportion of simultaneous excitation of two phases. Such stepping is referred to as **micro stepping.**

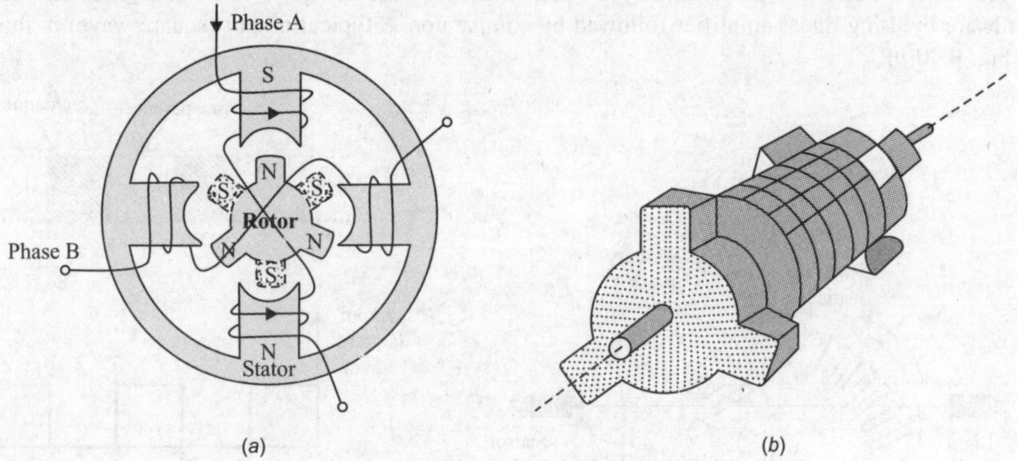

Fig. 9.29: Hybrid Stepper Motor (schematic views)

9.8 OPTICAL ENCODER

The optical encoders are very often used in modern (digital) control system design, particularly robots for the purpose of translating linear and rotational displacement into digitally coded or pulse signals. The two types of encoders: **absolute** and **incremental** are of current interest. They are different in the sense that they provide output signals in different forms.

The output from **absolute encoder**, is a **distinct digital code** corresponding to each particular least significant increment of resolution while the output from **incremental encoder**, is a **pulse** for each increment of resolution without making distinction between increments. Note that the signals emerging from both encoders, are compatible with digital logic hardware.

The choice out of these two, depends on economy and control design goal. The **primary concern** in use of **absolute encoder**, is that there is **data loss** during **power failure.** However, the options are available to outwit this situation. The **incremental encoders** are **far more widely** used in control systems by virtue of their **simple construction, low cost, ease in application and versatility**.

Incremental encoder

The linear and rotary both types are incremental encoders are found in real practice.

Figure 9.30(*a*) shows that a typical optomechanical arrangement of an incremental encoder has **four basic parts:** a **light source,** a **rotary (or translatory) disc,** a **stationary mask** and a **sensor.** As shown in Fig. 9.30(*b*), the disc has alternate opaque and transparent sectors of equal width. The photo sensor is located behind the mask. The light beam originating from light source, is either passed through the mask and reaches the sensor generating a pulse or blocked by the mask. In fact, as the disc rotates, during half the increment cycle the transparent sectors of rotating disc and stationary mask, are aligned to pass the light beam and during another half the light is blocked by the mask. Thus, for each increment one pulse appears at the output. If design demands very low resolution, (say few thousands of increments per revolution) multiple slit mask is often used in order to enhance the reception of light at photo sensor.

The wave form appearing at sensor output, is **sinusoidal** or **triangular** depending on demand of resolution. These wave forms may be converted into square wave form as needed in digital hardware by using linear amplifier followed by comparator. A typical output (square wave) is shown in Fig. 9.30(*b*).

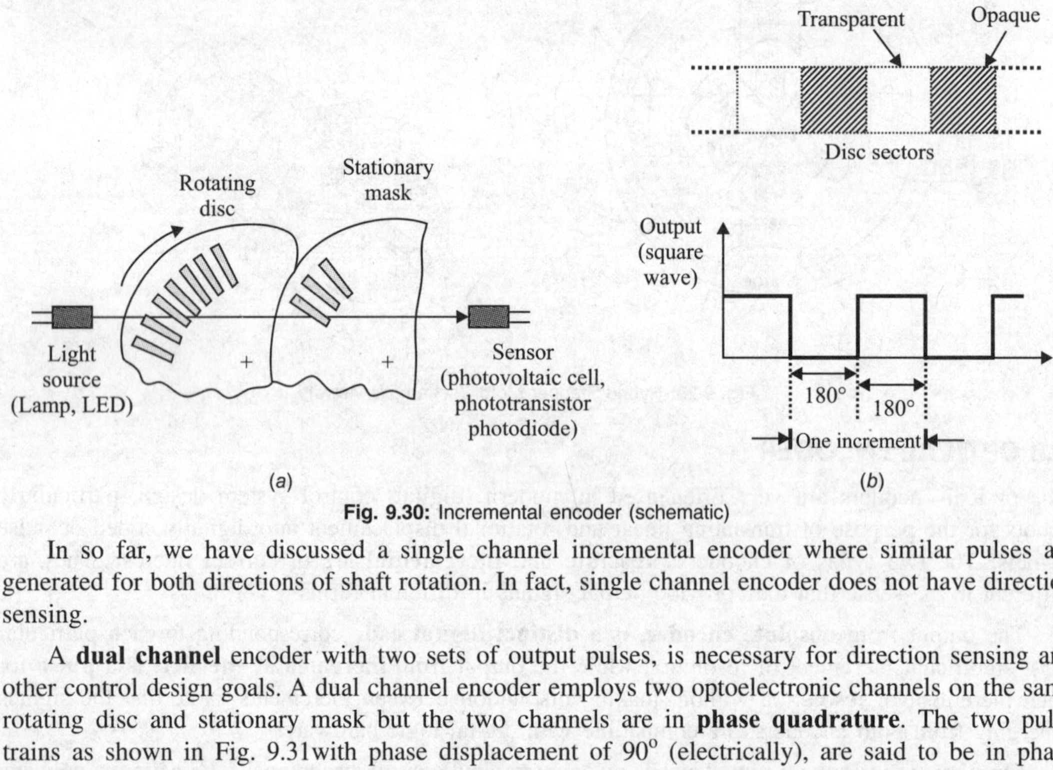

Fig. 9.30: Incremental encoder (schematic)

In so far, we have discussed a single channel incremental encoder where similar pulses are generated for both directions of shaft rotation. In fact, single channel encoder does not have direction sensing.

A **dual channel** encoder with two sets of output pulses, is necessary for direction sensing and other control design goals. A dual channel encoder employs two optoelectronic channels on the same rotating disc and stationary mask but the two channels are in **phase quadrature**. The two pulse trains as shown in Fig. 9.31with phase displacement of 90° (electrically), are said to be in phase quadrature. Note that the two signals have unique 0 to 1 and 1 to 0 logic transitions with respect to

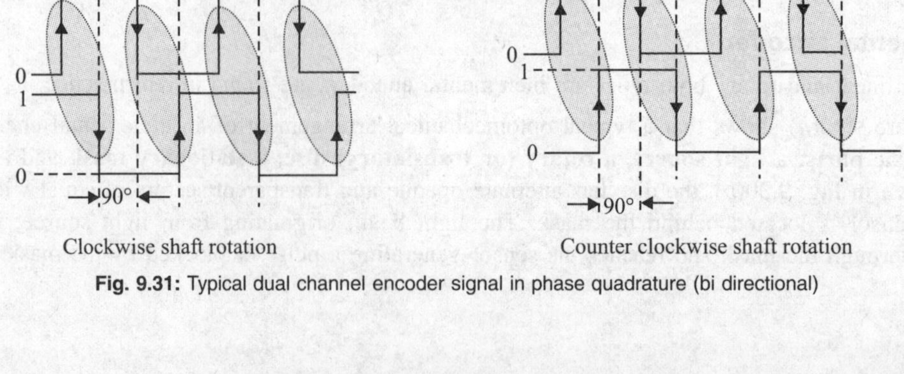

Fig. 9.31: Typical dual channel encoder signal in phase quadrature (bi directional)

direction (clockwise and counter clockwise) of shaft rotation. Based on unique 0 to 1 and 1 to 0 logic transition, a logic circuit can be designed to sense the direction.

In addition to direction sensing property, the dual channel encoder also exhibits improvement in resolution. Passing the quadrature signals through a differentiating circuit and reversing the polarity of negative pulses (full wave rectifying) as shown in Fig. 9.32, we obtain four impulses(short duration pulses) against each increment. A single channel incremental encoder has resolution = 360°/N where N = number sectors (in fact each sector is half transparent and half opaque). A dual channel encoder will have resolution = 360°/4N, thus exhibiting four fold improvement in resolution. If the encoder output (pulses) are applied to a counter which counts the number of pulses, it is also

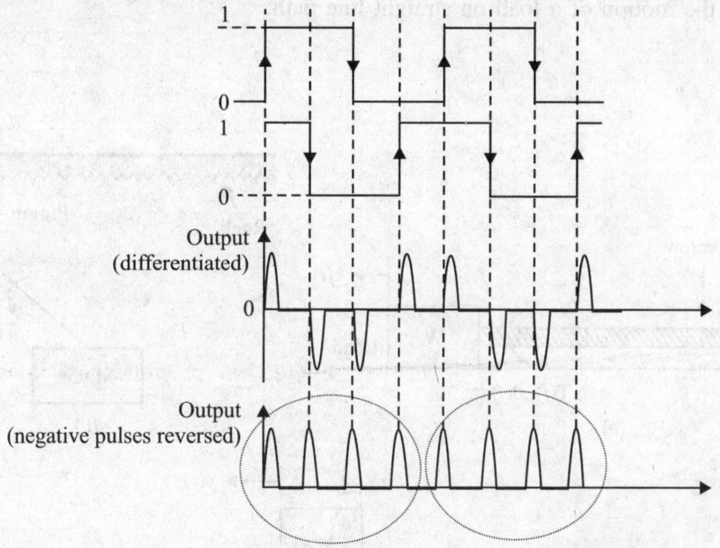

Fig. 9.32: Generating four impulses for each increment from quadrature pulse trains

possible to determine the speed of encoder shaft as each count (pulse) corresponds to a definite angle through which the shaft rotates. In addition to speed determination, these pulses may also be used for position indication or position control.

Absolute encoder

In the discussion just above, we have understood that determination of speed and position, is possible by means of incremental encoder, but it needs additional hardware for signal conditioning and counting the pulses. The absolute encoder does it without any additional hardware. It provides a distinct digital code for each increment. In this sense, it is truly a digital transducer.

The distinct binary codes are etched on the rotating disc that has as many tracks as the number bits in each digital code. In the digital code, the transparency corresponds to 1 and opaqueness to 0. The physical arrangement requires as many photo diode –LED pairs as the number of tracks on the disc. The photo diode –LED pairs are required to be carefully spread round the tracks so as to ensure that the signals do not interfere with each other.

9.9 MECHANICAL ARRANGEMENTS TO CONVERT ROTATIONAL MOTION INTO CORRESPONDING LINEAR MOTION

The design of some position control systems, often demand conversion of motion from rotary into linear and vice versa. The methods that are mostly used in practice to meet this demand, are discussed below:

(*i*) As shown in Fig. 9.33(*a*), the **motor and screw** assembly may be used to control the motion of a load on a straight line path. For example, load may be wiper arm of potentiometer in position control system.

(*ii*) A **rack and pinion** assembly driven by a motor as shown in Fig. 9.33(*b*), is another way to control the motion of a load on straight line path.

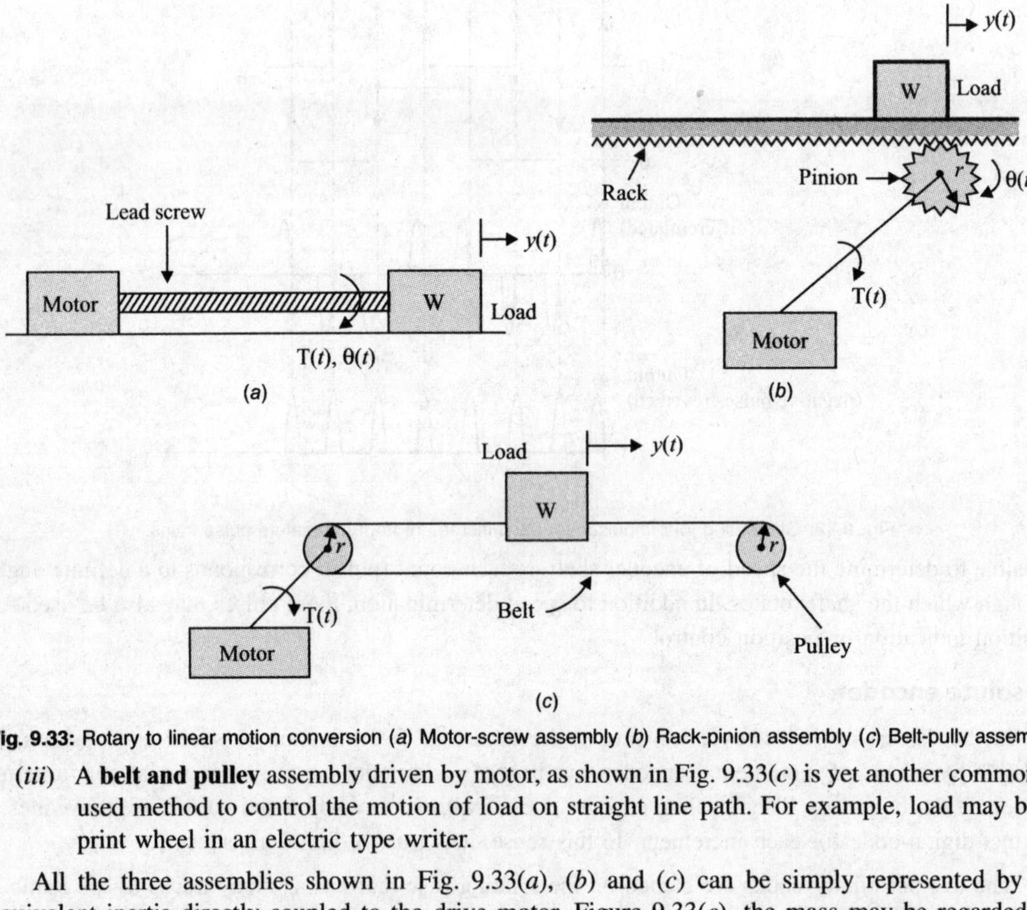

Fig. 9.33: Rotary to linear motion conversion (*a*) Motor-screw assembly (*b*) Rack-pinion assembly (*c*) Belt-pully assembly

(*iii*) A **belt and pulley** assembly driven by motor, as shown in Fig. 9.33(*c*) is yet another commonly used method to control the motion of load on straight line path. For example, load may be a print wheel in an electric type writer.

All the three assemblies shown in Fig. 9.33(*a*), (*b*) and (*c*) can be simply represented by an equivalent inertia directly coupled to the drive motor. Figure 9.33(*c*), the mass may be regarded as point mass moving about pulley of radius *r*. Ignoring inertia of the pulley, the equivalent inertia (J) seen by the drive motor, will be

$$J = Mr^2 = \frac{W}{g} r^2 \qquad ...(9.43)$$

where M is mass of the load.

Figure 9.33(*b*), the equivalent inertia given by (9.43) also holds true but in this case r = radius of pinion. The linear distance travelled by the mass is $2\pi r$ when pinion makes one complete revolution. In Fig. 9.33(*a*), let us define the **lead** (L) of the screw as the linear distance travelled by mass per revolution of the screw. If r is radius of screw, $L = 2\pi r$ and $r = L/2\pi$. Now, equivalent inertia seen by motor becomes

$$J = \frac{W}{(g)}\left(\frac{L}{2\pi}\right)^2 \quad ...(9.44)$$

9.10 BELT OR CHAIN DRIVE AND LEVER SYSTEM

We have already discussed the gear trains in Section 1.6. Let us recall that a gear train transmits energy from one part of mechanical system to another with flexibility that torque, speed and displacement may be altered. The **belt or chain drives** also serve the same purpose as gear trains except that they allow the transfer of energy over a longer distance without using large number of gears. A typical diagram of belt or chain drive between two pulleys of radius r_1 and r_2, is shown in Fig. 9.34. With the assumption that there is no slippage between the belt and the pulleys, the relation (1.44) discussed in case of gear train in Section 1.6, holds true in this case as well. However, for quick reference, it is again written in case of belt or chain drive as:

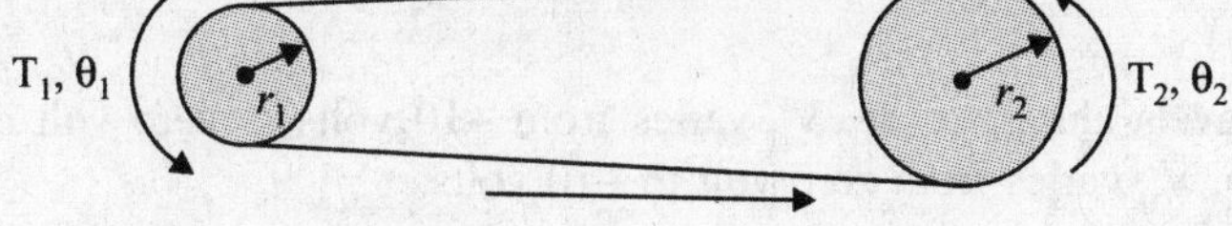

Fig. 9.34: Belt or chain drive

$$\frac{T_1}{T_2} = \frac{\theta_2}{\theta_1} = \frac{\omega_2}{\omega_1} = \frac{r_1}{r_2} \quad ...(9.45)$$

The **lever system** as shown in Fig. 9.35, is used to transmit force and translational motion in the same way as gear train transmits rotational motion. The relation between forces and distances governing the dynamics of the lever system, is:

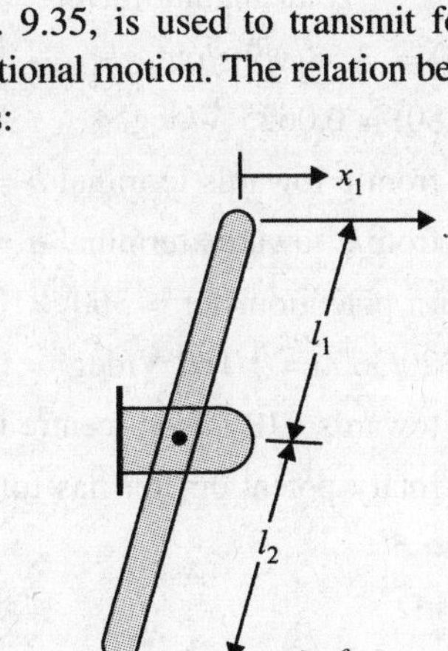

Fig. 9.35: Lever system

$$\frac{f_1}{f_2} = \frac{l_2}{l_1} = \frac{x_2}{x_1} \quad ...(9.46)$$

PROBLEMS AND SOLUTIONS

P 9.1: A single turn 10 K potentiometer with rotation angle of 320°, is shown in Fig. P9.1. The fixed terminals ***a*** and ***b*** are connected to + 10 V and – 10 V DC supply respectively. The centre tap ***c*** is grounded.

(*a*) Find gain constant (sensitivity) of potentiometer in V/rad.

(*b*) What is open circuit output voltage V_0 if wiper is rotated through an angle of 30° towards fixed terminal ***b*** from centre tap ***c***?

(*c*) What is open circuit output voltage if wiper is rotated through an angle 60° towards fixed terminal ***a*** from centre tap ***c***?

(*d*) If this potentiometer has 10 turns instead of single turn, what is new gain constant (volts/rad) and what is output voltage (open circuit) when wiper is rotated 64° from centre tap ***c*** towards +10 V.

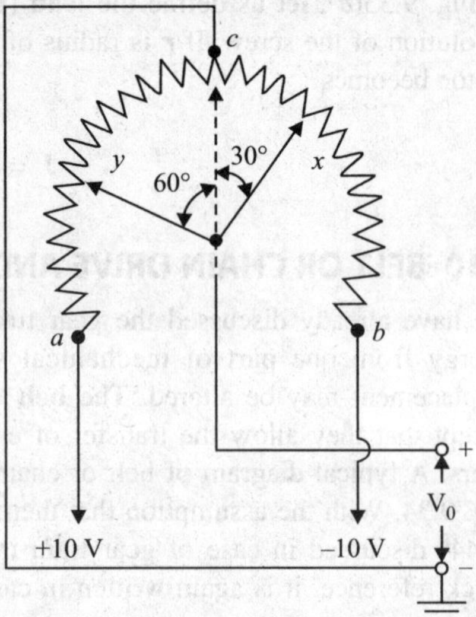

Fig. P 9.1

Solution:

(*a*) As wiper travels along arc **ac**, V_0 varies from +10 volts to zero volt and as wiper travels along arc **cb**, V_0 varies from zero volt to –10 volts.

$$320° = \frac{\pi}{180} \times 320 = 5.59 \text{ rad.}$$

So, gain constant $$K_s = \frac{\text{Total variation in } V_0}{\text{Total angular travel}} = \frac{10-(-10)}{5.59}$$

$$= 3.58 \text{ volts/rad.}$$

(*b*) K_s =3.58 V/rad = 3.58 ×(π/180) = 0.0625 V/deg.

V_0 for angular travel of 30° from ***c*** towards terminal ***b*** = –0.0625 × 30 = –1.875 V

(*c*) V_0 for angular travel of 60° from ***c*** towards terminal ***a*** = + (60 × 0.0625) = +3.75 V.

(*d*) Total angular travel for 10 turn potentiometer = 360° ×10 = 3600°

and new gain constant K_s = 20/3600 = 1/180 V/deg. = 0.318 V/rad

V_0 for angular travel of 64° towards +10V from centre tap = +64/180 = +0.36 volts

P 9.2: A multi turn wire wound rotary potentiometer has following specifications:

Total number of turns, N = 5

Total resistance, R_a = 10 kΩ

Total number of winding turns = 8000

Fixed reference voltage, V_R = 50 V

(*a*) With the movable arm set in the middle, the resistance measured is 5.05 kΩ. Express linearity as percentage change in resistance.

(*b*) With the movable arm set at quarter point, the resistance measured is 2.6 kΩ. What is corresponding linearity in terms of percentage change in resistance?

(*c*) What is resolution of the potentiometer?

(*d*) Find gain constant (K) of the potentiometer in volts/turn, volts/rad and Volts/deg.

Solution:

(*a*)
$$\% \text{ Linearity} = \frac{\text{Deviation from nominal mid-value}}{\text{Total resistance}}$$

$$= \frac{\text{Measured mid value} - \text{Nominal mid-value}}{R_a} \times 100$$

$$= \frac{5.05 - (1/2) \times 10}{10} \times 100 = 0.5\%$$

(*b*)
$$\% \text{ Linearity} = \frac{\text{Deviation at quarter point from nominal value}}{\text{Total resistance}} \times 100$$

$$= \frac{\text{Measured value at quarter point} - \text{Nominal value at quarter point}}{R_a} \times 100$$

$$= \frac{2.6 - \frac{1}{4} \times 10}{10} \times 100 = 1\%$$

(*c*)
$$\text{Resolution of potentiometer} = \frac{V_R}{\text{Number of winding turns}}$$

$$= \frac{50}{8000} = 6.25 \text{ mV}$$

(*d*) Gain constant of potentiometer $K = \frac{50}{5} = 10$ V/turn

or $K = \frac{20}{2\pi N} = \frac{50}{10\pi} = 1.59$ Volts/rad

or $K = \frac{50}{360 \times 5} = 0.0278$ V/deg. $= 27.8$ mV/deg

P 9.3: A helical multi turn potentiometer has following specifications:

Total resistance, R_a = 20 kΩ

Total number of turns, N = 10

Fixed reference voltage, V_R = 80 V

% linearity = 1%

(*a*) Find the range of voltage at mid-point setting that will appear at open circuited output terminals.

(*b*) Find output voltage when potentiometer is set at mid-point and output is terminated in load R_L= 50 K. Also find loading error percent.

Solution:

(*a*) Nominal voltage at mid-point setting = (1/2)× 80 = 40 V

Range of voltage = 40 ± 1% of 80

= 40 ± 0.8

The open circuit voltage will lie in the range from 39.2 V to 40.8 V.

(*b*) The potentiometer with wiper at mid-point and terminated in R_L = 50 K, is shown in Fig. P 9.3.

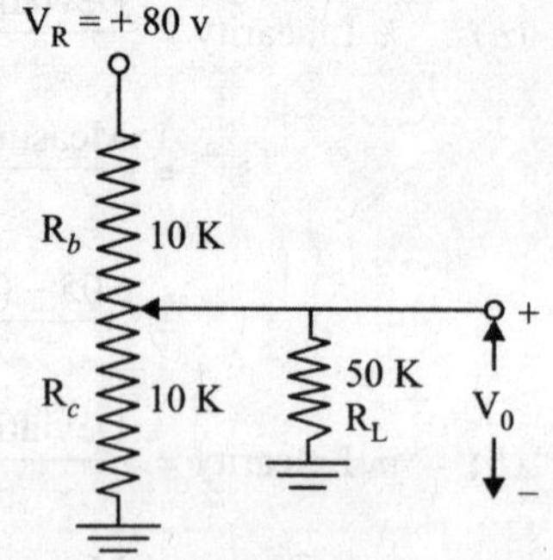

Fig. P 9.3

$$V_0 = \frac{V_R \times R_C || R_L}{R_b + R_c || R_L} = \frac{80 \times \dfrac{10 \times 50}{10 + 50}}{10 + \dfrac{10 \times 50}{10 + 50}}$$

= 36.36 V

Loading error = (1/2)80 – 36.36 = 3.64 V

$$\% \text{ loading error} = \frac{3.64 \times 100}{40} = 9.1\%$$

P 9.4: The straight line approximated torque-speed curve of AC servo motor for rated control voltage 115 V, 50 Hz is shown in Fig. P9.4. The moment of inertia of motor is 10^{-5} kg.m^2. Neglect the friction and determine transfer function model that relates shaft position θ_0 to control voltage V_C.

Solution: We have already discussed the dynamics of AC servo motor in Section 9.4 and developed signal flow graph as shown in Fig. 9.17. This is sketched, again here as Fig. P9.4(*a*) for quick reference. Note that minus sign appears in feedback path due to slope being negative.

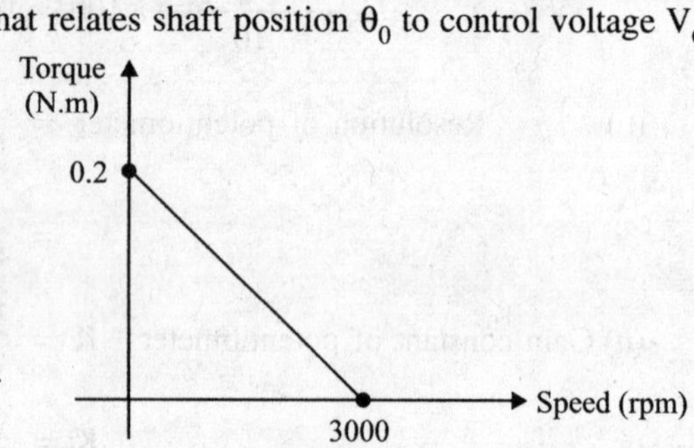

Fig. P 9.4

Let us find constant **K** and ***m*** from given torque-speed curve.

$$K = \frac{T_0}{V_{C_{rated}}} = \frac{0.2}{115} = 1.74 \times 10^{-3} \text{ N.m/volt}$$

Translating rpm into rad/sec, we have

$$3000 \text{ rpm} = \frac{3000 \times 2\pi}{60} = 100\,\pi \text{ rad/sec}$$

and slope

$$m = \frac{0.2}{100\pi} = 6.37 \times 10^{-4} \text{ N.m/rad.sec}^{-1}$$

$$\frac{K}{J_0} = \frac{1.74 \times 10^{-3}}{10^{-5}} = 174$$

$$B_0 = 0$$

$$\frac{m}{J_0} = \frac{6.37 \times 10^{-4}}{10^{-5}} = 63.7$$

Substituting the values computed just above, the signal flow graph of Fig. P9.4(*a*), takes the form as shown in Fig. P9.4(*b*).

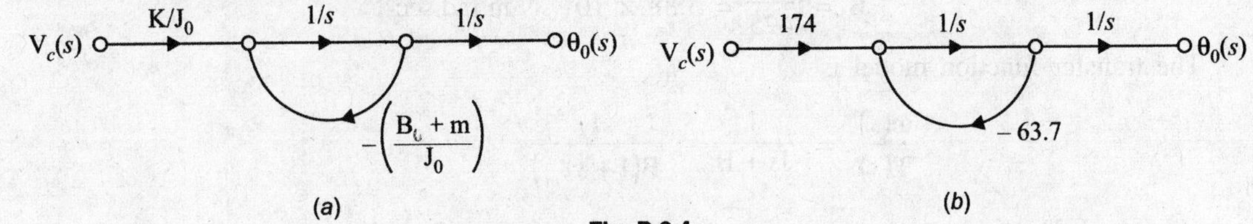

Fig. P 9.4

Now,
$$\frac{\theta_0(s)}{V_c(s)} = \frac{174/s^2}{1+\dfrac{63.7}{s}} = \frac{174}{s(s+63.7)}$$

P 9.5: The torque-speed curve of a DC servo motor is shown in Fig P 9.5. Determine mechanical time constant of motor if moment of inertia of rotor is 1.2×10^{-4} kg.m^2.

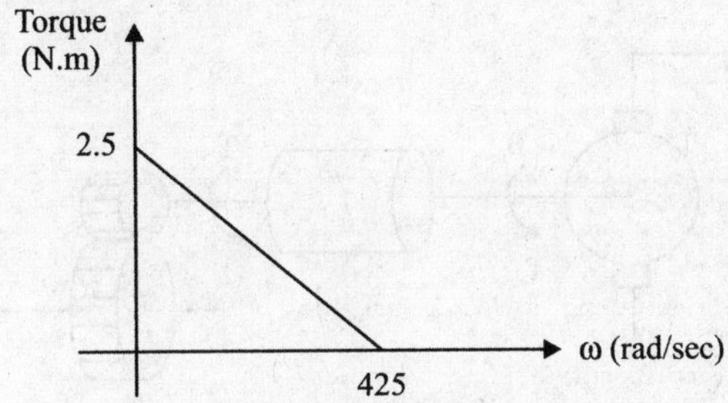

Fig. P 9.5

Solution: The equation governing the motor dynamics, is

$$T = \frac{Jd^2\theta}{dt^2} + B\frac{d\theta}{dt}$$

or
$$T = J\frac{d\omega}{dt} + B\omega$$

And its Laplace transform with zero initial conditions, is

$$T(s) = Js\,\omega(s) + B\omega(s)$$

or
$$T(s) - B\omega(s) = Js\,\omega(s)$$

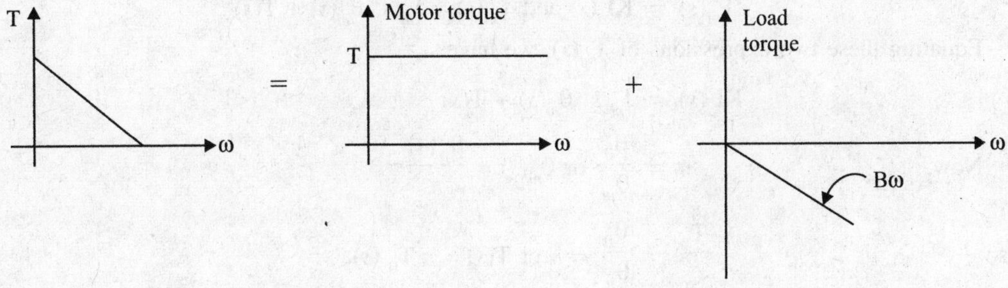

Thus, B can be determined from torque-speed curve. In fact, B equals slope of curve.

$$B = \frac{2.5}{425} = 5.88 \times 10^{-3} \text{ N·m/rad.sec}^{-1}$$

The transfer function model is

$$\frac{\omega(s)}{T(s)} = \frac{1}{Js + B} = \frac{1}{B(1 + s\tau_m)}$$

where $\tau_m = \frac{J}{B}$ is mechanical time constant and $\tau_m = \frac{1.28 \times 10^{-4}}{5.88 \times 10^{-3}} = 21.8$ m sec.

P 9.6: As shown in Fig. P9.6, an armature controlled DC servo motor drives a load with moment of inertia J_L. The torque developed by motor is T. The moment of inertia of motor rotor is J_m. The angular displacement of motor rotor and load element are θ_m and θ_0, respectively. The gear ratio is $n = \theta_0/\theta_m$. Develop transfer function model $\theta_0(s)/V_c(s)$.

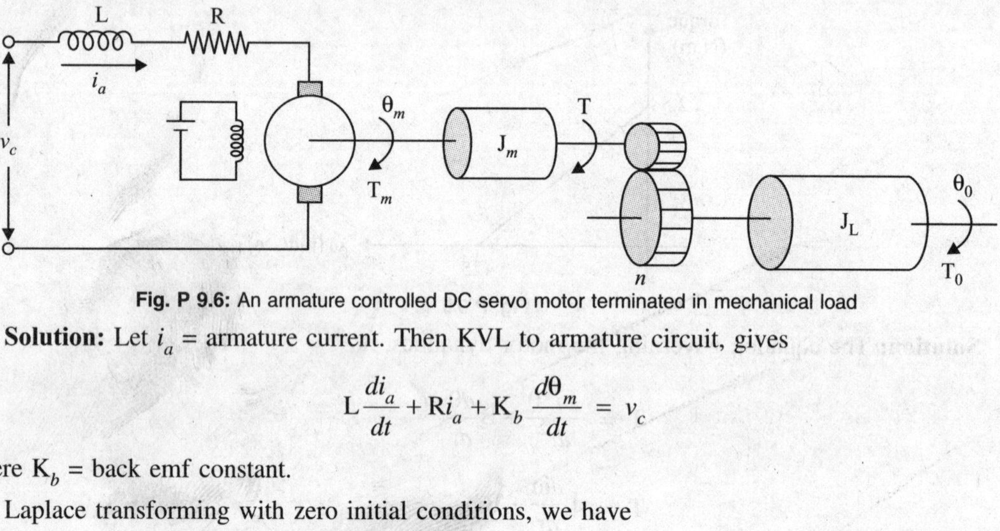

Fig. P 9.6: An armature controlled DC servo motor terminated in mechanical load

Solution: Let i_a = armature current. Then KVL to armature circuit, gives

$$L\frac{di_a}{dt} + Ri_a + K_b\frac{d\theta_m}{dt} = v_c$$

where K_b = back emf constant.

Laplace transforming with zero initial conditions, we have

$$(sL + R)I_a(s) + K_b s\,\theta_m(s) = V_c(s)$$

Also, $$T_m = J_m\ddot{\theta}_m + T$$

and $$T_m = Ki_a\ ; \ (K = \text{motor torque constant})$$

or $$T_m(s) = KI_a(s) \text{ and } T_m(s) = J_m s^2\theta_m(s) + T(s)$$

Equating these two expressions of $T_m(s)$, we have

$$KI_a(s) = J_m s^2\theta_m(s) + T(s)$$

Now, $$n = \frac{\theta_0}{\theta_m} \text{ or } \theta_m(s) = \frac{\theta_0(s)}{n}$$

also $$\frac{T}{T_0} = \frac{\theta_0}{\theta_m} = n \text{ or } T(s) = n\,T_0(s)$$

Substituting $\theta_m(s)$ and $T(s)$ in the expression for $KI_a(s)$, we get

$$KI_a(s) = J_m s^2 \frac{\theta_0(s)}{n} + nT_0(s)$$

Also substituting $\theta_m(s)$ in expression for $V_c(s)$, we obtain

$$V_c(s) = (sL + R)I_a(s) + K_b s \frac{\theta_0(s)}{n}$$

But $$I_a(s) = \frac{J_m s^2\theta_0(s)}{nK} + \frac{nT_0(s)}{K}$$

Thus, $$V_c(s) = (sL+R)\left[\frac{J_m s^2\theta_0(s)}{nK} + \frac{nT_0(s)}{K}\right] + K_b s\frac{\theta_0(s)}{n}$$

Now, $$T_0 = J_L \ddot{\theta}_0$$

or $$T_0(s) = J_L s^2\theta_0(s)$$

Substituting in the expression for $V_c(s)$, we get

$$V_c(s) = (sL+R)\left[\frac{J_m s^2\theta_0(s)}{nK} + \frac{n}{K} J_L s^2\theta_0(s)\right] + K_b s\frac{\theta_0(s)}{n}$$

or $$V_c(s) = (sL+R)\left[\frac{J_m s^2 + n^2 J_L s^2}{nK}\right]\theta_0(s) + K_b s\frac{\theta_0(s)}{n}$$

or $$nK V_c(s) = [s^2 (sL + R)(J_m + n^2 J_L) + sK_b K]\theta_0(s)$$

Hence, the transfer function model

$$\frac{\theta_0(s)}{V_c(s)} = \frac{nK}{s\left[s(sL+R)\left(J_m + n^2 J_L\right) + K K_b\right]}$$

P 9.7: The print wheel control system of a word processor, is shown in Fig P9.7. The control system consists of a DC motor driving belts and pulleys. Assume that belts are rigid. $T_m(t)$ is motor torque, $\theta_m(t)$ is motor displacement (angular), $y(t)$ is linear displacement of print wheel, J_m is motor inertia, B_m is motor friction, r is the pulley radius and M is the mass of print wheel.

(*a*) Write the differential equation describing system dynamics.

(*b*) Obtain transfer function model of system, $Y(s)/T_m(s)$.

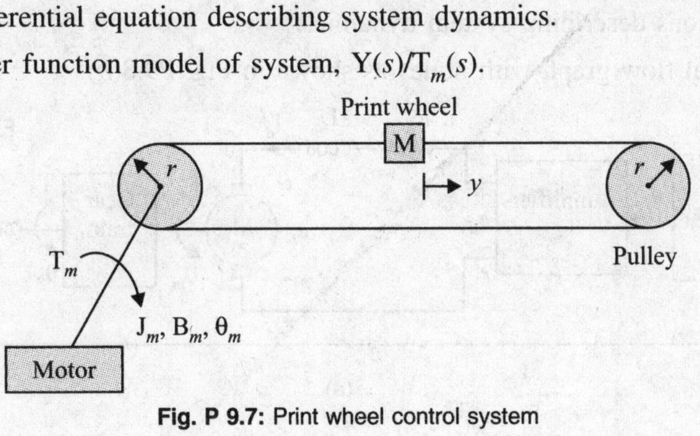

Fig. P 9.7: Print wheel control system

Solution:

(*a*) The equivalent inertia of print wheel seen by motor is $J_p = M_{r^2}$. So, the total inertia in the system will be $J_p + J_m$. The differential equation describing the control system can be written as

$$(J_P + J_m)\ddot{\theta}_m(t) + B_m\dot{\theta}_m(t) = T_m(t)$$

and
$$y(t) = r\,\theta_m(t)$$

(*b*) Laplace transforming with zero initial conditions the equations just above, we get

$$\theta_m(s)\,[(J_P + J_m)s^2 + sB_m] = T_m(s)$$

and
$$Y(s) = r\,\theta_m(s)$$

Thus,
$$\frac{Y(s)}{r}\left[\left(M_{r^2} + J_m\right)s^2 + sB_m\right] = T_m(s)$$

or
$$\frac{Y(s)}{T_m(s)} = \frac{r}{s\left[\left(M_{r^2} + J_m\right)s + B_m\right]}$$

P 9.8: Figure P9.8(*a*) shows the schematic diagram of a DC motor control system for controlling the print wheel as load in an electronic word processor. The control system variables and parameters are defined as follows:

K_s = gain of error detector (V/rad)
K_i = torque constant
K_a = amplifier gain
K_b = back emf constant
n = gear turn ratio = $\theta_2/\theta_m = T_m/T_2$
B_m = motor viscous-friction coefficient
J_m = motor inertia
K_L = torsional spring constant of the motor shaft
J_L = load inertia

(*a*) Write equations describing system dynamics.

(*b*) Sketch signal flow graph with nodes as shown in Fig. P9.8(*b*).

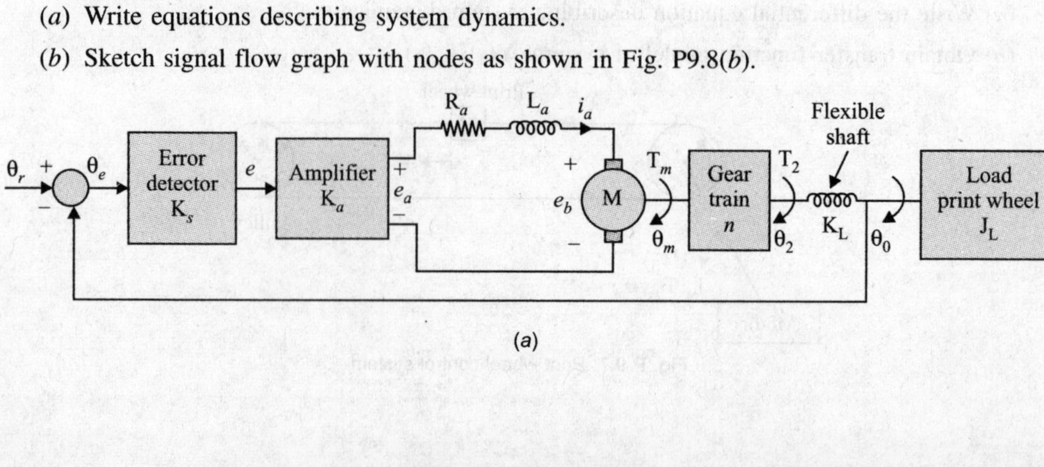

(*a*)

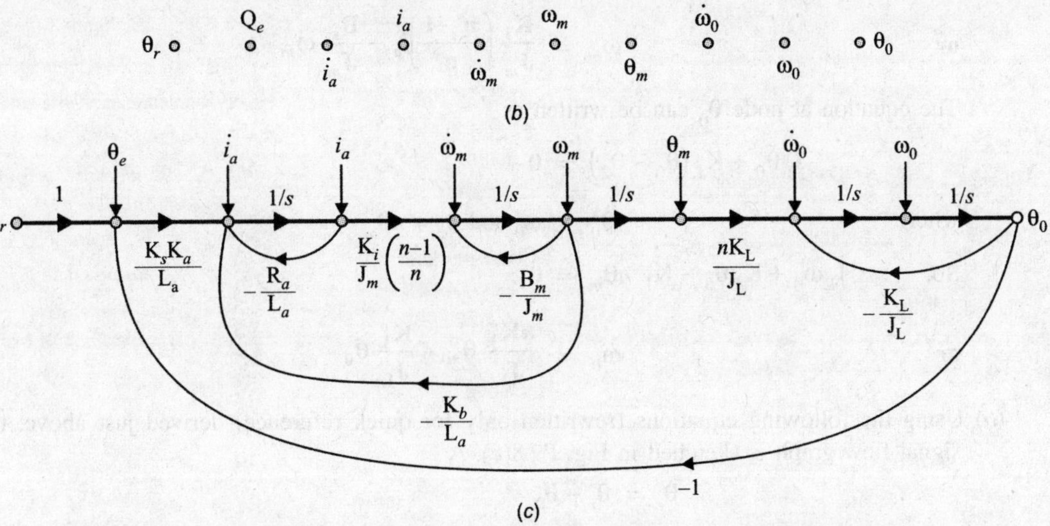

Fig. P9.8 (*a*) Print wheel control system, (*b*) Nodes of signal flow graph, (*c*) Complete signal flow graph

Solution:

(*a*) The equations describing dynamics of system shown in Fig. P9.8(*a*), can be written as follows:

$$\boldsymbol{\theta_e = \theta_r - \theta_0}$$

$$e = K_s\,\theta_e$$

$$e_a = e\,K_a = K_s\,K_a\,\theta_e$$

Also, KVL applied to armature circuit, gives

$$e_a = R_a i_a + L_a \dot{i}_a + e_b$$

where $e_b = K_b\,\omega_m$

So,
$$e_a = R_a i_a + L_a \dot{i}_a + K_b \omega_m$$

or
$$\dot{i}_a = \frac{e_a}{L_a} - \frac{R_a}{L_a} i_a - \frac{K_b}{L_a}\omega_m$$

or
$$\dot{i}_a = \frac{K_s K_a}{L_a}\theta_e - \frac{R_a}{L_a} i_a - \frac{K_b}{L_a}\omega_m$$

Now,
$$T_m = K_i\,i_a$$

and also,
$$T_m = J_m \ddot{\theta}_m + B_m \dot{\theta}_m + T_2$$

where
$$T_2 = \frac{T_m}{n} \text{ and } \dot{\theta}_m = \omega_m$$

So,
$$K_i\,i_a = J_m \dot{\omega}_m + B_m \omega_m + \frac{K_i\,i_a}{n}$$

or
$$K_i\left(1 - \frac{1}{n}\right) i_a = J_m \dot{\omega}_m + B_m \omega_m$$

or $$\dot{\omega}_m = \frac{K_i}{J_m}\left(\frac{n-1}{n}\right)i_a - \frac{B_m}{J_m}\omega_m$$

The equation at node θ_0 can be written as

$$J_L\ddot{\theta}_0 + K_L(\theta_0 - \theta_2) = 0$$

where $$\dot{\theta}_0 = \omega_0 \text{ and } \theta_2 = n\theta_m$$

So, $$J_L\dot{\omega}_0 + K_L\theta_0 - K_L\, n\theta_m = 0$$

or $$\dot{\omega}_0 = \frac{nK_L}{J_L}\theta_m - \frac{K_L}{J_L}\theta_0$$

(*b*) Using the following equations (rewritten only for quick reference) derived just above, the signal flow graph is sketched in Fig. P9.8(*c*).

$$\theta_e = \theta_r - \theta_0$$

$$i_a = \frac{K_s K_a}{L_a}\theta_e - \frac{R_a}{L_a}i_a - \frac{K_b}{L_a}\omega_m$$

$$\dot{\omega}_m = \frac{K_i}{J_m}\left(\frac{n-1}{n}\right)i_a - \frac{B_m}{J_m}\omega_m$$

$$\dot{\omega}_0 = \frac{nK_L}{J_L}\theta_m - \frac{K_L}{J_L}\theta_0$$

P 9.9: The linear model of a robot arm system being driven by a DC motor, is shown in Fig. P 9.9 (*a*). The system variables and parameters are given below:

DC motor	**Robot arm**
T_m = motor torque = $K_i\, i_a$	J_L = arm inertia
K_i = torque constant	B_L = arm friction coefficient
i_a = armature current of motor	θ_L = arm displacement
J_m = motor inertia	K = spring constant of shaft between the motor and arm
B_m = motor friction coefficient	B = friction coefficient of shaft between motor and arm
θ_m = motor shaft displacement	T_L = disturbance torque on arm

(*a*) Write equations describing dynamics of system with $i_a(t)$ and $T_L(t)$ as inputs and $\theta_m(t)$ and $\theta_L(t)$ as outputs.

(*b*) Sketch signal flow graph choosing $I_a(s)$, $T_L(s)$, $\theta_m(s)$ and $\theta_L(s)$ as node variables.

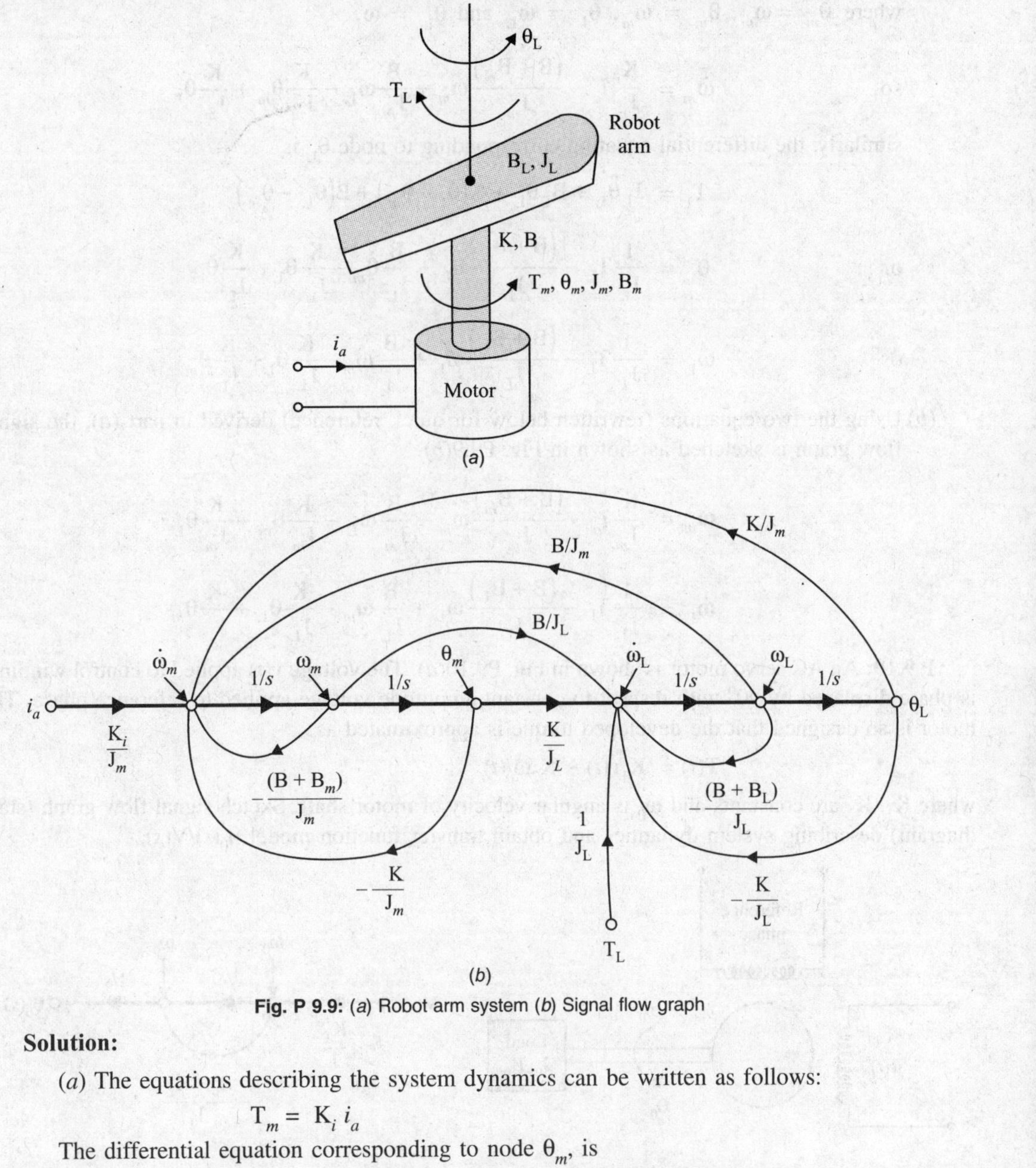

Fig. P 9.9: (*a*) Robot arm system (*b*) Signal flow graph

Solution:

(*a*) The equations describing the system dynamics can be written as follows:

$$T_m = K_i\, i_a$$

The differential equation corresponding to node θ_m, is

$$T_m = J_m \ddot{\theta}_m + B_m \dot{\theta}_m + K(\theta_m - \theta_L) + B(\dot{\theta}_m - \dot{\theta}_L)$$

or

$$\ddot{\theta}_m = \frac{K_i}{J_m} i_a - \frac{(B + B_m)}{J_m}\dot{\theta}_m - \frac{K}{J_m}\theta_m + \frac{B}{J_m}\dot{\theta}_L + \frac{K}{J_m}\theta_L$$

where $\ddot{\theta}_m = \dot{\omega}_m$, $\dot{\theta}_m = \omega_m$, $\ddot{\theta}_L = \dot{\omega}_L$ and $\dot{\theta}_L = \omega_L$

so, $$\dot{\omega}_m = \frac{K_i}{J_m} i_a - \frac{(B + B_m)}{J_m}\omega_m + \frac{B}{J_m}\omega_L - \frac{K}{J_m}\theta_m + \frac{K}{J_m}\theta_L$$

similarly, the differential equation corresponding to node θ_L is

$$T_L = J_L\ddot{\theta}_L + B_L\dot{\theta}_L + K(\theta_L - \theta_m) + B(\dot{\theta}_L - \dot{\theta}_m)$$

or $$\ddot{\theta}_L = \frac{1}{J_L}T_L - \frac{(B + B_L)}{J_L}\dot{\theta}_L + \frac{B}{J_L}\dot{\theta}_m - \frac{K}{J_L}\theta_L + \frac{K}{J_L}\theta_m$$

or $$\dot{\omega}_L = \frac{1}{J_L}T_L - \frac{(B + B_L)}{J_L}\omega_L + \frac{B}{J_L}\omega_m - \frac{K}{J_L}\theta_L + \frac{K}{J_L}\theta_m$$

(*b*) Using the two equations (rewritten below for quick reference) derived in part (*a*), the signal flow graph is sketched as shown in Fig. P9.9(*b*).

$$\dot{\omega}_m = \frac{K_i}{J_m} i_a - \frac{(B + B_m)}{J_m}\omega_m + \frac{B}{J_m}\omega_L - \frac{K}{J_m}\theta_m + \frac{K}{J_m}\theta_L$$

$$\dot{\omega}_L = \frac{1}{J_L}T_L - \frac{(B + B_L)}{J_L}\omega_L + \frac{B}{J_L}\omega_m - \frac{K}{J_L}\theta_L + \frac{K}{J_L}\theta_m$$

P 9.10: An AC servo motor is shown in Fig. P9.10(*a*). The voltage $v(t)$ applied to control winding, is phase displaced by 90° with respect to constant amplitude voltage applied to reference phase. The motor is so designed that the developed torque is approximated as

$$T(t) = K_1 v(t) - K_2\omega_0(t)$$

where K_1, K_2 are constants and ω_0 is angular velocity of motor shaft. Sketch signal flow graph (state diagram) describing system dynamics and obtain transfer function model $\theta_0(s)/V(s)$.

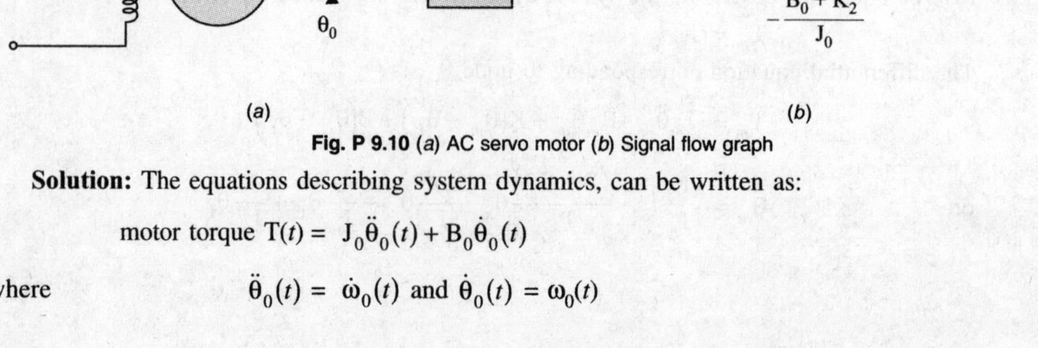

Fig. P 9.10 (*a*) AC servo motor (*b*) Signal flow graph

Solution: The equations describing system dynamics, can be written as:

motor torque $T(t) = J_0\ddot{\theta}_0(t) + B_0\dot{\theta}_0(t)$

where $$\ddot{\theta}_0(t) = \dot{\omega}_0(t) \text{ and } \dot{\theta}_0(t) = \omega_0(t)$$

and $\quad T(t) = K_1 v(t) - K_2\omega_0(t) = K_1 v(t) - K_2\dot{\theta}_0(t)$

So, Laplace transformation with zero initial conditions, gives

$$T(s) = K_1 V(s) - K_2\, s\theta_0\,(s) = J_0\, s^2\,\theta_0(s) + B_0\, s\,\theta_0\,(s)$$

and little algebraic manipulation gives the transfer function model

$$\frac{\theta_0(s)}{V(s)} = \frac{K_i}{s\left[Js + \left(B_0 + K_2\right)\right]}$$

Using the equation (derived just above)

$$\ddot{\theta}_0(t) = \frac{K_1}{J_0} v(t) - \left(\frac{B_0 + K_2}{J_0}\right)\dot{\theta}_0(t)$$

the signal flow graph is sketched as shown in Fig. P9.10(*b*).

P 9.11: An electromechanical system is shown in Fig. P 9.11. On energising the solenoid with $e(t)$, the lower arm of lever P moves to the left and upper arm to the right. The coil has resistance R and inductance L. The solenoid pull F = K $i(t)$ where $i(t)$ is current through solenoid. Obtain transfer function model X(*s*)/E(*s*) of the system. Ignore all inertia and consider effects of spring K and viscous damper B only. The back emf constant is K_b. The displacement of upper arm is *x*.

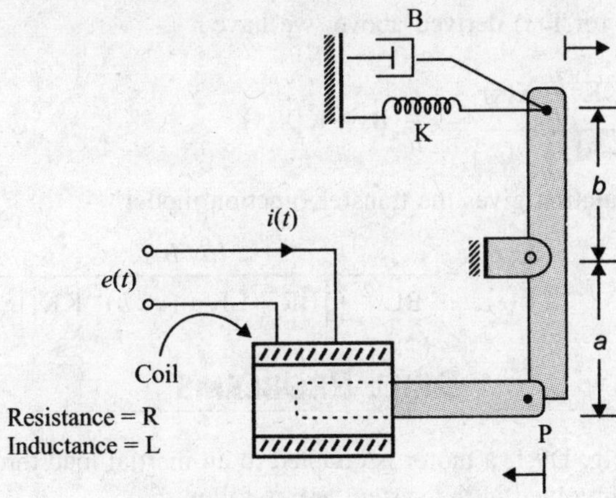

Fig. P 9.11: Electromechanical system

Solution: A back emf is generated when lower arm moves to the left and it is given by:

$$E_b = K_1\frac{a}{b}\frac{dx}{dt}; \; x = \text{displacement of upper arm}$$

The coil circuit equation can now be written as

$$e(t) = Ri(t) + L\frac{di(t)}{dt} + K_1\frac{a}{b}\frac{dx}{dt}$$

Taking Laplace transform with zero initial conditions, we have

$$E(s) = RI(s) + sLI(s) + K_1\frac{a}{b}\, sX(s)$$

or
$$I(s) = \frac{E(s) - K_1 \frac{a}{b} X(s)}{R + sL}$$

But the force developed by energised solenoid, is
$$f(t) = Ki(t)$$
or
$$F(s) = KI(s)$$
or
$$F(s) = \frac{KE(s) - KK_1 \frac{a}{b} X(s)}{R + sL}$$

The force on upper arm is $\frac{a}{b} f(t)$ and equation describing dynamics of upper arm is
$$\frac{a}{b} f(t) = B\frac{dx}{dt} + Kx$$

Laplace transforming with zero initial condition, we have
$$F(s) = \frac{b}{a}\left[Bs\, X(s) + KX(s)\right]$$

Equating expressions for F(s) derived above, we have
$$\frac{KE(s) - KK_1 \frac{a}{b} X(s)}{R + sL} = \frac{b}{a}\left[Bs + K\right]X(s)$$

and little algebraic manipulation gives the transfer function model
$$\frac{X(s)}{E(s)} = \frac{(a/b)K}{BLs^2 + \left[BR + LK + (a/b)^2\, KK_1\right]s + KR}$$

DRILL PROBLEMS

D 9.1: As shown in Fig. D9.1, a motor is coupled to an inertial load through a shaft. Significant variables and parameters involved in the system are as follows:

$T_m(t)$ = motor torque
J_m = motor inertia
B_m = motor friction coefficient
J_L = Load inertia
K = spring constant of shaft
$\theta_m(t)$ = motor displacement
$\theta_L(t)$ = load displacement

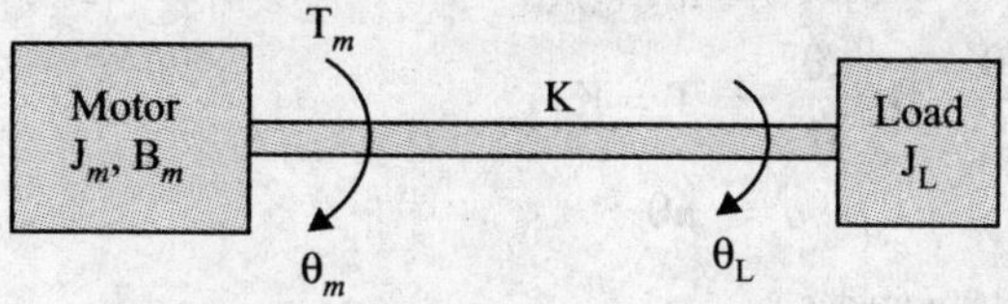

Fig. D 9.1: Motor coupled to load (system)

(*a*) Write torque equations describing system dynamics.

(*b*) Sketch state diagram (signal flow graph). Choose state variables

$$x_1(t) = \theta_m(t) - \theta_L(t)$$
$$x_2(t) = \dot{\theta}_L(t)$$
$$x_3(t) = \dot{\theta}_m(t)$$

(*c*) Obtain the transfer function model $\dfrac{\theta_m(s)}{T_m(s)}$ and $\dfrac{\theta_L(s)}{T_m(s)}$

Ans.

(*a*)
$$\ddot{\theta}_m(t) = -\frac{B_m}{J_m}\dot{\theta}_m(t) - \frac{K}{J_m}\left[\theta_m(t) - \theta_L(t)\right] + \frac{1}{J_m}T_m(t)$$

$$\ddot{\theta}_L(t) = \frac{K}{J_L}\left[\theta_m(t) - \theta_L(t)\right]$$

(*c*)
$$\frac{\theta_m(s)}{T_m(s)} = \frac{J_L s^2 + K}{s\left[J_m J_L s^3 + B_m J_L s^2 + K(J_m + J_L)s + B_m K\right]}$$

$$\frac{\theta_L(s)}{T_m(s)} = \frac{K}{s\left[J_m J_L s^3 + B_m J_L s^2 + K(J_m + J_L)s + B_m K\right]}$$

D 9.2: A position control system is shown in Fig. D9.2(*a*). The angular position ***r*** is the reference input to the system. The output shaft position determines the angular position ***c*** of wiper arm of the output potentiometer. The error voltage $e_v = e_r - e_c$ where $e_r = K_0 \,.\, r$ and $e_c = K_0 . c$ where K_0 is proportionality constant. K_1 is amplifier gain. The amplified output drives armature circuit of the DC motor. A fixed voltage is applied to field winding. The torque developed by motor is

$$T = K_2\, i_a\,;\; K_2 = \text{motor torque constant and}$$
$$i_a = \text{armature current}$$

K_3 is back emf constant. Gear ratio ***n*** of gear train, is so chosen that $\boldsymbol{c = n\theta}$. The overall system is designed to reduce error $\boldsymbol{e = r - c}$ to zero if it exists. J_0 is the inertia of the motor plus load plus gear train referred to motor shaft. B_0 is viscous friction coefficient of motor plus load plus gear train referred to motor shaft. Show that system of Fig. D9.2(*a*) and block diagram shown in Fig. D9.2(*b*) are equivalent.

Hint:

$$L_a \frac{di_a}{dt} + R_a i_a + K_3 \frac{d\theta}{dt} = K_1 e_v$$

$$J_0 \frac{d^2\theta}{dt^2} + B_0 \frac{d\theta}{dt} = T = K_2\, i_a$$

$$c = n\theta$$

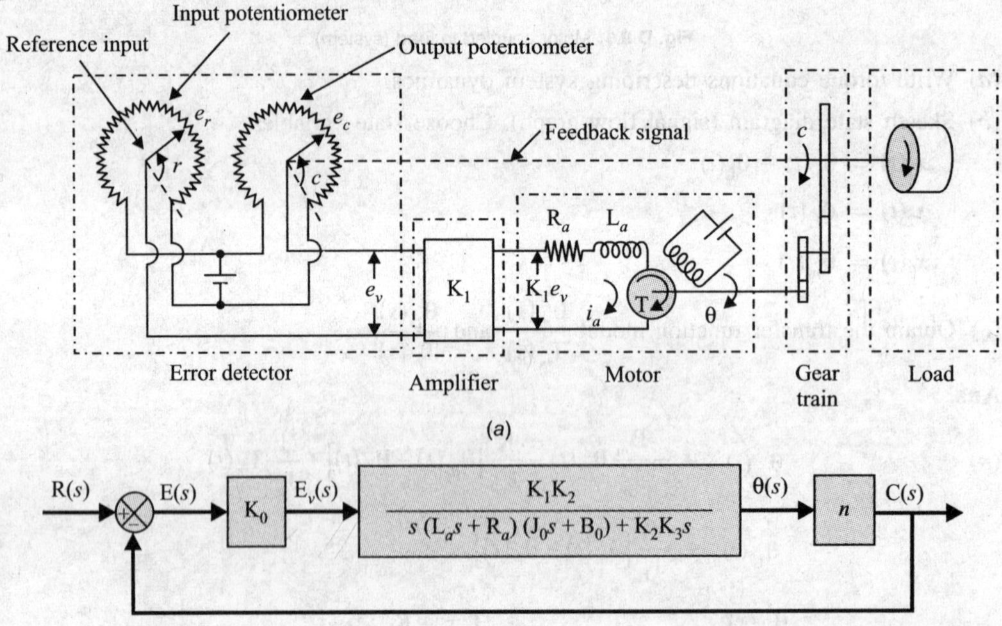

Fig. D 9.2(*a*) Position control system (*b*) Block diagram

D 9.3. The position control system is shown in Fig. D9.3(*a*). The control parameters and variables are defined as

e = error voltage	K_i = motor torque constant
θ_r = reference position	T_m = motor torque
θ_L = load position	J_m = motor inertia
K_s = potentiometer gain constant	B_m = motor frictional coefficient
K_A = amplifier gain	K_L = torsional spring constant
e_a = motor input voltage	J_L = load inertia
e_b = back emf	
K_b = back emf constant	
i_a = motor armature current	

Write state equations of the system and draw state diagram (signal flow graph) using the nodes as shown in Fig. D9.3(*b*).

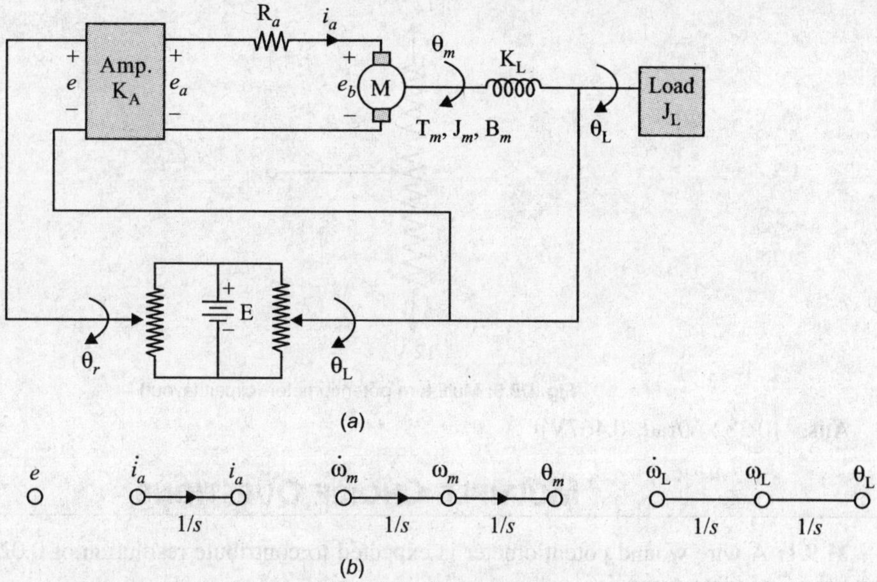

Fig. D9.3: (*a*) Position control system (*b*) Nodes of signal flow graph

D 9.4: As shown in Fig. D 9.4, the amplidyne is considered as two stage cascaded generator. Obtain the transfer function model $E_d(s)/V_f(s)$. The symbols used in the circuit have their usual meaning.

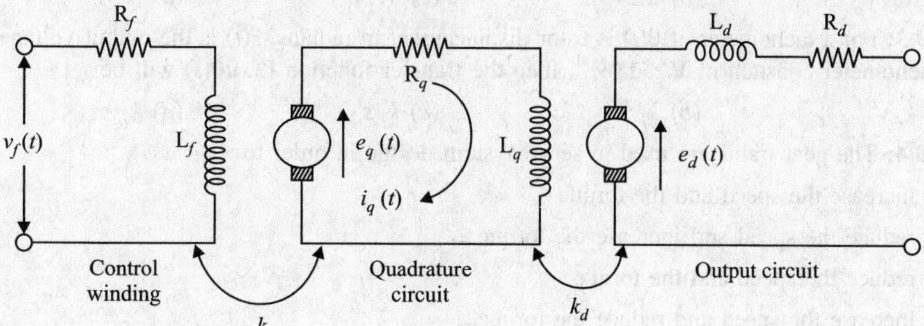

Fig. D9.4: An amplidyne circuit

Ans. $$\frac{E_d(s)}{V_f(s)} = \frac{k_d\, k_q}{R_q\, R_f\left(1+s\tau_f\right)\left(1+s\tau_q\right)};\ \tau_f = \frac{L_f}{R_f},\ \tau_q = \frac{L_q}{R_q}$$

D. 9.5: A 50 KΩ, 10 turn potentiometer has ±12 V supply connected to fixed terminals as shown in Fig. D9.5.

(*a*) Determine gain constant of potentiometer in V/rad.

(*b*) Find open circuit (unloaded) output voltage when the shaft is rotated 70° from mid-point towards + 12 V terminal.

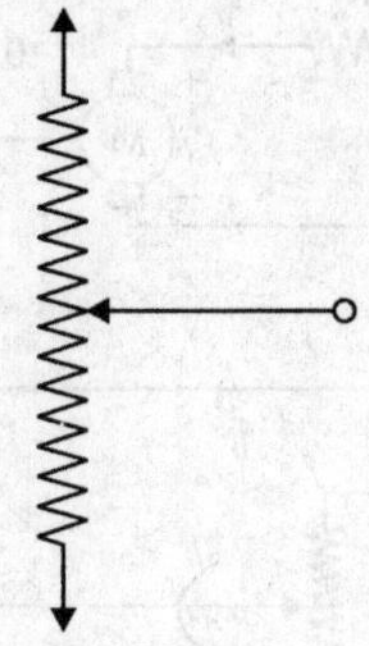

Fig. D9.5: Multi turn potentiometer (circuit layout)

Ans. [0.382 V/rad, 0.467V]

MULTIPLE CHOICE QUESTIONS

M 9.1: A wire wound potentiometer is expected to contribute resolution of 0.02 %. The required number of winding turns, is

(*a*) 4000 (*b*) 5000 (*c*) 2000 (*d*) 1500.

M. 9.2: A tachometer has sensivity of 4V/1000 rpm. The output voltage for shaft speed of 30 rad/sec, will be approximately

(*a*) 12 V (*b*) 2.24 V (*c*) 1.14 V (*d*) 3 V.

M 9.3: For a tachometer, if $\theta(t)$ is rotor displacement in radians, $e(t)$ is the output voltage and k_t is the tachometer constant in V/rad.sec^{-1}, then the transfer function $E(s)/\theta(s)$ will be

(*a*) $k_t\, s^2$ (*b*) k_t /s (*c*) $k_t\, s$ (*d*) k_t

M 9.4: The gear trains are used in servo system design in order to

(*a*) increase the speed and the torque

(*b*) reduce the speed and increase the torque

(*c*) reduce the speed and the torque

(*d*) increase the speed and reduce the torque.

M 9.5: In design of two phase AC servo motor, special care should be taken such that

(*a*) the inertia is high (*b*) the friction is low

(*c*) torque-speed curve has positive slope (*d*) torque-speed curve has negative slope.

M 9.6: The conventional two phase induction motor cannot be used in design of servo system because

(*a*) its cost is high (*b*) it is less accurate

(*c*) it is noisy (*d*) it destabilises the servo system.

M. 9.7: The purpose of the series quadrature windings in an amplidyne is to

(*a*) neutralize the effect of armature reaction (*b*) reduce commutation difficulties

(*c*) increase gain (*d*) increase the response time.

M 9.8: In position control system, the device used for providing rate feedback voltage, is called

(*a*) potentiometer (*b*) synchro transmitter

(*c*) synchro transformer (*d*) tacho generator

M 9.9: A synchro transmitter-receiver unit is a

(*a*) two phase AC device (*b*) three phase AC device

(*c*) DC device (*d*) single phase ac device.

M 9.10: Match the control system components in List I with their function in List II and select the correct answer using the codes given below the lists:

List I	**List II**
A. Servo motor	**1.** Error detector
B. Amplidyne	**2.** Rate feedback
C. Potentiometer	**3.** Actuator
D. Tacho generator	**4.** Power amplifier

Codes:

	A	**B**	**C**	**D**		**A**	**B**	**C**	**D**
(*a*)	3	1	2	4	(*b*)	3	4	2	1
(*c*)	4	3	2	1	(*d*)	3	4	1	2.

M 9.11: For a two phase servo motor, which one of the following statements is not true?

(*a*) The rotor diameter is small. (*b*) The rotor resistance is low.

(*c*) The applied voltages are seldom balanced. (*d*) The torque-speed characteristics are linear.

M. 9.12: In case of synchro error detector, the electrical zero position of control transformer is obtained when angular displacement between rotors, is

(*a*) 0° (*b*) 45° (*c*) 90° (*d*) 180°.

M 9.13: Consider the following for a variable reluctance stepper motor used in control system.

1. The static torque acting on the rotor is a function of angular misalignment between stator and rotor teeth.
2. There are two positions of zero torque: $\theta = 0^o$ and $180^o/T$ (T = number of rotor teeth).
3. Both zero torque positions are stable.
4. As the stator is excited, the rotor is pulled into the nearest minimum reluctance position.

Of these statements

(*a*) 2, 3 and 4 are correct (*b*) 1, 2 and 3 are correct

(*c*) 1, 2 and 4 are correct (*d*) 1, 3 and 4 are correct.

M 9.14: In a two phase AC servo motor, the rotor has resistance R and reactance X. The torque speed characteristics of the servo motor will be linear provided that

(*a*) $\frac{X}{R} << 1$ (*b*) $\frac{X}{R} >> 1$ (*c*) $X^2 = R$ (*d*) $\frac{X}{R} = 1$

M 9.15: Which of the following can work as error detecting devices?

1. A pair of potentiometers
2. A pair of synchros
3. A differential transformer
4. An amplidyne
5. A control transformer

Select the correct answer using the following codes:

(*a*) 1, 2 and 5 (*b*) 2, 3, 4 and 5 (*c*) 1, 3, 4 and 5 (*d*) 1, 2, 3 and 4.

M 9.16: The block diagram shown below, represents a hybrid servo system.

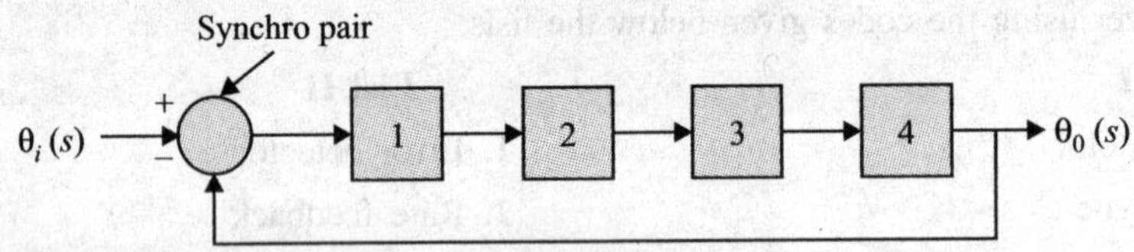

The blocks labelled 1, 2, 3 and 4 are respectively

(*a*) amplifier, demodulator, DC servo motor and load

(*b*) demodulator, amplifier, DC servo motor and load

(*c*) amplifier, DC servo motor, demodulator and load

(*d*) demodulator, DC servo motor, amplifier and load.

M. 9.17: Which of the following components can be used as a rotating amplifier in a control system?

1. An amplidyne
2. A separately excited DC generator
3. A self-excited DC generator
4. A synchro

Select the correct answer using the codes given below:

Codes:

(*a*) 3 and 4 (*b*) 1 and 2 (*c*) 1, 2 and 3 (*d*) 1, 2, 3 and 4.

M 9.18: In the field controlled DC motor, the entire damping is contributed by

(*a*) the armature resistance (*b*) the back emf

(*c*) the motor friction and load (*d*) feed resistance.

M 9.19: Which of the following rotors are used in a two phase AC servo motor?

1. solid iron 2. squirrel cage 3. drag cup.

Select the correct answer using the codes given below:

Codes:

(*a*) 1, 2 and 3 (*b*) 1 and 2 (*c*) 2 and 3 (*d*) 1 and 3.

M 9.20: For two phase AC servo motor, if R is resistance of rotor, X is reactance of rotor, L is length of rotor and D is diameter of rotor then

(*a*) X/R and L/D are both small. (*b*) X/R is large but L/D is small.

(*c*) X/R is small but L/D is large. (*d*) X/R and L/D are both large.

M 9.21: Match List I with List II and select the correct answer using the codes given below the lists:

List I	List II
A. Synchro transmitter	**1.** Dumb-bell rotor
B. Control transformer	**2.** Drag cup rotor
C. AC servo motor	**3.** Cylindrical rotor
D. Stepper motor	**4.** Toothed rotor
	5. Phase wound rotor

Codes :

	A	B	C	D		A	B	C	D
(*a*)	1	3	2	4	(*b*)	2	4	3	1
(*c*)	1	5	3	2	(*d*)	3	2	1	5.

M 9.22: Which of the following can be included in a control system such that total number of poles and zeros, does not change?

(*a*) Amplidyne (*b*) DC servo motor (*c*) tachometer (*d*) AC servo motor.

M 9.23: In a two phase AC servo motor with small X/R ratio, maximum torque occurs at

(*a*) synchronous speed (*b*) low speed (*c*) high speed (*d*) rated speed.

M 9.24: A stepper motor has stator wound for two phase and rotor consisting of 24 teeth. The step movement θ is

(*a*) 7.5° (*b*) 15° (*c*) 3.75° (*d*) 11.25°

M 9.25: A stepper motor has step angle of 1.8° and the pulse frequency is 300 pulses per second. Now consider the following statements.

1. The resolution of motor is 200 steps/revolution.
2. The speed of motor is 90 rpm.
3. The number of steps, the motor moves in 15 revolutions is 3000 steps.

Of these

(*a*) only 1 and 2 are correct (*b*) only 2 and 3 are correct

(*c*) only 3 is correct (*d*) 1, 2 and 3 all are correct.

ANSWERS

M 9.1. (*b*)	M 9.2. (*c*)	M 9.3. (*c*)	M 9.4. (*b*)	M 9.5. (*d*)
M 9.6. (*d*)	M 9.7. (*c*)	M 9.8. (*d*)	M 9.9. (*d*)	M 9.10. (*d*)
M 9.11. (*b*)	M 9.12. (*c*)	M 9.13. (*c*)	M 9.14. (*a*)	M 9.15. (*a*)
M 9.16. (*a*)	M 9.17. (*b*)	M 9.18. (*c*)	M 9.19. (*a*)	M 9.20. (*c*)
M 9.21. (*a*)	M 9.22. (*a*)	M 9.23. (*b*)	M 9.24. (*a*)	M 9.25. (*d*)

Important Hints

M 9.1:

$$\%\text{ resolution} = \frac{1}{N} = \frac{0.02}{100} \Rightarrow N = 5000$$

M 9.2:

$$1000\text{ rpm} = \frac{1000 \times 2\pi}{60}\text{ rad/sec}$$

$$K_s = \frac{4 \times 60}{2000\pi} = 0.038\ \text{V/rad.sec}^{-1}$$

$$V_0 = 0.038 \times 30 = 1.14\ \text{V}$$

M 9.24:

$$\theta = \frac{360}{m \cdot Nr} = \frac{360}{2 \times 24} = 7.5°$$

M 9.25:

$$\text{resolution} = \frac{360}{1.6} = 200\ \text{steps/revolution}$$

$$\text{speed} = \frac{1.8 \times 300 \times 60}{360} = 90\ \text{rpm}$$

number of steps in 15 revolutions = 200 × 15 = 3000.

□□□

INDEX

Absolute
- encoder, 673, 675
- stability, 166

AC
- servo motor, 647, 653
- tachometer, 657

Acceleration error constant, 76
Accuracy/precision/linearity, 641
Addition of
- a pole to open loop transfer function, 82
- a zero to closed loop transfer function, 81
- a zero to open loop transfer function, 82
- finite non-origin poles, 304
- pole to closed loop transfer function, 83
- poles at origin, 302
- zeros at origin, 305

All pass systems, 349
Alternative method to resolve the left column zero situation, 170
Amplidyne, 660
Analogous systems, 18
Angle criterion, 208
Angles of departure
- and approach, 220
- and arrival, 231

Angular potentiometer, 639
Arbitrary pole placement, 484
Armature controlled DC servo motor, 647
Associative law for summing points, 117

Asymptotic
- angles, 215, 231
- plot, 335
- stability, 163

Asymptotically stable system, 462
Autonomous, 1
Auxiliary polynomial, 173

Band width, 284, 286
Belt, 676, 677
Block, 115
Block diagram(s), 115
- development, 116
- reduction rules, 117

Blocks in
- cascade, 117
- tandem, 117

Bode
- phase plot, 330
- plot, 335

Bounded input-bounded output (BIBO) stability, 164
Branches, 119
Break away and break in (entry) points, 218
Break frequency, 332, 333

Canonical
- forms of state model, 448
- variable, 441, 455

Cascade
- compensator, 601
- compensator design for improving steady performance, 559
- compensator design for improving transient response, 564
- controller, 533

Causal system, 33
Causes of uncontrollability, 473
Centroid of the asymptotes, 216
Chain drive, 677
Changing a summer sign, 117
Characteristic
- equation, 27, 66, 208
- equation roots, 65
- polynomial, 27

Characteristics of phase lead compensator, 550
Closed loop
- control system (feedback control system), 2
- frequency response of unity feedback system, 355
- transfer function, 4

Coefficient test for stability, 166
Co-factor of *i*th forward path, 125
Companion matrix, 448
Comparing armature controlled DC servo-motors, 653
Compensating winding, 661
Compensator, 598, 601
- elements, 546

Completely
- observable system, 469
- state controllable system, 469
Complex
- conjugate poles, 336
- dominant poles, 80
Conditional frequency, 62
Conditionally stable systems, 425
Conjugate symmetry, 218
Constant
- amplitude oscillation, 164
- damping factor loci, 67
- magnitude loci, 357
- phase loci, 360
- ω_d loci, 67
- ω_n loci, 67
Construction of signal flow graph
- for electrical network, 123
- from block diagram, 123
Continuous and discrete systems, 33
Control adjustment, 2
Control system
- components, 636
- design, 532
Controllability, 456, 468
- matrix, 470
Controllable canonical form, 486
Controller, 532
- configurations, 532
Convolution, 464
Corner
- frequency, 343, 344
- plot, 327
Correlation between time response and frequency, 282
Corresponding time response, 65
Critical point, 362
- $-1 + j0$, 312
Critically damped, 68

D'Alembert's principle, 13
Damped frequency, 62
Damper, 11
Damping
- coefficient, 19
- constant, 62
- factor, 62
- ratio, 62
DC
- attenuation, 548
- servo motor, 647
- tachometer, 657
Decoupled form, 476
Delay time, 65
Demerits of potentiometer, 642
Deposit film potentiometer, 642
Derivative
- control, 237
- time, 539
Design procedure of lead compensator, 569
Designing
- lag-compensator, 575
- lead compensator, 574
- PD compensator, 571
- PI compensator, 572
Detectability of these system, 527
Determinant of a signal flow graph, 125
Deterministic and stochastic systems, 33
Diagonal
- matrix, 462
- system, 472
Diagonalisation, 476
Differential gap, 534
Disadvantage of passive lead network, 569
Disturbances on system response, 8
Divisor polynomial, 171
Dominant
- closed loop poles, 558
- poles, 80
Dual
- channel encoder, 674
- phase variable form of state model, 447
Dummy nodes, 121
Dynamic
- behaviour, 53
- error coefficients, 78
Dynamics of observer, 485

Effect of
- adding poles, 81, 235
- addition of poles, 302
- delay on root locus, 237
- lag-compensator on transient response, 561
- presence of delay in system on Bode plot, 351
- variation in gain K on Bode plot, 350
- varying one on system stability, 176
Eigen
- values, 462, 466, 477
- vector, 477
Electrical
- analog of gear train, 24
- elements, 11
- systems, 10
- time constant, 649, 652
- zero, 644
Error
- at break frequency, 337
- detectors, 636
- series, 78
- signal, 3
Errors due to approximation for transmittances, 333
Evaluation of
- K_a from Bode plot, 343
- K_p from Bode plot, 341
- K_v from Bode plot, 342

Features of Bode plot, 327
Feed forward compensation, 533
Feedback
- compensation, 532, 546, 590
- compensator design, 590
- control system, 72
- gains, 483
- matrix, 482
- path transfer function, 4
f-i analogy, 21
Field controlled DC servo motor, 651, 653
Finding
- decoupled state equations, 476
- time response from state model, 464
- transfer function from state space model, 461
- transfer function models, 352
First order
- system, 55
- system in state variable form, 464
Force current analogy, 21
Forced response, 30, 467
Force-position relationship, 12
Force-voltage (*f-v*) analogy, 18
Forward path, 122
- transfer function, 4
Frequency
- characteristics, 642
Frequency response, 335, 337
- analysis, 335
- design, 577
- of second order system, 285
- specifications, 362
Full order state observer, 485
f-v analogy, 20

Gain
- adjustment, 559
- and phase characteristics of phase lead, 554
Gain cross over
- frequency, 317, 349
- point, 317
Gain margin, 315, 319
- from Bode plot, 339
Gear train and its electrical analog, 22
Generating
- complete Nyquist plot, 321
- hardware for industrial controllers, 543
Graphical evaluation of
- angle and magnitude of $G(s)\ H(s)$, 210
- frequency response, 281
Graphical representation of frequency response, 287

Half stepping, 666, 667
High
- frequency asymptote, 332
- pass filter, 550
- velocity noise, 641, 642
Higher order systems, 79, 467
Homogeneity, 32
Hybrid stepper motor, 671

Identity
- matrix, 448
- observer, 484, 485
- system, 33
Imaginary axis
- crossing points, 223
- zeros and poles, 354
Impulse
- function, 54
- response stability, 163
- responses, 28
Incremental encoder, 673
Industrial automatic controllers, 534
Inherent design, 642
Initial condition (I.C.), 1
Input
- node (source), 121
- vector, 441
Insertion/removal of unity gain, 117
Insignificant poles, 80
Integral
- control, 236, 305, 536, 542
- time, 537
Interpreting
- magnitude response of closed loop system from constant M circles, 359
- stability from Bode plot, 349
Irrational transmittances, 349

Jordan canonical form, 458

Kalman's
- controllability test, 470
- observability test, 472

Lag-compensator design, 561, 578, 581
Lag-lead compensator, 555
- design, 574
- transmittance, 556
Lead
- compensator design, 568, 582, 584
- network, 550
Left column zero of array, 169
Lever system, 677
Limitations of classical transfer function model, 439
Linear system, 32
Loading noise, 642
Loci branches, 213
Loop, 122

Low frequency asymptote, 332
Low pass filter, 554

M circles, 356
Magnitude
- criterion, 208
- plot, 331
Major loop, 590
Mapping, 306
Marginally
- stable, 164, 173, 360
- unstable, 165
Mason's gain
- formula, 119
- rule, 124, 447
Mass, 11
Mathematical model of a continuous system, 53
Matrix exponential e^{At}, 466
Maximum
- overshoot, 63
- (peak) phase lag, 553
- phase lead, 547
- pull in, 664
- pull out, 664
Mechanical
- elements, 11
- lag network, 552
- lead network, 547
- system(s), 11, 459
- time constant, 649, 652
Merits of potentiometer, 642
Micro stepping, 673
MIMO system, 441
Minimum phase, 312
Minor loop, 590
- feedback compensation, 593
Modal matrix, 477
Modelling electrical, 459
Motor, 676
- and screw assembly, 676
- gain constant, 656
- time constant, 656
Moving a
- summing point, 118
- take off point, 118
Multi
- outputs, 30
- stack stepper motor, 668
- turn potentiometer, 640
Multiple
- gain cross over frequencies, 318
- outputs, 448, 451
- phase cross over frequencies, 316

N circles, 356, 360
Natural response, 30
- terms, 530
Natural undamped frequency, 62
Necessary and sufficient condition for complete state observability, 472
Need for the buffer amplifier, 586
Negative
- damping, 66
Nichols chart, 362
Nodes, 119
Noise, 641
Non
- autonomous, 1
- linear system, 32
- minimum phase, 312
Non-anticipative systems, 33
Nyquist
- criterion, 279
- plot and stability, 321
- procedure for minimum phase system, 312
- stability criterion, 305, 310, 311

Observability matrix, 472, 487
Observable canonical form, 486
Observer
- design, 484
- dynamics, 487
- error matrix, 486
- gain, 485
- gain matrix, 486
Octave, 329
Open loop
- control system, 2
- transfer function, 4
Optical encoder, 673
Order system, 27
Other ways of modelling, 451
Output
- matrix, 443
- node (sink), 121
- vector, 441
Overall gain, 4
Overdamped, 68
Overshoots, 63

Padee
- approximation, 238
- approximant, 278
Parabolic error constant, 76
Passive
- lag network, 568
- lead network, 568
PD compensator, 484, 564
- design, 564, 565, 591
Peak
- overshoot, 64, 82
- phase lead, 548
- resonance, 283
- time, 62
Perfect integrater, 165
Permanent magnet stepper motor, 670
Phase
- contribution of lag-compensator, 578

– cross over frequency, 339, 360
– lag-compensator, 551, 554
– lag-lead compensator design, 586
– lead compensator, 546
– plane, 440
– point, 315
– variable form of state model, 445
– *vs* log ω plot, 333
Phase margin, 315, 317
– from Bode plot, 339
PI compensator design, 559
– controller, 543
Pick off point, 116
PID controller design, 571
– controller, 545
Pinion, 676
Plot of
– magnitude, 327
– phase angle, 327
Polar plot(s), 287, 312, 547
– construction, 287
– of transfer functions, 300
– open loop frequency response, 359
Pole(s), 354
– pitch, 672
– placement, 481
– placement design, 483
Pole-zero
– cancellation, 469, 473, 561
– form, 73
Position error constant, 74
Positive damping, 66
Potentiometer, 637
Power handling capability, 641
Premature termination of array, 171
Principle of
– arguments, 306
– duality, 475, 476
Product terms of G(*j*ω) H(*j*ω), 327
Proper, rational transfer function, 445
Properties of STM, 466
Proportional
– band, 535
– control, 535
– gain, 535, 537, 539
– *plus* derivative control (PD), 539
– *plus* integral control, 537
– *plus* integral *plus* derivative control, 541
Pulley, 676

Quadruplet, 444

Rack, 676
Ramp error constant, 75
Rate
– control, 539
– feedback compensator design, 590
– gyro, 590
– sensor, 590
Rational transfer function, 445
Read out function, 444
Real axis
– locus segment, 231
– poles, 336
– segments, 214
Reduced order state observer, 484
Redundancy in state variable assignment, 474
Region of stability, 190
Relationship between Bode magnitude (dB) plot and number type of a system, 341
Relative stability, 166
– analysis, 175
– using Nyquist procedure, 313
Residual noise, 642
Resolution, 641
– noise, 642
– of the stepper motor, 672
Resolvent matrix, 465
Resonant frequency, 283, 285, 577
Resonant peak, 577
– magnitude, 360, 363
RHP boundary, 310
Rise time, 65, 82
Robust control, 440
Root contours, 239
Root loci for systems with
– other forms, 228
– positive feedback, 230
Root locus, 206, 278
– construction, 213
– construction rules, 225
– design, 557
– for feedback systems, 208
– of a G(*s*) H(*s*) product with pole-zero cancellation, 234
Root locus plots of
– negative feedback system, 233
– positive feedback system, 233
Roots forming symmetry about the origin, 171
Rotary potentiometer, 639
Rotating amplifier, 659
Routh array, 167
– properties, 169
Routh stability, 278
Routh's stability
– criterion, 168
– test, 167

Scalar multiplicativity, 32
Second order system, 60
Self loop, 122

Selsyn synchros, 643
Semigraphical procedure, 505
Sensitivity, 4, 535
– of error detector, 646
Series
– compensation, 546
– feedback compensation, 533
Servo, 636
– amplifier, 636, 659
– motors, 636, 647
– system, 3
Settling time, 64
Shorting noise, 642
Signal
– flow graph, 119
– multiplier, 119
Similarity transformation, 477
Simulation diagram, 441, 443
Simultaneous improvement in
– both transient as well as steady state response, 574
– transient as well as steady state performance, 571
Singular points, 307
Sinusoidal transmittance
– G_{lead} ($j\omega$), 548
– of lag-lead compensator, 557
Sketching
– Bode plot, 344
– magnitude (dB) plot, 336
– phase plot, 337
Some root locus plots, 226
Spring, 11
– constant, 19
Stability, 4, 456
– analysis, 163
– investigation, 305
Stabilizability, 531
Stall torque, 655
Standard test signals, 54
State, 440
– feedback, 481, 483
– plane, 440
– response, 467
– variables, 440
– vector, 440, 441
State model for systems with
– multiple inputs, 451
– single input, 448
State space, 440
– analysis and design, 439
– approach, 439
– model using canonical variables, 455
– representation, 440
State transition, 466
– matrix, 465
Steady state
– behaviour, 53
– performance, 72
– sinusoidal response, 279
Steady state error, 57, 73, 537
– in terms of gain, 77
– performance, 561
Steady state error due to
– parabolic input (acceleration input), 76
– ramp input (velocity input), 75
– step input, 74
Step
– acceleration input (parabolic function), 54
– angle, 667
– displacement input (step function), 54
– error constant, 74
– velocity input (Ramp function), 54
Stepper motor, 663
Strengths of frequency response, 278
Strictly proper, 445
Structure of full order state observer, 485
Summary of Nyquist procedure, 311
Summing point, 115
Superposition (additivity), 32
Sustained oscillation, 68
in system, 186
Synchro, 643
– receiver, 643
– transmitter, 643
System, 1
– stability, 163, 462
– symmetry, 473
– with memory and without memory, 33
Systems modelling, 10

Take off point, 116
Technique to cast away insignificant poles, 80
Telsyn synchros, 643
Temperature control system, 116
Test of
– observability for diagonal systems, 472
– output controllability, 472
– state controllability for diagonal systems, 469
Testing the largest time constant, 193
Three
– stack motor, 668
– term control, 543
Time constant, 56
– form, 73
Time invariant system, 32
Time varying system, 59

Tolerance band, 64
Tooth pitch, 672
Torque-position relationship, 12
Trajectory, 444
Transfer function, 26, 27, 445
- of a system with multi inputs, 30
Transient response, 53
- specifications, 62
Translatory potentiometer, 639
Transmission matrix, 444, 449
Transmittance, 27, 546
- factor, 119
- of PI controller, 537
Transportation lag, 349
Two blocks in feedback configuration, 117
- position/on-off control, 534
- stage power amplifier, 662
- system parameters on system stability, 176
Types of potentiometer, 642
Typical step response, 69

Undamped, 68
Underdamped, 68
Undershoots, 63
Unit impulse response of
- a general second order system, 70
- first order system, 58
Unit ramp response of
- first order system, 57
- PD controller, 540
- PID controller, 541
Unit step response of
- a general second order control system, 60
- first order system, 55
- P and PI controller, 538

Vander Monde matrix, 479
Variable reluctance stepper motor, 664
Variation of
- M_r with ξ, 285
- ω_r with ξ, 285
Velocity error constant, 75
Vibrational noise, 641
Voltage current relations, 11

Weaknesses of frequency response, 279
Weighing function, 26
Wire wound, 642

X/R ratio, 653

Zero
- input response, 30, 467
- state response, 29, 467
Zeros, 331, 333
Zeros to
- G(s) on the shape of polar plots, 302
- the product G(s) H(s), 235
- to transfer function, 81